CLINICAL LABORATORY CHEMISTRY

Robert L. Sunheimer, MS, MT(ASCP)SC, SLS

Upstate Medical University
Syracuse, New York

Linda Graves, EdD, MT(ASCP)

University of Maine at Presque Isle
Presque Isle, Maine

Pearson

Boston Columbus Indianapolis New York San Francisco Upper Saddle River
Amsterdam Cape Town Dubai London Madrid Milan Munich Paris Montréal Toronto
Delhi Mexico City São Paulo Sydney Hong Kong Seoul Singapore Taipei Tokyo

Library of Congress Cataloging-in-Publication Data
Sunheimer, Robert L.
  Clinical laboratory chemistry / Robert L. Sunheimer,
Linda Graves.—1st ed.
    p. ; cm.
  Includes bibliographical references and index.
  ISBN-13: 978-0-13-172171-5 (alk. paper)
  ISBN-10: 0-13-172171-2 (alk. paper)
  1. Clinical chemistry.   I. Graves, Linda.   II. Title.
  [DNLM: 1. Chemistry Techniques, Analytical.   2. Chemistry,
Clinical—methods.   3. Clinical Chemistry Tests.   4. Clinical Laboratory
Techniques.   QY 90 S958c 2011]
  RB40.S85 2011
  616.07'56—dc22
                          2010014917

Publisher: Julie Levin Alexander
Assistant to Publisher: Regina Bruno
Editor-in-Chief: Mark Cohen
Executive Editor: John Goucher
Associate Editor: Melissa Kerian
Assistant Editor: Nicole Ragonese
Editorial Assistant: Rosalie Hawley
Development Editor: Cathy Wein
Senior Media Editor: Amy Peltier
Media Project Manager: Lorena Cerisano
Managing Production Editor: Patrick Walsh
Production Liaison: Christina Zingone
Production Editor: Laura Lawrie
Manufacturing Manager: Ilene Sanford
Manufacturing Buyer: Pat Brown

Art Director: Christopher Weigand
Cover Designer: Kevin Kall
Cover Image: Richard N. Whelsky/
  Medical Photographer
Director of Marketing: David Gesell
Executive Marketing Manager: Katrin Beacom
Marketing Specialist: Michael Sirinides
Marketing Assistant: Judy Noh
Manager, Rights and Permissions: Zina Arabia
Manager, Visual Research: Beth Brenzel
Image Permission Coordinator: Richard Rodrigues
Composition: MPS Limited, A Macmillan Company
Printer/Binder: Quebecor World Color, Versailles
Cover Printer: Lehigh-Phoenix Color/Hagerstown

www.pearsonhighered.com

10 9 8 7 6 5 4 3 2
ISBN-13: 978-0-13-172171-5
ISBN-10: 0-13-172171-2

DEDICATION

Dedicated to my wife, Bonny; my daughter, Kristin; and my son, Mark.

Robert L. Sunheimer

Dedicated to Peetie Charette, my former Co-Director of the Medical Laboratory Technology Program of Maine, a valued colleague and friend who will be missed; and my husband, Bill Graves, who supported me through yet another writing project.

Linda Graves

CONTENTS

FOREWORD

Clinical Laboratory Chemistry is part of Pearson's Clinical Laboratory Science series of textbooks, which is designed to balance theory and practical applications in a way that is engaging and useful to students. The authors of and contributors to *Clinical Laboratory Chemistry* present highly detailed technical information and real-life case studies that will help learners envision themselves as members of the health-care team, providing the laboratory services specific to clinical chemistry that assist in patient care. The mixture of theoretical and practical information relating to clinical chemistry provided in this text allows learners to analyze and synthesize this information and, ultimately, to answer questions and solve problems and cases. Additional applications and instructional resources are available at www.myhealthprofessionskit.com.

We hope that this book, as well as the entire series, proves to be a valuable educational resource.

Elizabeth A. Zeibig, PhD, MLS(ASCP)^{CM}
Clinical Laboratory Science Series Editor
Pearson Health Science

Associate Dean for Graduate Education
Associate Professor
Department of Clinical Laboratory Science
Doisy College of Health Sciences
Saint Louis University

PREFACE

Clinical chemistry has been one of the fastest-developing disciplines in the clinical laboratory. While working on this book, we found that even from one rewrite to the next we sometimes had to add a section or revise a segment based on new information. The ever-evolving nature of clinical chemistry is one of the things we love about the discipline and one of its greatest challenges.

We have developed a textbook with page-to-page formatting that is aesthetically pleasing to the eyes and conducive to visually oriented learning. Several learning features are included that will help readers digest the large amount of material presented. We have also placed text content in special formats, including a Website for students and instructors and an instructor's guide to facilitate presentation of text material.

TARGET AUDIENCE

We designed this textbook, *Clinical Laboratory Chemistry*, especially for students in clinical laboratory technician (CLT), medical laboratory technician (MLT), and clinical laboratory technologist/medical technologist (MT) education programs. We also believe this book to be useful for practicing technicians and technologists who will find it helpful to review topics of interest or to look up the answer to a specific question. *Clinical Laboratory Chemistry* is readable, up-to-date, and concentrates on clinically significant analytes that students are likely to encounter. We focused on trying to balance the material that students at both levels require without making it overwhelming. Students will not read or use a textbook they have difficulty understanding.

TEXTBOOK FORMAT

Several approaches have been used in the past by authors of clinical chemistry textbooks to present the voluminous amount of material that spans the breadth of services provided by a clinical chemistry laboratory. We choose a fundamental approach that is suitable for both CLT/MLT and CLS/MT students. The two principle authors represent both academic levels.

The first four chapters present topics that focus on chemistry laboratory operations, laboratory data, and equipment commonly found in clinical chemistry laboratories. Chapter 5 reviews the principles of several immunoassays that are used to measure hormones, vitamins, and drugs. Diagrams are provided to aid in the understanding of the reactions involved in each immunoassay. Particular attention is paid to the label or detector compound used to facilitate the quantitative determination of analyte concentration.

The remaining chapters discuss individual organ systems and selected specialties (e.g., tumor markers and porphyrins) offered in many clinical chemistry laboratories. A brief review of the anatomy and physiology precedes a discussion of analytes associated with their respective organs and selected specialties. An explanation of laboratory tests associated with each analyte is provided. This is followed by a discussion of the appropriate pathophysiology that details the association of abnormal chemistry laboratory results with disease states.

SPECIAL FEATURES

Within each chapter are several unique pedagogical features that will assist the student in developing a complete understanding of the material presented. These features are designed to facilitate the flow of material through the chapters and to avoid disrupting the student's focus.

- **Objectives** are presented at two levels: Level I focuses on fundamentals of clinical chemistry and basic assimilation of laboratory data with disease; Level II focuses on advanced applications of clinical chemistry. These objectives have been written and reviewed by faculty representing both baccalaureate and associate degree programs. Instructors at both levels of education may use these objectives at their discretion.

- **Key Terms** represent the vocabulary or terminology used throughout the chapter that the authors believe to be significant to the complete understanding of the material. Many terms represent fundamental scientific terminology, whereas on many occasion the reader will encounter terms that are unique to clinical chemistry laboratories. These terms represent the language of the laboratory and should be learned before entering a clinical laboratory.

- A **Case in Point** is presented at the beginning of each chapter to provide the reader with an actual medical case or situation that can occur in a clinical laboratory. The case may be continued throughout the chapter, emphasizing concepts presented in that particular section. Each medical case begins with a presentation of the patient's history, physical findings, and laboratory results. Several issues and questions are presented that reflect the material presented in the textbook and serve to emphasize points the authors consider significant. The response to the questions and issues are presented at the end of the book.

- **What's Ahead** presents a bulleted summary of the main topics or issues the authors believe to be important for the reader to focus on.

- **Checkpoints** are questions that appear throughout the chapter. The purpose of these questions is to allow the reader to reflect on the material presented in order to answer questions or apply information covered in the preceding section.

- The Summary of each chapter concludes the text portion and serves to bring all the material together.
- Review Questions are included at the end of the chapter. There are two sets of review questions—Level I and Level II—that are referenced to their respective objective set. The answers to the review questions are located at end of the textbook.

A COMPLETE TEACHING AND LEARNING PACKAGE

The book is complemented by a variety of ancillary materials designed to help instructors be more effective and students more successful.

- The Instructor's Resource Manual is designed to equip faculty with necessary teaching resources regardless of the level of instruction. Features include suggested learning activities for each chapter, and classroom discussion questions.
- The Test Bank includes more than 600 questions to allow instructors to design customized quizzes and exams using our award-winning TestGen 7.0 test-building engine. The TestGen wizard guides you through the steps to create a simple test with drag-and-drop or point-and-click transfer. You can select test questions either manually or randomly and use online spellchecking and other tools to quickly polish your test content and presentation. You can save your test in a variety of formats both locally and on a network, print as many as 25 variations of a single test, and publish your tests in an online course. For more information, please visit www.pearsonhighered.com/testgen.
- PowerPoint lectures contain key discussion points, along with color images, for each chapter. This feature provides dynamic, fully designed, integrated lectures that are ready to use and allows instructors to customize the materials to meet their specific course needs. These ready-made lectures will save you time and ease the transition into the use of *Clinical Laboratory Chemistry*.

PEARSON myhealthprofessionskit™

MyHealthProfessionsKit is an online study guide that is completely unique to the market. It provides an array of assessment quizzes that have been developed within an automatic grading system that provides users with instant scoring once users submit their answers. Each student's quiz results can me e-mailed directly to the educator, if desired, as part of a homework assignment.

Go to www.myhealthprofessionskit.com and select Clinical Laboratory Science. Next, find this book and then click to log in. To obtain a username and password, please register using the scratch-off access code on the inside front cover of this book.

ACKNOWLEDGMENTS

The completion of a project as overwhelming as a textbook is impossible without the assistance of many individuals. Our contributors provided us with chapters in their areas of expertise. The editors from Pearson Health Science guided us through the various steps in the process and offered encouragement along the way. Reviewers gave us extremely useful feedback and suggestions on our various chapters.

Special thanks are extended to the staff at Pearson Health Science. Mark Cohen, Editor-in-Chief, has always been supportive throughout this long process. Associate Editor Melissa Kerian kept us on track and answered all our questions. Cathy Wein, a freelance development editor, helped with assembly of the manuscript and formatting guidelines.

EDITORIAL DEVELOPMENT TEAM

REVIEWERS

Charity Accurso, PhD
University of Cincinnati
Cincinnati, Ohio

Hassan Aziz, PhD
Armstrong Atlantic State University
Savannah, Georgia

Beth Blackburn, MT(ASCP)
Orangeburg-Calhoun Technical College
Orangeburg, South Carolina

David W. Brown, PhD
Florida Gulf Coast University
Ft. Myers, Florida

Karen M. Escolas, EdD, MT(ASCP)
Farmingdale State College
Farmingdale, New York

Candance Hill, BS, MAEd, MT(ASCP), CLS(NCA)
Jefferson State Community College
Birmingham, Alabama

Robert L. Klick, MS, MT(ASCP)
University at Buffalo
Buffalo, New York

Robin Krefetz, MEd, MT(ASCP), CLS(NCA), PBT(ASCP)
Community College of Philadelphia
Philadelphia, Pennsylvania

Pamela B. Lonergan, MS, MT(ASCP) SC
Norfolk State University
Norfolk, Virginia

Nadia Rajsz, MS, MT(ASCP)
Orange County Community College
Middletown, New York

Robert J. Sullivan, PhD, MT(ASCP)
Marist College
Poughkeepsie, New York

Debbie Wisenor, MA, CLS(NCA), MT(ASCP)
University of Louisiana at Monroe
Monroe, Louisiana

Patricia Wright, BSMT(ASCP)
Southeastern Community College
Whiteville, North Carolina

CONTRIBUTORS

Cheryl Haskins MSMT(ASCP) SC
Laboratory Alliance
Syracuse, New York

Linda E. Miller, PhD, MP(ASCP) SI
SUNY Upstate Medical University
Syracuse, New York

Jannie Woo, PhD
SUNY Upstate Medical University
Syracuse, New York

SECTION ONE
FUNDAMENTAL PRINCIPLES AND TECHNIQUES USED IN A CLINICAL CHEMISTRY LABORATORY

1

Laboratory Basics

■ OBJECTIVES—LEVEL I

Following successful completion of this chapter, the learner will be able to:

1. Identify two methods used to produce clinical laboratory-grade water for use in the clinical laboratory.
2. List three items that should be monitored during the water-purification process.
3. Identify four types of glassware available for laboratory use.
4. Identify four types of plastics used in laboratory plasticware.
5. Define the following terms: to contain (TC) and to deliver (TD) in reference to types of pipettes, molarity, molality, normality, thermocouple, percent solution, and hydrates.
6. Cite three types of balances used to weigh substances in the laboratory.
7. Complete the mathematical calculations presented in this chapter correctly.
8. Convert results from one unit format to another.
9. Calculate the volumes required to prepare a 1:2, 1:5, and 1:10 dilution.

■ OBJECTIVES—LEVEL II

Following successful completion of this chapter, the learner will be able to:

1. Explain the difference between air-displacement and positive-displacement micropipettes.
2. Distinguish swinging-bucket, fixed-angle-head, and ultra centrifuges from one another.
3. Explain the usefulness of liquid-in-glass, total-immersion, and partial-immersion thermometers.
4. Identify three alternative thermometers that do not contain mercury.
5. Distinguish density, specific gravity, and assay by weight from one another.
6. Identify a source of calibration material for balances, thermometers, and pipettes.

KEY TERMS

Atomic mass unit (amu)
Avogadro's number
Density
Gram molecular weight (GMW)
International unit (IU, U)

Molarity (M)
Normality (N)
Pascal
Relative centrifugal
   force (RCF)

Resistivity
Reverse osmosis
Revolution per minute
Thermocouple
Torr

A CASE IN POINT

The laboratory supervisor discovered the presence of a shift in quality-control values for serum creatinine in both levels of serum-based quality-control material. The quality-control data were consistently lower than previous results. The quality-control material is lyophilized (freeze-dried) and requires reconstitution with 5.0 mL of the diluent provided by the manufacturer. An investigation into the cause of this shift in quality-control results was initiated.

Issues and Questions to Consider

1. Identify several causes of a shift in quality-control data.
2. Indicate the type of pipette(s) that is suitable for reconstituting quality-control material.
3. Are the procedure and the techniques required to use pipettes the same for every pipette?

WHAT'S AHEAD

▶ Discussion of equipment commonly used in clinical chemistry laboratories.
▶ Identification of reliable resources for operating, calibrating, and documenting use of basic equipment in the laboratory.
▶ Identification of laboratory mathematics associated with several functions performed in the laboratory.

▶ INTRODUCTION

Laboratory testing requires technical skills and the ability to follow procedures. The technologist needs to be familiar with fundamental practices and the equipment used in laboratory testing so that he or she may successfully complete a testing procedure. These fundamental practices include selecting the proper types of pipettes, pipette use, centrifugation, laboratory calculations, temperature monitoring, and others that will be presented in this chapter.

▶ WATER

Water has numerous uses in the clinical laboratory, including the following:

• preparation of reagents,
• as a diluent for controls and calibrators,

• to flush and clean internal components of analyzers,
• to serve as a heating bath for cuvettes, and
• to wash and rinse laboratory glassware.

For many of these uses, the source of water must be of the highest purity, whereas the water required for rinsing general laboratory glassware may be less pure.

CLINICAL LABORATORY REAGENT WATER

Clinical and Laboratory Standard Institute (CLSI), formerly the National Committee for Clinical Laboratory Standard (NCCLS) has developed specifications for laboratory water quality. The criteria for *clinical laboratory reagent water* (CLRW) are outlined in CLSI, C3-A4 guidelines.[1] When selecting a water-purification system the purchaser must review the criteria established by CLSI so that all of the appropriate filters and components necessary to produce CLRW are included. Also, the source of the feedwater should be determined; it is usually laboratory tap water. The feedwater may contain unique contaminants, or may have a high mineral content (hardness), which will often require the inclusion of additional components in the water-processing system.

PURIFICATION

Many laboratories produce or purify their own water. There are several water-purification systems available to produce CLRW. Most water-filtration systems use a prefilter to begin the process. This prefilter has feedwater running through it to trap any particulates before the water is sent on to the next process, which is either distillation or reverse osmosis. Distillation is the process by which a liquid is vaporized and condensed to purify or concentrate a substance or separate volatile substances from less-volatile substances. Water that is only distilled does not meet the specific resistivity requirements for CLRW. To produce CLRW water using distillation, additional filters must be used—for example, ion exchange and carbon material. **Reverse osmosis** (RO) is a process by which water is forced through a semipermeable membrane that acts as a molecular filter. The RO filter removes 95–99% of organic compounds, bacteria, and other particulate matter and approximately 95% of all ionized and dissolved minerals, but not as many gaseous impurities. RO alone does not result in the production of CLRW, but—as with distillation—if additional filters are added to the system, CLRW water may be produced.

Ion-exchange filters remove ions to produce mineral-free deionized water. Deionization is accomplished by passing water through insoluble resin polymers that contain either anion- or cation-exchange resins. These exchange resins replace hydrogen ions (H^+) and hydroxyl (OH^-) ions for impurities present in ionized forms in the water. Another type of material used is a mixed bed resin that contains both anion and cation-exchange materials. Deionizers are capable of producing water that has specific resistance exceeding 1 to 10 Mohm·cm ($M\Omega \cdot cm$). Carbon filters containing activated charcoal may be added to the water-purification system to help remove several types of organic compounds that may be in the water.

A particulate filter may be added at the end of the system. This filter with a mean pore size of about 0.22 μm will serve to trap any remaining particulates as large as or larger than the pore size.

Monitoring Water Purity

Because water is such an integral part of laboratory analysis, its purity must be monitored on a consistent basis. The frequency of water testing depends on many factors, including the composition of the feedwater, availability of staff to perform the water testing, and the amount of water the laboratory uses during a given period of time. At the very least, resistivity and bacterial content of the water should be monitored on a regular basis. In addition, pH, silica content, and organic contaminants may be determined. Depending on the laboratory resources, some or all of these parameters should be checked on a periodic basis.

Most water-filtrations systems will have an in-line resistivity meter available. **Resistivity** is the electrical resistance in ohms measured between opposite faces of a 1.00-cm cube of an aqueous solution at a specified temperature. Resistivity measurements are used to assess the ionic content of purified water. An inverse relationship exists between the ion concentrations in water and the resistivity value. The higher the ion concentration in water, the lower the resistivity value will be. CLSI requires that CLRW water have a resistivity greater than 10 $M\Omega \cdot cm$.

Monitoring bacterial contamination can be accomplished quite easily. The water should be allowed to run for at least one minute to flush the system. Next, an aliquot of water is obtained and plated onto an appropriate growth media. After an appropriate incubation time, the number of colony-forming units on the agar plate is determined. Gram-negative rods are the most commonly found organisms in water after the purification process.

Water Use

Most water-purification systems are designed for easy access to end-product CLRW, so it is advisable to use only CLRW water for most laboratory procedures. When a procedure such as heavy metal testing or high-performance liquid chromatography (HPLC) requires the use of specially prepared water, then CLRW should not be used. The clinical laboratory should specify *special reagent water* (SRW) for that procedure. The criteria used to specify CLRW should be included in SRW specifications. These special applications may require different limits, so additional parameter may be added if necessary. For example, the water may have to be passed through more than one carbon filter or the pore size of the bacteria filter may have to be smaller.

► CHEMICALS

The chemicals used to prepare reagents for chemical testing exist in varying degrees of purity. Proper selection of chemicals is important so that the desired results may be attained. Chemicals acquired for reagent preparation are characterized by a grading system. The grading of any chemical is greatly influenced by its purity. The type and quantity of impurities are usually stated on the label affixed to the chemical container. Less-pure grades of chemicals include practical grade, technical grade, and commercial grade; all of them are unsuitable for use in most quantitative assays.

Most qualitative and quantitative procedures performed in the clinical laboratory require the use of chemicals that meet the specifications of the American Chemical Society. These chemicals are classified as either analytical grade or reagent grade. Examples of other designations of chemicals that meet high standards of purity include spectrograde, nanograde, and HPLC grade. These are often referred to as *ultrapure chemicals*.

Pharmaceutical chemicals are produced to meet the specifications defined in *The United States Pharmacopeia*, *The National Formulary*, and *The Food Chemical Index*. The specifications define impurity tolerances that are not injurious to health.

The International Union for Pure and Applied Chemistry (IUPAC) has developed standards and purity levels for certain chemicals. These include atomic weight standard (grade A), ultimate standard (grade B), primary standard (grade C), working standard (grade D), and secondary substances (grade E).

A very good source of highly purified chemicals, especially reference materials, is the National Institute of Standard and Testing (NIST) (Gaithersburg, MD). NIST defines its chemical and physical properties for each compound and provides a certificate documenting their measurements. NIST also provides standard reference materials (SRMs) in solid, liquid, or gaseous form. The solids may be crystalline, powder, or lyophilized.

CLSI and the College of American Pathologists (CAP) are two professional organizations that can provide laboratory staff with guidelines for proper chemical selection and reagent preparation.

► LABORATORY GLASSWARE AND PLASTICWARE

TYPES OF GLASSWARE

The most common type of glassware encountered in volume measurements is borosilicate glass. This glass is characterized by a high degree of thermal resistance; it has a low alkali content and is free of heavy metals. Commercial brands are known as Pyrex (Corning; Corning, NY) and Kimax (Kimble; Vineland, NJ). The caustic properties of concentrated alkaline solutions in borosilicate glass will etch or dissolve the glass and destroy the calibration. Borosilicate glassware with heavy walls—such as bottles, jars, and even larger beakers—should not be heated with a direct flame or hot plate. Glass should not be heated above its strain point; for example, the temperature for Pyrex is 515°C because rapid cooling strains the glass, which will crack easily when heated again. In the case of volumetric glassware, heating can destroy the calibration.

Corex* (Corning, NY) brand glassware is a special alumina-silicate glass strengthened chemically rather than thermally. Corex is six times stronger than borosilicate glass (e.g., Corex pipettes have a typical strength of 30,000 psi, compared with 2000 to 5000 psi for borosilicate pipettes) and will outlast conventional glassware tenfold. Corex also resists clouding and scratching better than other types of glassware.

Low actinic glassware is a glass of high thermal resistance with an amber or red color added as an integral part of the glass. The density of the red color is adjusted to permit adequate visibility of the contents yet give maximum protection to light-sensitive materials such as bilirubin standards. Low actinic glass is commonly used in containers for control material, calibrators, and reagents. A comparison between clear glass and low actinic glass is shown in Figure 1-1 ■.

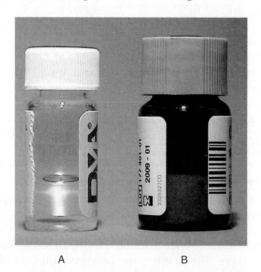

■ FIGURE 1-1 Types of glass bottle containers. A. Transparent glass; B. low actinic glass.

*Corning Glass no longer manufactures Corex glassware.

TYPES OF PLASTICWARE

Several types of plastics are used in clinical laboratories—for example, polypropylene, polyethylene, Teflon, polycarbonate, and polystyrene. Plastics are used for pipette tips, beakers, flasks, cylinders, and cuvettes.

Plastic pipette tips are made primarily of polypropylene. This type of plastic may be flexible or rigid, is chemically resistant, and can be autoclaved. These pipette tips are translucent and come in a variety of sizes. Polypropylene is also used in several tube designs, including specimen tubes and test tubes. Specially formulated polypropylene is used for cryogenic procedures and can withstand temperatures as low as −190°C.

Polyethylene is widely used in plasticware for test tubes, bottles, graduated tubes, stoppers, disposable transfer pipettes, volumetric pipettes, and test-tube racks. Polyethylene may bind or absorb proteins, dyes, stains, and picric acid, so care must be taken before selecting polyethylene.

Polycarbonate is used in tubes for centrifugation, graduated cylinders, and flasks. The usable temperature range is broad: −100 to +160°C. It is a very strong plastic but is not suitable for use with strong acids, bases, and oxidizing agents. Polycarbonate may be autoclaved but with limitations (refer to furnished instructions from the manufacturer).

Polystyrene is a rigid, clear type of plastic that should not be autoclaved. It is used in an assortment of tubes, including capped graduated tubes and test tubes. Polystyrene tubes will crack and splinter when crushed, thus care must be taken when handling damaged tubes. This type of plastic is not resistant to most hydrocarbons, ketones, and alcohols.

Teflon is widely used for manufacturing stirring bars, tubing, cryogenic vials, and bottle-cap liners. Teflon is almost chemically inert and is suitable for use at temperatures ranging from −270 to +255°C. This type of plastic is resistant to a wide range of chemical classes, including acids, bases, alcohols, and hydrocarbons.

VOLUMETRIC LABORATORYWARE

Pipettes

Many types of pipettes are available for use in a clinical laboratory, and each is intended to serve a specific function. Pipettes are used to reconstitute lyophilization controls and calibrators and prepare serum, plasma dilutions, and aliquoting specimens. Thus, a high degree of accuracy and precision is required. Manual pipettes fall into two general categories: transfer (volumetric) and measuring. Three subclassifications include *to contain* (TC), *to deliver* (TD), and *to deliver/blowout* (TD/blowout).

Class A Designation

Class A glassware, including pipettes, is manufactured and calibrated to deliver the most accurate volume of liquid. Class A specifications are defined by NIST. CAP specifies that volumetric pipettes must be of certified accuracy (class A) or the volumes of the pipettes must be verified by calibration techniques—for example, gravimetric or photometric. The letter A appears on all pipettes that conform to the standards of class A glassware. Volumetric glassware designated as class A has been manufactured to class A tolerances as established by American Standards and Testing Materials (ASTM) E 694 (West Conshohocken, PA) for volumetric ware. Other standards related to volumetric glassware include ASTM E542 for calibration of volumetric ware and ASTM E288 for volumetric flasks.[2-4]

Types of Pipets Pipettes designed to contain are often referred to as *rinse-out* pipettes because they must be refilled or rinsed out with the appropriate solvent after the initial liquid has been drained from the pipette. TC pipettes contain an exact amount of liquid that must be completely transferred for accurate measurement. Examples of TC pipettes are Sahli hemoglobin and Lang-Levy pipettes. These pipettes do not meet class A certification criteria.

TD pipettes are designed to drain by gravity pipettes and must be held vertically with the tip placed against the side of the container but without touching the liquid. The stated volume is obtained when draining stops. This type of pipette should not be blown out. Examples of TD pipettes include Mohr, serological, and volumetric transfer pipettes. These pipettes are designed to meet the requirements of class A type pipettes.

Volumetric TD pipettes have an open-ended bulb that holds the bulk of the liquid. On one side of the pipette is a long glass tube with a line indicating the extent to which the pipette is to be filled. The other end is tapered for smooth delivery of liquid. These pipettes should be allowed to drain freely without being shaken or hit against the container. Any disruption of the free-flowing liquid may result in an inaccurate delivery of the liquid.

Some TD pipettes are designed so that most of the contents are allowed to drain freely, after which the remaining fluid in the tip is blown out. These pipettes are not rinsed out. Examples of pipettes designed to be blown-out–type pipettes include Ostwald-Folin and serological pipettes. A TD/blowout pipette is identified by the presence of one or two frosted bands near the mouthpiece of the pipette. Never pipette or blow out solutions by mouth! The user must perform the procedure using an appropriate pipetting aid such as a safety bulb.

Serologic glass or plastic pipettes are long tubes with uniform diameters. They have volume graduations extending to the delivery tip of the pipette. The last volume of liquid blown out is included in the delivery volume. The design of the Mohr TD pipette is different from the serological. Mohr pipettes are not graduated to the tip. The accuracy of the Mohr pipettes is valid only when the pipette is filled. If smaller volumes are dispensed, the accuracy decreases proportionally. Several examples of glass pipettes used in clinical laboratory are presented in Figure 1-2 ■.

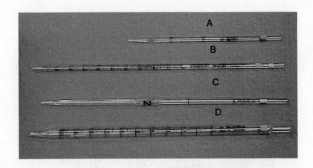

■ FIGURE 1-2 Several examples of glass pipettes. A. 0.2 mL TC; B. 1.0 mL TD serologic (blowout); C. 2.0 mL TD volumetric; and D. 10.0 mL TD Mohr. Note the two frosted- or etched-glass rings on pipette B

✓ **Checkpoint! 1–1**

Identify which type of glass pipette would be the best to use to reconstitute lyophilized, serum–based, quality-control material.

Micropipettes Two examples of commonly used micropipettes are air-displacement and positive-displacement micropipettes. These pipettes are capable of delivering liquid volumes from 1–1000 μL. Some micropipettes are designed to deliver a fixed volume, while others can deliver variable amounts of liquid.

An air-displacement micropipette uses a piston device to facilitate aspiration and ejection of liquids. A disposable, one-time-use polypropylene tip is attached to the pipette barrel. The pipette tip is placed into the liquid to be aspirated and drawn into and dispensed from this tip. A positive-displacement micropipette uses a capillary tip made of glass or plastic to transfer liquids. A Teflon-tipped plunger fits tightly inside the capillary. Liquid solutions are drawn up the capillary and pushed out with a squeegee effect, thus limiting the amount of carryover. These capillary tips are reusable and suited for rinsing out solutions. Some procedures require a washing or flushing step between samples.

Several examples of micropipettes commonly used in clinical laboratory are shown in Figure 1-3 ■. These micropipettes may be fixed volume or adjustable to specific volumes.

Pipette Calibration Monitoring the performance of pipetting devices is not only mandatory in laboratories licensed by their respective states but also very wise to do. Micropipettes should be verified for accuracy and precision before they are put into use and monitored during the course of the year. The frequency of verification depends in part on how extensively they are used and requirements by the licensing or accrediting agency. Proper maintenance of air-displacement pipettes is very important. This type of pipette has a fixed stroke length that must be maintained. These pipettes also have seals to prevent air from leaking into the pipette when the piston is moved. These seals require periodic

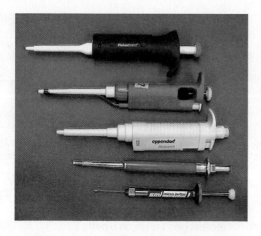

■ FIGURE 1-3   Several examples of micropipettes used in clinical laboratories.

lubrication to maintain their integrity. Positive-displacement micropipettes need to have their spring checked and the Teflon tip replaced periodically. A slide wire is used to quickly check the plunger setting. This check does not replace the scheduled precision and accuracy checks.

Several procedures are used by laboratories to verify precision and accuracy of micropipettes. Most of these procedures are time consuming, especially the procedures that require the weighting of water. It may take several hours to properly calibrate pipettes because of multiple weightings and monitoring environmental factors such as temperature, humidity, and atmospheric pressure. No matter; this verification procedure must be done to ensure proper performance of laboratory micropipettes.

CLSI has provided a gravimetric procedure that is acceptable for determining pipette accuracy and precision.[5] This gravimetric procedure is labor intensive but does provide a low-cost means of complying with the regulations set forth by the various accrediting agencies.

More expensive procedures for calibrating micropipettes include:

• commercial photometric pipette-calibration products,
• calibration-services providers, and
• Pipette Tracker (Labtronics Inc., Canada).

One major concern when considering the cost attributed to pipette-calibration procedures is technologist time. The technologist time required for the photometric procedures is often 50–60% less than the inexpensive manual-weighing techniques.

Volumetric Flasks

Volumetric flasks are a special type of glassware in the laboratory. These flasks are often used to prepare standards for quantitative procedures, so their accuracy is critical. Volumetric flasks used to prepare standards and other solutions require optimal accuracy and must meet class A specifications as defined by NIST. These specifications are imprinted on the flasks. Volumetric flasks are used to contain an exact volume when the flask is filled to the *mark*. A Teflon or ground glass stopper should be used to seal the flask. Volumetric flasks should not be used to store reagents.

Calibration of Volumetric Glassware

According to the strictest of standards, every piece of volumetric glassware in the clinical laboratory should be coded, and a record should be maintained of its calibration. Any piece of glassware that does not meet class A tolerance should be rejected. To prepare a piece of glassware for calibration, it should be thoroughly washed and dried using appropriate cleaning procedures. CLSI can provide the laboratory with an appropriate procedure to calibrate volumetric flasks.[5]

▶ MEASUREMENT OF MASS

Measuring mass is a fundamental process in the preparation of standards, reagents, gravimetric analysis, and the calibration of volumetric equipment, and it requires the use of an analytical balance. *Mass* is the quantity of matter contained, and masses are compared by their inertias; equal masses acquire equal velocities when acted on by equal forces for the same length of time. The weight of a body is the gravitational force exerted on it; unlike mass, weight varies with geographical position. *Weight* is a function of mass under the influence of gravity and is equal to mass multiplied by gravity. In the laboratory, the weight of an object is determined. To obtain the object's mass, we would have to divide the weight by the local acceleration resulting from gravity.

TYPES OF BALANCES

Several different types of balances are available, depending on what needs to be weighed. For example, to weigh a fecal fat specimen, an appropriate balance would be a top-loading precision balance capable of accurately weighing kilogram amounts. If a standard solution needs to be prepared for a toxicology assay that requires microgram quantities, then a single-pan microbalance should be used.

Unequal-Arm Substitution Balances

Unequal-arm substitution balances are typically a single-pan design and are commonly used in laboratories, though electronic balances are replacing almost all of these types of balances. This single-pan, mechanical, unequal-arm balance operates on the principal of removing weights rather than adding them. A fixed-mass counterweight is used to balance the combined mass of the pan and the removable weights across two arms of unequal weight. When a sample is placed on the weighting pan, the operator turning a set of knobs moves the internal weights in 1-g or 10-g increments one at a time. This is continued until the system returns to equilibrium, at which time the sum of the weights removed is equal to the weight of the object.

Magnetic Force-Restoration Balance

Another commonly used balance is the single-pan balance that relies on magnetic force restoration. *Restoring force* is the force required to put the balance back into equilibrium. The unknown mass is placed on the pan, and this system goes out of equilibrium. The operator adjusts the internal weights and restores partial equilibrium. A null-detector optics circuit senses when equilibrium is near and provides a signal to the sensor motor to generate a restoring current until equilibrium is reached. At this time, the unknown mass is equal to the mass of the weights removed plus the value of the restoring current.

Top-Loading Balances

Single-pan, top-loading balances operate on the same principle as single-pan analytical balances (i.e., weighing by substitution). Damping or the release of the pan is accomplished by magnetic rather than air release. These balances are especially suitable for quickly weighing larger masses (as much as 10,000 *g*) that do not require as much analytical accuracy, such as the preparation of large volumes of reagent.

Electronic Balances

There are several electronic balance designs. One design uses a strain-gauge load cell. This small, thin device changes electrical resistance when it is stretched or compressed. Typically, several strain gauges are used in a Wheatstone bridge arrangement and are glued onto the load cell in a protected location. A load cell is usually in the shape of a beam or a plate. When the beam or plate is displaced, it bends a tiny amount; this tiny bending is detected by the strain gauges. The amount of bending might be only a thousandth of an inch, but that is enough for a strain gauge to measure.

Another electronic balance design operates on the principle of electromagnetic force compensation. A coil placed between the poles of a cylindrical electromagnet is mechanically connected to a weighing pan. Mass placed on the pan produces a force that displaces the coil within the magnetic field. A regulator generates a compensation current just sufficient to return the coil to its original position. The more mass placed on the pan, the larger the deflecting force, and the stronger the current required to correct the deflection of the coil. The measuring principle is based on a strict linear relationship between compensation current and force produced by the load placed on the pan.

Several additional features may be available on some models of electronic balances. For example, some electronic balances include an electronic vibration damper. Any excess vibration can be detected when any variation of the pointer or oscillation of the number in the last decimal place of the digital display is observed. Another feature available in some models is built-in *taring*. This allows the weight of the weighing container to be automatically subtracted from the total weight of the sample. Also electronic balances can be interfaced with computers to provide calculations such as weight averaging and statistical analysis of multiple weighing. The fundamental design of electronic balances allows for faster weighing, which is advantageous when doing multiple readings—for example, during a pipette calibration.

Calibration of Balances

Laboratory balances require calibration at regular intervals. There is no fixed calibration interval for scientific applications, according to NIST. Calibration intervals should coincide with the requirements of the laboratory's licensing and accrediting organizations.

The mass-standard and test-weight-accuracy classes for weights used in calibrating balances have been updated and replace the older requirements specified by National Bureau of Standards classes S and S1 weights. The new mass-standards and test-weight-accuracy classes appropriate for laboratory balances include ASTM classes 1 and 2. Refer to ASTM E 617-97 for specific information regarding range, readability, and best uncertainly applicable to these classes.[6]

NIST class 1 weights (extra-fine accuracy) are available up to 250 mg and may be used for high-precision balances such as a single-pan and electronic balances that are precise to four decimal places. The range of weight for class 2 balances may be in excess of 1000 *g*. Meticulous care must be used when handling class 1 or 2 weights. The operator must avoid direct contact with the weights by using clean gloves or special lifting tools (for example, forceps). Hand contact with the weights can cause corrosion. The weights should not be dragged across any surface, including the stainless steel weighting pan. Usually the weights are sent in a specially designed covered box and should always be stored in that box.

ASTM provides calibration weights that range from a few milligrams to larger weights (for example, 10 g). An example of a standard weight set is provided in Figure 1-4 ■.

Several factors may affect the performance of a laboratory balance, including temperature, air drafts, floor vibrations, table instability, and static electricity. Minimizing the effects

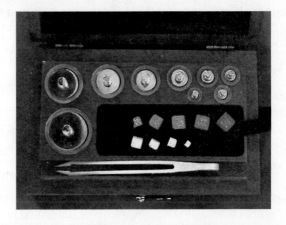

■ FIGURE 1-4 ASTM standard weight set used to calibrate laboratory balances.

of these factors on your weighing procedures can often be done quite easily. For example, if there are air drafts in the room a shroud or enclosure can be placed around the balance. A marble table can be used to reduce table vibrations or instability.

Balance Specifications

The laboratory staff should be familiar with the specifications attributable to the respective balances. Examples of several important balance specifications that an operator should be knowledgeable about include:

- *capacity*, or the maximum load one can weigh;
- *accuracy*, which is dependent on the smallest mass that will be weighed;
- *linearity*, or the ability of a balance to provide accurate output over its full range; and
- *resolution (readability)*, the smallest increment of weight that may be discernible.

Laboratory-accrediting agencies require that the accuracy of balances be verified at various time intervals. Consult your accrediting agency for specific information regarding your equipment.

▶ CENTRIFUGES

Centrifuges serve an important role in preparing specimens for analysis. Improper centrifugation of specimens often leads to erroneous data. Assays, especially immunoassays, have very low detectable levels for analytes in biological fluids and are prone to error caused by the presence of small fibrin clots and cells.

The main components of a centrifuge are the motor, drive shaft, and rotor assembly as illustrated in Figure 1-5 ■. An electromagnetic drive motor is used to provide the speeds required to separate particulates from samples. These motors use carbon brushes to facilitate creation of electromagnetic fields that ultimately makes the drive shaft turn; this in turn spins the rotor assembly. The buckets hold the tubes containing the samples.

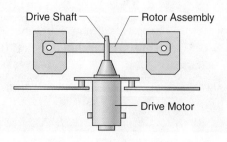

■ FIGURE 1-5   Schematic showing the three major components of a typical laboratory centrifuge.

RELATIVE CENTRIFUGAL FORCE AND REVOLUTIONS PER MINUTE

Relative centrifugal force (RCF) is defined as the weight of a particle in a centrifuge relative to its normal weight. It is the appropriate term for describing the force required to separate two phases being separated by centrifugation. Relative centrifugal force is expressed as some number times gravity (*g*). RCF is calculated using equation 1.1.

$$RCF = (1.118 \times 10^{-5})\ (r)\ (rpm)^2 \qquad (1.1)$$

where

1.118×10^{-5} = empirical factor;

r = radius in centimeters from the center of rotation to the bottom of the tube in the rotor cavity or bucket during centrifugation; and

rpm^2 = total number of revolutions per minute squared.

Example

Calculate the RCF of a centrifuge whose *r* is 10 cm from the center of rotation when the centrifuge is operated at a speed of 3000 rpm. (Answer = ~1000 × *g*)

Converting **revolutions per minute** to RCF can also be derived from a *nomogram* that is usually included in the manufacturer's manual or found in many clinical chemistry textbooks. A nomogram is defined as a representation by graphs, diagrams, or charts of the relationship between numerical variables.

RPM is a unit for expressing the number of complete rotations of a rotor occurring per minute. It is a measure of speed. Laboratory centrifuges are equipped with a speed dial that allows the user to set the RPMs. Therefore, if a procedure requires the samples to be centrifuged at 1000 × *g*, simply use the formula shown above or consult a clinical chemistry reference or textbook for a nomogram to make the conversion.

Several types of centrifuges are available to process specimens, separate low-density particles for analysis, and clear specimens of potential interfering compounds (lipids, for example). Examples of the types of centrifuges include:

- swinging-bucket rotor (also swing-out rotor with buckets),
- fixed-angle rotor,
- air-driven ultracentrifuge,
- ultracentrifuge, and
- refrigerated.

TYPES OF CENTRIFUGES

The swinging-bucket centrifuge shown in Figure 1-6 A ■ is routinely used to separate cells from serum or plasma. Both plain red-top tubes without serum-separator gel and serum-separator tubes can be centrifuged. The required relative centrifuge force is 1000 − 1200 × *g*, and centrifuge times are between 5 and 10 minutes. The swinging-bucket design allows

TOP VIEW

BOTTOM VIEW

SIDE VIEW

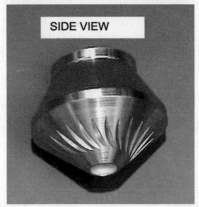

B

■ FIGURE 1-6 Examples of two types of laboratory centrifuges.
A. Swing-out rotor with buckets and B. fixed-angle rotor.

the tubes to assume a horizontal position when the centrifuge is at maximum *g* force. During centrifugation, particulates (cells, for example) constantly move along the tube while it is in the horizontal position. This movement of particles spreads the sediment uniformly against the bottom of the tube. When the centrifuge rotor comes to a complete stop, the surface of the sediment is flat with a column of serum or plasma above it.

Fixed-angle rotors allow tubes to be centrifuged at angles ranging from 25° to 52°, depending on the design. This rotor is shown in the top view in Figure 1-6 B ■. The sample holder is placed in one of the six positions in the rotor. During centrifugation, particles move along the side of the tube to form sediment that packs against the side and bottom of the tube. The surface of the sediment is parallel to the centrifuge shaft. Fixed-angle rotors are aerodynamically designed to yield much faster rotational speeds or greater *g* forces than swinging-bucket rotors. A microcentrifuge used to prepare pellets of DNA and RNA is equipped with a 25° or 45° fixed-angle rotor that can achieve RCFs of $18,000 \times g$ (14,000 RPMs). Another type of fixed-angle centrifuge that is widely used in the laboratory allows for a quick 2-minute spin at nearly $4400 \times g$ (8,500 RPM) for 7.0-mL blood tubes. This type of centrifuge is being used for preparation of *stat* samples. The purpose of a 2-minute spin procedures is to reduce turnaround time for certain critical laboratory tests.

Most centrifuges use an electromagnetic motor to rotate the rotors. One exception to this design uses air to spin the rotor. This centrifuge functions by directing compressed air onto grooves that are etched into the outer surface of the fixed-angle rotor. The rotor begins to move as the air blows across the grooves, which are shown in the bottom and side views of Figure 1-6 B. The maximum RCF is about $178,000 \times g$. This type of centrifuge is often used to "clear" or remove lipid particles from lipemic specimens.

Ultracentrifuges are much larger than regular laboratory centrifuges. They are often floor-model types and generate very high centrifugal forces—for example, $800,000 \times g$ (100,000 RPM). These high-speed centrifuges are used to fractionate lipoproteins, perform drug-binding assays, and prepare tissue for hormone-receptor assays.

A refrigerated centrifuge is used routinely in laboratories for separations requiring colder temperatures. Temperature ranges from –15° to 25°C are achievable with this type of centrifuge. Specimens for lactic-acid and plasma-ammonia determinations require the use of refrigerated centrifuges. Refrigerated centrifuges come equipped with either swinging buckets or fixed-angle rotors.

MAINTENANCE AND CALIBRATION

Routine maintenance of centrifuges includes cleaning the interior and exterior surfaces, the rotor, and the buckets with an appropriate disinfectant such as a 10% bleach solution. Any debris inside the centrifuge (for example, broken glass and stoppers) should be carefully removed.

Centrifuge timers and speeds should be checked periodically for proper function. The timers can be checked against an accurate timepiece. Centrifugation speeds can be checked using a strobe tachometer. If the results of either of these checks are outside tolerance, then the centrifuge requires service. Many centrifuges in use today have carbon brushes located within the housing of the motor. When these brushes wear down, the centrifuge can no longer maintain speed and will eventually fail to start. An appropriately trained staff technologist can replace these carbon brushes.

Refrigerated centrifuges must have temperature checks performed periodically. A NIST-certified thermometer should be used to verify the temperature of the refrigeration unit. If the temperature check falls outside acceptable tolerance, then the unit should be serviced.

▶ WATER BATHS

Water baths are routinely used to incubate or warm solutions for a specified period of time. For general clinical laboratory use, water baths must offer variable temperature control from +5°C above ambient temperature to 100°C, with accurate control to ±0.2°C. Water baths can be circulating or noncirculating. Circulating water baths provide the best temperature control. Another important consideration when selecting a constant-temperature bath is that the model be large enough to accommodate the desired working volume.

MAINTENANCE AND QUALITY CONTROL

Maintenance of a water bath is improved by filling it with clean, reagent-grade water. This prevents excess mineral deposits from building up if regular tap water is used. The accumulation of mineral deposits can affect the temperature-sensing element and generally leads to poor heat transfer. However, if an accumulation of these minerals does occur, then a weak hydrochloric acid solution will dissolve the deposits. Frequent cleaning and the use of freshwater will help prevent the overgrowth of bacteria and algae. Also, a 1:1000 dilution of thimerosal (Merthiolate) can be added to help prevent bacterial growth. Overheating and subsequent damage can occur if the water in the bath is allowed to completely evaporate. At higher temperatures, the bath should be covered, both to maintain proper temperature control and to prevent rapid evaporation of water into the atmosphere.

A thermometer calibrated against a certified NIST thermometer must be a component of any water bath. The temperature should be noted and recorded each day. Monitoring thermometer temperature is a significant function and directly affects the performance of the assay.

▶ HEATING BLOCKS, DRY-BATH INCUBATORS, AND OVENS

Heating blocks and dry-bath incubators are commonly used for incubating liquids at higher temperatures. Most incubators are constructed of an aluminum alloy that is capable of uniformly distributing heat. Their heating efficiency is less than a circulating water bath but will maintain a constant temperature within ±0.5°C. A certified thermometer or NIST-calibrated thermometer must be present in the heating block to monitor the temperature.

Heating ovens are used in chromatography procedures to dry chemicals, assist in organic extractions, and dry membranes or gels in electrophoresis. Several different designs are available, depending on the desired temperature and purpose. Oven designs include programmable, vacuum, and standard laboratory types. Temperature control is usually within ±1°C. The oven must have a certified thermometer or NIST-calibrated thermometer available to monitor the interior temperature.

▶ MIXING

Mixing is an operation intended to form a homogeneous mass or create a uniform homogeneous system. Mixing is used to bring solids into solution, to bring phases into intimate contact (for instance, in extraction procedures), to wash suspended solids, and to homogenize liquid phases, among many other operations. Improper mixing can result in the following problems.

• Invalid data may result if proteins settle out during long-term frozen storage of serum controls.

• Excessive mixing may cause denaturation of protein or hemolysis.

• A phase separation occurs when serum (or plasma) specimens stand for a period of time; they must be thoroughly mixed before analysis. The concentration of even small molecules in such a system will be heterogeneous as proteins settle and become more concentrated, decreasing the effective water concentration in this layer. This produces a water concentration gradient throughout the system and consequently a concentration gradient of all components.

SINGLE-TUBE MIXERS

A vortex mixer is an example of a single-tube mixer and is capable of variable-speed oscillations that result in a swirling motion of the liquid contents in a test tube or other container. The angle of contact and degree of pressure can be regulated for optimal mixing action. A very effective mixing action is created by using a multiple-touch sequence (i.e., touching and withdrawing the tube from the mixer's

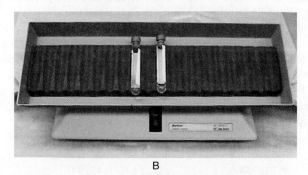

■ FIGURE 1-7   Two types of tube mixers. A. vortex mixing;
B. rocking motion of single and multiple tubes.

neoprene oscillating cup). The operator must be careful not
to completely fill the container or mix the liquid contents
too quickly because spillage can occur. To prevent spillage,
cover the tube with parafilm or a plastic cap.

MULTIPLE-TUBE MIXERS

Several types of mixers are available that can accommodate a
number of tubes, tube sizes, and provide different types of
mixing motions. For example, a Thermolyne Maxi-Mix
(Sybron Corporation, Dubuque, IA) (shown in Figure 1-7 A ■)
can conveniently be used for vortex mixing of one or several
tubes at one time. Changing the pressure applied to the tube
against the foam-rubber top alters the mixing action.

Rotary mixers use a circular motion on a tilted disk to provide continuous inversion of contents in tubes, which are
clip-mounted at the circumference of the rotating disk. Rotational speed can be varied to provide gentle or more vigorous
mixing. Control sera are conveniently reconstituted on this
type of mixer.

A commonly used multiple-tube mixer similar to that
shown in Figure 1-7 B ■ is suitable for specimen tubes and
uses a uniform, gentle rocking motion to maintain a homogeneous mixture of tube contents. This rocker-type mixer operates by tilting back and forth at variable speeds to provide
thorough mixing of whole blood samples.

▶ THERMOMETRY

Thermometers and other types of temperature-sensing
devices are used in the laboratory to monitor temperatures
in refrigerators, freezers, water baths, heating blocks, and
incubators. Special applications of thermometry include osmometry, refrigerated centrifuges, refrigerated reagent compartments of automated analyzers, warming compartments
of automated analyzers, and circulating waters baths for
cuvette compartments in automated analyzers. All of these
temperature-monitoring applications have the same requirements, including accurate measurements and maintaining a
constant temperature.

Appropriate quality-control procedures must be carried
out and documented routinely for all of these temperature-
monitoring devices. Any temperature-sensitive device that
fails to perform within established tolerances must be replaced. Because many assays performed in the laboratory use
enzymes, even the slightest deviation from the optimal temperature required to perform the assay may result in an erroneous result.

TYPES OF THERMOMETERS

The two types of liquid-in-glass thermometers most widely
used are total immersion and partial immersion. A total-
immersion thermometer requires that the bulb and entire
column of liquid be immersed into the medium measured.
These thermometers are used to monitor freezers and refrigerators. Partial-immersion thermometers must have the bulb
and stem immersed to the immersion line or to the defined
depth on the thermometer. This type of thermometer is often
used for water baths and heating blocks. Examples of these
thermometers are presented in Figure 1-8 A, B, and C ■.

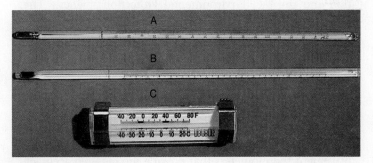

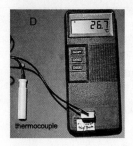

■ FIGURE 1-8   Four examples of commonly used temperature monitoring
devices. A. Mercury-filled partial-
immersion thermometer;
B. red-spirit–filled partial-immersion
thermometer; C. red-spirit–filled,
total-immersion refrigerator thermometer;
and D. digital type with thermocouple.

SPECIAL APPLICATIONS OF TEMPERATURE-SENSING DEVICES

Thermistors are used in several types of instruments found in the laboratory, including freezing-point–depression osmometers. A thermistor is a transducer that converts changes in temperature (heat) to resistance. It consists of a small bead constructed of a fused mixture of metal oxides that is attached to two leads and encapsulated in glass. The metal-oxide mixture has a large negative-temperature coefficient of resistance. Thus, a small decrease in temperature causes a relatively large increase in the resistance of the thermistor.

A **thermocouple** similar to the one shown in Figure 1-8D ■ is a sensor that consists of two dissimilar metals joined together at one end. When the junction of the two metals is heated or cooled, a voltage is produced that is calibrated to a temperature. Thermocouples come in several designs, including beaded wire, probes, and surface probes. An important feature of most thermocouples used in laboratory analyzers is their fast response times. The response time of a thermocouple is defined as the time required by a sensor to reach 63.2% of a step change in temperature under a specified set of conditions. Five time constants (the time required for a sensor to reach 63.2% of a step change) are required for the sensor to approach 100% of the step-change value. In the laboratory, thermocouples are used in gas and liquid chromatography, heating compartments in automated analyzers, thermo cuvettes, and circulating water baths used in automated chemistry analyzers.

MERCURY-FREE LABORATORIES

Three national initiatives now seek to remove all mercury from laboratories. For example, a June 1998 landmark agreement was reached between the American Hospital Association and the U.S. Environmental Protection Agency.[7] The memo of understanding between the two organizations seeks to decrease and eventually eliminate hospital pollution practices over a 5- to 10-year period. One goal is to eliminate mercury waste.

Mercury is contained in chemical reagents used by the laboratory as well as in mercury thermometers. The cost associated with the proper disposal of mercury and the environmental effects of mercury make replacing mercury thermometers a sound idea. Several alternatives exist for replacement thermometers that provide the required accuracy for laboratories procedures. They include:

- thermometers containing an organic red spirit that are pressurized with nitrogen gas,
- thermometers containing blue biodegradable liquid (isoamyl benzoate and dye),
- a red-liquid thermometer filled with kerosene,
- bimetal digital thermometers, and
- digital thermometers with stainless steel stems.

THERMOMETER CALIBRATION

Monitoring the accuracy of thermometers is necessary to ensure the reliability of procedures requiring temperature regulation. Thermometers may be purchased with a certificate to indicate traceability to standards provided by NIST. Also, many commercially available thermometers meet or exceed the tolerances of accuracy of NIST, the American National Standards Institute (New York, NY), and the Scientific Apparatus Makers Association (United Kingdom).

Noncertified thermometers can be calibrated by using a NIST SRM 934 thermometer or a NIST SRM 1968 gallium melting-point cell. The SRM 934 clinical laboratory thermometer is calibrated per specifications by the International Temperature Scale of 1990 (ITS-90) at 0°, 25°, 30°, and 37°C. A gallium melting-point cell consists of about 25 g of very pure (99.99999%) gallium metal that has a single, fixed melting point at 29.7646°C (as defined by ITS-90). The gallium is sealed in an inert plastic crucible and surrounded in a stainless steel envelope.

Temperature-monitoring devices should be verified for accuracy at 6 or 12-month intervals. Guidelines and procedures for proper monitoring and tolerances are available.[8]

▶ LABORATORY MATHEMATICS

SYSTEMS OF MEASUREMENT

Most of the U.S. scientific community switched to the metric system years ago, but some areas have not. The U.S. customary system of measures for length, area, liquid or dry volume, and mass can be found in Appendix A. Quite often we are faced with the task of converting these units to metric units during the course of everyday life and in the workplace. Medications, for example, are often prescribed in metric volume units such as milliliters (mL), but the utensils used to prepare the doses use U.S. units (e.g., teaspoon). Conversions of units for some commonly used measures are found in Appendix B.

Metric

Although metrication across the United States has not been entirely successful, it does exist within the scientific community. This system is based on fixed standard and on a uniform scale of 10. There are three basic units of measurements for length, mass, and volume. The basic units are as follows:

- length = meter
- mass   = gram
- volume = liter

A meter is defined as the distance traveled by light in vacuum during a time interval of 1/299,792,458 second. The gram is 1/1000 the quantity of matter in the international prototype kilogram. A kilogram (the SI* unit of mass), or

*SI = International System of Units (from Le Système International d' Unitès).

1000 grams, is defined as the mass of water contained by a cube whose sides are 1/10th the length of a meter, or 1 decimeter in length. The liter is defined as the volume of liquid contained within that same cube. The liter is also used as a name for exactly 1 cubic decimeter, 1000 cubic centimeters, or 0.001 cubic meters.

An additional measurement used within the clinical laboratories is area. This is a derivation of the measurement of length. The area of a surface is determined by multiplying the length times the width of a surface (for example, the area of a floor). Thus, area is measured in squared units. Common laboratory metric area measures are mm^2, cm^2, and m^2.

Prefixes are often used to express units of measurement. For example, the blood concentration of glucose for a sample may be reported in units of milligrams per deciliter (mg/dL). The prefix *milli* (m) is 1000th of a liter, and *centi* (c) 100th of a liter. A list of several prefixes and their abbreviations are shown below in Appendix C.

It is often necessary to convert units among different measurements within the metric system (for example, converting milliliters to liters). A common occurrence in the laboratory is the need to convert analytes from one unit to another. An example would be to convert milligrams per deciliter (mg/dL) to micrograms per milliliter (µg/mL). Usually, a simple ratio and proportion calculation will suffice, but knowledge of the prefixes and their corresponding multipliers is essential. Several examples of conversion among different measurements within the metric system are listed in Appendix D.

Le Système International d' Unitès

In 1960, the 11th General Conference of Weights and Measures adopted the name International System of Units (from *Le Système International d' Unitès*) (abbreviated SI) and all of the three classes of SI units: base, derived, and supplementary units.[9] The base units include seven fundamental but dimensionally independent physical quantities; these are listed in Appendix E. The derived units shown in Appendix F include units that can be formed by combining base units according to algebraic relationships among the corresponding quantities. Supplementary units—for example, the radian—conform to SI but have not been classified as either a base or a derived unit.

Conversion Between the Metric System and SI

The use of SI units is found primarily in the chemistry laboratory. SI units are used throughout the literature but have not been completely incorporated into the clinical laboratories. Two common applications of the SI units are in reporting analyte concentrations such as glucose in millimole per liter (mmol/L) and enzyme measurements in terms of mass rather than activity.

 ## Checkpoint! 1–2

What is the value in SI units for a blood glucose of 100 mg/dL?

Temperature The Celsius temperature scale (also called the *centigrade* scale) is widely used in clinical laboratories in the United States. On this scale, the interval between the freezing and boiling points of water is divided into 100 equal divisions. Water freezes at a temperature of 0°C, and boils at 100°C. Room temperature is around 20° to 25°C. Normal body temperature is approximately 37°C. Temperature measurements using the Fahrenheit temperature scale are also used. The interval between freezing and boiling of water is divided into 180 equal divisions. Thus, water freezes at 32°F and boils at 212°F. Room temperature is about 68°F, and normal body temperature is approximately 98.6°F.

A third temperature unit is Kelvin. The designation *absolute zero* is taken to be the zero point of the Kelvin temperature scale. Temperatures below 0°C and 0°F can be expressed in terms of Kelvin. Kelvin represents the theoretically lowest possible temperature and is equivalent to –273°C or –459°F. Therefore, no negative Kelvin temperatures are possible because, theoretically, no temperature is possible below the zero on this scale. Neither the term *degree* nor is the degree symbol are used with the Kelvin scale. Thus, 100K is 100 Kelvin, not 100 degrees Kelvin.

Equations 1.2–1.4 represent formulas that can be used to convert between temperature units.

$$T_C = (T_F - 32)/1.8 \tag{1.2}$$
$$T_F = 1.8\,T_C + 32 \tag{1.3}$$
$$T_K = T_C + 273 \tag{1.4}$$

 ## Checkpoint! 1–3

Convert the following:

1. *98°F to °C*
2. *10° C to °F*
3. *1°C to Kelvin*

Pressure Blood-gas measurements are routinely performed in the clinical laboratories and include PO_2 and PCO_2. The units associated with blood-gas measurements include mmHg and Pascal (Pa). Two additional units often used for blood-gas results are atmosphere (atm) and torr. An *atmosphere* is defined as the pressure exerted by a column of mercury 760 mm high with the mercury at 0°C. A **torr** is the pressure exerted by 1 mm of mercury at certain specified conditions, notably 0°C. A **Pascal** is the MKS unit (SI) of pressure and is equal to 1 newton meter squared ($1\ N/m^2$), and 1 Pa is equal to $1\ m^{-1}\ kg\ s^{-2}$. The following equalities relate each of these terms and can be used for conversions among them:

- 1 atm = 760 mmHg
- 1 torr = 1/760 atm (1.3158×10^{-3} atm)
- 760 mmHg = 760 torr
- 1 torr = 1 mmHg
- 760 mmHg = 101,325 Pa
- 1 mmHg = 0.133 kPa

Checkpoint! 1–4

The partial pressure of oxygen (PO_2) reported in a European journal of medicine article is 13 kPa. What is the PO_2 in mmHg?

Conversions

On occasion, a laboratory staff may be asked to convert one set of units to another. There are several approaches to these types of problems. Some individuals may remember the appropriate conversion factor to use while others may resort to clinical chemistry textbooks for the correct factor. Most calculators have programs and functions that allow for quick solutions to conversion problems. The Internet has sites with calculators available to perform the conversion. Three examples of different types of conversion are shown below:

1. Converting from different but comparable values.

   Milligrams per milliliters to milligrams per deciliter

2. Converting from one system to another.

   Gallons to liters

3. Interconversion between units.

   milligrams/deciliter to milliequivalents/liters

 Checkpoint! 1–5

1. *Convert 100 milligrams/deciliter to milligrams/liter.*
2. *Convert 2.5 quarts to liters.*
3. *Convert 140 milliequivalents of sodium/liter to milligrams of sodium/deciliter.*

Ratios

A ratio represents the relationship of one value to another. Ratios are widely used throughout the clinical laboratory. A 1:1 ratio represents an equal relationship, whereas a 1:2 ratio means that one value is twice that of the other. Some examples of ratios used in laboratories are shown below:

The serum to saline ratio is 1:9.

The saline to serum ratio is 9:1.

The serum to total volume ratio is 1:10.

Note that a colon is used to designate a ratio.

Dilutions

Manual dilution procedures are not performed as frequently as in the past because of improvements in computers and instrumentations in which a system performs the dilution automatically. There are occasions, however, when a dilution has to be prepared, so a brief review follows. For example, in chemistry, the specimen may have a concentration that lies outside of the linear range of the method. Immunology procedures often require the preparation of serial dilutions to determine titers of antibodies. Dilutions must be properly prepared and interpreted so that serious laboratory errors do not occur that will compromise good patient care.

There are two parts of a dilution. The first part is the sample to be diluted; the second is the diluents used to perform the dilution. The choice of diluents is significant, and substitutions should never be made without proper evaluation. Examples of diluents that are used include water, saline, buffers, alcohols, and proteins solutions.

Usually, dilutions are described by laboratory personnel as a "1 to 4" or "1 to 2" for example. This means that for a 1:4 dilution, for every 1 part of sample, there is a total of 4 *parts* of the solution. Therefore, in a 1:4 dilution, 1 part sample is used and 3 parts—*not* 4 parts—of the diluent is used. Examples of dilution nomenclature are shown below:

Prepare a 1 to 10 dilution of serum in saline.

Prepare a 1 in 10 dilution of serum in saline.

Prepare a 1/10 dilution of serum with saline.

 Checkpoint! 1–6

Describe the proper preparation of a 1 ml to 10 ml dilution of a serum sample with saline.

In the laboratory, technologists often experience the following situation: The chemistry analyzer prints a result that exceeds the upper limit of linearity. The resolution of this problem may be the preparation of a dilution followed by reanalyzing the dilution and then multiplying the new instrument result times the dilution factor.

 Checkpoint! 1–7

A patient's glucose result is "flagged" with the comment "Results exceed the upper limit of linearity." The method's upper limit of linearity is 700 mg/dL. The technologist prepares a 1 to 2 dilution of the serum specimen with saline and reanalyzes the diluted sample. A result, 400 mg/dL, is displayed. What is the concentration of the glucose in this patient's serum sample?

Solution Concentrations

The concentration of a solution may be expressed as molarity (M), normality (N), and, less frequently, molality (m). Accurate preparation of reagents requires fundamental knowledge of solution chemistry, basic mathematics, and techniques.

Density, Specific Gravity, and Assay by Weight

Density is defined as the amount of matter (weight) per unit volume of substance. This is a property of all substances, not just solutions. The units for density in the centimer-gram-second system are grams per cubic centimeter (cc). A cc is equal to 1 milliliter. Water has a density of 1.000 g per 1.00 mL at standard conditions for temperature and pressure (STP).

Specific gravity is a ratio of mass to volume and is the density of chemical in terms of grams per milliliter compared to pure water at STP. By definition, specific gravity is a dimensionless quantity. For example, if the specific gravity of a substance is 3.0, its density is 3.0 times (1.0 g per 1.00 mL) = 3.0 g/mL.

Specific gravity is a method of measuring density. When working with concentrated liquids, it is cumbersome to measure volume in grams. Converting grams to milliliters can be done using specific gravity and therefore making the preparation of the solutions easier. When a bottle label lists the specific gravity of that solution, the number indicates the mass of 1 mL of that solution. For example, a bottle containing a solution of nitric acid (HNO_3) is labeled with specific gravity 1.4, meaning that 1.0 mL of this solution weights 1.4 g.

Another factor to consider when making solutions is the percent assay by weight (%w/w) which also represents the purity of the solute contained in the solution. For example, if a bottle of nitric acid is 70% assay weight, there is 0.70 g of HNO_3 in 1.0 g solution.

For example, if the number of grams of HNO_3 per mL solution needs to be determined, simply multiply the specific gravity by the percent assay weight. Using the data just presented for nitric acid, the number of grams of HNO_3 per milliliter is calculated using the following proportion:

$$\frac{1.40 \text{ g } HNO_3 \text{ solution}}{1.00 \text{ mL solution}} \times \frac{0.700 \text{ g } HNO_3}{1.00 \text{ g } HNO_3 \text{ solution}}$$

$$= 0.980 \text{ g } HNO_3/\text{mL}$$

Atomic Mass Unit

Atomic mass unit (amu) is defined as the unit of mass used by chemists and physicists for measuring the masses of atoms and molecules. It is equivalent to 1/12th the mass of the most common atoms of carbon, known as carbon-12 atoms. Careful experiments have measured the size of this unit; the currently accepted value (1998) is $1.660\ 538\ 73 \times 10^{-24}$ grams. [This number equals 1 divided by Avogadro's number. In biochemistry, the atomic mass unit is called the dalton (Da)].

Dalton (Da)

Before 1961, physicists and chemists used the acronym *amu* for their respective uses of atomic mass units. The physicists' *amu* was defined as 1/16th of the mass of one atom of oxygen-16, while the chemists' *amu* was defined as 1/16th of the average mass of an oxygen atom. After 1961, the International Union of Pure and Applied Physics (IUPAP), in conjunction with IUPAC, agreed to a unified *atomic mass unit* (symbol μ) with 1μ equal to 1.000317 amu (physical scale) and 1.000043 amu (chemical scale). Thus, the accepted standard definition of the unified *amu* is equal to 1/12th of the mass of a carbon-12 atom.

As determined experimentally, the mass on one unified amu is $1.660538782 \times 10^{-27}$ kilogram. *Molar mass* (symbol M) is defined as the mass of 1 mole of a substance—for example,

a chemical element or compound. Molar masses are always quoted in grams per mole (g/mol or gmol^{-1}) in chemistry. The Dalton (Da) is used as a unit of molar mass, where 1 Da is equal to 1g/mol and thus 1 Da is equivalent to $1.660538782 \times 10^{-27}$ kg.

Molar mass is related to the molecular weight of compound and to the standard atomic weight of its constituent's elements. The dalton is often used in biochemistry to state the masses of large organic molecules such as proteins. These measurements are typically designated as kilodaltons (kDa). For example, albumin is a 65-kDa protein (more accurately, 66.438 kDa).

 ## Checkpoint! 1–8

The molecular mass of albumin, a protein found in significant amounts in the human body, is 66.438 kDa. How many grams does this represent?

Moles

The SI base unit of the amount of a substance (as distinct from its mass or weight) is the *mole*. Moles represent the actual number of atoms or molecules in a substance. An alternate designation is **gram molecular weight (GMW)**, because 1 mole of a chemical compound is the same number of grams as the molecular weight of a molecule of that compound measured in atomic mass units. The official definition, adopted as part of the SI system in 1971, is that 1 mole of a substance contains just as many elementary entities (atoms, molecules, ions, or other kinds of particles) as there are atoms in 12 grams of carbon-12 (the most common atomic form of carbon, consisting of atoms with 6 protons and 6 neutrons). The actual number of "elementary entities" in a mole is called **Avogadro's number**, named after Italian chemist and physicist Amedeo Avogadro (1776–1856). Careful measurement determines Avogadro's number to be approximately 6.02×10^{23}.

A mole of any substance is the amount of a substance that has a mass in grams numerically equivalent to its molecular weight in daltons.

 ## Checkpoint! 1–9

Determine 1 mole of sucrose ($C_{12}H_{22}O_{11}$) in daltons and grams.

Molarity

The GMW or gram formula weight of a chemical substance is the molecular (or formula) weight expressed in grams. The weights of sodium chloride (NaCl) and glucose ($C_6H_{12}O_6$), for example, are 58.3 daltons and 180 daltons, respectively. Their GMWs are 58.3 grams and 180 grams, respectively. Thus, 58.3 g of NaCl contains 6.02×10^{23} molecules of NaCl, and 180 g of glucose also contains 6.02×10^{23} glucose molecules. Thus, a mole of a substance is equal to its molecular weight in grams.

In the laboratory, solutions may be expressed in terms of molarity. **Molarity (M)** is equal to the number of moles of solute per liter of solution (solvent). Therefore, a 1M solution of sodium hydroxide (NaOH) whose GMW is 40g will contain 40 g of NaOH in 1 liter of diluents.

✓ Checkpoint! 1–10

How many grams of NaCl are required to prepare 1 liter of a 0.5-M solution? The GMW of NaCl is 58.5.

Millimoles

When small concentrations are used, they are frequently expressed in millimoles per liter (1000 mmol = 1 mol). For example, to prepare 10 mL of a 10-mmol (0.01 moles) NaOH solution (GMW = 40), 4 mg NaOH are diluted to 10 mL.

Normality

Normality (N) is equal to the number of gram equivalents of solute per liter of solution; it is dependent on the type of reaction involved (e.g., acid–base, oxidation). One gram equivalent weight (GEW) of an element or compound equals the GMW divided by the number of replaceable hydrogen ions or hydroxyl ions (also valence). Therefore, a one normal (1N) solution represents the number of gram equivalents of solute per liter of solution.

✓ Checkpoint! 1–11

1. *What is the gram equivalent weight of $Ca(OH)_2$ (GMW = 74)?*
2. *What is the GEW of H_2SO_4 (GMW = 98)?*
3. *How many milliliters of concentration H_2SO_4 (specific gravity 1.84, percent purity 96.2%) are required to prepare one liter of a one normal solution?*

Molal

In the laboratory, we sometimes measure the physical properties of solutions; for example, when we measure the osmolality of serum or urine. The molality of the solution is determined instead of the molarity. A molal solution is one mole of solute in one kilogram of solvent. Molal solutions are based on weight, not volume. Because the density of water at room temperature is approximately 1 gram per milliliter, 1000 grams of water occupies about 1 liter. Therefore, a 1 molal aqueous solution is approximately the same as a 1 molar solution.

✓ Checkpoint! 1–12

How many grams of NaOH are required to prepare a 2.00 molal solution?

Percent Solutions

Percent concentration is described as the number of parts of solute in 100 parts of solution. This type of solution is used commonly in the laboratories and may be expressed in terms of weight to volume or volume-to-volume. Percent solutions are often used as an ancillary reagent in a procedure or as a rinse solution for automated analyzers. If a percent weight to volume solution is required, the preparer must remember to dilute up to a total volume—for example, 100 mL. When preparing a percent weight-to-weight (w/w) solution, the weight of the solute and solvent are represented as a sum. For example, to prepare a 25.0% w/w solution of NaOH in distilled water, one must weigh out 25 g of NaOH into 75.0 g of water. Therefore, the 25g NaOH plus the 75g of water equals 100g.

✓ Checkpoint! 1–13

How many grams of solid NaOH are required to prepare 100 mL of a 10% solution in water?

Hydrates

The molecules of certain salts can chemically combine with one or more molecules of water to form hydrates. These water molecules must be considered when determining the amount of salt to be used in the preparation of a solution. Molecules having no water attached are termed *anhydrous*, a molecule with one water attached is termed a *monohydrate*, molecules with two waters attached are termed *dihydrates*, and so on. A common procedure in which anhydrous and hydrate compounds are required is the preparation of buffer solutions.

✓ Checkpoint! 1–14

A buffer solution requires that 3.00 grams of anhydrous Na_2HPO_4 be dissolved into 100 mL of reagent-grade water. The laboratory has $Na_2HPO_4 \cdot 7\,H_2O$. How much of the $Na_2HPO_4 \cdot 7\,H_2O$ should be used to prepare the solution?

Enzymes

The unit of enzyme activity used in the United States is the **International Unit (IU or U)**. This unit was part of the effort of the Enzyme Commission of the International Union of Biochemistry (circa 1961) to standardize the unit of enzyme activity. It is defined as that amount of enzyme that will catalyze the reaction of 1 μmol of substrate per minute (μm/min) of reaction under defined conditions of temperature, pH, and substrate concentration.

As part of its system of units, the World Health Organization adopted the katal (kat) as the unit of enzyme activity. One katal is defined as the activity of the enzyme that changes the substrate by 1 mole per second under defined conditions of temperature, pH, and substrate concentration. Thus, 1.0 nanokatal/L = 0.06 U/L.

Determination of Enzyme Activity Using Molar Extinction Coefficient Several clinical chemistry enzyme assays use the oxidation or reduction of coenzymes NADH or NADPH as part of a coupled indicator reaction. The increase or decrease in coenzyme activity is monitored at 340 nm. The molar extinction coefficient of NADH is 6.22×10^3 M^{-1}cm^{-1} at 340 nm. The formula for calculating enzyme activity using the molar extinction coefficient of NADH is:

$$[(\Delta \text{ absorbance of unknown})/\varepsilon \times d](106)(1/T)(V_t/V_s) = \text{IU/L} \quad (1.5)$$

where

ε = molar extinction coefficient

d = cuvette path length

T = time

V_t = volume of test

V_s = volume of sample

Derivation of Enzyme Factor Many of the enzyme methods used do not utilize calibrator solutions to quantitate enzyme activity in samples. Instead, the activity of an enzyme in a sample is determined by using an enzyme factor. The enzyme factor is an experimentally derived number based on the analytical procedure used. The origin of the enzyme factor is shown in equation 1.6:

$$[1/(\varepsilon)(d)][10^{-6}][1/T][V_t/V_s] = \text{enzyme factor} \quad (1.6)$$

 ### Checkpoint! 1–15

Determine the enzyme activity (concentration) of LDH in a patient specimen using a spectrophotometric assay. The method uses NADH as the reaction indicator. The reaction volume is 300 µL, and the sample volume is 50 µL. Absorbance readings are taken at 30-second intervals for 2.5 minutes. The light path for the spectrophotometer is 1.0 cm, and the delta absorbance for the sample is 0.020.

▶ ACIDS AND BASES

An acid is a species that produces hydrogen ions (H$^+$) and serves as a proton donor in aqueous solutions. There are two types of acids, strong and weak, which differ in the extent of their ionization in water. Strong acids—for example, hydrochloric acid—ionize completely, forming H$^+$ and Cl$^-$. Weak acids are only partially ionized to H$^+$ ions in water.

A base is a species that produces hydroxyl ions (OH$^-$); these are proton acceptors in aqueous solutions. A strong base such as NaOH in water solution is completely ionized to OH$^-$ and Na$^+$. Most strong bases are the hydroxides of the Group 1 and Group 2 metals. These are typical ionic solids, completely ionized both in the solid state and in water solutions. Weak bases produce OH$^-$ ions. They react with water molecules, acquiring H$^+$ ions and leaving OH$^-$ ions behind.

A list of acids and bases with their physical characteristics (for example, specific gravity, molecular weight, and approximate molarity and normality) is presented in Appendix G.

pH

The hydrogen ion activity of a solution is usually shown in shorthand notation as *pH*. By definition, pH is the negative logarithm of the hydrogen ion activity or concentration. Similarly, pOH is the negative logarithm of the hydroxyl ion activity.

Thus,

$$pH = -\log_{10}[H^+] = \log 1/[H^+]$$

and

$$pOH = -\log_{10}[OH^-] = \log 1/[OH^-]$$

In all aqueous solutions, the equilibrium for the ionization of water (K_w) must be satisfied—that is, $K_w = [H^+][OH^-] = 10^{-14}$. Therefore, if [H$^+$] is known, then the [OH$^-$] can easily be calculated. The data presented in Appendix H illustrate the association of molar concentrations [H$^+$] and [OH$^-$] in relation to pH.

▶ BUFFER SOLUTIONS

A buffer solution is defined as one that resists changes in pH. Most buffers consist of either a weak acid and its conjugate base or a weak base and its conjugate acid. Laboratorians use buffers whenever they need to maintain the pH of a solution at a constant and predetermined level. The preparation of buffers in the clinical laboratory requires quality reagents such as acids and bases, an accurate balance, and a pH meter. For most commonly used buffers, the amounts of salts of the acids and bases have been predetermined and may be found in reference books (for example, that by R. G. Bates).[10] This reference contains a very good discussion of the theoretical aspects of buffers and extensive information of how to prepare several buffer solutions. Proper preparation of buffers known as "Good" buffers for clinical and molecular diagnostic procedure have also been published.[11]

SUMMARY

This chapter presented several fundamental topics related to the clinical laboratory. Description of the equipment, measurements, and calculations were provided. Included in the discussion where appropriate was information regarding calibration and standardization of laboratory apparatus. Selected references to organization that may provide guidance on the use, care, and maintenance of basic laboratory equipment were also provided. A brief review of laboratory mathematics also was provided.

The ultimate purpose of this chapter was to provide readers with examples of good laboratory practices that will enable them to provide caregivers with the highest quality laboratory data.

REVIEW QUESTIONS

LEVEL I

1. Reverse osmosis is a suitable process to complete which of the following? (Objective 1)
   a. Remove particulates from the air.
   b. Purify water.
   c. Purify organic solvents used in liquid chromatography.
   d. Separate erythrocytes from serum.

2. The *to deliver* (TD) pipette is designed to drain by: (Objective 5)
   a. tapping it against the container.
   b. gravity.
   c. forcing air into the opening.
   d. rapidly moving it up and down.

3. One technique for monitoring water purity is to measure the _____ of an aliquot of water. (Objective 2)
   a. voltage
   b. partial
   c. absorbance
   d. resistivity

4. Polypropylene, polyethylene, polycarbonate, and polystyrene beakers and flasks are all examples of: (Objective 4)
   a. glassware.
   b. plasticware.
   c. semiconductor materials.
   d. types of micropipettes.

5. Low actinic glassware is: (Objective 3)
   a. clear and colorless.
   b. black and nontransparent.
   c. amber or red in color, and the contents are visible.
   d. brown, and the contents are not visible.

6. How many grams of H_2SO_4 are present in 1.5 mL of a concentrated acid solution with a specific gravity of 1.80 and an assay purity of 95.0% w/w? (Objective 9)
   a. 0.88
   b. 1.00
   c. 1.14
   d. 2.56

7. A serum lipase is diluted 1/20 with a result of 30 U/L. What is the patient's actual lipase result? (Objective 9)
   a. 1.5 U/L
   b. 30 U/L
   c. 300 U/L
   d. 600 U/L

8. A patient's calcium result is 7.8 mg/dL. Convert this result to mmol/L. (Objective 8)
   a. 0.00195
   b. 1.95
   c. 3.90
   d. 78.0

LEVEL II

1. A thermistor is an example of: (Objective 3)
   a. a potentiometer.
   b. an amperometer.
   c. a device in which resistance changes as temperature changes.
   d. a calorimeter.

2. Swinging-bucket and fixed-angle-head rotors are found in which of the following? (Objective 2)
   a. water-purification systems
   b. centrifuges
   c. multiple-tube mixers
   d. electronic balances

3. Which of the following represents an alternative to mercury-filled thermometers? (Objective 4)
   a. thermometers containing an organic red spirit and pressurized with nitrogen gas
   b. thermometers containing blue biodegradable liquid (isoamyl benzoate and a dye)
   c. digital thermometers with stainless steel stems
   d. All of the above.

4. Which of the following calibration devices is used to verify the accuracy of laboratory thermometers? (Objective 6)
   a. refractometer
   b. tachometer
   c. SRM 1968 gallium melting-point cell
   d. holmium oxide filter

PEARSON
myhealthprofessionskit™

Use this address to access the interactive Companion Website created for this textbook. Simply select "Clinical Laboratory Science" from the choice of disciplines. Find this book and log in using your user name and password.

REFERENCES

1. National Committee for Clinical Laboratory Standards. *Preparation and testing of reagents water in the clinical laboratory,* 4th ed. Approved Guideline. NCCLS Document C3-A3 (Wayne, PA: National Committee for Clinical Laboratory Standards, 2006).

2. American Society for Testing Material. ASTM E288-03. *Standard specification for laboratory glass volumetric flasks* (Philadelphia: ASTM, 2003).

3. American Society for Testing Material. ASTM E694-99. *Standard specification for laboratory glass volumetric apparatus* (Philadelphia: ASTM, 2005).

4. American Society for Testing Material. ASTM E542-01. *Standard practice for calibration of laboratory volumetric apparatus* (Philadelphia: ASTM, 2003).

5. National Committee for Clinical Laboratory Standards. *Determining performance of volumetric equipment.* Proposed Standard. NCCLS Document I8-P (Wayne, PA: NCCLA, 1996).

6. American Society for Testing Material. ASTM Standard E 617-97. *Standard specification for laboratory weights and precision mass standards* (Philadelphia: ASTM, 2003).

7. U.S. Environmental Protection Agency. *Eliminating mercury in hospitals: Environmental best practices for health care facilities* (Washington, DC: U.S. Environmental Protection Agency, November 2002).

8. National Committee for Clinical Laboratory Standards. *Temperature calibration of water baths, instruments, and temperature sensors,* 2nd ed. Approved Standard. NCCLS document I2-A2 (Wayne, PA: NCCLS, 1990).

9. The International System of Units (SI). BIPM.org (www.bipm.org/utils/common/pdf/s:_brochure_8_en.pdf); 8th ed. (Sèvres, Cedex, France: Bureau International des Poids et Mesures, 2006) (accessed August 2008).

10. Bates RG. *Determination of pH—Theory and practice,* 2nd ed. (New York: John Wiley and Son, 1973).

11. Good NE, Winget GD, Winter W, et al. Hydrogen ion buffers for biological research. *Biochemistry* (1966) 5: 467–477.

2

Instrumentation

CHAPTER OUTLINE

■ OBJECTIVES—LEVEL I

Following successful completion of this chapter, the learner will be able to:

1. Identify two physical properties of light.
2. Define the following wave parameters: amplitude, period, frequency.
3. Define the factors that characterize the energy of a photon.
4. Identify three types of light scatter.
5. List several major instrument components for the following analyzers:
   a. Spectrophotometer
   b. Fluorometer
   c. Nephelometer
   d. Mass spectrometer
   e. Gas chromatograph
   f. Densitometer
6. Identify the significant regions of the electromagnetic spectrum (EMS) from lowest energy to highest energy levels.
7. List four spectrophotometric function checks.
8. Define the following terms associated with electrochemical methods: potentiometer, amperometry, coulometry, conductance, resistivity, and voltammetry.
9. Write the Nernst equation.
10. Write the chemical reactions for the PO_2 and PCO_2 electrode.
11. Identify four examples of separation techniques used in the clinical laboratory.
12. List four examples of transducers used in biosensor devices.
13. Identify three factors that affect chromatographic resolution.
14. Define the following terms: diffuse reflection, retention time, R_f, fluorescence, chemiluminescence.
15. Identify four colligative properties of solutions.
16. Identify specific analyte(s) that are measured by each device or instrument.

■ OBJECTIVES—LEVEL II

Following successful completion of this chapter, the learner will be able to:

1. Diagram the correct sequence of significant components of a spectrophotometer.
2. Determine which component of a spectrophotometer malfunctioned, given the failure of a specific photometric function check.
3. Calculate the concentration of a solution given the absorbance values for tests and standards.
4. Convert between units used to describe wavelengths.

■ OBJECTIVES—LEVEL II (continued)

5. Predict the shape of the line when given the following x- and y-axis parameters:
   a. Absorbance (y) versus concentration (x) on linear graph paper
   b. Percent transmittance (y) versus concentration (x) on linear graph paper
6. Explain the fundamentals principles of selected instruments described in this chapter.
7. Explain how the absorbance and transmittance of light are related.
8. Calculate the wavelength, frequency, and photon energy of electromagnetic radiation (EMR).

KEY TERMS

Amperometry	Lab-on-a-chip	Resolution
Beer–Lambert law	Mass-to-charge ratio	Retention time
Chemiluminescence	Monochromator	Spectral bandwidth
Electromagnetic radiation	Nernst equation	Transducer
Fluorescence	Nominal wavelength	Turbidimetry
Heat of fusion	Osmometry	Wavelength
Ion-selective electrode	R_f	

 **A CASE IN POINT**

A staff technologist discovered that the quality-control (QC) results for her photometric assays, which include tests for glucose, urea nitrogen, and creatinine, were out of range. Further investigation revealed the presence of a shift in QC results that resulted in a QC failure according to Westgard rules. The shift in QC was below the mean for all photometric assays.

The technologist initiated the laboratory QC protocol for out-of-control situations. None of the procedures, which included reassaying the QC serum and recalibrating the assay, resolved the problem.

Issues and Questions to Consider

1. Identify components within the spectrometer that may result in a shift in QC results to below the mean.
2. Identify specific photometric function check(s) that can be completed and may provide evidence that a component(s) is not functioning properly.
3. Indicate which function check has the highest probability of being the one that failed and is responsible for the downward shift of QC results.
4. Provide a solution for the problem that would result in an upward movement of the QC data toward the mean.

WHAT'S AHEAD

▶ Explanation of the underlying electrical, mechanical, and optical principles of several analyzers found in clinical chemistry laboratories today
▶ Identification of the major components of selected analyzers
▶ Discussion of the theoretical aspects of the interaction of light with matter
▶ Definition of terms associated with instrumentation

▶ ABSORPTION SPECTROSCOPY

The interaction of EMR with matter provides the principle means of measuring analytes in biological fluids. These techniques include absorption spectroscopy, reflectometry, refractometry, and emission spectroscopy. It is interesting to note that as much as the analytical systems have changed, the "black box" or measuring device within the system has fundamentally remained the same. For example, the largest automated clinical chemistry system having test throughput rates exceeding 3000 tests per hour uses a spectrophotometer as the measuring device.

Before proceeding to instrument components that are the objects and devices within the analyzer, a review of the properties of light and the laws applicable to the interaction of light with matter will be provided. It should be noted that many of the observations presented in this section are applicable to other section topics (e.g., fluorescence and atomic absorption).

PROPERTIES OF LIGHT

Electromagnetic radiation exists both as Maxwell's waves and as streams of particles called photons. Photons are exchanged whenever electrically charged subatomic particles interact. Other manifestations of EMR include gamma rays, X-rays, microwaves, radio frequency radiation, and ultraviolet radiation. The energies involved with the specific regions of the EMS and their corresponding wavelengths change dramatically from radio waves to gamma radiation.

Some properties of EMR can be described by means of a classical sinusoidal wave model. Parameters associated with this waveform include wavelength, frequency, velocity, and amplitude. EMR requires no supporting medium for its transmission and passes readily through a vacuum.

The sine wave model does not provide the total picture when discussing the absorption and emission of radiant energy. EMR also exists as a stream of discrete particles, or packets (also bundles) of energy called photons. Photon energy is proportional to the frequency of the radiation. This dual nature of EMR is considered complementary and applies to the behavior of streams of electrons, protons, and other elementary particles.

The wave model allows us to represent EMR as both an electric and magnetic field that can undergo in-phase, sinusoidal oscillation at right angles to both each other and to the direction of propagation. The electric and magnetic fields for a monochromatic beam of plane-polarized light (with oscillation of either the electric or magnetic fields within a single plane) in a specific direction of propagation are shown in Figure 2-1 ■.

The electric vector of the waveform is shown in two-dimensional format. Remember that a vector has both magnitude and direction. The electric field vector at a certain point in time and space is proportional to its own magnitude. Time or distance of wave travel is plotted on the abscissa. Many of the instrument principles widely used in the laboratory involve the electric component of radiation and will represent the focus of discussion throughout this chapter. An exception will be nuclear magnetic resonance (NMR). In this technique, the magnetic component produces the desired effect.

Electromagnetic waveforms are characterized by several parameters. *Amplitude* of the sine wave is shown as the length of the electronic vector at maximum peak height. A *period, p,* is defined as the time in seconds required for the passage of successive maxima or minima through a fixed point in space. The number of oscillations of the waveform in a second is termed *frequency, ν*. The unit of frequency is hertz (Hz), which corresponds to one cycle per second. Frequency is also

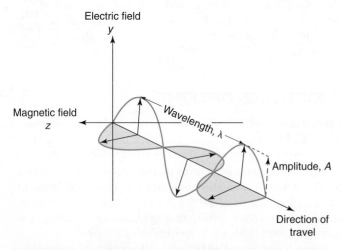

■ FIGURE 2-1 Diagram of a beam of monochromatic, plane-polarized radiation. The electric and magnetic fields are at right angles to one another and direction of travel. Wavelength, λ, from peak to peak and amplitude A, length of electric waveform at maximum peak height.

BOX 2-1 Conversion between units used to describe wavelengths.

1 nm = 10^{-9} m = 10^{-7} cm; other units sometimes used include:
1 μm = 10^{-6} m = 10^{-4} cm
1 Å = 10^{-10} m = 10^{-8} cm

equal to $1/p$. A **wavelength**, λ, is the linear distance between any two equivalent points on a successive wave. A widely used unit for wavelength in the visible spectrum is the nanometer, nm (10^{-9} m). EMR in the X-ray or gamma region may be expressed in terms of angstrom units, Å (10^{-10} meters). Finally, because of its much longer wavelength, EMR in the infrared region may have units corresponding to micrometer, mμ (10^{-6} m).[1] Examples of units used for wavelength measurements are shown in Box 2-1 ■.

VELOCITY OF PROPAGATION

Velocity of propagation, v_i, in meters per second is determined by multiplying frequency by wavelength shown in Equation 2.1.

$$v_i = \nu\lambda_i \qquad (2.1)$$

The frequency of light is determined by the source and does not change whereas the velocity depends on the composition of the medium through which it passes. Therefore, Equation 2.1 implies that wavelength of radiation is also dependent on the medium.

The velocity of light traveling through a vacuum is independent of wavelength and is at its maximum. This velocity is represented by the symbol c, and is equivalent to 2.99792×10^8 m/s (rounded to 3.00×10^8 m/s). Substituting in Equation 2.1, then

$$c = \nu\lambda = 3.00 \times 10^8 \,\text{m/s} = 3.00 \times 10^8 \,\text{m}\cdot\text{s}^{-1}$$
$$= 3.00 \times 10^{10} \,\text{cm/s} \qquad (2.2)$$

✓ **Checkpoint! 2–1**

What is the wavelength in nanometers for EMR having a frequency of 1.58×10^{15} Hz?

In any medium containing matter, the propagation of radiation is slowed by the interaction between the EMR field of the radiation and the electrons bound in the atoms of that matter. Because the radiant frequency does not vary and is fixed by the source, the wavelength must decrease as radiation passes from air to a slower medium.

ENERGY OF ELECTROMAGNETIC RADIATION

What about the energy of EMR? It should be remembered for safety concerns that optical devices emitting high-frequency EMR generate very high energies that may have deleterious

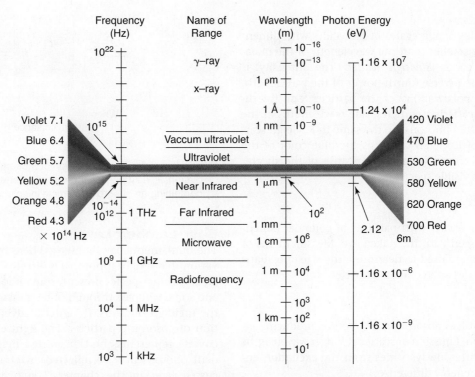

■ **FIGURE 2-2** The relationship among frequency, wavelength, and photon energy throughout the electromagnetic spectrum.

effects on ocular tissue. Examples include deuterium and xenon lamps. Wavelength and frequency are related to the energy of a photon as shown in Equation 2.3.

$$E = h\nu = hc/\lambda \qquad (2.3)$$

where

E = the energy of a photon in Joules or eV

h = Planck's constant $(6.626 \times 10^{-34} \text{ J} \cdot \text{s})$*

ν = frequency in Hz (cycles/s)

c = the velocity of light in a vacuum $(3.00 \times 10^8 \text{ ms}^{-1})$

The energies of photons are often stated in terms of an electron volt (eV). An eV is defined as the energy acquired by an electron that has been accelerated through a potential of one volt. The conversion factors between Joules and eV are as follows:

$$1 \text{ J} = 6.24 \times 10^{18} \text{ eV}$$

$$1 \text{ eV} = 1.602 \times 10^{-19} \text{ J}$$

To illustrate the difference in energies of photons in the EMS, compare the energy of photons in the ultraviolet (UV) region versus the visible region of the EMS.

*Also expressed in units 6.626×10^{-27} erg · s

✓ **Checkpoint! 2–2**

What is the photon energy (E) in eV of EMR of (a) 190 nm? (b) 520 nm?

The relationship of frequency, wavelength, and photon energy throughout the EMS can be seen in Figure 2-2 ■. This graphic illustrates, for example, that very-high-energy gamma photons have extremely short wavelengths and very high frequencies. The converse is true for TV and radio wave parameters.

SCATTERING OF RADIATION

Transmission of radiation in matter can be viewed as a momentary retention of the radiant energy by atoms, ions, or molecules followed by reemission of the radiation in all directions as the particles return to their original state. Destructive interference removes most but not all of the reemitted radiation involving atomic or molecular particles that are small relative to the wavelength of the radiation. The exception is the radiation that travels in the original direction of the beam; the path of the beam appears to be unaltered because of the interaction. It has been shown that a very small fraction of the radiation is transmitted at all angles from the original path and that the intensity of this scattered radiation increases with particle size.

Rayleigh Scatter

Scatter by molecules or aggregates of molecules with dimensions significantly smaller than the wavelength of the radiation is referred to as *Rayleigh scatter*. The intensity is proportional to the inverse fourth-power of the wavelength, the square of the polarizability of the particles, and the dimensions of the scattering particles. The wavelengths of the absorbed and emitted photons are the same in Rayleigh scatter. An example of Rayleigh scatter is the blue color of the sky, which results from the increased scatter of the shorter wavelength of the visible spectrum.

Tyndall Effect

The Tyndall effect occurs with particles of colloidal dimensions and can be seen with the naked eye. Measurements of scattered radiation are used to determine the size and shape of polymer molecules and colloidal particles.

Raman Scatter

Raman scatter involves absorption of photons producing vibrational excitation. Emission or scatter at longer wavelengths occurs. Raman scatter always varies from the excitation energy by a constant energy difference.

PRACTICAL ASPECTS OF LIGHT

Spectrophotometric techniques used in clinical laboratory measurements use either ultraviolet or visible light. In the ultraviolet and visible regions of the EMS, absorption is because of changes in the energy levels of electrons associated with certain types of chemical groups. Electromagnetic energy sources, such as tungsten halogen lamps or deuterium lamps, emit polychromatic light. The color of objects that we perceive is the result of light being transmitted or reflected from them. Thus, a substance appears red because it transmits or reflects red light, and absorbs bluish green. An object appears black (lacks color) because it absorbs all colors; conversely objects that appear white reflect all colors.

Chemical analyses that rely on the interaction of light with matter are used to measure a majority of the analytes in the clinical laboratory. For example, measurement of creatinine in blood is routinely performed using an alkalinized solution of picric acid, and the orange-red color produced is measured at about 500 nm. The final product appears orange-red in color because of absorption of greenish-blue color. The filter used to isolate visible light of 500 nm transmits the greenish-blue color, and the filter absorbs all other colors.

 ## Checkpoint! 2–3

What wavelength should be selected for a filter that will be used to measure a solution that appears purple in color?

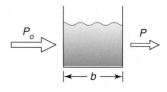

Where:

$$\text{Transmittance} = \frac{P}{P_o}$$

■ FIGURE 2-3   The attenuation of a beam of monochromatic radiation, P, as it passes through a cuvette with a light path, b, containing an absorbing particle in solution.

Beer–Lambert Law

The fundamental laws presented here are based on the observation that if light (photons) is directed into a container with an absorbing species present, some photons will be absorbed and some will pass through or be transmitted. In Figure 2-3 ■ the incident photons, P_o, enter a cuvette containing a solution of absorbing particles. The light energy, P, leaving the cuvette represents the transmitted light. Some of the incident photons may be reflected or refracted and will be discussed later in the chapter. The overall behavior of the incident photons forms the basis of the theories proposed by Bouguer, Lambert, and Beer.

In 1729 Bouguer published his finding regarding light transmittance through a thickness of absorbing material. He observed that if a unit thickness of material has transmittance, T, then thickness, x, of the material has the transmittance shown in Equation 2.4:

$$T = \varepsilon^{-ax} \qquad (2.4)$$

where

ε = the base of the natural system of log (2.71828)

a = the absorption coefficient

Unfortunately for Bouguer, his publications and findings were not generally known. In 1760 Lambert rediscovered the work of Bouguer and his work became know as Lambert's law.

Lambert's law states that for parallel monochromatic radiation that passes through an absorber material of constant concentration, the radian power decreases logarithmically as the light path increases arithmetically. Thus if a certain thickness of absorbing material absorbs 60% of the incoming radiation energy, then an equal thickness of absorbing material that follows will absorb 60% of the remainder. Also, equal thicknesses of an absorbing material will absorb a constant fraction of the energy incident upon the equal thicknesses. This principle is illustrated in Figure 2-4 ■. If a series of identical cells, each containing a portion of the same solution, is placed in the light beam, the results shown at the top of this illustration are produced. This shows the effect of increasing the solution light path on the transmittance. The percent transmittance (%T) of each cell is 60% of the previous cell and a plot of %T values against the number of cells or solution

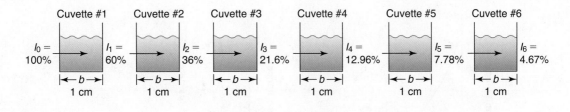

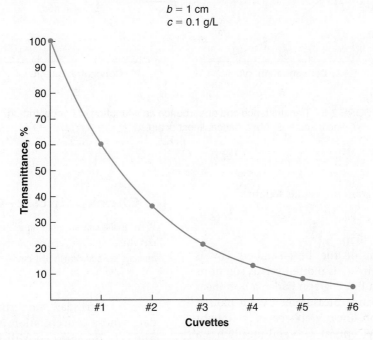

$b = 1$ cm
$c = 0.1$ g/L

■ **FIGURE 2-4** A plot of percent transmittance versus cuvettes 1 through 6 for a series of solutions of identical concentrations and cuvette path length. This illustrates the nonlinearity of the relationship between percent transmittance and concentration.

thickness is shown in Figure 2-4. The results reveal that %T does not vary linearly with increased light paths. If concentration is used as the variable, a similar curve will be created.

Lambert proved that for monochromatic radiation that passes through an absorber of constant concentration, there is a logarithmic decrease in the radiant power as the light path increases arithmetically. The absorbance (A) of a solution was determined to be equivalent to

$$A = -\log P/P_o \qquad (2.5)$$

The relationship between transmittance, t, or percent transmittance ($P/P_o \times 100$) and absorbance is shown in Equation 2.6. Absorbance and percent transmittance are inversely related as given by

$$A = \log 1/T = \log 100 \text{ percent}/\%T = \log 100 - \log \%T$$
$$= A = 2 - \log_{10}\%T \qquad (2.6)$$

✓ **Checkpoint! 2-4**

What is the absorbance of a solution whose percent transmittance is 10%?

Beer followed with studies (circa 1850) on the relationship between radiant power and concentration. Beer's approach was to keep the light path and wavelength constant while determining the relationship of radiant power, P, and concentrations of the absorbing species. Based on previous work by Lambert, Beer discovered that for monochromatic radiation, absorbance is directly proportional to the light path, b, through the medium and the concentration, c, of the absorbing species. The work culminated in the Beer–Lambert law or simply **Beer's law**.

The following represents the principles established:

$$A = -\log t = \log P/P_o = abc \qquad (2.7)$$

where

a = absorptivity in (L g^{-1} cm^{-1})

b = light path of 1 cm

c = concentration in units of g/L

When the concentration in Equation 2.7 is expressed in moles per liter and the light path is in centimeters, the term applied is *molar absorptivity* (also *extinction coefficient*) and

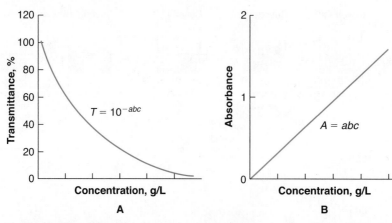

■ FIGURE 2-5   Transmittance and absorbance as a function of concentration. A. %*T*, linear scale. B. Absorbance, linear scale.

given the symbol ε (*a* times gram molecular weight):

$$A = \varepsilon bc \qquad (2.8)$$

where units for ε are (L mol^{-1} cm^{-1})

A graphic representation of the Beer–Lambert law is shown in Figure 2-5 ■. When %*T* is plotted versus solutions of known concentrations on linear graph paper, a nonlinear plot is produced. If absorbance measurements are plotted versus the same solutions on linear graph paper, a straight line is created. This line intercepts at zero and ideally has a slope of 1.0. These mathematical variables, slope and intercept, serve as valuable tools that laboratories use for method evaluations, calibrations, and troubleshooting problems with colorimetric techniques.

 Checkpoint! 2–5

What is the molar absorptivity of a 9.8×10^{-6} M solution of bilirubin dissolved in methanol having an absorbance of 0.600 when measured in a 1.0 cm cuvette at 435 nm?

The concentration of an analyte in solution can be determined by several different methods based on the Beer-Lambert law. The following are three examples of how concentrations of a known analyte can be determined:

1. Rearranging either Equation 2.7 or 2.8 and solving for concentrations *c*, we have

$$c = A/ab \qquad (2.9)$$

or

$$c = A/\varepsilon b$$

Absorbance (*A*) is determined by experimentation. The light path (*b*) of the cuvette is known and is usually 1 cm, and *a* or ε must be known. The concentration of analyte can be derived after all values are substituted in either equation shown above.

 Checkpoint! 2–6

What is the concentration in g/L of a solution of uric acid with an absorbance of 0.250, an absorptivity of 0.0625 L g^{-1}cm^{-1} at 570 nm, and cuvette path length of 1 cm?

2. Beer–Lambert's law states that $A = abc$. If a reference standard whose concentration is known is used and its color intensity is compared with the color intensity of the test samples, then

$$A_t = a \times C_t \times b \text{ and}$$
$$A_s = a \times C_s \times b$$

where

A_t = Absorbance of the test

A_s = Absorbance of the standard

C_t = Concentration of test analyte

C_s = Concentration of the standard

Because the light path is constant in the photometer (1 cm), *b* is constant. Concentration of the standard C_s is known; therefore

$$C_t = \frac{A_t}{A_s} \times C_s \qquad (2.10)$$

 Checkpoint! 2–7

What is the concentration of creatinine in a serum sample if given the following?

$$A_t = 0.140$$
$$A_s = 0.125$$
$$C_s = 0.90 \text{ mg/dL}$$

3. The concentrations of a known analyte in solution can be derived from a graph of absorbance measurements of a series of reference standards or calibrators versus their concentrations. The measured absorbance values are plotted on the *y*-axis (absorbance) and concentrations of standards on the *x*-axis. The absorbance values for the corresponding standards are plotted in the body of the graph, and a best-fit line is drawn to connect the data points. Next the absorbance value for the unknown sample is located on the *y*-axis and a perpendicular line is drawn to the best-fit line. Finally a perpendicular line is drawn from the point on the best-fit line to the *x*-axis and the numerical value for concentration is interpreted from the *x*-axis.

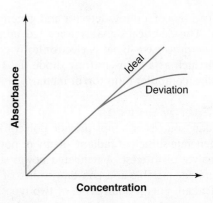

■ FIGURE 2-6 A plot of absorbance versus concentration that illustrates deviation from the Beer–Lambert law.

 Checkpoint! 2–8

What is the concentration of glucose in an unknown serum sample derived from the standard curve produced from the data shown in the following table?

Standard (mg/dL)	Absorbance
50	0.100
100	0.200
250	0.500
500	1.000
Sample:	
Patient	0.250

Linearity and Deviation from Beer's Law

The graph of absorbance versus concentration shows the intercept at zero and a linear plot with a slope equivalent to *ab*. Absorption spectroscopy is used best for solutions whose absorbance values are less than 2.0. Measured absorbance values greater than 2.0 may result in deviations from Beer's law because of interactions of light with a larger number of particles. This may cause a deviation from Beer's law resulting in a bending of the linear plot as shown in Figure 2-6 ■.

Deviations from Beer's law are caused by changes in instrument functions or chemical reactions. Instrument deviation is commonly a result of the finite bandpass of the monochromator. Beer's law assumes monochromatic radiation, but truly monochromatic radiation is achieved using only unique line emission sources. If absorptivity is constant over the instrument bandpass, then Beer's law is followed within close limits.

INSTRUMENTATION

A typical photometer or spectrophotometer contains five significant components in either a single- or double-beam configuration. The five components include (1) a stable source of radiant energy, (2) a device that isolates a specific region of the electromagnetic spectrum, (3) a sample holder, (4) a photo detector, and (5) a readout device. Each component of a typical photometer is shown in Figure 2-7 ■.

The following steps outline the function of each component in any absorption-type photometer as it detects light and provides information to the operator. The light source provides the energy that the sample will modify or attenuate by absorption. The light is polychromatic and all wavelengths are present. A wavelength selector isolates a portion of the spectrum emitted by the source and focuses it on the sample. The sample in a suitable container such as a cuvette absorbs a fraction of the incident light and transmits the remainder. The light that passes through or is transmitted

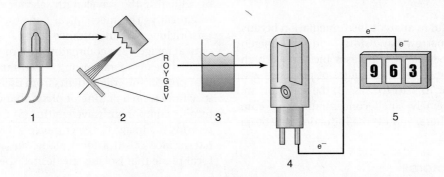

■ FIGURE 2-7 Schematic representation of five fundamental components of a spectrophotometer: 1. radiant energy source, 2. wavelength selector, 3. sample holder (cuvette), 4. photodetector, and 5. readout device.

strikes the cathode of a photodetector and generates an electrical signal. The electrical signal is processed, amplified, and digitized. The processed signal is electronically coupled to a display unit such as a light-emitting diode (LED), *X-Y* strip chart recorder, computer monitor, or meter.

Radiant Energy Sources

Radiant energy sources or lamps provide polychromatic light and must generate sufficient radiant energy or power to measure the analyte of interest. A regulated power supply is required to provide a stable and constant source of voltage for the lamp. Radiant energy sources are of two types: *continuum* and *line*. A continuum source emits radiation that changes in intensity very slowly as a function of wavelength. Line sources emit a limited number of discrete lines or bands of radiation, each of which spans a limited range of wavelengths.

Continuum sources find wide applications in the laboratory. Examples of continuum sources include tungsten, deuterium, and xenon. For testing in the visible region of the EMS, tungsten or tungsten-halogen lamps are widely used. Introducing a halogen gas into the lamp envelope counteracts the problem of increased atom vaporization from the high-temperature filament.

A deuterium lamp provides UV radiation in analytical spectrometers. The voltage applied is typically about 100 volts, which gives the electrons enough energy to excite the deuterium atoms in a low-pressure gas to emit photons across the full UV range.

The strong atomic interaction in a high-pressure xenon discharge lamp produces a continuous source of radiation, which covers both the UV and visible range. The discharge light is normally pulsed for short periods with a frequency that determines the average intensity of the light from the source and the source's lifetime.

In atomic emission line sources (e.g., mercury and sodium vapor lamps), the electrons move between atomic energy levels. If the atom is free of any interaction with other atoms, the amount of energy liberated can be precise, and all of the photons share a clearly defined wavelength. This is the characteristic sharp "line" emission from an electronic discharge in a low-pressure gas. Line sources that emit a few discrete lines find wide use in atomic absorption spectroscopy and fluorescent spectroscopy.

A laser source is useful in analytic instrumentation because of its high intensity, narrow bandwidth, and the coherent nature of its output. Several specific uses include (1) high-resolution spectroscopy, (2) kinetic studies of processes with lifetimes in the range of 10^{-9} to 10^{-12} s, (3) the detection and determination of extremely small concentrations of compounds in the atmosphere, and (4) the induction of isotopically selective reactions.

Wavelength Selectors

An important component of all photometers is the device used to select the appropriate wavelength. These devices are collectively referred to as **monochromators**. A monochromator is

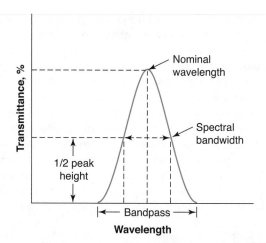

■ **FIGURE 2-8**   Spectral percent transmittance characteristics of a wavelength selector.

a spectroscope modified for selective transmission of a narrow band of the spectrum. There are several types of monochromators, including filters, prisms, grating monochromators, and, recently, holographic gratings. The quality of these selectors is described by their nominal wavelength, effective bandwidths, and bandpass. **Nominal wavelength** represents the wavelength in nanometers at peak transmittance. **Spectral bandwidth** is the range of wavelengths at a point halfway between the baseline and the peak. It is sometimes called the half-power point or *full width at half maximum* (FWHM). The light seen within the spectral bandwidth represents what your eye would see as the output from a monochromator projected onto a screen. Effective bandwidths represent the inverse measure of the quantity of the wavelength selector and should be approximately 1.5% of the nominal wavelength. The total range of wavelengths transmitted is the *bandpass*. The wavelength spectrum shown in Figure 2-8 ■ serves to identify the specific regions defined in the text.

Monochromators   In many spectrophotometric applications it may be necessary to vary or scan the wavelength of EMR while monitoring absorbance. This process of scanning the EMS is used to evaluate amniotic fluid samples for the presence of bilirubin. The full range of EMS can be scanned by adjusting the wavelength selector either manually or automatically. The mechanical construction of monochromators includes (1) an entrance slit that provides a rectangular optical image, (2) a collimating lens or mirror that produces a parallel beam of radiation, (3) entrance and exit windows that protect the components from dust and corrosive laboratory fumes, (4) a grating or prism that disperses the radiation into its component wavelengths, (5) a focusing element that reforms the image of the entrance slit and focuses it on a planar surface called a focal plane, and (6) an exit slit in the focal plane that isolates the desired spectral band.

Filters   There are two types of filters used in most spectrometers. They are absorption filters and interference filters. Absorption filters are used primarily for work in the visible region of the spectrum. These filters function by absorbing certain portions of the spectrum. The most common types consist of colored glass or dyes suspended in gelatin and sandwiched between glass plates. The spectral bandwidth of absorption filters range from about 30 to 50 nm. These types of filters are commonly used in photometers because they provide a less expensive alternative for the laboratory.

Interference filters are available for the ultraviolet, visible, and well into the infrared region of the EMS. This type of filter relies on optical interference to provide narrow bands of radiation. Specific examples of interference filters are Fabry-Perot, wedge, multilayer, and dichroic mirrors. Interference-type filters typically have spectral bandwidths of approximately 1.5% of the wavelength at peak transmittance. A comparison of spectral bandwidth of interference and absorption filters is shown in Figure 2-9 ■.

Prism Monochromators Several types of prism monochromators are used, and the choice depends on the wavelength region to be used. Two common types of prisms designs include 60-degree and 30-degree prisms. Among these types are the Cornu, Bunsen, and Littrow mounts. Each specific type of prism provides unique advantages and quality of light dispersion. Selection of the appropriate prism depends the specific application(s) of the analyst.

The action of a prism depends on the refraction of radiation by the prism material. The dispersive power depends on the variation of the refractive index with wavelength. A ray of radiation that enters a prism at an angle of incidence is bent toward the normal (vertical to the prism face), and at the prism–air interface it is bent away from the vertical. The dispersed light appears as the spectrum of colors that make up the incident radiation.

Grating Monochromators The surface of a grating monochromator is characterized as an array of a very large number of slits spaced equidistant from each other that reflect or transmit radiation. Only at certain definite angles is

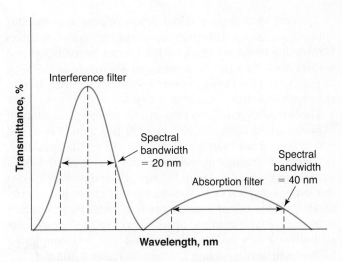

■ FIGURE 2-9 Spectral percent transmittance characteristics of an interference filter versus an absorption filter.

radiation of any given wavelength in phase. At other angles, the waves from the slits destructively interfere.

Most gratings used are of a reflective type. Referred to as diffraction gratings, they consist of parallel, equally spaced grooves ruled by a properly shaped diamond tool directly into a highly polished surface. The quality of the grating depends on (1) the straightness of the grooves, (2) the degree of parallelism, and(3) the equidistance of the grooves to each other. A specific type of diffraction grating is the echellette-type grating shown in Figure 2-10A ■. This grating is grooved or *blazed* such that it has relatively broad faces from which reflection occurs and narrow unused faces. This design provides highly efficient diffraction of radiation.

Three specific types of nonconcave diffraction gratings include Ebert, Littrow, and Czerny–Turner mounts. Each mounting surface is designed to produce the highest quality monochromatic radiation with the least amount of noise or unwanted radiation.

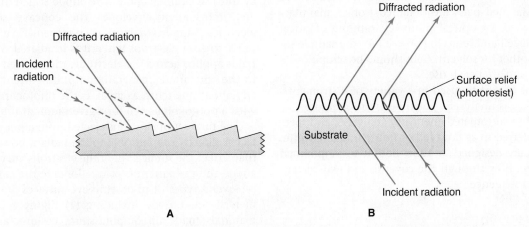

■ FIGURE 2-10 Two types of grating monochromators. A. Echellette-type grating. B. Holographic grating patterned in photoresist.

Concave grating permits the design of a monochromator without auxiliary collimating and focusing mirrors or lenses because the concave surface both disperses the radiation and focuses it on the exit slit. A principal advantage of this type of grating is the reduced expense because of the reduction in the number of mirrored surfaces used.

Holographic gratings represent the newer generation of light-dispersing devices. The surface of this grating is etched with a laser rather than a mechanically driven diamond tool. Holographic gratings provide a greater degree of perfection with respect to line shape and dimension and provide spectra that are affected less by stray radiation and ghosts (double images). The number of grooves per millimeter may approach 6000 using a laser source as opposed to only 2000 grooves per millimeter with a mechanical diamond tool. An example of a holographic grating design is shown in Figure 2-10B ■.

Monochromator Slits The slits of a monochromator are important because they play a significant role in determining the monochromator performance characteristics and quality. In practical spectrophotometry, the monochromator unit is not capable of isolating a single wavelength of radiation from the continuous spectrum emitted by the source. Rather, a finite band of radiation is passed by the monochromator. This finite band arises from the slit distribution. Slit jaws are formed by carefully machining two pieces of metal to give sharp edges. Meticulous care is taken to assure that the edges of the slit are exactly parallel to one another and that they lie on the same plane. In some monochromators, the openings of the two slits are fixed; more commonly, the space can be adjusted with a micrometer device.

Cuvettes

Cuvettes or cells hold samples and reagents that are made of material transparent to radiation in the spectral region of interest. In the UV region (below 350 nm) cuvettes are made from fused silica or quartz. Silicate glasses can be used in the region between 350 and 2000 nm. Plastic cuvettes have also found application in the visible region.

To obtain good absorbance data, the cuvettes used should be matched. Matched pairs are obtained from a manufacturer, and checked for optical purity by running identical samples in each cell and reserving one cell for the sample solution and the other for solvent. Cells should be scrupulously cleaned before and after each use.

Another type of cuvette used in a variety of instruments, including blood-gas analyzers, cooximeters to measure carboxy hemoglobin, automated chemistry analyzers, and spectrometers, is referred to as flow cell or flow through cuvette. These cuvettes are designed to allow liquids to be pumped into the cuvette, flow through the cuvette, and exit the cuvette in a timed sequence.

Radiation Transducers

The transmitted signal in the form of photons passes through the exit slit and collides with the cathode within the radiation transducer or photodetector. A **transducer** is a device that converts one form of energy to another. Photodetectors are transducers that convert EMR into an electron or photocurrent that is passed to a readout circuit. This photocurrent may require amplification, particularly when measuring low levels of radiant energy.

There are several types of photon transducers used. Each photon transducer has an active or photoemissive surface that is capable of absorbing radiation and generating a proportionate signal. In some types, the absorbed energy causes emission of electrons and the development of a photocurrent.

Photovoltaic or Barrier Layer Cell The photovoltaic cell is a basic phototransducer that is used for detecting and measuring radiation in the visible region. This cell typically has a maximum sensitivity at about 550 nm and the response falls off to about 10% of the maximum at 350 and 750 nm.

A photovoltaic cell has a flat copper or iron electrode upon which is deposited a layer of semiconductor material, such as selenium. The outer surface of the semiconductor is coated with a thin transparent metallic film of gold or silver, which serves as the second or collector electrode. When radiation of sufficient energy reaches the semiconductor, covalent bonds are broken, with the result that conduction electrons and holes or voids left by departed electrons are formed. The electrons migrate toward the metallic film and holes toward the base on which the semiconductor is deposited. The liberated electrons are free to migrate through the external circuit to interact with these holes. The result is an electrical current of a magnitude that is proportional to the number of photons that strike the semiconductor surface and a photocurrent that is directly proportional to the intensity of the radiation that strikes the cells.

These types of photodetectors are rugged, low cost, and require no external source of electrical energy. Low sensitivity and fatigue are two distinct disadvantages of these cells. For routine analyses at optimum wavelengths, these photocells provide reliable analytical data.

Vacuum Phototubes A vacuum phototube has a semicylindrical cathode and a wire anode sealed inside an evacuated transparent envelope. The concave surface of the electrode supports a layer of photoemissive material that tends to emit electrons when it is irradiated. When a potential is applied across the electrode, the emitted electrons flow to the wire anode, generating a photocurrent that is generally about one-tenth as great as the photocurrent associated with a photovoltaic cell for a given radiant intensity.

The number of electrons ejected from a photoemissive surface is directly proportional to the radiant power of the beam that strikes the surface. All of the electrons are collected at the anode and the current is proportional to the radiant power.

Several types of photoemissive surfaces are used in commercial phototubes, including (1) highly sensitive bialkali materials made up of potassium, cesium, and antimony; (2) red sensitive material using multialkalis (for example, Na/K/Cs/Sb); (3) ultraviolet sensitive composites with UV

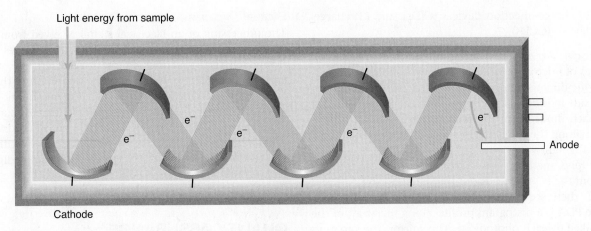

Light energy from sample

Cathode

Anode

■ FIGURE 2-11   Diagram of a photomultiplier tube (PMT)

transparent windows; and (4) flat response-type substances using Ga/As compositions (gallium/arsenic).

Photomultiplier Tubes   Photomultiplier tubes (PMTs) are commonly used when radiant power is very low, which is characteristic of very low analyte concentrations. The operating principle is similar to the phototube with one significant exception—the output signal from the PMT is amplified up to approximately a million fold.

The response of the PMT begins when incoming photons strike a photocathode source similar in composition to the surface of the phototube described earlier. Electrons are ejected from the surface of the photocathode. A PMT has nine additional electrodes called dynodes each having a potential applied to it that is approximately 90 V. Upon striking a dynode, each photoelectron causes emission of several additional electrons; these, in turn, are accelerated toward a second dynode, with an applied voltage of about 90 V, which further amplifies the incident signal. This process shown in Figure 2-11 ■ continues within the PMT until all the electrons are collected at the anode, where the resulting current is passed to an electronic amplifier.

Photomultiplier tubes are highly sensitive to UV and visible radiation. They also have very fast response time. These tubes are limited to measuring low-power radiation because intense light causes irreversible damage to the photoelectric surface.

Silicon Diode Transducers   Silicon diode transducers (SDTs) are more sensitive than vacuum phototubes but less sensitive than PMTs. The SDT has a spectral range from about 190 to 1100 nm. These devices consist of a reverse bias *p-n* junction formed on a silicon chip. The *p-n* junction is constructed by the fusing *p*-type and *n*-type semiconductive materials. The reverse bias configuration has the negative charge on the *n*-type material and the positive charge on the *p*-type with the power supply polarity arranged with positive pole to the *n*-type and negative pole to the *p*-type. The reverse bias creates a depletion layer that reduces the conductance of the junction to nearly zero. If radiation is allowed to impinge on the chip, holes and

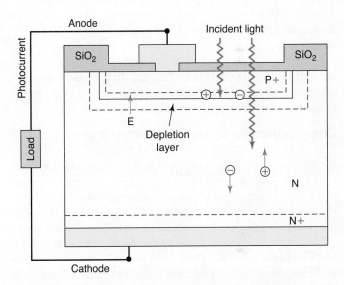

■ FIGURE 2-12   Diagram of a semiconductor diode. (P+ is *p*-type semiconductor material, N is *n*-type semiconductor material, E is emitter electrode, and SiO$_2$ is silicon dioxide.

electrons formed in the depletion layer are swept through the device to produce a current that is proportional to radiant power. This process is shown in Figure 2-12 ■.

Multichannel Photon Transducers   A multichannel transducer consists of an array of small photoelectric-sensitive elements arranged either linearly or in a two-dimensional pattern on single semiconductor chip. The chip, which is usually silicon and typically has dimensions of a few millimeters, also contains electronic circuitry that makes it possible to determine the electrical output signal from each of the photosensitive elements either sequentially or simultaneously. The alignment of a multichannel transducer is generally in the focal plane of a spectrometer so that various elements of the dispersed spectrum can be converted and measured simultaneously. There are several types of multichannel transducers currently used, including (1) photodiode arrays

(PDAs), (2) charge-injection devices (CIDs), and (3) charge-coupled devices (CCDs).

Photodiode Arrays By using the modern fabrication techniques of microelectronics, it is now possible to produce a linear (one-dimensional) array of several hundred photodiodes set side-by-side on a single integrated circuit (IC), or "chip." Each diode is capable of recording the intensity at one point along the line, and together they provide a linear profile of the light variation along the array.

A multiplex method is used to sort all of the signals received from the PDA. It reads out the signal one by one, and then feeds them sequentially to a single amplifier. The output of the PDA is a histogram profile, along the array, of the charge leaked by each photodiode. This mirrors the variation of light intensity across the array. Thus, the PDA detection occurs in three main stages:

1. Initialization
2. Accumulation of charge at each pixel
3. Readout signals

In comparison to the PMT, the PDA has a lower dynamic range and higher noise. Its great value lies in it use as a simultaneous multichannel detector.

Charge-Transfer Devices Recent developments in solid-state detection techniques have now produced very effective two-dimensional array detectors that operate on a charge-transfer process, as an alternative to photodiodes. The term *charge-transfer device* (CTD) is a generic term that describes a detection system in which a photon, striking the IC semiconductor material, releases electrons from their bound state into a mobile state. The released charge, consisting of negative electrons and positive holes, then drifts to and accumulates at surface electrodes. An array of these surface electrodes divides the detector into separate, light-sensitive "pixels." The charge that accumulates at each electrode is proportional to the integrated light intensity falling on that particular pixel.

There are two distinct classes of CTDs: charge-coupled devices (CCDs) and charge-injection devices (CIDs). In a CCD, all of the charge packets are moved "in step" along the array row from one pixel to the next as in a "bucket chain." At the end of the row, the charge packets are fed sequentially into an *on-chip* low-noise amplifier, which then converts the charge into a voltage signal. The overall signal profile across the two-dimensional array is recorded one row at a time, thus giving a series of voltage signals corresponding to all of the pixels in the detection area.

In a CID, the charge accumulated in each pixel can be measured independently and non-destructively by using a network of "sensing" electrodes, which can monitor the presence of the accumulated charge. This is an important factor that differentiates CID systems from CCD and PDA systems, in which the whole of the detection area is "read' destructively in a single process.

Signal Processors and Readout

The processing of an electrical signal received from a transducer is accomplished by (1) a device that amplifies the electronic signal, (2) rectifies alternating current (ac) to direct current (dc) or the reverse, (3) alters the phase of the signal, and (4) filters it to remove unwanted components. In addition, the signal processor may require performing such mathematical operations on the signal such as differential calculations, integration, or conversion to a logarithm. Several readout devices have been used and include digital meters, d'Arsonval meters, recorders, LEDs, cathode-ray tubes (CRTs), and liquid crystal displays (LCDs).

QUALITY ASSURANCE IN SPECTROSCOPY

There are several photometric parameters that must be monitored periodically to ensure optimal performance. Monitoring these parameters is required by laboratory regulatory agencies and accrediting organizations. The parameters routinely monitored include

- Wavelength accuracy
- Bandwidth
- Photometric accuracy
- Linearity
- Stray lights

Accuracy infers the closeness of a measurement to its true value. Wavelength accuracy implies that a photometer is measuring at the wavelength that it is set to. Wavelength accuracy can be assessed quite easily using special glass-type optical filters. Two examples of commonly used filters include didymium and holmium oxide. Didymium glass has a broad absorption peak around 600 nm, and holmium oxide has multiple absorption peaks with a sharp peak occurring at 360 nm. Failure of wavelength accuracy occurs when the spectrometer measurement exceeds the stated tolerance at a specified wavelength. For example, if the tolerance at 360 nm is ±1% and the spectrometer measures 365 nm, then the instrument failed this function check. Wavelength accuracy failures are usually because of a problem with the monochromator.

A mercury vapor lamp is used to measure spectral bandwidth. The lamp has several sharp, well-defined emission lines between 250 and 580 nm. Interference filters with very narrow bandwidths may be purchased to check the laboratory's photometers.

Assessment of photometric accuracy is performed using glass filters or solutions that have known absorbance values for a specific wavelength. The operator simply measures the absorbance of each solution at a specified wavelength and compares their his with the stated values. Each user should establish a tolerance for the measurements based on accepted criteria. Failure of a photometric accuracy check may be because of a failure of the polychromatic light source.

Linearity is defined as the ability of a photometric system to yield a linear relationship between the radiant power incident on its detector and concentration. The linearity of a spectrometer can be determined using optical filters or solutions that have known absorbance values for a given wavelength. Linearity measurements should be evaluated for both slope and intercept. Deviation from linearity may be caused by problems with the light source, monochromator, or detector.

Stray light is described as any light that impinges on the detector that does not originate from the light source. Stray light can have a significant impact on any measurement made. Stray light effect can be evaluated by using special cutoff filters. Stray light measurements that exceed accepted tolerance may caused by the presence of excess room light or sunlight. Stray light can also reach the detector if there is an opening anywhere in the protective "housing" of the monochromator. This opening or crack may occur if the spectrometer is dropped. Therefore, a stray light check should be done anytime the spectrometer is moved from one location to another.

 Checkpoint! 2–9

What would a laboratory's course of action be if its spectrophotometer failed a stray light check?

Types of Photometric Instruments

There are several instrument designs and configurations used for absorption photometry. Each has unique terminology associated with its design. The terminology is not universal among users but is presented here as a guide. A *spectroscope* is an optical instrument used for visual identification of atomic emission lines. It has a monochromator; usually a prism or diffraction grating in which the exit slit is replaced by an eyepiece that can be moved along the focal plane. The wavelength of an emission line can then be determined from the angle between the incident and dispersed beam when the line is centered on the eyepiece.

A *colorimeter* uses the human eye as the detector. The user compares the observed color of the unknown sample to a standard or a series of colored standards of known concentrations. *Photometers* consist of a light source, monochromatic filter and photoelectric transducer, signal processor, and readout. Some manufacturers use the term *colorimeter* or *photoelectric colorimeter* for photometers. These photometers use filters for isolation of specific wavelengths and not gratings or prisms.

A **spectrometer** is an instrument that provides information about the intensity of radiation as a function of wavelength or frequency. Spectrophotometers are spectrometers equipped with one or more exit slits and photoelectron transducers that permit determination of the ratio of the power of two beams as a function of wavelength as in absorption spectroscopy. Most spectrophotometers use a grating monochromator to disperse the light into a spectrum.

A single-beam instrument represents the simplest type of spectrometer (see Figure 2-13 ■). This instrument is designed to make one measurement at a time at one specified wavelength. In using a single-beam instrument, the absorption maximum of the analyte must be known in advance. The wavelength is then set to this value. The reference material (solvent blank) is positioned into the radiation path, and the instrument is adjusted to read $0\%T$ when the shutter is placed so it blocks all radiation from the detector, and to read $100\%T$ when the shutter is removed. After these adjustments have been made, the sample is placed into the path of light, the absorbance is measured, and the concentration is determined by the use of either a calibration curve for the substance or proper algebraic equations.

A double-beam instrument splits or chops the monochromatic beam of radiation into two components. One beam passes through the sample and other through a reference solution or blank. In this design, the radiant power in the reference beam varies with the source energy, monochromator transmission, reference material transmission, and detector response, making the difference between the sample and reference beam largely a function of the sample. The output of the reference beam can be kept constant, and the absorbance of the sample can be recorded directly as the electrical output of the sample beam.

There are two fundamental instrument designs for double-beam spectrophotometers: (1) double beam in space and (2) double beam in time. A double-beam-in-space design uses two photodetectors, one for the sample beam and the other for the reference beam. The two signals generated are directed to a differential amplifier, which then passes on the

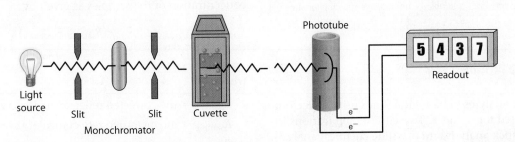

■ **FIGURE 2-13** Schematic of a single-beam spectrophotometer. Symbol e⁻ represents the electrical signal created by photomultiplier tube.

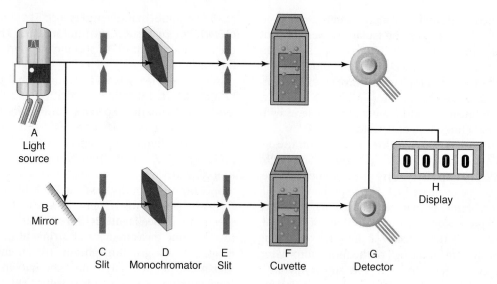

FIGURE 2-14 Schematic of a double-beam spectrophotometer.

difference between the signals to the readout device. A schematic of the double-beam-in-space system is shown in Figure 2-14 ■.

A double-beam-in-time instrument uses one photodetector and alternately passes the monochromatic radiation through the sample cuvette and then to the reference cuvette using a *chopper*. A chopper is the term used for a device, such as rotating sector mirrors, that breaks up or rotates the radiations beams. Each beam, consisting of a pulse of radiation separated in time by a dark interval, is then directed onto an appropriate photodetector.

A scanning double-beam system includes a double-beam spectrophotometer and a recorder that can provide an *x-y* plot of wavelength versus absorbance for a given test sample. This type of configuration is ideal for determining the wavelength spectrum of an analyte in solution. The spectrophotometer has an automatically driven wavelength cam that can rotate at a predetermined speed. The recorder can be calibrated to the wavelength of the monochromator to facilitate the identification of each peak maxima.

 Checkpoint! 2–10

How many detectors does a double-beam-in-space spectrophotometer have?

▶ REFLECTOMETRY

Measurement of analytes in biological fluids using reflectometry has been used for decades. Two clinical applications include urine dipstick analysis and dry-slide chemical analysis. A **reflectometer** is a filter photometer that measures the quantity of light reflected by a liquid sample that has been dispensed onto a nonpolished surface.

There are two types of reflectance: (1) specular and (2) diffuse. Specular reflectance occurs on a polished surface where the angle of incidence of the radiant energy is equal to the angle of reflection. Polished surfaces (e.g., a mirror) are used to direct and manage radiant energy but are not used to determine concentration. Diffuse reflectance occurs on nonpolished surfaces (e.g., grainy or fibrous surfaces). The reflected radiant energy tends to go in many directions. Diffuse reflection occurs within the layers and depends on the properties and characteristics of the layers themselves. A colored substance absorbs the wavelength of its color and reflects all other wavelengths at many different angles. Therefore the amount of a substance present can be measured as an indirect function of the reflected light.

A typical reflectometer used in a clinical laboratory detects only a constant fraction of the diffuse reflected light. Thus the reflectance of a sample is represented by

$$R_{(\text{diffuse reflectance})} = \frac{|R'|_{(\text{fraction of diffuse reflectance of sample})}}{R'_{(\text{fraction of diffuse reflectance of a standard})}} \quad (2.11)$$

The amount of light reflected by a solution dispensed onto a white granular or grainy surface is inversely related to the concentration of the samples as given by:

$$R_{\text{density}} = -\log\left(\frac{(R_{\text{sample}} - R_{\text{black}})}{R_{\text{standard}}}\right) \times R_{\text{white}} \quad (2.12)$$

where

R_{density} = the corrected reflectance density of the sample

R_{sample} = the measured reflectance of the sample

R_{black} = the reflectance of a black reference

R_{white} = the reflectance of a white reference

R_{standard} = the reflectance of a standard solution

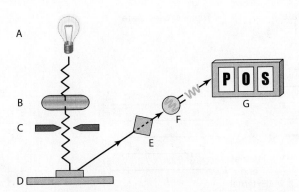

FIGURE 2-15 Diagram of a typical reflectometer. A. Polychromatic light source. B. Monochromator. C. Slit. D. Diffuse reflective surface. E. Lens. F. Photodetector. G. Readout device.

The ideal reflectance value of a pure white standard of ceramic material is *one*, signifying that all light is reflected; and conversely the ideal reflectance value of pure black material is *zero* where all light is absorbed.

The relationship of reflectance and concentration of an analyte is nonlinear. Therefore a series of standards must be assayed to correct for the inherent nonlinearity in reflectance measurements.

REFLECTOMETERS

The components of a reflectometer are very similar to a photometer, as shown in Figure 2-15 ■. A tungsten-quartz halide lamp serves as a source of polychromatic radiation. The light passes through a slit and is directed onto the surface of a urine dipstick "pad" or dry slide. A wavelength selector, such as a stationary filter or filter wheel for multiple analytes, is used to isolate the wavelength of interest. Solid-state photodiodes are typically used to detect the reflected radiant energy. Special optical devices (e.g., fiber optics or ellipsoidal mirrors) may be used to direct radiant energy onto the detector. A computer or microprocessor is used to convert nonlinear reflectance signals into direct readout concentration units.

▶ ATOMIC ABSORPTION SPECTROMETRY

Atomic absorption spectrometry (AAS) is used for quantitative analysis of metals such as calcium, lead, copper, and lithium. This method of analysis provides the laboratory with a very sensitive and specific means for measuring metals. Clinical laboratories have been slow to adopt this technique because of the complexity associated with the procedures and instrumentation. Also, AAS is time consuming to perform, labor intensive, and requires meticulous laboratory techniques.

The fundamental principle of AAS involves the absorptions of monochromatic EMR by an element in its ground state with a net zero charge. A sample containing the metal

in question is aspirated into a flame. The heat energy causes the metal to disassociate from any complexes (e.g., protein and salts). It also converts the element to ground state. The metal in its elemental or ground state is able to absorb EMR unique to that metal. The source of EMR for a specific metal comes from a hollow cathode lamp. This lamp contains a small amount of the pure metal to be measured in the sample. When the lamp is heated, the metal in the cathode emits EMR, which is directed into the flame and is absorbed by the same metal present in the sample. A detector, usually a PMT, is placed in close proximity to the flame and any unabsorbed monochromatic EMR is directed toward its cathode. The amount of absorbed radiation is directly proportional to the concentration of the metal in solution.

> ### ✓ Checkpoint! 2–11
>
> *What is the source of EMR required for a metal to be measured in a sample by atomic absorption spectroscopy?*

▶ MOLECULAR LUMINESCENCE SPECTROSCOPY

FLUOROMETRY

Clinical laboratory science has used the natural ability of compounds to fluoresce as a means to quantitate their concentrations in biological fluids for decades. In the 1970s there was a concerted effort to find a ligand label to replace radioactive isotopes in competitive immunoassays. Abbott Diagnostics (Abbott Park, Chicago, IL) showed that fluorescein served as a suitable fluorescent label to measure therapeutic drugs in serum. This sparked advancements in the use of novel labels such as biological or synthetic, fluorescent or nonfluorescent labels that could be bonded to either antigens or antibodies.

Fluorescent spectroscopy is widely used because of its inherent high sensitivity and high specificity. Highly sensitive data can be obtained as a result of the difference in wavelengths between the excitation energy and the emission energy of the fluorescent compound. With appropriate filtering, the result is a signal contrasted with nearly zero background. High specificity results from dependence on two spectra, the excitation and emission spectra, and the possibility of measuring the lifetimes of the fluorescent state. Two compounds that are excited at the same wavelength but emit at different wavelengths are readily differentiated without the use chemical separation techniques.

Principles of Fluorometry

Luminescence is based on an energy exchange process that occurs when valence shell electrons absorb EMR, become excited, and return to an energy level lower than their original level. Because some energy is lost before emission from the excited state by collision with the solvent or other molecules,

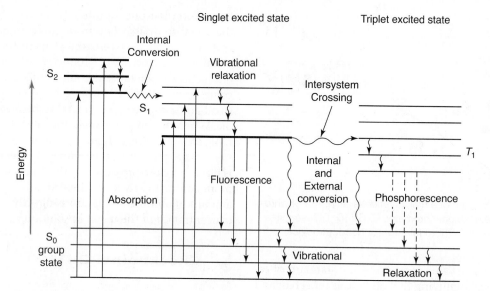

■ **FIGURE 2-16**   Mechanism of fluorescent and phosphorescent energy production. S_0 is ground-state electrons, S_1 is first singlet state, S_2 is second singlet state, T_1 is first triplet state.

the wavelength of the emitted light is longer than that of the exciting light. Most uncharged molecules contain even numbers of electrons in the ground state. The electrons fill molecular orbits in pairs, with their spins in opposite directions. No electron energies can be detected by application of a magnetic field to such spin patterns, and this electronic state is called the singlet state. Similarly, if an electron becomes excited by EMR and its spin remains paired with ground state, it becomes an excited singlet state electron. The lifetime of the excited state is the average length of time the molecule remains excited before emission of light. For a singlet-excited state, the lifetime of the excited state is on the order of 10^{-9} to 10^{-6} seconds. The light emission from a singlet-excited state is called **fluorescence**. When the spin of the electrons in the excited state is unpaired, the electron energy levels will be split if a magnetic field is applied, and this electronic state is called a triplet state. Triplet state lifetimes range from 10^{-4} to 10 seconds[1]. Light emission from an excited triplet state is called phosphorescence. The energy transitions resulting in fluorescence and phosphorescence is presented in Figure 2-16 ■.

Instrumentation

The conventional design of fluorometers place the detector at a 90° angle to the polychromatic light source and is shown in Figure 2-17 ■. This design is optimum because none of the sample cuvette surfaces directly illuminated by the excitation beam are directed toward the emission monochromator; therefore no cuvette fluorescence or light reflected from these surfaces enters the emission monochromator.

Sources   Several factors should be considered regarding the source of EMR, including (1) intensity, (2) wavelength distribution of emitted radiation, and (3) stability. Sources of EMR

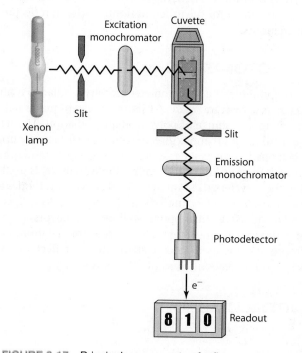

■ **FIGURE 2-17**   Principal components of a fluorometer.

typically include high-pressure dc xenon arc lamps, low-pressure mercury vapor lamps equipped with a fused silica window, and tunable dc lasers* pumped by a pulsed nitrogen gas laser or Nd:YAG laser (neodymium:yttriumaluminum garnet).

Filters and Monochromators   Both interference and absorption filters have been used in fluorometers for wavelength selection of both the excitation beam and emission

*Laser is an acronym for light amplification stimulated emission of radiation.

beam. Most spectrophotofluorometers use at least one and sometimes two grating monochromators.

Transducers The luminescent signal in most fluorescent assays is low intensity. Therefore, large amplifier gains are needed. PMTs are the most common transducers found in fluorescent instruments. Newer fluorometers on the market today use diode-array and CTDs, which allow the rapid recording of both excitation and emission spectra and are particularly useful in chromatography and electrophoresis.

Cuvettes or Cells Cuvettes or cells used for fluorescent measurement may be rectangular or cylindrical. They may be made of glass or quartz depending on the region of the EMS used in the assay. Most cell compartments use baffles to reduce the amount of scattered radiation reaching the detector. Fluorescent assays require meticulous handling of cells or cuvettes. It is very important to avoid leaving fingerprints on their surfaces because some skin oils often fluoresce when exited at specific wavelengths.

Applications of Fluorescent Spectroscopy

Fluorescent Polarization Immunoassay A unique application in fluorescent spectroscopy developed in the early 1970s and still used today is fluorescent polarization immunoassay (FPIA). The advantage of this application is that the measurement of the fluorescent-labeled bound fraction is determined in the presence of the fluorescent-labeled free fraction. This is referred to as a homogenous immunoassay. Thus, there is no need for physical separation of bound fraction from free fraction and results in reduced time for analysis.

This technique requires a fluorometer capable of providing linearly polarized light to excite sample molecules. A quartz halogen lamp provides the polychromatic light. An excitation filter isolates the monochromatic radiation that will be used to excite the fluorophore. The next component is a dichroic polarizer that absorbs the horizontal component of the beam and allows the vertical component to pass through. A liquid crystal serves to rotate the plane of polarization of the exciting light from vertical to horizontal, by applying an alternating voltage. The sample cuvette containing the sample and reagent is placed in the path of the oscillating polarized beam of light. The next array of internal components is aligned to receive the reemitted radiation from the fluorophore. An emission filter transmits the photons to be detected and a-vertical polarizer filter allows only vertically polarized light to pass through to the PMT detector.

Time-Resolved Fluorescent Immunoassay Time-resolved fluorescent assays are available for clinical use. A specific application that resulted in the development of automated instrumentation is time-resolved fluorescence immunoassays (TR-FIA). These assays are very sensitive and tend to minimize problems inherent with other fluorescent assays, such as overlapping excitation or emission spectra of compounds present in the sample with the fluorophore. The label most commonly used is a chelate of europium (Eu^{3+}). Energy is absorbed by the organic ligand, leading to an excited state as the electrons migrate from the ground-state singlet to the excited singlet state. The excitation may lead to any vibrational multiplet of the excited state S_1. The molecule rapidly returns to the lowest energy levels in S_1 by a nonradiative process.

The energy is transferred to the metal ion, which becomes excited and subsequently emits characteristic radiation. The radiative transition of excited Eu^{3+} after an energy transfer from triplet state results in an emission wavelength of 613 nm. The fluorescence lifetime of Eu^{3+} chelate is 10 to 1000 μs compared to nanoseconds for the most commonly used fluorophore. Therefore, Eu^{3+} with its longer emission lifetime makes it more attractive to use over a fluorophore like fluorescein, which has a lifetime of 4.5 ns.

Time-resolved fluorescence (TR-F) instruments are similar to a typical fluorometer except the time-resolved fluorometer uses a time-gated measure on only a portion of the total emission spectra. Equipment for TR-F measurement can be broken down into the following components:

- An excitation source that provides pulsed signals
- A sample chamber with optics for focusing the exciting light and fluorescence
- A detector for converting fluorescence into an electronic signal
- The recording electronics for processing the detector output

Front-Surface Fluorometry Front-surface fluorometry is used for immunoassay and non-immunoassay techniques. For example, the measurement of zinc protoporphyrins as free erythrocyte protoporphyrin is done using front-surface fluorescence. The unique instrument design allows the emitted radiation to be detected off a surface or (face) containing molecules that fluoresce.

The sample of blood is placed onto a thin glass slide. A quartz halogen lamp and excitation filter are placed at an angle less than 90° to the glass slide. Monochromatic light used to excite the compound to be measured is directed onto the surface of the glass slide where the sample has been applied. Light that is reemitted (fluorescence) is passed through an emission filter to isolate the specific wavelength. A photodetector is used to capture the photons and generate a signal proportional to the concentration of analytes in the sample. The emission filter and PMT are placed directly above the glass slide.

A second example represents a more sophisticated system to excite and detect emitted photons. The sample containing the fluorophore is placed onto a pure white surface. A high-energy polychromatic light source provides the photons for excitation of the fluorophore. An excitation filter is used to isolate the wavelength required by the fluorophore. The emitted light from the fluorophore is directed toward an emission filter, which isolates the fluorophore's primary emission wavelength. A photodetector is used to transform the photon energy to electrical energy. A unique component of this instrument design is the dichroic filter, which reflects a specific wavelength region, directs the excitation energy to the sample, and does not allow any unwanted photons to strike the photodetector.

 Checkpoint! 2-12

True or false? The excitation wavelength of a fluorophore has a higher energy value and a shorter wavelength than the emission wavelength.

► CHEMILUMINESCENCE

Chemiluminescence applications have increased dramatically because of the increased sensitivity over fluorescence. The primary application has been in the area of immunoassays, where several chemiluminescence compounds have been used as antigen labels.

Chemiluminescence differs from fluorescence and phosphorescence in that the emission of light is created from a chemical or electrochemical reaction and not from electromagnetic energy stimulation of electrons resulting in emission of photons. Chemiluminescence involves the oxidation of an organic compound, such as dioxetane, luminol, or acridinium ester, by an oxidant (hydrogen peroxide, hypochlorite, or oxygen). These oxidation reactions may occur in the presence of catalysts, such as enzymes (alkaline phosphatase, horseradish peroxidase, or microperoxidase), metal ions (Cu^{+2} or Fe^{+3} phthalocyanine complex), or hemin. The excited products formed in the oxidation reaction produce chemiluminescence on return to the singlet state.

The basic components of a chemiluminescent analyzer are shown in Figure 2-18 ■. The detector is a luminometer that contains a PMT and provides a very strong electrical output signal. A typical signal from a chemiluminescent compound as a function of time rises rapidly to a maximum as mixing of reagent and analyte is complete. An exponential decay of the signal follows until baseline is reached.

 Checkpoint! 2-13

Explain the mechanism of light production by chemiluminescence.

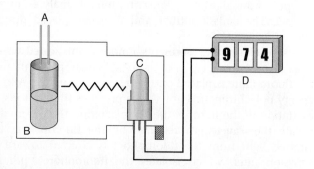

■ FIGURE 2-18   Diagram of the principal components of a luminometer. A. Reagent probes. B. Sample and reagent cuvette. C. Photomultiplier tube. D. Readout device.

► LIGHT SCATTER TECHNIQUES

Two useful techniques available for measuring the concentration of a solution that contains particles too large for absorption spectroscopy are nephelometry and turbidimetry. These methods are suitable for quantitative assays using antigen–antibody complexes or measuring the amount of proteins in fluids. In this section the focus is on the light scattered by the particles in solution.

Nephelometry is the measurement of the light scattered by a particulate solution. Three types of light scatters occur based on the relative size of the light wavelength.[2] If the wavelength (λ) of light is much larger than the diameter (*d*) of the particle, where $d < 0.1$, the light scatter is symmetric around the particle. Light scatter is minimally reflected at 90° to the incident beam and was described by Rayleigh.[3] If the wavelength of light is much smaller than the particle diameter, where $d > 10\lambda$, the light scatters forward because of the destructive out-of-phase backscatter, as described by the Mie theory. If the wavelength of light is approximately the same as the particle size, more light scatters in the forward direction than in other directions, as defined by the Rayleigh–Debye theory. A common application of nephelometry is the measurement of antigen–antibody reactions. Because most antigen–antibody complexes have a diameter of 250 to 1500 nm and the wavelengths used are 320 to 650 nm, the light scatter is essentially of the Rayleigh–Debye type.*

INSTRUMENTATION

A typical nephelometer consists of a light source, a collimator, a monochromator, a sample cuvette, a stray light trap, and a photodetector. Light scattered by particles is measured at an angle, typically 15° to 90°, to the beam incident on the cuvette. Figure 2-19 ■ shows two possible optical arrangements

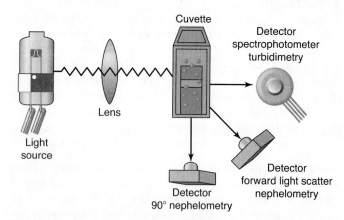

■ FIGURE 2-19   Light scatter measurements using nephelometry and turbidimetry.

*Mie theory is an analytical solution of Maxwell equations for the scattering of EMR by spherical particles.

for a nephelometer. Light scattering depends on the light wavelength and the particle size. For macromolecules whose sizes are close to or larger than the light wavelength, measurement of the forward light scatter increases the sensitivity of the measurement. Light sources include a mercury-arc lamp, a tungsten-filament lamp, a light-emitting diode, and a laser.

TURBIDIMETRY

Turbidimetry is the measurement of the reduction in light transmission caused by particle formation. Light transmitted in the forward direction is detected. The amount of light scattered by a suspension of particles depends on the specimen concentration and particle size. Solutions requiring quantitation by turbidimetry are measured using photometers or spectrophotometers. Higher sensitivity has been achieved using photodetectors that can detect small changes in photon signals. Sensitivity comparable to nephelometry can be attained using low wavelengths and high-quality spectrophotometers. Many clinical applications exist for turbidimetry. Various microbiology analyzers measure turbidity of samples to detect bacterial growth in broth cultures. Turbidimetry is routinely used to measure the antibiotic sensitivity from such cultures. In coagulation analyzers, turbidimetric measurements detect clot formation in the sample cuvettes. Turbidimetric assays have long been available in clinical chemistry to quantify protein concentration in biological fluids, such as urine and cerebrospinal fluid (CSF).

▶ REFRACTOMETRY

Refractometry has been applied to various measurements (e.g., total serum protein concentration, specific gravity of urine, and column effluent from high-performance liquid chromatography analysis). The principle of refractometry is based on the refraction of light as it passes through a medium such as glass or water. When light passes from one medium into another, the light beam changes its direction at the boundary surface if its speed in the second medium is different from that in the first. The angle created by the bending of the light is called the *critical angle*. The ability of a substance to bend light is called *refractivity*. The refractivity of a liquid depends on the wavelength of the incident light, temperature, nature of the liquid or solid medium, and concentration of the solute dissolved in the medium. If the first three factors are held constant, the refractivity of a solution is an indirect measurement of total solute concentration.

To measure the specific gravity or protein concentration of a sample, the user places a sample on the glass surface of the refractometer or total solids (TS) meter. The sample moves along the glass surface, creating a thin film over the glass. Polychromatic light is directed onto the meter. Inside the meter is an Amici compensator (prism) that is used to isolate a monochromatic beam of radiation, usually the sodium D line (about 589 nm). The image seen within the meter represents the critical angle created by the sample. This critical angle forms a line of demarcation (i.e., where the dark area meets the lighter area). A scale is devised that correlates a critical angle value with specific gravity or protein concentration.

Calibration of a TS meter is accomplished using an appropriate reference solution of pure water. An adjustment of the calibration is accomplished using the small screw embedded within the housing of the meter.

> ### ✓ Checkpoint! 2–14
>
> *What does the line of demarcation seen within the refractometer after a sample is applied to the glass surface represent?*

▶ OSMOMETRY

Osmometry is the measurement of the osmolality of an aqueous solution such as serum, plasma, or urine. As osmotically active particles (e.g., glucose, urea nitrogen, and sodium) are added to a solution the osmolality increases, and four other properties of the solution are also affected. These properties are osmotic pressure, boiling point, freezing point, and vapor pressure. They are collectively referred to as colligative properties of the solution because they can be related to each other and to the osmolality. As the osmolality of a solution increases, (1) the osmotic pressure increases, (2) the boiling point is elevated, (3) the freezing point is depressed, and (4) the vapor pressure is depressed. Osmometry is based on measuring changes in the colligative properties of solutions that occur because of variations in particle concentration. Freezing-point depression osmometry is the most commonly used method in the clinical laboratory for measuring the changes in colligative properties of a solution.

PRINCIPLES OF FREEZING-POINT DEPRESSION OSMOMETRY

A 1.0 mOsm solution will result in a depression in the freezing point of 0.00186°C when compared with the pure solvent (usually water). The measurement of this temperature change and its application to sample osmolality is found in the following thermodynamic equation:

$$\Delta T = RT_0^2 M_1 W_2 |H_f| W_1 M_2 \qquad (2.13)$$

where

ΔT = change in freezing-point temperature

R = gas constant, 1.987 calories/mole

T_0 = freezing point of the solvent in K

M_1 = the molecular weight of the solvent

M_2 = the weight of the solute in grams

W_1 = the weight of the solvent in grams

W_2 = the weight of the solute in grams

H_f = the heat of fusion of the solvent, calories/mole

Example:

What is the change in the freezing point (ΔT) for an aqueous solution where (the heat of fusion of water is 1436 calories per mole and the freezing point of water is 273.1 K) 60 grams of urea is dissolved in 1000 gram of water?

$$\Delta T = RT_0^2 M_1 W_2 / H_f W_1 M_2$$

$$\Delta T = (1.987)(273.1)^2(18.02)(60)/(1436)(1000)(60)$$

$$\Delta T = 1.86°$$

From these data the molality (m) may be calculated from the simplified equation:

$$m = \Delta T / 1.86$$

$$m = 1.86 / 1.86$$

$$m = 1$$

The thermodynamics of freezing-point osmometry are such that if one carefully measures the temperature of a solution as it is slowly cooled to a temperature below freezing, known as supercooling, and then initiates freezing, a uniform freezing curve results from the process of crystallization. As freezing is initiated, the temperature will rise rapidly to a point just below the freezing point of the solution, where it will then stabilize for a short period of time. The temperature rise is because of the release of latent **heat of fusion** as the sample crystallizes. During the phase of equilibrium, the freezing-point depression is read. It should be noted that what is being observed is the temperature change because of heat released by the sample as a phase change from liquid to solid is taking place.

FREEZING-POINT OSMOMETER

A freezing-point osmometer consists of a sample chamber containing a stirrer and a thermistor (temperature-sensing device) connected to a readout device. The sample is rapidly supercooled to several degrees below its freezing point in a refrigeration chamber usually containing ethylene glycol. The sample is then agitated with the stirrer to initiate freezing. As the ice crystals form, heat of fusion is released from the solution. The rate at which this heat of fusion is released from the ice being rapidly formed reaches equilibrium with the rate of heat removed by the colder temperature of the sample chamber. This equilibrium temperature, known as the freezing point of the solution, stays constant for several minutes once it is reached. This freezing point is detected by the thermistor, and the osmolality of the sample is converted to units of milliosmoles per kilogram of water.

A graph of the events that occur using a freezing-point depression osmometer is provided in Figure 2-20 ■. The y-axis represents temperature (°C) and the x-axis is time (minutes). Each significant point along the freezing-point curve is identified in the figure.

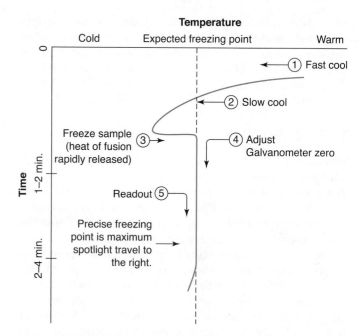

■ **FIGURE 2-20**   Diagram of a uniform freezing point depression curve illustrating the entire process which begins at number 1, fast cool, and ends with a direct readout in units of mOsmol/kg.

 Checkpoint! 2-15

If the number of solute particles in an aqueous solution is increased, would the change in freezing-point temperature increase or decrease?

▶ ELECTROCHEMISTRY

Electrochemistry involves measuring the current or voltage generated by the activity of specific ion species. Analytical electrochemistry for the clinical laboratory includes, for example, potentiometry, coulometry, voltammetry, and amperometry.

POTENTIOMETRY

The measurement of potential (voltage) between two electrodes in a solution forms the basis for a variety of procedures for measuring analyte concentration. Electrical potentials are produced at the interface between a metal and ions of that metal in a solution. Such potentials also exist when a membrane semipermeable to that ion separates different concentrations of an ion. To measure the electrode potential, a constant-voltage source is needed as the reference potential. The electrode with a constant voltage is called the *reference electrode*, whereas the measuring electrode is termed the *indicator electrode*. Concentration of ions in a solution can be calculated from the measured potential difference between the two electrodes. The measured cell potential is related to the molar concentration by the **Nernst equation**:

$$E = E° - RT/nF \ln a_{red}/a_{ox} = E° - 2.302 \, RT/nF \log a_{red}/a_{ox}$$

(2.14)

where

$$E = \text{cell potential measured}$$
$$E^O = \text{standard reduction potential}$$
$$N = \text{number of electrons involved in the reaction}$$
$$a_{red} = \text{activity of the reduced species}$$
$$a_{ox} = \text{activity of the oxidized species}$$
$$F = \text{faraday (96,487 coulombs)}$$
$$T = \text{absolute temperature}$$
$$R = \text{molar gas constant}$$

Substituting molar concentration for activity and common logarithm for natural log:

$$E = E^O - (0.0592/n) \log C_{red}/C_{ox} \qquad (2.15)$$

where

$$C_{red} = \text{concentration of reduced species}$$
$$C_{ox} = \text{concentration of oxidized species}$$

The Nernst equation is useful for predicting the electrochemical cell potential given the concentrations of oxidized and reduced species for a given electrode system.

Reference Electrodes

In most electroanalytical applications (e.g., blood pH) it is desirable that the half-cell potential of one electrode be known, constant, and completely insensitive to the composition of the solution under study. An electrode that fits this description is called a reference electrode.

Important attributes of a reference electrode include the following: (1) Potential is reversible and obeys the Nernst equation, (2) the electrode exhibits a potential that is constant with time, (3) the electrode returns to its original potential after being subjected to small currents, and (4) the electrode exhibits little hysteresis or lag with temperature cycling.[4] The standard hydrogen electron (SHE) exhibits many of these qualities but is not practical to use in clinical laboratory instrumentation. Thus, two commonly used reference electrodes, the calomel and silver/silver chloride electrodes, provide many of the analytical qualities necessary for reliable measurements yet are designed for clinical use.

The calomel electrode (saturated calomel electrode, SCE) consists of a mercury/mercury (I) chloride (calomel) paste in saturated potassium chloride located in an inner tube, which is positioned within another tube containing a saturated potassium chloride solution. The inner tube has an opening in the tip which allows the two solutions to come into contact with each other. Contact with the indicator electrode system is made by means of a fritted porcelain plug, sealed in the end of the outer tubing.

The silver/silver chloride electrode consists of a silver electrode immersed in a solution of potassium chloride that has been saturated with silver chloride. Silver/silver chloride electrodes have the advantage in that they can be used at temperatures greater than 60°C, whereas calomel electrodes cannot.

TABLE 2-1

Commonly used ion-selective electrodes in clinical laboratories.

Ion for which the electrode is selective	Key membrane component
Sodium (Na$^+$)	Silicate in glass
Lithium (Li$^+$)	Dodecylmehtyl-14-crown-4 (ether)
Calcium (Ca^{++})	Calcium di-(n-decyl)phosphate in di-(n-octylphenyl)-phosphate
Magnesium (Mg$^+$)	N'N"N'''-imin-6,1-hexandiyl-tris(N-heptyl-N-methyl-malonamide
Hydrogen (H$^+$)	Neutral H+ carrier tridodecylamine
Chloride (Cl$^-$)	Solvent polymeric membranes
Potassium (K$^+$)	Valinomycin

Ion-Selective Electrodes

An **ion-selective electrode** (ISE) is a membrane-based electrochemical transducer capable of responding to a specific ion. A potential difference or electron flow is created by selectively transferring the ion to be measured from the sample solution to the membrane phase. ISEs measure ion activities, specifically free ion concentration. The selectivity of these transducers is determined by the nature and composition of the membrane material used. These membranes are very complex and consist of several compounds, including plasticizers, organic solvents, inert polymers, and ionophores. An ionophore is defined as any molecule that increases the permeability of a membrane to a specific ion. Several examples of ISEs and their respective membrane components are shown in Table 2-1 ✪.

ISEs provide several advantages over "wet chemistry" and photometric techniques. These advantages include

- No reagent preparation
- No standard curve preparation
- Direct measurement (i.e., no dilutions)
- Cost effective (i.e., can be reused)
- Fast analysis times
- Precise measurements
- Very sensitive
- Very selective or specific for the analyte
- 24-hour state of readiness
- Easy to maintain
- Easily adapted to automation

pH Electrodes

Glass electrodes are the most common electrodes for measuring hydrogen ion activity. A pH electrode consists of a small bulb located at the tip of the electrode made of layers of hydrated and nonhydrated glass. Inside the electrode is a chloride ion buffer solution. A diagram of a pH electrode and its

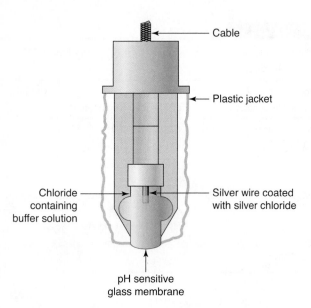

■ FIGURE 2-21   Diagram of a potentiometric pH electrode.

significant components is shown in Figure 2-21 ■. The buffer has a known hydrogen ion concentration. An internal electrode, usually silver/silver chloride, serves as a reference electrode. A SCE is used as an external reference electrode. One theory suggests that the sodium ions in the hydrated glass layer drift out. Sodium ions have a large ionic radius. Specimens containing hydrogen ions, which have a smaller ionic radius, replace the sodium ions. The result is a net increase in the external membrane potential difference. This potential propagates through the thin, dry membrane to the inner hydrated surface of the glass. Chloride ions in the inner buffer solution respond by migrating to the internal glass layer. The potential difference created at the pH electrode is referenced to the external reference electrode, and the difference or change is displayed as a numerical value for pH.

PCO$_2$ Electrodes

The PCO_2 electrode is a pH electrode contained within a plastic "jacket." This plastic jacket is filled with a sodium bicarbonate buffer and has a gas-permeable membrane (Teflon or silicone) across its opening. When whole blood containing dissolved CO_2 comes into contact with the Teflon membrane, CO_2 from the blood passes through and mixes with the buffer. The chemical reaction shown below occurs and results in a change in pH. The hydrogen ion activity is measured by a potentiometric pH indicator system. The basic design of a PCO_2 electrode is shown in Figure 2-22 ■.

$$CO_2 + H_2O \Leftrightarrow H_2CO_3 \Leftrightarrow HCO_3^- + H^+$$

COULOMETRY

Coulometry is an analytical method that involves measuring the quantity of electricity (in coulombs) needed to convert the analyte quantitatively to a different oxidation state. By

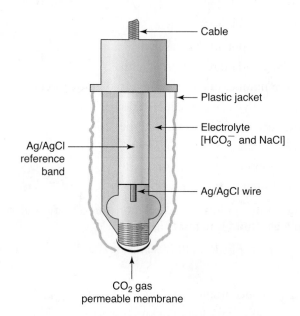

Reaction occuring in the electrolyte solution:

$$CO_2 + H_2O \longleftrightarrow H_2CO_3 \longleftrightarrow H^+ + HCO_3^-$$

■ FIGURE 2-22   A gas-sensing PCO_2 electrode.

definition a coulomb is the quantity of electricity or charge that is transported in one second by a constant current of one ampere. This relationship is shown in Equation 2.16.

$$Q = It \qquad (2.16)$$

where

Q = amount of electricity (coulombs = ampere/s)

I = current in amperes

t = time in seconds

A Faraday is the charge in coulombs associated with one mole of electrons. The charge of the electron is 1.60218×10^{-19} C; thus one Faraday is equal to $96,487$ C $\cdot$ mol^{-1} of electrons.

The relationship between the charge and the amount of substance electrolyzed, Faraday's Law, is expressed in Equation 2.17.

$$Q = Fnz \qquad (2.17)$$

where

F = Faraday constant (96,487 C $\cdot$ mol^{-1})

z = the number of electrons involved in the reduction (or oxidation) reaction

n = number of moles of analytes in the solution

Laboratory application of coulometry includes the measurement of chloride ions in serum, plasma, CSF, and sweat samples. The amount of chloride can be determined by coulometric titration using a chloridometer. The active electrode process involves generation of the titrating reagents, which are silver ions produced at a silver anode. The current is carefully maintained at a constant and accurately known level by means of an amperostat. The product of this current in amperes and the time in seconds required to reach an endpoint yields the number coulombs, which is proportional to the quantity of chloride in the sample.

AMPEROMETRY

Amperometry is the measurement of the current flow produced by an oxidation-reduction reaction. Amperometric methods have been used in clinical laboratories for decades and include immobilized enzyme electrodes, the Clark PO_2 electrode, and electrodes for measuring heavy metals such as lead.

The measurement of chloride in samples involves the use of two electrochemical methods: coulometry (discussed above) and amperometry. The chloride titrator includes a pair of silver electrodes that serve as the indicator electrodes. When all of the chloride in the sample has been consumed, silver ions appear in excess, which cause an increase in current. At this point all of the chloride is bound to free Ag^+ and the endpoint has been reached.

PO_2 Gas-Sensing Electrode

A widely used application of amperometry is the determination of the partial pressure of oxygen in blood samples using an oxygen-sensing electrode. The Clark PO_2 electrode consists of a gas-permeable membrane, usually polypropylene, which allows dissolved oxygen to pass through.[4] This membrane also prevents other blood constituents from passing through that may interfere with the electrode. Once the oxygen passes through the membrane, it mixes in a phosphate buffer solution and reacts with the polarized platinum cathode. The cathode has an applied voltage of approximately 630 mV, which results in the reduction of oxygen according to the following reaction:

$$\text{Cathode: } O_2 + 4e^- \rightarrow 2O^-$$
$$2O^- + 2H_2O \rightarrow 4OH^-$$

Reduction of molecular oxygen produces a flow of electrons (electrical current), and the magnitude of current is proportional to the amount of oxygen present in the sample.

Elemental silver is oxidized at the anode to an ionized form plus four electrons as shown below:

$$\text{Anode: } 4\,Ag \rightarrow 4\,Ag^+ + 4e^-$$
$$4\,Ag^+ + Cl^- \rightarrow 4\,AgCl$$

A diagram of the basic components of the PO_2 electrode is shown in Figure 2-23 ■. The gas-permeable membrane is secured at the tip of the electrode.

VOLTAMMETRY

Voltammetry comprises a group of electroanalytical methods in which information about the analyte is derived from the

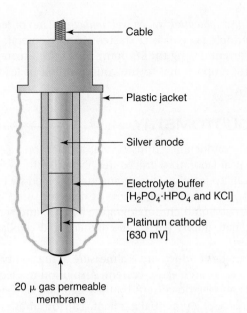

Reactions occuring in the cathode and anode:

Anode

$$4Ag \longleftrightarrow 4Ag^+ + 4e^-$$
$$4Ag^+ + 4Cl^- \longleftrightarrow 4Agcl$$

Electrolyte

$$KCl \longleftrightarrow K + Cl^-$$
$$H_2PO_4^- \longleftrightarrow H_2PO_4 + H^+$$

Cathode

$$O_2 + 4e^- \longleftrightarrow 2O^{--}$$
$$2O^{--} + 2H_2O \longleftrightarrow 4OH^-$$

■ **FIGURE 2-23** Schematic of a gas-sensing, amperometric PO_2 electrode.

measurement of current as a function of an applied potential obtained under conditions that promote polarization of an indicator, or working, electrode. Voltammetry is based on the measurement of a current that develops in an electrochemical cell under conditions of complete concentration polarization. Also in voltammetry a minimal consumption of analyte takes places, unlike coulometric assays, where essentially all of the analyte is converted to another state.

Anodic Stripping Voltammetry

The measurement of lead in whole blood samples can be performed in the clinical laboratories using anodic stripping voltammetry (ASV). This technique consists of three major steps to directly determine the concentration of lead in whole blood. The first step involves the reduction of lead and deposition of the lead onto the electrode. This deposition occurs at an applied voltage of about –0.5 V. Time, temperature, and stirring of the sample must be kept constant. The next step is a "resting period" in which stirring is halted but the potential remains on the electrode. Finally the lead is stripped from the electrode back into the solution by oxidation to the ionic form. The potential is scanned in the positive direction using *linear potential sweep voltammetry*. The potential at which stripping occurs is characteristic of the metal. The observed current during the stripping process is integrated and is directly proportional to the concentration of lead in the sample.

CONDUCTOMETRY

Electrical conductance finds several applications associated with clinical laboratory procedures. Examples include monitoring water purity; measuring analytes in blood such as urea nitrogen; serving as components of detectors used in high-performance liquid chromatography (HPLC); and serving in gas chromatography (GC), cell counters, and capillary electrophoresis.

Electrolytic conductivity is a measure of the ability of a solution to carry an electric current. A solution of electrolytes conducts an electric current by the migration of ions under the influence of a potential gradient. The ions move at a rate dependent on their charge and size, the microscopic viscosity of the medium, and the magnitude of the potential gradient. Thus, for an applied potential that is constant but at a value that exceeds the deposition potential of the electrolyte, the current that flows between the electrodes immersed in the electrolyte varies inversely with the resistance of the electrolytic solution. The reciprocal of resistance, $1/R$, is called the conductance, given the symbol G, and is expressed in reciprocal ohms, or mhos.

A conductivity meter is used to monitor water purity in many clinical laboratories. These meters may be used external to the water system or be added in-line with the water stream for continuous monitoring. A high meter reading indicates that more ions are present in the solution.

RESISTIVITY

Laboratories that produce their own water are required to monitor its purity. A convenient electrochemical technique that is supported by organizations such as Clinical and Laboratory Standards Institute (CLSI) for measuring water purity is resistivity. Resistivity is the electrical resistance in ohms measured between opposite faces of a 1.00-centimeter cube of an aqueous solution at a specific temperature. This measurement is accomplished by using a resistivity meter. Water is passed through the meter and its resistivity in Mohm-cm (MΩ-cm) is determined. CLSI requires that all water used for analytical procedures in the laboratory must have a resistivity of greater than 10 Mohm-cm.[5] Therefore the fewer ions present in the water, the greater the resistivity.

IMPEDANCE

Electrical impedance measurement is based on the change in electrical resistance across an aperture when a particle in conductive liquid passes through this aperture. Electrical impedance is used primarily in the hematology laboratory to enumerate leukocytes, erythrocytes, and platelets. In a typical electrical impedance instrument by Coulter, (Beckman Coulter, Brae, CA) aspirated blood is divided into two separate volumes for measurements. One volume is mixed with diluent and delivered to the cell bath, where erythrocyte and platelet counts are performed. As the blood passes through the aperture, the electric current between the electrodes changes each time a cell passes through. This produces a voltage pulse, the size of which is proportional to the cell size. The number of pulses is directly related to the cell count. Particles measuring between 2 and 20 fL are counted as platelets, whereas those measuring greater than 36 fL are counted as erythrocytes. The other blood volume is mixed with diluent and a cytochemical-lytic reagent that lyses only the red blood cells. A leukocyte count is performed as the remaining cells pass through an aperture. Particles greater than 35 fL are recorded as leukocytes.

▶ SEPARATION TECHNIQUE

ELECTROPHORESIS AND DENSITOMETRY

Electrophoresis is the separation of charged compounds in a liquid medium under the influence of an electrical field. When a voltage is applied to a salt solution (usually sodium chloride), an electrical current is produced by the flow of ions: cations toward the cathode, and anions toward the anode. Conductivity of a solution increases with its total ionic concentration. The greater the net charge of a dissolved compound, the faster it moves through the solution toward the oppositely charged electrode. The net charge of a compound, in turn, depends on the pH of the solution. Electrophoresis separations often require high voltages (50 to 200 V dc); therefore, the power supply should supply a constant dc voltage at these levels.

The buffer solution must have a carefully controlled ionic strength. A dilute buffer causes heat to be generated in the cell, whereas a high ionic strength does not allow good separation of the fractions. Common support media for electrophoresis in clinical work include cellulose acetate, agarose, and polyacrylamide gels. Total volume of specimen applied depends on the sensitivity of the detection method. For clinical work, 1 μL of serum may be applied. Once the electrophoresis is completed, the support medium is treated with a dye that colorizes the separated compounds. The most common dyes used for the visualization step include Amido Black, Ponceau S, Fat Red 7B, and Sudan Black B. To obtain a quantitative profile of the separated fractions, densitometry is performed on the stained support medium.

Densitometry is basically an absorbance measurement. A densitometer measures the absorbance of the stain on a support medium. The basic components of a densitometer include a light source, monochromator, and movable carriage to scan the medium over the entire area; an optical system; and a photodetector. Signals detected by the photodetector

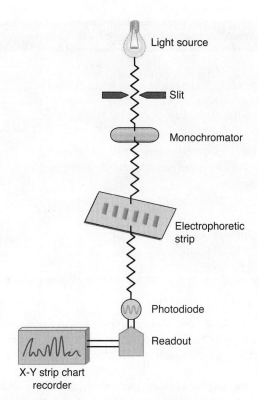

Light source

Slit

Monochromator

Electrophoretic strip

Photodiode

Readout

X-Y strip chart recorder

FIGURE 2-24 Diagram of a densitometer.

are related to the absorbance of the sample stain on the support, which is proportional to the specimen concentration. The support medium is moved through the light beam at a fixed rate so that a graph may be constructed that represents multiple density readings taken at different points. Most densitometers have a built-in integrator to find the area under the curve so that all sample fractions can be quantified. A diagram of a typical densitometer design is shown in Figure 2-24 .

Capillary Electrophoresis

A typical capillary electrophoresis (CE) system consists of a fused silica capillary, two electrolyte buffer reservoirs, a high-voltage power supply, and a detector linked to a data acquisition unit. The sample is introduced into the capillary inlet. When a high voltage is applied across the capillary ends, the sample molecules are separated by electroosmosis. Electroosmosis is the motion of a liquid when a voltage is applied between the ends of an insulating tube that contains that liquid. This liquid will migrate toward the cathode or anode. Positively and negatively charged ions will migrate toward the capillary surface and the anode or cathode depending on the pH of the solution. The positive ions in the specimen emerge early at the capillary outlet because the electroosmotic flow and then to their respective electrodes (anode or cathode). Negative ions in the specimen also move toward the capillary outlet but at a slower rate. They are detected using an appropriate detector. Examples of the different types of detectors are optical, conductivity, and electrochemical. Several advan-

tages of CE over conventional electrophoresis and HPLC include short analytical time, improved resolving power, smaller sample volumes. Complex mixtures of molecules can be separated with very high *theoretical plate* values (see discussion on chromatography resolution below). Separations may be completed in less than 10 minutes with very high applied voltage. The application of high voltage is made possible by the capillary's high surface-to-volume ratio, which allows for efficient heat transfer through the capillary wall.

Isoelectric Focusing

Proteins are polymers of amino acids that can be anions or cations depending on the pH environment. At a specific pH, a protein will have a net charge of zero when the positive charges and the negative charges of its amino acids cancel each other out. This pH value is referred to as the isoelectric point (pI). Isoelectric focusing (IEF) techniques are similar to electrophoresis except that the separating molecules migrate through a pH gradient. This pH gradient is created by adding acid to the anodic area of the electrolyte cell and adding base to the cathode area. A solution of ampholytes (mixtures of small amphoteric ions with different pIs) is placed between the two electrodes. These ampholytes have high buffering capacity at their respective isoelectric points. The ampholytes close to the anode carry a net positive charge, and those close to the cathode carry a net negative charge. When an electrical voltage is applied, each ampholyte will rapidly migrate to the area where the pH is equal to its isoelectric point. With their high buffering capacity, the ampholytes create stable pH zones for the slower migrating proteins. The advantage of isoelectric focusing techniques lies in their ability to resolve mixtures of proteins. Using narrow-range ampholytes, macromolecules differing in isoelectric point by only 0.02 pH units can be identified. Isoelectric focusing has been useful in measuring serum isoenzymes of acid phosphatase, creatine kinase, and alkaline phosphatase. IEF is also used for detecting oligoclonal immunoglobulin bands in CSF for evaluation of patients with multiple sclerosis.

✔ **Checkpoint! 2-16**

Which electrode would albumin migrate to if the buffer pH is 8.6 and its pI is 4–5.8?

CHROMATOGRAPHY

Chromatography is a separation technique based on physical and chemical interactions of compounds in a sample with a mobile phase and stationary phase, as the compounds travel through a support medium. The compounds interacting more strongly with the stationary phase are retained longer in the medium than those that favor the mobile phase.

Chromatographic techniques may be classified according to their mobile phase (e.g., gas chromatography and liquid

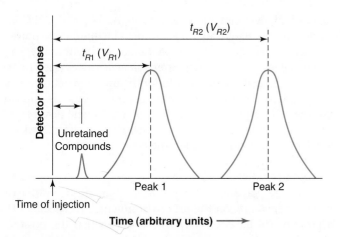

■ **FIGURE 2-25** Chromatogram showing the parameters that are used to characterize a separation of a multicomponent mixture. The letter *t* represents time, and *V* represents volume.

BOX 2-2 Factors controlling the resolution of multicomponent mixtures of analytes.

k', Strength of solvent: polarity
Strength of packing material: surface area or amount of stationary phase
Temperature

α Chemistry of solvent: functional groups
Chemistry of packing material: functional groups
Chemistry of samples: presence of hydrogen bonds or derivatization

N Flow rate: linear velocity
Column length
Average particle size of packing material
Viscosity of solvent
Mass of injection

chromatography). Figure 2-25 ■ shows a typical chromatogram representing the concentration of each detectable compound eluting from the column as a function of time.

Retention time (R_T) is the time it takes a compound to elute off the column once it has been injected. This value is characteristic of a compound and is related to the strength of its interaction with the stationary phase and the mobile phase. The retention time therefore can be used to determine a compound's identity.

Resolution (R_s) is a measure of the ability of a column to separate two or more analytes in a sample. Several factors have a significant impact on the ability of the chromatographic system to separate compounds:

• Column retention factor (k')
• Column efficient (N)
• Column selectivity (α)

Column retention factor describes the distribution of solutes between stationary phase and mobile phase. It is a measure of the time the sample component resides in the stationary phase relative to the time it resides in the mobile phase.[6]

Column efficiency describes the ease of physical interaction between solute molecules and column packing material. Column efficiency is expressed in terms of "theoretical plates." A theoretical plate is an abstract concept that describes the equivalent to the length of a column necessary to allow equilibrium of the solute to occur between the stationary phase and the mobile phase.

The selectivity factor is a measure of the relative separation between the peak centers of two solutes. It is equivalent to the ratio of the retention factors (k') of the second eluting compound and the first eluting compound. By definition (α) is always greater than 1.

The association of these three factors to resolution can be shown by Equation 2.18:

$$R = 4N^{1/2}[(\alpha - 1/\alpha)][(k/1 + k)] \qquad (2.18)$$

In Box 2-2 ■ several examples of specific factors that affect α, k', and N discussed above are listed. Changing anyone of these specific factors will affect the resolution of the separation.

Thus for any laboratory evaluation involving chromatographic separations, careful monitoring of all these three factors is important to the successful separation of multicomponent mixtures.

Historically, chromatographic separation techniques and associated instrumentation have been time consuming, labor intensive, and complex. Thus clinical laboratories were hesitant to incorporate them. In an effort to bring these techniques back into the laboratory, instrument manufacturers have designed instrumentation to be more user friendly, less complex, and more automated. Also, manufacturers have streamlined the techniques so that overall analysis times are significantly reduced. Three separation techniques will be presented in this chapter: thin-layer chromatography (TLC), gas chromatography (GC), and HPLC.

Thin-Layer Chromatography

Thin-layer chromatography is used in many laboratories as an initial screening technique for the detection of drugs of abuse in urine (DAU). TLC may also be used as a confirmatory method in conjunction with an immunoassay screening method for DAU testing. The stationary phase was manufactured as a thin layer or coating of adsorbent that is bonded to a solid support such as glass or plastic. The solid adsorbent may consist of a basic silica material or a more complex adsorbent with functional groups such as cyanide or fluorine attached to the silica.

The sample is processed so that the compounds to be separated are extracted out of the specimen matrix, usually urine. The extracted material is applied or "spotted" onto the stationary phase bonded to a glass plate or plastic support. A container such as a glass jar is partially filled with a liquid mobile phase. Next the chromatographic "plate" is coated with stationary phase and an applied extracted sample is placed into the glass jar containing the liquid mobile phase. The movement of the mobile phase and separation of compounds in the samples occur because of select factors shown in Box 2-2. Compounds in the samples are identified and characterized based on the distance they traveled through the stationary phase and their degree of coloration with specific colorizing solutions. Most compounds have a characteristic identifier known as R_f. **R_f** is a ratio used in TLC that represents the distance the compound migrated from the origin to the center of the "spot" or separated zone divided by the distance traveled by the solvent from the origin, represented as the solvent front.

Gas Chromatography

Gas chromatography (GC) is a separation technique that uses a "carrier" gas to move compounds through a stationary phase located within a column. This technique is useful to separate compounds that are naturally volatile or can be easily converted into a volatile form. GC has been a widely used technique for decades because of its high resolution, low detection limits, accuracy, and short analytical times. Clinical applications include separating and identifying various organic molecules, including many therapeutic and toxic drugs. Retention of a compound in GC is determined by its vapor pressure and volatility, which, in turn, depend on its interaction with the stationary phase.

Gas Chromatographs The basic design of a gas chromatograph GC consists of the following components which are shown in Figure 2-26 ■ :

• Carrier gas supply

• Sample injection device and GC inlet

• Column

• Detector

• Data system

Carrier gases (mobile phase), which must be chemically inert, include helium, hydrogen, and nitrogen. These carrier gases must be of high purity, and the flow must be tightly controlled to ensure optimum efficiency and precise test results.

Samples are injected manually using a microliter syringe or automatically using a programmable automated sampler. The injected sample is directed into a heated sample "port" whose entrance is described as a self-sealing, silicone rubber diaphragm. The sample is vaporized and moves into the head of the analytical column. If the molecules of interest are not volatile enough for direct injection, it may be necessary to prepare a derivative of the compounds into a more volatile form. Most derivatization reactions belong to one of three groups: silylation, alkylation, and acylation. Silylation is the most common technique and replaces active hydrogen(s) on the compounds with alkylsilyl groups. This substitution results in a more volatile form that is also less polar and more thermally stable. Examples of other sampling techniques are headspace sampling and pyrolysis.

The main purpose of the inlet is to provide accurate and reproducible transfer of the sample to the column without diminishing the integrity of the samples or the efficiency of the separation. Several types of capillary column inlets are available for separation analysis. A common inlet used for GC separation of drugs of abuse is the *split/splitless* injector. The split injector is used to introduce only a small amount of the sample vapor onto the analytical column. The splitless injector is designed to transfer the entire samples to the column. Split injection result in narrow inlet bands because of the rapid sample vapor transfer, whereas the splitless injector creates high-sensitivity analysis.[7]

Two types of columns generally used for clinical applications in GC are packed columns and open tubular, or capillary, columns. Capillary columns are commonly used because of their high efficiency and larger theoretical plate number.

Chromatographic columns are constructed of stainless steel, glass, fused silica or Teflon and come in various sizes. The columns may be U-shaped or coiled to conserve space.

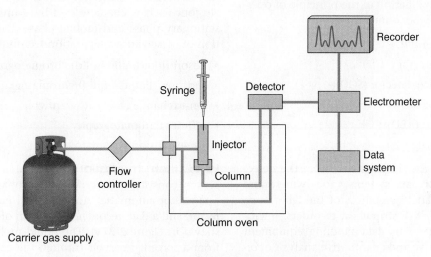

■ FIGURE 2-26   Schematic of a typical gas chromatograph.

Gas chromatography columns are housed in an oven that provides the high temperatures required to separate multi-component mixtures. Two types of stationary phases commonly used in GC are solid absorbent (gas-solid chromatography [GSC]) and liquids coated on solid supports (gas-liquid chromatography [GLC]). In GSC, the same material (usually alumina, silica, or activated carbon) acts as both the stationary phase and the support phase. Although this was the first type of stationary phase developed, it is not as widely used as other types primarily because of the strong retention of polar and low volatile solutes by the column. GLC uses liquid phases such as polymers, hydrocarbons, fluorocarbons, liquid crystals, and molten organic salts to coat the solid support material. Calcine diatomaceous earth graded into appropriate size ranges is often used as a stationary phase because it is a stable inorganic substance. The use of fused silica capillary columns in which the stationary phase is chemically bonded onto the inner surface of the column has become very popular with chromatographers. The advantage of this type of column is that the stationary phase does not leave the solid support and bleed into the detector, and a uniform monomolecular layer of the stationary phase is obtained through the bonding procedure.

The detectors used in a gas chromatography must provide the analyst with optimum performance characteristics that include:

- High sensitivity
- Good stability and reproducibly
- Linear response over a wide concentration range
- Expanded temperature range from room temperature to between 300 and 400°C
- Short response times
- Nondestructivity of the samples

There are many unique detector designs used in gas chromatographs GC. Several examples are listed below; the flame ionization detector is commonly used for clinical chemistry applications. A detailed explanation of the principle of operation for each detector may be found in Skoog.[8]

- Flame ionization detector (FID)
- Thermo conductivity detector (TCD)
- Sulfur chemiluminescence detector (SCD)
- Electron capture detector (ECD)
- Atomic emission detector (AED)
- Thermionic detector (TID)

A data system receives the output signal from the detectors. The configuration of data systems varies widely and often depends on the financial resources of the laboratory. Data systems may include X-Y strip chart recorders, microprocessors, and computers. This data-handling equipment will facilitate both qualitative and quantitative analysis of biological samples. Quantitative analysis may be accomplished by assaying a series of standards, integrating the area of the peaks, or calculating peak height ratios.

 Checkpoint! 2–17

What is the difference between a split and splitless injector?

Liquid Chromatography (High-Performance Liquid Chromatography)

GC as a separation technique has some restrictions that make liquid chromatography (LC) a suitable alternative. Many organic compounds are too unstable or are insufficiently volatile to be assayed by GC without prior chemical derivatization. Liquid chromatography techniques use lower temperatures for separation, thereby achieving better separation of thermolabile compounds. These two factors allow LC to separate compounds that cannot be separated by GC. Finally, it is easier to recover a sample in LC than in GC. The mobile phase can be removed, and the sample can be processed further or reanalyzed under a different set of conditions.

There are many types of LC available, and the selection of an appropriate type depends on a variety of factors. These factors include analysis time, type of compounds to be separated, and detection limits. Paper, thin-layer, ion-exchange, and exclusion LC often result in poor efficiency and a very long analysis time because of slower mobile phase flow rates. HPLC emerged in the late 1960s as a viable type of LC that provided advantages over other types of LC and GC. HPLC uses small, rigid supports and special mechanical pumps producing high pressure to pass the mobile phase through the column. HPLC columns can be used many times without regeneration. The resolution achieved with HPLC columns is superior to that of other types of LC, analysis times are usually much shorter, and reproducibility is greatly improved. All of these attributes of HPLC render it a better method of separation over other types of LC.

There are five commonly used separation techniques in LC: adsorption, partition, ion exchange, affinity, and size exclusion. Each is characterized by a unique combination of stationary phase and mobile phase. For a detailed explanation of these various separation techniques, refer to Skoogs.[9]

- Adsorption (liquid–solid) chromatography
- Partition (liquid–liquid) chromatography
- Ion-exchange chromatography
- Affinity chromatography
- Size-exclusion chromatography

HPLC Instrumentation   A typical liquid chromatography system consists of a liquid mobile phase, a sample injector (manual or automatic), a mechanical pump, a column, a detector, and a data recorder. A diagram of these components is shown in Figure 2-27 ■. The liquid mobile phase is pumped from a solvent reservoir through the column. There are two methods for delivery of mobile phase(s), *isocratic* or *gradient*.

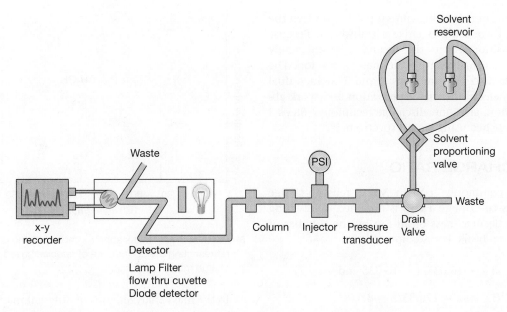

■ **FIGURE 2-27** Schematic of a typical high-performance liquid chromatograph.

Isocratic LC uses one mobile phase. Gradient LC involves the use of two or more mobile phases that are automatically programmed to pump for a specific interval of time. A mechanical pump must provide precise and accurate flow, often working at high pressures (up to 6000 psi). The pump must have low internal volume and be constructed of material that does not react with the solvent. Sample injection is achieved using a syringe or an automated sampler and depositing the sample into a loop. Most analytical separations are performed using packed column. There are many types of packing material available. Selection of the appropriate packing material is largely dependent on the type of compound(s) to be separated. In liquid chromatography, the physical properties of the sample and mobile phase are often very similar. Two basic types of detectors have been developed. One is based on the differential measurement of a physical property common to both the sample and mobile phase; examples include refractive index, conductivity, and electrochemical detectors. The other is based on the measurement of a physical property that is specific to the sample, either with or without the mobile phase; examples include absorbance and fluorescent detectors.

A widely used clinical application of HPLC is the separation and quantitation of specific hemoglobins; for example, hemoglobin A_{1c} (H_{A1c}). The systems available for this type of analysis are automated and minimize many of the laboratory tasks inherent with HPLC techniques. The method for delivery of mobile phase is gradient elution. Most methods use two phosphate-containing buffers at different pH values for mobile phases and the column consists of packing material that is a nonspherical copolymer with carboxylic acid functional groups attached. The separated analytes are detected using a flow-through curvet within a photometer. The separations are completed in less than five minutes and the resolution of the hemoglobin compounds is very good.

✓ Checkpoint! 2–18

Compare these two solvent delivery methods for liquid chromatography: isocratic and gradient elution.

▶ MASS SPECTROMETRY

Mass spectrometry (MS) is a powerful analytical technique that is used to identify unknown compounds, determine concentrations of known substances, and study the molecular structure and chemical composition of organic and inorganic material.

Mass spectrometry is used to

- Determine how drugs are used in the body.
- Detect and identify illegitimate steroids in athletes.
- Determine damage of human genes because of environmental causes.
- Detect dioxin in food and humans.
- Identify and characterize proteins involved in biological processes.
- Detect the presence of metabolic disorders in infants.
- Search for unique proteins in biological samples for use as therapeutic or diagnostic targets.

Atomic mass spectrometry analysis involves the following steps: (1) atomization; (2) conversion of a substantial fraction of the atoms formed in step 1 to stream of ions (usually singly charged positive ions); (3) separation of the ions formed in the second step on the basis of their mass-to-charge ratio (m/z), where m is the mass of the ion in atomic mass units and z is its charge; and (4) counting the number of ions of each type or measuring current produced when the ions formed from the samples strike a transducer. Because most of the ions formed in the second step are singly charged, the m/z is usually simply the mass of the ion. The techniques required to perform steps 1 and 2 include dual sample insertion, electrothermal vaporization and spark ablation, laser ablation, and glow discharge techniques. Steps 3 and 4 are accomplished using a mass spectrometer.[10]

MASS-TO-CHARGE RATIO

A measurement commonly used in MS is mass-to-charge ratio (m/z). **Mass-to-charge ratio** is obtained by dividing the atomic or molecular mass of an ion by the number of charges that the ion bears. For example,

$$12C^1H_4^+, m/z = 16.035/1 = 16.035 \text{ and for}$$

$$^{13}C^1H_4^{2+}, m/z = 17.035/2 = 8.518$$

Many ion masses have a single charge; thus the term *mass-to-charge ratio* is often shortened to the more convenient term *mass*. Strictly speaking, this abbreviation is incorrect, but it is widely used in MS literature. Mass-to-charge also represents the *x*-axis for MS spectrums of molecules plotted against their relative abundance (*y*-axis).[11]

MASS SPECTROMETERS

Mass spectrometers have three fundamental components: an ion source, a mass analyzer, and an ion detector. These are shown in Figure 2-28 ■. The purpose of the inlet system is to introduce a micro amount of sample into the ion source, where the components of the sample are converted into gaseous ions by bombardment with electrons, photons, ions, or molecules. Ionization may also be created by thermal or electrical energy. The output of the ion source is a stream of positive or negative ions in a gaseous medium that are then accelerated into the mass analyzer.

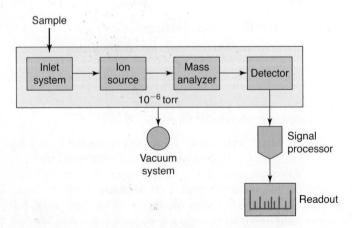

■ FIGURE 2-28 Block diagram of a mass spectrometer.

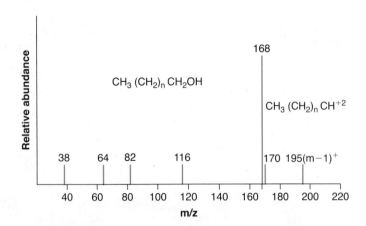

■ FIGURE 2-29 Mass spectrograph showing discrete lines representing the abundance of species based on their mass-to-charge ratio.

In the mass analyzer, ions of different masses are separated into species, so the different species of ions strike the detector or transducer at different times. This small ion current is amplified by the detector. The output of the electronics is a compilation of the abundance of the ions (*y*-axis) versus m/z (*z*-axis) as shown in Figure 2-29 ■.

Ion Sources

Three sources commonly used for MS include:

1. Electrospray ionization (ESI)
2. Matrix-assisted laser desorption ionization (MALDI)
3. Surface-enhanced laser desorption ionization (SELDI)

Electrospray Ionization The first step in ESI is to dissolve the analyte in a mixture of organic and aqueous solvents containing an acid or base. The analyte is then introduced into the electrospray tip and moves along a short length of stainless steel capillary tube, to the end of which is applied a high positive or negative electrical potential, ~3–5 kV. When the solution reaches the end of the tube, a powerful eclectic field causes it to be nearly instantaneously vaporized into a spray of very small droplets of solution in solvent vapor. Prior to entering the mass spectrometer, this mist of droplets flows through an evaporation chamber that can be heated to prevent condensation. As the droplets move through this area, solvent evaporates rapidly from the surfaces and the droplets get smaller and smaller.[12]

Matrix-Assisted Laser Desorption Ionization The MALDI source consists of a solid mixture of analyte and matrix on a sample plate, along with a laser light and ion optics, as shown in Figure 2-30 ■.[13] The matrix for positive ions consists of an organic chromophore molecule such as sinapinic acid. The peptide or protein is dissolved in a mixture of organic and aqueous solvents and mixed with matrix solution. The final matrix-sample mixture is placed on a sample plate and is allowed to dry to form crystals. Once the sample plate is inserted into the mass spectrometer through a vacuum

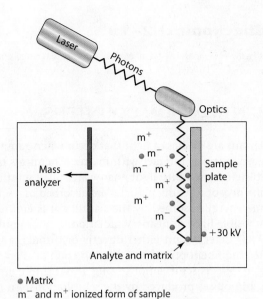

FIGURE 2-30   Representation of matrix-assisted laser desorption ionization (MALDI).

positive ions are pushed from the region above the plate and into the analyzer. The MALDI technique is considered an off-line ionization technique because the sample is purified, deposited, and dried on the sample plate before analysis.[14]

Surface-Enhanced Laser Desorption Ionization

Surface-enhanced laser desorption ionization (SELDI) is a technique that allows measurement of proteins from complex biological specimens, including serum, plasma, intestinal fluids, urine, cell lysates, and cellular secretion products. Proteins are captured using techniques such as adsorption, partition, electrostatic interaction, or affinity chromatography on a solid-phase protein chip surface. A laser ionizes samples that have been co-crystallized with a matrix on a target surface. The protein chip chromatographic surfaces in SELDI are uniquely designed to retain proteins from complex mixtures according to their specific properties. After the addition of a matrix solution, proteins can be ionized with a nitrogen laser and their molecular masses measured by time of flight (TOF) MS (Figure 2-31 ■).

The protein chip arrays are the heart of the SELDI-TOF MS technology and distinguish it from other MS-based systems. Each array is composed of different chromatographic surfaces that, unlike HPLC or GC, are designed to retain, not elute, proteins of interest. The protein chip arrays have an aluminum base with several specific areas composed of a chemical (anionic, cationic, hydrophobic, hydrophilic, or metal ion) or biochemical (immobilized antibody, receptor, DNA, enzymes, etc.) active surface. Each surface is designed to retain protein according to a general or specific physicochemical property unique to that protein. Chemically active surfaces retain whole classes of proteins, whereas surfaces to which biochemical agents, such as an antibody or other types of affinity reagent, are coupled are designed to interact specifically with a single target protein.[15]

interlock, it is positioned so that the pulsed laser light strikes a small portion of the sample. A nitrogen laser creating photons at 337 nm is commonly used in MALDI. The neodymium:yttrium aluminum garnet (Nd:YAG) laser has also been used successfully for specific MS applications.

When the organic chromophore absorbs the photons, it vaporizes and lifts the analyte ions from the surface and into a gas phase plume directly above the target plate. The gaseous plume is believed to consist of a conglomerate of neutral, metastable ions; positive ions; negative ions; ion clusters; and compound fragments. After the laser fires, the sample plate is charged to 20 to 30 kV for positive ions. The

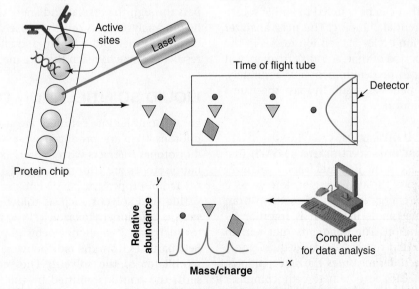

FIGURE 2-31   Surface-enhanced laser desorption ionization (SELDI) and time of flight (TOF) analysis.

Mass Analyzers

ESI, MALDI, and SELDI sources have been coupled to many different types of mass analyzers. These include the following:

- ESI-QqQ (triple quadrupole)
- ESI-QIT (quadrupole ion trap)
- MALDI-ToF-MS (time of flight)
- ESI-QqToF (quadrupole time of flight)
- ESI-FTMS (Fourier transform)

Three components that are generic to these systems are time of flight (TOF), quadrupole, and Fourier transform ion cyclotron and will be presented briefly in this section. For a more detailed discussion of the specific mass spectrometers, refer to Bier, 2002.[16]

Time of Flight   The TOF analyzer consists of a metal flight tube, and the *m/z* ratios of the ions are determined by accurately and precisely measuring the time it takes the ions to travel from the MALDI or SELDI sources to the detector. Given that all ions of different *m/z* receive the same kinetic energy, low *m/z* ions will reach the detector sooner than high *m/z* ions.

Quadrupole Ion Trap   Specifically the three-dimensional quadrupole field ion trap is a three-electrode device. Ions are injected into the device and collected in packets from a selected source. The ion packets enter through an entrance-end cap and are analyzed by scanning the radio frequency (RF) amplitude of the ring electrode. The ions resonate sequentially from low to high *m/z* and are ejected from the ion trap through the exit-end cap electrode to a detector. An advantage to this design is that this system can perform tandem mass-spectrometry scan modes in the same analyzer.

Fourier Transform Ion Cyclotron Mass Spectrometer   A Fourier transform ion cyclotron mass spectrometer (FTMS) is a trapping MS that can be coupled to an ESI source with many stages of differential pumping. This mass analyzer measures the number of ion cycles that an ion makes in the trapping cell for a given period of time. An ion's frequency, if measured long enough, can be determined very accurately, and therefore this instrument can result in mass spectrums with extremely high resolution.

Tandem Mass Spectrometers

Following ESI, the tandem mass spectrometer (MSMS) first sorts molecular ions by *m/z* in the first quadrupole analyzer (Q1). These sorted ions pass through a region known as a collision cell (Q2), where energy is imparted to them from high concentrations of inert molecules such as argon or nitrogen, causing fragmentation. The fragments that remain ions are then sorted a second time by the next quadrupole analyzer (Q3). A signal recorder measures the detected ions. Highly selective analysis can be obtained with computer control and communication between the first and final quadrupole analyzers.

 Checkpoint! 2–19

Identify the two measured parameters of a compound using a mass spectrometer.

▶ SCINTILLATION COUNTERS

Scintillations are flashes of light that occur when gamma rays or charged particles interact with matter. Chemicals used to convert their energy into light energy are called scintillators. If gamma rays or ionizing particles are absorbed in a scintillator, some energy absorbed by the scintillator is emitted as a pulse of visible light or near-UV radiation. A photomultiplier tube (PMT) detects light either directly or through an internal reflecting fiber optic. A scintillation counter is an instrument that detects scintillations using a PMT and counts the electrical impulses produced by the scintillations. An important application of scintillation counting is radioimmunoassay (RIA). Two types of scintillation methods exist: crystal scintillation and liquid scintillation.

CRYSTAL SCINTILLATION COUNTERS

A crystal scintillation counter is generally is used to detect light (scintillations) created by the interaction of gamma particles from radioisotopes with matter. Examples of gamma-emitting radioisotopes include ^{125}I and ^{137}Cs. The sample containing the radioisotope is placed into a sodium iodide (NaI) crystal, which contains 1% thallium. When gamma radiation from the sample penetrates the sodium iodide crystal, it excites the electrons of the iodide atoms and raises them to higher energy states. When the electrons return to ground state, energy is emitted as UV radiation. The UV radiation is promptly absorbed by the thallium atoms and emitted as photons in the visible or near-UV range. The photons pass through the crystal and are detected by a PMT. A pulse-height analyzer sorts out the pulse signals from the PMT according to their pulse height and allows only those within a restricted range to reach the rate meter for counting.

LIQUID SCINTILLATION COUNTERS

Liquid scintillation counters are used to detect and count photons that are produced when beta radiation from radioisotopes interacts with matter. Examples of beta-emitting radioisotopes are tritium (^3H) and $^{14}carbon$. A sample with the radioisotope is suspended in a solution or "cocktail" consisting of a solvent such as toluene; a primary scintillator such as 2,5-diphenyloxazole (PPO); and a secondary scintillator such as 2,2′-*p*–phenylenebis (5-phenyloxazole) (POPOP). Beta particles from the radioactive sample ionize the primary scintillator of the solvent. The secondary scintillator absorbs the photons emitted by the primary scintillator and reemits them at a longer wavelength. The secondary scintillator facilitates more effective energy transmission from the beta particles, especially when a large amount of quenching is present. Quenching is a process that results in a reduction of the photon output from the sample. This phenomenon may be because of chemical quenching, in which impurities in the sample compete with the scintillators for energy transfer, or color quenching, in which colored substances such as hemoglobin absorb the photons produced by the scintillator. The photons produced in the sample are detected and amplified by the PMT in the same manner as described in the section on crystal scintillation counters.

▶ NUCLEAR MAGNETIC RESONANCE SPECTROSCOPY

New technology using nuclear magnetic resonance (NMR) spectroscopy has become available to allow lipoprotein particles of different size to be directly quantified. The data produced are in the form of a lipid profile or panel and reveals information that has not been available to the clinicians in the past.

Nuclear magnetic resonance is a phenomenon that occurs when the nuclei of certain atoms are immersed in a static magnetic field and exposed to a second oscillating magnetic field, the magnetic component of EMR. Some nuclei experience this phenomenon and other do not, depending on whether they possess a property called "spin." The spin of a proton causes NMR signal. The nucleus of a hydrogen atom does spin, and because hydrogen atoms occur very frequently, they are useful in determining structure.

When EMR is used to bombard molecules, the hydrogen atoms present will absorb photons of different energies depending on the environment of the hydrogen atom. For example, in the compounds containing hydrogens and chlorides, the proton near two chloride atoms will produce a different NMR than the proton located near a single chloride and single hydrogen. By plotting the amount of energy absorbed by the spinning nuclei versus the frequency of the EMR applied to the molecules, we obtain a NMR spectrum of the molecule.

The actual source of the NMR signal used for subclass quantification is the protons of the terminal methyl groups of the various types of lipid carried in the particles, primarily the cholesterol ester and triglyceride of particle core, and the phospholipids of the particle shell. The signals from all these different lipids combine to produce a bulk, lipid signal that has a characteristic frequency and shape that is directly dependent on the size of the particle and the diameter of the phospholipids shell, excluding the influence of the apolipoprotein attached to the particle.[17]

Basic components of an NMR spectrometer consists of the magnet that is used to separate the nuclear spin energy state, and a transmitter that supplies the radio frequencies (RFs) or irradiating energy. A sample probe contains coils for coupling the sample with the RF field(s), and a computer is used to evaluate the data.

▶ FLOW CYTOMETRY

A flow cytometer is an instrument that measures multiple cell parameters and other types of particles as they flow individually in front of a light source. It detects and correlates the signals from multiple detectors at various angles. These signals are triggered by physical factors and fluorescent emissions from those cells as a result of illumination and staining procedures. The applications of flow cytometry have expanded to counting and sorting of cells, viral particles, DNA fragments, bacteria, and latex beads.

KEY FEATURES OF A FLOW CYTOMETER

The key features of a typical flow cytometer are (1) the cells or particles, (2) a method to illuminate the cells, (3) the use of fluidics to ensure the cells flow individually past the illuminating beam, (4) a detector to measure the signals coming from the cells and (5) a computer to correlate the signals after they are stored in a data file.

Cells (Particles)
The term *particle* is used as a more general term for any of the objects flowing through a flow cytometer. An *event* is a term that is used to indicate anything that has been interpreted by the instrument to be a single particle. An event may be determined correctly or incorrectly by a flow cytometer. Methods have been developed to compensate for measurement of unwanted events. An example is the correction techniques for measuring the simultaneous movement of two particles past the detector.

Illumination
A laser light serves as a source of illumination for most flow cytometers. Laser light is used because it provides intense light in a narrow beam. The particle in a stream of fluid can move through this light beam rapidly and in most instances one particle at a time. The resulting illumination from the particle is bright enough to produce scattered light or fluorescence of detectable intensity.

Several types of lasers are currently used and include argon, red and green helium, and neon. The light produced by these lasers generates a specific wavelength, which is defined and inflexible, based on the characteristics of the lasing medium.

Fluidics
In flow cytometry the particles need to be suspended in a fluid. Each particle is then analyzed as it passes through the analysis point over a brief but defined period of time. The fluidics in a cytometer decrease the probability that multiple cells will group together at the analysis point. This is accomplished by injecting the sample cell suspension into the center of a wide, rapidly flowing stream (referred to as the sheath

stream), where, according to the principles of hydrodynamics, the cells will remain confined to a narrow core at the center of the wider stream. The term *hydrodynamic focusing* is applied to this phenomenon and the result is coaxial flow or a narrow stream of cells flowing in a core within a wider sheath stream.

Detector

Light (the signal) from a particle is collected in both the forward direction along the path of the laser beam as well as at a right angle (orthogonal) to the laser beam. Lenses are used to collect the light and focus the beam of radiation onto a photodiode. Across the forward lens is an obscuration bar that serves to block the laser beam itself after it passes through the stream. Only light from the laser that has been refracted or scattered as it strikes a particle in the stream will be diverted enough from its original direction to avoid the obscuration bar and strike the forward-positioned lens and the photodiode behind it.

Detecting fluorescent light is similar to detecting side-scatter light, but with the addition of wavelength-specific mirrors and filters. These mirrors and filters reflect and transmit light at well-defined wavelengths. The wavelengths selected are optimized to detect the appropriate emission spectra of the fluorescent dye used. The change in wavelength between the illuminating laser and detected fluorescence is referred to as the Stoke's shift.

Data

In a conventional flow cytometer, where analog signals require processing, the current from each photometer will be converted to a voltage, amplified and processed for ratio calculation or spectral cross-talk correction, and ultimately be digitized by an analog-to-digital converter (ADC). The final output number will have required binning of the analog amplified and processed values into (digital) channels.

Newer technology incorporated into flow cytometers digitizes the voltage signal coming from the photodetector immediately without prior amplification and processing. This improves both the process and the signal linearity.

At this point all the data about each of a group of cells is stored in data files in an array where each cell has the associated output from each detector. This array is called "list mode." The remaining process involves computers only. The software packages available will produce histograms and scattergrams of any set of parameter values for the cells in the data file, provide a display of correlated data between any two parameters' values, and restrict the display of information of certain cells in the data file (referred to as gating).

✓ Checkpoint! 2–20

Identify the term used to describe the injection of a sample cell suspension into the center of a rapidly flowing stream or sheath.

▶ MICROSCALE TECHNOLOGIES

The miniaturization of analytical devices has been ongoing for several decades. Clinical laboratories have benefited from this technology in several ways, including

- Improved turnaround time (TAT) and testing efficiency
- Improved cost efficiency
- Improved analytical performance of laboratory tests
- More efficient operation of laboratories
- Better utilization of space and other laboratory resources

Miniaturization of devices begins with the analytical device fabricated onto a small chip usually made of silicon, plastic, or glass. This microchip is etched with small channels and outfitted with small laboratory components, including pumps, valves, filters, and electronic circuitry to control fluid flow through the chip. This elaborate microchip configuration is also referred to as a **lab-on-a-chip**. Lab-on-a-chip represents a total microanalysis system (μTAS) incorporating sample preparation, separation, detection, and quantification on a microchip surface. The μTAS requires only a minute amount of sample and reagents. Micromachines are microchips with miniaturized mechanisms such as gears and motors. The process of fabricating these labs-on-a-chip and micromachines is called micro machining. Micromachine fabrication involves the formation of a labyrinth of very small channels leading to an array of diaphragms, valves, motors, lasers, LEDs, ISEs, and other mechanical microstructures on a silicon wafer.

The applications of microscale technologies are abundant across the whole scientific community. Several examples of these applied technologies and their fabrication are presented in this chapter.

OPTODES

Optical sensors (optodes) are used in clinical laboratory instrumentation designed to measure blood gases and electrolytes. An optode designed to measure PO_2, for example, uses a dye that is immobilized within a polymer film. The polymer film must allow molecular oxygen to dissolve within its structure. The oxygen within the polymer film oxidizes the dye compound. If the dye compound is a fluorophore, the intensity of the fluorescence energy will be decreased or "quenched" by the oxygen. Therefore the higher the PO_2 in the blood sample, the greater the quenching of the fluorescent dye.

Optical pH sensors use immobilized pH indicators (e.g., fluorescein) to determine the pH of a blood specimen. As the hydrogen ion concentration changes in a sample, the amounts of protonated or deprotonated forms change also. The PCO_2 in a blood sample can be measured using a modification of the pH optode. The pH optode is covered by an outer gas-permeable hydrophobic film to allow CO_2 to equilibrate with the pH-sensing layer. As the PCO_2 level in the sample increases, the pH of the bicarbonate layer decreases, and the corresponding decrease in the concentration of the deprotonated form of the indicator is detected using optical sensors.

BIOSENSORS

The development of miniaturized sensing devices for biomedical use has also created applications in clinical laboratories—specifically near-bedside testing and critical-care testing. Continued development of biosensors will provide the needed improvements in patient care, convenience, cost, and turnaround time that may result in a significant impact on health care in the future. During the past several years, new technologies have resulted in the following: (1) the integration of sensors with miniaturized (micro machined) analyzers, (2) rapid growth of genosensors, (3) ultra-fast monitoring of dynamic events in the microscopic environments, (4) molecular-sized electrodes, and (5) introduction of advanced sensing materials. These developments, along with continued experimentation with novel biocatalytic and affinity biosensors, will improve the capabilities of biosensor and electroanalysis in the clinical laboratory.

There have been almost as many operational definitions of a biosensor as there have been authors on the subject. Thus, only a fundamental description of a biosensor is presented here. A biosensor comprises a biologically sensitive material (a biocatalyst) in contact with a suitable transducing system that converts the biochemical signal into an electrical signal.

Biocatalysts include enzymes, multienzyme systems, antibodies, membrane components, organelles, bacteria, and mammalian or plant tissues. These biocatalysts are responsible for the sensitivity and specificity of the biosensors. Some of the newer biosensors do not use biological molecular recognition agents; instead, synthetically produced molecules or abiotic molecules (i.e., not of biological origin), such as crown ethers, cryptands, calixarenes, and molecularly imprinted polymers, are being used.

Analytes approach the assembly by one of several mass transport processes (e.g., stirring, flowing stream, or diffusion). The biocatalyst must be close to a suitable transducer to produce an electrical signal. Intimate contact between the biocatalyst and the transducer is achieved by immobilization of the biocatalyst at the device surface. Examples of methods to accomplish this include (1) *adsorption* of proteins to the metal or metal oxide surfaces used for transducers; (2) chemically *cross-linking* the biocatalyst with an inert, generally proteinaceous, material to form intermolecular bonds; (3) physically restraining the biocatalyst at the transducer surface by *entrapment* in polymer matrices such as polyacrylamide or agarose or by retention with a polymer membrane comprising cellophane, polyvinyl alcohol, or polyurethane; and (4) *covalently attaching* the biocatalyst directly to the surface of the transducer.[18]

✪ TABLE 2-2

Types of biosensors used in clinical laboratories.

Transducer	Measurement	Applications
Ion-selective electrodes	Potentiometric	Na^+, K^+, Ca^{++}, Li^+
Gas-sensing electrodes	Potentiometric	PCO_2
Enzyme electrodes	Amperometric	Glucose
Conductimeter	Conductance	Urea nitrogen
Piezoelectric crystals, acoustic waves	Mass change	Immunosensor, volatile gases and vapors
Thermistor	Calorimetric	Enzymes
Luminescence	Optical	pH, enzyme substrates

Once the analyte is bound to the biocatalyst or receptor, the transducer detects a change in one or more physicochemical properties, including the following:

- Electrochemical
- Mass
- Heat
- Optical properties
- pH
- Specific ions

Several examples of types of biosensors used in laboratories and their respective transducer systems are shown in Table 2-2 ✪.[19]

Electrochemical

Electrochemical transducers (potentiometric or amperometric-type biosensors) are routinely used in clinical laboratories. The potentiometric transducer is based on the principle of accumulation of a membrane potential as a result of the selective binding of ions to a sensing membrane. The change in potential is measured. Amperometric transducers measure current flow through an electrochemical cell held at a constant voltage. The current generated by the redox reaction of the analyte at the sensing electrode is directly proportional to the analyte concentration at the electrode surface.

Conductimetric

Conductimetric measurements are useful because many chemical reactions produce or consume ionic species and thus alter the overall electrical conductivity of a solution. In the urea nitrogen electrode the enzyme urease is immobilized in the membrane and converts urea to its ionic product ammonium. The increase in solution conductance measured is proportional to urea concentration.

Piezoelectric

The piezoelectric principle is applied using crystals of quartz coated with an adsorbent. The absorbent selectively binds to the analyte of interest, which increases the mass of the coated crystal and alters its basic frequency of oscillation. A change in mass would be detected by monitoring the oscillation frequency. This change is proportional to the analyte concentration.

Acoustic wave techniques require that the oscillation of the piezoelectric crystals be at a higher frequency (30 to 200 MHz), and an acoustic wave is generated by application of an alternating voltage across a layer of interlaced gold or titanium electrodes. A transducer situated a few millimeters away detects the acoustic signal produced.

Calorimetric

A biological component is attached to a heat-sensing transducer or thermistor. Alternatively, the biological component is immobilized on a column with an embedded thermistor. The reaction between these two components generates a specific amount of heat, which can be detected and measured. For example, most enzyme-catalyzed reactions are accompanied by heat production of 25 to 100 kJ/moles, and applications have been developed for measurement of cholesterol, glucose, urea, and triglycerides.

Optical

This type of transducer uses fiber-optic technology to measure the reflected fluorescence light from immobilized chemicals at the end of small fiber-optic probes. Specific examples include the determination of hydrogen ion activity using miniaturized probes in which pH-sensitive dyes are immobilized at fiber termini.

BIOSENSOR TECHNOLOGIES

Four specific biosensor designs are presented here as examples of devices that have been in use for several years. These biosensors have been included in instruments designed for central laboratory use and POCT.

1. Enzyme-based biosensors with amperometric detection were first developed for the measurement of glucose in blood. Modifications of the original design led to the development of electrodes to measure cholesterol, pyruvate, alanine, and creatinine. The design of a "glucose electrode" is based on immobilizing glucose oxidase on the surface of an amperometric PO_2 sensor. A solution of glucose oxidase is deposited between the gas-permeable membrane of the PO_2 electrode and an outer semipermeable membrane. Glucose in the blood diffuses through the semipermeable membrane and reacts with the glucose oxidase. The consumption of oxygen occurs near the surface of the PO_2 sensor. The rate of decrease in PO_2 is a function of the glucose concentration and is monitored by the PO_2 electrode. The amperometric measurement of oxygen (Clark electrode) was discussed previously.

2. Enzyme-based biosensors with potentiometric and conductometric detection have been developed for the measurement of urea nitrogen (blood urea nitrogen, BUN),

glucose, creatinine, and acetaminophen. For example, the urea nitrogen biosensor uses a polymembrane ion selective electrode (ISE) for measuring ammonium ions. The enzyme urease is immobilized at the surface of the ammonium ISE. Urease catalyzes hydrolysis of urea to NH_3 and CO_2. The ammonia produced dissolves to form ammonium, which is detected by the ISE. The signal created by the ammonium produced is proportional to the logarithm of the concentration of urea in the sample.

3. Enzyme-based biosensors with optical detection are used to measure analytes such as glucose, cholesterol, and bilirubin. These sensors use immobilized enzymes and indicator dyes. A method of detection includes absorbance, reflectance, fluorescence, and luminescence. The operating principle of the cholesterol biosensor is based on fluorescence quenching (i.e., a reduction in fluorescence intensity of an oxygen-sensitive dye that is coupled to consumption of oxygen resulting from the enzyme-catalyzed oxidation of cholesterol by the enzyme cholesterol oxidase).[20]

4. Affinity biosensors use binding proteins, antibodies, or oligonucleotides (e.g., DNA) as an immobilized biological recognition element. These proteins have a high binding affinity and high specificity toward a clinically significant analyte. Affinity sensors detect the coupling reaction between the selective binding unit (SBU), such as avidin, antibody, single-stranded DNA, lectin, and host artificial molecular recognition species, and its complementary component, including compounds such as biotin, antigen, complementary single-stranded DNA, sugar sequence, and guest target compound. The design of an affinity biosensor ensures that binding of SBU and complement takes place on the transducer surface. The transducer converts the binding event into a measurable response. Affinity biosensors are divided into two categories, nonlabeled and labeled. Nonlabeled affinity biosensors directly detect the affinity complex by measuring physical changes resulting from the formation of the affinity complex at the transducer interface. In contrast, labeled affinity sensors incorporate a detectable label, and the presence of the affinity complexes is then determined through measurement of the label.

 Checkpoint! 2–21

Identify the major components of a biosensor.

Point of Care (POC)

Point-of-care testing (POCT) is also referred to as "bedside," "near patient," "physician's office," "off-site" and "ancillary" testing and is a rapidly growing aspect of health care. There are several definitions of POCT. The definition that embodies the content of this chapter is "any test that is performed at the time at which the test result enables a decision to be made and an action taken that leads to an improved health outcome."[21] POCT is used in many health-care settings, including intensive care units, surgical wards, emergency departments, coronary care units, and pediatric units. Two major advantages of POCT are (1) reduced turnaround-times (TATs) and (2) improved patient management.

The devices or analyzers used for POCT should possess the following attributes: (1) are easy to use, (2) are durable and reliable over a long period of time, (3) produce results that are in close agreement with a central laboratory, and (4) are safe to operate.

Several criteria that distinguish POCT devices from conventional laboratory testing equipment are listed below:
POCT devices should

- Be portable
- Have consumable reagent cartridges
- Generate results within minutes
- Require minimum operating steps
- Have the capability to perform tests on whole blood specimens
- Have flexible test menus
- Contain built-in/ integrated calibration and quality control
- Require ambient temperature storage for reagents

POCT devices are designed to provide both qualitative and quantitative measurements. The analytical principles used and examples of analytes tested are shown in Table 2-3 ✪. The design features that enable these devices to generate reliable results based on these analytical principles include[21]

- Operator interface
- Bar-code identification
- Sample delivery method
- Reaction cell
- Sensors
- Control and communication systems
- Data management and storage

Types of Technologies   There are many POCT devices on the market today, and a detailed explanation of each is beyond the scope of this textbook. Instead a few examples of specific technologies will be presented here; the reader is encouraged to review the references cited for a more detailed explanation.

Single-use qualitative or semiquantitative strips have been used for urine analysis and more recently for measuring cardiac markers and hCG. More complex strips have been developed to perform quantitative analysis for numerous analytes using a variety of analytical principles shown in Table 2-3. Basically each strip contains pads consisting of several layers of porous material. The uppermost layer is a semipermeable membrane that prevents red cells from entering the matrix. Sample is added to the pad and begins to move down through the layers. The layers of porous material contain reagents specific for the analyte being tested. The pads will change color if the analyte is present, and the more analyte in the sample the more intense will be the color. Most strip methods use a reflectometer to measure the color developed. Other analytical principles have been used with the strips, including electrochemistry, immunoturbidimetry, and light scattering.

Biological sensors (e.g., immunosensors) are used to measure more complex molecules such as troponin, various infectious agents, and antibodies. A typical immunosensor using flow-through technology has an antibody covalently coupled to the surface of a porous matrix. When the patient sample is added to the matrix, the analyte of interest binds to the detector antibody (e.g., gold-labeled antibody). A sandwich is formed after the addition of a second antibody, which may be a biotinylated antibody, and traps the label at the position of the first antibody. The antibody analyte complex then flows in a lateral direction along the cellulose nitrate test strip until it reaches the capture zone, which contains streptavidine bound to a solid phase. The biotin in the antibody analyte complex binds to the streptavidin and immobilizes the complex. The complex is then visualized as a purple band by the gold particles attached to one of the antibodies. A reflectometer is used to provide quantitative results.

✪ TABLE 2-3

Analytical principles and analytes measured using POCT devices.

Analytical Principle	Analyte
Reflectance	Urine and blood chemistries (e.g., glucose)
Lateral-flow immunoassay	Infectious disease agents, cardiac markers, hCG
Electrochemistry	Glucose, pH, blood gases, electrolytes, metabolites (e.g., creatinine and urea nitrogen)
Light scattering	Coagulation
Immunoturbidimetry	HbA1c, urine albumin
Spectrophotometry	Blood chemistry
Fluorescence	pH, blood gases, electrolytes, metabolites
Multiwavelength spectrophotometers	Hemoglobin species, bilirubin
Time-resolved fluorescence	Cardiac markers, drugs, CRP
Electrical impedance	Complete blood count

✔ **Checkpoint! 2–22**

List several advantages for patients and caregivers when using POCT.

Nanotechnology

"Nanotechnology has given us the tools . . . to play with the ultimate toy box of nature-atoms and molecules.

Everything is made from it. . . . The possibilities to create new things appear limitless."[22] Horst Stormer, 1998 Physics Nobel Prize Winner

Nanotechnology is defined in part as the manipulation of living and nonliving matter at the level of the nanometer. It is at this scale that quantum physics takes over from classic physics and the properties of elements change character in a novel and unpredictable way.

To get a sense of size, consider these comparisons:

- Ten atoms of hydrogen lined up side by side stretch to one nanometer.
- A DNA molecule is about 2.5 nm wide (25 times bigger than a hydrogen atom).
- A red blood cell is about 5000 nm in diameter, about one-twentieth the width of a human hair.
- A nanometer is 10^{-9} meters in length, and 10^{-12} puts us in the realm of the nucleus of an atom.

Laboratory applications are not described in the true sense of nanotechnology but rather as miniaturization. This means that laboratory devices are not micromachined to between 1 and 100 nanometers but rather scaled down to approximately 100 μm (micrometers). One example of an analyzer that fits the miniaturization models is the iSTAT handheld system (Abbott Point of Care, Inc., Princeton, NJ). It was the first device to include mirofabricated electrodes and chemical sensors on a silicon chip.

The process of fabricating miniature devices is very complex and involves many manufacturing steps. A brief description of manufacturing techniques and key engineering concerns (e.g., microfluidic and valving systems) will be presented.

Several manufacturing techniques are used to produce miniature devices and include photolithography on silicon, glass and plastic surfaces, and molding of polymers. Photolithography is the process of transferring geometric shapes on a mask to the surface of silicon wafer. Once the wafer has been designed, the drilling process can be started to fabricate chambers in the microdevice. Mechanical, ultrasonic, and laser methods are used for drilling holes and creating chambers. Laser drills are used to produce very small holes approximately 2 μm in diameter. Bonding and sealing structures to create liquid-tight chambers and microchannels is necessary in the fabrication of a microdevice.

"Microfluidics" is a term associated with the movement of fluids through the microchannels. The volume of fluid that typically flows through the nanochip ranges from nanoliters to picoliters. The mechanism to move or transport fluids in a nanochip is called "valving." Chip-based microvalve systems are designed as active microvalves (with an actuator) or passive (check) microvalves (without an actuator).[23] Examples of actuators include solenoids, plungers, bimetallic actuators, and piezoelectric actuators.

Atom Technology

Atom technology refers to a spectrum of new technologies that operate at the nanoscale and below. "Below" is described as the manipulation of molecules, atoms, and subatomic particles to create new products. Some people believe that the nanoscale is not the final frontier.

Atom technology is transdisciplinary. It borrows from physics, engineering, molecular biology, and chemistry. The ultimate goal is to manipulate the fundamental building blocks of matter. Once the tools are developed to control and manipulate matter, researchers will be positioned to exploit and integrate technologies, including biotechnology, informatics, cognitive sciences, and more.

Atomtech are technologies converging at the nanoscale, and the etc [group] published a lengthy dissertation outlining issues, impact, risks, and polices associated with this emerging technology.[24] This organization is an international civil society dedicated to the conservation and sustainable advancement of cultural and ecological diversity and human rights. The organization supports socially responsible developments in technologies useful to the poor and marginalized, and it addresses governance issue affecting the international community.

SUMMARY

The applications of laboratory instrumentation broaden as the rate of technology development accelerates. Clinical instruments range from the hand-held analyzers used at the patient's bedside to the extremely sophisticated systems located in large, centralized laboratories. No field in medicine has expanded more rapidly with technology than laboratory medicine, especially in the area of instrument automation. With the nearly constant and wide-ranging advances in technology, it is a challenge for this chapter or any text on the topic to be comprehensive and current on all laboratory instrumentation. What may be state-of-the-art technology today, seen only as ideas on drawing boards or prototypes in research laboratories, can become reality in clinical laboratories in a matter of months.

Clinical laboratories are expected to be scrutinized closely during the current health-care initiatives, especially with regard to finances. At this time, it is still not possible to forecast the impact of new technologies on lowering the operating costs while still providing timely laboratory data to clinicians in this era of cost containment.

Federal regulations on clinical laboratories (Clinical Laboratory Improvement Act, 1988) impose quality control for all clinical laboratory procedures. The impact on POC testing has been tremendous, as seen in decreased availability of accurate instruments, inadequate use of quality control, and insufficient laboratory expertise. POC analyzers are expected to evolve and become even smaller, easier to operate, and more accurate to conform to the new laboratory regulations. One cannot predict the future, but this brief review of instrumentation shows that change will be a constant in the development and use of instrumentation in the clinical laboratory. These changes will inevitably support and promote both decentralization, as more hand-held analyzers are made available to satellite units, and centralization in laboratory medicine and thus help maintain the speed and efficiency for large test volume that health-care providers demand.

REVIEW QUESTIONS

LEVEL I

1. Which of the following is an example of a photodetector? (Objective 5)
   a. diffraction grating
   b. photomultiplier tube
   c. prism
   d. tungsten halogen lamp

2. A holmium oxide filter is used to determine which of the following? (Objective 7)
   a. stray light of a spectrophotometer
   b. linearity of a spectrophotometer
   c. concentration of an unknown solution
   d. wavelength accuracy of a spectrophotometer

3. Diffuse reflection occurs on which type of surface? (Objective 14)
   a. polished
   b. grainy or fibrous
   c. mirror like
   d. transparent

4. "Light energy that occurs when a specific compound absorbs EMR, becomes excited, and returns to an energy levels lower than the original levels" describes which of the following? (Objective 14)
   a. fluorescence
   b. electrophoresis
   c. absorption spectrometry
   d. nuclear magnetic resonance

5. The release of light energy by a molecule after reacting with chemical is termed: (Objective 14)
   a. mass spectroscopy
   b. osmometry
   c. potentiometry
   d. chemiluminescence

6. The total solute concentration of a urine samples can be determined by: (Objective 16)
   a. UV-visible spectrophotometry.
   b. flow cytometry
   c. refractometry
   d. nephelometry

7. "Osmometry is based on measuring changes in the colligative properties of solutions." Which of the following is an example of a colligative property of a solution? (Objective 15)
   a. molar absorptivity
   b. conductivity
   c. specular reflectance
   d. freezing point

8. Which of the following is identified with the functional principle of a pH electrode? (Objective 8)
   a. coulometry
   b. anodic stripping voltammetry
   c. potentiometry
   d. vapor pressure

LEVEL II

1. What is the energy in Joules of an X-ray photon with a wavelength of 2.70 angstroms? (Objective 8)
   a. 6.96×10^{-25}
   b. 7.36×10^{-16}
   c. 1.36×10^{-15}
   d. 1.36×10^{15}

2. What is the percent transmittance of a solution whose absorbance is 0.375? (Objective 3)
   a. 42.0
   b. 16.25
   c. 4.22
   d. 1.625

3. What is the molar concentration of solution whose molar absorptivity is 2.17×10^3 L cm^{-1} mol^{-1}, %T is 10, and the cuvette light path is 1.0 cm? (Objective 3)
   a. 2.17×10^{-3} M
   b. 4.61×10^{-3} M
   c. 4.61×10^{-4} M
   d. 4.61×10^4 M

4. Determine the concentration of a solution in mg/dL given the following: (Objective 3)
   Absorbance of the unknown: 0.200
   Absorbance of the standard: 0.100
   Concentration of the standard: 50 mg/dL
   a. 1.0
   b. 10.0
   c. 25.0
   d. 100.0

5. Blood lead is measured using an electrochemical technique known as anodic stripping voltammetry. Identify, in order of occurrence, the three significant steps in anodic stripping voltammetry. (Objective 6)
   a. Reduction and deposition of lead onto an electrode, resting step, stripping lead off the electrode.
   b. Stripping lead off the electrode, reduction and deposition of lead onto an electrode, resting step.
   c. Resting step, reduction and deposition of lead onto an electrode, stripping lead off the electrode.
   d. Oxidation and deposition of lead onto an electrode, resting step, stripping lead off the electrode.

6. What shape line is created when the absorbance of a series of solutions of known concentration is plotted on the y-axis and the concentrations are plotted on the x-axis? (Objective 5)
   a. sigmoid shape
   b. hyperbole
   c. straight line (linear)
   d. circle

REVIEW QUESTIONS (continued)

LEVEL I

9. Retention time in gas chromatography is defined as: (Objective 14)
   a. the time it takes after a sample injected onto a column from the column
   b. the time it takes after a sample elutes from the column and is finally detected
   c. the time it takes after a sample enters a column to when the sample has traveled half the length of the column
   d. the time it takes after a sample injected onto a column is collected in the waste container

10. What are the three major components of a mass spectrometer? (Objective 5)
   a. phototube, monochromatic filer, tungsten lamp
   b. flame ionization detector, packed separation column, oven for the column
   c. light source, samples, cuvette, stray light trap, and phototube detector
   d. ion source, mass analyzer, ion detector

11. Which of the following is an example of a type of transducer used in biosensor devices? (Objective 12)
   a. reflectometer
   b. osmometer
   c. optical
   d. nephelometer

12. Alpha (α), k prime (k'), and N are three factors that: (Objective 13)
   a. affect chromatographic resolution
   b. affect mass/charge ratios
   c. affect colligative properties of solutions
   d. affect the refractive index of a solution

LEVEL II

PEARSON myhealthprofessionskit™

Use this address to access the interactive Companion Website created for this textbook. Simply select "Clinical Laboratory Science" from the choice of disciplines. Find this book and log in using your user name and password.

REFERENCES

1. Willard HH, Merritt LL, Dean JA, et al. *Instrumental methods of analysis.* Belmont, CA: Wadsworth, 1988: 197–201.

2. Pesce A, Frings CS, Gauldie J. Spectral techniques. *In* Kaplan LA, Pesce AJ, Kazmierczak SC. *Clinical chemistry theory, analysis, correlation.* 4th ed. St Louis, MO: Mosby, 2003: 97–99.

3. Rayleigh, Lord B. On waves propagated along the plane surface of an elastic solid. *ProcLondonMathSoc* (1885) xviv: 4–11.

4. Hitchman ML. *Measurement of dissolved oxygen.* New York: Wiley, 1978.

5. National Committee for Clinical Laboratory Standards. *Preparation and testing of reagents water in the clinical laboratory,* 4th ed. Approved Guideline. NCCLS Document C3-A3. Wayne, PA: National Committee for Clinical Laboratory Standards, 2006.

6. Gocan S. Stationary phase for thin-layer chromatography. *JChromotogSci* (2002) 40: 538-49.

7. Taylor T. Sample injection systems. *In* Handley AJ, Adlard ER eds. *Gas chromatographic techniques and applications.* Boca Raton, FL: CRC Press LLC, 2001: 52–70.

8. Skoog DA, Holler FJ, Nieman TA. *Principles of instrumental analysis,* 5th ed. Philadelphia, PA: Saunders, 1998: 705–711.

9. Skoog DA, Holler FJ, Nieman TA. *Principles of instrumental analysis,* 5th ed. Philadelphia, PA: Saunders; 1998: 739–58.

10. Rubinson KA, Rubinson JF. *Contemporary instrumental analysis.* Upper Saddle River, NJ: Pearson Education, 2000: 515–21.

11. Rubinson KA, Rubinson JF. *Contemporary instrumental analysis.* Upper Saddle River, NJ: Pearson Education, 2000: 523–28

12. Herbert CG, Johnston R. *Mass spectrometry basics*. Boca Raton, FL? CRC Press LLC, 2003: 55–60.

13. Rubinson KA, Rubinson JF. *Contemporary instrumental analysis*. Upper Saddle River, NJ: Pearson Education,, 2000: 533–35.

14. Herbert CG, Johnston R. *Mass spectrometry basics*. Boca Raton, FL: CRC Press LLC, 2003: 7–12.

15. Herbert CG, Johnston R. *Mass spectrometry basics*. Boca Raton, FL: CRC Press LLC, 2003: 153–98.

16. Bier ME. Analysis of proteins by mass spectrometry. *In* Howard GC, Brown WE, eds. *Modern protein chemistry—practical aspects*. Boca Raton, FL: CRC Press, 2002: 76 –80.

17. Rubinson KA, Rubinson JF. *Contemporary instrumental analysis*. Upper Saddle River, NJ: Pearson Education, 2000: 477–90.

18. Cunningham A. *Introduction to bioanalytical sensors*. New York: John Wiley & Sons, 1998.

19. Morgan CL, Newman DJ, Price CP. Immunosensors: technology and opportunities in laboratory medicine. *ClinChem* (1996) 42: 193–206.

20. Trettnak W, Wolfbeis OS. A fiber-optic cholesterol biosensor with an oxygen optode as the transducer. *AnalBiochem* (1990) 2(184): 124–27.

21. Price CP, St. John A. Point of Care testing. In: Burtis CA, Ashwood ER, Bruns DE, eds. *Tietz textbook of clinical chemistry and molecular diagnostics*, 4th ed. Philadelphia, PA: W. B. Saunders, 2006: 299–305.

22. Amato I. Nanotechnology-shaping the world atom by atom (NSTC report) at http://www.wtec.org/loyola/nano/IWGN.Public.Brochure/IWGN.Nanotechnology.Brochure.pdf (accessed September 1, 2008).

23. Shoji S, Esashi M. Microflow devices and systems. *JMicromech-Microeng* (1994) 4: 157–71.

24. The big DOWN: from genomes to atoms. ETC[group] publications, January 2003. (http://www.etcgroup.org/en) (accessed September 1, 2008).

3

Laboratory Automation

■ OBJECTIVES—LEVEL I

Following successful completion of this chapter, the learner will be able to:

1. List four advantages of automated chemical analysis.
2. Define the following terms: throughput, test menu, carryover, discrete testing, random-access testing, open-reagent analyzer, and closed-reagent analyzer.
3. Identify five laboratory tasks associated with the preanalytical stages of laboratory testing.
4. Identify three reasons why automation is necessary.
5. Give examples of how automated analyzer performs the following functions:
   a. mixing
   b. incubating
   c. transfer reagents
6. List four tasks associated with the analytical stage of laboratory testing.
7. Identify five demands placed on the laboratory that serve to drive automation.
8. List three techniques used to mix samples and reagents in an automated system.
9. Identity three techniques used to incubate samples and reagents.
10. List three drawbacks of total laboratory automation.
11. Identify three tasks associated with the postanalytical stage of laboratory testing.

■ OBJECTIVES—LEVEL II

Following successful completion of this chapter, the learner will be able to:

1. Explain the concept of total laboratory automation.
2. Distinguish the three stages of laboratory testing from one another.
3. Differentiate between proportioning reagent by volumetric addition and by continuous flow.
4. Explain the operating principle of a Peltier thermal electric module.
5. Distinguish between workstation and work cell.
6. Explain the principle used for clot detection in automated analyzers.

KEY TERMS

Aliquot	Discrete testing	Test menu
Batch analysis	Liquid-level sensor	Throughput
Bulk reagents	Middleware	Total laboratory automation (TLA)
Carryover	Modular instruments	Unit test reagents
Closed-reagent analyzer	Open-reagent analyzer	Work cell
Clot detector	Random-access testing	Workstation

A CASE IN POINT

The supervisor of a laboratory has been given the task of updating the core laboratory, including processing, instrumentation in chemistry, hematology, and urinalysis.

Issues and Questions to Consider

1. What are several options available to accomplish this reorganization of the core laboratory?

2. What factors should be considered before making a final decision on types of equipment to purchase?

WHAT'S AHEAD

▶ Discussion of automation within the three stages of laboratory testing.

▶ Explanation of automated techniques used to complete the various tasks required during the analytical stage of testing.

▶ Overview of automated systems designs.

BOX 3-1 Laboratory Demands that Drive Automation

▶ Reduce turnaround times (TATs)
▶ Staff shortages
▶ Economic factors
▶ Less maintenance
▶ Less down time
▶ Faster start-up times
▶ 24/7 uptime
▶ Increase throughput
▶ Computer and software technology
▶ Primary tube sampling
▶ Increase number of different analytes on one system
▶ Increase number of different methods on one system
▶ Reduce lab errors
▶ Increased number of specimens
▶ Improve safety of laboratory staff
▶ Environmental concerns such as biohazard risks

▶ INTRODUCTION

Automated methods of analysis have been available in the laboratory since the mid-1950s when Leonard Skeggs released the first single-channel autoanalyzer.[1] This analyzer was designed to automate the analytical phase of analyte testing. The pre- and postanalytical stages still required considerable manual work, and it would be several decades before these two stages achieved the level of automation that exists today.

The demands placed on the laboratory by clinicians, the public, government, insurance carriers, and laboratorian have served to change the way laboratory tests are performed. Laboratory automation is a facet of the clinical laboratory that has responded the most as a result of these demands. Aided by the rapid advancement in technologies and computers, the changes in automation can be described as exponential or linear with a steep slope. Several specific examples of demands that drive automation are listed in Box 3-1 ■. Many of these demands will persist for years to come and continue to provide the laboratory with new and innovative ways of performing laboratory tests.

The reasons then for automating chemical analyses are somewhat different than today. To remain competitive, clinical laboratories now face many challenges in addition to those listed in Box 3-1. These challenges are a result of the continued reduction of government reimbursement rates for laboratory tests, cost-restraint measures, the managed-care industry, and the movement toward containment of national health-care costs.[2]

The main impetus behind automation has been the need to create automated systems capable of reducing or eliminating the many manual tasks required to perform analytical procedures. Continued development has led first to the consolidation of most high-volume chemistry measurements onto a single platform and more recently to the consolidation of chemistry and immunoassay systems onto a single platform. By eliminating manual steps, the opportunity to reduce error is enhanced because there is less potential for error resulting from technologist fatigue or erroneous sample identification.

The introduction and implementation of the laboratory information system (LIS) that occurred in the 1970s automated the process of information flow in the clinical laboratory. One significant result brought about by the use of LIS was a decrease in the expected 5% transcription error rate seen when laboratory results were manually transcribed into various medical-record formats. In the 1980s, intralaboratory transport systems were designed to include specimen carriers and a conveyor belt or track that would take samples between instruments. Preanalytical processing stations for centrifugation, decapping and capping, and storage helped lead to a total laboratory automation approach to testing laboratory specimens.

This chapter will explore the automation of the three stages of laboratory testing and provide the reader with a scenario that puts it all together.

▶ AUTOMATED ANALYSIS

One means of gaining an understanding and appreciation of automation is to identify what advantages automation provides to chemical analysis, especially in clinical laboratories. A partial listing of these advantages is shown in Box 3-2 ■. Another way to interpret the information in the box is to ask, How can automation create these advantages? To appreciate how tests are performed in modern clinical laboratories, we can review the progression of instrument designs, computers, and technologies that have brought us to where we are today.

The measurement of samples using automated instrumentation has undergone an evolutionary process since the Technicon AutoAnalyzer (Technicon Instrument Corp., Tarrytown, NY). It began with a single-channel analyzer using continuous-flow analysis and measured one analyte on each sample. These samples were measured in a sequential manner, or one sample after another. The specimen **throughput**, or numbers of tests performed per hour was approximately 40 to 60. Technicon continued to develop their systems, which evolved into multiple-channel instruments. A good example is the SMAC II, which produced specimen-throughput rates as high as 150 per hour with test throughput of approximately 3750 per hour, depending on test configuration. One major disadvantage to this type of instrument design was that all testing was performed in a parallel fashion. This resulted in the measurement of every analyte configured on the system for every sample. This inflexibility in testing led to the development of analyzers that

provided **discrete testing**, which measured only the test requested on a sample.

The next generation of automated analyzers included centrifugal analyzers and modular analyzer configurations. Centrifugal analyzers were discrete, batch-type systems. **Batch analysis** means that a group of samples is prepared for analysis and a single test is performed on each sample in the group. A significant limitation of these discrete analyzers is their reduced throughput rates as compared to other measuring techniques—for example, random-access testing. Because the analyzers were configured to measure one analyte at a time, the only way to improve throughput using discrete analyzers was to purchase multiple analyzers so that several tests could be run simultaneously, depending on how many analyzers the laboratory purchased.

The solution to the limitations associated with centrifugal analysis was to design a modular system that could be configured to measure any specimen by a command to the processing systems, as well as be analyzed by any available process, in or out of sequence with other specimens and without regard to their initial order. This is the essence of **random-access testing**. Modular instruments also allow the user to include additional modules (e.g., ion-selective electrodes or immunoassays).

The ultimate result of combining modular design with random-access testing was to increase specimen-throughput rates to hundreds per hour and test-throughput rates to thousands per hour.

 Checkpoint! 3–1

Identify several areas of concern in the laboratory that were improved using automated chemistry analyzers.

▶ PREANALYTICAL STAGE

The three stages of laboratory testing are preanalytical, analytical, and postanalytical. Improving efficiency and productivity during the preanalytical stage of laboratory testing was not the main focus of the laboratory staff at the outset. Likewise, the postanalytical stage received little attention. This lack of attention to improvements for these two stages was partly because technologies had yet to be developed that would serve to change the scope of each stage.

The preanalytical stage involves primarily sample or specimen processing. Mechanisms for transporting specimens to the laboratory and within the laboratory include:

• human carriers or runners,
• pneumatic-tube delivery systems,
• electric-track–driven vehicles,
• mobile robots, and
• conveyors or track systems.

For decades, specimens drawn within a facility were brought to the laboratory, usually by those who drew the blood, or "runners." If the specimens were obtained from outside the laboratory facility—for example, at a clinic or a physician's office—a courier service was often used. A courier service is described as a specimen-batching process in which an individual picks up specimens at an off-site location and then delivers them to the testing laboratory by some schedule. These individuals were the first link between patient and laboratory and were also a source of problems that resulted in some remarkable changes and innovations to the process of laboratory testing as a whole. One early solution to replacing humans as specimen couriers was the introduction of pneumatic tubes.

Pneumatic-tube delivery systems were installed to provide point-to-point delivery of specimens to the laboratory and offered several advantages over human transport: quicker delivery of specimens, no delays during transit, and 24-hour-a-day, 7-days-a-week availability. The tubes are sent very quickly to the laboratory encased in a carrier lined with a foam-type material to reduce breakage. Pneumatic-tube systems are designed to prevent hemolysis by avoiding significant elevations of g forces during acceleration and deceleration.

Electric-track vehicles can transport a larger number of specimens than pneumatic tubes. The electric tracks require a station for loading and unloading specimens, which may pose a problem in facilities with limited space. Like couriers, electric tracks allow specimens to be batched.

Later, mobile robots of many designs were used by laboratories to transport specimens from within and outside the facility. Samples are usually batched for pickup and loaded onto the mobile robot for movement to their destination.

Conveyors or track systems are used in some laboratory facilities, especially if the laboratory receives large numbers of specimens. These systems transport specimens in a horizontal fashion as well as vertically to another floor.

A **workstation** refers to a designated area in which a limited number of specific tasks are completed. The processing workstation serves as an area where specimens requiring laboratory testing are received. Once the specimens arrive at the laboratory processing workstation, several tasks are required to be completed, examples of which are listed in Box 3-3 ■.

BOX 3-3 Examples of Sample-Processing Tasks

► Identify specimens
► Label specimens using bar-code labels
► Sort and route
► Centrifuge sample tubes
► Decap tubes
► Prepare sample aliquots
► Recap, store, retrieve
► Transport
► Sample level detection
► Store and retrieve

Some novel approaches have been used, culminating in what is termed *preanalytical modules*. These modules are available from several instrument manufacturers. There were many earlier attempts to process specimens with minimal human involvement before the development of preanalytical modules. Several of these preanalytical modules are presented in this chapter.

Labeling specimens by hand is time consuming and proved to be a large source of laboratory error. Labeling went beyond the specimen tube and included pour-off tubes, sample cups, dilution cups, and send-out containers. The use of printed bar-code labels facilitated this process tremendously. Later, as computers became more sophisticated and communication between computers improved, the bar-code label system improved processing time and reduced preanalytical errors.

Manual sorting or separating of samples was needed because of the many types of testing that most laboratories did. Specimen tubes of all shapes and sizes would be received in the processing area of a clinical laboratory, and the technologist would have to sort tubes by stopper color, size, tests ordered, instrument-design requirements, and tube destinations.

In the earlier days of clinical laboratories, each red-top tube was double spun so that clot removal was optimal. This step required that specimen tubes be decapped by hand. The invention of the serum separator tube eliminated the need for this double-spin technique so that decapping at this stage was not necessary. A specimen would have to be decapped to process **aliquots**—that is, a portion of a specimen—poured into sample cups or introduced into the analyzer. The decapping of tubes posed health hazards to the technologist via aerosols that leave the specimen tubes or by direct contact with the blood.

Centrifugation of blood-collection tubes required the technologist to manually load tube carriers and place them into the centrifuge. The tubes would then be removed from the centrifuge and resorted, aliquots would be processed, and the samples distributed to their destinations or target area. This whole process was fraught with potential safety hazards, opportunities for mistakes, and large increases in sample-processing time.

Many specimens require aliquots to be poured off, a process also known as *splitting* the samples. The aliquots are used by instrument operators, sent to other laboratory sections, sent out to references laboratories, and used for dilutions. Like other manual-processing steps, aliquoting blood samples exposes the laboratory staff to the sample and can pose a potential health hazard. The chance for errors also increases, and processing time may be prolonged.

When samples are no longer needed for testing, they are stored in a refrigerator or freezer. All of the samples are stored in an organized fashion in case they are needed again for repeat testing. Manual storage and retrieval of samples may result in problems for some laboratories, including losing samples, improper storage of samples, and difficulty in locating samples.

The Abbott Diagnostics ACCELERATOR Stand-Preanalytics (Abbott Diagnostics, Abbott Park, IL) is a front-end, preanalytical laboratory automation solution. Automation of the preanalytical process is of increasing interest to laboratory managers all over the world. The ACCELERATOR combines preanalytical functions—including presorting, centrifugation, volume check and clot detection, decapping, secondary tube labeling, aliquotting, and destination sorting into analyzer racks—on a small-footprint instrument.

The Abbott ACCELERATOR automates labor-intensive tasks such as sample identification, decapping, aliquotting, sorting, and archival preparation of sample tubes. Automated processing units such as the ACCELERATOR also lower staff exposure to biohazards, decrease preanalytical errors, and reduce repetitive motion injuries associated with manual sample processing.

The Beckman Coulter Power Processor (Bechman Coulter, Brea, CA) automated sample-processing system has many of the same features as other preanalytical specimen-processing units and is designed to sort samples into discrete *personality racks* for use with most Beckman Coulter automated analyzers and generic output racks for other instruments.

 Checkpoint! 3–2

The preanalytical stage of laboratory testing remains the greatest source of laboratory error. Identify several examples of preanalytical tasks that may lead to laboratory error.

AUTOMATED SPECIMEN PROCESSING

Automated specimen processing (also known as front-end sample processing) represents the most cost effective automation strategies for the clinical laboratory.[3] Two goals for automating specimen processing are (1) to minimize non–value-added steps in laboratory process for example sorting tubes and (2) increase available time for value-added steps in the tasks that technologists perform, which will help make a difference in the quality of the test results and, ultimately, patient care.

Several front-end sample-processing systems are available to improve on all of the shortcomings associated with manual sample processing. The system designs may be either integrated or modular specimen processing. Some modular systems are designed to exist as stand-alone, front-end processors.

Integrated specimen-processing systems allow the user to perform some or all specimen-handling tasks. These systems, however, only process certain types of samples and specimen containers. (This inflexibility with specimen containers is the result of many laboratories, especially those in hospitals, pursuing modular specimen-processing systems.) Each module has it own on-board computer that is linked to a master-controller computer system. Modular systems also can accommodate several different types of specimens—for example, whole blood, serum, and plasma—with their respective specimen containers.

Most manufacturers of preanalytical systems attempt to provide the user with some or all of the tasks needed to prepare samples for testing, including:

• presorting,

• centrifugation,

• volume checks,

• clot detection,

• decapping,

• secondary tube labeling,

• aliquoting, and

• destination sorting into analyzer racks.

The *stand-alone* system automates one portion of front-end processing. A stand-alone system has no analyzer or track system attached to it. The Tecan FE500 is an example of a stand-alone processor. Stand-alone systems automate the sample sorting, sample uncapping, and aliquot functions of the front-end samples processing. A centrifuge is not included in this design. If serum or plasma is required, then the sample must be carried to the centrifuge by the technologist.

Archiving and retrieving specimens in an automated fashion is also available in stand-alone designs. Automated sample-archiving systems use bar-coded specimens that are scanned and placed in numbered positions in numbered racks. Retrieval of a specimen is initiated by entering the patient's sample-accession number or a medical record number into the archival systems database. The rack number and position in the rack are determined and displayed for the user. Some systems include a refrigerator for sample storage and automatic disposal of samples at predetermined times. Box 3-4 ■ shows several examples of preanalytical processing units.

BOX 3-4 Examples of Automated Sample-Processing Systems

▶ A&T CliniLog System (Kanagawa, Japan)
▶ Bayer Diagnostic ADVIA LabCell (Tarrytown, NY)
▶ Beckman Coulter Power Processor System (Brea, CA)
▶ Olympus OLA-2500 Decapper, Sorter, Archiver, Aliquotter (Melville, NY)
▶ Roche Diagnostic PSD1 Decapper and Sorter (Indianapolis)
▶ Roche Diagnostic VSII Aliquotter and Sorter
▶ Tecan Genesis FE-500 Workcell (Durham, NC)
▶ Thermo Clinical Labsystem's TC Automation Systems (Thermo Electron, Waltham, MA)

► Sample introduction and transport to cuvette or dilution cup
► Reagent measurements, transport, and introduction to cuvette
► Mixing of sample and reagent
► Incubation
► Detection
► Calculations
► Readout and result reporting

► ANALYTICAL STAGE

The analytical stage of testing has evolved to a very sophisticated level because of progress in technology, improvements in computer technology, and many of the drivers of automation listed earlier. Box 3-5 ■ lists many of the tasks required in the analytical stage of laboratory testing.

SAMPLE INTRODUCTION

Automatic sampling may be accomplished using several different physical mechanisms. Peristaltic pumps and positive liquid-displacement pipettes are two examples. Peristaltic pumps are examples of older technology but are still used in some instrument designs that measure electrolytes. Positive liquid-displacement pipettes are usually a single pipette that transfers samples from cups or tubes to the next analytical process. Most positive-displacement pipettes function in one of two ways: either dispensing aspirated samples into the reaction container or flushing out samples together with diluent.

Transferring the sample from the sample cup or tube to its destination via the sample probe is accomplished in several different ways. Some analyzers use a robot-like arm with an aspiration probe attached that pivots back and forth, picking up a sample and depositing it into a reaction vessel or onto the surface of a porous pad. Other systems may use a worm gear device that pulls the sample probe from one point to another.

In most analyzers, samples are transferred using a thin, stainless steel probe. The probe may be required to pierce a rubber stopper or pass directly into a test tube or cup. A given quantity of sample is aspirated into the probe, and the probe is moved toward an appropriate container for dispensing. A potential source of problems with this type of sample probe is the formation of a clot in a sample that subsequently attaches to the probe. These clots may plug the probe, making continued use impossible. Also the clot may occupy sample volume and thus cause an error in the measurement. Because of

the sticky nature of serum or plasma, the clot may adhere to the sample probe; as the sample probe swings toward it next destination, the whole clot and sample vessel may move along with it. This could result in an instrument malfunction or sample-probe misalignment. Several sample probe designs have **clot detectors**. Clot detection is accomplished by using a pressure transducer. As the pressure transducer comes into contact with the sample, the analyzer measures the difference in pressure between air and the surface of the sample. If the pressure created is greater than a specified cutoff value, then the sample is not aspirated.

Another feature associated with sampling is the ability of the sampler to detect the presence or a liquid. A **liquid-level sensor** is designed to detect the presence of a sample by measuring the electrical capacitance of the surrounding area. The pipette and liquid-level sensor travel a specified distance into the sample container to determine if liquid is present or not.

One problem with reusable sample probes is **carryover**, or the contamination of one sample by the previous sample. This contamination may cause serious variation in results for subsequent tests. Several instrument modifications have been used to reduce carryover. One method used is to aspirate a wash solution between each pipetting. Another technique is to back flush the probe using a wash solution. The wash solution flows through the probe in a direction opposite to that of the aspiration and into a waste container. This technique also tends to minimize the risk of forcing a small clot further into the system.

Many samplers use disposable plastic pipette tips to transfer samples. This has the distinct advantage of eliminating contamination from carryover within the sample probe and from sample to sample. A downside to the use of disposable tips is the increased cost of performing the assays.

 Checkpoint! 3-3

Identify two techniques that can be used in automated analyzers to reduce carryover of samples.

REAGENTS

Reagents used in automated analyzers required attention to several concerns that include:

• preparation,
• storage,
• proportioning, and
• dispensing.

Most laboratories use **bulk reagents** that represent large quantities of reagents or solutions that are used for most analyses, flushing, and priming. Bulk reagents are ready for use with little or no preparation. If the reagent is lyophilized, most analyzers will automatically dispense the proper diluent to dissolve the dried reagents. Chemistry analyzers that use **unit test** reagents (i.e., where there is only enough reagent for single test) may require some preparation. For dry-slide analysis in which a thin film is impregnated with the appropriate reagent, preparation consists of wetting the reagent with water, buffer, or sample. Another type of unit test reagent is a container or test tube consisting of premeasured liquid or powdered material to which water, buffer, or sample is added.

Reagents that are packaged either wet or dry are maintained within the reagent compartments, and a complete inventory is established on a real-time basis within the computer. Most of the methodologies used in the laboratory require only a single reagent, but several require two or more. As a reagent becomes depleted, the computer signals the operator that the reagent container is empty and that a new one should be added. The amount of inventory for reagents that needs to be available within the analyzer depends on the number of tests completed for any given analyte. On-board reagent storage compartments are refrigerated to maintain reagent stability.

Reagent identification and inventory processes are accomplished by use of bar-coded labels. The bar-code label may also contain additional information such as expiration dates, lot numbers, and the number of tests the contents of the container may provide. Some analyzers may couple a liquid-level sensor onto the reagent probe, which will alert the operator as to whether a sufficient quantity of reagent exists to complete the tests.

For immunoassay tests, the bar-coded reagent label stores critical information about calibrators. Examples of stored information include but are not limited to the concentration of calibrant, expected detector responses, calibration curve algorithms, and tolerances for acceptability of calibration. This information is often referred to as *master lot* or *master calibration*.

An important classification category for all automated analyzers is based on reagents. Automated analyzers are categorized as either an *open* or *closed*. This distinction is often a key determinant as to whether a laboratory will select an analyzer. An **open-reagent analyzer** is a system in which reagents other than the instrument manufacturer reagents can be used. The operator also may have more flexibility to change the parameters necessary to run the particular test. Open-reagent analyzers provide users with more flexibility by easily adapting new methods and analytes. In a **closed-reagent analyzer**, on the other hand, the operator can only use the instrument manufacturer's reagents. Reagents for closed-reagent analyzers can be more expensive, and it may not be possible to introduce new tests that are not already performed on the closed system. There are several advantages to both categories of reagent systems, so users should thoroughly investigate their requirements.

The correct proportion of reagents and samples must be constant to achieve precise and accurate results. For unit test applications, the reagents are already proportioned in the appropriate amounts, so only the sample needs to be proportioned. Methods that require the addition of bulk reagents pose additional risks of increasing imprecision. When bulk reagents are used, proportioning is accomplished by volumetric addition.

The delivery of bulk reagents requires automated, volumetric dispensing devices. Random-access analyzers use syringes or volumetric overflow devices that volumetrically proportion reagent and sample into a test tube or other type of container.

Another mechanism used for to proportion reagents and samples is the continuous-flow technique. The sample and reagents are proportioned by their relative flow rates. Peristaltic pumps are used for continuous-flow proportioning. Many instrument designs use electronic valves to control reagent flow time. The flow rate is controlled by the air pressure applied to the reagent container, the flow resistance in the tubing connected to the reaction vessel, and the internal diameter of the tubing.

Liquid reagents are aspirated, delivered, and dispensed into mixing chambers or reaction vessels by pumps or positive-displacement syringes. These pumps are connected to the reagent containers using plastic tubing. On command from the computer, each pump draws a given amount of reagent or diluent out of the container and transports it via the tubing to its destination where it is dispensed.

Syringe devices are widely used in automated systems for both reagent and sample delivery. Most are positive-displacement devices, and the volume of reagent delivered is controlled by a computer. If the reagent syringe is to be used for more than one reagent, then adequate flushing between sampling is essential to reduce carryover of reagent.

Direct tube sampling is offered on many of the new models lines of instruments. Direct tube sampling along with bar-code identification has eliminated pouring off into another container, thus reducing errors and minimizing technologist exposure to patient's samples. Direct tube sampling can be performed after the stoppers have been removed or in systems such as the Beckman Coulter SYNCHRON LX i725 Clinical System while the stoppers is on. The LX i725 incorporates cap-piercing technology into its module. A blade is used to slit the stopper, and the sample probe pierces the stopper to withdraw an aliquot of sample.

✓ Checkpoint! 3–4

Indicate whether the following situation represents a closed- or an open-reagent analyzer:

A staff technologist is interested in adding a new assay to the laboratory's chemistry analyzer. The technologist contacts the instrument manufacturer and is told that the assay is not currently available and that no competitor's reagents can be used on this chemistry analyzer.

light-emitting diode produces light that is directed into a cuvette to measure hs-CRP.

▶ POSTANALYTICAL STAGE

The postanalytical stage begins after the electrical signal is generated by the detector, which represents analyte concentration. This electrical signal is directed into the analyzer's microprocessor or computer. The instrument computer represents a means to accomplish several tasks, which include signal processing, data handling, and process control.

Signal processing involves the conversion of an analog signal derived from the detector to a digital signal that is usable by all communication devices. The processing of data by computers has allowed the automation of nonisotopic immunoassays, reflectance photometry, and other nonlinear assays because computer algorithms can transform nonlinear standard input signals into linear-calibration plots.

Data processing by computers includes data acquisition, calculations, monitoring, and displaying data. In addition to transforming data into linear-calibration plots, computers can perform statistics on patient and control values. They can perform corrections on data, subtract blank responses, and determine first-order linear regression for slope and intercept. Computers can monitor patient results against reference values. They can also test control data against established quality-control (QC) protocols. Computer monitors can display all types of information including patient results, QC data, maintenance, and instrumentation operation checks.

The computer has profoundly affected the entire process of automated laboratory instruments. Within the analyzer, the computer commands and times the electromechanical operations so that they can be done in a uniform manner, in repeatable fashion, and in the correct sequence. These operations include activating pipetting devices, moving cuvettes from one point to another, moving sample tubes, and dispensing reagents, to name a few.

A computer provides a means of communication between the analyzer and operator. Instrument computers can display information usable by the operator such as warning that something may not be working properly or that a specific reaction has exceeded method-defined parameters.

Chemistry analyzer computers can display graphical information such as Levy-Jennings QC charts and calibration curves. They can also "flag" data that does not meet predefined criteria. The operator can reprogram the computer to meet a specific need such as adding a new test or changing an operating parameter.

Computers have the ability to be linked to other computers, which has drastically improved automation efforts. Instrument computers can be linked via interfaces such as RS-232 to an LIS to provide a means of transmitting information in either unidirectional or bidirectional format. Instrument computers are now being equipped with the means to link to the Internet via TCP/IP (transmission control protocol/

Internet protocol). Instrument manufacturers have designed analyzer computers to link the analyzer to the manufacture's site. This linkup is in real time and serves to monitor the instruments performance at all times. If a problem with the analyzer does develop, the manufacturer can see real-time data to help the laboratory resolve the problem in the shortest time possible.

Several other features of note are available in newer instrument designs. On-board troubleshooting is available on many systems through the analyzers' computers. In the event a problem occurs, technologists may access a system's help protocols, which guides them through a step-by-step procedure to resolve the problem. Some of these on-board troubleshooting programs are quite sophisticated and include video and graphics. An on-board training program is also available in some systems, effectively augmenting staff training for new users of the system.

▶ AUTOMATED SYSTEM DESIGNS

TOTAL LAB AUTOMATION

A **total laboratory automated (TLA)** approach can be described as the combination of several instruments, consolidated instruments, work cells, integrated work cells, or integrated modular works cells that are coupled to a specimen-management and transportation system as well as a process-control software component to automate a large percentage of laboratory work.

The concept of total laboratory automation in a clinical laboratory has its roots in Japan, where a prototype was first evaluated in the early 1980s.[4] This TLA design used one-arm robots, conveyors belts, and modifications to existing chemistry analyzers to perform as many of the preanalytical and analytical tasks as possible without human intervention. Each laboratory workstation was coupled to the conveyor belts so that samples could be moved from one workstation to another. Continued research and modifications to these earlier systems led to the development of commercial TLA designed for hospital-based laboratories.

An example of a TLA is the Roche Diagnostics system, which includes a modular preanalytics feature and platform C (i.e., the chemistry analyzer). It also includes an integrated tract device that connects all of the laboratory workstations including front-end processing, instrumentation, and archiving together to create a continuous, inclusive network that serves to automate nearly every step involved in the testing of each sample. TLA can incorporate testing specimens for chemistry, hematology, coagulation, and immunochemistry.

The advantage of TLA includes a decrease in labeling errors, reduced turnaround times, and a potential reduction in full-time equivalents (FTEs). Some laboratories using a high degree of automation choose not to reduce FTE but instead bring new assays into the laboratory and use the staff to perform the added tests.

MIXING

There are many examples of unique mixing devices and techniques used in automated systems, including:

- magnetic stirring,
- rotating paddles,
- forceful dispensing,
- the use of ultrasonic energy, and
- vigorous lateral displacement (e.g., centrifugal analyzers).

Dry-slide analyzers do not require mixing of sample and reagents. The sample is allowed to flow through the layers containing the reagents.

INCUBATION

Warming of instrument components or solutions in automated analyzers is accomplished by heating air, water, or metal. The warming process must be constant and accurate. Electronic thermocouples and thermistors are used to monitor and maintain required temperatures in the analyzer. Circulating water baths are commonly used to warm solutions—reaction mixtures in cuvettes, for example—so these analyzers require a water-purification and delivery system, which is usually external to the analyzer and an additional cost. In some analyzers, the cuvettes or reaction vessels are allowed to incubate within a chamber containing circulating air. Heated metal blocks are widely used devices for incubating cuvettes, test tubes, or plastic pouches containing solutions. The timing for each incubation period is monitored by the instrument's computer system and represents an extremely complex process given the throughput for these systems.

Two novel approaches for incubating reactions mixtures have been developed and released on currently used automated chemistry analyzers. Bayer Diagnostics (Tarrytown, NY) uses an elongated cuvette path length and a fluorocarbon-oil incubation bath to maximize the accuracy of results by enhancing absorbance values while using microvolume technology for samples and reagents. This design feature is found in the company's model ADVIA chemistry systems.

Beckman Coulter uses a Peltier thermoelectric module in the shape of a ring to maintain a constant temperature for analysis. A Peltier module is a small solid-state device that functions as a heat pump. The Peltier thermal ring is made of copper. The cuvettes, usually quartz glass, are placed within the Peltier thermal ring so that each cuvette is surrounded on three sides by copper. Temperature is maintained by using heating and sensing elements in physical contact with a copper core filled with Freon 134A, and it is controlled by the reaction-heat-controller board assembly mounted in the thermal ring handle.

Checkpoint! 3–5

Give three examples of how liquid solutions are warmed in automated analyzers.

DETECTION

In automated analyzers, absorption spectroscopy remains the principle means of measuring a wide variety of compounds. Reflectance photometry has been adapted to dry-slide analysis and has been used in chemistry laboratories for decades. The use of fluorescent compounds such as fluorescein as signal generators have been used to measure drugs, hormones, and vitamins in several immunoassays analyzers. In the past decade, chemiluminescence compounds such as acridinium have replaced fluorescent compounds because of improvements in sensitivity.

Electrochemiluminescent methods have also been incorporated into automated systems. Automated electrolyte measurements have been accomplished using ion-selective electrodes (ISEs). The actual detectors used are discussed in Chapter 2, Instrumentation. This section will focus on new approaches to measure compounds with automated analyzers.

Novel approaches to measurement designs include not only addition of new measurement principles but also inclusion of two or more unique detectors in one analytical system. Most of the integrated chemistry analyzers being marketed today incorporate several measuring platforms. Each platform requires a distinct detector. The Roche COBAS Integra 800 (Indianapolis) incorporates a spectrometer, ISE, fluorescence polarization immunoassay optics, and turbidimetric optics. The interior of the COBAS Integra is shown in Figure 3-1 ■.

The Beckman Coulter LX i725 Clinical System includes a luminometer, spectrometer, electrochemical detectors, and near-infrared detector. Infrared detection is used for the near-infrared particle immunoassay method to measure high-sensitivity C-reactive protein (hs-CRP). A 940-nm

■ FIGURE 3-1   COBAS Integra 800
Courtesy of Roche Diagnostics

The major drawbacks of TLA are the needs for substantial financial investment and increased floor space.[5] Initial investments may be millions of dollars, and space requirements may exceed 4000 square feet. Another factor that requires attention by the planners because of the complex nature of TLA is the need for highly technical personnel to operate and troubleshoot the system. Other challenges to TLA include infrastructure remodeling, personnel team building, and software interfacing. In addition, several of these systems do not allow the work flow to be interrupted so that emergency (*stat*) samples can be analyzed.[6]

INTEGRATED MODULAR SYSTEMS

In the United States, only some 7% of the laboratories are considered liable to benefit from TLA. A hospital with fewer than 600 beds may not be suitable for TLA. Therefore, modular automation provides a more attractive approach for hospital laboratories and physician group laboratories because the systems are smaller, require less initial capital investment, and require less planning than TLA.[2] Modular systems can be configured to include several different platforms such as hematology and immunochemistry. The combination of modules also can include multiple identical models of analyzers and pre- and postanalytical modules. These modules are linked into a single testing platform that interconnects by use of a track or other connector-type device. Individual modules can be added to the entire system to reflect changes in either workload or testing patterns.

The Bayer IMS integrated modular system's use of Smart Access software determines the most efficient and productive route for each individual tube to take while incorporating automatic dilutions and reruns of samples. The Bayer IMS combines clinical chemistry with immunoassay and without limiting future menu expansion.

Workstations

Workstations represent a unique environment within a laboratory facility dedicated to one type of testing—for example, hematology or immunoassays. All of the stages of specimen testing are carried out for that particular testing platform at its respective workstation. Each workstation functions as its own separate entity. Integrating workstations into one common unit close to specimen processing can improve laboratory operations such as turnaround time and error reduction. The Beckman Coulter DL 2000 Data Management System provides workstation connectivity. This system connects chemistry, immunochemistry, electrophoresis, and immunoassay workstations to provide improved workflow management.

Work Cells

A **work cell** is a combination of a specimen manager with instruments or consolidated instruments of chemistry and immunoassay reagents that provide a broad spectrum of analytical tests. A *specimen manager* is a mechanical device that

■ FIGURE 3-2   Siemens StreamLAB
Courtesy of Siemans Healthcare Diagnostics

allows the storage and buffering of specimens before and after analysis and may include pre- and postanalytical specimen-processing capabilities such as centrifugation and decapping. Modular work cells are work cells where the instruments used are designed to interface directly with the specimen manager. The Siemens StreamLAB Analytical Workcell is an example of this type of work cell and is shown in Figure 3-2 ■. before and after analysis. *Modular work cells* are work cells in which the instruments are configured to interface directly with the specimen manager. The Dade Behring StreamLAB is an example of this type of work cell (see Figure 3-2).

The StreamLABsystem integrates pre and postanalytical specimen-processing capabilities with multiple analytical components, including the Siemens Dimension systems, via a single operator interface. Centrifugation, tube sealing, and refrigerated storage are optional modules.

Work-cell technology has been advanced by the new ADVIA WorkCell CDX Automation Solution, which provides enhanced computer operations through hardware upgrades and a new operating system. The system also has the ability to add more ADVIA Centaur Immunoassay Systems or additional ADVIA Chemistry Systems with smart algorithms implemented in the ADVIA CentraLink Networking Solution. A full view of the ADVIA WorkCell CDX system is shown in Figure 3-3 ■.

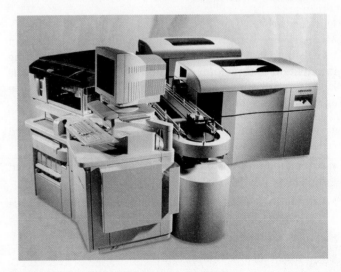

■ FIGURE 3-3   Siemens Work Cell CDX
Courtesy of Siemans Healthcare Diagnostics

Checkpoint! 3–6

Identify two advantages of integrated modular systems over single-batch, discrete analyzers.

FULLY INTEGRATED SYSTEMS

The trend in automation design is to integrate several modules into one continuous system that will allow the user to assay photometric, immunoassay, and electrochemistries. The **test menu**, a listing of all of the tests or analytes that the instrument is capable of measuring, can include as many as 150 analytes. All modular integrated systems use random-access technology to allow the analysis of several different types of chemistry assays.

The Olympus AU5400 Integrated Chemistry-Immunoassay analyzer is an example of this trend toward combining instrument platforms. This system is a true random-access analyzer with test throughput rates exceeding 3000 photometric tests/hour. For ISE measurements, the throughput rate is approximately 600 samples/hour. A distinct advantage with these modular systems is the ability to be able to link two or more modules and thereby increase throughput.

OPTIONS FOR INTEGRATED AUTOMATED SYSTEMS

Instrument Connectors

Olympus developed a unit it calls the AU-Connector that enables all types of clinical chemistry analysis to be consolidated into a single workstation without compromising the connected systems' function or performance. The AU-Connector uses intelligent sample management and tube-presorting capabilities to keep all the analyzers working at full potential. The AU-Connector does not use a tract device, which can create additional throughput bottlenecks and increase turnaround times.

Middleware

Middleware is software that allows a laboratory to connect its existing LIS and instrumentation to facilitate automating information and perform tasks not currently done with the laboratory's existing hardware and software. Middleware packages provide several features and functionality, including:

- automatic verification of test results through rules-based decision processing,
- automated and customized work and information based on a laboratory's specific needs,
- automatic tracking of data and location of samples requiring storage,
- automated sample interference testing and detection,
- real-time reflexive testing, and
- automatic comparison of current results with previous results on a patient's test (delta checking).

▶ PUTTING IT ALL TOGETHER

There are several manufactures of highly automated chemistry analyzers. These companies have been producing automated systems for several decades and include the following:

- Abbott Diagnostics, (Irving, TX)
- Beckman Coulter, Inc. (Brea, CA)
- Olympus (Center Valley, PA)
- Ortho-Clinical Diagnostics (Raritan, NJ)
- Roche Diagnostics Corp. (Indianapolis)
- Siemens Healthcare Diagnostics Inc. (Deerfield, IL)

Because instrumentation changes every several years, we will not review each system in detail. To provide the reader with an example of many of the features described earlier in a samples-in, samples-out format, we offer a detailed description of one system, the Roche Modular Analyzer (Figure 3-4 ■). The process begins at the analytical stage of testing and all preanalytical tasks have been completed.

The core unit transports all samples from the loader through each analytical module to the rerun buffer and unloader. Sample racks are designed to accommodate a wide variety of test tubes, samples cups, and other containers. Each rack is identified for its designated purposes (e.g., calibration, quality controls, and patient samples). A cup-detection mechanism is used to ensure that cups are present. A bar-code reader scans each sample, and software programs determine the most efficient route to complete the sample processing. The sample racks are placed into the loader and moved throughout the transport assembly. The racks will stop at each appropriate analytical module, depending on the test requested.

The first analytical module in this system is the modular plug-in-type ISE cartridge and associated reference electrode cartridge. Three measuring ISEs—sodium, potassium, and chloride—are included. Samples are pipetted into an ISE dilution vessel, and a sonic mixer is used to mix samples and diluent. A sample probe is rinsed and cleaned both inside and outside to reduce carryover. A syringe-type pipetting

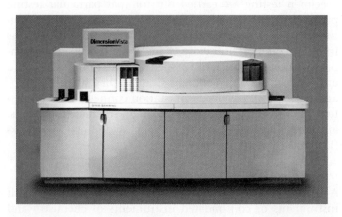

■ FIGURE 3-4   Roche Modular Analytics Serum Work Area
Courtesy of Roche Diagnostics

device is used to delivered reagents to the dilution vessel. Once the measurements are made, a vacuum system is used to aspirate diluted sample and send it to waste.

The second analytical module is the photometric module, which provides measurements of many analytes of several different samples types, including serum, plasma, urine, and spinal fluid. Sample racks are transported to the photometric module. Aliquots of samples are transferred from a sample tube or cup to a reaction vessel using a sample pipetter or syringe and a sample probe. Reagents necessary for the each measurement are contained in a refrigerated reagent compartment. Many slots or positions are available to accommodate both one- and two-reagent methods. A reagent-dispensing probe is used to transfer reagent from the container to the reaction vessel. A rinse station is used to clean the reagent probes after each delivery. A reaction disk with reaction cells or cuvettes are reusable and partially immersed in a heated water bath for temperature control. Once the sample and all reagents are dispensed into the reaction cell, a mechanical stirrer mixes the reaction mixture. The photometric measuring system includes a tungsten halogen lamp, diffraction grating monochromator, and photodetector. After the absorbance measurements are made, a vacuum system is used to aspirate the reaction mixture and transfer the contents to waste. Each cuvette is washed, rinsed, dried, and optically checked. If an optical check for a cuvette fails, then the cuvette is not used again.

The third measurement module provides automated immunoassay determinations for analytes, including therapeutic drugs, hormones, vitamins, and urine drugs of abuse. Sample racks are automatically moved to the immunoassay module where a sample probe transfers aliquots of specimens to assay cups to accommodate the immunoreactions. A "gripper" is used to grasp tips or assay cups and then transport them to various stations or sites in the analyzers. Reagents used for the immunoassays are placed in a reagent disk that accommodates several reagent containers. Reagent probes are used to pipette reagents and transfer them to an assay cup. A cap open-and-close mechanism for automatic opening and closing of the reagent containers is used to reduce evaporation. A vortex-type mixer is used for the particulate reagent (antibody-coated metal particles) so that settling does not occur. Rinse and wash cycles are initiated after each transfer of reagents. Sipper probes aspirate the reaction mixture from an assay cup located in the incubator and transfer it to the measuring channels. The measuring unit consists of (1) detection cells with a magnet, (2) a photomultiplier tube, (3) Peltier thermoelectric module, and (4) flow-through measuring channel.

The final destination of each sample rack is the re-run buffer and unloader. Samples that need to be re run can be diluted to satisfy one of two needs; 1) If the original measurement produces a value that exceeds the upper technical limit, 2) if the original measurement produces a value that is below the lower technical limit. Dilution can be performed automatically or requested manually by the operator. When all testing is completed the sample racks are moved into the unloader segment of the system and can then be archived into the laboratories specimen storage facility.

Several ancillary functions are associated with this example of an automated system. A complete inventory of all reagents, including bulk reagents and some consumables, is provided via the computer and is constantly updated for the user. Waste generated during the operation of the analyzer is also monitored. The system provides automatic dilution using first-run or rerun dilution parameters that are programmed into the computer. Automatic calibration is carried out using two-point or full calibration protocols and is user defined in terms of frequency. Maintenance of the system can be established using maintenance "pipes" that provide a single request option that represents a series of system-controlled maintenance items performed one after the other. For example, a maintenance pipe for a photometric module may consist of the following:

• photometer check,
• cell blank measurement,
• cell wash, and
• cell prime.

▶ FUTURE TRENDS

The demands placed on the laboratory are continually changing and evolving, and this will lead to new approaches to the delivery of health care. Test menus will continue to increase, resulting in multiple detectors and platforms being incorporated into a single automated system. For example, the Siemens Dimension Vista 1500 Intelligent Lab System shown in (Figure 3-5 ■) with integrated platforms incorporates five measuring techniques in one system. They include:

• photometry,
• turbidimetry,

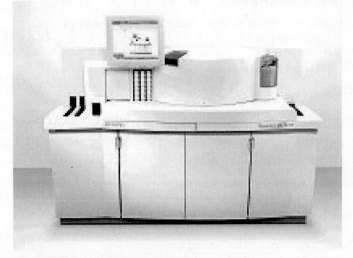

■ FIGURE 3-5   Siemens Dimension Vista 1500
Courtesy of Siemans Healthcare Diagnostics

- nephelometry,
- integrated sensor technology, and
- chemiluminescence for luminescent oxygen-channeling immunoassay (LOCI) technology.

A continuing trend toward modular instrument design will provide users with more flexibility. Computers that serve the laboratory will continually be upgraded to accommodate the various analytical configurations of these modules. Employee-safety concerns will result in modifications to instruments that will reduce toxic and biological exposures to technologists.

Increased interest in proteomics will eventually bring this discipline into the clinical laboratory. The instrumentation required will need to meet the demands placed on the laboratory, which means automated systems will have to be manufactured. Currently, several manufacturers are marketing automated polymerase chain reaction (PCR) analyzers, including:

- Biomek (Beckman Coulter Inc., Brea, CA)
- DNAtrak (Perkin Elmer, Boston)
- Evolution P3 precision pipetting platform with modular dispense technology (Perkin Elmer, Shelton, CT)
- Sciclone ALH 3000 (Caliper Life Sciences, Hopkinton, MA)
- Roche COBAS Amplicor Analyzer (Indianapolis)

The COBAS Amplicor Analyzer (Figure 3-6 ■) is a benchtop system that fully automates the amplification and detection steps of the PCR testing process on a single instrument. It combines five instruments into one: thermal cycler, automatic pipetter, incubator, washer, and reader. This real-time PCR analyzer provides reliable testing for analytes including the human immunodeficiency–1, hepatitis C, and hepatitis B viruses.

 Checkpoint! 3–7

List five examples of detectors that can be found in automated clinical chemistry analyzers.

■ **FIGURE 3-6** Roche COBAS Amplicor Analyzer Automated Real-Time PCR System
Courtesy of Roche Diagnostics

SUMMARY

The drivers of automation discussed in this chapter are continually changing and evolving, and this will lead to new approaches to the delivery of health care. The test menus will grow even larger, and new technologies will be needed to measure these analytes. These new technologies will need to be incorporated into existing systems.

Another demand that systems manufacturers will need to address is the ability to link different instruments or modules to their systems. A standardized interface will be required to accomplish this, and laboratory information systems will need to adjust their processors as well. Increased emphasis on employee safety, supported by legislative efforts, will lead instrument manufactures to continue developing their products so that they will reduce the exposure of technologist to body fluids. As the field of proteomics evolves, so will the methods and instruments required for measurement. Proteomic systems are currently coming to the marketplace; in time, this type of testing will find a place in the clinical laboratories. Automated systems manufacturers—Roche Diagnostic, for example—are currently marketing automated sample processing systems for PCR analysis. The automated sample processing is accomplished using the MagNA Pure LC (MP, Roche Applied Science, Indianapolis) instrument. The assay is completed using the Roche COBAS analyzer. This analyzer can function as a stand-alone or linked to Roche Pre-Analytics.

REVIEW QUESTIONS

LEVEL I

1. Automating laboratory testing has resulted in which of the following? (Objective 1)
   a. increased numbers of repetitive-stress injuries
   b. a reduction in the number of testing errors
   c. increased turnaround times
   d. increased safety-risk factors

LEVEL II

1. Which of the following devices are used to proportion reagents by the continuous-flow technique? (Objective 3)
   a. peristaltic pumps
   b. syringes or volumetric overflow devices
   c. conveyor belts
   d. vortex mixers

LEVEL I

2. Which of the following is an example of a preanalytical sample-processing task? (Objective 6)
   a. sorting samples
   b. evaluating patient results
   c. calculating patient results
   d. interpreting quality-control data for an assay

3. Which of the following are considered demands place on the laboratory that serve to drive automation? (Objective 7)
   a. reduced turnaround time
   b. increased throughput
   c. compensates for staff shortages in the laboratory
   d. All of the above.

4. Laboratory test throughput is defined as: (Objective 2)
   a. the number of tests performed per minute.
   b. the number of tests performed per hour.
   c. the number of tests performed per hour divided by 60.
   d. the number of tests performed on each sample.

5. An open-reagent analyzer allows the user to do which of the following? (Objective 2)
   a. Keep all of the reagent containers open or uncapped throughout the analysis.
   b. Buy and use reagents from a vendor other than the instrument manufacturer.
   d. Operate the analyzer without keeping the reagent cover over the reagent compartment.
   d. Develop additional assays on the instrument's open channel.

6. Sample and reagent can be incubated using which of the following? (Objective 9)
   a. water
   b. Peltier thermoelectric modules
   c. heating air
   d. All of the above.

7. The postanalytical stage of testing begins: (Objective 6)
   a. after the electrical signal is generated by the detector.
   b. as soon as the specimen is sent to a reference laboratory
   c. as soon as the specimen is received in the laboratory.
   d. after the sample and reagent have been incubated.

8. The term *test menu* refers to a list of: (Objective 2)
   a. analytical parameters for completing a test.
   b. calibration factors for the test.
   c. tests offered by the clinical laboratory.
   d. automated analyzers used by the clinical laboratory.

LEVEL II

2. Total laboratory automation is described as: (Objective 1)
   a. a combination of several instruments, consolidated instruments, and a work cell that are coupled to a specimen-management and transportation system as well as process-control software to automate a large percentage of laboratory work.
   b. a single analyzer connected to a computer via an RS-232 interface.
   c. a laboratory in which all of the analyzers are configured as automated systems.
   d. a combination of a preanalytical specimen-processing unit and an automated analyzer, either chemistry or hematology.

3. The difference between work cells and workstations is that a work cell: (Objective 5)
   a. provides an area in which all stages of specimen testing are carried out for that particular laboratory section.
   b. does not require any automated analyzer in its design.
   c. incorporates one type of testing platform thus provides fewer analytical tests.
   d. incorporates more than one type of testing platform, thus providing a broad spectrum of analytical tests.

4. Automated analyzers detect clots in samples by: (Objective 6)
   a. measuring the electrical capacitance of the samples.
   b. measuring the electrical resistivity of the samples.
   c. using a pressure transducer.
   d. measuring the absorbance of the samples.

REFERENCES

1. Skeggs LT Jr. An automatic method for colorimetric analysis. *Am J Clin Pathol.* (1957) 28: 311.

2. Sarkozi L, Simson E, Ramanathan L. The effects of total laboratory automation on the management of a clinical chemistry laboratory. Retrospective analysis of 36 years. *Clinica Chimica Acta* (2003) 329: 89–94.

3. Felder RA. Modular workcells: modern methods for laboratory automation. *Clin Chem Acta* (1998) 278(2): 256–267.

4. Sasaki M. Total laboratory automation in Japan: Past, present and the future. *Clin Chim Acta* (1998) 278: 217–227.

5. Wilson LS. New benchmarks and design criteria for laboratory consolidations. *Clinical Leadership & Management Review* (2003) 17(2): 90–98.

6. Battisto DG. Hospital clinical laboratories are in a constant state of change. *Clinical Leadership & Management Review* (2004) 18(2): 86–99.

4

Laboratory Operations

■ OBJECTIVES—LEVEL I

Following successful completion of this chapter, the learner will be able to:

Statistics

1. Compare and contrast descriptive statistics and inferential statistics.
2. Explain a Gaussian distribution.
3. Define *accuracy* and *precision*.
4. Identify three types of errors.
5. Identify five factors to consider when selecting quality-control material.
6. Explain the characteristics of a Levey–Jennings chart and include *x*- and *y*-axis labels.
7. Explain each Westgard rule violation.
8. Characterize proportional and systematic error.

Safety

9. Provide the correct words that correspond to the following abbreviations: OSHA, MSDS, NFPA, HEPA, JCAHO, RACE, PASS, and CFR.
10. List several responsibilities of employers and employees in maintaining a safe work environment.
11. Identify the classes of fires and the appropriate type of fire extinguisher to use.
12. List five important safety procedures to follow when handling electrical equipment.
13. Identify three hazards related to handling biological specimens.
14. List five examples of personal protective equipment (PPE) and engineered controls used to protect laboratory staff.
15. Define *teratogens, carcinogens*, and *transplacental carcinogenesis*.
16. List five examples of risk factors for cumulative trauma disorders.

Informatics

17. State the two items required to interface computers.
18. State two ways of entering laboratory data into a computer.

■ OBJECTIVES—LEVEL II

Following successful completion of this chapter, the learner will be able to:

Statistics

1. Interpret the results of selected laboratory statistics.
2. Select the appropriate statistic(s) for a given set of measurements.
3. Interpret a Levey–Jennings quality-control plot.
4. Identify Westgard rule violations and determine a course of action.
5. Identify three sources of errors associated with linear regression by least squares.

■ OBJECTIVES—LEVEL II (continued)

6. Outline a protocol used for selection of a laboratory method.
7. Identify and explain four factors that must be addressed before determining a reference interval.
8. Explain the differences between parametric and nonparametric statistics.
9. Calculate selected statistics.

Safety

10. Select the appropriate Codes of Federal Regulations (CFR) to review for a specific safety issues.
11. Identify specific elements of a chemical hygiene plan.
12. Interpret a selected NFPA 704-M warning label.
13. Explain the proper storage and inventory practices for chemicals.
14. Identify elements of an exposure-control plan.
15. Identify several risk factors associated with CTD.

Informatics

16. Outline the functions of a laboratory information system (LIS) in the clinical environment.
17. Explain how data are transferred from one computer system to another computer system.

KEY TERMS

Accuracy
Bias
Bidirectional interface
Bloodborne pathogen
Calibration
Carcinogen
Chemical hygiene plan
Codes of Federal Regulations
Cumulative trauma disorder
Descriptive statistics
Discordant results
Engineered controls

Ergonomic
Exposure-control plan
Hazard identification system
Inferential statistics
Laboratory information system
Limit of detection
Linearity
Local area network
Material safety data sheets
Nonionizing radiation
Occupational and Safety Health
    Administration

Online data entry
Peripheral device
Personal protective equipment
Precision
Range
Reference interval
Quality control
Skewness
Universal precautions
Xenobiotic

@ A CASE IN POINT

A member of the laboratory staff was asked to clean out a closet storage area where chemicals had been stored and, while cleaning, found a bottle of picric acid.

Issues and Questions to Consider

1. What is the NFPA704–M placard designation for health hazard, fire/flammability, reactivity, and special hazard?
2. What is the physical appearance of the chemical?
3. Where can you look for information regarding the safety issues for this chemical?
4. On which important information about the chemical should laboratory workers focus?
5. What specific steps should be taken to properly handle, use, and dispose of this chemical?

WHAT'S AHEAD

▶ Identification and explanation of laboratory statistics commonly used in clinical laboratories
▶ Discussion of appropriate utilization of selected statistics
▶ Interpretation of selected statistics for a given set of data
▶ Definition and explanation of quality control and quality assurance
▶ Discussion of a protocol used to establish quality-control ranges.
▶ Explanation and interpretation of Westgard rules for quality control
▶ Discussion of types of laboratory errors
▶ Explanation of the origin of laboratory safety and the organizations that monitor its compliance
▶ Identification and discussion of specific safety issues relevant to clinical laboratories
▶ Identification of computer hardware used in laboratory information systems
▶ Discussion of laboratory information system components used to manage data in the laboratory

► LABORATORY STATISTICS

Statistics (e.g.,arithmetic mean, standard deviation, and variance) should be viewed as tools that are available for the laboratory staff to use. These tools will assist the user in making important decisions about data collected during a study. Knowing which tool or statistic to use is important. Selecting the wrong "stat" will most likely result in a wrong assumption being made or an inappropriate action taken. A discussion of the tasks required to produce reliable laboratory data will be presented in this chapter. Procedures for doing these tasks will be described. Guidelines for selecting the proper statistic or "tool" to use to help make decisions will be included. Appropriate interpretation of data will also be presented.

Computation of most statistics is performed using any of several computer software packages or calculators. Therefore, derivation and memorization of formulas is not necessary. Most textbooks on clinical chemistry include formulas; thus only a limited few will be included in this textbook. What is most important is the selection of appropriate statistics and accurate interpretation of the statistical calculations.

BASIC LABORATORY STATISTICS

Data Presentation

Data can be presented in a variety of formats. Most data generated in chemistry or hematology laboratory are numeric in nature (e.g., 100 mg/dL or 4.0×10^6 red blood cells/mm^3). But for some laboratories, including blood bank and microbiology, the data are mostly nonnumeric. Does this mean that these laboratories' methods and instruments cannot be evaluated using statistics? No. It would be impractical to attempt to include examples of statistics that could be applied to all sorts of nonnumeric data. Instead a brief discussion of examples of how data can be presented and an introduction to several types of data will be presented. The reader can review any of the statistic references listed at the end of the chapter for specific applications.

Three examples of how data can be presented include rank, continuous, and discrete. *Ranked* data are arranged from highest to lowest according to magnitude and then assigned numbers that correspond to each observation placed in the sequence. For example, suppose a researcher is looking at the leading causes of death in the United States and the results of a literature search reveal the following:

1st = heart disease

2nd = cancer

3rd = cerebrovascular disease

Based on the results, found the data are ranked such that heart disease is shown to be the leading cause of death in the United States, cancer is second, and cerebrovascular diseases third.

Continuous data represent a continuous variable that can assume any value within the range of scores that define the limits of the variable. Examples of continuous data include temperature, time, and the cholesterol level in a subject. Remember that there are no gaps in continuous data.

Discrete data are a subset of continuous data. A characteristic of discrete data is that discrete variables assume a defined set of integers. Both order and magnitude are important. For example, age recorded in integral years is reported as 62 or 63 years old but not 62.491 years old.

Types of Data

Nominal (categorical) is a type of data in which the values fall into unordered categories or classes. Nominal data are numbers applied to nonnumeric variables. For example, a group of subjects' blood types could be coded as: group O = 1, group A = 2, group B = 3, and group AB = 4. No individual or variable can be assigned to more than one group, and order or sequence is not important.

Ordinal data (rank order) are numbers that are discrete and ordered (ranked). The order or ranking is important. For example, injuries may be classified according to their level of severity as follows:

4 = the most severe or even fatal injury

3 = a severe injury

2 = moderately severe

1 = a minor injury

The intervals are not usually known and are often unequal.

Descriptive statistics include the mean, range, variability, and distribution of a data set. They represent the commonly used statistical computations in the clinical laboratory.

Inferential statistics are concerned with the relationship among different sets or samples of data. For example, is the mean of one set of data significantly different from another? If we find a difference between the means of the two samples, what are the probabilities associated with this difference? Examples of inferential statistics include the following:

- *F*-test
- *t*-test
- *Z*-test
- chi square

Two additional important terms are **population** and **sample**. A population refers to the universe of values or attributes, such as all of the fasting serum cholesterol levels of all apparently healthy males in the United States.

Samples refer to a portion or subset of a population, such as the serum cholesterol levels of all males in the laboratory. Often the distinction between population and sample is not made very clear (e.g., a sampling could be argued to be a population).

Parameters and *statistics* are terms associated with population and samplings, respectively. A parameter describes a quantitative feature of a population. Statistics describes a quantitative feature of a sample.

✪ TABLE 4-1

Terms and symbols associated with population and samples statistics.

Parameter		Statistic
μ	Arithmetic mean or average	$\bar{x}$
σ	Standard deviation	s
σ^2	Variance	s^2
Z	Distribution of means	t
$\mu \pm 1.96\,\sigma$	95% confidence interval	$\bar{x} \pm t_{1-\alpha}s$

Symbols are used throughout statistics. Table 4-1 ✪ compares symbols used for parameters versus statistics. Parameter statistics tend to use Greek symbols whereas statistics use Roman or italicized characters.

Measures of Central Tendency

The **arithmetic average** or **mean** is commonly used to describe data. For example, what is the average glucose value for all samples in a study? Summing all data and dividing the sum by the number of data represented by n or N determines arithmetic means. The **median** in a sample set is the middle value or the 50th percentile value when the data are rank ordered by magnitude. Close observation of the data will show that half the data points are above and half are below the median. The **mode** is the data point that occurs most frequently. Data for a sampling can be distributed within one mode (unimodal), two modes (bimodal), several modes (polymodal), or have no distinct mode at all. Evaluating mode may uncover a population with two or more distinct modes and lead to the conclusion that there are two or more subsets within the large population. Measures of central tendency are illustrated in Table 4-2 ✪. Calculations of central tendency allow the derivation of graphical plots that can clearly show modal distributions and skewness. Skewness is described as not symmetrical (asymmetrical). Data that are unimodal are symmetrical and the mean, median, and mode coincide. When data are skewed, neither the mean nor the mode estimate central tendency well, although the median retains it distinction. Examples of various modes of distribution are shown in Figure 4-1 ■.

✪ TABLE 4-2

Data for 10 replicate measurements of calcium in serum and results of selected descriptive statistics.

Replicate Number	Results
1	9.9
2	9.8
3	9.7
4	9.2
5	9.5
6	9.6
7	9.9
8	9.3
9	9.6
10	9.6
Sum	96.1
Arithmetic mean	9.61
Median	9.6
Mode	9.6
Range	9.2 – 9.9
Standard deviation	0.2331
Variance	0.0543
CV	2.4%

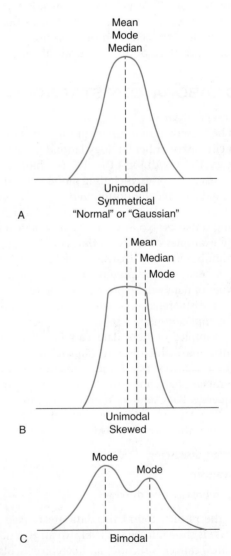

■ FIGURE 4-1   Modes of distribution. **A.** Unimodal symmetrical "normal Gaussian." **B.** Unimodal skewed. **C.** Bimodal.

Measures of Dispersion

Range   Range is a measure of spread or variation in a set of data. The range of data is the difference between the largest and the smallest numbers of the data set or the smallest to the largest number. For example, the data in Table 4-2 show a range of 9.2–9.9.

Standard Deviation (*s*)   Standard deviation is a commonly used estimator of dispersion because it is predictably related to a common type of data distribution, the Gaussian or normal distribution. **Standard deviation** is a measure of the dispersion of a group of values around a mean. It is derived from the curve of normal distribution and bears a meaningful relation to the area under the normal distribution curve so that, for example, the area under the center of the curve represented by the mean $\pm 1.96s$ is ~95% of the whole area. A plot of a normal curve distribution is shown in Figure 4-2 ■.

Coefficient of Variation (CV)   **Coefficient of variation (CV)** is another way of expressing standard deviation and is also referred to as *relative standard deviation*. It is defined as 100 times the standard deviation divided by the mean. CV is expressed as a percentage. It relates the standard

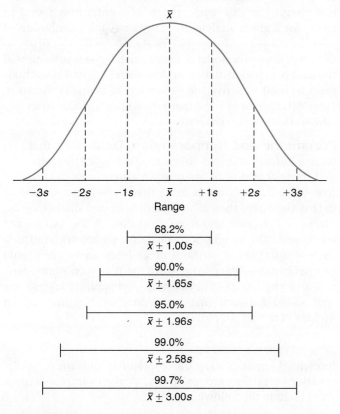

■ **FIGURE 4-2** A normal bell-shaped curve showing $\pm 1, 2, 3s$ and the relative percentages associated with each area under the curve.

✪ TABLE 4-3			
Precision data for the measurement of glucose in pooled serum samples measured by four different methods.			
	Pooled Sample Mean (mg/dL)	Pooled Sample SD (mg/dL)	Pooled Sample %CV
Method 1	240	12.2	5.1
Method 2	180	8.1	4.5
Method 3	120	6.5	5.4
Method 4	70	5.0	7.1

deviation to the level at which the measurements are made. For example, from the data in Table 4-2 the mean and *s* of the measurements are 9.61 and 0.2331, respectively. The CV for the data set is 2.4% ($9.61/0.2331 \times 100$).

Many laboratories find CV a useful statistic. Because the numbers are in percentage format, it is often easier to relate to CV than to standard deviation. Remember that the lower the %CV the better the performance of the assay. For example, from the data shown in Table 4-3 ✪ it would appear that the precision based on standard deviation of Method 4 is better than Methods 1, 2, and 3. But if the standard deviation is related to the mean concentrations levels, Method 4 has the highest %CV and therefore is the poorest performing method.

Variance (*s*²)   When the values of a set of observations lie close to their mean, the dispersion is less than when they are scattered over a wide range. If the laboratory needs to know the measure of dispersion of data relative to the scatter of the values about the mean, a statistic is available to provide this information. The statistic is called variance. Variance is calculated by squaring the standard deviation. It can also be derived by subtracting the mean from each of the values, squaring the resulting differences, and then adding up the squared differences. This sum of the squared deviations of the values from their mean is divided by the sample size minus one ($N - 1$) to obtain the sample variance. The variance of the set of data shown in Table 4-2 is 0.0543.

Calculated variances are also used in hypothesis testing such as *F* test. The *F* test calculation shown later in this chapter uses a ratio of the larger variance to the smaller variance for two sets of data.

Outliers   An outlier in a sample set is a measurement that belongs to a population other than the one to which most of the measurements belong. Outliers can distort the computed values of statistics and cause incorrect inferences to be made about the population parameters of interest. There are tests that can be used to determine whether a value is an outlier or not. The gap test[1] and Prescott Test for Outliers[2] are two examples used for clinical laboratory statistics.

✓ Checkpoint! 4–1

A laboratory technologist was asked to determine the descriptive statistics for the data presented in the following table. The data represent replicate measurements of sodium in serum samples. Twenty measurements were completed on one sample. Determine the statistics listed for the data shown in the table:

- *Arithmetic mean*
- *Standard deviation*
- *Variance*
- *Percent coefficient of variation*
- *Median*
- *Mode*
- *Range*

Data Number	Result	Data Number	Result
1	140	11	139
2	141	12	138
3	139	13	140
4	140	14	140
5	140	15	141
6	138	16	139
7	142	17	139
8	141	18	141
9	141	19	140
10	140	20	140

Population Distributions

Normal Distribution   The normal distribution of an analytes in a selected population is a continuous distribution, symmetric around the mean, predictably related to sigma (σ) and variance (σ^2). Gauss explained concepts associated with normal distribution when he studied errors of replicate measurement. Graphical plots of the patterns of clustering about the mean with the symmetric distribution of outlying values produced a bell-shaped curve. This normal distribution of data characterized many biological attributes, such as height, weight, and biochemical measurements.

When studying measurements of an analyte in large populations, the parameters listed in Table 4-1 should be used. Thus if an analyte concentration such as glucose is determined in a population of adults, the plot of frequency versus arithmetic mean (μ) or sigma (σ) as presented in Figure 4-3 ■ will be created. The formula for Z answers the question of how many sigmas away from the population mean is the value in question. Thus,

$$Z = (x - \mu)/\sigma \qquad (4.1)$$

Most laboratorians are involved with smaller samplings that require a different approach than for population studies. Therefore, *statistics* terms and symbols must be used (see Table 4-1). Additional questions must be asked when evaluating distribution data using small samplings. For example, is this small sample likely to be from a normally distributed population? Is the sample distribution symmetrical or is it non-Gaussian?

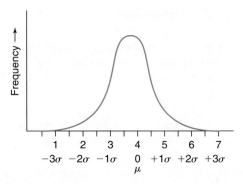

■ FIGURE 4-3   A normal distribution curve with a mean and ±1, ±2, and ±3 sigmas (σ).

✓ Checkpoint! 4–2

A normal population study for uric acid in serum samples reveals that the mean of the group of samples is 6.0 mg/dL and the standard deviation is 0.5 mg/dL. What is the probability of a serum uric acid value of 6.9 or greater occurring in this series?

Nonnormal Distribution   Most laboratory data involving the evaluation of methods, quality control, and proficiency surveys follow a normal distribution and lend themselves to calculations using fundamental descriptive statistics. This is not always the case with analyte concentrations found in blood for a given analyte. In many cases the distribution of data is **nonnormal** (asymmetrical or skewed). The distribution may also be bimodal or log normal. If the distribution of the data is non-Gaussian, then alternative statistical methods must be used to derive the parameters or statistics shown in Table 4-1. Several of these methods will be introduced in specific sections of this chapter.

Parametric and Nonparametric Data   Data that presume a normal Gaussian distribution are called **parametric**, and those that do not are **nonparametric** or distribution free. Once the decision is made that a given distribution of data is Gaussian, then all statistical tests and methods (e.g., means, s, t-test) are said to be parametric. If the data do not conform to the assumptions made for parametric statistics, then nonparametric methods of statistical inference should be used instead. Two examples are the Wilcoxon signed-rank test and the Mann–Whitney test. Nonparametric tests are well suited for data that are ordinal or nominal, which include data that are ranked.

Inferential Statistics

Inferential statistics allow the laboratorian to address specific questions regarding groups or sets of data. Examples of questions include the following:

- Is the observed difference between two sample or population means due to chance alone?

- Is the standard deviation (s) of one set of data significantly different from another?

Two commonly used tests that may provide answers to such questions are the *t*-test and *F*-test or *F*-ratio. These tests will be discussed in the next section, on hypothesis testing.

Hypothesis Testing Hypothesis testing is useful if the laboratorian needs to draw some conclusion about a population parameter (e.g., a laboratorian needs to make an inference about the mean of a continuous random variable, using the information contained in a sample of observations).

To perform such a test, we begin by claiming that the mean of the population is equal to some postulated value μ_0. This statement about the value of a population parameter is called the *null hypothesis*, or H_0. If the laboratorian needs to test whether the mean serum troponin I concentration of the subpopulation of patients with hypertension is equal to the mean of the general population of 40- to 60-year-old males, for instance, the null hypothesis would be

$$H_0: \mu = \mu_0$$

The alternative hypothesis, represented by H_A, is a second statement that contradicts H_0. This is shown as

$$H_A: \mu \neq \mu_0$$

Together, the null and the alternative hypothesis cover all possible values of the population mean (μ); thus one of the two statements must be true.

After formulating the hypothesis, we next draw a random sample of size n from the population of interest. The sample size should be as large as possible. The mean of the samples is computed.

Another decision that needs to be made is which statistical test and formula to use. For example, should an *F*-test or *t*-test (Student's *t*-test, paired or unpaired) be used? The answer to these questions goes back to the *null hypothesis*. Another way of stating this is to answer the question, "What significance is being tested?" It is beyond the scope of this text to present the numerous scenarios a laboratorian may face. Example 4-1 discusses a problem that occurs commonly in laboratories.

Once the appropriate statistical test is selected, all the variables of the formula must be addressed. For example, the formula may require a standard deviation, bias, square root of n, or standard deviation of the difference to be determined.

Another important aspect of hypothesis testing is *significance* and *significance level*. Significance as used in statistical inference refers to the probability statement (p), which implies that chance and random distortion of samplings have played only a small part in the observed differences. Common significant levels for clinical applications are $p < 0.01$ or $p > 0.05$, which means that there is only a 1% or 5% chance, respectively, that the observed difference is due to chance alone.

Statistic tables may often use the symbol alpha (α) to represent significance levels. The area under the curve or $1 - \alpha$ is considered part of the sample population, and alpha is the area under the curve that may represent the second overlapping population.

One-Tailed and Two-Tailed Tests Another criterion that requires a decision while formulating a hypothesis is whether to evaluate the data using one-tailed or two-tailed tests. If the question concerns the probability of observing a difference in only one direction (i.e., does $\bar{x}_1$ exceed $\bar{x}_2$?), then select one tailed *t*-test. The one-tailed *t*-test describes a distribution extending from minus infinity to positive *s*- or σ-values that includes 95% of the area under the curve, for $t_\alpha = 0.05$. If the question concerns only magnitude but not the direction (i.e., is $\bar{x}_1$ different from $\bar{x}_2$?), then select two-tailed *t*-test and the area under the curve is the central 95% excluding 2.5% from each tail.

Most clinical laboratory applications use the two-tailed test. A two-tailed test decreases the chance of falsely rejecting the null hypothesis. Conversely, the two-tailed convention may cause failure to accept the difference when one exists.

Paired *t*-Test and Unpaired *t*-Test Two frequently used variations of the *t*-test are the following:

* The unpaired *t*-test (Student's *t*-test or *t*-independent), in which the means and standard deviations of two independent samples are compared and each standard deviation is presumed to be from the same population

* The paired *t*-test or *t*-dependent, in which the average difference between a series of paired observations is analyzed

The unpaired *t*-test or *t*-independent is used to study the difference between two groups in which each subject is tested only once. The question is whether or not there is a difference in the data of one or more variable(s) between two groups that were independent of one another. By *independent* we mean that the two groups were not related in any way. Clinical laboratories rarely need to evaluate unpaired or independent data; rather, most clinical laboratory studies involve paired or dependent data.

Suppose the laboratory is implementing a second method for measuring sodium. A comparison of the two methods is mandatory. The *t*-test can provide valuable information about differences between the means of both sets of data. Two additional statistics necessary to calculate the *t*-value for paired data are bias and the standard deviation of differences (S_d). The data in Table 4-4 ✪ indicate that the mean difference (bias) between paired data is statistically different (see also Example 4-1). The *t*-test has provided a probability regarding whether or not there is a statistical difference between the two methods under the conditions of the evaluation.

Degrees of Freedom Degrees of freedom (**df, DF**) is described as the number of data points that affect the statistical analysis. For example, suppose 30 measurements of glucose are obtained. The series has 30 degrees of freedom because no single measurement affected the other 29. However, if we calculate the mean of the 30 values, the series now has ($n - 1$) or 29 df, because the 30th value in the series cannot be changed without altering the mean. For small samplings of data the df is calculated using $n - 1$.[3] For larger samplings and population studies the number of

TABLE 4-4

A two-tailed paired-sample *t*-test for serum sodium measurements of 10 different patient samples by two different methods (Biosensor vs. ISE).

Sample Number	Biosensor	ISE
1	141	138
2	140	136
3	144	147
4	144	139
5	142	143
6	146	141
7	149	143
8	150	145
9	142	136
10	148	146

Bias = 3.2

$t = 3.402$

$p = 0.008$

degrees of freedom may be calculated using other formulas such as $n - 2$.

Bias

Bias is described and used in several different ways in laboratory statistics. Bias is the difference between two means, or the mean difference. The value of the bias is used in the *t*-test and in protocols for method comparisons. Bias is also defined as the presence of nonrandom events (e.g., estimating the serum cholesterol levels in apparently normal healthy individuals when it was not realized that they were on cholesterol-lowering medications). Finally, bias may be described as lack of accuracy. For example, a laboratory is attempting to determine the accuracy of a method by doing an analytical recovery experiment and does not realize that there is a reducing substance in the samples that constantly affects the assay.

Example 4-1

A laboratory purchased a new biosensor device to measure sodium in blood; the new device will replace the currently used ion selective electrode. A split sample study was carried out, and the data are shown in Table 4-4. A two-tailed paired-sample *t*-test was selected as a means to evaluate the two devices. The following information is included:

$H_0: \mu_1 = \mu_2$

$H_A: \mu_1 \neq \mu_2$

$\alpha = 0.05$

$n = 10$

df $= 9$

Bias $= 3.2$

$t = 3.402$

$t_{0.05} = 2.262$

Conclusion:

Reject H_0

$0.005 < p < 0.01$

$p = 0.008$

Accept the alternative hypothesis, which is $\mu_1 \neq \mu_2$.

Standard Error of the Mean

The measurement of serum sodium in either of the two methods shown in Table 4-4 is characterized by a certain degree of imprecision. The error of the mean measurement of the set of values is smaller than that of a single measurement, so the more times a measurement is made, the more certain you can be of its true value. If several means are calculated from different groups of measurements of this population, the individual means are distributed about the actual population mean. The random variation in this group of means is distributed by the standard error of the mean.

Standard Deviation of Differences The standard deviation of the differences (SD_d) is a measure of the variability in the calculation of the difference between each member of a set of paired data. In a comparison of methods the SD_d indicates the degree of variability of the difference between the values of the new method and the values of the accepted method. The standard deviation assumes a normal distribution for these differences that is equally distributed on both sides of the bias. A large SD_d may be due to the following:

- The distribution of the difference about the bias is wide, indicating a large amount of random error in the measurement on one or both methods.

- The distribution may not be normally distributed.

Standard Deviation of the Duplicates

A standard deviation of the duplicates (also, standard deviation of differences of paired data) is useful to estimate the variability of measurements of the same sample in duplicate. It allows estimation of the minimum change expected from duplicate analytical error alone and permits estimation of whether a real change in the patient's status is likely to have occurred. The calculation involves taking the square root of the sum of the differences of duplicate data, divided by two, times the number of data pairs.

Standard Error of the Mean of Differences

The standard error of the mean of differences quantifies the uncertainty in the estimate of the mean of a population. This uncertainty in the estimate is related to the random error that is present with all sampling techniques. Just as it is impossible to repeat exactly a value for a single sample, it would be impossible to obtain the exact mean for the difference of paired data if the samples were repeated.

It must be remembered that there is a difference between the standard deviation and the standard error. The standard deviation quantifies the variability in the entire population of data, whereas the standard error quantifies the variability in the estimate of the mean of the population. Therefore, the value for the standard error will always be less than the value for the standard deviation.

Confidence Intervals

A confidence interval (CI) can be defined as a range around an experimentally determined statistic that has a known probability of including the true parameter. Applications of confidence interval include the following:

- Estimating the range of values that include a specified proportion of all members of a population, such as the "normal" or "reference interval" of values for a laboratory test
- Hypothesis testing

The confidence interval (e.g., 95%) associated with a given set of data will or will not actually include the true size of the sample population, but in the long run 95% of all possible 95% CI will include the true difference of mean values associated with the sample population. As such, CI describes not only the size of the sample population but quantifies the certainty with which you can estimate the size of the sample population.

The size of the interval depends on the level of confidence you want that the sample studied will actually include the true sample population. Therefore, it can be assumed with 95% confidence that the true population mean lies within the range of $\bar{x} - 1.96\ (\sigma/\sqrt{N})$ to $\bar{x} + 1.96\ (\sigma/\sqrt{N})$. This expression is termed the 95% CI.

Although a 95% CI is used most often in clinical practice, the laboratory is not restricted to this choice. The laboratory may prefer to have a greater degree of certainty regarding the value of the population mean; in this case, it could choose to use a 99% CI instead of a 95% interval. Because 99% of the observations in a normal distribution lie between –(2.58) and (2.58), a 99% CI for μ is $-2.58\ (\sigma/\sqrt{N})$ and $2.58\ (\sigma/\sqrt{N})$. Therefore, approximately 99 out of 100 confidence intervals obtained from 100 independent random samples of size n drawn from this population would cover the true mean μ. As is expected, the 99% CI is wider than the 95% CI; the smaller the range of values considered, the less confident the interval covers μ.

In some cases, sigma for a given population is known and one may be interested in determining a mean for a smaller sample group and assessing whether its sample mean falls within the population confidence interval. Example 4-2 illustrates this application of confidence interval for a cholesterol study in women.

Example 4-2   Estimating a 95% CI for μ when σ is known

The laboratory staff is interested in the distribution of serum cholesterol levels for all females in the United States who are hypertensive and smoke. The distribution is approximately normal with an unknown mean, μ, and the standard deviation, σ, is 50 mg/dL. The laboratory staff is interested in estimating the mean serum cholesterol for this population.

Before the staff members go out and select a random sampling, the probability that the interval, $\bar{x} - 1.96\ (50/\sqrt{N})$, $\bar{x} + 1.96\ (50/\sqrt{N})$, includes the true population mean, $\mu = 0.95$. The staff draws 14 samples from the population of hypertensive smokers, and these women have a mean cholesterol level of 220 mg/dL. Based on this sampling, a 95% CI for μ is

$$220 \text{ mg/dL} - 1.96\ (50/\sqrt{14}),\ 220 \text{ mg/dL} + 1.96\ (50/\sqrt{14}) = 199 - 246 \text{ mg/dL}$$

Therefore, the group is 95% confident that the limits of 199 – 246 mg/dL cover the true mean, μ.

✓ Checkpoint! 4–3

A clinical laboratory is required to correlate its new biosensor method for whole blood potassium concentrations with results obtained from its existing method, which uses an ion selective electrode analyzer. The following data were obtained on paired samples (note: the number of sample pairs has been reduced below the recommended number for comparison of methods to simplify the problem):

1. *Determine the following statistics from the data below:*
   - *Bias*
   - *Standard deviation of the difference*
   - *Standard error of the mean of differences*
   - *p-value*
2. *Assuming $t_{\alpha = 0.05}$ two-tailed distribution, should the null hypothesis, $H_0: \mu_1 = \mu_2$, be accepted or rejected?*

Sample Number	K^+ meq/L Biosensor Technique	K^+ meq/L ISE Technique
1	4.5	4.6
2	4.0	4.0
3	3.9	4.0
4	3.8	3.9
5	5.0	5.1
6	5.5	5.7
7	3.0	3.2
8	3.4	3.6
9	3.1	3.3
10	4.2	4.2
11	4.4	4.7
12	5.2	5.3
13	5.1	5.0
14	3.7	3.9
15	4.4	4.5
16	3.2	3.3
17	2.8	3.0
18	6.0	6.3
19	5.8	6.2
20	4.6	4.6

Errors of Inference

A null hypothesis can be accepted or rejected. Either there is no difference between groups or there is really an inequality (such as the difference between two groups.) There are two types of errors, identified as Type I (alpha) and Type II (beta), associated with an inappropriate conclusion of hypothesis testing.

A type I error is defined as erroneously rejecting the null hypothesis or proclaiming a difference to exist when one does not. Type I error is also known as the *level of significance*. It is often created when the level of significance (*p*-value) or the amount of risk that one is willing to take in any test of the null hypothesis is inappropriate.

A type II error is defined as erroneously accepting the null hypothesis, with the result that a real difference is not detected. In other words,, there may actually be a difference between the populations represented by the sample group, but you mistakenly conclude that there is not.

F-Test

The *F*-test or *F*-ratio is the ratio of the variance (s^2) calculated for both sets of data. A table of critical *F*-ratio values is used to determine if the null hypothesis should be accepted or rejected. Or, simply stated, is the difference between the variance such that chance alone could well account for it?

The statistical procedure associated with the *F*-test is similar to that for the *t*-test. The differences include the following: (1) the nature of the null hypothesis differs between the *F*-test and the *p*-test, (2) each set of data must account for numbers of degrees of freedom, and (3) a unique probability table must be used.

LINEAR REGRESSION AND CORRELATION

Regression

Regression analysis is useful in assessing specific aspects of the relationship between variables, and the ultimate objective is to predict or estimate the value of one variable based on a given value of the second variable. Several examples of regression analysis are discussed in this chapter.

Correlation

Correlation statistics measure the strength of the relationship between variables. When measures of correlation are computed from a set of data, we are interested in the degree of correlation between the variables.

Linear Regression by Least Squares

Regression analysis is commonly used in the comparison of two methods or two instruments and to evaluate the linearity of an instrument or method. Regression calculations provide a simple and general descriptive statement relating one set of observations to another.

Linear regression by the method of least squares positions a straight line among the points on the graph in such a way that the sum of the squares of the vertical distances from

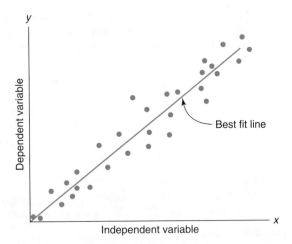

■ FIGURE 4-4   Plot of independent variable and dependent variable using linear regression by least squares.

each point to the fitted line is the smallest value possible. In Figure 4-4 ■ two sets of data representing variables, *x* and *y*, are plotted on a graph. The data sets represented by closed circle forms a scatterplot. The *x*-variable is referred to as the *independent variable* because the investigator frequently controls it. The *y*-variable is called the *dependent variable*, and the experimenter measures it to determine the effect of the independent variable; thus the terminology refers to the regression of *y* on *x*. The following are the assumptions underlying the simple linear regression model.[4]

- Values of the independent variable *x* are said to be "fixed."
- The variable *x* is measured without error.
- For each value of *x* there is a subpopulation of *y*-values.
- The variances of the subpopulation of *y* are all equal.
- The means of the subpopulation of *y* all lie on the same straight line.
- The *y*-values are statistically independent.

A least squares or "best-fit" straight line is included in Figure 4-4. The graph shows that there is an association between the two variables. In such cases, it is often desirable to use a mathematical description of such an *x*-*y* relationship. The simplest such expression is the general equation for a straight line:

$$Y = a + bX \tag{4.5}$$

This linear model or simple linear regression formula contains two parameters: the intercept *a* and slope *b*. The data presented in Table 4-5 ✪ show that a nearly linear relationship between glucose oxidase (*X*) and glucose hexokinase (*Y*) exists and is expressed in the following equation for a straight line: $Y = -0.0067 + 0.952X$

where

a (intercept) $= -0.0067$

b (slope) $= 0.952$

TABLE 4-5

Least squares regression analysis of glucose measurements in serum by two different methods.

Sample Number	Glucose Oxidase (x-axis)	Glucose Hexokinase (y-axis)
1	0.9	0.9
2	1.1	1.0
3	1.0	0.8
4	1.3	1.2
5	0.4	0.3
6	0.1	0.2
7	1.5	1.6
8	5.2	4.8
9	0.8	0.8
10	0.4	0.4
11	4.4	4.2
12	3.5	3.9
13	1.0	0.8
14	2.8	2.6
15	2.5	2.2
16	2.4	2.0
17	0.7	0.8
18	1.1	0.9
19	1.1	1.0
20	0.5	0.6
Slope	0.952	
Intercept	−0.0066	
$S_{y.x}$	0.186	
Bias $(y - x)$	−.085	

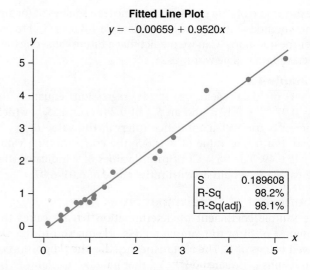

Fitted Line Plot
$y = -0.00659 + 0.9520x$

S	0.189608
R-Sq	98.2%
R-Sq(adj)	98.1%

■ FIGURE 4-5   Fitted line plot for data from Table 4-5.

- If the range of data is too narrow, random error will exert a larger influence than when calculations are based on a wider range of data.

- There may be a tendency to assume that the accepted method is "true," when in fact errors that are detected by regression statistic may actually be from the accepted method rather than the new method.

If the data being evaluated exhibit any of the above limitations, then an alternative regression method should be explored. Two commonly used regression analysis are the methods of Deming and Passing and Bablok, both of which are discussed below.

Deming's Regression Analysis
Regression analysis by the method of Deming is used with increasing frequency by clinical chemists.[5] Deming's method attempts to overcome shortfalls in the least squares method. The least squares method assumes there is <u>no imprecision</u> in the measurements of either the independent or dependent variable. In Deming's regression analysis, random error from *both methods* is taken into account. This random error can be determined by calculating a λ, which is a ratio that represents the variances of repeated measurements of x- and y-values.

Passing and Bablok Regression Analysis
The method of linear regression developed by Passing and Bablok makes no assumption regarding the distribution of the samples and the measurement errors.[5] The result does not depend on the assignment of the methods (or instruments) to X and Y. The slope and intercept are calculated within a 95% confidence interval. The CI is used to determine whether there is only a chance difference between a slope and 1 and between intercept and 0.

Standard Error of the Estimate ($S_{y.x}$)
The standard error of the estimate ($S_{y.x}$) measures the variability about the regression line. The calculated value of ($S_{y.x}$) describes

This equation reveals that the slope (*b*) is positive, the best-fit line crosses the Y-axis just below the origin at −0.0067, and the fitted line shown in Figure 4-5 ■ extends from the lower left-hand corner of the graph to the upper right-hand corner. Also, for each unit increase in X, $\hat{y}$ (calculated value for Y on the line at each X value) increases by an amount equal to [0.952(X) + (−0.0067)]. For example, if a given value of X = 3.0, then $\hat{y}$ = [0.952(3.0) + (−0.0067)] = 2.8.

Limitations of Least Squares Linear regression analysis by the method of least squares is acceptable for many clinical evaluations. This method is affected by several limitations that will significantly alter the results and lead to erroneous conclusions. Examples of these limitations include the following:

- Nonlinearity of the data will affect both the slope and the intercept.

- Outliers at either the high or low end of the data will cause the line to pivot on the mean of the data and move toward the directions of the outliers.

the dispersion of data points around the least squares fit (or any other method of fitting the line.) Example 4-3 uses $(S_{y\cdot x})$ to answer the question, "Can we predict that a future value of $\hat{y}$ for a given value x will lie within $\pm 2S_{y\cdot x}$?"

Example 4-3

A set of data yields a linear regression equation of $Y = 0.972X + 0.171$, and an $S_{y\cdot x}$ of 0.240. For $\pm 2S_{y\cdot x}$, which represents the 95% confidence interval, the value of $S_{y\cdot x}$ is 0.480. If a future value of X is,6.5, the corresponding value of $\hat{y}$ is 6.49. The 95% CI of future values of $\hat{y}$ indicates that the value of y would lie within the range of 6.01–6.97.

Coefficient of Determination

The sample coefficient of determination (r^2) measures the closeness of fit of the sample regression equation to the observed values of Y. The r^2-statistic measure the proportion of the Y sums of squares $\Sigma Y - \overline{Y}^2$ that may be "explained" by the regression formula $Y = a + bX$. That is, if all the data points on a scattergram lie on a straight line, all the variability as measured by the Y sums of squares may be accounted for (or explained by) knowing X.[6]

In Figure 4-6(A) ■ we see that the observations all lie close to the regression line and we would expect r^2 to be close to 1.0. For example, the computed r^2 for these data is 0.985, indicating that about 99% of the total variation in the y_i is explained by the regression. In Figure 4-6(B) ■ the y_i are widely scattered about the regression line, and we suspect that r^2 is closer to 0.0. The computed r^2 for the data is 0.400; that is, less than 50% of the total variation in the y-values is explained by the regression.

The largest value that r^2 can assume is 1, a result that occurs when all the variation in the y_i is explained by the regression. When $r^2 = 1$, all the observations fall on the regression line. This situation is shown in Figure 4-6(C) ■.

The lower limit of r^2 is 0. In this situation none of the variation in the y_i is explained by the regression. Figure 4-6(D) ■ illustrates a situation in which r^2 is close to zero.

Correlation Coefficient

There are two different correlation coefficients. The first, called the Pearson product-moment correlation coefficient (r), quantifies the strength of association between two variables that are normally distributed.[6] The second is Spearman's (r_s) and is a nonparametric rank-order correlation coefficient. It is used to quantify the strength of a trend between two variables that are measured on an ordinal scale.[7]

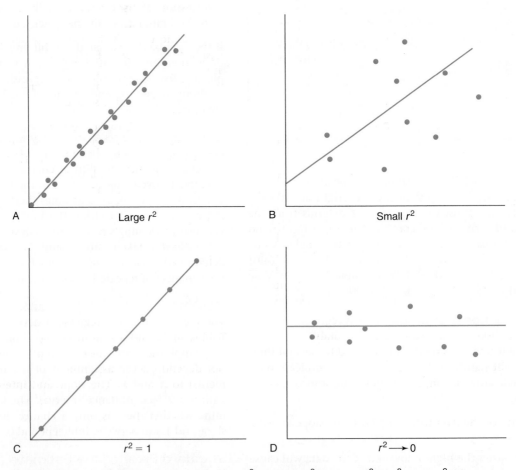

A   Large r^2

B   Small r^2

C $r^2 = 1$

D $r^2 \longrightarrow 0$

■ FIGURE 4-6   Coefficients of determination (r^2). A. Large r^2. B. Small r^2. C. $r^2 = 1$. D. r^2 approaches zero.

✪ TABLE 4-6

Relative intervals for interpretation of correlation coefficient, *r*.

Interval	Interpretation
0.76–1.0	Strong
0.55–0.75	Moderate
0.26–0.5	Fair
0.0–0.25	Weak or no relationship

The Pearson product-moment correlation coefficient is calculated by setting up a ratio the sums over all the observed (X, Y) points. Its value does not depend on which variable we identify as X and which we identify as Y. The magnitude of r describes the strength of the association between two variables, and the sign of r indicates the direction of this association: $r = +1$ when the two variables increase together and $r = -1$ when one decreases as the other increases. Descriptive interpretation for values of r is presented in Table 4-6 ✪. These intervals present general agreement for r-values.

✓ Checkpoint! 4–4

Calculate the following from the data presented in Checkpoint 4-3:

- *Linear regression by least squares, including slope and intercept*
- *Correlation coefficient*
- *Coefficient of determination*

ANALYSIS OF VARIANCE

Analysis of variance (ANOVA) is a statistic that can be used to compare data in experiments where there are more than two variables. For example, suppose a laboratory has three methods for measuring sodium in blood. The laboratory would like to compare whole blood values to serum and plasma. One approach is to develop several different studies to compare two variables at a time using a *t*-test, or one can simply use ANOVA in a single experimental design. It is not within the scope of this text to provide detailed examples and explanations of ANOVA. There are many references, including computer software programs, that provide detailed instructions for performing and interpreting ANOVA.

EVALUATION AND SELECTION OF METHODS

Introduction

Clinical laboratories frequently change methodologies, add new laboratory tests, and replace or add new instruments. The implementation of any of these changes requires proper evaluation and validation. Centers for Medicare and Medicaid Services (CMS) through the Clinical Laboratory Improvement Act (CLIA '88) require that certain tasks be completed to ensure the quality of laboratory tests resulting from these changes. The quality of laboratory testing is monitored via quality-control procedures and performance on proficiency tests.

Clinical laboratories today are not required to complete an exhaustive evaluation of new assays as was required in the past. Instead, if the assay is approved by the U.S. Food and Drug Administration (FDA), then an abbreviated evaluation and a validation protocol is required. The protocols can be reviewed in the regulations provided by CLIA.

Issues to Consider Before Evaluating a New Assay

Before adding new instrumentation, changing instruments, or adding new assays, several issues should be addressed. A brief discussion of selected issues is presented here.

Need The need for implementing a new method may be the result of (1) a clinician's inquiry; (2) a new method that is more accurate or precise than the existing one; (3) a new method that is more cost effective, provides faster analysis time, or is less labor intensive than an exixting method; and/or (4) a new method that provides more clinically useful information (e.g., better specificity) than provided by the current method.

Literature/Consultation with Colleagues A thorough review of the literature and conversations with colleagues and users of the method in question can provide a wealth of information that will save the laboratory staff time and possibly money. Information on method performance parameters, ease of performance, and clinical utility may be ascertained from these sources.

Cost Issues Cost per test is an important issue and requires close scrutiny. A detailed discussion of cost accounting a new procedure is not within the scope of this text. There are many resources available to the laboratory to assist in determining cost, profit, and fees associated with the test. Some items that require attention and input from the laboratory are as follows:

- Capital outlay
- Direct and indirect cost
- Supplies
- Consumables

Types of Specimens Several issues relevant to biological specimens need to be addressed, including (1) knowledge of the type(s) of specimens for chemical analysis, (2) which specimen collection containers are appropriate, (3) specimen procurement requirements, and (4) the volume of specimen required.

Safety Issues In most cases all of the appropriate safety practices are already in place, but the evaluator of the new method or instrument must ensure that any new or different safety considerations are identified and handled in the proper fashion. For example, the new method may use a chemical not currently in use by the laboratory. An material safety data sheet (MSDS) form must be obtained and reviewed for any special handling requirements. Questions concerning the use of personal protective equipment and disposal of waste must be addressed. Safety in the laboratory is discussed later in this chapter.

Additional Equipment During the planning stages of method selection and instrument evaluation, it should be determined whether new equipment is required (e.g., a −70°C freezer, centrifuges, or a personal computer). In many cases the added equipment may have to be in place before the evaluation begins.

Facilities Requirements Laboratory facility issues are not usually a concern unless a new piece of equipment or an analyzer is required. If new equipment or an analyzer will be needed, then the laboratory should address the following requirements:

- Space
- Electrical system
- Plumbing/water
- Temperature control
- Waste disposal
- Storage (reagents, consumables)

Manufacturers' Claims The FDA requires manufacturers and developers of methods and analyzers to document claims about the analytical performance of the method. These should be reviewed in advance so that the laboratory can make any necessary accommodations.[8]

Clinician Requirements The decision to accept or reject a new method should be based on the ability of the method to meet the requirements of the physician who will be using the result for patient care. Careful evaluation of error must be undertaken because if the error of the test result is excessive, patient injury may result. The greatest chance for misdiagnosis caused by an analytical error in a test result occurs at the concentration of analyte at which a medical decision is to be made. See also the subsequent section on performance standards for further discussion of medical decision levels.

Quality-Control Issues A significant part of a method evaluation is establishing quality control (QC) for the analyte. Several issues need to be addressed before determining a QC range. An abbreviated list of issued includes the following:

- What type of QC material is appropriate (e.g. lyophilized or liquid)?
- What should be the proper matrix substance (e.g., whole blood, urine, or cerebrospinal fluid [CSF])?
- How many concentration levels are required (e.g., two or three)?
- What should be the concentrations be for each level?
- How frequently should the controls be assayed?

Once the QC material is selected, the evaluator can establish QC ranges using accepted protocols. These ranges will be incorporated into Levey-Jennings QC charts and subject to the laboratory's QC rules.

Training Proper training of laboratory staff is essential. The staff must be given enough time to learn the method and associated instrumentation (if it is a new piece of equipment). All aspects of the assay should be taught, not just the procedure.

Staff The laboratory must have staff that is competent to perform the assay and make critical decisions regarding procedure and results. Additional training may be required if the methodology and/or instrumentation are overly complex and beyond the capabilities of the laboratory staff.

ANALYTICAL PERFORMANCE PARAMETERS

Accuracy

The International Federation of Clinical Chemistry (IFCC) defines **accuracy** as the closeness of the agreement between the measured value of an analyte to its "true" value.[9] Several options are available to the laboratory staff to ascertain the accuracy of their method. A brief description of a select few follows.

Definitive and reference methods established by the National Reference System for Clinical Laboratories (NRSCL) may be used. NRSCL is part of the CLSI and provides definitive and reference methods for analytes (e.g., glucose, cholesterol, sodium total protein, and others.)[10]

Commercially available reference materials may be used to monitor the accuracy of a method. A list of these materials, identified as standard reference materials (SRMs), is found in NIST Special Publications 260 and can be obtained from that company (National Institute of Standards and Testing, 100 Bureau Drive, Stop 1070, Gaithersburg, MD 20899-1070). **http://www.nist.gov/public_affairs/contact.htm**

Accuracy can be evaluated using proficiency testing samples. An estimation of a system's bias can be made from proficiency testing performance. The laboratory must evaluate the specific test method's observed value against a comparison value, which is the mean value reported for all similar methods (peer group means), the mean value for all methods, or the definitive method value. Bias is calculated by subtracting the comparison value from the laboratory method value. The algebraic sign indicates whether the laboratory value is higher (positive bias) or lower (negative bias) than the group mean. This procedure does not establish accuracy but rather bias from the comparison value.

Recovery experiments using the method of addition are also used to determine the accuracy of a method. This technique is commonly used to evaluate the accuracy of therapeutic and abused drug assays. The method of addition tests whether the method can measure the analyte in the presence of all other compounds that are contained in the matrix of the sample.

Precision

The **precision** of a method is its ability to produce the same value for replicate measurements of the same sample. Precision is also described as the random variation in a population of data. Estimates of precision are determined for within-run and between-run analysis using mean, standard deviation, and %CV.

Within-run repeatability is described as the closeness of agreement between results of successive measurements carried out under the same conditions. Between-run reproducibility is the closeness of agreement between results of measurements performed under changed conditions of measurement (e.g., time, operator, and calibrator).[11]

Types of Errors

Random Random analytical error occurs without prediction or regularity. Factors contributing to random error are those that affect the precision (reproducibility) of the measurement. Examples of factors affecting the precision of an assay include (1) temperature fluctuations; (2) unstable instrumentation; (3) changes in reagents; (4) variation in manual techniques, such as pipetting, mixing, and timing; and (5) operator variation.

Systematic Systematic analytical error is defined as error that is consistently low or high. If the error is consistently low or high by the same amount over the entire concentration range, it is called *constant* systematic error. Constant systematic error may be caused by the presence of contaminants in the reagent that affect the results by the same amount in all samples. Interference studies can be used to ascertain the amount of constant error present in the method.

If the error is consistently low or high by an amount proportional to the concentration of the analyte, it is termed *proportional* systematic error. Proportional systematic error may be caused by the incorrect assignment of the amount of analyte in the calibrator. Proportional error is also caused by side reactions of the analyte, where the percentage of analyte that undergoes a side reaction will be the percentage of error in the method. A distinguishing feature of proportional error is that the magnitude of the error increases as the concentrations of analyte increase. The amount of proportional error can be determined by a recovery study.

The presence of constant and/or proportional error can be seen easily if a graph of comparative data is created by measuring the analyte concentration of two different methods, as shown in Figure 4-7 ■.

Total error of an analytical method represents the overall effect of the components of error. The relationship of total error to constant and proportional error is shown in Figure 4-8 ■. The distribution of values around a central value represents random error. The shift of the central value of the distribution from the true value represents systematic error.

Discordant Results **Discordant result** is a term commonly used to describe laboratory results that do not agree. Other terms associated with discordance are *false negative,*

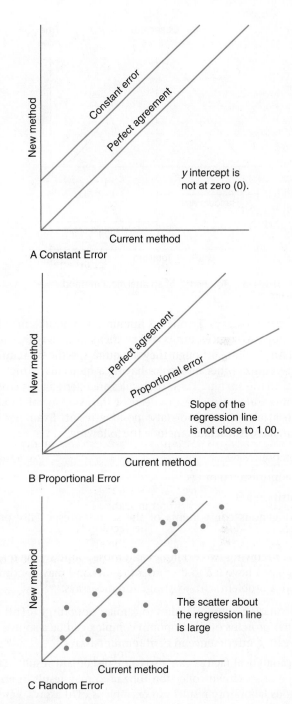

■ FIGURE 4-7   Plot of errors. **A.** Constant error. **B.** Proportional error. **C.** Random error.

false positive, errors, and *inaccurate results.* The consequence of discordant results may lead to incorrect diagnosis and therapy, thereby increasing morbidity or mortality.[12] Discordant results provide laboratorians with one of their greatest challenges because such results often occur sporadically and may occur with some specimens but not others.

Discordant results are caused by a variety of factors, some of which are associated with the preanalytical stage of testing. Several incidents of discordant results have been associated

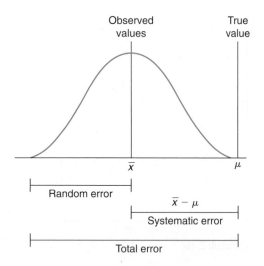

■ FIGURE 4-8   Total error of an analytical method.

with immunoassays. Erroneous immunoassay results may be caused by two sources: antibody specificity and assay interference. Antibodies, although they are quite specific by nature, may recognize other molecules having a molecular structure similar to the analyte. These molecules are referred to as cross-reactants and may increase or decrease the assay signal, resulting in a false negative or false positive result. Examples of potential cross-reactants include the following:

- Metabolites
- Coadministered drugs
- Nutrition aids
- Endogenous components of the samples, especially proteins

The following two examples and those found in the references listed have led to discordant results and may continue to pose a problem with existing immunoassays.[13,14]

1. Circulating autoantibodies to cardiac troponin I (cTnI) or other proteins of the troponin complex can be a source of negative interference in cTnI immunoassays.[15]

2. Preanalytical factors can also cause discordant results. The type of specimen-collection tube used may result in erroneous laboratory results. An example is cTnI assays, where Ca^{2+} ions are needed to maintain the complex forms of the analyte. Anticoagulants, like ethylenediaminetetraacetic acid (EDTA) or heparin, bind Ca^{2+} ions; thus some cTnI assays show different results for serum versus plasma.[16]

Discordant results may also occur in specimens that are lipemic, icteric, or hemolyzed. Lipemic specimens contain an abnormal amount of lipids, which can interfere with light-measuring techniques, or the lipids can create a dilutional effect, resulting in a falsely low value for the analyte being measured. A specimen that is icteric contains a large amount of bilirubin, which may interfere with light-measuring techniques. Hemolysis is defined as the disruption of the red cell membranes and results in the release of hemoglobin. The hemoglobin molecule may cause interference with both immunoassays and non-immunoassays.

Linearity

The word *linear* means "of or relating to a line or lines," and **linearity** is the quality or state of being linear. Linearity experiments are done to verify the expected range of analyte concentration over which the assay is linear. The experiment requires that reference solutions containing a wide range of specific quantities of an analyte be analyzed by the candidate method. The concentrations of analytes are plotted on the *x*axis versus the instrument response (*y*-axis) and a best-fit line is drawn. Ideally the plot should be linear and pass through zero. A slope and intercept may be calculated from the data. If the plot is linear, the range tested is termed the linear range of the method. CLIA '88 regulations require that the reportable range (see below) be verified and do not require that a linearity experiment be performed.

Analytical Range

The range of concentration in the sample over which the method is applicable without "modification" is termed analytical range.[17] Analytical range is tested by a linearity experiment.[18] The analytical range should be wide enough to include 95 to 99% of the expected samples without predilution. CLIA'88 states that once the analytical range of a method has been validated, it is termed the *reportable range* of the method.[19]

Analytical Sensitivity

The International Union for Pure and Applied Chemistry (IUPAC) defines analytical sensitivity as the slope of the calibration curve and the ability of an analytical procedure to produce a change in the signal for a defined change of the analyte quantity.[20] The terms *analytical sensitivity* and *detection limit* are often used interchangeably because they are interrelated in that they both are attributes of a sensitive method. Ideally a method should have a high level of analytical sensitivity and a low detection limit. Another representation of a methods sensitivity is referred to as "limit of absence" and is defined as the lowest concentration of analyte that a method can differentiate from zero.[21]

Analytical Specificity

The ability of a method to measure only the analyte it claims to measure without reacting with other related substances is termed analytical specificity.[21] For example, an assay that is specific for glucose measures only glucose and not other hexose sugars such as galactose. Analytical specificity is a very significant issue in immunoassays. The question of how specific the antibody is for the analytes to be measured requires investigation. Substances commonly found in blood samples (e.g., bilirubin, lipids, and hemoglobin) may also compromise analytical specificity. These compounds may affect specificity by virtue of their color, turbidity, or other chemical properties.

Limit of Detection

Limit of detection represents the lowest concentration or quantity of an analyte that significantly exceeds the measurement of a blank sample.[22] The detection limit depends on the amplitude of the blank value and must be precise at that level. A laboratory measurement less than the detection limit should be reported as "less than the detection limit value."

Interferences

The effects of a compound(s) other than the analyte being measured are termed interference. An interfering compound adversely affects the accuracy of the method. An example of interference is the effect of the presence of other reducing substances in blood on a test that measures a specific compound that is a reducing substance itself. Information regarding potential interferences on laboratory test may be found in the literature and dedicated publications.[23] Protocols that test for the presence of interfering substances are provided by CLSI (NCCLS).[24]

Stability

Stability of reagents, calibrators, and controls must be investigated. Fortunately in today's clinical laboratories most of these materials are prepared commercially and the use of laboratory prepared reagents is minimal. Reagent stability impacts the laboratory in both cost and efficiency. Therefore, prior knowledge of these issues will help the laboratory make sound decisions in implementing a method. In-house preparation of reagents requires a more labor-intensive scrutiny and evaluation process. Assaying samples with new or fresh reagents against older reagents can be done to evaluate reagent stability. The observed difference between the old and new reagents can be tested by use of a *t*-test.

Ruggedness

The ruggedness of an assay refers to the ability of the assay to perform in a consistent, reliable fashion when used by different operators and with different batches of reagents over an extended period of time. A "rugged" assay gives the laboratory staff confidence that the data being generated are accurate and precise over time.

Robustness

Robustness is defined as the capacity of a method to remain unaffected by small, deliberate variation in method parameters; thus it is a measurement of the reliability of a method. A robust assay provides the clinical laboratory with results that instill confidence in both laboratorians and clinicians.

Performance Standards

Performance standards need to be established before any analytical experiments are begun. The focus of performance standards is error. The laboratory should establish performance goals that specify the total error allowed at a specific concentration of analyte. The concentrations are selected at medical decision levels at which test results are most critically interpreted by clinicians for diagnosis, monitoring, or therapeutic decision. For example, glucose would require critical concentrations for hypoglycemia and hyperglycemia. Therefore, any evaluation of glucose methods would require a study of error at these two medical decision levels. CLIA '88 provides recommended allowable error data for several laboratory analytes.[19] Several individuals, including Barnett,[25] Tonks,[26] Cotlove,[27] and Elevitch,[28] have provided additional criteria for performance standards.

Reference Interval

Determining the reference interval of an analyte is an issue that requires close attention. CLSI has published a document that provides guideline on how to define and determine reference intervals in the clinical laboratory.[29] A more detailed discussion of reference interval is provided below.

Experimental Phase of Method Evaluation

Once the issues discussed above have been addressed, the experimental phase can be started. The actual experiments or tasks performed and the order in which they are performed vary considerably from laboratory to laboratory. Therefore, it is imperative that the evaluator be familiar with the requirements of CLIA'88, as well as state and federal regulations pertaining to method performance criteria. A brief discussion of several tasks necessary to evaluate methods is presented here. For more detailed information regarding this topic, the reader is encouraged to review the reference provided.[30] The reader must be aware that this list may not represent all of the tasks appropriate for evaluating a method. Also, the order in which these tasks are completed may be altered to suit the particular evaluator's philosophy.

1. Write a protocol for the evaluation.
2. Write the procedure for performing the test.[31]
3. Document the validation of the method.
4. Determine within-run precision by analyzing serum pools at two to three levels 20 times or more. The concentrations of the serum pools should be near the medical decision levels. Calculate the mean and standard deviation and compare them to the allowable standard deviation or calculate random error appropriate for the test and compare with the allowable total error. If the within-run precision data are acceptable, then determine between-run precision. This can be accomplished by analyzing the serum pools each working day for at least 20 days.
5. Determine the reportable range by measuring a set of reference solutions or a series of dilutions either of control material or specimen pools. Plot the observed values against the relative dilutions of the series of samples.[32] Estimate the analytical range by visual inspection of the graph. Compare your data with the manufacturer's data. It is recommended that you do not extend your range beyond the available calibrators.

6. Determine the accuracy of the method by comparing the performance of the candidate or new method with that of a reference method or definitive method. Accuracy of a new method may also be evaluated by comparing the performance of the candidate method with that of the method being replaced. The assumption is that the current method is accurate.

7. Determine the sensitivity of the new method by determining the slope of the calibration curve. Data can be extrapolated from an analytical range study.

8. Estimate a detection limit by adding the mean of blank measurements to the product of the standard deviation of the blank measurement and a numerical factor selected based on the confidence level desired.

9. Assess the test specificity by determining the interference due to substance(s) regularly found in blood samples (e.g., hemoglobin, lipids, and bilirubin) and commonly used drugs.

10. Establish the reference interval for the analyte. CLIA '88 Amendments require the laboratory to verify that the manufacturer's reference interval is appropriate for the patient population that is served by the laboratory.[19]

There are several additional tasks that should be included along with the performance evaluation before the method is implemented into the laboratory. These include the following:

• Prepare a document to add to the laboratory procedure manual describing the step-by-step performance of the assay.

• Develop in-service training materials and instruct staff on the proper performance of the test.

• Establish quality-control ranges and implement procedures for documenting and interpreting quality-control data for the test.

• Prepare instructions for the laboratory staff that include instructions on specimen collection, test availability, reference intervals, turnaround time, possible interferences, and clinical use of the test.

Method Comparisons of Qualitative Tests (Paired Proportions)

Several assays performed routinely in the clinical laboratory provide qualitative results (e.g., pregnancy tests, where the results of the test are reported as positive or negative and urine dipstick chemistries may be reported as 1+, 2+, etc. Evaluation of qualitative assays cannot be carried out in the same manner as quantitative assays; therefore, a different approach and statistics must be used. An example of a statistical test that is useful for comparing two qualitative assays with nominal-type data is the McNemar test.[33] The results of a study assessing concordance of data between qualitative assays are entered into a 2 × 2 contingency table similar to that of Table 4-7. This test compares paired proportions and focuses on the pairs' discordant data. An example of an application of a McNemar test is shown in Example 4-4 . The data in Example 4-4 represent a

study comparing the results of two test kits for qualitative determination of β-hCG in urine. A contingency table is developed and reveals the presence of 46 discordant results (or 7.7%). The McNemar test will be useful to determine if the number of discordant results is significant.

Example 4-4

Consider the following information taken from a correlation study between two different β-hCG assays. The data presented below show that there are a total of 594 patient specimens that are assayed by both β-hCG kits. The results are posted in a 2 × 2 contingency table, and there are 46 sets of results that are discordant. A determination of significance is assessed using hypothesis testing and a chi square table of probability. The null hypothesis and alternate hypothesis must be stated:

	Test Kit 2 Positive	Test Kit 2 Negative	Total
Test Kit 1 Positive	257	1	258
Test Kit 1 Negative	45	291	336
Total	302	292	594

H_o = There is no significant difference in response of both assay kits to patient specimens.

H_A = There is a significant difference in the response of both assay kits to patient specimens.

The level of significance or α for this evaluation is 0.05.

Number of degrees of freedom = $(2 \text{ rows} - 1) \times (2 \text{ rows} - 1) = 1$

Calculate the McNemar statistic using the following formula:

$$\chi^2 = \frac{[|O - E|]^2}{E} + \frac{[|O - E|]^2}{E}$$

where

O = the observed value

E = the expected value = total discordant data / 2

$$\chi^2 = \frac{[|45 - 23|]^2}{23} + \frac{[|1 - 23|]^2}{23}$$

$$\chi^2 = \frac{[|22|]^2}{23} + \frac{[|22|]^2}{23}$$

$$\chi^2 = \frac{484}{23} + \frac{484}{23}$$

$$\chi^2 = 21 + 21$$

$$\chi^2 = 42.0$$

The chi square value for one degree of freedom (from a table of chi square) and $\alpha = 0.05$ is 3.84. The calculated McNemar statistic (χ^2) is equal to 42.0; thus the p-value is less than 0.001. Because the p-value is less than α, we reject the null hypothesis and conclude that the proportion of discordant values is significant.[7]

Note: The McNemar statistic is similar to a chi square statistic. The chi square sums the observed and expected proportions for every cell in the table, with the expected value for each cell based on a calculation of probability of results. Chi square should only be used to test the significance of differences when the rows and columns of a contingency table are independent of each other. In comparison of method experiments, the rows and columns are not independent because they represent the response of the same samples to two different tests. Thus the McNemar test can be thought of as a chi square statistic for paired data. Therefore, caution is advised when using the McNemar statistic, and the McNemar statistic should not be used as the sole reason for accepting or rejecting an experiment.

► CLINICAL DECISION LIMITS

DIAGNOSTIC TESTS

In the health science field, the application of probability laws and concepts is intended to evaluate screening tests and diagnostic criteria. Clinicians are interested in increasing their ability to correctly predict the presence or absence of a particular disease from knowledge of test results (positive or negative) and/or the status of presenting symptoms (present or absent). Clinicians may also be interested in information regarding the likelihood of positive and negative test results and the likelihood of the presence or absence of a particular symptom in patients with or without a particular disease.

Bayes' theorem is often employed in issues of diagnostic testing or screening. Screening is the application of a test to individuals who have not yet exhibited any clinical symptoms in order to classify them with respect to their probability of having a particular disease. Those who test positive are considered to be more likely to have the disease and are usually subjected either to further diagnostic procedures or to treatments. Diagnostic screening tests are often employed by health-care professionals in situation where the early detection of disease would contribute to a more favorable prognosis for the individual or for the population in general. Bayes' theorem allows health-care professionals to use probability to evaluate the associated uncertainties.

Screening tests are not always infallible. That is, a testing procedure may yield a *false positive* or a *false negative* result. Therefore, the following issues of probability should be considered in evaluating the usefulness of test results:

1. Given that a subject has the disease, what is the probability of a positive test result (or the presence of symptoms)?

2. Given that a subject does not have the disease, what is the probability of a negative test result (or the absence of symptoms)?

3. Given a positive screening test (or the presence of a symptom), what is the probability that the subject has the disease?

4. Given a negative screening test result (or the absence of a symptom), what is the probability that the subject does not the have the disease?

The probability values for the four situations noted above are derived from experimental data and formatted in a 2 × 2

✪ TABLE 4-7

McNemar test for comparing two qualitative β-assays.

	Test Kit 2 Positive	Test Kit 2 Negative	Total
Test Kit 1 Positive	Number of positive results for test Kits 1 and 2	Number of positive results for test kit 1 and negative results for test kit 2	Total tests results for row
Test Kit 1 Negative	Number of negative results for test kits 1 and 2	Number of negative results for test kit 1 and positive results for test kit 2	Total test results for row
Total	Total of number of results in column	Total of number of results in column	Total test results for row

contingency table, as shown in Table 4-7 ✪. Every probability estimate can be derived by applying Bayes' theorem and associated calculations shown in the table.

A laboratory may be interested in knowing the diagnostic utility of its cardiac troponin I assay used to evaluate patients suspected of having an acute myocardial infarction. A study can be created in which blood samples are taken from patients admitted to the emergency department complaining of chest pain and troponin I is measured on all specimens. The diagnostic parameters associated with the predictive value are presented in Table 4-8 ✪ and are calculated based on the information provided by the emergency department clinician and laboratory staff. An example of data collected for

✪ TABLE 4-8

Diagnostic parameters associated with predictive value theory.

	Number of Subjects with Positive Test Results	Number of Subjects with Negative Test Results	Total
Number of Subjects with Disease	TP	FN	TP + FN
Number of Subjects without Diseases	FP	TN	FP + TN
Totals	TP + FP	FN + TN	TP + FP + TN + FN

TP, True positives (number of diseased patients correctly classified by the test)
FP, False positives (number of patients without the disease misclassified by the test)
FN, False negative (number of diseased patients misclassified by the test)
TN, True negative (number of patients without the disease correctly classified by the test)
Diagnostic sensitivity = TP / TP + FN
Diagnostic specificity = TN / FP + TN
Predictive value of positive test, PV+ = TP / TP + FP
Predictive value of negative test, PV− = TN / TN + FN
Efficiency of the test (number of patients correctly classified) = TP + TN / TP + FP + TN + FN
Prevalence = TP + FN / TP + FP + TN + FN

⊗ TABLE 4-9

Data representing a study to determine the clinical utility of a laboratory test for measuring serum cardiac troponin I.

	AMI Present	AMI Absent	Total
Troponin I			
Positive	400 (TP)	4 (FN)	404 (TP + FN)
Negative	2 (FP)	200(TN)	202 (FP + FN)
Total	402 (TP + FN)	204 (FN + TN)	606 (TP + FP + FN)

Diagnostic sensitivity = 99%
Diagnostic specificity = 99%
Predictive value of positive test = 99.5%
Predictive value of negative test = 98%
Efficient of the test = 99%
Prevalence = 66.7%

such a study is shown in Table 4-9 ⊗. The diagnostic sensitivity of a test is the probability of a positive test result given the presence of the diseases. In this example the sensitivity of the test is 99%. The diagnostic specificity of a test is the probability of a negative test result given the absence of the disease. Specificity for data presented is 99%. The predictive value of a positive test is the probability that a subject has the disease, given that the subject has a positive screening test result and is computed to be 99.5%. Lastly, the predictive value of a negative test is the probability that a subject does not have the disease, given that the subject has a negative screening test result and is computed to be 98%. Test efficiency is the fraction, usually in percentage, of all tested individuals who were correctly classified as either having or not having the disease. The diagnostic efficiency is 99%.

The prevalence of the disease in the population being evaluated is important in determining predictive values. By rearranging the calculation for predictive value of a positive test (PV+) and the predictive value of a negative test (PV−) to include prevalence, the following formulas are obtained:

$$PV+ = \frac{[\text{prevalence} \times \text{sensitivity}] \times 100\%}{[\text{prevalence} \times \text{sensitivity}] + [(1 - \text{prevalence})(1 - \text{specificity})]}$$

$$PV- = \frac{[\text{prevalence} \times \text{specificity}] \times 100\%}{[\text{prevalence} \times \text{specificity}] + [(1 - \text{prevalence})(1 - \text{sensitivity})]}$$

Therefore, as the prevalence of the disease changes so will the PV+ and PV− values. For example, what would be the PV+ if the prevalence of the disease were 60% or 5%? Based on the data shown in Table 4-9, the adjusted PV+ for a prevalence of 60% is 99.3% and for 5% the PV+ is 83.9%.

RECEIVER OPERATOR CHARACTERISTICS (ROCS)

The relationship between sensitivity and specificity may be illustrated using a graph identified as a receiver operator characteristics (ROC) curve. A ROC curve is a line graph that plots the probability of a true positive result or the sensitivity of the test (on the vertical axis) against the probability of a false positive result or specificity of the test (on the horizontal axis) for a range of different cutoff points. Each point on the ROC plot represents a sensitivity/specificity pair corresponding to a particular decision threshold. A test with perfect discrimination (i.e., no overlapping the two distributions of results) has a ROC plot that passes through the upper left corner, where the true-positive fraction is 1.0 or 100% (perfect sensitivity), and the false-positive fraction is zero (perfect specificity). The theoretical plot for a test with no discrimination (identical distribution of results for the two groups) is a 45° diagonal line extending from the lower left corner to the upper right corner. Most plots fall in between these two extremes.[34]

Figure 4-9(A) ■ shows the ROC curve for biopsy results as predicted by prostate-specific antigen (PSA) levels. Each point on the curve represents a different cutoff value for the disease of interest. The critical value selected is the value at which the curve deviation from the diagonal line from (0, 0) or bottom left to (1, 1) top right is the greatest. The strongest predictor of a biopsy result will be indicated where the perpendicular distance from the line of equality is a maximum, which is about at the point of the coordinates (0.36, 0.75). Using the coordinates and appropriate software, the cutoff point value for PSA concentration can be determined.[35]

Diagnostic accuracy is also determined from a ROC analysis. An example of this application is shown in Figure 4-9(B) ■. The ROC shown compares NT-ProBNP versus clinically estimated likelihood for the emergency department clinician to diagnose acute congestive heart failure (CHF). The results of the NT-Pro BNP test were superior to those of clinical judgment alone, with significantly greater area under the curve (0.90). The area under the curve derived solely from clinical judgment was 0.86. Finally, the area under the curve derived from NT-ProBNP testing plus clinical judgment (0.93) was superior to each diagnostic modality alone.[36]

RELATIVE RISK AND THE ODDS RATIO

Scientific investigations include *designed experiments*, in which at least one variable is manipulated in some way, and *observational studies*, in which neither the subjects under study nor any of the variables of interest are manipulated in any way.[37] Thus an observational study may be described as an investigation that is not an experiment. A basic form of observational study is one in which there are only two variables of interest. One of the variables is termed the *risk factor*, or independent variable, and the other variable is referred to as the *outcome* or dependent variable. The term *risk factor* is used to designate a variable that is thought to be related to some outcome variable. For example, cigarette smoking (risk factor) may be associated with a subject suffering from cancer (outcome variable).

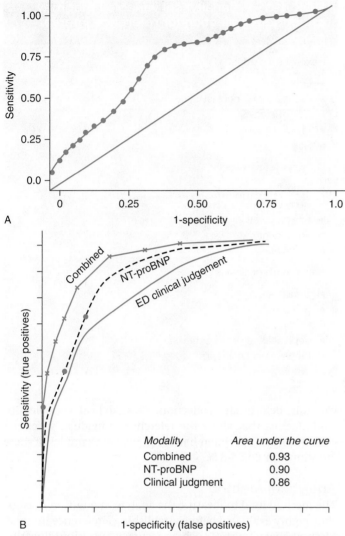

A

B

■ FIGURE 4-9 A. Receiver operator characteristics curve for biopsy results as predicted by prostate-specific antigen level. Area under ROC curve = 0.71. **B.** Receiver operating characteristic curve of NT-proBNP.

Two types of observational studies that appear frequently in the literature are *prospective studies** and *retrospective studies*. A prospective study is one in which two random samplings of subjects are selected. One sampling consists of subjects possessing the risk factor, and the other sampling consists of subjects who do not possess the risk factor. The subjects are followed into the future and a record is kept on the number of subjects in each sample who, at some point in time, are classifiable into each of the categories of the outcome variables. A 2 × 2 contingency table is used for analysis of the data. A retrospective study is the reverse of a prospective study. The samples are selected from those falling into the categories of the outcome variable. The researcher then looks back at the subjects and determines which ones have and which ones do not have the risk factors.

*Prospective epidemiological studies are also called cohort studies.

Measured data from a prospective study are used to determine *relative risk*. Relative risk is defined as the ratio of the risk of developing a disease among subjects with the risk factor to the risk of developing the disease among subjects without the risk factor.

The appropriate measure for comparing cases (having the disease) and controls (not having the disease) in a retrospective study is the *odds ratio*. The odds ratio is described as the ratio of the probability of success to the probability of failure.

The range of values for odds ratio is between zero and infinity. A value of one indicates no association between the risk factor and disease condition. A value less than one indicates reduced odds of the disease among subjects with the risk factor. An odds ratio greater than one indicated increased odds of having the disease among subject in whom the risk factor is present.

▶ REFERENCE RANGE (NORMAL RANGE)

A reference interval (also called reference range) is the interval between and including two reference limits. Reference limit is a numerical value(s) derived from the reference distribution. The key word thus far is *reference*, which denotes a well-defined selection of subjects used to mathematically determine the numerical values equivalent to reference limits and thus reference interval. Before the use of the word *reference* as the modifier for interval or range, the comparator used for clinical decision making was "normal" range. The use of the word *normal* was often imprecise and confusing. The limits and intervals derived from apparently "normal" healthy subjects may not have represented a certain segment of the population. For many analytes, the usual test result distribution of healthy and diseased population overlaps. Thus normal value does not always indicate a lack of disease.

Reference intervals are derived from a reference individual. A reference individual is selected on the basis of well-defined criteria. It is usually important to define the individual's health, age, sex, and race. An adequate number of reference individuals must be selected to determine a reference interval. This is the reference sample group and is representative of the reference population.

The recommendation for clinical laboratories to determine reference intervals is provided by CLIA'88 (§493.1213), Federal Register 57(40) Feb 28, 1992. This section outlines the protocol for obtaining reference values and reference intervals. To assist the laboratory in determining reference intervals, CLSI has published a document that establishes guidelines and procedures for determining valid reference intervals for quantitative procedures.[19, 29]

There are several factors that must be addressed before determining a reference interval:

- Selection of reference individuals
- Preanalytical variables

- Analytical methods
- Statistical applications

Selection of Reference Individuals

The selection of reference individuals must be determined before any data are collected. Data should be collected from "normal," healthy individuals. Defining what is to be considered healthy becomes the initial problem in a reference interval study. Therefore, a list of *selection criteria* must be established. In addition, the criteria used to exclude nonhealthy subjects from the reference samples must be established. Several examples of possible *exclusion criteria* are shown in Box 4-1 ■.[38]

Another set of criteria that require attention is referred to as *partitioning criteria*. Examples of possible partitioning criteria are listed in Box 4-2 ■.

If the reference interval data shows a significant difference between any of these factors, then the laboratory may have to establish different reference intervals to reflect these differences. For example, if the reference interval for serum creatinine kinase is significantly different between males and females, then the laboratory is compelled to show two different reference ranges, one for the males and one for the females.

Preanaltytical Variables

Preanalytical variables can have a profound effect on establishing reference intervals and therefore must be studied vary

BOX 4-1 Possible exclusion criteria for reference interval studies.

- Recent illness
- Alcohol consumption
- Abnormal blood pressure
- Occupation
- Drug abuse
- Obesity
- Prescription drugs
- Oral contraceptives
- Pregnancy
- Tobacco use
- Recent transfusion

BOX 4-2 Partitioning criteria for reference interval studies.

- Race
- Sex
- Age
- Blood group
- Ethnic background
- Exercise
- Fasting or nonfasting
- Tobacco use
- Circadian variation

BOX 4-3 Preanaltytical variables that may affect data collection for reference interval studies.

Reference individual

- Diet
- Fasting versus nonfasting
- Drug therapies
- Physical activity
- Stress

Collection of specimens

- Time
- Body posture
- Site preparation
- Equipment
- Technique

Handling of specimens

- Transport
- Storage
- Clotting
- Separation of serum or plasma
- Preparation for analysis

carefully before data collection. Preanalytical variables include factors that affect the reference individual, specimen collection, and specimen handling. Several examples of these are shown in Box 4-3 ■.

Analytical Method

The method used to measure the analyte in question must be thoroughly evaluated before establishing a reference interval. Method characteristics such as accuracy, precision, linearity, interferences, and minimum detection limits must be described in detail. Factors that affect analytical performance, including equipment, reagents, calibrators, and calculations, require control and documentation. Variability between reagent lots, technologists, and instrument-to-instrument differences (if the analyte is to be measured on different analyzers) must be determined. The gathering of data should include values derived from different technologists and from more than one lot of reagents. Quality-control materials should be assayed throughout the reference interval study to monitor the analytical procedure. Finally, samples used for establishing the reference interval should be measured over several days to reflect average between-run variations.

Statistical Applications

The reference interval may be described as the interval between the 2.5th (lower reference limit) and 97.5th (upper reference limit) percentiles of a group of data obtained from a reference population. Before selecting upper and lower percentile limits, it must be determined whether the data conform to a Gaussian distribution. If the data are Gaussian, then the parametric method may be used. For non-Gaussian distributions the evaluator must use nonparametric methods, which make no specific assumptions concerning the mathematical form of the probability distribution represented by the observed reference values.[39]

CLSI guidelines recommend that a minimum of 120 reference values be used and the reference interval be determined by the nonparametric method.[29] In this method the reference values are ranked by frequency of occurrence at various concentrations and then the 2.5th and 97.5th percentiles values are selected. Further refinement of a reference interval includes the use of confidence intervals (CIs) to determine the 90% CI of the upper and lower percentiles values. Determining confidence intervals is useful because they remind the evaluator of the variability of estimate and provide a quantitative measure of this variability. Also, confidence intervals tend to become narrow as the size of the sample increases. This will allow the evaluator to see the improved precision, in an estimated 95% reference interval, that would be obtained from a larger sampling of reference individuals.

▶ QUALITY ASSURANCE AND QUALITY CONTROL

Quality control has its roots in the early automotive industry. Shewhart developed a type of control chart for use in industry.[40] Since that time quality control has evolved to a level that has served to minimize error, especially in the analytical stage of testing. For a review of the origin of present-day quality-control procedures, refer to the references cited.[41–47]

QUALITY ASSURANCE

Total quality management (TQM) is a management concept that began in industry and provides a management philosophy for organizational development and a management process for improving the quality of workmanship.[48] Clinical laboratories have incorporated many of the aspects of TQM, especially quality assurance and quality control.[49] TQM programs accomplish the following:

- Monitor and evaluate the ongoing and overall quality of the total testing process and the effectiveness of its policies and procedures.

- Identify and correct problems and ensure the accurate, reliable, and prompt reporting of test results.

- Ensure the adequacy and competency of the staff.

Quality control refers to the procedures for monitoring and evaluating the quality of the analytical testing process of each method to ensure the accuracy and reliability of patient test results and reports.

Quality assurance (QA) in a health-care facility represents global issues and is the responsibility of everyone involved in the care of patients. There are many essentials for a quality assurance program, including the following:

- Commitment
- Facilities
- Resources
- Competent staff
- Reliable procedures, methods, and instrumentation

The clinical laboratory is an important component of the overall QA within a facility due in part to the volume of testing and the clinical importance of the laboratory data being generated. If an error occurs in any one step during the acquisition, processing, analysis, and reporting of a laboratory test result, it will invalidate the quality of the analysis and the laboratory will not realize its QA goals.

QUALITY CONTROL

The purpose of assaying control material is to verify the stability and accuracy of calibration and testing systems. This is an extremely important function supported by several organizations and regulatory agencies.

The International Organization for Standardization (ISO) 9000 series identified nine quality-ensuring items that are common to all quality-control systems:[50]

- Instituting an effective quality-control system
- Ensuring valid and timely measurements
- Using calibrated measuring and testing equipment
- Using appropriate statistical techniques
- Developing a product identification and traceability system
- Maintaining adequate record-keeping systems
- Ensuring an adequate product handling, storage packaging, and delivery system
- Maintaining an adequate inspection and testing system
- Ensuring adequate personnel training and experience

CLIA'88 embraces most of these items in its regulations under *subpart K* and *subpart P*, and therefore it is incumbent on clinical laboratories to do the same.[29]

Quality-Control Materials

The best practice when selecting the appropriate material for QC is to use a matrix that is similar to the test specimens. For example, if the analyte to be measured in a patient specimen requires whole blood, then the QC material should be whole blood. Unfortunately, matrixes such as whole blood are difficult to stabilize; therefore, a suitable alternate must be found.

There are several factors to consider when selecting quality-control materials:

- The materials must be stable.
- The materials must be available in aliquots or vials.

- The materials can be analyzed periodically over a long span of time.
- There is little vial-to-vial variation.
- The concentration of analyte should be in the normal and abnormal ranges.

Commercially prepared control materials are manufactured in three different forms: (1) lyophilized or freeze dried pooled material, (2) liquid pooled material, and (3) frozen pooled material. Lyophilized materials are processed so that the water content is minimal, thus requiring reconstitution and special mixing procedures. This process often results in error because of incorrect pipetting and/or failure to following mixing procedures exactly as written. A means to alleviate these problems is to acquire control material that is already liquefied and may be stored frozen or refrigerated.

Target Values

Commercial QC material may be assayed or unassayed. Assayed QC material has concentrations or *target values* that are determine by the manufacturer. Unassayed QC material has no predetermined target values and its ranges must be determined by the laboratory. The target values for the QC material include the mean and standard deviations of the analyte for the particular control material. Each target value must represent a concentration in the normal range and abnormal range for bilevel QC material and an additional abnormal level for trilevel QC material. The decision whether to assay two levels or three levels of QC material depends on the analyte to be measured, the method of measurement, and the clinical significance of the analyte to disease. For example, blood-gas analytes require three levels of controls; general chemistries (e.g., glucose and urea nitrogen) require at least two levels; and immunoassays usually require three levels.

Quality-Control Limits

Establishing QC limits, represented by ± 1, 2, and 3 standard deviation(s), is not an easy task. One must be knowledgeable of the analyte method, instrumentation, allowable error, CLIA'88 regulations, and other significant aspects of laboratory testing. The control limits must fall within the total allowable error of the method and allow for successful completion of proficiency testing required by state and/or federal regulatory agencies. Recommended procedures for establishing QC limits may be found in several clinical chemistry textbooks.[51,52]

Levey–Jennings Control Charts

Levey–Jennings (L-J) charts show the difference between the observed values and the expected mean. The charts are created by calculating the mean concentration and up to ±3s for a pool of QC material. These data are plotted on the y-axis. The x-axis is divided into days, usually in 30-day intervals. As quality-control data are generated during the month, the

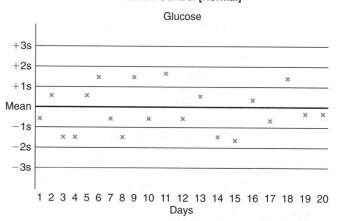

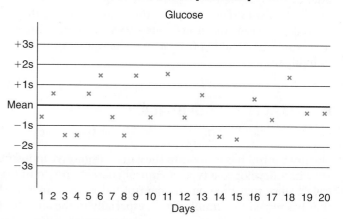

■ FIGURE 4-10   Levey–Jennings quality-control charts for normal and abnormal concentrations of glucose in blood.

values are plotted on the L-J chart. L-J charts are very useful for observing patterns of data (for instance, *trends* and *shifts*). A trend is a pattern of data in which all of the QC values continue to increase or decrease over a period time. QC trends may occur, for example, when reagents begin to deteriorate or a polychromatic light source continually diminishes in luminescent intensity or when a monochromatic filter becomes delaminated. If a trend is allowed to continue, the QC results will eventually exceed ± 2s. A shift in QC values is described as consecutive data that remain on one side of the mean for a period of time. The presence of a shift in QC results may be due to a change in incubation temperature, contamination of reagents, changes in calibrator values, or changes in pipet volumes. A typical L-J QC chart for blood glucose measurement using bilevel (normal and abnormal) QC material is shown in Figure 4-10 ■.

Power Functions

Power functions (also called power curves) are useful for (1) evaluating the performance capabilities of individual control procedures, (2) comparing the performance of different control procedures, and (3) designing a new procedure with improved performance characteristics. Refer to the references cited for details on the use of power functions.[53,54]

Westgard Multirule Procedures

A multirule procedure developed by Westgard and colleagues uses an assortment of control rules for interpreting QC data.[55] The procedure requires a chart in which lines for control limits are drawn at the mean; this chart can be adapted to existing L-J charts by adding one or two sets of control limits. This system provides a more structured use than that proposed by Levey–Jennings and shows either random or systematic errors. A summary of the rules is shown in Table 4-10 ✪ and are diagrammed in Figure 4-11 ■.

Table 4-11 ✪ summarizes the interpretation of QC data in the examples provided in Figure 4-12 ■. The table provides the day, the decision to accept or reject, the control rule violated, and the type of error that may have caused the questionable control value.

Sources of Random and Systematic Errors

Once the possibility of an error has been detected, the next step is to determine the cause of the error. Sources of error vary depending on whether they are random or systematic. Examples of sources of error are shown in Table 4-12 ✪.

Detecting Quality-Control Problems

Computers Computer software applications for quality control are available for both PCs and laboratory information systems (LISs). QC results can be entered into these computers either manually by the operator or automatically through data transmission ports. Target values and control limits are established for each analyte and QC level. Levey–Jennings plots and QC rules such as Westgard are also available. QC data can be reviewed by the operator to determine acceptability of the result. If Westgard rules are available, the computer will determine whether any test results in a rule violation and alert the operator. The advantages of using computers for QC include (1) real-time review, (2) early detection of QC problems, and (3) documentation of the QC process.

Patient results Monitoring patient results, especially serial results on a single patient, can alert the technologist to quality issues that require attention. Just because the quality-control values are acceptable does not mean that a change in the performance of an assay did not occur soon after the controls were run. The data shown in Table 4-13 ✪ represent serial determinations at 1-hour time intervals of glucose on the same patient over a period time. The results of bilevel control testing in the morning were acceptable. However, beginning at the third glucose measurement, the patient's results increased steadily before returning to normal at the sixth measurement. This pattern of results should cause the technologist to question the results. For example,

1. Does this deviation from normal represent a bias in the assay?

2. Is this shift in glucose results for this patient a result of a treatment or procedure?

3. Do these results correlate with other glucose testing (e.g., portable glucose meter [glucometer])?

This example clearly demonstrates that technologists must use every available resource to verify patient data and evaluate quality-control data.

Steps to Remedy Westgard Rule Violations

If QC results violate any Westgard rules, the laboratory must have a procedure that will resolve these violations. This procedure includes the following: (1) steps to take toward problem resolution, (2) proper documentation of the problem and solution(s), and (3) indications of whom to notify of the rule infractions (e.g.,a supervisor or QC officer). The following example identifies steps that may be taken if a problem occurs while assaying quality-control materials.

- If a control value is outside the acceptable range, then reassay a fresh aliquot of the same control material.

- If the control value is still outside the acceptable range, then reconstitute a new vial of control material and reassay.

- If the control value is still outside the acceptable range, then look for obvious problems such as clots in the specimen, low reagent levels, and instrument-related mechanical faults. This may be a good time to replace the reagents.

- If the control value is still outside the acceptable range, then consider recalibrating the assay.

- If the control value is still outside the acceptable range, then the technologist may be directed to consult the QC officer and/or supervisor for further direction.

✪ TABLE 4-10

Westgard quality control rules and interpretation.
Note: standard deviation (s).

Control Rule	
Designations	Interpretation
1_{2s}	One control observation exceeding the mean ±2s—this rule may be used as a **warning** rule that initiates testing of the control data by the other control rules.
1_{3s}	One control observation exceeding the mean ±3s—recommend rejecting patient results and this rule is sensitive to random error.
2_{2s}	Two consecutive control observations exceeding the same mean plus 2s or mean minus 2s limit—recommend rejecting patient results; this rule is sensitive to systematic error.
R_{4s}	One observation exceeding the mean plus 2s and another exceeding the mean minus 2s—recommend rejecting patient results; this rule is sensitive to random error.
4_{1s}	Four consecutive observations exceeding the mean plus 1s or the mean minus 1s—recommend rejecting patient results; this rule is sensitive to systematic error.
$10\overline{X}$	Ten consecutive control observations falling on one side of the mean (above or below, with no other requirement on size of the deviations)—recommend rejecting patient results; this rule is sensitive to systematic error.

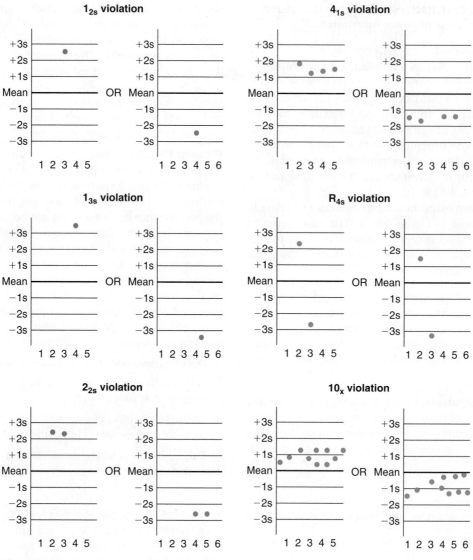

■ FIGURE 4-11   Quality-control rule violations.

Evaluation of quality-control data shown in Figure 4-12 using the Westgard multirule procedure.

Day	Level 1	Level II	1_{2s}	1_{3s}	2_{2s}	R_{4s}	4_{1s}	$10\bar{x}$	RE	SE	Accept	Reject	Warning
2	X			X					X			x	
5	X	X			X					x		x	
8	X		X							x			x
9	X				X					X		x	
12	X	X				X				X		x	
16	X						X			X		x	
23	X							X		X		x	

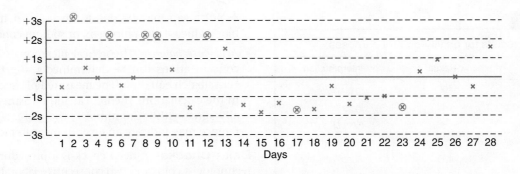

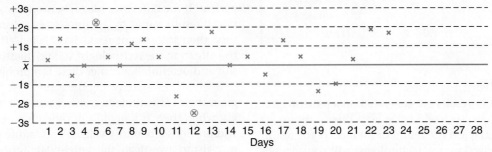

■ FIGURE 4-12   Examples of data plotted Levey–Jennings graphs with the rule interpretations presented in Table 4-11.

⊗ TABLE 4-12

Sources of random and systematic error.

Random Error	Systematic Error
Operator technique	Improper alignment of sample or reagent pipettes
Use of non-reagent-grade water	Unstable incubator chambers
Incorrect reconstitution of control material	Change of reagent lot
Power supply	Change in calibrator lot
Pipetting mistakes	Deterioration of reagents in use
Automated pipette problems	Deterioration of control material while in use
Air bubbles in tubing	Evaporation of sample during analysis
	Dirty filter or gradual delaminating of monochromatic filter
	Change in test operator
	Recent calibration
	Deteriorating light source
	Incorrect handling of control material

✓ Checkpoint! 4–5

The mean and standard deviation for creatinine level I and II quality control pool are shown in the following table:

Creatinine	Level I	Level II
Mean	0.8	6.0
Standard deviation	0.1	0.3

The technologists in the chemistry laboratory recorded the following quality-control values for a 10-day period:

	Level I	Level II
Day 1	0.9	6.1
Day 2	1.15	6.2
Day 3	0.8	5.9
Day 4	0.7	6.7
Day 5	0.7	6.8
Day 6	0.8	5.5
Day 7	0.9	6.0
Day 8	1.3	6.5
Day 9	0.8	6.2
Day 10	0.9	6.1

Using the data presented, do the following:

1. *Determine all of the Westgard rule violations.*
2. *Identify the type of errors associated with each rule violation.*
3. *Determine a course of action for each rule violation.*

Lesser-Used Quality-Control Applications

There are several QC applications available for use by the laboratory staff, and these can be reviewed in the references cited: (1) cumulative sum control chart (cusum), (2) Shewhart mean and range control charts, and (3) moving averages and standard deviations.[56–58] These applications can augment QC protocols currently used in the laboratory.

⊗ TABLE 4-13

Use of patient results for quality-control purposes.

Sample	Result	Interpretation
Control I (normal)	100	In control
Control II (abnormal)	250	In control
1	98	Normal
2	100	Normal
3	135	Abnormally high
4	200	Abnormal high
5	290	Abnormal high
6	100	Normal

Evaluating Patient Results

Whenever laboratory results are generated on a patient specimen, the technologist is faced with at least three options: (1) accept the result, (2) reject the result, or (3) modify the result. Before making a selection, the technologist will consider additional information, as discussed next.

The technologist can determine whether the laboratory result correlates with the patient's condition. This information is available to hospital-based laboratories but usually not to reference or other "outside" laboratories. But if the clinical information is available, it can prove to be invaluable. Examples of clinical correlation of laboratory results are as follows:

- If the patient is known to be jaundiced, then a normal total bilirubin should be questioned.

- If a specimen is labeled as a trough collection and the results generated by the laboratory are in the toxic range, the technologist should question the validity of the result and/or specimen.

- If the total thyroxine level is normal for a patient who is diagnosed as having hyperthyroidism, the technologist should question the laboratory results.

- If the technologist has a positive result for a β-hCG pregnancy test on a male patient, then the results and specimen should be questioned.

Laboratory results generated on a patient specimen may be correlated with other measured and/or calculated laboratory tests. The following examples involve commonly ordered laboratory tests.[59]

- *Anion gap*: Very high (>20 mmol/L) or very low (<10 mmol/L) results may indicate an error in the measurement of one or more of the electrolytes, which include sodium, potassium, chloride, and bicarbonate (TCO_2).

- *Osmolar gap*: Normally the osmolar gap is nearly zero; thus if the patient's result is high, additional investigation into the measured values and the patient's condition may be warranted. Normally, high osmolar gaps are seen in patients who have ingested alcohol; therefore a phone call to the clinician may confirm the patient's condition. An elevated

osmolar gap may be due to an error in measurement of blood glucose, urea nitrogen, or sodium concentration.

- *Acid–base balance*: Theoretical bicarbonate and total CO_2 concentrations can be determined using the Henderson-Hasselbalch equation. The measured pH and pCO_2 are used in the calculation. Theoretical and measured results generally agree within 2.0 mmol/L; therefore, a significant difference may indicate an error in measurement.

Delta Checks A delta check is a procedure in which the technologist compares two consecutive laboratory results on a patient. If the difference (delta) between two consecutive laboratory results varies by an amount established by the laboratory, then the patient result has "failed" the delta check. In this situation the technologist should question the patients result and determine whether there is a problem with the assay or analyzer before releasing the results. Most laboratories program their LIS with delta check limits that alert the technologist when there is a significant difference between the current and previous laboratory values. This failure prompts the technologist to investigate the patient data further.

For example, suppose the laboratory's delta check limit for serum potassium is 20%. The previous potassium level for a patient was 4.5 mmol/L, and the current potassium result is 5.8 mmol/L. This difference exceeds the delta limit of 20% and therefore may represent a significant change is blood potassium levels. The operator then must try and determine what may have caused this change.

Limit Checks Limit checks (often referred to as critical or panic values) represent laboratory results that may represent serious conditions relative to the patient. These results, if valid, must be brought to the attention of the caregiver immediately. Limit checks are useful for detecting clerical errors, such as transposed digits, and specimen integrity problems, such as hemolysis. The following are examples of limit check ranges:

- Potassium, < 3.0 and > 6.0 mmol/L
- Calcium, < 6.5 and > 13.0 mg/dL
- Sodium, < 120 and > 150 mmol/L

Autoverification Autoverification is described as verification and release of patient results using software-based algorithms with decision-making logic via the LIS. Laboratories are adopting autoverification to make the process of accepting or rejecting laboratory results on patient specimens more efficient and to reduce turnaround time. The basic tenet is that if all laboratory results are within a set of parameters or rules established by the laboratory, then the technologist does not have to review the results and physically perform the keyboard strokes to accept the results. Thus the computer determines that the data are acceptable using the preprogrammed algorithms and accepts the data.

Proficiency Testing CLIA '88 mandates that proficiency testing (PT) be conducted in clinical laboratories. Proficiency tests, also referred to as *surveys*, are an example of external quality control wherein an agency or organization provides biological samples whose concentrations are unknown to the testing clinical laboratory. The clinical laboratory submits its data for the unknown sample to the PT testing service. The results are returned to the laboratory with an evaluation of the laboratory's performance. A commonly used criterion for evaluation of PT results for a given clinical laboratory has been comparing PT test results with the results of peer groups and considering all values that exceed 2s to be "unacceptable." PT samples are routinely sent several times a year for analysis, and the overall performances from all the PT samples are used to determine whether a laboratory performance is acceptable or unacceptable. If the laboratory performance is unacceptable, then the laboratory is placed on a probationary status for that category of tests (e.g. electrolytes or blood gases) until the laboratory successfully passes subsequent PT. If the laboratory continues to perform unacceptably, then it will not be allowed to report patient results for that test(s) until the lab is reinstated.

CLIA'88 regulations specify that PT specimens should be treated the same as patient specimens. Therefore, "special handing" of PT specimens is discouraged. Except for the special instructions for preparing PT specimens, they must be included with the laboratory's daily assay regimens.

Calibration The purpose of **calibration** as defined by CLIA '88 is to "substantiate the continued accuracy of the test system throughout the laboratory reportable range of the test results for the test system." [19] Materials used to perform a calibration are called *calibrators*. Calibration material or calibrators can be described as "a material—either solutions or a device of known, or assigned quantitative or qualitative characteristics (e.g., concentration, activity, intensity, reactivity)—used to calibrate, graduate, or adjust a measurement procedure or to compare the response obtained with the response of a test specimen and /or sample." [60] Controls and calibrators are different. Calibrators are prepared differently from controls; therefore, controls must not be used as calibrators. Calibrators should not be used as controls unless there are no suitable control materials for a particular test. CLIA'88 will allow the use of calibrators as controls if that situation occurs. Commercially prepared calibrators have assigned values for each analyte that are determined by a definitive or reference method. The calibrator value assigned for the material is then programmed into the analyzer's computer for use as a comparator in measuring an unknown sample.

▶ LABORATORY SAFETY

Safe laboratory practices are the responsibility of everyone who enters the facility. Knowledge and common sense are key to ensuring that accidents are kept to a minimum. All laboratory staff should develop an appreciation for safety because working safely is as much an attitude as it is a practice.

✪ TABLE 4-14

Organizations actively involved in providing safety standards, guidelines, education, and procedures for clinical laboratories.

AEC	Atomic Energy Commission
CAP	College of American Pathologists
ASCP	American Society for Clinical Pathology
AACC	American Association for Clinical Chemistry
CDC	Centers for Disease Control
EPA	Environmental Protection Agency
OSHA	Occupational and Safety Health Administration
JC	Joint Commission (formerly JCAHO, Joint Commission on Accreditation of Healthcare Organizations)
NRC	National Regulatory Commission
CLIS (NCCLS)	Clinical and Laboratory Standards Institute (formerly National Committee for Clinical Laboratory Standards)
NIOSH	National Institute for Occupational Safety and Health
NFPA	National Fire Protection Agency

There are several organizations actively involved in providing a safe environment for laboratory personnel. Examples of these organizations include state, federal, professional, and accrediting agencies. Specific examples are listed in Table 4-14 ✪.

OCCUPATIONAL SAFETY AND HEALTH ADMINISTRATION

The Occupational Safety and Health Administration (OSHA) is a federal agency within the Department of Labor that was created by Congress through public law 91-596 in 1970 and provides federally mandated policies and procedures to ensure safety in the workplace, including laboratories and health-care facilities. OSHA standards are found in the Codes of Federal Regulations (CFR). The standard most associated with clinical laboratories is identified as Standard 29 CFR. This standard is divided into several sections that are identified using a part number (e.g., 1910. 1450.65) titled "Occupational Exposure to Hazardous Chemicals in the Laboratory." [61] A list of clinically relevant sections and their respective part numbers is shown in Table 4-15 ✪. Detailed information for each part may be found in the Federal Register, Document 29 CFR.

Four major OSHA programs have made a significant impact on the safety practices in clinical laboratories:

• Occupational exposure to hazardous chemicals in the laboratory, 29 CFR 1910.1450.

• *Hazard Communication* (29 CFR 1910.1200 for laboratories), which includes *Right to Know*, OSHA poster 2203, and MSDS. In 1988 OSHA expanded this regulation to include hospital workers.

⊕ TABLE 4-15

Relevant OSHA standards, including their respective parts and titles, for clinical laboratories.

1910.77	Toxic and chemical hazards
1910.94	Environment
191.0.101	Compressed gases
1910.120	Hazardous waste operations and emergency response (HAZMAT teams)
1910.134	Respirators
1910.155	Fire protection
1910.157	Portable fire extinguishers
1910.301	Electrical
1910.1030	Bloodborne pathogens
1910.1048	Formaldehyde
1910.1096	Ionizing radiation
1910.1200	Hazard communications
1910.1450	Occupational exposure to hazardous chemicals in laboratories

BOX 4-4 Specific elements included in the OSHA chemical hygiene plan.

▶ Glossary of terms
▶ A description of standard operating procedures
▶ An inventory of all chemicals
▶ Material safety data sheets (MSDSs)
▶ Proper labeling and storage of chemicals
▶ An inventory of personal protective equipment
▶ A description of engineering controls
▶ Procedures for waste removal and disposal
▶ Requirements for employees' physical and medical consultations
▶ Training requirements
▶ Procedures for proper record keeping
▶ Designation of a chemical hygiene officer and safety committee

- Occupational exposure to bloodborne pathogens. (29 CFR 1910.1030) in 1991. OSHA issued the *Bloodborne Pathogens Standard* to protect workers from this risk.

- In 2001, in response to the Needlestick Safety and Prevention Act, OSHA revised the Occupational exposure to bloodborne pathogens 29 CFR1910.1030 to include needlestick safety.

CHEMICAL SAFETY

Chemical safety awareness is of paramount importance in the clinical laboratory. The staff technologist may handle several different types of chemicals in a working day. Therefore, knowledge of the compounds, their associated hazards, and proper handling is very important in reducing the extent of injury in case of an accidental exposure.

Chemical Hygiene Plan
OSHA mandated that each laboratory establish a chemical hygiene program.[61] This plan is designed to provide laboratory staff with information necessary to handle chemicals. Several specific elements included in the hygiene plan are listed in Box 4-4 ■.

Operating Procedures
Specific operating protocols should be developed for proper disposition of accidents in laboratories involving chemical spills. Accidents that involve chemicals splashed into the eyes require that the eyes be flushed immediately with copious amounts of water from an eyewash fountain. Chemical exposure to other body tissues also requires that the affected body tissue be flushed with water using a faucet or shower. After that a physician should evaluate the victim. Flushing

should last for at least 15 minutes. Chemical spills should be handled appropriately. For small chemical spills there are commercially available chemical spill kits that may be used. If the spill is a large one or if the chemical is very noxious, then the facility's chemical spill response team should be notified. Also, a drench-type safety shower should be available that can deliver several gallons of water in a short period of time.

Policies for avoiding unnecessary chemical exposure must be defined. Activities such as smoking, eating, drinking, and applying cosmetics must be prohibited in all workstations. Wearing of proper footwear must be enforced. No sandals, canvas shoes, or any open-toed foot ware should be allowed. Staff with long hair or loose clothing and jewelry should be reminded that these items must be secured. Contact lenses should not be worn in the laboratory because they prevent proper washing of the eyes in case of a chemical splash. Also, plastic contact lenses may be damaged by organic vapors, which may lead to chronic eye infections. Handwashing after handling chemicals and before leaving the laboratory should be emphasized.

Material safety data sheets (MSDSs) consist of 16 sections, including identification, composition, hazards, first aid, firefighting concerns, and safe handling, that are relevant to the chemical being used. MSDSs must be obtained for all chemicals and must be located in an area that is available to all laboratory personnel at any time. The format of the MSDS has been standardized by the American National Standards Institute (ANSI), which makes the document easier to interpret.[62]

Identifying and Labeling Chemicals
Proper labeling of chemicals is described in OSHA regulation 29 CFR 1910.1450. The actual labeling system used is adopted from the Hazard Identification System developed by the National Fire Protection Agency (NFPA).[63] This identification

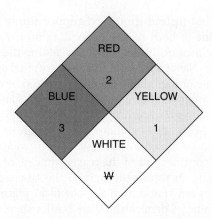

■ FIGURE 4-13 NFPA 704-M. Identification system using diamond shaped symbol.

system, referred to as 704-M Identification System, covers the nine classes of hazardous materials, as follows:

- Explosives
- Compressed gases
- Flammable liquids
- Flammable solids
- Oxidizer materials
- Toxic materials
- Radioactive materials
- Corrosive materials
- Miscellaneous materials not classified by any other means

Materials are identified using four small, diamond-shaped symbols grouped into a large diamond shape, as shown in Figure 4-13 ■. The smaller diamonds are color coded to represent a specific health hazard, as shown in Table 4-16 ✪. The degree of hazard is rated using a scale of 0 to 4, with 4 indicating the most serious risk. Chemical labels on the original containers must not be removed or defaced. For chemicals not in their original container, the labeling information must include at least the following:

- Identification of the hazardous chemical
- Route of body entry
- Health hazard
- Physical hazard
- Target organ(s) affected

Storing and Inventorying Chemicals

A chemical inventory should be conducted on an annual basis. An accurate listing of all chemicals and their hazards should be available at all times. During the inventory process the technologist should pay particular attention to the following:

- Keep only the amount of chemicals that is needed. Less is better.

✪ TABLE 4-16

NFPA 704-M identification system of warning labels for chemical hazards.

Color Codes	Interpretations	
Blue—Health Hazard	4—deadly	
	3—extreme danger	
	2—hazardous	
	1—slightly hazardous	
	0—normal material	
Red—Fire Hazard		
Flash points	4—below 73°F	
	3—below 100°F	
	2—below 200°F	
	1—above 200°F	
	0—will not burn	
Yellow—Reactivity	4—may detonate	
	3—shock and heat may detonate	
	2—violent chemical change	
	1—unstable if heated	
	0—stable	
White—Special Hazards	oxidizer	ox
	Corrosive	Cor
	Acid	ACID
	Alkali	ALK
	Use no water	☇
	Radiation	☢

- If possible, purchase chemicals in plastics containers to avoid breaking glass.
- Rotate your chemical inventory and note expiration dates.
- Dispose of chemicals if not used within a year—especially peroxide-forming compounds.
- Make sure all secondary containers are properly labeled.
- If possible, relocate corrosive, flammable, and reactive chemicals to below eye level.

Proper storage of hazardous materials in the laboratory is very important. Several examples of poor storage practices are shown in Box 4-5 ■. Examples of problems with storing chemicals alphabetically are shown in Table 4-17 ✪.[67]

Categorical storage of chemicals is a safe approach for separating potentially reactive chemicals. Each chemical must be stored in its respective hazardous category. For example, inorganic acids such as hydrochloric acid (HCL) and sulfuric acid (H_2SO_4) should be stored as a group and should not be stored with caustic bases, such as sodium hydroxide (NaOH) and ammonium hydroxide (NH_4OH). Chemicals classified as oxidizing compounds, such as potassium dichromate and silver nitrate, can be stored together, but away from other compounds. Chemical compatibility charts are available that outline general classes of incompatible chemicals.[64]

BOX 4-5   Examples of poor chemical storage practices.

► Chemicals stored in random order
► Chemicals stored in alphabetical order
► Chemicals stored by poorly chosen categories
► Chemicals stored in a hood while the hood is in use for other purposes
► Flammables stored in domestic or household-type refrigerators
► Food stored beside chemicals in refrigerator
► Chemicals stored on shelves above eye level
► One bottle of a chemical is sitting on the top of a second bottle of a chemical
► Overcrowded shelves
► Shelving on which chemicals are stored is not strong enough to support chemicals
► Shelves not securely fastened to a permanent structure
► Inventory control is poor or nonexistent
► Containers with no labels or inappropriate labels
► Containers stored on the floor
► Caps on containers are missing, not on properly, or badly deteriorated

Flammable chemicals must be stored in an appropriate safety cabinet (see section safety cabinets below) and should not be mixed in with other classes of compounds. Volatile or flammable liquids should never be stored in refrigerators that are not designed to accommodate hazards the solvents may create. If the laboratory has corrosive or flammable chemicals that require refrigeration, there are several manufacturers that distribute explosion-proof and corrosion-proof refrigerators. These units are designed to comply with NFPA and OSHA specifications.

Transferring chemicals from one container to another can result in an accident or overexposure. Therefore, care must be taken to avoid these situations. A bottle should never be held

✪ TABLE 4-17

Problems associated with storing chemicals alphabetically.

Chemical Combinations	Problems
Acetic acid + acetaldehyde	Small amounts of acetic acid will cause the acetaldehyde to polymerize, releasing a large amount of heat.
Ammonium nitrate + acetic acid	Mixture will ignite especially if acids is concentrated.
Hydrogen peroxide + ferrous sulfide	Vigorous reaction, highly exothermic.
Lead perchlorate + methanol	Explosive mixture if agitated.
Potassium cyanide + potassium nitrite	Potentially explosive mixture if heated.

by its neck, but instead should be gripped firmly around its body with one or both hands, depending on the size of the bottle. Acids must be diluted slowly by adding them to water while mixing. Never add water to a concentrated acid. Transferring volatile or flammable chemicals should be carried out in an OSHA approved fume hood. Also, safety glasses should be worn whenever transferring chemicals from one container to another. Acids, caustic materials, and strong oxidizing agents should be mixed in the sink. This provides water for cooling and confinement of the reagent in case of a spill.

Transporting chemicals throughout the laboratory can be hazardous. For safe transport of any chemical, place it in a secondary container. Chemicals in glass bottles should be transported in rubber or plastic containers that protect them from breakage and, in the event of breakage, help to contain the spill. To move heavy containers or multiple numbers of containers from one area of the laboratory to another, use a cart.

Potentially Explosive Compounds

Laboratory staff must be aware of any chemicals that have the potential to explode or detonate. An *explosive* chemical is a chemical that causes a sudden release of pressure, gas, or heat when subjected to shock, pressure, or high temperatures. Dried picric acid may explode if its container is dropped. Lead azide, which is often found in drainpipes, is potentially explosive. For example, concentrated perchloric acid (90% assayed weight) is potentially explosive. Information regarding the explosive nature of chemicals may be found in their accompanying MSDS.

Cryogenic Material

Cryogenic liquids (cryogens) are liquefied gases that are kept in their liquid state at very low temperatures. The word *cryogenic* means "producing, or related to, low temperatures," and all cryogenic liquids are extremely cold. Cryogenic liquids have boiling points below –150°C (–238°F). Carbon dioxide and nitrous oxide, which have slightly higher boiling points, are sometimes included in this category. All cryogenic liquids are gases at normal temperatures and pressures. These gases must be cooled below room temperature before an increase in pressure can liquefy them. Various cryogens become liquids under different conditions of temperature and pressure, but all have two properties in common: (1) They are extremely cold, and (2) small amounts of liquid can expand into very large volumes of gas.

The vapors and gases released from cryogenic liquids also remain very cold. They often condense the moisture in air, creating a highly visible fog. In poorly insulated containers, some cryogenic liquids actually condense the surrounding air, forming a liquid-air mixture. Cryogenic liquids are classified as "compressed gases" according to OSHA Standard Industrial Class 2813. Everyone who works with cryogenic liquids must be aware of their hazards and know how to handle them safely.

Cryogenic liquids such as liquid nitrogen, helium, and oxygen are, by definition, extremely cold. Contact of exposed skin to cryogenic liquids can produce a painful burn. A splash of cryogenic liquid in the eye can cause loss of vision. Whenever handling cryogenic liquids, always wear proper personal protective equipment, including a lab coat that is buttoned, heavy gloves, and a face shield or safety goggles. Hazards associated with cryogens include the following:

- Fire or explosion
- Asphyxiation
- Pressure buildup
- Embrittlement of material
- Tissue damage

Formaldehyde

Formaldehyde (formalin) is a colorless liquid with a characteristic pungent odor and is routinely used in laboratories for tissue processing. The primary deleterious health effects of formaldehyde exposure are respiratory cancer, dermatitis, sensory irritation (eye, nose, and throat), and sensitization. In addition to inhalation hazards, solutions of formaldehyde can damage skin and eye tissue immediately on contact. OSHA and specific state guidelines have been established to assist laboratories in handling and disposing of this chemical.

Safety Cabinets

Limit the amount of flammable materials that you need to store. OSHA defines the maximum amount of flammable and combustible liquids that can be stored in any laboratory with NFPA-approved flammable storage cabinets. These limits may be modified by local fire departments. The regulations are defined by the solvents'classification. The classification of flammable solvents is defined in terms of flash points, with Class IA and IB solvents being the most combustible. Large amounts of volatile solvents must be stored in a safety cabinet approved by NFPA and meet OSHA standards. The cabinet should be properly vented, and the doors should be self-closing.

Chemical Waste

Disposal of chemical waste is the responsibility of individual laboratories. The responsibility and liability start when the original chemical is brought into or made in the laboratory and continues until the waste has been completely destroyed. The Resource Conservation and Recovery Act (RCRA) dictates this "cradle-to-grave" responsibility and liability. Congress passed the RCRA in 1976. RCRA established a system for managing nonhazardous and hazardous solid wastes in an environmentally sound manner. Specifically, it provides for the management of hazardous wastes from the point of origin to the point of final disposal. RCRA also promotes resource recovery and waste minimization.[65] In addition, each laboratory must comply with all local, state Department of Environmental Conservation (DEC), and U.S. Environmental Protection Agency (EPA) mandates.[66]

Laboratories are identified by RCRA as "waste generators" and thereby require a permit for proper waste disposal.

Laboratories are considered small-quantity waste generators and are therefore governed by the policies and procedure appropriate to the corresponding permit. Pouring chemicals down the sink drain should not be done without authorization from the laboratory's environmental health and safety officer. There are several chemicals that must not be disposed of by pouring down sink drains:

- Organic solvents with a boiling point of less than 50°C
- Hydrocarbons
- Halogenated hydrocarbons
- Nitro compounds
- Mercaptan
- Freon
- Azides and peroxides
- Concentrated acids and bases

BIOHAZARDS: UNIVERSAL PRECAUTIONS

Exposure Control Plan

OSHA requires all laboratories to develop and implement an exposure control plan.[67] This plan is designed to help prevent accidental exposure of laboratory personnel to bloodborne pathogens. **Bloodborne pathogens** are pathogenic microorganisms that are present in human blood and can cause disease in humans. These pathogens include, but are not limited to, hepatitis B virus (HBV) and human immunodeficiency virus (HIV).The exposure control plan also includes procedures for the proper handling and disposal of all medical waste produced by the laboratory. The exposure control plan should include sections on the following: (1) purpose, (2) scope, (3) references, (4) definition of terms, (5) delineation of responsibilities, and (6) detailed procedural steps.

OSHA guidelines for this plan require that the employer place each employee into one of three groups defined as follows:

Group I: A job classification in which all employees face occupational exposure to blood or other potentially infectious materials.

Group II: A job classification in which some employees face occupational exposure to blood or other potentially infectious materials. The job description for this group represents duties that put the employee at risk for possible exposure to human blood or other potentially infectious materials.

Group III: A job classifications in which employees will not face any occupational exposure to blood or other potentially infectious materials.

Employees in groups I and II require OSHA bloodborne pathogen training at the outset of their employment and annually thereafter. Also, these employees must be offered vaccination against hepatitis B within 10 working days of initial assignment to a position involving possible exposure.

Group III employees do not require OSHA training. However, if a group III employee assumes a job assignment that will result in possible exposure to bloodborne pathogens, then he or she must be offered the hepatitis vaccination and is required to attend an OSHA training session before beginning job duties.

Biological Hazards

Laboratory personnel must be aware of potential sources of exposure to infectious agents such as HBV and HIV. Examples of exposure sources include the following:

- Centrifuge accidents
- Needle punctures
- Spilling infectious material on bench surfaces
- Cuts and scratches from contaminated glassware
- Removing stoppers from blood drawing tubes

Universal precautions is a practice described by the National Institute for Occupational Safety and Health (NIOSH) in which every clinical laboratory should treat all human blood and other potentially infectious materials as if they were known to contain infectious agents such as HBV, HIV, and other bloodborne pathogens.[68] Examples of infectious materials include blood, serum, plasma, blood products, vaginal secretions, semen, cerebrospinal fluid, and synovial fluid. Hands and other skin surfaces should be washed immediately and thoroughly if contaminated with blood or other body fluids. Hands should be washed immediately after gloves are removed. Although saliva has not been implicated in HIV transmission, to minimize the need for emergency mouth-to-mouth resuscitation, mouthpieces, resuscitation bags, and other ventilation devices should be available for use in areas in which the need for resuscitation is predictable. Health-care workers who have exudative lesions or weeping dermatitis should refrain from all direct patient care and from handling patient-care equipment until the condition resolves.

Pregnant health-care workers are not known to be at greater risk of contracting HIV infection than health-care workers who are not pregnant; however, if a health-care worker develops HIV infection during pregnancy, the infant is at risk of infection resulting from perinatal transmission. Because of this risk, pregnant health-care workers should be especially familiar with and strictly adhere to precautions to minimize the risk of HIV transmission.[69]

Needlestick Regulations

OSHA updated 29 CFR 1910.1030 to include the Needlestick Safety and Preventions Act,[70,71] which requires that needles used for withdrawing blood must have a "built-in safety feature or mechanism that effectively reduces the risk of an exposure incident." All laboratories must provide the appropriate devices and monitor their use as part of a quality-assurance program.

The Centers for Disease Control (CDC) reported that implementing the following measures might prevent needlestick injuries:[72]

- Use safe and effective alternatives to needles.
- Participate in the selection and evaluation of needle safety devices.
- Use only devices equipped with safety mechanisms.
- Do not recap needles.
- Implement a plan that will ensure safe handling and disposal of needles.
- Immediately dispose of used needles in the proper sharps disposal containers.
- Communicate needlestick hazards to your employers.
- Participate in infection-control training.

Personal Protective Equipment

Personal protective equipment (PPE) is specialized clothing or equipment worn by an employee for protection against a hazard. General work clothes (e.g., uniforms, pants, shirts or blouses) are not intended to function as protection against a hazard and are not considered to be personal protective equipment.

Gloves The routine use of gloves to protect laboratory staff against exposure to bloodborne pathogens was mandated by OSHA in 1991.[67] As laboratory staff began wearing gloves on a daily basis for prolonged periods of time, reports of skin sensitivities traceable to the plastic gloves increased. Skin sensitivities may be due to either protein from the rubber tree or to the chemicals used in the production of latex. NIOSH has recommended that employers provide gloves with reduced protein content (e.g., latex-free and powder- or cornstarch-free gloves). Also, every employee should be provided with continuing education and training materials about latex allergies, and high-risk employees should be periodically screened for allergy symptoms.

Gloves are also available to protect employees from exposure to chemicals (e.g., acids, bases, and solvents). Glove materials highly rated for handling acids and bases include nitrile, neoprene, and natural rubber. Nitrile- and neoprene-based gloves are also very good for handling organic solvents. For guidance on proper selection of gloves, refer to the MAPA Spontex, Inc. (Columbia, TN) chemical resistance guide at the following Web site: www.spontexusa.com.

Eyewear Laboratory staff often do not properly protect their eyes while handling hazardous materials. Workers who wear glasses often feel "safe" and do not believe that additional eyewear is necessary. Conventional or prescription eyeglasses are not impervious to chemicals, and therefore proper eyewear is recommended. Proper eyewear is described as glasses or goggles that provide protection to the sides of the face as well as the front. The lenses must also be impervious to chemicals. Another acceptable face-and-eye protection device is the face shield.

Respirators Respirators are required for selected laboratory procedures. Most notable is the handling of pathogenic microorganisms such as strains of *Mycobacterium tuberculosis*.[73]

Respirators should contain high efficiency particulate air (HEPA) filters if no other engineering controls are available. Handling of certain chemicals may require the use of a respirator. The MSDS accompanying the chemical should indicate whether a respirator is required. Filtration respirators will require the use of one or two cartridges designed to filter out hazardous chemical vapors. Strict adherence to all safety recommendations is necessary when handling hazardous materials requiring the use of any type of respirator.

Laboratory Coats and Footwear Employers must provide proper laboratory coats to each employee. Characteristics of proper laboratory coats include the following:

- They must have cuffed sleeves and be full length.
- They should remain buttoned to avoid contaminating street clothes.
- The material should be resistant to liquids.
- The construction of lab coats resistant to liquids includes layers of polypropylene, spun bonded filaments, and melt-blown polypropylene microfibers.

A spun bonded filaments refers to process of producing filaments or webs from polypropylene and other organic substances that will possess positive features such as restricting fluids from passing through. Meltblown polypropylene is a thermally bonded ultrafine fiber that has been self-bonded and forms a three-dimennsional random microporous structure. This material is impervious to the passage of liquids through the fibers.

Laboratory staff should wear footwear that is comfortable and safe. The shoe material should be nonporous. No open-toed shoes, sandals, or flip-flops should be worn in the laboratory.

Engineered Controls An **engineered control** is defined as safety equipment that isolates or removes the blood-borne pathogen hazard from the workplace and represents the preferred method for controlling hazards. Several examples of engineered controls are listed in Box 4-6 ■.

Glass versus Plastic Exposure to broken glass in the laboratory poses a risk similar to that posed by a needlesick. Therefore, clinical laboratories should use products that reduce the risk of exposure to broken glass. Many glass products, including specimen collection tubes, capillary tubes, pipettes,

---

BOX 4-6 Examples of safety-engineered controls.

- ► Splash shields
- ► Biosafety cabinets
- ► Secondary containers
- ► Impervious needle boxes
- ► Automatic pipettes
- ► Centrifuge caps
- ► Self-sheathing needles

⊗ TABLE 4-18

Biosafety levels 1–4 with associated agents.

BSL	Agents
BSL 1	Not known to consistently cause disease in healthy adults
BSL 2	Associated with human disease, hazard by percutaneous injury, ingestion, mucous membrane exposure
BSL 3	Indigenous or exotic agents with potential for aerosol transmission; disease may have serious or lethal consequences
BSL 4	Dangerous/exotic agents, which pose high risk of life-threatening disease, aerosol-transmitted lab infections; or related agents with unknown risk of transmission

and slides, can be replaced by their plastic counterparts. Benefits of converting to plasticware are as follows: (1) reduced cost is associated with biohazard waste disposal because plastic is lighter than glass; and (2) plastics can be incinerated, thus reducing waste "downstream." Laboratory staff must remember that when switching to plasticware, especially blood collection tubes, they must review the manufacturer's data and verify method performance characteristics (e.g., accuracy, precision and interference) related to the plasticware.

Biosafety Levels
The CDC and NIH have developed criteria for handling infectious materials. These criteria are classified into four biosafety levels: biosafety level 1 (BSL1) through BSL4.[74] Table 4-18 ⊗ provides a portion of the information contained in the reference cited. Knowledge of BSLs is required when questions arise regarding proper use and functional parameters of laboratory hoods.

Toxic Substances
Toxic substances pose a potentially significant and long-lasting risk to all laboratory personnel. Chemicals in this group have the potential to affect the offspring of female workers. The term **xenobiotic** refers to a substance that is foreign to a living organism and is usually harmful. The routes of entry for xenobiotic substances include the following:

- Pulmonary
- Oral
- Transcutaneous absorption
- Percutaneous (ocular contact)

Once a xenobiotic enters the body, it circulates until a suitable cellular receptor is located. Examples of receptors include the following:

- Membranes
- Antibodies
- Circulating carrier proteins
- Water-soluble intracellular proteins
- Intracellular cytoskeletal proteins

A xenobiotic substance that results in the production of a cancer-producing tumor is termed a *carcinogen*. Examples of compounds currently classified as carcinogens that may be found in laboratories are as follows:

- Asbestos
- Arsenic
- Benzene
- Benzidine
- Nickel

Teratogens

A **teratogen** is a substance that acts preferentially on an embryo at precise stages of its development, thereby leading to possible anomalies and malformations. *Organogenesis* is a term that describes the 15th to 60th days of fetal development, in which the embryo is most sensitive to the action of a teratogenic substance. Examples of teratogens include thalidomide, polyaromatic hydrocarbons, dimethymercury, diethylstilbestrol (DES), and vinyl chloride.

Transplacental Carcinogenesis Transplacental carcinogenesis refers to a condition in which a carcinogenic substance such as DES induces cancer in the descendents of females who have had contact with this carcinogen. The presence of DES in the mother may lead to genital cancers in her daughters.

FIRE SAFETY

OSHA and the NFPA have combined their resources to provide policies, procedures, and standards for fire safety. State and local government agencies have adopted these standards for laboratories and businesses and are authorized to fine laboratories for noncompliance.

Laboratories should have available the means to extinguish small fire in a room, confine fire, and extinguish clothing that has caught on fire. Common sense, personal safety, and the safety of others must always be considered before attempting to extinguish a fire. Examples of fire safety devices that are used for extinguishment in the laboratory include the following: (1) fire blankets, which can be used to smother fires and are especially effective for a situation when someone's clothes catch fire; (2) safety showers, which are designed to provide nearly 50 gallons of water that can be used for extinguishment; and (3) fire extinguishers.

Several types of extinguishers are available for fire suppression. Laboratories should have an extinguisher available near each door and on opposite sides of the laboratory. Laboratory staff should know the location of these extinguishers and be trained in the proper operation of each type of extinguisher available in the laboratory.

Fire extinguishers are designed to suppress and extinguish different classes of fires. The fire suppression agents within the extinguishers characterize the various classes of extinguishers. The classes of fires are designated by capital letters A, B, C, and D. The classes of fires, recommended extinguishing

⊗ TABLE 4-19

List of classes of fires, associated hazards, and proper types of extinguisher to use.

Fire Class	Hazard	Extinguisher Type
A	Ordinary combustible material (e.g., paper, wood, cloth and some rubber and plastic material)	Water or dry chemical
B	Flammable combustible liquids, gasses, grease, and similar materials	Dry chemical or carbon dioxide
C	Energized electrical equipment	Carbon dioxide, dry chemical or Halotron (Halon)
D	Combustible metals such as magnesium, titanium, sodium, and potassium	Specific for the type of metal in question

agents, and types of hazards suitable for each type of extinguisher are shown in Table 4-19 ⊗.

The appropriate extinguishing agent to use for computer fires is Halon 1301 (bromotrifluoromethane) or Halon 1211 (bromochlorodifluoromethane), which is being replaced with Halotron (American Pacific Corporation Halotron Division, Las Vegas, Nevada USA 89109).[75]

Sprinkler systems are located in most buildings according to fire code regulations and OSHA mandates.[76] NFPA 13 discusses the policies and procedures relevant to sprinkler systems. Another important safety issue involves the storage of materials near sprinkler heads. NFPA guidelines state that "the minimum vertical clearance between sprinkler and material below shall be 18 inches."[76]

Two commonly used acronyms relevant to fire-related emergencies are defined in Box 4-7 ■. Individuals who are in a situation to initiate RACE should exercise extreme caution and common sense before commencing this procedure.

BOX 4-7   RACE and PASS for fire-related emergencies.

In case of fire:
RACE

- Rescue
- Activate alarm call (phone number)
- Contain fire
- Evacuate area

How to use a fire extinguisher properly:
PASS

- Pull the pin
- Aim extinguisher
- Squeeze handle
- Spray back and forth

Every hospital, outpatient clinic, and other medical-related facility has instituted procedures for alerting occupants of a fire-related emergency. Some facilities use a public announcement system of codes (e.g., *code Elmer* or *code red*) to alert staff that there is a fire emergency. Automatic alarm systems use flashing lights and loud, high-pitched sounds to warn staff that there may be a fire emergency. Larger institutions may have a fire brigade of employees who will respond immediately to fire-related emergencies. Whatever the alarm method used, every laboratory staff member should know what to do when the alarm sounds.

COMPRESSED GASES

The Department of Transportation (DOT) regulates the labeling of gas cylinders that are transported by interstate carriers. NFPA diamond-shaped labels, discussed previously, are used on all large cylinders and boxes containing small cylinders. The OSHA standards regarding compressed gases are based on information from the Compressed Gas Association, Inc.[77]

Gases may be

- Flammable or combustible
- Corrosive
- Explosive
- Poisonous
- Inert
- A combination of the aforementioned hazards

There are two important aspects of compressed gases of which laboratory staff must be aware. First, staff should know the dangers, if any, associated with the gas(es) they are using. Second, staff must know the proper procedure for securing gas cylinders (see last paragraph below).

The contents of the cylinder must be clearly identified. Such identification must be stamped or stenciled onto the container or label. Never rely on the color of a cylinder for identification. Color coding is not reliable because the same color may be used for different types of gases. Always read the label.

Signs should be posted conspicuously in the area where flammable gases are used. These signs must identify the gas and appropriate precautions to take if a problem should arise. For example, if a laboratory has a tank of hydrogen gas, a sign should be posted that indicates that it is a flammable gas and no smoking or open flames are allowed.

Because gases are contained in highly pressurized metal containers, the large amount of potential energy resulting from compression of the gas makes the cylinder a potential rocket or fragmentation bomb. Cylinders must be kept in an upright position and secured to the wall using a chain, which will prevent the cylinder from tipping. Cylinders may also be attached to bench tops, placed in a holding cage, or have a nontip base attached. If a gas cylinder does fall, the immediate concern is that the on–off valve or regulator located on top of the cylinder may break off. If the value breaks off, the

gas may be released at a very high velocity, causing the cylinder to propel like a rocket.

ELECTRICAL SAFETY

Working with electricity can be dangerous. Laboratory technologists work with electricity directly (e.g., when troubleshooting electrical components) and indirectly (e.g., by operating analyzers). Technologists are often exposed to electrical hazards.

Handling "live" or charged electrical devices requires users to be completely focused on the task and aware of possible hazards. Several examples of what to do and what not to do when working with electrical equipment are listed in Box 4-8 ■.

RADIATION SAFETY

Clinical laboratory personnel may be exposed to two types of radiation sources: nonionizing and ionizing. Both types can pose a considerable health risk to exposed workers if not properly controlled. **Nonionizing radiation** sources emit electromagnetic radiation ranging from extremely low frequency (ELF) to ultraviolet (UV). *Ionizing* radiation sources emit particulate (alpha, beta, neutrons) and electromagnetic (X-rays, gamma rays) radiation.

Sources of nonionizing radiations commonly found in clinical laboratories include the following:

- Microwaves
- Infrared lamps and lasers

BOX 4-8 Electrical safety recommendations.

- ► Use properly grounded electrical circuits and devices.
- ► Electrical wires and cords should have no frayed edges, cuts, or exposed wiring.
- ► Use of extension cords should be discouraged.
- ► Multiplug adapter and "cheater" plugs (two-prong to three-prong adapters) should not be used.
- ► Use only Underwriter Laboratories (UL)–approved, fuse-protected, multiple-socket surge protectors.
- ► Use ground fault interrupters (GFIs) in areas where there is a source of water (e.g. sink faucets and water filtration systems).
- ► Keep heat sources and liquids away from outlets.
- ► Do not handle any electrical devices with damp or wet hands.
- ► Train employees in how to handle shock injuries.
- ► Electrical panels should be clear of obstacles for easy access (allow at least 3 feet from the panel).
- ► Never work on exposed electrical devices alone.
- ► All shocks should be reported immediately, including small "tingles."
- ► Do not work on or attempt to repair any instrument while the power is still on. Unplug the equipment.

- Visible lamps (tungsten-halogen)
- Ultraviolet lamps (xenon and deuterium)

Protective measures usually include some type of shielding and / or containment devices that will protect the user from direct exposure. The tissues of the eyes are the most vulnerable to radiation exposure; therefore, appropriate eyewear is recommended.[78]

Laser light is used in several commercial brands of office copiers and pointers used for making presentations. Lasers typically emit optical (UV, visible, infrared [IR]) radiations and are primarily an eye and skin hazard. Common lasers include CO_2 IR laser, helium-neon, neodymium, yttrium aluminum garnet (YAG), ruby visible lasers, and the nitrogen UV laser. Special precautions must be followed when maintaining these light sources, and it is critical that the user avoid direct exposure to laser beams. These precautions are provided by the manufacturer of the equipment and are outlined in OSHA standard 1910.1096.

WORKPLACE SAFETY ISSUES

Ergonomics
Ergonomics is described as the study of problems related to people adjusting to their environment. Employers and employees must work together to adapt working conditions to suit the worker. Laboratory work involves a myriad of tasks that are challenging to the mind and body. Over time, serious complications can develop from performing these tasks and may render the staff member nonfunctional and perhaps cause the worker to leave the work force. Every effort must be made to seek better ways to perform daily laboratory work and thus ensure a long and healthy work life for the employee.

Cumulative Trauma Disorders
One way to identify tasks that may place employees at risk of developing problems is to perform a cumulative trauma disorders (CTD) risk factor survey. **Cumulative trauma disorders**, also called "cumulative trauma syndrome," "repetitive motion disorder," and "overuse syndrome," refers to disorders associated with overloading particular muscle groups from repeated use or maintaining a constrained posture. The disorders typically develop over weeks, months, or years and may involve tendons, as in de Quervain's disease; nerve entrapment, as in carpal tunnel syndrome; muscles, as in tension neck syndrome; or blood vessels, as in vibration white-finger disease or occupational Raynaud's phenomenon.[79,80] Risk factors that are associated with CTD are shown in Box 4-9 ■.

Noise and Hearing Conservation
Clinical laboratories—especially highly automated, heavily populated facilities—may generate a substantial amount of low-level noise due to the high amount of activity and communication. Low-level noise or "white noise" can result in fatigue, irritation, and headaches. OSHA standard 1910.95,

---

> **BOX 4-9 List of cumulative trauma disorder (CTD) risk factors**
>
> ▶ Repetitive and prolonged activity
> ▶ Forceful exertions, usually with the hands
> ▶ Prolonged static postures
> ▶ Awkward postures of the upper body (e.g., reaching above the shoulders or behind the back)
> ▶ Continued physical contact with work surfaces, such as contact with edges
> ▶ Excessive vibrating from equipment
> ▶ Working in cold temperatures
> ▶ Inappropriate or inadequate hand equipment
> ▶ Poor body mechanics (e.g., continued bending at the waist, continued lifting from below the knuckles or above the shoulder, or twisting at the waist)
> ▶ Lifting or moving objects of excessive weight or asymmetric sizes
> ▶ Prolonged sitting, especially with poor posture
> ▶ Lack of adjustable chairs, footrest, body supports, and work surfaces at workstation
> ▶ Slippery footing
> ▶ Noise (low-level noise > 85 decibels/8 hours can result in fatigue, irritability, and headaches)
> ▶ Lifting
> ▶ Repetitive stress issues
> ▶ Standing for long periods of time

Subpart App G, allows a *time weighted average* (TWA) of >85 decibels/8 hours.[81] This revised amendment requires that employees be placed in a hearing conservation program if they are exposed to noise levels that exceed this TWA. A hearing conservation program is designed to provide an employee with proper equipment to reduce exposure to high levels of noise. Examples of equipment used are hearing plugs or muffs that are fitted to each employee.

Laboratory Hoods (Ventilation)/Biosafety Cabinets
Laboratory hoods, also called fume hoods, are used to ventilate unwanted fumes from chemical reagents. According to OSHA guidelines outlined in 1910.1450 App A, a laboratory hood with 2.5 linear feet of hood space per person should be provided for every two workers if they spend most of their time working with chemicals.[82] Each hood should have a continuous monitoring device to allow convenient confirmation of adequate hood performance.

More sophisticated biohazard hoods or biological safety cabinets that can remove particulates are available depending on the needs of the laboratory. An abbreviated listing of specifications for Class I–III biosafety cabinets is provided in Table 4-20 ✪. Classes I and II biosafety cabinets are appropriate for BSL II and III, whereas class III cabinets must be used for BSL III and IV.[74]

TABLE 4-20

Three classes of biosafety cabinets and associated descriptors.

Type	Description
Class I	Face velocity, 75 feet per minute
	Open front, exhaust only through HEPA filter
Class II	Face velocity, 100 feet per minute
	Recirculation through HEPA filters and exhaust via HEPA filter
Class III	Supply air inlets and exhaust through two HEPA filters

Note: HEPA = high-efficiency particulate air

All ventilation devices should be checked periodically for proper flow rates, and HEPA filters should be changed according to the manufacturer's recommendations. Proper procedures and documentation of all maintenance should be described in the laboratory's chemical hygiene plan.

Waste Management Issues

The three types of waste generated in most laboratories include chemical, radioactive, and biohazardous. OSHA, CDC, and the Nuclear Regulatory Commission (NRC) provide specific guidelines on proper disposal of all types of waste materials. In addition, each state defines specific procedures to follow that are appropriate for the state's resources (e.g.,some states may have a radioactive dump site or large landfill sites whereas other states may not have these). A few issues for laboratories to consider if they are creating waste materials are as follows:

- Create a recycling program.
- Dispose of the waste according to guidelines provided by federal, state, and local authorities.
- Create a waste management plan.
- Instruct employees on the types of liquids that can be discharged into sewers.
- Obtain a discharge permit for the laboratory.
- Monitor sewer discharge.
- Properly label waste containers.
- Separate incompatible waste and/or materials.

▶ INFORMATICS

Computers, specifically personal computers (PCs), can be used as stand-alone units to provide the laboratory with a means of compiling laboratory data for quality control and statistics compilation. Computers are also incorporated into most laboratory analyzers and provide a variety of functions to assist the technologist. For example, instrument computers are used for calibration, quality control, diagnostics, maintenance, and inventory control. Newer analyzers include tutorials for training, operating, and troubleshooting. These tutorials may include video and sound, which enhance the learning experience. More sophisticated configurations and uses of computers incorporate laboratory information systems (see the following discussion) and hospital information systems with other hospital units (e.g., pharmacy, radiology, business offices, intensive care units, and emergency departments).

LABORATORY INFORMATION SYSTEMS

A **laboratory information system (LIS)** can be described as a group of microprocessors and computers connected together to provide management and processing of information. The functions of an LIS span all three phases of laboratory testing. LISs have served to reduce laboratory errors, reduce turnaround times, and improve the overall efficiency of clinical laboratories. LIS functions associated with the three stages of testing are shown in Box 4-10 ■.

There are several features and functions associated with LISs that deserve additional discussion and clarification. They will be presented in the following paragraphs.

Patient demographics refers to the patient-specific information that must be entered into the LIS. This information includes but is not limited to the following:

- Patient's name
- Sex

BOX 4-10 Laboratory information system functions that support the three stages of laboratory testing.

Preanalytical
- Test ordering
- Phlebotomy draw lists
- Specimen accessioning
- Specimen tracking
- Blood drawing labels

Analytical
- Instrument worklist
- Manual result entry
- Automated or on line entry mode
- Patient delta checks
- Quality control
- Result calculation
- Result verification
- Instrument interfaces

Postanalytical
- Patient cumulative reports
- Result inquiries
- Historical patient archiving
- Workload recording
- Billing
- Result correction
- Data interfaces to other computer systems

- Age/date of birth (DOB)
- Patient number (assigned by health-care facility)
- Referring physician
- Admitting diagnosis

Once the patient demographics are entered into the system, laboratory staff must enter all the test-request information. This process is referred to as *order entry*. Examples of information required at order entry include the following:

- Patient identifier
- Ordering physician
- Test-request time and date
- Test name
- Test priority (e.g., stat, routine)
- Special instructions pertaining to the request
- Specimen draw time

Bar-code labels are created via the LIS and used as identifiers in both automated and manual laboratory procedures. Bar codes consist of a series of lines or squares of differing widths that represent numbers and/or letters. These bar codes are "read" or deciphered by a device called a *bar-code reader*. All of the information represented by a given bar code is transmitted to a computer for further processing. Bar-code identification is used for patient specimens, reagents, calibration parameters, quality-control purposes, and inventory procedures. A significant advantage of using bar-code labels is the reduction in errors associated with transmitting information. Bar-code labeling also speeds up the process of entering information into a computer.

Connecting computers to other computers or instruments is called *interfacing*. Interfacing computers requires two items to be in place. First, there must be a physical hook-up, usually via an RS-232 port, between each computer. Newer, more sophisticated instruments may use an Ethernet connector over a local area network (LAN) to complete an interface. Second, the LIS must have an appropriate interface software program to allow it to receive and transmit data with the instrumentation in the laboratory. ASTM and CLSI provide standards for proper interfacing.[82,83]

LIS interface with laboratory instrumentation requires a means of transmitting information between and among computers. There are two modes of transmission associated with an LIS–instrument interface: (1) a unidirectional interface and (2) a bidirectional interface. A unidirectional interface only transmits or uploads results to the LIS computer. A **bidirectional interface** allows the LIS to simultaneously transmit or download information to and receive uploaded information from the instrument. Specific examples of information transmitted between LIS and instruments are specimen number, tests to be performed, and test results.

LISs are now capable of interfacing with point-of-care testing (POCT) devices within the facility or offsite (e.g.,in doctors'

offices). Uploading information from the POC device to the LIS has circumvented problems (e.g., the amount of time required to document testing events or test results in the patient's record and transcription errors involving patient test results). Many health-care facilities use PCs with special software to assist in transmitting information from POC devices to LISs.

Entering patient results into an LIS can be accomplished in one of two ways. First, data can be transmitted automatically from instrument to LIS; this is often referred to as online data entry. Second, data can be entered manually; that is, the operator physically enters the results of patient tests into the computer. Manual entry mode is necessary if the procedure is nonautomated or if the analyzer cannot be connected to the LIS. Unfortunately, manual entry mode inherently increases the mistake rate for laboratories; thus online data entry is the preferred means for entering patient data. In an effort to reduce mistake rates in either entry mode, laboratory staff have used LIS features such as displaying flags that signify results outside reference ranges, results that are life threatening, results outside technical ranges, or results that fail delta checks. LISs can also display prior results for a patient, and these results can be used to verify current laboratory results.

SUMMARY

Laboratory operations encompass several distinct areas of responsibility, and due to the limited scope of this textbook only three were presented in this chapter. The three areas—statistics and quality control, safety and computers—are all within the purview of each member of the laboratory staff. This chapter focused on the working concepts and practices of selected topics in an effort to serve as a resource for students of laboratory science and technical staff members in clinical laboratories.

Statistics should be looked at as tools that can be used to make good decisions about laboratory data. Most computers are equipped with software that is capable of performing all of the necessary calculations so the laboratorian need not be concerned with that part of the process. Laboratorians can instead focus on interpreting statistics, which is a critically important task.

Most of the statistics and quality-control topics and issues discussed in this chapter were selected because they represent what is currently being practiced in most clinical laboratories. A significant amount of the information presented is used on a daily basis by practicing clinical laboratory scientists.

Working safely and in a safe working environment is of utmost importance. Employers are obligated to provide a safe working laboratory, and in return employees are responsible for adhering to all safety regulations. The safety topics presented represent what is currently of interest in clinical laboratories.

Computers play a significant role in laboratory medicine. They are used in all facets of laboratory operations, from handling scientific data to managing all of the documents generated by each laboratory. Fundamental knowledge of computer principles and operations is of paramount importance in clinical laboratories today.

REVIEW QUESTIONS

LEVEL I

1. OSHA standards are found in which of the following documents? (Objective 9)
   a. Codes of State Regulations (CSRs)
   b. Codes of Federal Regulations (CFRs)
   c. material safety data sheets (MSDSs)
   d. National Fire Protection Agency (NFPA)

2. Liquid nitrogen is an example of which of the following? (Objective 15)
   a. teratogen
   b. mutagen
   c. halogen
   d. cryogen

3. A class A fire involves which of the following? (Objective 11)
   a. combustible metals such as sodium
   b. flammable liquids (e.g., ethanol)
   c. ordinary combustible materials, including paper and wood
   d. electrical equipment

4. The letters MSDS are defined as which of the following? (Objective 9)
   a. manual safety devices schedule
   b. material safety data sheet
   c. material safety design sheet
   d. medical systems data sheets

5. Proportional error is characterized by which of the following? (Objective 8)
   a. The magnitude of the error increases as the concentration of analyte increases.
   b. The magnitude of the error stays the same over a wide concentration range.
   c. The magnitude of the error is inversely proportional to concentration.
   d. The magnitude of the error decreases as the concentration of analyte increases

6. Which of the following statements best defines the term *accuracy*? (Objective 3)
   a. a test's ability to produce the same value for replicate measure of the same sample
   b. a test's ability to correlate to another method
   c. the closeness of the agreement between the measured value of an analyte to its true value
   d. laboratory results that do not agree

7. Which of the following is an example of an inferential statistic? (Objective 1)
   a. mean
   b. standard deviation
   c. Student's *t*-test
   d. correlation coefficient

LEVEL II

1. The NFPA 704-M identification system is associated with which of the following warning labels? (Objective 12)
   a. warning labels for mechanical devices
   b. warning labels for biohazardous materials
   c. warning labels for chemical hazards
   d. warning labels for laboratory safety hoods

2. Which of the following statements best describes the proper method of storing chemicals in the laboratory? (Objective 13)
   a. according to their chemical properties and classifications
   b. alphabetically, for easy accessibility
   c. inside a chemical fume hood
   d. inside a walk-in-type refrigerator

3. Which of the following Codes of Federal Regulations (CFR) would be appropriate to review if there was a concern about hazardous chemicals in the laboratory? (Objective 10)
   a. 29 CFR 1910.1450
   b. 29 CFR 1910.1030
   c. 29 CFR 1910.1096
   d. 29 CFR 1910.1200

4. A plan designed to help prevent accidental exposure of laboratory personnel to bloodborne pathogens is called a(n): (Objective 11)
   a. chemical exposure plan.
   b. carcinogen exposure plan.
   c. exposure-control plan.
   d. universal precautions plan.

5. Cumulative trauma disorders (CTDs) are described as which of the following? (Objective 15)
   a. disorders due to blunt trauma injuries to the head caused by an automobile accident
   b. disorders associated with the overloading of particular muscle groups from repeated use
   c. posttraumatic stress disorders associated with soldiers returning from battle
   d. trauma due to emotional stress and anxiety

6. What is the mean, standard deviation(s), and percent coefficient of variation (%CV) for the data shown in the following table? (Objective 9)

Sample Number	Potassium (mmol/L)
1	5.2
2	5.0
3	4.8
4	5.1
5	4.9
6	5.2
7	5.0
8	5.3
9	4.9
10	4.8

REVIEW QUESTIONS (continued)

LEVEL I

8. Which of the following statements defines a 2_{2s} quality-control rule violation? (Objective 7)
   a. Two consecutive control values exceed $\pm 2s$.
   b. Two nonconsecutive control values exceed $\pm 2s$.
   c. Four consecutive controls values exceed $\pm 2s$.
   d. Two control values exceed $\pm 4s$.

9. Bidirectional interface with a laboratory information system (LIS) means that: (Objective 17)
   a. the LIS can simultaneously transmit data in one direction only.
   b. the LIS cannot simultaneously transmit or download information to a laboratory instrument.
   c. the LIS can simultaneously transmit or download information to a laboratory instrument and the instrument can transmit data to the LIS.
   d. the LIS can only upload information from an instrument.

LEVEL II

   a. mean = 4.80 mmol/L, s = 3.5 mmol/L , %CV = 72.9
   b. mean = 5.20 mmol/L, s = 0.0175 mmol/L , %CV = 0.34
   c. mean = 5.02 mmol/L, s = 0.175 mmol/L , %CV = 3.5
   d. mean = 3.50 mmol/L, s = 0.175 mmol/L , %CV = 5.0

7. Linear regression by least squares is useful for identifying which of the following? (Objective 5)
   a. systematic and proportional error
   b. quality-control failures
   c. variation among three or more methods
   d. cutoff values that maximize both sensitivity and specificity

8. Which of the following statistics is used to qualify the strength of the relationship between two variables? (Objective 1)
   a. t-test
   b. F-test
   c. confidence interval
   d. correlation coefficient

9. A laboratory measurement that is less than the detection limit should be reported as which of the following? (Objective 1)
   a. greater than the detection limit value
   b. the actual concentration value given by the analyzer or as calculated by the operator
   c. less than the detection limit value
   d. zero

10. Establishing a reference interval requires which of the following? (Objective 7)
    a. a random selection of individuals
    b. a well-defined group of individuals
    c. an identical number of subjects representing various groups of individuals
    d. no criteria at all with regard to subjects used for data collection

11. Which of the following actions represents the best course of action for a 1_{3s} quality-control failure? (Objective 4)
    a. Release all patient results but rerun the control material that failed.
    b. Release all patient results and do not rerun control material that failed.
    c. Do not release patient results and rerun the control material that failed.
    d. Do not release patient results, recalibrate the assay, and then rerun all levels of control material.

12. The presence of a cross-reacting compound in a serum sample may result in which of the following? (Objective 1)
    a. an accurate result
    b. a precise result
    c. a discordant result
    d. a reliable result

REVIEW QUESTIONS (continued)

LEVEL I

LEVEL II

13. The linearity of a method is characterized by which of the following? (Objective 1)
    a. slope and bias
    b. slope and intercept
    c. intercept and p-value
    d. a large correlation coefficient

14. The null hypothesis is represented by which of the following? (Objective 1)
    a. $\mu_1 = \mu_0$
    b. $\mu_0 = \mu_0$
    c. $\mu_0 \neq \mu_0$
    d. $\mu \neq \mu_0$

15. Which of the following represents the correct conclusion based on the data shown below? (Objective 1)
    $$0.005 < p < 0.01$$
    $$p = 0.007$$
    $$\alpha = 0.05$$
    a. Accept the null hypothesis.
    b. Reject the null hypothesis.
    c. The method is linear.
    d. Cannot be interpreted.

16. Laboratory information systems provide useful functions in all three stages of laboratory testing, including: (Objective 16)
    a. preanalytical, analytical, and postanalytical
    b. interfacing, networking, and Internet
    c. hypothesis testing, patient correlation, and linearity
    d. manual procedures, automated procedures, and semiautomated procedures

PEARSON
myhealthprofessionskit™

Use this address to access the interactive Companion Website created for this textbook. Simply select "Clinical Laboratory Science" from the choice of disciplines. Find this book and log in using your user name and password.

REFERENCES

1. Hollander M, Wolfe DA. *Nonparametric statistical methods*, 2nd ed. New York: Wiley-Interscience, 1999.

2. Prescott P. An approximate test for outliers in linear model. *Technometrics* (1975) 17:129–32.

3. Daniel WW. *Biostatistics: A foundation for analysis in the health sciences*, 8th ed. Hoboken, NJ: Wiley, 2005: 40–41.

4. Zar JH. *Biostatistical analysis*, 4th ed. Upper Saddle River, NJ: Prentice Hall, 1999: 332–33.

5. Linnett K, Boyd JC. Selection and analytical evaluation of methods with statistical techniques. *.In*: Burtis CA, Ashwood ER, Bruns DE (eds.). *Tietz textbook of clinical chemistry and molecular diagnostics*, 4th ed. Philadelphia: W. B. Saunders, 2006: 388–90.

6. Daniel WW. *Biostatistics: A foundation for analysis in the health sciences*, 8th ed. Hoboken, NJ: Wiley; 2005 426–28.

7. Daniel WW. *Biostatistics: A foundation for analysis in the health sciences*, 8th ed. Hoboken, NJ: Wiley; 2005: 730–33.

8. Labeling for in-vitro diagnostic products. US Food and Drug Administration (FDA).Title 21 CFR, Part 809.10 *Federal Register*, 2009.

9. Buttner J, Broth R, Boutwell JH et al. (IFCC Committee on Standards). Provisional recommendations on quality control in clinical chemistry: General principles and terminology. *ClinChem* (1976) 22: 532–39.

10. National Committee for Clinical Laboratory Standards. *Source book of reference methods, materials and related information for the clinical laboratory: Proposed guidelines*. NRSCL 12-P. Wayne, PA: National Committee for Clinical Laboratory Standards, 1994.

11. Linnett K, Boyd JC. Selection and analytical evaluation of methods with statistical techniques. *In*: Burtis CA, Ashwood ER, Bruns DE

(eds.). *Tietz textbook of clinical chemistry and molecular diagnostics*, 4th ed. Philadelphia: W. B. Saunders, 2006. 357–58.

12. Kroll MH, Elin RJ. Interference with clinical laboratory analyses. *ClinChem* (1994) 40(11 part 1): 1996–2005.

13. Taylor AE, Khoury RH. Interference in immunometric assays for gonadotropins. *JClinLigandAssay* (1997) 20: 190–99.

14. Datta P, Hinz V, Klee G. Comparison of four digoxin immunoassays with respect to interference from digoxin-like immunoreactive factors. *ClinBiochem* (1996) 29: 541–47.

15. Eriksson S, Halenius H, Pulkki K et al. Negative interference in cardiac troponin I immunoassays by circulating troponin autoantibodies. *ClinChem* (2005) 51(5): 839–47.

16. Desgupta A, Chow L, Wells A et al. Effect of elevated concentrations of alkaline phosphatase on cardiac troponin I assays. *JClinLabAnalysis* (2001) 15: 175–77.

17. Buttner J, Borth R, Boutwell JH et al. International Federation of Clinical Chemistry. Committee on Standards. Expert Panel on Nomenclature and Principles of Quality Control in Clinical Chemistry. Approved recommendation (1978) on quality control in clinical chemistry. Part 1. General principles and terminology. *ClinChemActa* (1979) 98 (1–2):129F–143F.

18. National Committee for Clinical Laboratory Standards. *Evaluation of the linearity of quantitative analytical methods: Approved guidelines*. NCCLS Document EP6-P. Wayne, PA: National Committee for Clinical Laboratory Standards, 2003.

19. Health Care Financing Administration (42CFR part 493, et al) the Public Health Service, U.S. Department of Health and Human Services: Clinical Laboratory Improvements Amendments of 1988. Final rule. *Federal Register* (1992) 57:7002–7288.

20. Currie LA. Nomenclature in evaluation of analytical methods including detection and quantification capabilities (KUPAC Recommendation 1995). *PureApplChem* (1995) 67:1699–1723.

21. Garber CC, Carey NR. Evaluation of methods. *In* Kaplan LA, Pesce AJ, Kazmierczak SC (eds.). *Clinical chemistry—theory, analysis, and correlation*, 4th ed. St. Louis, MO: Mosby, 2003: 414–15.

22. Linnet K, Boyd JC. Selection and analytical evaluation of methods—with statistical techniques. *In*: Burtis CA, Ashwood ER, Bruns DE (eds.). *Tietz textbook of clinical chemistry and molecular diagnostics*, 4th ed. Philadelphia: W. B. Saunders, 2006. 359–60.

23. Young DS. *Effects of drugs on clinical laboratory tests*, 5th ed. Washington, DC: AACC Press, 2000.

24. National Committee for Clinical Laboratory Standards. *Interference testing in clinical chemistry: Approved guideline*. NCCLS Document EP7-P. Wayne, PA: National Committee for Clinical Laboratory Standards, 2002.

25. Barnett RN. Medical significance of laboratory results *AmJClinPath* (1968) 50: 671–72.

26. Tonks D. A study of the accuracy and precision of clinical chemistry determinations in 170 Canadian laboratories. *ClinChem* (1963) 9: 217–18.

27. Cotlove E, Harris E, William G. Biological and analytic components of variations in long-term studies of serum constituents in normal subjects. III. Physiological and medical implications. *ClinChem* (1970) 16: 1028–29.

28. Elevitch FR (ed.).*CAP Aspen conference 1976: Analytical goals in clinical chemistry*. Skokie, IL: College of American Pathologists, 1977.

29. National Committee for Clinical Laboratory Standards: *How to define and determine reference intervals in the clinical laboratory: Approved guidelines*, 2nd ed. Document C28-A2. Wayne, PA: National Committee for Clinical Laboratory Standards, 1996.

30. Linnett K, Boyd JC. Selection and analytical evaluation of methods with statistical techniques. . *In*: Burtis CA, Ashwood ER, Bruns DE (eds.). *Tietz textbook of clinical chemistry and molecular diagnostics*, 4th ed. Philadelphia: W. B. Saunders, 2006, 353–96.

31. National Committee for Clinical Laboratory Standards. *Clinical laboratory technical procedure manual: Approved guideline*, 3rd ed. NCCLS Document GP2-A3. Wayne, PA: National Committee for Clinical Laboratory Standards, 1996.

32. Linnett K, Boyd JC. Selection and analytical evaluation of methods with statistical techniques. *In*: Burtis CA, Ashwood ER, Bruns DE (eds.). *Tietz textbook of clinical chemistry and molecular diagnostics*, 4th ed. Philadelphia: W. B. Saunders, 2006: 359–60.

33. Dawson B, Trapp RG. *Basics and clinical biostatistics*, 4th ed. New York: McGraw-Hill, 2004, 119–21.

34. Zweig MH, Campbell G. Receiver-operating characteristic (ROC) plots: A fundamental evaluation tool in clinical medicine. *ClinChem* (1993) 39(4): 561–77.

35. Riffenburg RH. *Statistics in medicine*, 2nd ed. Boston: Elsevier Academic Press, 2006: 256–59.

36. McCullough PA, Nowak RM, McCord J et al. B-type natriuretic peptide and clinical judgment in emergency diagnosis of heart failure. *Circulation* (2002) 106: 416–22.

37. Daniel WW. *Biostatistics: A foundation for analysis in the health sciences*, 8th ed. Hoboken, NJ: Wiley, 2005: 634–36.

38. Sasse EA. Reference interval and clinical decision limits. *In*: Kaplan LA, Pesce AJ, Kazmierczak SC (eds.). *Clinical chemistry—theory, analysis, and correlation*, 4th ed. St. Louis,MO: Mosby, 2003: 366–67.

39. Solberg HE. Establishment and use of reference values. *In*: Burtis CA, Ashwood ER, Bruns DE (eds). *Tietz textbook of clinical chemistry and molecular diagnostics,* 4th ed. Philadelphia: W. B. Saunders, 2006: 425–30.

40. Shewhart WA. *Economic control of quality of the manufactured product*. New York: Van Nostrand, 1931.

41. Wernimont G. Use of control charts in the analytical laboratory. *IndEngChem AnalEd* (1946) 18: 587–92.

42. American Society for Testing and Materials. *ASTM Report of committee E-11 on quality control of material*. Philadelphia: American Society of Testing and Materials, 1947.

43. American Society of Testing and Materials. *ASTM Special Technical Publication #15-C: Manual on QC of material*. Philadelphia: American Society of Testing and Materials, 1950.

44. Levey S, Jennings ER. The use of control charts in the clinical laboratory. *AmJClinPath* (1950) 20: 1059–66.

45. Henry R, Segalove M. The running of standards in clinical chemistry and the use of control charts. *JClinPath* (1952) 5: 305–311.

46. Bull BS, Elashoff RM, Heilbron DC et.al. A study of various estimators for the derivation of quality control procedures from patient erythrocyte indices. *AmerJClinPath* (1974) 61: 473–81.

47. Westgard JO, Groth, T, Aronsson T et al. Performance characteristics of rules for internal quality control: Probabilities for false rejection and error detection. *ClinChem* (1977) 23(10): 1857–67.

48. Berwick DM, Godfrey AB, Roessner J. *Curing health care: New strategies for quality improvement*. San Francisco: Jossey-Bass, 1990.

49. Westgard JO, Burnett, RW, Bowers GN. Quality management science in clinical chemistry: A dynamic framework for continuous improvement of quality. *ClinChem* (1990) 36: 1712–16.

50. Breitenberg M. Questions and answers on quality, the ISO 9000 standard series, quality systems registration and related issues, U.S. Department of Commerce, National Institute of Standards and Technology Publication NISTIR 4721, Gaithersburg, MD, 1991, USDC.

51. Westgard JO, Klee GG. Quality management. *In*: Burtis CA, Ashwood ER, Bruns DE (eds.). *Tietz textbook of clinical chemistry and molecular diagnostics*, 4th ed. Philadelphia: W. B. Saunders, 2006: 502–504.

52. Blick KE, Passey RB. Quality control for clinical chemistry laboratory. *In*: Kaplan LA, Pesce AJ, Kazmierczak SC (eds.). *Clinical chemistry—theory, analysis, and correlation*, 4th ed. St. Louis, MO: Mosby, 2003: 389–91.

53. Westgard JO, Klee GG. Quality management. *In*: Burtis CA, Ashwood ER, Bruns DE (eds.), *Tietz textbook of clinical chemistry and molecular diagnostics*, 4th ed. Philadelphia: W. B. Saunders, 2006: 499–501.

54. Westgard JO, Groth, T. Power functions for statistical control rules. *ClinChem* (1979) 25: 863–64.

55. Westgard JO, Barry PL, Hunt MR et.al. A multi-rule Shewhart chart for quality controlling clinical chemistry. *ClinChem* (1981) 27: 494–501.

56. Westgard JO, Klee GG. Quality management. *In*: Burtis CA, Ashwood ER, Bruns DE (eds.), *Tietz textbook of clinical chemistry and molecular diagnostics*, 4th ed. Philadelphia: W. B. Saunders, 2006: 506–507.

57. Westgard JO, Klee GG. Quality management. *In*: Burtis CA, Ashwood ER, Bruns DE (eds.). *Tietz textbook of clinical chemistry and molecular diagnostics*, 4th ed. Philadelphia: W. B. Saunders, 2006: 508–509.

58. Westgard JO, Klee GG. Quality management. *In*: Burtis CA, Ashwood ER, Bruns DE (eds.). *Tietz textbook of clinical chemistry and molecular diagnostics*, 4th ed. Philadelphia: W. B. Saunders, 2006: 509–12.

59. Whitehurst P, DiSilvio TV, Boyadjian G. Evaluation of discrepancies in patients' results: An aspect of computer-assisted quality control. *ClinChem* (1975) 21: 87–92.

60. National Committee for Clinical Laboratory Standards. *Nomenclature and definitions for use in the NRSCL and other NCCLS documents: Proposed standard*. 3rd ed. NCCLS document NRSCL 8-P3. Wayne, PA: National Committee for Clinical Laboratory Standards, 1996.

61. Occupational exposure to hazardous chemicals in the laboratory. Occupational Safety and Health Administration (OSHA) Final Rule. Document 29 CFR, Part 1910.1450.65. *Federal Register*, 1991; 55:3300-35.

62. Turk, AR. New MSDS could make safety data easier to understand. *Material Management in Health Care* (1993) 2: 14–15.

63. NFPA 704: Standard for the Identification of the Fire Hazards of Materials for Emergency Response. National Fire Protection Agency, 2001 edition, Quincy Mass.

64. Bretherick L, Urben P, Pitt M. *Bretherick's handbook of reactive chemical hazards*, 6th ed. Oxford: Butterworth Heinemann, 1999.

65. Resource Conservation and Recovery Act (RCRA, Pub. L. 94-580). Dept. of Energy (DOE), Washington, DC, 1976.

66. Protection of the environment, Codes of Federal Regulations, Environmental Protection Agency, 40: parts, 260–299. Rockville, MD: Government Institutes, Inc., 1990.

67. Occupational exposure to bloodborne pathogens. Occupational Safety and Health Administration (OSHA) Final Rule. *Federal Register* (1991) 56(235): 64004–64182.

68. National Institute for Occupational Safety and Health (NIOSH). *Guidelines for prevention of transmission of human immunodeficiency virus and hepatitis B virus and health-care and public safety workers*. Centers for Disease Control (NIOSH) publication No. 89-107. Atlanta, GA: Centers for Disease Control, July 1989.

69. United States Department of Health and Human Services: Recommendations for prevention of HIV transmission in health care settings. MMWR 36 no. SU02; 001, 1987.

70. The Needlestick Safety and Prevention ACT, Nov. 6, 2000; H.R. 5178.

71. Occupational exposure to bloodborne pathogens; Needlesticks and other sharps injuries. Occupational Safety and Health Administration (OSHA) Final Rule. Document 29 Part 1910.1030. *Federal Register* (2001) 66: 5318–25.

72. National Institute for Occupational Safety and Health (NIOSH). *Alert: Preventing needlestick injuries in health care settings*. Centers for Disease Control (NIOSH) publication No. 2000-108. Atlanta, GA: Centers for Disease Control, 1999.

73. Respiratory protection. Occupational Safety and Health Administration (OSHA). Document 29 CFR, Part 134. *Federal Register*, 2006.

74. *Primary containment for biohazards: Selection, installation and use of biological safety cabinets*, 2nd ed., Centers for Disease Control publication No. 89-107. Atlanta, GA: Centers for Disease Control, 2000.

75. National Fire Protection Association (NFPA) 13: Standard for installation of sprinkler systems, 2001 edition, Quincy, Mass.

76. Automatic sprinkler systems. Occupational Safety and Health Administration (OSHA). *Federal Register* (2005) 5: 501–502.

77. Handbook of Compressed Gases by Compressed Gas Association, 2nd ed. New York: Van Nostrand Reinhold, Co.1981.

78. Non-ionizing radiation. Occupational Safety and Health Administration (OSHA). Document 29 CFR, Part 1910. 67. *Federal Register* (2005) 6: 450–58.

79. National Institute for Occupational Safety and Health (NIOSH). *Elements of ergonomics programs. A primer based on workplace evaluation of musculoskeletal disorders*. Centers for Disease Control (NIOSH) publication No. 97-117. Atlanta, GA: Centers for Disease Control, July 1997.

80. National Institute for Occupational Safety and Health (NIOSH). *Musculoskeletal disorders and workplace factors: A critical review of epidemiologic evidence for work-related musculoskeletal disorders of the neck, upper extremities, and low back*. Centers for Disease Control (NIOSH) publication No. 97-141. Atlanta, GA: Centers for Disease Control, July 1997.

81. Monitoring noise levels non-mandatory informational appendix. Occupational Safety and Health Administration (OSHA). Document 29 CFR, part 1910.95 subpart G. *Federal Register* (2005) 5: 211–33.

82. American Society for Testing and Materials. ASTM standard E1238-97. *Specification for transferring clinical laboratory data messages between independent computer systems*. Philadelphia: American Society for Testing and Materials, 2001.

83. National Committee for Clinical Laboratory Standards. *Laboratory automation standards: Auto-1-A (specimen container/specimen carriers), Auto-2-A (bar codes for specimen containers identification), Auto-3-A, Auto-4-A (communications with automated clinical laboratory systems, instrument, devices and information systems)-A (systems operation requirement, characteristics, and information elements), and Auto-5-A (electromechanical interfaces)*. Wayne, PA: National Committee for Clinical Laboratory Standards, 2001.

5

Immunoassays

■ OBJECTIVES—LEVEL I

Following successful completion of this chapter, the learner will be able to:

1. Explain the fundamental differences among enzyme immunoassays, fluorescent immunoassays, and chemiluminescent immunoassays.
2. Identify five specific examples of labels used in immunoassays.
3. Identify five specific examples of solid-phase material used to bind antibodies.
4. Identify three advantages of monoclonal antibodies used in immunoassay systems.
5. Define the terms *antigen*, *antibody*, *immunogen*, and *hapten*.

■ OBJECTIVES—LEVEL II

Following successful completion of this chapter, the learner will be able to:

1. Contrast immunogen and antigen.
2. Explain the differences between competitive immunoassays and noncompetitive (sandwich) immunoassays.
3. Outline the procedural steps involved in a typical heterogeneous enzyme immunoassay.
4. Explain the mechanism involved to produce photons of light energy from a chemiluminescent compound.
5. Distinguish homogeneous immunoassays from heterogeneous immunoassays.
6. Explain the principles of the reactions for the following immunoassays:
   a. EMIT
   b. FPIA
   c. CEDIA
   d. SLFIA
   e. CLIA
   f. LOCI

KEY TERMS

Acridinium ester	Competitive immunoassay	Immunogen
Affinity	Fluorescein	Luminescence
Antibody	Hapten	Paramagnetic particle
Antigen	Heterogeneous immunoassay	Sandwich immunoassay
Avidity	Homogeneous immunoassay	Solid phase
Chemiluminescence	Hook effect	

 A CASE IN POINT

A 21-year-old female (gravida 0, para 0)* presents to the emergency department complaining of nausea and abdominal pain over the past couple of weeks. The patient states that she shows vaginal spotting, is sexually active, admits to drug and alcohol use, and has an intrauterine device. An attending physician conducts a physical examination that reveals lower-leg edema, and a palpable lower abdominal mass. The physician orders an ultrasound and a urine pregnancy test followed by a serum quantitative human chorionic gonadotropin, beta polypeptide (β-hCG) measurement.

Diagnostic Test Results

Results of the ultrasonography: The ultrasound revealed a "snowstorm" appearance that may be the result of a molar pregnancy (for example, hydaditiform mole).

Urine qualitative pregnancy test: Negative

Serum quantitative β-hCG: 700 IU/L (Reference interval = <0.5 − 2.9 IU/L)

Issues and Questions to Consider

1. What is the clinical significance of a serum quantitative β-hCG value of 700 IU/L?

2. Why is the urine qualitative pregnancy test negative, whereas the serum quantitative β-hCG is 700 IU/L?

3. What methodologies were used for both the qualitative and quantitative tests?

*Gravida is the total number of times a women has been pregnant, and para indicates the number of viable births.

WHAT'S AHEAD

- An overview of the fundamental of immunologic reactions.
- Explanation of the principles of selected immunoassays.
- Description of immunoassay techniques and their respective labels or detectors.

Yalow asserted, "We never thought of patenting RIA. . . . Patents are about keeping things away from people for the purpose of making money. We wanted others to be able to use RIA."[1]

▶ INTRODUCTION

Immunoassays are testing methods used to detect and measure either antigens or antibodies in a solution. An antigen may be a hormone, a vitamin, or drugs found in biological samples such as blood. Antibodies are proteins that are produced in response to antigenic stimulators. When antigen and antibodies are used in a testing system, they provide the laboratory with a very sensitive, specific, and reliable method.

What were some of the factors that stimulated scientists to investigate and develop immunoassays as a means of measuring compounds in biological samples? Before she was awarded the Nobel Prize in medicine in 1977, Rosalyn Yalow and colleague Solomon Berson in the mid-1950s observed that most assays for complex compounds such as proteins, steroids, and vitamins were fraught with limitations.[2] These included lack of sensitivity, specificity, useful detection limits, poor precision, and poor accuracy. Also, the assays themselves were labor intensive, time consuming, and inefficient.

"Yalow and Berson had provided a means to observe the previously invisible world of antigen–antibody reactions that take place in solution. Before their ideas and methods, scientists were restricted in the analysis of reactions between antigens and antibodies to those that produced visible precipitation or other evidence, such as the clumping of red blood cells."[3]

Since Yalow and Berson released their radioimmunoassay (RIA) for insulin measurement in1960, there have been hundreds of immunoassays developed and marketed using several unique labels, assay platforms, and applications for clinical chemistry and other disciplines within clinical laboratories. For example, in the early 1970s SYVA Company (Dade Behring Inc., Cupertino, CA) introduced the homogeneous enzyme multiplied immunoassay technique (EMIT) to measure therapeutic drugs in serum and drugs of abuse in urine. This was followed with the development of fluorescence polarization immunoassay (FPIA) by researchers at Abbott Diagnostics (Abbott Park, Chicago). In the late 1970s, Burd published his work with substrate-labeled fluorescent immunoassay (SLFIA) and provided the laboratory with a technique to measure drugs and various immunoglobulins.[4] The use of chemiluminescent compounds provided an improvement in assay sensitivity over existing techniques and is the predominant immunoassay used today. (Note: All of the specific immunoassays discussed in this chapter are still used today. Although some of the methods developed during the 1960s and 1970 may be used less frequently now, they still provide valuable and reliable data to the clinician.)

Immunoassay testing platforms exist for point-of-care technology (POCT) products. Several devices are currently on the market for use in small laboratories and over the counter for consumers to purchase. There are issues with POCT assays, including sensitivity and specificity, that are currently being addressed by the manufacturers.

► BASICS OF IMMUNOCHEMICAL REACTIONS

ANTIBODIES

Immunochemical reactions involve the binding of an **antibody** to its complementary antigen. Antibodies are immunoglobulins that are capable of binding specifically to both natural and synthetic antigens. These antigens can be proteins, carbohydrates, nucleic acids, and lipids, among other types of molecules. The immunoglobulins are grouped into five classes designated as IgG, IgA, IgM, IgD, and IgE. Immunoglobulin G is the predominant class used as an immunochemical reagent. The other classes are not used in any significant quantity for analytical purposes. More detailed information of immunoglobulin structures and functions can be found in the reference section at the end of the chapter.[5,6]

Polyclonal antibodies are derived from different plasma cell lines or clones, and *monoclonal* antibodies are derived from essentially one plasma cell line or clone. Monoclonal antibodies are further described as *uniform homogeneous antibodies* directed to specific *epitopes* (antigenic sites). A specific cell line allows for the secretion of reactive immunoglobulins specific to a single epitope. Monoclonal antibodies allow for analysis of molecules on an epitope-to-epitope basis because of their narrow limits of specificity. Earlier immunoassays used mainly polyclonal antisera in their reagent systems, which resulted in problems with analytical specificity. For example, polyclonal antisera used in earlier reagent systems for human chorionic gonadotropin (hCG) (pregnancy testing) were capable of detecting not only hCG but also luteinizing hormone and follicle-stimulating hormone because of their structural similarities. Currently, most manufactures of immunoassay reagents use monoclonal antisera with significantly improved binding specificity. In contrast to polyclonal antibodies, a disadvantage of some monoclonal antibodies is the inability to recognize the entire or intact molecule. An example of this is problem occurs when different antigens with a common epitope appear to be the same antigens. The monoclonal antibody may cross-react with different antigens. This cross-reactivity may be attributed to the probable existent of the same amino acid sequences, carbohydrates, or lipids on different molecules.[7]

The advantages of using monoclonal antibodies in immunoassays include the following:[7]

• Monoclonal antibodies provide a well-defined reagent.

• Monoclonal antibody production can yield an unlimited quantity of homogeneous reagent with highly consistent affinity and specificity.

• Monoclonal antibodies can be prepared through immunization with a nonpurified antigen.

There are certain limitations to the use of monoclonal antibodies in immunoassays. Two significant examples are (1) possible insufficient reactivity in precipitation or agglutination because of weak or nonexistent network formation in the immunocomplex when single monoclonal antibodies are used, and (2) antigens with several heterogeneous epitopes are more difficult to characterize immunochemically with a single monoclonal antibody.[7]

ANTIGENS, IMMUNOGENS, AND HAPTENS

An **immunogen** is a chemical substance capable of inducing an immune response. The term *immunogen* is used when referring to material capable of eliciting antibody formation when injected into a host. The term **antigen** is used to describe any material capable of reacting with an antibody without necessarily being capable of inducing antibody formation. The following are examples of the difference between the two terms. Egg albumin is an immunogen because it is capable of inducing formation of anti–egg albumin antibody. The drug secobarbital is an antigen because it will react with antisecobarbital antibodies, but it is not an immunogen because it does not induce antibody formation unless first conjugated with a protein.

An immunogen may be a protein or a substance coupled to a carrier compound, usually a protein, that is capable of inducing the formation of an antibody when introduced into a foreign host. If the antibodies produced are circulating in the vascular system it is referred to as *humoral*; if the antibody is bound to tissue, it is called *cellular*.

Haptens are an example of another substance that is capable of producing antibodies. Haptens are chemical determinants that, when conjugated to an immunogenic carrier, are able to stimulate the synthesis of an antibody specific for the hapten. Therefore, haptens are capable of binding a protein to create antibodies but cannot by themselves stimulate an immune response. One example of the unique property of haptens involves the commonly used compound dinitrophenol (DNP). Injected into a host, DNP cannot by itself produce an antibody. When conjugated to bovine serum albumin (BSA), however, and subsequently injected into a host, several different antibodies are produced, including anti-DNP, anti-BSA, and anti-DNP/BSA.

 Checkpoint! 5–1

Distinguish *immunogen* from *antigen*.

ANTIGEN–ANTIBODY REACTION KINETICS

The kinetics for the reversible Ag–Ab reactions is as follows:[8]

$$\text{Ag} + \text{Ab} \underset{k_2}{\overset{k_1}{\rightleftarrows}} \text{AgAb} \qquad (5.1)$$

Where Ag represents free antigen, Ab represents free antibody sites, AgAb is the antigen–antibody complex, and k_1 and k_2 are the association and dissociation rate constants, respectively.

The strength of binding between an antigen and antibody is influenced by the following:

1. Cation salts present in a solution of antigen and haptens will inhibit the binding of antibody with a cationic hapten. For example, salts of ammonia, potassium, and sodium will affect the degree of hydration of the salt with an anionic group located in the antibody binding site. A similar event occurs for anionic haptens and anionic salts, including iodine, chloride, and fluoride.

2. The presence of linear polymers such as dextran 500 or polyethylene glycol in a mixture of antigens and antibodies can cause an increase in the rate of immune complex growth.

The energy and strength of interaction between the antibody and antigen is described by their affinity and avidity. **Affinity** refers to the thermodynamic energy of interaction of a single antibody binding site and its corresponding epitope on the antigen. **Avidity** refers to the overall strength of binding of an antibody to an antigen and includes the sum of the binding affinities of all the individual binding sites on the antibody. Therefore, affinity is a property of the substance bound (antigen), and avidity is a property of the binder or antibody.[9]

▶ GENERAL PRINCIPLES OF IMMUNOASSAYS

Several variations of immunoassays exist to measure the concentration of hormones, therapeutic drugs, drugs of abuse, and vitamins in biological fluids. A discussion of the fundamentals of immunoassay reactions will be presented first, followed by the presentation of specific immunoassays used in clinical chemistry laboratories.

IMMUNOASSAY LABELS

Several compounds or substances have been used as labels for antigens or antibodies in immunoassay procedures. A list of labels and their method of detection is presented in Table 5-1 ✪. The first labels used were radioactive isotopes and the technique was called radioimmunoassay (RIA). These RIAs required specialized instrumentation called gamma or beta

TABLE 5-1

Labels Used for Immunoassays and Method of Detection

Labels	Method
Enzymes	Absorbance spectroscopy
Peroxidase	
β-galactosidase	
Alkaline phosphatase	
Glucose-6-phosphate dehydrogenase	
Radioisotopes	Scintillation counting
^{125}Iodine	
^{131}Iodine	
Tritium	
Fluorescein	Fluorescence
Acridinium esters	Chemiluminescence
Dioxetane	Chemiluminescence
Luminol	Chemiluminescence
Ruthenium III	Electrochemiluminescence
Europium	Time-resolved fluorescence

scintillation counters to detect and measure the amount photons produced due to the interaction of radiation with matter. RIAs are still used primarily in reference laboratories and rarely available in hospital based clinical laboratories. Radioactive isotopes labels were replaced by enzymes and fluorescent labels. The enzyme and fluorescent labels were eventually replaced by chemiluminescent labels. Fluorescent and chemiluminescent labels have the ability to produce light or **luminescence**. Luminescence can be detected with a photodetector—for example, a PMT.

The conjugated antigens used are prepared by the manufacturers using a variety of techniques. Conjugation is accomplished via chemical reactions that complex antigens to labels or other materials required to complete the reaction. An example of a coupling reaction sequence that attaches an enzyme label to a protein antigen is shown in Figure 5-1 ■. In this reaction, *m*-maleimidobenzoyl-N-hydroxysuccinimide ester (MBS), a bivalent reagent consisting of an activated ester and maleimide, can react with an NH_2 group and SH group. MBS reagent allows the attachment of both the protein moiety and enzyme label to its structure.[7]

The instrumentation used to detect the label is dependent on the type of labels used—for example, if a fluorescent label is used then a fluorometer is required. A computer will determine

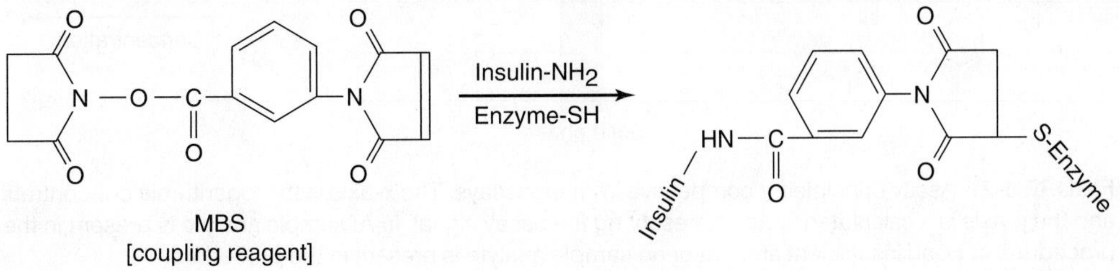

FIGURE 5-1 Preparation of insulin–enzyme conjugate using the heterobifunctional reagent MBS.

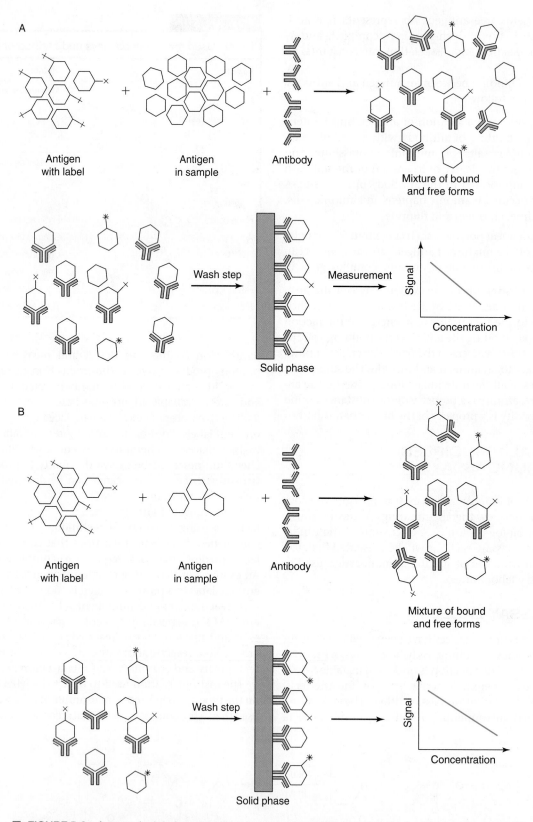

■ FIGURE 5-2 Assay principle for competitive immunoassays. The *x*-axis is the logarithmic concentration, and the *y*-axis is a calculated factor representing the assay signal. In A. sample analyte is present in the procedure. In B. an insufficient amount or no sample analyte is present in the procedure.

the concentration of analyte from data gathered by measuring calibration material and relating it to instrument or detector response. The instrument response is converted to a calculated parameter—for example, net polarization as in FPIA.

COMPETITIVE IMMUNOASSAYS

A **competitive immunoassay** begins with the addition of a patient sample containing the analyte to be measured (antigen) to a mixture of competing antigen with a "label" attached (antigen–label conjugate) and the corresponding antibody as shown in Figure 5-2 ■. The labeled antigen conjugate competes with the patient's antigen for binding sites available on the antibody. When the competition phase is complete, there are patient analyte antigens and labeled antigens that did not bind to the antibody. These antigens are referred to as *free* or *unbound antigens*. The proportion of unbound antigens depends on the kinetics of the immunoassay reaction. The reaction mixture will now consist of a population of free and bound antigens. Some of the antigens are labeled (reagent), and some are not (patient). The goal of an immunoassay is to measure the label on one of the forms of antigens, either the bound or free, and the concentration of analyte is directly proportional to the assay signal.

At this juncture in the assay, one of two procedures will be performed. If the assay is designed to be a **homogeneous immunoassay**, there will be no physical separation of the bound and free forms or phases. If the assay is a **heterogeneous immunoassay**, there will be a physical separation of bound from free forms. A detailed explanation of homogeneous immunoassays (e.g., EMIT and FPIA) will be presented later in this chapter under "Specific Immunoassay Techniques." We will proceed with this example using the heterogeneous technique, which is how the assays were first introduced into the clinical laboratory.

All heterogeneous immunoassays require a separation step to distinguish the reacted immunocomplex (bound) from unreacted antigen (free). Many substances have been and are currently being used to facilitate this separation. Examples of substances used to separate bound from free fractions include the following:

- double antibodies,
- solid-phase antibodies,
- polyethylene glycol,
- talc,
- solvent or salt precipitates, and
- charcoal.

A commonly used technique that facilitates separation of bound fractions incorporates a **solid-phase** material. In this technique, usually an antibody is immobilized onto a solid-phase material via covalent bonding or physical adsorption through noncovalent interactions. Examples of solid-phase materials are gel particles made of agarose or polyacrylamide, plastic beads, particles coated with iron oxide, and **paramagnetic particles**.

Once the bound and free fractions have been separated, one of the two fractions needs to be retained for quantitative or qualitative analysis. Most immunoassays retain the bound fraction and discard the free fraction.

NONCOMPETITIVE SANDWICH IMMUNOASSAYS

A noncompetitive **sandwich immunoassay** technique (also referred to as an immunometric assay) uses a label that is attached to a second antibody. Also, the antigen must have two binding sites to complex both antibodies.

The procedure is outlined below and shown in Figure 5-3 ■:

- The reaction commences with the addition of the patient sample to an antibody affixed to a solid phase.
- The antibody on the solid phase captures the antigen (analyte) in the specimen.
- A washing step serves to separate bound from free fractions.
- A second antibody with the label attached binds to the antigen (analyte)–antibody–solid-phase complex.
- The signal from the label can then be determined after the elimination of free or unbound conjugates through a washing step.
- Sample analyte concentration versus antibody label signal is nonlinear.

 Checkpoint! 5–2

Which of the following types of immunoassays—homogeneous or heterogeneous immunoassays—requires a physical separation of bound from free fractions?

► SPECIFIC IMMUNOASSAY TECHNIQUES

MICROPARTICULATE ENZYME IMMUNOASSAY

Microparticulate enzyme immunoassay (MEIA) uses a solid-phase material described as latex microparticles. The latex microparticles are coated with an analyte-specific antibody. A second antibody labeled with alkaline phosphatase (ALP) is added along with the patient sample. The analytes in the patient sample binds to both the enzyme-labeled antibody and the antibody coated onto the microparticles. An aliquot of the antibody–analyte–antibody complex is transferred to the glass-fiber matrix, which serves to anchor the complex as illustrated in Figure 5-4 ■. A wash buffer is added to remove

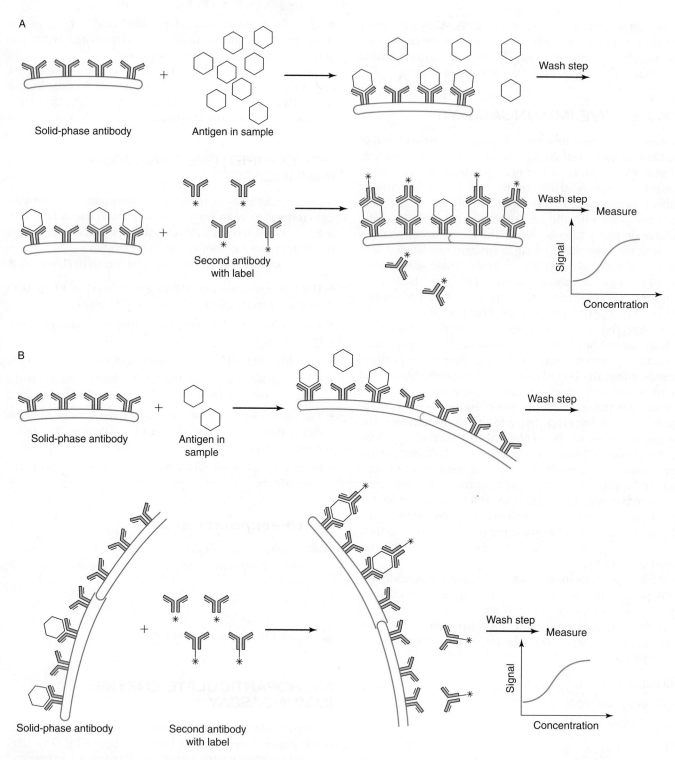

■ **FIGURE 5-3**   Assay principle for noncompetitive (sandwich) immunoassays. The *x*-axis is the logarithmic concentration; the *y*-axis is a calculated factor representing the assay signal. In A. sample analyte is present. In B. there is an insufficient amount of or no sample analyte present.

any unbound material. The final step in the assay is the addition of the substrate 4-methylumbelliferyl phosphate (MUP). Antibody-bound ALP will react with the substrate MUP to produce a fluorescent product methylumbelliferone (MU).

The excitation and emission wavelengths for MU are 380 nanometer (nm) and 450 nm, respectively. The rate at which 4-MUP is generated on the matrix is proportional to the concentration of analyte in the test samples.

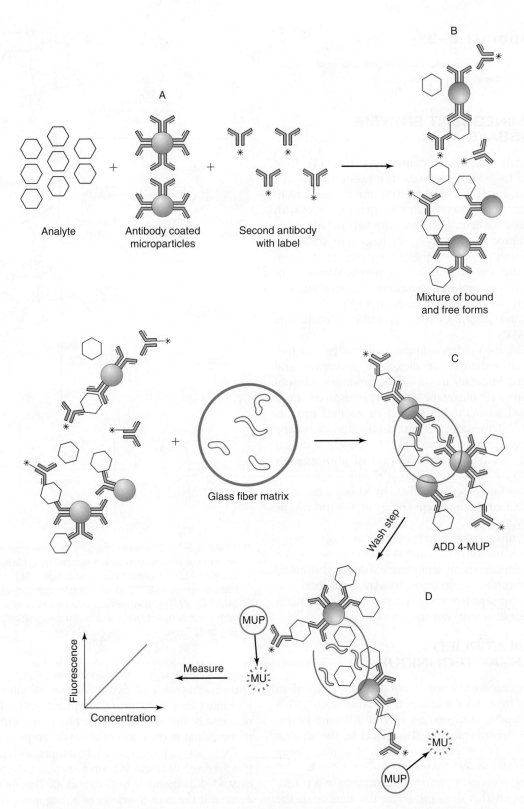

■ **FIGURE 5-4** Microparticulate enzyme immunoassay (MEIA). A. Sample, microparticles coated with antianalyte antibodies and antianalyte: ALP conjugates are incubated together. B. The analyte in the patient's samples binds to both the enzyme-labeled antibody and antibody-coated microparticles, forming a "sandwich." C. An aliquot of the reaction mixture is transferred to the glass-fiber matrix and washed. D. The substrate 4-methylumbelliferyl phosphate (MUP) is added to the matrix. The fluorescent product, methylumbelliferone (MU), is measured.

✓ **Checkpoint! 5–3**

How is fluorescent electromagnetic radiation produced in the microparticle enzyme immunoassay method?

CHEMILUMINESCENT ENZYME IMMUNOASSAYS

Chemiluminescent enzyme immunoassays (CL-EIA) have been developed to provide methods to measure therapeutic drugs, hormones, and tumor markers. This immunoassay technique uses a chemiluminescent substrate that can react with an enzyme to produce light. Therefore, **chemiluminescence** is described as a process whereby light energy is produced via a chemical reaction. Chemiluminescence differs from bioluminescence in that chemiluminescence uses synthetic compounds—for example, luminol, dioxetane, and acridinium—to produce light. Bioluminescence, on the other hand, incorporates natural substrates such as luciferin adenosine triphosphate (ATP).

Earlier, CL-EIA used either a luminol derivative and peroxidase with an enhancer or dioxetane derivative and ALP.[10] A later modification to CL-EIA incorporates the substrate adamantyl 1,2-dioxetane phenyl phosphate (also phenylphosphoyl adamantyldioxetane) or AMPPD into its reagent system.[11] This substrate is an adamantyl dioxetane derivative. The partial reaction sequence for CL-EIA is shown in Figure 5-5 ■ and begins with cleavage of phosphoester bond of AMPPD by ALP and triggers a chemically initiated electron-exchange luminescence (CIEEL) by releasing electron-rich dioxetane. Chemiluminescent energy at 477 nm can be detected quickly.[12]

This assay results in improved sensitivity over other types of immunoassays including EIAs and FIAs and can be applied to automated immunoassay analyzers. "Among the many methods, chemiluminescent immunoassay and their enzymes are the most sensitive assay systems capable of detecting analytes present at very low concentrations."[7]

ENZYME MULTIPLIED IMMUNOASSAY TECHNIQUE

The first homogeneous EIA was developed and released in the early 1970s by SYVA Company (Palo Alto, CA).[13] This assay is still available and used on several different instrument platforms. An enzyme is used as a label on the antigen and is referred to as the enzyme multiplied immunoassay technique. The EMIT assay is illustrated in Figure 5-6 ■.

The enzyme glucose-6-phosphate dehydrogenase is conjugated to an antigen that does not affect the activity of the enzyme. An antigen-specific antibody is added, which results in the inhibition of enzyme activity. Addition of patient sample with antigen analyte results in a competition between antigen–enzyme and sample antigen for binding sites on the antibody. The antibody that binds to the antigen–enzyme inhibits the enzyme by inducing or preventing conformational

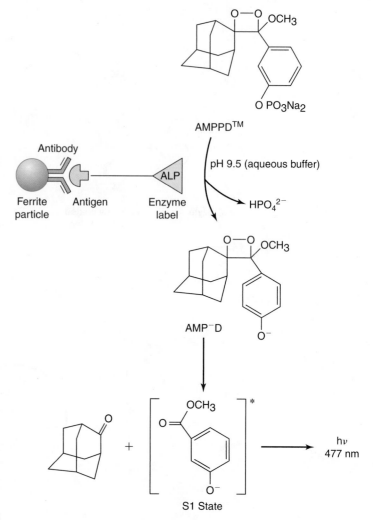

■ **FIGURE 5-5** Chemiluminescent enzyme immunoassay. AMPPD, a chemiluminescent substrate, is cleaved by the enzyme ALP attached to analyte antigen. ALP catalyzes the conversion of AMPPD to an electron-rich dioxetane phenolate (AMP-D). AMP-D further decomposes to phenolate (S1 state) with a subsequent release of electromagnetic radiation (EMR) (*hv*) at 477 nm.

changes necessary for enzyme activity.[14] The unbound antigen–enzyme is left free to react with a substrate to produce a product that is measured colorimetrically. Thus, the assay response is modulated by steric hindrance. Enzyme activity is proportional to the concentration of sample antigen.

EMIT assays were initially developed to screen urine samples for drugs of abuse. Within a few years, the SYVA Company expanded its menu of tests to include therapeutic drugs with detectable limits in the range of nanograms per milliliter.

 Checkpoint! 5–4

In the EMIT assay, which form—bound or free—is left to react with a substrate?

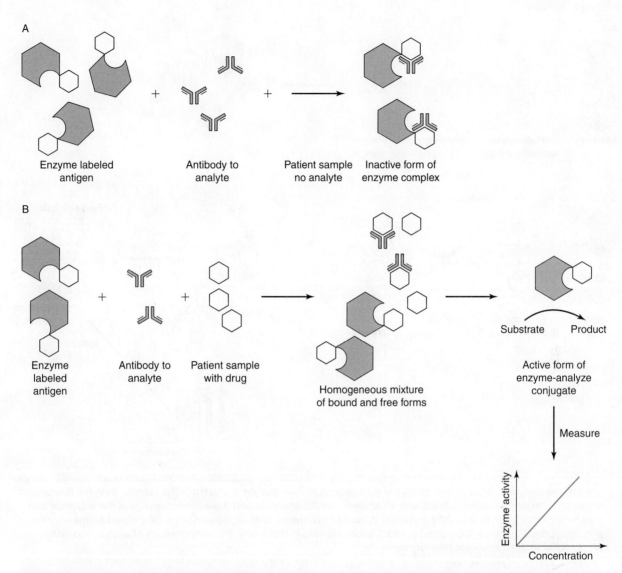

■ FIGURE 5-6   The enzyme multiplied immunoassay technique. In **A.** if no analyte is present, the activity of the enzyme, glucose- 6-phosphatase dehydrogenase, is inhibited by the binding of antibody to the antigen conjugated with enzyme. In **B.** if analyte is present in the sample, the enzyme will be available to convert the substrate (glucose-6-phosphatase dehydrogenase) to product. The enzyme activity is proportional to the concentration of the analyte.

SUBSTRATE LABEL FLUORESCENT IMMUNOASSAY

SLFIA can be used to measure therapeutic drugs, hormones, and immunoglobulins (e.g., IgG and IgA). It is not as analytically sensitive as other FIAs because the amplification properties of the enzyme are not enhanced.[4] The reaction is illustrated in Figure 5-7 ■. A patient sample is added to reagents that include the enzyme β-galactosidase, analyte-specific antibody, and a substrate, β-galactosylumbeliferone, conjugated to the antigen. The substrate–antigen complex is nonfluorescent. A fluorescent product, umbelliferone is formed by the cleavage of substrate by the enzyme β-galactosidase. However when the substrate–antigen conjugate is allowed to react with a specific antibody to the antigen, there is no cleavage of the substrate

complex with the enzyme. Patient antigen (analyte) competes with fluorescent substrate–antigen (a reagent) for binding sites on the antibody. Analyte concentration in the sample is directly proportional to the fluorescent intensity measured.

CLONED ENZYME DONOR IMMUNOASSAY

Cloned enzyme donor immunoassay (CEDIA) (Microgenics Corporation, part of Thermo Fisher Scientific Incorporated, Fremont, CA) uses a unique application of recombinant DNA technology to homogeneous immunoassay techniques.[15] A β-galactosidase protein was engineered into a large polypeptide (an enzyme acceptor) and a small polypeptide (an enzyme donor). A unit of enzyme acceptor and a unit of enzyme

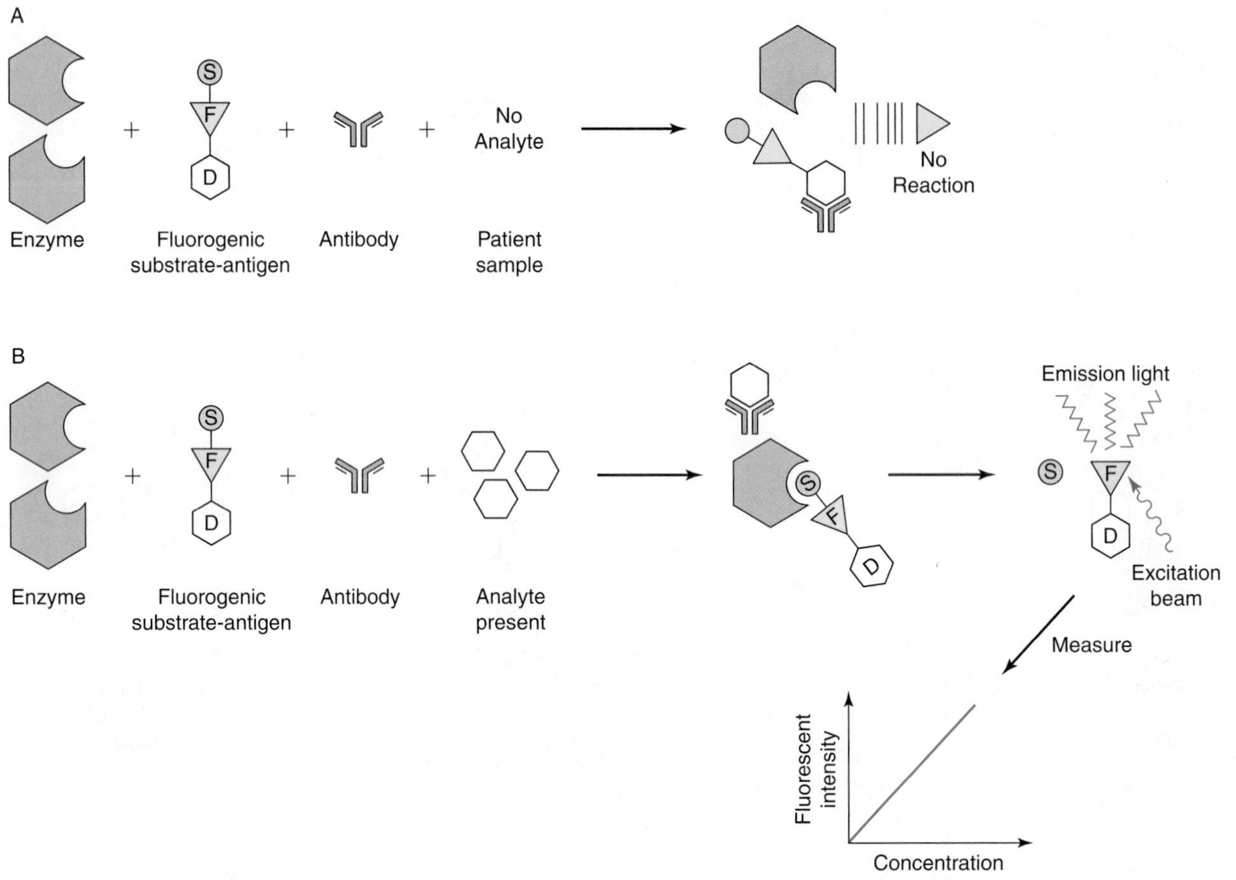

■ FIGURE 5-7   Substrate label fluorescent immunoassay. The substrate β-galactosylumbelliferone is conjugated with the analyte (reagent) and forms a nonfluorescent substrate. In A. if no analyte is present in the sample, then the fluorogenic substrate–antigen is allowed to react with an antigen-specific antibody, and there is no cleavage of the substrate complex by the enzyme. In B. if analyte is present in the sample, then substrate is available to be cleaved by the enzyme β-galactosidase to form a fluorescent product, umbelliferone. In this assay, the concentration of analyte is directly proportional to the fluorescent intensity measured.

donor were assembled to form an enzymatically active tetra-meric molecule of β-galactosidase. The β-galactosidase catalyzes galactopyranoside substrate to produce a colored product that can be measured by spectrophotometry.

As illustrated in Figure 5-8 ■, the assay is initiated when a patient sample is added to a reagent containing an antian-alyte antibody, an enzyme donor–analyte conjugate, and an enzyme acceptor. If the analyte is present in the patient samples, then the antibody will bind to this analyte and the enzyme donor (ED) will be free to form active enzyme with the enzyme acceptor (EA). This reaction serves to modulate the amount of active β-galactosidase formed. The signal generated by the conversion of substrate to product by the enzyme is directly proportional to the analyte concentra-tion in the patient serum. If no analyte is present in the samples, then the antibody will bind to ED and inhibit the reassociation of ED and EA, so no active enzyme will be formed. CEDIA assays are linear because the amount of enzyme formed is directly proportional to the amount of analyte present. The menu of tests using CEDIA includes

therapeutic drugs, drugs of abuse, immunosuppressant drugs, and several hormones.

✓ Checkpoint! 5–5

What type of technology is used to engineer the creation of large polypeptides of β-galactosidase in the CEDIA?

FLUORESCENCE POLARIZATION IMMUNOASSAY

Abbott Diagnostics combined light polarization (reviewed in Chapter 2, Instrumentation), fluorometry, and immunoas-say techniques to provide a homogeneous immunoassay for the measurement of therapeutic and abused drugs.[16] Abbott's fluorescence polarization immunoassay provided an alternative to EMIT, which was being used at that time. This new homogeneous FIA used **fluorescein**, which ex-cites at 490 nm and reemits fluorescence at 520 nm as the label or detector compound.

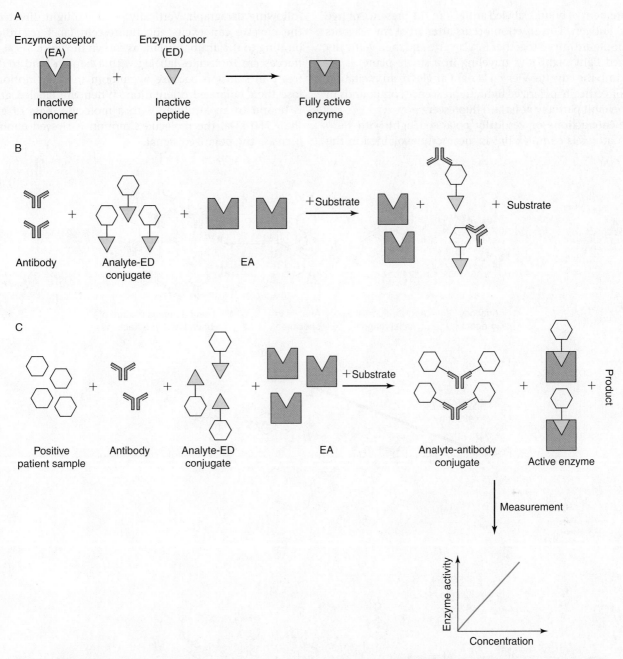

FIGURE 5-8 Cloned enzyme donor immunoassay. In A. an enzyme acceptor (EA) reacts with an enzyme donor (ED) to form a β-galactosidase tetramer, a fully active enzyme. In B. if no analyte is present in a sample, then the antibody will bind to the ED and inhibits association of ED with EA. In C. if analyte is present in a sample, it binds to the antibody, which leaves the analyte–ED conjugate available to attach to the enzyme. The enzyme catalyzes a reaction with the substrate, producing a product.

The assay begins with the addition of the patient sample to an analyte-specific antibody. If analyte is present in the sample, it will bind to the antibody. Next, the fluorescent antigen–tracer (the fluorescent molecule) is added to the mixture, and any unbound antibody will bind to the antigen–tracer complex. Finally, the amount of polarized fluorescence emitted is measured. The quantity of analyte in the serum samples is inversely proportional to the amount of polarized fluorescence emitted.

In summary, when there is a high concentration of analyte, most of it binds to the antibody, leaving a high concentration of antigen–label complex that is free to rotate in solution and emit nonpolarized fluorescence. Where there is very little analyte in the sample, most of the antibody is bound to the antigen–label, reducing the rotation of the free antigen–label complex and causing it to emit polarized fluorescence.

Vertically polarized light is used to excite the electrons in fluorescein molecules and allows for the detection and measurement of bound labeled antigen in the presence of free labeled antigen. This reaction occurs after all of the reagents and sample are mixed together in a cuvette and then vertically polarized light (light rays traveling in a single plane) is directed into the cuvette. Viewed at right angles to an excitation beam of vertically polarized light, fluorescence compounds in solution emit partially polarized fluorescence.[17]

The interactions of vertically polarized light with fluorescein atoms is complex but is succinctly explained in the following paragraph. Vertically polarized light directed into the cuvette can excite the fluorescein label regardless of binding to the antibody. But, as shown in Figure 5-9 ■, small fluorescein molecules labeled with a hapten tend to "spin" freely in solution because of random thermal motion and lose their polarized orientation. When the labeled antigen is bound to an antibody with a molecule mass of greater than 160 kDa, the molecule's motion is slowed enough to increase the polarized signal.

A

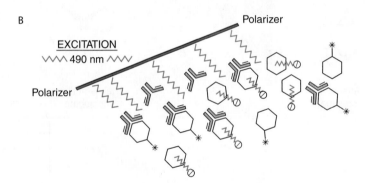

B

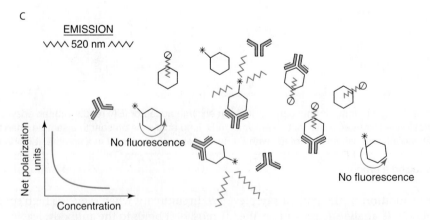

C

■ FIGURE 5-9   Fluorescent polarization immunoassays. In A. all reactants are added to a cuvette, creating a homogenous mixture of bound and free fractions. In B. monochromatic light (excitation) is polarized and enters the cuvette and excites fluorescein molecules (label) attached to both the bound and free fractions. In C. small fluorescein-labeled analyte free from antibody fails to emit fluorescent and net polarization is decreased. However, the fluorescein-labeled analyte bound to the antibody remains at the same angle, because of its very slow movement, and emits fluorescence; the net polarization is increased.

Abbott Diagnostics marketed this technique in several of its instrument platforms beginning with the Abbott TDx chemistry analyzer (see Chapter 2, Instrumentation). The TDx is described as a discrete-batch analyzer that can measure only one analyte at a time in a fixed number of samples. It is a closed-reagent system that provides the laboratory with a relatively fast, reliable, and uncomplicated means of measuring analytes in patient samples.

 Checkpoint! 5–6

Explain why a positive result in a fluorescent polarization immunoassay produces a low polarized signal.

CHEMILUMINESCENT IMMUNOASSAYS

A commonly used label for a chemiluminescent immunoassay (CLIA) is **acridinium ester** (AE). The assay formats include sandwich, competitive, and antibody-capture, which gives this method added utility. The AE is conjugated to a protein moiety (antibody) and serves as a reagent.[18] The procedure for the "sandwich" format begins [Figure 5-10(A) ■] with the addition of a sample to a reagent that contains AE-labeled antibody. The AE-labeled antibody binds specifically with the analyte-specific antigen in the samples. Next, a solid phase containing paramagnetic particles (PMP) that are coated with antibody specific for the antigen in the sample is added. PMP binds to the antigens that are bound to AE-labeled antibody. A "sandwich" complex is formed on the magnetic microparticle. The reaction cuvette is exposed to a magnetic field, which draws PMP toward the magnet. While the magnets hold PMP in place, sample and reagent not bound to PMP are washed away. The cuvette now contains AE bound to antigen, which is bound to PMP by an antibody. Finally pretrigger and trigger reagents are added to facilitate the creation of photons of energy characteristic of AE.

In a sandwich format, the analyte-specific antigen concentration in the samples and the light emission in relative luminescent units have a direct relationship. If more analyte-specific antigen molecules are present in the samples, then more AE is present, and therefore light emission is greater.

The trigger reagents are typically acids and bases that produce a pseudobase compound that is nonluminescent and N-methyl acridone, which spontaneously creates characteristic electromagnetic radiation (EMR. This method and associated chemical processes are illustrated in Figure 5-10(B) ■. The sensitivity of this assay is greater than FIAs and is adaptable to several analyzer platforms.

ELECTROCHEMILUMINESCENT IMMUNOASSAY

A unique modification of CLIA applied to the measurement of hormones and therapeutic drugs is used by Roche Diagnostics. The technique is referred to as *electrochemiluminescence immunoassay* (ECLIA) and uses a chemical label that can generate light via an electrochemical reaction. A commonly used label is ruthenium $Ru(bpy)_3^{2+}$ with tripropylamine (TPA), which produces light as illustrated in Figure 5-11 ■.[19] Ruthenium III complex tris(2,2'-bipyridyl)ruthenium III has a reaction site for the conjugation of analytes using the activation reagent N-hydroxysuccinimde. Ruthenium III that is conjugated to an antibody can be applied to sandwich-type assays for large molecules. The conjugate generates light on the surface of a gold electrode. Ruthenium III on a solid phase and TPA are oxidized on the surface of electrodes to form $Ru(bpy)_3^{2+}$ and TPA^+, respectively. The ionic form of TPA^+ will spontaneously lose electrons. The excited state decays to the ground state through a normal fluorescence mechanism, emitting a photon with a maximum wavelength of 620 nm. Emission light (620 nm) is created as $Ru(bpy)_3^{3+}$ converts to $Ru(bpy)_3^{2+}$ coincident with the reduction of TPA.

The competitive immunoassay procedure is outline below:

• A patient sample is added to an analyte-specific biotinylated antibody. An immunocomplex is formed, and the amount formed is dependent on the analyte concentration in the sample.

• Streptavidin-coated microparticles and an analyte derivative labeled with ruthenium complex are added to the mixture. Any vacant sites on the biotinylated antibodies become occupied and form an antibody–antigen complex. The entire complex becomes bound to the solid phase via interaction of biotin and Streptavidin.

• The reaction mixture is aspirated into the measuring cell, where the microparticles are magnetically captured onto the surface of the electrode. Unbound fractions are then removed with a wash solution.

 Checkpoint! 5–7

What chemicals are used to facilitate the production of EMR in a chemiluminescent immunoassay that incorporates the label acridinium ester?

LUMINESCENT OXYGEN-CHANNELING IMMUNOASSAY

Luminescent oxygen-channeling immunoassay (LOCI) is an example of a novel technique applied to the measurement of chemical analytes in body fluids and is offered on the Dade Behring Dimension VISTA Intelligent Lab System (Dade Behring, Newark, DE). Examples of analytes that can be measured by this technique include thyroid compounds, cancer markers, fertility markers, cardiac markers, and anemia-related compounds.

LOCI is a homogenous immunoassay that provides highly sensitive real-time monitoring of particle–particle interactions. The two particles include *Chemibeads* and *Sensibeads*, which represent key reagents in this procedure. This technique

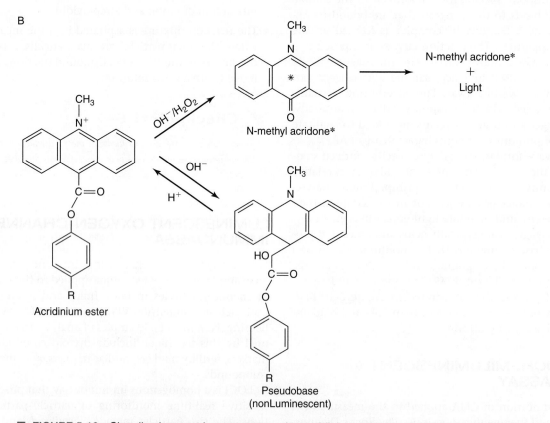

■ **FIGURE 5-10** Chemiluminescent immunoassay using acridinium ester for a label. In A. a complete immunoassay sequence generates light energy by the addition of pretrigger and trigger reagents. In B. acridinium ester undergoes chemical transition to N-methyl acridone with chemically produced light energy.

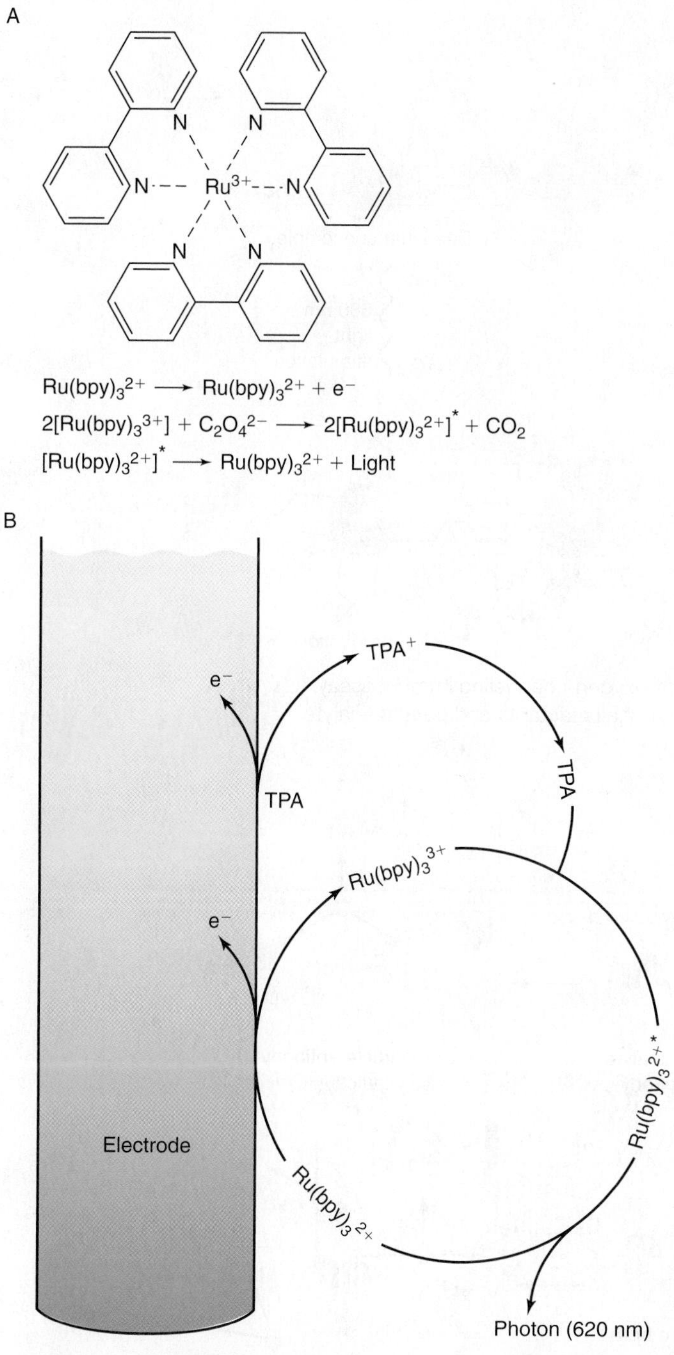

A

$$Ru(bpy)_3^{2+} \longrightarrow Ru(bpy)_3^{2+} + e^-$$
$$2[Ru(bpy)_3^{3+}] + C_2O_4^{2-} \longrightarrow 2[Ru(bpy)_3^{2+}]^* + CO_2$$
$$[Ru(bpy)_3^{2+}]^* \longrightarrow Ru(bpy)_3^{2+} + Light$$

B

■ FIGURE 5-11 Electrochemiluminescent immunoassay using ruthenium III as a label. Part A. shows the structure of the ruthenium III label and the chemical reaction of $Ru(bpy)_3^{2+}$ with oxalate to produce light energy. In B. the reaction of $Ru(bpy)_3^{2+}$ and TPA with a gold-plated electrode generates light energy at 620 nm.

uses the short diffusion distance of $^1\Delta_gO_2$ (first electronic excited state of oxygen) produced in one particle to cause preferential reaction within a particle pair. The resulting chemiluminescent reaction in a neighboring particle allows for ready monitoring of particle pair formation without

interference from particles in the sample. A distinct advantage of LOCI over other immunoassays is that it is relatively insensitive to interferences by particles or other substances present in biological samples.[20,21]

The reaction principle of the Dade Behring LOCI method is illustrated in part in Figure 5-12 ■. Exact substances used are not entirely known because they are proprietary information and not released to the public. The assay uses two synthetic beads: a Sensibead coated with Streptavidin that contains photosensitive dye and a Chemibead coated with a binding partner specific for the method and which contains a chemiluminescent dye. A third component of the system is a biotinylated receptor. This receptor is prepared by chemically attaching biotin (vitamin B₇) to a protein, usually an antibody. The process is termed *biotinylation* and results in the creation of larger complexes, thereby increasing the sensitivity of an assay. The reactants combine with analyte to form a bead-aggregated immunocomplex. Illumination of this complex by light at 680 nm releases singlet oxygen from the Sensibead, which diffuses into the bound Chemibead and triggers chemiluminescence. A blocking layer surrounding each bead minimizes the potential for nonspecific binding. The final measurement of the LOCI signal is taken at 612 nm. The resulting signal is an inverse function of the analyte concentration.

✓ Checkpoint! 5–8

Identify the element and its electronic transition state that is responsible for triggering chemiluminescence in the LOCI.

▶ OTHER IMMUNOASSAYS

KINETIC INTERACTION OF MICROPARTICLES IN A SOLUTION

The principle of *kinetic interaction of microparticles in a solution* is based on measurable changes in light transmission related to the interaction of microparticles in a solution and the analyte of interest. The analyte as shown in Figure 5-13 ■ is conjugated to microparticles in solution, and analyte polyclonal antibody is solubilized in buffer. In the absence of analyte, free antibody binds to analyte–microparticle conjugates, causing the formation of particle aggregates. As the aggregation reaction proceeds in the absence of sample drug, the absorbance increases.[22]

When the urine sample contains the analyte is question, it competes with the particle-bound drug derivative for free antibody. Antibody bound to sample analyte is no longer available to promote particle aggregation, and subsequent particle lattice formation is inhibited. The presence of sample analyte reduces the increasing absorbance in proportion to the concentration of analyte in the sample. Sample analyte content is determined relative to the value obtained for a known amount of concentration of the drug.

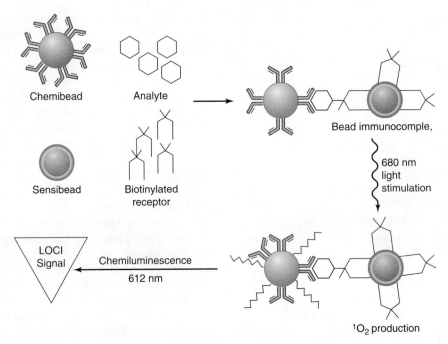

■ **FIGURE 5-12**   Homogeneous luminescent oxygen-channeling immunoassay. Solid-phase beads facilitate the attachment of all reactants and patient analyte.

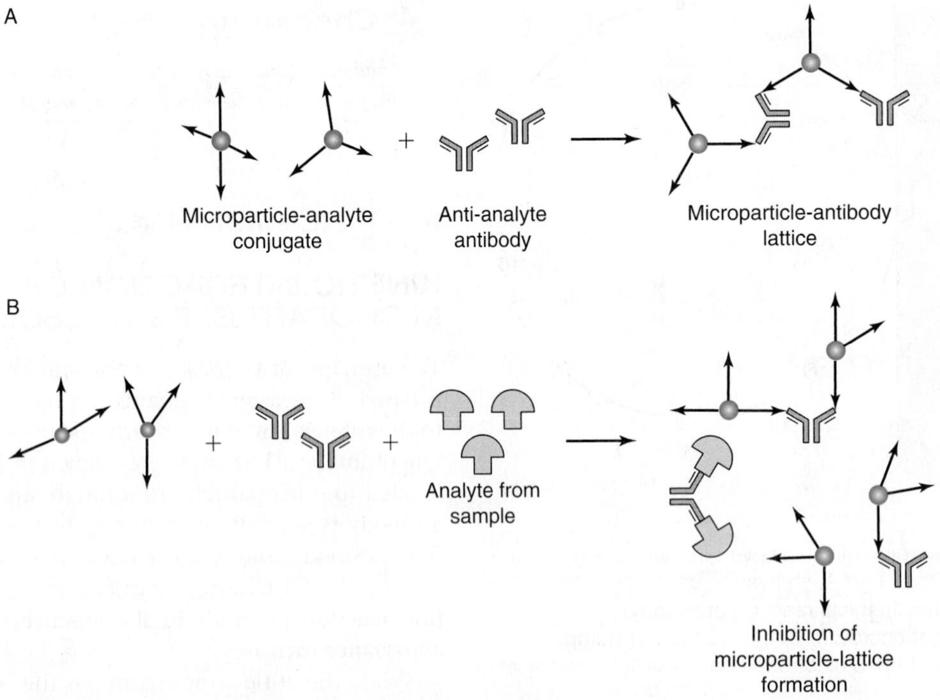

A

Microparticle-analyte conjugate   +   Anti-analyte antibody   →   Microparticle-antibody lattice

B

+   +   Analyte from sample   →   Inhibition of microparticle-lattice formation

■ **FIGURE 5-13**   Methodology for kinetic interaction of microparticle in a solution. Part A. shows a sample with no analyte. Part B. shows a sample containing analyte to be measured.

Roche Diagnostics provides drugs of abuse screening assays on several different platforms that include:

- Abuscreen ONLINE,
- COBAS Integra Chemistry Systems, and
- Roche Modular Analytical Systems.

LIMITATION OF IMMUNOASSAYS

Hook Effect

There must be an excess of antibodies, both capture and enzyme conjugates, relative to the analyte being detected to produce accurate results by immunoassays. It is only under the conditions of antibody excess that the dose-response curve is positively sloped and the assay provides accurate quantitative results. As the concentration of analyte begins to exceed the amount of antibody present, the dose-response curve will flatten (plateau); with further increases in analyte, the curve may become negatively sloped in a phenomenon termed **hook effect** or high-dose hook effect. Because the possibility exists that some samples may have analyte concentrations in excess of the amount of antibody present, questionable sample results must be validated using a dilution protocol to test the linearity. This protocol will establish whether the patient's results lie on the positively sloped region of the linearity plot or on the negatively sloped hook region of the linearity plot. Failure to validate the potential for hook effect can result in severe underestimation of the true analyte concentration.

✓ **Checkpoint! 5–9**

Identify the test protocol used to determine the presence of hook effect in a patient sample measured by an immunoassay.

▶ NEW DEVELOPMENTS IN IMMUNOASSAYS

The development and release of point-of-care technology devices is on the rise partly because of the shortage of qualified laboratory professional and the demands by the clinician for shorter turnaround times. Thus, a plethora of POCT devices are currently on the market to measure a variety of analytes in blood and urine.[7] Many of these devices are located in emergency departments, operating rooms, clinics, and doctor's offices. Table 5-2 lists several examples of POCT instruments categorized by their separation technique with detected analytes.

⊘ TABLE 5-2

Four Examples of Separation Techniques Used in Selected POCT Instruments

Product Name/Measured Analyte	Specimen	Sensitivity	Manufacturer
Dip-Strip Immunochromatographic Methods			
AimStick protein band dipstick human chorionic gonadotropin	Urine	20 mIU/mL	Craig Medical
Daina Screen hepatitis B surface antigen	Serum/plasma	3.1 ng/mL	Dainabot
Immunochromatographic Methods			
PointCare human chorionic gonadotropin	Urine	25 mIU/mL	Orgenics
Biocard troponin I	Serum	0.1 ng/mL	Ani Biotech OY
Clearview human chorionic gonadotropin II	Urine	25 mIU/mL	Inverness Medical Group
QuickVue one-step human chorionic gonadotropin	Urine	25 mIU/mL	Quidel
CARD-I-KIT troponin I	Serum	0.1 ng/mL	AboaTech Ltd
TROPT troponin T	Whole blood	0.1 ng/mL	Roche
Prostate-specific antigen rapid screen test	Whole blood	>4 ng/mL	Craig Medical
Triage Cardiac:	Whole blood		Biosite Diagnostics
Myoglobin		0.6 ng/mL	
Troponin I		10 ng/mL	
Creatine kinase		170 ng/mL	
Isoenzyme MB			
Flow-Through Methods			
ICON-II human chorionic gonadotropin	Serum	20 mIU/mL	Beckman Coulter
NycoCard C-reactive protein	Whole blood	8 mg/L	Axis-Shield PoC AS
Combined Flow-Through and Immunochromatography			
RapidSignal hepatitis B surface antigen	Serum/plasma	5 ng/mL	Orgenics

Flow-through or immunodiffusion POCT devices have been available for several years. They are commonly used for detecting hCG in urine (pregnancy testing). A patient sample is placed onto a nylon membrane or mesh containing immobilized anti-hCG. Below the nylon membrane is an absorbent pad that becomes saturated with sample. A washing buffer is applied followed by a solution of ALP-labeled anti-hCG antibody. The endpoint of the reaction is observed in the center of the absorbent pad. For most methods, a color develops if analyte is present in the sample.

A second type of POCT immunoassay technique is the "dipping strip" or immunochromatographic test. In this method, the test strip contains a spotting area with antibody at the end of the strip. The strip is dipped into a samples cup containing the patient's specimen and then dipped into a washing cup to remove excess substance. Finally, the strip is dipped into a labeling solution and into a color developer, which may be colored latex or colloidal gold.

Analytes in the sample form a complex with the conjugate and migrate to the detection zone, thereby immobilizing the capture antibody to form a positive colored line. Excess amount of the conjugate and samples migrate to the other end of the strip.

Newer POCT devices are currently being developed that will allow the user to measure multiple analytes from one sample simultaneously. Two types of devices based on solid phases are available:

1. microsectioning that uses a microchip as a solid phase, and

2. microparticles as a solid phase with two technologies available to recognize different assay parameters; one uses a fluorolabeled latex particle, and the other involves the use of particles of different sizes.

SUMMARY

Radioimmunoassay ushered in a whole new paradigm in laboratory measurements of complex structures such as proteins and hormones with improvements in sensitivity. From the earliest work of Yalow and Berson to the inclusion of immunoassays into point-of-care technology devices and other miniaturized analyzers, clinical laboratories have seen an accelerated growth pattern that may parallel computerization.

Clinicians now have the opportunity to receive laboratory data that is reliable and timely and has clinically useful sensitivities and specificities. These improvements in test performance occurred over several decades. With each new assay type, greater strides were made in their overall performance. For example, the assay sensitivity of fluorescent immunoassays was less than that of enzyme immunoassays. Similarly, the analytical sensitivity of chemiluminescent immunoassays was lower than that of fluorescence immunoassays.

Mastery of immunoassays must begin with a fundamental knowledge of antigens and antibodies. Once that knowledge is obtained, the next step is to gain an understanding of the nuances of each different type of immunoassay. So many variables are associated with each immunoassay that it is incumbent on the analyst to study these in order to be able to provide reliable and meaningful results to the clinician.

This chapter has provided the reader with fundamental knowledge and understanding of immunoassays to which more information can be added from other more detailed resources. Understanding all aspects of immunoassays can help ensure that the technologists performing the test will provide care givers with the best possible laboratory data.

REVIEW QUESTIONS

LEVEL I

1. Monoclonal antibodies are derived from: (Objective 5)
   a. different cell lines.
   b. different clones.
   c. a single cell line.
   d. a heterogeneous mixture of clones.

2. Which of the following best describes haptens? (Objective 5)
   a. Haptens are substances that are capable of binding an antibody but cannot by themselves stimulate an immune response.
   b. Haptens are immunoglobulins capable of binding specifically to both natural and synthetic antigens.
   c. Haptens can be any substance capable of reacting with an antibody but without necessarily being capable of inducing antibody formation.
   d. Haptens are compounds that can be attached to antigens or antibodies and used as labels for immunoassays.

LEVEL II

1. Which of the following best describes a significant difference between homogeneous and heterogeneous immunoassays? (Objective 2)
   a. There is a physical separation of bound from free fractions in a homogeneous immunoassay.
   b. There is no physical separation of bound from free fractions in a homogeneous immunoassay.
   c. A homogeneous immunoassay usually takes longer to perform because bound fractions must be physically separated from free fractions.
   d. Separation of bound and free fractions is optional for either type of immunoassays.

2. In fluorescence polarization immunoassay, the small fluorescein molecules chemically attached to the hapten reacts in which of the following manner? (Objective 6)
   a. They don't spin at all.
   b. They spin 100 times faster than normal.

REVIEW QUESTIONS (continued)

LEVEL I

3. A competitive immunoassay includes an antibody, patient antigen, and: (Objective 5)
   a. a paramagnetic particle to which an antibody attaches.
   b. an antibody to the antibody–antigen complex.
   c. a label that is not attached to either an antibody or antigen.
   d. an antigen with a label attached to its structure.

4. Acridinium ester and dioxetane are both examples of labels used in which of the following immunoassays? (Objective 2)
   a. fluorescence polarization immunoassay
   b. enzyme-multiplied immunoassay technique
   c. chemiluminescent immunoassays
   d. substrate-labeled fluorescent immunoassay

5. Polyclonal gels, particles coated with iron oxide, and plastic beads are all examples of: (Objective 3)
   a. a solid-phase material.
   b. antigen labels.
   c. immunogens.
   d. double antibodies.

LEVEL II

   c. They spin more slowly in solution.
   d. They spin freely in solution.

3. Which of the following immunoassays uses a gold electrode to facilitate production of light? (Objective 4)
   a. fluorescence polarization immunoassay
   b. substrate-labeled fluorescent immunoassay
   c. electrochemiluminescence immunoassay
   d. enzyme-multiplied immunoassay technique

4. Luminescent oxygen-channeling immunoassay takes advantage of which of the following distance factors to initiate a chemiluminescent reaction? (Objective 4)
   a. long diffusion distance of oxygen molecules in the first electronic state
   b. short diffusion distance of oxygen molecules in the first electronic state
   c. short absorption distances of oxygen
   d. short diffusion distance of hydrogen molecules in the first electronic state

5. The antibody that binds to hapten-enzyme in the EMIT assay: (Objective 6)
   a. inhibits enzyme activity.
   b. enhances enzyme activity.
   c. facilitates the enzyme catalytic properties.
   d. has no effect on the reaction.

PEARSON myhealthprofessionskit™

Use this address to access the interactive Companion Website created for this textbook. Simply select "Clinical Laboratory Science" from the choice of disciplines. Find this book and log in using your user name and password.

REFERENCES

1. Straus E. *Rosalyn Yalow, Nobel Laureate: Her life and work in medicine* (Cambridge, MA: Perseus Books, 1998).

2. Yalow R, Berson S. Assay of plasma insulin in human subjects by immunological methods. *Nature* (1959) 184: 1648–1649.

3. Pizzi RA. Rosalyn Yalow: Assaying the Unknown. *Modern Drug Discovery* (2001) 4, 9: 63–64.

4. Burd JF, Wong RC, Feeney JE et al. Homogeneous reactant-labeled fluorescent immunoassay for therapeutic drugs exemplified by gentamicin determination in human serum. *Clin Chem* (1977) 23: 1402–1408.

5. McPherson RA, Massey HD. Laboratory evaluation of immunoglobulin function and humoral immunity. In McPherson RA, Pincus MR (Eds.), *Henry's clinical diagnosis and management by laboratory methods*, 21st ed. (Philadelphia: Saunders Elsevier, 2007): 836–840.

6. Johnson AM. Amino acids, peptides, and proteins. In Burtis CA, Ashwood ER, Bruns DE (Eds.), *Tietz textbook of clinical chemistry and molecular diagnostics*, 4th ed. (Philadelphia: WB Saunders, 2006): 570–574.

7. Ashihara V, Kasahara Y, Nakamura RM. Immunoassay and immunochemistry. In McPherson RA, Pincus MR (Eds.), *Henry's clinical diagnosis and management by laboratory methods*, 21st ed. (Philadelphia: Saunders Elsevier, 2007): 793–800.

8. Steward MW. Overview: Introduction methods used to study the affinity and kinetics of antibody–antigen reaction. In Weir M. (Ed.), *Handbook of experimental immunology, vol I: Immunochemistry*, 4th ed. (Oxford, UK: Blackwell Scientific, 1986): 25.

9. Kricka LJ, Phil D, Chem C et al. Principles of immunochemical techniques. In Burtis CA, Ashwood ER, Bruns DE (Ed.), *Tietz textbook of clinical chemistry and molecular diagnostics*, 4th ed. (Philadelphia: WB Saunders, 2006): 219–240.

10. Thorpe GH, Kricka LJ, Moseley SB et al. Phenols as enhancer of the chemiluminescent horseradish peroxidase-luminol-hydrogen peroxide reaction; application in luminescence-monitored enzyme immunoassays. *Clin Chem* (1985) 31: 1335–1341.

11. Bronstein I, Edwards B, Voyta JC. 1, 2-dioxetanes: novel chemiluminescent enzyme substrates. Applications to immunoassays. *J Biolumin Chemilumin* (1989a) 4: 99–111.

12. Bronstein I, Juo RR, Voyta JC. Novel chemiluminescent adamantyl 1, 2-dioxetane enzyme substrate. In Stanley P, Kricka LJ (Es.), *Bioluminescence and chemiluminescence* (Chichester, UK: John Wiley & Sons, 1991): 74–82.

13. Rubenstein KE, Schneider RS, Ullman EF. Homogeneous enzyme immunoassay. A new immunological technique. *Biochim Biophys Res Commun* (1972) 47: 846–851.

14. Rowley GL, Rubenstein JE, Huisjen J, et al. Mechanism by which antibodies inhibit hapten–malate dehydrogenase conjugates. *J Biol Chem* (1975) 250: 3759–3766.

15. Henderson DR, Freidman SB, Harris JD, et al. CEDIA, a new homogenous immunoassay system. *Clin Chem* (1986) 32: 1637–1641.

16. Dandliker WB, Kelly RJ, Dandliker J, et al. Fluorescence polarization immunoassay. Theory and experimental method. *Immunochemistry* (1973) 10: 219–227.

17. Jolley ME. Fluorescence polarization immunoassay for the determination of therapeutic drug levels in human plasma. *J Anal Toxicol* (1981) 5: 236–240.

18. Weeks I, Beheshti I, McCapra F, et al. Acridinium esters as high specific activity labels in an immunoassay. *Clin Chem* (1983) 29: 1472–1479.

19. Blackburn GF, Shah HP, Kenten JH, et al. Electrochemiluminescence detection for development of immunoassays and DNA probe assays for clinical diagnostics. *Clin Chem* (1991) 37, 9: 1534–1539.

20. Ullman EF. Kirakossian H, Singh S, et al. Luminescence oxygen channeling immunoassay: Measurement of particle binding kinetics by chemiluminescent. *Proc Natl Acad Sci* (1994) 91: 5426–5430.

21. Poulsen F, Jensen KB. A luminescent oxygen channeling immunoassay for the determination of insulin in human plasma. J of Bimolecular Screening 2007; 12; 2: 240–47.

22. Armbruster DA, Schwarzhoff RH, Hubster EC, et al. Enzyme immunoassay, kinetic microparticle immunoassay, radioimmunoassay, and fluorescence polarization immunoassay compared for drugs-of-abuse screening. *Clin Chem* (1993) 39, 10: 2137–2146.

SECTION TWO
ANALYTES

6

Carbohydrates

■ OBJECTIVES—LEVEL I

Following successful completion of this chapter, the learner will be able to:

1. List and define the major classes of carbohydrates and give examples of each.

2. Recognize the structure of the common stereoisomers of carbohydrates.

3. Review the digestion of carbohydrates from the role of salivary amylase to the final CO_2 and H_2O.

4. Define the most common terms associated with carbohydrate metabolism.

5. Briefly summarize the three major biochemical pathways associated with carbohydrate metabolism.

6. List the various hormones that affect carbohydrate metabolism and summarize their functions (gland and action).

7. Compare and contrast the main characteristics of the two major types of diabetes mellitus (DM) (type 1 and type 2).

8. Review other carbohydrate disorders, including

   i. Gestational diabetes

   ii. Other types of diabetes (secondary diabetes mellitus)

   iii. Impaired glucose tolerance

9. List and explain the changes that occur in the body with hyperglycemia (complications of DM).

10. List the American Diabetes Association criteria for the diagnosis of DM, impaired glucose tolerance, and impaired fasting glucose.

11. Define *hypoglycemia* and discuss the common causes of drug-induced, reactive, and fasting hypoglycemia.

12. List the three factors in Whipple's triad.

13. Summarize the common enzymatic glucose methodologies: glucose oxidase and hexokinase.

14. Review urine and cerebrospinal fluid glucose clinical significance and methodologies.

■ OBJECTIVES—LEVEL II

Following successful completion of this chapter, the learner will be able to:

1. Summarize the inborn errors of metabolism affecting carbohydrate metabolism, including glycogen storage diseases and galactosemia.

2. Analyze the tests used to diagnose DM, including the criteria and factors that affect those tests.

3. Interpret laboratory tests used to classify patients as normal, impaired glucose tolerance, or diabetic.

4. Explain the various tests (e.g. ketones, glycosylated hemoglobin, and microalbumin) and their importance in monitoring DM.

KEY TERMS

Aldehyde	Glycogen	Ketose
Aldose	Glycogen storage diseases (GSDs)	Microalbumin
Carbohydrates	Glycogenesis	Monosaccharide
C-peptide (connecting peptide)	Glycogenolysis	Oral glucose tolerance test (OGTT)
Counterregulatory hormones	Glycolysis	Pancreatic amylase
Diabetes mellitus	Glycosylated hemoglobin	Polydipsia
Disaccharide	Hexokinase	Polyphagia
Drug-induced hypoglycemia	Hypoglycemia	Polysaccharide
Galactosemia	Impaired fasting glucose (IFG)	Polyuria
Gestational diabetes	Impaired glucose tolerance (IGT)	Reactive hypoglycemia
Glucagon	Insulin	Salivary amylase
Gluconeogenesis	Insulinopenia	Stereoisomers
Glucose oxidase	Ketones	Whipple's triad

A CASE IN POINT

Heidi, an 18-year-old white woman, was brought to the hospital emergency room in a comatose state. Her roommates stated that she had been nauseated earlier in the day. Upon physical examination, it was noted that she was breathing deeply and rapidly, her breath had a fruity odor, and her skin and mucous membranes were dry.

Heidi's physician received the following laboratory data.

Laboratory Results
Reference Range

Na$^+$	128	135–145 mmol/L
K$^+$	5.7	3.4–5.0 mmol/L
pH	7.12	7.35–7.45
Serum glucose	750	70–99 mg/dL
Serum acetone	3+	Neg

Urinalysis

Color/Appearance	Pale Yellow/Clear	
Glucose	4+	Neg
Ketones	2+	Neg

Issues and Questions to Consider

1. Identify all abnormal laboratory values.

2. On the basis of Heidi's history, clinical findings, and laboratory data, Heidi would be classified as having what type of hyperglycemia? Why?

3. Which of the laboratory findings is *most* valuable in establishing the diagnosis?

4. Explain the presence of glucose in the urine.

5. What type of antibodies (autoimmune) are often found in type 1 diabetics?

6. What is causing the "fruity odor" of Heidi's breath?

7. The nitroprusside chemical reaction with glycine, used to indicate a positive acetone reaction, reacts with which ketone bodies?

8. Why are the ketone bodies increased in diabetes mellitus?

9. Heidi is prescribed a daily regimen of insulin. Which of the laboratory procedures discussed in this chapter would be of most value in determining the degree of glucose control over a two-month period? What would be Heidi's target value?

WHAT'S AHEAD

▶ A review of carbohydrate biochemistry and metabolism.

▶ An analysis of the hormonal regulation of glucose.

▶ A summary of the clinically significant carbohydrate disorders focusing on types 1 and 2 diabetes mellitus and hypoglycemia.

▶ A description and analysis of the most common glucose methodologies: glucose oxidase and hexokinase.

▶ A review of the guidelines for diagnosing diabetes mellitus and other disorders.

▶ An examination of the other carbohydrate analytes.

▶ INTRODUCTION

Carbohydrates are the most abundant organic molecules found in nature. They are stored in the liver as glycogen and serve as our primary source of food and energy. Glucose provides building blocks for many metabolic processes, especially in the brain and red blood cells. Regulatory hormones such as insulin and glucagon maintain plasma glucose within a very narrow range even after a large meal or an overnight fast. Laboratory tests are critical in determining how efficient the body is in preventing hyperglycemia (abnormally high plasma glucose levels) or hypoglycemia (abnormally low glucose levels).

▶ BIOCHEMISTRY OF CARBOHYDRATES

Carbohydrates are a group of organic compounds including sugars, glycogen, and starches that contain only carbon, oxygen, and hydrogen. The ratio of hydrogen to oxygen is usually 2 to 1 and the general formula is $C_x(H_2O)_y$. Carbohydrates can be classified using four criteria: (1) number of carbons in the chain, (2) size of the carbon chain, (3) location of the carbonyl (CO) group, and (4) stereoisomers.

The number of carbons in the monosaccharide molecule is used to categorize the simple sugars. **Trioses**, which have

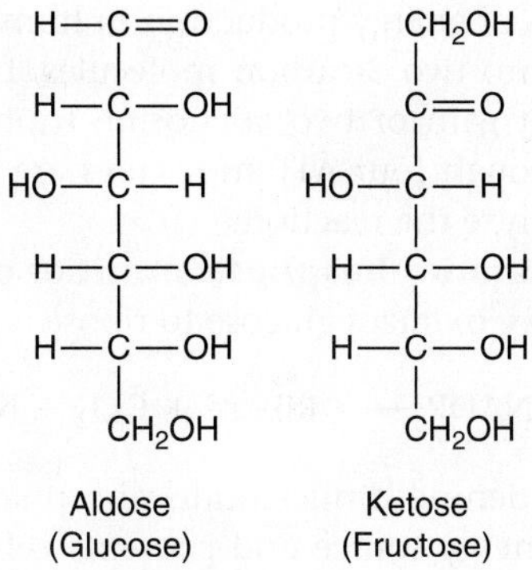

FIGURE 6-1   Aldose and ketose

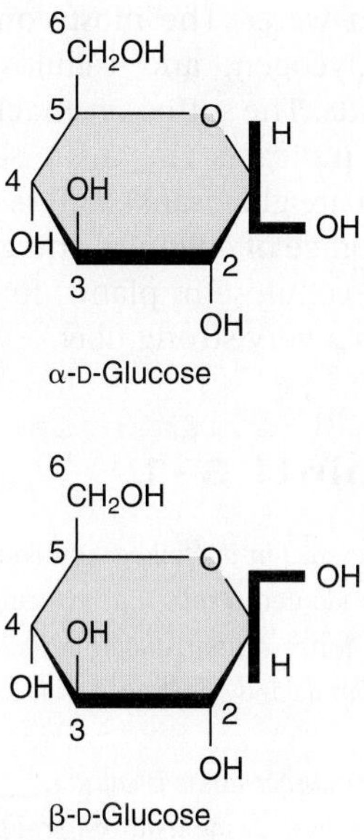

FIGURE 6-3   Glucose stereoisomers.

three carbons, are the smallest carbohydrates, followed by tetroses, pentoses, and hexoses with four, five, and six carbons, respectively. Glucose, the monosaccharide of interest in this chapter, is a hexose.

One of the carbon atoms in a monosaccharide is double bonded to an oxygen molecule to form a carbonyl group (CO), and the location of the carbonyl group is another characteristic used to describe sugars. If the carbonyl group is at the end of the carbon chain (first or last), the molecule is an **aldehyde**, and the monosaccharide is called an **aldose**. A molecule containing the functional group on an internal carbon is called a **ketone** and the monosaccharide a **ketose**. Figure 6-1 ■ illustrates an aldose and a ketose. Carbon atoms in hexoses are numbered from one to six starting with the aldehyde or ketone group.

Stereoisomoers, molecules that have the same empirical or chemical formula but have mirror image structural formulas, are also used to describe monosaccharides. D and L isomers refer to the position of the hydroxyl group on the carbon atom next to the last (bottom) CH_2OH group, which is carbon 5 in a six-carbon sugar. The D or **dextro** isomer has the hydroxyl group on the right, and in the L or **levo** isomer the hydroxyl group is on the left, as illustrated in Figure 6-2 ■. Most sugars in the body are D isomers.

Another stereoisomer of glucose (α and β) illustrated in Figure 6-3 ■ is based on the position of the –OH group on the anomeric carbon (C1), which is the C atom bonded to an

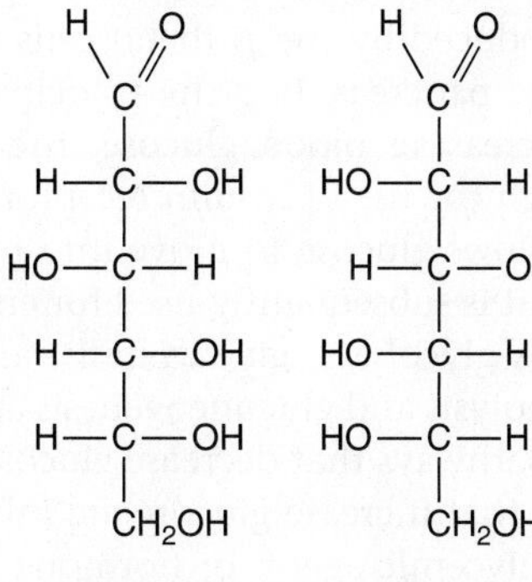

OH group and an O. In a ring structure the hydroxyl group is below the plane of the ring or on the right hand side in the α form. In the β isomer the OH group is above the plane of the ring or on the left side.

Carbohydrates can be represented by a number of models. Sugars that contain four or more carbons are usually found in cyclic forms, with the aldehyde or ketone reacting reversibly with the hydroxyl group present in the monosaccharide. In the Fisher projection the aldehyde or ketone group is at the top of the structure. It can be represented by a straight chain or cyclic. The **Haworth projection** is formed when the aldehyde or ketone reacts with the alcohol group on carbon 5, which results in a symmetrical ring structure. (See Figure 6-3.)

Disaccharides are formed by the interaction of two monosaccharides with the loss of a water molecule. Three of the most common disaccharides are maltose (two glucose molecules), lactose (glucose and galactose), and sucrose (glucose and fructose). The chemical bond between the monosaccharides always involves the aldehyde or ketone group of one monosaccharide joined to the alcohol (OH–functional group) or the aldehyde or ketone group of the second sugar. If the linkage between the two monosaccharides is between the aldehyde and ketone groups of both sugars, the disaccharide has no free aldehyde or ketone group and is a nonreducing sugar. Sucrose is an example of a nonreducing sugar. If the linkage is between the aldehyde or ketone group of one sugar and the hydroxyl group of the second monosaccharide, there is a potentially free aldehyde or ketone group and the disaccharide is a reducing sugar (e.g., glucose).

Polysaccharides are formed by glycosidic linkages of many monosaccharides, and they can be exogenous, found in most foods, or endogenous, formed in the body from monosaccharides. They are a group of complex carbohydrates composed of more than 20 monosaccharides and are

FIGURE 6-2   D- and L-Glucose

usually insoluble in water. The most common polysaccharides are starch, glycogen, and cellulose, which contain 25–2500 glucose units. The suffix *-an* attached to the name of a monosaccharide indicates the main sugar present (e.g., starch and glycogen are glucosans). Polysaccharides have two major functions: storage of chemical energy and structural. A structural example, cellulose in plants, forms a long straight chain that results in a very strong fiber.

✓ Checkpoint! 6–1

1. *Explain the difference in the linkage of two monosaccharides to form a disaccharide that results in a reducing carbohydrate and the linkage that forms a nonreducing carbohydrate. Name one reducing disaccharide and one disaccharide that is a nonreducing carbohydrate.*
2. *A five-carbon monosaccharide is called a _____.*
3. *A glucose molecule with the hydroxyl group on the left on the next to the last carbon atom is a(n) _____-glucose.*

▶ CARBOHYDRATE METABOLISM

Metabolism of carbohydrates begins in the mouth with the digestion of polysaccharides, including glycogen and starch, by **salivary amylase**. Salivary amylase, an enzyme produced by the salivary glands, catabolizes polysaccharides to intermediate-sized glucosans called "limit dextrins" and maltose. Salivary amylase is inactivated by the acid pH when it arrives in the stomach.

The food reaches the intestine, where the pH is increased by alkaline pancreatic juices and where **pancreatic amylase** completes the digestion of starch and glycogen to limit dextrins and maltose. The intestinal mucosa secretes a group of three disaccharidases (maltase, lactase, and sucrase) that digest maltose, lactose, and sucrose to monosaccharides (glucose, galactose, and fructose). The monosaccharides are absorbed by the intestinal mucosa: glucose and galactose by active transport and fructose by passive transport.

The monosaccharides are then transported by the portal vein to the liver. Monosaccharides other than glucose (e.g., galactose and fructose) are converted to glucose and utilized for energy. The ultimate goal is to convert glucose to CO_2 and H_2O. Depending on the body's needs, carbohydrates are channeled into four pathways: (1) they are converted to liver glycogen and stored; (2) they are metabolized completely to CO_2 and H_2O to provide immediate energy; (3) they are converted to keto acids, amino acids, and proteins; and (4) they are converted to fats (triglycerides) and stored in adipose tissue.

Three major biochemical pathways are involved: Embden-Meyerhoff pathway, hexose monophosphate shunt pathway (HMP shunt), and glycogenesis. The **Embden–Meyerhoff pathway** results in the glycolysis of glucose into pyruvate or lactate and can occur with or without oxygen. It is the principal means of energy production in humans. Glucose is broken down into two 3-carbon molecules, lactate or pyruvate, with a net gain of two adenosine triphosphate (ATP) molecules. Although four ATP molecules are produced, two are need to catalyze the reaction.

The **hexose monophosphate shunt** or **pentose phosphate pathway** oxidizes glucose to ribose and CO_2.

$$\text{Glucose} + \text{NADP} \longrightarrow \text{Ribose} + CO_2 + \text{NADPH} \qquad (6.1)$$

Nicotinamide adenine dinucleotide phosphate (NADPH) is an important energy source and prevents cell damage from oxidation and free radicals.

Four major terms are used to describe what occurs in glucose metabolism. **Glycogenesis** is the conversion of glucose to glycogen for storage and often occurs after a heavy meal. **Glycogenolysis** involves the breakdown of glycogen to form glucose and other intermediate products and is the process that regulates glucose levels between meals. **Gluconeogenesis** is the formation of glucose from noncarbohydrate sources such as amino acids, glycerol, or lactate and occurs during long-term fasting. **Glycolysis** is the conversion of glucose or other hexoses into 3-C molecules (lactate or pyruvate). For a more detailed discussion of carbohydrate metabolism, the reader is referred to the many excellent biochemistry textbooks.

▶ HORMONE REGULATION

Glucose levels are normally kept within a narrow range despite changes in eating and fasting. Several hormones from different glands work together to regulate blood glucose levels. Insulin is the most important hormone that lowers blood glucose when glucose levels are elevated. A group of **counterregulatory hormones**, including glucagon, increases levels when blood glucose concentrations are too low, as occurs during fasting or between meals. These two groups of hormones are counterregulatory, meaning they prevent the other group from getting out of control. For example, insulin prevents glucagon from increasing the glucose levels to dangerously high levels, and glucagon prevents insulin from decreasing glucose to unsafe levels. Table 6-1 ✪ summarizes the principal glucose regulatory hormones.

Insulin is produced by the β-(beta) cells of the islets of Langerhans in the pancreas. It is the principal hormone responsible for decreasing blood glucose, the hypoglycemic hormone. Through the use of insulin receptors on cell membranes, insulin allows glucose to move into muscle and adipose cells, where it is subsequently used for energy or stored. Insulin stimulates glycolysis, glycogenesis, and lipogenesis, whereas glycogenolysis and gluconeogenesis are inhibited. In other words, all pathways that decrease glucose are increased and all pathways that increase glucose are inhibited. Insulin is the only hypoglycemic agent, or hormone, that decreases blood glucose.

TABLE 6-1

Principal glucose regulatory hormones.

Hormone	Gland	Principal Effects
Insulin	B cells, islets of Langerhans—Pancreas	↑glycolysis, glycogenesis, and lipogenesis ↓glycogenolysis and gluconeogenesis
Glucagon	α cells, islets of Langerhans—Pancreas	↑glycogenolysis, gluconeogenesis, lipolysis ↓glycolysis, glycogenesis, and lipogenesis
Epinephrine	Adrenal medulla	↑glycogenolysis, gluconeogenesis, lipolysis ↓insulin (glucose uptake)
Glucocorticoids (Cortisol)	Adrenal cortex	↑glycogenolysis, gluconeogenesis, lipolysis ↓insulin (glucose uptake)
Adrenocorticotropic Hormone (ACTH)	Anterior pituitary	↑glycolysis, gluconeogenesis, lipolysis ↓insulin (glucose uptake)
Growth Hormone	Anterior pituitary	↑glycolysis, gluconeogenesis, lipolysis ↓insulin (glucose uptake)
Thyroxine Triiodothyronine	Thyroid	↑glycogenolysis, gluconeogenesis
Somatostatin	δ cells, islets of Langerhans—Pancreas	Inhibit secretion of glucagon and insulin

Insulin is synthesized from a precursor called proinsulin, which is composed of two amino acid chains connected by two disulfide bonds and linked by an area called "**connecting peptide**" or **C-peptide**. The α chain is 21 amino acids in length, and the β chain consists of 30 amino acids. When insulin is required, the pancreatic β-cells activate proinsulin by converting it into equimolar amounts of insulin and C-peptide. C-peptide has no biological activity, but it appears necessary to ensure the correct structure of insulin. Therefore, endogenous insulin levels have a strong correlation with C-peptide levels and both increase and decrease proportionally if the insulin is endogenous or synthesized by the person's pancreas.

Insulin and C-peptide determinations can distinguish between endogenous and exogenous or injected insulin. Injected or exogenous insulin will result in high insulin levels without an increase in C-peptide because injectable insulin is already active and does not contain C-peptide. For example, if a patient's insulin level is high and the C-peptide level is normal or low, the insulin is exogenous (injected).

COUNTERREGULATORY HORMONES

Another pancreatic hormone, **glucagon**, produced by the α cells of the islets of Langerhans, has the opposite effect and is known as a hyperglycemic agent. The release of glucagon is stimulated by a decrease in glucose. It is the principal hormone that increases glucose levels during the fasting state by stimulating glycogenolysis, gluconeogenesis, and lipolysis. Glucagon inhibits glycolysis, glycogenesis, and lipogenesis. Diabetes mellitus sometimes results in impaired secretion of glucagon, resulting in hypoglycemic episodes. Stress and exercise can also induce glucagon secretion.

Other hormones that cause an increase in glucose levels are secreted by the adrenal gland, anterior pituitary gland, and thyroid gland. The adrenal medulla synthesizes and secretes **epinephrine**, and the adrenal cortex synthesizes and secretes cortisol. Epinephrine is stimulated in the "fight or flight" syndrome and is secreted during periods of physical or emotional stress. Epinephrine serves as a very important backup when glucagon secretion is impaired. **Glucocorticoids**, primarily **cortisol**, are secreted by the cortex in response to adrenocorticotropic hormone (ACTH) from the anterior pituitary. Epinephrine and glucocorticoids (cortisol) stimulate glycogenolysis, gluconeogenesis, and lipolysis and inhibit insulin secretion, which decreases the entry of glucose into the cells.

The anterior pituitary secretes **adrenocorticotropic (ACTH)** and **growth hormones**, which are also antagonistic to insulin. Like glucagon and the adrenal hormones, they increase glycolysis, gluconeogenesis, and lipolysis and inhibit insulin and thereby glucose uptake by the cells.

The thyroid hormones, **thyroxine (T_4)** and **triiodothyronine (T_3)**, stimulated by thyroid stimulating hormone (TSH) from the anterior pituitary, have a minimal impact on glucose levels by increasing glycogenolysis and gluconeogenesis. **Somatostatins** are secreted by the delta (δ) cells of the islets of Langerhans in the pancreas. Although they do not appear to have a direct effect on carbohydrate metabolism, they do indirectly inhibit the secretion of insulin, growth hormone, glucagon, and other counterregulatory hormones.

✓ Checkpoint! 6–2

1. Two hormones regulating glucose metabolism are secreted by the pancreas. The hypoglycemic hormone is ___Insulin___ and the hyperglycemic hormone is ___glucagon___.
2. List two glands, other than the pancreas, that regulate glucose metabolism and the hormones they secrete. Salivary
3. A patient's insulin level is twice her C-peptide level (in moles). Provide a brief explanation for this scenario.

► CLINICAL SIGNIFICANCE

DIABETES MELLITUS

Diabetes mellitus (DM) is a common disorder, the seventh leading cause of death in the United States and a primary cause of blindness and end-stage renal disease. The American Diabetes Association (ADA) statistics report that a total of 20.8 million children and adults or 7.0% of the population have diabetes. One and a half million new cases of diabetes were diagnosed in 2005. Forty-one million people are classified as prediabetes or impaired glucose tolerance, which will be discussed later in this chapter.[1] About one in every 400 to 600 children under the age of 20 (176, 500, or 0.22%) are type 1 diabetics. Two million adolescents (or 1 in 6 overweight adolescents) aged 12 to19 have prediabetes.[2] The emergence of type 2 diabetes in this younger population is a major world health problem because these children are at increased risk of disabling chronic complications of DM. It is estimated that 30% of type 2 diabetics in the United States are undiagnosed.

Diabetes mellitus is a group of diseases characterized by hyperglycemia due to defects in insulin production, insulin action, or both. Although it can be associated with serious complications and premature death, diabetics can take steps to control the disease and prevent, postpone, or lower the risk of complications.

The ADA classifies diabetes mellitus into four categories: two major classes (type 1 and type 2), Other (secondary diabetes), and Gestational. Type 1 diabetes, immune-mediated diabetes, or insulin-dependent diabetes mellitus (IDDM), is caused by destruction of the β-cells of the islets of Langerhans in the pancreas, resulting in absolute insulin deficiency. Type 2 diabetes mellitus, previously called non-insulin-dependent (NIDDM), is characterized by insulin resistance with relative insulin deficiency. In other words, the insulin levels are decreased and the insulin that is present is not effective because it can't get into the cells. The third category, Other, is hyperglycemia related to secondary conditions, and the fourth category is gestational diabetes (GDM) associated with pregnancy.[3]

Type 1 diabetes mellitus accounts for about 5 to 10% of diabetics. It was previously called juvenile-onset diabetes and insulin-dependent diabetes, but the terminology of choice is now type 1. It is most commonly diagnosed in childhood and adolescence, but it has been found in adults. A cellular-mediated autoimmune response attacks the β cells of the islets of Langerhans, leading to absolute deficiency of insulin or **insulinopenia,** a decrease in insulin production. Islet cell cytoplasmic antibodies (ICA), autoantibodies to insulin (IAA), autoantibodies to glutamic acid decarboxylase (anti-GAD), and autoantibodies to tyrosine phosphatases IA-2 and IA-2β have been implicated in type 1, and one or more these autoantibodies are present in 80 to 90% of patients with type 1. These occur because the body doesn't recognize the islet cells or insulin as self and forms autoantibodies.[4] Symptoms of type 1 develop when the β-cell volume has been decreased by 80 to 90% and only 10 to 20% of the β cells are still active. Type 1 diabetics are also prone to other autoimmune diseases, such as pernicious anemia, myasthenia gravis, Hashimoto's thyroiditis, and Addison's disease.

An initiating event (e.g., a viral infection with coxsackie virus A or B, mumps, rubella, Epstein Barr virus, retrovirus) or an environmental factor leads to the production of the autoantibody in the genetically predisposed individual.

BOX 6-1 Signs and Symptoms of type 1 diabetes mellitus (3 Ps).

► **Polydipsia**: excessive thirst
► **Polyuria**: increased secretion and discharge of urine
► **Polyphagia**: increased appetite, eating large amounts of food

---

Autoimmunity to a β-cell protein initiated by a viral protein that is similar in amino acid structure to a β-cell protein appears to be a likely explanation. Type 1 typically begins with a one- to two-week progressive weight loss, irritability, and respiratory infection progressing to more severe symptoms.

Type 1 diabetes is associated with the inheritance of particular human leukocyte antigen (HLA) alleles: HLA-DR, HLA-DQ, and HLA-D, which code for the class II major histocompatability complex (MHC). Type 1 diabetes is characterized by rapid onset and insulin dependence, and type 1 diabetics are prone to go into ketosis when glucose levels are out of control. Three symptoms usually associated with type 1 are the 3 Ps: polydipsia, polyuria, and polyphagia (defined in Box 6-1 ■). Other signs and symptoms include rapid weight loss, nausea, vomiting, mental confusion, dehydration, stupor, and, if undiagnosed and untreated, coma and death. Diabetics are in a semistarvation state; although their blood glucose levels are very high, glucose cannot enter the cells to provide needed energy.[6]

Type 1 diabetics are more likely to go into **diabetic ketoacidosis (DKA)** due to the tendency to produce high levels of ketones when their glucose levels are out of control. Acidosis, high ketones, and elevated glucose levels are the three characteristics of DKA, which occurs as a result of insulin deficiency and counterregulatory hormone, especially glucagon, excess. Low insulin levels lead to decreased uptake of glucose by the cells, and glucagon stimulates gluconeogenesis, lipolysis, and glycogenolysis, leading to an even more severe hyperglycemia.

Ketones are formed by the incomplete catabolism of fatty acids from lipid stores. Because diabetics cannot utilize glucose for energy, they turn to the next source of energy, lipids. The body can only metabolize a certain quantity of lipids, and once the threshold is reached, incomplete fat metabolism results in the formation of ketones. The high glucose levels cause an increased urine output or polyuria because of osmotic diuresis. Symptoms of diabetic ketoacidosis are polyuria, polydipsia, Kussmaul respiration (deep, rapid breathing), nausea and vomiting, dehydration, weight loss, and generalized weakness.

Type 2 diabetes mellitus is the most common and milder form and accounts for 90 to 95% of diabetics. It has been referred to as adult-onset and non-insulin-dependent, but the preferred terminology is type 2. The main defect of type 2 diabetes is insulin resistance and dyslipidemia, a risk factor for cardiovascular disease (CVD). These individuals have a relative rather than an absolute lack of insulin. Their

β-cells remain intact and they usually do not require insulin to maintain adequate glucose control.

Diet and exercise are very important determinants in the pathogenesis of type 2, and the risk of developing type 2 diabetes increases with age, obesity, and lack of physical activity.[7] Type 2 is associated with a much stronger genetic predisposition than type 1. Eighty percent of type 2 patients are obese at the time of diagnosis, and type 2 is 10 times more likely to occur in an obese individual with a diabetic parent than one without a family history. Obesity is associated with insulin resistance or a decreased response to insulin. The primary defect is a decrease in the number of insulin cell receptors or an increase in cell receptors that are ineffective, meaning that they do not allow insulin and thereby glucose into the cell. No relationship to viruses and no autoantibodies are associated with type 2. Risk factors for type 2 diabetes are listed in Box 6-2 ∎.

Out-of-control type 2 diabetes will lead to a nonketotic hyperosmolar state and not ketoacidosis. Type 2 diabetes mellitus patients have a relative lack of insulin, but the counterregulatory hormones, especially glucagon, are decreased or normal, therefore leading to an inhibition of β oxidation and no ketone production. Hyperglycemia (>300–500 mg/dL), serum osmolality greater than 315 mOsm/kg, and severe dehydration are characteristic of this state. Dehydration complicates the condition by not allowing the excretion of glucose in the urine.

The characteristics of type 2 diabetes mellitus are insidious (gradual) onset, not necessarily insulin dependent, and not ketosis prone. Early diagnosis is difficult because the gradual onset of symptoms delays an appointment with a physician; therefore, it may take weeks or months before diagnosis; often type 2 is detected during a routine physical.[8] Differentiation between type 1 and type 2 diabetes mellitus is summarized in Table 6-2 ✪.

BOX 6-2 Risk factors for type 2 diabetes.

► Age ≥45 years
► Overweight (BMI ≥25 kg/m^2)^a
► Family history of diabetes (i.e., parents or siblings with diabetes)
► Habitual physical inactivity
► Race/ethnicity
► Previously identified IFG or IGT
► History of GDM or delivery of a baby weighing > 9 lb
► Hypertension (≥140/90 in adults)
► HDL cholesterol ≤35 mg/dL and/or a triglyceride level ≥250 mg/dL
► Polycystic ovary syndrome
► History of vascular disease

Source: American Diabetes Association. Screening for type 2 diabetes. *Diabetes Care* (2004) 27(Supplement 1): S11–S14.

^{a} Basal metabolic index (BMI) may not be correct for all ethnic groups.

✪ TABLE 6-2

Type 1 and type 2 diabetes mellitus.

	Type 1 Diabetes Mellitus	Type 2 Diabetes Mellitus
Onset	Abrupt	Insidious
Insulin	Absolute insulin deficiency	Relative deficiency or insulin resistance
Ketosis	Ketosis prone	Non–ketosis prone
Autoantibodies	80–90% autoantibodies • Islet cell cytoplasmic antibodies (ICAs) • Insulin • Glutamic acid decarboxylase (GAD) • Tyrosine phosphatases IA-2 & IA-20β	

Complications of diabetes mellitus, especially in uncontrolled or poorly managed diabetes, are very serious; diabetics are at increased risk for a number of life-threatening health problems. Heart disease is the leading cause of death in diabetics. Heart disease and strokes are two to four times higher in diabetics than nondiabetics due to arteriosclerosis as a result of hyperlipidemia. Abnormalities in fat metabolism associated with diabetes, increased lipids, and poor circulation lead to arteriosclerosis and associated problems. Increased triglycerides and reduced high-density lipoproteins (HDLs) are the key characteristics of dyslipidemia in type 2 diabetes.[3, 5, 9]

High blood pressure is present in about 73% of diabetics, and many require medication to control their hypertension. Patients with diabetes are diagnosed with hypertension twice as often as nondiabetic individuals. Diabetes is the leading cause of blindness in adults in the 20 to 74 age group. **Diabetic retinopathy** accounts for 12,000 to 24,000 new cases of blindness each year.[3, 9]

Forty-four percent of new cases of **end-stage renal disease** are associated with diabetes, and diabetic nephropathy occurs in 20 to 40% of diabetics. Kidney damage is caused by glucosuria and degenerative changes in the glomerular arterioles and capillaries. The first sign of renal damage is proteinuria. **Nervous system damage** is present in 60 to 70% of diabetics, including impaired sensation in the feet or hands, carpal tunnel syndrome, and other nerve problems. Peripheral neuritis as a result of degenerative damage to peripheral nerves can cause pain and loss of sensation. Diabetics are more likely to require amputation of lower limbs due to an increased **susceptibility to infection** (e.g., gangrene, due to hyperglycemia and poor circulation). Bacteria thrive in the presence of high levels of glucose, and due to poor circulation antibiotics are not as effective.[3, 9] Complications of diabetes mellitus are summarized in Box 6-3 ∎.

Gestational diabetes(GDM) is diagnosed when abnormal glucose concentrations are discovered for the first time during pregnancy. It occurs in 1 to 5% of pregnancies due to

BOX 6-3   Complications of diabetes mellitus.

▶ Heart disease and strokes
▶ Diabetic retinopathy
▶ End-stage renal disease
▶ Nervous system damage (e.g., peripheral neuritis)
▶ Susceptibility to infections (amputation of lower limbs)

BOX 6-4   Low risk for gestational diabetes.

▶ Age <25 years of age
▶ Weight normal before pregnancy
▶ Have no family history (i.e., first-degree relative) of diabetes
▶ Have no history of abnormal glucose tolerance
▶ Have no history of poor obstetric outcome
▶ Are not members of an ethnic/racial group with a high prevalence of diabetes (e.g. Hispanic American, Native American, Asian American, African American, Pacific Islander)

Source: American Diabetes Association. Standards of medical care in diabetes. *Diabetes Care* (2005) 18: S4–S53

BOX 6-5   Etiologic classification of diabetes mellitus.

1. Type 1 diabetes
2. Type 2 diabetes
3. Other specific types
   a. Genetic defects of β cell function
   b. Diseases of the pancreas
      i. Pancreatitis
      ii. Trauma/pancreatectomy
      iii. Pancreatic cancer
      iv. Cystic fibrosis
      v. Hemochromatosis
   c. Endocrinopathies
      i. Acromegaly
      ii. Cushing's syndrome
      iii. Glucagonoma
      iv. Hyperthyroidism
      v. Pheochromocytoma
      vi. Aldosteronism
   d. Drug- or chemical induced
      i. Vacor
      ii. Nicotinic acid
      iii. Thyroid hormone
      iv. Thiazides
      v. Dilantin
   e. Genetic syndromes with a higher than normal incidence of diabetes
      i. Down's syndrome
      ii. Turner's syndrome
      iii. Huntington's chorea
4. Gestational diabetes

metabolic and hormonal changes, and it must be diagnosed early so patients can be monitored and treated. Pregnant women 25 years of age and older should be screened between 24 and 28 weeks of gestation, and younger women if they are in a high-risk category. The ADA recommends that women who are at low risk for developing gestational diabetes not be screened for GDM.[3] However, if a woman does not meet all of the criteria for low risk outlined in Box 6-4 ■, she should be screened.

Most women revert to normal glucose metabolism after delivery, but there is a high probability (30 to 60%) that a woman with gestational diabetes will develop diabetes later in life. One characteristic of gestational diabetes is a larger than normal baby, usually over 9 pounds. Another complication of which the physician must be aware is that before birth the baby's insulin secretion was stimulated by the high glucose levels of the mother, but once it is born, the high insulin levels may continue, resulting in severe hypoglycemia in the newborn.

Other specific types, often referred to as **secondary diabetes mellitus**, are associated with secondary conditions. **Endocrinopathies** such as Cushing's syndrome, hyperthyroidism, glucagonoma, and acromegaly are associated with secondary diabetes. Certain drugs or chemicals, including nicotinic acid, glucocorticoids, thyroid hormone, diazoxide, and dilantin, have been known to induce diabetes in susceptible individuals. Genetic syndromes (e.g., Down's syndrome, Turner's syndrome, and Huntington's chorea) are also associated with a high incidence of diabetes.[4] For a complete summary of the etiologic classification of diabetes mellitus, see Box 6-5 ■.

CRITERIA FOR THE DIAGNOSIS OF DIABETES MELLITUS

The ADA's three criteria for the diagnosis of diabetes, involving casual (random) glucose, fasting plasma glucose, and 2-hour postload glucose, are listed in Box 6-6 ■.

Impaired Glucose Tolerance (IGT) and Impaired Fasting Glucose (IFG)

The Expert Committee of the American Diabetes Association identified a group that is in the gray area—not abnormal enough to be diagnosed diabetics but not normal—which the Committee calls impaired glucose tolerance (IGT) and impaired fasting glucose (IFG). These are defined as having a fasting plasma glucose (FPG) $\geq$100 mg/dL but <126 mg/dL. In the oral glucose tolerance test (OGTT), the 2-hour values are $\geq$140 mg/dL but <200 mg/dL.[4]

These patients are considered "prediabetics" and are at increased risk of developing diabetes and cardiovascular disease in the future. Some patients with IGT will revert to normal, but many will be diagnosed as diabetics later in life. The criteria determined by the ADA are outlined in Box 6-7 ■.

1. Symptoms of diabetes plus casual plasma glucose concentration ≥200 mg/dL (11.1 mmol/L). *Casual* is defined as any time of day without regard to time since last meal. The class symptoms of diabetes include polyuria, polydipsia and unexplained weight loss.

Or

2. FPG ≥126 mg/dL (7.0 mmol/L). *Fasting* defined as no caloric intake for at least 8 hours.

Or

3. 2-hour postload glucose ≥200 mg/dL (11.1 mmol/L) during an OGTT. The test should be performed as described by the World Health Organization (WHO), using a glucose load containing the equivalent of 75 g anhydrous glucose dissolved in water.

In the absence of unequivocal hyperglycemia, these criteria should be confirmed by repeat testing on a different day. The third measure (OGTT) is not recommended for routine clinical use.

Source: American Diabetes Association. Standards of medical care in diabetes. *Diabetes Care* (2005) 28: S4–S36.

BOX 6-7   Categories for fasting plasma glucose and 2-hour postload glucose.

Fasting Plasma Glucose

▶ Normal: FPG <100 mg/dL
▶ Impaired fasting glucose: FPG 100–125 mg/dL
▶ Diabetes mellitus: FPG ≥126 mg/dL

2-Hour Postload Glucose (OGTT)

▶ Normal: 2-hour postload glucose <140 mg/dL
▶ Impaired glucose tolerance: 2-hour postload glucose 140–199 mg/dL
▶ Provisional diabetes mellitus: 2-hour postload glucose ≥200 mg/dL (must be confirmed)

Source: American Diabetes Association. Diagnosis and classification of diabetes mellitus. *Diabetes Care* (2005) 28: S37–S42.

Checkpoint! 6–3

1. Identify the following conditions or symptoms as type 1 or type 2 diabetes mellitus.
   a. ketosis prone _____
   b. non-insulin-dependent _____
   c. obesity_____
   d. autoantibodies_____
2. List three risk factors for diabetes mellitus.

BOX 6-8   Whipple's triad

1. Signs and symptoms of hypoglycemia
2. Documentation of low plasma glucose at the time patient is experiencing the signs and symptoms
3. Alleviation of symptoms with the ingestion of glucose and an increase in plasma glucose

HYPOGLYCEMIA

Hypoglycemia is an abnormally low plasma glucose level, usually defined as below 50 mg/dL in men and below 45 mg/dL in women, accompanied by symptoms associated with hypoglycemia.[10] Three factors, called **Whipple's triad**, used to diagnose hypoglycemia, are listed in Box 6-8 ■.

Drug-induced hypoglycemia is the most common cause and accounts for over 50% of patients who are hospitalized for hypoglycemia. Insulin, alcohol, sylfonylureas, salicylates, propranolol, and quinine can lead to hypoglycemia in susceptible patients. The remaining group is non-drug-induced and is divided into fasting hypoglycemia and reactive hypoglycemia.

Fasting hypoglycemia occurs after an overnight fast or a fast of more than 8 hours. Fasting hypoglycemia diagnosed in infancy or childhood can be caused by various liver enzyme deficiencies, listed in Box 6-9 ■. Other causes of fasting hypoglycemia are islet cell adenoma or carcinoma (insulinomas), severe liver disease, and severe renal disease. Symptoms of fasting hypoglycemia are usually **neuroglycopenic** and involve the central nervous system (CNS), including confusion, inappropriate behavior, visual disturbance, stupor, seizures, and coma. These can be mistaken for inebriation or psychological problems. Fasting hypoglycemia is less

BOX 6-9   Non-drug-induced hypoglycemia.

1. Fasting hypoglycemia
   a. Inherited hepatic enzyme deficiencies (e.g., glucose-6-phosphatase; phosphorylase)
   b. Inherited defects in ketogenesis
   c. Islet cell adenoma or carcinoma (insulinoma)
   d. Nonpancreatic neoplasms
   e. Insulin-receptor antibody hypoglycemia
   f. Surreptitious insulin injection/accidental insulin overdose
   g. Severe liver disease
   h. Severe renal disease
   i. Septicemia
2. Reactive hypoglycemia
   a. Hereditary fructose intolerance
   b. Galactosemia
   c. Alimentary hypoglycemia (gastrointestinal surgery: gastrectomy, gastrojejunostomy, pyloroplasty)
   d. Early onset type 2 diabetes mellitus
   e. Idiopathic

common than reactive, but it is more severe and can be life threatening.[11]

A supervised fast is the most reliable test for the diagnosis of fasting hypoglycemia. The fast is continued for as long as 72 hours or until symptoms develop in the presence of hypoglycemia. The patient is allowed to drink calorie-free and caffeine-free liquids. Plasma glucose, insulin levels, and C-peptide levels are drawn every 6 hours until the glucose level falls below 60 mg/dL and then they are drawn every one to two hours. The test is discontinued when the glucose level falls to ≤45 mg/dL and the patient exhibits signs of hypoglycemia.[12] The fast is discontinued after 72 hours if hypoglycemia and associated symptoms do not develop. A low plasma glucose with an elevated insulin level is highly suggestive of an insulinoma.

Reactive hypoglycemia occurs within 2 to 4 hours after eating due to a delayed and exaggerated increase in plasma insulin, and it can be a predictor of early onset type 2 diabetes mellitus. Alimentary hypoglycemia is found in patients who have had gastrointestinal surgery. Reactive hypoglycemia is usually associated with milder and briefer decreases in plasma glucose. **Adrenergic symptoms**, including sweating, nervousness, faintness, palpitations, and hunger, are associated with reactive hypoglycemia. These are caused by increased sympathetic activity and release of epinephrine. Reactive hypoglycemia is more common and usually not life threatening.[11]

 Checkpoint! 6–4

1. *List the three components of Whipple's triad.*
2. *Linda has a glucose level of 50 mg/dL, but no symptoms (no weakness, shaking, or dizziness). Does she suffer from hypoglycemia?*

INBORN ERRORS OF CARBOHYDRATE METABOLISM

Glycogen storage diseases (GSDs) are caused by genetic defects resulting in the deficiency of a specific enzyme in the glycogen metabolic system. Symptoms vary and can include muscle cramps and wasting, hepatomegaly, and hypoglycemia. The most common forms of GSD are types I, II, III, and IV, which account for 90% of all cases.

Type I, **glucose-6-phosphatase deficiency**, also known as **von Gierke disease**, is the most common form. It involves the last step in glycogenolysis, which is the conversion of glucose-6-phosphate to glucose. Von Gierke is an autosomal recessive disease characterized by hypoglycemia, metabolic acidosis, ketonemia, and ketouria. Glycogenolysis cannot occur, which leads to a buildup of glycogen in the liver and hepatomegaly. Severe hypoglycemia, hyperlipidemia, uricemia, increased lactate, and growth retardation are characteristic of von Gierke patients. Hypoglycemia

is caused by the body's inability to convert glycogen to glucose. Primary symptoms improve as the patient ages, but liver tumors, liver cancer, chronic renal disease, and gout may develop.[13]

Type II or **Pompe's disease** (*acid maltase deficiency*) is caused by the lysosomal alpha-D-glucosidase deficiency in skeletal and heart muscles. It is divided into two forms based on age of onset. The infantile form is seen within a few months of birth and is characterized by muscle weakness, dyspnea, and cardiomegaly. Cardiac failure and death usually occur before age 2 despite medical treatment. Juvenile and adult forms of GSD II affect the skeletal muscles in the body's limbs and torso. Treatment can extend life; however, there is no cure. Respiratory failure is the primary cause of death.[13]

Type III or **Cori's disease** is caused by a glycogen debrancher enzyme deficiency in the liver, muscles, and some blood cells. About 15% of patients have only liver involvement. Debrancher enzymes unlink the small branches of the glycogen molecule. GSD III characteristics include hypoglycemia, retarded growth, and hepatomegaly. GSD III also causes muscle wasting, cardiomegaly, and hyperlipidemia.[13]

Type IV or **Anderson's disease** is also associated with a glycogen brancher enzyme deficiency in the liver, brain, heart, skeletal muscles, and skin fibroblasts. The glycogen that is produced is abnormal and accumulates in the cells, resulting in organ damage. Again, infants with type IV appear normal at birth but are diagnosed with hepatomegaly within a few months, develop cirrhosis of the liver by age 3 to 5, and die as a result of chronic liver failure.[13]

Another inborn error or metabolism involving the lack of an enzyme, galactose-1-phosphate uridyl transferase, prevents the breakdown of galactose and results in **galactosemia**. This is a autosomal recessive trait, which means that the infant must inherit the abnormal gene from both parents. Galactose, a major monosaccharide formed in the breakdown of lactose found in milk, cannot be utilized and is excreted in the urine of affected newborns.[14]

Galactose in large amounts can cause serious complications. Galactosemia usually causes no symptoms at birth, but jaundice, diarrhea, and vomiting soon develop. If untreated, symptoms are failure to thrive, cataracts, hepatic dysfunction, extreme mental retardation, and death.[14]

Diagnosis of galactosemia is made usually with a blood test from a heel stick as part of the mandatory newborn screening. Another screening test is the presence of reducing carbohydrates in the urine, which will be discussed in the section on urine glucose. All newborns and infants under three years of age should be screened for the presence of reducing sugars using Clinitest tablets. Galactose is a reducing sugar and will reduce copper, resulting in a positive test. Strict dietary restrictions can control the condition, but there continues to be a high incidence of long-term complications involving speech and language and learning disabilities.

▶ SPECIMEN COLLECTION AND HANDLING

Glucose levels can be performed on serum, plasma, or whole blood. Most clinical chemistry instruments use serum or plasma to monitor patient values. Instruments for the self-monitoring of blood glucose (SMBG) for diabetics who monitor their glucose levels at home without access to a centrifuge use whole blood. Whole-blood measurements are about 15% lower than serum or plasma because the space occupied by the red blood cells results in less space for plasma and therefore glucose in whole blood. If serum or plasma is used, it should be separated from the red cells within an hour after collection to prevent decreased glucose due to glycolysis by the cells. Glycolysis decreases blood glucose by 5 to 7% per hour in uncentrifuged coagulated blood at room temperature.

Sodium fluoride is the anticoagulant of choice if the specimen will be delayed in reaching the lab for processing and analysis. Sodium fluoride is a preservative and anticoagulant that prevents glycolysis.

▶ GLUCOSE METHODOLOGIES

The most common methodologies used to measure glucose are enzymatic glucose oxidaseandhexokinase. Although there is a long and varied history of glucose methodologies, the other procedures are of historical interest only and beyond the scope of this textbook.

GLUCOSE OXIDASE

In the **Trinder reaction** (glucose oxidase/peroxidase—GOD/POD), glucose oxidase catalyzes the oxidation of glucose to gluconic acid and hydrogen peroxide (H_2O_2).

$$Glucose + O_2 + 2H_2O \xrightarrow{Glucose\ Oxidase} Gluconic\ acid + 2H_2O_2 \quad (6.2)$$

The indicator reaction uses peroxidase and a chromogenic oxygen acceptor. *O*-dianisidine, 3-methyl 2-benzothiazolinone hydrazone, and *N,N*, dimethylaniline are examples of three oxygen acceptors resulting in color formation, which can then be measured.

$$H_2O_2 + Reduced\ chromagen \xrightarrow{Horseradish\ peroxidase} Oxidized\ chromagen + H_2O \quad (6.3)$$

Glucose oxidase is highly specific for β-D glucose. Glucose in solution is approximately 36% α and 64% β; therefore, many companies add a mutarotase to convert the α to β for a complete reaction. The disadvantage of the Trinder reaction is that the indicator or secondary peroxidase reaction is much less specific for glucose than the glucose oxidase. Uric acid, ascorbic acid, bilirubin, and glutathione inhibit the reaction by competing with the chromagen in the peroxidase reaction,

resulting in lower glucose values. The advantage of glucose oxidase is the lower cost. It can be used for cerebrospinal glucose but is not suitable for urine glucose due to many interfering substances in urine. Many home monitoring devices utilize GOD/POD.

HEXOKINASE

Hexokinase is the initial enzyme used in a second coupled enzyme reaction for glucose. It catalyzes the phosphorylation of glucose by adenosine triphosphate (ATP) to form glucose-6-phosphate and adenosine diphosphate (ADP). Hexokinase is the reference method for glucose methodologies.

$$Glucose + ATP \xrightarrow[Mg^{++}]{Hexokinase} Glucose\text{-}6\text{-}phosphate + ADP \quad (6.4)$$

The indicator/secondary reaction involves the oxidation of glucose-6-phosphate by nicotinamide adenine dinucleotide phosphate ($NADP^+$) to form NADPH in proportion to the glucose and the glucose-6-phosphate produced in the first reaction.

$$Glucose\text{-}6\text{-}phosphate + NADP^+ \xrightarrow{Glucose\text{-}6\text{-}phosphate\ dehydrogenase\ (G6PD)} 6\text{-}phosphogluconolactone + NADPH + H^+ \quad (6.5)$$

The reduction of NADP to NADPH results in a decrease in absorbance at 340 nm. The sources of G6PD can be yeast, with the cofactor NAD, or bacterial (Leuconostoc mesenteroides), with NADP as the cofactor.

Interfering substances in the hexokinase reaction include some drugs, hemolysis, bilirubin, and lipemia. Unlike GOD/POD,, hexokinase is not affected by ascorbic acid or uric acid. The disadvantage is cost; hexokinase is more expensive than glucose oxidase.

REFERENCE INTERVAL

Fasting serum and plasma glucose levels are approximately 70–99 mg/dL. Many variables will affect this range, including methodology, laboratory, and instrumentation. Fasting specimens should be collected after a fast of at least 8 hours but not greater than 16 hours.

URINE AND CSF GLUCOSE

Urine glucose levels can be measured semiquantitatively by the GOD/POD reaction using urinalysis reagent strips. Results are reported as negative, 1+, 2+, 3+, and 4+ corresponding to concentrations of 100 mg/dL to 2 g/dL.

False positive results are found when the container is contaminated with oxidizing agents (bleach) and detergents. High levels of ascorbic acid (vitamin C) and ketones, high specific gravity, low temperatures (refrigerated urine), and improperly preserved specimens result in false negative or lower glucose values.

A nonspecific method using copper reduction methodology using Clinitest tablets is also available. A color change from a negative blue color through green, yellow, and orange/red occurs when the cuprous ions are oxidized by a reducing substance (e.g., glucose). Clinitest methodology is also used to determine the presence of non-glucose-reducing substances (e.g., galactose).

CSF glucose can be performed using the glucose oxidase or hexokinase serum methodologies. The reference range of CSF glucose is 40–70 mg/dL. The CSF glucose is approximately two-thirds to three-quarters of the plasma glucose. For example, if the serum glucose is 100 mg/dL, you would expect the CSF glucose to be between 66 and 75 mg/dL. A plasma glucose drawn at the same time is critical in interpreting the CSF glucose result, because the serum and CSF glucoses should be compared. For example, if the serum glucose is abnormally high or low, the CSF glucose can be abnormal and fall outside the reference range but be within the 66–75% of serum glucose.

▶ LABORATORY DIAGNOSIS

FASTING GLUCOSE

Fasting plasma glucose (FPG) is the preferred screening tests for nonpregnant adults. The criteria for diagnosis of diabetes, impaired fasting glucose and impaired glucose tolerance, are outlined in Box 6-7.

GLUCOSE TOLERANCE AND 2-HOUR POSTLOAD TESTS

A 2-hour postprandial glucose may be performed two hours following a meal. A more standardized 2-hour glucose can be obtained by performing a 2-hour postload glucose level when a patient is given a 75-g glucose drink and a glucose level is drawn 2 hours later. A glucose level over 200 mg/dL is indicative of diabetes mellitus, but it must be confirmed by an abnormal fasting glucose or random glucose on a subsequent day.[3]

The ADA no longer recommends the oral glucose tolerance test (OGGT) as a screening test because it is inconvenient for the patient as well as unnecessary. It is still recommended for the diagnosis of gestational diabetes. If it is performed, it should be carried out to obtain the most reproducible results possible.[15] The patient should have fasted for 10 to 16 hours and the OGGT should be performed in the morning. Factors that affect OGTT and need to be taken into consideration, controlled, or eliminated before the test are listed in Box 6-10 ■. Medications known to affect OGTT should be discontinued if possible, and a fasting glucose level should have been performed previously. The drink containing 75 g of glucose should be consumed within 5 minutes, which should be timed when the patient begins to drink. Glucose levels are drawn fasting, ½ hour, 1 hour, 2 hours, and 3 hours.

BOX 6-10 Factors that affect glucose tolerance tests.

- ▶ Gastrointestinal surgery and malabsorption
- ▶ Carbohydrate intake (at least 150 g for 3 days before the OGTT)
- ▶ Inactivity (nonambulatory patients)
- ▶ Weight (obesity)
- ▶ Stress (surgery, infection, hospitalization)
- ▶ Nausea
- ▶ Anxiety
- ▶ Caffeine (coffee, tea)
- ▶ Cigarette smoking
- ▶ Time of day
- ▶ Amount of glucose ingested (75 g)

▶ OTHER CARBOHYDRATE-RELATED ANALYTES

KETONES

Uncontrolled or out-of-control type 1 diabetics are susceptible to diabetic ketoacidosis, which is characterized by hyperglycemia and excess ketones. Ketones are produced by the liver from the catabolism of fatty acids from lipid stores when diabetics cannot use carbohydrates (glucose) for energy. The fatty acids are not metabolized completely, resulting in the presence of ketones. Three major ketone bodies and approximate percentages are acetone (2%), acetoacetic acid (20%), and β-hydroxybutyric acid (78%). **Ketonemia** is the presence of excess ketones in the blood and causes the characteristic fruity odor of the breath of diabetics in ketoacidosis. **Ketonouria** is the presence of excess ketones in the urine.[16]

Ketones are present in other instances of carbohydrate deprivation, including starvation, anorexia,[17] glycogen storage disease, and prolonged vomiting and diarrhea (gastrointestinal disturbances). Diabetics should be tested for urine ketones during acute illness, stress, persistent hyperglycemia, pregnancy, or symptoms consistent with diabetic ketoacidosis (nausea, vomiting, or abdominal pain).

Ketones in the serum or urine are most commonly detected using the sodium nitroprusside reaction in urine reagent strip tests and Acetest tablets. Sodium nitroprusside reacts with acetoacetic acid in an alkaline medium, yielding a purple color. If glycine is added, acetone will also react; however, β-hydroxybutyric acid will not react in any routine screening test for ketones. This is not a problem because ketones are always formed in the same percentages, so if the urine is positive for acetone and acetoacetic acid, β-hydroxybutyric acid is also present.

Testing for β-hydroxybutyrate is becoming more widespread in the last few years. β-hydroxybutyrate in the presence of NAD is converted to acetoacetate and NADH by β-hydroxybutyrate dehydrogenase at pH 8.5, as illustrated in Equation 6-6.

$$\beta\text{-hydroxybutyrate} + \text{NAD} \xrightleftharpoons{\beta\text{-hydroxybutyrate dehydrogenase}} \text{Acetoacetate} + \text{NADH} + \text{H}^+ \quad (6.6)$$

The NADH produced reduces nitroblue tetrazolium (NBT) to a purple compound that is read spectrophotometrically.

$$\text{NADH} + \text{NBT} \xrightleftharpoons{\text{Diaphorase}} \text{NAD} + \text{Reduced NBT} \quad (6.7)$$

GLYCOSYLATED HEMOGLOBIN (GLYCATED HEMOGLOBIN)

Glycation is the nonenzymatic addition of a sugar residue to amino groups of proteins. **Glycosylated hemoglobin** is formed by the glycation of glucose to the N-terminal valine of one or both β chains of hemoglobin A to eventually form HbA_{1c}. Glycation occurs over the 120-day life span of red blood cells (RBCs) and is proportional to the glucose levels during that period.[18]

Glycosylated hemoglobin is an index of the patient's average blood glucose over a 2- to 3-month period. It can be used to determine compliance with therapy and the extent to which diabetic control has been achieved; in other words, how well the patient controlled his or her glucose levels over the past 3 months. Methodologies are based on either charge difference between glycated hemoglobin and nonglycated hemoglobin or structural characteristics of glycogroups on hemoglobin. The preferred method is affinity chromatography, which is neither temperature dependent nor affected by the presence of hemoglobin F, S, or C. Box 6-11 ■ lists the most common methodologies for glycosylated hemoglobin. Immunoturbidometric and ion-exchange chromatography are the methods adopted by most clinical laboratories.[19]

Every 1% change in HbA_{1c} is equivalent to a 30-mg/dL change in glucose. For example, a 4% HbA_{1c} corresponds to an average glucose of 60 mg/dL, and a 5% HbA_{1c} is equivalent to 90 mg/dL. The ADA recommends HbA_{1c} as the primary target for glycemic control and a HbA_{1c} less than 7 %. HbA_{1c} should

BOX 6-11 Glycosylated hemoglobin (HbA_{1c}).

Based on charge differences between glycated and nonglycated hemoglobin

▶ Cation-exchange chromatography
▶ Agar gel electrophoresis

Based on separation of components based on structural differences between glycated and nonglycated

▶ Boronate affinity chromatography
▶ Immunoassay

be performed at least biannually on all diabetics and more often with those exhibiting poor glycemic control. Studies have shown that glycemic control reduces the development and progression of micro- and macrovascular complications of diabetes mellitus, including retinopathy, nephropathy, and neuropathy.[18]

A significant advantage of glycosylated hemoglobin is that it does not have to be done fasting; therefore, it can be drawn during an office visit. It is also not affected by diet immediately preceding the time it was drawn. Therefore, unlike glucose levels, when the patient can be "good" for a day or two before the visit and lower his or her glucose levels, glycosylated hemoglobin provides a true picture of how well the patient is controlling glucose levels with diet and/or medication.

MICROALBUMIN

Diabetic nephropathy is one of the most serious complications of diabetes mellitus. Over a period of years, the kidneys are damaged by the excess glucose in the urine of uncontrolled diabetics. Two mechanisms leading to albuminuria have been discussed. One mechanism is the general vascular dysfunction that allows albumin to pass through the glomerular membrane. A second factor is vascular inflammation caused by the leakage that results in further damage to the blood vessels.

Microalbumin can be used to diagnose or monitor diabetic nephropathy before it causes further renal damage. In type 1 diabetics, microalbuminuria is usually discovered five years or more after diagnosis, because it takes a few years for the damage to occur. In type 2 diabetics, it is usually present at the time of diagnosis because the kidneys have already been damaged (the diabetes has usually been present for a few years before diagnosis). In 80% of type 1 diabetics, urinary albumin excretion increases at a rate of 10–20% per year, with development of clinical proteinuria in 10 to 15 years.[20]

Microalbumin is positive before overt proteinuria and is defined as 30–300 mg of albumin/24 hours or 30–300 μg albumin/mg creatinine on two of three urine collections. Normal proteinuria detected by urine chemistry reagent strips do not detect protein levels below 500 mg/dL. The ADA defines *microalbuminuria* as follows: At least two of three urine specimens collected within a period of three to six months test positive as diagnostic for microalbuminuria (MA). Type 1 diabetics should be screened annually beginning five years after diagnosis, and type 2 diabetics should be tested annually beginning at the time of diagnosis.[3]

The ADA recommends optimizing glucose control to reduce the risk and/or slow the progression of nephropathy. Also, in hypertensive patients, optimizing the control of blood pressure through angiotensin-converting enzyme (ACE) inhibitors or angiotensin- reducing blockers (ARBs) will help to reduce and/or slow the progression of nephropathy.[3]

Checkpoint! 6–5

1. *John, a diabetic, has a glycosylated hemoglobin of 8%. According to the ADA, is this acceptable? Approximately what would be his average glucose level?*

2. *Why is microalbuminuria usually present at the time of diagnosis in type 2 diabetics whereas it takes several years to develop in type 1?*

SUMMARY

Glucose is the major carbohydrate, a source of energy for the body that provides building blocks for many metabolic processes. The structural formula for carbohydrates is $C_x(H_2O)_y$. Carbohydrates, including glucose, can be classified by the number of carbons in the chain, the size of the carbon chain, the location of the carbonyl group (aldose or ketose), and the various stereoisomers. The metabolism of carbohydrates begins in the mouth with salivary amylase; depending on the body's requirements, carbohydrates are channeled into four pathways: converted to liver glycogen and stored; metabolized completely to CO_2 and H_2O for energy; converted to keto acids, amino acids, and proteins; or converted to fats and stored in adipose tissue.

Glucose levels are controlled by a number of hormones. One major hormone, insulin, lowers glucose when glucose levels are increased. Insulin stimulates glycolysis, glycogenesis, and lipogenesis and inhibits glycogenolysis and gluconeogenesis. Counterregulatory hormones, including glucagon, epinephrine, glucocorticoids, ACTH, growth hormone, thyroid hormones, and somatostatin, increase glucose levels. The counterregulatory hormones increase glycogenolysis, gluconeogenesis, and lipolysis and inhibit insulin secretion.

Diabetes mellitus is classified into two major categories type 1 and type 2. Type 1 is characterized by rapid onset, insulin dependence, and the tendency to go into ketosis when glucose levels are out of control. Type 1 is usually juvenile onset, autoimmune (antibody formation), and often associated with an initiating event like a viral illness. Type 2 is the most common form (90 to 95%) and is usually diagnosed in adults. The characteristics of type 2 are insidious (gradual) onset, not necessarily insulin dependent, and not ketosis prone. Type 2 is associated with obesity and lack of physical exercise and has a strong genetic link (runs in families).

Diagnosis of diabetes mellitus is obtained by FPG ≥126 mg/d, 2-hour postload ≥200 mg/dL, or symptoms and a casual (random) glucose ≥200 mg/dL. Impaired glucose tolerance is defined as FPG ≥100 mg/dL but <126 mg/dL and 2-hour values ≥140 mg/dL but <200 mg/dL.

Glucose oxidase/peroxidase and hexokinase are the two major glucose methodologies. Other procedures important in monitoring DM are ketones, glycosylated hemoglobin, and microalbumin.

Hypoglycemia is usually defined as below 50 mg/dL in men and 45 mg/dL in women. Whipple's triad must be present: signs or symptoms of hypoglycemia, a low plasma glucose at the same time, and alleviation of symptoms with the ingestion of glucose. Drug-induced hypoglycemia is the most common cause. Non-drug-induced hypoglycemia is divided into two categories: fasting hypoglycemia and reactive hypoglycemia. Fasting hypoglycemia can be caused by various liver enzyme deficiencies, insulinomas, severe liver disease, and severe renal disease. Reactive hypoglycemia is associated with patients who have undergone gastrointestinal surgery and early onset diabetes mellitus

REVIEW QUESTIONS

LEVEL I

1. Which of the following is a glucose molecule with the hydroxyl group on the right side on C1? (Objective 2)
   a. D-glucose
   b. L-glucose
   c. α-glucose
   d. β-glucose

2. Formation of glucose from non carbohydrate sources describes which of the following? (Objective 4)
   a. glycogenesis
   b. glycogenolysis
   c. gluconeogenesis
   d. glycolysis

3. In what form is glucose stored in the liver? (Objective 3)
   a. maltose
   b. fructose
   c. glycogen
   d. starch

LEVEL II

1. Which of the following OGTT results indicates impaired glucose tolerance? (Objective 3)

One-hour glucose	Two-hour glucose
a. 120 mg/dL	190 mg/dL
b. 115 mg/dL	105 mg/dL
c. 215 mg/dL	150 mg/dL
d. 212 mg/dL	205 mg/dL

2. Which test may be performed to assess the average blood glucose level that an individual maintained during the previous 2- to 3-month period? (Objective 4)
   a. plasma glucose
   b. two-hour postload glucose
   c. oral glucose tolerance
   d. glycosylated hemoglobin

REVIEW QUESTIONS *(continued)*

LEVEL I

4. Breakdown of glycogen to form glucose and other intermediate products is called: (Objective 4)
   a. glycolysis.
   b. gluconeogenesis.
   c. lipogenesis.
   d. glycogenolysis.

5. The conversion of glucose into lactate or pyruvate and then CO_2 and H_2O is called: (Objective 5)
   a. glycogenesis.
   b. glycogenolysis.
   c. gluconeogenesis.
   d. glycolysis.

6. Which of the following inhibits glycolysis and glucose uptake by muscle cells and causes a rise in blood glucose levels? (Objective 6)
   a. parathyroid hormone
   b. glucagon
   c. insulin
   d. gastrin

7. Which of the following hormones has the ability to decrease blood glucose concentration? (Objective 6)
   a. glucagon
   b. epinephrine
   c. insulin
   d. thyroxine

8. Decreased blood glucose, increased insulin and low C-peptide is seen in: (Objective 6)
   a. Addison's disease.
   b. injection of exogenous insulin.
   c. administration of sulfonyl urea.
   d. glucagonoma.
   e. none of the above

9. Which of the following is the primary hyperglycemic hormone? (Objective 6)
   a. insulin
   b. glucagon
   c. growth hormone
   d. thyroid hormone

10. Which form of diabetes usually manifests itself later in life and is associated with obesity, physical inactivity, and glucosuria? (Objective 7)
    a. congenital
    b. gestational
    c. type 1
    d. type 2

11. Which of the following is characteristic of type 1 diabetes mellitus? (Objective 7)
    a. low insulin levels
    b. ketosis often accompanied by hyperglycemia
    c. high frequency of autoantibodies to islet cells
    d. all of the above

LEVEL II

3. Which statement regarding glycosylated hemoglobin is true? (Objective 4)
   a. It has a sugar attached to the N-terminal valine of the β-chain.
   b. It is dependent on the time averaged blood glucose over the RBC life.
   c. Levels below 7% indicate adequate regulation for 8 to 12 weeks prior to sampling.
   d. all of the above

4. What is cutoff value for HbA_{1c} recommended by the American Diabetes Association for adequate control of blood glucose? (Objective 4)
   a. 5%
   b. 6%
   c. 7%
   d. 8%

5. What enzyme is responsible for von Gierke disease (type 1 glycogen storage disease)? (Objective 1)
   a. B-glucosidase
   b. glycogen synthetase
   c. glucose-6-phosphatase
   d. glycogen phosphorylase

REVIEW QUESTIONS (continued)

LEVEL I

12. All of the following are confirmatory of diabetes mellitus *except*: (Objective 10)
    a. fasting blood glucose of 130 mg/dL.
    b. two-hour postload of 250 mg/dL.
    c. urine glucose in excess of 300 mg/dL.
    d. one- and two-hour glucose tolerance values greater than 200 mg/dL.

13. In preparing a patient for an oral glucose tolerance test (OGTT), all of the following are done *except*: (Objective 10)
    a. patient must be ambulatory for 3 days prior to the test.
    b. carbohydrate intake must be below 150 g/day prior to the test.
    c. no food, coffee, tea or smoking 8 hours before and during the test.
    d. administration of 75 g of glucose to adults after a 10- to 12-hour fast.

14. Which of the following may result in hypoglycemia? (Objective 11)
    a. insulinoma
    b. early type 2 diabetes
    c. ethanol
    d. all of the above

15. What type of hypoglycemia follows a meal? (Objective 11)
    a. reactive
    b. fasting
    c. alimentary
    d. none of the above

16. All of the following statements about clinical hypoglycemia are true *except*: (Objective 11)
    a. neuroglycopenic symptoms must be present at the time of the low blood sugar.
    b. symptoms can be relieved by ingestion of carbohydrates.
    c. high fasting insulin levels must be present to make a diagnosis.
    d. C-peptide levels are normal or elevated.

17. In the glucose oxidase/peroxide methodology, which product reacts with the chromagen? (Objective 13)
    a. nitroprusside
    b. hydroxide
    c. hydrogen peroxide
    d. tartrate

18. The hexokinase reaction for serum glucose: (Objective 13)
    a. measures the amount of hydrogen peroxide produced.
    b. produces a green condensation product with o-toluidine.
    c. reduces cupric ions to cuprous ions.
    d. uses a glucose-6-phosphate dehydrogenase (G-6-PD) catalyzed indicator reaction.

19. In the glucose oxidase methods, a mutarotase is added to: (Objective 13)
    a. remove inhibitors to glucose oxidase.
    b. convert α-glucose to β-glucose.

LEVEL II

REVIEW QUESTIONS *(continued)*

LEVEL I	LEVEL II

LEVEL I

c. convert alpha glucose to gluconic acid.

d. convert fructose to glucose.

e. measure the amount of hydrogen peroxide produced.

20. Glucose concentration in whole blood is: (Objective 13)
   a. less than the concentration in plasma or serum.
   b. greater than the concentration in plasma or serum.
   c. equal to the concentration in plasma or serum.
   d. meaningless because of its instability.

21. If a patient's serum glucose is 85 mg/dL, what would you expect the cerebrospinal fluid glucose to read? (Objective 14)
   a. 90 mg/dL
   b. 80 mg/dL
   c. 70 mg/dL
   d. 60 mg/dL

22. A patient with an insulinoma may exhibit dizziness and fainting attributable to: (Objective 11)
   a. acidosis.
   b. ketosis.
   c. hypoglycemia.
   d. hyperglycemia.

23. The following glucose tolerance test results are indicative of what state? (Objective 10) Glucose values are assayed by the glucose oxidase method.

Fasting:	130 mg/dL
½ h:	165 mg/dL
1 h:	225 mg/dL
2 h:	150 mg/dL

   a. normal
   b. diabetes mellitus
   c. Addison's disease
   d. hyperinsulinism
   e. hypothyroidism

PEARSON
myhealthprofessionskit™

Use this address to access the interactive Companion Website created for this textbook. Simply select "Clinical Laboratory Science" from the choice of disciplines. Find this book and log in using your user name and password.

REFERENCES

1. National Diabetes Information Clearinghouse. *National Diabetes Statistics.* http://diabetes.niddk.nih.gov/dm/pubs/statistics. (accessed 9/12/2007).

2. American Diabetes Association. *Diabetes Statistics.*www.diabetes.org/diabetes-statistics.jsp (accessed 2/6/2008).

3. American Diabetes Association. *Standards of Medical Care in Diabetes* (2005) 28: S4–S36.

4. American Diabetes Association. Diagnosis and classification of diabetes mellitus. *Diabetes Care*(2005) 28: S37–S42.

5. Chiasera JM. Diabetes mellitus: A growing healthcare concern. *Advance* (2005) 17(17): 22–28.

6. Ogedegbe HO. Diabetes mellitus: A clinical laboratory perspective. *LabMedicine* (2006).37(5): 292–97.

7. American Diabetes Association. Screening for type 2 diabetes. *Diabetes Care*(2004) 27: S11–S14.

8. Winter WE. Diabetes disease management: Differentiating type I and type 2 diabetes. *Clinical Laboratory News* (2005) 34(7): 14–16.

9. National Center for Chronic Disease Prevention and Health Promotion. National Diabetes Fact Sheet United States 2005. www.cdc.gov/diabetes/pubs/ndfs_2003.pdf (accessed 2/6/2006).

10. Raghavan VA, Srinivasan AR, Snow, KF. Hypoglycemia. www.emedicine.com/med/topic1123.htm (accessed 9/12/2007)

11. The Merck Manual. Hypoglycemia. www.merck.com/mrkshared/mmanual/sctin2/chapter13/13/13e.jsp (accessed 2/6/2008).

12. Snow KJ. Hypoglycemia. www.emedicine.com/med/topic1123.htm (accessed 2/6/2008).

13. Barrett J.. Glycogen storage diseases. *Gale Encyclopedia of Medicine*. *www.findarticles.com/p/articles/mi_g2601/is_0005/ai_2601000598* (accessed 9/12/2007).

14. Galactosemia. www.galactosemia.org/galactosemia.asp (accessed 2/6/2008).

15. Sacks D, Burns DE, Goldstein DE, Maclaren NK, McDonald JM, Parrott M. Guidelines and recommendations for laboratory analysis in the diagnosis and management of diabetes mellitus. *Clinical Chemistry* (2002) 48: 436–72.

16. Goldstein DE, Little RR, Lorenz R et al. Tests of glycemia in diabetes. *Diabetes Care* (2004) 27: 1761–73.

17. Daee A, Robinson P et al. Psychological and physiological effects of dieting in adolescents. *SouthMedJ* (2002) 95(9): 1032–1041.

18. Krishnamurti U, Steffes MW. Glycohemoglobin: A primary predictor of the development or reversal of complications of diabetes mellitus. *Clinical Chemistry* (2001). 47: 1157–65.

19. Molinaro RJ. Targeting HbA1c: Standardization and clinical laboratory measurement. *MLO* (2008) 40(1): 10–19.

20. Busby DE, Atkins RC. The detection and measurement of microalbuminuria: A challenge for clinical chemistry. *MLO* (2005) 37(2): 8–16.

Suggested Readings

1. Lerardi-Curto L. Glycogen storage disease type 0. www.emedicine.com/PED?topic873.htm. *The Merck Manual*. Hypoglycemia. www.merck.com/mrkshared/mmanual.section2/chapter13/ 13e.jsp (accessed 9/12/2007).

2. Smeeks FC. Hypoglycemia. www.emedicine.com/emerg/TOPIC272.htm (accessed 2/12/2008).

3. Goldstein DE, Little RR, Lorenz, Rodney A et al. Tests of glycemia in diabetes. *Diabetes Car.* (2004) 27(7): 1761–73.

7

Lipids, Lipoproteins, and Cardiovascular Disease

■ OBJECTIVES—LEVEL I

Following successful completion of this chapter, the learner will be able to:

1. Review cholesterol metabolism, absorption, synthesis, and catabolism.
2. Outline and describe classes of clinically significant lipids.
3. Define unsaturated and saturated fatty acids.
4. List and explain the role of the major apolipoproteins.
5. Summarize lipid metabolism, including exogenous, endogenous, and reverse cholesterol pathways.
6. List the major components and the percentage composition of the major lipoproteins—for example, apoproteins, cholesterol, and triglycerides.
7. Review the four major lipoproteins and their density and function.
8. List conditions associated with hypercholesterolemia and hypocholesterolemia.
9. Summarize the major cholesterol methodologies.
10. Identify causes of hypertriglyceridemia and hypotriglyceridemia.
11. Review triglyceride methodologies.
12. Summarize HDL-C methodologies.
13. Calculate LDL-C using the Friedewald formula.
14. Examine Lipoprotein (a) [Lp(a)] and its clinical significance.

■ OBJECTIVES—LEVEL II

Following successful completion of this chapter, the learner will be able to:

1. Review metabolic syndrome and its identification and clinical significance.
2. Summarize the recommendations of the National Cholesterol Education Program (NCEP).
3. Identify the risk factors for coronary heart disease.

KEY TERMS

Acyl-cholesterol acyltransferase (ACAT)
Apolipoprotein
Atherogenic
Cholesterol
Cholesterol oxidase
Chylomicrons
Coronary artery disease (CAD)
Coronary heart disease (CHD)
Endogenous cholesterol
Essential fatty acids
Exogenous cholesterol
Fatty acids

Friedewald formula
Glycerol esters
Glycerol kinase
High-density lipoprotein (HDL)
Hypercholesterolemia
Hypocholesterolemia
Hyperlipidemia
Hypertriglyceridemia
Hypotriglyceridemia
Lecithin-cholesterol acytransferase (LCAT)
Lipase
Lipids

Lipoprotein (a) [Lp(a)]
Lipoprotein lipase (LPL)
Lipoproteins
Low-density lipoprotein (LDL)
Metabolic syndrome
Monounsaturated fatty acids
Phospholipids
Polyunsaturated fatty acids
Prostaglandins
Saturated fatty acids
Triglycerides
Very-low-density lipoprotein (VLDL)

 A CASE IN POINT

Robert R., a 47-year-old male, went to his physician for an "annual" physical that had been postponed for more than 3 years. He had started his own business 4 years ago and had been extremely busy getting it established. Robert's medical history indicated he was a nonsmoker, and his father and grandfather had histories of myocardial infarctions before age 55. Because he hadn't seen this patient in a few years, the physician decided to order routine screening tests.

Chemistry Results

		Reference Range
Sodium	143	136–145 mEq/L
Potassium	4.6	3.6–5.0 mEq/L
Chloride	109	101–111 mEq/L
CO_2	29.0	24.0–34.0 mEq/L
Glucose	95	70–99 mg/dL
BUN	16	7–24 mg/dL
Creatinine	1.0	0.5–1.2 mg/dL
Bilirubin total	0.7	0.2–1.2 mg/dL
AST	32	5–40 IU/L
ALP	80	30–157 IU/L
Protein	7.5	6.0–8.4 g/dL
Albumin	4.6	3.5–5.0 g/dL
Calcium	8.5	8.5–10.5 mg/dL

Miscellaneous Chemistry

		Reference Range
Cholesterol	355	Recommended (Desirable): <200 mg/dL

Urinalysis
Macroscopic

		Reference Range
Color	Yellow	Colorless to amber
Appearance	Clear	Clear
Specific gravity	1.014	1.001–1.035
pH	6.0	5–7
Protein	Neg	Neg
Glucose	Neg	Neg
Ketones	Neg	Neg
Bilirubin	Neg	Neg
Blood	Neg	Neg
Urobilinogen	Normal	Normal
Nitrite	Neg	Neg
Leukocyte esterase	Neg	Neg

Microscopic: Not indicated

Issues and Questions to Consider

1. Circle or highlight the abnormal result(s).
2. List eight secondary conditions or disorders associated with the abnormal result(s) in question 1.
3. Which can the physician rule out with Robert's medical history, physical exam, or current laboratory results?
4. What is the most probable cause of the abnormal result in this patient?

The following week, a lipid profile was performed on a 12-hour fasting specimen:

Lipid Profile

Cholesterol	350	Recommended (desirable): <200 mg/dL
High-density lipoprotein cholesterol	40	Recommended (desirable): >60 mg/dL
Triglycerides	340	Recommended (desirable): <150 mg/dL

5. Given the above information, what is Robert's LDL-cholesterol? Based on this, is he at high risk, moderate risk, or within the recommended (desirable) range for coronary heart disease?
6. If Robert's triglycerides were 450 mg/dL, could the LDL-C be calculated? Why or why not? What would be the next step?
7. List eight risk factors associated with CHD as determined by the National Cholesterol Education Program (NCEP) Adult Treatment Panel.
8. How many other risk factors, outlined in question 7, does Robert currently have given the information provided? Is he at high risk for CHD?
9. What is the follow-up testing and treatment decision recommended by the NCEP Adult Treatment Panel?

WHAT'S AHEAD

► A review of the classification and biochemistry of lipids.
► A summary of the main lipoproteins.
► An overview of lipoprotein metabolism.
► A review of the major causes of hyperlipidemia.

WHAT'S AHEAD (continued)

▶ A summary of analytical procedures for measuring cholesterol, triglycerides, HDL, and LDL.
▶ An examination of metabolic syndrome.
▶ A report of the major findings and recommendations of the National Cholesterol Education Program (NCEP).

▶ INTRODUCTION

Lipids are found in all tissues of the body and play a vital role in virtually all aspects of biological life. They have many functions, including being one of the body's main sources of energy and energy storage. Lipids are converted to hormones or hormone precursors, and they also serve as structural and functional components of cell membranes. Lipids also provide insulation for nerve conduction and heat retention.

Lipids are a class of organic compounds that are actually or potentially esters of fatty acids. Lipids are soluble in organic solvents and nearly insoluble in water because they are extremely hydrophobic. The major lipids are cholesterol, triglycerides, phospholipids, and glycolipids. This chapter will concentrate on the most common and clinically significant lipids: sterol derivatives (cholesterol), glycerol esters (triglycerides), and lipoproteins.

▶ CLASSIFICATION AND BIOCHEMISTRY OF LIPIDS

Lipids are classified into five classes: sterol derivatives, fatty acids, glycerol esters, sphingosine derivatives, and terpenes. A more comprehensive list is found in Box 7-1 ■.

BOX 7-1 Classification of Clinically Important Lipids

1. Sterol derivatives
   a. Cholesterol and cholesteryl esters
   b. Steroid hormones
   c. Bile acids
   d. Vitamin D
2. Fatty acids
   a. Short chain (2–4 C atoms)
   b. Medium chain (6–10 C atoms)
   c. Long chain (12–26 C atoms)
   d. Prostaglandins
3. Glycerol esters
   a. Triglycerides
   b. Phospholipids
4. Sphingosine derivatives
   a. Sphingomyelin
   b. Glycosphingolipids
5. Terpenes
   a. Vitamin A
   b. Vitamin E
   c. Vitamin K

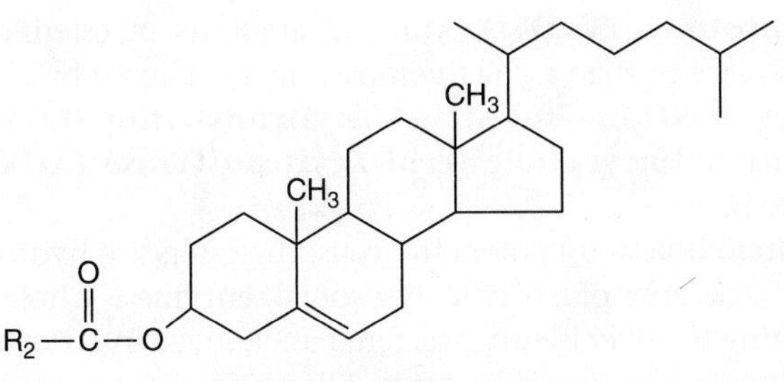

Cholesterol ester

■ FIGURE 7-1   Cholesterol ester

CHOLESTEROL

Cholesterol is found exclusively in animals and humans and is the primary sterol derivative. Cholesterol contains 27 carbon atoms and four fused rings (A, B, C, and D) called a *perhydrocyclopentanophenanthrene nucleus* (illustrated in Figure 7-1 ■). Virtually all cells in the body and body fluids contain some cholesterol. Cholesterol has three major functions: a structural component of cell membranes, a precursor of bile acids, and a precursor of steroid hormones.

Exogenous cholesterol is absorbed in the diet, bile, intestinal secretions, and cells. It is found in animal products, especially meat, egg yolk, seafood, and whole-fat dairy products. The average American diet contains 100 to 700 mg of cholesterol per day; the recommended level is less than 300 mg. **Endogenous cholesterol** is produced by the liver and is made from simpler molecules, particularly acetate. Daily synthesis of endogenous cholesterol, 500 to 1000 mg, is regulated by feedback from the levels of exogenous cholesterol. The higher the level of exogenous cholesterol, the less the body manufactures to a point. However, that doesn't mean lowering dietary cholesterol is not a good idea: it is the first line of defense against hypercholesterolemia.

ABSORPTION

To be absorbed in the intestine, cholesterol is solubilized by the formation of mixed micelles. Practically all cholesterol in the intestine is present in the unesterified (free) form because esterified cholesterol is rapidly hydrolyzed in the intestine by cholesterol esterases secreted by the pancreas and small intestine. Cholesterol is solubilized by the formation of mixed micelles containing unesterified cholesterol, fatty acids, triglycerides, diglycerides, monoglycerides, phospholipids, and conjugated bile acids. Micelles are important not only in solubilizing cholesterol but also in facilitating its transport to the intestinal luminal cell wall where it is absorbed. Maximum absorption of fat and cholesterol occurs in the middle and terminal ileum of the small intestine. As absorption of fat and cholesterol occurs, the micelles break up, reducing further cholesterol absorption.

CHOLESTEROL SYNTHESIS AND CATABOLISM

Once cholesterol is synthesized, it is released into circulation in complexes called **lipoproteins**, primarily **very-low-density** lipoproteins (VLDLs). Minimal amounts of esterification takes place in the vascular compartment catalyzed by two enzymes: **lecithin-cholesterol acyltransferase (LCAT)** in plasma and **acyl-cholesterol acyltransferase (ACAT)** in the cells.

After cholesterol enters the cells, the esters are hydrolyzed by the action of specific lysosomal enzymes. Cholesterol reaching the liver is either secreted unchanged into the bile or metabolized to form bile acids. Bile acids play an important role in cholesterol and fat absorption in the small intestine.

The cholesterol secreted directly into the biliary system is solubilized by mixed micelles of bile acids (glycholic acid, taurocholic acid, taurochenodeoxycholic acid) and phospholipids. If the amount of cholesterol exceeds the capacity of the solubilizing agents, then a supersaturated state can occur that results in the formation of gallstones.

FATTY ACIDS

Fatty acids are the simplest form of lipids with the chemical formula RCOOH. R stands for an alkyl group with a straight chain of an even number of carbon atoms. An alkyl group is any hydrocarbon radical with the general formula C_nH_{2n+1}. Fatty acids are classified primarily on their chain length (short, medium, and long) as described in Box 7-1. Fatty acids are very important to human nutrition and metabolism, serving as components of more complex lipids found in membranes, hormones, and vitamins. They store large amounts of energy, especially in the form of triglycerides.

Fatty acids are also classified according to the degree of saturation. A **saturated fatty acid** contains an alkyl chain without a double bond between C atoms. A **monounsaturated fatty acid** contains one double bond. **Polyunsaturated fatty acids** have more than one double bond. Double bonds are usually *cis,* meaning the hydrogen atoms are on the same side of the molecule. *Trans* double bonds have hydrogen atoms on opposite sides of the molecule and do not bend. They are synthesized by catalytic hydrogenation, which hardens fats. This can be seen, for example, in the production of polyunsaturated plant oils and margarine.

Fatty acids are categorized by the number of carbons and the number of double bonds. For example, linoleic acid has 18 carbons and two double bonds and is 18:2. The carbon atoms are numbered from the carboxyl end of the molecule, and the double bonds are usually three carbon molecules apart. The body can synthesize most fatty acids, except linolenic and linoleic acids, which are referred to as **essential fatty acids.** These are found in plants and are vital for health, growth, and development. They must be ingested in the diet. Most fatty acids are found in glycerol esters, including triglycerides.

PROSTAGLANDINS

Prostaglandins are derivatives of fatty acids comprising 20 C atoms, including a five-carbon cyclopentane ring. Although they are extremely potent, their full physiological

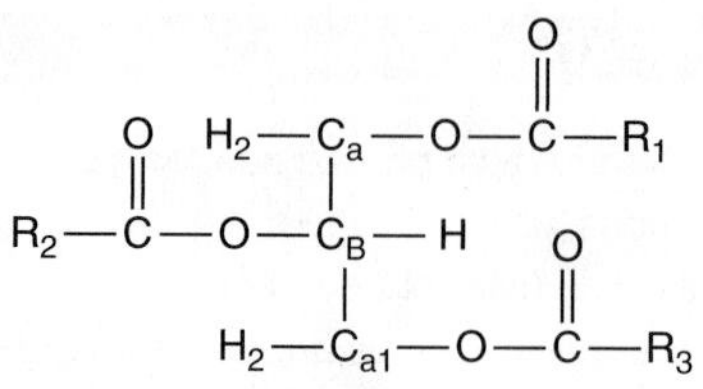

■ FIGURE 7-2   Triglyceride

function is unknown. They appear hormone-like in action, but they are different from hormones in that they are synthesized at the site of action and in almost all tissues. Prostaglandins are short lived with a half-life of seconds.

GLYCEROL ESTERS

Glycerol esters are one of the common alcohols found in human metabolism. Glycerol is a three-carbon molecule containing three hydroxyl groups. Esterification of glycerol with fatty acids produces acylglycerols, which are classified as *mono-, di-,* or *tri-* according to the number of fatty acids.

Triglycerides

Triglycerides (TGs) are the most common glycerol esters in plasma, comprising glycerol and three fatty acids (see Figure 7-2 ■). They make up approximately 95% of fat stored in tissue. Triglycerides from plants tend to have large amounts of linoleic residues and are called *polyunsaturated* fats that are liquid even at 4°C. Triglycerides from animal sources tend to have different residues and are saturated fats that are solid even at room temperature. Triglycerides can also be *exogenous* (from the diet) or *endogenous* (synthesized in the body).

Triglycerides are digested in the duodenum and proximal ileum. Pancreatic **lipase** hydrolyzes the triglyceride molecule to glycerol and fatty acids. Following absorption, the triglycerides are resynthesized in the intestinal epithelium and combined with cholesterol and apolipoprotein B-48 to form chylomicrons, which will be discussed later in this chapter.

PHOSPHOLIPIDS

Phospholipids are basically triglycerides with an additional group (hydrogen, choline, and serine) that form diacylphosphoglyceride, phosphatidylcholine, and phosphatidylserine, respectively. Phosphotidylcholines are often referred to collectively as *lecithins.* Phosphatidylserine is part of a group of phospholipids called *cephalins.* Phospholipids are synthesized in the liver and are an important part of the outer shell of lipoproteins, which are discussed in the next section.

▶ LIPOPROTEINS

Lipids are synthesized in the intestine or liver but must be transported to many distant organs and tissues to carry out their metabolic functions. Neutral fats—for example,

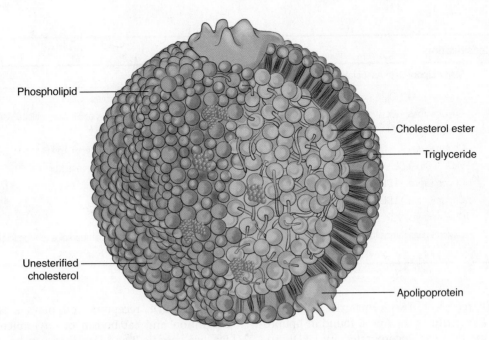

Phospholipid

Cholesterol ester

Triglyceride

Unesterified cholesterol

Apolipoprotein

■ FIGURE 7-3   Lipoprotein molecule

triglycerides and cholesteryl esters—are *hydrophobic* (repel water molecules). Lipid transport and delivery via plasma would not be possible without some form of *hydrophilic* (attracting water molecules) adaptation.

Lipids are transported by means of complex micellar structures called *lipoproteins* that consist of an outer layer of proteins, **apolipoproteins**, polar lipids (phospholipids), and unesterified cholesterol, as well as an inner core of neutral lipids (phospholipids and esterified cholesterol). The core neutral lipids are largely inactive components or passengers, and the apolipoproteins in the outer layer interact with a series of enzymes or tissue receptors. These are mainly responsible for further metabolism and catabolism of the particle. Figure 7-3 ■ illustrates a generic lipoprotein molecule.

Lipoproteins are classified into five categories:

1. chylomicrons,
2. very-low-density lipoproteins,
3. intermediate-density lipoproteins,
4. low-density lipoproteins, and
5. high-density lipoproteins.

Ultracentrifigation was the methodology used historically to differentiate the five lipoprotein classes, but other procedures have been developed to measure specific lipoproteins.

✓ **Checkpoint! 7-1**

1. *Differentiate between exogenous cholesterol and endogenous cholesterol.*
2. *Why is it difficult to lower cholesterol by only limiting dietary cholesterol?*
3. *Which lipid class is responsible for 95% of fat stored in tissue?*

APOLIPOPROTEINS

Apolipoproteins are proteins found as an integral part of the lipoprotein molecules described in the following section. They are classified in five major categories—Apo A, B, C, D, and E—with several subcategories under each. Apolipoproteins activate enzymes in the lipoprotein metabolic pathways, bind to specific cell receptors to allow the uptake of lipoproteins into the cell, and help maintain the structural integrity of the lipoprotein molecule.

Apolipoprotein A

Apolipoprotein A (Apo A) forms the major protein found primarily in **high-density lipoproteins (HDL)**, but it is also found in chylomicrons. Approximately 50% of HDL is protein, and Apo A-I and A-II make up 90% of the HDL protein in a 3:1 ratio. Apolipoprotein classification is summarized in Table 7-1 ○.

Apo A proteins originate in the intestine or liver. After they enter circulation as a component of chylomicrons, Apo A-I and Apo A-II transfer to and accumulate in HDL. Apo A-I has a role in the activation of lecithin-cholesterol acyltransferase, which esterifies free cholesterol from the outer layer of the lipoprotein molecule and shifts the esterified cholesterol into the core of the chylomicron. Apo A-I is the major protein in HDL and a measure of antiatherogenic HDL. Apo A-II plays a structural role in HDL and may inhibit LCAT and activate hepatic lipase. Apo A-IV is found in newly secreted chylomicrons but is not present in any significant amount in chylomicron remnants.

Apolipoprotein B

Apolipoprotein B (Apo B) makes up the major protein part of all lipoproteins other than HDL and is responsible for

⚙ TABLE 7-1

Apolipoprotein Classification		
Apolipoprotein	Major Lipoprotein Association	Function
Apo A-I	Chylomicrons, HDL*	Major protein of HDL, activates LCAT
Apo A-II	Chylomicrons, HDL	Primarily in HDL, activates hepatic lipase, inhibits LCAT
Apo A-IV	Chylomicrons, HDL	
Apo B-48	Chylomicrons	Solely in chylomicrons, formed from Apo B-100 in the intestinal epithelium
Apo B-100	VLDL, LDL$_1$, LDL*	Major LDL protein; binds to LDL receptor
Apo C-I	Chylomicrons, VLDL, LDL$_1$, and HDL	May activate LCAT
Apo C-II	Chylomicrons, VLDL, LDL$_1$, and HDL	Activates lipoprotein lipase
Apo C-III	Chylomicrons, VLDL, LDL$_1$, and HDL	Inhibits lipoprotein lipase
Apo E	Chylomicrons remnants, VLDL, LDL$_1$, and HDL	Mediates the uptake of remnant particles (chylomicrons, VLDL, or LDL$_1$ remnants)

*Primary lipoprotein

binding LDL to LDL receptors. Two forms exist in humans. *Apo B-100* or *large B* constitutes the Apo B found in lipoproteins synthesized in the liver, incorporated into LDL and VLDL, and secreted. There is a 1:1 ratio of VLDL to Apo B-100; in other words, there is one molecule of Apo B-100 in each VLDL molecule. The higher the concentration of Apo B-100, the greater the risk of cardiovascular disease.

Apo B-48 or *small B* is believed to be synthesized in the intestinal wall and associated with chylomicrons. In normal patients, little if any Apo B-48 exists in fasting plasma. An increase in Apo B-48 may indicate a defect in the clearance of Apo B-48 and the presence of chylomicron remnants, which can occur in renal failure.

Apolipoprotein C
Apolipoprotein C (Apo C) is a low-molecular-weight apolipoprotein synthesized primarily in the liver. Three categories of Apo C are C-I, C-II, and C-III. Apo C-I, the smallest of the C apolipoproteins, has been associated with activation of LCAT in vitro. Apo C-II activates extrahepatic **lipoprotein lipase (LPL)**, which breaks down triglycerides at the cellular level, releasing fatty acids for metabolism and energy. LPL is responsible for the hydrolysis of triglyceride-rich lipoproteins, chylomicrons, and VLDL.

Most Apo C from HDL is transferred to newly secreted chylomicrons and VLDL. Most Apo C is in VLDL and HDL, and it is associated with all lipoproteins except LDL.

Apolipoprotein D
Nothing is known about the synthesis or catabolism of apolipoprotein D (Apo D). It may function as a transfer protein, moving cholesterol or triglyceride among various lipoproteins.

Apolipoprotein E
Apolipoprotein E (Apo E) is synthesized in the liver and released as part of nascent (newly formed) HDL. It also binds

to hepatic LDL receptors and plays a significant role in recognition and catabolism of chylomicron remnants and LDL$_1$ via specific B and E receptors in hepatic cells. Apo E is incorporated into HDL, where it is transferred to VLDL and chylomicrons.

✓ **Checkpoint! 7–2**

1. Which of the following apolipoproteins is described by the following?
   a. Major protein found in HDL: _____
   b. Associated with a high risk of cardiovascular disease: _____
   c. Associated with chylomicron remnants and renal failure: _____
   d. Activates lipoprotein lipase: _____

▶ LIPOPROTEIN METABOLISM

Lipoproteins undergo a series of complex metabolic processes in which changes and exchanges occur continuously in and between the various lipoproteins. Lipid metabolism occurs in four pathways: lipid absorption (discussed earlier under cholesterol), exogenous, endogenous, and reverse cholesterol. The composition, density, and function of the five major lipoproteins are summarized in Table 7-2 ⚙.

EXOGENOUS PATHWAY

The *exogenous pathway* is made up of the metabolism of **chylomicron** (CM), the largest and least dense (<0.94 g/mL) of the lipoprotein classes. Chylomicrons are formed from lipids absorbed in the intestines and are responsible for transporting dietary or exogenous fat, mostly triglycerides, from the intestines to the liver and peripheral cells.[1]

⊘ TABLE 7-2

Lipoprotein Composition, Density, and Function

Lipoprotein	Composition	Density (g/mL)	Function
Chylomicrons	Triglycerides: 84% Apolipoproteins: 1–2% Cholesterol: 7% Phospholipids: 6%	<0.94	Transport dietary or exogenous triglycerides from the intestines to the liver.
Very-low-density lipoproteins	Triglycerides: 44–60% Cholesterol: 16–22% Apolipoproteins: 2–8% Phospholipids: 18%	0.94–1.006	Transports endogenous triglycerides from the liver to muscle and adipose tissue.
LDL$_1$ (intermediate-density lipoproteins)	Triglyceride: 25% Cholesterol: 38% Apolipoproteins: 11% Phospholipids: 26%	1.006–1.019	
Low-density lipoproteins	Triglyceride: 11% Cholesterol: 62% Apolipoproteins: 20% Phospholipids: 23%	1.019–1.063	Principal lipoprotein carrier of cholesterol—2/3s of the cholesterol in plasma—to the body's tissues.
High-density lipoproteins	Triglyceride: 3% Cholesterol: 19% Apolipoproteins: 50% Phospholipids: 25–30%	1.063–1.210	Cholesterol scavenger; returns cholesterol to the liver to form bile acids and to be excreted in bile.

Apolipoproteins—primarily A-I, A-II, and B-48—are incorporated into the membrane, and triglycerides and cholesterol are packed into the core. Chylomicrons contain mostly triglycerides, and the percentage composition of the major components is:

a. triglycerides, 84%;

b. apolipoproteins, 1–2%;

c. cholesterol, 7%; and

d. phospholipids, 6%.

Once formed in the intestine, chylomicrons are transported through exocytosis and absorbed into the bloodstream through the lymphatics, where they travel to the various body organs and tissues. Apo A-I activates LCAT, which esterifies free cholesterol from the outer layer, and shifts esterified cholesterol to the core of the chylomicron. During the lipolytic process, chylomicrons make temporary contact with HDLs and exchange material. Some of the Apo A-I, A-II, and A-IV apoproteins, as well as phospholipids and free cholesterol, pass from the chyomicron to the HDL in exchange for Apo E and C-II.

Apo C-II activates lipoprotein lipase on the surface of the endothelial cells, which hydrolyzes the triglycerides on the surface of the chylomicrons to glycerol and free fatty acids. The free fatty acids are bound to albumin and stored in adipose tissue or used for energy. Each particle gets progressively smaller as the fatty acids and glycerol are released to the tissues.

The Apo B-48 and Apo E proteins in the membrane of the 90% delipidated chylomicron recognize and attach to hepatic cell B and E receptors. The hepatic cells engulf and catabolize the *chylomicron remnant*, and the residues of triglycerides and cholesterol esters are (1) used to synthesize VLDLs, (2) released to form bile acids, or (3) stored as cholesteryl esters. Chylomicrons transport dietary or exogenous triglycerides from the intestine to hepatic or peripheral cells, where they are stored or used for energy. Figure 7-4 ■ illustrates the exogenous pathway.

ENDOGENOUS PATHWAY

Very-Low-Density Lipoproteins

The endogenous pathway involves the metabolism of VLDLs to low-density lipoproteins. Hepatocytes can synthesize endogenous triglycerides from carbohydrates and free fatty acids. VLDLs have a density of 0.94–1.006 and are primarily a transporter of endogenous triglycerides from the liver to muscle and adipose cells.[1] It has been compared to a tanker truck carrying oil or gas. VLDL comprises approximately half triglyceride and approximately 5% apolipoproteins Apo B-100, Apo E, and a small amount of Apo C. The percent composition of the major components is:

a. triglycerides, 44–60%;

b. cholesterol, 16–22%;

c. apolipoproteins, 2–8%; and

d. phospholipids, 18%.

VLDL originates in the liver from chylomicron remnants and endogenous triglycerides that are used to synthesize and release nascent VLDL. It is transported through exocytosis into the extracellular space, sinusoids, and bloodstream. As

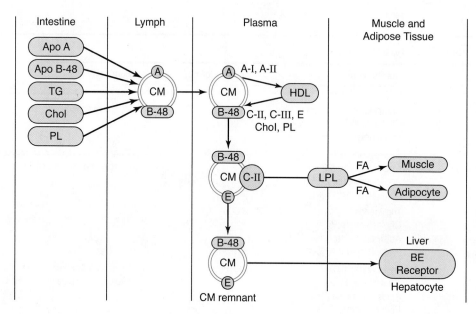

FIGURE 7-4   Exogenous lipoprotein pathway

occurs in chylomicrons, C-II activates LPL on the capillary cell walls and hepatic cells, which metabolizes the triglycerides and releases free fatty acids and glycerol. The fatty acids are taken up by muscle cells for energy or by adipose cells for storage. Apo B-100 is the primary apolipoprotein on formation, but like chylomicrons, VLDL also receives Apo C-II and Apo E from HDL. The primary protein is Apo B-100 with a small amount of Apo E.

During the metabolism of VLDL, apolipoprotein C-II is transported back to HDL, and the lipoprotein is now LDL$_1$. Half of the VLDL becomes LDL$_1$ (formerly *intermediate-density lipoprotein* or IDL), and the other half is taken up by remnant receptors or LDL receptors in the liver. During fasting, production of VLDL decreases; conversely, when there is an excess dietary intake of carbohydrates, VLDL production increases.

Low-Density Lipoproteins L

Low-density lipoprotein (LDL) is the principal lipoprotein carrier of cholesterol, accounting for approximately 70% of total cholesterol in the plasma. It arises primarily from the degradation of VLDL to LDL$_1$.[1] LDL has a density of 1.019–1.063, and the percent composition of the major components is:

a. triglyceride, 11%;

b. cholesterol, 62%;

c. apolipoproteins, 20%; and

d. phospholipids, 23%.

Acyl-cholesterol acyltranferase converts the free cholesterol to esterified cholesterol for storage. The LDL membrane is almost exclusively Apo B-100 and a small amount of Apo E. After binding to tissue receptors, the LDL is engulfed by the cell and degraded by lysosomal enzymes (endocytosis). Normally, two-thirds of the LDL is removed by LDL receptors and the remaining third by scavenger receptors in extrahepatic tissues.

If there is an excess of free cholesterol, three systems can be activated. In the first system, LDL receptors that regulate cellular cholesterol synthesis are inhibited by suppressing a gene that codes for the receptor protein. In other words, LDL receptors are decreased on cells that have an excess of cholesterol. In the second, the synthesis of endogenous cholesterol is decreased when the metabolic pathway is inhibited. In the third, the formation of cholesteryl esters that are catalyzed by ACAT is increased.

LDL is the major transporter of cholesterol in plasma to peripheral tissues for use as a structural component of cell membranes, as a precursor of steroid hormones, or as storage as cholesteryl esters. It is the most atherogenic (causing degeneration or thickening of the walls of the larger arteries in arteriosclerosis) lipoprotein and a major risk factor for **coronary heart disease (CHD)**. Figure 7-5 ■ summarizes the endogenous pathway.

REVERSE CHOLESTEROL TRANSPORT

High-Density Lipoprotein

High-density lipoprotein, the smallest and most dense lipoprotein (1.063–1.210), is secreted from the liver and intestinal mucosal cells. It contains primarily Apo A-I as well as some Apo A-II and C. Free cholesterol from cell membranes is also added to the nascent HDL. The percent composition of the major components is:

a. triglycerides, 3%;

b. cholesterol, 19%;

c. apolipoproteins, 50%; and

d. phospholipids, 25–30%.

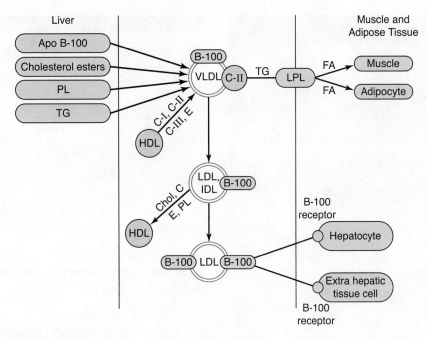

■ **FIGURE 7-5**   Endogenous lipoprotein pathway

LCAT and its cofactor, Apo A-I, esterify the cholesterol in HDL, and the cholesterol esters are then transferred to the hydrophobic core, increasing the density of the HDL molecule. The size of the HDL is determined by the level of cholesteryl esters and the activity of LCAT. By esterifying cholesterol, LCAT promotes the uptake of free cholesterol by HDL.

HDL is a cholesterol scavenger that removes cholesterol from tissues, esterifies it, and carries it to the liver for disposal.[1] Cholesteryl esters are delivered to the liver by three methods. In the first method, they are selectively taken up by HDL, and the HDL particles are returned to circulation for further transport. In the second, cholesteryl esters are transferred from HDL to Apo B-100 containing LDLs. In the third, HDL Apo E is recognized and taken up by hepatic remnant receptors. Cholesterol can then be converted to bile acids and excreted in the bile. HDL is called the "good" lipoprotein, and it provides protection against CHD.

✓ Checkpoint! 7–3

1. *Which lipoprotein has the lowest ratio of lipid to protein (the highest percentage of protein)?*

2. *The exogenous pathway involves primarily which lipoprotein?*

3. *Chylomicron remnants are catabolized and channeled in what three pathways?*

4. *Which lipoprotein class is described as*

   a. *the "good" lipoprotein: _____*

   b. *containing almost exclusively Apo B-100: _____*

   c. *the transporter of endogenous triglycerides from the liver to muscle and adipose cells: _____*

▶ HYPERLIPIDEMIA

Total cholesterol is the most often ordered screening test for **hyperlipidemia** (an excessive concentration of lipids in the blood) followed by a fractionated lipid profile: triglycerides, *HDL cholesterol* (HDL-C), and calculated *LDL cholesterol* (LDL-C). Approximately 70–75% of total cholesterol in plasma is carried by LDL, 15–20% by HDL, and 5–10% by VLDL and chylomicrons. An increase in total cholesterol can result in the development of atherosclerotic plaques, which have been shown to be a risk factor in the development of **coronary artery disease (CAD)** by many studies,[2] including the Framingham Heart Study.

HYPERCHOLESTEROLEMIA

Primary **hypercholesterolemia** can be caused by a genetic defect: familial hypercholesterolemia (FH). FH leads to an increase in LDL because of decreased or complete absence of LDL receptors, which results in the body's inability to metabolize LDL.[3]

Secondary causes (caused by other conditions) of hypercholesterolemia include hypothyroidism, uncontrolled diabetes mellitus, nephrotic syndrome, and extrahepatic obstruction of the bile duct.[4] Drugs (steroids, for example) and stress may also be responsible for a high cholesterol. Postmenopausal women who are not on hormone-replacement therapy may have elevated levels of cholesterol as a result of low estrogen levels. Box 7-2 ■ lists the causes of hypercholesterolemia and hypocholesterolemia.

BOX 7-2 Hypercholesterolemia and Hypocholesterolemia

Hypercholesterolemia

► Primary: Congenital
  • Various hyperlipidemias
  • Familial hypercholesterolemia (FH)
► Secondary (caused by other conditions)
  • Hypothyroidism
  • Uncontrolled diabetes mellitus
  • Nephrotic syndrome
  • Extrahepatic bile duct obstruction
  • High-fat diet, obesity
  • Drugs (e.g., steroids)
  • Decreased estrogen
  • Hepatitis
  • Bacterial and viral infections
  • Stress: physical or emotional
    o Acute trauma, surgery
    o Myocardial infarction
    o Burns

Hypocholesterolemia

► Hyperthyroidism
► Hepatocellular disease
► Anemias
► Starvation
► Anorexia
► Genetic defects (e.g., abetalipoproteinemia)

BOX 7-3 Hypertriglyceridemia

► Hypothyroidism
► Nephrotic syndrome
► Acute alcoholism
► Obstructive liver disease
► Acute pancreatitis
► Uncontrolled diabetes mellitus
► Glycogen storage disease
► High-fat diet
► Drugs (e.g., steroids)
► Low estrogen levels

HYPOCHOLESTEROLEMIA

Hypocholesterolemia or decreased cholesterol has been associated with hyperthyroidism and hepatocellular disease, which results in a decrease in the synthesis of endogenous cholesterol. Malnutrition, starvation, and eating disorders (e.g., anorexia and bulimia) may cause a decrease in cholesterol. Hypocholesterolemia is also found in certain genetic defects (e.g., abetalipoproteinemia).

TRIGLYCERIDES

Hypertriglyceridemia

Hypertriglyceridemia is secondary to many conditions almost identical to those listed for hypercholesterolemia. Hypothyroidism, nephrotic syndrome, acute pancreatitis, and acute alcoholism are examples of conditions associated with hypertriglyceridemia. See Box 7-3 ■ for more detail.

The appearance of the plasma or serum can serve as a valuable clue to triglyceride levels. If the plasma is clear, then the triglycerides are probably less than 200 mg/dL. A hazy or turbid appearance suggests a triglyceride level between 200 and 300 mg/dL. Opaque, milky, or lipemic plasma is indicative of a triglyceride of more than 600 mg/dL.

Hypotriglyceridemia

Hypotriglyceridemia is found in rare diseases such as abetalipoproteinemia.

Reference Range

The reference range for triglycerides is 60–150 mg/dL.

Lipoprotein (a)

Lipoprotein (a) [Lp(a)] is another lipoprotein particle that resembles LDL with the addition of apolipoprotein (a). Lp(a) contains a carbohydrate-rich protein—Apo (a)—that is bound to Apo B-100, with each molecule containing one molecule of Apo (a) and one molecule of Apo B-100. Lp(a) links lipid metabolism with blood coagulation. It is similar to both LDL and plasminogen, resulting in both atherogenic and thrombogenic potential. Lp(a) may inhibit thrombolysis, and high levels are associated with increased risk of CHD as supported by many studies that have shown a correlation between an elevated Lp(a) and an increased risk of myocardial infarction. The concentration of Lp(a) is influenced by genetics, which may explain the genetic predisposition to CHD.[5,6]

► CHOLESTEROL METHODOLOGIES

Enzymatic cholesterol methodologies have replaced chemical methods. For historical purposes, chemical methods were based on the reaction of cholesterol as a typical alcohol with a strong concentrated acid. The Lieberman–Burchard reaction measured cholesterol extracted into cold chloroform and then treated with acetic anhydride, acetic acid, and concentrated sulfuric acid. The green complexes formed—cholestapolyenes and cholestapolyene carbonium ions—were measured spectrophotometrically.

The reference method developed by Abell et al. reacted 0.5 mL of serum in 5.0 mL of alcoholic KOH to hydrolyze the cholesterol esters. Ten mL of hexane were added to extract the total cholesterol. An aliquot was dried in a vacuum and then redissolved in 3.2 mL of acetic acid, acetic anhydride, and sulfuric acid as in the Liebermann–Burchard reaction.

Cholesterol oxidase is the most common enzymatic procedure. The first step in the reaction involves the hydrolysis of cholesteryl esters by cholesterol esterase (cholesteryl ester hydrolyase):

$$\text{Cholesterol esters} + H_2O \xrightarrow{\text{Cholesterol esterase}}$$
$$\text{Free cholesterol} + \text{Fatty acids} \quad (7.1)$$

In the second reaction, free cholesterol is oxidized by cholesterol oxidase, yielding hydrogen peroxide (H_2O_2), which can be quantified by a procedure similar to the glucose oxidase–peroxidase assay:

$$\text{Free cholesterol} + O_2 \xrightarrow{\text{Cholesterol oxidase}}$$
$$\text{Cholestenone} + 4\,H_2O_2 \quad (7.2)$$

Hydrogen peroxide generated by cholesterol oxidase is decomposed by horseradish *peroxidase* in the presence of 4-aminoantipyrine and phenol to yield a quinoneimine dye:

$$H_2O_2 + \text{4-Aminoantipyrine} + \text{Phenol} \xrightarrow{\text{Peroxidase}}$$
$$\text{Quinoneimine} + 4\,H_2O \quad (7.3)$$

Quantitation of the colored oxidation product—in this case, quinoneimine—is directly proportional to the cholesterol.

The enzymatic reaction is subject to several interfering substances that compete with the oxidation reaction, including bilirubin, ascorbic acid, and hemoglobin. Reagents and methodologies have been refined to minimize interferences. For example, bilirubin oxidase has been added to some reagents to eliminate bilirubin interference, and reading at dual wavelengths eliminates hemoglobin interference. Most cholesterol methodologies are linear to 600–700 mg/dL.

Reference Values

The desirable range for cholesterol is <200 mg/dL. The specimen does not have to be drawn fasting because the postprandial increase is not clinically significant (<3%). Serum cholesterol is low at birth but increases approximately 40% by the third day of life and is fairly stable until age 20. Cholesterol levels increase 30–30 mg/dL by age 55 in women and age 60 in men. Cholesterol levels are similar in males and females from 30 to 35 years of age. At this point, male levels increase more quickly than females. At age 55, male levels decline and female levels increase until age 60 when female cholesterol levels are higher than males. In summary, the concentration for women is slightly lower than men until menopause and then it increases, especially if the woman is not on hormone-replacement therapy.

▶ TRIGLYCERIDE METHODOLOGIES

Triglycerides are usually quantitated by enzymatic methods, either colorimetric or fluorometric, measuring the glycerol part of the lipid molecule. The first step in the reaction involves hydrolysis of the triglyceride molecule to glycerol and three fatty acids by a triacylhydrolase (lipase):

$$\text{Triglyceride} \xrightarrow{\text{Lipase}} \text{Diglyceride} + \text{Fatty acid} \quad (7.4)$$

$$\text{Diglyceride} \xrightarrow{\text{Lipase}} \text{Monoglyceride} + \text{Fatty acid} \quad (7.5)$$

$$\text{Monoglyceride} \xrightarrow{\text{Lipase}} \text{Glycerol} + \text{Fatty acid} \quad (7.6)$$

See Chapter 9 (Enzymes) for further information on the lipase reaction. The glycerol is converted to glycerol-1-phosphate by **glycerol kinase**:

$$\text{Glycerol} + \text{Adenosine triphosphate (ATP)} \xrightarrow{\text{Glycerol kinase}}$$
$$\text{Glycerophosphate} + \text{Adenosine diphosphate (ADP)} \quad (7.7)$$

Glycerol kinase also reacts with the free glycerol present in normal serum (10–20 mg/dL). The correction for free glycerol can be accomplished in two ways: (1) double cuvette blanking using a blank without lipase (two cuvettes, one without lipase measuring free glycerol and one with lipase measuring the total glycerol) or (2) designated calibration blanking of the average free glycerol (subtracting the absorbance of the average free glycerol found in serum from the total glycerol).

The fourth reaction is catalyzed by *glycerophosphate oxidase*, yielding dihydroxyacetone and hydrogen peroxide:

$$\text{Glycerophosphate} + O_2 \xrightarrow{\text{Glycerophosphate oxidase}}$$
$$\text{Dihydroxyacetone} + H_2O_2 \quad (7.8)$$

The fifth or indicator reaction utilizes peroxidase and a dye, resulting in formation of a chromagen (e.g., 4-aminoantipyrene) to a quinoneimine dye:

$$H_2O_2 + \text{Dye} \xrightarrow{\text{Peroxidase}} \text{Color} \quad (7.9)$$

A modification of this procedure proceeds from equation 7.6. The fourth reaction, *pyruvate kinase*, transfers the phosphate of phosphoenolpyruvate to the ADP formed in the second reaction.

$$\text{Phosphoenolpyruvate} + \text{ADP} \xrightarrow{\text{Pyruvate kinase}} \text{ATP} + \text{Pyruvate} \quad (7.10)$$

The fifth or indicator reaction, lactate dehydrogenase, converts pyruvate to lactate, oxidizing nicotinamide adenine dinucleotide plus hydrogen (NADH) to nicotinamide adenine dinucleotide (NAD) and resulting in a decrease in absorbance proportional to the triglyceride concentration:

$$\text{Pyruvate} + \text{NADH} + H^+ \xrightarrow{\text{Lactate dehydrogenase}} \text{Lactate} + \text{NAD}^+ \quad (7.11)$$

Most triglyceride methodologies are linear up to 700 mg/dL. The enzymatic procedures are fairly specific and do not react with glucose or phospholipids.

REFERENCE RANGE

The recommended level for triglycerides is less than 150 mg/dL; borderline high is 150–199, high 200–499, and very high ≥500 mg/dL. A baseline triglyceride is obtained after a 10- to 14-hour fast. Concentration of triglyceride is increased moderately after a meal, with the peak elevation 4 to 5 hours after ingestion. As indicated earlier in this section, plasma or serum is usually turbid or milky when triglycerides are grossly elevated.

 Checkpoint! 7–4

1. *Mary has diabetes mellitus and is hypothyroid. Would she be at higher or lower risk of cardiovascular disease? Explain.*
2. *An opaque or lipemic plasma would indicate elevated levels of which lipid? What would be the approximate level?*
3. *Lp(a) contains one molecule of which two proteins?*

► HIGH-DENSITY LIPOPROTEIN METHODOLOGIES

Total cholesterol is of limited value when assessing the risk of coronary artery disease. As noted earlier, approximately 70% of the body's cholesterol is transported by LDL and 20% by HDL. High-density lipoprotein competes with LDL for binding to tissue receptors and reduces cholesterol levels in blood vessel walls by transporting cholesterol from tissues to the liver for catabolism and excretion in bile. LDL transports cholesterol from the site of origin to deposition in the tissues, including blood vessels. HDL appears to be inversely related to the risk of cardiovascular disease—in other words, the higher the HDL, the lower the risk.

HDL is measured in samples in which the LDL and VLDL have been precipitated and removed. Ultracentrifugation is the gold standard to which all other methods are compared. Most HDL methodologies use a polyanion that reacts with the positively charged groups of lipoproteins combined with divalent cations, resulting in the precipitation of LDL and VLDL. The extremely small amount of cholesterol in chylomicrons, not normally present in fasting serum, does not affect the final result. Phosphotungstate or dextran sulfate with $MgCl_2$ are the most common precipitants. The specimen is then centrifuged to accelerate the precipitation of the lipoproteins rich with Apo B-100; the cholesterol remaining or not precipitated in the supernatant is HDL-C, which is measured by the normal cholesterol methodology.

An automated homogeneous assay (requiring no separation step) uses sulfated α-cyclodextrin with dextran sulfate and $MgCl_2$ to reduce the reaction of cholesterol from VLDL, LDL, and chylomicrons with cholesterol esterase and cholesterol oxidase. This eliminates the precipitation step.

The desirable range for HDL-C is ≥60 mg/dL. The gray area is 35–59 mg/dL, and high risk for CAD is <35 mg/dL.

► LOW-DENSITY LIPOPROTEIN: CALCULATED AND DIRECT

LDL-C can be calculated using the **Friedewald formula** if the total cholesterol, HDL-C, and triglycerides have been determined. It assumes that most of the cholesterol is found in LDL, VLDL, and HDL because chylomicrons are not present in normal fasting serum. An estimate of the VLDL-C can be calculated by dividing the triglycerides by 5, based on the average ratio of triglycerides to cholesterol in VLDL.

Total cholesterol = VLDL-C + LDL-C and HDL-C

The Friedewald formula (substituting TG/5 for VLDL-C) is

LDL cholesterol (LDL) = (Total cholesterol) − (HDL-C + TG/5)

Paul's lipid panel results are as follows:

Total cholesterol: 210 mg/dL

TG: 100 mg/dL

HDL-C: 23 mg/dL

His calculated LDL is 210 − (23 + 100/5) = 167 mg/dL.

The Friedewald formula is not valid if the patient's triglyceride is more than 400 mg/dL. A triglyceride above 400 indicates the presence of chylomicrons, which means the assumption of no or extremely low chylomicrons is not valid.

Five LDL-C homogeneous assays are available. These are recommended by the NCEP when the triglyceride is more than 400 mg/dL and the Friedewald formula is not valid. The assays contain different detergents and other chemicals, which results in specific blocking or solubilization of lipoprotein classes to achieve specificity for LDL-C. In these automated procedures, the LDL-C is measured enzymatically in the same cuvette.[7]

One example of a homogeneous LDL-C assay is from Daiichi Pure Chemicals Company and distributed by Genzyme Diagnostics. A 3-μL serum sample is incubated with 300 μL of reagent 1 [ascorbic acid oxidase, 4-aminoantipyrene, peroxidase, cholesterol oxidase, buffer (pH 6.3) and a detergent that solubilizes non-LDL lipoproteins] for 5 minutes at 37°C. The cholesterol reacts with cholesterol esterase and cholesterol oxidase to form hydrogen peroxide. The hydrogen peroxide is consumed by 4-aminoantipyrene and yields no color production. Reagent 2 (100 μL) is then added which contains N-N-bis-(4 sulfobuytl)-m-toluidine disodium salt, buffer (pH 6.3), and a specific detergent that releases cholesterol from LDL. The hydrogen peroxide in this step reacts with N-N-bis-(4 sulfobuytl)-m-toluidine disodium salt to generate a colored product measures at 546 nm (main) and 660 nm (subsidiary), which is proportional to LDL-C. For a more comprehensive review of LDL methodologies, see the references at the end of this chapter.[8]

The desirable range for LDL is 100–130 mg/dL. LDL from 130 to 159 is borderline high risk of CAD, and high risk is >160 mg/dL.

► METABOLIC SYNDROME

The **metabolic syndrome** is a group of interrelated metabolic risk factors that appear to directly promote the development of atherosclerotic cardiovascular disease (ASCVD). These risk factors are also related to the development of type 2 diabetes mellitus. The most commonly recognized metabolic risk factors are atherogenic dyslipidemia (low HDL, high LDL, and elevated triglycerides), elevated blood pressure, and elevated blood glucose. Metabolic syndrome is also associated with a prothrombotic and proinflammatory state. It is classified as a syndrome, which suggests that metabolic syndrome has more than one cause; no single pathogenesis or cause has been identified.[9]

CLINICAL IDENTIFICATION

The criteria for clinical diagnosis of metabolic syndrome are described in Table 7-3 ☻. They include waist circumference, elevated triglycerides, low HDL-C, elevated blood pressure, and elevated fasting glucose. The predominant underlying risk factors for metabolic syndrome are abdominal obesity and insulin resistance (type 2 diabetes), which are associated with physical inactivity, aging, and hormonal imbalance.[10]

Other proinflammatory and prothrombotic factors related to metabolic syndrome can be measured. Elevated levels of C-reactive protein (CRP) is one of the easiest measurements to detect acute-phase proteins or proinflammatory state. A CRP of >30 mg/dL in a patient without any other detectable causes can identify metabolic syndrome. An average CRP of 19 was found in patients with one risk factor compared to 68 when all five criteria were present. Prothrombotic states can be identified with elevations of fibrinogen, plasminogen activator inhibitor-1, and other coagulation factors. Abdominal obesity (visceral fat) carries a higher risk for metabolic syndrome than fat around the hips and thighs. Waist-circumference criteria are given, although waist-to-hip ratio and body mass index

(BMI) can be used. In fact, BMI is less prone to error and can be calculated from data already available (height and weight). Abdominal obesity has a much stronger correlation with insulin resistance than lower-body obesity; for example, lower-body obesity, has not been linked with insulin resistance.[10]

Dyslipidemia is defined by low HDL, high LDL, increased small, dense LDL (sdLDL) and high triglycerides. Circulating LDL particles are of two distinct phenotypes: Type A are large, buoyant LDL particles; type B are small LDL particles. Elevated VLDL is associated with an increased exchange of VLDL triglycerides for cholesteryl esters in HDL. This results in a decrease in the cholesterol content of HDL. It is also associated with the exchange of cholesteryl esters from LDL for VLDL triglycerides, resulting in an increase in sdLDL. An increase in sdLDL is associated with increased risk of metabolic syndrome and is also an independent predictor of coronary heart disease. Small, dense LDL is thought to be more susceptible to oxidative modification and therefore may be more toxic to the vascular endothelium found on the walls of blood vessels. The sdLDL particles are also taken into the vessel wall 50% faster than larger LDL particles, contributing to a more serious prognosis.[10]

Insulin resistance is difficult to measure. In insulin-resistant patients, insulin loses its ability to suppress mobilization of free fatty acids, resulting in postprandial increases in triglycerides, VLDL, LDL, and sdLDL and a decrease in HDL. Measurement of insulin levels is not recommended for the diagnosis or evaluation of metabolic syndrome because of such problems as assay standardization, cross-reactivity, and interlaboratory variability. A measure of glucose intolerance such as impaired fasting glucose is used as a surrogate marker of insulin resistance.

The first line of treatment of metabolic syndrome is to address the major risk factors: LDL-C above goal, hypertension, and diabetes. If type 2 diabetes is not present, then the goal would be to prevent or delay the onset. Studies have demonstrated that metabolic syndrome patients with type 2 diabetes are at much higher risk for ASCVD. Lifestyle intervention is the first step, including a 7% to 10% weight reduction over 6 to 12 months through reduced caloric intake and increased physical activity. If lifestyle changes are not adequate, then drug therapies for the individual risk factors are recommended.[10]

▶ NATIONAL CHOLESTEROL EDUCATION PROGRAM (NCEP)

Atherosclerosis is the leading cause of death (morbidity and mortality) in the United States. The relationship between elevated lipids, deposits, or plaques containing especially esterified cholesterol in the arterial walls and CAD has been well established. The increased awareness and treatment of abnormal lipid levels has helped decrease the mortality rate. The NCEP has identified algorithms for the assessment of patients for the risk of coronary artery disease. NCEP's Adult Treatment Panel III (ATP III) Report *Detection, Evaluation, and Treatment of High Blood Cholesterol in Adults* provides updated

---

☻ TABLE 7-3

Criteria for Clinical Diagnosis of Metabolic Syndrome

Measure*	Categorical Cutpoints
Elevated waist circumference	≥102 cm in men ≥88 cm in women
Elevated triglycerides	≥150 mg/dL (1.7 mmol/L) or On drug treatment for elevated triglycerides
Reduced HDL-C	≤40 mg/dL (0.9 mmol/L) in men ≤50 mg/dL (1.1 mmol/L in women) or On drug treatment for reduced HDL-C
Elevated blood pressure	≥130 mm Hg systolic blood pressure ≥85 mm Hg diastolic blood pressure or On antihypertensive drug treatment (for a patient with a history of hypertension)
Elevated fasting glucose	≥100 mg/dL or On drug treatment for elevated glucose

*Any of three of five constitute a diagnosis of metabolic syndrome.
Source: Grundy SM, Cleeman JI (Co-chairs, Diagnosis and Management of the Metabolic Syndrome, American Heart Association–National Heart, Lung, and Blood Institute Scientific Statement. *Circulation* (2005) 112: 1–18

TABLE 7-4

ATP III Classification of LDL, Total, and HDL Cholesterol (mg/dL)

LDL cholesterol	<100	Optimal
	100–129	Near optimal, above optimal
	130–159	Borderline high
	160–189	High
	≥190	Very high
Total cholesterol	<200	Desirable
	200–239	Borderline High
	≥240	High
HDL cholesterol	<40	Low (High risk)
	≥60	High (Desirable)

Source: National Cholesterol Education Program. *Detection, Evaluation, and Treatment of High Blood Cholesterol in Adults* (Adult Treatment Panel III) 2001, Executive Summary, p. 3.

guidelines for cholesterol testing and management. ATP III recommends intensive lowering of LDL-C in several groups of patients: The higher the risk for CHD, the lower the goal—the lower the LDL. A fasting lipoprotein profile including cholesterol, LDL cholesterol, HDL cholesterol, and triglyceride is recommended for all adults age 20 and older every 5 years.[2] The risk values for LDL cholesterol, total cholesterol and HDL cholesterol are found in Table 7-4 ✪.

Box 4-4 ■ shows the major risk factors that must be considered in addition to lipid levels. ATP III also combines two criteria—lipid levels and number of risk factors—to determine

BOX 7-4 Major Risk Factors (Exclusive of LDL Cholesterol) That Modify LDL Goals

▶ Cigarette smoking
▶ Hypertension (blood pressure ≥140/90 mmHg) or on anti-hypertensive medication
▶ Low HDL cholesterol (<40 mg/dL)
▶ Family history of premature CHD (CHD is male first-degree relative < 55 years; CHD in female first-degree relative <65 years)
▶ Age (men ≥45 years, women ≥55 years)

Other secondary causes of elevated LDL cholesterol or other forms of hyperlipidemia that should be assessed or ruled out before initiation of therapy:

▶ Diabetes mellitus
▶ Hypothyroidism
▶ Obstructive liver disease
▶ Chronic renal failure
▶ Drugs that increase LDL cholesterol and increase HDL (steroids, progestins, etc.)

Source: National Cholesterol Education Program. Detection, Evaluation, and Treatment of High Blood Cholesterol in Adults (Adult Treatment Panel III) 2001, Executive Summary. *JAMA* (2001) 285: 2486–2497.

TABLE 7-5

Three Categories of Risk that Modify LDL Cholesterol Goals

Risk Category	LDL Goal (mg/dL)
CHD and CHD risk equivalents	<100
Multiple (2+) risk factors	<130
Zero to one risk factor	<160

Source: National Cholesterol Education Program. Detection, Evaluation and Treatment of High Blood Cholesterol in Adults (Adult Treatment Panel III) 2001, Executive Summary. *JAMA* (2001) 285: 2486–2497.

LDL goals.[2] These are defined in Table 7-5 ✪. CHD risk equivalents are:

• other clinical forms of atherosclerotic disease (e.g., peripheral arterial disease, abdominal aortic aneurysm, and symptomatic carotid artery disease),

• diabetes mellitus, and

• multiple risk factors that confer a 10-year risk for CHD of >20%.

For patients at the highest risk for CHD or with the highest risk equivalents listed above, the goal is an LDL-C below 100 mg/dL. For patients with two risk factors—for example, cigarette smoking and hypertension—the goal is to decrease the LDL-C below 130 mg/dL. The goal for patients with no or only one risk factor is an LDL-C below 160 mg/dL. For additional information, see the references listed at the end of this chapter.[2]

✓ Checkpoint! 7–5

1. Bill's lipid profile is

   Total cholesterol: 290 mg/dL

   TG: 140 mg/dL

   HDL-C: 40 mg/dL

   What is his calculated LDL-C?
2. List the five criteria for the clinical diagnosis of metabolic syndrome.
3. Describe dyslipidemia as defined for the metabolic syndrome.

SUMMARY

Lipids are vital in all aspects of biological life, including roles as energy sources, hormones and hormone precursors, and structural and functional components of cell membranes. They are a class of organic compounds characterized by solubility in organic solvents and insolubility in water.

Lipids are divided into five classes: sterol derivatives (cholesterol), fatty acids, glycerol esters (triglycerides), sphingosine derivatives, and terpenes. Cholesterol is the most commonly ordered screening test for hyperlipidemia, followed by fractionation into HDL-C, LDL-C, and triglycerides.

Lipoproteins are classified into five categories: chylomicrons, very-low-density lipoproteins (VLDLs), intermediate-density lipoproteins (IDLs or LDL$_1$), low-density lipoproteins (LDLs), and high-density lipoproteins (HDLs). Lipoproteins undergo a series of complex metabolic processes occurring in four pathways: lipid absorption, endogenous pathway, exogenous pathway, and reverse cholesterol pathway. The exogenous pathway is based on the metabolism of chylomicrons that transport dietary or exogenous triglycerides from the intestine to hepatic or peripheral cells, where they are stored for energy. The endogenous pathway involves the metabolism of very-low-density lipoproteins to low-density lipoproteins. LDL is the major transporter of cholesterol in plasma to peripheral tissues for use as a structural component of cell membranes, a precursor of steroid hormones, and storage as cholesteryl esters. The reverse cholesterol transport is made up of high-density lipoproteins, a cholesterol scavenger that removes cholesterol from tissues, esterifies it, and carries it to the liver for disposal.

Total cholesterol comprises 70–75% LDL-C, 15–20% HDL, and 5–10% VLDL. HDL, the "good" cholesterol, is inversely related to the risk of coronary heart disease: The higher the HDL-C, the lower the risk. LDL-C is directly related to the risk of CHD: The higher the LDL-C, the greater the risk.

Metabolic syndrome is a group of interrelated risk factors that appear to directly promote the development of atherosclerotic cardiovascular disease. The most commonly recognized risk factors are atherogenic dyslipidemia, elevated blood pressure, and elevated blood glucose. Dyslipidemia is defined by low HDL, high LDL, increased sdLDL, and high triglycerides. The first line of treatment is to address the major risk factors: Lower LDL, decrease blood pressure, and control diabetes mellitus or impaired glucose tolerance.

The National Cholesterol Education Program (NCEP) Adult Treatment Panel III (ATP III) provides updated guidelines for cholesterol testing and management. It also recommends lowering of LDL-C to lower levels than the previous report. ATP III also proposes a fasting lipoprotein profile that includes cholesterol, LDL-C, HDL-C, and triglyceride on all adults over 20 years of age every 5 years. The goals are total cholesterol <200 mg/dL, LDL-C <100 mg/dL (optimal), and HDL-C ≥60 mg/dL.

REVIEW QUESTIONS

LEVEL I

1. Lipids are characterized by: (Objective 2)
   a. insolubility in water and solubility in organic solvents.
   b. solubility in water and organic solvents.
   c. solubility in saline.
   d. solubility in weak acids.
   e. solubility in water and insolubility in organic solvents.

2. Which is the most commonly found sterol in humans? (Objective 1)
   a. cholesterol
   b. cortisol
   c. estriol
   d. pregnanediol

3. The cyclopentanoperhydrophenanthrene nucleus is found in: (Objective 1)
   a. nucleic acids.
   b. lecithin.
   c. cephalin.
   d. cholesterol.
   e. triglycerides.

4. Cholesterol functions in all of the following *except*: (Objective 1)
   a. bile acid precursor.
   b. steroid hormone precursor.
   c. prostaglandin synthesis.
   d. membrane structure.

5. Cholesterol esters are formed through the esterification of cholesterol with: (Objective 1)
   a. protein.
   b. digitonin.
   c. fatty acids.
   d. Apo A-I.

LEVEL II

1. Which of the following is a variant form of LDL associated with an increased risk of coronary heart disease? (Objective 2)
   a. HDL
   b. Apo A-I
   c. Lp(a)
   d. VLDL

2. The Friedewald formula for estimating LDL cholesterol should *not* be used when the: (Objective 2)
   a. HDL cholesterol is greater than 40 mg/dL.
   b. triglyceride is greater than 400 mg/dL.
   c. plasma shows no visible evidence of lipemia.
   d. total cholesterol is elevated based on the age and sex of the patient.

REVIEW QUESTIONS *(continued)*

LEVEL I

6. Which of the following is *false* concerning bloodcholesterol concentrations? (Objective 8)
   a. Increased cholesterol is associated with diabetes mellitus.
   b. Increased cholesterol is associated with hyperthyroidism.
   c. Increased cholesterol is associated with nephrotic syndrome.
   d. Cholesterol levels decrease following strenuous exercise.
   e. Cholesterol levels are unchanged when comparing results from fasting and nonfasting blood.

7. Which enzyme is used in enzymatic cholesterol determinations to form hydrogen peroxide? (Objective 10)
   a. lactate dehydrogenase
   b. cholesterol oxidase
   c. cholesterol kinase
   d. cholesterol dehydrogenase

8. Which enzyme catalyzes the following reaction:

   glycerol + ATP $\longrightarrow$ glycerophosphate + ADP (Objective 10)
   a. glycerol phosphate dehydrogenase
   b. glycerol kinase
   c. hexokinase
   d. pyruvate kinase
   e. glycerol phosphate oxidase

9. The *desirable* range for cholesterol levels as determined by the NCEP is: (Objective 18)
   a. less than 200 mg/dL.
   b. 200–240 mg/dL.
   c. more than 240 mg/dL.
   d. more than 300 mg/dL.

10. Serum lipase catalyzes the: (Objective 13)
    a. hydrolysis of triglyceride to fatty acids and glycerol.
    b. hydrolysis of triglyceride to fatty acids and water.
    c. synthesis of triglyceride from fatty acid and glycerol.
    d. synthesis of cholesterol from fatty acids.

11. The enzymatic determination of glycerol in triglyceride analysis usually involves conversion of glycerol to: (Objective 13)
    a. oleic acid.
    b. phospholipid.
    c. formaldehyde.
    d. CO_2 and H_2O.
    e. glycerol phosphate.

12. Which of the following lipid results would be expected to be falsely elevated on a serum specimen from a nonfasting patient? (Objective 13)
    a. triglyceride
    b. cholesterol
    c. HDL
    d. LDL

13. Blood collected about an hour after the patient has eaten a heavy meal often has elevated: (Objective 5)
    a. chylomicrons.
    b. globulin.

LEVEL II

REVIEW QUESTIONS (continued)

LEVEL I

    c. thymol turbidity.
    d. calcium.

14. From 60% to 75% of the plasma cholesterol is transported by: (Objective 6)
    a. chylomicrons.
    b. very-low-density lipoproteins.
    c. low-density lipoproteins.
    d. high-density lipoproteins.

15. Triglyceride is the main constituent of: (Objective 6)
    a. chylomicrons.
    b. VLDL.
    c. LDL.
    d. HDL.
    e. a and b.

16. The lipoprotein that is responsible for the transport of cholesterol from peripheral cells to the liver for excretion is: (Objective 5)
    a. HDL.
    b. LDL.
    c. VLDL.
    d. ILDL.

17. After a meal, the bloodstream transports chylomicrons and VLDL to all tissues of the body. The principal site of uptake is the: (Objective 5)
    a. liver.
    b. intestine.
    c. kidney.
    d. brain.

18. An increase of 10 mg/dL of HDL cholesterol would: (Objective 14)
    a. increase the risk of myocardial infarction.
    b. decrease the risk of myocardial infarction.
    c. have no effect.

19. Which apolipoprotein has the ability to *increase* the risk of coronary heart disease? (Objective 14)
    a. Apo A-I
    b. Apo B-100
    c. Apo B-48
    d. Apo E-11

20. A 52-year-old man went to his doctor for a physical examination. The patient was overweight and had missed his last two appointments because of business dealings. His blood pressure was elevated, his cholesterol was 210 mg/dL, and his triglyceride was 150 mg/dL. The result of an HDL cholesterol test was 23 mg/dL (20–60 mg/dL) Which of the following would be this patient's calculated LDL cholesterol value? (Objective 15)
    a. 157 mg/dL
    b. 137 mg/dL
    c. 55.4 mg/dL
    d. Cannot determine from the information given.

LEVEL II

REVIEW QUESTIONS (continued)

LEVEL I

21. The results on a lipid profile on a 40-year-old patient with a history of alcoholism would be: (Objective 19)
    a. elevated.
    b. decreased.
    c. normal.
    d. Unable to determine from the information.

22. To produce the most reliable results, a specimen for a lipid profile should be drawn: (Objective 10)
    a. immediately after eating.
    b. fasting, 12 to 14 hours after eating.
    c. fasting, 4 to 5 hours after eating.
    d. randomly, or any time without regard to the last meal.

23. Which of the following is a reagent commonly used to precipitate VLDL and LDL, thereby allowing the measurement of HDL-C? (Objective 14)
    a. zinc sulfate
    b. isopropanol
    c. sulfosalicylic acid
    d. dextran sulfate and $MgCl_2$

LEVEL II

PEARSON
myhealthprofessionskit™

Use this address to access the interactive Companion Website created for this textbook. Simply select "Clinical Laboratory Science" from the choice of disciplines. Find this book and log in using your user name and password.

REFERENCES

1. Kingsbury KJ, Bondy G. Understanding the essentials of blood lipid metabolism. *Prog Cardiovasc Nursing* (2003) 18, 1: 13–18.

2. National Institutes of Health–National Heart, Lung, and Blood Institute. *Third report of the National Cholesterol Education Program (NCEP) Detection, Evaluation, and Treatment of High Blood Cholesterol in Adults (Adult Treatment Panel III), Executive Summary* (May 2001) (www.nhlbi.nih.gov/guidelines/cholesterol/index.htm), accessed September 14, 2007.

3. Naoumova RP, Soutar AK. *Mechanisms of disease: Genetic causes of familial hypercholesterolemia* (www.medscape.com/viewprogram/6817_pnt), accessed February 15, 2008.

4. Fpnotebook. Hypercholesterolemia (www.fpnotebook.com/CV149.htm), accessed February 15, 2008.

5. Fogelman A, Superko HR. Lipoproteins and atherosclerosis the role of HDL cholesterol, Lp(a), and LDL particle size. *American College of Cardiology 48th Annual Scientific Session* (www.medscape.com/viewarticle/439375), accessed September 14, 2007.

6. Holmes DT, Schick BA et al. Lipoprotein (a) is an independent risk factor for cardiovascular disease in heterozygous familial hypercholesterolemia. *Clin Chem* 51, 11: 2067–2073.

7. National Institutes of Health–National Heart, Lung, and Blood Institute. *Third report of the National Cholesterol Education Program (NCEP): Implications of recent clinical trials for the National Cholesterol Education Program Adult Treatment Panel III guidelines 2004* (www.nhlbi.nih.gov/guidelines/cholesterol/index.htm), accessed September 14, 2007.

8. Nauck M, Warnick RG, Rafai N. Methods for measurement of LDL-cholesterol: A critical assessment of direct measurement by homogeneous assay versus calculation. *Clin Chem* (2002) 48, 2: 236–254.

9. Grundy SM, Brewer B, et al. Definition of metabolic syndrome: Report of the National Heart Association–National Heart, Lung, and Blood Institute–American Heart Association Conference on Scientific Issues Related to Definition. *Circulation* (2004) 109: 433–438.

10. Grundy SM, Cleeman JI, et al. Diagnosis and management of the metabolic syndrome. American Heart Association–National Heart, Lung, and Blood Institute. *Circulation* (2005) 112: 1–18.

Suggested Readings

1. Harris, NS, Winter, WE. The chemical pathology of insulin resistance and the metabolic syndrome. *MLO* (October 2004) 36, 10: 20–25.

2. Hoefner DM. The ruthless malady: Metabolic syndrome. *MLO* (October 2003) 35, 10: 12–23.

3. Krauss, RM. Triglyceride-rich lipoproteins: LDL particle size and atherogenesis (www.medscape.com/viewarticle/420330), accessed July 12, 2007.

4. Ogedegbe, HO. Apolipoprotein A-1/B ratios may be useful in coronary heart disease risk assessment. *Lab Med* (2002) 10, 23: 790–793.

5. Reaven, GM. The metabolic syndrome: Requiescat in pace. *Clin Chem* (2005) 51, 6: 931–938.

8

Amino Acids and Proteins

■ OBJECTIVES—LEVEL I

Following successful completion of this chapter, the learner will be able to:

1. Describe protein structure.
2. List the major functions of protein.
3. Discuss clinically significant proteins, including function, clinical significance, and protein band in electrophoresis.
4. Discuss causes of hyperproteinemia.
5. Discuss causes of hypoproteinemia.
6. Explain the principle of major protein methodologies.
7. Describe urinary protein screening, clinical significance, and methodologies.
8. Describe cerebrospinal fluid protein, clinical significance, and methodologies.
9. List major functions of albumin.
10. List causes of hypoalbuminemia.
11. Discuss the major cause of hyperalbuminemia: dehydration.
12. Explain major albumin methodologies.
13. Discuss the major components of protein electrophoresis.
14. List in order the protein electrophoresis bands and approximate percentages of total protein.
15. Explain changes in the protein electrophoresis associated with the more common causes of abnormal patterns.
16. Calculate A/G ratio.

■ OBJECTIVES—LEVEL II

Following successful completion of this chapter, the learner will be able to:

1. Discuss more specific plasma proteins, including the zone of electrophoresis, and their clinical significance.

KEY TERMS

Acute phase reactions
α_1-acid glycoprotein
α_1-antichymotrypsin
α_1-antitrypsin
α_2-macroglobulin
A/G ratio
Albumin
Alkaptonuria
Alpha-fetoprotein
Amino acids
Ampholyte
Amphoteric
Analbuminemia
Anencephaly
Apoprotein
β_2-microglobulin
Biuret reaction
Bromcresol green
Bromcresol purple
Ceruloplasmin
Conjugated protein
C-reactive protein
Cystinuria

Denaturation
Essential amino acids
Fusion of the β–γ bands
  (bridging)
γ-globulins
Globular protein
Globulin
Glycoprotein
Haptoglobin
Hemopexin
Homogentisic acid
Hyperalbuminemia
Hyperproteinemia
Hypoalbuminemia
Hypoproteinemia
Immunoglobulin
Immunoglobulin A
Immunoglobulin D
Immunoglobulin E
Immunoglobulin M
Isoelectric point (pI)
Lipoprotein
Maple syrup urine disease

Metalloprotein
Mucoprotein
Multiple myeloma
Nucleoprotein
Ochronosis
Peptide
Peptide bond
Phenylketonuria
Primary structure
Protein electrophoresis
Protein-energy malnutrition
  (PEM)
Proteinuria
Relative hyperproteinemia
Relative hypoproteinemia
Secondary structure
Secretory IgA
Simple protein
Spina bifida
Transferrin
Tertiary structure
Transthyretin
Wilson's disease (WD)

A CASE IN POINT

Judy, a 40-year-old woman with a past history of kidney infections, was seen by her physician because she had felt lethargic for a few weeks. She also complained of decreased frequency of urination and a bloated feeling. The physician noted periorbital swelling and general edema including a swollen abdomen.

Urinalysis

Macroscopic:

Color	Yellow
Appearance	Cloudy/frothy
Specific gravity	1.022
pH	7.0
Protein	3+ (500 mg/dL) (SSA: 4+)
Glucose	Neg
Ketones	Neg
Bilirubin	Neg
Blood	Neg
Urobilinogen	Normal
Nitrite	Neg
Leukocyte Esterase	Neg

Microscopic:

WBCs	0–3/HPF
RBCs	0–1/HPF
Epithelial Cells	Rare squamous/HPF
	Rare renal tubular epithelial/HPF
Casts	0–3 Hyaline/LPF
	0–1 Renal tubular epithelial/LPF
	0–1 Granular/LPF
	0–1 Waxy/LPF
	0–1 Fatty/LPF
Other	Occasional oval fat bodies

Chemistry

		Reference Range
Protein	5.0 g/dL	6.0–8.4 g/dL
Albumin	2.4 g/dL	3.5–5.0 g/dL
Cholesterol	370 mg/dL	<200 mg/dL
BUN	33 mg/dL	7–24 mg/dL
Creatinine	2.1 mg/dL	0.5–1.2 mg/dL

Issues and Questions to Consider

1. Circle or highlight the abnormal value(s) or discrepant result(s) in the urinalysis.

2. What type of disease/condition would be characterized by the urinalysis and chemistry results?

3. What urinalysis result(s) led to your probable diagnosis?

4. Are the abnormal chemistry tests consistent with the probable diagnosis? Explain why or why not.

5. Discuss the physiological cause of the edema.

6. What is Judy's A/G ratio? Is it within the reference range? Is it consistent with the probable diagnosis?

7. Describe what you would expect to see in this patient's protein electrophoresis.

8. Which specific proteins would be decreased? Which specific proteins would be increased?

lysine; and in infants, two additional amino acids: arginine and histidine.

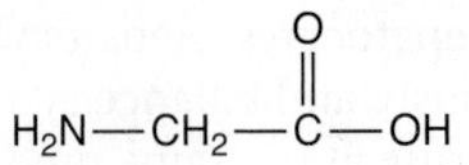

The structure of amino acids is **amphoteric**—containing two ionizable sites, a proton accepting group (NH_2), and a proton donating group (COOH). At physiological pH, approximately 7.4, the COOH easily loses a hydrogen ion and becomes COO^-, and NH_2 gains the hydrogen ion and becomes NH_3^+. When both are ionized, the amino acid is called an **ampholyte**, dipolar ion, *or* zwitterion *(*the older nomenclature).

The **isoelectric point (pI)** is the pH at which the amino acid or protein has no net charge and the positive charges equal the negative charges. The pI can vary from pH 3 to 10, and there is no pH at which all 20 amino acids are neutral. At a pH greater than the pI the protein carries a negative charge and at a pH less than the pI the protein carries a positive charge. For example, if the pI of a protein is 7.8 and the pH is 8.6, the protein will be negatively charged.

Amino acids in proteins are linked to each other through **peptide bonds**. A molecule of water is split between the carboxyl group of one amino acid and the amino group of another, and a covalent bond called a peptide bond is formed, as seen in Figure 8-2 ■. The end of the protein that has the amino free group is called the N-terminal end, and the opposite end that has the carboxyl group free is called the C-terminal end.

A **peptide** contains two or more amino acids. Dipeptides, tripeptides, tetrapeptides, and ogliopeptides contain two, three, four, and up to five amino acids, respectively. Polypeptides contain more than five amino acids. When the number of amino acids exceeds 40, the molecule takes on the properties of a protein chain. In serum, proteins average 100 to 150 amino acids.

Protein structures can be described using four structural categories—primary, secondary, tertiary, and quaternary. **Primary structure** is determined by the sequence of amino acids in the polypeptide chain, the identity and specific order of the amino acids. This sequence is found in the genetic DNA coding, as discussed later in this chapter. Peptide bonds are the primary bonds between atoms in the primary structure, and the molecule is one-dimensional.

Secondary structure is determined by the interaction of adjacent amino acids. The winding of the polypeptide chain, the formation of hydrogen bonds between the NH and CO groups of the peptide bonds, and the occasional disulfide bonds, affect its secondary structure. It is the regular, recurring arrangement of the primary structure in one dimension. Three possible conformations—α-helix, β-pleated

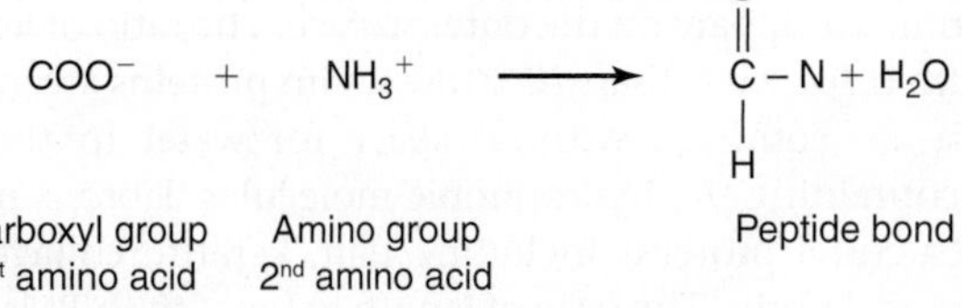

▶ INTRODUCTION

Proteins are complex polymers of α-amino acids that are produced by living cells in all forms of life. Each protein is composed of a maximum of 20 different amino acids in varying numbers and sequences. All proteins contain carbon, hydrogen, oxygen, and nitrogen, and some contain sulfur. The average content of nitrogen is 16%, and its presence differentiates proteins from carbohydrates and lipids.

Proteins are involved in many cellular processes, including maintenance of colloidal osmotic pressure, coagulation, and transport of various molecules, which will be discussed under protein functions. This chapter will concentrate on the proteins, protein methodologies, and aminoacidopathies that are most clinically significant or that you are more likely to encounter in the clinical laboratory setting.

▶ PROTEIN STRUCTURE

Amino acids contain an amino group (NH_2), carboxyl group (COOH), hydrogen, and an R group (radical or side chain) with the formula $RCH(NH_2)COOH$. The nucleus of the amino acid is the α-carbon to which the carboxylic and **amino** group are attached. The radical group is what distinguishes one amino acid from another. The simplest amino acid is glycine, illustrated in Figure 8-1 ■, which has a hydrogen atom as its R group.

Although all amino acids can be synthesized by some animals, in higher life forms, including humans, some amino acids called **essential amino acids** are not synthesized and must be ingested in the diet. In human adults there are eight essential amino acids: valine, leucine, isoleucine, phenylalanine, tryptophan, methionine, threonine, and

$$H_2N-CH_2-\overset{\overset{\displaystyle O}{\|}}{C}-OH$$

■ FIGURE 8-1   Glycine

$$COO^- \quad + \quad NH_3^+ \quad \longrightarrow \quad \overset{\overset{\displaystyle O}{\|}}{\underset{\underset{\displaystyle H}{|}}{C}}-N + H_2O$$

Carboxyl group   Amino group   Peptide bond
1st amino acid   2nd amino acid

■ FIGURE 8-2   Peptide bond

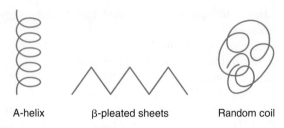

A-helix   β-pleated sheets   Random coil

■ FIGURE 8-3   Secondary structure of proteins

sheets, and random coils—are illustrated in Figure 8-3 ■. Secondary structure remains one-dimensional.

Tertiary structure is the way in which the chain folds back upon itself to form a three-dimensional structure. The bonds responsible for this structure are covalent bonds, for example disulfide bonds, and noncovalent bonds such as hydrogen, hydrophobic, electrostatic, and Van der Waals. These are mainly interactions of amino acids with the R-groups of more distant amino acids. Tertiary structure determines the chemical and physical properties of the protein.

Quaternary structure is the arrangement of two or more polypeptide chains to form a protein. Only proteins with more than one polypeptide chain have quaternary structure. The number of polypeptide chains and type (identical or different) determines the specific properties of the complex. For example, creatine kinase is an enzyme that is dimeric, or consists of two polypeptide chains, M or B, resulting in three possible combinations, MM, MB, and BB.

Proteins are classified into two major groups—simple proteins and conjugated proteins. **Simple proteins** are those composed only of amino acids, for example, albumin. **Conjugated proteins** are proteins which have nonprotein groups attached to them, providing certain characteristics to the protein. **Metalloproteins** have a metal ion attached to the protein, like ceruloplasmin, which has copper. **Lipoproteins** contain lipids such as cholesterol, triglycerides, and phospholipids. **Glycoproteins** are compounds consisting of a simple protein and carbohydrates, which make up less than 4% of the total weight. **Mucoproteins** are linked with large, complex carbohydrates (>4% of the total weight). **Nucleoproteins** are a combination of a simple protein and nucleic acids (DNA, RNA). Conjugated proteins without their nonprotein groups or ligands are called **apoproteins**; for example, when lipids are freed from a lipoprotein, the remaining molecule is called an *apolipoprotein*.

Proteins can also be categorized by shape into two major groups: globular and fibrous. **Globular proteins** are compact, folded, and coiled chains that are relatively soluble. The hydrophobic groups are folded within the protein, and the hydrophilic groups are on the outer surface. The ratio of length to breadth (L/B) is less than 10. Most serum proteins are globular, which are compact, without space for water in the inner core containing the hydrophobic molecules. Fibrous proteins are structural proteins including hair, keratin, collagen, troponin, and fibrin. The ratio of length to breadth (L/B) is greater than 10. Fibrous proteins are insoluble in water or saline.

DENATURATION

Denaturation is the disruption of the bonds holding the secondary, tertiary, or quaternary structures together. If the bonds are broken, the polypeptide chains unfold or their quaternary structure is lost. This results in the loss of activity and also the functional and structural characteristics of the protein molecule. Denaturation can occur as a result of heat, changes in pH, mechanical forces, exposure to chemicals (solvents, detergents, metals), and exposure to ultraviolet light. The bonds holding the quaternary and tertiary structures together are weak, and if they are destroyed, proteins—for example, enzymes—will lose their activity.

✓ Checkpoint! 8–1

1. *Define amphoteric.*
2. *At a pH above its isoelectric point, a protein carries a* _____ *(positive or negative) charge.*
3. *A protein with three polypeptide chains are arranged to forms its* _____ *structure.*
4. *Conjugated proteins that contain cholesterol and triglycerides are called* _____.

▶ PROTEIN METABOLISM

Digestion of dietary proteins by proteolytic enzymes originates in the gastrointestinal tract. Amino acids are released and absorbed in the jejunum and transported through the portal circulation system to amino acid pools, where they are stored. The amino acids pools are especially important in conserving the essential amino acids discussed earlier. The liver and other organs utilize the amino acid pools to synthesize the body's proteins, as described in the following section.

In the kidneys, amino acids are filtered through the renal glomeruli but are subsequently reabsorbed by the renal tubules. Although the details of the process of reabsorption are not known, it is an active transport system that is based on membrane-bound carriers and intraluminal Na^+ concentration. Increased plasma levels of amino acids, as occurs in aminoacidopathies, covered later in this chapter, results in increased renal excretion of amino acids.

▶ PROTEIN SYNTHESIS

Most plasma proteins—except for immunoglobulins (antibodies), some coagulation factors, protein hormones, and hemoglobin—are synthesized in the liver and secreted into circulation by the hepatocytes. Hepatocytes synthesize many proteins simultaneously and balance synthesis with degradation, which is occurring at the same time. This balance maintains plasma protein levels within a fairly narrow range.

The primary structure or sequence of amino acids is determined by the sequence of purine and pyrimidine (adenine, guanine, cytosine, and pyrimidine) bases in the DNA molecule, which codes for the particular protein. The double-stranded DNA molecule unfolds and one strand serves as a template for the messenger RNA (mRNA). The information on the mRNA also has the initiation and termination codes to begin and end the specific protein molecule.

The code is carried by the mRNA from the nucleus to the cytoplasm of the cell, where it attaches to a ribosome receptor protein on the ribosome. The code contains codons, or sequences of three bases specific for a particular amino acid. The next step in the process is getting the amino acids in the genetic code to the mRNA. The amino acid linked to another RNA, called transfer RNA (tRNA) that corresponds to the specific codon, is carried to the ribosome and is attached to the matching codon. The free tRNA returns to the cytoplasm, where it can bind to another amino acid. The next amino acid in the sequence is added and the cycle repeats itself until the protein is completed when the terminal codon is reached. When this occurs, the mRNA and the ribosome dissociate. The proteins are then secreted into the space of Disse and move through the hepatic sinusoids into the bloodstream.

PROTEIN FUNCTIONS

Proteins serve many functions in the body. One of the major roles is maintenance of water distribution between cells and tissue. When protein levels are decreased, the osmotic pressure is also decreased, allowing more water into the interstitial fluid, resulting in edema. Plasma colloidal osmotic pressure (COP) provided by the proteins tends to retain water in the vascular space. Structural proteins also provide support for the body, tissues, or cells. Keratin, found in nails and hair, and collagen are examples of two structural proteins. Collagen is a strong, fibrous, insoluble protein found in connective tissue and makes up about 25% of the body's weight. Proteins also function as enzymes (biological catalysts).

Coagulation proteins are important in maintenance of hemostasis or blood coagulation. Many proteins function as transport vehicles to move various ligands (an ion or molecule that reacts to form a complex bond with another molecule) to where they are needed or stored in the body. Examples of molecules that are transported by protein are iron carried by transferrin, bilirubin linked to albumin, and thyroid hormones bound to thyroxine-binding globulin (TBG). Peptide hormones, for example, insulin, serve critical functions in the body. Hemoglobin, the major protein in red blood cells, carries oxygen throughout the body. Immunoglobulins (antibodies) are very important constituents of the body's immune system. See Box 8-1 ■ for a list of protein functions.

BOX 8-1 Protein Functions

▶ Maintenance of colloidal osmotic pressure and water distribution
▶ Structural—support for the body, tissue or cell
  • Collagen
  • Keratin—hair, nails
▶ Transport molecule, examples
  • Transferrin-Fe^{+3}
  • Albumin-bilirubin
  • Thyroid binding globulins
  • Hormones
▶ Enzymes
▶ Peptide hormones, insulin
▶ Coagulation
▶ Hemoglobin
▶ Antibodies

▶ AMINOACIDOPATHIES

Aminoacidopathies are inherited disorders of amino acid metabolism. The disorder can be a specific enzyme in the metabolic pathway or in the membrane transport system for amino acids. Over 100 aminoacidopathies have been identified, including alkaptonuria, cystinuria, phenylketouria, and maple syrup urine disease, which we will focus on in this chapter.

ALKAPTONURIA

Alkaptonuria is a rare inherited disease that results from the deficiency of the enzyme homogentisic acid oxidase in the catabolic pathway of tyrosine. This deficiency leads to a buildup of **homogentisic acid** (HGA) in the tissues of the body. It is an autosomal recessive condition, which means both parents have one normal gene and one alkaptonuric gene that they pass on to their offspring.

Ochronosis, one of the characteristics of alkaptonuria, is the darkening of the tissues of the body because of the excess homogentisic acid in alkaptonurics. This occurs later in the disease, usually when patients are in their 40s, when the pigments accumulate and cause slate blue, gray, or black discoloration of the cartilage in the joints and ears, skin, and sclerae (whites) of the eyes. It also causes bluish discoloration of the nails in some patients. The deposition in joints can lead to arthritis-like degeneration of the large joints (hips) as well as intervertebral discs at the thoracic and lumbar levels, leading to back pain.

One of the earliest signs of the disorder is the tendency for diapers to stain black due to the oxidation of HGA in the urine. Homogentisic acid can be identified in the urine, using gas chromatography and mass spectroscopy. Reduction of phenylalanine and tyrosine in the diet is thought to reduce or minimize complications later in life and although this cannot be proven, it is a reasonable approach to treating patients with alkaptonuria.[10]

CSYTINURIA

Cystinuria is not a metabolic enzyme deficiency but a defect in the amino acid transport system. Normally, amino acids are freely filtered by the renal glomeruli and then actively reabsorbed by the proximal convoluted tubules. In patients with cystinuria, cystine is not reabsorbed and the concentration excreted in urine is increased 20 to 30 times normal. Cystine is not the only amino acid affected by the defect in the transport mechanism. Other diamino acids, including lysine, arginine, and ornithine, are also excreted in significant amounts; however, cystine appears to cause the most problems.

Cystine is somewhat insoluble, resulting in its precipitation in the renal tubules and the formation of urinary calculi (kidney stones). Symptoms suggestive of kidney stones may result in the diagnosis of cystinuria. Patients report flank pain or pain in the side or back, which may be progressive, getting increasingly worse, and it may radiate to the lower flank, pelvis, groin, or genitals. The urinalysis may also indicate blood in the urine.

Increased fluid intake (a minimum of 6 to 8 glasses of water per day) and alkalizing the urine with sodium bicarbonate, sodium citrate, or other medications will help prevent the formation of kidney stones. Penicillamine may also be used to increase the solubility of cystine. This is a chronic, lifelong condition, but it is not life threatening and does not affect other organs.

MAPLE SYRUP URINE DISEASE

Maple Syrup Urine Disease (MSUD) is named for the characteristic maple syrup or burnt sugar odor of the urine of persons with this condition. MSUD is caused by the absence or very low levels of the enzyme α-ketoacid decarboxylase, which results in the abnormal metabolism of three essential amino acids: leucine, isoleucine, and valine. These are converted to toxic ketoacids that cannot be oxidized, leading to their accumulation in the serum, urine, and spinal fluid.

The first symptoms of MSUD in the newborn are poor appetite, irritability, and the characteristic odor of their urine. MSUD should be considered when an infant presents with severe acidosis within the first 10 days of life, and screening should be done 12 hours after birth or later. Infants also lose their sucking reflex and become listless, have a high-pitched cry, and become limp, with episodes of rigidity. If treatment is not initiated quickly, the symptoms progress rapidly to seizures, coma, and death. It is usually lethal within the first month of life if unrecognized and untreated. The earlier the diagnosis and treatment, the lower the risk of permanent damage.

Treatment involves a special, very carefully controlled diet requiring careful monitoring of protein intake. It centers around a synthetic formula or "medical food" that provides all essential nutrients and all amino acids except leucine, isoleucine, and valine. These are added in very controlled amounts to provide the necessary levels for normal growth and development without exceeding levels that would lead to a build up of these amino acids. The diet requires lifelong restriction of branched-chain amino acids.

PHENYLKETONURIA

Phenylketonuria (PKU) is another inborn error of metabolism that results in the inability to metabolize the essential amino acid phenylalanine. It is an autosomal recessive trail, occurring in approximately 1 in 15,000 births in the United States. The biochemical defect is a deficiency of the enzyme phenylalanine hydroxylase (PAH) that converts phenylalanine to tyrosine.

PKU screening is one of the mandatory tests performed on newborns. Classic PKU is diagnosed when phenylalanine levels exceed 20 mg/dL without treatment. Other characteristics of PKU are a "mousy" urine odor due to the breakdown products of phenylalanine, including phenylpyruvic acid, found in the urine; eczema; and fair coloring due to the tyrosine deficiency.

Pregnant women known to be carriers of the PKU gene or definitely carrying a PKU fetus should also be maintained on a phenylalanine-restricted diet from conception to birth. Studies have reported that women who followed a normal diet while carrying a PKU fetus always delivered a baby who was microcephalic and mentally retarded.

Elevated phenylalanine levels are toxic to developing brain tissue and negatively impact brain function. If not diagnosed early, patients with PKU are usually mentally retarded, but the affect of PKU can be controlled through dietary treatment. If appropriate treatment is initiated early, the patient's IQ should be within 5–8 points of their siblings'.

The most important treatment is phenylalanine restriction and supplementation of diet with essential amino acids, minerals, and vitamins. Aspartame, one of the primary sweeteners in foods and soft drinks, should also be avoided. Formerly, the diet was discontinued at 5 or 6 years of age, but it has now been determined that slight brain damage occurs after the discontinuation of the diet, and most physicians no longer recommend suspending the program.

✓ **Checkpoint! 8–2**

1. *Most plasma proteins are synthesized in the* _____.
2. *List three major protein functions.*
3. *Identify the aminoacidopathy associated with the following:*
   a. *ochronosis*_____
   b. *"mousy" urine odor*_____
   c. *formation of kidney stones* _____
   d. *deficiency of homogentisic acid oxidase* _____

▶ SPECIFIC PLASMA PROTEINS

TRANSTHYRETIN (TTR) (PREALBUMIN)

Transthyretin (TTR) binds with thyroxine and triiodothyronine (thyroid hormones) and retinol (vitamin A) and serves as a transport protein. It is a nonglycosylated, tetrameric serum protein consisting of four identical subunits of 127 amino acids each, which are synthesized in the liver, and the choroid plexus of the brain. TTR is rarely seen on cellulose acetate electrophoresis and is more likely detected in high-resolution electrophoresis (HRE).

The main clinical significance of TTR is its role as a sensitive marker of poor nutritional status such as protein-energy malnutrition (PEM). Decreased TTR indicates dietary intake of protein is not adequate, resulting in decreased synthesis of transthyretin by the liver. People at risk for PEM are the elderly and those who are hospitalized or in a nursing home. Chronic illnesses (diabetes, arthritis), increased nutritional losses, open wounds, burns, and malabsorption (gastrointestinal protein-losing diseases) are other causes of PEM. TTR is also decreased in acute inflammatory response (acute phase reactant-APR), liver disease, nephrotic syndrome, and other protein-losing renal diseases.

TTR has a high ratio of essential-to-nonessential amino acids, making it a sensitive indicator of the quality of protein intake. It can detect PEM earlier because of its short half-life of 2 to 8 days. TTR methodologies include immunonephelometry, immunoturbidimetry, and radial immunodiffusion. The reference range for transthyretin is 190–359 mg/dL, but as with most analytes it is method dependent. Mild PEM would be a TTR of 151–200 mg/dL, moderate 100–150 mg/dL, and severe <100 mg/dL.

ALBUMIN

Albumin is synthesized in the liver and comprises approximately 60% of total serum protein. The rate of synthesis is dependent on protein intake and is subject to feedback regulation by the plasma albumin level and plasma colloidal

---

BOX 8-2 Functions of Albumin

- ▶ Maintain plasma colloidal osmotic pressure
- ▶ Transport and store a wide variety of ligands
  - Bilirubin
  - Long chain fatty acids
  - Therapeutic drugs (e.g., warfarin, phenylbutazone, salicylate, penicillin)
  - Calcium
  - Magnesium
  - Hormones (e.g., thyroxine, triiodothyronine, cortisol)
- ▶ Serve as an endogenous source of amino acids

BOX 8-3 Hypoalbuminemia

- ▶ *Increased catabolism: tissue damage and inflammation
- ▶ Impaired or decreased synthesis
  - Primary: liver disease
  - Secondary: diminished protein intake, malnutrition, malabsorption
- ▶ Increased loss of protein
  - Nephrotic syndrome
  - Chronic glomerulonephritis
  - Diabetes mellitus/diabetic nephropathy
  - Extensive burns
  - Acute viral gastroenteritis

*Chief biological function

---

osmotic pressure (COP). The chief biological function of albumin is to maintain plasma COP. The high concentration of albumin makes up approximately 80% of the total pressure, which keeps intravascular fluid inside the blood vessels and out of the interstitial fluid, which would result in edema. Another function of albumin is to transport and store a wide variety of ligands, such as bilirubin, listed in Box 8-2 ■. A ligand is an ion or molecule that reacts to form a complex with another molecule.

The most common cause of decreased albumin levels, *hypoalbuminemia*, is increased catabolism due to tissue damage and inflammation. The liver is too busy synthesizing proteins to repair the body; therefore, it is unable to keep up with albumin production. Impaired or decreased synthesis can be divided into two categories: primary, associated with liver disease, and secondary due to diminished protein intake, malabsorption, or malnutrition. The third category is increased protein loss, which is mainly renal. See Box 8-3 ■ for more detail.

Hyperalbuminemia is of little diagnostic significance except in dehydration. The increase in albumin is usually artifactual due to a decrease in plasma volume; as a result, normal albumin levels are diluted in a larger volume of plasma.

▶ GLOBULINS

α_1-GLOBULINS

α_1-Antitrypsin (AAT)

α_1**-antitrypsin** is the major α_1-globulin making up approximately 90% of α_1 proteins. AAT is a glycoprotein synthesized by the liver and released into the plasma. It is an acute phase reactant (APR) with antiprotease activity resulting in neutralizing leukocyte elastase and collagenase. Levels may double in acute and chronic infections and other conditions resulting in the release of APRs. AAT deficiency, one of the most common genetically lethal diseases in Caucasians (1:4000), is associated with lung and liver disease.

AAT is one of a family of serum proteins called serpins (serine proteinase inhibitors). Examples of other proteins in this group are α_1-antichymotrypsin, antithrombin, and antithrombin. α_1-antitrypsin is actually a misnomer. Although 90% of antitrypsin activity is attributed to AAT, plasma contains very little trypsin, and most of the AAT function is to inhibit nonprotein proteinases, especially collagenase and elastase.

AAT deficiency is related to early onset emphysema because particles and bacteria are continually removed from the lungs by polymorphonuclear neutrophils (PMNs) that release elastase. Normally, sufficient AAT is present to bind the free elastase released from the neutrophils. If AAT levels are not adequate, proteases, including elastase from the neutrophils, attack lung tissue. Elastase reacts with elastin in the vascular endothelium, destroying the lung tissue. AAT is very important in prevention of loss of elastin lung recoil, which results in emphysema. Emphysema with onset at 45 years of age of earlier and emphysema occurring in the absence of smoking are common features of AAT deficiency. The most common symptoms are dyspnea on exertion (84%), wheezing (74%), cough (42%), and chronic bronchitis (8–40%).

Although AAT deficiency is associated with early onset emphysema, a congenital deficiency can also result in juvenile hepatic cirrhosis, where AAT is synthesized by the hepatocytes but not released. Pulmonary emphysema is the major cause of disability and death; however, liver cirrhosis and/or cancer are present in 30–40% of patients over age 50 with AAT deficiency. Increased levels of AAT are present in inflammatory reactions (APR), pregnancy, and women on estrogen.

Quantitative assays, including immunoturbidimetry and immunonephelometry, are available to determine levels of AAT that have a reference range of 90–190 mg/dL. They do not, however, provide information on the particular genetic defect responsible for the deficiency. Detection of AAT deficiencies is critical because effective replacement therapy is available and treatment should be initiated as early as possible. If the level is below 50 mg/dL, a phenotyping should be performed to classify the specific genetic defect. Genetic defects (Pi, Z, or null phenotypes) are beyond the scope of this book and the reader is referred to the references at the end of the chapter.

α_1-Acid Glycoprotein (Orosomucoid) (AAG)

α_1-**acid glycoprotein** (AAG) is the major glycoprotein increased during inflammation (APR). It was one of the first glycoproteins to be isolated in its pure state. Elevated levels are found in rheumatoid arthritis, cancer, pneumonia, and other conditions resulting in an increased APR.

Alpha-Fetoprotein (AFP)

Alpha-fetoprotein (AFP) is the first α_1-globulin from the fetus to appear in the mother's serum during pregnancy. AFP is the principal fetal protein (fetal albumin-like protein) in maternal serum used to screen for the antenatal diagnosis of neural tube defects, including spina bifida and anencephaly. **Spina bifida** is a congenital defect in the walls of the spine that allows a protrusion of the spinal cord or meninges. **Anencephaly** is congenital absence of the brain or cranial vault, resulting in a disorganized mass of neural tissue. These conditions allow increased passage of fetal protein into amniotic fluid and indicate fetal distress. AFP is also increased in some abdominal wall defects in the fetus.

Alpha-fetoprotein levels peak in the fetus at 13 weeks gestation (end of first trimester) and decrease at 34 weeks. In maternal serum, the level peaks at 30 weeks gestation. Screening is usually performed at 16 to 18 weeks gestation, and normal levels are determined using such variables as the weight of the mother, the race of the mother (higher in African Americans), presence of Type 1 diabetes mellitus, and multiple births. For example, if the mother is having twins or triplets, the AFP would increase proportionally.

Multiple of the median (MoM) is calculated by dividing the patient's AFP by the median reference value for the gestational age. Elevated AFP leading to incorrect interpretation of the MoM can be caused by fetal demise, incorrect gestational age, multiple fetuses, and fetomaternal bleeds.

Alpha-fetoprotein is decreased in Down's syndrome and Trisomy 18. AFP can also be used as a tumor marker, with increased levels found in 80% of patients with heptocellular cancer, 50% of germ cell tumors (gonadal), and all children with hepatoblastoma.

α_1-Antichymotrypsin

α_1-**antichymotrypsin** (ACT) is a serine proteinase which catalyzes chymotrypsin and mast cell chymase. ACT is found between the α_1 and α_2 zones. It is elevated in inflammation and indicates APR protein synthesis. Decreased levels of ACT are associated with asthma, emphysema, and liver disease.

α_2-GLOBULINS

Haptoglobin (Hp)

Haptoglobin is an α_2-globulin and an acute phase reactant that binds free hemoglobin in plasma. It consists of two nonidentical chains, α and β, $[(\alpha\beta)_2]$ linked by disulfide bonds, and its primary function is irreversible binding with the α-chains of the globin portion of free oxyhemoglobin A, F, S, and C in plasma. The complex is then removed from the plasma within minutes by the mononuclear-phagocyte system (spleen, thymus, lymph nodes), where the components are metabolized to free amino acids and Fe within hours. Haptoglobin prevents loss of hemoglobin through the renal glomeruli. Approximately 1% of red blood cells are removed from circulation each day, and the body normally synthesizes enough Hp to take care of a normal load. Hp's primary function was originally through to be the preservation of iron and prevention of renal tubular damage by hemoglobin excretion. However, another important role of Hp is the control of local inflammatory response through a number of processes. For example, the Hp-hemoglobin complex is a complex peroxidase that can hydrolyze peroxidases released during phagocytosis and catabolism by polymorphonuclear leukocytes at the site of inflammation. It is also a natural bacteriostatic agent in infections with iron-requiring bacteria, for example, *Escherichia coli*.

Hp depletion is the most sensitive indicator of intravascular hemolysis. In transfusion reactions and certain hemolytic disorders (hemolytic anemias), Hp levels are not sufficient to take care of the increased load and Hp is decreased. Hp levels are also decreased in severe burns, acute and chronic hepatocellular disease, malaria, and disseminated intravascular coagulation (DIC). Epstein-Barr and cytomegalovirus infections and high estrogen levels (oral contraceptives or pregnancy) may result in low levels of Hp. RBC destruction during strenuous exercise can also cause a temporary decrease in Hp.

Haptoglobin is an APR that is synthesized late and weak reacting. As with all APRs, Hp is increased in conditions involving inflammation, infection, tissue necrosis, or malignancy.

Haptoglobin can be measured by immunochemical methods, including immunonephelometry and immunoturbidimetry. The reference range is 16–199 mg/dL.

Ceruloplasmin (Cp)

Ceruloplasmin (Cp) is the principal copper (Cu)-containing protein in plasma containing 95% of the total serum copper. It is not, however, a transport protein (it does not gain or lose copper), but it does prevent copper toxicity. It is now thought that Cp plays a role in copper metabolism by releasing copper to key copper-containing enzymes. The primary role of Cp seems to be in plasma redox reactions, and it can be an oxidant or antioxidant, depending on various factors such as the presence of ferric ions and ferritin-binding sites. For example, Cp oxidizes Fe^{+2} to Fe^{+3} allowing incorporation of iron into the transferrin molecule without the formation of toxic products. Cp does play a role in preventing lipid oxidation and free radical formation, which is damaging to cells.

Although copper is an essential nutrient, it is very toxic to cells in high concentrations. The primary storage site is the liver, and the principal site of excretion is the biliary tract. Ceruloplasmin binds most of the copper released into plasma. Cp carries 6 to 8 Cu atoms per apoceruloplasmin that are half cuprous (Cu^+) and half cupric (Cu^{+2}). Pure Cp has a blue color. Copper homeostasis is based on the balance between absorption in the intestine and excretion in the biliary tract.

Wilson's disease (WD) is a rare autosomal recessive trait where Cp levels are reduced and the dialyzable Cu concentration is increased. A mutation in the gene for a copper-transporting ATPase (ATP7B) results in decreased movement of copper into bile and accumulation of copper in the liver

and other tissues. This enzyme is also responsible for ensuring apoceruloplasmin has its full complement of copper. Patients usually develop symptoms in their 20s or 30s, although it may be earlier or later. Excessive accumulation of Cu in the liver, kidney, and brain can lead to degenerative cirrhosis, chronic active hepatitis, renal tubular acidosis, and neurological damage (clumsiness, tremors) unless treated with a copper chelator, for example, penicillamine or trientine. Although the incidence of WD is low, it is one of the more common causes of chronic liver disease in children. Copper also deposits in the eyes, resulting in the characteristic Kayser-Fleischer rings, pigmented rings at the outer margins of the cornea and the sclera. In other tissues, excessive copper can cause renal tubular damage, kidney stones, osteoporosis, arthropathy, cardiomyopathy, and hypoparathyroidism.

Diagnosis of Level 1 WD is defined as low serum ceruloplasmin (<20mg/dL) and the presence of Kayser-Fleischer rings. If treated before the onset of cirrhosis and neurological symptoms, patients can lead a normal lifespan. Penicillamine can be used to chelate copper and promote its excretion in the urine. Because WD is an autosomal recessive trait, following diagnosis all siblings and parent's siblings should be screened.

Increased levels are also detected during pregnancy and in women on oral contraceptives. Cp is an acute phase reactant that increases late in the condition. Decreased levels are found in liver disease because of impaired synthesis, malnutrition, and conditions resulting in increased protein loss.

Cp can be measured by immunonephelometry, immunoturbidimetry, or radial immunodiffusion. The reference range is 20–60 mg/dL.

α_2-Macroglobulin (AMG)

α_2-**macroglobulin** (AMG) is one of the largest plasma proteins, consisting of four identical subunits that are actually two dimer subunits. AMG is a protease inhibitor that inhibits trypsin, pepsin, and plasmin. Although it appears to be an important antiproteinase, plasma concentrations of AMG are only 1/10 those of AAT. AMG is involved in the primary or secondary inhibition of enzymes in the complement, coagulation, and fibrinolytic pathways.

In nephrotic syndrome, AMG is characteristically increased up to 10 times normal, because it is retained whereas smaller proteins are excreted in the urine. Increased synthesis of all proteins, including AMG, by the liver to compensate for the urinary loss of smaller proteins also contributes to the elevation. AMG makes up for the decrease in smaller proteins. α_2-macrogloculin is also elevated in liver disease and estrogen [oral contraceptives or hormone replacement therapy (HRT)], and slightly increased in diabetes mellitus. AMG may be increased up to 70% in cirrhosis despite the decrease in hepatic mass.

Immunonephelometry, immunoturbidimetry, and radial immunodiffusion assays for AMG are available, although of very limited clinical use. The reference range for AMG is 30–300 mg/dL.

β-GLOBULINS

Transferrin (TRF)

Transferrin (TRF) is the major component of β-globulins and the principal plasma protein for transport of iron (Fe^{+3}—ferric ion) from the intestine, where it is absorbed by apotransferrin, to red cell precursors in the bone marrow or to the liver, bone marrow, or spleen for storage. TRF transports iron to storage sites, where it is bound to apoferritin and stored as ferritin. It also prevents excretion of Fe through the kidney.

Adults have 3–5 grams of body iron; however, only 0.1% or 3 mg circulates in the plasma bound to transferrin. Each transferrin molecule can carry two ferric ions but normally only one-third of the sites are occupied. TRF is responsible for most of the total iron-binding capacity of plasma.

Transferrin is important in the differential diagnosis of anemias and in monitoring the treatment of iron deficiency anemia when the transferrin level is increased but the percent saturation is decreased. TRF is also increased in hepatitis, pregnancy, and women on oral contraceptives or HRT. It is a negative acute phase reactant, with low levels occurring in inflammation and malignancy as well as nephrotic syndrome, and hemochromatosis. TRF levels are also decreased in starvation, anorexia, and malnutrition and can serve as a marker of protein-energy malnutrition (PEM).

The reference range for transferrin is 191–365 mg/dL. TRF can be measured by various immunochemical methods, including rate nephelometry and RID, and indirectly with total iron-binding capacity (TIBC). TIBC ($\mu g/dL$) × 0.70 will give an approximate concentration of transferrin (mg/dL).

Hemopexin

Hemopexin removes heme from circulation. When red blood cells are destroyed, hemopexin transports heme to the liver, where is catabolized by the reticuloendothelial system. It also removes heme from the breakdown of myoglobin or catalase.

Increased levels are found in pregnancy and in diabetes mellitus. Decreased levels are found in some malignancies, hemolytic anemias, and Duchenne-type muscular dystrophy.

β-Lipoproteins (LDL)

β-Lipoproteins are a class of lipoproteins that travel with the β-globulins. Lipoproteins are complexes of proteins and lipids whose function is to carry lipids in the plasma. β-lipoproteins are classified as low-density lipoproteins (LDLs), which transport the majority of cholesterol in the body from the liver to the tissues.

High levels of LDLs are a risk factor for atherosclerosis and heart disease, and will be covered in Chapter 7, Lipids and Lipoproteins. Elevated LDLs are also seen in nephrotic syndrome, hepatobiliary disease, diabetes mellitus, and hypothyroidism.

β₂-Microglobulin (BMG)

β₂-microglobulin (BMG) is a protein on the cell membrane of most nucleated cells and is found in especially high levels in lymphocytes. BMG is a low molecular weight protein that comprises the common light chain of Class I major histocompatability complex (MHC) antigens found in all nucleated cells. It is a small protein that is filtered by the renal glomeruli but is reabsorbed and catabolized by the proximal convoluted tubules.

β₂-microglobulin is increased in renal failure, inflammation, and neoplasms, especially those associated with β-lymphocytes. The plasma level of BMG is a good indicator of glomerular filtration rate. It is used primarily to test for renal tubular function in renal transplant patients when decreased renal tubular function indicates early rejection. BMG levels are also elevated in patients with renal tubulointerstitial disorders, including upper urinary tract infections, anticancer drug toxicity, and heavy metal toxicity.

BMG elevations are also associated with a number of inflammatory conditions such as hepatitis, rheumatoid arthritis, systemic lupus erythematosus (SLE) and acquired immunodeficiency syndrome (AIDS). The elevation of BMG in these conditions may be due to increased lymphocyte turnover. BMG can be measured by enzyme immunoassay and radioimmunoassay, with an upper limit of 2.8 mg/dL in serum and 0.2 mg/L in urine.

C-Reactive Protein (CRP)

C-reactive protein is an acute phase reactant and a nonspecific indicator of bacterial or viral infection, inflammation, and tissue injury or necrosis. It is called *C-reactive* because it was first discovered in the serum of patients with *Streptococcus pneumoniae* infections resulting in precipitation with C-substance, a polysaccharide of *Streptococcus pneumoniae*. CRP reacts with proteins present in many bacteria, fungi, and protozoal parasites.

CRP is composed of five identical subunits, nonglycosylated polypeptides, synthesized in the liver. When CRP is bound to bacteria, it activates complement, neutrophils, and monocyte-macrophages and leads to phagocytosis. These all play a role in CRP's recognition of microorganisms and its role as an immunomodulator in the body's defense. It binds to tissue breakdown products formed during the inflammatory process and activates the complement cascade.

CRP was one of the first APRs to be discovered and one of the most sensitive. Levels increase rapidly within 24–48 hours up to 2000-fold, but it has a short half-life of 19 hours. Once the stimulus is eliminated, the levels return to normal very quickly. Levels rise dramatically following myocardial infarction, trauma, psychological or physical stress, infection, inflammation (e.g., rheumatic fever, rheumatoid arthritis), surgery, and various cancers.

Normally CRP levels in serum are <0.8 mg/dL, and most healthy patients have levels below the detection limit of many procedures. The most dramatic increases in concentration are following a myocardial infarction (MI). Increased CRP predicts risk of an MI or coronary event. Sensitive CRP levels of <1, 1–3, and >3 mg/dL are indicative of low, moderate, and high risk of MI. The new immunoassays, including laser nephelometry, immunoturbidimetry, and enzyme immunoassays, are much more sensitive, which has led to an increased interest in CRP determinations.

Fibrinogen

Fibrinogen is an acute phase reactant and one of the coagulation factors required for the clotting of blood. Its major function is the formation of a fibrin clot when it is activated by thrombin. Therefore, fibrinogen is consumed in the clotting process and is not found in serum. Fibrinogen levels are increased in pregnancy and in women on oral contraceptives. Decreased levels are found in patients with conditions resulting in extensive coagulation.

Complement

Complement is a cascade system of at least 20 glycoproteins, identified by C and a number, for example, C3, that interact to provide many of the effector functions of humoral immunity and inflammation including facilitating phagocytosis and lysis of certain foreign cells, for example bacteria. Once activated the complement cascade results in lysis of the cell.

Complement proteins, especially C3 and C4, are acute phase reactants (APRs) and increased in inflammatory states. They are however, late reacting and weak. C3 and C4 are also elevated in biliary obstruction. Variant and genetic deficiencies of C2 and C4 are associated with autoimmune, immune complex diseases including lupus erythematosus and glomerulonephritis. C3 deficiencies are found in infections especially with encapsulated bacteria. Decreased levels are found when the coagulation proteins have been consumed or used up such as disseminated intravascular coagulation, and in conditions when the body cannot produce adequate proteins, for example, malnutrition.

γ-GLOBULINS

γ-globulins are immunoglobulins, or humoral antibodies. Each immunoglobulin (Ig) molecule consists of two or more basic units consisting of two identical heavy (H) chains and two identical light chains (L). The variable region at the N-terminal end of the four chains provides the antigen-binding sites, and their amino acid sequences determine the antigenic specificity characteristics of the Ig molecule. The remainder of the molecule, the constant region, is identical for every Ig molecule of a given class (Figure 8-4 ■).

Classes and heavy chains (idiotypes) are

IgM = μ (mu)
IgG = γ (gamma)
IgA = α (alpha)
IgD = δ (delta)
IgE = ε (epsilon)

The light chains are produced independently and are then assembled with the heavy chains. All light chains are of two types—Kappa, κ, and Lambda, λ. They occur in all classes in

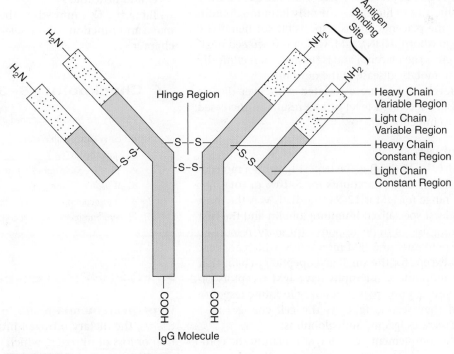

■ FIGURE 8-4 Immunoglobulin

the proportion of $\kappa:\lambda$ of 2:1, and the two halves of a given molecule always have the same type of light chain, kappa or lambda.

Immunoglobulins (Igs) are unique in that they are not synthesized in the liver where the majority of proteins are produced, but by mature B-lymphocytes and plasma cells. Upon encountering a foreign antigen, B-lymphocytes develop and secrete a highly specific antibody capable of binding the antigen. A B-lymphocyte produces only one antibody against a single antigen with a variable region on the heavy and light chains specific for the antigen. As they mature, B-lymphocytes develop into plasma cells, and a clone or cell line of plasma cells specific for the antigen develops. Different clones are produced for every antigen the body encounters that stimulates an immune response.

Immunoglobulin G (IgG)

Immunoglobulin G (IgG) has a molecular weight of 150,000 and is the most abundant Ig in serum (70–75% of Igs). The reference range for IgG is 565–1765 mg/dL. IgG antibodies are produced in response to the antigens of most bacteria and viruses. They are also soluble enough to travel to extracellular spaces and neutralize toxins. IgG is also a complement activator capable of catalyzing the complement cascade, which is how it destroys bacteria and other infectious agents.

IgG globulins are produced by clones or subsets of plasma cells generated in response to different antigens. Each clone produces one specific antibody based on the variable region of the IgG molecule. IgG production for a specific antigen follows IgM, which is the first antibody produced. IgG has four subclasses identified as IgG_1, IgG_2, IgG_3, and IgG_4. IgG_1 is the principal Ig to cross the placenta and protect the baby during the first three months of life. IgG is the only immunoglobulin that can cross the placenta and is responsible for hemolytic disease of the newborn (HDN). IgM, which is discussed next, is too large to cross the placenta and is therefore not clinically significant in hemolytic disease of the newborn.

Increased levels of IgG are associated with liver disease and infections. A decreased level of IgG results in increased susceptibility to infections.

Immunoglobulin M

Immunoglobulin M (IgM) is the largest Ig, with a molecular weight of 900,000, and accounts for 5–10% of total Igs. The reference range for IgM is 12–21 mg/dL. IgM is the most primitive and least specialized immunoglobulin and the first produced during an immune response (primary response). Most IgM is a pentamer of 5 IgM monomers, which are very similar to IgG except for the small glycopeptide J-chain that links the five monomers. B-lymphs have IgM receptors on their surfaces and in primary or secondary immune response to an antigen they secrete IgM. As the cell changes to a plasma cell, it secretes IgG immunoglobulins.

IgM is also a complement activator, stimulating the complement cascade. It is the only immunoglobulin that newborns can produce.

Immunoglobulin A (IgA)

Immunoglobulin A (IgA) has a molecular weight of 160,000 and comprises 10–15% of Ig. The reference range for IgA is 7–13 mg/dL. IgA can also activate complement, but its exact role in serum as a monomer is not clear. A second form, **secretory IgA**, is found in secretions including tears, sweat, saliva, milk as well as gastrointestinal and bronchial secretions. Secretory IgA is a dimer (2 molecules) linked by a J-piece (MW ~380,000) and is synthesized mainly by the plasma cells in the mucous membranes of the gut and bronchi and ductules of lactating breasts. Secretory components provides IgA with resistance to enzymes which allows it to protect the body's mucosal membranes from bacteria and viruses. IgA in breast milk provides the newborn protection from intestinal infections.

Immunoglobulin D

Immunoglobulin D (IgD) has a molecular weight of 184,000 and makes up <1% of serum Igs. The reference range is 0–8 mg/dL. Its primary function is unknown.

Immunoglobulin E

Immunoglobulin E (IgE) has a molecular weight of 180,000 and concentrations are very low (0.3 μg/mg). The reference range for IgE is 3–423 IU/mL. Increased levels of IgE are produced in allergic reactions, urticaria, hay fever, and asthma. IgE levels are increased in allergies especially during acute episodes. When an antigen or allergen cross links two IgE molecules, the mast cells are stimulated to release histamine and other vasoactive amines that are responsible for the symptoms associated with allergies, for example, hay fever and urticaria.

Table 8-1 ✪ summarizes the reference values, molecular mass, and functions of the selected proteins discussed in this chapter.

✓ Checkpoint! 8–3

1. Identify the protein described by the following:
   a. early-onset emphysema _____
   b. spina bifida _____
   c. sensitive indicator of intravascular hemolysis_____
   d. Wilson's disease_____
2. The most common cause of hypoalbuminemia is _____.
3. Hyperalbuminemia is most commonly found in _____.

▶ HYPERPROTEINEMIA

Hyperproteinemia results in a positive nitrogen balance; that is, the dietary nitrogen intake is greater than the excretion or loss of nitrogen, which occurs mainly in the urine. It usually occurs as a result of hemoconcentration or dehydration, where the plasma volume is decreased, causing all of

✪ TABLE 8-1

Characteristics of Selected Proteins

	Reference Value g/dL	Molecular Mass (D)	Function
Transthyretin (Prealbumin)	0.01–0.04	55,000	Marker of poor nutritional status; transport protein— binds thyroid hormones and retinol (Vitamin A)
Albumin	3.5–5.5	66,300	Maintain plasma colloidal pressure; transport ligands, e.g., bilirubin
α_1-Globulin			
α_1-antitrypsin	0.09–0.19	52,000	Acute phase reactant protease inhibitor; associated with early onset emphysema, juvenile hepatic cirrhosis
α_1-acid glycoprotein (orosomucoid)	0.04–0.14	44,000	Acute phase reactant
α_1-fetoprotein		70,000	Antenatal diagnosis of neural tube defects
α_1-antichymotrypsin	0.03–0.06	68,000	Proteinase inhibitor
α_2-Globulins			
Haptoglobin	0.03–0.20	85,000–1 million	Acute Phase reactant; binds free hemoglobin
Ceruloplasmin	0.02–0.60	133,000	Copper metalloprotein; Peroxidase activity; oxidant-antioxidant
α_2-macroglobulin	0.13–0.30	725,000	Protease inhibitor (trypsin, pepsin, and plasmin)
β-Globulins			
Transferrin	0.20–0.36	78,000	Transports iron (Fe^{+3})
Hemopexin	0.05–0.10	60,000	Binds heme
β-lipoproteins	0.25–0.44	3,000,000	Transports lipids, especially cholesterol
β-microglobulin	0.0001–0.0002	11,800	Cellular membrane of nucleated cells, especially lymphocytes; renal tubular function test
C-reactive protein	0.001	118,000	Acute phase reactant; Indicator of risk of myocardial infarction
Fibrinogen	0.20–0.45	341,000	Coagulation factor
Complement			Immune response resulting in cell lysis
γ-globulins			
IgG	0.7–1.6	144,000–150,000	Antibody, increased in immune reactions
IgA	0.090–0.410	~160,000	Antibody
IgM	0.03–0.36	970,000	Antibody

the proteins to be increased proportionally. It is a **relative hyperproteinemia** because the protein concentration is usually normal, but it is dissolved in less plasma.

Another much less common cause is hyperproteinemia due to an increase in globulins such as **multiple myeloma**.

▶ HYPOPROTEINEMIA

Hypoproteinemia causes a negative nitrogen balance, when excretion of nitrogen exceeds intake or synthesis of protein. The most common cause is an increase in plasma water volume, or hemodilution, which results in a decrease in the concentration of all proteins and a **relative hypoproteinemia**. This occurs in water intoxication, salt retention syndromes, and as a result of massive IV infusions and administration of volume expanders (e.g., dextran).

A second major category is an increase in protein loss through the kidneys, gastrointestinal tract, and skin. Examples of excessive loss are nephrotic syndrome, blood loss

after trauma, and burn patients. A third category is decreased intake, as in malnutrition, starvation, or malabsorption. The last category is decreased synthesis in liver disease and immunodeficiency disorders. See Box 8-4 ■ for additional information.

▶ TOTAL PROTEIN METHODOLOGIES

BIURET

Total serum protein concentration is a routine procedure performed in most clinical laboratories. The most common protein methodology is the **Biuret reaction** based on the presence of peptide bonds found in all proteins. When a solution of protein (serum or plasma) is treated with cupric (Cu^{+2}) divalent ions in a moderately alkaline medium, a violet-colored chelate, which absorbs light at 540 nm, is formed between the cupric ion and carbonyl oxygen and the amide nitrogen atoms of the peptide bond. The compound has to contain at least two H_2N-C-, H_2N-CH_2, CH_2-H_2N-CS- or other

BOX 8-4 Hyperproteinemia and Hypoproteinemia

Hyperproteinemia

▶ Hemoconcentration/dehydration
- Inadequate intake of H_2O
- Excessive sweating
- Vomiting
- Salt-losing syndromes
- Diarrhea
- Addison's disease
- Diabetic ketoacidosis

▶ Increase in globulins—multiple myeloma

Hypoproteinemia

▶ Increase in plasma water volume—hemodilution (relative hypoproteinemia)
- Water intoxication
- Salt retention syndromes
- Massive IV infusions
- Volume expanders (e.g., dextran)

▶ Increase in protein loss (kidneys, gastrointestinal tract, and skin)
- Nephrotic syndrome
- Blood loss after trauma
- Burn patients
- Trauma
- Inflammatory bowel disease

▶ Decreased intake
- Malnutrition
- Malabsorption
- Starvation/anorexia

▶ Decreased synthesis
- Liver disease
- Immunodeficiency disorders

similar groups joined together or directly through a carbon or nitrogen atom.

Biuret reagent contains (1) sodium potassium tartrate, which complexes with cupric ions to prevent their precipitation in alkaline solution; (2) copper sulfate, the major reactant providing the Cu^{+2} ions; (3) potassium iodide, an antioxidant that stabilizes the cupric ions; and (4) NaOH to provide the alkaline pH.

The Biuret reaction requires at least two peptide bonds to react, and the color produced is proportional to the number of peptide bonds that have reacted and thereby the concentration of protein in the system. The reaction is simple, can be automated, and is precise enough for clinical use. Serum protein is stable for a week or more at room temperature and a month at 2–4°C. The main interfering factor is hemolysis, which can falsely elevate protein levels due to RBC proteins. Other common causes of interference are lipemia, bilirubinemia, and turbidity. Serum blanks may be used to correct for these interfering substances.

REFRACTOMETRY

Refractometry is also used for a rapid, approximate measure of total serum protein concentration within the range of 3.5–10 g/dL. Velocity of light is changed as it passes through the boundary between two transparent layers, for example air and water, causing light to be bent or refracted. In plasma the major solute protein is diluted in water and the refractive index of the solution increases in proportion to the concentration of the solute, protein. The higher the protein concentration, the greater the refraction or bending of light. Other nonprotein compounds are present in serum, but they do not contribute significantly to the refractive index.

Clinical refractometers used in urinalysis to measure specific gravity are also calibrated for protein in g/dL.

REFERENCE RANGE

The serum protein in healthy, ambulatory adults is 6.0–8.3 g/dL. A physiological decrease of approximately 0.5 g/dL occurs in bedridden patients due to a shift in water distribution into the extracellular compartments. Tourniquet application over 1 minute can result in hemoconcentration and elevate protein by 0.5 g/dL.

▶ A/G RATIO

Total protein and albumin determinations are routinely performed in all clinical laboratories. Globulin methodologies are not routinely available; however, globulin concentration can be calculated by subtracting the albumin from total protein.

$$Globulin = Total\ Protein\ (g/dL) - Albumin\ (g/dL)$$

The **A/G ratio** can then be determined by dividing the albumin concentration by the calculated globulin. For example, a patient with a total protein of 7.0 g/dL and an albumin of 4.0 g/dL has a globulin level of 3.0 g/dL (7.0 − 4.0). The A/G ratio would be 4.0 ÷ 3.0, or 1.33.

The reference range for A/G ratio is approximately 1.0–1.8 with albumin levels normally higher than globulins. Low protein levels, as discussed in a previous section, can be a result of liver disease, renal disease, or decreased intake or malabsorption. A low A/G ratio may be found in overproduction of globulins, for example, multiple myeloma and other autoimmune diseases; underproduction of albumin in liver disease; or selective loss of albumin as occurs in nephrotic syndrome. A high A/G ratio suggests underproduction of globulins found in some leukemias and genetic deficiencies.

▶ URINARY PROTEINS

Urinary proteins originate mostly from the blood and filtration through the renal glomeruli. Proteins in the urine have been filtered by the glomeruli and have not been reabsorbed by the renal tubules. Normally, the higher molecular weight proteins are not present in the glomerular filtrate, and only smaller proteins, for example, albumin, are present. Also renal tubular reabsorption is inversely related to molecular size; the smaller the protein, the more likely it is to be reabsorbed. Other proteins in urine are secreted by the kidney and also from the vaginal and prostatic secretions.

The most common method of screening for urinary protein is by urine reagent test strips based on the protein error of indicators. An indicator, tetrabromphenol blue, at a pH of 3.0 is yellow in the absence of protein and turns green and finally blue as the concentration of protein increases. The protein accepts ions from the indicator, resulting in the change in color due to the protein and not the pH. The indicator changes color because of the loss of hydrogen ions from the indicator to the proteins and not because of a change in pH. This is a semiquantitative assay with results reported as negative, trace, 1+, 2+, 3+, and 4+, corresponding to certain levels of protein in mg/dL. The chemical reagent strips are more sensitive to albumin than globulins; therefore, if globulins are present, the reaction may be negative or falsely decreased. It is also important to correlate the protein level with the specific gravity of the urine, because a trace of protein in a dilute urine is more clinically significant than a trace in a concentrated urine.

Turbidometric methods are based on the precipitation of protein with anionic acids, including sulfosalicylic acid (SSA), trichloracetic acid (TCA), or benethonium chloride. Various concentrations of SSA can be used and methods do vary among laboratories.

SSA will precipitate substances other than protein, resulting in a false positive turbidity. The most commonly encountered substances are radiographic (X-ray) dyes, tolbutamide metabolites, cephalosporins, penicillins, and sulfonamides. Patient history and medical records will confirm the possibility of a false positive.

► CEREBROSPINAL FLUID PROTEIN

Cerebrospinal fluid (CSF) is a clear, colorless fluid that contains small amounts of glucose and protein. CSF protein levels as well as glucose, red and white blood cell count and differential, and culture and sensitivity are routinely performed on CSF. The reference range for CSF protein is 15–45 mg/dL.

CSF protein levels are determined to detect increased permeability of the blood–brain barrier to plasma proteins or increased production of immunoglobulins in the spinal canal (intrathecal). Increased levels of CSF proteins are found in various types of meningitis, cerebral hemorrhage, spinal cord tumors, neurosyphilis, multiple sclerosis, and brain abscesses.

► ALBUMIN

Albumin is a small protein with a molecular weight approximately 66,000 that comprises approximately 60% of all serum proteins. It is synthesized in the liver, rate dependent on protein intake, and subject to feedback regulation by the plasma albumin level and the colloidal osmotic pressure (COP). Albumin first appears in plasma as proalbumin. The chief biological function is to maintain plasma COP, as discussed earlier in this chapter.

► HYPERALBUMINEMIA

Hyperalbuminemia is of little diagnostic significance except in **dehydration**. Most increases in albumin are artifactual due to decreased plasma volume. Artifactual increases in albumin also occur after prolonged tourniquet application.

► HYPOALBUMINEMIA

The most common cause of **hypoalbuminemia** is increased catabolism, tissue damage, and inflammation. A second category is impaired or decreased synthesis, with the primary cause being liver disease and secondary causes including diminished protein intake, malnutrition, and malabsorption. The third category is increased loss of protein in the urine in nephrotic syndrome, chronic glomerulonephritis, diabetes, and extensive burns. A rare cause is **analbuminemia**, the absence of albumin, a genetic autosomal recessive trait. Hypoalbuminemia decreases plasma colloidal osmotic pressure, resulting in edema, leakage of water from the intravascular fluid into the body's tissues.

► ALBUMIN METHODOLOGIES

Albumin methodologies are based on binding of albumin with anionic dyes. **Bromcresol green (BCG)** and **bromcresol purple (BCP)** are the two most common dyes. Bromcresol green is not as specific for albumin and will bind to globulins, albeit more slowly. However, this is not a problem if the procedure is performed promptly. Methyl orange, another anionic dye, was also used but has lost popularity because of nonspecific binding. The pH in albumin methodologies is adjusted lower than the isoelectric point so albumin is positively charged (pH < pI).

Four main requirements of dye-binding methods are as follows: (1) specific binding of the dye to albumin in the presence of serum or plasma proteins; (2) high binding affinity between the dye and albumin, so small changes in ionic strength and pH will not break the dye-protein complex; (3) a substantial shift in the absorption wavelength of the dye in the bound form so that it will be spectrally distinct from the free form present in excess; (4) an absorption maximum for the bound form at a wavelength distinct from those where bilirubin and hemoglobin, the main interfering chromogens, can interfere.

REFERENCE RANGE

The reference range for albumin is 3.4–5.0 g/dL or 34–50 g/L.

▶ PROTEIN ELECTROPHORESIS

Protein electrophoresis is the migration of charged solutes or particles in a liquid medium under the influence of an electrical field. Proteins are ampholytes or zwitterions and can move toward the anode or cathode, depending on the charge. Protein electrophoresis can be performed on cellulose acetate or agarose gel. Agarose gel is preferred because of its higher resolution and efficiency, and it does not have to be cleared or presoaked. Most laboratories purchase the apparatus, materials, and reagents from a single supplier to ensure compatability and the best results.

Protein electrophoresis is performed on serum to avoid complication of the fibrinogen band in the β–γ region. A standard barbital buffer with ionic strength of 0.05 and pH 8.6 is used. At this pH the serum proteins have a net negative charge (pH > pI) and migrate toward the anode. The sample size, 3–5 μL, is applied to the support media using a mechanical device. The support medium is interposed between two electrodes immersed in the buffer, and a constant current and voltage is applied for 40–60 minutes, depending on the system used. Migration distance varies directly with the charge carried by the protein.

The rate of migration is affected by five factors. The net electrical charge based on the pH of the buffer and the pI of the proteins will increase or decrease mobility. The size and shape of the molecule, the strength of the electrical field, the properties of the supporting medium, and the temperature will also affect the distance traveled by the protein.

After electrophoresis, the bands are fixed by immersing the strip in an acid medium, usually acetic acid and methanol, which denatures and immobilizes the proteins. The strip is then stained so the bands can be visualized. Coomassie Brilliant Blue is the most sensitive and most widely used stain. Amido black, Ponceau S, and Bromphenol blue are also utilized in some systems.

The protein strips are then passed through a densitometer, where an optical beam is passed through the strip and absorbance reading of each fraction is measured and displayed on a recorder chart as a series of peaks. The strip is placed in a holder and is slowly moved through a beam of light of the correct wavelength for the dye used. A certain amount of light is absorbed depending on the concentration of the protein and the dye bound to it. The larger and darker the band, the greater the protein concentration and the higher the absorption. As with all laboratory procedures a reference serum is run along with the patients' for comparison.

The readout is a trace that relates to the amount of light absorption. Data are further analyzed electronically to provide a percentage distribution of each fraction of the strip.

$$\text{The relative } \% = \frac{\text{Abs of the Specific Band}}{\text{Absorbance of all Bands}}$$

A normal serum protein electrophoresis (SPE) has five bands. The % distribution × total protein = g/dL. For example, a

⚙ TABLE 8-2

Protein Electrophoresis Bands

	Relative %	g/dL
Albumin (anode—fastest)	53–65	3.5–5.0
α_1-globulin	2–5	0.1–0.3
α_2-globulin	7–13	0.6–1.0
β-globulin	8–14	0.7–1.1
γ-globulin (cathode—slowest)	12–20	0.8–1.6

total protein of 7.5 g/dL and an α_2-globulin of 10% calculates to an α_2-globulin of 0.75 g/dl (7.5 g/dL × 0.10) rounded off to 0.8 g/dL. The bands, relative percentages, and ranges in g/dL are listed in Table 8-2 ⚙.

COMMON ABNORMAL PROTEIN ELECTROPHORESIS PATTERNS

Common abnormal serum protein electrophoresis patterns are classified by an abnormal band or bands. Abnormal increases in α_2-globulins are found in diseases associated with vasculitis (inflammation of blood vessel walls), and immune complex diseases, for example, rheumatoid arthritis. β_1-globulins increases are found in iron deficiency anemia due to increases in transferrin or high levels of estrogen.

Fusion of the β–γ bands, or **bridging**, is a result of fast moving γ-globulins that prevent resolution of β- and γ-globulins. Cirrhosis is the most common cause of β–γ bridging, although it can also be found in respiratory or skin infections and rheumatoid arthritis. Increases in γ-globulins are associated with immune reactions and chronic inflammatory disease, for example, rheumatoid arthritis, metastatic cancer, bacterial infections, and liver disease.

Nephrotic syndrome is characterized by a decrease in albumin and γ-globulin bands in conjunction with an increase in α_2-globulins that suggests selective proteinuria. In selective **proteinuria** increased numbers of smaller molecular weight proteins pass through the glomerular membranes of the renal nephron and are excreted, but the larger proteins are retained. Proteins like albumin are excreted by the kidney and larger molecular weight molecules like α_2-macroglobulin are retained. It is important to note that albumin must be decreased by at least one-third before it is evident on the electrophoresis.

A diffuse increase in the γ-globulin band is found in polyclonal gammopathy associated with chronic liver disease, chronic inflammatory disease, immune reaction, or metastatic cancer. Absence or decrease in the γ-globulin band is indicative of acquired or congenital immunodeficiency.

In **acute phase reactions**, both of the major α_1 proteins, α_1-antitrypsin (AAT) and α_1-glycoprotein (AAG); α_2-band proteins, ceruloplasmin and haptoglobin; and the β-globulins, C3, C4, and C-reactive protein, are increased and albumin is decreased. This pattern suggests an acute phase reaction. Acute phase reactants are increased in infection,

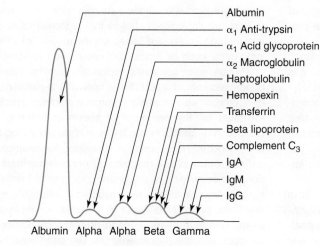

1. Normal serum protein electrophoresis

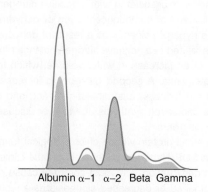

2. Nephrotic Syndrome (↓ albumin, α₁, β, and γ globulins and ↑ α₂ globulin)

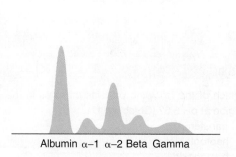

3. Acute Phase Reactants (↑ α₁ and α₂ globulins)

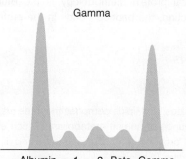

4. Multiple Myeloma (↑ γ globulin-M spike)

■ FIGURE 8-5   Common electrophoresis patterns

tumor growth or malignancy, rheumatoid arthritis, hepatitis, surgery, trauma, burns, and myocardial infarction. The APRs play an important role in protecting the body from inflammation, especially in complement activity and controlling enzyme activity.

Positive acute phase reactants that increase during APR are α_1-antitrypsin, α_1-acid glycoprotein, haptoglobin, ceruloplasmin, α_2-macroglobulin, fibrinogen, and C-reactive protein. The negative APRs are albumin, transferrin, and transthyretin (prealbumin). APR levels increase at different rates but all reach their maximum within 2–5 days. The first to elevate is C-reactive protein, followed by AAG, AAT, Hp, C4, and fibrinogen. The final APRs are C3 and Cp. See Figure 8-5 ■ for common electrophoresis patterns.

SUMMARY

Proteins are composed of 20 different amino acids in varying numbers and sequences and contain carbon, hydrogen, oxygen, and nitrogen; some also contain sulfur. The secondary, tertiary, and quaternary structures of proteins are determined by the amino acid sequence that is the primary structure.

Proteins are involved in many cellular processes, including maintenance of colloidal osmotic pressure, coagulation, and transport of various materials. For example, plasma colloidal osmotic pressure (COP) provided by proteins tends to retain water in the vascular space. Structural proteins including collagen provide support for the body, tissues, or cells.

Aminoacidopathies are inherited disorders of amino acid metabolism. For example, alkaptonuria is a rare inherited disease that results from the deficiency of the enzyme homogentisic acid oxidase in the catabolic pathway of tyrosine, which leads to a buildup of homogentisic acid (HGA) in the tissues of the body. Another aminoacidopathy, phenylketonuria (PKU), is an inborn error of metabolism that results in the inability to metabolize the essential amino acid phenylalanine.

Total protein levels can be measured as well as methodologies for specific proteins. Transthyretin (TTR) binds with thyroxine and triiodothyronine (thyroid hormones) and retinol (vitamin A) and serves as a transport protein. The main clinical significance of TTR is its role as a sensitive marker of poor nutritional status such as protein-energy malnutrition (PEM). α_1-antitrypsin is the major α_1-globulin making up approximately 90% of α_1-proteins. AAT deficiency is associated with early onset emphysema; however, a congenital deficiency can also result in juvenile hepatic cirrhosis, in which AAT is synthesized by the hepatocytes but not released. Examples of other specific proteins that are clinically significant and can be measured are alpha-fetoprotein, haptoglobin, ceruloplasmin, C-reactive protein, and immunoglobulins.

Hyperproteinemia is associated with a positive nitrogen balance that occurs as a result of hemoconcentration or dehydration. It is usually a relative hyperproteinemia as a result of dehydration. Hypoproteinemia is related to a negative nitrogen balance often occurring as a result of an increase in water volume, which leads to a relative hypoproteinemia. A second category is increased loss of protein through the kidneys, gastrointestinal tract, and skin. The third category is decreased synthesis due to liver disease and immunodeficiency disorders.

Albumin is a small protein whose chief biological function is to maintain plasma COP. Hyperalbuminemia is of little clinical significance except in dehydration. The most common cause of hypoalbuminemia is impaired or decreased synthesis as a result of liver disease. Other causes include diminished protein intake, malnutrition, and malabsorption.

The most common total protein methodology is the Biuret reaction. In the Biuret reaction, the peptide bonds in the protein molecules bind with cupric ions in a moderately alkaline solution to form a violet colored chelate that is measured at 540 nm. Albumin methodologies are based on the binding of albumin with anionic dyes, such as bromcresol green and bromcresol purple.

Protein electrophoresis separates proteins into five separate bands: albumin, α_1-globulin, α_2-globulin, β-globulin, and γ-globulin. Common abnormal serum protein electrophoresis patterns are identified by the presence of abnormal band(s). Abnormal increases in α_2-globulins are found in diseases associated with vasculitis and immune complex diseases, for example, rheumatoid arthritis. Acute phase reactants or proteins are increased in infection, tumor growth or malignancy, rheumatoid arthritis, hepatitis, surgery, trauma, burns, and myocardial infarction. In acute phase reactions, both of the major α_1 proteins, α_1-antitrypsin (AAT) and α_1-glycoprotein (AAG); α_2-band proteins, ceruloplasmin and haptoglobin; and the β-globulins, C3, C4, and C-reactive protein are increased and are called positive AAPs (acute phase proteins), and albumin, transthyretin, and transferrin are decreased or negative AAPs.

REVIEW QUESTIONS

LEVEL I

1. Proteins, carbohydrates, and lipids comprise the three major biochemical compounds of human metabolism. Which element distinguishes proteins from carbohydrate and lipid compounds? (Objective 1)
   a. carbon
   b. oxygen
   c. nitrogen
   d. hydrogen

2. What is the common property of amino acids? (Objective 1)
   a. They contain sulfur.
   b. They have –OH and –COOH groups on adjacent carbon.
   c. They reduce copper.
   d. They are obtained on hydrolysis of starch.
   e. NH_2 and COOH groups attached to the same carbon.

3. Amino acids are joined to form proteins through: (Objective 1)
   a. hydrogen bonds.
   b. peptide bonds.
   c. disulfide bonds.
   d. ionic bonds.

4. At its isoelectric point a protein is: (Objective 1)
   a. negatively charged.
   b. neutral (no net charge).
   c. positively charged.

5. The chemical bond between amino acids in the primary structure of proteins are: (Objective 1)
   a. hydrogen bonds.
   b. disulfide bonds.
   c. peptide bonds.
   d. hydrophilic bonds.

LEVEL II

1. Which of the following proteins migrates in the α_2-globulin region at pH 8.6? (Objective 17)
   a. transferrin
   b. haptoglobin
   c. hemopexin
   d. fibrinogen

2. Elevated serum protein and decreased albumin are associated with: (Objective 15)
   a. glomerulonephritis.
   b. cirrhosis.
   c. starvation/anorexia.
   d. multiple myeloma.

3. What is the clinical significance of testing for prealbumin? (Objective 17)
   a. Levels correlate with diabetic nephropathy
   b. High levels are associated with acute inflammation
   c. A series of low levels indicate poor nutritional status
   d. Low levels are related to decreased cortisol

4. Which of the following conditions is associated with an acute inflammatory pattern on protein electrophoresis? (Objective 15)
   a. cirrhosis
   b. rheumatoid arthritis
   c. myocardial infarction
   d. malignancy

5. The screening test for antenatal detection of neural tube defects is: (Objective 17)
   a. alpha fetoprotein.
   b. ceruloplasmin.
   c. β_2-microglobulin.
   d. α_1-antichymotrypsin.

REVIEW QUESTIONS (continued)

LEVEL I	LEVEL II

6. α-helixes and β-pleated sheets describe the _____ structure of proteins. (Objective 1)
   a. primary
   b. secondary
   c. tertiary
   d. quaternary

7. The primary function of albumin is to: (Objective 9)
   a. enhance the ionization of calcium.
   b. regulate colloidal osmotic pressure.
   c. promote the antigenic activity of plasma proteins.
   d. provide a neutral binding site for all receptors.

8. Albumin methodologies include all of the following *except*: (Objective 12)
   a. bromcreol purple.
   b. bromcresol green.
   c. phosphotungstic acid.
   d. methyl orange.

9. A protein which is conjugated with cholesterol and triglyceride is called a: (Objective 1)
   a. glycoprotein.
   b. lipoprotein.
   c. metalloprotein.
   d. mucoprotein.

10. What is the most common cause of hyperproteinemia? (Objective 4)
    a. nephrotic syndrome
    b. malnutrition
    c. dehydration
    d. liver disease

11. Transferrin is associated with the transport of: (Objective 3)
    a. ferritin.
    b. ferric ions.
    c. ferrous ions.
    d. heavy metals.
    e. hemoglobin.

12. Which of the following proteins transports thyroxine and Vitamin A in the blood and serves as a marker of nutritional status? (Objective 3)
    a. α-1 antitrypsin
    b. hemopexin
    c. ceruloplasmin
    d. transthyretin

13. Wilson's disease is associated with a deficiency of: (Objective 3)
    a. hemoglobin.
    b. α_1-antitrypsin.
    c. ceruloplasmin.
    d. haptoglobin.

REVIEW QUESTIONS *(continued)*

LEVEL I

14. Early onset emphysema and juvenile cirrhosis are associated with a deficiency of: (Objective 3)
    a. α-1 antitrypsin.
    b. hemopexin.
    c. α-2 macroglobulin.
    d. C-reactive protein.
    e. haptoglobin.

15. Hypoproteinemia would be found in cases of: (Objective 5)
    a. multiple myeloma.
    b. dehydration.
    c. vomiting.
    d. nephrotic syndrome.

16. Globulin in serum is usually determined by: (Objective 13)
    a. calculation as difference between total protein and albumin.
    b. Biuret reaction.
    c. $MgSO_4$ fractionation.
    d. ethanol fractionation.
    e. paper chromatography.

17. Examples of negative acute phase reactants are: (Objective 15)
    a. albumin, transthyretin, and IgG.
    b. ceruloplasmin, haptoglobin, and fibrinogen.
    c. albumin, transthyretin, and transferrin.
    d. C-reactive protein, fibrinogen, and haptoglobin.

18. pH 8.6 is used for serum protein electrophoresis so that: (Objective 14)
    a. all serum proteins will have a net negative charge.
    b. all serum proteins will have a net positive charge.
    c. electroendosmosis is avoided.
    d. heat production is minimized.

19. Which of the following is associated with "beta-gamma bridging"? (Objective 15)
    a. multiple myeloma
    b. malignancy
    c. hepatic cirrhosis
    d. rheumatoid arthritis

20. How many heavy chains and how many light chains make up an IgG molecule? (Objective 3)
    a. two heavy, two light
    b. two heavy, four light
    c. four heavy, two light
    d. four heavy, four light

LEVEL II

REFERENCES

1. American Academic of Pediatrics. Maternal serum α-fetoprotein screening. Committee on Genetics. *Pediatrics* (2001) 88, 6: 1282–1283.

2. Anderson SA, Cockayne S. *Clinical chemistry: Concepts and applications.* (New York, NY: McGraw Hill, 2003).

3. Arnold G, Renner CJ, Konop R, et al. Phenylketonuria (www.emedicne.com/ped/topic.1787.htm), accessed September 14, 2007.

4. Bishop ML, Fody EP, Schoeff L. *Clinical chemistry: Principles, procedures, correlations*, 5th ed. (Philadelphia, PA: Lippincott Williams & Wilkins, 2005).

5. Burtis CA, Ashwood ER, Bruns DE. *Tietz textbook of clinical chemistry and molecular diagnostics*, 4th ed. (Elsevier Saunders, 2006): 533–595.

6. Demling RH, DeSanti L. Involuntary weight loss and protein-energy malnutrition: Diagnosis and treatment (www.medscape.com/viewprogram/713), accessed September 15, 2007.

7. Eden, Edward MD. Alpha1-antitrypsin deficiency in COPD: Clinical implications. *Medscape* (www.medscape.com/viewarticle/408728/).

8. Graves C, Miller KE, Sellers AD. Maternal serum triple analyte screening in pregnancy. *American Family Physician* (2002) 65, 5: 915.

9. Harvard. Iron Transport and cellular uptake (2001) (Sickle.bwh.harvard.edu/iron_transport.html).

10. Roth KS. Alkaptonuria (www.emedicine.com/ped/topic64.htm), accessed September 14, 2007.

11. Lewandrowski K. Clinical chemistry: Laboratory management and clinical correlations. (Philadelphia, PA: Lippincott Williams & Wilkins, 2002): 531–560.

12. Schilsky ML, Oikonomou I. Inherited metabolic liver disease (www.medscape.com/viewarticle/502872_print), accessed September 14, 2007.

13. Tavill AS. Wilson's disease. *The Cleveland Clinic: Disease management project.* (2004) (www.clevelandclinicmeded.com/medicalpubs/diseasemanagement/gastro/wilsons/wilsons1.htm), accessed September 14, 2007.

14. WebMD. New guidelines for alpha-1 antitrypsin deficiency. (2003) (www.medscape.com/viewarticle/462776).

9

Enzymes

■ OBJECTIVES—LEVEL I

Following successful completion of this chapter, the learner will be able to:

1. Define enzyme and list general functions of enzymes.
2. Write the formula for enzyme-catalyzed reactions.
3. List the six major groups of enzymes and the reactions catalyzed by each group.
4. Review enzyme catalysis, including the role of enzymes in decreasing activation energy.
5. Define apoenzyme, prosthetic groups, and holoenzyme.
6. Define cofactor, coenzyme, and metalloenzyme and give examples of each.
7. Interpret the formula $Q = K \times E \times t$ and define first-order and zero-order reactions.
8. Explain how various factors affect enzyme reactions—for example, pH, temperature, and substrate concentration.
9. Examine the differences among competitive, noncompetitive. and uncompetitive inhibition.
10. Summarize the reaction catalyzed (including activators and coenzymes), methodologies, clinical significance, and reference ranges for the following cardiac enzymes: creatine kinase (CK), creatine kinase isoenzymes, and lactate dehydrogenase (LD).
11. Review the clinical significance of the three major CK isoenzymes (heart, muscle, and brain), including their dimeric composition and major sources.
12. Identify a normal CK isoenzyme pattern and the typical pattern following a myocardial infarction (MI).
13. Briefly examine other CK isoenzyme procedures—for example, electrophoresis and immunoinhibition.
14. Summarize the reaction catalyzed (including activators and coenzymes), methodologies, clinical significance, and reference ranges for the following liver enzymes: aspartate aminotransferase (AST), alanine aminotransferase (ALT), and alkaline phosphatases (ALP).
15. Differentiate the five major LD isoenzymes, including their tetrameric composition and the major tissue(s) involved.
16. Summarize the reaction catalyzed (including activators and coenzymes), methodologies, clinical significance, and reference ranges for the following biliary tract enzymes: gammaglutamyl transferase (GGT) and 5′-nucleotidase (5′-NT).
17. Relate the reaction catalyzed (including activators and coenzymes), methodologies, clinical significance, and reference ranges for the following pancreatic and liver enzymes: amylase (AMY), lipase (LPS), trypsin (TRY), and chymotrypsin (CHY).

■ OBJECTIVES—LEVEL II

Following successful completion of this chapter, the learner will be able to:

1. Examine urinary amylase, including the clinical significance and calculation of the amylase-to-creatinine clearance ratio.

2. Calculate the CK relative index (RI).

3. Review and differentiate the reactions catalyzed (including activators and coenzymes), methodologies, clinical significance, and reference ranges for the following miscellaneous enzymes: acid phosphatase (ACP), aldolase (ALD), and cholinesterase (CHE).

KEY TERMS

Absolute specificity
Acid phosphatase (ACP)
Activators
Active site
Alanine aminotransferase (ALT)
Aldolase
Alkaline phosphatase (ALP)
Allosteric sites
Amylase creatinine clearance ratio (ACCR)
Apoenzyme
Aspartate aminotransferase (AST)
Bond specificity
Cholinesterase
Chymotrypsin (CHY)
Coenzymes
Cofactor

Competitive inhibition
Creatine kinase (CK)
Denaturation
De Ritis ratio
Energy of activation (EA)
Enzyme–substrate (ES) complex
First-order kinetics
Gamma glutamyl transferase (GGT)
Group specificity
Holoenzyme
Hydrolases
Inhibitors
International unit (IU)
Isoenzymes
Isomerases
Lactate dehydrogenase (LD)
Ligases

Lipase
Lyases
Macroamylasemia
Metalloenzymes
Michaelis–Menten constant (Km)
Noncompetitive inhibition
5′-Nucleotidase (5′-NT)
Oxidoreductases
pH
Prosthetic group
Relative index (RI)
Stereospecificity
Transferases
Trypsin (TRY)
Uncompetitive inhibition
V max
Zero-order kinetics

 A CASE IN POINT

Doreen T., a 60-year-old woman, was seen in the emergency room complaining of chest pain, which was moderate to severe. She had experienced substernal pain for the previous 6 to 7 weeks with dyspnea on exertion. The pain, however, had become more fre-

quent and severe with constant pain and pressure for the last 2 to 3 days. Doreen appeared anxious and complained of weakness, sweating, and nausea. Her blood pressure was 110/66.

Table 1 Chemistry Results

	Day 1	Day 4	Day 6	Day 9	Day 11	Reference Range or Desired Range
Sodium	136	130	139	143	153	135–145 mEq/L
Potassium	3.7	3.0	4.3	3.9	3.8	3.6–5.0 mEq/L
Chloride	94	103	107	113	114	98–107 mEq/L
CO_2		25.0	24.0	23	30.0	24.0–34.0 mEq/L
Anion Gap		2.0	8.0	7.0	9.0	10–20 mmol/L
Glucose	319	519	379	310	234	70–99 mg/dL
BUN	23	53	81	99	79	7–24 mg/dL
Creatinine	0.9	1.4	2.3	2	1.9	0.5–1.2 mg/dL
Calcium	9.7					8.5–10.5 mg/dL
Magnesium		1.1	1.7	1.9	2.0	1.3–2.5 mEq/L
Digoxin			2.60	1.12		0.80–2.00 ng/mL
Cholesterol	350					<200 mg/dL
Triglyceride	275					<150 mg/dL
Bilirubin	0.2					0.2–1.2 mg/dL
AST	76					5–40 IU/L
ALP	84					30–157 IU/L
Protein	7.3					6.0–8.4 g/dL
Albumin	4.1					3.5–5.0 g/dL
TSH				0.72		0.3–3.0 uIU/mL

Table 2 Cardiac Profile

	Day 1	Day 2	Day 2	Day 2	Day 3	Day 4	Reference Range
	20:30	5:06	12:25	20:35	12:15	7:15	
CK	668	1383	3461	3743	2117	973	24–170 IU/L
CK–MB	47.1		146.6	93.0	24.8	12.1	0.0–3.8 ng/mL
Troponin I	36.6	184.0	5745.0	926.1			0.0–0.4 ng/mL

CK–MB Reference Range		Troponin I Reference Range	
0–3.8:	Normal	0–0.4 ng/mL:	No evidence of myocardial injury
3.9–10.4:	Borderline	0.5–2.0 ng/mL:	Mild elevation, possible myocardial injury
>10.4:	Significantly elevated	>2.0 ng/mL:	Significantly elevated, consistent with myocardial injury

Doreen was given Nitrostat 1/150 grain PRN for pain, and Inderal was administered. She was also taking digoxin. An electrocardiogram was performed and revealed an atrial flutter and the possibility of a true posterior infarct and lateral ischemia.

Issues and Questions to Consider

1. (A) After reading Doreen's initial patient history, what chemistry profile would the emergency room physician order on this patient? (B) What tests are included in this profile in your laboratory, and what are the collection times?

2. What laboratory results in Tables 1 and 2 are abnormal?

3. The laboratory results in Tables 1 and 2 indicate what condition?

4. Heart muscle contains which creatine kinase (CK) isoenzyme(s)?

5. How many hours after a myocardial infarction (MI) would you find an elevated CK and CK-MB? How long would they remain elevated?

6. Define CK relative index (RI). Calculate the CK-MB RI for all specimens with total CK and CK-MB.

7. What is the cause of the elevated AST?

8. What other medical problem or condition does this patient have? What laboratory values support your decision?

9. Do patients with the condition you described in the previous questions have a higher risk of myocardial infarction or stroke than the general population? Why or why not?

WHAT'S AHEAD

▶ A summary of enzyme function and biochemistry.

▶ A review of the kinetics of enzyme reactions.

▶ Identification of factors influencing enzyme reactions—for example, substrate concentration, temperature, and pH.

▶ A discussion of enzymes used in the diagnosis of cardiac disease.

▶ A review of enzymes included in liver function panels.

▶ Differentiation of enzymes associated with hepatobiliary disorders or obstructive liver disease from those associated with hepatocellular disorders.

▶ An examination of digestive and pancreatic enzymes and their role in the treatment of acute pancreatitis.

▶ An overview of other clinically significant enzymes.

▶ INTRODUCTION

Enzymes are biological catalysts present in all the cells of the body; each enzyme catalyzes a specific reaction. Biological systems developed enzymes or catalysts to enable metabolic reactions to occur at body temperature. Enzymes enhance reaction rates from 10^6 to 10^{12} times the speed of those reactions without enzymes.

Enzymes supply the energy and or chemical changes necessary for vital activities. These include muscle contraction, nerve conduction, respiration, digestion or nutrient degradation, growth, reproduction, and maintenance of body temperature. Enzymes are proteins; all enzymes are synthesized by the body under the control of a specific gene in the same way other proteins are formed.

▶ ENZYME BIOCHEMISTRY

Enzymes have three properties. First, they are not altered or consumed during the reaction. Second, only small amounts of them are required because they are used over and over again. Third, they accelerate the speed at which a chemical reaction reaches equilibrium, but they do not alter the equilibrium constant (the amount of product formed from the substrate).

As proteins, enzymes have a unique sequence of amino acids that provides the primary structure. The twisting and shape from the interaction of adjacent amino acids form the secondary structure: α-helix, β-pleated sheet, and random coil. The tertiary structure is determined by the interaction of the R groups of more distant amino acids. The quaternary structure is the arrangement of two or more polypeptide chains to form a protein. For example, creatine kinase is an enzyme that is dimeric: it consists of two polypeptide chains, M or B or both, resulting in three possible combinations: MM, MB, and BB. See Chapter 8 for a more complete description of protein structure.

Enzymes are composed of a heat-labile protein portion called the **apoenzyme**, which requires a coenzyme for full catalytic activity. The **prosthetic groups** are the bound coenzymes discussed later in this section. The **holoenzyme** is the apoenzyme and cofactor or coenzyme, which forms the catalytically active unit.

Enzymes catalyze reactions in which one or more substrate (S) molecules are converted to a product (P). A catalyst increases the rate of a particular chemical reaction without itself being consumed or permanently altered.

$$\text{Substrate} \xrightarrow{\text{Enzyme}} \text{Product} \qquad (9.1)$$

Enzymes have an **active site** or *center*, the actual place where particular bonds in the substrate are strained and ruptured and which closely fits the substrate. Think of the analogy of a lock and key with the substrate being the key. The active sites may involve only 5 to 10 amino acids out of a total of 200 to 300 amino acids in the enzyme molecule. The rest of the enzyme molecule comprises **allosteric sites** (also called *regulatory sites*).

Certain enzymes require the presence of a particular ion to induce the necessary configuration for proper binding to the substrate. **Cofactors** are organic or inorganic compounds that are required for enzyme function. *Inorganic cofactors*, or **activators**, include metallic ions—for example, Fe^{+2}, Ca^{+2}, Mg^{+2}, Zn^{+2}, Cu^{+2}, and Mn^{+2}. These are required in **metalloenzymes**, which have metal ions bound tightly to the molecule. **Coenzymes** are organic cofactors that commonly have a structure related to vitamins (usually vitamin B). Coenzymes include nicotinamide adenine dinucleotide (NAD^+) and pyridoxal-5-phosphate (P-5-P).

Enzymes are classified by the degree of substrate specificity. Enzymes with **absolute specificity** catalyze only one specific substrate and one reaction. **Group specificity** indicates an enzyme that catalyzes substrates with similar structural groups—for example, phosphatases that catalyze reactions containing organic phosphate esters. **Bond specificity** describes the catalyzing of a reaction with a certain type of bond—for example, the peptide bonds in proteins. Enzymes with **stereospecificity**, or *stereoisomer specificity*, catalyze reactions with only certain optical isomers. An example is glucose oxidase, which reacts only with β-D glucose.

▶ ENZYME KINETICS

An understanding of enzyme kinetics is critical to the comprehension of enzyme reactions and methodologies. As stated in the previous section, the enzyme catalyzes the conversion of substrate to product. The **energy of activation (EA)** is the amount of energy required to energize one mole of the

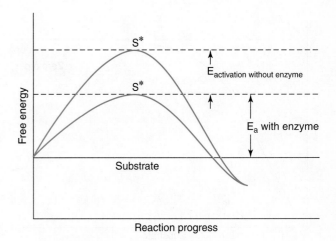

Enzymes lower the amount of energy required to reach the activation state S.

■ **FIGURE 9-1** Energy of Activation

substrate to form the activated complex. The enzyme binds with the substrate to form an **enzyme–substrate (ES) complex** that provides the free energy required for the reaction. This energy allows the reaction to proceed without the addition of external energy. In other words, the energy barrier of the reaction is decreased, and the breakdown of substrate to product is increased. See Figure 9-1 ■ for a graph that illustrates the energy of activation. This complex brings the substrate molecule into proper alignment with the enzyme so the catalytic reaction can take place at the active site.

$$E + S \rightarrow ES \longrightarrow P + E \qquad (9.2)$$

V max occurs when the substrate concentration is high enough that all enzyme molecules are bound to the substrate and all active sites are engaged. The **Michaelis-Menten constant (Km)** is the substrate concentration in moles per liter when the initial velocity is ½ Vmax. Km is a constant and remains the same for a given enzyme–substrate pair under given conditions. It is an expression of the relationship between the velocity of an enzyme reaction and the substrate concentration.

In a graph of substrate concentration versus velocity in the initial section, the velocity is almost linear with respect to substrate concentration, which indicates a reaction in **first-order kinetics**. The velocity is directly proportional to the substrate concentration. In other words, the more substrate, the higher the velocity (see Figure 9-2 ■).

Enzyme reactions in the clinical laboratory are measured in **zero-order kinetics**. A plateau is reached where the reaction rate is independent of substrate concentration. The enzyme cannot work any faster, all of the enzyme is bound to substrate, and the reaction rate is based on enzyme activity only. Laboratories measure enzymes using zero-order kinetics when the reaction is based only on enzyme concentration.

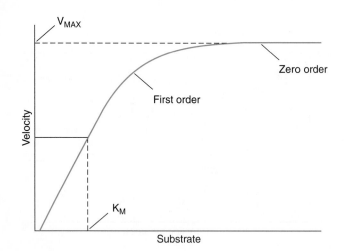

■ FIGURE 9-2 First- and Zero-Order Kinetics

▶ FACTORS INFLUENCING ENZYME ACTION

Many factors influence enzyme action, including substrate concentration, pH, temperature, and activators or cofactors. Enzymes are measured by activity and not by mass because of the small amounts of enzyme normally present in serum. The unit of measurement, the **International unit (IU)**, is the amount of enzyme that produces 1 μmol of product per minute under standardized conditions of temperature, pH, substrate, and activators.

Substrate concentration increases the rate of enzymatic reaction to a certain point (first-order kinetics) as described in the previous section. The higher the concentration of substrate molecules present, the more likely they are to bind to an enzyme until all of the enzyme molecules are bound to substrate in zero-order kinetics. All reactions must be in zero-order kinetics for accurate enzyme measurement; substrate must be in sufficient excess that no more than 20% is converted to product in a normal reaction. This means that the substrate concentration in reagents must be at least 10 times—preferably 100 times—greater than the normal enzyme levels can metabolize during that period. This allows measurement of abnormally elevated enzymes without dilution.

When the substrate concentration is high and essentially constant during an enzyme reaction, the velocity is based solely on the *enzyme concentration*. The higher the enzyme concentration (which is what is being measured), the more substrate is converted to product and the higher the velocity. As discussed previously, the enzyme concentration in the patient's serum is beyond our control; therefore, sufficient substrate must be in the reagent to result in zero-order kinetics.

Temperature increases the velocities of most chemical reactions, doubling for each 10°C increase up to a point when protein denaturation increases. The increase in temperature increases the movement of molecules, increasing the likelihood of collision of the enzyme and substrate molecules. Most enzymes have an optimal temperature close to 37°C

(physiological temperature); activity decreases as temperature rises above 40°C to 50°C because of denaturation of the enzyme. **Denaturation** is the alteration of the secondary, tertiary, or quaternary structure of a protein that results in inactivity. Although the primary structure is not altered, the active site is no longer functional. See Chapter 8 for a more detailed discussion of denaturation.

The optimal **pH** range for enzymes is relatively small, and the curve produced by plotting enzyme activity versus pH is usually a steep, bell-shaped curve with rapid falloffs before and after the optimal point. Most enzymes have an optimal pH from 7.0 to 8.0, or physiological pH. At extremes of pH—acid or alkaline conditions—enzymes are denatured. There are exceptions—for example, acid phosphatase and alkaline phosphatase—where enzymes do function at extremes of pH.

Cofactors, activators, or *coenzymes* increase the rate of an enzymatic reaction. Activators (for example, Mg^{+2}) provide an electropositive site that attracts the negatively charged groups of the substrate. Coenzymes are electron acceptors or donors in various enzymatic reactions. Activated coenzymes should be added to the enzyme reagent in excess.

Inhibitors are various substances that can decrease the rate of reaction. For some enzymes, a product of the reaction may inhibit enzyme activity. Various drugs or medications the patient is taking may be competitive or noncompetitive inhibitors of certain enzymes. In **competitive inhibition**, the substance is similar to the normal substrate and competes with the substrate for the binding or active site of the enzyme. Competitive inhibition can be reversed by increasing the concentration of the substrate. For example, if a therapeutic drug (e.g. sulfonamide) is competing with the substrate, then increasing the substrate concentration will increase the probability that the enzyme will bind to the substrate rather than the drug.

In **noncompetitive inhibition**, the inhibitor is structurally different than the substrate and binds to an allosteric site on the enzyme molecule. There is no competition between the inhibitor and substrate, and the affinity of the enzyme for the substrate is not affected. The addition of substrate will not overcome noncompetitive inhibition.

Uncompetitive inhibition occurs when an inhibitor binds to the ES complex to form an enzyme–substrate–inhibiting complex that does not yield product. It is more common in reactions involving two substrates. In this case, the addition of more substrate will actually increase the inhibition.

✓ Checkpoint! 9–1

1. *The apoenzyme and the cofactor or coenzyme that form the catalytically active unit is called the ————.*
2. *The actual place in the enzyme where the substrate is converted to product is the ———— ————.*
3. *Nicotinamide adenine dinucleotide (NAD^+) is an example of a(n) ————.*
4. *How do enzymes affect the energy of activation of a reaction?*
5. *Define zero-order kinetics.*
6. *List four factors that influence enzyme reactions.*

▶ DIAGNOSTIC ENZYMOLOGY

Common enzyme measurements are measurements of either product formation or substrate depletion over time and not the actual enzyme concentration. Activity, not the mass of the enzyme, is measured. Hundreds of different enzymes are present in the human body, and they function primarily on or within cells, including cell membrane, cytosol, nuclei, and organelles (e.g., mitochondria and ribosomes). Enzyme concentrations in serum are usually from the disintegration of cells during the normal process of breakdown and replacement. A certain percentage of cells in all tissues of the body are constantly being catabolized and replaced.

Abnormal serum enzyme levels are found in various diseases and in inflammation when more cells than normal are being destroyed. Panels of certain enzymes and associated enzyme levels provide characteristic patterns for specific tissues or organs. Concentrations of certain isoenzymes also provide evidence of a particular disease. For example, heart, brain, and muscle tissues are rich in creatine kinase. When an increased breakdown occurs in any of those tissues, the levels of CK, especially the associated isoenzyme, increase in the serum.

▶ ENZYME CLASSIFICATION

Enzymes are given a systematic name that clearly identifies the nature of the reaction catalyzed with a unique numerical designation. The Commission on Enzymes (Enzyme Commission of the International Union of Biochemistry) has established six classes. The individual enzymes are identified by the name of the substrate or group on which the enzyme acts with the addition of the suffix–ase. The six groups are as follow.

1. **Oxidoreductases** catalyze the oxidation–reduction reactions between two substrates—for example, lactate dehydrogenase (LD).

2. **Transferases** catalyze the transfer of a group other than hydrogen between two substrates—for example, aspartate aminotransferase (AST).

3. **Hydrolases** catalyze the hydrolytic (addition of water) cleavage of compounds—for example, amylase.

4. **Lyases** catalyze the removal of groups from substrates without hydrolysis (addition of water), leaving double bonds in the product.

5. **Isomerases** catalyze the interconversion of isomers—for example, phosphohexose isomerase converts α-glucose to β-glucose.

6. **Ligases** catalyze the joining of two molecules coupled with the hydrolysis of a pyrophosphate bond in adenosine triphosphate (ATP) or similar compounds.

▶ ISOENZYMES

Isoenzymes are multiple forms of an enzyme that can catalyze the enzyme's characteristic reaction. They are structurally different because they are encoded by distinct structural genes; therefore, isoenzymes can be measured to identify the organ or tissue of origin. For example, lactate dehydrogenase-1 (LD-1) is found in highest concentration in cardiac muscle, red blood cells (RBCs), and kidney tissue. LD-5 levels are highest in the liver and skeletal muscle.

The existence of multiple forms can be used to determine the origin of a disease by ascertaining the particular isoenzyme present. An increase in LD-1, for example, indicates heart, RBCs, or kidney as the source. Organ-specific patterns can be important in the diagnosis of disease.

Isoenzymes can be differentiated by various physical properties, including electrophoresis and resistance to chemical or heat inactivation. They also may have significantly measurable differences in their catalytic properties. Immunochemical methodologies have been developed for the measurement of specific isoenzymes and are critical in the diagnosis of disease.

▶ CARDIAC ENZYMES

The major cardiac enzymes are creatine kinase and lactate dehydrogenase. Other proteins—including troponin, C-reactive protein, and myoglobin—will be discussed in Chapter 18, Cardiac Function.

CREATINE KINASE

Creatine kinase is a cytoplasmic and mitochondrial enzyme that catalyzes the reversible phosphorylation of creatine by ATP:

$$\text{Creatine} + \text{ATP} \xrightarrow[\text{Mg}^{+2}]{\text{CK}} \text{Creatine phosphate} + \text{ADP} \quad (9.3)$$

The equilibrium of the CK reaction is dependent on pH. At a pH greater than 9.0, the forward reaction is favored; a pH less than 6.8 favors the reverse reaction. Mg^{+2} is an obligate activating ion that functions with adenosine diphosphate (ADP) and ATP in all kinases.

Creatine kinase is very important in muscle tissue, allowing high-energy phosphate to be stored in a more stable form than ATP. When muscle contracts, the phosphate group is used to form ATP, a source of energy for muscle contraction. Creatine phosphate is converted to creatine and ATP, the reverse reaction, to provide energy for muscle contraction.

Creatine kinase levels are highest in three major tissues: muscle, brain, and heart, in descending order. It is also found in placenta, gastrointestinal tract, thyroid, liver, bladder, prostate, small intestine, and kidney, although in much lower quantities than the three main sources.

Creatine kinase is a dimer consisting of two subunits: B for brain, and M for muscle. This results in three possible combinations: MM, MB, and BB. On electrophoresis, MM is found near the cathode (the negative electrode) and BB travels the farthest to the anode (the positive electrode).

CK-1 (BB), the most anodal isoenzyme, travels the farthest on electrophoresis and is found predominantly in the brain and central nervous system and in small amounts in other tissues including the lung and prostate. It rarely crosses the blood–brain barrier because of its large size. CK-2 (MB) is found predominantly in heart muscle and to a minor degree in skeletal muscle. It is clinically important in the diagnosis of myocardial infarctions (MIs). CK-3 (CK-MM) is found in skeletal and cardiac muscle and is the major isoenzyme in normal serum. More on these can be found in the isoenzyme section that follows.

CK CLINICAL SIGNIFICANCE

Elevations of CK are found primarily in conditions affecting brain, muscle, or cardiac muscle. Increases in both MM and MB isoenzymes indicates damage that is cardiac in origin. Following a myocardial infarction, the total CK begins to elevate in 4 to 6 hours, peaks at 24 hours, and returns to normal within 2 to 4 days.[1] Myocardial infarction is the only condition with this typical pattern. Examples of other cardiac conditions that result in an increase in MM and MB without the typical rise and fall are myocarditis, severe angina, cardiac catheterization, tachycardia, cardioversion, and congestive heart failure.

Skeletal muscle diseases or conditions that are associated with an elevated MM fraction include muscular dystrophy (MD) especially the Duchenne type. Muscular dystrophy is a sex-linked genetic disease that is progressive. CK activity in MD is highest in infancy and childhood (7–10 years), and it may be increased before the disease is apparent. The elevation may be as much as 50 times the upper limit of normal; however, CK decreases as the patient gets older because of a decrease in muscle mass as the disease progresses.

Other conditions associated with elevations of CK-MM are convulsive seizures (epilepsy), muscle trauma, extreme physical exercise, and viral myositis.[2] CK is elevated following physical exercise, especially in individuals not in good physical shape. Numerous studies[3] have reported increased CK and CK-MB values with postmarathon values more than 6 times higher than the premarathon measurements. This indicates exertional rhabdomyolysis and leakage from skeletal muscle during the race. CK activity has an inverse relationship with thyroid function; in other words, hypothyroidism is associated with elevated CK and hyperthyroidism with decreased CK.

Creatine kinase (BB) found in brain and central nervous system disorders is associated with brain injury, including cerebral ischemia, acute cerebrovascular accident (CVA, or stroke), and cerebral trauma. In summary, CK is a sensitive indicator of myocardial infarction and muscle diseases, especially muscular dystrophy. CK-BB is also a tumor marker associated with prostate cancer and small-cell carcinoma of the

---

BOX 9-1 Creatine Kinase: Clinical Significance

Major Sources of Elevation

Myocardial Tissue (Cardiac) MM and MB

- Myocardial infarction
- Myocarditis
- Cardiac trauma
  - Open-heart surgery
  - Cardiac catheterization
- Severe angina
- Congestive heart failure
- Tachycardia
- Cardioversion (paddles)
- Cardiopulmonary resuscitation
- Acute cocaine abuse

Skeletal Muscle Diseases (CK-2) (MM)

- Muscular dystrophy, especially Duchenne type
- Convulsive disorders: e.g., epilepsy, grand mal seizures
- Muscle trauma
  - Crush injuries
  - Surgery
- Extreme physical exercise
- Viral myositis
- Intramuscular injections
- Rhabdomyolysis

Brain (CK-1) (BB)

- Cerebral ischemia
- Acute CVA (stroke)
- Head injuries, cerebral trauma
- Brain tumor
- Cerebral thrombosis
- Reye's syndrome
- Hypothyroidism

Miscellaneous elevations of CK-1 BB

- Normal childbirth (muscle contraction)
- Lung tumor
- Hypothermia
- Prostate, bladder, kidney, and breast tumors

---

lung. Although its clinical usefulness has not been proven, it is also elevated in other cancers such as breast, colon, ovary, and stomach. See Box 9-1 ■ for a more complete list of sources of CK elevation.

CK METHODOLOGIES

Numerous photometric, fluorometric, and coupled-enzyme methodologies have been developed to measure CK activity using the forward (Cr → PCr) or the reverse (PCr → Cr). The most common CK methodology is the reverse reaction *Oliver and Rosalki* because it proceeds 2 to 6 times more rapidly than the forward reaction. In the more complex methodologies involving multiple reactions a shortened form will precede the more detailed version: CK → Hexokinase → G-6-phospiate dehydrogenase.

$$\text{Creatine phosphate} + \text{ADP} \xrightarrow[\substack{\text{Mg}^{+2} \\ \text{pH } 6.7}]{\text{CK}} \text{Creatine} + \text{ATP} \quad (9.4)$$

The second reaction involves the ATP in the initial reaction as a substrate in the hexokinase reaction:

$$\text{ATP} + \text{Glucose} \xrightarrow{\text{Hexokinase}} \text{Glucose-6-phosphate} + \text{ADP} \quad (9.5)$$

The indicator reaction is catalyzed by glucose-6-phosphate dehydrogenase and involves the reduction of $NADP^+$ to NADPH.

$$\text{Glucose-6-phosphate} + \text{NADP}^+ \xrightarrow{\text{Glucose-6-Phosphate dehydrogenase}}$$
$$\text{6-Phosphogluconate} + \text{NADPH} + \text{H}^+ \quad (9.6)$$

The rate of NADPH formation is proportional to the CK activity, provided the concentrations of all other components in the enzyme system are present in sufficient excess that the CK activity is the only limiting factor—in other words, it is a zero-order reaction.

Endogenous inhibitors including Ca^{+2} compete with the obligate activator Mg^{+2}. Ca^{+2} can be nullified by adding ethylenediaminetetraacetic acid (EDTA) to bind the calcium, or excess Mg^{+2} can be added to the reagent. Sulfhydryl groups at active sites can also inactivate CK. Addition of sulfhydryl compounds to the reagent—for example, N-acetylcysteine, monothioglycerol, dithiothreitol, or glutathione—can restore CK activity. N-acetylcysteine is the most common activator or sulfhydryl compound added to CK reagents.

Serum is the specimen of choice. CK activity is unstable and rapidly lost during storage. It is stable for 4 hours at room temperature, 48 hours at 4°C, and 1 month at –20°C. CK is light sensitive and inactivated by daylight. It is not affected by slight hemolysis, but enzymes and intermediates released in moderate hemolysis may affect the lag phase.

Tanzer-Gilvarg (CK → PK → LD → NADH) uses the forward reaction.

$$\text{Creatine} + \text{ATP} \underset{\substack{\text{Mg}^{+2} \\ \text{pH } 9.0}}{\overset{\text{CK}}{\rightleftharpoons}} \text{Creatine phosphate} + \text{ADP} \quad (9.7)$$

$$\text{ADP} + \text{Phosphoenolpyruvate} \overset{\text{Pyruvate kinase}}{\rightleftharpoons} \text{Pyruvate} + \text{ATP} \quad (9.8)$$

$$\text{Pyruvate} + \text{NADH} + \text{H}^+ \overset{\text{Lactate dehydrogenase}}{\rightleftharpoons} \text{Lactate} + \text{NAD}^+ \quad (9.9)$$

CK REFERENCE RANGE

The reference range is influenced by gender, with males having the higher muscle mass, which results in a slightly higher reference range.

Males: 46–180 U/L

Females: 15–171U/L

CK activity is also affected by several physiological variables: age (children slightly higher than adults), physical activity, and race (African Americans higher than Caucasians). Bed rest, even just overnight, can decrease CK by 20%; therefore, hospitalized patients who are bedridden would have a lower CK than when ambulatory. Elderly patients with decreased muscle mass would also have a lower CK; on the other hand, exercise, especially long jogs, can elevate CK.

CK ISOENZYME METHODOLOGIES

CK isoenzymes are measured by immunoinhibition, mass assay, and electrophoresis. Immunoinhibition with mass assay, which can be automated, has gained popularity and has replaced electrophoresis in most clinical laboratories.

Immunoinhibition requires a specific antibody against the M subunit that reacts with the M on the MB isoenzyme and completely inhibits the MM isoenzyme. The only CK activity remaining is the B subunit of the MB fraction, assuming the absence of BB in normal serum. The CK activity is measured and multiplied by 2 to correct for the inhibited M portion of the CK-MB molecule. The most important caution in using immunoinhibition is the assumption that the serum contains no CK-BB or abnormal macro forms, which may not be valid in rare cases.

Immunoinhibition has been replaced in many laboratories by *mass assay*. A sandwich technique is used with two antibodies: one against the M subunit (anti-M) and one against the B subunit (anti-B). The first antibody is bound to a solid surface, and the second antibody is labeled with an enzyme or other label. Only CK-MB is measured, because neither MM or BB will react with both antibodies. In some assays, a third antibody that reacts only with CK-MB is added. The measurement of the activity of the enzyme label is proportional to the mass of CK-MB. CK-MB mass assays are more sensitive than immunoinhibition and will detect an elevated CK-MB earlier.

CK electrophoresis is semiquantitative and can be performed on agarose or cellulose acetate. CK isoenzymes are separated by electromotive force at a pH of 8.6. At this pH, CK-BB (CK-1) migrates most rapidly toward the anode or positive pole, CK-MB migrates midway, and CK-MM remains near the point of origin. The bands are visualized by incubating the support medium with a concentrated CK assay mixture that uses the Oliver and Rosalki reverse reaction. NADPH formed in this reaction may be detected by observing fluorescence after excitation with UV light at 360 nm or NADPH reduction of a tetrazolium salt, nitro blue tetrazolium, to form a colored formazan.

The advantage of electrophoresis is the detection of CK-BB and abnormal bands or macro forms. It can also determine if the separation of the three forms is satisfactory.

CK ISOENZYME REFERENCE RANGE AND RELATIVE INDEX

In a normal gel electrophoresis CK-BB is absent or trace; CK-MB is ≤6% of the total, and CK-MM is 94–100%. A CK-MB ≥ 6%

is the most specific indicator of a myocardial infarction. Both skeletal muscle and cardiac injuries may increase the CK-MB, but the CK-MB in muscle is not high enough to increase the percentage more than 6%. CK-MB never exceeds 40% of the total CK activity—this is physiologically impossible! In immunochemical or immunoinhibition, the reference range of CK-MB is 0–10 U/L.

The **relative index (RI)** is calculated by divided the CK-2 mass (μg/L) by the total CK activity (U/L).

$$\text{Relative index (RI)} = \frac{\text{CK-2 } (\mu g/L \text{ or } ng/mL) \times 100}{\text{Total CK (U/L)}}$$

$$(9.10)$$

For example, a CK-2 mass of 20 μg/L and a total CK activity of 500 U/L results in an RI of 4%. The reference range for the relative index is less than 3%. Ratios greater than 5 are indicative of a cardiac source; between 3 and 5 is a gray zone that requires serial determination to diagnose or rule out myocardial infarction.[1]

LACTATE DEHYDROGENASE

Lactate dehydrogenase is an oxidoreductase, a hydrogen-transfer enzyme that catalyzes the oxidation of L-lactate to pyruvate with NAD^+ as a hydrogen acceptor. LD will not act on D-lactate, and only NAD^+ will serve as a coenzyme.

$$\text{L-Lactate} + NAD^+ \xrightleftharpoons[\text{pH 7.4 to 7.8}]{\text{pH 8.8 to 9.8}} \text{Pyruvate} + NADH + H^+$$

$$(9.11)$$

The forward LD reaction is favored when the pH is 8.8 to 9.8 and the reverse reaction is at a pH of 7.4 to 7.8 (physiological pH). LD is an important enzyme in the Embden–Meyerhoff metabolic pathway of glycolysis for tissues that metabolize glucose for energy.

Lactate dehydrogenase is found in the cytoplasm of most cells in the body, so elevations of LD without any other information is nonspecific for any disease or disorder. LD concentration in tissues is approximately 500 times higher than serum levels; therefore, damage to most organs will increase serum levels. The highest levels are found in liver, heart, and skeletal muscle. LD is a tetramer consisting of four polypeptide chains. Two subunits H (heart) and M (muscle) form five isoenzymes.

1. LD-1 (HHHH): heart, RBCs, kidney
2. LD-2 (HHHM)
3. LD-3 (HHMM): spleen, lungs and many tissues
4. LD-4 (HMMM):
5. LD-5 (MMMM): liver and skeletal muscle

In healthy individuals, the concentration of LD isoenzymes are LD-2 > LD-1 > LD-3 > LD-4 > LD-5.

LD Clinical Significance

Increases in LD may be absolute (total LD) or relative (increase in one of the isoenzymes). Cardiac isoenzymes are elevated in myocardial infarction, myocarditis, shock, and congestive heart failure. In myocardial infarctions, more LD-1 is released than LD-2, so a characteristic "flipped" pattern is observed where LD-1 > LD-2. LD rises between 12 and 24 hours, peaks at 48 to 72 hours, and returns to normal within 7 to 10 days.[4] LD levels are usually normal in angina and pericarditis.

The highest levels of LD are found in pernicious anemia or megaloblastic anemia. Almost any condition that increases erythrocyte breakdown is associated with an elevated LD that shows a characteristic flipped pattern but without the typical rise-and-fall pattern associated with an MI. Liver disease, including cirrhosis and obstructive jaundice, will increase the liver isoenzyme (LD5). Elevations of LD in hepatic disease are not as great as with aspartate aminotransferase and alanine aminotransferase, which are discussed later in this chapter. The muscle isoenzyme is elevated in muscle disorders, especially muscular dystrophy. See Box 9-2 ■ for a more comprehensive list.

BOX 9-2 Lactate Dehydrogenase: Clinical Significance

Cardiac or Heart
► Myocardial infarction
► Myocarditis
► Shock
► Congestive heart failure

Hemolytic Disease
► Hemolytic anemias
► Pernicious anemia: highest levels

Liver Disease
► Cirrhosis
► Obstructive jaundice
► Viral hepatitis
► Toxic hepatitis
► Infectious mononucleosis
► Hepatic cancer or metastasis

Muscle
► Muscular dystrophy
► Muscle trauma
► Physical exercise
► Crush injuries

Miscellaneous
► Pulmonary infarct or embolism
► Renal disease: glomerulonephritis, pyelonephritis
► Renal and bladder malignancies
► Malignancies
► Hodgkins
► Abdominal cancer
► Lung cancer

LD Methodologies

Most current methods measure the interconversion of NAD^+ to NADH at 340 nm. The suggested reference method is the kinetic reaction of pyruvate to lactate, although lactate to pyruvate is the most often used. The *Wacker procedure* is associated with the forward reaction, lactate to pyruvate.

$$\text{L-Lactate} + NAD^+ \xrightarrow{\text{Lactase dehydrogenase}} \text{Pyruvate} + \text{NADH} + H^+ \quad (9.12)$$

LD is specific for L-Lactate, and substrate inhibition by lactate is less than that produced by pyruvate. Also, NAD^+ preparations contain fewer endogenous LD inhibitors than NADH used in the reverse reaction.

An advantage of the reverse reaction, *Wroblewski and LaDue* ($P \rightarrow L$), is the lower cost.

$$\text{Pyruvate} + \text{NADH} + H^+ \xrightarrow{\text{Lactase dehydrogenase}} \text{L-Lactate} + NAD^+ \quad (9.13)$$

LD Reference Range

The LD reference range varies considerably by methodology (forward or reverse) and age (children higher than adults). There are no clinically significant gender differences.

- $L \rightarrow P$ 100–224 U/L @ 37°C
- $P \rightarrow L$ 80–300 U/L @ 37°C

Hemolysis results in higher LD because of increased RBC LD-1 and LD-2 isoenzymes, with RBCs having 150 times the concentration of LD than serum. LD specimens can be stored at room temperature for 3 days with no loss of activity. LD5 will be decreased at cold temperatures (–20°C).

ENZYMES IN CARDIAC DISEASE

The two major cardiac enzymes are CK and CK-MB. LD and AST were historically on cardiac panels but have been replaced by more sensitive and specific tests that are discussed in Chapter 18, Cardiac Function.

CK increases to above the upper reference limit 4 to 6 hours after onset of a myocardial infarction, peaks at 24 hours, and returns to normal within 2 to 4 days. CK-MB follows a similar pattern.[1]

A negative CK-MB over a period of 6 to 8 hours has a 95% negative predictive value, indicating that the patient did not have an MI. Total CK has a 98–100% diagnostic sensitivity and 85% diagnostic specificity, and CK-MB has a 98–100% sensitivity and specificity.

Specimens are drawn on admission 6, 12, 24, and 48 hours after the infarct or onset of pain. Specimens should be obtained at least on admission, and 6 to 9 hours and 12 to 24 hours after the onset of symptoms. Diagnosis is usually made within 12 hours of admission, meaning an MI is confirmed or ruled out.

> ## ✓ Checkpoint! 9–2
>
> 1. CK-3 (MM) is found primarily in what tissue?
> 2. List the CK isoenzyme associated with the following:
>   a. tumor-associated marker _____
>   b. muscular dystrophy _____
>   c. cerebrovascular accident _____
>   d. rhabdomyolysis _____
>   e. myocardial infarction _____
> 3. John's CK-2 is 30 µg/L, and total CK is 400 U/L. What is his relative index, and what does it indicate?
> 4. Describe the "flipped" pattern in LD electrophoresis and what condition(s) are associated with it.
> 5. Bill has a negative or normal CK-MB at 8 hours. Can his physician confirm or rule out a myocardial infarction?

▶ ENZYMES IN LIVER DISEASE

Liver enzymes discussed in this section include aspartate aminotransferase (AST), alanine aminotransferase (ALT), and alkaline phosphatase (ALP). Marked elevations of AST and ALT are associated with hepatocellular disease or damage to hepatocytes. Marked elevation of ALP is associated with hepatobiliary disease or obstructive liver disease. Although all three enzymes are elevated in both types of liver disease, marked elevations of AST And ALT are associated with hepatocellular disease and marked increases in ALP with hepatobiliary. Additional enzymes used in hepatobiliary disease will be discussed in the next section.

ASPARTATE AMINOTRANSFERASE

Aspartate aminotransferase is one of a group of enzymes that catalyze the interconversion of amino acids and α-oxoacids by the transfer of amino groups. Transamination is important in the intermediary metabolism in the synthesis and degradation of amino acids. The ketoacids formed by the transamination are ultimately oxidized by the tricarboxylic acid cycle to provide a source of energy. AST catalyzes the deamination of aspartate.

$$\text{L-Aspartate} + \text{2-Oxoglutarate} \underset{\text{P-5-P}}{\overset{\text{Aspartate aminotransferase}}{\rightleftarrows}} \text{Oxaloacetate} + \text{L-Glutamate} \quad (9.14)$$

The α-oxoglutarate and L-glutamate couple serves as an amino acid acceptor and donor pair in all amino transfer reactions. The specificity of the individual enzymes is derived from the particular amino acid that serves as the other donor of an amino group—in this case, aspartate. The required coenzyme, pyridoxal-5-phosphate (P-5-P), serves as a necessary prosthetic group for full catalytic transfer.

AST is found predominantly in the cytoplasm of the cells of the heart, liver, skeletal muscle, and kidney. A *mitochondrial* form (mAST) is found chiefly in the liver and is only present in serum in cases of severe liver damage.

AST Clinical Significance

Elevations of aspartate aminotransferase occur mostly in myocardial (cardiac), liver, and muscle diseases or conditions. Cardiac conditions resulting in elevated AST are myocardial infarctions, congestive heart failure, pericarditis, and myocarditis.[5]

Viral hepatitis, cirrhosis, alcoholic hepatitis, and Reye's syndrome are examples of hepatic elevations of AST.[6] Following viral and alcoholic hepatitis, nonalcoholic steatohepatitis is the most common cause of AST elevation in liver disease.[7] Elevations of AST and ALT, discussed later in this section, are present even before symptoms of liver disease are present (e.g., jaundice).

Muscle conditions associated with elevated AST are muscular dystrophy and skeletal muscle injury. Box 9-3 ■ provides a more complete list. In summary, elevations in AST are most commonly associated with myocardial infarction, hepatocellular disease, and skeletal muscle disorders.

BOX 9-3 Aspartate Aminotransferase: Clinical Significance

Cardiac and Heart

▶ Myocardial infarction
▶ Congestive heart failure
▶ Pericarditis
▶ Myocarditis

Liver

▶ Viral hepatitis
▶ Alcoholic hepatitis
▶ Nonalcoholic steatohepatitis
▶ Toxic hepatitis (drugs or chemicals)
▶ Cirrhosis
▶ Bile duct obstruction
▶ Hepatic carcinoma
▶ Reye's syndrome
▶ Infectious mononucleosis

Muscle

▶ Muscular dystrophy
▶ Dermatomyositis
▶ Skeletal muscle injury
▶ Gangrene

Miscellaneous

▶ Pulmonary infarct
▶ Pulmonary embolism
▶ Acute pancreatitis

AST Methodologies

AST is routinely measured using a modification of the *Karmen method* with the addition of coenzyme P-5-P to ensure full catalytic activity. Malate dehydrogenase is the indicator reaction that measures the decrease in absorbance at 340 nm as NADH is oxidized to NAD^+.

$$\text{L-Aspartate} + \text{2-Oxoglutarate} \underset{\text{Pyridoxal-5-phosphate}}{\overset{\text{Aspartate aminotransferase}}{\rightleftharpoons}}$$
$$\text{Oxaloacetate} + \text{L-Glutamate} \quad (9.15)$$

$$\text{Oxaloacetate} + \text{NADH} + \text{H}^+ \xrightarrow{\text{Malate dehydrogenase}}$$
$$\text{Malate} + \text{NAD}^+ \quad (9.16)$$

AST is stable at room temperature for 48 hours and 3 to 4 days at refrigerator temperature.

AST Reference Range

The reference range at 37°C is 5–30 U/L. There are no clinically significant gender differences. Hemolysis should be avoided because AST is 10 to 15 times higher in RBCs than in serum.

ALANINE AMINOTRANSFERASE

Alanine aminotransferase is an aminotransferase which catalyzes the deamination of alanine instead of aspartate.

$$\text{Alanine} + 2 - \text{oxoglutarate} \underset{\text{Pyridoxal-5-phosphate}}{\overset{\text{Alanine aminotransferase}}{\rightleftharpoons}}$$
$$\text{Pyruvate} + \text{L-glutamate} \quad (9.17)$$

Alanine aminotransferase is found in greatest concentration within the cytoplasm of the liver and is predominantly a liver specific transaminase. It is present in smaller amounts in the kidney, the heart, and skeletal muscle. P-5-P is a required coenzyme as with AST and other aminotransferases.

ALT CLINICAL SIGNIFICANCE

ALT's diagnostic significance is confined mainly to hepatocellular disorders, and it is more specific than AST. Acute viral hepatitis and toxic hepatitis result in the highest elevation. Obstructive liver disease, hepatic cancer, and cirrhosis are associated with moderate elevations of ALT. As noted earlier, ALT and AST may be increased before the patient exhibits symptoms of liver disease (e.g., liver cancer.)

The **De Ritis ratio** (AST/ALT) is helpful in determining the cause of the liver disease. The normal ratio ranges from 0.7 to 1.4, depending on methodology. In alcoholic liver disease and cirrhosis AST > ALT; therefore, the ratio of AST/ALT is greater than 1.0. The additional AST is mitochondrial AST (mAST), which is increased in patients with extensive liver degeneration and necrosis. Although the exact mechanism has not been determined, it appears that the production of ALT decreases in patients with cirrhosis. A ratio greater than 2.0 is usually indicative of alcoholism or alcoholic hepatitis. Multiple explanations are proposed for the higher AST in alcoholic hepatitis. For example, alcohol increases the mitochondrial AST, which is not elevated in other causes of hepatitis. In most types of liver injury, hepatocyte activity of both the cytoplasmic and mitochondrial AST is decreased, but alcohol produces a decrease in only the cytoplasmic AST.[8] Viral hepatitis, acute inflammatory disease, and obstructive liver disease are associated with ALT > AST; therefore, the ratio of AST/ALT is < 1.0. The reader is cautioned to confirm the ratio used in various sources because the De Ritis ratio is also calculated as ALT/AST.

ALT METHODOLOGIES

ALT is routinely measured with a modification of Wroblewski and LaDue using lactate dehydrogenase as the indicator reaction (AST → LD).

$$\text{Alanine + 2-Oxoglutarate} \xrightarrow[\text{Pyridoxal-5-phosphate}]{\text{Alanine aminotransferase}}$$

$$\text{Pyruvate + L-Glutamate} \quad (9.18)$$

$$\text{Pyruvate + NADH} \xrightarrow{\text{Lactase dehydrogenase}}$$

$$\text{L-Lactate + NAD}^+ \quad (9.19)$$

The decrease in absorbance is measured at 340 nm as NADH is oxidized to NAD^+. ALT should be measured within 24 hours because activity decreases even at refrigerator temperature (4°C) or frozen at –25°C. ALT is stable at –70°C.

ALT REFERENCE RANGE

The reference range for ALT at 37°C is 6–37 U/L. Hemolysis should be avoided because ALT levels in RBCs are 5 to 8 times higher than serum levels.

ALKALINE PHOSPHATASE

Alkaline phosphatase is a generic name for a group of enzymes with maximum activity in the pH range of 9.0 to 10.0. ALP frees inorganic phosphate from an organic phosphate monoester, resulting in the production of an alcohol.

$$\text{Phosphomonoester + H}_2\text{O} \xrightarrow[\text{Zn}^{+2},\ \text{Mg}^{+2}]{\text{Alkaline phosphatase}}$$

$$\text{Alcohol + HPO}_4^- \quad (9.20)$$

ALP is nonspecific, is able to catalyze reactions with many phosphomonoesters, and requires the correct ratio of Mg^{+2} and Zn^{+2} as activators.

Alkaline phosphatase is present is practically all tissues of the body especially at or near the cell membranes. ALP has multiple forms including isoenzymes, which are encoded at separate genetic loci. The *primary sources* or the richest sources of ALP are the cell membranes of hepatocytes lining the sinusoidal border of the parenchymal cells and the bile canaliculi, and osteoblasts in the bone. It is also found in decreasing concentrations in placenta, intestinal epithelium,

spleen, and kidney. As with CK and LD, different isoenzymes are produced by the various tissues.

ALP CLINICAL SIGNIFICANCE

As discussed in the previous paragraph, ALP is clinically significant in disorders involving the hepatocytes (liver) and osteoblasts (bone). The highest levels of ALP are found in hepatobiliary disease involving biliary tract disorders and hepatobiliary obstruction. The liver responds to any type of biliary obstruction by stimulating the production of ALP by the hepatocytes. It is considerably higher in obstructive disease than with hepatocellular or parenchymal cell disorders where AST and ALT are extremely elevated.[6] Therefore, a markedly elevated ALP and a slightly or only moderately elevated AST and ALT is suggestive of obstructive liver disease. A markedly elevated AST and ALT with a moderate or slight elevation of ALP points to a hepatocellular origin. Extrahepatic obstruction results in higher elevations of ALP than intrahepatic obstruction; the greater the obstruction, the higher the ALP.

Elevations because of osteoblastic activity are associated with Paget's disease (osteitis deformans) in which there is excessive bone destruction followed by osteoblastic stimulation to rebuild bone. ALP is also elevated in osteomalacia and rickets, but it is normal in osteoporosis. Physiological increases are related to healing bone fractures when osteoblastic activity is increased to synthesize new bone, in pregnancy because of placental ALP, and in infants and children during growth spurts to stimulate the production of new bone.

Separation of ALP isoenzymes led to the discovery of an abnormal isoenzyme, carcinoplacental or Regan, which is associated with malignant disease (cancer). It is nearly identical to the placental isoenzyme found in the serum of pregnant women. Box 9-4 ■ provides further detail on conditions associated with an elevated ALP.

ALP METHODOLOGIES

Several ALP procedures have been developed using a variety of substrates. *Bowers and McComb* is the most common procedure. It uses 4-nitrophenol phosphate (4-NPP) [formerly P-nitrophenylphosphate (P-NPP)] as the substrate, and the yellow product, 4-nitrophenoxide, is measured at 405 nm.

$$\text{4-Nitrophenyphosphate (4-NPP)} \xrightarrow[\substack{\text{Mg}^{+2}\\ \text{pH 10.3}}]{\text{Alkaline phosphatase}}$$

$$\text{4-Nitrophenoxide} \quad (9.21)$$

Other substrates include phenolphthalein monophosphate, thymolphthalein phosphate, and α-naphthyl phosphate. The preferred buffer to maintain the pH at 10.3 is 2-amino-2-methyl-1-propanol (AMP). As with all transferases, the correct ratio of Zn^{+2} and Mg^{+2} is required for optimal activity.

BOX 9-4   Alkaline Phosphatase: Clinical Significance

Elevated ALP

Hepatic

- ▶ Hepatobiliary disease
- ▶ Biliary tract disorders
- ▶ Hepatobiliary obstruction
- ▶ Liver cancer
- ▶ Infectious and viral hepatitis
- ▶ Alcoholic cirrhosis

Bone (Osteoblastic)

- ▶ Paget's disease (osteitis deformans)
- ▶ Osteomalacia
- ▶ Rickets
- ▶ Osteogenic sarcoma
- ▶ Hyperparathyroidism

Carcinoplacental (Regan)

Physiological increase

- ▶ Healing bone fractures (osteoblastic activity)
- ▶ Pregnancy (placental ALP)
- ▶ Infants and children (growth spurts)

ALP REFERENCE RANGE

The reference range for ALP is 30–90 U/L, but it is dependent on the specific procedure or modification used and the instrument. Pediatric and adolescent ALP levels are 2 to 3 times higher than the adult levels because of physiological increases, and they are highly variable because of growth spurts. Pregnancy is also associated with an elevated ALP as a result of placental ALP.[6]

Serum or heparinized plasma are the specimens of choice. EDTA and other complexing anticoagulants are not acceptable because they bind to Mg^{+2} and Zn^{+2} in addition to Ca^{+2}. ALP specimens should be kept at room temperature and analyzed within 4 hours. ALP levels decrease slowly at refrigerator temperature (~2%/day).

✓ Checkpoint! 9–3

1. *Which enzyme is most specific for hepatic disease: AST or ALT?*

2. *Linda's AST is 190 U/L and her ALT is 90 U/L. What is the De Ritis ratio, and what does it indicate?*

3. *Tom is a 15-year-old boy with an ALP of 160 U/L.*
   a. *Should his physician be concerned?*
   b. *What is the most likely explanation?*

4. *The primary sources or the tissues richest in the following enzymes are:*
   a. *AST* _____ _____ _____
   b. *ALT* _____ _____
   c. *ALP* _____ _____

5. *Bill's AST is 50 U/L, and his ALP is 250 U/L. Is his problem hepatic or obstructive liver disease?*

▶ BILIARY TRACT ENZYMES

In addition to alkaline phosphatases, two enzymes are associated with hepatobiliary disorders or hepatic obstruction: gamma glutamyl transferase and 5′-nucleotidase.

GAMMA GLUTAMYL TRANSFERASE

Gamma glutamyl transferase (GGT) is a transferase, a membrane-associated enzyme that transfers the γ-glutamyl group from glutathione and other γ-glutamyl peptides to amino acids or small peptides to form the γ-glutamyl amino acids and cysteinyl-glycine. Peptidases catalyze hydrolytic cleavage of peptides to form amino acids or smaller peptides or both.

$$\text{Glutathione + Amino acid} \xrightarrow{\text{Gamma glutamyltransferase}}$$
$$\text{γ-Glutamyl-peptide + L-Cysteinylglycine} \quad (9.22)$$

GGT is present in serum and all cells except muscle. GGT in the serum is mainly from the kidney, liver, pancreas, and intestine, where it is found primarily in the cell membrane. Most serum activity is from the liver, therefore GGT is used to evaluate liver function, especially hepatobiliary tract disorders.

GGT Clinical Significance

The primary role of GGT is the detection and differential diagnosis of hepatobiliary disease. Cholestasis or biliary obstruction is associated with the highest elevation: 5 to 30 times the upper reference limit. GGT is higher in extrahepatic or posthepatic cholestasis than in intrahepatic cholestasis. Examples of extrahepatic obstruction include a gallstone or tumor blocking the bile duct. Intrahepatic obstruction occurs in liver cancer and biliary cirrhosis.[6] In hepatocellular disease (e.g., hepatitis and cirrhosis), the elevation is 2 to 5 times the upper reference limit. Marked elevations of GGT, an average of 12-fold, are seen in obstructive liver disease and moderate increases in hepatocellular disease.[6]

GGT is also used to detect injury caused by chronic alcoholism or drug ingestion.[9,10] GGT is considered the enzyme of choice in detecting alcoholism and is used by drug treatment centers to detect relapses.

GGT is also associated with acute pancreatitis, diabetes mellitus, and myocardial infarctions. One useful role of GGT is to differentiate liver disease from bone disease. It is not elevated in bone disease, so if a patient has an elevated ALP and a normal GGT, then the ALP would most likely be of bone origin. However, it does not completely rule out coexisting bone disease. GGT is very important when testing children or pregnant women to verify physiological increases in bone and placental ALP, respectively.

GGT Methodology

GGT methodologies were historically based on the reaction with γ-glutamyl-P-nitroanilide (GGPNA) with the measurement of yellow product, P-nitroaniline, at 405 nm.

Answer pg A-51

$$\gamma\text{-Glutamyl-P-nitroanilide} +$$
$$\text{Glycylglycine} \xrightarrow{\text{Gamma glutamyltransferase}}$$
$$\text{P-Nitroaniline} + \gamma\text{-Glutamylglycylglycine} \quad (9.23)$$

Newer substrates have been adopted, including L-γ-glutamyl-3-carboxy-4-nitroaniline.

5'-NUCLEOTIDASE

5'-nucleotidase (5'-NT, NTP) hydrolyzes the phosphate group from nucleoside-5'-phosphates—for example, 5'-adenosine monophosphate (AMP):

$$\text{Adenosine-5'-Monophosphate} + H_2O \xrightarrow{\text{5'-Nucleotidase}}$$
$$\text{Adenosine} + H_2PO_4 \quad (9.24)$$

5-nucleotidase is a microsomal and membrane-associated enzyme found in a variety of tissues, most specifically in liver tissue.

5'-NT Clinical Significance

The presence of 5'-NT is useful with ALP and GGT results in determining whether ALP elevation is from bone or liver disease. It is predominantly elevated in diseases of the biliary tract where the presence of bile salts stimulates its release from hepatocytes. A hepatobiliary panel including ALP, GGT, and 5-NT in which ALP is elevated and GGT and 5'-NT are normal would provide evidence of an osteoblastic source of ALP. It would also be useful in growing children and pregnant women to rule out liver as the source of elevated ALP, and provides evidence of osteoblasts or placenta as the source. Unlike GGT, 5'-NT is not elevated with drugs or alcohol.

5'-NT Methodology

The two most common substrates are AMP and inosine-5'-phosphate (INP). A major disadvantage of these two substrates is reaction with nonspecific phosphatases, including alkaline phosphatase.

In the INP methodology, 5'-NT activity catalyzes INP, which yields inosine. In the second reaction, inosine is converted to hypoxanthine catalyzed by purine-nucleoside phosphorylase, and the hypoxanthine is oxidized to urate by xanthine oxidase in the third reaction. Each mole of hypoxanthine produces two moles of hydrogen peroxide (H_2O_2). The H_2O_2 is measured spectrophotometrically at 510 nm through the oxidation of a dye. Various manufacturers use inhibitors of alkaline phosphatase to reduce interference.

5'-NT Reference Range

The reference range for 5'-NT is 3–9 U/L, with no clinically significant gender differences.

 Checkpoint! 9–4

Robert has an elevated ALP and an elevated GGT. What is the most likely source?

▶ DIGESTIVE AND PANCREATIC ENZYMES

Pancreatic function is ascertained using two major screening enzymes: amylase and lipase. In today's clinical laboratory, amylase levels are often confirmed or supported with lipase measurements. Historically, lipase methodologies were too time consuming, and their sensitivity left something to be desired. Newer methodologies have renewed interest in lipase testing in the clinical laboratory.

AMYLASE

Amylase (AMY) is a hydrolase that catalyzes the hydrolysis of complex carbohydrates including starch, amylopectin, glycogen, and their partially hydrolyzed products. *Amylose* is a long, unbranched chain of glucose molecules linked by glucosidic bonds. *Amylopectin* is a branched and chained polysaccharide with α-1,6 linkages at branch points. Human amylase is an α-amylase that catalyzes the hydrolysis of internal α-1,4 glycosidic linkages in starch and glycogen, but it cannot hydrolyze α-1,6 bonds; hydrolysis ceases when a branch point is reached. This results in the formation of *limit dextrins*. Plant and bacterial amylases are β-amylases that can act on the terminal reducing end of carbohydrates.

The main sources of AMY in human serum are the salivary glands and the acinar cells of the pancreas. AMY is also found in smaller amounts in skeletal muscle, intestinal epithelium, and fallopian tubes. Two distinct isoenzymes of amylase are pancreatic amylase (P-AMY) and salivary amylase (S-AMY). It is the smallest of the enzymes discussed to this point and is readily filtered by the glomeruli into the urine.

Clinical Significance

The pancreas and salivary glands are the two richest sources of AMY. AMY is increased in acute pancreatitis, obstructive liver disease, acute alcoholism, and other conditions that affect the pancreas. In acute pancreatitis, AMY activity increases within 5 to 8 hours following the onset of symptoms, peaks at 12 to 72 hours, and decreases to normal levels by the third or fourth day.[11] The higher the elevation of AMY, the greater the likelihood the patient has acute pancreatitis. Urine AMY peaks at higher concentrations and persists longer than serum AMY, which is consistent with the body's clearing of AMY through the urine.

Box 9-5 ■ lists other intraabdominal conditions (e.g., perforated ulcer and acute appendicitis) that are associated with elevated AMY and must be ruled out by the physician. Increases in salivary AMY are associated with mumps, parotitis, and salivary gland lesions.

MACROAMYLASEMIA

Macroamylasemia is an artifactual increase in serum AMY found in 1% to 2% of the population. In these individuals, AMY binds with IgG or IgA to form a complex too large to be

BOX 9-5   Amylase: Clinical Significance

Hyperamylasemia

Pancreas

- ► Acute pancreatitis
- ► Chronic pancreatitis
- ► Alcoholic liver disease
- ► Obstructive liver disease, cholecystitis
- ► Pancreatic cancer
- ► Pancreatic trauma

Salivary Gland

- ► Mumps
- ► Parotitis
- ► Salivary gland lesions
- ► Maxillofacial surgery

Intraabdominal Conditions

- ► Perforated peptic ulcer
- ► Intestinal obstruction
- ► Acute appendicitis
- ► Ruptured ectopic pregnancy
- ► Cholecystitis
- ► Peritonitis
- ► Diabetic ketoacidosis
- ► Gastritis, duodenitis

Miscellaneous

- ► Septic shock
- ► Cardiac surgery
- ► Tumors

Macroamylasemia

filtered by the kidney; therefore, it remains in circulation. The serum AMY is elevated as much as 6 to 8 times the upper reference limit; however, macroamylasemia is not associated with any disease and patients are asymptomatic. As you would expect, the urine AMY is decreased. This condition can be diagnosed with an amylase:creatinine clearance ratio (discussed later in this chapter). Macroamylasemia is not clinically significant, but it should be identified to help the physician rule it out as a cause of hyperamylasemia.

AMY Methodologies

Amylase methodologies are classified into four categories: saccharogenic, amyloclastic (iodometric), chromogenic, and enzymatic. As with most analytes, enzymatic is the method of choice. The other three will be described briefly for historical purposes.

Saccharogenic methods measure the enzyme activity by quantitating the reducing substances formed (sugars, dextrins) by their reducing properties. In other words, it measures the products formed by the reaction. The chief reducing substance present is maltose, a disaccharide consisting of two glucose molecules.

Amyloclastic methods determine the decrease in substrate (starch) concentration by the addition of iodine, which turns blue when it binds to starch. When the starch is hydrolyzed to maltose and dextrins (fewer than 6 glucose units), the blue color disappears. The disappearance of the blue color is associated with a high AMY.

Chromogenic assay involves the use of a dye-labeled AMY substrate (amylose or amylopectin). As the AMY hydrolyzes the starch substrate, smaller water-soluble dye fragments are produced. The increase in color intensity of the soluble dye substrate is proportional to the AMY activity.

Enzymatic AMY procedures are based on the amylolytic hydrolysis of small oligosaccharides, which results in better controlled and more consistent reactions. The *maltotetraose* reaction is used in many instruments (AMY → Maltose phosphorylase → β-Phosphoglucose mutase → Glucose-6-phosphate dehydrogenase):

$$\text{Maltotetraose} + H_2O \xrightarrow[Ca^{+2},\,Cl^-]{\alpha\text{-Amylase}} 2\text{ Maltose} \qquad (9.25)$$

$$\text{Maltose} + P_i \xrightarrow{\text{Maltose phosphorylase}} \text{Glucose} + \beta\text{-Glucose-1-P} \qquad (9.26)$$

$$\beta\text{-Glucose-1-P} \xrightarrow{\beta\text{-Phosphoglucose mutase}} \text{Glucose-6-P} \qquad (9.27)$$

$$\text{Glucose-6-P} + NAD^+ \xrightarrow{\text{Glucose-6-phosphate dehydrogenase}}$$
$$\text{Gluconate-6-P} + NADH + H^+ \qquad (9.28)$$

The amount of AMY is proportional to the rate of production of NADH and the increase in absorbance at 340 nm. For each bond hydrolyzed by AMY, two molecules of NADH are produced.

Maltopentose is a second enzymatic methodology (AMY → α-Glucosidase → Hexokinase → G6PD).

$$\text{Maltopentaose} \xrightarrow{\alpha\text{-Amylase}} \text{Maltotriose} + \text{Maltose} \qquad (9.29)$$

$$\text{Maltotriose} + \text{Maltose} \xrightarrow{\alpha\text{-Glucosidase}} 5\text{ Glucose} \qquad (9.30)$$

$$\text{Glucose} + ATP \xrightarrow{\text{Hexokinase}} \text{Glucose-6-phosphate} + ADP \qquad (9.31)$$

$$\text{Glucose-6-phosphate} + NAD^+ \xrightarrow{\text{Glucose-6-phosphate dehydrogenase}}$$
$$\text{6-P-gluconolactone} + NADH + H^+ \qquad (9.32)$$

Reference Range

The reference range for AMY is very dependent on methodology and the instrument, but it is approximately 30–100 U/L for serum and 1–7 U/L for urine.

AMYLASE CREATININE CLEARANCE RATIO

The **amylase creatinine clearance ratio (ACCR)** compares the renal clearance of AMY to the clearance of CR on the same urine and serum.

$$\text{ACCR (\%)} = \frac{\text{Urine AMY (U/L)} \times \text{Serum creatinine (mg/dL)}}{\text{Serum AMY (U/L)} \times \text{Urine Creatinine (mg/dL)}} \qquad (9.33)$$

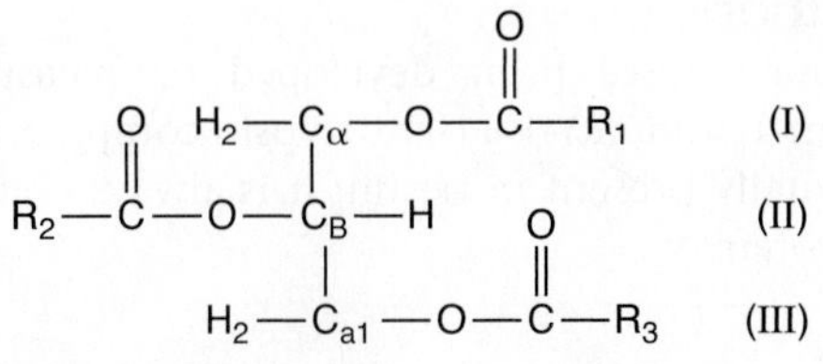

$$O$$
$$O \quad H_2—C_\alpha—O—C—R_1 \quad (I)$$
$$R_2—C—O—C_B—H \quad O \quad (II)$$
$$H_2—C_{a1}—O—C—R_3 \quad (III)$$

■ FIGURE 9-3   Triglyceride

Because volume and time are identical, they cancel out in a random or 2h specimen. The reference range is 2–5%, but it is affected by the AMY assay used. The ACCR is elevated in acute pancreatitis (>8%) because the renal clearance of AMY is greater than that of CR. The ratio returns to normal levels after the AMY is cleared from the serum. In macroamylasemia, the ACCR is <2% because the large complex cannot be filtered by the glomeruli, and the decreased ACCR differentiates macroamylasemia from other causes of hyperamylasemia.[11]

LIPASE

Lipase (LPS) hydrolyzes glycerol esters of long-chain fatty acids (triglycerides) to produce alcohol and fatty acids. Figure 9-3 ■ shows the structure of the triglyceride molecule. Only the ester bonds at carbon 1 and 3 (α and $\alpha1$ positions) are attacked, not β. However, through a process of isomerization, the fatty acid on the β carbon is moved to an α carbon. Then LPS can hydrolyze the third fatty acid on the triglyceride molecule, although at a much slower rate. The following three reactions result in the hydrolysis of triglyceride to glycerol and three fatty acids:

$$\text{Triglyceride} + H_2O \xrightarrow{\text{Lipase}} \alpha, \beta\text{-Diglyceride} + \text{Fatty acid I}$$
$$(9.34)$$

$$\alpha, \beta\text{-Diglyceride} + H_2O \xrightarrow[\text{Isomerase}]{\text{Lipase}}$$
$$\beta\text{-Monoglyceride} + \text{Fatty acid III} \quad (9.35)$$

$$\beta\text{-Monoglyceride} \rightarrow \alpha\text{-Monoglyceride} \xrightarrow{\text{Lipase}}$$
$$\text{Glycerol} + \text{Fatty acid II} \quad (9.36)$$

Lipase acts only at the interface between water and the substrate when the substrate is present in emulsified form; the rate of LPS action depends on the surface area of the dispersed substrate. Bile salts, synthesized by the liver and stored in the gall bladder, ensure that the surface of the dispersed substrate remains free of other proteins. Bile salts and a cofactor, collipase, are very important in the digestion of lipids, especially triglycerides. Collipase is added to LPS reagents as an activator.

Pancreatic LPS is the only LPS of clinical significance. Lipase is found in smaller amounts in intestinal mucosa, stomach leukocytes, and other tissues. LPS is also a small enzyme that is filtered by the glomeruli like AMY. It is completely reabsorbed in the tubule, however, so none is found in the urine.

LPS Clinical Significance

LPS is produced by the acinar cells of the pancreas. In acute pancreatitis, LPS rises within 4 to 8 hours, peaks at 24 hours, and remains elevated for 8 to14 days. It rises sooner and stays elevated longer than does serum AMY. It is also less affected by intraabdominal conditions described under AMY, making it more specific for acute pancreatitis but less sensitive. The panel of AMY and LPS can be used to differentiate pancreatic from salivary AMY. If AMY is increased and LPS is normal, then the AMY is probably of salivary origin (e.g., mumps) or an intraabdominal condition.

Although LPS is more specific than AMY for acute pancreatitis, other conditions can be associated with an elevated LPS, including chronic pancreatitis, duodenal ulcer, peptic ulcer, intestinal obstruction, acute cholecystitis, acute alcohol poisoning, and trauma to the abdomen as a result of surgery or accident. LPS levels more than 5 times the upper reference limit are convincing evidence of acute pancreatitis and rule out other intraabdominal conditions.

LPS Methodologies

Historically, lipase methodologies were time consuming and not especially accurate. Turbidimetric and titrimetric assays will be reviewed briefly, but enzymatic LPS is the methodology of choice in today's clinical laboratories. *Titrimetric (Cherry–Crandall)* involved a buffered stabilized 50% emulsion of olive oil incubated at 37°C for 24 hours. The free fatty acids released from the triglycerides in the olive oil were titrated with NaOH with phenolphthalein as an indicator to a purple endpoint. The higher the patient's LPS, the higher the level of fatty acids released.

Turbidimetric methodologies added the patient's serum to an emulsion of triglycerides, which had a milky appearance because the micelles absorb and scatter light. Clearing of the solution occurs as LPS hydrolyzes triglycerides in the micelles and micelle disruption takes place. The rate of micellular disintegration is measured by the decrease in turbidity in the reaction mixture.

Enzymatic LPS reactions have largely replaced titrimetric and turbidimetric methodologies (LPS → Glycerol kinase → L-α-Glycerophosphate kinase → Peroxidase).

$$\text{1,2-Diacylglycerol} + H_2O \xrightarrow[\text{pH 8.7}]{\overset{\text{Pancreatic lipase}}{\text{Collipase}}}$$
$$\text{2-Monoacylglycerol} + \text{Fatty acid} \quad (9.37)$$

$$\text{2-Monoacylglycerol} + H_2O \xrightarrow{\text{Monoglyceride lipase}}$$
$$\text{Glycerol} + \text{Fatty acid} \quad (9.38)$$

$$\text{Glycerol} + \text{ATP} \xrightarrow{\text{Glycerol kinase}}$$
$$\text{L-}\alpha\text{-Glycerophosphate} + \text{ADP} \quad (9.39)$$

$$\text{L-}\alpha\text{-Glycerophosphate} + O_2 \xrightarrow{\text{L-}\alpha\text{-Glycerophosphate kinase}}$$
$$\text{Dihydroxyacetone phosphate} + H_2O_2 \quad (9.40)$$

$$2 H_2O_2 + 4 \text{ Aminoantipyrene} + \text{Dye} \xrightarrow{\text{Peroxidase}}$$
$$H_2O + \text{Quinonemine dye} \quad (9.41)$$

Various dyes are used in the indicator–peroxidase reaction, including sodium N-ethyl-N-(2-hydroxyl-3-sulfopropyl)-m-toluidine (TOOS).

LPS Reference Range

The reference range depends on the procedure used. The upper reference limit for the TOOS methodology is 45 U/L at 37°C. LPS activity is stable at room temperature for a week, at refrigerator temperature for 3 weeks, and frozen for several years.

PANCREATIC ENZYMES

Serum amylase and lipase and urine amylase provide valuable information that is useful in differentiating acute pancreatitis from other intraabdominal conditions, acute from chronic pancreatitis, and stages of acute pancreatitis. As noted, LPS is not as affected by intraabdominal conditions as AMY. Serum LPS rises slightly sooner (4 to 8 hours) than AMY (5 to 8 hours), and LPS remains elevated longer (8 to14 days) than AMY (3 to 4 days). Urine AMY remains elevated longer than either serum AMY or LPS. A ratio of lipase to amylase may correlate with the cause of acute pancreatitis. Ratios greater than two may suggest alcoholic pancreatitis, but studies have shown it not to be a reliable indicator.[12]

TRYPSIN

Trypsin (TRY) is a proteinase that hydrolyzes the peptide bonds formed by the carboxyl groups of lysine or arginine with other amino acids. Acinar cells of the pancreas synthesize two different forms of trypsin [trypsin-1 (TRY-1) and trypsin-2 (TRY-2)] as inactive proenzymes (zymogens) trypsinogen-1 and -2, which are stored in zymogen granules until they are secreted into the intestine. In the intestine, trypsinogen is converted to the active enzyme trypsin by the intestinal enzyme enteropeptidase (enterokinase) or by active trypsin through autocatalysis.

TRY Clinical Significance

Trypsin measurements are important in screening for cystic fibrosis and chronic pancreatitis. In healthy individuals, TRY-1 is the major form found in serum. In acute pancreatitis, the increase in the levels of TRY-1 parallels amylase values. The form of TRY-1 indicates the severity of acute pancreatitis. In milder forms, more than 80% of TRY-1 is free trypsinogen-1. In more severe cases of acute pancreatitis with a mortality rate ~20%, the percentage of free trypsinogen-1 drops to 30% of the total TRY 1.

In newborns with cystic fibrosis, the TRY 1 levels are often high, but they fall dramatically as the disease progresses. Trypsin is not clinically important in treating patients with cystic fibrosis or acute pancreatitis, so it is usually only performed in reference or research laboratories.

TRY Methodologies

Immunoassays have been developed to measure TRY-1, trypsinogen-1, and TRY-1-α1-antitrypsin complex. Free TRY-1 is not normally present in serum; it is always complexed to another protein.

CHYMOTRYPSIN

Chymotrypsin (CHY) is a serine proteinase that hydrolyzes peptide bonds connecting the hydroxyl group of tryptophan, leucine, tyrosine, or phenylalanine. CHY prefers the carboxyl group of the aromatic amino acids residues as opposed to trypsin, which hydrolyzes bonds involving the amino group of the aromatic amino acids.

The pancreatic acinar cells synthesize two different chymotrypsins (1 and 2) as inactive proenzymes or zymogens, chymotrypsinogen-1 and -2. The zymogens are secreted into the intestine, and chymotrypsinogen is converted to the active CHY by trypsin as described in the previous section.

CHY Clinical Significance

The major application of CHY measurement is to investigate chronic pancreatic insufficiency. In patients with steatorrhea, CHY levels are decreased below the lower reference limit. In patients with chronic pancreatic insufficiency, CHY levels can determine whether oral pancreatic enzyme supplements are sufficient and whether the dosage needs to be adjusted. CHY is more resistant to catabolism in the intestine than TRY; therefore, it is the enzyme of choice for detecting pancreatic enzymes in the feces.

✓ Checkpoint! 9–5

1. *Mary has an elevated amylase and a normal lipase. What is the most likely source of the amylase? What are other possible explanations?*

2. *Amy has an amylase creatinine clearance ratio of 2%. What is the most likely condition associated with the low ACCR?*

3. *Which of the following pancreatic enzymes is elevated first in acute pancreatitis: serum amylase, serum lipase, or urine amylase? Which would remain elevated the longest?*

▶ MISCELLANEOUS CLINICALLY SIGNIFICANT ENZYMES

ACID PHOSPHATASE

Acid phosphatase (ACP) is a group of hydrolases similar to alkaline phosphatases; the major difference is the pH of the reaction. The optimal pH of ACP is 5.0–6.0.

$$\text{Phosphomonoester} + H_2O \xrightarrow[\text{pH 5.0}]{\text{Acid phosphatase}}$$

$$R\text{-OH} + HPO_4^- \quad (9.42)$$

The tissue richest in ACP is the prostate, where levels are 1000 times greater than in other tissues. ACP is also present in bone, liver, spleen, kidneys, RBCs, and platelets.

ACP Clinical Significance

Total ACP was historically the screening test for prostate cancer, especially metastatic prostate cancer. This was a somewhat insensitive and nonspecific test with a high percentage of false positives and false negatives. Today, newer markers, including prostate-specific antigen (PSA), are more useful tests to screen and diagnose prostate cancer.

Other conditions involving the prostate that are associated with elevated ACP are benign prostate hypertrophy (BPH) and prostate surgery. Although the studies report conflicting results, elevation of ACP for 24 hours following a prostate examination [digital rectal examination (DRE)] would indicate waiting at least 24 hours after a DRE before drawing a specimen for ACP.

ACP was also used in forensics in the investigation of rape. Seminal fluid in vaginal washings will result in ACP activity for 4 days following rape (intercourse).

Bone disease is a third category of elevated ACP. Osteoclasts are rich in ACP, so Paget's disease, breast cancer with metastases to the bone, and Gaucher's disease are associated with elevated ACP. Tartrate-resistant ACP (TR-ACP), described in the following section, is also increased physiologically in growing children, as is ALP.

ACP Methodology

Acid phosphatase is a group specific enzyme which will catalyze reactions with most phosphomonoesters. Total ACP methodologies use the same substrates as ALP but at an acid pH.

$$\text{P-Nitrophenolphosphate} \xleftarrow[\text{pH 5.0}]{\text{Acid phosphatase}}$$
$$\text{P-Nitrophenol + Phosphate ion} \quad (9.43)$$

The reaction products are colorless at an acid pH, but with the addition of alkali to end the reaction they are changed to chromagens that can be measured spectrophotometrically. The reaction can be modified to measure prostatic ACP with the addition of tartrate, which inhibits prostatic ACP (TR-ACP). The reaction is performed with and without tartrate.

Total ACP − ACP after tartrate inhibition = Prostatic ACP

Most of the ACP in serum is usually from osteoclasts.

Immunochemical methodologies, including radioimmunoassay and immunoprecipitation, have been developed. An immunological procedure using antibodies against TR-ACP has been introduced. The monoclonal antibody is bound to a solid phase and reacted with the patient's serum. After the reaction with the patient's TR-ACP, the ACP is measured at pH 6.1 with P-nitrophenolphosphate.

ALDOLASE

Aldolase (ALD) catalyzes the cleavage of D-fructose-1,6-diphosphate to D-glyceraldehyde-3-phosphate (GLAP) and dihydroxyacetone phosphate (DAP). ALD is an important enzyme in the glycolytic breakdown of glucose to lactate.

ALD Clinical Significance

Serum ALD measurements have been clinically significant in diagnosing and monitoring skeletal muscle diseases. Increases in ALD in combination with the CK:AST ratio can be useful in differentiating neuromuscular atrophies from myopathies. However, in most cases, the measurement of ALD does not provide sufficient additional information that is not provided by other enzymes (including CK, AST, and LD) to be useful clinically.

CHOLINESTERASE

Cholinesterase (CHE) is a hydrolyase enzyme that catalyzes the hydrolysis of choline esters to form choline and the corresponding fatty acid. The acetylcholine reaction is found in equation 9.44.

$$\text{Acetylcholine + H}_2\text{O} \xrightarrow{\text{Acetylcholinesterase}}$$
$$\text{Choline + Acetic acid} \quad (9.44)$$

Cholinesterases are divided into two groups: acetylcholinesterase (true cholinesterase) and pseudocholinesterase (acylcholine acylhydrolase).

Acetylcholinesterase (ACHE) is found primarily in red cells, lungs, spleen, and the central nervous system (gray matter of the brain and nerve endings). It is responsible for the prompt hydrolysis of acetylcholine, which is released at the nerve endings and mediates transmission of the neural impulse across the nerve synapse. Degradation of acetylcholine is necessary to depolarize the nerve so that is can be repolarized in the next contraction. ACHE is thus vital for the transmission of nerve impulses.

Acylcholine acylhydrolase or pseudocholinesterase is found in the serum, liver, and white matter of the brain. Although its biological function is unknown, it comprises most of the cholinesterase found in serum.

CHE Clinical Significance

The clinical significance of cholinesterase levels can be divided into three categories: detection of pesticide poisoning, liver function test, and detection of abnormal genetic variants. Exposure to some organophosphate compounds, including the pesticides Parathion and Sarin, decreases enzyme activity by binding to CHE and inhibiting the reaction. Chronic exposure to organophosphates by inhalation or through the skin results in a decrease of both acetylcholinesterase and pseudocholinesterase. Pilots who spray pesticides over crop fields have mandatory cholinesterase levels drawn before and after the spraying season. Symptoms such as headache and fatigue usually do not occur until there is a 40% drop in levels. After an 80% drop, neuromuscular effects are noted, including tremors, difficulty talking, paralysis, twitching, cramps, and weakness.[13] Inactivation of all ACHE will result in death.[14]

As a liver function test, CHE can serve as a measure of the synthetic ability of the liver. In acute hepatitis and chronic hepatitis, the levels of CHE may decrease 30% to 50%. In liver cancer and advanced cirrhosis, the levels fall 50 to 70%. CHE may be a useful monitor for patients with liver transplants or in the prognosis of patients with liver disease.

Abnormal genetic variants are found in patients who cannot hydrolyze succinyldicholine (Suxemethonium), a muscle relaxant administered during surgery. Normally, the succinyldicholine that is administered can be hydrolyzed by CHE. In patients with low levels of CHE, hydrolysis of the drug occurs much more slowly, resulting in prolonged apnea as a result of respiratory muscle paralysis. Rarely, death can occur if a respirator is not available.

Genetic variants can be detected by resistance to inhibition by *Dibucaine*, and the degree of drug resistance varies with the genetic genotype. Genetic testing of patients with a family history of the genetic variant (or a history of family members who have had trouble coming out of anesthesia) before surgery is recommended. Genetic variants are beyond the scope of this textbook; the reader is referred to the references at the end of this chapter.

CHE Methodology

CHE methodologies use various substrates including butyrylthiocholine. Butyrylthiocholine is hydrolyzed to butyrate and thiocholine, followed by reacting thiocholine with a chromogenic disulfide agent including 5,5′-dithio-bis (2-nitrobenzoate) (DTNB).

$$\text{Butyrylthiocholine} \xrightarrow{\text{Cholinesterase}}$$
$$\text{Butyrate + Thiocholine}\quad(9.45)$$

$$\text{Thiocholine + DTNB} \rightarrow \text{Mixed disulfide +}$$
$$\text{5-Mercapto-2-nitrobenzoic acid (5-MNBA)}\quad(9.46)$$

The colored 5-MNBA is measured at 410 nm. Other substrates include the iodide salts of acetylthiocholine, proprionylthiocholine, and succinylthiocholine.

CHE Reference Range

The reference range for CHE depends on the methodology. Using succinylthiocholine, the range for women is 33–76 U/L and 40–78 U/L for men in patients with the normal CHE genotype.

SUMMARY

Present in all cells of the body, enzymes are biological catalysts that enhance metabolic reactions from 10^6 to 10^{12} times the speed of those reactions without enzymes. Enzymes are not altered or consumed during reactions, small concentration of enzymes can be used over and over again, and they accelerate the speed at which a chemical reaction reaches equilibrium, but they do not alter the equilibrium constant.

Many factors influence enzyme reactions, including substrate concentration, pH, temperature, and activators or cofactors. Substrate concentration increases the rate of enzymatic reaction until it reaches a certain point when all of the enzyme molecules are bound to substrate in zero-order kinetics. Temperature also increases the velocity of most chemical reactions, doubling for each 10°C rise, up to the temperature when denaturation of the enzyme occurs. Optimal pH is usually from 7.0–8.0, or physiological pH. Cofactors or activators increase the rate of enzyme reactions with different activators or activators required for various groups of enzymes.

The two major cardiac enzymes are creatine kinase (CK and CK-MB) and lactate dehydrogenase. CK is a dimmer with three isoenzymes of two subunits M (muscle) and B (brain) with three possible combinations: MM, MB, and BB. Elevations of CK are found primarily in the brain, skeletal muscle, and cardiac muscle. Elevations of the CK-MB and CK-MM isoenzymes is indicative of cardiac damage. A CK-MB ≥6% is a specific indicator of a myocardial infarction. Lactate dehydrogenase is found primarily in cardiac, hemolytic, liver, and muscle diseases. LD is a tetramer consisting of combinations of two subunits: H and M. Following a myocardial infarction a characteristic "flipped" pattern is observed in which LD-1 isoenzyme levels are greater than LD-2.

Liver panels include two groups of enzymes. Aspartate aminotransferase and alanine aminotransferase are associated with hepatocellular injury. Alkaline phosphatase, gamma glutamyl transferase, and 5′-nucleotidase are suggestive of hepatobiliary disorders. For example, a marked increase in AST and ALT and a normal or mildly elevated ALP and GGT is indicative of hepatocellular injury (e.g., hepatitis).

Pancreatic enzyme panels include amylase and lipase, which are useful clinically to differentiate acute pancreatitis from other intraabdominal conditions. The highest levels of AMY are found in the salivary glands and the acinar cells of the pancreas. Amylase is not as specific for pancreatic conditions as lipase; for example, an elevated AMY and a normal LPS could indicate a salivary gland disorder (e.g. mumps, parotitis).

Trypsin and chymotrypsin are pancreatic enzymes rarely used to investigate pancreatic insufficiency in chronic pancreatitis and to determine whether enzyme supplements in cystic fibrosis patients are sufficient. Acid phosphatase was historically the screening test for prostate cancer, but with prostate-specific antigen methodologies it is not the test of choice. Cholinesterases are clinically significant in three categories: detection of pesticide poisoning, liver function test, and detection of abnormal genetic variants of cholinesterase.

REVIEW QUESTIONS

LEVEL I

1. Factors governing the rate of enzyme reactions are: (Objective 8)
   a. pH.
   b. temperature.
   c. substrate concentration.
   d. cofactors.
   e. All of the above.

2. In the assay of an enzyme, zero-order kinetics is best described by which of the following statements? (Objective 7)
   a. Enzyme is present in excess; rate of reaction is variable with time and dependent only on the concentration of the enzyme in the system.
   b. Substrate is present in excess; rate of reaction is constant with time and dependent only on the concentration of enzyme in the system.
   c. Substrate is present in excess; rate of reaction is constant with enzyme concentration and dependent only on the time in which the reaction is run.
   d. Enzyme is present in excess; rate of reaction is independent of both time and concentration of the enzyme in the system
   e. None of the above.

3. The protein part of the enzyme molecule with the cofactor is called: (Objective 5)
   a. an apoenzyme.
   b. a zymogen.
   c. a holoenzyme.
   d. a coenzyme.

4. The Michaelis–Menten (Km) constant is: (Objective 7)
   a. the concentration of substrate at which the reaction proceeds times one-half the maximum enzyme velocity.
   b. the temperature of optimal enzyme velocity.
   c. the pH of optimal enzyme activity.
   d. the concentration of substrate giving twice maximum enzyme velocity.

5. An international unit of enzyme activity is the quantity of enzyme that: (Objective 8)
   a. converts 1 micromole of substrate to product per liter.
   b. forms 1 mg of product per dL.
   c. converts 1 micromole of substrate to product per minute.
   d. forms 1 millimole of product per liter.

6. Enzymes that exist in multiple forms that are molecularly different but can act on the same substrate are termed: (Objective 11)
   a. apoenzymes.
   b. isoenzymes.
   c. coenzymes.
   d. holoenzymes.

LEVEL II

1. Levels of this enzyme are useful in the diagnosis of a genetic defect that leads to increased apnea (difficulty breathing) after surgery using succinyl choline. (Objective 3)
   a. cholinesterase
   b. aldolase
   c. leucine aminopeptidase
   d. 5′-nucleotidase
   e. acid phosphatase

2. A patient's CK-MB is reported as 17 μg/L, and the total CK is 300 IU/L. What is the CK relative index? (Objective 2)
   a. 0.56%
   b. 3.2%
   c. 4.6%
   d. 5.6%

3. Which of the following chemical determinations may be of help in establishing the presence of seminal fluid? (Objective 3)
   a. lactate dehydrogenase
   b. isocitric dehydrogenase
   c. acid phosphatase
   d. alkaline phosphatase

REVIEW QUESTIONS (continued)

LEVEL I

7. Creatine kinase catalyzes the reversible phosphorylation of creatine by: (Objective 10)
   a. AMP.
   b. ATP.
   c. pyrophosphate.
   d. orthophosphate.
   e. FADP.

8. Serum CK activity is increased in myocardial infarction and in all of the following except: (Objective 10)
   a. liver disease.
   b. skeletal muscle disorders.
   c. cardiac surgery.
   d. vigorous exercise.
   e. heart disease other than MI.

9. In the Oliver and Rosalki method, the reverse reaction is used to measure CK activity. The coupling reactions use the enzyme(s): (Objective 10)
   a. hexokinase and G-6-PD.
   b. pyruvate kinase and LD.
   c. luciferase.
   d. adenylate kinase.

10. Which of the following is a required activating ion for creatine kinase reactions? (Objective 10)
    a. magnesium ions
    b. chloride ions
    c. potassium ions
    d. zinc ions

11. Which CK isoenzymes(s) will be elevated following a myocardial infarction? (Objective 11)
    a. MM
    b. MB
    c. BB
    d. MM and MB
    e. MB and BB

12. All of the following statements regarding CK-MB (CK-2) are true except: (Objective 11)
    a. CK-MB >6% or 10 IU/L followed by an LD flip is specific evidence of AMI.
    b. CK-MB peaks before total CK after an acute myocardial infarction.
    c. CK-MB can be elevated after AMI without an increase in total CK.
    d. Levels are normal in cardiac ischemia.

13. Lactate dehydrogenase catalyzes the reaction of: (Objective 10)
    a. lactate → aspartate.
    b. lactate → pyruvate.
    c. lactate → oxaloacetate.
    d. lactate → ketoglutarate.

LEVEL II

REVIEW QUESTIONS (continued)

LEVEL I

LEVEL II

14. Increased LD because of the elevation of LD isoenzymes 1 and 2 is caused by: (Objective 15)
    a. muscular dystrophy.
    b. myocardial infarction.
    c. pancreatitis.
    d. hepatic damage.

15. The products formed by the forward reaction of aspartate aminotransferase are: (Objective 14)
    a. alanine and alpha ketoglutarate.
    b. oxalacetate and glutamate.
    c. aspartate and glutamine.
    d. glutamate and NADH.

16. Which enzyme catalyzes this reaction? (Objective 14)
    a. L-alanine $+ \alpha$-ketoglutarate $\rightarrow$ glutamate $+$ pyruvate
    b. alkaline phosphatase
    c. alanine aminotransferase
    d. aspartate aminotransferase
    e. gamma-glutamyl transpeptidase

17. Aspartate aminotransferase is *markedly* elevated in: (Objective 14)
    a. pyelonephritis.
    b. viral hepatitis.
    c. intestinal epithelium.
    d. Paget's disease.
    e. thyroiditis.

18. ALT is an enzyme that is *most* specific for the: (Objective 14)
    a. heart.
    b. pancreas.
    c. liver.
    d. kidney.

19. Alkaline phosphatase levels may be increased in: (Objective 14)
    a. pancreatic disease.
    b. liver or bone disease.
    c. kidney or bone disease.
    d. kidney or liver disease.

20. A 14-year old child is found to have a mildly elevated alkaline phosphatase. This: (Objective 14)
    a. indicates bone disease.
    b. is normal for growing children.
    c. indicates kidney disease.
    d. is an indicator of a mild hemolytic disorder.
    e. indicates liver disease.

21. Given the following results:

    Alkaline phosphatase: marked increase
    Aspartate transferase: slight increased
    Alanine transferase: slight increase
    γGT: marked increase

REVIEW QUESTIONS (continued)

LEVEL I

This is most consistent with: (Objective 16)
a. acute hepatitis.
b. osteitis fibrosa.
c. chronic hepatitis.
d. obstructive liver disease.

22. Which of the following is associated with a high serum amylase? (Objective 17)
a. mumps
b. intestinal obstruction
c. alcoholic liver disease
d. All of the above.

LEVEL II

PEARSON
myhealthprofessionskit

Use this address to access the interactive Companion Website created for this textbook. Simply select "Clinical Laboratory Science" from the choice of disciplines. Find this book and log in using your user name and password.

REFERENCES

1. Schreiber D, Miller S. Use of cardiac markers in the emergency department (www.emedcine.com/emerg/topic9332.htm), accessed November 8, 2007.

2. Garry J, McShane JM. Postcompetition elevation of muscle enzyme levels in professional football players (www.medscape.com/viewarticle/408045_print), accessed November 8, 2007.

3. Krantz A, Lewandrowski K, et al. Effect of marathon running on hematologic and biochemical laboratory parameters, including cardiac markers. *Amer J Clin Path* (2002) 118 6: 856–863.

4. Ogedegbe HO. Biochemical markers in risk stratification and diagnosis of acute coronary syndromes. *Lab Med* (2002) 1, 33: 42–53.

5. Harris EK, Wang ET, Shaw ST Jr. Statistical criteria for separate reference intervals: Race and gender groups in creatine kinase. *Clin Chem* (1991) 39, 9: 1580–1582.

6. Dufour RD, Lott JA, Nolte FS, et al. Diagnosis and monitoring of hepatic injury 1. Performance characteristics of laboratory tests. *Clin Chem* (2000) 46, 12: 2027–2049.

7. Mendler M. *Fatty liver: Nonalcoholic fatty liver disease (NAFLD) and nonalcoholic steatohepatitis (NASH)* (www.medicinenet.com/script/main/art.asp?articlekey=1909&pf=3&page=1), accessed October 26, 2006.

8. Nyblom H, Berggren U, Balldin J, Olsson R. High AST/ALT ratio may indicate advanced alcoholic liver disease rather than heavy drinking. *Alcohol & Alcoholism* (2004) 39, 4: 336–339.

9. Hietala J, Pukka K, Koivisto H, et al. Serum glutamyl transferase in alcoholics, moderate drinkers and abstainers: Effect on GT reference intervals at population level. *Alcohol & Alcoholism* (2005) 40, 6: 511–514.

10. Gamma-glutamyl transferase levels vary with weight, sex, and alcohol use (www.medscape.com/viewarticle/536567_print0), accessed November 2, 2007.

11. LeClerc P, Forest J-C. Variations in amylase isoenzymes and lipase during acute pancreatitis, and in other disorders causing hyperamylasemia. *Clin Chem* (1983) 29, 6: 1020–1023.

12. Chiemprabha A. Hyperamylasemia (www.emedicine.com/med/TOPIC3409.HTM), accessed October 31, 2007.

13. Tietz N, Shuey DF. Lipase in serum—the elusive enzyme: An overview. *Clin Chem* (1983) 39, 5: 746–756.

14. Chang K-C, Changchien C-s, Kuo C-M, et al. Clinical analysis of the efficacy in lipase/amylase ratio for acute pancreatitis. *J. Intern Med Taiwan* (2005) 16: 113–120.

Suggested Readings

1. Corestti JP, Cox C, Schultz TJ, Arvan DA. Combined serum amylase and lipase determinations for diagnosis of suspected acute pancreatitis. *Clin Chem* (1993) 39, 12: 2495–2499.

2. Dufour R. Laboratory approach to acute and chronic hepatitis. *MLO* (September 2003) 35, 9:10–16.

3. Dufour R. Detecting liver disease complications. *Advance* (April 21, 2003) 15, 9.

4. Falck-Ytter Y, McCullough AJ, et al. Clinical features and natural history of nonalcoholic steatosis syndromes. *Seminar Liver Disease* (2001), 21, 1: 17–26.

5. Allen LA, O'Donnell CJ, Camargo CA, et al. Comparison of long-term mortality across the spectrum of acute coronary syndromes. *Am Heart J* (2006) 151, 5: 1065–1071.

6. Sougioultzis S, Dalaskas E. et al. Alcoholic hepatitis: From pathogenesis to treatment. *Curr Med Res Opin* (2005) 212, 9: 1337–1346.

10

Tumor Markers

■ OBJECTIVES—LEVEL I

Following successful completion of this chapter, the learner will be able to:

1. List five roles of tumor markers in the assessment of cancers.
2. Describe four different methodologies that may be used to detect markers associated with malignancy.
3. List commonly used tumor markers and state their clinical significant in relation to cancer.
4. Identify several laboratory tests used to evaluate the following:
   a. Prostate disease
   b. Ovarian cancer
   c. Breast cancer
   d. Bladder cancer
   e. Pancreatic cancer

■ OBJECTIVES—LEVEL II

Following successful completion of this chapter, the learner will be able to:

1. Describe the difference between diagnostic sensitivity and specificity as related to biomarkers for cancer.
2. Calculate diagnostic sensitivity and specificity for a given set of data.
3. Interpret percent diagnostic sensitivity and specificity relative to predicting the presence of disease or no disease.
4. Define each of the following examples of clinical applications of tumor markers: screening, diagnosis, monitoring treatment, detection of recurrence, and prognosis (staging).
5. Discuss the impact of emerging technologies on clinical applications for tumor markers.
6. Explain the difference in molecular configuration between tPSA, fPSA, and cPSA.
7. Explain the clinical use of serum HER-2/neu levels in patients with breast cancer and on adjuvant therapy, for example, with Herceptin.
8. Discuss the clinical usefulness of the following biochemical tumor markers:
   a. CA 15-3
   b. CA 125
   c. CA 549
   d. CA 27.29
   e. CA 72-4
   f. CEA

KEY TERMS

Benign	Neoplasia	Screening
Carbohydrate antigen	Oncofetal	Staging
Epitope	Antigen	Tumor
Malignant	Oncogene	
Metastases	Prognosis	

A CASE IN POINT

A 38-year-old woman presented for biopsy and excision of a recently discovered lump in her right breast. Along with her routine preoperative blood work, her physician requested a blood test for CA 27.29. Subsequent requests for CA 27.29 levels were made for her two-week post-op follow-up visit and at one year and two years following her surgery. Results are shown in the table below. All testing was performed by the same laboratory, using the same immunoassay instrumentation and reagent systems. Reference range for CA 27.29 using this test system is 0–40 U/mL. Between-run precision of the assay is 10% at a level of 30 U/mL and 5% at a level of 90 U/mL.

Sample	Result (U/mL)
Pre-op	602
2 weeks post-op	38
1 year post-op	41
2 years post-op	60

Issues and Questions to Consider

1. Why did the values fall after surgery?

2. Which value(s) indicate significant changes from the previous value?

3. How would interpretation of these results be affected if the laboratory had changed instrument manufacturers during the time period represented?

WHAT'S AHEAD

▶ Description of the molecule characteristics of tumor markers.

▶ Discussion of methods of detection and/or quantitation of those molecules.

▶ Discussion of tumor markers currently used as markers of malignancy.

▶ Identification of additional tumor markers for each specific cancer discussed.

▶ TUMOR MARKERS

After heart disease, cancer is the second leading cause of death in the United States.[1,2] Because malignancy is not considered "curable" by standards of contemporary medicine; it becomes increasingly important to improve our means of prevention, early detection, and treatment of the disease.

Any molecule could function as a **tumor** marker providing that changes in circulating levels of the molecule correlate with tumor presence, growth, or activity. The molecule may be a product of the **malignant** tissue, or it may be a substance produced by the body in response to the malignancy. It should be measurable in body fluids. Ideally, it should be present when malignancy exists and not present in healthy people. However, because most tumor markers are molecular forms that exist at some level in normal, **benign**, and malignant **neoplasia**, they are not specific enough to be used as a means of **screening** an asymptomatic population. Tumor markers are rarely diagnostic for cancer. With the exception of PSA, most markers are elevated in more than one type of cancer.

HISTORICAL BACKGROUND

Historically, Bence Jones protein in the urine of patients with multiple myeloma was the first tumor marker. It was identified in 1847 but not characterized as monoclonal immunoglobulin light chains until the 1950s. Many molecules were associated with tumors and malignancies throughout the 20th century, but it was not until the mid-1960s that these tumor-associated molecules were used for monitoring patients with specific cancers. In the last quarter of the 20th century, advances in antigen/antibody research and molecular genetics were used to identify, characterize, and apply a wide spectrum of molecules that can be used as tumor markers.

▶ CLINICAL APPLICATIONS: THE ROLE OF LABORATORY TESTS

SCREENING

Ideally, a marker could be used to identify every individual who has a malignancy (100% sensitivity) and only those individuals who have that malignancy (100% specificity) at a stage early enough in the course of the disease to begin effective treatment. If the cost of performing an assay for that marker was such that the marker could be used to screen the general population, screening for and then treating any identified malignancies would be a tremendous public health advantage and realize substantial savings in health care costs. Gold and Freeman attempted to use serum carcinoembryonic antigen (CEA) levels to screen for colorectal cancer in 1965.[3] They found that the CEA as a tumor marker did not have the sensitivity and specificity necessary to successfully identify those individuals who have the disease. Unfortunately, this is true for many of the currently identified biomarkers. Most do not have the sensitivity and specificity necessary to screen the general asymptomatic public. However, there are a few biomarkers that have shown clinical utility for screening, especially within a smaller subset of the population. Lok et al. had success in using alpha-fetoprotein (AFP) to screen for hepatoma in a Chinese population that had a high prevalence of that disease.[4] The specificity of the PSA (prostate-specific antigen) makes it useful in screening men over 50 years of age, another population with a high prevalence of a certain specific malignancies.[5,6,7] National Academy of Clinical Biochemistry (NACB) clinical practice guidelines published in 2005 recommend screening for all men beginning at age 50.[8]

DIAGNOSIS

The same issues of sensitivity and specificity also make a serum test unsuitable for **diagnosis** of cancer. One cannot use a cutoff value for a tumor marker diagnostic for cancer if some individuals who have the disease exhibit values below the cutoff or if some individuals with elevated levels do not have the disease. However, if a patient has signs and symptoms of the disease, risk factors for the disease, and an elevation in a tumor marker associated with the disease, this elevation may confirm the clinical diagnosis. Clinically, a diagnosis of cancer is made based on a combination of physical examination, imaging, and laboratory tests.

It is also possible that a combination of elevated levels of multiple markers may confirm a clinical diagnosis of malignancy, or that elevation of the marker combined with other empirical clinical data. For example, a clinician may be interested in assessing a patient for prostate disease and will request a serum PSA levels in combination with prostate gland volume or enlargement as determined by transrectal ultrasound measurements to confirm the diagnosis.

MONITORING TREATMENT

By contrast, serum levels of tumor markers are widely accepted as useful for monitoring treatment of diagnosed disease states. The problems associated with lack of sensitivity or lack of specificity does not preclude this type of application because the patient has already been determined to have the disease in question. If a biomarker is elevated at diagnosis, and levels fall over the course of treatment for the disease, it can be presumed that the treatment is effective. If levels do not fall as expected, it may be an indication that the selected course of therapy is not effective and point the clinician to opt for a different treatment regimen. Once suitably low levels or negligible levels are attained, periodic measurements with consistent low values would indicate that the progression of the disease has been halted.

DETECTION OF RECURRENCE

Once a tumor has been reduced in size, removed, or otherwise inactivated, and tumor marker levels have been reduced, any subsequent elevation of the same marker can be attributed to recurrence, either by regrowth of a tumor or possibly by a metastatic process. Most of the commonly measured biomarkers will present with elevations in the circulating blood significantly earlier than the tumor can be detected by physical procedures. The relative ease of obtaining a blood sample for detection of recurrence over time-consuming or invasive physical procedures has made this application of tumor markers to clinical diagnosis and management of disease widely accepted.

PROGNOSIS

Prognosis is the prediction of the course and end of a disease, and the estimate of chance for recovery and it can be generally stated that the most dramatic elevations of the tumor marker are associated with the most widespread or aggressive tumors. Biomarkers generally increase with progressive disease and decrease with remission.

It is important for both clinicians and laboratorians to understand that several considerations must be addressed in assessing the significance of laboratory results for tumor markers. First, a single result should not be used to make or confirm a diagnosis. Some benign conditions produce transient elevations in some biomarkers. In contrast, a malignant condition is more likely to produce consistent elevation or increasing elevations.

Second, values obtained from different manufacturers' test kits cannot be used interchangeably. Differences in affinities of the antibodies used in immunoassay kits may produce variability of the results among commercial applications. Differences in the sensitivities of the measuring systems used in automated systems may produce variability even among laboratories using the same commercial system. In most laboratories, tumor marker assay procedures are appended with a comment to each result, stating which kit was used to generate the result and that results from different kits cannot necessarily be used interchangeably.

Third, the physiology of the body must be taken into account when evaluating the significance of a tumor marker result. Questions regarding, for example, the half-life of the marker in a normal individual, the ability of the biomarkers to be cleared from the circulation, and the effects of liver or renal disease on the marker metabolism must be addressed in an effort to better understand the data provided by the measurement.

In summary, the criteria for an ideal tumor marker should include the following:[9]

1. Be easy and relatively inexpensive to measure in easily obtainable body fluids

2. Be specific to the tumor under investigation and commonly associated with it

3. Have a direct relationship between blood levels of the marker and tumor mass

4. Have an abnormal plasma level, urine level, or both, in the presence of metastases at a stage at which no clinical or presently available diagnostic methods reveal their presence

5. Have plasma levels, urine levels, or both, that are stable and not subject to aberrant fluctuations

6. If present in the plasma of healthy individuals, exist at a much lower concentration than that found in association with all stages of cancer

► CLINICAL MEASUREMENT

ASSESSMENT OF THE PREDICTIVE VALUE OF A TEST

The predictive value of a method refers to its ability to detect patients with disease (sensitivity) and its ability to exclude patients without disease (specificity).

A test that is designed to diagnose a specific disease can have four possible outcomes:

1. It can test positive for patients who have the disease, (i.e., true positive, TP)

2. It can test negative for patients who do not have the disease, (i.e., true negative, TN)

3. It can test positive for patients who do not have the diseases, (i.e., false positive, FP)

4. It can test negative for patients who do have the diseases, (i.e., false negative, FN).

An ideal situation would be for all positive and negative values of the test to be true positives and true negatives. But this is not always possible, and the predictive value of a test is an indication of the balance between the true positives and true negatives for a test versus the false positives and false negatives.

DIAGNOSTIC SENSITIVITY AND SPECIFICITY

The clinical usefulness of any biomarker is dependent on both its diagnostic sensitivity and diagnostic specificity. **Diagnostic sensitivity**, as related to tumor markers, is a measure of how often the assay system detects the biomarker when the disease is present (i.e., positivity in disease). An assay with 100% sensitivity can detect the biomarker in question in every patient who has the respective disease (True Positive). **Diagnostic specificity** is used to describe the probability that a laboratory test will be negative in the absence of disease. An assay with 100% specificity will always be negative when the patient does not have the disease in question (True Negative). An assay system with less than 100% specificity will have some instances of false-positive results, that is, positive results in the absence of the disease in question, or positive results when a tumor is benign or nonmalignant.

The predictive value for a test is the probability that a disease is present when the test is positive or that a disease is absent when a test is negative. The formulas for predictive value are similar to the formulas for sensitivity and specificity but with a different emphasis. The predictive value of a positive test relates the number of true positives for a test to the total number of positive tests. Conversely, the predictive value for a negative test relates the true negative to the total number of negative tests.

Assessing the predictive value for a test is important but may be impractical for most clinical laboratories. However, many manufacturers will list the predictive value as well as the sensitivity and specificity of a test, and thus it is important to understand the principles of theses terms. A true table representing a fundamental matrix for estimating positivity and negativity in malignancy and nonmalignancy is shown in Table 10-1 ✪.

The values in percent for sensitivity, specificity, and predictive value are calculated by the following formula:

$$\text{Predictive value of a positive test} = \frac{TP}{TP + FP} \times 100\% \quad (10.1)$$

$$\text{Predictive value of a negative test} = \frac{TN}{FN + TN} \times 100\% \quad (10.2)$$

$$\text{Sensitivity} = \frac{TP}{TP + FN} \times 100\% \quad (10.3)$$

$$\text{Specificity} = \frac{TN}{TN + FP} \times 100\% \quad (10.4)$$

✪ TABLE 10-1

Truth Table for Positivity and Negativity in Malignancy and Nonmalignancy

	Positive	Negative
Malignancy	True Positive	False Negative
No Malignancy	False Positive	True Negative

The following example illustrates the use of assessing the predictive value of a test: A clinical chemistry laboratory is required to assess the predictive value for a nest test for cardiac troponin I (cTnI). The laboratory supervisor needs to know the predictive value of this test for diagnosing myocardial infarction (MI) in his or her own patient population. Data were collected and are shown in the following table.

Results (present or absent)	Number of Patients with MI	Number of Patients without MI	Total
Positive	250 (TP)	10 (FP)	260
Negative	50 (FN)	490 (TN)	540
Total	300 (TP + FN)	500 (FP + TN)	

$$\text{Predictive value of a positive test} = \frac{250}{250 + 10} \times 100\% = 96\%$$

$$\text{Predictive value of a negative test} = \frac{490}{50 + 490} \times 100\% = 91\%$$

$$\text{Sensitivity} = \frac{250}{250 + 50} \times 100\% = 83\%$$

$$\text{Specificity} = \frac{490}{490 + 10} \times 100\% = 98\%$$

The data show that sensitivity and specificity alone do not necessarily give an indication as to the performance of the test in the presence or absence of disease.

 Checkpoint! 10–1

Define the following terms: diagnostic sensitivity, diagnostic specificity, predictive value of a positive test, and predictive value of a negative test.

▶ METHODOLOGIES

IMMUNOASSAYS

Historically, immunoassays have been the primary methodology used to identify and measure proteins used as tumor markers. Immunoassays are very versatile and may be used to quantitate either antigen or antibody molecules. Because of the specificity associated with the antigen/antibody binding, immunoassays offer a versatile assay with the specificity necessary for measuring only the desired molecule with few cross-reacting substances. For detailed descriptions of theory and applications of immunoassay techniques, refer to Chapter 5, Immunoassays.

It should be noted that the antibodies used in immunoassays can be directed against different **epitopes** of a biomarker molecule. Even antibodies against the same epitope may vary in their specificity or affinity for the antigen. For this reason, immunoassays produced by different manufacturers may produce different results for the same sample. Results from different manufacturers' reagent systems cannot be used interchangeably. Laboratories must report which assay was used to determine results so that clinicians can make appropriate decisions when evaluating a patient's results over time.

MASS SPECTROMETRY

Mass spectrographic techniques used to detect biomarkers include matrix-assisted laser desorption ionization–time of flight (MALDI-TOF) and surface-enhanced laser desorption ionization–time of flight (SELDI-TOF). Review text in Chapter 2, Instrumentation, for principles of operation.

These relatively new technologies can generate proteomic patterns that can be compared with those of normal healthy individuals. If cancer cells manufacture proteins that are different from those proteins produced by normal cells, it can be theorized that detecting those different proteins could be used to diagnose malignancy. Using sophisticated software algorithms, one can identify patterns that are present only when malignancy is present.

MICROARRAYS

Cancer is a highly complex disease, which can include multiple genomic alterations on the pathway from a normal cell to a diseased state. Continued research during the past decade has allowed for distinguishing how these genomic modifications drive cancer cell survival by altering mechanisms for cell cycle control, differentiation, apoptosis, tumor vascularization, and metabolism.

Understanding complex diseases such as cancer require a universal overview of the genomic changes implicated in its progression. Researchers are uncovering the mechanisms and pathways involved in cancer by applying whole genome products for gene expression, SNP genotyping, and copy number analysis. The results of this type of research reveal new targets for drug development and disease biomarkers for diagnostic applications.

Currently, new tools are providing a larger view of the cancer genome, including alternative splicing levels and transcription regulation. Also, an increasing number of publications demonstrate the value of combining multiple modes of genomic analysis, including gene- and exon-levels expression, DNA variation, and copy number data, to fundamentally improve our understanding of the disease. New technologies are available to provide a comprehensive platform to accomplish the following (see Chapter 25, *Molecular Diagnostic*, for additional discussion of these topics):

• Point mutations

• Translocation

• Gene amplification/loss

• Epigenetic modifications

• Deletions

• Altered gene expression.

One example of a product currently available is the Affymetrix GeneChip (Santa Clara, CA), which is designed to investigate the molecular mechanisms of cancer. The results of testing samples using this microarray may lead to improving cancer diagnosis, for patient stratification in prognostic and therapeutic decision making, and for developing safer and more efficient treatments.

▶ SELECTED TUMOR MARKERS

There are several hundred examples of biochemical tumor markers being evaluated for clinical use. It is well beyond the scope of this clinical chemistry textbook to even list most of them. This section will focus on the tumor markers that are currently available in clinical laboratories or are currently being considered for use by clinicians. Emphasis will be placed on molecular characteristics, clinical application(s), and methodologies where appropriate. Many of the technique discussed in this chapter are described in both Chapter 2, *Instrumentation*, and Chapter 25, *Molecular Diagnostics*, in this textbook. For additional details regarding specific information of selected tumor markers, refer to the reference cited, particularly the reference texts edited by Carl Burtis, Edward Ashwood, and David Bruns, as well as Richard A. McPherson and Mathew Pincus.

The approach used in this chapter to facilitate discussion of individual tumor markers is to organize the tumor markers by neoplasms (e.g., prostate cancer or breast cancer, with the exceptions of enzymes and hormones as tumor markers). A rather lengthy listing of tumor markers and cancer types is provided in Table 10-2 ✪. This table clearly demonstrates an example of the vast numbers of biomarkers related to

✪ TABLE 10-2

Biochemical Grouping of Tumor Markers and Associated Cancer Types

Tumor Markers	Cancer Types
Enzyme:	
Alcohol dehydrogenase	Liver
Alkaline phosphatase	Bone, liver, leukemia, sarcoma
Alkaline phosphatase-placental	ovarian, lung trophoblastic, GI, Hodgkin's
Amylase	Pancreatic
Creatine kinase BB	Prostate, lung (small cell), breast, colon, ovarian
Galactosyltransferase	Colon, bladder, GI
Gamma-glutamyltransferase	Liver
Hexokinase	Liver
Lactate dehydrogenase	Liver, lymphomas, leukemia
Leucine aminopeptidase	pancreatic, liver
5'-nucleotidase	Liver
Prostatic acid phosphatase	Prostate
Hormones:	
Adrenocorticotropic (ACTH)	Cushing's syndrome, lung (small cell)
Antidiuretic (ADH)	Lung (small cell), adrenal cortex, pancreatic
Calcitonin	Medullary thyroid
Gastrin	Glucagonoma
Growth (GH)	pituitary adenoma, renal, lung
hCG	choriocarcinoma, embryonal, testicular
Human placental lactogen	trophoblastic, gonads, lung, breast
Neurophysins	breast
Parathyroid (PTH)	liver, renal, breast, lung
Prolactin	Pituitary adenoma, renal, lung
Vasoactive intestinal peptide (VIP)	panaceas, bronchogenic, pheochromocytoma, neuroblastoma
Oncofetal antigens:	
Alpha fetal protein (AFP)	Hepatocellular, germ cell
Carcinoembryonic cancer (CEA)	colorectal, GI, pancreatic, lung, breast
Pancreatic oncofetal	Pancreatic
Tennessee antigen	Colon, GI, bladder
Tissue polypeptide antigen	breast, colorectal, ovarian, bladder
Cytokeratins:	
Tissue polypeptide antigen	breast, colorectal, ovarian, bladder
Tissue polypeptide-specific antigen (TPS)	lung
Cytokeratin 19 fragments (CYFRA 21-1)	lung
Squamous cell carcinoma antigen (SCCA)	cervix, lung, skin, head neck, GI

⊕ TABLE 10-2 *(Continued)*

Tumor Markers	Cancer Types
Carbohydrates:	
Mucin-	
CA 15-3	breast, ovarian
CA 549	breast
CA 27.29	breast
DU-PAN-2	pancreatic, ovarian, GI, lung
CA 125	ovarian
Blood group related	
CA 19-9	Pancreatic, GI, hepatic
CA 19-5	GI, pancreatic, ovarian
CA 50	Pancreatic, GI, colon
CA 72-4	Ovarian, breast, GI, colon
CA 242	GI, pancreatic
Proteins:	
β_2-microglobulins	multiple myeloma, β-cell lymphoma, Waldenstroms macroglobulinemia, chronic lymphocytic leukemia
C-peptide	insulinoma
Ferritin	liver, lung, breast, leukemia
Immunoglobulin	multiple myeloma, lymphomas
Pancreas-associated antigen	trophoblastic, germ cell
Pregnancy-specific protein I	trophoblastic, germ cell
Prothrombin precursor	hepatocellular
Nuclear matrix proteins (NMP22)	bladder cancer
Complement factor-H	bladder cancer
Fibronectin	bladder cancer
Immnocyt	bladder cancer
Bladder tumor associated (BTA)	bladder cancer
Heat shock protein 27(HSP 27)	breast, endometrial, leukemia
S-100 proteins	melanoma
Thyroglobulin (Tg)	thyroid
Prostate specific antigen	prostate cancer
Receptors and other markers:	
Estrogen and progesterone	breast
Androgen receptor	prostate
Hepatocyte growth factor or c-Met	heptocellular, prostate, colorectal
Epidermal growth factor receptor (EGFR)	Head, neck, ovarian, cervical, bladder, esophageal
Genetic markers:	
N-ras mutation	acute myeloid leukemia, neuroblastoma
c-myc gene	β- and T-cell lymphoma, small cell lung
c-erb B-2 (HER-2/neu)	breast
bcl-2	leukemia, lymphoma
BCR-ABL	chronic myelogenous leukemia
Tumor suppressor genes	
BRCA1	neurofibromatosis 1, melanoma, breast
BRCA2	breast
p53	breast, colorectal, lung, liver

various cancer diseases under investigation. A second box, Box 10-1 ■, identifies the markers that are made available to care givers in a majority of hospital-based clinical laboratories and larger physician office laboratories (POLs).

ENZYMES AS TUMOR MARKERS

Prior to the discovery of radioimmunoassay in the mid-1950s, serum enzyme levels were the only tumor makers. Elevation of serum activity suggested the presence of a cancer. Measurement of enzymes did not provide the sensitivity or specificity required for identifying the type of cancer or the specific organ involved. The separation and identification of isoenzymes did improve upon the organ specificity. Unfortunately, reliable quantitative measurements of the isoenzyme fractions were not available. Three enzymes—alkaline phosphatase, creatine kinase, and lactate dehydrogenase—are discussed in more detail below.

BOX 10-1 Examples of tumor markers available in hospitals and physician office laboratories

▶ Alpha fetal protein (AFP)
▶ CA 15-3
▶ CA 125
▶ Carcinoembryonic cancer (CEA)
▶ Ferritin
▶ Growth (GH)
▶ hCG
▶ β_2-microglobulins
▶ Parathyroid (PTH)
▶ Prolactin
▶ Prostate specific antigen
▶ Thyroglobulin (Tg)

Once immunoassays were developed and used by the clinical laboratories, the concentration of an enzyme could be measured as a protein antigen instead of its catalytic activity. Also the analytical sensitivity and specificity improved, thus the clinicians were able to receive reliable quantifiable data for selected tumors markers that would aid in the overall care of their patients.

ALKALINE PHOSPHATASE

Principle sources of alkaline phosphatase (ALP) are liver, bone, and intestine (see Chapter 9, Enzymes). In serum from normal adults, most ALP comes from the liver or biliary tract. Patients with primary or secondary liver cancer, osteoblastic lesions associated with prostate cancer with bone metastases, and breast cancer with bone metastases will have elevated ALP.

Measurement of serum ALP in patients with liver metastases can provide a more accurate correlation with the extent of liver involvement than those of other liver function tests. Serum determination of 5'nucleotidase, γ-glutamyltransferase or ALP isoenzymes may assist the clinician in differentiating the source of elevated ALP. Other cancers, such as leukemia, sarcoma, and lymphoma with hepatic involvement, may show elevated ALP levels.

Another source of ALP is the placenta. Placental ALP (PALP) is elevated in sera of pregnant women. It represents one of the first oncodevelopmental markers along with AFP and CEA. PALP is elevated in several types of cancer such as ovarian, lung, trophoblastic, GI, and seminoma.

CREATINE KINASE

Creatine kinase (see Chapter 9, Enzymes) is a dimer consisting of two subunits designated M and B and three isoenzymes CK-1 (CK-MM), CK-2 (CK-MB), and CK-3 (CK-BB). CK-1 is elevated in small cell carcinoma of the lung and prostate cancer. Creatine kinase isoenzyme measurements

are included in a prostate cancer panel, identified as ProstAsure.[11,12] ProstAsure is a medical device that employs a nonlinear classification technique based on artificial neural networks (ANN) to simultaneously analyze multiple prostate tumor markers. Six input vectors are included in the algorithm, tPSA, prostatic acid phosphatase (PAP), and three creatine kinase isoenzymes. The goal of ProstAsure is to detect the presence of prostate cancer and benign prostate hypertrophy in men over 50 years of age.

LACTATE DEHYDROGENASE

Lactate dehydrogenase (LD) is an enzyme (see Chapter 9, Enzymes) that is elevated in a variety of malignancies but is somewhat nonspecific. It has been shown to be elevated in cancers including liver, non-Hodgkin's lymphoma, acute leukemia, nonseminomatous germ cell testicular cancer, seminoma, neuroblastoma, and others. LD is useful because it correlates well with tumor mass in solid tumors, and it helps predict survival and response to various treatments. LD isoenzymes provide very little assistance because of lack of organ specificity.

HORMONES AS TUMOR MARKERS

Tumors of specific endocrine glands may produce significant elevations of the hormones that are produced as part of that gland's normal physiological function. Measurement of these hormones may be useful in monitoring therapy or recurrence of disease. Three hormones, β-human chorionic gonadotropin, calcitonin, and adrenocorticotropin hormone, are discussed in more detail below.

β-HUMAN CHORIONIC GONADOTROPIN (β-hCG)

β-human chorionic gonadotropin is a glycoprotein hormone secreted by the syncytiotrophoblastic cells of the normal placenta and consists of two subunits (alpha and beta). The α-subunit is structurally different from the β-subunit and is common to several other hormones, including LH, FSH, and TSH. However, the β-subunit is unique to hCG, and the 28–30 amino acids located on the carboxyl terminus are antigenically unique. Serum hCG concentration in men and nonpregnant women is normally less that 5.0 U/L. In early pregnancy, the free β-subunit is produced together with intact or whole (molecule) hCG. In later trimesters, the free α-subunit predominates. Most cancer patients produce a greater amount of both free β-subunit and intact molecules than free subunits alone.

Serum levels of hCG are usually greater than a million units per liter in patients with trophoblastic tumors, a majority of those with nonseminomatous testicular tumors, and less frequently in those with seminoma. Moderate elevations are seen in cases of melanoma and cancer of the breast, GI, lung, and ovary.

Measurement of hCG is useful in detecting trophoblastic tumors together with AFP in detection of nonseminomatous testicular tumors. Serum levels of hCG correlate with the tumor volume and disease progression. hCG can be used to screen for choriocarcinoma (a rare cancer of the uterus) in women who are at high risk for the disease.

A clinically useful application of hCG measurement is for monitoring the treatment and progression of trophoblastic disease.[13] Based on the knowledge that serum levels of hCG correlate with tumor volume, extremely high initial hCG levels (e.g., >200,000–400,000 U/L) is considered to put patients at high risk for treatment failure. Serum hCG levels are expected to decline after surgical removal of the tumor. A normal serum half-life of hCG is ~12–20 hours. If hCG levels decrease slowly or remain elevated, it may indicate the presence of residual disease. An important assay parameter for serum hCG measurement is its detection limit because any residual hCG activity may indicate the presence of a tumor.

The current immunoassays using monoclonal antibodies detect various forms of hCG, including intact hCG or whole molecule, "nicked" hCG or hCGn (which is a partially degraded hCG missing a peptide bond), hCG α-subunit, hCG β-subunit and hCG core (residual hCG β).[14] Patients with malignancies may produce various forms of hCG, therefore knowledge of the assays analytical specificity is important.

Currently, most hCG assays are immunometric (sandwich) and measure primarily intact molecule when an antibody for the α-subunit and an antibody for the β-subunit are used in the sandwich format. These intact hCG assays do not measure free α-subunit or β-subunit because free subunits cannot form a sandwich with both antibodies. An assay designated "total βhCG" measures both the intact hCG and free β-subunits. In many cases, the tumor of the patient produces a large amount of free β-hCG, therefore a total β-hCG may be preferred.

In most clinical laboratories, the quantitative β-hCG immunoassay used as a pregnancy test also serves as the tumor marker assay. Some laboratories may provide reference intervals or interpretive information to delineate results between the two clinical applications, whereas other laboratories allow requesting clinicians to interpret their own tumor marker assay results.

 Checkpoint! 10–2

Why is knowledge of analytical specificity important for quantitative β-hCG assays when used as a tumor markers assay?

CALCITONIN

Calcitonin is produced by the C-cells of the thyroid and is a polypeptide with 32 amino acids. It is normally secreted in response to an increase in serum calcium from bone, thereby lowering the serum calcium level. Patients with familial medullary carcinoma of the thyroid, which is an autosomal dominant disorder, will present with elevated serum levels of calcitonin. Calcitonin levels correlate with tumor volume and extent of tumor involvement in local and distant metastases. Calcitonin is also clinically useful for monitoring treatment and detecting the recurrence of disease.

Serum levels of calcitonin are elevated in patients with carcinoid and cancer of the lung, breast, kidney, and liver. It is also elevated in benign conditions such as pulmonary diseases, pancreatitis, hypertension, pernicious anemias, Paget's bone disease, and pregnancy.

ADRENOCORTICOTROPIN HORMONE

Adrenocorticotropin hormone (ACTH) is a polypeptide hormone with 39 amino acids and molecular weight of 4,500. It is produced by the corticotrophic cells of the anterior pituitary gland (see Chapter 15, Endocrinology). Patients with small cell carcinoma (SCC) of the lungs may produce proACTH, the precursor to ACTH. The molecular weight of ProACTH is 22,000. It has a 5% bioactivity and contributes most of the immunoactivity of ACTH. Many of the immunoassays available measure both the hormone and precursor hormone.

Adrenocorticotropin hormone can be produced outside the pituitary (i.e., ectopic production). This is the case in patients with SCC of the lungs. Serum levels often exceed 200 ng/L in these patients. Many other conditions result in elevated serum levels of ACTH, including pancreatic, breast, gastric, and colon cancers and benign conditions such as chronic pulmonary lung disease, mental depression, obesity, hyperthyroidism, diabetes mellitus, and stress.

▶ GERM CELL TUMORS

ALPHA FETOPROTEIN

Alpha fetoprotein and **oncofetal antigen (protein)** is a 70 kDa glycoprotein that is normally produced by the fetal yolk sac and fetal hepatocytes. It is genetically and structurally similar to albumin. Production of AFP declines rapidly at birth and healthy adults and children have negligble or undetectable levels in serum. It is useful as a tumor marker in cases of hepatocellular carcinoma but also may be elevated during pregnancy and in cases of benign liver diseases, for example, hepatitis and cirrhosis. In healthy individuals, serum AFP levels are less than 10 μg/L.[15] During pregnancy maternal AFP levels increase starting at 12 weeks of gestation to a peak of about 500 μg/L during the third trimester. The fetal AFP reaches a peak of 2 g/dL at 14 weeks and then declines to about 70 mg/dL at term.[16]

Alpha fetal protein is useful for detecting hepatocellular carcinoma especially in the earlier stages when the tumor is small. Serum levels have been shown to correlate with the size of the tumor.[17] The concentration of AFP may exceed 300 μg/L in a patient with hepatocellular carcinoma. AFP has been used in determining prognosis and in the monitoring of therapy for hepatocellular carcinoma. Elevated levels of AFP and serum bilirubin levels are associated with shorter survival time.

In many instances, using more than one tumor marker provides more accurate and definitive information regarding the medical condition of a patient. Measuring both AFP and hCG can be useful in classifying and **staging** germ cell tumors. Germ cells tumors are often one cell type but may be a mixture of seminoma, yolk sac, choriocarcinomatous elements (embryonal carcinoma, or teratoma).[18] Patients with yolk sac tumors will have an elevated AFP, whereas in choriocarcinoma hCG is elevated. Both tumor markers are elevated in embryonal carcinoma. Neither marker is elevated in teratoma.

Neural tube defects (NTD) represent some of the most common, severe congenital malformation. They occur when there is failure of the neural plate and its coverings to fuse properly by the 27th day after conception. NTD screening consists of measuring serum AFP concentration in maternal blood. In the second trimester, when maternal screening is performed, maternal serum AFP levels increase approximately 15% per gestational week. The maternal serum AFP increases while amniotic fluid levels are decreasing, due to the combined changes in transfer to maternal serum and maternal clearance.[19] (See reference cited for additional information on this subject.)

Several immunoassays are available for measuring AFP using antigen or antibody labels that provide low detection limits, acceptable precision, and adaptability to automated platforms.

Checkpoint! 10–3

Briefly describe the rise and fall pattern of serum alpha fetal protein in a normal progressing pregnancy.

▶ PROSTATE DISEASE

PROSTATE-SPECIFIC ANTIGEN

Prostate-specific antigen (PSA), also referred to as human kallikrein 3 (hK3), is a serine protease (34kDa) consisting of 237 amino acid residues and four carbohydrate side chains. PSA is found in the epithelial cells of the acini and ducts of the prostate gland. It serves to lyse seminal coagulum and catalyzes fibronectin, semonogolin I, and semonogolin II. The origin of PSA is shown in Figure 10-1 ■. The relative distribution of PSA in body fluids is as follows:

In seminal fluid 60–70% exists as free (uncomplexed) PSA (f PSA) and <5% is complex with protease inhibitors such as protein C inhibitor (PCI). PCI is identified as the major complexing inhibitor.

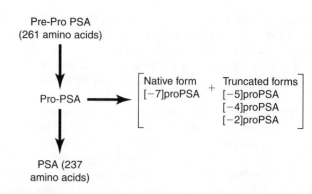

■ FIGURE 10-1   Origin of PSA

- In blood 55–95% of immunodetectable PSA is bound to α1-antichymotrypsin (ACT) and 5–45% of immunodetectable PSA is free.*
- In serum PSA forms stable 1:1 complexes with ACT, PCI, α-1-protease inhibitor (API), and α_2-macroglobulin (AMG). Neither PCI nor AMG are immunodetectable in serum, and ACT (98%) and API (0.5–2%) comprise the majority of total PSA (tPSA).

The assays currently available for routine use include (1) tPSA, (2) free PSA (fPSA), and (3) complex PSA (cPSA). These assays can be utilized by the clinician for the following:

- Detecting early stage prostate cancers
- Evaluating disease progression
- Assessing therapeutic response
- May also be useful in identifying postsurgical residual disease or tumor recurrence

Other approaches to improve diagnostic performance in evaluating patients with prostate related symptoms are measuring PSA density, PSA transitional density, fPSA/total PSA (tPSA ratio), or PSA velocity.

Measuring serum biomarkers for early detection of prostate cancer is a highly controversial topic in medicine. Much has been written about PSA testing, including questions regarding the level of tPSA that should prompt the clinician to require a follow up biopsy, the appropriate cutoff values, and the frequency with which to test a patient, as well as whether the entire male population should be screened for prostate cancer. The consensus seems to be that PSA testing by itself is not effective in the screening or detection of early prostate cancer because PSA is specific for prostate tissue but not for prostate cancer.

A condition known as benign prostate hypertrophy (BHP) is very prevalent in men over the age of 50 years. Serum tPSA values in patients with BPH are similar yet significantly different from those associated with early prostate cancer.[21] Thus a significant overlap in PSA values exists between these two

---

*The majority of uncomplexed or fPSA in serum appears to be an inactive form that cannot complex with protease inhibitors and may be either a PSA zymogen or an enzymematically inactive, cleaved portion of PSA.[20]

conditions, and selecting an appropriate cutoff value has been difficult. If a clinician uses a combination of diagnostic techniques including serum tPSA and digital rectal examination (DRE) followed by transurethral ultrasonography, a more accurate assessment of the presence or absence of cancer is possible.

A widely used cutoff value representing low risk of disease for tPSA is 4.0 ng/mL. There are studies that show the following cutoffs may provide the clinician with information regarding the appropriate cutoff value for recommending a biopsy.[22]

2.6 ng/mL	low risk—a biopsy may not be necessary
4–10 ng/mL	diagnostic grey zone—a biopsy may be necessary
>10 ng/mL	greater likelihood of cancer

FREE PSA

Free PSA can be measured by immunoassays techniques, and there are several manufacturers that have developed FDA-approved automated methodologies. Measurement of fPSA should be used only with the total PSA immunoassay to calculate the ratio (% fPSA) of free PSA (fPSA) to total PSA (tPSA). This calculated ratio is derived by dividing the measured fPSA by the measured tPSA and multiplying by 100. The interpretation of % fPSA is based on work done by Luderer and Chen et al., who determined that a % fPSA ≤ 25% detected 95% of prostate cancer when the tPSA is between 4 and 10 ng/mL.[23]

COMPLEX PSA

The molecular forms included in the fraction of complex PSA (cPSA) have been identified above. The AMG-PSA is not detectable in serum because of the engulfment and subsequent masking of PSA epitopes by the α_2-macroglobulin molecule.[20] One significant advantage of measuring cPSA is that there is minimal effect on the levels of cPSA after prostate manipulation either by flexible cystoscopy or digital rectal examination (DRE). Studies have shown that cPSA has slightly higher diagnostic specificity for patients with tPSA in the range of 4.1–10.0 ng/mL.[24] Thus the number of biopsies and associated cost performing them will be reduced.

There are FDA-approved immunoassays for measuring serum cPSA, and the accepted cutoff is up to 3.6 ng/mL. A two-site chemiluminescent immunoassay is available on the ADVIA Centaur (Bayer Corp., Tarrytown, NY)

 Checkpoint! 10–4

Explain the differences among tPSA, fPSA, and cPSA.

FUTURE SERUM BIOMARKERS FOR PSA

Human glandular kallikrein 2 (hK2) and PSA (human kallikrein 3) are serine proteases and are very similar in structure. They are almost exclusively found in prostate epithelium.[25] The amount of hK2 in seminal fluid is about 100,000-fold higher than in serum.[26] hK2 exists as free and complexed to several proteins, including ACT and AMG.

Tissue expression of hK2 using immunohistochemical techniques is most frequently used in high-grade cancers and lymph node metastases, and much less often for low-grade cancer and benign prostate hypertrophy (BPH).[27] Measuring serum hK2 alone or in combination with tPSA/fPSA has been shown to be a significantly better predictor of organ confined (pT2a/b) prostate cancer versus nonorgan-confined prostate cancer (pT3a) than measuring only tPSA.[28]

ProPSA (pPSA) is a proenzyme or precursor form of PSA and is associated with prostate cancer. Studies have shown that this biomarker may be better at detecting prostate cancer and reducing the number of unnecessary biopsies than fPSA. This markers is useful when tPSA is in the range of 2.5–4.0 ng/mL.[29]

The primary form of pro PSA found in tumor extracts is (–2) ProPSA. It has two amino acid propeptide leaders, hence (–2) proPSA. It has been shown to demonstrate a greater histochemical reactivity in malignant prostate tumors than in benign tissue.[30] Currently, a serum assay is available on the Beckman Access (p2PSA).

Prostate cancer gene 3 (PCAs) encodes a prostate specific mRNA that is highly overexpressed in prostate cancer tissue. Compared with benign prostate tissue, urinary PCA3 can add specificity to a diagnostic algorithm for prostate cancer in men with a negative biopsy but elevated serum tPSA.[31] PCA is determined using molecule diagnostic techniques and the value is incorporated into a *PCA3 score*. A PCA3 score is determined by dividing the PCA3 results by PSA mRNA. The cutoff value for PCA3 is 35.

Currently Gen Probe (Gen Probe Inc., San Diego, CA) has an assay available and marketed as the APTIMA PCA3 assay to determine PCA3. Studies show that PCA3/PSA mRNA correlation with biopsy resulted in ROC of 0.746, diagnostic sensitivity of 69%, and diagnostic specificity of 79%.[32]

Early prostate cancer antigen (EPCA-2) is a prostate cancer–associated nuclear protein. Studies have shown that this marker is useful for early detection of prostate cancer. An ELISA assay is available that uses anti-epitope 2.22. The recommended cutoff value is 30 ng/mL. Data show that the diagnostic specificity in men with tPSA of <2.5 ng/mL is 92–97%.[33]

Another approach to evaluate patients with prostate disease is to measure multiple biomarkers especially molecular diagnostic markers. For example by measuring or detecting GalNAc-T3, PSMA, Hepsin and DD3/PCA3 by molecular diagnostic techniques the ability of the clinician to diagnosis cancer of the prostate is greatly enhanced.[34]

 Checkpoint! 10–5

Explain why PCA3 may be a better biomarker for prostate cancer than tPSA.

► BREAST CANCER

CA 15-3

Carbohydrate antigen 15-3 (CA 15-3) is a glycoprotein with a molecular mass of 400 kDa. It is a type of mucin often overexpressed on the cell surfaces of malignant glandular cells for example breast cancer. This mucin is released into circulation where it can be detected, thus making it a useful tumor marker.

Measuring serum levels of CA 15-3 provides a more sensitive and specific biomarker for monitoring the clinical course of patients with metastatic breast cancer. Serum levels increase with higher stages of breast cancer. CA 15-3 levels can be used to predict adverse outcomes in breast cancer patients. In auxiliary node-negative patients, pre-op CA15-3 concentrations are a significance prognostic factor.[35]

The upper limit of CA 15-3 in healthy subjects is 25 kU/L. Elevated levels of CA 15-3 are also found in other malignancies including: pancreatic, lung, ovarian, liver and colorectal cancer. It is also elevated in benign diseases (e.g., benign liver and benign breast disease). The U.S. Food and Drug Administration (FDA) have approved CA 15-3 for monitoring therapy for advanced or recurrent breast cancer. Thus serum concentrations >30 U/L reflect worse overall survival pattern.

HER-2/*neu*

HER-2/neu (also c-erb B2 gene) is an **oncogene** identified as a 185 kDa transmembrane protein of the tyrosine kinase receptor family. The *neu* portion of this oncogene is referenced to its association with neural tumors. It belongs to a family of EGFR structurally related to HEGFR. HER-2/neu plays a key role in the regulation of normal oncogene cell growth. In cancer this gene may become amplified and creates excess copies of itself (over expression). It is normally expressed on the epithelia of several organs including bladder, lungs, pancreas, breast and prostate. Amplification is found in breast, ovarian and GI tumors.

In patients with breast cancer serum HER-2/neu is more important as a prognostic and predictive marker. Serum HER-2/neu is elevated in only ~5–10% of newly detected breast cancers. An increase in serum HER-2/neu before adjuvant therapy has been shown to be associated with increased tumor size, tumor grade and positive lymph nodes. In patients undergoing therapy, pre and post chemotherapy HER-2/neu levels may serve as a prognostic marker for disease-free survival and overall survival.

A useful clinical application of serum HER-2/neu measurement is monitoring the patient response to chemotherapy. Patients, who undergo treatment with Herceptin (also trastuzumab) (registered trademark of Genentech Inc., South San Francisco, CA), which is a humanized monoclonal antibody against HER-2/neu receptor, can be followed to evaluate patient's cancer state. For example, if the serum HER-2/neu level is high while the patient is on Herceptin therapy, the outcome usually results in a shorter treatment response. Conversely, low levels suggest longer or complete treatment response. HER-2/neu is also over expressed and/or amplified in other cancers including ovarian, bladder and prostate so it can be used as a therapeutic agent in theses cancers.[36]

Currently there are two FDA approved assays for HER-2/neu. One is a microtiter plate sandwich immunoassay using two monoclonal antibodies: one to capture human HER-2 ECD (extracellular domain) and the second is biotinylated designed to detect the complex. The assay is HER-2 ELISA from (Siemens Medical Solutions Diagnostics- Oncogene Science Biomarker Group, Cambridge, MA). The second assay is HER-2/neu [H2N] on the ADVIA Centaur (Bayer Diagnostics). It is a two-site sandwich immunoassay using direct, chemiluminescent technology. Heterophile antibodies and therapeutic Trastuzumab do not interfere with this assay because different antigen epitopes are targeted.[37]

Methods using immunohistochemical (IHC) staining of breast cancer tissue are available. One example is the InSite HER-2/neu kit (BioGenex Lab, San Ramon, CA). This method is used to detect increased levels of HER-2/neu proteins in cancer cells. The staining of breast tissue using IHC techniques is relatively easy to perform but suffers from interanalyst variation. A second technique used for detection of HER-2/neu gene amplification in breast tissue specimens is direct fluorescent in situ hybridization assay (FISH). A manufacture's kit version of this assay is the HER2 FISH pharmDx Kit for determination of HER2 gene amplification (Dako No. American Inc., Carpinteria, CA). Dako also has an IHC technique, the Herceptest for determination of HER2 protein overexpresson. A FISH based DNA probe assay is available from Vysis Inc. (Abbott Molecular Inc., Des Plaines IL) using the product name, PathVysion HER2 DNA probe kit. The salient features of these and other assays are shown in Table 10-3 ✪.

BRCA1 AND BRCA2

There is a group of patients with breast cancer that have an inherited predisposition to develop breast and ovarian cancer

✪ TABLE 10-3

Features of HER-2/neu test methods for breast cancer

Method	Target	Slide	Specimen Type	Storage	FDA Approval
IHC	Protein	Yes	Paraffin, frozen	Yes	Yes
FISH	Gene	Yes	Paraffin	No	Yes
CISH	Gene	yes	Paraffin	Yes	No
RT-PCR	Gene	No	Fresh, (frozen)	Yes	No
So. Blot	Gene	No	Paraffin, frozen, fresh	No	No
Serum ELISA	Protein	No	Fresh, (frozen)	No	Yes
Tissue ELISA	Protein	No	Fresh (frozen)	No	No

that is inherited as an autosomal dominant trait. This subset of breast and ovarian cancer patients usually has a family history where mother, aunts and grandmother may have these cancers.

Two unique genetic loci have been identified in this group of cancer patients: BRCA1 on chromosome 17q and BRCA 2, localized to 13q 12-13. BRCA1 encodes for an 1863 amino acid protein that can act as a transcription factor.[38] Both are involved in DNA repair. The mutated allele for BRCA1 or BRCA 2 gene in somatic cells may be inherited from either parent. Also certain ethnic groups, such as Ashkenazi Jews have an increase probability of harboring germ-line BRCA1 and BRCA 2 **mutations**. Women who carry BRCA1 gene mutations have a life time risk of breast cancer that ranges from 56–87%.[39] Similar risks is associated with BRCA2 gene mutations. Men with BRCA1 or BRCA2 mutations are at risk for male breast cancer and may also have an increased risk for developing pancreatic cancer as compared with men who do not have the mutation.[40] Women carriers of BRCA1 or gene mutations also have an increased risk for developing ovarian cancer.

CA 549

CA 549 is an acid glycoprotein. It is comprised of two separate entities with molecular masses of 400 and 512 kDa. One antibody is a murine IgG identified as BC4E 549 and the other antibody is a murine IgM, named BC4N 154.

Most healthy women have a CA 549 value below 11 kU/L. Serum levels may be elevated in pregnancy and benign breast disease and benign liver disease. It may also be elevated in non breast metastatic carcinoma including ovarian, prostate and lung cancer.

CA 549 is not useful in detecting early breast cancer because of the high proportion of women with breast cancer having low CA 549 levels. It is useful in detecting recurrence of breast cancer in patients after initial therapy followed by adjuvant therapy. In women who experience an increasing CA 549 value after an initial decrease or stabilization may have developed a metastatic disease. Monitoring serum CA 549 in patients with advance breast cancer shows a correlation with disease progression and regression and aids in the detection of metastases.[41,42]

CA 27.29

CA 27.29 is a mucin biomarker (MUC1)–associated antigen. It is detected by a monoclonal antibody, B27.29, that is produced against an antigen in ascites of patients with metastatic breast cancer. The reactive portion of the amino acid sequence of B27.29 overlaps with a segment of DF3 used in the CA 15-3 assay.[43]

The FDA has approved CA 27.29 for clinical use in detection of recurrent breast cancer in patients with stage II or stage III disease. Data show that at a cutoff value of 37.7 kU/L

BOX 10-2   Biomarkers used primarily for breast cancer detection

- ▶ HER-2/neu also c-erb B2 gene
- ▶ CA 15-3
- ▶ CA 549
- ▶ CA 27.29
- ▶ BRCA1 and BRCA2
- ▶ BCA225 (breast cancer antigen)
- ▶ NCC-ST439 (national cancer center)
- ▶ TPA (Tissue polypeptide antigen)
- ▶ IAP (Immunosuppressive acidic protein)
- ▶ PTEN (Phosphatase and tensin homolog)
- ▶ *c-myc* gene (gene product p62)
- ▶ EGFR (epidermal growth factor receptor)
- ▶ Estrogen and progesterone receptors
- ▶ S-100A4
- ▶ HSP27 (heat shock protein)
- ▶ MCA (mucin like carcinoma-associated antigen)

for CA 27.29 resulted in a positive predictive value of 83.3% and a negative predictive value of 92.6% or the detection of recurrent breast cancer. CA 27.29 is thought to be a better marker for recurrence of breast cancer than CA 15-3.[44]

An extended listing of biomarkers used almost exclusively for breast cancer detection is presented in Box 10-2 ■. These biomarkers show promise in facilitating the appropriate management of both male and female patients suspected of having a breast related disease.

 Checkpoint! 10–6

What is the relationship of serum HER-2/neu levels to treatment with Herceptin (trastuzmab)?

▶ OVARIAN CANCER

A very large study titled PLCO screening trials (prostate, lung, colon ovarian) in the United States with approximately 78,000 female's subjects is currently in progress to evaluate ovarian cancer.[45] One specific analytes currently available in most clinical laboratories, CA 125 is being studied in regards to its clinical usefulness as a biomarker for ovarian cancer.

CA 125

CA 125 is a 200 kDa glycoprotein that is 24% carbohydrate and is expressed by epithelial ovarian tumors and other pathological and normal tissues of mulerian duct origin. In healthy individuals, the consensus upper limit of CA 125 levels is 35 kU/L. CA 125 is very useful for epithelial ovarian cancer (EOC) and endometrial carcinomas. CA 125 measurement is most often utilized for monitoring recurrence of disease, and for differential diagnosis of pelvic masses in postmenopausal women.

Chambers et al. published findings for diagnostics sensitivity and specificity and positive predictive values for CA 125 and other potential biomarkers for ovarian cancer. They also included multiple biomarkers testing.[46] Their data showed that measuring only CA125 resulted in a sensitivity of 65%, specificity of 97–99, and a positive predictive value of 4.6%. Measuring CA125 and performing a trans vaginal sonography improve sensitivity (70%), specificity (99.9%), and positive predictive value (26.8). One significant discovery was that measurement of serum proteomic-based markers provides the highest percent for each diagnostic parameter; that is, they were all greater than 94%.

Two examples of specific proteomic base markers that may prove useful to clinicians relative to ovarian cancer are eosinophil derived neurotoxin (EDN) and COOH-terminal osteopontin fragments. Eosinophil-derived neurotoxin is a major secretory product of human eosinophilic leukocytes and COOH-terminal osteopontin fragments consists of arginine-glycine aspartic acid glycoprotein. These two analytes correlate well with various stages of ovarian cancer with a 93% diagnostic specificity and 72% diagnostic sensitivity for early sates ovarian cancer.[47]

Other potential biomarkers for ovarian cancer are cytokines, which include IL-6, IL-8, vascular endothelial growth factor (VEGF), and epidermal growth factor (EGF). One study shows that simultaneous measurement of a panel of serum cytokines and CA 125 using LABMap technology may present a promising approach for ovarian cancer detection.[48]

Kallikrein are a subgroup of serine protease enzymes and the human kallikrein (hK) gene locus spans a region of ~300 kb of chromosomes 19q13.4. Kallikreins are expressed in a wide variety of tissues including prostate (see section on prostate disease), breast (see section on breast cancer), ovarian and testes. hK6 has been used as a serum marker for diagnosis, prognosis and monitoring of ovarian cancer.[49] Other kallikreins associated with ovarian cancer includes hK5, hK6, hK10, and hK11.

Plasmalogens are a class of phospholipids possessing an ether lipid, with an ether-link also known as vinyl ether, at the first carbon position of the glycerol. The second carbon has a typical ester-linked fatty acid and the third carbon usually has a phospholipid head group like choline or ethanolamine. Two specific examples relevant to ovarian cancer biomarkers are plasmenylphosphatidic acid (PPA) and plasmenyethanolamine (PPE). In cardiac tissue nearly 50% of phosphatidylcholine contains the alkenyl ether at carbon number one. Nervous tissues, testes and kidney also contain significant amounts of plasmalogens. These compounds can protect cells against the damaging effects of single oxygen, which at high concentrations can kill cells. Thus plasmalogens are currently being studied relevant to cardiac dysfunction.

Plasmalogens have been isolated from colon tumors.[50] Recently a study was conducted to determine the clinical usefulness of measuring plasmalogens in patient with ovarian cancer. Measurement of these markers has shown very good sensitivities and specificity depending upon the cutoff value selected. This study shows that blood levels of plasmalogens are greatly decreased in ovarian cancer patients.[51]

Lysophophatidic acid (LPA, 1-acyl-2-hydroxy-sn-glycero-3-phosphate) is a form of glycerophospholipid consisting of various species with both saturated and unsaturated fatty acid tails. It is a normal component of serum that is released by activated platelets during platelet aggregation.[52]

Studies have shown that LPA is produced by ovarian cancer cells and that LPA itself also act as an ovarian cancer activating factor.[53] Elevated levels of LPA were found in ascites of ovarian cancer patients but also in the corresponding plasma samples.

 Checkpoint! 10–7

What is the principle clinical usefulness of CA 125 measurement in patients with ovarian cancer?

▶ COLORECTAL CANCER

CARCINOEMBRYONIC ANTIGEN

Carcinoembryonic antigen (CEA) is 22 kDa glycoprotein comprised of ~50% carbohydrate. It is a single polypeptide chain consisting of 641 amino acids. CEA consists of a large family of related cell surface glycoprotein and has been describe as part of an immunoglobulin gene "superfamily." It is found in embryonic tissue hence its name.

Serum levels of CEA are elevated in a variety of cancers, such as colorectal (70%), lung (45%), gastric (50%), breast (40%), pancreatic (55%), ovarian (25%) and uterine (40%) carcinomas. Elevated blood levels are also associated with benign disease (false positive results) and a number of tumors that do not produce CEA which represent false negative results. Therefore CEA measurement should not be used for screening purposes.[54]

The clinical usefulness of CEA is primarily to monitor therapy. Elevated blood levels of CEA over a prolonged period of time may strongly suggest the presence of colon cancer but may also be associated with other cancer. CEA is useful for detecting recurrence of colon cancer. In the healthy population the upper limit of CEA is 3 μg/L (3 ng/mL) in nonsmokers and ~5.0 μg/L (5 ng/mL) in smokers.

GUANYLYL CYCLASE C

Guanylyl cyclase C (GC-C) is a receptor that meditates fluid and electrolyte secretion and is expressed in brush border membrane of intestinal mucosa cells. GC-C is expressed only in intestinal mucosal cells lining the intestine, from the duodenum to the rectum. It is not expressed by other extraintestinal tissues.[55]

The expression of GC-C remains after intestinal mucosal cells transforms into neoplastic tissue. This expression occurs in both primary and metastatic colorectal tumors. Thus data from a study by Fava et al. suggest that GC-C may be a unique marker for detection of colorectal cancer cells in blood during post operative surveillance.[55]

An assay has been developed and manufactured by Targeted Diagnostics Technologies and Therapeutics Inc. (West Chester PA), Whole Blood GCC-B1 that detects a receptor GCC-B1 that is found on colorectal cancer cells. This technique can find one cancer cell in 10,000,000 mononuclear blood cells using RT-PCR. It is useful for detection recurrent metastatic colorectal cancer.

 Checkpoint! 10–8

What is the clinical usefulness of serum carcinoembryonic antigen measurements?

▶ PANCREATIC CANCER

Several blood group antigens (CA 19-9, CA 50, and CA 242) have been identified as useful markers for pancreatic cancer. Theses carbohydrate related tumor markers either are, (1) antigens on the tumor cell surface, or (2) secreted by the tumor cell.

CA 19-9

CA 19-9 is a glycoprotein (sialylated lacto-N-fucopenteose II ganglioside) that is a sialylated derivative of the Le^a blood group antigen. CA 19-9 is produced by normal pancreatic and biliary ductular cells and by colon, endometrial, gastric and salivary epithelia. Serum CA 19-9 exists as a mucin, which is a high molecular weight glycoprotein. Patients who are genotypically Le^{a-b-} do not express CA 19-9.[56]

The clinical usefulness of CA19-9 is in monitoring pancreatic cancer and colorectal cancer. Elevated blood levels of CA 19-9 can indicate recurrence one to seven months prior to detection by radiograph or the onset of clinical symptoms.

Several immunoassays are available to measure serum levels of CA 19-9. Most are radioisotopic immunoassays that are not practical for clinical laboratories today. Roche Diagnostic Corporation has introduced an electrochemiluminscent immunoassay using biotinylated monoclonal CA 19-9 antibody and monoclonal CA19-9 specific antibody labeled with a ruthenium complex to form a sandwich complex. This assay is available on the Elecsys and MODULAR analytics E 170 module.

CA 50

CA 50 is a monoclonal antibody developed against the human colon adenocarcinoma cell line COLO 205. The antigens to CA 50 are sialylated Le^a and sialylated Le^a lacking fucose. They exist in serum as glycoprotein and gangliosides in tissues. In epithelial carcinoma the principle form of CA 50 is sialytated Le^a and is also recognized by CA 19-9. Elevated levels of CA 50 occur in approximately 80–97% patients with pancreatic cancer. It is also elevated in numerous other cancers and benign disease but to a lesser amount.

CA 242

CA 242 is a monoclonal antibody developed from a human colorectal carcinoma cell line COLO 205. CA 242 recognizes the epitopes of CA 50 and CA 19-9. Locations of CA 242 includes apical border of ductoral cells of the human pancreas and in the epithelial and goblet cells of the colonic mucosa.

DU-PAN-2

DU-PAN-2 is an antibody that recognized an epitope that is mucin. The molecule mass is between 100–500 kDa and is 80% carbohydrate. The antigen is located primarily in the glandular epithelia of the pancreatic and biliary system. It is also found in several other tissue sources throughout the body.[57] Elevated serum levels of DU-PAN-2 are seen in patients with pancreatic (~60%), biliary tract (~45%) and hepatocellular (~44%) carcinomas.

 Checkpoint! 10–9

On a molecular level CA 19-9 is associated with which blood group antigen?

▶ BLADDER CANCER

Over 500,000 people in the United States are affected by bladder cancer and ~60,000 new cases were reported for 2006.[2] The most common type is transitional cell carcinoma (TCC). Current biomarkers are useful for monitoring recurrence. Detection of bladder cancer is usually accomplished by either cystoscopy or cytology of shed cells. It can also be detected by monitoring the presence of non cellular markers.

Several biomarkers have been evaluated for use in bladder disease but only a few are commercially available. Four biomarkers will be discussed in more detail below. An extended listing of bladder biomarkers is presented in Box 10-3 ■.

BLADDER TUMOR ANTIGEN

Bladder tumor antigen (BTA) is a high molecular weight polypeptide composed of complexes of basement membrane

BOX 10-3   Extended listing of bladder biomarkers

- ▸ Telomerase
- ▸ EGFR (epidermal growth factor receptor
- ▸ BLCA-4 (bladder cancer 4)
- ▸ Cytokeratin
- ▸ Soluble Fas
- ▸ p53
- ▸ pRb
- ▸ P21 gene
- ▸ p16 gem
- ▸ c-ras
- ▸ mdm-2
- ▸ Ki67
- ▸ α-FGF (fibroblast growth factor)
- ▸ β-FGF (fibroblast growth factor)
- ▸ E-cadherin
- ▸ P120
- ▸ MMP (matrix metalloproteinase)
- ▸ Fibrin-fibrinogen degradation product
- ▸ Hyaluronic acid (hyaluroindase)
- ▸ Lewis X antigen
- ▸ Nuclear shape, DNA content
- ▸ Survivin
- ▸ Bbb α_1 B-glycoprotein
- ▸ Vascular Epidermal growth factor (VEGF)
- ▸ Proliferating cell nuclear antigen (PCNA)
- ▸ Fibrinopeptide
- ▸ CD44
- ▸ Psoriasin

proteins. BTA can be found in urine due to invasion of the basement membrane by tumors or production by the tumor itself or a combination of these processes. Tumor antigens in urine are the easiest to analyze but cannot be used as the sole mechanism for tumor detection. They should be used along with cystocopy and cytology.

BTA Stat (Alidex Inc. Redmond WA) is a qualitative point-of-care assay that is designed to detect the presence of human complement related H factor in urine. It is an agglutination assay performed on a test strip. A visible color change represents a positive result. The specificity of the test varies from 68–72%.[58]

A quantitative test, BTA TRAK (Alidex Inc.) is also available. This test uses a standard enzyme immunoassay format to detect human complement factor H related protein. The manufacturer recommended cutoff for detection bladder cancer is 14 U/mL. At this cutoff the diagnostic sensitivity is 67–77% and specificity is 50–75%.[58]

NUCLEAR MATRIX PROTEIN-22

Nuclear matrix proteins (NMP) are part of the internal structural framework of the nuclei of the cells. Their functions are to support the nuclear shape, organize DNA and participate in DNA replication, transcription and gene expression. NMP-22 is a nuclear mitotic apparatus involved in the proper distribution of chromatin to daughter cells during cellular replication. It may be released from the nucleus of the tumor cells during apoptosis. NMP-22 concentration are ~25-fold greater in bladder cancer than in mean levels isolated from normal bladder.

NMP-22 can be assayed in urine using a qualitative urine POC proteome assay. This assay has been approved by the FDA to be used to detect NMP-22 as a monitoring test for recurrence of bladder cancer. The product is the NMP-22 BladderCheck test (Matritech, Newton, MA). It is a lateral flow immunochromatographic assay that detects elevated amount of nuclear mitotic apparatus protein. The cutoff is 10 U/mL and when used in conjunction with cytology it increased the detection of cancer to 99% compared with cystocopy alone at 91.3%.[59]

Fluorescence in situ hybridization (FISH) is a diagnostic technique that can be used to screen bladder carcinoma. This technique relies on chromosomal alterations associated with bladder cancer. Loss or part of chromosome 9 is the most common genetic alteration in bladder cancer. Alterations of the other chromosomes including 1,3,7,11,17 and loss of the 9p21 locus can also appear in bladder cancer. Interphase FISH applies labeled DNA probes to chromosomal centromeres or unique loci to detect cells with numerical or structural abnormalities.[60]

The UroVysion Bladder Cancer Kit (UroVysion Kit) from Abbott Diagnostics is an FDA approved FISH assay for detecting bladder cancer. This assay detects aneupoloidy for chromosomes 3,7,17 and loss of the 9p21 locus. The test requires urothelial cells, a fluorescent microscope, computer assisted image analyzer and specially trained laboratory personnel. FISH technique has shown diagnostic sensitivity of ~84% and specificity of ~95%. Thus FISH assays have the ability to detect tumor recurrence before any evidence of urothelial carcinoma on biopsy.[61]

ImmunoCyt (DiagnoCure Quebec Canada) is an FDA approved test that uses a combination of cytology with an immunofluorescence assay. This test detects cellular markers specific for bladder cancer in exfoliated cells using a panel of three fluorescent monoclonal antibodies to mucin and CEA associated with bladder cancer. Antibody 19A21 labeled with Texas red detects a high molecular weight form of CEA and fluorescein labels monoclonal antibody detect mucin which are expressed on most bladder cancer cells but not in normal transitional epithelia cells. The diagnostic sensitivity is 86% and specificity is 79%.[58,60] This assay may have a role in decreasing the frequency of cystoscopic exams for monitoring patients with low risk bladder cancer.[62]

 Checkpoint! 10–10

Identify the factor detected by the BTA test.

► LUNG CANCER

NEURON-SPECIFIC ENOLASE

One specific enzyme classified as an enolase, a glycolytic enzyme (phosphopyruvate hydratase), which may be useful to clinicians to evaluate lung cancer is neuron-specific enolase (NSE). This enzyme is the form of enolase located in neuronal tissue and in the cells of the diffuse neuroendocrine system. NSE is present in tumors including small cell lung carcinoma (SCLC), neuroblastoma, pheochromocytoma, carcinoid, medullary carcinoma of the thyroid and pancreatic endocrine tumors.[63]

Serum levels of NSE can be determined using several different immunoassay methods. The upper limit is 12 μg/mL. Diagnostic sensitivity and specificity in patients with SCLC are 80% and 80–90% respectively. The NSE levels correlate to a degree with stages and may provide a useful prognosis for disease progression. NSE has been recommended for the differential diagnosis of small cell carcinoma (SCC).[64]

► CYFRA 21-1

Cytokeratins are intermediate filament keratins found in the intracytoplasmic cytoskeleton of epithelial tissue. There are two types of cytokeratins: the low molecular weight, acidic type I cytokeratin and the high molecular weight, basic or neutral type II cytokeratin. Cytokeratins are usually found in pairs comprising a type I cytokeratin and a type II cytokeratin. One specific protein in this group that provides clinically useful information relative to lung cancer is CYFRA 21-1 or cytokeratin 19 fragments.

Data have shown that CYFRA 21-1 can provide useful information in diagnosing non small cell lung carcinoma (NSCLC). It is the most sensitive biomarker in NSCLC particularly SCC. A positive correlation exists with increase levels of CYFRA 21-1 and in stages of NSCLC and SCC. CYFRA 21-1 is also useful for therapeutic monitoring, recurrence and prognosis.[64]

SQUAMOUS CELL CARCINOMA ANTIGEN

Squamous cell carcinoma antigen (SCCA) is a glycoprotein (formerly tumor-associated antigen 4), approximately 45 kDa. Isoelectricfocusing techniques have revealed the presence of two subfractions of SCCA, neutral and acidic. The neutral fraction is found in both malignant and non malignant squamous cells, whereas the acidic fraction is located primarily in malignant cells. The acidic fraction is the one that is released into the blood circulation. Although SCCA has a lower sensitivity than CYFRA 21-1 it has superior specificity for SCC and can be used for histology sub typing. Patients with serum concentration of SCCA of >2 μg/mL have a 95% probability of being diagnosed with NSCLC and 80% probability of being diagnosed with SCC. SCCA may be used in the differential diagnosis of NSCLC, especially when used in combination with CYFRA 21-1 and CEA.[65]

Serum levels of SCCA can be measured using IRMA or microparticulate enzyme immunoassay (MEIA) on the Abbott IMX or AXsym analyzers (Abbott Diagnostics). The Abbott assay high limit for healthy subjects is 1.5 μg/L. Elevated levels of SCCA may be seen in several benign conditions including, liver diseases, renal failure and pulmonary infection.

OTHER LUNG CANCER BIOMARKERS

Two biomarkers, thyroid transcription factor-1 (TTF-1) and p63 have been demonstrated to be useful markers for distinguishing between SCLC and poorly differential pulmonary squamous cell carcinoma.[66] TTF-1 is a 38 kDa member of the NKx2 family of homeodomain transcription factors. It has been shown to be a sensitive marker for certain types of lung carcinomas including SCLC and to be nonimmunoreactive in pulmonary squamous cell carcinoma.[67] The second biomarker p63 is a homologous nuclear protein and is expressed in most poorly differentiated pulmonary squamous cell carcinomas.[68] Immunohistochemical staining techniques are available to detect immunoreactivity of both of these markers simultaneously to assist with differentiation between SCLC and poorly differentiated pulmonary squamous cell carcinoma.[66]

 Checkpoint! 10–11

Which biomarker is reported to be the most sensitive for non small cell lung carcinoma?

► GASTRIC CANCER

The mortality associated with gastric cancer has decreased steadily over the past several decades. Unfortunately, the likelihood that an individual patient will survive five years after being diagnosed with gastric cancer is low.

Several biomarkers have been evaluated for clinical use as biomarkers for gastric cancer. These include CEA, CA 19-9, CA 242, CA72-4, and β-hCG. CA 72-4 is the biomarker that has come to the forefront as a potentially useful biomarker for gastric cancer.

Tumor-associated glycoprotein (TAG-22), also called CA 72-4, is a mucin with high molecular weight (~200–400 kDa). It can be identified by using monoclonal antibody B72-3.[69] This biomarker has been used to diagnose cancer and monitor immunotherapy. Studies have shown that CA 72-4 specificity is 92% and has a positive predictive value 86%.[70]

Studies have shown that the free beta subunits of hCG have been reported to be expressed in various digestive tract malignancies. Data from the study resulted in a sensitivity of 41% for β-hCG in serum in gastric cancer patients.[71] Measurement of multiple biomarkers improved the prognostic value of serum biomarkers. For example, preoperative measurement of β-hCG and CA 72-4 as well as stage and histological type of tumor provided independent prognostic information for patients with gastric cancer.[72]

▶ GALL BLADDER CANCER

Carcinoma of the gall bladder is a common malignant tumor of the biliary tract. The initial clinical presentation is nonspecific and makes early diagnosis difficult. Detection of cancer of the gall bladder in the early stage may improve patient survival. Unfortunately, this is not usually the case and gall bladder cancer is generally diagnosed at an advanced stage. Tumors are often detected by routine computer tomography and ultrasonography. Several serum biomarkers have been investigated for use in evaluating patients with suspected cancer of the gall bladder and to differentiate cancer of the gall bladder from cholelithiasis.

Studies have shown that measurement of CA 242, CA 125, CA 15-3, and CA 19-9 may be useful to clinicians who treat patients with gall bladder disease.[73] The best clinical use of these biomarkers is in combination rather than measuring a single marker. For example, a combination of CA 242 and CA 125 can be useful as a diagnostic measure in preoperative evaluation of patients suspected of having gall bladder cancer. CA 19-9 and CA 125 used in combination may be useful to differentiate gall bladder cancer from cholelithiasis.

SUMMARY

Biochemical tumor markers have the potential to provide valuable information to the clinician with regard to screening, diagnosis, prognosis, monitoring, and detecting recurrence. The factors that determine which of these clinical applications are appropriate for a given tumor marker include diagnostic sensitivity, diagnostic specificity. Broadly speaking, more biochemical tumor markers are used for monitoring patient condition and detecting recurrence. The reader must also keep in mind that laboratory measurement of biochemical tumor represents only one of the diagnostic tests that a patient is subjected to while being evaluated for the presence or absence of disease. For example a patient that is being evaluated for breast cancer may require imaging tests, histological tests, and cytological tests that will be used in conjunction with results of serum biochemical tumor markers tests to afford the clinician a better opportunity to maker the right medical decision.

Currently, the laboratory tests available in hospital based laboratories and physician officer laboratories are primarily immunoassays that use a variety of antigen or antibody labels. These assays are designed form automated high throughput analysis. They provide a wealth of data for the clinician but often lack the specificity and/or sensitivity that could make them a better diagnostic test. Newer technologies are emerging that may make their way into clinical laboratories and provide hospital-based physicians with more powerful assays. Some examples of these techniques are molecular diagnostic techniques and instrumentation, mass spectroscopy and MALTI-TOF, and SELDI-TOF.

This chapter is limited in scope because of the overwhelming number of potentially useful biochemical tumor markers. The focus of the discussion was on biochemical tumor markers currently measured in most clinical laboratories and emerging biochemical tumor markers that may soon be available in these laboratories. For those initials who continue to work in the field of laboratory science, you will experience a paradigm shift very similar to the one that occurred in the 1970s when immunology laboratories replace serology. Biochemical tumor marker measurements will evolve from immunoassays to molecular diagnostic assay and techniques in the next five years!

REVIEW QUESTIONS

LEVEL I

1. Which of the following blood group antigens is CA 19-9 derived from? (Objective 3)
   a. Le^a
   b. A
   c. Kell
   d. Duffy

2. Human complement factor H related protein is associated with which of the following biochemical tumor markers? (Objective 3)
   a. DU-PAN-2
   b. CA 125
   c. HER-2/neu
   d. Bladder tumor antigen

LEVEL II

1. Squamous cell carcinoma antigen and CYFRA 21-1 can be used in combination to provide reliable laboratory data for the differential diagnosis of which disease? (Objective 8)
   a. liver cancer
   b. pancreatic cancer
   c. nonsquamous cell lung carcinoma
   d. pulmonary embolism

2. Diagnostic sensitivity represents which of the following conditions? (Objective 1)
   a. identifies only patients who have the diseases
   b. smallest concentration of an analyte that can be detected from zero.
   c. negativity in disease
   d. positively in disease

REVIEW QUESTIONS (continued)

LEVEL I

3. Which of the following enzymes is a useful biomarker for small cell lung carcinoma? (Objective 3)
   a. acid phosphatase
   b. alkaline phosphatase
   c. neuro-specific enolase
   d. creatine kinase

4. CA 72-4 is a useful biochemical marker for which of the following cancers? (Objective 3)
   a. gastric
   b. bladder
   c. liver
   d. prostate

5. Alpha fetal protein is useful in detecting which of the following carcinomas? (Objective 4)
   a. prostate
   b. breast
   c. hepatocellular
   d. brain

6. Measuring serum HER-2/neu is clinically useful for which of the following? (Objective 4)
   a. diagnosing breast cancer
   b. staging breast cancer
   c. determining the method of treatment for breast cancer
   d. monitoring the patients response to therapy, for example, with Herceptin (trastuzmab)

7. The biochemical marker CA 125 is expressed by which of the following tissues? (Objective 3)
   a. epithelial cells in the ovaries
   b. squamous cells in the lungs
   c. chromaffin cells in the adrenal gland
   d. heptocytes in the liver

8. Biomarkers designed with the prefix letters CA contain a significant amount of which of the following compounds? (Objective 1)
   a. carbamino
   b. carbohydrate
   c. carbon dioxide
   d. calcium

9. Which of the following biochemical tumor markers is useful in screening patients for neural tube defect? (Objective 3)
   a. CA 125
   b. alpha fetal protein
   c. DU-PAN-2
   d. NMP-22

LEVEL II

3. Which is the cause of increased alkaline phosphatase in a male with metastatic prostate carcinoma? (Objective 8)
   a. bone cancer
   b. brain cancer
   c. gall bladder cancer
   d. bladder cancer

4. Which of the following best describes the molecular form(s) measured by an intact β-hCG assay? (Objective 3)
   a. α-hCG and free β-hCG subunits
   b. only "nicked" hCG or hCGn
   c. the whole molecule
   d. hCG core and "nicked" hCG or hCGn

5. Which of the following is the best response to the scenario described below? (Objective 8)

   The laboratory performs an ACTH measurement using an assay that is specific for ACH only. The results of the laboratory tests are abnormally high. The clinician states that there is no evidence of a problem with the patient's pituitary.
   a. The possible cause of an elevated ACTH is ectopic production from a tumor, for example, lung.
   b. The possible cause of an elevated ACTH is ectopic production from a bone disease.
   c. The possible cause of an elevated ACTH is the presence of high levels of Pro-ACTH.
   d. The possible cause of an elevated ACTH is the presence of a tumor on the posterior pituitary.

6. Which of the following statements best reflects the production of alpha fetal protein (AFP) in a healthy individual? (Objective 8)
   a. The production of AFP declines rapidly at birth, and healthy adults and children have negligible or undetectable levels in serum.
   b. The production of AFP increases rapidly at birth, and healthy adults and children have very high levels in serum.
   c. The production of AFP declines rapidly at birth, and healthy adults and children have higher than normal levels in serum.
   d. The production of AFP remains the same through life.

7. The molecular difference between free PSA (fPSA) and complex PSA (cPSA) is best described by which of the following statement? (Objective 6)
   a. Free PSA is an active form that can complex with protease inhibitors.
   b. Free PSA is bound by compounds such as ACT, PCI, API, or AMG
   c. Free PSA is not bound by compounds such as ACT, PCI, API, or AMG.
   d. Free PSA cannot be detected in serum, whereas complex PSA can.

REVIEW QUESTIONS (continued)

LEVEL I

LEVEL II

8. Measuring serum levels of CA 15-3 is clinically useful for which of the following? (Objective 8)

    a. monitoring the clinical course of patients with metastatic breast cancer

    b. diagnosing patients with breast cancer

    c. monitoring the clinical course of patients with prostate cancer

    d. screening patients with breast cancer

9. Which of the following statements is true about HER-2/neu? (Objective 7)

    a. It is rarely expressed on the epithelia of any other organs.

    b. It amplifies and creates excess copies of itself (over expression)

    c. It does not amplify or create excess copies of itself.

    d. The neu portion of this oncogene is referenced to its association with neutrophils.

PEARSON
myhealthprofessionskit™

Use this address to access the interactive Companion Website created for this textbook. Simply select "Clinical Laboratory Science" from the choice of disciplines. Find this book and log in using your user name and password.

REFERENCES

1. Kung HC, Hoyert DL, Xu J, et al. Deaths: Final data for 2005. CDC National Vital Statistics Reports (2008) 56, 10: 31–35.

2. Jemal A, Siegel R, Ward E. Cancer statistics, 2008. *CA A Cancer Journal for Clinicians* (2008) 58: 71–96.

3. Gold P, Freeman SO. Demonstration of tumor-specific antigens in human colonic carcinomata by immunological tolerance and absorption techniques. *J Exp Med* (1965) 121:439–462.

4. Lok AS, Lai CL. Alpha-fetoprotein monitoring in Chinese patients with chronic hepatitis B virus infection: Role in the early detection of hepatocellular carcinoma. *Hepatology* (1989) 10, 3: 398–399.

5. Prostate specific antigen (PSA) best practice policy. American Urological Association (AUA). *Oncology* (2000) 14: 267–272.

6. Coley CM, Barry MJ, Fleming C, et al. G. Clinical guidelines part II. Early detection of prostate cancer: Estimating the risks, benefits, and costs [position paper]. *Ann Intern Med* (1997) 126: 468–479.

7. Coley CM, Barry MJ, Fleming C, et al. Clinical guidelines part III: Screening for prostate cancer [Position Paper]. *Ann Intern Med* (1997) 126: 480–484.

8. Lilja H, Semjonow A, Sibley P, et al. National Academy of Clinical Biochemistry guidelines for the use of tumor markers in prostate cancer (2005).

9. Coombes RC, Neville AM. Significance of tumor-index substance in management. In Stoll BA, ed., *Secondary spread in breast cancer* (Chicago: William Heinemann Medical Books, 1978).

10. Semmes OJ, Feng Z, Adam BL, Banez LL, Bibbee WL, et al. Evaluation of serum protein profiling by surface-enhanced laser desorption ionization time-of-flight mass spectrometry for the detection of prostate cancer; I. Assessment of platform reproducibility. *Clin Chem* (2005) 51: 102–112.

11. Stamey TA, Barnhill SA, Zhang Z, et al. Effectiveness of ProstAsure™ in detecting cancer (PCa) and benign prostatic hyperplasia (BPOH) in men age 50 and older. *J Urol* (1996) 155 (suppl): abstract 504.

12. Stamey TA, Barnhill SA, Zhang Z, et al. Comparison of a neural network with high sensitivity and specificity to free/total serum PSA for diagnostic prostate cancer in men with a PSA < 4.0 ng/mL. *Monogr Urol* (1998) 19: 21–32.

13. Berkowitz RS, Goldstein DP. Gestational trophoblastic disease. In Mossa AR, Schimpff SC, Robson MC (Eds.), *Comprehensive textbook of oncology.* (Baltimore: Williams & Wilkins, 1991): 1046–1051.

14. Birkin S, Yershova O, Myers RV, et al. Analysis of human choriogonadotropin core 2 o-glycan isoforms. *Mol Cell Endocrinol* (2003) 204: 21–30.

15. Chan DW, Kelsten M Rock R, et al. Evaluation of monoclonal immunoenzymometric assay for alpha-fetoprotein. *Clin Chem* (1986) 32: 1318–1322.

16. Chan DW, Booth RA, Diamandis EP. Tumor markers. In Burtis CA, Ashwood ER, Bruns DE (Eds.), *Tietz textbook of clinical chemistry and molecular diagnostics*, 4th ed. (Philadelphia: WB Saunders, 2006): 745–795.

17. Kelsten ML, Chan DW, Bruzek DJ, et al. Monitoring hepatocellular carcinoma by using monoclonal immunoenzymometric assay for alpha-fetoprotein. *Clin Chem*, (1988) 34: 76–81.

18. Bartlett NL, Freiha FS, Torti FM. Serum markers in germ cell neoplasm. *Hematol Oncol Clin North Am* (1991) 5: 1245–1260.

19. Ashwood ER, Knight GJ. Clinical Chemistry of pregnancy. In Burtis CA, Ashwood ER, Bruns DE (Eds.), *Tietz textbook of clinical chemistry and molecular diagnostics*, 4th ed. (Philadelphia: WB Saunders, 2006): 2167–2168.

20. Christensson A, Laurell CB, Lilja H. Enzymatic activity of prostate-specific antigen and its reactions with extracellular serine proteinase inhibitors. *Eur J Biochem* (1990) 194, 3: 755–763.

21. Oesterling JE, Chan DW, Epstein JI, et al. Prostate specific antigen in the preoperative and postoperative evaluation of localized prostatic cancer treated with radical prostatectomy. *J Urol* (1988) 1, 39: 766–772.

22. Punglia RS, D'Amico AV, Catalona WJ. Effect of verification bias on screening for prostate cancer by measurement of prostate-specific antigen. *N Engl J Med* (2003) 349, 4: 335–342.

23. Luderer AA, Chen Y, Soriano TF, et al. Measurement of the proportion of free to total prostate-specific antigen improves diagnostic performance of prostate-specific antigen in the diagnostic gray zone of total prostate-specific antigen. *Urology* (1995) 46, 2: 178–194.

24. Okegawa T, Noda H, Nutahara K, et al. Comparison of two investigative assays for the complexed prostate specific antigen in total prostate-specific antigen between 4.1 and 10 ng/ml. *Urology* (2000) 55: 700–704.

25. Schedlich LJ, Bennetts BH, Morris BJ. Primary structure of a human glandular kallikrein gene. *DNA* (1987) 6: 429–437.

26. Lovgren J, Valtonen-Andre C, Marsal K, et al. Measurement of prostate-specific antigen and human glandular kallikrein 2 in different body fluids. *J Androl* (1999) 20: 348–355.

27. Darson MF, Pacelli A, Roche P, et. al. Human glandular kallikrein 2 expression in prostate adenocarcinoma and lymph node metastases. *Urology* (1999) 53: 939–944.

28. Haese A, Graefen M, Steuber T, et al. Human glandular kallikrein 2 levels in serum for discrimination of pathologically organ-confined from locally-advanced prostate cancer in total PSA-levels below 10 ng/mL. *Prostate* (2001) 49: 101–109.

29. Baumgart Y, Otto A, Schafer A, et al. Characterization of novel monoclonal antibodies for prostate-specific antigen (PSA) with potency to recognize PSA bound to (2-macroglobulin. *Clin Chem* (2005) 51, 1:84–92.

30. Weinzierl CF, Susx, Pierson TB, et al. Measuring [-2] proPSA in serum: Analytical performance of the Access p2PSA assay from Beckman Coulter. *Clin Chem* (2007) 53, 6: A115.

31. Marks LS, Fradet Y, Deras IL, et al. PCA3 molecular urine assay for prostate cancer in men undergoing repeat biopsy. *Urology* (2007) 69: 532–535.

32. Groskopf J, Aubin S, Deras IL, et al. APTIMA PCA3 molecular urine test: Development of a method to aid in the diagnosis of prostate cancer. *Clin Chem* (2006) 52, 6: 1089–1095.

33. Leman ES, Cannon GW, Trock BJ, et al. EPCA-2: A highly specific serum marker for prostate cancer. *Urology* (2007) 69: 714–720.

34. Landers KA, Burger MJ, Tebay MA, et al. Use of multiple biomarkers for a molecular diagnosis of prostate cancer. *Int J Cancer* (2005) 114, 6: 950–956.

35. Duffy MJ, Duggan C, Keane R, et al. High preoperative CA 15-3 concentrations predict adverse outcome in node-negative and node-positive breast cancer: Study of 600 patients with histologically confined breast cancer. *Clin Chem* (2004) 50: 559–563.

36. Luftner D, Henschke, P, Flath B et al. Serum HER-2/new as a prediction and monitoring parameter in a phase II study with weekly paclitaxel in metastatic breast cancer. *Anticancer Res* (2004) 24: 895–906.

37. Luftner D, Luke C, Possinger K. Serum HER-2/neu in the management of breast cancer patients. Clin Biochem, (2003) 36, 4:233–240.

38. King MC, Rowell S, Love SM. Inherited breast and ovarian cancer. *JAMA* (1993) 269: 1975–1980.

39. Ford D, Easton DF, Bishop DT et al. Risks of cancer in BRCA1-mutation carriers. Breast Cancer Linkage Consortium. *Lancet* (1994) 343: 692–695.

40. Frank TS, Deffenbaugh AM, Reid JE, et al. Clinical characteristics of individuals with germline mutations in BRCA1 and BRCA2: analysis of 10,000 individuals. *J Clin Oncol* (2002) 20, 6: 1480–1490.

41. Chan DW, Beveridge RA, Bhargava A, et al. Breast cancer marker CA 549: A multicenter study. *AJCP* (1994) 1010: 465–470.

42. Chan DW, Beveridge RA, Bruzek DJ, et al. Monitoring breast cancer with CA 549. *Clin Chem* (1988) 34: 2000–2004.

43. Reddish MA, Helbrecht N, Almedia AF, et al. Epitope mapping of Mab within the peptide core of the malignant breast carcinoma associated mucin antigen coded for the human MCU 1 gene. *T Tumor Marker Oncol* (1992) 7: 19–27.

44. Chan DW, Beveridge RA, Muss H, et al. Use of TRUQUANT BR RIA fore early detection of breast cancer recurrence in patients with stage II and stage III disease. *J Clin Oncol* (1997) 15: 2322–2328.

45. Munkarah A, Chatterjee M, Tainsky MA. Update on ovarian cancer screening. *Am J Obstetrics and Gyn* (2007): 1922–1926.

46. Chambers AF, Vanderhyden BC. Ovarian cancer biomarkers in urine. *Clin Cancer Res* (2006) 12, 2: 323–327.

47. Ye Bin, Skates S, Mok SC, et al. Proteomic-based discovery and characterization and glycosylated eosinophil-derived neurotoxin and COOH-terminal osteopontin fragments for ovarian cancer in urine. *Clin Cancer Res* (2006) 12(2): 432–440.

48. Gorelik E, Landsittel DP, Marrangoni AM, et al. Multiplexed immunobead-based cytokine profiling for early detection of ovarian cancer. *Cancer Epidemiology, Biomarkers & Prevention* (2005) 14, 4: 981–987.

49. Diamandis EP, Scorilias A, Fracchioli S, et. al. Human kallikrein 6 (hK6): A new potential serum biomarkers for diagnosis and prognosis of ovarian carcinoma. *J Clin Oncol* (2003) 21: 1035–1043.

50. Dueck DA, Chan M, Tran K, et al. The modulation of choline phosphoglyceride metabolism in human colon cancer. *Molecular & Cellular Biochemistry* (1996), 162, 2: 97–103.

51. Shan L, Davis L, Hazen SL. Plasmalogens, a new class of biomarkers for ovarian cancer detection. *Clin Chem* (2007) 53, 6: A110.

52. Meleh M, Pozlep B, Mlaka A, et al. Determination of serum lysophosphatidic acid as a potential biomarker for ovarian cancer. *J Chrom Bio* (2007) 858: 287–291.

53. Fang X, Gaudette D, Furui T, et al. Lysophospholipid growth factors in the initiation, progression, metastases, and management of ovarian cancer. *Acad Sci* (2000) 905: 188–208.

54. Sikorska HM, Fuks A, Gold P. Carcinoembryonic antigen. In Sell S (Ed.), *Serological cancer markers*. (Totowa, NJ: Humana Press, 1992): 47–97.

55. Fava TA, Desnoyers R, Schulz S, et al. Ectopic expression of guanylyl cyclase C in CD34+ progenitor cells in peripheral blood. *J Clin Oncol* (2001) 19, 19: 3951–3959.

56. Lee P, Pincus MR, McPherson RA. Diagnosis and management of cancer using serologic tumor markers. In McPherson RA, Pincus MR (Eds.), Henry's clinical diagnosis and management by laboratory methods, 21st ed. (Philadelphia: WB Saunders, 2007): 1353–1366.

57. Metzgar RS, Sawabu N, Hollingsworth MA. DU-PAN-2: A clinically useful mucin marker of differentiation of pancreatic and other ductal cells and their tumors. In Sell S, ed., *Serological cancer markers*. (Totowa, NJ: Humana Press, 1992): 355–374.

58. Dey P. Urinary markers of bladder carcinoma. *Clinica Chima ACTA* (2004) 340: 57–65.

59. Grossman HB, Soloway M, Messing E. et al. Surveillance for recurrent bladder cancer using a point-of-care proteomic assay. *JAMA* (2006) 295, 3: 299–305.

60. Nielsen ME, Schaeffer EM, Veltri RW, et al. Urinary markers in the detection of bladder cancer: What's new? *Current Opinion in Urology* (2006) 16: 350–355.

61. van Rhijn BW, van de Poel HG, van de Kwast TH. Urine markers for bladder cancer surveillance: A systematic review. *Eur Urol* (2005) 47: 736–748.

62. Messing EM, Teot L, Korman H, et al. Performance of urine test in patients monitored for recurrence of bladder cancer: A multicenter study in the United States. *J Urol* (2005) 174: 1238–1241.

63. Chan DW, Booth RA, Diamandis EP. Tumor Markers. In Burtis CA, Ashwood ER, Bruns DE (Eds.), *Tietz textbook of clinical chemistry and molecular diagnostics*, 4th ed. (Philadelphia: WB Saunders, 2006): 745–795.

64. Cho WC. Potentially useful biomarkers for the diagnosis, treatment and prognosis of lung cancer. *Biomed and Parmacotherapy*, (2007) 61, 9: 515–519.

65. Molina X, Filella, JM, Auge R, et al. Tumor markers (CEA, CA 125, CYFRA 21-1, SCC and NSE) in patients with non-small cell lung cancer as an aid in histological diagnosis and prognosis. Comparison with the main clinical and pathological prognostic factors. *Tumor Biol* (2003) 24: 209–218.

66. Kalhor N, Zander DS, Lie J. TTF-1 and p63 for distinguishing pulmonary small-cell carcinoma from poorly differentiated squamous cell carcinoma in previously pap-stained cytologic material. *Modern Pathol* (2006) 19: 1117–1223.

67. Kaufmann O, Fietze E, Mengs J, et al. Value of p63 and cytokeratins 5/6 as immunohistochemical markers for the differential diagnosis of poorly differentiated and undifferentiated carcinomas. *Anotomic Pathol* (2001) 116: 823–830.

68. Wang BY, Gil J, Kaufman D, et al. P63 in pulmonary epithelium, pulmonary squamous neoplasm, and other pulmonary tumors. *Hum Pathol* (2002) 33: 921–926.

69. Sheng SL, Wang Q, Huang G. Development of time-resolved immunofluorometric assays for CA 72-4 and application in sera of patients with gastric tumors. *Clinica Chimica Acta* (2007) 380: 106–111.

70. Filella X, Molina R, Jo J, et al. Tumor associated glycoprotein-72 (TAG-72) levels in patients with non-malignant and malignant disease. *Bull Cancer* (1992;) 79: 271–277.

71. Louhimo J, Finne P, Alfthan H, et al. Combination of HCGβ, CA 19-9 and CEA with logistic regression improves accuracy in gastrointestinal malignancies. *Anticancer Res* (2002) 22: 1759–1764.

72. Louhimo J, Kokkola A, Alfthan H, et al. Preoperative hCG and CA 72-4 are prognostic factors in gastric cancer. *Int J Cancer* (2004) 111: 929–933.

73. Shukla VK, Gurubachan D et al. Diagnostic value of serum CA242, CA 19-9, CA 15-3 and CA 125 in patients with carcinoma of the gallbladder. *Tropical Gastroenterology* (2006) Oct–Dec, 27, 4: 160–165.

11

Nonprotein Nitrogen and Renal Function

■ OBJECTIVES—LEVEL I

Following successful completion of this chapter, the learner will be able to:

1. List and briefly describe the major parts of the urinary system.
2. Trace the ultrafiltrate (urine) flow through the major parts of the nephron.
3. Trace the blood flow in the kidney from the renal artery to the renal vein.
4. Summarize the three major renal processes; glomerular filtration, tubular reabsorption, and tubular secretion, including where they occur in the nephron and constituents involved.
5. Explain the difference between active transport and passive transport in relation to renal concentration.
6. List the major components of nonprotein nitrogen (NPN).
7. Identify the source of blood urea nitrogen (BUN) and the major organ of the urea cycle.
8. Review the most common BUN methodologies including chemical reactions and specificity.
9. State the reference range for BUN.
10. Convert BUN to urea and urea to BUN.
11. Define azotemia and uremia.
12. Outline common causes of prerenal, renal, and postrenal azotemia.
13. Identify causes of a decreased BUN.
14. Explain the source of creatinine (CR).
15. Review the Jaffe reaction and creatinase procedures.
16. Cite the reference range for creatinine.
17. Classify sources of increased creatinine.
18. Calculate the BUN:CR ratio and discuss its clinical significance.
19. Summarize the formation and excretion of uric acid.
20. Review the major uric acid methodologies.
21. Explain primary hyperuricemia (gout), including causes (precipitating factors) and treatment.
22. Outline causes of secondary hyperuricemia.
23. Review the renal clearance tests, including creatinine, the protein:creatinine ratio, and inulin clearance.
24. Calculate a creatinine clearance given the relevant data.
25. Summarize the etiology and clinically significant laboratory findings of major renal diseases.

■ OBJECTIVES—LEVEL II

Following successful completion of this chapter, the learner will be able to:

1. Review cystatin C and its role in assessing renal function.
2. Examine other screening tests for renal disease, including total protein, β_2-microglobulin, low-molecular-weight proteins, urinalysis, and microalbuminuria.
3. Explain dialysis and its role and complications in patients in renal failure.

KEY TERMS

Acute glomerulonephritis
Azotemia (prerenal, renal
 and postrenal)
Blood urea nitrogen (BUN)
BUN:creatinine (BUN:CR) ratio
Chronic glomerulonephritis
Creatininase (creatinine
 amidohydrolase)
Creatinine (CK)
Creatinine clearance (CrCl)
Cystatin C
Diabetic nephropathy
Dialysis

Glomerular filtration
Glomerular filtration rate (GFR)
Glomerulus
Gout
Hemodialysis (HD)
Hyperuricemia
Jaffe reaction
Lesch–Nyhan syndrome
Microalbuminuria
Nephron
Nephrotic syndrome
Nonprotein nitrogen (NPN)
Peritoneal dialysis (PD)

Phosphotungstic acid (PTA)
Primary gout
Pyelonephritis (acute and chronic)
Renal clearance
Tubular reabsorption
Tubular secretion
Urea
Urea reduction ratio (URR)
Urease
Uremia
Uricase

@ A CASE IN POINT

Dave C., an 80-year-old male, was admitted through the emergency room with difficulty breathing, coughing, and chest pain.

Chemistry Results

	2/18	2/19	Reference Range
Sodium	141	140	136–145 mEq/L
Potassium	4.4	4.7	3.6–5.0 mEq/L
Chloride	107	106	101–111 mEq/L
CO_2	27.0	30.0	24.0–34.0 mEq/L
Anion gap	5.0	5.0	10–20 mmol/L
Glucose	95	112	70–99 mg/dL
Bilirubin total	0.3		0.2–1.2 mg/dL
Troponin I	<0.3	<0.3	0.0–0.4 ng/mL
AST	10		5–40 IU/L
ALP	42		30–157 IU/L
Protein	6.5		6.0–8.4 g/dL
BUN	48	49	7–24 mg/dL
Creatinine	1.7	1.8	0.5–1.2 mg/dL
Calcium	8.8		8.5–10.5 mg/dL
Albumin	3.6		3.5–5.0 g/dL
Digoxin	1.06	0.82	0.00–2.00 ng/mL
CK-MB	3.1	3.2	*

* 0–3.8 ng/mL = Normal   3.9–10.4 = Borderline
>10.4: Significantly elevated

Issues and Questions to Consider

1. What are Dave's abnormal test results?
2. Do the laboratory results rule out a myocardial infarction?
3. What is Dave's BUN:creatinine ratio? What is a normal BUN:creatinine ratio?
4. What type of azotemia does this indicate?
5. (A) Where does this type of azotemia originate? (B) List five conditions associated with this type of azotemia.
6. What is the most likely diagnosis in this case?

WHAT'S AHEAD

▶ An overview of renal anatomy and physiology.
▶ A discussion of the clinical significance, most common methodologies, and reference ranges of BUN, creatinine, and uric acid.
▶ A review of the most common screening tests for renal disease.
▶ A summary of analytical procedures for measuring glomerular filtration.
▶ A comparison and contrast of renal pathophysiology and the laboratory findings of the most common renal diseases.
▶ An overview of dialysis treatment for patients with chronic renal disease.

▶ INTRODUCTION

The kidneys play a vital role in maintaining levels of many substances in the human body, retaining critical components and eliminating what is not essential. The most crucial roles are removal of waste, toxic, and surplus products from the body; homeostasis of the body's water; regulation of acid–base levels and electrolytes; and hormonal regulation. Renal function tests—for example, blood urea nitrogen (BUN), creatinine, and electrolytes—are included in chemistry screening profiles to screen for renal disease, water balance, and acid–base disorders.

Nonprotein nitrogen was the original test of renal function. NPN comprises products from the catabolism of proteins and nucleic acids that contain nitrogen but are not part of a protein molecule. Earlier NPN testing required a protein-free filtrate (PFF): precipitating protein from the serum and then centrifuging and using the PFF supernatant for testing. This procedure was time consuming and technically difficult. This chapter will concentrate on procedures developed to measure individual NPN compounds, procedures that are much simpler and can be automated.

▶ RENAL ANATOMY

The urinary system consists of two kidneys, two ureters, a bladder, and urethra as illustrated in Figure 11-1 ■. The kidneys are located in the posterior abdominal wall and are approximately 12 cm long, 6 cm wide, and 2.5 cm in depth; each weighs approximately 140 g. The kidneys are divided into two distinct areas: the outer layer or *cortex* and the inner layer or *medulla*. Each kidney contains approximately 1 million to 1.5 million **nephrons**, the functional unit of the kidney (see Figures 11-2 ■ and 11-3 ■).

Renal blood flow is vital to renal function. Blood is supplied to the kidney by the renal artery from the abdominal

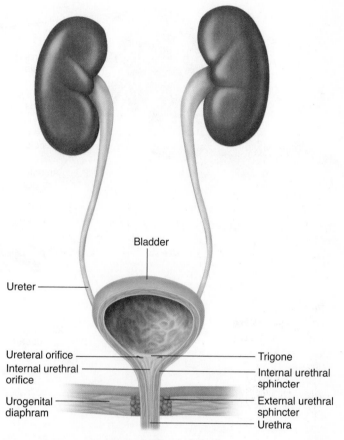

■ **FIGURE 11-1** Urinary System

From Colbert, Ankey, and Lee, *Anatomy and Physiology for Health Professions* (Upper Saddle River, NJ: Prentice Hall, 2007), p. 417.

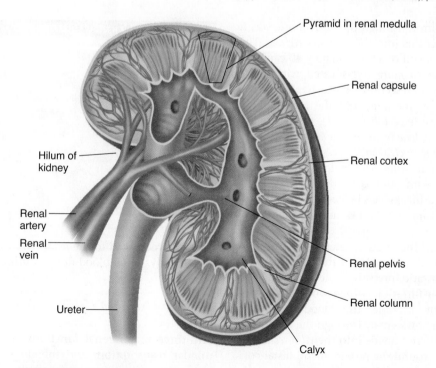

■ **FIGURE 11-2** Kidney

From Colbert, Ankey, and Lee, *Anatomy and Physiology for Health Professions* (Upper Saddle River, NJ: Prentice Hall, 2007), p. 404.

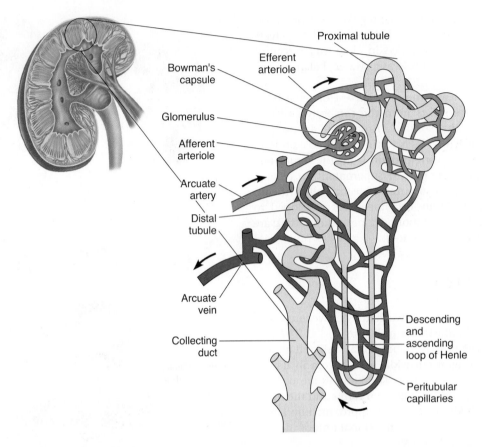

■ FIGURE 11-3 Nephron
From Colbert, Ankey, and Lee, *Anatomy and Physiology for Health Professions* (Upper Saddle River, NJ: Prentice Hall, 2007), p. 406.

aorta and enters the nephron through the *afferent arteriole*. It flows through the glomerulus into the *efferent arteriole*. The **glomerulus** consists of a coil of approximately 40 capillary loops referred to as the *capillary tuft* located within the *Bowman's capsule*, the initial section of the nephron. The blood is filtered in the glomerulus, and the filtrate flows through the *proximal convoluted tubule (PCT)*, the *descending loop of Henle*, the *ascending loop of Henle*, the *distal convoluted tubule (DCT)*, the *collecting duct, renal calyces,* and *the ureters, bladder,* and *urethra* in that order. The collecting duct from each nephron combines with other collecting ducts to form the renal calyces, where urine collects before passing into the ureters, bladder, and urethra. Box 11-1 ■ outlines the flow of urinary filtrate.

Renal blood flows from the afferent arteriole to the efferent arteriole, and the smaller diameter of the efferent arteriole results in a hydrostatic pressure differential that is important for glomerular filtration, which will be discussed later. From the efferent arteriole, the blood enters the *peritubular capillaries* and flows slowly through the cortex and medulla, where the capillaries divide into the *vasa recta*. The peritubular capillaries surround the proximal and distal convoluted tubules and are responsible for the immediate reabsorption of essential substances from the fluid in the PCT.

BOX 11-1   Urinary Filtrate Flow
1. Bowman's capsule
2. Proximal convoluted tubule (PCT)
3. Descending loop of Henle
4. Ascending loop of Henle
5. Distal convoluted tubule (DCT)
6. Collecting duct
7. Renal calyces
8. Ureter
9. Bladder
10. Urethra

The vasa recta lead to the *renal vein,* where blood is returned to the body. Renal blood flow is outlined in Box 11-2 ■.

▶ RENAL PHYSIOLOGY

The three major renal functions are glomerular filtration, tubular reabsorption, and tubular secretion. **Glomerular filtration** occurs in the glomerulus, which is the first part of the nephron. Filtration is enhanced by several factors.

First, the pressure in the glomerular capillaries is high because of the difference in size between the afferent and efferent arterioles described earlier. Second, the semipermeable glomerular basement membrane allows low-molecular-weight molecules less than 66,000 daltons to pass through into the filtrate. Albumin and other low-molecular-weight proteins, glucose, amino acids, urea, and creatinine are freely filtered and proceed to the proximal convoluted tubules. Third, the basement membrane is negatively charged, so large, negatively charged molecules (e.g., proteins) are repelled.

Tubular reabsorption occurs through *active transport* and *passive transport*. Every minute, 1200 mL of ultrafiltrate is filtered through the 2 million nephrons. Obviously, the body cannot lose 1200 mL of water containing essential nutrients every minute. The PCTs reabsorb approximately 80% of the fluid and electrolytes filtered by the glomerulus, including 70% of the sodium and chloride and most of the potassium, phosphate, and sulfate.

In active transport, the substance to be reabsorbed must be combined with a carrier protein contained in the membranes of the renal tubular cells. It is transported against a concentration gradient and requires the expenditure of energy from adenosine triphosphate (ATP). Sodium is moved through active transport in the proximal convoluted tubule, the ascending loop of Henle, and the collecting tubules. When the concentration of the substance (e.g., glucose) exceeds the capacity of the active transport system, then the substance is excreted or "spilled into" the urine. The *renal threshold* is the concentration above which the substance cannot be totally reabsorbed and is excreted in the urine. For example, the renal threshold of glucose is 160 to 180 mg/dL. If an individual's serum glucose exceeds the renal threshold—for example, with diabetes—the urine glucose will be positive.

Passive transport requires no energy and is characterized by the movement of a substance from an area of high concentration to one of lower concentration. The mode of transport also depends on the location in the nephron; for example, chloride is absorbed actively in the ascending loop of Henle and passively in the proximal convoluted tubule. Water and urea are always reabsorbed through passive transport.

Tubular secretion is the opposite of tubular reabsorption and involves the passage of substances from the peritubular capillaries into the tubular filtrate. Tubular secretion serves two major functions: elimination of waste products not filtered by the glomerulus and regulation of acid–base balance in the body through secretion of 90% of the hydrogen ions excreted by the kidney. Many foreign substances (e.g., medications) cannot be filtered by the glomerulus because they are bound to proteins that make them too large to pass through the glomerular pores. However, when they pass through the peritubular capillaries, they develop a strong bond for the tubular cells and dissociate from their carrier protein, resulting in their transportation into the filtrate by the tubular cells. The major site for removal of these nonfiltered substances (H^+, NH_3, K^+, and weak acids and bases) is the PCT.

ANALYTES ASSOCIATED WITH RENAL FUNCTION

NONPROTEIN NITROGEN

Nonprotein nitrogen (NPN) comprises the products of the catabolism of proteins and nucleic acids that contain nitrogen but are not part of protein molecules. The kidneys remove excess NPN from the body; therefore, one of the major reasons for measuring NPN is to evaluate renal function. Approximately 15 NPN compounds are found in plasma. The clinically significant NPN compounds discussed in this chapter are urea nitrogen, uric acid, and creatinine. Ammonia will be covered in Chapter 19, Liver Function. The most common NPN compounds and their relative percentage concentrations are listed in Table 11-1.

UREA OR BLOOD UREA NITROGEN

Urea or **blood urea nitrogen (BUN)** is the major nitrogen-containing metabolic product of protein catabolism in humans. It is formed from exogenous protein (protein in the diet) or endogenous protein from the breakdown of cells in the body. Approximately 75% of the NPN in the body is eventually excreted. A large percentage of protein is converted to urea during the breakdown of protein. Figure 11-4 ■ shows the chemical structure of urea.

The synthesis of urea is carried out exclusively by the hepatic enzymes of the *Krebs and Henselheit urea cycle*. In the

TABLE 11-1

Nonprotein Nitrogen Compounds and Relative Concentration (%)	
Blood urea nitrogen (BUN)	45%
Amino acids	20%
Uric acid	20%
Creatinine	5%
Creatine	1–2%
Ammonia	0.2%

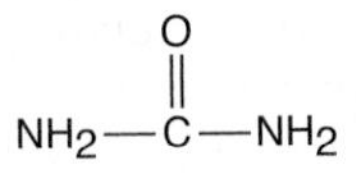

$$NH_2 - \underset{\underset{Urea}{\|}}{\overset{\overset{O}{\|}}{C}} - NH_2$$

■ FIGURE 11-4   The Chemical Structure of Urea

intestine, urea is metabolized by intestinal bacteria to form ammonia and carbon dioxide. The ammonia is reabsorbed through the portal system (a venous system comprising the hepatic portal vein and its tributaries) and carried to the liver, where it is reconverted to urea, which is less toxic. More than 90% of the urea is excreted by the kidneys, and the remainder is lost through the gastrointestinal tract and skin.

There is a direct relationship between urea and the **glomerular filtration rate (GFR)**. Urea is freely filtered in patients with normally functioning kidneys. In a patient with a normal to increased GFR, approximately 40% of the BUN is reabsorbed and 60% excreted; therefore, a well-hydrated patient excretes more BUN, resulting in a lower serum BUN. In a dehydrated patient, 70% of the BUN is reabsorbed and 30% excreted; as a result, the patient's serum BUN is increased and urine BUN is decreased. BUN is dependent on three variables: urea concentration, glomerular filtration rate, and level of hydration.

BUN Clinical Significance

Many different renal diseases are associated with an elevated BUN; however, there are many nonrenal factors that also result in high BUNs. In **azotemia**, there is an increase in blood urea and other NPN compounds. **Uremia** is an increase in urea and BUN. Azotemia is classified in three categories: prerenal, renal, and postrenal. **Prerenal azotemia** occurs before the kidney, usually because of a decreased renal blood flow. Examples of conditions associated with prerenal azotemia are congestive heart failure, shock, advanced cirrhosis, and mild dehydration.[1]

Renal azotemia is associated with kidney disease: for example, glomerulonephritis, nephrotic syndrome, and acute renal failure. In renal disease, the GFR is decreased and less BUN is excreted in the urine.[2]

Postrenal azotemia occurs after the urine has left the kidney and results from an obstruction of urine flow through the kidneys, bladder, or urethra. Tumors of the bladder or prostate gland, prostatic hypertrophy, nephrolithiasis (kidney stones), and severe infections are examples of postrenal azotemia.[3] Box 11-3 ■ has a more complete listing of prerenal, renal, and postrenal causes of azotemia.

Decreased BUN is found in five conditions: (1) decreased protein intake, (2) severe liver disease, (3) overhydration, (4) in the third trimester of pregnancy (because of increased plasma volume), and (5) a syndrome of inappropriate antidiuretic hormone secretion (SIADH).[4] In liver failure, the Krebs and Henselheit cycle is no longer effective, so BUN is not synthesized. During pregnancy, the fetus is using maternal

BOX 11-3   Prerenal, Renal, and Postrenal Azotemia

Prerenal azotemia
► Congestive heart failure
► Dehydration
► Shock (from blood loss)
► Advanced cirrhosis
► Septic states
► Increased protein catabolism
  • Muscle wasting (starvation)
  • Gastrointestinal hemorrhage
  • Steroids
  • Uncontrolled diabetes mellitus
  • High fever

Renal Azotemia
► Uremia
► Uremic syndrome
► Glomerulonephritis
► Nephrotic syndrome

Postrenal Azotemia
► Tumors of the bladder or prostate gland
► Prostatic hypertrophy
► Gynecologic tumors
► Nephrolithiasis
► Severe infections

amino acids, so less protein is available to be catabolized to BUN.

The **BUN:creatinine (BUN:CR) ratio** can be used to distinguish between the three major types of azotemia. The normal ratio is between 12:0 and 20:1. In renal disease, because the BUN and CR are both elevated proportionally, the ratio will fall within the normal range. A high ratio of >20:1 to 30:1 with a high BUN and a normal or only slightly elevated CR is associated with prerenal azotemia.[1] High ratios with an elevated CR suggests postrenal obstruction (azotemia) or prerenal azotemia in addition to renal disease. For example, a patient who has a BUN of 45 mg/dL and a CR of 1.3 mg/dL would have a ratio of 35 (45/1.3) (a moderately elevated BUN and only slightly elevated CR), which would be indicative of prerenal azotemia—for example, from congestive heart failure. A ratio >36 is suggestive of upper gastrointestinal bleeding.

Decreased ratios are much less common but can be seen in renal dialysis patients because BUN is more dialyzable than CR. Other conditions associated with a low ratio are acute tubular necrosis, low-protein diets, starvation, severe diarrhea, vomiting, SIADH, and severe liver disease.

BUN Methodologies

BUN methodologies can be categorized as enzymatic urease reactions and the diacetyl or Fearon reaction. The most common BUN methodologies utilize **urease** in the initial reaction.

$$Urea + 2H_2O \xrightarrow{Urease} 2NH_4^+ + CO_3^{-2} \quad (11.1)$$

The reaction then proceeds to Bertholot's reaction, Nessler's reaction, or glutamate dehydrogenase.

In *Berthelot's reaction,* the ammonium ion is reacted with phenol and hypochlorite in an alkaline medium to form indophenol blue, the chromagen that is measured. Sodium nitroprusside is added to catalyze the reaction.

$$NH_4^+ + 5NaOCl + Phenol \xrightarrow[\text{Na nitroprusside}]{NaOH}$$

$$Indophenol\ blue + 5NaCl + 5H_2O \quad (11.2)$$

In *Nessler's reaction,* the addition of a double iodide compound $(2HgI_2 + 2KI)$ results in the formation of a yellow to orange-brown compound with NH_4^+.

$$2HgI_2 + 2KI + NH_4^+ \longrightarrow NH_2Hg_2I_3 + 4KI + NH_4I \quad (11.3)$$

The *glutamate dehydrogenase* procedure is the most commonly used. The disappearance of NADH is measured as a decrease in absorbance as NADH is oxidized to NAD^+.

$$NH_4^+ + 2\text{-oxoglutarate} + NADH \xrightarrow{\text{Glutamate dehydrogenase}}$$

$$NAD^+ + Glutamate + H_2O \quad (11.4)$$

The *diacetyl* or *Fearon reaction* is a colorimetric reaction based on the condensation of diacetyl with urea to form the chromogen diazine.

$$Urea + Diacetyl + H_2O \xrightarrow[\text{Strong acid}]{H^+} Diazine + 2H_2O \quad (11.5)$$

The reaction of diacetyl and urea results in a diazine derivative that absorbs strongly at 540 nm. Thiosemicarbazide and Fe(III) ions are added to stabilize and enhance the reaction.

BUN Reference Range

Blood urea nitrogen is the term still used, although laboratories no longer test whole blood, and in the United States it is reported out as *urea nitrogen* not *urea.* The reference range is 7–18 mg/dL. BUN is stable at room temperature for 24 hours, several days at refrigerator temperature (4°C), and 2 to 3 months frozen. A high-protein diet increases BUN, although a single meal would have negligible effect. Although BUN is slightly higher in males, the gender difference is not clinically significant.

A BUN of <8–10 mg/dL probably results from overhydration. A BUN between 50–150 mg/dL is clearly abnormal and beyond variation from urine flow or nitrogen load and indicates impairment of glomerular filtration rate. A BUN from 150–250 mg/dL is conclusive evidence of severe renal impairment.

It is useful to know how to convert from urea to urea nitrogen (BUN) and vice versa (to understand journal articles in European publications, for example). The conversion factor is calculated using the molecular weight of BUN (60) and nitrogen (28). To convert from urea to urea nitrogen, divide the molecular weight of nitrogen by the molecular weight of urea (28/60) for a factor of 0.467. Therefore, a urea of 56 mg/dL × 0.467 is equivalent to a urea nitrogen (BUN) of 26 mg/dL (rounding off to the nearest whole number).

To convert urea nitrogen to urea, the factor is (60/28) or 2.14.

✓ Checkpoint! 11–1

1. *A journal article reports a urea nitrogen of 10 mg/dL. What would be the equivalent urea concentration?*
2. *Classify the following as prerenal, renal, or postrenal azotemia.*
   a. *Dehydration* _____
   b. *Glomerulonephritis* _____
   c. *Congestive heart failure* _____
   d. *Nephrolithiasis* _____
   e. *Shock* _____
3. *A BUN:CR ratio of 15 with a moderately elevated BUN and creatinine can be classified as prerenal, renal, or postrenal azotemia?*

▶ CREATININE

Creatine is synthesized in the liver from three amino acids: arginine, glycine, and methionine. **Creatinine (CR)** is a waste product derived from creatine and creatine phosphate. Creatine is produced when creatine phosphate (phosphocreatine) loses a phosphoric acid molecule during the process of muscle contraction, and creatinine is an anhydride formed when creatine loses a water molecule:

$$Phosphocreatine + ADP \xrightarrow{Creatine\ kinase} Creatine + ATP \quad (11.6)$$

$$Creatine \longrightarrow Creatinine + H_2O \quad (11.7)$$

The constancy of endogenous creatinine production is proportional to the muscle mass of the individual, and creatinine is released into the body fluids at a constant rate. The constant plasma levels over a 24-hour (24h) period makes creatinine a good endogenous substance to use as an indicator of glomerular filtration. Creatinine is readily filtered by the glomeruli and does not undergo any significant tubular reabsorption.

Creatinine levels are affected by three main variables: relative muscle mass, creatine turnover, and renal function.

CREATININE CLINICAL SIGNIFICANCE

Creatinine is primarily an index of renal function and measures the GFR. Increased serum creatinine is present when the formation or excretion of urine is impaired because of prerenal, renal, or postrenal causes. Creatinine values are usually not above the upper reference limit until one-half to two-thirds of renal function is lost.

Low values are not clinically significant. Serum CR levels and urinary CR excretion are functions of muscle mass in normal individuals and show little response to dietary changes. High levels of CR in urine is unique to the kidney. CR levels in urine can be used to detect if a urine has been diluted (e.g., drug testing). CR and BUN levels can be utilized to identify a fluid as urine because urine has the highest concentrations of these two substances.

CREATININE METHODOLOGIES

The **Jaffe reaction** is a reaction between creatinine and picric acid in an alkaline medium, yielding a red-orange compound of creatinine and a picrate ion. Concentration or alkalinity of reagent is critical in the Jaffe reaction. Interfering substances include protein, glucose, uric acid, ascorbic acid, acetone, ketoacids, and medications (cephalosporins and other antibiotics).

A kinetic method was developed to reduce the affect of interfering substances. The absorbance at 520 nm is measured between 20 and 80 seconds. Certain interfering substances—for example, acetoacetate—react faster and others such as protein are slower; therefore, creatinine is the main reactant between 20 and 80 seconds.

An enzymatic methodology using **creatininase (creatinine amidohydrolase)** is a recent advance. Creatininase is followed by creatinase, sarcosine oxidase, and peroxidase reactions.

$$\text{Creatinine} + H_2O \xrightarrow{\text{Creatininase}} \text{Creatine} \quad (11.8)$$

$$\text{Creatine} + H_2O \xrightarrow{\text{Creatinase}} \text{Sarcosine} + \text{Urea} \quad (11.9)$$

$$\text{Sarcosine} + H_2O + O_2 \xrightarrow{\text{Sarcosine Oxidase}} \text{Glycine} + \text{Formaldehyde} + H_2O_2 \quad (11.10)$$

$$H_2O_2 + \text{Reduced indicator} \xrightarrow{\text{Peroxidase}} \text{Oxidized indicator} + H_2O \quad (11.11)$$

Creatininase may become method of choice with less interference, more accuracy, and greater precision.

A second enzymatic procedure also uses creatinase, followed by creatine kinase, pyruvate kinase (PK), and lactate dehydrogenase (LD) (creatinase → CK → PK → LD).

$$\text{Creatinine} + H_2O \xrightarrow{\text{Creatinase}} \text{Creatine} \quad (11.12)$$

$$\text{Creatine} + \text{ATP} \xrightarrow{\text{Creatine kinase}} \text{Creatine phosphate} + \text{ADP} \quad (11.13)$$

$$\text{ADP} + \text{Phosphoenolpyruvate} \xrightarrow{\text{Pyruvate kinase}} \text{ATP} + \text{Pyruvate} \quad (11.14)$$

$$\text{Pyruvate} + \text{NADH} \xrightarrow{\text{Lactate dehydrogenase}} \text{Lactate} + \text{NAD}^+ \quad (11.15)$$

This procedure requires a larger sample size and is not as popular as the peroxidase reaction. The National Kidney Disease Education Program has made recommendations for improving serum creatinine measurements.[5]

Creatinine Reference Range
The creatinine reference range for men is 0.9–1.2 mg/dL, and slightly lower for women (0.6–1.1 mg/dL) because of lower muscle mass. Intraindividual variation is small, and creatinine is not affected by diet unless it ultimately affects muscle mass.

► URIC ACID

Uric acid is the major product of nucleoprotein catabolism in humans and higher primates. The breakdown of adenine and guanine, purine nucleosides found in nucleic acids (DNA and RNA), results in the formation of uric acid . Adenine and guanine are catabolized to xanthine, and uric acid is produced in the liver from xanthine by the action of the enzyme xanthine oxidase.

$$\text{Xanthine} \xrightarrow{\text{Xanthine oxidase}} \text{Uric acid} \quad (11.16)$$

Uric acid is formed from both exogenous (dietary) and endogenous nucleotides, but most uric acid is derived from endogenous nucleic acids, cells breaking down, and cells being replaced. Seventy percent of uric acid is excreted in the urine, and the remainder is degraded by bacteria in the gastrointestinal tract.

In lower mammals, uric acid is further reduced by uricase to form allantoin, which is water soluble and mammals' major product of purine catabolism.

$$\text{Uric acid} \xrightarrow{\text{Uricase}} \text{Allantoin} \quad (11.17)$$

URIC ACID CLINICAL SIGNIFICANCE

Hyperuricemia is defined as a serum or plasma uric acid concentration of >7.0 mg/dL in men and >6.0 mg/dL in women. Causes of hyperuricemia are divided into four categories: increased dietary intake, overproduction of uric acid, underexcretion of uric acid, and specific enzyme defects. Most are caused by a combination of overproduction and underexcretion of uric acid.

Primary gout is associated with overproduction and essential hyperuricemia. It is an inborn error of metabolism found predominantly in men 30 to 50 years of age and is 7 times more common in men than women. Symptoms include arthritis (pain, inflammation of the joints), nephropathy, and nephrolithiasis.[6]

The patient often presents with an inflamed big toe, usually the first joint affected in gout. **Gout** occurs when monosodium urate precipitates in supersaturated body fluids;

deposits of urates are responsible for the symptoms. Acute attacks of gout can be precipitated by alcohol, high-protein diets, stress, acute infection, surgery, and certain medications. Treatment of gout includes a diet adequate but not high in protein, no alcohol, normal weight maintenance, and drug therapy if required. Patients should avoid or limit organ meats or meats with a high purine content (e.g., liver, kidney, salmon, haddock, scallops, and sardines). Allopurinol is one of the medications that inhibits xanthine oxidase, resulting in decreased levels of serum and urine uric acid.[6]

Other causes of increased uric acid production include cytotoxic chemotherapy and radiation therapy in cancer patients, which results in increased cell destruction and increased endogenous nucleic acids.

Secondary hyperuricemia can be attributed to several causes, including renal retention of uric acid in acute or chronic renal disease and renal failure. Toxemia of pregnancy, severe exercise, poisons, and drug therapy are examples of other conditions associated with secondary hyperuricemia.

Enzyme deficiencies including **Lesch-Nyhan syndrome** can also result in hyperuricemia. In Lesch-Nyhan, a deficiency of hypoxanthine-guanine phosphoribosyl transferase (HGPRT) produces an elevated uric acid. It is an X-linked genetic disorder characterized by mental retardation, abnormal muscle movements, and behavioral problems such as pathological aggressiveness and self-mutilation. See Box 11-4 ■ for a more complete list of conditions associated with hyperuricemia.

Hypouricemia, defined as a uric acid <2.0 mg/dL, is much less common than hyperuricemia. It can be secondary to severe hepatocellular disease or defective renal tubular reabsorption of uric acid (congenital or acquired, Fanconi's syndrome, or Wilson disease).

URIC ACID METHODOLOGIES

The two most common uric acid methodologies are phosphotungstic acid (Carraway) and uricase. **Phosphotungstic acid (PTA)** measures the development of a blue color (tungsten blue) when phosphotungstic acid is reduced by uric acid in an alkaline medium. Sodium carbonate is added to maintain the alkaline pH.

$$\text{Uric acid} + \text{Phosphotungstic acid} \xrightarrow{\text{Na}_2\text{CO}_3}$$

$$\text{Allantoin} + CO_2 + \text{Tungsten blue}   (11.18)$$

Interference by substances that reduce PTA is similar to those that affect the Jaffe creatinine reaction. They include endogenous compounds (glucose and ascorbic acid) and exogenous compounds (acetaminophen, acetylsalicylic acid, and caffeine). **Uricase** is more specific, does not require a protein-free filtrate, and is used by 99% of the labs. Uricase catalyzes the oxidation of uric acid to allantoin.

$$\text{Uric acid} \xrightarrow{\text{Uricase}} \text{Allantoin} + H_2O_2 + CO_2   (11.19)$$

The decrease in absorbance is measured at 293 nm, which is a peak absorbance for uric acid and one at which allantoin does not absorb. The decrease in absorbance is thus inversely proportional to the uric acid concentration; the lower the absorbance, the higher the uric acid concentration.

A second indicator reaction utilizing peroxidase and a dye (4-aminoantipyrene) is a second modification and the most common automated method.

$$H_2O_2 + \text{4-Aminoantipyrine} \xrightarrow{\text{Peroxidase}} \text{Chromagen}   (11.20)$$

Ascorbate is added to minimize ascorbic acid interference.

Uric Acid Reference Range

The reference range for the uricase methodology is 3.5–7.2 mg/dL for males and 2.6–6.0 mg/dL for females. Uric acid is susceptible to bacterial action, so specimens should be refrigerated. Increased hemolysis and bilirubin decrease uric acid using the peroxidase reaction. Uric acid is not affected by recent diet, but it can be influenced by a long-term purine-rich diet.

► ANALYTICAL PROCEDURES FOR ASSESSMENT OF GLOMERULAR FILTRATION

CLEARANCE TESTS

Renal clearance is defined as the rate at which the kidneys remove a substance from the plasma or blood or a quantitative expression of the rate at which a substance is excreted by the kidneys in relation to the concentration of the same substance in the plasma, which is usually expressed as mL cleared per minute. The best markers are freely filtered by the glomeruli, produced at a constant rate, not reabsorbed or secreted by tubules, and present in stable concentrations in the plasma. They have an inexpensive and rapid assay for detection.[7]

Creatinine Clearance

Creatinine clearance (CrCl) is the most popular and practical method for estimating the GFR. It is easily measured,

---

BOX 11-4 Hyperuricemia

Primary Hyperuricemia
► Gout
► Cytotoxic chemotherapy
► Radiation therapy (leukemia, lymphoma)
► Malignancy (cancer)

Secondary Hyperuricemia
► Acute or chronic renal disease, renal failure
► Glycogen storage disease
► Toxemia of pregnancy
► Severe exercise
► Poisons (lead, alcohol)
► Drug therapy (diuretics, salicylates)
► Lesch-Nyhan syndrome

and extensive data is available for all age groups. Creatinine is a very good indicator of glomerular filtration rate for three reasons. First, it is freely filtered by the glomeruli. Second, it is not reabsorbed by the tubules to any significant extent. Third, creatinine is released into the plasma at a constant rate, resulting in constant plasma levels over 24 hours. A linear decrease in creatinine clearance over time as renal function fails has been documented for different diseases (e.g., chronic glomerulonephritis).

Creatinine levels are measured on serum and urine specimens. A 24-hour urine is usually collected. The CrCl is calculated using the serum and urine creatinine levels and the urine volume.

$$\text{Clearance (X)} = \frac{U \times V}{P} \qquad (11.21)$$

U = urine concentration in mg/dL

P = plasma concentration in mg/dL

V = urine flow in mL/ minute (1440 min/24h) $\longrightarrow$

$$\frac{\text{24h volume (mL/day)}}{\text{1440 min/day}}$$

The volume of urine in mL/minute is calculated by divided the 24h volume in mL/day by 1440 minute/day, which results in mL/min. Plasma creatinine is inversely proportional to the clearance: The higher the plasma creatinine, the lower the clearance.

Creatinine clearance has to be corrected to an adult body surface area (BSA) of 1.73 m², which is especially important for small or pediatric patients and obese patients. This can be done in two ways using the Dubois formula or a nomogram. The Dubois formula is:

SA (surface area in m²)

$$= W \text{ (kg)}^{0.425} \times H \text{ (cm)}^{0.725} \times 0.007184 \qquad (11.22)$$

The nomogram is much easier. See Appendix I.
The correction or normalization factor for BSA is added to the CrCl equation.

$$\frac{(U \times V)}{P} \times \frac{1.73 \text{ m}^2}{\text{BSA m}^2 \text{ (normalization factor)}}$$

$$= \text{CrCl (mL/minute/1.73 m}^2) \qquad (11.23)$$

The corrected CrCl for a larger or obese individual will be lower because the normalization factor will be less than one. For a pediatric patient or small adult, the corrected CrCl will be increased.

The patient should not drink any caffeinated beverages (tea, coffee) the day of the test and urine collection. The clearance can be performed on a 4-, 12-, or 24h urine with the important variable being a well hydrated patient. The patient should drink 500 mL of water 10 to 15 minutes before the test to ensure proper hydration and a urine flow rate of more than 2 mL/min, which is critical to an accurate CrCl.

The patient's instructions should stress the importance of complete collection because the largest source of error is incomplete urine collection. The bladder should be completely emptied before timing is initiated. Vigorous exercise and muscle mass (as discussed under serum creatinine) can affect creatinine values. Proteinuria also results in increased creatinine clearance. The reference range for males is 97–137 mL/minute; for females, 88–128 mL/minute. Creatinine clearance decreases with age approximately 1 mL/min/year. Increased CrCl is not clinically significant and probably an error in specimen collection. Decreased CrCl indicates decreased GFR as a result of acute or chronic damage to the glomeruli. Mild renal impairment is indicated by a CrCl of 50–79 mL/min, moderate 10–49 mL/min, and severe <10 mL/min.

✓ Checkpoint! 11–2

1. *Why is creatinine clearance the most widely used test for estimating the glomerular filtration rate?*
2. *List three reasons why creatinine is described as a good indicator of the glomerular filtration rate.*
3. *What is the most common source of error in calculating the creatinine clearance?*
4. *A creatinine clearance was ordered on an obese patient with kidney disease.*

   Weight: 350 lb
   Height: 5'4''
   24-hour urine volume: 1850 mL
   Plasma creatinine: 6.5 mg/dL
   Urine creatinine: 120 mg/dL
   BSA: 2.30 m²
   What is this patient's creatinine clearance?

---

eGFR

Many studies have recommended using an estimating or prediction equation—eGFR—to estimate glomerular filtration rate from the serum creatinine level in patients with chronic renal disease and those at risk for chronic kidney disease (CKD)—for example, those with diabetes, hypertension, cardiovascular disease, or family history of kidney disease.[8] The primary reasons are that GFR and creatinine clearance are not as accurate as using creatinine alone, and creatinine is more often measured than urinary albumin. Also, the Modification of Diet in Renal Disease (MDRD) Study equation has been thoroughly validated and is superior to other methods of approximating GFR. It does not require weight or height variables because it is normalized to 1.73 m² body surface area, which is the accepted BSA as discussed in the section on creatinine clearance.[7]

In patients 18 years of age and older, the MDRD equation is the best means currently available to use creatinine values as a measure of renal function. The equation has been validated in the Caucasian and African American populations with impaired renal function (eGFR < 60 mL/min/1.73 m^2) between the ages 18 and 70. It requires four variables: serum or plasma creatinine, age in years, gender, and race (African American or not).[8]

Two versions of the equation are available, depending on whether or not the serum creatinine has been calibrated to be traceable using an isotope dilution mass spectrometry (IDMS) method. The following MDRD equations are recommend by the National Kidney Disease Education program[9]:

Original MDRD Study equation
When S_{cr} is in mg/dL (conventional units):

$$\text{eGRF (mL/min/1.73 m}^2) = 186 \times (S_{cr})^{-1.154} \times (\text{Age})^{-0.203} \times$$
$$(0.742 \text{ if female}) \times (1.210 \text{ if African American}) \qquad (11.24)$$

IDMS-traceable MDRD Study equation
When S_{cr} is in mg/dL (conventional units):

$$\text{eGRF (mL/min/1.73 m}^2) = 175 \times (S_{cr})^{-1.154} \times (\text{Age})^{-0.203} \times$$
$$(0.742 \text{ if female}) \times (1.210 \text{ if African American}) \qquad (11.25)$$

Laboratories have elected to routinely report both the eGRF and the creatinine.

PROTEIN:CREATININE RATIO

Proteinuria as determined by a 24h urine protein and creatinine clearance is subject to several sources of error, including incompleteness of collection, as discussed in a previous section. A protein:creatinine ratio for urine collected over a shorter period of time (e.g., 4h or random urine) has been determined as a result of many studies to predict the presence of significant proteinuria. Normal protein excretion is <100–150 mg/24h, and creatinine excretion is fairly constant at 15–20 mg of creatinine per kilogram of body weight per day. A normal protein:creatinine ratio is <0.1 (100–150 mg protein/1000–1500 mg creatinine). A metaanalysis of 16 studies by Price et al.[10] concluded that a protein:creatinine ratio performed on a random urine has a strong correlation with 24h protein excretion.

INULIN CLEARANCE

Inulin, an endogenous, naturally occurring polysaccharide found in artichokes, is the gold standard for measuring glomerular filtration rate. Inulin is injected and measured in serum and urine in the appropriate time frame. Major disadvantages are that, unlike creatinine, there is no easily performed methodology for inulin, and it an invasive procedure requiring injection of inulin.

CYSTATIN C

Cystatin C, a single-chain, nonglycosylated, low-molecular-weight protein synthesized by all nucleated cells, is a cysteine protein inhibitor. Its most important characteristics—small

size and high isoelectric point (pI = 9.2)—enable it to be freely filtered by the glomeruli and catabolized in the proximal convoluted tubules. Cystatin C is produced at a constant rate, and serum concentrations are not affected by muscle mass, diet, race, age, or gender.[11]

The most practical cystatin C methodologies for use in the clinical laboratory are latex particle–enhanced turbidimetric and nephelometric immunoassay.

Studies comparing cystatin C and creatinine have found that cystatin C is superior to creatinine for the detection of renal disease, and it is especially useful in detecting mild to moderate impairment of renal function. Serum cystatin C levels increase when the GFR is less than 88 mL/min, while creatinine remains normal until the GFR decreases to 75 mL/min.[11] In a study by Villa et al.,[12] of 25 patients with renal dysfunction, five (20%) had elevated serum creatinine levels whereas 19 (76%) had elevated cystatin C.

▶ SCREENING FOR RENAL DISEASE

SERUM PROTEIN

Severe renal disease is characterized by decrease in total protein, especially the smaller-molecular-weight proteins such as albumin. Both serum total protein and albumin levels therefore will be decreased in some renal diseases, which will be discussed later in this chapter. In nephrotic syndrome, a typical protein electrophoresis will be characterized by a decrease in albumin, α_1-globulin, and gamma globulin and an increase in α_2-globulin (due to increased α_2-macroglobulin). Chronic renal disease is associated with a similar electrophoretic pattern with a decreased total protein, albumin, α_1-globulin, and gamma globulin.

β_2-MICROGLOBULIN

β_2-microglobulin (BMG) is a small, nonglycosylated protein found on the cell membrane of most nucleated cells, and it is present in especially high levels in lymphocytes. β_2-microglobulin is increased in renal failure, and the plasma level of BMG is a good indicator of renal tubular function.[13,14] It is used primarily to test for renal tubular function in renal transplant patients when decreased tubular function indicates early rejection. BMG has been found in some studies to be an earlier indicator of rejection than CR because is it not influenced by lean muscle mass or variations in daily excretion. (See Chapter 8, Amino Acids and Proteins, for additional information.)

LOW-MOLECULAR-WEIGHT PROTEINS

Low-molecular-weight proteins such as α_1-microglobulin, α_2-microglobulin, β-trace protein, and cystatin C are cleared from the plasma by glomerular filtration and can be considered freely filtered at the glomerular filtration barrier.

These proteins are filtered and metabolized in the proximal convoluted tubules and eliminated in the urine. A disadvantage of all of the proteins listed, except for cystatin C, is the influence of nonrenal factors, including inflammation and liver disease. Cystatin C, however, appears to be more specific for measuring the GFR than the previously discussed proteins and creatinine clearance.

URINALYSIS

A routine urinalysis may be the first indication of renal disease. The *appearance* (color, turbidity) of the urine can provide valuable clues. For example, an amber or darker yellow urine is indicative of a more concentrated urine or the presence of bilirubin or both, pink or red is associated with hemoglobin or myoglobin, and a white foam is suggestive of proteinuria.

The *reagent strip tests* provide valuable information regarding renal function. The cellulose pads on a plastic strip are impregnated with various reagents with as many as 10 pads or tests measuring different constituents on one strip. The reader is referred to several excellent urinalysis textbooks in the reference list at the end of this chapter for more detailed information.

The *protein* test mat utilizes tetrabromphenol blue and is based on the "protein error of indicators." Proteinuria is often the first symptom of kidney disease. Many biological variables affect protein excretion, including upright position, exercise, fever, heart failure, and, of course, kidney disease. Urine protein is discussed in more depth in Chapter 8, Amino Acids and Proteins.

Reagent strip methodologies that are more specific for albumin, myoglobin, and other low-molecular-weight proteins are available. Chromogenic reaction of albumin with bis-(3′,3′-diiodo-4′,4′-hydroxy-5′,5′-dinitrophenyl)-3,4,5,6-tetrabromosulonephthalein (DIDNTB) at pH 1.5 produces a measurable color reaction.[15]

Hemoglobin testing is also part of the routine urinalysis, with the possible sources of hemoglobin being glomerular, tubulointerstitial, or postrenal. The latter two causes are the most common. The detection of hemoglobin is based on the peroxidase activity of hemoglobin using a peroxide substrate and an oxygen acceptor—for example, tetramethyl benzidine, a buffer, and an organic peroxide. The color change (depending on the oxygen acceptor) is usually from orange to pale to dark green, depending on the free red blood cells or hemoglobin present. Free hemoglobin will result in a uniform green color, and intact red cells will produce a speckled pattern on the test mat.[15]

Red cells or hemoglobin in the urine are indicative of renal or bladder disease. Hematuria indicates trauma or irritation, including from (1) renal disease (e.g., glomerulonephritis, pyelonephritis), (2) cystitis, (3) tumor, (4) trauma, (5) exposure to toxic chemicals or drugs, (6) strenuous exercise, (7) acute febrile episodes, (8) appendicitis, (9) smoking and (10) menstrual contamination. Hemoglobinuria may also be associated with (1) hemolytic transfusion reaction (HTR), (2) hemolytic anemia, and (3) severe burns.[15]

The presence of *leukocytes* is usually associated with a bacterial infection in the bladder or kidney. The reagent strip reaction utilizes the hydrolysis of an indoxylcarbonic acid ester to indoxyl and a diazonium salt, resulting in the production of a purple azo dye.

Nitrite detection is used to determine the presence of nitrate-reducing bacteria in the bladder or kidney, cystitis or pyelonephritis, respectively. The presence of nitrites and leukocyte esterase is diagnostic of a bacterial infection.

Specific gravity provides an indication of the kidney's ability to selectively reabsorb chemicals and water from the glomerular filtrate, one of the body's most important functions. Specific gravity is the density of a substance (urine) compared with the density of a similar volume of distilled water at a similar temperature. It is a measurement of the density of dissolved particles in the urine as influenced by the size and number of particles. As renal disease progresses and more nephrons are damaged, the specific gravity becomes isothenuric or fixed at 1.010. At this point, the kidney is no longer able to dilute or concentrate urine.

Reagent strip specific gravity is measured using a polyelectrolyte (methylvinyl ether or malic anhydride), a pH indicator (bromthymol blue), and an alkaline buffer. Ionic solutes cause H^+ ions to be released by the polyelectrolyte, decreasing the pH of the test pad causing it to become more acid. As the pH decreases, the indicator changes color from blue to green to yellow, indicating low to high specific gravities, respectively.

MICROALBUMIN

Albumin may be the first sign of renal disease and is acknowledged to be an independent predictor of nephropathy in diabetes mellitus.[16] The National Institute of Diabetes and Digestive and Kidney Diseases recommends a microalbumin on a spot, untimed urine. Using the ratio of urine albumin to creatinine, **microalbuminuria** is defined as a level of 30–300 mg albumin/g creatinine.[17] See Chapter 6, Carbohydrates, for additional information.

▶ RENAL PATHOPHYSIOLOGY

GLOMERULAR DISEASE

Glomerular diseases are associated primarily with damage to the glomeruli of the renal nephron. Various renal diseases are characterized by distinctive chemistry and urinalysis profiles. The reader is referred to the excellent urinalysis textbooks in the references at the end of this chapter for more in-depth information.

GLOMERULONEPHRITIS

Acute glomerulonephritis (AGN) is characterized by a rapid onset of symptoms that indicate damage to the glomeruli. It is most often associated with children and young adults following a group A streptococcal infection (e.g., strep throat or respiratory infection). AGN is not caused by a bacterial infection but by circulating immune complexes (antigen–antibody) that trigger an inflammatory response in the glomerular basement membrane. The precipitation of immune complexes results in glomerular damage. Nonstreptococcal glomerulonephritis can be caused by other bacteria, viruses, parasites, and chemicals that are toxic to the kidney. Systemic diseases (e.g., systemic lupus erythematosus, SLE) are also associated with glomerulonephritis.[18]

Symptoms include rapid onset, fever, malaise, nausea, oliguria, hematuria, and proteinuria. Secondary complications include edema, especially periorbital (around the eyes) and of the knees and ankles; hypertension from mild to moderate; and electrolyte imbalance (Na^+ and K^+).[18]

The urinalysis exhibits marked hematuria, increased protein, and dysmorphic RBCs as well as hyaline, red blood cells (RBCs), and granular casts in the sediment. RBC casts are pathognomonic for glomerulonephritis. Renal function tests are abnormal with elevated BUN and creatinine and decreased GFR. As toxicity subsides, the urinalysis and renal function tests return to normal.[18]

Chronic glomerulonephritis is associated with the end stage of persistent glomerular damage with irreversible loss of renal tissue and chronic renal failure. Eighty percent of patients have had some other form of glomerulonephritis, and 20% are unrecognized or have subclinical symptoms. Symptoms include edema (usually the presenting symptom), fatigue, hypertension, anemia, metabolic acidosis, proteinuria, and decreased urine volume from oliguria to anuria. Death from uremia can occur without dialysis or renal transplant.[19]

NEPHROTIC SYNDROME

Nephrotic syndrome may occur as a complication of glomerulonephritis or as a result of circulatory disorders that affect blood pressure and flow of blood to the kidney—for example, diabetes mellitus and lupus erythematosus. Nephrotic syndrome can also be associated with a variety of conditions: minimal change nephropathy, focal segmental glomerulosclerosis (FSGS), carcinoma, drugs, and infection. A change in the permeability of glomerular membrane allows high-molecular-weight proteins and lipids into the glomerular filtrate, resulting in glomerular and tubular damage. Symptoms include massive proteinuria (>3 g/day), albuminuria (>1.5 g/day), pitting edema, hyperlipidemia, and hypoalbuminemia. Low albumin and high lipids are hallmarks of nephrotic syndrome.

The urinalysis is characterized by marked proteinuria; hematuria; a sediment containing urinary fat droplets, oval fat bodies, and renal tubular cells; and renal epithelial, waxy, and fatty casts. The acute symptoms are shock and decreased renal blood flow.

TUBULAR DISEASE

Tubular disease affects the renal tubules and often occurs during the progression of renal disease and decreasing GFR. It results in decreased reabsorption and excretion of certain substances and a reduction in the concentrating function of the kidney.

PYELONEPHRITIS

Acute pyelonephritis is most often seen in women as a result of untreated cases of cystitis or lower urinary tract infections. It is an ascending infection and usually does not cause permanent damage to the renal tubules. **Pyelonephritis** is an inflammatory process involving a bacterial infection of the renal tubules by gram negative bacteria—for example, *Escherichia coli*, *Klebsiella*, *Proteus*, and *Enterobacter*. It is associated with an acute onset with presenting symptoms of urinary frequency and burning and lower back pain.[20]

The causes or conditions associated with pyelonephritis are incomplete emptying of bladder, vesicoureteral reflux, diabetes mellitus, pregnancy, urinary obstruction, and catheterization. Vesicoureteral reflux is the flow of urine up the urethra during urination instead of downward to the bladder. This allows bacteria to migrate upward, resulting in frequent urinary tract infections. The urinalysis results will indicate positive leukocyte esterase and nitrite, positive proteinuria, and decreased specific gravity. The sediment exhibits white blood cells (WBCs), often in clumps; WBC casts; microscopic hematuria; and granular, renal tubular, and waxy casts. It may resemble cystitis except for the presence of casts, especially WBC casts.[20]

Chronic pyelonephritis causes permanent scarring of renal tubules and can lead to renal failure. The most common cause is vesicoureteral reflux nephropathy described in the previous section. From 10% to 15% of chronic pyelonephritis patients progress to chronic renal failure and end-stage renal disease (ESRD).

The urinalysis in chronic pyelonephritis exhibits alkaline pH, positive leukocyte esterase and bacteria (+/−), proteinuria (moderate <2.5 g/day), decreased specific gravity, and granular, waxy, broad, WBC, and renal cell casts in the sediment. Polyuria and nocturia are also characteristic of chronic pyelonephritis.

CYSTITIS

Cystitis is a bladder infection characterized by dysuria, which is more common in females than males because of the shorter urethra in women. It is caused by the same intestinal flora associated with pyelonephritis and can progress to pyelonephritis if untreated or if treatment guidelines have not been adhered to (antibiotic therapy).

The urinalysis indicates small protein (<0.5 g/day), hematuria, leukocyte esterase and nitrite positive, and increased WBCs, bacteria, and transitional epithelial cells. The absence of cellular (WBC or renal epithelial) casts differentiates cystitis (lower urinary tract infection) from pyelonephritis (upper urinary tract infection).

CHRONIC KIDNEY DISEASE

The number of people with kidney failure requiring dialysis is expected to double to 630,000 by 2010. CKD is underdiagnosed and undertreated, but if it is diagnosed and treated early, the rate of progression to renal failure can be delayed or stopped.[13] The National Institute of Diabetes and Digestive And Kidney Diseases defined CKD as either kidney damage or a GFR <60/min (1.73 m^2) for three months or more.[21]

Chronic kidney disease is associated with hypertension and diabetes. Autoimmune diseases (e.g., lupus), urinary tract and systemic infections, nephrolithiasis, and some medications are also risk factors for CKD. The National Kidney Foundation's Kidney Disease Outcomes Quality Initiative (K/DOQI) developed guidelines that are summarized in Table 11-2 ✪ and can be found in its entirety at www.kidney.org/PROFESSIONALS/kdoqi/guidelines.cfm. Recommended screening tests are basic metabolic profile, calculated GFR, urinalysis, and albumin- or protein-to-creatinine ratio on a spot (random) urine.[21]

RENAL FAILURE

Acute renal failure is an acute increase in the serum creatinine level of 25% or more and a GFR of less than 10 mL/min. The incidence and prevalence of end-stage renal disease—kidney failure treated by dialysis and transplantation—have more than quadrupled in the last two decades. Renal failure results from acute tubular necrosis from vasoconstriction, nephrotoxic agents, and hemorrhaging. The urinalysis findings will depend on the cause of renal failure. Prerenal causes include decreased renal blood flow and hypotension, traumatic or surgical shock, burns, transfusion reactions, congestive heart failure, and septicemia. Renal causes are associated with any progressive glomerular, tubular, or vascular disease. Postrenal failure results from a problem in urine flow after it exits the kidney (e.g., obstruction of urine flow by calculi, blood clots, tumors, or an enlarged prostate).

Symptoms of acute renal failure are decreased urine production, oliguria (<400 mL/day), and anuria. Characteristics of acute renal failure are peripheral edema, hypertension, and congestive heart failure. Uremic syndrome or ESRD is diagnosed in a patient with the preceding symptoms and increased BUN and CR levels.

RENAL CALCULI

Renal calculi are solid aggregates of chemical or mineral salts formed in the renal tubules, renal pelvis, ureters, or bladder. They are more common in males than females. Of all stones, 67% of stones are calcium oxalate with or without phosphate, 12% magnesium ammonium phosphate, 8% calcium phosphate, 8% urate (uric acid), 1–2% cystine, and 2–3% mixtures.[22] Four factors influence the formation of kidney stones: (1) increase in concentration of chemical salts as a result of dehydration or increase in salts in the diet (e.g., vegetarians ingest more oxalates), (2) change in urinary pH (more alkaline pH results in less solubility of Ca and oxalate), (3) urinary stasis, and (4) presence of a foreign body. Urine contains a number of promoters of stone formation as well as an equal concentration of inhibitors. When the promoters are increased, stone formation ensues.[23]

Renal calculi can cause ulceration and bleeding. Symptoms include nausea, vomiting and intense pain from the kidney, forward and downward toward the abdomen, genitalia, and legs. Small stones described as <5 mm in diameter can pass spontaneously through the urine as "gravel." Treatment includes allowing stones to pass, lithotripsy (sound waves to break up stones into smaller pieces that can be eliminated), and surgery.[24]

DIABETES MELLITUS

Some 45% of type 1 diabetes mellitus patients develop **diabetic nephropathy** within 15 to 20 years of diagnosis. The damage is primarily to the glomerulus, although the rest of the nephron can be affected as well, and is a result of the hyperglycemic urine filtrate that passes through the nephron. Diabetic nephropathy is diagnosed when a patient

✪ TABLE 11-2			
Classification of CKD and Action Plan			
CKD Stage	**Description**	**GFR (mL/min/1.73 m^2)**	**Action**
At increased risk	Risk factors for CKS are present but without markers of kidney damage	>90	Periodically test for CKD; treat modifiable risk factors for CKS
1	Kidney damage with normal or increased GFR	>90	Diagnose and treat type of kidney disease
2	Kidney damage with mild reduction of GFR	60–89	Adjust drug dosages for level of GFR
3	Moderate reduction of GFR	30–59	Evaluate and treat complications of CKD; avoid nephrotoxic drugs
4	Severe reduction of GFR	15–29	Prepare for kidney-replacement therapy
5	Kidney failure	<15 (or on dialysis)	Start kidney-replacement therapy

Source: Woodhouse S, Batten W, Hendrick H, Malek PA. The glomerular filtration rate: An important test for diagnosis, staging and treatment of chronic kidney disease. *Lab Med* (2006) 37, 4: 244–246.

with diabetes presents with proteinuria with no evidence of a urinary tract infection. Overt nephropathy is characterized by a protein excretion greater than 0.5 g/day or albumin excretion of approximately 300 mg/day. Diabetes is the leading cause of renal failure.[25] See Chapter 6 (Carbohydrates) for further information.

✓ **Checkpoint! 11–3**

1. Which renal disease is associated with each of the following?
   a. Bacterial infection of the renal tubules: _____
   b. Circulating antigen–antibody complexes: _____
   c. Bladder infection: _____
2. What chronic disease is the leading cause of renal failure?

▶ DIALYSIS

Dialysis or kidney transplant may be the only treatment options available for patients with acute renal failure or ESRD. Their kidneys are no longer able to excrete the body's waste products (e.g., BUN), so dialysis is the only treatment available for reducing the levels of toxic products. Dialysis is a process whereby larger macromolecules are separated from low-molecular-weight compounds by their rate of diffusion through a semipermeable membrane.

Many forms of dialysis are available, but all forms use a semipermeable membrane surrounded by a dialysate bath. The dialysate is the fluid used to remove or deliver compounds or electrolytes that the kidney can no longer excrete or retain in the proper concentrations. In the traditional and most common method **hemodialysis (HD)**, the synthetic membrane is in a machine outside the body. The patient is connected to the dialysis machine and his arterial blood and dialysate are pumped at high rates in opposite directions, 150–250 mL/minute and 500 mL/minute, respectively. The patient's blood minus the toxic products is returned to his circulation and the dialysate that contains the waste products is discarded.

In **peritoneal dialysis (PD)**, the peritoneal wall acts as the dialysis membrane and the dialysate is introduced and removed through gravity. Two types are available: (1) continuous ambulatory peritoneal dialysis and (2) continuous cycling peritoneal dialysis. Both require continuous (24 hour a day, 7 days a week) treatment because they are not as effective as hemodialysis.[26]

The main disadvantage of PD is the risk of infection resulting in peritonitis. Peritonitis can cause decrease in fluid removal and ultimately in scarring and fibrosis, which result in permanent loss of filtration.[26]

EVALUATING DIALYSIS PATIENTS

Dialysis patients must be closely evaluated for many factors, including their well-being, cardiovascular risk, nutritional status, and degree of achievable ultrafiltration. Routine laboratory tests consist of hemoglobin, phosphate, and albumin and the clearance of BUN and creatinine.

The simplest calculation for determining the adequacy of dialysis is the **urea reduction ratio (URR)**. It determines the percentage fall in urea during a dialysis session.

$$URR = \frac{\text{Predialysis BUN} - \text{Postdialysis BUN} \times 100\%}{\text{Predialysis BUN}} \quad (11.26)$$

A URR greater than 60% indicates an adequate dialysis.[27]

A more complicated formula developed by the MDRD study is more accurate, but the URR is adequate for screening purposes. Urea kinetic modeling (UKM)—which also considers age, sex, presence of diabetes or cardiovascular disease, and creatinine clearance—is the most widely used method in the assessment of dialysis adequacy. These were found to be independent predictors of GFR.[21] The GFR calculation can be programmed into most laboratory information systems. See the references at the end of the chapter for further information on the MDRD study and formula.[27]

COMPLICATIONS OF DIALYSIS

Patients on dialysis are prone to a number of diseases—for example, cardiovascular disease, hypertension, β_2-microglobulin amyloidosis, and malnutrition. Patients with ESRD are at increased risk of cardiovascular disease: Approximately 40% show evidence of coronary artery disease and 75% for left ventricular hypertrophy.[26,27]

HD patients often exhibit hypertension following dialysis. The treatment is to ensure that the volume of excess fluid be kept to a minimum. This is monitored by setting a "target" or "dry" weight. At the target weight, there should be no apparent edema and blood pressure should be normal.[26]

Dialysis patients with ESRD also have poor appetite, and protein metabolism is decreased because of inflammation and chronic acidosis, which results in malnutrition. Nutritional status is evaluated by weight loss and measurement of albumin. Albumin <3.5 g/dL (bromcresol green) or <3.0 g/dL (bromcresol purple) are indicative of malnutrition or undernutrition.[27]

SUMMARY

Nonprotein nitrogen (NPN), the original test of renal function, comprises products of protein and nucleic acid catabolism that contain nitrogen but are not part of a protein molecule. Blood urea nitrogen (BUN) is the major NPN product of protein catabolism in humans. Ammonia is transported through the portal system to the liver, where it is converted to urea that is less toxic. BUN levels are directly related to the GFR and the level of hydration. In a dehydrated patient, 70% of the BUN is reabsorbed in the renal tubules; therefore, the serum BUN is increased and the urine BUN levels are decreased. The opposite occurs in a normal to well-hydrated patient.

Azotemia is defined as increased blood urea and other NPN compounds. Prerenal azotemia occurs because of a decreased renal blood flow from congestive heart failure, shock, and dehydration, among other causes. Renal azotemia is associated with kidney disease, including glomerulonephritis, nephrotic syndrome, and renal failure. Postrenal azotemia occurs after urine has passed the kidney as a result of obstruction of flow by tumors, nephrolithiasis, and severe infections.

The BUN:creatinine ratio can differentiate between the major types of azotemia. The normal ratio is 12:1 to 20:0, and in renal disease the ratio falls within the normal range but both BUN and creatinine are increased. An increased ratio (20:1 to 30:1) with a high BUN and a normal or only slightly elevated creatinine is associated with prerenal azotemia. A high ratio with an elevated creatinine suggests a postrenal azotemia or a combination of renal and prerenal azotemia.

BUN methodologies include urease reaction and the diacetyl or Fearon reaction. The ammonium ion in the urease reaction is measured using Bertholot's reaction, Nessler's reaction, or glutamate dehydrogenase. BUN levels are affected by a high protein diet over time but are not significantly altered by a single day's increase in protein.

Creatinine is synthesized in the liver from three amino acids: arginine, glycine, and methionine. Creatinine production is proportional to the individual's muscle mass and is released into body fluids at a constant rate. It is primarily an index of renal function and measures the glomerular filtration rate. Creatinine methodologies include the Jaffe reaction (alkaline picrate) and creatinase. Creatinine reference ranges are slightly higher for men than women because of greater muscle mass.

Uric acid is the major product of purine (adenine and guanine) in humans. Sources of uric acid are endogenous from the breakdown of cells in the body and exogenous from dietary protein degradation in the gastrointestinal tract. Causes of hyperuricemia are increased dietary intake, overproduction of uric acid, underexcretion of uric acid, and specific enzyme defects. Primary gout is associated with overproduction and is called *essential hyperuricemia*. Other causes of increased production include chemotherapy and radiation therapy in cancer patients. Decreased excretion is found in renal retention of uric acid in chronic renal disease and renal failure.

Two major uric acid methodologies are phosphotungstic acid (Carraway) and uricase. Phosphotungstic acid is based on the reduction of phosphotungstic acid to tungsten blue by uric acid. Uricase oxidizes uric acid to allantoin, which can be measured by a decrease in absorbance at 293 nm or by using a second indicator reaction peroxidase where a dye is oxidized to a colored chromagen that can be measured.

Creatinine clearance is the most popular method for estimating the GFR. Creatinine is a very good indicator of glomerular filtration rate because it is freely filtered by the glomeruli, is not reabsorbed, and is produced and released into the plasma at a constant rate. Creatinine levels are determined on serum and urine specimens, and the mL of plasma per minute cleared of creatinine is calculated. The clearance has to be corrected for adult body surface area of 1.73 m², especially for small or obese patients.

REVIEW QUESTIONS

LEVEL I

1. Each kidney is composed of more than a million urinary units. Each of these is called a: (Objective 2)
   a. glomerulus.
   b. nephron.
   c. photon.
   d. medulla.

2. The order of blood flow through the nephron is: (Objective 3)
   a. afferent arteriole, peritubular capillaries, vasa recta, efferent arteriole.
   b. efferent arteriole, peritubular capillaries, vasa recta, afferent arteriole.
   c. peritubular capillaries, vasa recta, afferent arteriole, efferent arteriole.
   d. afferent arteriole, efferent arteriole, peritubular capillaries, vasa recta.

3. Substances removed from the blood by tubular secretion include primarily: (Objective 4)
   a. protein, hydrogen, and ammonia.
   b. protein, hydrogen, and potassium.
   c. protein bound substances (i.e. penicillin), hydrogen, and potassium.
   d. amino acid, glucose, ammonia.

LEVEL II

1. Review the following serum test results:

Creatinine	2.5 mg/dL	(0.7–1.5 mg/dL)
Cholesterol	200 mg/dL	(<220 mg/dL)
Glucose	110 mg/dL	(70–110 mg/dL)
Urea (BUN)	40 mg/dL	(8–26 mg/dL)
Uric acid	6.9 mg/dL	(2.5–7.0 mg/dL)

   These results are *most* consistent with: (Objective 18)
   a. impaired renal function.
   b. impaired glucose metabolism.
   c. diagnosis of gouty arthritis.
   d. increased risk of coronary artery disease.

2. A technologist obtains a BUN value of 59 mg/dL and serum creatinine value of 3.1 mg/dL on a patient. These results indicate: (Objective 18)
   a. renal disease.
   b. liver failure.
   c. gout.
   d. prerenal azotemia.

REVIEW QUESTIONS *(continued)*

LEVEL I

4. Many compounds in human serum contain nitrogen. Some of these substances are protein in nature, whereas others are nonprotein nitrogen compounds. What is the compound that comprises the majority of the nonprotein-nitrogen fraction in serum? (Objective 6)
   a. creatine
   b. uric acid
   c. creatinine
   d. ammonia
   e. urea

5. Plasma concentration of urea is significantly elevated by all of the following *except*: (Objective 7)
   a. gastrointestinal bleeding.
   b. exogenous protein catabolism.
   c. dehydration.
   d. renal disease.
   e. severe liver disease.

6. The plasma urea concentration is increased by: (Objective 11)
   a. urinary stasis or dehydration.
   b. a vegetarian diet.
   c. an increase in diuresis (urine production).
   d. an increase in renal blood flow.

7. Postrenal azotemia is found in: (Objective 11)
   a. glomerulonephritis.
   b. congestive heart failure.
   c. high protein diets.
   d. tumors of the bladder obstructing urine flow.

8. Renal azotemia is found in: (Objective 11)
   a. prostate gland tumors.
   b. high protein diets.
   c. uremia or nephrotic syndrome.
   d. hemolytic anemia.

9. Prerenal azotemia is caused by: (Objective 11)
   a. congestive heart failure.
   b. chronic renal failure.
   c. renal tumors.
   d. glomerulonephritis.

10. A BUN of 9 mg/dL is obtained by a technologist. What is the urea concentration? (Objective 10)
    a. 18.3 mg/dL
    b. 19.3 mg/dL
    c. 10.3 mg/dL
    d. 9.3 mg/dL

11. Urea may be determined after reaction with urease by measuring _____ in an indicator reaction. (Objective 8)
    a. ammonia
    b. uric acid
    c. diacetyl monoxine
    d. creatinine

LEVEL II

3. All of the statements below regarding creatinine are true *except:* (Objective 14)
   a. the rate of formation per day is relatively constant.
   b. plasma levels are highly dependent on diet.
   c. it is completely filtered by the glomeruli.
   d. it is not reabsorbed by the renal tubules.

4. Which of the following serum results correlates *best* with the rapid cell turnover associated with chemotherapy treatment regimens? (Objective 22)
   a. creatinine of 2.5 mg/dL
   b. potassium of 5.0 mmol/L
   c. urea nitrogen of 30 mg/dL
   d. uric acid of 11.0 mg/dL

REVIEW QUESTIONS (continued)

LEVEL I

LEVEL II

12. Serum urea nitrogen and serum creatinine determinations are frequently requested together so that their ratio may be evaluated. What is the range of the normal ratio of urea nitrogen to creatinine? (Objective 18)
    a. 1/1 and 10/1
    b. 12/1 and 20/1
    c. 15/1 and 20/1
    d. 20/1 and 30/1
    e. 25/1 and 30/1

13. A high BUN:creatinine ratio with a significantly elevated creatinine is usually seen in: (Objective 18)
    a. liver disease.
    b. low protein intake.
    c. tubular necrosis.
    d. postrenal azotemia.

14. The amount of creatinine in urine is directly related to: (Objective 14)
    a. protein intake.
    b. urine color.
    c. body muscle mass.
    d. severity of renal disease.

15. The measurement of creatinine is based on the formation of a yellow-red color when creatinine reacts with: (Objective 15)
    a. alkaline picrate.
    b. Ehrlich's reagent.
    c. acetic anhydride–sulfuric acid.
    d. phosphomolybdate.
    e. Titan yellow.

16. Creatinine clearance is used to estimate the: (Objective 23)
    a. tubular secretion of creatinine.
    b. glomerular secretion of creatinine.
    c. renal glomerular and tubular mass.
    d. glomerular filtration rate.

17. Given the following data, the patient's creatinine clearance in mL/min is: (Objective 24)

urine creatinine	90 mg/dL
serum creatinine	0.90 mg/dL
patient's total body mass	1.73 m^2
24h urine volume	1500 mL

    a. 104 mL/min
    b. 117 mL/min
    c. 124 mL/min
    d. 140 mL/min

18. The creatinine clearance formula the term "1.73" is used to: (Objective 23)
    a. normalize clearance making it independent of muscle mass (size).
    b. correct clearance for creatinine that is secreted by the renal tubules.
    c. normalize clearance making it independent of filtrate flow.
    d. adjust clearance so that it is equal to inulin clearance.

REVIEW QUESTIONS (continued)

LEVEL I

19. Uric acid is derived from the: (Objective 19)
    a. oxidation of proteins.
    b. catabolism of purines.
    c. oxidation of pyrimidines.
    d. reduction of catecholamines.

20. Gout is a pathological condition that is characterized by the accumulation of which of the following in joints of the body tissues? (Objective 21)
    a. calcium
    b. BUN
    c. creatinine
    d. uric acid
    e. ammonia

21. John Doe is admitted to the hospital for a complete workup. His chemistry screening profile results are: (Objective 22)

BUN	12 mg/dL
Creatinine	1.5 mg/dL
Uric acid	60 g/dL

    What condition are these indicative of?
    a. glomerulonephritis
    b. hepatitis
    c. muscular dystrophy
    d. gout

LEVEL II

PEARSON
myhealthprofessionskit™

Use this address to access the interactive Companion Website created for this textbook. Simply select "Clinical Laboratory Science" from the choice of disciplines. Find this book and log in using your user name and password.

REFERENCES

1. Shaver M-J, Shah SV. Acute renal failure: Acute renal failure in the hospitalized patient. ACP Medicine Online (2002) (www.medscape.com/viewarticle/534676_print), accessed January 12, 2007.

2. Salifu MA. Azotemia. *Emedicine* (2007) (www.emedicine.com/med/TOPIC194.HTM), accessed November 12, 2007.

3. Gottlieb S. Acute renal failure (www.medscape.com/viewarticle/414667_print), accessed November 12, 2007.

4. RNCEUS. Blood urea nitrogen (www.rncaus.com/renal/renalbun.html), accessed November 12, 2007.

5. Meyers GL, Miller WG, Coresh J, et al. Recommendations for improving serum creatinine measurement: A report from the Laboratory Working Group of the National Kidney Disease Education Program. *Clin Chem* (2006) 52, 1: 5–18.

6. Edwards NL, Weaver A, Schumacher HR. Hyperuricemia and gout: New clinical insights and emerging treatments (www.medscape.com/viewprogram/4933_print), accessed February 6, 2008.

7. National Kidney Disease Education Program. Rational for use and reporting of estimated GFR (www.nkdep.nih.gov/professionals/estimated_gfr.htm), accessed February 14, 2008.

8. Woodhouse S, Batten W, Hendrick H, Malek PA. The glomerular filtration rate: An important test for diagnosis, staging, and treatment of chronic kidney disease. *Lab Med* (2006) 37, 4: 244–246.

9. National Kidney Disease Education Program. Suggestions for laboratories (www.nkdep.nih.gov/resources/laborator_reporting.htm), accessed February 14, 2008.

10. Price CP, Newell RG, Boyd JC. Use of protein: Creatinine ratio measurements on random urine samples for prediction of significant proteinuria. *Clin Chem* (2005) 51, 9: 1577–1586.

11. Seliger SL, DeFillippi C. Role of cystatin C as a marker of renal function and cardiovascular risk (www.medscape.com/viewprogram/6158_print), accessed February 8, 2008.

12. Villa P, Jiminez M, et al. Serum cystatin C concentration as a marker of acute renal dysfunction in critically ill patients. *Crit Care* (2005) 9: R139–R143.

13. Herrero-Morin JD, Malaga S, et al. Cystatin C and beta-2-microglobulin: Markers of glomerular filtration in critically ill children. *Crit Care* (2007) 11, 3: R59.

14. Filler G, Priem F, et al. β_2-microglobulin, and creatinine compared for detecting impaired glomerular filtration rates in children. *Clin Chem* (2002) 48, 5: 729–736.

15. Strasinger SK, Schaub DM. *Urinalysis and body fluids,* 5th ed. (Philadelphia: F.A. Davis Co., 2008).

16. National Kidney Foundation. Kidney Early Evaluation Program. *KKEP 2204 Report* (www.kidney.org/news/keep), accessed February 8, 2008.

17. American Diabetes Association. *Standards of medical care in diabetes* (2005) 28: S4–S36.

18. Parmar MS. Acute glomerulonephritis (www.emedicne.com/MED/topic879.htm), accessed February 11, 2008.

19. Salifu MA, Delano BG. Chronic glomerulonephritis (www.emedicine.com/med/TOPIC880.HTM), accessed February 11, 2008.

20. Shoff WH, Green-McKenzie J. Acute pyelonephritis (www.emedicine.com/TOPIC2843.HTM), accessed February 11, 2008.

21. National Kidney Foundation K/DOQI Guidelines. *K/DOQI clinical practice guidelines for chronic kidney disease: Evaluation, classification, and stratification. Part 4 Definition and classification of stages of chronic kidney disease* (www.kidney.org/professionals/kdoqi/guidelines_ckd/p4_class_g1.htm), accessed February 11, 2008.

22. Kidney stones: Types, causes, and treatment for renal calculi (www.kidney-pain.com/articles/kidney-stones/index.php), accessed February 11, 2008.

23. The many causes of kidney stones: From diet to medical conditions (www.kidney-pain.com/articles/kidney-stones/causes-of-kidney-stones.php), accessed February 11, 2008).

24. Ibid.

25. National Kidney Foundation. The problem of kidney and urologic disease (www.kidney.org/news/newsroom/printfact.cfm?id=11), accessed February 11, 2008.

26. Coppo R, Bargman JM. Initiation of dialysis from registry to practice (www.medscape.com/viewprogram/5835), accessed April 13, 2008.

27. Kloppenburg WD, Stegman CA, et al. Assessing dialysis adequacy and dietary intake in the individual hemodialysis patient. *Kidney Int'l* (1999) 55: 1961–1969.

Suggested Readings

1. Cohen EP, Lemann J Jr. The role of the laboratory in evaluation of kidney function. *Clin Chem* (1991) 37, 6: 785–796.

2. Leanos-Miranda A, Marquez-Acosta J, et al. Protein:creatinine ratio in random urine samples is a reliable marker of increased 24-hour protein excretion in hospitalized women with hypertensive disorders of pregnancy. *Clin Chem* (2007) 53, 9: 1623–1628.

3. Myers GL, Miller WG, et al. Recommendations for improving serum creatinine measurement: A report from the Laboratory Working Group of the National Kidney Disease Education Program. *Clin Chem* (2006) 52, 1: 5–18.

12

Body Water and Electrolyte Homeostasis

■ OBJECTIVES—LEVEL I

Following successful completion of this chapter, the learner will be able to:

1. List examples of electrolytes found in plasma water, interstitial fluid, and intracellular water.
2. Identify the analytes required to calculate anion gap and osmolality.
3. State the specific fluid compartments that make up total body water.
4. Distinguish between serum and plasma.
5. State the principle differences between interstitial fluid and plasma.
6. List five examples of body fluids that are assayed for electrolyte composition.
7. Select the electrolyte associated with each of the following:
   a. major intracellular cation
   b. major extracellular cation
   c. major extracellular anion
8. Identify four methods used to measure chloride in sweat.
9. Identify methods and instrument techniques used to measure electrolytes in body fluids.
10. Name the four colligative properties of solutions.

■ OBJECTIVES—LEVEL II

Following successful completion of this chapter, the learner will be able to:

1. Define anion gap and discuss its clinical significance. Calculate and interpret anion gap results from a given set of data.
2. Describe an appropriate course of action for the measurement of sodium ion in a sample that shows gross lipemia.
3. Describe the homeostatic regulation of sodium, potassium, and chloride in body water compartments.
4. Explain the concept of electrolyte exclusion effect.
5. Discuss the clinical aspects of increased and decreased plasma body fluid electrolyte concentrations.
6. Diagram the renin-angiotensin-aldosterone system for water and electrolyte regulation.
7. Discuss the role of atrial natriuretic peptide in response to an increased intravascular volume.
8. Interpret the clinical significance of serum and urine osmolality and osmol gap.

KEY TERMS

Anion gap
Colligative properties
Diuresis
Electrolyte
Extracellular water
Hyperkalemia
Hypernatremia

Hypovolemia
Interstitial fluid
Kalium
Natrium
Natriuresis
Natriuretic peptides
Osmolal Gap

Osmolality
Osmolarity
Plasma
Serum
Syndrome of inappropriate ADH
(SIADH)
Total body water

A CASE IN POINT

A 42-year-old Caucasian male presented to the emergency department, complaining of malaise, tiredness ("just no energy"), and mild pain described as coming from "my bones." The physician requested an initial set of laboratory tests based upon a complete history and physical examination. Results of the initial laboratory tests are shown below:

Serum Chemistries	Results (Conventional units)	Reference Interval (Conventional units)
*Sodium	122 mmol/L	136–145
*Potassium	3.5 mmol/L	3.5–5.1
Chloride	101 mmol/L	98–107
Total carbon dioxide	20 mmol/L	23–29
Anion gap	1 mmol/L	7–16
Glucose	125 mg/dL	74–106
Creatinine	0.8 mg/dL	0.9–1.3
Urea nitrogen	8 mg/dL	6–20
Calculated osmolality	254 mOsm/Kg	N/A
Calcium	6.0 mg/dL	8.8–10.2
Phosphate	1.3 mg/dL	2.4–4.4
Magnesium	1.8 mg/dL	1.6–2.6
Triglyceride	300 mg/dL	<150
HDL-C	15 mg/dL	27–67
LDL-C	50 mg/dL	87–186
Total protein	12.5 g/dL	6.4–8.4
Albumin	1.0 g/dL	3.5–5.2
A/G	0.09 mg/dL	1.0–2.2

*Sodium and potassium were measured using an indirect measuring ISE.

Hematology Results

WBC count	$12.5 \times 10^3/\mu L$	$4.0–11.0 \times 10^3/\mu L$
RBC count	$3.0 \times 10^6/\mu L$	$4.70–6.10 \times 10^6/\mu L$
Hemoglobin	11.0 g/dL	13.0–18.0
MCV	93.0 fL	82.0–99.0
MCH	33.0 pg	27.0–34.0
MCHC	36.0%	32.0–36.0
Platelet count	200,000 μL	140,000–440,000 μL
Neutrophils	68%	39–78
Bands	2%	0–12
Lymphocytes	20%	15–56
Monocytes	4%	2–14
Plasma cells	6%	0–1
Anisocytosis	Mildly elevated	
Rouleaux	Markedly elevated	

Issues and Questions to Consider

1. What is/are this patient's most striking laboratory result(s)?
2. What disease(s) or condition(s) could account for these abnormal results?
3. Why is the patient's anion gap decreased?
4. What additional test(s) should be ordered to support presumptions made in question 2?

Based on the results of the initial laboratory tests and history/physical, the clinician requested additional laboratory tests to confirm the diagnosis of multiple myeloma. The laboratory staff also completed confirmation tests of the patient's specimen to rule out the possibility of pseudohyponatremia. The results of the additional laboratory tests are shown below.

Serum Chemistries	Results (Conventional units)	RI (Conventional units)
Measured serum osmolality	310 mOsm/kg	275–295
Osmolal gap	56 mOsm/kg	
SPE	IgG kappa monoclonal gammopathy	
IgG	5000 mg/dL	700–1,600
IgA	56 mg/dL	70–400
IgM	39 mg/dL	40–230

Laboratory confirmation tests (sodium and potassium) were assayed using a direct ISE.

Sodium	141 mmol/L
Potassium	3.5 mmol/L
Chloride	100 mmol/L
Total CO$_2$	21 mmol/L
Anion gap	20 mmol/L
Recalculated osmolal	291 mOsm/kg
Recalculated osmolal gap	18 mOsm/kg

Issues and Questions to Consider

5. Was the laboratories course of action appropriate? Explain.
6. What is the explanation for the falsely low sodium value?

WHAT'S AHEAD

Water is an essential component of all living entities. It is the milieu within which lies all of the necessary components to sustain a living organism. These components include both organic and inorganic solutes. The discussion in this chapter focuses on the following:

▶ Compartmentalization of body fluid and its electrolytes
▶ Fundamental mechanisms involved in the maintenance of these specialized compartments
▶ Homeostasis and disorders of sodium, potassium, chloride, and bicarbonate
▶ Measurement of sodium, potassium, chloride, and bicarbonate in body fluids

▶ INTRODUCTION

In the adult human body, nearly 60% of its mass is water. Body water exists within compartments that contain most of the organic and inorganic substances required to maintain a living body. **Total body water** (TBW) is distributed into the intracellular fluid (ICF) compartment and the extracellular fluid (ECF) compartment in a ratio approximating 66.6%: 33.3%.[1]

The intracellular fluid compartment includes all water within cell membranes and provides the medium in which chemical reactions of cell metabolism occur. A plasma cellular membrane physically separates ICF and ECF compartments. The ECF compartment includes all water external to cell membranes and provides the medium through which all metabolic exchanges occur. The ECF may be further divided into **interstitial fluid** (ISF) and intravascular fluid (IVF), or plasma compartments, which are separated by the capillary endothelium. Intravascular fluid represents the volume of fluid accessible to direct measurement. The ISF allows ions and small molecules to diffuse freely from plasma and also directly bathes the cells of the body. The distribution of TBW into compartments is illustrated in Figure 12-1 ■.

Within the IVF (*whole blood*) compartment, plasma, the liquid fraction, constitutes about 3.5 L for the average adult, with a hematocrit of approximately 40% and 5L total blood volume. Most laboratory tests to determine hydration status and electrolyte and acid–base status are performed on samples from the intravascular (plasma, serum, or whole blood) compartments. Other clinically significant ECF compartments include cerebrospinal fluid (CSF) and urine, which are also used routinely for measurement of analytes for the same purposes stated above.

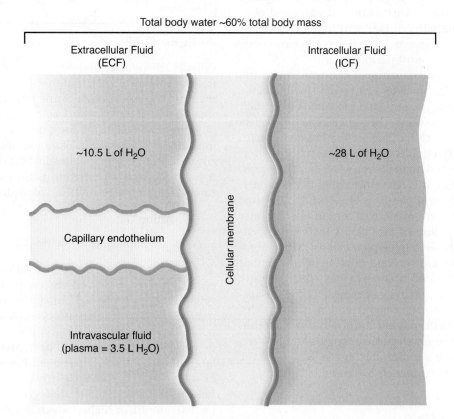

Total body water ~60% total body mass

Extracellular Fluid (ECF)   Intracellular Fluid (ICF)

~10.5 L of H_2O ~28 L of H_2O

Capillary endothelium

Cellular membrane

Intravascular fluid (plasma = 3.5 L H_2O)

■ FIGURE 12-1 Total body distribution between body fluid compartments. Water volume figures represent approximate values for a normal-size adult.

▶ BODY FLUID COMPOSITION

Electrolyte is a term that reflects the major ions of body fluids and constitutes the majority of osmotically active particles (see discussion below). The primary cation (positive charge) electrolytes include these:

- Sodium ion (Na^+)
- Potassium ion (K^+)
- Calcium ion (Ca^{2+})
- Magnesium ion (Mg^{2+})

The primary (negative charge) anions include these:

- Chloride ion (Cl^-)
- Bicarbonate (HCO_3^-)
- Phosphate (HPO_4^{2-}, $H_2PO_4^{2-}$)
- Sulfate (SO_4^{2-})
- Organic ions (e.g., lactate)
- Negatively charge proteins (albumin)

Hydrogen ion (H^+) is a cation with a concentration approximately 1 million times lower in plasma than most electrolytes listed above and therefore is negligible in terms of osmotic activity (see discussion below). The total number of positive ions (including H^+) *must* equal that of the negative ions for electrical neutrality. Distribution of water and electrolytes in a typical muscle cell is shown in Table 12-1 ✪.[2] Most electrolytes are measured from the plasma compartment because it yields the most reliable data.

Plasma is the primary compartment of interest and is the only compartment where analytes can be measured directly. Plasma is composed of water, ions, and macromolecules such as fibrinogen and other proteins. The term *plasma water* refers to the water phase in which ions are present but not protein, lipids, and the other macromolecules. Thus, the concentration of ions in plasma is lower than in plasma water due to the presence of macromolecules. It is relevant to note that although the concentration of ions in plasma is that portion measured and reported, it is the concentration of ions in plasma water that affects the distribution of ions across the capillary endothelium. Individuals with hyperproteinemia or hyperlipidemia will demonstrate a lower concentration of ions in their plasma even though the concentration of ions in plasma water and the resultant chemical activity of these ions may be normal. **Serum** is the liquid part of blood remaining after a clot has formed. Serum lacks the protein fibrinogen and other clotting factors. The difference in electrolyte composition between serum and plasma is not significant except for K^+, where the plasma value may be approximately 8% lower than serum.

✓ Checkpoint! 12-1

Compare and contrast serum versus plasma.

✪ TABLE 12-1

Water and Electrolyte Composition of Body Fluid Compartments in a Typical Muscle Cell (ISF, interstitial fluid, ICW, intracellular water)

	Plasma (mmol/L)	ISF	ICW
Total body water (L):			
Plasma	3.5		
Interstitial fluid (ISF)	10.5		
Intracellular fluid (ICW)	28		
Cations:			
Sodium	142	145	12
Potassium	4.0	4.0	156
Calcium	6.0	2–3	3
Magnesium	2.0	1–2	26
Trace elements	1		
Total Cations	155	153	197
Anions:			
Chloride	103	116	4.0
HCO_3^-)	27	31	8.0
Proteins	16	—	55
Organic acids	5.0 ⎱	6.0 ⎱	130
HPO_4^{3+}	2.0 ⎰	⎰	
SO_4^{2+}	1.0		
Total anions	154	153	197

ANION GAP

Clinical laboratories routinely measure only plasma concentrations of Na^+, K^+, Cl^-, and HCO_3^-. The sum of the measured cations exceeds that of the measured anions. Thus, the sum of unmeasured plasma anions (e.g., negatively charged proteins and phosphates) must be greater than that of the unmeasured cations (e.g., calcium and magnesium). The difference between the sum of measured cations and the sum of the measured anions is referred to as the **anion gap**. Anion gap can be determined using either of the following equations:

$$\text{Anion Gap} = [Na^+ + K^+] - [Cl^- + HCO_3^-] \quad (12.1)$$

$$\text{Anion Gap} = Na^+ - [Cl^- + HCO_3^-] \quad (12.2)$$

The latter equation is widely used because plasma K^+ concentrations are relatively constant and may be spuriously elevated due to hemolysis. Hemolysis occurs whenever there is trauma to the fragile erythrocytes. A significant amount of potassium resides inside the erythrocyte; therefore, if they are destroyed, the potassium will be transferred to the extracellular fluid. For example, if a patient has a measured plasma K^+ of 4.0 mmol/L, which is within its reference range, that same sample with a moderate degree of hemolysis may measure 6.0 mmol/L or higher, which is significantly abnormal.

Initially, the idea of the anion gap was used as a quality control tool. When the electrolytes were measured, an anion gap was calculated, and if the value were outside the interval of 7–16 mmol/L, this would alert the analyst that an error might have been made for one of the electrolyte measurements. However, abnormal anion gaps are seen in several clinical conditions that must be considered before attributing the abnormal calculated value to laboratory error.

The determination of anion gap can provide useful information to the clinician. For example, because the total plasma cation concentration must equal the total plasma anion concentration, and a decrease in unmeasured cations has little effect in the calculation, an increase in anion gap is usually caused by an increase in the concentration of one or more of the unmeasured anions (e.g., proteins, SO_4^{2-}, $H_2PO_4^{2+}$) that are present in plasma.

Changes in anion gap are often seen in patients with metabolic acidosis or metabolic alkalosis. An example of metabolic acidosis with an increase in anion gap and associated electrolytes is shown in the following example:

Example

The calculated anion gap using equation 12-2 for the laboratory results listed below is 16 and is within the reference interval, but the actual AG is 21.

Sodium	141 mmol/L
Potassium	5.0 mmol/L
Chloride	103 mmol/L
HCO_3^-	22 mmol/L
Anion Gap	21

This is due to an increase in the unmeasured anions in the blood derived from the substances identified by the mnemonic device MUDPILES. Most anion gap metabolic acidoses can be explained by one (or a combination) of eight underlying mechanisms described by this mnemonic, whose letters represent methanol, uremia, diabetes mellitus, paraldehyde, isoniazid, lactic acidosis, ethylene glycol, and salicylate.

 Checkpoint! 12–2

Identify examples of unmeasured anions that could cause an increase in the anion gap.

INTERSTITIAL FLUID

Interstitial fluid (ISF) can be described as an ultrafiltrate of plasma. The major difference between ISF and plasma is the presence of protein in the plasma and virtually none in ISF. Plasma is separated from the ISF within the endothelial lining of the capillaries, which acts as a semipermeable membrane that allows passage of water and diffusible solutes but not compounds of large molecular mass such as proteins. The ultrafiltration process is not absolute, thus small amounts of protein can be present. In certain pathological conditions such as bacterial sepsis, the permeability of the vascular endothelium increases, resulting in (1) leakage of albumin, (2) a reduction in circulating fluid volume (shock), and (3) hypotension.

INTRACELLULAR WATER COMPARTMENT

The intracellular water (ICW) compartment represents a heterogeneous mixture of solutes that are difficult to measure directly because of the lack of cells free of contamination. There are significant differences in intracellular solute concentrations between different cell types. However, some features of most cell fluids are quantitatively similar and distinguish ICW from ECW. The major cations of ICW are K^+ and Mg^{2+}, and the Na^+ concentration is always low. However, proteins, organic acids, phosphates, and sulfates represent the major anions in ICW, with chloride and bicarbonate present at low concentrations.

REASONS FOR DIFFERENCE IN BODY FLUID COMPARTMENTS

Differences in body fluid compartments both in solute concentration and liquid volume are a direct result of interactions of physical forces within and among compartments. The presence of polyanionic protein molecules in plasma, which cannot cross semipermeable membranes, creates an equilibrium referred to as *Gibbs-Donnan* equilibrium. This equilibrium results in a plasma water concentration slightly greater than that in ISF and a plasma water anion concentration slightly less than that in ISF. (See Kaplan and Pesce for a more detailed explanation of Gibbs-Donnan equilibrium, with examples.[3])

Ions are distributed between fluid compartments by both active and passive transport. The distribution of ions shown in Table 12-1 is similar between plasma and ISF but quite different in ICW. The major ions in plasma and ICF, for example, are Na^+, Cl^-, and HCO_3^-, whereas in ICW they are K^+, Mg^{2+}, proteins, and organic phosphates. This unequal distribution of ions is due to an active transport of sodium from inside to outside the cell, against an electrochemical gradient. Active transport requires energy derived from mechanisms such as glycolysis. In the membrane, an active sodium pump utilizing energy from ATP is coupled with the active transport of potassium into the cell.

The active transport process involves the Na^+/K^+–ATPase mechanism to move ions into and out of cells. Also working toward moving ions into and out of cells is a process utilizing Na^+–H^+ exchange that actively pumps H^+ out of ICF in exchange for Na^+. This process is vital for maintaining intracellular pH homeostasis and proper fluid volume in many types of cells.

▶ COLLIGATIVE PROPERTIES

An important factor that determines the distribution of water among body water compartments is osmotic pressure. Osmotic pressure is one of the four **colligative properties** of a solution, and the other three are vapor pressure, freezing point depression, and boiling point. When a solute is added to a solvent, the vapor pressure of the solution is lowered below that of the pure solvent. As a result of the change in vapor pressure, the boiling point of the solution is raised above that of the pure solvent, and the freezing point of the solution is lowered below that of the pure solvent. These four colligative properties are directly related to the total number of solute particles per mass of solvent. The osmotic pressure represents a hydrostatic pressure that develops and is maintained when two solutions of different concentrations exist on opposite sides of a semipermeable membrane. Osmotic pressure regulates the movement of solvent (water) across a membrane that separates two solutions. Different membranes may have larger or smaller pore sizes, thereby allowing the membrane to be selective for certain solutes of specific size or shape. For example, certain membranes may be more selective for small molecules or ions and allow them to pass through, but not allow larger particles such as proteins (macromolecules) through. A difference in concentration of osmotically active molecules, for example, glucose, protein, and urea nitrogen, which may not be able to cross a specific membrane, can cause those molecules that are able to cross the membrane to migrate and establish an osmotic equilibrium that then creates an osmotic pressure, as described above.

Osmolarity is a term that expresses concentration per volume of solution, where a 1 osmolar solution contains 1 osmol/L of solution. The term **osmolality** represents a more precise thermodynamic expression than osmolarity, because solution concentrations expressed on a weight factor are temperature independent, whereas those based on volume will vary with temperature. Osmolality is expressed as Osmol/kg H_2O and is used to identify the *number* of moles of a particle per kilogram of water, *not* the *kind* of particle.

The following example serves to illustrate this difference. A solution of NaCl that is 1.0 millimolal is equivalent to 2.0 milliosmolal because NaCl separates into Na^+ and Cl^-. Each kind of ion represents a particle that contributes to the osmolality. Another example is a 1.0 millimolal calcium chloride ($CaCl_2$) solution that is equivalent to 3.0 milliosmolal because each molecule ionizes to yield one Ca^{2+} and two Cl^-.

Measurement of fluid osmolality is accomplished in a clinical chemistry laboratory, using either freezing point depression osmometry or vapor pressure osmometry. Refer to Chapter 2, Instrumentation, for a description of both of these techniques. Osmometry measures the concentrations of molecules and ions in general and not the concentration of particular molecules. Clinical laboratories routinely provide measured serum and urine "osmolality" by either instrument technique. These measurements reflect the number of osmotically active particles in serum or urine of which sodium, urea nitrogen, and glucose are the major contributors.

There are a few calculations that have been developed to calculate serum osmolality. These calculations are presented in equations 12-3 to 12-6.

Calculated osmolality (mOsm/kg) = 2(Na$^+$, meq/L) +

glucose (mg/dL)/18 + Urea N (mg/dL)/2.8   (12.3)

Calculated osmolality (mOsm/kg) = 1.86(Na$^+$, meq/L) +

glucose (mg/dL)/18 + Urea N (mg/dL)/2.8 + 9   (12.4)

Calculated osmolality (mOsm/kg) = 1.86(Na$^+$, meq/L) +

0.056(glucose, mg/dL) + 0.36(Urea N, mg/dL) + 9   (12.5)

Calculated osmolality (mOsm/kg) = 2(Na$^+$, mmol/L) +

(glucose, mmol/L) + (Urea N, mmol/L)   (12.6 (SI Units))

Equation 12-6 uses SI units and is straightforward. In equations 12-3 and 12-6, the factor 2 counts the cation (sodium) once and the corresponding anion once. Glucose and blood urea nitrogen (BUN) are undissociated molecules and are counted once each. All other components are ignored. The dividing factors in equations equations 12-3 and 12-4 represent the respective molecular weights and conversion from deciliters to liters.

The measured osmolality can be compared with the calculated osmolality, and the difference, termed the **osmolal gap**, is shown in equation 12-7. A normal osmolal gap is approximately zero. Clinicians use osmolal gap to provide relevant information regarding their patients. An abnormal osmolal gap may indicate the presence of significant amounts of unmeasured substances in the blood.

Osmolal gap, Osm/kg = measured Osm/kg −

calculated Osm/kg   (12.7)

Examples of clinically significant unmeasured substances in the blood that result in an abnormal osmolal gap include alcohols (ethanol, isopropanol, and methanol), acetone (ketone bodies), and other organic solvents.

✓ **Checkpoint! 12–3**

Given the following data, calculate the serum osmolality using equation 12-3 and osmolal gap equation 12-7. Also, list two examples of substances that can result in an increased osmolal gap.

Sodium	135 mEq/L
Potassium	4.5 mEq/L
Chloride	108 mEq/L
Carbon dioxide	7 mEq/L
Anion gap	20 mEq/L
Glucose	162 mg/dL
Creatinine	1.5 mg/dL
Urea nitrogen	17 mg/dL
Serum Osmolality*	320 mOsm/kg

Measured by freezing point depression osmometry

Measurement of urine osmolality reflects the kidneys' ability to concentrate urine. If the random urine osmol is >600 mOsm/kg, it can be assumed that the kidneys' ability to concentrate urine is satisfactory. A patient with a very dilute urine specimen and a low urine osmolality is not presumptive for renal dysfunction; therefore, additional renal function tests should be considered.

One significant distinction that should be noted between measurement techniques for osmolality identified above is that a vapor pressure osmometer does not measure volatile solutes such as alcohols in serum. Therefore, vapor pressure osmometry may not be suitable for assessing the osmolality in patients who present to the emergency department (ED)with possible alcohol ingestion.

COLLOID ONCOTIC PRESSURE

Colloid oncotic pressure (COP) is the pressure created in a solution by the presence of large (>30 kDa) proteins, also called *colloids*. The measurement of COP reflects the osmolal contribution of a group of large molecules and may provide the clinician with useful information relevant to a patient's condition. For example, in conditions of reduced plasma proteins, such as malnutrition and proteinuria, excessive fluid may build up in tissue (edema), due to low oncotic pressure. A low oncotic pressure may also be caused by hypoalbuminemia.

▶ REGULATION OF WATER AND ELECTROLYTES

HYPOTHALAMUS AND WATER METABOLISM

Regulation of water and electrolytes in the body is under the influence of hormones and other substances. An abbreviated description of the principle mechanisms will be presented here and further details can be found in the references listed.

Water metabolism is under the influence of the hypothalamus. The hypothalamus is the central nervous system (CNS) structure that most affects the autonomic nervous system (ANS). Neurons in the hypothalamus respond to (1) increases in **extracellular water** (ECW) osmolarity, (2) decreases in intravascular volume, and (3) angiotensin II. Body water balance (i.e., a balance between free water output [water without solute] by the kidney and free water intake) is maintained in part by increases and decreases of ECW osmolarity. Hormones that serve to regulate salt and water balance are listed in Table 12-2 ✪.

In conditions where ECW is increased, neurons in the hypothalamus cause a decrease in intravascular volume, which then results in a stimulation of neurons in the water intake area and produces the sensation of thirst, thus stimulating water intake. Stimulation of neurons in the water output area results in the release of antidiuretic hormone (ADH) from the posterior pituitary gland. ADH stimulates water reabsorption in the collecting ducts of the kidneys, which results in formation of a hypertonic urine and decreased output of free water. Changes in blood influence circulatory levels of angiotensin II, which also serves to stimulate the hypothalamus areas (see additional discussion of angiotensin II below).[3] The mechanism involved with this process is shown in Figure 12.2 ■.

RENIN-ANGIOTENSIN-ALDOSTERONE SYSTEM

Body sodium and water content, arterial blood pressure, and potassium balance are regulated in part by the renin-angiotensin-aldosterone system. Renin is a proteolytic enzyme produced, stored, and secreted by cells located in the juxtaglomerular cells of the kidneys. These granular-type cells are located in the wall of afferent arterioles. Renin secretion is increased by a reduction in renal perfusion pressure, stimulation of sympathetic nerves to the kidneys, and reduced Na^+ concentrations in the fluid of the distal tubules. Renin catalyzes the conversion of angiotensinogen, produced in the liver, to angiotensin I. Angiotensin I is converted to angiotensin II by a converting enzyme, angiotensin-converting enzyme (ACE), found on the surface of vascular endothelial cells. This conversion takes place primarily in the lungs and kidneys. Angiotensin II is a potent vasoconstrictor and has several important physiological functions, including these:

- Stimulation of aldosterone secretion by the adrenal cortex
- Arteriolar vasoconstriction, which increases blood pressure
- Stimulation of ADH secretion and thirst
- Enhancement of NaCl reabsorption by the proximal tubule

The hormone aldosterone stimulates Na^+ reabsorption in the distal nephron, and the body retains water. The

✪ TABLE 12-2			
Hormones That Regulate Sodium Chloride (NaCl) and Water Reabsorption			
Hormone	Major Stimulus	Nephron Site of Action	Effect on Transport
Angiotensin II	↑ renin	proximal tubule	↑ NaCl and H_2O reabsorption
Aldosterone	↑ angiotensin II, ↑ plasma K^+	distal tubule, collecting ducts	↑ NaCl and H_2O reabsorption
ANP, BNP, CNP	↑ ECFV	collecting ducts	↓ NaCl and H_2O reabsorption
Urodilatin	↑ ECFV	collecting ducts	↓ NaCl and H_2O reabsorption
ADH	↑ plasma osmol ↓ ECFV	distal tubule and collecting duct	↑ H_2O reabsorption

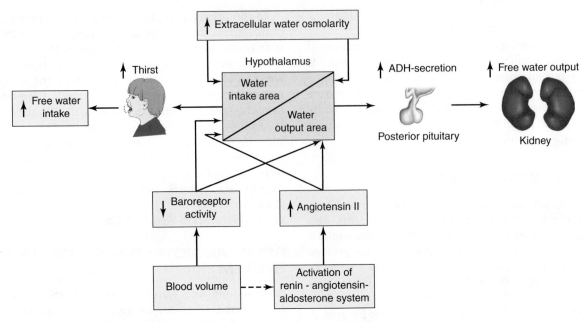

■ FIGURE 12-2   Water balance regulated through the hypothalamic axis.

renin-angiotensin-aldosterone system and its impact on water and sodium balance are summarized in Figure 12-3 ■.

In times of water deprivation (also dehydration), plasma sodium concentration increases (hypernatremia), and both total and intracellular water decreases. A low effective circulating volume, which can be sensed by baroreceptors located in the carotid sinus and aortic arch, signal medullary control centers in the brain to increase sympathetic outflow to the juxtaglomerular cells, thus increasing renin release. Renin converts angiotensinogen to angiotensin I. Angiotensin I is converted to angiotensin II in the lungs and kidneys. Angiotensin II is a potent vasoconstrictor. Also, angiotensin II stimulates aldosterone secretion by the adrenal cortex, thirsting behavior, and ADH secretion. Aldosterone stimulates sodium reabsorption in the distal nephron, and as a result of this sodium reabsorption, the body retains water. In addition, the thirsting mechanism results in an increase in water intake.

 Checkpoint! 12–4

Explain the body's response via the renin-angiotensin-aldosterone system to a situation of water deprivation.

NATRIURETIC PEPTIDES

Water and sodium metabolism is also regulated by other substances located in various organs and tissues in the body, including a family of peptides called **natriuretic peptides**. They include atrial (A)-type natriuretic peptide (ANP), brain (B)-type natriuretic peptide (BNP), and (C)-type natriuretic peptide (CNP) (see also Chapter 18, Cardiac Function). Each

natriuretic peptide is tissue specific and independently regulated. Natriuretic peptides are released in response to intravascular volume expansion, which tends to reduce blood pressure and plasma volume. This response is orchestrated between the brain, vasculature, adrenal glands, and kidneys. Natriuretic peptides play an important role in guarding the body against salt-induced hypertension and moderating congestive heart failure (CHF). The physiologic response to an increase in intravascular volume is shown in Figure 12-4 ■. Natriuretic peptides have reciprocal effects on the renin-angiotensin-aldosterone system.[4]

Atrial natriuretic peptide is a hormone produced primarily in the cardiac atria and is released in response to changes in pressure within the atria. Three functions of ANP are listed here:

1. Reducing the venous pressure that occurs with a given increase in blood volume

2. Increasing vascular permeability

3. Promoting **natriuresis** and **diuresis**

The mechanisms promoting natriuresis and diuresis are based upon the results of direct effects on the kidneys that serve to (1) increase GFR, (2) suppress renin-angiotensin-aldosterone system (which inhibits tubular sodium reabsorption), and (3) moderate the effects of antidiuretic hormone (ADH) in the collecting ducts (which inhibit water reabsorption). In the brain, ANP inhibits salt appetite, water intake, and secretion of ADH and corticotrophin.

BNP (see also Chapter 18, Cardiac Function) is a hormone produced in the cardiac ventricles and is released when the pressure within the ventricles increases due to an increase in volume expansion and pressure overload. Its effect on cardiovascular, natriuresis, and diuresis is similar to ANP.

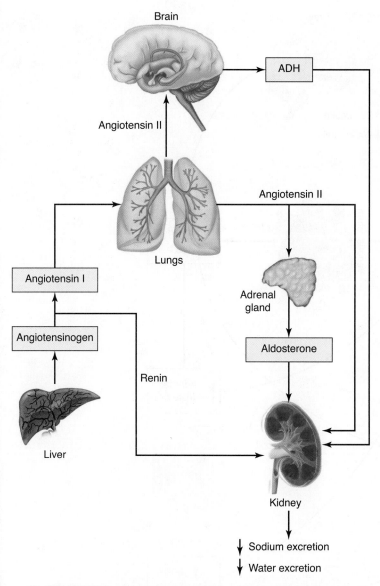

■ **FIGURE 12-3** Principle components of the renin-angiotensin-aldosterone system. Activation of this systems resulting in decreased excretion of Na$^+$ and water by the kidneys. Angiotensin I is converted to angiotensin II by angiotensin-converting enzyme (ACE), which is present on all vascular endothelial cells (see text for more details).

CNP is produced in the vascular endothelial cells, brain, and renal tubules. Circulating levels of CNP in plasma are minute, thus it is believed to act at the local level. Regulation of CNP is not completely understood at this time. It is a potent venous dilator but has no natriuretic effects.

Urodilatin is produced in the kidneys and is similar in structure to ANP. It is secreted by the renal distal tubules and collecting ducts and is not present in the systemic circulation. Urodilatin's principle influence is on the kidneys. Urodilatin secretion is stimulated by a rise in blood pressure and an increase in extracellular fluid volume (ECFV). It inhibits NaCl and water reabsorption across the medullary

portion of the collecting ducts. The regulation of urodilatin is not completely understood, but it is known to be a more potent diuretic and natriuretic than ANP.

 Checkpoint! 12–5

Describe the response of ANP in a patient with CHF and severe hypertension.

DISORDERS OF WATER IMBALANCE

Disorders of water imbalance include dehydration and overhydration, which result from an imbalance of water and sodium intake and output. Dehydration may result from a lack of water alone. Total body water is reduced but Na$^+$ levels are usually normal. This condition represents a simple form of dehydration, which may result from failure to replace water by not drinking it or failure of the regulatory (or effector) mechanisms that promote conservation of water by the kidneys.

Another condition related to water imbalance is dehydration, which may result from a deficit of water and sodium. Water balance may be more negative than, equal to, or less negative than sodium balance. If water balance is more negative than sodium balance, the result is *hypernatremic* or *hyperosmolar* dehydration; if it is equally negative, *normonatremic* or *isomolar* dehydration results; and if it is less negative, *hyponatremic* or *hyposmolar* dehydration results.

Overhydration may be due to water intoxication where total body water (TBW) is increased and total body sodium remains normal. It is rarely due to overconsumption of water (polydipsia). Usually water intoxication is a result of impaired renal free water excretion as the result of greater ADH secretion than is required to maintain normal ECW osmolarity. A condition associated with overhydration due to this mechanism is termed **syndrome of inappropriate ADH secretion** (SIADH). SIADH can occur due to increased secretion of ADH by hypothalamus secondary to decreased venous return to heart, with no decrease in total blood volume, brought about by any of the following:

- Asthma
- Pneumothorax
- Bacterial or viral pneumonia
- Positive pressure ventilation
- Chronic obstructive pulmonary disease (COPD)
- Right-side heart failure

SIADH may also develop due to increased secretion of ADH by hypothalamus in the absence of appropriate osmolar or volume stimuli due to any of the following:

- Hypothyroidism
- Pain, fear

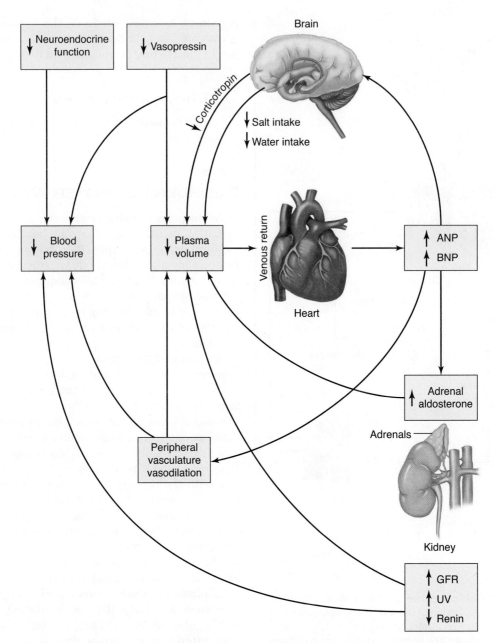

■ FIGURE 12-4 Natriuretic peptides (ANP and BNP) effects in response to an increase in intravascular volume. (ANP, atrial natriuretic peptide; BNP, brain natriuretic peptide; CNP, C-type natriuretic peptide; GFR, glomerular filtration rate; URO, urodilatin)

- Anesthesia or surgical stress
- Drugs, including morphine, barbiturates, and carbamazepine
- CNS disorders resulting from skull fractures and possible brain tumors

Retention of water and sodium results in the expansion of the extracellular compartment. This expansion can be caused by oliguric renal failure, nephritic syndrome, congestive heart failure, cirrhosis, or primary hyperaldosteronism. Excess TBW can be associated with normal or low serum sodium (hyponatremia) and osmolarity in these conditions.

► ELECTROLYTES (SODIUM, POTASSIUM, CHLORIDE, AND BICARBONATE)

SODIUM HOMEOSTASIS

Sodium is the major cation of ECF and comprises nearly 90% of the inorganic ions per liter of plasma. Thus, Na^+ is responsible for about one-half of the osmotic strength of plasma. Sodium plays a central role in maintaining the normal distribution of water and osmotic pressure in ECF compartment. Dietary

NaCl is almost completely absorbed from the GI tract. An adult requires only a fraction of ingested NaCl, and the excess is excreted by the kidneys, which serve as the ultimate regulator of Na^+ and water in the body.

Sodium is freely filtered by the glomerulus. Approximately 80% of the filtered Na^+ is then actively reabsorbed in the proximal tubules, whereas Cl^- and water are passively absorbed in an isoosmotic and electrically neutral fashion. About 20–25% of Na^+ is reabsorbed in the loop of Henle along with Cl^- and more water. In the distal tubule, aldosterone in concert with $Na^+ - K^+$ and $Na^+ - H^+$ exchange systems promotes reabsorption of Na^+ and indirectly of Cl^- from the remaining 5–10% of the sodium load. The amount of Na^+ in urine is ultimately determined by action in distal tubules. A more detailed description of these processes can be found in Chapter 11, Renal Function.

DISORDERS OF SODIUM HOMEOSTASIS

Hyponatremia

Hyponatremia (hypo [deficiency] + new Latin **natrium** [sodium] + emia [blood]) is clinically significant at plasma concentrations of <136 mmol/L. Patients with decreased plasma Na^+ levels may experience generalized weakness and nausea. Plasma Na^+ concentrations <120 mmol/L are associated with altered mental status in patients and may progress to a more severe altered mental status at concentrations <105 mmol/L. Because Na^+ contributes significantly to fluid osmolality, it is prudent to determine plasma or serum osmolality in patients showing signs and symptoms of hyponatremia.

A common condition associated with hyponatremia is hypoosmotic hyponatremia. This type of hyponatremia can be due to either excess loss of Na^+ (*depletional* hyponatremia) or increased ECF volume (*dilutional* hyponatremia). A clinician must complete a though history and physical exam in order to differentiate these two conditions.

Depletional hyponatremia is accompanied by a loss of ECF water, with some Na^+ loss. A physical examination often reveals signs of **hypovolemia** (e.g., orthostatic hypotension, tachycardia, and decreased skin turgo or elasticity). Hypovolemia may result from both renal and extrarenal loss of Na^+. If urine Na^+ is low (<10 mmol/L), the loss is extrarenal because the kidneys are properly retaining filtered Na^+ in response to increased aldosterone. Extrarenal loss of Na^+ is attributed to GI disorders such as vomiting and diarrhea or through the skin via burns and sweating.[5]

In patients whose urine Na^+ is >20 mmol/L, the cause is usually renal in nature. Examples of conditions resulting in renal loss of Na^+ include the following:

- Osmotic diuresis
- Adrenal insufficiency
- Administration of diuretics (e.g., spironolactone)

Dilutional hyponatremia is often seen in patients who retain excess water in association with weight gain or excess edema. Other conditions associated with dilutional hyponatremia are listed here:

- Liver failure
- Congestive heart failure (CHF)
- Renal failure
- Nephrotic syndrome
- Inappropriate ADH secretion

Hypernatremia

Hypernatremia, or increased blood levels of sodium, is associated with values >150 mmol/L and is always hyperosmolar, which means that there is a greater deficiency of ECW than of sodium. Patients who are hypernatremic usually present to the clinician with neurological signs and symptoms, for example, tremors, irritability, ataxia, confusion, and possibly coma. Causes of hypernatremia include the following:

- Ingestion of large amounts of sodium salts
- Administration of hypertonic NaCl or $NaHCO_3$
- Primary hyperaldosteronism
- Excessive sweating
- Diabetes insipidus
- Osmotic diuresis

 Checkpoint! 12–6

Indicate several conditions that would result in depletional-type hyponatremia.

SODIUM MEASUREMENT

In clinical laboratories, serum or plasma samples are routinely measured for sodium content, using either direct or indirect ion-selective electrodes (ISE). The difference between the two techniques is that a direct ISE does not require a dilution of the sample, whereas an indirect ISE does. The composition of membranes in these ISEs is highly selective for Na^+ over K^+ or H^+. Other techniques for measuring Na^+ include atomic absorption spectroscopy, spectrophotometry, and flame emission photometry. Capillary blood obtained from a finger stick and whole blood (arterial or venous) in a heparinized syringe (as part of electrolytes and a blood gas panel) are also common types of samples sent to the laboratory for analysis of sodium. Whole blood can be collected in tubes containing lithium or ammonium salts of heparin, but *not* sodium heparin. Other body fluids submitted to the laboratory by clinicians for Na^+ measurements include ocular, GI, and feces. Sodium ion concentration in urine is performed on 24-hour urine collections and random or "spot" urine specimens.

Grossly lipemic (high levels of lipids) blood samples may produce an error in measurement if an indirect ISE is used. The error is the result of the *electrolyte exclusion effect*, which refers to the exclusion of electrolytes from the fraction of the total plasma volume that is occupied by solids. For a complete explanation of this phenomenon, refer to Burtis, Ashwood, and Bruns.[5] An example of the effect and thus the magnitude of error is shown below.

If a serum or plasma sample is grossly lipemic, the sodium value may be falsely low due to *electrolyte exclusion effect*. The lipids present in the sample can be removed, or "cleared," by centrifuging the specimen at a much faster speed (greater g force) than a regular laboratory centrifuge used to separate cells from serum or plasma. A special type of centrifuge, called an *ultracentrifuge*, can attain the g forces necessary to clear the sample of lipids.

Example

A technologist assays a grossly lipemic serum sample for electrolytes using an indirect ISE (dilutional). The results are as follows:

Na	128 mmol/L
K	4.5 mmol/L
Cl	109 mmol/L
TCO_2	25 mmol/L
Anion Gap	−6

The same sample is "cleared" of the lipids by ultracentrifugation and reanalyzed by the indirect measuring ISE. The results are as follows:

Na	145 mmol/L
K	4.5 mmol/L
Cl	109 mmol/L
TCO_2	25 mmol/L
Anion gap	9

Notice that the anion gap was also corrected to a more meaningful value. An alternate solution to ultracentrifugation is to use a direct ISE method.

 Checkpoint! 12–7

Identify at least three techniques used to measure Na^+ in body fluids.

POTASSIUM HOMEOSTASIS

Potassium is the major intracellular cation. The concentration of K^+ in tissue cells is ~150 mmol/L and in erythrocytes the concentration is ~105 mmol/L (i.e., about 23 times its concentration in plasma). An energy-dependent system involving the Na^+/K^+–ATP pump continually transports K^+ into the cell against a concentration gradient. This pump is an important means of maintaining and adjusting the ionic gradients on which nerve impulse transmission and contractility of cardiac and skeletal muscles depend.

Potassium from ingested food is absorbed in the GI tract and is quickly distributed. A small amount is taken up by cells and most excreted by the kidneys. Potassium filtered through the glomeruli is almost completely reabsorbed in the proximal tubules and is then secreted in the distal tubule in exchange for Na^+ under the control of aldosterone. An overview of K^+ homeostasis is shown in Figure 12-5 ■. A more in-depth discussion of potassium relevant to renal function may be found in Chapter 11.

Excess amounts of K^+ in the body occur when intake exceeds output due to an abnormality of the potassium homeostatic mechanism. Causes of potassium retention (excess) include the following:

- High-potassium diet
- Oral potassium supplements
- IV potassium administration
- Renal failure
- Hypoaldosteronism

Potassium depletion occurs when K^+ output exceeds intake. There are several reasons for low body K^+, including these:

- Low K^+ diet
- Alcoholism
- Anorexia nervosa
- Increase GI loss due to vomiting, diarrhea, or laxative use
- Urinary loss due to high aldosterone levels or renal disease
- Administration of diuretics

DISORDERS OF POTASSIUM HOMEOSTASIS

Hyperkalemia

Elevated or depressed plasma levels of K^+ often result in serious medical conditions for patients. **Hyperkalemia** (hyper [increase] + new Latin **kalium** [potassium] + emia [blood]) or hypokalemia (decreased blood levels) are terms associated with these medical conditions. Several clinical conditions associated with hyperkalemia are listed here:

- Pseudohyperkalemia
  - Hemolysis
  - Leukocytosis
  - Vigorous arm exercise
  - Tight application of a tourniquet
  - Squeezing the area around a venipuncture site
- High-potassium uptake
- Decreased K^+ excretion
- Intracellular to extracellular shift due to metabolic acidosis
- Crush injuries
- Digitalis (cardiac glycoside drug) overdose
- Tissue hypoxia
- Metabolic acidosis

Hyperkalemia has the potential of producing serious problems for a patient, which can ultimately result in death. The

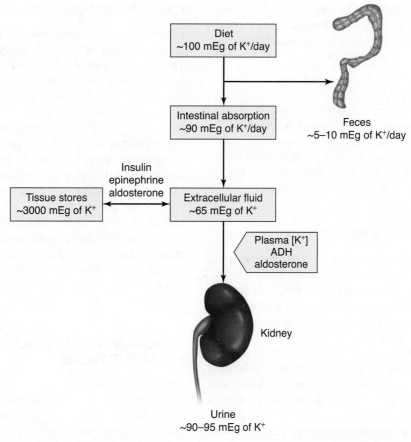

■ FIGURE 12-5 Overview of K$^+$ homeostasis (see text for additional details).

reason is due in part to the profound effects potassium has on cardiac tissue. General symptomologies of hyperkalemia include those listed here:

- Mental confusion
- Weakness
- Tingling sensations
- Flaccid paralysis of the extremities
- Respiratory muscle weakness

Cardiac-associated effects of hyperkalemia include brady-cardia (slowed heart rate) and conduction defects apparent on an electrocardiogram (EKG or ECG). Hyperkalemia may result in several modifications of the electrical conductivity of cardiac myocytes. The electrical changes will adversely affect the mechanical functions of the heart tissues. For example, the EKG may no longer show a P wave (atrial contraction), and the QRS complex (ventricular contractions) will be prolonged. The mechanical cardiac functions associated with these electrical stimuli will be inadequate. Moderate to severe hyperkalemia may lead to ST elevations on the EKG, often an early sign of cardiac ischemia. Prolonged severe hyperkalemia (potassium concentration is >7.0 mmol/L) can lead to peripheral vascular collapse and cardiac arrest. Plasma levels of potassium >10.0 mmol/L are almost always associated with death.

Increased levels of K$^+$ in blood can occur in metabolic acidosis. The mechanism involves the redistribution or transfer of intracellular K$^+$ into ECF as H$^+$ shifts intracellularly and K$^+$ shifts outward to maintain electrical neutrality.

Hypokalemia

Hypokalemia (plasma K$^+$ concentration <3.5 mmol/L) can be caused by movement of K$^+$ into the cell from the ECW space, increased output, or decreased intake. Examples of conditions that may cause hypokalemia include these:

- Metabolic alkalosis
- Diuretic administration
- Increased GI loss
- Increased urinary loss

In metabolic alkalosis, K$^+$ moves from ECF into the cells as H$^+$ moves in the opposite direction. Also, renal conservation of H$^+$ in the distal tubules occurs at the expense of K$^+$ ions.

Potassium measurement in urine is useful to the clinician for the differential diagnosis of hypokalemia. A urine K$^+$ concentration <20 mmol/L with hypokalemia may reflect inadequate intake of K$^+$ or nonurinary losses, whereas a urine K$^+$ > 20 mmol/L is consistent with urinary loss.

Hypokalemia often results in ST depression (EKG) with flattened T waves. This may lead to serious cardiac conditions (e.g., tachyarrhythmias, which are abnormally fast hearts rates with irregular heartbeats).

 Checkpoint! 12–8

Plasma potassium concentration may be elevated or decreased in acid–base disturbances, depending upon the conditions associated with the patient. Identify the plasma K^+ levels as either hyperkalemia or hypokalemia in metabolic acidosis and metabolic alkalosis. Also, briefly discuss the mechanism for these acid–base conditions.

POTASSIUM MEASUREMENT

The information presented earlier in the chapter regarding the measurement techniques and specimen types for Na^+ is similar for K^+, with a few exceptions that will be presented here. There is a difference between K^+ concentration in serum and that found in plasma and whole blood. Potassium ion concentration in plasma and whole blood is ~0.1–0.7 mmol/L lower than that in serum. Serum K^+ concentration is ~0.2–0.5 mmol/L higher than that for plasma K^+. The magnitude of this difference depends on the platelet count, because the extra K^+ in serum is mainly a result of platelet rupture during coagulation.

Specimens used for K^+ measurement must be free from hemolysis, because release of K^+ from cells will falsely increase the measured values. It has been estimated that slight hemolysis of a specimen can raise K^+ values ~3%, marked hemolysis about 12%, and gross hemolysis as much as 30%.[5]

If a specimen shows any degree of hemolysis, a note or comment should append the results, indicating that a specified degree (slight, moderate, or gross) of hemolysis was present and that the K^+ result may be falsely elevated.

Specimens for K^+ measurement should be processed expeditiously. Potassium ions begin to leak out of erythrocytes and other cells, which will result in spuriously elevated plasma or serum K^+ levels. Finally, common problems that occur during blood drawing, and which may alter blood levels of K^+, involve the inappropriate time of release of the tourniquet. A phlebotomist may not release the tourniquet before beginning to draw blood after a patient clenches his or her fist repeatedly. The plasma K^+ values can spuriously increase because of muscle activity where K^+ leaves muscle cells and enters the plasma compartment.

Measurement of serum, plasma, and whole blood K^+ is accomplished using an ion-selective electrode (ISE) that incorporates a neutral antibiotic valinomycin into its organic liquid membrane. Valinomycin has a strong attraction for K^+ and is highly selective for K^+ over Na^+. This electrode design is extensively used in clinical laboratories and has been adapted to several instrument platforms.

CHLORIDE HOMEOSTASIS

Chloride is the major extracellular anion and represents the largest fraction of the total inorganic anion concentration. Thus the majority of the osmotically active particles in plasma are comprised of $Cl^- \sim 103$ mmol/L and $Na^+ \sim 140$ mmol/L. Similar to Na^+, Cl^- plays an important role in the maintenance of water distribution, osmotic pressure, and anion–cation balance in the ECF. The concentration of Cl^- in ICF in contrast to ECF is very low (~50 mmol/L).

Chloride from ingested food is almost completely absorbed by the GI tract. In the glomerulus of the kidney, Cl^- is filtered from plasma and passively reabsorbed along with Na^+ in the proximal tubules. Chloride ion is actively reabsorbed in the ascending loop of Henle by the *chloride pump*, in concert with passive reabsorption of Na^+. Excess Cl^- is excreted in the urine and is also lost in sweat. In patients without acid–base disturbances, the Cl^- concentration in plasma usually follows that of Na^+. Therefore, if serum Na^+ concentration increases, then the serum Cl^- will increase.

DISORDERS OF CHLORIDE HOMEOSTASIS

Hyperchloremia

An increase in Cl^- in blood is termed *hyperchloremia*. This is often due to a situation where intake of Cl^- exceeds output due to an abnormality of a Cl^- homeostasis mechanism. For most cases, the cause of Cl^- retention is the same as those for Na^+ retention. Thus, the pathophysiology of Cl^- excess is similar to that of Na^+ excess.

There is one clinical situation where this is not true: metabolic acidosis. The two major extracellular anions are Cl^- and HCO_3^-. In metabolic acidosis, extracellular HCO_3^- is consumed by reaction with hydrogen ions that are generated in greater quantity in metabolic acidosis. If no organic anions are produced with the H^+, Cl^- is needed to replace the consumed HCO_3^- to maintain electrical neutrality. An increase in Cl^- concentration is caused by the reabsorption of a greater proportion of Na^+ with Cl^- than with HCO_3^- in the renal tubules of the kidney. Hyperchloremia (elevations of blood levels of Cl^-) parallel those of Na^+ with some exceptions; one of these is chronic metabolic alkalosis and acidosis.

Hypochloremia

Hypochloremia, or low blood Cl^-, occurs when the output of Cl^- exceeds intake. Similar to the causes of Cl^- excess, Cl^- depletion parallels Na^+ depletion. One notable exception exists and that is hypochloremia associated with metabolic alkalosis, where the amount of Cl^- in the body is low but Na^+ is not. This usually occurs due to GI loss of Cl^- and not Na^+.

The measurement of Cl^- in urine is useful in the differential diagnosis of metabolic alkalosis. If the metabolic alkalosis is due to a decrease in ECW volume, urinary Cl^- concentrations are usually <15 mmol/L. Administration of saline solution will correct this condition. The Cl^- concentration in a patient with metabolic alkalosis and a normal ECW volume will be >15 mmol/L, and administration of saline will not resolve this condition.

CHLORIDE MEASUREMENT

There are several methods available to measure Cl^- in body fluids. The majority of Cl^- determinations are performed in serum or plasma. Chloride is also measured in whole blood, urine, feces, gastric aspirate, and sweat. Chloride is stable in serum and plasma, and there are no significant preanalytical factors that will result in erroneous Cl^- measurements.

ISEs are commonly used to measure Cl^- in biological fluids. Spectrophotometric and coulometric-amperometric titrations techniques are also available. The ISEs used for Cl^- analysis utilize either direct or indirect methodologies. A solvent polymeric membrane using quaternary ammonium salt anion-exchange material, such as tri-n-octylpropylammonium chloride deconate, is used to make the electrode highly selective for chloride and reduces the potential for interference from other negatively charged ionic species in the sample.

A spectrophotometric method used to measure Cl^- is based on the reaction of Cl^- with mercuric thiocyanate. The Cl^- reacts with mercuric thiocyanate to form mercuric chloride and free thiocyanate (SCN^-). The SCN^- reacts with ferric ions (Fe^{3+}) to form a reddish complex of ferric thiocyanate. Many laboratories have discontinued the use of this method because of the toxic substances present, including mercury.

Sweat Chloride

The measurement of chloride in sweat is useful for evaluating patients with cystic fibrosis (CF). Cystic fibrosis is characterized as a syndrome with a broad spectrum of clinical presentations associated with a defect in the cystic fibrosis transmembrane conductance regulator (CFTR) protein. This protein functions to regulate electrolyte transport across epithelial membranes. Testing for CFTR and its many mutations is available but is often not informative in specific cases. Thus, sweat testing is the standard for diagnostic testing.[6]

Current recommendations by the United States Cystic Fibrosis Foundation (USCFF) for diagnosis of CF include (1) laboratory evidence of CFTR abnormalities documented by elevated sweat chloride concentrations or (2) identification of two CF mutations or (3) in vivo demonstration of characteristic abnormalities in ion transport across nasal epithelium.[6]

Sweat analysis can be accomplished by several different techniques:

- Chloride in sweat using a chloridometer[a]
- Direct skin electrode[b]

- ISEs test patch
- Manual titration of Cl^- using silver nitrate
- Measuring the conductivity of a sample using a sweat conductivity analyzer[c]
- Measuring sodium by ISE
- Measuring the osmolality of a solution using a vapor pressure osmometer[d] or freezing point depression osmometer[e]

Standardization of sweat analysis is an important concern for laboratories and CLSI formerly NCCLS, developed guidelines to facilitate this test.[7] There are three stages for sweat Cl^- testing:

1. Sweat stimulation by pilocarpine iontophoresis
2. Collection of the sweat onto support medium
3. Sweat analysis by one of the techniques listed above

Chloride titration analyzers can be used to measure Cl^- concentrations in biological samples. These instruments are not automated and therefore not suitable for high-volume test use. Laboratories performing sweat testing frequently use chloride "titrators," which have the distinct advantage of requiring small sample volumes ($\sim 20\ \mu L$). The electrochemical principles of the "Cl^- titrator," coulometry coupled with amperometry, are discussed in Chapter 2, Instrumentation.

Regardless of the measuring technique, CLSI guidelines require that the sweat test consist of (1) collecting sweat into a gauze, filter paper, or Macroduct[f] coils, (2) evaluating the amount collected either in weight (milligram) or volume (μL), and (3) then measuring the sweat chloride concentrations. Instrumentation used for sweat chloride analysis must be able to detect as low as 10 mmol/L. For diagnostic purposes, a sweat chloride >60 mmol/L is consistent with CF; concentrations between 40 and 60 mmol/L are considered borderline; and values <40 mmol/L may be considered normal.

BICARBONATE (HCO_3^-)

A portion of this topic relevant to acid–base balance is discussed in Chapter 13, Blood Gases, pH, and Acid–Base Balance. The focus of discussion for this chapter will be total carbon dioxide (TCO_2) that is measured on many automated analyzers by acidification of a serum or plasma sample and measurement of the CO_2 released during the reaction or by alkalinization and measurement of total HCO_3^- (see below). Results of this measurement under specified conditions correlate closely with calculated results for TCO_2 concentration by blood gas analyzers.

---

[a]Labconco Digital Chloridometer (Kansas City, MO)
[b]Wescor Nanoduct Neonatal Sweat Analysis System (Wescor Inc., Logan, UT)
[c]Wescor Sweat-Chek
[d]Wescor VAPRO Model 5520 vapor pressure osmometer
[e]Advanced Model 3320 Microosmometer (Advanced Instruments, Inc., Norwood, MA)
[f]Wescor Model 3700 Macroduct sweat stimulation and collection system

Total carbon dioxide measurements made using a photometric assay can be performed on either serum or heparinized plasma. Venous or capillary samples are often used too. Measurements should be carried out as soon as the stopper is removed from the tube to minimize changes in CO_2 concentrations due to the exchange of atmospheric gases with those in the sample.

A significant number of laboratories measure TCO_2 using an indirect electrode technique or a photometric assay that incorporates an enzyme system. The electrode-based method incorporates a pCO_2 electrode (see Chapter 2, Instrumentation) that detects gaseous CO_2 released after acidification of the sample. Direct measuring pCO_2 electrodes are available, but measurements may be affected by loss of specificity. Several coupled enzymatic assays available for measurement of TCO_2 use phosphoenolpyruvate carboxylase (PEPC). The reaction begins when carbon dioxide and carbonic acid (H_2CO_3) in the blood sample are converted to HCO_3^- via alkalinization. Phosphoenolpyruvate carboxylase catalyzes the conversion of HCO_3^- and PEP to oxaloacetate and inorganic phosphate. A second reaction sequence incorporates the enzyme malate dehydrogenase to convert oxaloacetate to malate. The coenzyme NADH is transformed to NAD^+ and the resultant consumption of NADH causes a decreased absorbance at a specified wavelength. The decrease in absorbance is proportional to TCO_2 content.

SUMMARY

Maintenance of appropriate water balance in the human body is a complex task. Water balance is affected by many organs via integrated biochemical and physiological mechanisms. Several hormones, enzymes, metabolics, and electrolytes are required for the entire process to function properly. Specific mechanisms such as the renin-angiotensin-aldosterone system interact with organs, including kidney, heart, lungs, and brain, to aid in the regulation of water and electrolyte homeostasis. Physical properties of solutions that are described as colligative properties are also important. The movement of particles within the various compartments of the body is accomplished by physical and biochemical mechanisms. These particles may be small ions or large proteins.

A succinct review of the four major electrolytes has been presented. Included in the discussion are physiology, methods of measurement, and disorders associated with electrolyte imbalance. Other electrolytes such as calcium, phosphorus, and magnesium will be discussed in subsequent chapters in this text.

REVIEW QUESTIONS

LEVEL I

1. The anion gap is determined from which of the following groups of electrolytes? (Objective 1)
   a. sodium, chloride, potassium, and calcium
   b. sodium, chloride, potassium, and phosphorus
   c. sodium, chloride, potassium, and TCO_2
   d. TCO_2, chloride, potassium, and magnesium

2. Which of the following groups of analytes are used to calculate serum osmolality? (Objective 2)
   a. sodium, chloride, potassium, and calcium
   b. aldosterone, antidiuretic hormone, and cortisol
   c. urea nitrogen, glucose, and creatinine
   d. glucose, urea nitrogen, and sodium

3. What is the cation with the highest extracellular concentration? (Objective 7)
   a. potassium
   b. sodium
   c. calcium
   d. magnesium

4. What is the anion with the highest extracellular concentration? (Objective 7)
   a. cyanide
   b. bicarbonate
   c. chloride
   d. phosphate

LEVEL II

1. An elevated anion gap usually represents an increase in which of the following? (Objective 11)
   a. measured cations
   b. unmeasured cations
   c. measured anions
   d. unmeasured anions

2. A patient with normal concentrations of electrolytes and a moderately elevated blood ethanol level will most likely have which of the following? (Objective 8)
   a. a lower than normal serum osmolal gap
   b. a higher than normal serum osmolal gap
   c. a lower than normal measured serum osmolality by freezing point depression osmometry
   d. the same results for both the calculated and measured serum osmolality

3. Which of the following sequences accurately reflects the renin-angiotensin-aldosterone system in response to hyponatremia? (Objective 6)
   a. Renin is released from the kidney; renin converts angiotensinogen to angiotensin I; angiotensin I is converted to angiotensin II.
   b. Angiotensinogen is released from the kidney; angiotensinogen converts to angiotensin I; angiotensin I is converted to renin.

REVIEW QUESTIONS (continued)

LEVEL I

5. What is the cation with the highest intracellular concentration? (Objective 7)
   a. potassium
   b. sodium
   c. calcium
   d. magnesium

6. The presence of slightly visible hemolysis can significantly increase the serum level of which of the following electrolytes? (Objective 1)
   a. sodium
   b. potassium
   c. chloride
   d. bicarbonate

7. A 1 millimolal solution of KCl is how many milliosmolal? (Objective 2)
   a. 0.5
   b. 1.0
   c. 2.0
   d. 4.0

8. Which of the following instrumentation techniques is used to measure sodium and potassium and serum or plasma? (Objective 9)
   a. ion-selective electrodes
   b. refractometry
   c. photometry
   d. gas chromatography

9. Which of the following contributes the most to the serum total CO_2? (Objective 1)
   a. partial pressure CO_2 (pCO_2)
   b. dissolved CO_2 (dCO_2)
   c. bicarbonate (HCO_3^-)
   d. carbonium ion

10. Which of the following measures chloride in sweat? (Objective 8)
    a. pH electrodes
    b. coulometry and amperometry
    c. photometry
    d. gas sensing electrodes

11. What is the difference between serum and plasma? (Objective 4)
    a. Serum has fibrinogen and plasma does not.
    b. Serum does not contain fibrinogen but plasma does.
    c. Serum is devoid of all proteins but plasma is not.
    d. Technically, there is no difference between serum and plasma.

12. Which of the following is not considered a colligative property of solution? (Objective 10)
    a. vapor pressure
    b. freezing point depression
    c. temperature coefficient
    d. boiling point

LEVEL II

   c. Angiotensin I is released from the kidney; angiotensin I converts angiotensinogen to angiotensin II; angiotensin II is converted to renin.
   d. Renin is released from the kidney; renin converts angiotensinogen to angiotensin II; angiotensin II is converted to angiotensin I.

4. Which of the following represents the correct responses to the analytes listed for a patient that is "overhydrated"? (Objective 6)
   a. elevated serum osmolality, elevated serum sodium, elevated serum ADH
   b. elevated serum osmolality, decreased serum sodium, decreased serum ADH
   c. decreased serum osmolarity, decreased serum sodium, increased serum ADH
   d. normal values for serum osmolality, sodium, and ADH

5. A patient who has an increase in intravascular fluid volume will experience which of the following? (Objective 7)
   a. no change in blood levels of cardiac natriuretic peptides
   b. a decrease in the release of cardiac natriuretic peptide
   c. an increase in the release of cardiac natriuretic peptide
   d. a suppression of the renin-angiotensin-aldosterone system

6. Hypoosmotic hyponatremia can be due to which of the following? (Objective 8)
   a. excess loss of sodium, or an increased extracellular fluid volume
   b. increase of serum sodium and increased extracellular fluid volume
   c. excess loss of serum sodium and decreased interstitial fluid volume
   d. no change in extracellular fluid volume

7. Syndrome of inappropriate ADH secretion (SIADH) is associated with which of the following? (Objective 6)
   a. decreased ADH and increased serum sodium
   b. increased ADH and increased serum sodium
   c. decreased ADH and decreased serum sodium
   d. increased ADH and decreased serum sodium

8. In most cases, hyperchloremia parallels which of the following conditions? (Objective 5)
   a. hypercalcemia
   b. hyperkalemia
   c. hypernatremia
   d. high levels of organic ions

9. Which of the following conditions is characterized by an increased anion gap? (Objective 1)
   a. salicylate intoxication
   b. diabetes mellitus
   c. lactate acidosis
   d. all of the above

REVIEW QUESTIONS (continued)

LEVEL I

13. What is a major difference between interstitial fluid and plasma? (Objective 5)
    a. Interstitial fluid contains almost no protein, whereas plasma contains a large amount protein.
    b. Interstitial fluid contains mostly proteins and plasma has none.
    c. Interstitial fluid contains no water but plasma does.
    d. Interstitial fluid is the same as plasma.

LEVEL II

10. Hyponatremia may be seen in which of the following conditions? (Objective 5)
    a. vomiting
    b. diarrhea
    c. cirrhosis
    d. edema

11. Hyperkalemia may be seen in which of the following? (Objective 5)
    a. increased urinary loss
    b. increased GI loss
    c. metabolic alkalosis
    d. metabolic acidosis

12. Which of the following conditions can result in inaccurate serum potassium levels (e.g., pseudohyperkalemia)? (Objective 5)
    a. digitalis overdose
    b. tissue hypoxia
    c. hemolysis
    d. high-potassium intake

13. An elevated chloride concentration in sweat is suggestive of which of the following diseases? (Objective 5)
    a. multiple sclerosis
    b. multiple myeloma
    c. cystic fibrosis
    d. Cushing's disease

PEARSON
myhealthprofessionskit™

Use this address to access the interactive Companion Website created for this textbook. Simply select "Clinical Laboratory Science" from the choice of disciplines. Find this book and log in using your user name and password.

REFERENCES

1. Levy MN, Koeppen BM, Stanton BA. *Berne & Levy principles of physiology,* 4th ed. (Philadelphia: Mosby, 2006): 525–526.

2. Klutts JS, Scott MG. Physiology and disorders of water, electrolytes, and acid base metabolism. In Burtis CA, Ashwood ER, Bruns DE (Eds.), *Tietz textbook of clinical chemistry and molecular diagnostics*, 4th ed. (Philadelphia: WB Saunders, 2006): 1747–1757.

3. Lorenz JM, Kleinman LI. Physiology and pathophysiology of body water and electrolytes. In Kaplan LA, Pesce AJ, Kazmierczak SC (Eds.), *Clinical chemistry—theory, analysis, and correlation*, 4th ed. (St. Louis: Mosby, 2003): 441–458.

4. Levin ER, Gardner DG, Samson WK. Natriuretic peptides. *NEJM* (2007) 339(5): 321–328.

5. Scott MG, LeGrys VA, Klutts JS. Electrolytes and blood gases. In Burtis CA, Ashwood ER, Bruns DE (Ed.), *Tietz textbook of clinical chemistry and molecular diagnostics*, 4th ed. (Philadelphia: WB Saunders, 2006): 983–998.

6. Grody WW, Cutting GR, Klinger KW, et al. Laboratory standards and guideline for population-based cystic fibrosis carrier screening. *Genetics in Medicine* (2001), 3(2): 149–154.

7. NCCLS. Sweat Testing: Sample collection and quantitative analysis; approved guideline-second ed. NCCLS document C34-A2 [ISBN 1-56238-407-4]. NCCLS, 940 West Valley Road, Suite 1400, Wayne Pennsylvania 19087-1898 USA, 2000.

13

Blood Gases, pH, and Acid–Base Balance

CHAPTER OUTLINE

■ OBJECTIVES—LEVEL I

Following successful completion of this chapter, the learner will be able to:

1. State the Henderson–Hasselbalch equation and identify the respiratory and metabolic components.
2. Calculate various blood gas parameters given the appropriate equation(s).
3. Identify the four major body buffer systems.
4. Identify the five ways in which carbon dioxide is carried in blood.
5. Identify appropriate calibration materials to use for pH, PCO_2, and PO_2 measurements.
6. Describe the proper control material to use for blood pH, PCO_2, and PO_2 measurements.
7. Identify preanalytical sources of errors in blood-gas analysis.
8. Identify the specimen of choice discuss the proper handling of specimen for blood-gas analysis.

■ OBJECTIVES—LEVEL II

Following successful completion of this chapter, the learner will be able to:

1. Explain how blood buffers contribute to the stabilization of blood pH.
2. Compare and contrast the external and internal convection system for respiration.
3. Identify the significant components of the respiratory apparatus.
4. Describe the effect of increasing and decreasing pH, PCO_2, temperature and 2,3-diphosphoglycerate (2,3-DPG) on dissociation of oxygen from hemoglobin.
5. Explain how changes in PO_2 and PCO_2 affect peripheral and central chemoreceptors.
6. Explain the role of the kidneys in acid–base balance.
7. Given a set of arterial blood gases (ABGs) and pH results, determine which acid–base disorder is most appropriate.
8. Explain how the body attempts to compensate for acid–base disorders; include kidneys, lungs, and body buffers.
9. Identify which analytes are measured and which parameters are calculated using a blood-gas analyzer.

KEY TERMS

Acidemia

Acidosis

Alkalemia

Alkalosis

Base excess

Carbamino compound

External convection

Henderson–Hasselbalch equation

Hypercapnia

Internal convection

Metabolic acidosis

Metabolic alkalosis

Oxygen saturation

Partial pressure

Respiratory acidosis

Respiratory alkalosis

Respiratory apparatus

A CASE IN POINT

A 42-year-old male Caucasian enters the emergency department of a local hospital. He is confused, disoriented, sleepy, and combative, and his breath has a fruity odor and smells like acetone. The emergency room staff establishes intravenous and intraarterial lines. An emergency room physician requests blood samples to be drawn for glucose, electrolytes, serum acetone, and arterial blood gas (ABG). Results of the laboratory tests follow.

Tests	Results	Reference Intervals
Routine Chemistries:		
Sodium	125 mmol/L*	135–145
Potassium	3.4 mmol/L	3.5–5.1
Chloride	90 mmol/L	98–107
Total CO_2	10 mmol/L	22–28
Anion Gap	25 mmol/L	7–16
Glucose	600 mg/dL	74–106
Urea nitrogen (BUN)	40 mg/dL	6–20
Creatinine	3.5 mg/dL	0.7–1.3
Acetone	Positive	Negative
ABGs:		
pH	7.30	7.35–7.45
PCO_2	25 mmHg	35–45
Bicarbonate (HCO_3^-)	16 mmol/L	18–23
Base excess (deficit)	–2	(–2)–(2)

*Millimoles per liter

Issues and Questions to Consider

1. What are the most striking abnormal test results?

2. Is this patient in acidosis or in alkalosis?

3. Is the acid–base imbalance metabolic or respiratory?

4. What is the possible cause of the acid–base imbalance based on the history provided and laboratory findings?

A second set of ABGs were drawn and the results follow.

pH	7.34	
PCO_2	34	mmHg
HCO_3^-	18	mmol/L
Base excess (deficit)	–1	mmol/L

5. Why have these results begun to change?

6. Specifically, what is the body doing to cause these results to shift?

7. What modified condition does this patient have?

A final set of ABGs were drawn from the patient with the following results.

pH	7.45	
PCO_2	40	mmHg
HCO_3^-	22	mmol/L
Base excess	0	mmol/L

8. What has happened to this patient by this point in time?

9. What is the patient's condition now?

WHAT'S AHEAD

▶ Description of pH, buffer systems, and the respiratory system.

▶ Discussion on the role of the kidneys in acid–base balance.

▶ Explanation of disorders of acid–base balance.

▶ Description of methods and instrumentation used to measure pH, blood gases, and hemoglobin.

▶ Discussion of preanalytical factors, maintenance, instrument calibration, quality control, and sources of error.

▶ INTRODUCTION

DESCRIPTION OF pH

Acid–base balance can be described as the maintenance of homeostasis of the hydrogen-ion concentration of body fluids. Any deviation from normal may cause significant changes in the rate of cellular chemical reactions. The outcome of these reactions may have a deleterious effect on an individual. Acid–base balance is defined by the degree of acidity or alkalinity of a body fluid. This is determined by the *pH* or negative log_{10} of the hydrogen-ion concentration [H^+] in moles/L. For example, a pH of 6 can be expressed as 10^{-6} or 0.000001 moles of H^+/L. The range of pH is between 1, which is highly acidic, and 14, which is strongly alkaline. A value of 7 represents a neural pH. Therefore, any solution with a pH below 7 is acidic, and any solution with a pH greater than 7 is a base.

The acidity of a solution is determined by the *concentration of hydrogen ions* (cH^+). An acid is a hydrogen donor. For example, carbonic acid (H_2CO_3) can donate one H^+ through dissociation, which is shown in equation 13.1:

$$H_2CO_3 \leftrightarrow H^+ + HCO_3^- \tag{13.1}$$

An alkaline substance, or base, is described as a substance that can yield hydroxyl ions (OH^-). A base for example bicarbonate (HCO_3^-) is an H^+ or proton acceptor as shown in equation 13.2:

$$HCO_3^- + H^+ \leftrightarrow H_2CO_3 \qquad (13.2)$$

If there are an equal number of H^+ and OH^- in solution, then the pH of the solution is 7. Equal numbers of H^+ and OH^- produce water, which is neutral—neither acidic nor alkaline. This reaction is shown in equation 13.3:

$$H^+ + OH^- \leftrightarrow H_2O \qquad (13.3)$$

A *buffer* is defined as a solution containing a weak acid and its conjugate base that resists changes in pH when a strong acid or base is added. The human body has several *buffer systems* of varying degrees of effectiveness. These buffers represent a potential first line of defense when the body's pH deviates from normal.

▶ BUFFERS AND ACID–BASE BALANCE

American biochemist L. J. Henderson expressed the relationship between H^+ and the acid from which it came and the base that accepted it as shown in equation 13.4:

$$H^+ = K \frac{[acid]}{[base]} \qquad (13.4)$$

In the equation, K represents a constant that is unique to the acid–base pair produced in the solution. The human body has four important blood buffers: bicarbonate, hemoglobin, phosphate, and protein. These buffers react quickly in an attempt to offset any small changes in body pH.

BICARBONATE BUFFER SYSTEM

The most important blood buffer is the H_2CO_3–HCO_3^- pair. This buffering system accounts for the majority of the total buffering capacity in the extracellular space. This is because a large amount of carbon dioxide (CO_2) is produced within the body as a whole, and the potential for large amounts of acid (particularly H_2CO_3) to build up is greatest. In equation 13.5, CO_2 and water react to form H_2CO_3, which in turn dissociates into H^+ and HCO_3^-. This equation is significant and will be referred to throughout the chapter.

$$H_2O + \text{dissolved } CO_2 \xrightleftharpoons[]{\overset{\text{Carbonic anhydrase}}{K_{hydration}}} H_2CO_3 \xrightleftharpoons[]{K_{dissociation}} H^+ + HCO_3^-$$
$$(13.5)$$

The total concentration of CO_2 (tCO_2), HCO_3^-, concentrations of dissolved CO_2 ($cdCO_2$) and cH^+ are interrelated. Each reaction has an equilibrium constant (K) and (pK). The K value for the hydration reaction is 2.29×10^{-3} ($pK = 2.64$) at 37°C, and the K value for the dissociation reaction is 2.04×10^{-4} ($pK = 3.69$).

In 1908, Henderson conducted studies in which he substituted concentrations for HCO_3^-, CO_2, and H^+ while assuming the concentration of water to be constant. He coupled these two reactions together and incorporated the constant K' with a value of 4.68×10^{-7}, a (pK' of 6.33) at 37°. Equation 13.6 represents this concept:

$$K' = \frac{cH^+ \times cHCO_3^-}{cdCO_2} \qquad (13.6)$$

where:

K' = the combined dissociation constant

cH^+ = the concentration of hydrogen ions

$cHCO_3^-$ = the concentration of bicarbonate ion

$cdCO_2$ = the concentration of dissolved carbon dioxide

The amount of $cdCO_2$ includes a small quantity of undissociated H_2CO_3 and can be expressed as $cdCO_2 = \alpha \times PCO_2$. The lowercase Greek letter alpha (α) represents the solubility coefficient for CO_2. It follows then that the $cHCO_3^-$ is equal to the $ctCO_2$ minus $cdCO_2$, which includes H_2CO_3. Thus, by definition the $cHCO_3^-$ includes undissociated sodium bicarbonate, carbonate ($NaCO_3$), and carbamate (carbamino—CO_2; $RCNHCOO^-$), which are present in minute amounts in plasma.

From equation 13.5 we can see that when CO_2, a gas, is dissolved in H_2O in the presence of carbonic anhydrase, one mole of H_2CO_3 is formed. The pressure exerted by the presence of CO_2 gas is referred to as the **partial pressure** of CO_2 (PCO_2). The units for partial pressures of gases can be expressed in mmHg, torr, or pascals. A list of the various units and their relationship to each other is shown in Box 13-1 ∎.

A rearrangement of the Henderson equation (equation 13.6) can allow for the calculation of cH^+. If $cdCO_2$ is replaced by $\alpha \times PCO_2$, then equation 13.7 results.

$$cH^+ = \frac{K' \times \alpha \times PCO_2}{cHCO_3^-} \qquad (13.7)$$

BOX 13-1 Physical Pressure Relationships and Units Often Used in the Literature for Blood Gases

1 atmosphere (atm) = 760 mmHg
*1 torr = 1/760 atm
760 mmHg = 760 torr
1 torr = 1 mmHg
1 mmHg = 0.133 kPa
1 kPa = 7.5 mmHg
760 mmHg = 101,325 pascals

*A torr is a unit of pressure equal to 1 millimeter of mercury rise in a barometer.

A pascal is the Systèm International unit of pressure and is equal to 1 N (newton) m^{-2} or 1 m^{-1} kg s^{-2}

Henderson's research was followed by K. A. Hasselbalch in 1916, who proceeded to show that a logarithmic transformation of equation 13.7 was a more practical form, substituting the symbol pH ($= -\log cH^+$) and pK' ($= -\log K'$). Therefore, pH is defined as the negative log of the activity of H^+ (aH^+), which is actually measured by pH meters or pH electrodes (in blood-gas analyzers). Rearrangement of equation 13.7 results in the **Henderson–Hasselbalch equation** (equation 13.8):

$$pH = pK' + \log \frac{cHCO_3^-}{(\alpha \times PCO_2)} \qquad (13.8)$$

The normal average value for pK' in blood at 37°C is equal to 6.103, and the solubility coefficient (α) for CO_2 gas in normal plasma at 37°C is 0.0306 mmol $\times L^{-1} \times mmHg^{-1}$. If these values are substituted into equation 13.8, the following expression is derived, allowing for the computation of pH:

$$pH = pK' + \log \frac{cHCO_3^-}{(\alpha \times PCO_2)}$$

$$pH = 6.103 + \log \frac{cHCO_3^-}{(0.0306 \times PCO_2)}$$

where:

PCO_2 = measured in millimeters of mercury (mmHg)

$cHCO_3^-$ = measured in mmol/L

$ctCO_2$ = measured in mmol/L

From equation 13.8, the ratio of HCO_3^- to H_2CO_3 can be deduced. This is an important ratio to use when evaluating acid–base disorders. The derivation of the 20:1 ratio is shown in Figure 13-1 ■. Substituting normal values for HCO_3^- and PCO_2, the product of α and PCO_2 is 1.2. This product represents 1.2 milliequivalents (mEq) H_2CO_3/L. Therefore, for each mmHg of PCO_2, there are 0.031 mEq/L of plasma H_2CO_3. A normal healthy individual with a PCO_2 of 40 mmHg will have a plasma carbonic acid level of 1.2 mEq/L.

A practical application of the above expression could be the calculation of hydrogen-ion concentration derived from

---

Assuming an individual is normal and healthy with properly functioning kidneys, lungs, and using normal arterial blood gas values:

$$pH = pK + \log$$

$$7.4 = 6.1 + \log \frac{24 \text{ mmol/L}}{0.031^* \times 40 \text{ mmHg}}$$

$$7.4 = 6.1 + \log \frac{24}{1.2}$$

$$7.4 = 6.1 + \log \frac{20}{1}$$

$$7.4 = 6.1 + 1.3$$

$$7.4 = 7.4$$

*The value 0.031 also allows for conversion of mmHg to millimole per liter.

■ **FIGURE 13-1**   Derivation of the 20:1 ratio of bicarbonate (HCO_3^-) to carbonic acid (H_2CO_3) in whole blood at 37°C.

---

measurements of any two of the four parameters. This calculation would require manipulation of the Henderson–Hasselbalch equation by taking the antilogarithm, combining the constants, and expressing hydrogen-ion concentration in nanomoles per liter (nmol/L) to produce the following expression:

$$cH^+ = 24.1 \times \frac{PCO_2}{cHCO_3^-}$$

If normal values are substituted into the expression for PCO_2 and $cHCO_3^-$ above, then the value for cH^+ is as shown in the following example.

Example

$$cH^+ = 24.1 \times \frac{40}{25.4} \text{ nmol/L} = 38.0 \text{ nmol/L}$$

> ✓ **Checkpoint! 13–1**
>
> *What is the calculated pH for a patient sample if the measured bicarbonate concentration is 22 mmol/L and pCO₂ is 57 mmHg?*

In live human tissue cells, CO_2 is produced and released into the blood. A pressure differential between body water compartments causes this movement. The regulation of H_2CO_3 is under the control of the pulmonary system. This control is immediate and influenced by the presence or absence of CO_2. Therefore, any change in plasma cH_2CO_3 (see equation 13.5), is respiratory in nature. Also, as the cH_2CO_3 in an aqueous medium (liquid) changes, so does the amount of dCO_2 gas. The lungs control the cH_2CO_3 in blood.

Plasma $cHCO_3^-$ is primarily under the control of the kidneys. If there is a change in blood $cHCO_3^-$, it is caused by a metabolic process and associated with the kidneys. Replacing the terms in equation 13.8 with *metabolic* and *respiratory* along with their respective organs, the kidneys and lung, we have equation 13.9:

$$pH = pK + \log \frac{\text{metabolic (kidneys)}}{\text{respiratory (lungs)}} \qquad (13.9)$$

The HCO_3^- buffer system is significant for several reasons: (1) H^+ is eliminated as CO_2 by the lungs (refer to equation 13.5), (2) a change in PCO_2 modifies the ventilation rate, and (3) the kidneys can alter the concentration of bicarbonate. This buffer system also counters the effects of increases in concentrations of fixed nonvolatile acids such as glycolic acid (a metabolite of ethylene glycol) by binding the dissociated H^+.

HEMOGLOBIN BUFFER SYSTEM

Hemoglobin (Hb) serves several roles in acid–base balance and the respiration process. These roles include (1) binding approximately 20% of CO_2 as a carbamino compound (Hb-NH–COO⁻), (2) binds and transports H^+ and O_2, and (3) participates in the *chloride shift* and *isohydric shift*. These two processes allow large amounts of CO_2 produced by metabolism to be carried in the blood with little or no change in pH.

PHOSPHATE BUFFER SYSTEM

The phosphate buffers are essential within the erythrocytes. Organic phosphate in the form of 2,3-diphosphoglycerate (2,3-DPG) accounts for <20% of the nonbicarbonate buffer value of erythrocyte fluid.

Phosphate buffers enable the kidney to excrete H^+. About 95% of phosphate is present as a dihydrogen phosphate, NaH_2PO_4, which reacts with strong acids such as hydrochloric acid to form sodium chloride and a weaker acid, H_3PO_4. This reaction results in the neutralization of strong acids. The remaining 5% of phosphate exists as monohydrogen phosphate (HPO_4^{2-}).

PROTEIN BUFFER SYSTEM

Most protein buffers are cellular buffers, with a small percentage occurring in the plasma. The buffering capacity of proteins comes from the terminal groups of amino acids. Some amino acids are basic (e.g., histidine and lysine), and others are acidic (e.g., glutamic acid and asparagine). The ionizable side chains of the amino acids allow them to pick up or release H^+. The protein albumin, with 16 histidine residues within its structure, contributes a significant percentage of buffering capability.

In summary, the relative buffering capacity (mEq/L) from highest to lowest is hemoglobin ($pK_a = 7.2$), protein, bicarbonate ($pK_a = 6.33$), and phosphate ($pK_a = 6.8$). Buffering capacity depends on the concentration of the buffer and the relationship between the apparent pK_a (pK_a) of the buffer and the desired pH. Hemoglobin is also present in highest concentration, $\sim$53 mmol/L; next is HCO_3^-, $\sim$25 mmol/L; and finally phosphate, $\sim$1.2 mmol/L. Therefore, the major buffer of blood is Hb, which is found primarily in erythrocytes. Hemoglobin has the highest relative buffering capacity and concentration (mmol/L) at physiological pH.

 Checkpoint! 13–2

Which buffer is the major buffer of blood and why?

▶ RESPIRATORY SYSTEM AND GAS EXCHANGE

DIFFUSION OF OXYGEN AND CARBON DIOXIDE

The most important fundamental mechanism of O_2 and CO_2 transport is *diffusion*. Diffusion is the movement of an uncharged, hydrophobic solute through a lipid bilayer. The random movements of molecules such as O_2 and CO_2, whether in a gaseous phase or dissolved in water, result in a net movement of the substance from regions of high concentration to regions of low concentration. No expenditure of energy is involved. The driving force for diffusion is the *concentration gradient*. A concentration gradient, where *gradient* refers to a change in value of a quantity, represents the difference in concentration on either side of a membrane, which allows for diffusion of solutes or gases from one side to another.

EXTERNAL AND INTERNAL CONVECTION SYSTEMS

Ventilation is the process of moving air into and out of the lungs. In larger organisms and mammals, simple diffusion of gases will often not be adequate. Thus, to facilitate this movement of gases, a mechanism called **external convection** is employed. In humans, the external convection system consist of the lungs, the airway, and respiration muscles. The external convection system serves to maximize gas exchange by continuously supplying atmospheric air to the external surface of the gas-exchange barrier (e.g., alveoli), thereby maintaining a high external PO_2 and low external PCO_2. The average human inspires about 4.0 liters of fresh air per minute with an alveolar PO_2 of approximately 100 mmHg and an alveolar PCO_2 of about 40 mmHg.

The circulatory system is an **internal convection** system that maximizes the flow of O_2 and CO_2 across the gas-exchange barriers by delivering blood that has low PO_2 and a high PCO_2 to the inner surface of the barrier. Perfusion is the process of delivering blood to the lungs.

Respiratory pigments such as hemoglobin improve the dynamics of O_2 uptake by blood passing through the lungs. Hb reversibly binds to about 96% of the O_2 that diffuses from the alveolar air spaces to the pulmonary capillary blood, thereby greatly increasing the carrying capacity of blood for O_2. Hemoglobin plays a significant role in the transport of CO_2 by reversibly binding CO_2 and by acting as a pH buffer as described above.

 Checkpoint! 13–3

Compare and contrast external and internal convection systems and include anatomical, physiological, and mechanical features.

THE RESPIRATORY APPARATUS

The respiratory system uses highly efficient convection systems (ventilatory and circulatory) for long-distance transport of O_2 and CO_2, and it reserves diffusion primarily for short-distance movements of O_2 and CO_2. The important components of the respiratory apparatus are the following[1,2]

1. It is a means of moving outside air to the alveolar air spaces. *Inspiration* occurs when the muscles of respiration increase the volume of the thoracic cavity, creating a partial vacuum in the alveolar air spaces, and causing the

alveoli to expand passively. *Fraction of inspired air* (FIO$_2$) is the percentage of oxygen in the ambient air, which is equal to 0.2093 (20.93%).

2. It carries O$_2$ and CO$_2$ in the blood. Erythrocytes transport O$_2$ from the lungs to the peripheral tissues and transport CO$_2$ in the opposite direction. Erythrocytes have especially high levels of Hb and other components such as 2,3-DPG, carbonic anhydrase, and Cl$^-$–HCO$_3^-$ exchanges that help to rapidly load and unload large amounts of O$_2$ and CO$_2$. In the pulmonary capillaries, Hb binds O$_2$, thus enabling the blood to carry about 65 times more O$_2$ than normal saline. Hemoglobin also reacts with some of the CO$_2$ produced by the mitochondria and carries this CO$_2$ back to the lungs.

3. It provides a surface for gas exchange. The alveoli are the gas-exchange barriers in human lungs. Alveoli provide an enormous surface area but are not actually thick. This large, thin surface area allows for passive diffusion of gases between the alveolar air spaces and the pulmonary capillaries.

4. It contains an internal convection system and a circulatory system that consists of a four-chamber heart. There is a separate systemic and pulmonary circulation system that serves to deliver oxygenated blood to peripheral tissues and returns CO$_2$-laden blood back to the lungs for removal (perfusion).

5. It provides A mechanism for locally regulating ventilation and perfusion. The entire process of bringing air into the body (inspiration) and removing CO$_2$ (*exhalation*) is not uniform throughout the lungs. However, efficient gas exchange requires, to the best possible extent, a ratio of ventilation to perfusion to be uniform for all alveoli. The

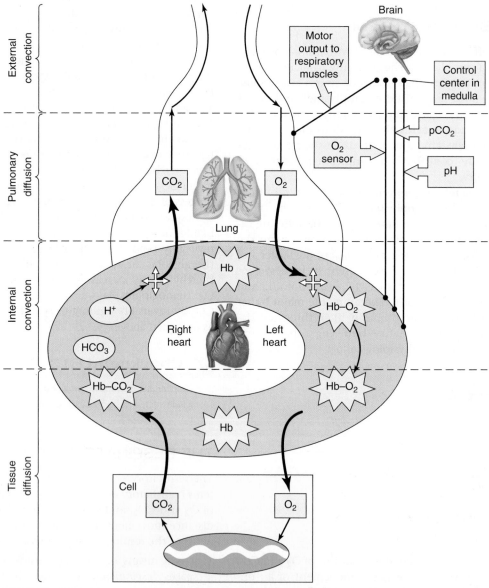

■ FIGURE 13-2   Summary of the respiratory apparatus used to move O$_2$ and CO$_2$ through the body.

body utilizes a sophisticated feedback-control mechanism to regulate this ratio.

6. It is a mechanism for centrally regulating ventilation. Respiratory control centers in the central nervous system (CNS), consisting of neurons within the medulla called *respiratory-related neurons*, rhythmically stimulate the muscles of inspiration. These respiratory control centers must also modify the pattern of ventilation during exercise or changes in physical or mental activity. Sensors[2] for arterial PO_2, PCO_2, and pH are part of feedback loops that stabilize these three blood-gas parameters. These sensors include peripheral chemorecptors located in the carotid bodies and aortic bodies in the thorax and central chemoreceptors located on the "brain" side of the blood–brain barrier (see section on chemorecptors later in the chapter).

A summary of the **respiratory apparatus** just described is shown in Figure 13-2 ■.

✓ Checkpoint! 13–4

Atmospheric oxygen needs to enter the body and be distributed to cells. Briefly outline the processes involving the respiratory apparatus that serve to accomplish this end.

▶ DISTRIBUTION OF GASES IN THE BODY

The distribution of gases, especially O_2 and CO_2 in atmospheric air, is shown in Table 13-1 ✪. The table shows that the total of all partial pressures equals 760 mmHg. These numbers represent what the total pressure would be under specified conditions for temperature and elevation. A more detailed representation of relative pressures in various compartments throughout the body is found in (Figure 13-3 ■).[3]

✪ TABLE 13-1		
The Percentage Composition and Partial Pressures (in mmHg) of Gases in a Dry Air Atmosphere		
	DRY AIR Atmosphere	
Gas	Fraction in Air (%)	Partial Pressure at Sea Level (mmHg)
Nitrogen	78.17	593.70
Oxygen	20.90	159.0
Carbon dioxide	0.03	0.23
Argon	0.90	7.07
Water	0	0
Total	100	760

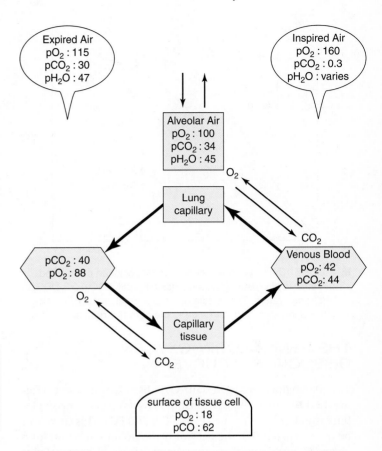

■ FIGURE 13-3 Partial pressures of oxygen and carbon dioxide in various tissue compartments throughout the body. (Note: Units for partial pressures are in mmHg.)

The mechanism by which O_2 and CO_2 are transported throughout the body contributes to the maintenance of normal body pH through the elimination or retention of CO_2.

TRANSPORT OXYGEN IN THE BLOOD

The complete hemoglobin molecule has the stoichiometry $[\alpha(heme)]_2 \, [\beta(heme)]_2$ and can bind as many as four O_2 molecules. Hemoglobin can bind O_2 only when the iron is in the ferrous (Fe^{2+}) state. The Fe^{2+} in Hb is oxidized to ferric iron (Fe^{3+}), either spontaneously or under the influence of compounds such as nitrites or sulfonamide. The result of this oxidation is methemoglobin, which is incapable of binding O_2.

The environment provided by the globin portion (polypeptide) is critical to the physiology of the O_2–heme interaction. This interaction is fully reversible, allowing repetitive capture and release of O_2. Thus, when oxygen binds with heme, Fe^{2+} is oxidized to Fe^{3+}. This reaction is *irreversible*. However, when heme is part of hemoglobin, interactions with about 20 amino acids cradle the heme in the globin so that O_2 loosely and *reversibly* binds to Fe^{2+}. The most important amino acid in this reaction is histidine, which binds the Fe^{2+}. Each histidine donates a negative charge that serves to stabilize the Fe^{2+}–O_2 complex.

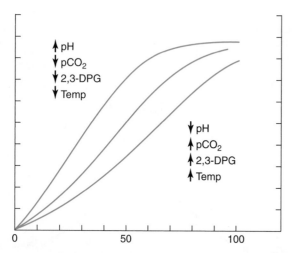

■ **FIGURE 13-4** Hemoglobin–oxygen dissociation curves and the effects of changing factors, including pH, PCO_2, 2,3-DPG, and temperature. The middle curve is normal, the curve to the left is a left shift, and the curve to the right is a right shift.

THE HEMOGLOBIN–OXYGEN DISSOCIATION CURVE

The hemoglobin–oxygen (Hb–O_2) dissociation curve (Figure 13-4 ■) shows the relationship between percent hemoglobin saturation with oxygen (%Hb–O_2) and PO_2. This curve can be used to determine the percent hemoglobin saturation for a given PO_2 and O_2 content. Superimposed onto the normal plot are curves that represent the effects of increased or decreased factors such as pH, temperature, PCO_2, and 2,3-DPG. The curves are shifted to the left or right, depending on the quantity of the respective factors.

Notice that the curve is sigmoid or S-shaped. The reason for this lies in the relationship of the four O_2-binding sites on the hemoglobin molecule. At low PO_2 values, increases in PO_2 produce relatively small increases in O_2 binding. This reflects the relatively low O_2 affinity of Hb in this molecular state. At moderate PO_2 values, the amount of bound O_2 increases more steeply with increases in PO_2, reflecting the increased O_2 affinity as more Hb molecules shift to a more relaxed state. Ultimately, the Hb–O_2 dissociation curve flattens out at high PO_2 values as Hb becomes more saturated.

The PO_2 at which the hemoglobin is half-saturated is knows as *P50*. A shift in the Hb–O_2 dissociation curve is detected by the P50. A normal P50 is approximately 26.5 mmHg. A decreased P50 means the Hb–O_2 dissociation curve has shifted to the left with an elevated percentage of hemoglobin saturation and a decrease in available O_2 to the tissue. An increased P50 means the Hb–O_2 dissociation curve has shifted to the right with a reduced percentage of Hb saturation and an increase in available O_2 to the tissue.

The presence of disease can result in significant changes in the Hb–O_2 dissociation curve. Anemia, which is characterized by a reduced Hb content of the blood, will cause a marked reduction for O_2 carried in the blood.

 Checkpoint! 13–5

What is the approximate percent hemoglobin saturation if a sample of whole blood has a PO_2 of 30 mmHg?

FACTORS AFFECTING THE HEMOGLOBIN–OXYGEN DISSOCIATION CURVE

Metabolically active tissues are characterized as (1) having a high demand for O_2, (2) being warm, (3) producing large amounts of CO_2, and (4) being acidic. Hemoglobin is intrinsically sensitive to high temperatures, high PCO_2, and low pH. These three factors tend to decrease the O_2 affinity of hemoglobin by acting at nonheme sites and shifting the equilibrium between molecular states of hemoglobin.

Temperature
Increasing the temperature causes the Hb–O_2 dissociation curve to shift to the right while decreasing the temperature has the opposite effect. At a PO_2 of 40 mmHg (mixed-venous blood), the amount of O_2 bound to hemoglobin becomes progressively less at higher temperatures. Thus, high temperature decreases the O_2 affinity of hemoglobin and O_2 is released. The effects of increasing or decreasing temperature on the Hb–O_2 are shown in Figure 13-4.

pH
The presence of excess H^+ in the blood—and hence a lowered pH—causes the hemoglobin dissociation curve to shift to the right. This will occur in conditions associated with respiratory acidosis and increased amounts of acid metabolites. The result of a shift to the right is a decrease in O_2 affinity for hemoglobin. The effects of changes in pH on the Hb–O_2 dissociation curve are shown in Figure 13-4.

The effect of pH on hemoglobin–oxygen affinity is known as the Bohr effect. When blood reaches tissue, the affinity of hemoglobin for oxygen is decreased by the high cH^+, thus allowing the more efficient unloading of O_2 at these sites. This equilibrium is shown in equation 13.10.[4]

$$Hb(O_2)_4 + 2H^+ \leftrightarrow Hb(H^+)_2 + 4O_2 \qquad (13.10)$$

PCO_2
A decrease in blood pH often is the result of an increase in extracellular PCO_2, which causes CO_2 to enter erythrocytes, leading to a decrease in intracellular pH. An increase in blood CO_2 concentration is referred to as **hypercapnia**. Hypercapnia results in a shift to the right of the Hb–O_2 dissociation curve as shown in Figure 13-4. The mechanism is as follows: As PCO_2 increases, CO_2 combines with unprotonated amino groups on Hb-NH_2 to form a **carbamino compound** (Hb–NH–COO^-). Only the four amino termini of the globin chains are susceptible to appreciable carbamino formation.

Amino groups exist in protonated form ($Hb-NH_3^+$) in equilibrium with unprotonated form $Hb-NH_2$. The reaction of CO_2 with $Hb-NH_2$ tends to shift hemoglobin away from $Hb-NH_3^+$ toward $Hb-NH-COO^-$ as shown in equation 13.11.[4]

$$Hb-NH_3^+ \leftrightarrow Hb-NH_2 \leftrightarrow Hb-NH-COO^- \quad (13.11)$$

The overall effect of carbamino formation is therefore a negative shift in the charge on one amino acid side chain, causing a shift in the formation of Hb, and reducing its O_2 affinity as shown in equation 13.12.

$$(O_2)_4Hb-NH_3^+ + CO_2 \leftrightarrow (O_2)_3Hb-NH-COO^- + O_2 + 2H^+ \quad (13.12)$$

Therefore, an increase in PCO_2 causes hemoglobin to release O_2, which is important for the systemic tissues. Conversely, an increase in PO_2 causes hemoglobin to unload CO_2, which is important in the lungs.

2,3-Diphosphoglycerate

The affinity of Hb for O_2 is highly sensitive to the presence of the glycolytic metabolite 2,3-DPG. The concentration of 2,3-DPG is about the same as that of hemoglobin. Diphosphoglycerate binds to hemoglobin in a 1:1 stoichiometry. It interacts with the central cavity formed by the two beta chains. When O_2 binds to Hb, the shape of the central cavity changes and the DPG-bound Hb becomes unstable. Therefore, the 2,3-DPG affinity of oxygenated hemoglobin is only about 1% as great as that of deoxygenated hemoglobin. Conversely, binding of 2,3-DPG to Hb destabilized the interaction of Hb with O_2. This shifts the following reaction in equation 13.13 to the right and O_2 is released as.

$$Hb(O_2)_4 + 2,3\text{-DPG} \leftrightarrow Hb(2,3\text{-DPG}) + 4O_2 \quad (13.13)$$

The impact on the Hb–O_2 dissociation curve is a shift to the right. The effects of any change in 2,3-DPG concentrations can be seen in Figure 13.4.

In tissue *hypoxia* seen in patients with anemia or in a high-altitude environment where a low PO_2 exists, glycolysis begins to increase, which leads to an increase in the levels of 2,3-DPG and a decrease in oxygen affinity for hemoglobin.

✓ Checkpoint! 13–6

Predict whether the hemoglobin–oxygen dissociation curve will shift to the right or left for each of the following:

1. *Increase blood pH.*
2. *Increase 2,3-DPG.*

Transport of CO_2 in the Blood

Carbon dioxide is transported in the blood primarily as HCO_3^- and is carried to a lesser extent by four other related compounds. The sum of all five forms constitutes in part the total carbon-dioxide content (tCO_2).

Bicarbonate HCO_3^- can form in three ways: (1) H_2CO_3 can dissociate into HCO_3^- and H^+, (2) HCO_3^- forms when carbonate combines with H^+, and (3) CO_2 can combine directly with hydroxyl ions (OH^-) to form HCO_3^-.

Carbonate HCO_3^- dissociates to form carbonate (CO_3^{2-}) plus H^+. The concentration of CO_3^{2-} is extremely low and is not quantitatively important for CO_2 transport.

Carbonic Acid H_2CO_3 can form from CO_2 and H_2O and from H^+ and HCO_3^-. Similar to CO_3^{2-}, H_2CO_3 is formed in low concentrations, so it is not an important transporter of CO_2.

Dissolved Carbon Dioxide The concentration of dCO_2 makes up only about 5% of tCO_2 of arterial blood.

Carbamino Compounds In arterial blood, carbamino compounds account for approximately 5% of total CO_2.

In the 1920s, Van Slyke introduced a method that measured all five of these compounds, and this method remains the basis for assaying blood HCO_3^- in clinical laboratories today.[4]

The transport of CO_2 depends on carbonic anhydrase, the Cl^-–HCO_3^- exchanger (chloride or *Hamburger shift*) and hemoglobin. The movement of CO_2 from the systemic capillaries to the lungs is illustrated in Figure 13-5 ■. While blood is flowing through the capillary beds, it picks up CO_2 produced from biological oxidation in the mitochondria and diffuses out of cells, through the extracellular space, across the capillary

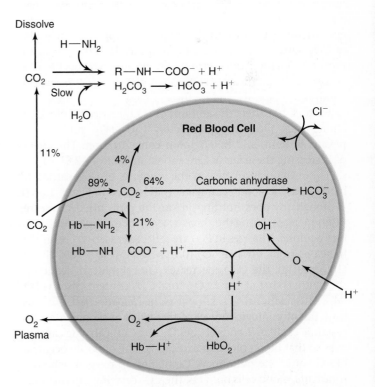

■ FIGURE 13-5   Movement of carbon dioxide from systemic capillaries to the lungs.

✪ TABLE 13-2

Blood Parameters Affecting the Amount of Total CO_2 Carried by Blood

Blood Parameters	Effects of Increasing the Parameter
PCO_2	• Increased concentration of dissolved CO_2 • Increased formation of HCO_3^- • Increased formation of carbamino compounds
Concentration of plasma protein	• Increased plasma buffering power, thus the capacity for consuming H^+ indirectly, promotes the formation of HCO_3^-.
plasma pH	• Increased formation of HCO_3^- in plasma • Increased pH inside red cell, thus promoting formation of HCO_3^- and carbamino hemoglobin
Concentration of hemoglobin	• Increased formation of carbamino hemoglobin • Increased buffering power inside erythrocyte
PO_2	• Tends to decrease buffering power of hemoglobin and the capacity for consuming H^+ decreases, thus indirectly restrains formation of HCO_3^- and carbamino hemoglobin • Decreased formation of carbamino hemoglobin

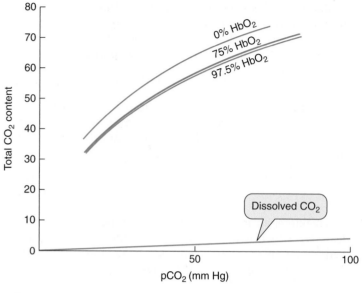

■ **FIGURE 13-6**   Carbon dioxide dissociation curves illustrate the Haldane effect.

✓ Checkpoint! 13–7

Carbon dioxide needs to be removed from cells, carried through the blood, and removed from the body. Identify the five different forms in which CO_2 is carried in the blood.

endothelium, and into the blood plasma. Approximately 11% of CO_2 remains in blood plasma and travels to the lungs, but most (89%) enters the erythrocytes. Inside erythrocytes, CO_2 follows three pathways: (1) becomes dissolved, (2) forms carbamino compounds, and (3) forms HCO_3^{-4}.

Factors that affect the amount of tCO_2 carried by blood include three blood-gas parameters (pH, PO_2, and PCO_2), plasma protein, and hemoglobin concentration. The effects of increasing these factors are shown in Table 13-2 ✪.

The relationship between the amounts of CO_2 carried by blood and the blood parameters is illustrated in Figure 13-6 ■. Although pH does not appear on the plot, pH decreases as PCO_2 increases along the *x*-axis as seen, for example, in respiratory acidosis. Changes in PO_2 are reflected as Hb–O_2, so as PO_2 increases from ~0% Hb–O_2 (0 mmHg PO_2) to ~98% Hb–O_2 (100 mmHg), the PCO_2 decreases. One significant feature of the figure is that at any PCO_2, total CO_2 contents rises as PO_2 falls. This relationship is known as the *Haldane effect*.[4] To illustrate the physiologic impact of this effect as blood enters the systemic capillaries and releases O_2, the CO_2-carrying capacity rises so that blood picks up extra CO_2. The opposite occurs as blood enters the pulmonary capillaries. In oxygen-rich environments, blood cells bind O_2, the CO_2 carrying capacity falls, and excess CO_2 is removed from the cells. The dissolved CO_2 (dCO_2) component is plotted at the bottom and shows that dCO_2 rises only slightly with increases in PCO_2.

▶ CONTROL OF VENTILATION

Breathing is something we rarely think about until something goes wrong. However, people with asthma, chronic obstructive pulmonary disease (COPD), or other respiratory conditions are very aware of their breathing. The feeling of dyspnea or shortness of breath is one of the most unpleasant feelings in life. The control of ventilation is one of the most important of all brain functions.

Two events must be accomplished for proper ventilatory control. First, there must be automatic rhythm in the contractions of the respiratory muscles. Second, the rhythm of contraction must accommodate changing metabolic demands—for example, changes in blood pH, PO_2, or PCO_2.

AUTOMATIC CENTERS IN THE BRAIN STEM

The output of the CNS to muscles of the ventilatory apparatus usually occurs automatically. There is no conscious effort involved. Respiratory-related neurons within the medulla of the brain generate signals that are distributed appropriately to various cranial and spinal motors neurons. These neurons directly innervate the respiratory muscle. The specific site containing the neurons that generate the respiratory rhythm is unknown, but their functions have been widely studied.

The most important respiratory motor neurons are those that send axons out from the phrenic nerve to innervate the diaphragm. The diaphragm is one of the primary muscles of inspiration. During exercise when respiratory output increases, activity also occurs in motor neurons that innervate a wide variety of accessory muscles of inspiration and expiration.

All of the muscles of respiration are active at different times within the respiratory cycle, and the brain can alter this timing depending on prevailing conditions. The pattern of alternating inspiratory and expiratory activity that occurs under normal conditions during sleep, at rest, and during moderate exercise is termed *eupnea*. During eupnea, neural output to respiratory muscles is highly regular, with rhythmic bursts of activity during inspiration only to the diaphragm and external intercostal muscles. Expiration occurs only because of the cessation of inspiration and passive elastic recoil of the lungs. The phrenic nerve activity is increased during exercise, and both amplitude and frequency of phrenic nerve activity changes to accommodate an increase in body activity. The activity of the nerves that supply accessory muscle of inspiration is increased. With this increased effort, the accessory muscles of expiration also become active, thus producing exhalation that is more rapid and permitting an increase in respiratory frequency.

PERIPHERAL AND CENTRAL CHEMORECEPTORS

Respiration is initiated by neurons that generate the respiratory rhythm functions like a clock that times the automatic cycling of inspiration and expiration. Sometimes the clock stops "ticking" in the absence of *tonic drive* inputs, resulting in the absence of ventilation, or *apnea*. This tonic drive develops from many sources, but the most important are the central and peripheral chemoreceptors that monitor O_2, CO_2, and pH. The respiratory neurons change with the strength of the drive from the chemoreceptors, resulting in changes in both depth and frequency of ventilation.

The peripheral chemoreceptors, located in the carotid and aortic bodies, are mainly sensitive to decreases in arterial PO_2. These two peripheral chemoreceptor sites are shown in (Figure 13-7 ■).[2] The sensitivity of the chemorecpetors is heightened by high PCO_2 and low pH. They relay their sensory information to the medulla via the glossopharyngeal nerve and the vagus nerve. The central chemoreceptors, located on the brain side of the blood–brain barrier, are sensitive only to increases in arterial PCO_2, and more slowly to a decrease in arterial pH, and not at all to changes in arterial PO_2.

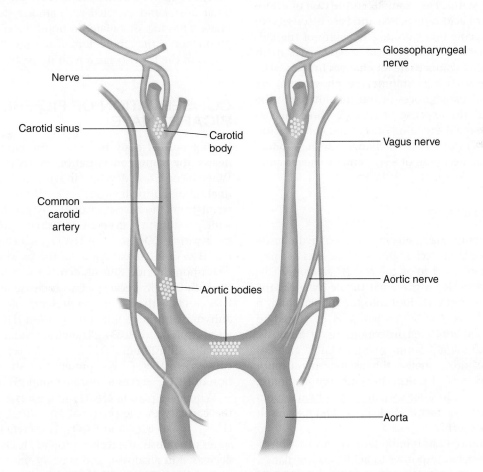

Nerve

Carotid sinus

Common carotid artery

Carotid body

Aortic bodies

Glossopharyngeal nerve

Vagus nerve

Aortic nerve

Aorta

■ FIGURE 13-7   Anatomic positions of the aortic and carotid bodies that represent two peripheral chemoreceptor sites.

Changes in these signals cause an increase in alveolar ventilation that tends to return these O_2 and CO_2 to normal. Therefore, the peripheral and central chemoreceptors, in addition to supplying tonic drive to the respiratory neurons, form the critical sensory end of the negative-feedback system that uses respiratory output to stabilize arterial PO_2, PCO_2, and pH.

✓ Checkpoint! 13–8

Explain how peripheral and central chemoreceptors respond to changes in blood PCO_2.

▶ ROLE OF THE KIDNEYS IN ACID–BASE BALANCE

The kidneys play an important role in the body's attempt to maintain a physiological pH. The average pH of plasma and of the glomerular filtrate is ~7.4, whereas the average urinary pH is ~6.0. These values represent the kidney's attempt to excrete nonvolatile acids that are produced by metabolic processes. Renal function tends to adjust to different alterations of acid–base status. For example, in the case of acidosis, renal excretion of acid is increased and base is conserved; just the opposite occurs in alkalosis. The ability of the kidneys to excrete variable amounts of acid or base makes them significant organs that compensate for changes in body pH.

Several different acids (e.g., sulfuric and phosphoric) are produced during metabolic processes and are buffered in the extracelluar fluid at the expense of HCO_3^-. Three mechanisms that facilitate renal excretion of acid and conservation of HCO_3^- include (1) the Na^+-H^+ exchange, (2) the production of ammonia and excretion of NH_4^+, and (3) the reclamation of HCO_3^-.

Na^+-H^+ EXCHANGE

The kidney cells contain a plasma-membrane, ATP-hydrolyzing protein that is capable of exchanging sodium ions for protons, a mechanism referred to as the *Na^+-H^+ exchange*. This process occurs principally in the renal tubule, whereas the Na^+-H^+ exchanger extrudes H^+ ions into the tubular fluid in exchange for Na^+ ions. Na^+-H^+ exchange is enhanced in states of acidosis and inhibited in conditions of alkalosis. Cells in the proximal tubules cannot maintain an H^+ gradient of more than ~1 pH unit, whereas the distal tubules cannot maintain more than ~3 pH units. Therefore, the maximum urine acidity is ~pH 4.4. In some specific forms of renal tubular acidosis, this exchange process is defective and may result in a decrease in blood pH.[5]

Potassium-ion homeostasis is important in acid–base physiology, especially in renal responses to acid–base imbalance. Potassium ions compete with H^+ in the renal tubular Na^+-H^+ exchanger mechanism. If the intracellular K^+ concentration of renal tubular cells is elevated, more K^+ and fewer H^+ are exchanged for Na^+. The result is less-acidic urine, thereby increasing the acidity of body fluids. If K^+ concentration is decreased, then more H^+ ions are exchanged for Na^+, and the urine pH becomes acidic and body fluids more alkaline. Therefore, hyperkalemia (increase in serum K^+) contributes to acidosis and hypokalemia (decrease in serum K^+) to alkalosis.

AMMONIA PRODUCTION AND AMMONIUM-ION EXCRETION IN THE KIDNEY

Ammonia is produced in the renal tubular cell through the reactions of glutamine and other amino acids that are derived from muscle and liver cells. The NH_4^+ produced from the deamination of amino acids dissociates into NH_3 and H^+, depending on the pH. A normal blood pH would create a ratio of NH_4^+ to NH_3 of about 100:1. Ammonia is a gas and readily diffuses across the cell membrane into the tubular lumen, where it combines with H^+ to form NH_4^+. The equilibrium between NH_4^+ and NH_3 at an acid pH shifts to the left at a ratio of ~10,000 to 1, thereby facilitating the formation of NH_4^+. Most of the NH_4^+ created in the tubular lumen cannot easily cross cell membranes and therefore is trapped in the tubular urines and excreted with anions—for examples, phosphate, chloride, or sulfate. In nondiseased individuals, NH_4^+ production is the tubular lumen accounts for the excretion of ~60% of the H^+ associated with nonvolatile acids.[5]

CONSERVATION OF FILTERED BICARBONATE

During normal body functions, the glomerular filtrate has nearly the same concentration of HCO_3^- as does plasma. When a condition develops in which the acidity in the proximal tubular urine increases, the $cHCO_3^-$ decreases. This may result from the excretion of H^+ by the Na^+-H^+ exchangers, which results in a decrease in urinary pH. The H^+ excreted react with HCO_3^- to form H_2CO_3, which dissociates to CO_2 and H_2O in the brush border of the proximal tubular cells.

Carbon dioxide diffuses across the tubular wall and into the tubular cells because of an increase in urinary CO_2. The CO_2 reacts with H_2O in the presence of cytoplasm carbonic anhydrase in the tubular cells to form H_2CO_3, which dissociates into H^+ and HCO_3^-. Therefore, reclamation of HCO_3^- is actually diffusion of CO_2 into tubular cells, and it eventually converts to HCO_3^-. The presence of an increase concentration of HCO_3^- serves to restore blood pH back to normal.[5]

Whenever plasma $cHCO_3^-$ increases above ~28 mmol/L, the capacity of the proximal and distal tubules to reclaim HCO_3^- is exceeded and HCO_3^- is excreted into the urine. In states of acidosis, the reclamation of HCO_3^- is enhanced (and decreased in alkalosis), and is likely the result of an increase in Na^+-H^+ exchange. This process of conservation or excretion of HCO_3^- in the kidney during acidosis or alkalosis serves to support other compensatory mechanisms that attempt to restore the normal $cHCO_3^-/cdCO_2$ ratio.[5]

▶ DISORDERS OF ACID–BASE BALANCE

Acid–base disorders often represent severe conditions for patients. The hallmark for these disorders is an abnormal blood pH. A patient's blood pH may be lower than the reference range, a condition termed **acidemia**. A blood pH higher than the reference range is referred to as **alkalemia**. The disorders associated with acidemia are termed **acidosis**; similarly, those associated with alkalemia are called **alkalosis**.

If the condition or disorder is the result of a problem associated with the respiratory system—specifically, PCO_2—then the imbalance is respiratory in nature. A disorder involving a change in $cHCO_3^-$ is metabolic. A disorder is *primary* if the imbalance results from a change in PCO_2 or $cHCO_3^-$. Therefore, a patient may have a primary metabolic acidosis because of a decrease in bicarbonate. Several examples of these types of disorders will be discussed in the following sections. Finally, *mixed acid–base disorders* are a result of more than one pathologic process occurring in an individual. Mixed acid–base disorders are more complex medically and clinically and are beyond the scope of this chapter; therefore, they will not be included in this chapter.

COMPENSATORY MECHANISMS

The body has several mechanisms that serve to counter any imbalance in pH, electrolytes, and gases. These mechanisms are designed to compensate for changes in ion concentrations, especially cH^+. The human body does not tolerate even small changes in blood pH and will attempt to bring the blood pH back to normal. This attempt of the body to restore itself back to normal is termed *compensation*. For example. if the body's blood pH becomes acidic because of a metabolic disturbance, then the lungs attempt to reduce the PCO_2. This compensation can be full or partial, depending on the amount of change in pH it takes to return to normal.

DETERMINING ACID–BASE STATUS

Measurement and interpretation of Na^+, Cl^-, and HCO_3^- levels is important early in the assessment of patients with suspected acid–base imbalance. Normally, with electrolyte disorders, Na^+ and Cl^- change in the same direction and by nearly the same amount, so Cl^- can be compared to Na^+. If the cCl^- changes independently of, or out of proportion to, changes in Na^+, it usually indicate an acid–base disorder.

Determinations of $cHCO_3^-$ are used to assess the metabolic component of acid–base disorders, whether a primary metabolic disorder or compensation for a respiratory disorder. The $cHCO_3^-$ (mmol/L) can be determined using a blood-gas analyzer. It is not directly measured but calculated based on the measurements of PCO_2 and pH. A bicarbonate value includes ions of hydrogen carbonate, carbonate, and carbamate in the plasma.

A calculated anion gap (discussed in Chapter 12, Body Water and Electrolyte Homeostasis) is a significant indicator for the presence of an acid–base disorder. It is recommended that anion-gap values be compared with the patient's previous results if they are available because of possible variations in method performance for each electrolyte included in the calculation.

Measurement of PCO_2 is essential to assess the respiratory component of acid–base disorders. Arterial blood samples are recommended to evaluate oxygenation, but venous blood gases can be used instead. When a patient is not in shock, venous PCO_2 is normally about 3–4 mmHg higher and venous pH 0.03–0.04 lower than arterial.

Determining the pH of a blood specimen is useful for assessing the magnitude of acidosis or alkalosis, but remember that pH is significantly affected by compensation or mixed disorders. Ion-selective electrodes are routinely used to measure pH in blood.

The normal or reference values for several blood parameters are presented in Table 13-3 ✪ for the reader's convenience. These reference values may vary, depending on the instrumentation and methods used to determine their concentrations.

Base excess is a calculated blood-gas parameter used to evaluate the metabolic components of the patient's acid–base status. The term *base excess* describes a clinical situation in which there is an excess of HCO_3^- (*positive* base excess) or a deficit of HCO_3^- (*negative* base excess). The base excess in a blood sample with a pH of 7.40, PCO_2 40 mmHg, and cHb of 15 mg/dL at a temperature of 37°C is zero. Thus, the measured hemoglobin (Hb) value, PCO_2, and $cHCO_3^-$ are essential for determining base excess or base deficit. The calculation for base excess is shown in the following example.

Example
Base excess = $(1.0 - 0.0143 \text{ Hb})(HCO_3^-) - (9.5 + 1.63 \text{ Hb})$ $(7.4 \text{ pH}) - 24$

A second method for determining base excess is to use a nomogram (one is provided in Figure 13-8 ■). Base excess is interpolated from the value of pH and PCO_2. To use the nomogram, measure the blood-gas parameters indicated and

✪ TABLE 13-3	
Reference Ranges for Arterial Blood pH, Gases, and Derived Analytes	
pH	7.35–7.45
PCO_2 (mmHg)	35–45
PO_2 (mmHg)	85–108
HCO_3^- (mEq/L and mmol/L)	18–23
TCO_2 (mEq/L and mmol/L)	22–28

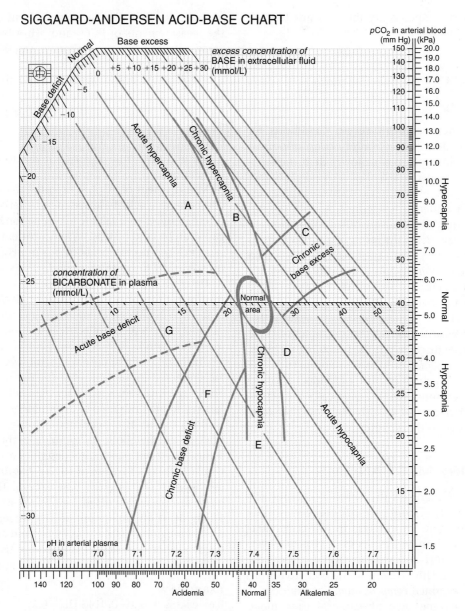

SIGGAARD-ANDERSEN ACID-BASE CHART

■ FIGURE 13-8   Siggaard–Andersen acid–base chart for determining base excess and assessing acid–base balance using measurements of pH and PCO_2.

find each result on the appropriate axis. Draw intersecting lines and identify the area or region on the chart. Refer to the following example.

Example

A patient's blood-gas results are as follows:

pH = 7.2

PCO_2 = 45 mmHg

In which area do the lines intersect and thus what conditions does this patient have? What is the base deficit?

Answer: The points on the chart fall into the area identified as "G," acute base deficit, and the patient is in acute metabolic acidosis. The base deficit is –10.

 Checkpoint! 13–9

Name the three most significant measured arterial blood-gas parameters that are required to evaluate acid–base status of a patient.

METABOLIC ACIDOSIS

Metabolic acidosis represents a base deficient disorder that results from either an accumulation of fixed acids or a loss of extracellular buffers. The blood pH decreases as a result of a decrease of $cHCO_3^-$. Accumulation of fixed acids occurs in several disorders including diabetes. In diabetic ketoacidosis (DKA) metabolic acids, for example acetoacetic acid and β-hydroxybutyric acid are produced in greater quantities. These acids enter the plasma and react with HCO_3^- to form H_2CO_3. Carbonic acid is converted to CO_2, which is eliminated by the lungs. The $cHCO_3^-$ decreases with very little lose of PCO_2. Therefore, the $cHCO_3^-/\alpha \times PCO_2$ ratio are decreased resulting in a lower blood pH.

A second mechanism resulting in metabolic acidosis involves lose of HCO_3^- from the body. This can occur in severe diarrhea and with abdominal fistulas. The $cHCO_3^-$ goes down, resulting in a decrease in the ratio of $cHCO_3^-/\alpha \times PCO_2$.

Other conditions such as renal failure can cause metabolic acidosis. The diminished number of functioning renal tubular cells results in a proportionate impairment of acid excretion, resulting in the accumulation of H^+ and other anions. These anions include phosphates, sulfates, and other products of protein metabolism. The causes of metabolic acidosis are listed in Box 13-2 ■.

Evaluating metabolic acidosis in a patient requires measurement of electrolytes, pH, blood gases, and calculating anion gap (see Chapter 12, Body Water and Electrolyte Homeostasis). Anion gap is increased because of renal failure, DKA, lactic acidosis, and ingestion of drugs or other chemicals that result in a decreased blood pH. Renal failure and DKA were discussed earlier. In lactic acidosis, the concentration of lactic acid increases during anaerobic metabolism. This is caused by strenuous muscular exercise or the presence of systemic infections. Lactic acidosis also results from tissue hypoxia (low PO_2) because of poor perfusion caused by cardiac failure.

Chemicals such as methanol, ethanol, ethylene glycol (a component of radiator fluid), and isopropyl alcohol (see Chapter 22, Toxicology) metabolize to acids, thus increasing the concentration of fixed acids. In addition, a compound such as salicylate (acetylsalicylic acid: aspirin) when taken in significant amounts initially causes a respiratory alkalosis (discussed later) but ultimately results in a decreased pH and metabolic acidosis.

BOX 13-2   Possible Causes of Metabolic Acidosis in Relationship to Normal and High Anion Gaps

Normal Anion Gap
▶ Severe diarrhea
▶ Pancreatitis
▶ Intestinal fistula
▶ Renal tubular acidosis

High Anion Gap
▶ Methanol toxicity
▶ Ethyl alcohol toxicity
▶ Ethylene glycol toxicity
▶ Lactic acidosis
▶ Uremia
▶ Renal failure
▶ Diabetes mellitus
▶ Salicylate toxicity

Anion-gap metabolic acidosis can be explained by one (or a combination) of eight factors identified by the mnemonic device MUDPILES:

Methanol

Uremia of renal failure

Diabetes or ketoacidosis

Paraldehyde toxicity

Isoniazid (or iron or ischemia)

Lactic acidosis

Ethylene glycol

Salicylate

A normal anion gap in a patient with metabolic acidosis usually presents with an increased concentration of Cl^- and thus hyperchloremic acidosis. A loss of HCO_3^- causes the kidneys to retain Cl^- in an effort to achieve electroneutrality. Conditions associated with a normal anion gap and hyperchloriemia include vomiting secondary to intestinal obstruction, diarrhea, and fistulas. This condition is also associated with renal tubular acidosis, which is an uncommon condition that may be inherited or acquired. The mechanism involves the cells of the proximal tubule. These cells are unable to reabsorb an appropriate amount of filtered HCO_3^-; instead, Cl^- is reabsorbed, raising the plasma Cl^- concentration.

Hyperchloremia will suppress renal HCO_3^- resorption. An endogenous cause of hyperchloremia is ureterosigmoidostomy, where the bowel mucosa that is in contact with urine selectively reabsorbs Cl^- and Na^+, thereby raising the plasma level of Cl^-. Exogenous causes are the ingestion or infusion of solutions such as intravenous sodium chloride. If this continues over a prolonged period, a mild acidosis will develop.

Compensation in Metabolic Acidosis
In metabolic acidosis, the increased cH^+ acts as a stimulant of the respiratory center. A deep, rapid, and gasping respiratory pattern (known as *Kussmaul breathing*) develops. This breathing pattern lowers PCO_2, thereby causing the blood pH to move toward normal. In addition, the kidneys begin to retain HCO_3^-. This compensatory mechanism occurs relatively fast in an effort to bring the blood pH to within normal limits.

METABOLIC ALKALOSIS

Metabolic alkalosis is characterized as a primary excess of HCO_3^-. Alkalosis can result from (1) the addition of base to the body, (2) a decrease in the amount of base leaving the body, and (3) the loss of acid-rich fluids. Laboratory findings include a plasma pH above the upper limit of normal; PCO_2 is normal, and plasma HCO_3^- is elevated. Severe cases of acute metabolic alkalosis may even result in death. Several additional causes of metabolic alkalosis are listed in Box 13-3 ■.

Metabolic alkalosis commonly arises from the loss of Cl^- from body fluids. This may be the result of vomiting or aspiration of gastric fluids, which leads to a loss of gastric hydrochloric acid that in turn leads to a rise in blood pH because of an increased renal retention of HCO_3^- to counter the Na^+ reabsorbed by the proximal tubules. Two terms associated with this condition are *hypochloremia alkalosis* and *Cl^--responsive metabolic alkalosis*. Urine Cl^- measurements may be <10 mmol/L.

Patients who have a condition in which their levels of adrenocortical hormones (mineralocorticoids or glucocorticoids) are elevated (such as Cushing disease, primary or secondary hyperaldosteronism, adrenal hyperplasia, and pituitary adrenocorticotropic-hormone–producing adenoma) may develop metabolic alkalosis. Elevated aldosterone or cortisol stimulates increased Na^+ reabsorption, which results in a state of increased K^+ and H^+ excretion. The resulting decreased tubular K^+ concentration stimulates NH_3 production and thus renal H^+ excretion as NH_4^+.

Compensation in Metabolic Alkalosis

The lungs will attempt to compensate by causing retention of CO_2 (hypercapnia), which in turns causes an increase in cH_2CO_3 and $cdCO_2$. In addition, the ratio of $cHCO_3^-/cdCO_2$ approaches its normal value, although the actual concentration of both $cHCO_3^-$ and $cdCO_2$ increases. Increases in PCO_2 are variable because of an erratic respiratory response to the metabolic alkalosis condition.

The kidneys respond to the alkalosis by decreasing Na^+–H^+ exchange, decreasing the formation of ammonia, and decreasing the conservation of HCO_3^-. This response by the kidney is also variable, depending on the blood cK^+ and actual blood volume.

RESPIRATORY ACIDOSIS

Respiratory acidosis is characterized by an increase of PCO_2 above normal limits. Blood pH is lower than normal, and the $cHCO_3^-$ may be normal or increased. Normally, the respiratory system, which is the sole route of CO_2 excretion, can remove all the CO_2 that is produced. The hallmark of respiratory acidosis is a defect in the excretion of CO_2. Thus, a person tends to retain CO_2 (hypercapnia).

Clinically, respiratory acidosis results from hypoventilation. Hypoxia is an associated finding unless the patient is receiving oxygen. The symptoms are nonspecific and are usually the result of hypoxia, high PCO_2, and acidosis. An acute rise in PCO_2 leads to drowsiness, restlessness, and tremors. A PCO_2 greater than 80 mmHg often results in stupor and coma. Brain function is modified, especially when PCO_2 rises acutely because the entry of CO_2 in the brain is not accompanied by a concurrent rise in the blood levels of HCO_3^-.

Emphysema causes a destruction of alveoli that reduces the surface area available for gas exchange. Diseases such as cancer result in the thickening of the alveolar membrane. Congestive heart failure that results in fluid accumulation in the alveoli impairs gas diffusion and results in alveolar collapse because of increasing surface tensions.

CNS disorders may impair respiratory drive. Examples include vascular disorders, trauma, epilepsy, hypoglycemia, and hypoxia or drug ingestion. Any one of these disorders can affect the respiratory center or its regulatory mechanisms (or both).

Airway obstruction is a common cause of respiratory acidosis. For example, large-diameter airways may be obstructed by food, vomitus, tumors, or other foreign objects. Smaller airways may be obstructed because of asthma, COPD, or the accumulation of mucus associated with cystic fibrosis. Pneumonia is an example of a cause of small-airway obstruction and alveolar obstruction.

An inability to move the chest wall to ventilate the lungs may result from disorders that affect the peripheral nerves (e.g., polio, respiratory muscles (dystrophy), or trauma to the ribs). A pneumothorax resulting from a penetrating wound in the chest wall will also limit inflation of the lungs.

Finally, abnormal ventilation–perfusion ratio because of emphysema or circulatory impairment from cardiac arrest usually results in severe respiratory acidosis. Stagnation of peripheral blood in the limbs from poor circulation also results in respiratory acidosis. Box 13-4 ■ lists the causes of respiratory acidosis.

Compensation in Respiratory Acidosis

Blood buffers, especially HCO_3^-, are the first attempt of the body to counter the accumulation of H^+ by forming H_2CO_3. The kidneys respond to retain HCO_3^- and secrete and excrete H^+ and Cl^-. The H^+ are excreted and eliminated in the form of hydrochloric acid (HCl) and ammonium chloride (NH_4Cl). Chloride is also reduced because of the chloride shift. Sodium is retained by the kidney to combine with the elevated HCO_3^-. In addition, ammonium-ion (NH_4^+) formation is increased and subsequently excreted over the next few days. These compensatory mechanisms represent a slower process than previously noted in other acid–base disturbances.

The lungs respond to the elevated PCO_2 by increasing the pulmonary rate and the depth of respiration if the primary defect is not in the respiratory center. The release of CO_2

BOX 13-4 Possible Causes of Respiratory Acidosis that Result in Retention of Carbon Dioxide

Suppression of Respiratory Center

▶ Drugs such as narcotics and barbiturates
▶ CNS tumors, trauma, and degenerative disorders
▶ Sleep apnea

Weakened Respiratory Muscle Function

▶ Myasthenia gravis
▶ Periodic paralysis
▶ Intaperitoneal aminoglycosides
▶ Guillain–Barré syndrome
▶ Botulism
▶ Poliomyelitis
▶ Amyotrophic lateral sclerosis (ALS)
▶ Myxedema

Airway Problems

▶ Aspiration of foreign body or vomitus
▶ COPD
▶ Adult respiratory distress syndrome
▶ Severe asthma or pneumonia
▶ Pneumothorax or hemothorax

BOX 13-5 Possible Causes of Respiratory Alkalosis

Hypoxemia

▶ Pulmonary disease
▶ Congestive heart disease
▶ Severe anemia
▶ High–altitude exposure

Stimulation of the Medullary Respiratory Center

▶ Hyperventilation syndrome
▶ Hepatic encephalopathy
▶ Salicylate intoxication
▶ Pregnancy (increased progesterone)
▶ Neurologic disorders (CVA)

Others

▶ Excessive mechanical ventilation

---

through the lungs results in a decreased $cdCO_2$. This results in a change in the ratio of $cHCO_3^-:cdCO_2$ and cH^+, causing the pH to move toward normal.

RESPIRATORY ALKALOSIS

Respiratory alkalosis is characterized by a reduction in PCO_2 below normal and an increase in blood pH. The major contributory cause of respiratory alkalosis is hyperventilation. Hyperventilation manifests itself with an increased depth or rate of breathing that leads to a lowering of PCO_2 and an increase in the ratio of PCO_2 to HCO_3^-.

The increase in pH does not usually exceed 7.60, and the $cHCO_3^-$ may only decrease minimally. In chronic respiratory alkalosis, the symptoms may be mild because the blood pH is so close to normal. Acute respiratory alkalosis presents a much different set of symptoms. These symptoms may include shortness of breath, a sense of choking, dizziness, nervousness, an altered level of consciousness, and eventually tetany. Some of these symptoms—tetany in specific—result from a decreased level of ionized calcium.

Respiratory alkalosis can be caused by many diverse conditions. Box 13-5 ■ lists several common causes of respiratory alkalosis. Several conditions are closely associated with other acid–base disturbances. For example, in salicylate intoxication, respiratory alkalosis is the initial disorder observed. If the intoxication is severe, then metabolic acidosis

develops. Another example is a patient who suddenly recovers from metabolic acidosis. If the extracellular $cHCO_3^-$ of the patient is suddenly increased, hyperventilation persists for 12 to 24 hours. Hyperventilation lowers the PCO_2, with a concurrent slight reduction of $cHCO_3^-$. The ratio of $HCO_3^-:H_2CO_3$ increases, thus returning the pH toward normal. This mechanism of changing the pH during acidosis is referred to as *compensatory respiratory alkalosis*. The result is a decrease in the PCO_2 and bicarbonate concentration and a pH value that is closer to the reference interval.

Compensation in Respiratory Alkalosis

The initial compensatory reaction to respiratory alkalosis is to make more H^+ available to combine with HCO_3^-. The H^+ ions come from the cells by a shift of the cellular reactions to the right. The more H^+ made available from the cell buffers, the more extracellular HCO_3^- that are consumed, and the greater degree of compensation. The loss of H^+ ions is balanced electrically by the entry of Na^+ and K^+ from the extracellular fluid.

Renal physiology plays a significant role in respiratory alkalosis. The kidney attempts to eliminate HCO_3^-. This is brought about by the decreased net acid excretion. A reduction in the excretion of titratable acid and NH_4^+ is the primary reason for this to occur. Therefore, HCO_3^- used for titration of metabolic acid is not recovered by renal HCO_3^- regeneration. Sodium and potassium salts are excreted instead of NH_4^+ salts. The kidney also retains and increases plasma cCl^-. Chloride ions are also increased because of the chloride shift. These compensatory mechanisms progress at a slow pace in an effort to bring the body pH back to within normal limits.

Table 13-4 ✪ provides a summary of each acid–base disorder and the corresponding pH and blood-gas measurements. In addition, Table 13-5 ✪ reviews the four major acid–base disorders, including their clinical causes.

⊗ TABLE 13-4

Summary of Laboratory Features for Simple Metabolic Acid–Base Disorders, Including pH, PCO_2, and $cHCO_3^-$*

	Primary Change	Compensation	pH	PCO_2	$cHCO_3^-$
Metabolic acidosis	↓ $cHCO_3^-$	↓ PCO_2	↓	↓ N	↓
Metabolic alkalosis	↑ $cHCO_3^-$	↑ PCO_2	↑	↑ N	↑
Respiratory acidosis	↑ PCO_2	↑ $cHCO_3^-$	↓	↑	↑ N
Respiratory alkalosis	↓ PCO_2	↓ $cHCO_3^-$	↑	↓	↓ N

*Also included are their respective principle compensatory actions.

⊗ TABLE 13-5

Summary of the Four Major Acid–Base Disorders, Including Examples of Immediate and Clinical Causes

Clinical Disorder	Immediate Cause(s)	Clinical Causes
Metabolic acidosis	1. Accumulation of acids other than CO_2 or H_2CO_3 2. Elimination of alkali (fixed PCO_2)	• Decreased urinary secretion of H^+ • Ketoacidosis • Lactic acidosis • Severe diarrhea (HCO_3^- loss)
Metabolic alkalosis	1. Accumulation of alkali 2. Removal of acids other than CO_2 or H_2CO_3 (fixed PCO_2 therapy)	• Decreased HCO_3^- (e.g., $NaHCO_3$) • Decreased H^+ (severe vomiting)
Respiratory acidosis	Increased PCO_2	• Decreased alveolar ventilation (e.g., drug overdose) • Decreased lung capacity (e.g., pulmonary edema)
Respiratory alkalosis	Decreased PCO_2	• Increased alveolar ventilation caused by hypoxia, anxiety, aspirin intoxication

▶ MEASUREMENT OF pH, PCO_2, AND PO_2

The measurement of blood gases are not always carried out in the clinical laboratory. Many institutions place blood-gas analyzers in departments located throughout the hospital—for example, operating rooms, neonatal intensive care units, emergency rooms, and adult intensive-care units. They may be placed onto mobile carts and transported to various units in the hospital. Operation and maintenance of blood-gas analyzers provides a unique challenge to each facility.

SPECIMENS

Specimen Types
Samples for blood-gas analysis may be drawn from several sites, depending on the patient's condition and the specific needs of the person requesting the sample. A widely used practice today is to establish an arterial line that will remain in a patient for the duration of his or her stay in the hospital. All ABG samples will be drawn from that line.

An arterial sample is preferred versus venous blood for blood-gas analysis. The radial artery is the recommended anatomical site. The femoral, brachial, or pedal arteries can also be used. For neonates, the temporal artery is another acceptable site. Because pH and blood gases must be measured in a whole-blood sample, an anticoagulant is necessary. The anticoagulant of choice is heparin. Thus, the specimen of choice for pH and blood-gas measurement is heparinized whole blood.

Venous samples are affected by peripheral circulatory effects and cellular metabolic needs. Veins may be used to reflect a patient's acid–base condition, particularly pH, PCO_2, and bicarbonate concentration. A convenient puncture site for venous blood collection is the brachial vein.

Capillary samples may be obtained from the earlobe, finger, toe, or heel. The heel should not be used after two to three months of age because the tissue may become sclerotic. The capillary sampling area must be hyperemic when the puncture is performed and during sample collection. This can be accomplished by massaging the sampling area until it is saturated with blood. If the sample area is not hyperemic, the results will only represent the local tissue and not the general status of the patient.

Transport and storage of samples for blood-gas measurements is a primary preanalytical concern. Improper handling of blood-gas samples can result in errors in measurement and lead to improper medical decisions. Samples for blood-gas measurements are usually transported to the laboratory by an appropriate hospital employee. This person will walk the specimen to the receiving area of the laboratory. Another method of transport that is used by many laboratories is a pneumatic-tube delivery system. The patient's specimen is placed into a carrier or tube and sent to the laboratory through a system of interconnecting plastic or metal pipes. A pneumatic pressure system is used to move the carrier containing the blood-gas sample to the laboratory. The result is a rapid and efficient means of delivering ABG specimens to the laboratory. The integrity of the sample is maintained, and the overall turnaround time is shortened using this system.

Storage time and transport time should be kept to a minimum because significant error may occur because of metabolism, the diffusion of gases through the plastic container, and elevated potassium values. Samples should not be kept for more than 10 minutes at room temperature. If a long transport time is anticipated, then the ABG specimen must be placed in a container with ice and water to maintain the temperature at 0–4°C (32–39.2°F). Samples should be kept at this temperature for no longer than 30 minutes. Avoid placing samples in ice without water because this may freeze and hemolyze the blood cells.

Preanalytical Handling of Specimens

Specimens for blood-gas analysis must be handled appropriately. After a sample of blood is removed from the patient, every effort must be made to remove air bubbles immediately and without agitation. An air bubble with relative volume of 0.5% to 1.0% or more of the blood in the syringe is a potential source of significant error. The sample should be mixed immediately to dissolve the heparin and prevent clotting, which could block the tubing in the analyzer and cause long and unnecessary downtime.

Samples for blood-gas measurements should be mixed thoroughly before they are injected into the blood-gas analyzer. Proper mixing ensures that all phases of the sample are homogenous and representative of the whole sample. The procedure to properly mix each sample requires simple inversion of the sample several times and avoidance of any excessive agitation of the sample. Failure to mix a sample properly may result in analysis of either the plasma phase or erythrocytes. If this occurs, the oximetry results such as carboxyhemoglobin will be of no value. In addition, inadequate mixing may also lead to erroneous pH, PCO_2, and PO_2 values.

> ✓ **Checkpoint! 13–10**
>
> *Identify the specimen of choice for blood-gas determinations.*

INSTRUMENTATION

The fundamental design of blood-gas analyzers has not changed significantly during the last two decades. A schematic of a typical blood-gas analyzer is shown in Figure 13-9 ∎. The electrodes used to measure pH, PCO_2, and PO_2 are discussed in detail in Chapter 2, Instrumentation. Electrolyte measurements are made using ion-selective electrodes, which are also discussed in Chapter 2. Total hemoglobin concentration is also measured using a spectrometer. Several other parameters are derived by calculations using measured parameters in various combinations (see the following discussion).

Measurement of blood gases begins with the introduction of a properly prepared specimen. Usually, only 50–120 μL of specimen is necessary. A specimen is injected into the inlet port and travels to the surface of each electrode through the actions of a peristaltic pump. The specimen is allowed to remain on the surface of the electrode for several seconds so

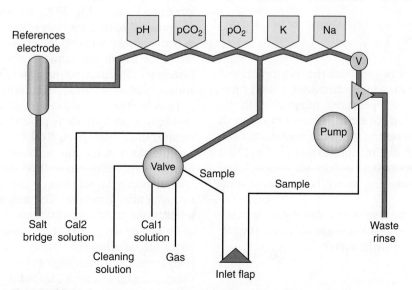

∎ FIGURE 13-9 Diagram of a typical blood-gas analyzer that includes ion-selective electrode for sodium and potassium.

BLOOD GAS ANALYZER MODEL XYZ

Date:		Time:		Sample volatile:

Identification:

Patient H>
Operator
Sample type
Temperature

Blood Gas values

pH	7.515	
pCO_2	25.8	mmHg
pO_2	127	mmHg
sO_2	99.1	%

Temperature Corrected Values

pH (T)	7.515	
pCO_2 (T)	25.8	mmHg
pO_2 (T)	127	mmHg

Electrolyte values

cNa^+	132	mmol/L
cK^+	4.0	mmol/L
cCa^{2+}	1.19	mmol/L

Metabolite values

cGlu	144	mg/dL
Lactic acid	1.0	mmol/L

Calculated values

$cHCO_3^-$	23.6	mmol/L
ABE	−0.9	mmol/L
$ctCO_2$	21.5	mmol/L

Oximetry values

ctHb	10.9	g/dL
FO_2Hb	97.3	%
FCOHb	0.9	%
FMetHb	0.9	%

■ FIGURE 13-10   Example of a reporting format for test results using a blood-gas analyzer.

that an electrical signal can be produced that is proportional to the relative amount of analyte present. After a short time, the microprocessor signals the peristaltic pump to flush the specimen to waste. An electrical signal is created from each measuring electrode and is sent to the microprocessor for calculation. The results from the microprocessor are sent to a printer or laboratory information system for review. Total time for the whole measurement process is less than 3 minutes for most analyzers.

Figure 13-10 ■ shows the results of a blood-gas measurement using a current model of a blood-gas analyzer. The following types of information are included:

- patient demographics,
- sample type,
- FIO_2 (%),
- patient temperature,
- blood-gas values,
- electrolytes values,
- metabolite values,
- calculated values,
- temperature-corrected values, and
- oximetry values.

CALIBRATION

Calibration of blood-gas analyzers is an essential and critical function. Without proper calibration, results for quality-control samples and patient specimens would be unacceptable. Calibration allows the laboratorian to quantitate blood-gas parameters and provide reliable measurements. All measured parameters require calibration.

Calibration of pH requires two electrolyte buffer solutions of known composition and pH. The source of the calibration buffers should be traceable to the National Institute of Standard and Technology. Usually, the pH values are about 6.8 and 7.4 at 37°C. The calibration solutions are pumped into the sample chamber where a pH measurement is taken. Blood gases are calibrated using calibration gases. Most blood-gas analyzers use commercially prepared CO_2 gas at two different concentrations to calibrate the PCO_2 electrode. The PO_2 electrode is calibrated using "no atmospheric" air as one calibrator value and "room air" as the other calibrator. The final low-gas mixture that is introduced to the electrodes yields a PO_2 and PCO_2 around 0 and 35 mmHg, respectively. High-gas mixtures yield a PO_2 and PCO_2 around 150 mmHg and 80 mmHg, respectively. Exact values for calibrators may vary with each blood-gas analyzer and tank of CO_2 gas.

A gas calibration begins when gas is released from the tanks and pumped into the calibration buffers. The solutions are mixed and warmed to 37°C. Microliter volumes of calibrant are transported to the surface of each electrode where a voltage or current measurement is taken. Each gas measurement is corrected for barometric (atmospheric) pressure using a built-in barometer. The data are manipulated in the microprocessor, and calibration information is derived for pH, PO_2, and PCO_2.

In a two-point calibration, the measured electrical response of the electrode is plotted against the known values for the buffers and gases. A calibration line (linear) is created, and the slope of the line is calculated. All subsequent measurements are compared to this calibration line.

Initiation of a calibration is usually done automatically via a programmed timetable. The frequency is dictated by recommendations from accreditation agencies and manufacturers as well as by daily laboratory operation demands. Operation protocols developed by state and federal agencies requires that a two-point calibration be performed for all analytes every eight hours and a one-point calibration for all analytes every four hours. In addition, a one-point calibration of PO_2 and PCO_2 should be initiated every 30 minutes.

> ✓ **Checkpoint! 13–11**
>
> *Explain why calibration of a blood-gas analyzer is so important.*

Quality-control material should be assayed after any of the following tasks: (1) a successful calibration is achieved, (2) maintenance procedures are completed, or (3) when troubleshooting steps are finished. Quality-control protocols developed by state and federal agencies require laboratories to run at least one level of control per shift with all three levels of controls assayed within a 24h period. Some laboratories may assay all three levels per 8h shift. Electrochemical devices, especially electrodes, have an inherent ability to drift over time. Therefore, it is important to monitor the analyzers' performance frequently. Assaying quality-control material and performing frequent calibration can minimize problems associated with electrode drift.

Laboratories are required to assay three levels of controls. Each level corresponds to three clinically significant acid–base levels: acidosis, alkalosis, and normal. Most laboratories use commercially prepared control materials. These control solutions are aqueous buffered fluids contained in sealed ampoules or vials. Each vial contains a known gas mixture appropriate for the level of control. Before injection, the vials are mixed vigorously to make a homogenous solution. The top of the vial is snapped off, and the solution is immediately injected into the sample port. Measurements are made for each analyte, and the results are documented. All quality-control data are evaluated for acceptability based on the quality-control protocol utilized by each laboratory.

Other materials and techniques may be used for quality control. Commercially available blood-based and fluorocarbon-based control materials may be used instead of aqueous-based controls. Both of these materials mimic the characteristics of human whole blood better than the aqueous-based controls. This is most notable when measuring PO_2.

Another technique used by laboratories to monitor the performance of their blood-gas analyzers is to prepare whole-blood controls using a tonometer. Whole-blood samples are tonometered (gas equilibrated) using known gas mixtures. The samples are then injected into the blood-gas analyzer, and measurements are made for PO_2 and PCO_2. The advantage to using tonometered blood is that it most closely resembles actual patient's blood, thereby interacting with the electrodes in a similar fashion. Several disadvantages to tonometered blood make this technique less popular in the clinical laboratory. Disadvantages include acquiring a suitable sample to tonometer, the cost of purchasing a tonometer, the time required to achieve equilibrium (often hours), and the inability to control pH measurements with tonometer blood samples.

MAINTENANCE

Proper maintenance of blood-gas analyzers is essential. Failure to routinely clean and perform function checks may result in extended downtime and increased turnaround time, cost, and errors. Maintenance schedules should be strictly followed. The analyzer should be monitored for any instrument "flags," abnormal noises produced by the analyzer (e.g., whining motors or squeaky pumps), and quality-control failures.

Many blood-gas analyzers have an internal barometer that is used to correct for changes in barometric pressure. The barometer should be checked for accuracy on a periodic basis. To verify the barometer in the laboratory blood-gas analyzer simply call the nearest airport or meteorological station to get its barometric measurement (use the Internet to find phone numbers and locations). Next, follow the manufacturer's recommendations for correcting the analyzer.

Accurate measurement of ABGs is temperature dependent and therefore requires verification. Critical functions performed automatically by most blood-gas analyzers include temperature sensing, monitoring, and correcting blood-gas measurements to temperature. The modern blood-gas analyzers are equipped with sophisticated thermosensors located in the proximity of the measuring chamber. If a temperature check exceeds acceptable tolerance, then an audible alarm and visible flags are shown on the monitor and printed on the results form. Acceptable tolerances for temperatures are usually ±0.1–0.2°C.

Calibration-verification procedures are required to confirm the calibration and linearity of pH, PO_2, and PCO_2. The calibration material can be purchased from a reliable manufacturer and usually consists of four or five vials of assayed material. Every vial has a known amount of analyte in a buffered solution that contains electrolytes and a gas mixture of CO_2, O_2, and N_2. All vials are prepared and analyzed according to manufacture instructions. Every measured result is compared to the manufacturer's range of values. If a measured value falls outside of the target range, then the laboratory must take corrective action. Linearity is evaluated by plotting the measured value on the *y*-axis and the expected or target value on the *x*-axis. Next, connect the plotted points and assess the linearity by evaluating the slope and intercept of the line. In addition, a calculated correlation coefficient *r* may be done (see Chapter 4, Laboratory Operations).

Many facilities have two or more blood-gas analyzers in their laboratories or off-site departments. The measured parameters should be cross-calibrated between all analyzers to ensure that data from each are similar. Several procedures may be used to validate the correlation between each analyzer. One procedure that is used by many laboratories involves collecting patient samples at three different levels of pH, PO_2, and PCO_2, followed by analysis of each sample on every blood-gas analyzer. The data can then be correlated using the appropriate statistics. A tolerance value should be determined in accordance with each laboratory's regulatory protocol.

SOURCES OF ERROR

Potential sources of error can occur from the time the physician requests an ABG be drawn on a patient to the time the physician looks at the results. Laboratory errors are typically categorized as preanalytical, analytical, and postanalytical. Specific examples of these types of errors relative to ABGs are found in Box 13-6 ■. The preanalytic phase is a frequent source of errors in blood-gas analysis. These errors usually occur during specimen collection or sample introduction.[6]

Samples containing air bubbles can result in a significant bias of blood-gas values. An air bubble left in a sample for a few minutes can affect the PO_2 values significantly, so it is important that samples be maintained under anaerobic conditions. If air bubbles do enter the sample, they should be removed before sample introduction. It has been shown that an air bubble with a relative volume of ~1% of the sample may introduce a bias of 20–30%, depending on the overall integrity of the sample.[7] The bias tends to increase the longer an air bubble remains in the syringe and the more the sample is agitated. The effect is particularly large if the sample is stored at a low temperature because of an increased affinity between hemoglobin and oxygen.[8] Because PO_2 in atmospheric air is about 150 mmHg, the bias will normally be positive because the normal PO_2 value in arterial blood is 90–100 mmHg. The bias could be negative in cases of a supranormal PO_2 value.

Another source of error with whole-blood samples is the bias from settling or separation of the phases of the blood.

Once whole blood is left in a syringe, it has the tendency to start separating into the main components (i.e., plasma and blood cells). Blood from patients suffering from disease may begin to separate immediately on entering the syringe. The bias in measurements results from a nonhomogeneous sample and does not reflect the true results. Blood-gas parameters that are most affected by this occurrence include total concentration of hemoglobin (tHb), PO_2, acid–base excess, and total oxygen content.[9]

Delays in measuring blood gases in blood samples may also introduce a bias. Blood cells continue to metabolize after blood has left the body. This changes the values of blood gases, pH, and metabolites. A spurious decrease in bias may be seen in results for PO_2 and pH, whereas an increase in bias would be detected in PCO_2 measurements.

Blood collected in a syringe will begin to coagulate immediately. Microclots are produced in as little as 15 seconds. The National Committee for Clinical laboratory Standards recommends that samples be completely anticoagulated to avoid formation of microscopic clots.[10] Clots have the potential to bias patient's results without the analyst knowing. These erroneous results can lead to inappropriate diagnosis and treatment of the patient. A clot may also block or plug tubing, resulting in erroneous values or not allow the aspiration of another sample. Clots may also result in nonhomogeneous samples, which may bias the measurement of several blood-gas parameters.

Finally, bias of PCO_2 and hemoglobin concentrations can occur because of dilution from saline in arterial lines and liquid heparin. The diluting of samples results in falsely low values for these analytes. Biases as high as 40% for these measurements have been reported in studies.[10]

✓ Checkpoint! 13–12

Identify several sources of preanalytical errors and indicate briefly an example of a negative effect on patient care.

TEMPERATURE CORRECTION

Many blood-gas analyzers manufactured within the last several years allow the users to correct for abnormal body temperatures. This feature is useful for patients who have been exposed to low temperatures for a prolonged period. *Hypothermia* is the term used when the body temperature falls below normal. Exposure to prolong cold temperatures can occur to people who remain outdoors—for example, hikers, sportspersons who hunt or fish, and mountain climbers.

When a specimen is analyzed for ABGs that requires a temperature correction, the operator selects the temperature for the sample. The microprocessor will apply the appropriate calculation to correct for the pH, PO_2, and PCO_2. Both corrected and uncorrected results will be displayed on the computer monitor. Blood-gas analyzers may be able to correct down to 15°C. Consult the instrument's manufacturer for the exact temperature-correction range.

► HEMOGLOBIN MEASUREMENT

The optical system of most blood-gas analyzers can measure some or all of the parameters listed in Table 13-6 ✪. The measurement of tHb is necessary because this value is used for the calculation of several other blood-gas parameters. Several of these will be discussed in the following paragraphs.

The concentration of tHb represents the sum of all hemoglobin fractions detected by the spectrometer. This determination is accomplished in the following manner. A blood sample is introduced into the analyzer and transported to a hemolyzer unit that ruptures or lyses erythrocytes. An aliquot of the hemolyzed sample is moved into a cuvette. Light from a polychromatic light source, usually a tungsten halogen lamp, is directed toward the cuvette. The light transmitted through the cuvette is guided to the spectrometer. A grating monochromer isolates specific wavelengths of light unique to each hemoglobin compound in the sample. Transmitted light strikes photodetectors that generate a voltage signal relative to the intensity of photons. All voltage signals are sent to the microprocessor, where the calculations of the hemoglobin fractions are made.

For most blood-gas analyzers, the concentration of hemoglobin is equaled to the sum of oxy-, carboxy-, meth-, and reduced hemoglobin. Corrections may be made for the presence of bilirubin pigments, abnormal hemoglobins, and nonhemoglobin substances that may absorb light within the same wavelength range used to measure the hemoglobin fractions.

Calibration of hemoglobin is performed using a colored dye material with a known total hemoglobin concentration.

The absorbance of the calibrant is measured within the oximeter (spectrometer). A current is generated as monochromatic light strikes a photodiode. The current is sent to the analyzer's computer, and a calibration factor is calculated. If the calibration factor is within tolerance, then the calibration is acceptable. If not, the sample must be reanalyzed and troubleshooting techniques perhaps initiated.

Quality control for total hemoglobin is accomplished by measuring two different levels of quality-control material with known hemoglobin composition. Several commercially available quality-control materials are available. Most of these materials are colored dyes that mimic the photometric characteristics of human types of hemoglobin.

► DERIVED PARAMETERS

Derived parameters are calculated based on measured data. Calculations are made using equations programmed into the analyzer. The accuracy of the calculations depends on data transmitted to the microprocessor. Table 13-7 ✪ shows several examples of derived parameters available on many models of currently used blood-gas analyzers.

Hemoglobin-dependent *base excess* can be described as the mEq/L base needed to change plasma pH by 0.01 units. An approximate average of base required is 0.7 mEq/L. An increased base is called *base excess*. A decreased base is called base deficit or negative base excess. The normal base excess is between +2 to –2 mEq/L. The hemoglobin concentration is important because the blood-buffering capacity is greatly dependent on this quantity. The addition of a base, such as HCO_3^-, raises the buffer content of the blood and results in positive base excess. The loss of base from diarrhea, for example, lowers the blood buffer content and results in a base deficit. Calculation of base excess is dependent on the hemoglobin measurement, bicarbonate concentration, and pH.

The Siggaard–Andersen alignment nomogram can be used for determining several blood-gas parameters including base excess, total carbon dioxide and bicarbonate concentration. These parameters require the measurements of pH, PCO_2 and hemoglobin. A Siggaard–Andersen nomogram is shown in (Figure 13-11 ■).[11]

✪ TABLE 13-6

A List of Parameters Either Measured or Calculated that Are Commonly Available in Blood Gas Analyzers that Include Cooximeters

Parameter	Description
ctHb	Concentration of total hemoglobin
sO$_2$	Oxygen saturation
FO$_2$Hb	Fraction of oxyhemoglobin
FCOHb	Fraction of carboxyhemoglobin
FHHb	Fraction of deoxyhemoglobin
FMetHb	Fraction of methemoglobin
FHbF	Fraction of fetal hemoglobin
ctBil	Concentration of total bilirubin (sum of unconjugated and conjugated bilirubin)

✪ TABLE 13-7

Examples of Derived and Measured Blood-Gas Parameters

Derived Parameter	Measured Analyte Parameter
Bicarbonate	PCO_2
Total carbon dioxide	PCO_2
Total oxygen content	Total hemoglobin
Oxygen saturation	PO_2
Actual base excess	

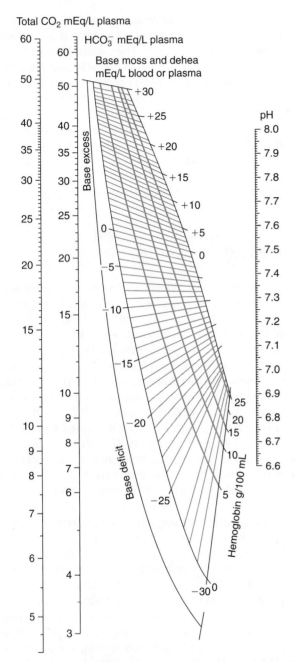

Total CO₂ mEq/L plasma

Oxygen saturation (O_2 sat or sO_2) describes the amount of oxygen bound to hemoglobin and is used to determine the effectiveness of oxygen therapy and respiration. Estimation of the percent sO_2 (%sO_2) is derived from the measured pH and PO_2 and the equation for a normal oxygen-dissociation curve. Oxygen saturation can also be derived from a published nomogram. Finally, sO_2 can be measured by most cooximeters using the difference in the wavelength of maximum absorbance value for oxyhemoglobin and deoxyhemoglobin.

SUMMARY

The buffers systems play an integral role in attempting to maintain normal acid–base status. In this chapter, each buffer is discussed along with the significant equations relevant to understanding how the buffers function. Knowledge of the human respiration system is important to the complete understanding of acid–base physiology. The mechanisms available to move gases and ions through the various barriers and phases within the body are important to understand.

The hemoglobin–oxygen dissociation curve for each component is affected by many factors such as temperature and pH. These factors and their effects on the affinity of oxygen for hemoglobin are presented. The role of the lungs and kidney in maintaining acid–base balance and appropriate gas exchange is discussed.

Knowledge of pH, blood gases, and other derived parameters is an important part of overall quality assurance in measuring ABGs. An understanding of metabolic and respiratory acidosis and alkalosis provides laboratory personnel with the means to produce reliable results. This chapter attempts to equip the reader with this knowledge and understanding.

Measurement of pH, PCO_2, and PO_2 involves the use of sophisticated analyzers, specimens that require extra effort in maintaining sample integrity, and low tolerance for error. In this chapter, preanalytical, analytical, and postanalytical concerns have been presented and an emphasis placed on calibration, quality control, and instrument maintenance. Finally, a discussion about potential sources of errors is presented. Unfortunately, not every potential error could be presented here, but the reader has been informed about several important sources of potential error.

■ **FIGURE 13-11** Siggaard–Andersen alignment nomogram for determining unmeasured blood gas parameters.

REVIEW QUESTIONS

(Refer to Table 13-4 for appropriate reference intervals.)

LEVEL I

1. What is the pH of a sample with a PCO_2 of 40 mmHg and bicarbonate concentration of 26 mmol/L? (Objective 2)
   a. 5.4
   b. 6.4
   c. 7.4
   d. 8.4

LEVEL II

1. Which of the following is a compensatory mechanism for a patient in metabolic acidosis? (Objective 8)
   a. metabolic alkalosis
   b. hyperventilation
   c. renal regulation of PCO_2
   d. renal regulation of carbonic acid

LEVEL I

2. Which of the following is not an example of a major body buffer? (Objective 3)
   a. bicarbonate
   b. hemoglobin
   c. protein
   d. sulfate

3. Carbonic acid concentration in plasma usually equals: (Objective 1)
   a. bicarbonate concentration.
   b. total CO_2 minus 0.03 times bicarbonate concentration.
   c. PCO_2.
   d. 0.03 times PCO_2.

4. Which of the following is the preferred sample type for pH and blood-gas measurement? (Objective 8)
   a. serum
   b. plasma
   c. arterial whole blood
   d. venous blood

5. Exposing a whole-blood sample to the atmosphere for a prolong period may cause: (Objective 7)
   a. positive bias in PCO_2.
   b. positive bias in PO_2.
   c. negative bias in PO_2.
   d. no change in PO_2.

6. Which of the following is the primary carrier of carbon dioxide? (Objective 4)
   a. dCO_2
   b. carbonate
   c. bicarbonate
   d. carbamino compound

7. Which of the following techniques is recommended just before introducing a patient specimen into a blood-gas analyzer? (Objective 8)
   a. Expose sample to air to allow it to equilibrate.
   b. Push out any air bubbles and clots from the syringe
   c. Warm the sample in your hand for several minutes.
   d. Draw as much as 0.5 mL of saline into the syringe to ensure that any clot in the sample will be dissolved.

8. Which of the following is an appropriate source of calibrating gas(es) used to calibrate PCO_2 in a blood-gas analyzer? (Objective 5)
   a. sealed ampoules of aqueous buffer with know amounts CO_2
   b. atmospheric air (room air)
   c. commercially prepared CO_2 gas at various concentrations
   d. a highly carbonated aqueous solution of water

LEVEL II

2. A patient suffering from COPD has the following laboratory ABGs: pH = 7.24, PCO_2 = 55 mmHg, and plasma bicarbonate concentration [HCO_3^-] = 25 mEq/L. Which of the following describes this patient? (Objective 7)
   a. respiratory acidosis
   b. respiratory alkalosis
   c. metabolic acidosis
   d. metabolic alkalosis

3. A patient presents to the emergency department showing signs of hypoxia. The patient also was anxious and nervous. ABGs were requested, and the laboratory results were pH = 7.55, PCO_2 = 30 mmHg, and plasma [HCO_3^-] = 22 mEq/L. Which of the following describes this patient? (Objective 7)
   a. respiratory acidosis
   b. respiratory alkalosis
   c. metabolic acidosis
   d. metabolic alkalosis

4. Which of the following would occur to the affinity of oxygen for hemoglobin in a patient suffering hypothermia? (Objective 4)
   a. no change to the normal hemoglobin–oxygen dissociation curve
   b. a shift to the right in the hemoglobin–oxygen dissociation curve
   c. a shift to the left in the hemoglobin–oxygen dissociation curve
   d. a decrease in the oxygen affinity for hemoglobin

5. Which of the following devices located within a blood-gas analyzer is used to measure carboxyhemoglobin (CO–hemoglobin)? (Objective 9)
   a. a spectrometer
   b. a carbon-monoxide electrode
   c. a carbon-monoxide meter
   d. a carbon-monoxide pressure transducer

6. Which of the following effects would pH have on the hemoglobin–oxygen dissociation curve for a patient who is in respiratory acidosis? (Objective 4)
   a. no change to the normal hemoglobin–oxygen dissociation curve
   b. a shift to the right in the hemoglobin–oxygen dissociation curve
   c. a shift to the left in the hemoglobin–oxygen dissociation curve
   d. a decrease in the oxygen affinity for hemoglobin

7. A base excess of +6 mEq/L suggests a base excess change above normal by the: (Objective 6)
   a. lungs.
   b. heart.
   c. kidneys.
   d. erythrocytes.

REVIEW QUESTIONS (continued)

LEVEL I

LEVEL II

8. The movement of uncharged, hydrophobic solutes through a lipid bilayer and against concentration gradients is termed: (Objective 2)
   a. perfusion.
   b. external convection.
   c. internal convection.
   d. diffusion.

9. To which one of the following conditions in the human body are the peripheral chemoreceptors sensitive? (Objective 5)
   a. decrease in arterial PO_2
   b. decrease in arterial PCO_2
   c. decrease in arterial bicarbonate
   d. decrease in arterial carbon-dioxide levels

10. As a patient's PO_2 increases, the amount of total carbon dioxide carried in blood: (Objective 8)
    a. remains the same.
    b. increases.
    c. decreases.

11. Which of the following is the oxidation state for iron in hemoglobin that will allow hemoglobin to bind oxygen? (Objective 4)
    a. +1
    b. +2
    c. +3
    d. +4

12. What effect does an increase in PCO_2 have on the hemoglobin–oxygen dissociation curve? (Objective 4)
    a. a shift to the left of the hemoglobin–oxygen dissociation curve
    b. a shift to the right of the hemoglobin–oxygen dissociation curve
    c. no change in the hemoglobin–oxygen dissociation curve
    d. does not affect the binding or dissociation of hemoglobin with oxygen

PEARSON
myhealthprofessionskit

Use this address to access the interactive Companion Website created for this textbook. Simply select "Clinical Laboratory Science" from the choice of disciplines. Find this book and log in using your user name and password.

REFERENCES

1. Boron WF. Organization of the respiratory system. In Boron WF, Boulpaep EL (Eds.), *Medical physiology: A cellular and molecular approach*, 2nd ed. (Philadelphia: Saunders, 2009): 613–623.

2. Ibid., 725–737.

3. Klutts JS, Scott MG. Physiology and disorders of water, electrolyte, and acid–base metabolism. In Burtis CA, Ashwood ER, Bruns DE (Eds.), *Tietz textbook of clinical chemistry and molecular diagnostics*, 4th ed. (St Louis: Saunders, 2006): 1764–1765.

4. Boron WF. Transport of oxygen and carbon dioxide in the blood. In Boron WF, Boulpaep EL (Eds.), *Medical physiology: A cellular and molecular approach*, 2nd ed. (Philadelphia: Saunders, 2009): 681–685.

5. Klutts JS, Scott MG. Physiology and disorders of water, electrolyte, and acid–base metabolism. In Burtis CA, Ashwood ER, Brun DE (Eds.), *Tietz fundamentals of clinical chemistry*, 6th ed. (St. Louis: Saunders, 2008): 667–668.

6. Lock R. The whole-blood sampling handbook. *Radiometer Copenhagen* (2000): 8–20.

7. Biswas CK, Ramos JM, Agroyannis B, et al. Blood gas analysis: Effect of air bubbles in syringe and delay in estimation. *Brit Med J* (1982) 284: 923–927.

8. Harsten A, Berg B, Inerot S, et al. Importance of correct handling of samples for the result of blood gas analysis. *Acta Anaesthesiol Scand* (1988) 32: 365–368.

9. Muller-Plathe O. Preanalytical aspects in STAT analysis. *Blood Gas News* (1998) 7: 4–11.

10. National Committee for Clinical Laboratory Standard. *Blood gas preanalytical considerations. Specimen collection, calibration and controls.* NCCLS document C27-A (Wayne, PA: National Committee for Clinical Laboratory Standards, 1993).

11. Sherwin JE. Acid–base control and acid–base disorders. In Kaplan LA, Pesce AJ, Kazmierczak SC (Eds.), *Clinical chemistry theory, analysis, correlation* (4th ed.) (St. Louis: Mosby, 2003): 471–472.

14

Mineral and Bone Metabolism

CHAPTER OUTLINE

■ OBJECTIVES—LEVEL I

Following successful completion of this chapter, the learner will be able to:

1. Identify three forms of calcium as they exist in circulation.
2. List three distinct methods for measuring total serum calcium.
3. Identify two chemical compounds used to measure inorganic phosphate in serum.
4. Identify three compounds used to measure magnesium in serum.
5. Identify two main causes of hypercalcemia.
6. Indicate the source of parathyroid hormone.
7. Discuss the feedback effects of PTH on calcium and phosphorus levels in circulation.
8. List three functions of vitamin D.
9. Describe the structure, source, and function of calcitonin.
10. Identify biochemical markers specific for bone formation and resorption.
11. List several methods used to measure biochemical markers for bone.

■ OBJECTIVES—LEVEL II

Following successful completion of this chapter, the learner will be able to:

1. Discuss the clinical usefulness of measuring ionized magnesium.
2. Compare the structures and biological functions of intact PTH and PTH fragments.
3. Discuss the importance of intraoperative PTH during surgical procedures for primary hyperparathyroidism.
4. Compare and contrast bone modeling and remodeling.
5. Discuss several reasons why biochemical markers of bone turnover are not diagnostic markers for bone disorders, whereas "gold" standard techniques are.
6. Identify several bone-related diseases and the appropriate biochemical marker(s) for bone that may be measured to assess a patient.
7. Identify the appropriate specimen of choice for selected markers.
8. Review sources of interferences for methods used to measure calcium, phosphorus, and magnesium in blood.
9. Predict the laboratory findings for serum total calcium, phosphorus, and urinary phosphate in patients with primary hyperparathyroidism.
10. Correlate laboratory data for minerals and bone markers with disease.

KEY TERMS

Bone alkaline phosphatase
Bone modeling
Bone remodeling
Bone resorption
Bone turnover
Collagens
Hydroxyapatite

Hypercalcemia
Hyperparathyroidism
Intraoperative PTH
Osteoblast
Osteocalcin
Osteoclast
Osteoporosis

Osteoprotegerin
Paget's disease
Parathyroid hormone (PTH)
Pyridinium
RANKL
Telopeptide
Type I collagen

☯ A CASE IN POINT

A 55-year-old Caucasian female underwent a routine physical that was unremarkable in all aspects. Her clinician requested several laboratory tests, and the results of key tests are shown below.

Tests	Patient's results	RI (Conventional Units)	RI (SI Units)
Sodium	142 mEq/L	136–145 mEq/L	136–145 mmol/L
Potassium	4.0 mEq/L	3.8–5.0 mEq/L	3.8–5.0 mmol/L
Chloride	108 mEq/L	98–110 mEq/L	98–110 mmol/L
TCO$_2$	23 mEq/L	24–30 meq/L	24–30 mmol/L
Glucose	96 mg/dL	70–110 mg/dL	3.9–6.1 mmol/L
Urea nitrogen	22 mg/dL	8–23 mg/dL	2.9–8.2 mmol/L
Creatinine	1.3 mg/dL	0.6–1.2 mg/dL	53–106 μMol/L
Calcium, total	11.5 mg/dL	9.2–11.0 mg/dL	2.3–2.7 mmol/L
PTH, intact	110 ng/L	10–65 ng/L	10–65 pg/mL

Issues and Questions to Consider

1. Which test(s) are outside the reference interval (RI)?
2. What are some plausible explanations for these abnormal results?
3. What is the probable diagnosis for this patient?
4. How can this laboratory facilitate the management of this patient?
5. What is the outcome for this patient?

WHAT'S AHEAD

▶ Discussion of calcium, phosphorus, and magnesium homeostasis
▶ Identification of methods used in laboratories to measure calcium, phosphorus, and magnesium
▶ Discussion of parathyroid hormone, vitamin D, and calcitonin
▶ Description of clinical disorders of associated with calcium, phosphorus, magnesium, parathyroid hormone, calcitonin, and vitamin D
▶ Review of bone physiology
▶ Identification of biomarkers of bone formation and bone resorption
▶ Overview of selected bone disorders and biomarkers available to assess the patient

▶ MINERALS, PARATHYROID HORMONE, VITAMIN D, AND CALCITONIN

CALCIUM HOMEOSTASIS

Calcium is the fifth most abundant mineral element in the human body. Approximately 98% of the 1000 to 1200 g of calcium in the adult is present in the skeleton, primarily as **hydroxyapatite**, which is a crystal lattice composed of calcium, phosphorus, and hydroxide. The remaining fraction of total body calcium is present in extracellular fluid and in a variety of tissues, particularly skeletal muscle. Only about 1% of the total skeletal reservoir of calcium is readily exchangeable with extracellular fluid (ECF).

Calcium in the blood is present almost exclusively in the plasma; its concentration is maintained within a narrow range of 8.8–10.3 mg/dL (2.2–2.58 mmol/L) in healthy individuals.[1] Quoted reference intervals vary among laboratories, partly as a result of different analytical methods. Calcium in plasma exists in three distinct forms:

1. Free, or ionized, calcium, the physiologically active form, accounting for about 50% of total calcium
2. Forming complexes with a variety of anions, including bicarbonate, lactate, phosphate, and citrate (about 10%)
3. Bound to plasma proteins (40%) where ~32% is primarily attached to albumin, ~8% to globulins, and an unknown percentage to α_2-Heremann-Schmid globulin (α_2HSG, Fetuin-A)

The binding of calcium to plasma proteins is a freely reversible process, which is governed by dissociation constant (*k*). This was recognized in 1935 by McLean and Hastings, who represented the relationship as follows:[2]

$$[Ca^{2+}][Pr^{2-}]/[CaPr] = K \qquad (14.1)$$

where

$[Ca^{2+}]$ = the concentration of ionized calcium.

$[CaPr]$ = the concentration of protein-bound calcium.

$[Pr^{2-}]$ = the concentration of free protein capable of binding calcium.

K is a constant, which is specific for each protein species and varies with pH and temperature.

Simple transposition of this equation serves to emphasize that the concentration of ionized calcium will remain constant

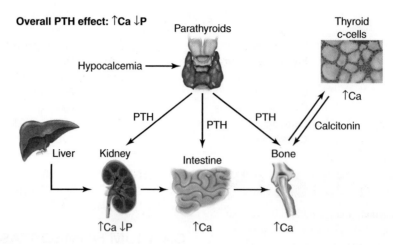

Overall PTH effect: ↑Ca ↓P

■ FIGURE 14-1   Calcium homeostasis.

and is thus independent of the absolute value of bound calcium as long as the ratio of the concentrations of protein-bound calcium to free protein remains constant:

$$[Ca^{2+}] = K[CaPr]/[Pr^{2-}] \qquad (14.2)$$

This concept and relationship is used to construct various equations and nomograms for estimating ionized calcium from measurements of total calcium and protein concentrations. The estimation of ionized calcium has been replaced with direct determination of ionized calcium and can now be performed conveniently using ion-selective electrode technology. When properly performed, ionized calcium is distinctly superior to total calcium for establishing the diagnosis of hyperparathyroidism and other causes of hypercalcemia.

Approximately 80% of calcium is bound to albumin and the remaining 20% is bound to globulins. Calcium binds to the negatively charged sites on proteins, therefore its binding is pH dependent. In a state of alkalosis, there is an increase in negative charges and binding and a decrease in free calcium; conversely, states of acidosis lead to a decrease in negative charge and binding and an increase in free calcium. In vitro, for each 0.1-unit change in pH, approximately 0.2 mg/dL of inverse change occurs in the serum free calcium concentration. Patients who may have a disease that results in a high globulin protein concentration, such as multiple myeloma, will usually bind sufficient calcium to produce an increase in the total serum calcium concentration.

In patients who are not severely ill, a total serum calcium measurement along with an albumin and total protein is usually sufficient. The total protein and albumin can be used to calculate corrected serum calcium. If a patient has a low albumin (hypoalbuminemia), the calcium should be corrected. A calculation that is often used is shown below:[1]

Total calcium (mg/dL) corrected for hypoalbuminemia

= Total calcium (measured) + [(Normal albumin* −

Patient albumin) × 0.8]   (14.3)

*Usually 4.4 g/dL

Maintenance of calcium homeostasis involves the participation of three major organs—the small intestine, the kidneys, and the skeleton. In the adult, there is no persistent net gain or loss of calcium in health. During growth and pregnancy, a positive calcium balance must be maintained. Calcium homeostasis is regulated by various hormones that act principally upon the major organs involved in calcium metabolism. The major hormones are parathyroid hormone (PTH) and the hormones derived from renal metabolism of vitamin D_3, notably 1,25-dihydroxycholecalciferol. Quite possibly, calcitonin plays a role in the regulating process, although its significance in humans is controversial. Other hormones that affect calcium metabolism, but whose secretions are not primarily affected by changes in plasma calcium and phosphate, include thyroid hormones, growth hormone, adrenal glucocorticoids, and gonadal steroids. Present concepts of the major hormonal regulations of calcium metabolism are summarized in Figure 14-1 ■.

✓ **Checkpoint! 14–1**

Identify the three forms of calcium that exists in plasma. Which form is the biologically active form?

IONIZED CALCIUM

The status of calcium in humans is more accurately determined by measuring ionized calcium (or free), which is tightly regulated and is the biologically active form.[3] As stated above, the interpretation of total serum calcium value is complicated by its association with protein and inorganic and organic ions. Evaluation of ionized calcium concentration is less complicated provided the specimen has been properly obtained, handled, and analyzed. Ionized calcium concentration in plasma or serum is sensitive to both pH and temperature. Therefore, it is recommended that the pH of the sample be measured with the ionized calcium. Also, the specimen should be analyzed immediately after the stopper in the blood drawing tube is removed. There are several preanalytical variables that should be addressed and are discussed in the reference cited.[4]

The clinical usefulness of measured ionized calcium is often greater than the measurement of total calcium in hospitalized patients, especially patients undergoing major surgery who have received citrated blood, platelets, heparin, bicarbonate, intravenous solutions, or calcium. Patients having liver transplant operations and other major operations that require cardiopulmonary bypass require expeditious measurement of ionized calcium, potassium, pH, and arterial blood gases for proper maintenance of cardiac function.

Patients with renal disease often have comorbidities with bone- and mineral-related disorders. Ionized calcium measurements are more advantageous than total calcium for evaluating therapy and calcium metabolism because of fluctuations in protein concentrations, pH, protein binding of calcium, and calcium complexes with organic and inorganic ions.

In patients diagnosed with hypercalcemia, it has been shown that ionized calcium provides better clinical information than total calcium. If these patients have subsequent surgically proven primary hyperparathyroidism, they usually have increased ionized calcium rather than serum total calcium. Ionized calcium is more sensitive than total calcium in detecting hypercalcemia associated with malignancy possibly due to the presence of hypoalbuminemia.

URINARY CALCIUM

The concentration of calcium in urine reflects intestinal absorption, skeletal resorption, and renal tubular filtration and reabsorption (see below). The intestinal and renal components are relatively fixed; calcium excretion in the fasting state is used to assess the skeletal component. An increased urinary calcium concentration may be associated with an increase in osteoclastic bone resorption. Measuring urinary calcium is useful in assessing renal stone disease and high-turnover osteoporosis.

CALCIUM METHODS

There are basically three methods commonly used to measure total serum calcium and ionized calcium: (1) colorimetric analysis, (2) atomic absorption , and (3) indirect potentiometry. Among the colorimetric assays, the two widely used metallochromatic indicators are orthocresolphthalein complexone (CPC or OCPC), which is chemically (3'3"-bis [[bis-(carboxymethyl) amino]-methyl]-5', 5"-dimethylphenolphthalein, and arsenazo III (1, 8-dihydroxynaphthalene-3, 6-disulfonic acid-2, 7-bis [azo-2]-phenylarsonic acid). The colorimetric methods are not as accurate as atomic absorption spectrometry but are easily adapted to automated chemistry analyzers.

Atomic absorption spectrometry has been selected as the reference method by NCCLS to measure total serum calcium.[5] This method has been compared with the definitive method using isotope dilution-mass spectrometry developed by NIST. Although atomic absorption spectrometry provides better accuracy and precision, it is not routinely used in clinical chemistry laboratories. Several possible reasons are the preanalytical tasks often required to perform assays by atomic absorption; increased time of analysis; and the ease with which samples, controls, and calibrators may be contaminated by dirty glassware, pipette tips, and laboratory bench surfaces.

Measurement of ionized calcium can be accomplished using a stand-alone ionized calcium analyzer equipped with a calcium-specific electrode or using a blood gas analyzer that has a calcium-specific ion-selective electrode (ISE) (see Chapter 2, Instrumentation). These ISEs are capable of measuring ionized calcium in whole blood, plasma, or serum. Ionized calcium membranes are unique and consist of a liquid membrane containing the ion-selective calcium sensor dissolved in inorganic liquid trapped in a polymeric matrix. Newer electrodes use a neural carrier ionophore, for example, ETH 1001, to selectively measure ionized calcium.

PHOSPHORUS HOMEOSTASIS

Phosphorus is also an abundant element in the body and is omnipresent in its distribution. About 85% of the 500–600 g of phosphorus (measured as inorganic phosphorus) in the adult is present in bone as hydroxyapatite crystals. The remaining phosphorus is mostly combined with lipids, proteins, carbohydrates, and other organic substances to fill vital roles as phospholipids, nucleic acids, nucleotides, constituents of cell membranes and cell cytoplasm, and compounds that are important in biochemical energy storage and exchange.

Most of the phosphorus in ECF is inorganic, predominantly as two species: HPO_4^{2-} and $H_2PO_4^-$. Negligible amounts of HPO_4^{2-} exist in the physiologic pH range. The relative amounts of the two phosphate ions are pH dependent. In health, serum phosphorus varies over a rather wide range of 2.8–4.5 mg/dL (0.89–1.44 mmol/L).[6] Higher phosphorus levels occur in growing children. Ingestion of food can significantly alter serum phosphorus concentration; whereas phosphate-rich food can increase serum phosphorus concentration, a high-carbohydrate meal can cause a significant decrease. Adult values are lower than normal during menstruation.

Three major organs involved in phosphorus homeostasis are the small intestine, kidneys, and skeleton. Because phosphorus is present in virtually all foods, dietary deficiencies do not usually occur. The average dietary intake for adults is about 800–1000 mg, most of which is derived from milk and dairy products. About two-thirds of ingested phosphate is absorbed, mostly in the jejunum. The remaining dietary phosphate is excreted in the feces, mostly as insoluble calcium compounds. Intestinal absorption of phosphate is an active, energy-dependent process. Absorption is increased in association with decreased dietary calcium and increased acidity of the intestinal contents. Absorption is also augmented by the action of vitamin D and growth hormone; vitamin D tends to increase intestinal absorption and renal reabsorption, whereas growth hormone tends to decrease renal excretion.

Most of the phosphorus absorbed from the intestines of adults is excreted in the urine. About 90% of plasma phosphorus is filterable by the glomeruli and about 80% of the filtered phosphate is reabsorbed. Parathyroid hormone inhibits renal tubular reabsorption of phosphate.

PHOSPHORUS METHODS

Nearly all colorimetric assays used to measure serum inorganic phosphate are based on the reaction of phosphate ions with ammonium molybdate to form a phosphomolybdate complex. The colorless phosphomolybdate complex is measured directly by ultraviolet (UV) at 340 nm or reduced to molybdenum blue and measured at 600–700 nm.

Several reducing agents have been used for producing the blue phosphomolybdate complex, including stannous chloride, ferrous ammonium sulfate, ascorbic acid, sardine hydrochloric acid, and aminonaphtholsulfonic acid. Each reagent system exhibits advantages over the others, including improved stability, increased color stability, lowered detection limits, or reduced hydrolysis of organic esters. There are other methods for quantitative analysis of serum, including enzymatic assays and vanadate-molybdate. Examples of enzymes used are glycogen phosphorylase, purine nucleosides phosphorylase, and sucrose phosphorylase. All of these enzymes assays are coupled enzymatic assays incorporating additional coenzymes and cofactors.

 Checkpoint! 14–2

Identify the three major organs involved in both calcium and phosphorus homeostasis.

MAGNESIUM HOMEOSTASIS

Magnesium is a divalent cation (Mg^{2+}). It is the second most abundant intracellular divalent cation, and the fourth most prevalent cation in the body. In a normal adult, the total body magnesium is ~25 g, of which 50–60% is in bone and the remaining 40–50% is in the soft tissues. The location of the skeletal component of magnesium is on the surface of bone and is readily available for exchange with other ions if necessary. The skeleton also serves as a reservoir for maintaining a normal extracellular magnesium concentration. Approximately, 1% of the total body magnesium is extracellular. Similar to calcium, magnesium in plasma or serum circulates in three forms: free or ionized (~55%); bound to proteins (~30%); and forming complexes with phosphate, citrate, and other ions (~15%). Measurement of serum or plasma magnesium may not accurately reflect the level of total body magnesium stores because only ~1% of total body magnesium resides in the extracellular fluid (ECF). Therefore,

some clinicians suggest measuring magnesium in erythrocytes, mononuclear blood cells, or muscle biopsies. In current clinical practice, clinicians request serum magnesium levels on their patients.

Magnesium serves several very important roles in the human body. It is an essential element for the function of more than 300 cellular enzymes. Many reactions requiring ATP, replication and transcription of DNA, and translation of mRNA must have magnesium present. The hexokinase method used to measure glucose in blood is an example of an analogous bodily reaction that requires the presence of magnesium.

Magnesium is required for cellular energy metabolism and membrane stabilization, nerve conduction, ion transport, and Ca^{2+} channel activity. Magnesium plays an important role in the maintenance of intracellular potassium concentration by regulating potassium movement through membranes of the myoctyes. A deficiency of Mg^{2+} can result in refractory plasma electrolyte abnormalities and cardiac arrhythmias, especially after cardiac surgery. Magnesium and calcium both regulate the secretion of PTH.

An average normal dietary intake of magnesium ranges form 6–15 mmol/day (140–160 mg). Approximately 30–40% is absorbed, primarily in the jejunum and ileum. Calcitriol ($1, 25(OH)_{2D3}$) improves magnesium absorption in the intestine. Magnesium is excreted in the urine and under normal conditions may match the intestinal absorption. Reabsorption of filtered magnesium occurs in the following sections of the kidneys: (1) proximal tubules (20%), (2) thick ascending loop of Henle (60%), and (3) distal convoluted tubule (5–10%). Several factors, both hormonal and nonhormonal, influence both loop of Henle and distal tubule reabsorption. These factors include PTH, insulin, magnesium restriction, potassium depletion, and changes in acid–base status. Magnesium in plasma is the primary regulatory factor. In states of hypermagnesemia, transport in the loop of Henle is inhibited, whereas hypomagnesemia stimulates transport whether or not there is magnesium depletion. Other factors that could influence reabsorption are hypercalcemia and the rate of sodium chloride reabsorption.

Magnesium deficiency leads to a rapid decrease in serum magnesium concentration. This results in a reduction of magnesium excretion by the kidneys. It may take several weeks to reestablish equilibrium with bone stores.

URINARY MAGNESIUM

Urinary magnesium levels provide the clinician with valuable data to evaluate patients who present with hypomagnesaemia that develops late in the course of magnesium deficiency. The clinician may be able to differentiate between renal magnesium wasting and extrarenal loss in patients with hypomagnesemia by measuring urinary magnesium excretion. Patients with hypomagnesemia can have 24-hour urine magnesium excretion levels of below 1.0 mmol/L (RI = 3.0–5.0 mmol/L).

The gold standard for determining body magnesium status is the *parenteral magnesium loading test* (MLT).[7] This test relies on uncompromised renal function to derive the retention of an IV magnesium load. Elemental magnesium is given to the patient by IV and a 24-hour urine collection is started. Excretion of <70% of the infused magnesium is considered indicative of magnesium deficiency.

MAGNESIUM METHODS

The most commonly used methods for measuring total serum magnesium are colorimetric assays. Other techniques include atomic absorption spectrometry, flame emission spectrometry, and fluorometry. There are a variety of metallochromic indicators or dyes used that change color upon selective binding of magnesium. Examples of compounds used to measure serum magnesium are calmagite, methylthymol blue, formazan dye, mango/xylidyl blue, chlorophosphan aso II, and arsenazo. Individual methods principles can be found in references cited at the end of the chapter.[8]

Enzymatic methods have been developed using hexokinase or other enzymes that use Mg^{2+}–ATP as a substrate. A coupling reaction with glucose-6-phosphate dehydrogenase allows for monitoring the change of absorbance at 340 nm with the formation of NADPH.

Ionized magnesium measurements in serum are available using ISE incorporating a specific ionophore, for example, a 14-crown ether-containing compound. This neutral carrier ionophore is selective for Mg^{2+}, although with some ISEs ionized calcium is also detected and measured. Thus, a chemometric correction is required to calculate the true magnesium levels in a sample.

Ionized magnesium measurements are affected by pH. The rate of change of ionized magnesium measurements is not as significant as that seen in ionized calcium determinations. As the pH increases, ionized magnesium decreases, and conversely, with a decrease in pH, ionized magnesium increases.

 ### Checkpoint! 14–3

Provide examples for each of the following: (1) Two metallochromatic indicators used for measuring total serum calcium, (2) the ammonium complex widely used to measure serum inorganic phosphorus, and (3) two compounds used to measure serum total magnesium.

PARATHYROID HORMONE (PTH)

Parathyroid hormone (PTH) is secreted primarily as a single-chain polypeptide consisting of 84 amino acids with a molecular mass of 9.5 kDa. It is derived from a larger precursor, pre-proPTH (115 amino acids), which undergoes two successive cleavages, both at the amino-terminal sequences, to yield, first, an intermediate precursor, proPTH, and then the hormone itself.

PTH circulates in heterogeneous forms, whereas it secretes as an intact hormone but subsequently breaks up into fragments in circulation. The major circulating forms are principally the middle and carboxyl terminal fragments. Biological activity resides in the first 34 amino acids, and both the intact and the amino terminal fragments have short half-lives. By contrast, the carboxyl fragment has a long half-life. They are cleared by the kidneys but accumulate easily in the event of renal dysfunction.

The primary physiologic function of PTH is to regulate the concentration of ionized calcium in ECF. PTH secretion causes a rise in serum ionized calcium concentration and a fall in phosphorus concentration. By way of an effective negative feedback mechanism, hypercalcemia leads to PTH suppression. PTH secretion may also be mediated by the magnesium concentration. Patients with low serum magnesium concentration often require magnesium intake to increase the serum PTH levels before the serum calcium concentration can be restored to the desired interval.

A significant hormonal effect of PTH is bone resorption, which serves to restore calcium concentration in ECF. The end result of PTH action on bone is *true* bone resorption and not simply demineralization. PTH-mediated bone resorption is affected by increased activity of osteoclasts (see cells of bone below). Increased conversion of osteoprogenitor cells to osteoclasts occurs as a consequence of more prolonged PTH stimulation. Additional effects of PTH on bone are increased formation of collagenase, which degrades the matrix of bone and increased breakdown of the ground substance of bone.

The major effects of PTH on the kidneys are the simultaneous reduction in reabsorption of sodium, phosphorus, calcium, and bicarbonate ions in the proximal tubule and the enhanced reabsorption of calcium at the distal tubule. The net effect is a rise in serum calcium concentration and phosphaturia and mild metabolic acidosis.

The effect of PTH on intestinal absorption of dietary calcium is indirect. PTH stimulates the renal synthesis of the active vitamin D metabolite $[1,25(OH)_2D_3]$, which in turn acts as a regulator of intestinal absorption of calcium.

INTRAOPERATIVE PTH[(9–13)]

Primary hyperparathyroidism (PHPT) has become a common disease, affecting an estimated 28 per 100,000 people each year in the United States. Increased recognition of PHPT—resulting from advances in screening tests—has produced a clinical profile of **hyperparathyroidism** characterized by mild hypercalcemia with absent or subtle symptoms. The number of parathyroidectomies performed for PHPT has also increased dramatically since 1996.

In the surgical management of PHPT, **intraoperative PTH** assays have been shown to improve the success of parathyroid gland surgery. Minimally invasive parathyroidectomy has replaced the traditional four-gland bilateral exploration as the procedure preferred by many institutions. Intraoperative PTH assays have been used by many surgeons to detect decreases in plasma PTH levels after all hypersecreting tissue has been excised. When combined with accurate preoperative localization using a nuclear imaging technique, technetium-99m sestamibi (^{99m}Tc-sestamibi) scanning, the rapid PTH assay has been reported to prevent dissection of previously operated-on tissue in cases of recurrent hyperparathyroidism and to allow targeted unilateral surgery with shortened operative times, same-day hospital discharge, and potential cost savings in cases of single parathyroid adenomas.

PARATHYROID HORMONE METHODS

Radioimmunoassay provided the clinician with the first test that lowered analytical sensitivity but presented problems with analytical specificity. The early assays were limited in that they used polyclonal antibodies directed against epitopes that were located mostly within the mid- or carboxyl terminal portion of the PTH molecule. These assays detected primarily carboxyl-terminal PTH fragments that do not activate the PTH/PTHrP receptors. These assays were an unreliable predictor of biologically active PTH in the circulation. To improve detection of intact PTH rather than hormonal fragments, noncompetitive immunoassays were developed. The newer noncompetitive immunoassays provided improved sensitivity and specificity for intact PTH due to (1) advances in amino acid sequence of human PTH; (2) a better understanding of the secretion metabolism, clearance, and circulation form of PTH; and (3) the synthesis of fragments of human PTH for use as immunogens, tracers, and calibrators.

 Checkpoint! 14–4

Indicate the principle effect of PTH on the kidneys, intestine, and bone.

VITAMIN D AND METABOLITES

Vitamin D_3, or cholecalciferol, is a prohormone because it can be transformed into physiologically active compounds by irradiation with ultraviolet light. It is found in certain animal tissues and products, particularly fish livers, the livers of fish-eating mammals, and irradiated milk. Almost all of vitamin D_3 in plasma is bound to serum alpha globulin with a molecular mass of 60 kDa. The hormonal form of vitamin D_3, 1,25-dihydroxyvitamin D_3, [1,25(OH)$_2D_3$], acts through a nuclear receptor to carry out its many functions, including calcium absorption, phosphate absorption in the intestine, calcium mobilization in bone, and calcium reabsorption in the kidney. It also has several noncalcemic functions in the body.[14]

In addition to dietary sources, cholecalciferol is synthesized in the skin by ultraviolet irradiation of 7-dehydrocholesterol. Vitamin D_3 is biologically inert and is transported to the liver, where it undergoes hydroxylation to produce 25-hydroxycholecalciferol [25(OH)D_3]. Although 25(OH)D_3 has limited biological activity, it is the major circulating metabolite of vitamin D_3. In the kidney, 25(OH)D_3 undergoes further hydroxylation to form 1,25(OH)$_2D_3$ and other dihydroxy metabolites; the most important are 24,25(OH)$_2D_3$ and 1,24,25(OH)$_3D_3$. The structural changes of vitamin D metabolites discussed are presented in Figure 14-2 ■.

The production of 1, 25(OH)$_2D_3$ is regulated by a negative feedback mechanism, depending on the need for calcium and phosphorus in the circulation. This compound is now currently monitored in serum to indicate the vitamin D status of patients.[14] Decreased blood calcium stimulates the parathyroid glands to secrete PTH, which in turn increases production of 1,25(OH)$_2D_3$ in the renal proximal tubules. Conversely, a rise in blood calcium suppresses PTH secretion, which lowers the production of 1,25(OH)$_2D_3$. Although PTH is required for the mobilization of calcium from bone and for the renal conservation of calcium, stimulation of intestinal reabsorption of calcium is achieved indirectly on the intestine through 1,25(OH)$_2D_3$. The ability of the vitamin D hormone to facilitate calcium transport across the intestinal membrane provides a successful means of administering exogenous vitamin D hormone and adequate dietary calcium to patients with hypoparathyroidism and

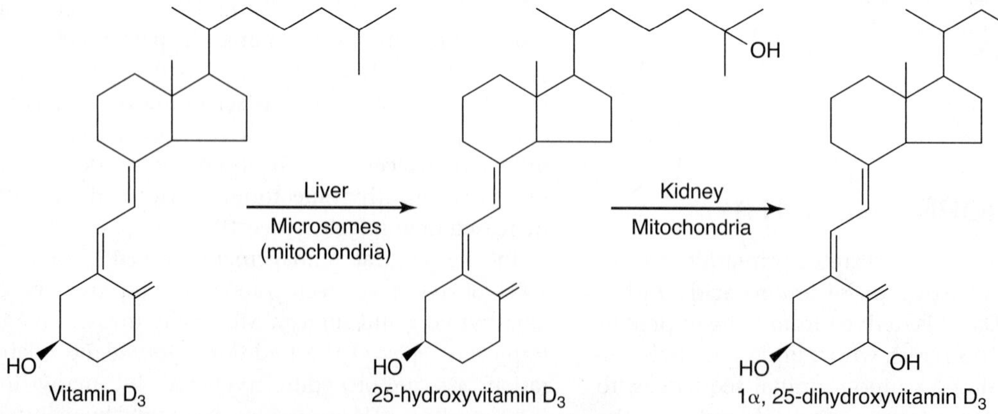

FIGURE 14-2   Metabolism of vitamin D in the liver and kidneys.

pseudohypoparathyroidism. Administration of the vitamin D hormone has also been shown to be effective in the therapeutic management and/or prevention of postmenopausal and age-related osteoporosis. Production of $1,25(OH)_2D_3$ is also stimulated by hypophosphaturia. The demonstration that the locations of $1,25(OH)_2D_3$ are not limited to its target tissues, namely the intestine, bone, and kidney, has expanded the therapeutic function of vitamin D. There is evidence that besides its calciotropic (relating to calcium) properties, vitamin D may also be a developmental hormone with differentiative activity. The therapeutic usefulness of treating psoriasis, female reproduction, and certain malignant diseases with use of $1,25(OH)_2D_3$ is entirely possible if the hypercalcemic activity can be suppressed. The development of analogues of vitamin D compounds to achieve such properties is in progress.

Vitamin D deficiency, an unrecognized epidemic among both children and adults in the United States, not only causes rickets among children but also precipitates and exacerbates osteoporosis among adults and causes the painful bone disease osteomalacia. Vitamin D deficiency has been associated with, but does not necessarily cause, cancers,[15] cardiovascular disease,[16] multiple sclerosis,[17] rheumatoid arthritis,[17] and type 1 diabetes mellitus.[17] Maintaining blood concentrations of 25-hydroxyvitamin D above 80 nmol/L (approximately 30 ng/mL) is not only important for maximizing intestinal calcium absorption but also may be important for providing the extrarenal 1-alphahydroxylase that is present in most tissues to produce $1,25(OH)_2D_3$.

VITAMIN D METHODS

Vitamin D assays that are useful for clinical evaluation of patient status focus on measuring the metabolites, $25(OH)D_3$ and $1,25(OH)_2D_3$. Several of the techniques used to measure these metabolites require samples to be deproteinized or extracted to free the metabolites from *D-binding protein* (DBP). This procedure is followed by a purification technique to separate the various forms of vitamin D, lipids, and interfering substances. This can be accomplished by column chromatography. Once the samples have been purified, the metabolites can then be quantitated. The method of quantitation depends on the metabolites being measured.[18]

Quantitative methods for measuring $25(OH)D_3$ include immunoassays, LC-MSMS, and HPLC separation followed by UV absorption spectroscopy. Most of these assays are labor intensive and some require very expensive and complex instruments such as the LC-MSMS and are not suited for hospital-based clinical laboratories. Automated analyzers are currently available that provide random access technology for measurement of $25(OH)D_3$. The methods include two site chemiluminescent assays that provide adequate sensitivity and specificity.

Measurement of $1,25(OH)_2D_3$ is accomplished primarily using RIA techniques. Radioreceptor assays are available but are not often performed in hospital-based clinical laboratories.

CALCITONIN

Calcitonin is a 32-amino acid peptide that is synthesized and secreted by parafollicular cells of the thyroid glands and is active in the bones, kidneys, and gastrointestinal tract. Although calcitonin has been viewed as a major calcium-regulating hormone because it tends to lower calcium and phosphorus, the precise physiological role of calcitonin is unclear. Although it inhibits osteoclastic bone resorption, the effect is transient and is unlikely to affect calcium homeostasis.

Calcitonin is useful in the evaluation of patients with nodular thyroid disease because it is elevated in medullary thyroid carcinoma. Calcitonin is also valuable therapeutically for the treatment of osteoporosis and Paget's disease.

▶ MINERAL DISORDERS

HYPERCALCEMIA

Hypercalcemia is a common disorder and affects about 1% of the general population. Clinicians are concerned with patients that develop hypercalcemia because one of the two main causes of hypercalcemia is malignancy and the other is primary hyperparathyroidism (PHPT). Together they account for about 80–90% of the hypercalcemic patients. Malignancy as the cause occurs more frequently in hospitalized patients, whereas PHPT appears more in ambulatory patients. Additional causes of hypercalcemia are included in Table 14-1 ✪.

HYPERCALCEMIA ASSOCIATED MALIGNANCY

There are two subtypes of *hypercalcemia-associated malignancy*. The first, known as local osteolytic hypercalcemia, occurs in patients with bone metastases. The malignancy is

✪ TABLE 14-1	
Causes of Hypercalcemia and Hypocalcemia	
Hypercalcemia	**Hypocalcemia**
Primary hyperparathyroidism	Hypoparathyroidism
Renal failure	Chronic renal insufficiency
Renal transplantation	Malignancy
Malignancy	Extreme physical activity (ionized CA^{2+} decreases)
Immobilization	Calcimimetic drugs
Thiazide diuretics	Vitamin D deficiency
Lithium	Septic shock
Vitamin D or A intoxication	Pancreatitis
Thyrotoxicosis	Rhabdomyolysis
	After thyroidectomy
	Calcium deficiency rickets

described as an accelerated bone resorption that exceeds the capacity of body control system to excrete or deposit calcium. Examples include malignant melanoma, breast carcinoma, and lymphoma. The other type, known as *humeral hypercalcemia of malignancy* (HHM), occurs in patients without bone metastases. The tumor produces a humoral factor that acts distally to cause bone resorption and on the kidney to inhibit calcium excretion. Prevalence of HHM is high, nearly 80%. It is often found in patients with squamous cell carcinoma of the lung and other epithelial cancers. The humoral factor was first suspected to be PTH because afflicted patients share clinical presentations of patients with PHPT. The humeral factor, now called parathyroid hormone-related peptide (PTH-rP), is shown to share the n-terminal sequence homology with PTH, and binds to PTH receptors. It exhibits PTH-like effects such as hypophosphatemia, hypercalcemia, and elevated cAMP, but it should suppress levels of PTH and $1,25(OH)_2D_3$.

Although immunoassays for intact PTH are specific enough to show high values in PHPT and low values in HHM, it may still be equivocal because intact PTH may not be elevated in all patients with PHPT. Measuring $1,25(OH)_2D_3$ in blood samples may aid in the proper assessment of patients. The most specific confirmatory test is the determination of PTH-rP. PTH-rP assays with acceptable specificity are now available commercially.

HYPOCALCEMIA

Patients with low blood calcium levels exhibit several adverse symptoms, which may include muscle spasms, carpopedal spasm, cardiac arrhythmias, and peripheral paresthesia. Complications that are more serious may develop such as laryngeal spasm, convulsion, respiratory arrest, and tetany. The causes of hypocalcemia can be categorized and include specific conditions, as listed in Table 14-1. These categories are interpreted as (1) deficiency in PTH production and secretion, (2) resistance to PTH secretion, (3) deficiency in vitamin D or vitamin D metabolites, and (4) deficiencies in bone mineralization with normal metabolism of PTH and vitamin D. Common causes of hypocalcemia are chronic renal failure, hypomagnesemia, hypoparathyroidism, vitamin D deficiency, and acute pancreatitis. For a more extensive review of the causes of hypocalcemia, refer to the reference indicated.[1]

PRIMARY HYPERPARATHYROIDISM

Primary hyperparathyroidism (PHPT) is caused by a disorder of calcium metabolism in relationship to excessive PTH release. It is characterized by excessive secretion of calcium in the absence of an appropriate physiologic stimulus. Hypersecretion may occur unilaterally in an adenoma (frequency ~80%), bilaterally in hyperplasia (frequency ~15–20%), or very rarely in carcinoma (frequency ~3–5%). Common symptoms involve lethargy, coma, psychosis,

anorexia, and hypertension. Over 50% of patients with PHPT are asymptomatic; their calcium elevation is normally detected from biochemical screening for other diseases. When symptoms are present, they usually involve the skeleton, kidney, gastrointestinal tract, and the central nervous system. Major laboratory findings of the patients include elevated calcium accompanied by elevated PTH. Because PTH inhibits reabsorption of phosphorus and bicarbonate, phosphaturia and hyperchloremic acidosis may develop; measurement of electrolytes, pH, and arterial blood gasses are additional tests that help to confirm PHPT. McLeod et al. correlated the blood levels of total calcium and ionized calcium in patients with PHPT and showed that the ionized calcium levels were elevated in a greater number of PHPT patients than total serum calcium, as shown in Figure 14-3 ■.[19]

A summary of results of serum total calcium and PTH measurements for selected diseases is presented in Table 14-2 ✪. It must be remembered that these results are variable depending upon the total physiological state of the patient.

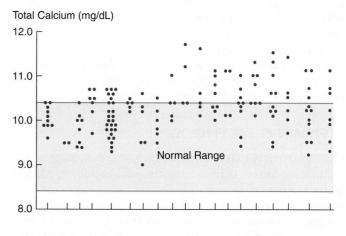

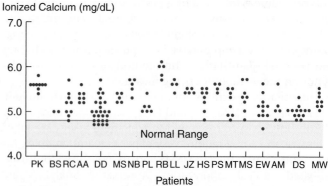

Total Ca and Ionized Ca concentrations in patients with primary HPT (McLeod et al., *Surgery* (1984) 95: 667–673)

■ FIGURE 14-3   Distribution of total calcium and ionized calcium measurement in patients with primary hyperparathyroidism. The results show that ionized calcium measurement provides the clinician with better diagnostic information than total calcium.

TABLE 14-2

Usual Findings for Serum Total Calcium and PTH Concentrations in Selected Disease States		
Disease States	Total [Ca^{2+}]	[PTH]
Primary hyperparathyroidism	increased	increased
Hypercalcemia of malignancy	Increased	decreased
Hypoparathyroidism	Decreased	decreased

✓ **Checkpoint! 14–5**

Predict the serum concentration of total calcium and PTH in a patient with PHPT.

HYPERPHOSPHATEMIA

There are no direct symptoms that result from hyperphosphatemia. However, when high levels of phosphate are maintained for a prolonged period of time, mineralization is increased, and the result is that calcium phosphate may be deposited in abnormal sites within the body. This is referred to as *ectopic calcification.*

Common causes of hyperphosphatemia include decreased renal excretion in acute and chronic renal failure; increased intake with excessive oral, rectal, or IV administration; or increased extracellular load due to transcellular shift that occurs in acidosis. A list of causes of both hyperphosphatemia and hypophosphatemia is presented in Table 14-3 ✪.[1]

HYPOPHOSPHATEMIA

Symptoms of hypophosphatemia may not become evident until serum inorganic phosphorus levels decrease to about one-half the lower limit of normal. These symptoms include proximal weakness, anorexia, dizziness, myopathy, respiratory failure, impairment of cardiac contractility, and metabolic encephalopathy.

A small percentage of hospitalized patients may exhibit slight hypophosphatemia. Severe hypophosphatemia is often found in patients who abuse alcohol and may be due to poor food intake, vomiting, and antacid use. Hypophosphatemia can occur when extracellular phosphate shifts into the cells by respiratory alkalosis associated with sepsis, salicylate poisoning, alcohol withdrawal, hepatic coma, increased insulin during glucose administration, and recovery from diabetic ketoacidosis. Lower blood levels of inorganic phosphorus may be the result of increased renal excretion secondary to hyperparathyroidism, renal tubular defects, and diuretic therapy.

HYPERMAGNESEMIA

Hypermagnesemia is a relatively rare condition where the serum magnesium levels exceed about 1.0 mmol/L (RI = 0.66 − 1.0 mmol/L). The symptoms of hypermagnesemia

✪ **TABLE 14-3**

Causes of Hyperphosphatemia and Hypophosphatemia	
Hyperphosphatemia	Hypophosphatemia
Increased exogenous load or increased intestinal absorption	*Decreased exogenous load or decreased intestinal absorption*
IV infusion	Dietary phosphate restriction
Oral supplementation	Antacid abuse
Vitamin D intoxication	Chronic diarrhea
Phosphate-containing enemas	Vitamin D deficiency
	Steatorrhea
Increased endogenous load	Alcohol abuse
Tumor lysis syndrome	
Rhabdomyolysis	*Altered internal redistribution*
Bowel infarction	Recovery from malnutrition
Hemolysis	Recovery from diabetic ketoacidosis
Acid-base disorders	Hormonal other agents (glucose
	Insulin, glucagons, epinephrine, cortisol
Reduced urinary excretion	Sepsis
Renal failure	Hungry bone syndrome
Hypoparathyroidism	Respiratory alkalosis
Magnesium deficiency	
Acromegaly	*Increased urinary excretion*
	Primary hyperparathyroidism
Pseudohyperphosphatemia	Secondary hyperparathyroidism
Multiple myeloma	High sodium load
Hemolysis in vitro	Metabolic or respiratory acidosis
Hypertriglyceridemia	
Sample drawn from indwelling catheters.	*Pseudohypophosphatemia*
	Multiple myeloma
	Monoclonal gammopathy
	Bilirubin interference
	Mannitol interference
	Phenothiazine interference

include hypotension, bradycardia, respiratory depression, depressed mental faculties, and electrocardiogram abnormalities. This condition is often caused by inappropriate treatment or ingestion of medications prescribed by a physician. Elderly individuals and patients with bowel disorders or renal insufficiency are at the highest risk.

Another example of hypermagnesemia may be found in obstetric patients who develop preeclampsia or eclampsia (hypertensive disorders of pregnancy) and are given a magnesium supplement, usually by IV. These patients must have their serum magnesium levels monitored periodically to ensure that the magnesium concentration in blood does not exceed the therapeutic level of 2.3–3.1 mmol/L. This level is much higher than the reference interval stated previously and would be flagged as a panic or critical value in most laboratories. Patients whose serum magnesium levels are higher than the upper therapeutic levels will develop some of the symptoms described above.

HYPOMAGNESEMIA

The effects on humans of low serum levels of magnesium (<0.6 mmol/L) are usually associated with hypocalcemia. Signs and symptoms are quite similar to those previously listed for hypocalcemia, with the addition of depression and psychosis.

Magnesium deficiency is not uncommon in hospitalized patients. Approximately 10% of the patients admitted to hospitals and ~65% of patients in intensive care units may present with hypomagnesemia. The most common causes of hypomagnesaemia in these patients are the result of conditions associated with the GI tract and kidney, as shown in Table 14-4 ✪.

A substantial loss of magnesium via the kidneys represents a significant cause of magnesium deficiency. Clinically important causes shown in Table 14-4 include alcohol, aminoglycosides antibiotics (e.g., gentamycin), diabetes mellitus (e.g., through osmotic diuresis), and diuretics (e.g., furosemide).

✪ TABLE 14-4

Causes of Hypermagnesemia and Hypomagnesemia

Hypermagnesemia	Hypomagnesemia
Excessive magnesium intake	*Impaired intake or intestinal absorption*
Magnesium containing enemas	*Malnutrition*
Excessive intake of antacids	Malabsorption syndromes
Intestinal obstruction following Mg ingestion	Vitamin D deficiency
Parenteral Mg administration	*Increased intestinal losses*
Tocolytic therapy with Mg-sulfate	Prolonged vomiting, diarrhea
	Intestinal drainage
Kidney: Impaired Mg excretion	*Kidney*
Renal failure	Genetic Mg wasting syndromes
	Bartter's syndrome
Endocrine	Autosomal dominant, with low bone mass
Adrenal insufficiency	Tubulointerstitial disease
Hypothyroidism	Acute tubular necrosis
	Renal transplantation
Redistribution	*Drugs and toxin acting on the kidney*
Trauma	Ethanol
Extensive burns	Diuretic
Shock	Cisplatin
Sepsis	Cyclosporine
Post cardiac ARREST	Aminoglycosides
Others	*Endocrine*
Applied tourniquet	Diabetes mellitus
Hemolysis	Hypercalcemia
	Metabolic acidosis
	Other
	Citrate in blood transfusion
	Pregnancy
	Preeclampsia
	Pancreatitis, burns

▶ BONE STRUCTURE, PHYSIOLOGY, AND METABOLISM

A review of bone structure, physiology, and metabolism is presented here so that reader have an immediate source of information that will assist them as they review the individual biomarkers and disease.

COLLAGENS

Collagens provide the structural framework of bones and cartilages and also provide shape and most of the biomechanical properties, for example, resistance to pressure, torsion, and tension. There are approximately 27 genetically distinct collagen types in vertebrates. They possess diverse structural and biochemical features, but only about half of them occur in cartilage and bone. Their specific functions in the tissues are only partially known.

The structural features of all collagens, illustrated in Figure 14-4 ■, is the triple helix, a coiled-coil component in the form of a right-handed helix of 1.5 nm diameter composed of three polypeptides chains (α-chains). Another feature of the collagen triple helix is the abundance of proline and hydroxyproline.[21]

In summary, bone consists of a large amount of densely packed cross-linked heterofibrils of types I and V collagen that provide the architectural structure and the substrate for mineralization. A more detailed discussion of collagens can be found in the reference cited.

FIBRILLOGENESIS

The formation of fibrils (fibrillogensis), shown in Figure 14-5 ■, and maturation of collagens is an important process leading to bone formation. Of the 27 forms of collagen referenced to previously, type I represents the major component of bone and collagen II, and is also the main component of cartilage. The focus of most of this discussion will be on type I collagen and bone mineralization.[22]

The short, nonhelical structures at the end of the molecule in Figure 14-6 ■, the telopeptide (both N and C telopeptide) which contain the cross-linking sites have a profound effect on the kinetics of fibril formation. They may possibly serve as a catalyst during fibrillogenesis.

Structural stability to newly formed collagen fibrils is accomplished by the formation of intermolecular cross-links.

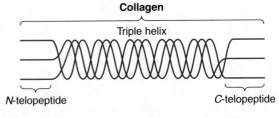

■ FIGURE 14-4 The structural features of collagens.

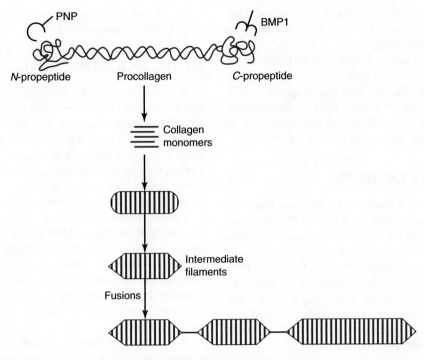

■ FIGURE 14-5   Formation of fibrils termed fibrillogenesis.

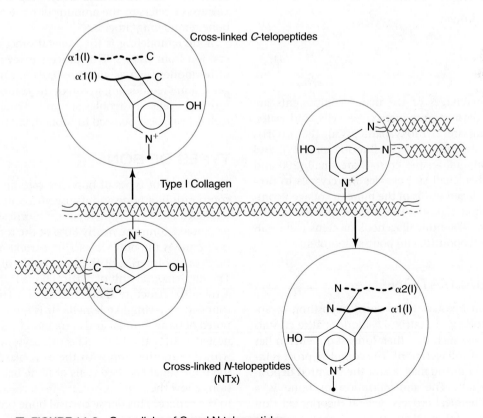

■ FIGURE 14-6   Cross links of C and N telopeptides.

The process begins with the formation of intermediate, chemically reducible cross-links followed by maturation to more stable, nonreducible bonds.

Cross-linking in bone tissue involves hydroxylation of telopeptide lysine residues, which gives rise to primarily keto-imine forms of cross-link. This reaction results in the formation of **pyridinium** and pyrrolic cross-links on maturation. Two specific pyridinium cross-links are pyridinoline (Pyr or PYD) and an analog named deoxypyridinoline (DPD).[23]

THE PHYSIOLOGY OF BONE

Bones serve many purposes, including (1) providing mechanical protection for internal organs, (2) allowing the direction of motion, (3) facilitatating the locomotion process, (4) providing a protective housing for blood-forming marrow, and (5) serving as a reservoir for mineral ions (e.g., calcium, phosphorus, and magnesium). Bones are organized optimally to resist loads imposed by functional activities. With growth and development, the bone tissue is constantly reshaped and remodeled to maintain this maximization and to maintain a form appropriate to its mechanical function.

The composition of bone is described as a composite tissue consisting of the following material in decreasing order of amount:

- Minerals
- Collagen
- Water
- Noncollagenous protein
- Lipids
- Vascular elements
- Cell

The relative percentages of the major components are ~60% minerals, ~35% organic matrix, ~5 cells, and water. The mineral is impure hydroxyapatite, $Ca_{10}(PO_4)_6(OH)_2$ containing carbonate, citrate, Mg^{2+}, fluoride ion (Fl^-), and strontium. Bone mineral apatite crystals are small (200 and 400 angstroms). This small size enables the crystals to provide the flexible collagen fibrils of the mineral collagen composite with structural rigidity. The organic matrix is 90% collagen and about 10% noncollagenous proteins (osteocalcin, osteonectin, osteopontin, and bone sialoprotein).

BONE MINERALIZATION

Bone mineralization involves the ordered deposition of apatite on a **type I collagen** matrix. The bone apatite crystals are deposited in a way such that their longest dimension lies parallel to the axis of collagen fibril. The complete process includes the formation of the matrix and the oriented deposition of these crystals. The mineralization of bone is a complex physical chemical process. Several theories exist on the exact mechanisms involved and therefore are beyond the scope of this textbook. For further insight into this process, refer to the reference cited.[22]

BONE MODELING AND REMODELING

Once mineralization of calcified cartilage or osteoid begins, the mineralized matrix remains in a dynamic state. It is continually being reshaped as the bones grow and as the loads applied to the bone change. The periosteum, where new bone formation starts, is responsive to mechanical load, allowing bone modeling to occur. Mineralize matrix or crystals grow, collect into a rounded mass, dissolve, and change in composition during bone remodeling.

During **bone modeling** (endochrondral ossification), the calcified cartilage matrix serves as a site for deposition of mineral in the form of the primary spongiosa. Mineralized crystals develop into woven and compact bone. The calcified cartilage and woven bone is removed and replaced by woven bone alone. This modeling is accomplished by the resorption of bone on the endosteal surface and the formation of periosteal bone.[24]

Bone remodeling is a process that involves formation of bone on surfaces previously containing bone. There are two phases to bone remodeling: (1) Bone breakdown is started by the osteoclasts that adhere to the bone surface and create a secondary lysosome with an acid pH to dissolve the hydroxyapatite mineral; and (2) the rebuilding phase depends on the recruitment of osteoblasts that lay down an osteoid, and becomes mineralized in a tightly regulated fashion. **Bone turnover** refers to the amount of bone renewed during the bone remodeling process.[24]

Bone remodeling is the skeletal process that allows mineral ion homeostasis. This process is regulated by a number of hormones such as vitamin D, PTH, calcitonin, and estrogens. Bone remodeling involves the complex action of bone-forming cells (osteoblasts) and bone resorption cells (osteoclasts), as discussed in more detail below.

TYPES OF BONE

The two major types of bone are *cortical* and *trabecular*. The composition of bone and its organization is illustrated in the reference cited for your review.[25] Cortical bones (also compact bone) comprise nearly 80% of the total bone mass. Cortical bone is the outer layer (the cortex) of bones and forms most of the interior of the long bones of the body (femur). The fundamental unit of cortical bone is the osteon, which is a tube-like structure that consists of a Haversian canal surrounded by a ring-like lamella. It is quite dense and is composed of bone mineral and extra blood vessels and osteocytes nested within the bone. The osteocytes are interconnected with one another and with the osteoblasts on the surface of the bone canaliculi (see Cells of Bone below). These connections allow the transfer of Ca^{2+} from the interior of the bone to the surface. This dense cortical bone provides much of the strength for weight bearing by the long bones.

Trabecular (also cancellous or medullary) bone makes up the rest of the total bone mass. It is located in the interior of bones and is especially prominent within the vertebral bodies. Trabecular bone is composed of thin spiculaes of bone that extend from the cortex into the medullary cavity. Both osteoblasts and osteoclasts line the bone spicules and are involved in bone remodeling.

CELLS OF BONE

There are several different types of cells associated with bone growth and development throughout life, and they require an organized and synchronized interaction for these processes to occur. These cells include the following:

- Osteoclasts
- Osteoblasts
- Osteocytes
- Bone-lining cells
- Precursor of specialized cells
- Mesenchymal stem cells
- Chondrocyte lineage cells

A brief discussion of osteoclasts, osteoblasts, and osteocytes will be presented here, and details of each can be found in the reference cited.[26]

OSTEOCLASTS

Osteoclasts promote bone-resorption and are found on the growth surfaces of bone. They are characterized as multinuclear giant cells (~100 μm) with ruffled borders that secrete products leading to bone destruction. There are several substances that stimulate and inhibit the actions of these cells; they can be found in Box 14-1 ■.[25]

BOX 14-1 Examples of Substances that Promote and Inhibit Osteoclasts Formation

The following are known to *promote* osteoclasts formation:

▶ PTH
▶ 1,25-dihydroxy vitamin D_3
▶ Osteoprotegerin ligand (OPG-L)
▶ Receptor activator of NF$_k$B ligand (RANKL)
▶ Interleukin-1,3,6,11
▶ Transforming growth factor-β (TGF-β)

The following agents are known to **inhibit** osteoclasts formation:

▶ Calcitonin
▶ Bisphosphonate
▶ Osteoprotegerin (OPG)
▶ 17β-estradiol

The morphological features of the osteoblast facilitate its unique function to resorb bone and calcified cartilage. Characteristic features of the osteoclast that result from attachment to the mineralized surface are (1) appearance of the sealing zone, (2) formation of the "ruffled border" on the apical bone surface, (3) a distinct organization of the cytoskeleton, (4) projections of the basal lateral membrane for secretion of mineral into the vascular space, (5) polarization of the osteoclast nuclei at the basilar surface, and (6) several mitochondria to support cellular activities.

OSTEOBLASTS

Osteoblasts promote bone formation and they function to (1) synthesize and secrete unmineralized bone matrix (the osteoid), (2) participate in the calcification and resorption of bone, and (3) regulate the flux of Ca^{2+} and PO_4^{2-} in and out of bone. Osteoblasts occur as a layer of contiguous cells that in their active state are cuboidal. Regulation of osteoblasts and bone mineralization involves alkaline phosphatase (maturation state), osteocalcin (OC), and bone sialoprotein (BSP), which regulate ordered deposition of mineral. Osteoblast and osteocytes have receptors for key regulators of bone turnover, which include PTH, cytokines, $1,25(OH)_2D_3$, and sex steroids. These factors provide mechanisms for mediating the coupling of osteoblast and osteoclast activity.

OSTEOCYTES

Osteocytes are found within the bony matrix and are derived from osteoblasts that have encased themselves within bone. They represent the most abundant cell in the skeleton. Osteocytes are in direct communication with the bone-lining cells, and with each other within the mineralized matrix through cellular processes that lie within, and are tethered to the canaliculi channels.

SIGNALING

The activity of osteoblast and osteoclasts to form and resorb bone is controlled by circulating hormones, cytokines, and growth factors. Both cytokines and growth factors are involved in a complex signaling system in bone that determines how bone is formed and resorbed in the remodeling process. Osteoblasts serve to regulate osteoclast formation from hemopoietic precursors through contact-dependent mechanisms that are controlled by hormones and by the production of locally generated inhibitors. The significant effectors are products of the tumor necrosis factor ligand and receptor family.

A major stimulator of both the differentiation of preosteoclasts to osteoclasts, and the activity of the mature osteoclast, is a specific protein substance called *receptor activator of nuclear factor (NF-$_k$B) ligand* (**RANKL**). RANK ligand is expressed by osteoblasts, bone marrow stromal cells, and activated T cells. It occurs in circulation as a soluble molecule, making it suitable for laboratory assessment. RANK ligand

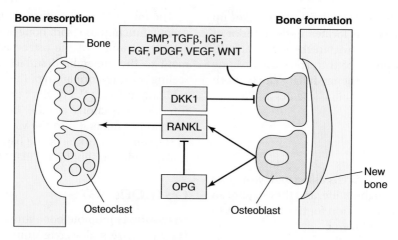

■ FIGURE 14-7 The interaction of RANKL and OPG in bone formation and resorption.

binds to and stimulates a membrane-bound receptor of the osteoclast called RANK, a member of the TNF receptor family. Osteoblastic and stromal cells also produce a soluble substance of the TNF receptor families called osteoprotegerin (OPG). **Osteoprotegerin** protects the bone from osteoclastic activity by binding RANK ligand. This process is shown in Figure 14-7 ■. The exact role of these proteins in the development of various forms of osteoporosis and osteopetrosis is only beginning to be understood. A consensus, however, believes that the balance between the amount of OPG and RANKL produced by the osteoblast/stromal cell appears to be a very important factor.[27]

BONE RESORPTION AND FORMATION

Bone resorption and formation, as illustrated in Figure 14-8 ■, is an intricate process involving bone cells, tissue factors, cytokines, hormones, minerals, and other bone-related substances. A summary of the process will be presented here and a detailed discussion may be found in the reference provided.[25]

Osteoblastic cells are stimulated to secrete factors such as macrophage colony-stimulating factor (M-CSF) by PTH and vitamin D. These and other substances induce stem cells to differentiate into osteoclast precursors, mononuclear osteoclasts, and finally mature, multinucleated osteoclasts. Osteoblasts also secrete ionized calcium and inorganic phosphate, which nucleate on the surface of bone. PTH indirectly stimulates bone resorption by osteoclasts. Osteoclasts are devoid of PTH receptors. Therefore, the PTH binds to a receptor on an osteoblast and stimulates the release of factors, such as IL-6 and soluble RANK ligand, and the expression of membrane-bound RANK ligand. These factors promote bone resorption by osteoclasts.[25]

Checkpoint! 14–6

Describe the difference between bone modeling and remodeling.

► BIOMARKERS OF BONE FORMATION AND RESORPTION

GENERAL FEATURES OF BONE MARKERS

A good biochemical marker of bone should be specific for bone tissue and reflect only formation and not resorption. It should correlate well to the gold standard techniques and procedures currently utilized. The bone marker method should be easy to perform and provide information on the dynamic process of bone turnover.

The measurement of biochemical bone markers has several useful features in clinical practice. It may assist the clinician in understanding the various aspects of osteoporosis, including (1) pathogenesis, (2) mechanism of therapies action, (3) selecting optimal doses in clinical trials, (4) familiarity with the time course of onset and resolution of treatment effects in clinical trials, and (5) its relationship to bone mineral density.

Biochemical bone markers may be useful in providing input to clinical questions involving the following:

• Diagnosing osteoporosis

• Identifying "fast bone losers" and patients at high risk for fractures

• Selecting the best treatment of osteoporosis

• Providing any early indication of the response of treatment

Finally bone markers are considered valid tools for assessing skeletal response to the following:

• Ophorectomy

• Physical exercise

• Immobilization

• Alcoholism

• Smoking

TABLE 14-5

Biochemical Markers of Bone Formation

Marker	Tissue of Origin	Sample Type
Total alkaline phosphatase (ALP)	Bone, liver, intestine, kidney, placenta	Serum
Bone-specific alkaline phosphatase (BAP)	Bone	Serum
Osteocalcin (OC)	Bone, platelets	Serum
Carboxy terminal propeptide of type I procollagen (PICP)	Bone, soft tissue, skin	Serum
Amino-terminal propeptide of type I procollagen (PINP)	Bone, soft tissue, skin	Serum

- Vitamin D deficiency
- Chronic inflammatory bowel disease
- Chronic starving
- Hyperthyroidism

BIOMARKERS OF BONE FORMATION

Biomarkers of bone formation have been available to the clinician for many years. Some of the earlier biomarkers used are slowly being replaced with better biomarkers as technology changes, which will provide the clinician with valuable information in the future. Several of these biomarkers are listed in Table 14-5 ✪. An ideal biomarker for bone formation should satisfy the criteria listed here:

- It should be a structural protein released into the blood in a rate proportional to its incorporation into bone, and the fraction released should be unchanged by disease.
- It should have a well-characterized function and should not be released unaltered during bone resorption.
- Its metabolic pathway and serum half-life should be known.

Currently, no serum biomarker of bone formation is ideal, but several of them possess characteristics that make them valuable to the clinician and therefore to their patients.

BONE ALKALINE PHOSPHATASE

The enzyme alkaline phosphatase (ALP) is found in many tissues, including liver, bone, intestine, placenta, and kidney (see Chapter 9, Enzymes). There are four different genes code for the tissue-nonspecific, intestinal, placental, and germ-cell isoenzyme. The ALPs from liver, bone, and kidney are isoforms of the same gene product, the tissue-nonspecific gene.

Alkaline phosphatase assays routinely available in the clinical chemistry laboratory measure total ALP activity and do not distinguish the source of the isoenzymes. Clinicians usually rely on other laboratory tests to help determine the source of ALP elevations. For example, various liver function tests and hCG measurements may provide a key to the source

of ALP. Separation of ALP isoenzymes by chromatographic assays (see below) may be of value in determining the source of ALP. In normal individuals, ALP originates primarily from liver and bone. The bone isoenzyme, **bone alkaline phosphatase** (BAP or BALP), is the marker for bone formation and is found in osteoblast.

CLINICAL USEFULNESS

The exact function of BAP is unknown. It is thought to be involved in these:

- Promoting and supporting mineralization
- Modulating osteoclast activity
- Deposition of osteoid

BAP is increased in a variety of conditions associated with bone, including these:

- Osteoporosis
- Osteomalacia
- Rickets
- Hyperparathyroidism
- Renal osteodystrophy
- Thyrotoxicosis
- Acromegaly
- Bony metastases
- Hydroxycorticol excess
- Paget's disease

Measuring blood levels of BAP has several advantages over osteocalcin (OC) measurements (see below). The in-vivo half-life of BAP is from 1 to 3 days, much longer than that of OC, thus it is relatively unaffected by diurnal variation. BAP is more stable in vitro and does not require special specimen handling. In patients with impaired renal function, BAP is more useful than OC because BAP is not cleared by glomerular filtration.

Patients with liver disease, severe osteomalacia, or increased concentration of 1,25 $(OH)_2$ vitamin D may provide misleading BAP results. Patients with severe osteomalicia may show marked increases in BAP without an increase in bone mineralization because of mineralizing defects.

MEASUREMENT OF BAP

There have been many methods available to measure BAP. A list of representative methods is shown below. Refer to the reference cited for a detailed explanation of each method listed.[28]

- Heat inactivation
- Chemical inhibition
- Electrophoresis

- Isoelectric focusing
- Lectin precipitation and chromatography
- HPLC

Electrophoresis provides qualitative results only, whereas heat inactivation and chemical inhibition methods have been used with some success for determining isoenzyme activity.

OSTEOCALCIN

Osteocalcin (also bone Gla protein is a protein that is synthesized by mature osteoblasts, odontoblasts, and hypertrophic chondrocytes. Most of the protein found in the circulation reflects osteoblast activity because the number of osteoblasts exceeds the number of odontoblasts. Osteocalcin originates as a 75-amino-acid propeptide that is first carboxylated at one to three glutamic acid sites before further post-translational processing. It is characterized by the presence of three residues of calcium-binding amino acid, gamma-carboxyglutamic acid (Gla). The degree of carboxylation determines OC's affinity for Ca^{2+} and calcium-containing proteins and minerals and is influenced by vitamin K. After carboxylation, a 25-amino-acid sequence is removed and the protein is released.

A majority of OC produced is found in bone matrix and comprises nearly 15% of the protein of the extracellular matrix. The exact role of OC in bone metabolism has not been fully established. Osteocalcin may have a role in limiting the extent of bone mineralization within the matrix. It also may be involved in recruiting osteoclasts, regulating osteoblast maturation, and recruiting scavenger cells.

There are several immunoreactive forms of OC found in the circulation. Approximately one-third is the intact molecule, one-third is a large N-terminal mid-region fragment, and one-third is N-terminal mid-region and C-terminal fragments. The fragments are derived from proteolytic degradation of the intact molecule or are generated as breakdown products during bone resorption.

Most experts believe that OC is the best indicator for bone formation even though some OC originates from bone resorption.[29] Unlike with certain PTH fragments, there is no agreement regarding which form provides the best clinical information. Serum levels are usually elevated in most states of high bone turnover and decreased in conditions of low bone turnover.

CARBOXY-TERMINAL PROPEPTIDE OF TYPE I PROCOLLAGEN AND AMINO-TERMINAL PROPEPTIDE OF TYPE I PROCOLLAGEN

The following two biomarkers represent specific products of proliferating osteoblasts and fibroblasts: (1) amino-terminal propeptide of type I procollagen (PINP) and (2) carboxy-terminal propeptide of type I procollagen (PICP). Both of these biomarkers are cleaved by specific extracellular endoproteinases

from newly translated collagen type I polypeptides. They are considered to reflect the collagenous phase of bone formation. An important factor to be considered is that type I collagen is also a component of several soft tissues (e.g., fibrocartilage, skin, heart valves, intestine, and large blood vessels), and there is a possibility that circulating procollagen from soft tissue synthesis of type I collagen may contribute to the overall amount of these two biomarkers.

BIOMARKERS OF BONE RESORPTION

Several biomarkers of bone resorption have been made available due in part to the changes in technologies over the past several years. Examples of bone resorption biomarkers are presented in Table 14-6 ✪. An ideal marker for bone resorption should meet the criteria listed:

- It should be a degradation product of a matrix component not found in any other tissue.
- Its serum level should not be under separate endocrine control.
- It should not be reutilized in new bone formation.

Currently, no serum biomarker of bone resorption is "ideal," but several of them possess characteristics that make them valuable to the clinician and therefore to his or her patients.

✪ TABLE 14-6

Biochemical Markers of Bone Resorption

Marker	Tissue of Origin	Sample Type
Hydroxproline	Bone, cartilage, soft tissue, skin	Urine
Pyridinoline (Pyr or PYD)	Bone, cartilage, tendon, blood vessels	Urine, serum
Deoxypyridinoline (DPD)	Bone, dentin	Urine, serum
Carboxyl-terminal cross-linked telopeptide of type I collagen (CTX-I)	All tissues containing type I collagen	Urine, serum
Amino-terminal cross-linked telopeptide of type I collagen (NTX-I)	All tissues containing type I collagen	Urine, serum
Carboxyl-terminal cross-linked telopeptide of type I collagen(ICTP, CTX-MMP)	Bone and skin	Serum
Hydroxylysine glycosides	Bone, soft tissues, skin, serum complement	Urine
Bone sialoprotein (BSP)	Bone, dentin	Serum
Tartrate-resistant acid phosphatase (TRAP or TR-ACP)	Bone, blood	Plasma, serum
Collagen I alpha 1 helicoidal peptide (HELP)	All tissues containing type I collagen	Urine
Cathepsins (e.g., K, L)	K: primarily in osteoclasts	Plasma, serum
	L: macrophages, osteoclasts	

URINARY CALCIUM

An initial assessment of a patient can begin with a 24-hour, or fasting (morning), urinary calcium sample that is normalized to creatinine. This measurement provides a simple and inexpensive marker for patients suspected of having a bone-related condition. An increase in excretion indicates a condition of accelerated bone resorption but does not provide additional information. The test is not sensitive to smaller changes expected in osteoporosis of low bone turnover. Diet and several drugs, including calcium-regulating hormones, diuretics, estrogen, and vitamin D supplements, may influence excretion and should be considered by the clinician when interpreting results.

BONE SIALOPROTEIN

Bone sialoprotein (BSP) is a glycosylated protein approximately 70–80 kDa and synthesized by osteoblasts, osteoclasts, and odontoblasts. Its role includes (1) attachment of osteoblasts to surfaces, (2) binding to alpha 2 chains in type 1 collagen, and (3) nucleation of hydroxyapatite crystals. BSP accounts for about 5–10% of noncollagenous protein of the bone extracellular matrix.

Several immunoassays are available for measurement of BSP in serum or plasma. The clinical usefulness for BSP is in assessment of patients with cancer. The expression of BSP by myeloma and other tumors has been described in patients who eventually developed bone metastases or who had poorer outcomes.

TARTRATE-RESISTANT ACID PHOSPHATASE

A tartrate-resistant acid phosphatase (TRAP or TR-ACP) is expressed by osteoclast and also hepatic Kupffer cells and macrophages of the alveoli and spleen. There are at least two forms of TRAP in serum. They are TRAP 5a, which contains additional sialic acid residues and has an optimal pH of 5, and TRAP 5b, which has an optimal pH of 5.8. Osteoclast show greatest abundance of TRAP5b isoenzyme.

There are immunoassays and electrophoresis methods available to measure TRAP in serum and plasma. Immunoassays specific for TRAP 5b are currently being evaluated, but none have been approved by the FDA.

HYDROXYPROLINE

Hydroxyproline (Hyp) is an amino acid present in all fibrillar collagens and partly collagenous proteins, including C1q and elastin. It is also present in newly synthesized and mature collagen. It is released following the enzymatic breakdown of collagen, with approximately 10% of the total circulating Hyp pool excreted in the urine. The remaining Hyp is reabsorbed, further metabolized, or reused for collagen synthesis.

Measurement of urinary Hyp levels reflects both collagen synthesis and breakdown of all body collagens, thus represents a low degree of specificity. Urine levels of Hyp can be determined colorimetrically or by HPLC, but these have been replaced by other, more specific bone resorption markers.

GLYCOSYLATED HYDROXYLYSINE (GHL)

Galactosyl hydroxylysine (GHL) and glucosyl-galactosyl-hydroxylysine (GGHL) are both produced during procollagen synthesis. Bone collagen contains primarily GHL, whereas GGHL is more prevalent in skin collagen. Both compounds are not metabolized in the body to a large extent, and exogenous dietary sources are less pronounced than for Hyp. GHL can be measured in urine by HPLC. The assay shows good correlation with other bone resorption markers. The lack of an immunoassay for quantitative analysis has limited the clinical application of this analyte.

COLLAGEN CROSS-LINKS

Cross-links formed between adjacent collagen molecules strengthen and stabilize type I collagen of bone. During resorption, cross-links are released into the circulation and subsequently excreted. Cross-links found in blood and urine include free cross-linking amino acids and larger fragments containing specific NH_2-terminal telopeptide and COOH-terminal telopeptide sequences. A brief discussion of several examples of cross-links will be presented next.

PYRIDINOLINE AND DEOXYPYRIDINOLINE

Pyridinoline (Pyr or PYD) and deoxypyridinoline (DPD) are the primary cross-links in skeletal tissue and act as stabilizers of mature cross-links in types I, II, and III collagen of most major connective tissues (dentin, bone, ligaments, tendons, muscle, vascular walls, and intestines). Both Pyr and DPD are not found in skin. Pyridinoline is found in most tissues, whereas DPD is found in highest concentration in bone and is considered the more bone-specific marker. Measurement of Pyr or DPD provides an advantage over urinary Hyp because they are not influenced by dietary intake and are unaffected by the degradation of newly synthesized collagen. They are not further metabolized or reused in collagen biosynthesis.

TELOPEPTIDES

Cross-linked telopeptide is the term associated with the measurement of collagen degradation products connected to the cross-link regions in type I collagen, as illustrated in Figure 14-6. Telopeptide refers to the nonhelical regions of type I collagen. There are immunoreactive epitopes located on peptide fragments that are derived from the N-terminal (NH_2^-) and designed as amino terminal cross-linked telopeptide of type I collagen (NTX) and C-terminal (COOH–) designated carboxy terminal cross-linked telopeptide of type I collagen (CTX). A larger cross-linked fragment derived from the C-terminal, collagen type I cross-linked C telopeptide (ICTP) has been recognized and a serum immunoassay is available for quantitative analysis.

Neither CTX-I nor NTX-I is recycled or metabolized upon release into the circulation. Both are excreted into urine and can be measured in either urine or serum excretion. Serum excretion of analytes is highest during the early morning hours but is variable. If urine is used, the sample collection should be standardized to improve precision.

 Checkpoint! 14–7

Identify two reasons why measurement of pyridinoline and de-oxypyridinoline provides an advantage over measurement of urinary hydroxyproline for assessment of bone disorders.

APPLICATION OF BONE MARKERS

Evaluation of patients with bone-related diseases may include diagnostic tests for bone turnover and measurement of biochemical markers of bone formation and bone resorption. Bone markers measurements appear to be useful for monitoring patients during their drug therapy regimens and determining prognosis. The usefulness of assessing fracture risk by measuring bone biomarkers is questionable, but much research is being done, especially with RANK and RANKL.

The gold standard techniques currently being used to assess patients with suspected bone diseases are (1)isotope bone scan for bone mineral density (BMD), (2) bone biopsy (histomorphometry) for micro-architecture of bone tissue, and (3) whole body calcium kinetic studies. Some of the drawbacks of these techniques are listed here:

- Time-consuming
- Invasive
- Require specialized equipment
- Cannot be repeated frequently
- May only reflect bone turnover in local areas of skeleton

Measuring biomarkers of bone turnover does provide advantages over the standard techniques listed above, including these:

- Relatively inexpensive
- Noninvasive
- Requires no special patient preparation
- Can be repeated over short time intervals, thus providing quick results

There are several reasons why measuring biomarkers of turnover are not routinely used by the clinician:

1. The measurements are affected by several preanalytical variables, both controllable and uncontrollable, as shown in Box 14-2 ■.

> **BOX 14-2 Controllable and Uncontrollable Sources of Preanalytical Variability**
>
> Controllable
> ▶ Circadian
> ▶ Fasting
> ▶ Diet
> ▶ Seasonal
> ▶ Exercise
>
> Uncontrollable
> ▶ Age
> ▶ Gender
> ▶ Pregnancy
> ▶ Fracture
> ▶ Diseases
> ▶ Drugs
> ▶ Immobility

2. Fluctuation may be observed daily, seasonally, and intra-individually.
3. A variety of methods are used and there is a lack of standardization.
4. Problems exist with assay specificity and precision.
5. There are instrumentation issues; for example, the laboratory may have to purchase an HPLC system or resort to radioimmunoassay methods. Many laboratories select the assays to bring in-house based on existing instrumentation/automation, and most platforms have limited applications available.

ASSAYS FOR BONE MARKERS

Numerous assays are available to measure biochemical markers of bone formation and resorption. An extensive list of laboratory assays for biomarkers is presented in Table 14-7 ✪. Specimens for analysis include urine, serum, and plasma. Urine samples for measurement of bone biomarkers often require special handling procedures, and questions should be asked before sample procurement. Examples of questions to ask are presented here:

- Does the test require timed collection?
- Does the test require a preservative? If yes, which one?
- Should the sample be protected from sunlight?
- Is the presence of hemolysis or bilirubin acceptable?
- What are the specific temperature storage requirements?

Also, the urine assays must be standardized against creatinine concentrations for proper interpretation. Therefore, it is very important that when a laboratory receives a request for bone marker analysis, specimen collection requirements must be verified. Serum samples usually do not require any special handling procedures beyond the laboratory's routine

TABLE 14-7

Laboratory Assays for Biochemical Markers of Bone Diseases

Marker	Trade Name
Bone alkaline phosphatase(BAP)	Metra BAP
Deoxypyridinoline	Metra DPD
Pyridinoline	Metra PYD
Carboxy-terminal cross-linked telopeptide of type I collagen (CTX-1)	Serum CrossLaps (These are all ELISA)
Intact osteocalcin (AA 1-49) and large N-mid fragment (AA1-43)	N-MID Osteocalcin
Total procollagen type I amino terminal propeptide	Total PINP
β-CTX	β-Crosslaps
Tartrate-resistant acid phosphatase	Bone TRAP (EIA)
Cross laps	α-Crosslaps (ELISA)
RANKL	Ampli sRANKL (ELISA)
BAP	Ostase (EIA)
Osteocalcin	N-MID-OC (ELISA)
Osteoprotegerin	OPG (ELISA)
Intact and N-terminal propeptide type III procollagen	PIIINP (ELISA)
Osteoprotegerin	OPG (EIA)
RANKL	RANKL (EIA)
Osteocalcin	Osteocalcin (IRMA)
Amino-terminal cross-linked telopeptide of type I collagen	Osteomark (ELISA)
Intact Osteocalcin	N-tact Osteo (IRMA)
Deoxypyridinoline (urine)	Immunlite Pyrilinks (CL-ELIA)
N-telopeptide	NTX (enhanced CLIA)
BAP	Access Ostase BAP (CLIA)

Note: Several of these assays are for research only and not for use in diagnostic procedures. Consult manufacturer for appropriate application.

blood-handling protocol. Plasma samples are acceptable for certain assays, and the only variable that needs to be verified is the type of anticoagulant that can be used.

There is a variety of measurement techniques used to measure bone biomarkers, for example:

- Spectrophotometry
- RIA, immunoradiometric assay (IRMA)
- EIA
- Enzyme-linked immunosorbent assay (ELISA)
- CLIA
- HPLC

Spectrophotometric assays are popular because they can usually be adapted to automated chemistry analyzers, and use serum or plasma as the specimen of choice. Immunoassays, especially EIA or CLIA, provide another useful means to measure bone biomarkers in clinical laboratories. Most chemistry laboratories have automated immunoassay analyzers that have the potential for incorporating a bone biomarkers assay. Of course, there are questions of assay specificity that need to be investigated before a laboratory brings the assay in-house.

High-performance liquid chromatography is used to measure bone biomarkers, especially in urine samples. A significant disadvantage to chromatographic techniques on urine samples is that the sample usually requires pretreatment/extraction, which increases the assay time and may decrease analytical precision. HPLC shares a similar disadvantage to RIA in that most hospital-based clinical laboratories do not have an HPLC system.

▶ METABOLIC BONE DISORDERS

OSTEOPOROSIS

Osteoporosis is defined as "a progressive systemic skeletal disease characterized by low bone mass and micro-architectural deterioration of bone tissue, with a consequent increase in bone fragility and susceptible to fracture."[30] Osteoporosis is a disease associated with aging and, because of the number of individuals with the disease, it represents a significant expense in health care. Patients with osteoporosis experience fractures throughout the skeletal system, with the most common fractures occurring in the vertebral column, distal radius (forearm and hand), and proximal femur (hip fracture). Other fractures sites include pelvis, proximal humorous, distal femur, tibia, and ribs.

Osteoporosis is divided etiologically into primary and secondary types. In primary osteoporosis, there are typical associations with patient's age, but the exact etiology of bone loss is not known. The most common type of primary osteoporosis is menopausal osteoporosis, characterized by decreased amounts of hormones with maximal loss of bone mass is the first menopausal decades. It manifests itself primarily as a loss of trabecular (cancellous) bone.

The prognostic assessment of fracture risk is primarily based on the quantification of bone mineral density (BMD) and with certain cases on the rate of bone loss that can be assessed indirectly by measurement of biochemical markers of bone turnover. Measurement of PTH and 25(OH) vitamin D may be useful in the differential diagnosis of osteoporosis. Other biomarkers, including hormones, cytokines, and growth factor production, may provide important information regarding the pathogenesis of the disease. The assessment of BMD and measurement of biomarkers of bone turnover appear to be significant methods to diagnose osteoporosis, predict future fractures, and monitor therapeutic regimens.[31]

⊘ TABLE 14-8

Biochemical Markers of Bone Turnover Available to Provide the Clinician with the Most Reliable Information to Assess Patients with Osteoporosis

Formation	Resorption
Total bone alkaline phosphatase	Serum CTX
BAP	PYD
OC	DPD
PINP	Urinary: NTX
PICP	Urinary: CTX

Several biomarkers of bone turnover are currently available to assess patients for osteoporosis and are listed in Table 14-8 ⊘. The clinician will select the best combinations of biomarkers along with BMD to evaluate his or her patient's condition, treatment modality, and susceptibility to future fractures.

SECONDARY CAUSES OF OSTEOPOROSIS

The secondary causes of osteoporosis are categorized and listed in Box 14-3 ■. A discussion of their etiology, treatment modality, and diagnostic tests, including clinical laboratory blood and urine analytes, is beyond the scope of this textbook and can be found in the reference cited.[32]

TRANSPLANTATION OSTEOPOROSIS

Transplantation osteoporosis is shown to be a viable side effect following transplant of kidney, heart, liver, lung, and bone marrow. The pathogenesis of transplantation osteoporosis is complex and not completely understood. The many risk factors for transplantation osteoporosis are common among potential transplant organ recipients. It has been shown that posttransplantation immunosuppressant drugs significantly enhance the loss of bone and increase the incidence of bone fractures. An excellent discussion of transplantation osteoporosis is presented in the reference cited.[33] As with other bone-related diseases, the role of biomarkers of bone turnover will depend upon specifics of the diseases, treatment modalities, and prognostic criteria.

RICKETS AND OSTEOMALACIA

Rickets and osteomalacia are diseases associated with defective bone and cartilage mineralization in children and bone mineralization in adults. The calcification of cartilage does not manifest itself properly, especially at epiphyseal growth plates, and results in a delayed maturation of the cartilage cellular sequence and disorganization of cell arrangement.[33] This process leads to widening of the epiphyseal plate, skeletal abnormalities, and possible growth retardation. The abnormal

BOX 14-3   Causes of Secondary Osteoporosis

Hypogonadal condition

► Turner's syndrome
► Athletic amenorrhea
► Anorexia nervosa
► Hypeprolactinemia

Thyroid disorders
Additional endocrine conditions

► Hyperparathyroidism
► Diabetes mellitus
► Hemochromatosis
► Acromegaly

Medications

► Glucocorticoids
► Anticonvulsant drugs
► Cyclosporine
► Anticoagulants

Renal disorders
Gastrointestinal conditions

► Primary biliary cirrhosis
► Inflammatory bowel disease
► Hepatic diseases

Hematological disorders

► Multiple myeloma

Rheumatologic conditions
Immobilization
Genetic conditions

► Osteogenesis imperfecta
► Vitamin D–resistant osteomalacia
► Hypohatsia

calcification of bone is limited to the newly formed organic matrix deposited at the bone-osteoid surface of remodeling tissue. This abnormal process leads to an increase in the bone-forming surface covered by incomplete mineralized osteoid, an enlarged osteoid volume and thickness, a reduction of mineralized surface, and an increased susceptibility to fractures and/or bone deformation.

There are numerous causes of rickets and osteomalacia syndromes. A detailed list can be found in the reference cited.[35] The roles of biomarkers of bone turnover are varied depending upon the specific disorders. Urine hydroxyproline has been used for quite some time. More recently, Pyr and DPD have been used frequently for calcium- and phosphorous-related disorders.

PAGET'S DISEASE

Paget's disease of bone is a chronic, localized disease characterized by increased bone remodeling, bone hypertrophy, and abnormal bone structure. Patients often suffer bone deformity. Paget's disease often progresses to osteoarthritis, increased risk for fractures, nerve compression syndromes, and formation of neoplasm. This disease most frequently presents itself before the age of 40. It is believed that both genetic and environmental factors contribute to the pathogenesis of **Paget's disease**.[36]

Biochemical markers of bone turnover have been used with varying degrees of success for both diagnosis and monitoring of diseases activity. Biochemical markers of choice in assessing this disease include serum BAP, PINP, serum alkaline phosphatase, and NTX.

SUMMARY

Mineral and bone metabolism are an integral part of human development. Each mineral and its associated hormone work in concert to directly and indirectly affect hundreds of biochemical and physiological reactions and processes. These same minerals and hormones work along with several biomarkers associated with bone formation and resorption. The processes involved in bone formation and resorption are complex and represent a sophisticated orchestration of proteins, enzymes, cells, and minerals that last the lifetime of a human being.

Unfortunately there are numerous disorders that can develop during an individual's lifetime involving these biochemicals and their processes. The laboratory in concert with the clinician can aid those stricken with one of these disorders by measuring specific biomarkers and/or minerals associated with the disorder.

In many instances, the markers measured in the clinical laboratory are used in conjunction with other diagnostic tests performed in other areas of health care, for example, imaging techniques, thereby creating a partnership between allied health disciplines that further helps the patient during his or her time of ill health.

REVIEW QUESTIONS

LEVEL I

1. Ionized calcium is characterized as which of the following? (Objective 1)
   a. bound to protein
   b. forming complexes to phosphates
   c. free or unbound
   d. bound to a specific antibody

2. Which of the following is the physiologically active form of calcium? (Objective 1)
   a. ionized
   b. protein bound
   c. antibody bound
   d. calcium phosphate

3. Which of the following methods is not used to measure total calcium concentration in blood? (Objective 2)
   a. colorimetric
   b. atomic absorption
   c. immunoassays
   d. indirect potentiometry

4. Phosphomolybdate is used to measure which of the following? (Objective 3)
   a. inorganic phosphorus
   b. calcium
   c. magnesium
   d. telopeptides

5. The principle physiological function of PTH is to regulate which of the following? (Objective 6)
   a. the concentration of ionized calcium in cerebrospinal fluid
   b. the concentration of ionized calcium in extracellular fluid
   c. the inhibition of the renal synthesis of 1,25(OH)2D3
   d. the concentration of magnesium during bone formation

LEVEL II

1. Which of the following can affect the concentration of ionized calcium in serum? (Objective 8)
   a. pH and temperature
   b. barometric pressure
   c. partial pressure of carbon dioxide
   d. magnesium concentration

2. If the pH of a sample decreases, what is the predicable response to the serum magnesium concentration? (Objective 8)
   a. The concentration will increase 1 mmol/L for each whole unit of pH change.
   b. The concentration will decrease.
   c. The concentration will increase.
   d. The concentration will not be altered.

3. Which of the following statements best reflects the diagnostic criteria of primary hyperparathyroidism? (Objective 8)
   a. no serum PTH detected and no serum calcium detected
   b. detectable amounts of serum PTH concomitant with elevated total serum calcium
   c. elevated total serum calcium only
   d. a normal level of serum PTH and serum calcium

4. Which of the following represents an important clinical utility or usefulness of intraoperative PTH? (Objective 8)
   a. surgical management of primary hyperparathyroidism
   b. surgical management of prostate cancer
   c. surgical management of coronary heart disease
   d. surgical management of thyroid disease

REVIEW QUESTIONS *(continued)*

LEVEL I

6. Bone remodeling is a process that is best described as which of the following? (Objective 10)
   a. Involving the synthesis of collagen followed by secretion as a precursor procollagen molecule
   b. Involving the ordered deposition of apatite on a type I collagen matrix
   c. Involving formation of bone on surfaces without any bone material
   d. Involving formation of bone on surfaces previously containing bone

7. Osteoblast promotes which of the following? (Objective 10)
   a. fibrillogenesis
   b. bone resorption
   c. bone formation
   d. embryogenesis

8. Which of the following is a biochemical marker of bone formation? (Objective 10)
   a. telopeptides
   b. bone alkaline phosphatase (BAP)
   c. tartrate-resistant acid phosphatase (TRAP)
   d. bone sialoprotein

9. The primary cross-links in skeletal tissue include which of the following biochemical marker of bone metabolism? (Objective 10)
   a. pyridinoline and deoxypyridinoline
   b. osteocalcin
   c. tartrate-resistant acid phosphatase
   d. hydroxyproline

LEVEL II

5. The production of 1,25(OH)2D3 is regulated by which of the following? (Objective 10)
   a. positive feedback mechanism associated with circulating levels of calcium and inorganic phosphorus
   b. negative feedback mechanism associated with circulating levels of magnesium
   c. negative feedback mechanism associated circulating levels of calcitonin
   d. negative feedback mechanism associated circulating levels of calcium and inorganic phosphorus

6. A blood specimen is received in the laboratory from a pregnant patient in the obstetric unit with blood magnesium ordered by the attending physician. The serum magnesium levels is 6.0 mmol/L (RI = 3.0–5.0 mmol/L).). Which of the following rational is most probable? (Objective 10)
   a. The patient has Paget's disease.
   b. The patient has primary hyperparathyroidism.
   c. The patient is on magnesium sulphate intravenous infusion because she is experiencing preeclampsia.
   d. The patient has had a massive heart attack.

7. Which of the following is considered a major problem in using biomarkers of bone turnover in the management of patients with osteoporosis in a clinical setting? (Objective 5)
   a. They are noninvasive.
   b. The measurements are affected by several preanalytical variables, both controllable and uncontrollable.
   c. They require no special patient preparation.
   d. They can be repeated over short time intervals, thus providing quick results.

8. Which laboratory test listed would be useful to the clinician to assess osteoporosis? (Objective 6)
   a. serum NTX-I
   b. total serum calcium
   c. urinary magnesium
   d. serum PTH

9. A patient diagnosed with rickets due to a dietary deficiency of vitamin D would present with which of the following laboratory results? (Objective 6)
   a. increased serum PINP, increased serum TRAP, increased serum DPD
   b. decreased serum PINP, decreased serum TRAP, decreased serum DPD
   c. normal serum PINP, normal serum TRAP, normal serum DPD
   d. increased serum total calcium, decreased serum total alkaline phosphatase, increased serum inorganic phosphorus

REVIEW QUESTIONS *(continued)*

LEVEL I

LEVEL II

10. Which of the following laboratory results best represents a patient with Paget's disease? (Objective 9)
    a. decreased urinary hydroxyproline, decreased serum alkaline phosphatase
    b. decreased urinary hydroxyproline, increased serum alkaline phosphatase
    c. normal urinary hydroxyproline, normal serum alkaline phosphatase
    d. increased urinary hydroxyproline, decreased serum alkaline phosphatase

11. "A progressive systemic skeletal disease characterized by low bone mass and micro-architectural deterioration of bone tissue, with a consequent increase in bone fragility and susceptible to fracture" describes which of the following diseases? (Objective 6)
    a. rickets
    b. Paget's disease
    c. renal osteodystrophy
    d. osteoporosis

PEARSON myhealthprofessionskit™

Use this address to access the interactive Companion Website created for this textbook. Simply select "Clinical Laboratory Science" from the choice of disciplines. Find this book and log in using your user name and password.

REFERENCES

1. Klemm KM, Klein MJ. Biochemical markers of bone metabolism. In McPherson RA, Pincus MR, *Henry's clinical diagnosis and management by laboratory methods*, 21st ed. (Philadelphia: Saunders Elsevier, 2007): 171–172.

2. McLean FC, Hastings AB. The state of calcium in the fluids of the body. I. The conditions affecting the ionization of calcium. *J Biol Chem* (1935) 108: 285–300.

3. Endres DB, Rude RK. Mineral and bone metabolism. In Burtis CA, Ashwood ER, Bruns DE (Eds.), *Tietz textbook of clinical chemistry and molecular diagnostics*, 4th ed. (Philadelphia: WB Saunders, 2006): 1903.

4. Ibid., 1899.

5. Ibid.,1897–1899.

6. Op cit., Klemm KM, Klein MJ., 173.

7. Elin RJ. Magnesium: The fifth but forgotten electrolyte. *Am J Clin Path* (1994) 102: 616–622.

8. Op cit., Endres DB, Rude RK., 1911–1912.

9. Mandell DL, Genden EM., Mechanick JI, et al. The influence of intraoperative parathyroid hormone monitoring on the surgical management of hyperparathyroidism. *Arch Otolaryngol Head Neck Surg* (2001) 127: 821–827.

10. Nelson CM, Victor NS. Rapid intraoperative parathyroid hormone assay in the surgical management of hyperparathyroidism. *The Permanente Journal* (Winter, 2007) 11, 1: 3–5.

11. Westerdahl J, Bergenfelz A. Parathyroid surgical failures with sufficient decline of intraoperative parathyroid hormone levels. unobserved multiple endocrine neoplasia as an explanation. *Arch Surg* (2006) 141: 589–594.

12. Haustein SV, Mack E, Starling JR, et al. The role of intraoperative parathyroid hormone testing in patients with tertiary hyperparathyroidism after renal transplantation. Am Assoc of Endo Surgeons—Annual meeting abstract (www.ednocrinesurgery.org/mtgs/205/abstracts/paper13.html)

13. Kao PC, vanHeerden JA, Farley DR, et al. Intraoperative monitoring of parathyroid hormone with a rapid automated assay that is commercially available. *Annals of Clin & Lab Science* (2002) 32(3): 244–251.

14. DeLuca HF. Overview of general physiologic features and functions of vitamin D. *Am J Clin Nutr* (2004) 80(suppl): 1689S–1696S.

15. Pilz S, Dobnig H, Winklhofer-Roob B, et al. Low serum levels of 25-hydroxyvitamin D predict fatal cancer in patients referred to coronary angiography. *Cancer Epidemiology, Biomarkers & Prevention* (2008) 17: 1228–1233.

16. Dobnig H, Pilz S, Scharnagl H, et al. Independent association of low serum 25-hydroxyvitamin D and 1,25-dihydroxyvitamin D levels with all-cause and cardiovascular mortality. *Archives of Internal Medicine* (2008) 168: 1340–1349.

17. Holick MF. Sunlight and vitamin D for bone health and prevention of autoimmune diseases, cancers and cardiovascular diseases. *Am J Clin Nutr* (2004) 80: 1678S–1688S.

18. Op cit., Endres DB, Rude RK., 1921–1925.

19. McLeod MK, Monchik JM, Martin HF. The role of ionized calcium in the diagnosis of subtle hypercalcemia in symptomatic primary hyperparathyroidism. *Surgery* (1984) 95(6): 667–673.

20. Op cit., Endres DB, Rude RK., 1909.

21. Mark K. Structure, biosynthesis and gene regulation of collagens in cartilage and bone. In Seibel, MJ, Robins SP, Bilezikian JP (Eds.) *Dynamics of bone and cartilage metabolism—principles and clinical applications*, 2nd ed. (San Diego: Academic Press, 2006): 5–7.

22. Robins SP. Fibrillogenesis and maturation of collagens. In Seibel, MJ, Robins SP, Bilezikian JP (Eds.), *Dynamics of bone and cartilage metabolism—principles and clinical applications*, 2nd ed. (San Diego: Academic Press, 2006): 41–44.

23. Kraenzlin ME, Seibel MJ. Measurement of biochemical markers of bone resorption. In Seibel, MJ, Robins SP, Bilezikian JP (Eds.), *Dynamics of bone and cartilage metabolism—principles and clinical applications*, 2nd ed. (San Diego: Academic Press, 2006): 545.

24. Boskey AL. Mineralization, structure and function of bone. In: Seibel, MJ, Robins SP, Bilezikian JP Editors. Dynamics of Bone and Cartilage Metabolism—principles and clinical applications, 2nd ed. San Diego: Academic Press, 2006: 201–209.

25. Barrett EJ, Barrett P: The parathyroid glands and vitamin D. In Boron WF, Boulpaep EL (Eds.), *Medical physiology*, undated edition. (Philadelphia, Elsevier Saunders, 2005): 1089–1091.

26. Lian JB, Stein GS. The cells of bone. In Seibel MJ, Robins SP, Bilezikian JP (Eds.), *Dynamics of bone and cartilage metabolism—principles and clinical applications*, 2nd ed. (San Diego: Academic Press, 2006): 221–240.

27. Martin TJ, Sims NA. Signaling in bone. In Seibel, MJ, Robins SP, Bilezikian JP (Eds.), *Dynamics of bone and cartilage metabolism—principles and clinical applications*, 2nd ed. (San Diego: Academic Press, 2006): 262–264.

28. Endres DB, Rude RK. Mineral and bone metabolism. In Burtis CA, Ashwood ER, Bruns DE (Eds.), *Tietz textbook of clinical chemistry and molecular diagnostics*, 4th ed. (Philadelphia: WB Saunders, 2006): 1940–1941.

29. Naylor KE, Eastell R. Measurement of biochemical markers of bone formation. In Seibel, MJ, Robins SP, Bilezikian JP (Eds.), *Dynamics of bone and cartilage metabolism—principles and clinical applications*, 2nd ed. (San Diego: Academic Press, 2006): 535–536.

30. World Health Organization (WHO). Assessment of fracture risk and its application to screening for postmenopausal osteoporosis. Technical report series 843, (Geneva: WHO, 1994).

31. Garnero P, Delmas PD. Laboratory assessment of postmenopausal osteoporosis. In Seibel, MJ, Robins SP, Bilezikian JP (Eds.) *Dynamics of bone and cartilage metabolism—principles and clinical applications*, 2nd ed. (San Diego: Academic Press, 2006): 611–613.

32. Mulder JE, Kulak CM, Shane E. Secondary Osteoporosis. In Seibel, MJ, Robins SP, Bilezikian JP (Eds.) *Dynamics of bone and cartilage metabolism—principles and clinical applications*, 2nd ed. (San Diego: Academic Press, 2006): 717–718, 734.

33. Kulak CM, Shane E. Transplantation osteoporosis: biochemical correlation of pathogenesis and treatment. In Seibel, MJ, Robins SP, Bilezikian JP (Eds.), *Dynamics of bone and cartilage metabolism—principles and clinical applications*, 2nd ed. (San Diego: Academic Press, 2006): 701–711.

34. Drezner MK. Osteomalacia and rickets. In Seibel, MJ, Robins SP, Bilezikian JP (Eds.) *Dynamics of bone and cartilage metabolism—principles and clinical applications*, 2nd ed. (San Diego: Academic Press, 2006): 739.

35. Ibid., 740.

36. Grauer A, Siris E, Ralston S. Paget's disease of bone. In Seibel, MJ, Robins SP, Bilezikian JP (Eds.) *Dynamics of bone and cartilage metabolism—principles and clinical applications*, 2nd ed. (San Diego: Academic Press, 2006): 779–781.

15

The Endocrine System

■ OBJECTIVES—LEVEL I

Following successful completion of this chapter, the learner will be able to:

1. Identify three major types of hormones.
2. State which of the three classes of hormones characterizes the following compounds:
   a. thyroxine
   b. cortisol
   c. parathyroid hormone
   d. epinephrine
   e. estrogen
3. Define negative feedback.
4. List five examples of hormones found in the anterior pituitary gland.
5. List two examples of hormones found in the posterior pituitary gland.
6. Know the location of the thyroid gland, adrenal glands, pituitary, and hypothalamus.
7. Identify the hormones released by the thyroid gland.
8. Identify the mineralocorticoids and the glucocorticoids.
9. Identify the hormones produced by the adrenal medulla.
10. List example(s) of target tissues for each of the following hormones:
    a. thyroxine
    b. prolactin
    c. testosterone
    d. antidiuretic hormone
    e. oxytocin
    f. growth hormone
    g. aldosterone
    h. cortisol
    i. epinephrine
    j. luteinizing hormone
11. Associate abnormal laboratory results with a disease or syndrome.
12. Know the functions of the hormones presented.
13. State the methods used to quantitate the amount of hormones in blood.

■ OBJECTIVES—LEVEL II

Following successful completion of this chapter, the learner will be able to:

1. Explain the role of the hypothalamus in regulating pituitary control of endocrine function.
2. Compare positive and negative feedback-control mechanisms.

■ OBJECTIVES—LEVEL II *(continued)*

3. Correlate laboratory data on patients with Addison's disease or Cushing's syndrome.

4. Explain the purpose of stimulation and suppression tests.

5. Outline the biosynthesis, secretion, transport, and action of thyroxine, triiodothyronine, and thyroid-stimulating hormone.

6. Explain how the adrenal gland functions to maintain blood pressure, potassium and glucose homeostasis.

7. Recognize clinical features of pheochromocytoma and neuroblastoma.

8. Correlate laboratory data on patients with hypothyroidism or hyperthyroidism.

9. Define the role of the laboratory in the diagnosis and management of some common endocrine disorders.

10. Identify four factors that may affect the interpretation of laboratory results for hormone measurements.

11. Review the appropriate laboratory thyroid-function tests used to properly evaluate or monitor patients with suspected thyroid disease.

12. Identify laboratory tests that will provide useful information for assessing selected endocrine diseases or syndromes.

KEY TERMS

Adenohypophysis
Addison's disease
Adrenal cortex
Adrenal medulla
Adrenocorticosteroid
Androgen
Anterior pituitary
Basal level
Catecholamines
Circadian rhythm
Conn's syndrome

Cushing's syndrome
Diurnal rhythm
Euthyroid
Glucocorticoids
Goiter
Gonadotropins
Hormone
Hypothalamic–pituitary–thyroid axis
Hypothalamus
Mineralocorticoids
Negative feedback

Neurohypophysis
Pituitary gland
Posterior pituitary
Releasing factor
Sella turcica
Sex-hormone binding globulin (SHBG)
Steroids
Stimulation test
Suppression test
Syndrome

A CASE IN POINT

A 21-year-old white female involved in a motor-vehicle accident is admitted to the emergency department. The patient has multiple injuries to soft tissue, a fractured ankle, pneumothorax, subarachnoid hemorrhage, and brain contusion. A computerized tomography (CT) scan of the abdomen revealed a mass near the kidney. Fourteen days after the accident, the patient developed severe abdominal pain during a physical therapy session. Her family revealed that the patient had a previous history of hypertension. When questioned, the patient admitted to having attacks of intermittent headache, palpitation, and anxiety some months earlier, each lasting for a few minutes. Another CT scan showed the same mass in the same area. Laboratory results are shown in the following table.

Tests	Laboratory Results	Reference Interval
Plasma epinephrine	300 pg/mL	20–97 pg/mL
Plasma norepinephrine	10,285 pg/mL	125–310 pg/mL
Metanephrine	3850 μg/day	0.3–0.9 μg/day
VMA	15.5 mg/day	1.5–7.5 mg/day
Sodium	141 mEq/L	136–142 mEq/L
Potassium	4.3 mEq/L	3.5–5.0 mEq/L
Cortisol (8 a.m.)	16 μg/dL	6–21 μg/dL

Issues and Questions to Consider

1. Based on the patient's symptoms and available history, what is your presumptive diagnosis? What tests would you order to confirm your diagnosis?

2. Comment on the relative usefulness of catecholamines and metabolites in the diagnosis of this patient's condition.

3. What would you recommend as the best management for this entity?

4. What is the prevalence of this condition as a cause of hypertension? How often is the tumor malignant? What are the associated familiar disorders?

▶ INTRODUCTION

BASIC CONCEPTS

An important objective of this chapter is to delineate the role of the clinical laboratory in the diagnosis of endocrine diseases. Endocrine diseases normally result from either an excess or a deficient production of one or more hormones. In severe cases, the clinical presentations of these endocrine diseases and syndromes are usually quite characteristic and obvious. The endocrinologist who assesses patients with such presentations forms a preliminary or presumptive diagnosis and then requests appropriate endocrine tests from the laboratory to confirm or rule out disease. These laboratory test results would be most useful in situations when the disease is mild and the presentation is not obvious. Examples of how test selections are made and how the results are interpreted to confirm or to rule out the differential diagnoses are found throughout this chapter.

The endocrine system comprises a group of ductless glands that secrete hormones directly into the bloodstream. By definition, a hormone is a chemical substance normally produced by specialized cells in some part of the body that is carried by the bloodstream to another part from which it exerts its regulatory effect. For example, adrenocorticotropic hormone (ACTH) is secreted from the pituitary, but it affects the functional activities of the adrenal cortex, which is the inner portion of the adrenal glands located atop the kidneys. Similarly, hormones from the adrenal cortex regulate the carbohydrate, fat, protein, and mineral metabolism of the body. The endocrine glands serve to maintain the circulatory status and homeostasis of the organism's proper functioning such as growth and development.

HORMONE CLASSIFICATION

There are three major classifications of hormones based on structure in humans; examples are shown in Figure 15-1 ■. They are steroids, polypeptides, and the amines. Several examples of specific hormones within each classification are shown in Box 15-1 ■. As discussed in the latter part of this chapter, hormone structures are important in deciding on what methodology to use for their measurement.

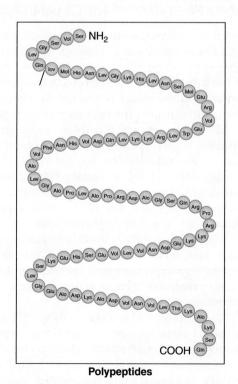

Polypeptides

Steroids

Amines

■ FIGURE 15-1   Structural representation of the three types of hormones: polypeptides (parathyroid hormone), steroids (cortisol), and amines (3,5,3′,5′-tetraiodothyronine).

BOX 15-1   Chemical Classification of Selected Hormones

Peptides

▶ Adrenocorticotropin hormone (ACTH)
▶ Antidiuretic hormone (ADH) (also known as arginine vasopressin, or AVP)
▶ Calcitonin
▶ Cholecystokinin
▶ Corticotropin-releasing hormone (CRH)
▶ Follicle-simulating hormone (FSH)
▶ Glucagon
▶ Gonadotropin-releasing hormone (GnRH)
▶ Growth hormone (GH)
▶ Insulin
▶ Luteinizing hormone (LH)
▶ Oxytocin
▶ Parathyroid hormone (PTH)
▶ Prolactin (PRL)
▶ Secretin
▶ Somatostatin
▶ Thyroid-stimulating hormone (TSH)
▶ Thyrotropin-releasing hormone (TRH)
▶ Vasoactive intestinal peptide

Amino Acids

▶ Dopamine
▶ Epinephrine (also adrenaline)
▶ Norepinephrine (noradrenalin)
▶ Serotonin (5-hydroxytryptamine, 5-HT)
▶ Thyroxine (T_4)
▶ Triiodothyronine (T_3)

Steroids

▶ Aldosterone
▶ Cortisol
▶ Estradiol (E_2)
▶ Progesterone
▶ Testosterone

Polypeptide Hormones

Polypeptides are composed of amino acid residues of varying lengths. Their biological activities usually reside in a small number of amino acids. Polypeptides or protein hormones are water soluble and circulate freely in plasma as a whole molecule or as fragments that may be either active or inactive. For example, parathyroid hormone (PTH) circulates as a whole-polypeptide chain (i.e., intact PTH) and as fragments that are designated as N-terminal (NH_2) and C-terminal (COOH).

Usually, the half-life of these hormones is short, often less than 30 minutes, and plasma concentrations fluctuate significantly in many physiological and pathological conditions. Polypeptide hormones initiate their response by binding to cell-membrane receptors and create a "second messenger" system that completes the specific action(s) of the hormone (see the following).

Steroid Hormones

The steroids are all derived from cholesterol and are hydrophobic and insoluble in water. The four-ring structure, a cyclopentanoperhydrophenanthrene ring, is the basic framework for this group of hormones, whose characteristics are usually conferred by the side chains. These hormones are usually bound reversibly in blood to specific carrier proteins. A small fraction of the total circulating steroid hormone is free or unbound; this is the fraction that is available to promote a physiological response. The half-life of steroid hormones is 30–90 minutes.

Amines

Hormones derived from amino acids—for example, tyrosine—include thyroxine (T_4) and catecholamines (epinephrine). They are water soluble and circulate in plasma either bound to proteins (T_4) or free (catecholamines). The half-life for protein-bound thyroxine is approximately 7–10 days, whereas free catecholamine half-life is less than a minute. These hormones interact in a similar fashion as polypeptides by binding to membrane receptors of target cells.

 Checkpoint! 15–1

Identify in which class each of the following hormones belongs: epinephrine, insulin, and testosterone.

HORMONE RECEPTORS AND FEEDBACK MECHANISMS

The mechanisms used to elicit hormone action on target tissues are complex, and a thorough discussion of this topic is beyond the scope of this book. Therefore, only an overview of hormone receptors and feedback mechanisms will be presented, and the reader is directed to references cited in the chapter for further explanations of these topics.

All hormones act on their respective target glands through highly specific binding proteins called *receptors* that are located either on the surface of the cell membrane or within the cytosol of the target cell. There are two major classes of hormone receptors: *membrane* and *nuclear*. Membrane receptors primarily bind protein hormones and catecholamines (e.g., epinephrine). Nuclear receptors bind smaller molecules that can diffuse across membranes such as steroids, vitamin D, and thyroid hormones. Examples of membrane receptors include cytokines and kinases, whereas nuclear receptors include estrogen receptors and thyroid-hormone receptors.

The binding of a hormone to its specific receptor serves as an initial *signal* to a cell. A signal may represent an entity created by a biochemical reaction involving a second messenger—for example, cyclic adenosine monophosphate (cAMP). The initial signal is amplified, usually several times, and ultimately affects the nucleus of the target cell so that it elicits an alternate gene expression. This process leads to the synthesis of a specific messenger ribonucleic acid message and new protein synthesis. The *new protein* leaves the cell to elicit its respective biological response. Each class of receptor responses are unique and may function quite differently than the examples here.

A fundamental feature of the endocrine system is *feedback control* and includes both **negative feedback** and positive feedback. Each major hypothalamic–pituitary–hormone axis is regulated by negative and positive feedback (loops), a process that serves to maintain a level of hormone concentration within a narrow range. Examples of these feedback loops include (1) cortisol on the axis of corticotrophin-releasing hormone (CRH) and ACTH (CRH–ACTH), (2) thyroid hormones on the axis of thyrotropin-releasing hormone (TRH) and thyroid-stimulating hormone (TSH) (TRH–TSH), and (3) gonadal steroids on the axis of gonadotropin-releasing hormone (GnRH), luteinizing hormone (LH), and follicle-stimulating hormone (FSH) (GnRH–LH–FSH) axis. These feedback loops contain both negative (e.g., T_4) and positive (e.g., TRH, TSH) components that allow for precise control of hormone levels. The following are two specific examples of feedback loops of the hypothalamic–pituitary–target organ axis using thyroid hormones and LH–FSH.

When circulating levels of thyroid hormones—T_4, for example—are low, the hypothalamus rapidly senses the decline in hormone output and increases its production of a specific hypothalamus-base **releasing factor** (hormone); in this example, it is TRH, which enters the portal circulation in the brain to stimulate pituitary hormone synthesis and secretion of TSH. TSH is transported to the thyroid gland to produce and release T_4. This is a *positive-feedback loop*. Conversely, when the circulating levels of T_4 are excessive, the high levels of T_4 feedback negatively to the hypothalamic–pituitary axis, resulting in reduced synthesis of TRH and subsequently pituitary TSH secretion. The reduced pituitary secretion of TSH in turn reduces the normal stimulation of the thyroid gland to maintain hormone levels.

Positive feedback is not as common, but it does occur. Biogenic amines and peptidergic neurons (neurons that secrete peptide hormones) in the hypothalamus respond to neural and sensory input from the brain to elicit the release of gonadotropin hormones (LH–RH). This input is *pulsatile*, which means that it occurs rhythmically in a series of high and low concentrations. LH–RH flows to the pituitary to synthesize and release FSH and LH. In the female, FSH causes ovarian follicular development and the production of estradiol, whereas LH causes corpus luteum development and increase secretion of progesterone. Estradiols released into circulation feed back both negatively and positively to the hypothalamic–pituitary axis to control the menstrual cycle and LH secretion.

There are feedback loops that do not involve the hypothalamic–pituitary axis such as calcium feedback to the parathyroid glands to reduce PTH secretion, glucose feedback control of insulin secretion from the pancreas, and leptin (a lipid) feedback on the hypothalamus. These examples are described in other chapters of this book. Examples of feedback mechanisms are shown in Figure 15-2 ■.

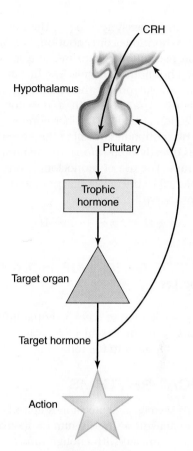

■ FIGURE 15-2   Feedback mechanisms.

✓ Checkpoint! 15–2

The circulating levels of thyroxin in a 42-year-old female patient are below the normal range. Explain how the body attempts to increase the blood levels of T_4 in this patient using negative feedback.

▶ TECHNIQUES OF HORMONE MEASUREMENT

Bioassays, chemical methods, and competitive-protein-binding assays have been used in hormone measurements in the past. These assay techniques have been replaced with immunoassays that are both sensitive and specific (see Chapter 5, Immunoassays).

Two types of immunoassay techniques are used. In one immunoassay technique, the antigen is labeled with a tracer or detector molecule (e.g., enzyme, fluorophore, or a chemiluminescent compound). The reaction is termed *competitive* when a labeled antigen competes with patient antigen for antibody-binding sites. A hormone (H) in serum competes for antibody sites (Ab) with a known quantity of labeled hormone (H*) to form hormone–antibody complex (H*—Ab and H—AAb). Hormone concentration is inversely proportional to the bound complex (H*—Ab).

$$H^* + H + Ab \longleftrightarrow H^*—Ab + H—Ab$$

A second immunoassay technique, the most widely used measurement of hormone concentration, is *immunometric*. In this technique, two antibodies are used to react with two antigenic sites on a hormone molecule. The first antibody is usually bound to a solid phase, the anchoring antibody. The second antibody is labeled with a detector molecule to form the detecting antibody. The hormone is "sandwiched" between the two antibodies, forming the Ab—H—Ab* complex, which is directly proportional to the hormone concentration in serum. The use of monoclonal antibodies labeled with tracer has led to the development of assays with greatly improved sensitivity.

$$Ab + H + Ab \longleftrightarrow Ab—H—Ab*$$

▶ INTERPRETATION OF HORMONE RESULTS

Several factors are known to have potential influence on the concentrations of hormones in biological fluids. These include but are not limited to the following.

BIOLOGICAL RHYTHMS

Certain glands secrete their hormones episodically in peaks and valleys, so drawing a single serum or plasma sample may not provide the clinician with enough information to arrive at a diagnosis. Several hormones may exhibit a **diurnal rhythm** or a **circadian rhythm**, which means that their concentrations change during a 24h interval. For example, plasma concentration of cortisol at 8 p.m. is approximately 50% of the plasma concentration at 8 a.m. Other hormones that show diurnal rhythm are growth hormone (GH), ACTH, TSH, PTH and catecholamines.

Autonomy
Autonomy applies to syndromes of hormone excess, which are usually caused by tumors of the endocrine gland and when hormone overproduction is not suppressed by feedback inhibition.

Protein Binding
Certain hormones are bound to carrier proteins in circulation such as T_4 and cortisol. Variation in binding-protein concentration affects total hormone concentration—for example, the effect of thyroxine-binding globulin (TBG) concentrations on total thyroxine discussed later in this chapter. Some hormones are protein bound once they are secreted from the gland, and they circulate in bound form. The bound hormone may not be biologically active; the free or unbound hormone is.

Heterogeneity
Some hormones—for example, PTH—circulate in blood as multiple distinct immunologic forms referred to as *heterogeneity*. The parathyroid gland secretes PTH as whole or intact hormone. Some of the intact hormone is then cleaved between specific amino acids and circulates in blood as hormone fragments. Heterogeneity presents difficulties in development of immunoassays because the antiserum recognizes only specific forms—for example, intact hormones or fragments in circulation.

Assay Limitations
Some assays do not have sufficient sensitivity to differentiate lower range of normal from the pathologically low or hormone-deficient state. The first-generation TSH assays developed in the 1970s are a good example of this. Since that time, newer TSH assays (fourth generation) have markedly improved analytical sensitivities that provide clinicians with reliable data to assess patients with unique thyroid conditions. In some assays, the antiserum for immunoassays may cross-react with other hormones in the specimen, causing overestimation of results.

▶ STRATEGY OF ENDOCRINE FUNCTION ASSESSMENT

Endocrine disorders are the consequences of hyperfunctioning or hypofunctioning of the target or the tropic glands (see Figure 15-3 ■). An example using ACTH (a tropic hormone) and cortisol, the target or glandular hormone of the adrenal cortex, illustrates this point.[1] What is the classification of a patient who presents with the following condition to his or her physician? The patient has a tumor in the adrenal cortex that releases a greater than normal amount of cortisol, and the blood-level ACTH is lower than normal. Thus, an increase in blood concentrations and decreased blood levels of ACTH place the patient condition in the box at the upper left side in Figure 15-3, so the interpretation or classification is primary hyperfunction. It follows then that if both the blood levels of

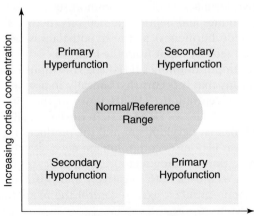

■ FIGURE 15-3 This diagram represents the classification of endocrine diseases as primary or secondary hyper or hypo-functions. The x-axis represents the tropic hormone concentrations for example ACTH and the y-axis is the target/glandular hormone concentrations for example cortisol.

cortisol and ACTH are lower than normal, then the patient is classified as having a secondary hypofunction as a result of a disease of the pituitary secondary to adrenal failure.

Measuring **basal levels** of hormone secretion may be adequate for diagnosis when the disorder is severe. However, when the excess or deficiency of hormone production is not definitive, stimulation or suppression tests are usually performed. **Stimulation tests** evaluate the secretory reserve of the gland when testing for hypofunction. For example, the ACTH stimulation test is used to assess the ability of the adrenal to secrete cortisol when adrenal insufficiency is suspected. A **suppression test** evaluates a hyperfunctioning gland by demonstrating its inability to suppress excessive hormone production. The dexamethasone suppression test (discussed later in the chapter) is used to suppress overproduction of cortisol in the adrenal gland when **Cushing's syndrome** is suspected.

> ### ✓ Checkpoint! 15–3
> *Distinguish between stimulation test and suppression test.*

▶ THE PITUITARY GLAND

The pituitary is a small gland that extends below the **hypothalamus** of the brain and lies within the **sella turcica** at the base of the skull. It is formed by the fusion of the neural tissue of the hypothalamus forming the infundibular stalk and the posterior lobe (**neurohypophysis**) and the glandular tissue of the Rathke's pouch forming the anterior lobe (**adenohypophysis**).

PHYSIOLOGY

The **pituitary gland** was once considered the "master gland." Current evidence indicates that hormone secretion from the anterior lobe of the pituitary is actually controlled by the hypothalamus, which manufactures small peptides called *releasing* and *inhibiting factors*, and physiologic feedbacks from the target organs. Thus, pituitary endocrinopathies may be the result of a disease or condition primary to the pituitary, secondary to the hypothalamus, or end-organ feedback mechanisms. Deficient production of the hypothalamic hormones is common in tumors, granulomatous disorders, and trauma and may lead to a variety of pituitary hormone deficiencies. Overproduction of hypothalamus hormones is rare.

Assessment of **anterior pituitary** lobe function in patients with pituitary tumors is important for the identification of clinically significant hormone deficiency states because of the tumor itself as well as in the reevaluation of patients after pituitary surgery or irradiation to detect hormone deficiencies that occur as a result of treatment. Testing of pituitary function is usually performed under basal conditions, but quite often it is performed under provocative conditions as a means of revealing the subtle or mild deficiencies

that may occur in the target glands. For example, to test the hypothalamic–pituitary–adrenal axis, a normal morning serum cortisol level is usually adequate to suggest that the axis is intact. Frequently, however, an ACTH stimulation test is used when the morning cortisol results are low or equivocal or when there is a strong clinical suspicion of adrenal insufficiency. Alternately, if there is a strong suspicion of Cushing's, a dexamethasone suppression test can be used to watch for pituitary gland suppression.

ANTERIOR PITUITARY HORMONES

The anterior pituitary gland secretes at least six principal hormones as listed in Table 15-1 . GH and prolactin (PRL) act primarily on diffuse target tissues, but the other tropic hormones ACTH, TSH, and the **gonadotropins** LH and FSH act primarily on specific target endocrine glands. How these pituitary hormones act on the respective target glands will be discussed here (GH and PRL) and later with their respective endocrine targets (e.g., ACTH with adrenal glands, TSH with thyroid gland, and LH and FSH with the reproductive system).

GROWTH HORMONE

Also known as somatotropin, GH is the most abundant hormone produced by the adenohypophysis. Structurally similar to PRL, GH is a polypeptide containing 191 amino acids. Its secretion is controlled by the hypothalamus, which secretes growth-hormone releasing hormone, a 44-amino acid hypothalamic peptide that stimulates GH synthesis and release. GH mediates its anabolic and metabolic actions through an intermediary—insulin-like growth factor 1 (IGF-l), which is also known as somatomedin C—thereby causing growth. IGF-l is a polypeptide similar in molecular structure to insulin and is hepatic in origin. GH secretion is pulsatile, with greatest levels at night, generally correlating with the onset of sleep. GH secretory rates decline markedly with age so that hormone production in middle age is about 15% of production during puberty. GH secretion is reduced in obese individuals, though IGF-l levels are usually preserved.

CLINICOPATHOLOGIC CORRELATIONS

Growth Hormone Deficiency

Serum GH is undetectable for most of the day in healthy, nonstressed individuals. This fact and the episodic nature of GH secretion make a single sampling difficult to interpret. In children with GH deficiency, short stature is commonly encountered. The diagnosis of GH deficiency is usually made by provocative stimuli, including exercise, insulin-induced hypoglycemia, and other pharmacologic tests that normally increase GH to >7 μg/L in children. Random GH measurements do not distinguish normal children from those with true GH deficiency. Adult GH deficiency is rare.

⊕ TABLE 15-1

Anterior and Posterior Pituitary Hormones and Their Actions

Hormones	Target Tissues	Action
Anterior Pituitary Hormones: Adenohypophysis		
Adrenocorticotrophic hormone	Adrenal cortex	Glucocorticoid synthesis and release
Follicle-stimulating hormone	Ovary	Estrogen synthesis
	Testis	Spermatogenesis
Growth hormone	Liver	Somatomedin synthesis
	Bone	Enhance growth stimulation
Luteinizing hormone	Ovary	Ovulation, corpus luteum, and progesterone production
	Testis	Testosterone synthesis
Prolactin	Breast	Lactation
Thyroid-stimulating hormone	Thyroid gland	Thyroid-hormone synthesis and release
Posterior Pituitary Hormones: Neurohypophysis		
Hormones	Target tissue	Action
Arginine vasopressin also antidiuretic hormone	Kidney	Maintain osmotic homeostasis
Oxytocin	Uterus	Uterine contraction, hemostatis at the placental site

A significant proportion of truly GH-deficient adults have low-normal IGF-l levels. The most validated test to distinguish pituitary-sufficient patients from those with adult GH deficiency is the *insulin-induced hypoglycemia stimulation test* (insulin tolerance test). In this test, patients are infused with insulin, which results in hypoglycemia. The stress of insulin-induced hypoglycemia triggers the release of GH and ACTH from the pituitary in normal patients. Laboratory measurement of serum GH, cortisol, and glucose are required. In a normal, nondiseased patient, both cortisol and GH should rise. No response or inadequate response may be a result of pituitary hormone deficiency or hypothalamic lesions.

Growth Hormone Excess

Acromegaly is a chronic disease of middle-aged persons marked by elongation and enlargement of bones of the extremities and certain head bones such as the frontal bone and jaws, and with enlargement of the nose and lips and thickening of the soft tissues of the face. GH hypersecretion is usually the result of somatotrope adenomas but is infrequently caused by extrapituitary lesions. Serum IGF levels are elevated in acromegaly, and they are matched to age and gender. Consequently, an IGF-l level provides a useful laboratory screening measure when acromegaly is suspected at clinical presentation. The diagnosis is confirmed by demonstrating the failure of GH suppression to <1 μg/L within 1 to 2 h of an oral glucose load (75 g).

Pituitary gigantism is associated with elevated levels of growth hormone occurring before long-bone growth is complete. Patients with this condition also experience a remarkable acceleration of linear growth. Usually, physical appearances provide the clinician with a clue to the presence of this condition, but if appearances are inconclusive then similar testing modalities that are used for acromegaly may be adopted.

Prolactin

Prolactin is a polypeptide produced by the lactotrophs of the pituitary. The major function of PRL is the initiation and maintenance of milk production. Its normal level is low because of the inhibitory actions of dopamine from the hypothalamus. Because PRL secretion is pulsatile and has a short half-life, more than one sample should be drawn at 30-minute intervals when screening a patient for hyperprolactinemia. The major circulating form of PRL is the nonglycosylated monomer. Falsely lowered results may occur because of assay artifacts; sample dilution is required to measure these high values accurately. Falsely elevated values may be caused by aggregated forms of circulating PRL that are biologically inactive as in macroprolactinemia.

Clinicopathologic Correlations

PRL is a sensitive indicator of pituitary dysfunction and the hormone most frequently produced in excess by pituitary tumors. It is also the first hormone to become deficient from infiltrative disease or tumor compression of the pituitary. Normal adult serum PRL levels are 3–14.7 ng/mL in men and 3.8–23 ng/mL in women. Hyperprolactinemia is the most common pituitary-hormone hypersecretion syndrome in both males and females, whereas pregnancy and lactation are the important physiologic causes.

Tumors arising from lactotrope cells, described as acidophilic cells in the pituitary that secrete PRL, account for approximately half of all functioning pituitary tumors. Hyperprolactinemia in women is a characteristic feature in women with amenorrhea, galactorrhea, and infertility. PRL levels can be increased by many physiologic and pathologic factors as well as medications; such elevations rarely exceed 200 ng/mL.

 Checkpoint! 15–4

Why is prolactin a sensitive indicator of pituitary dysfunction?

Posterior Pituitary Hormones

The two **posterior pituitary** hormones are arginine vasopressin (AVP), which is also known as antidiuretic hormone (ADH), and oxytocin. They are both small oligopeptides synthesized in the hypothalamus.

Antidiuretic Hormone

As one of the principal endocrine regulators of fluid and electrolyte balance, ADH regulates renal free-water excretion to maintain homeostasis. At low dose, ADH controls the resorption of water by the distal tubules of the kidneys and regulates the osmotic content of blood. At high doses, ADH causes contraction of arterials and capillaries, especially those of the coronary vessels, to produce localized increases in blood pressure.

CLINICOPATHOLOGIC CORRELATIONS

Diabetes Insipidus

Diabetes insipidus (DI) is characterized by copious production of urine (polyuria) accompanied by intense thirst (polydipsia) because of ADH deficiency. Urine output of water in excess of 2.5 L/day warrants investigation. DI may be classified as *central* (also hypothalamic DI) because of absence of deficient ADH secretion or as *nephrogenic* because of renal resistance to the action of ADH.

Central type DI is caused by a failure of the pituitary gland to secrete normal amounts of ADH in response to osmoregulatory factors. The loss of water triggers the thirst mechanism, which promotes increased water intake and prevents dehydration. Severe dehydration will occur if the thirst mechanism is compromised. Approximately 30% of patients who develop central type DI develop it for unknown reasons. Neoplasia, neurological surgery, head trauma, infections, autoimmune disorders, and hypoxic disorders account for the remaining 70% of patients with this condition.

Patients with nephrogenic DI are incapable of stimulating cAMP formation. There are two unique causes of this condition: (1) mutation in the vasopressin receptor and (2) mutations in the aquaporin-2 water channels. Acquired forms of nephrogenic DI include metabolic disorders (e.g., hypokalemia and hypercalcemia), drugs (e.g., lithium and barbiturates), and renal diseases (polycystic disease and chronic renal failure).[2]

Diagnosis of DI includes documentation of polyuria (water output >2.5L/day) and urine glucose testing to exclude glycosuria. Measurement of urine and serum creatinine, electrolytes (especially sodium), and osmolality may be useful. Further testing may include an overnight water-deprivation test or a hypertonic saline-infusion test.

A water-deprivation test is a diagnostic that distinguishes central DI from nephrogenic DI. The patient's blood levels of ADH are monitored along with urine and plasma osmolality. After a period of time, a dose of ADH is given and urine osmolality is measured. In central DI, low ADH causes the kidney to conserve water rapidly. In nephrogenic DI, however, the patient will show plasma and urine osmolalities that are similar to those of patients with central DI during water depravation. As the plasma osmolality exceeds approximately 300 mOsm/kg, the plasma ADH will begin to increase. After ADH injection, urine osmolality will not change significantly.

Patients who are subjected to the saline infusion test will be given a 3% solution of saline by intravenous injection. Their plasma osmolality and ADH are monitored for a 2-hour period. A normal patient shows a plasma ADH that is >7 pg/mL when plasma osmolality is ≥310 mOsm/kg.

Syndrome of Inappropriate Antidiuretic Hormone Secretion

Syndrome of inappropriate antidiuretic hormone secretion (SIADH) is described as the continuous production of ADH in the absence of known stimuli for its release. This syndrome is characterized by plasma ADH concentrations that are "inappropriately" increased relative to a low plasma osmolality, a normal to increased plasma volume, hypotonicity of the plasma, hyponatremia, and hypertonicity of the urine with continued sodium excretion. SIADH may be caused by one of the following: production of vasopressin by a malignancy, the presence of acute or chronic diseases of the central nervous system (CNS), pulmonary disorders, or side effects of certain drug therapies.[3]

Diagnostic protocols for SIADH require findings of plasma hypoosmolality (≤275 milliosmoles/kg) and hyponatremia (≤130 mmol/L). Quantitative urine sodium measurements and determination of serum and urine osmolality are also recommended. SIADH should be suspected if the urine osmolality is greater than plasma osmolality without correspondingly low urine sodium concentration (usually >60 mmol/L). In some patients, additional diagnostic tests are available, including a water-loading test, quantitative plasma ADH, plasma renin measurement, and estimation of plasma renin activity. SIADH is characterized by high ADH and low renin concentrations. If both quantitative assays are low, then a primary defect in renal water excretion may be the cause of the patient's symptoms.[2]

Oxytocin

Oxytocin is a nonapeptide that is secreted by the magnicellular neurons of the hypothalamus and is stored in the neurohypophysis along with AVP. It contributes to the contraction of uterine smooth muscle and stimulates the epithelial cells surrounding the mammary glands, resulting in milk ejection.

► THE ADRENAL GLAND

A cross-section of the adrenal gland is shown in Figure 15-4 ■. The outermost layer (zona glomerulosa) is a relatively narrow layer of small cells in poorly defined clusters that produces aldosterone. The broad lipid-filled zona fasciculata produces cortisol. The dense compact cells of the zona reticularis secrete **androgens**. The **adrenal medulla**, which lies below the zona reticularis, produces **catecholamines**.

ADRENAL MEDULLARY HORMONES

Three naturally occurring catecholamines are derived from the adrenal medulla: epinephrine, norepinephrine, and dopamine. Their structures are shown in Figure 15-5 ■. The main secretory products of the adrenal medulla are epinephrine and norepinephrine. Catecholamine production is not exclusive to the adrenal medulla; synthesis can take place in the neurons of the sympathetic and CNS and in chromaffin cells found in other areas of the abdomen and neck. Specifically, norepinephrine is the principal product synthesized in the CNS, and epinephrine is the principal catecholamine produced by the adrenal glands.

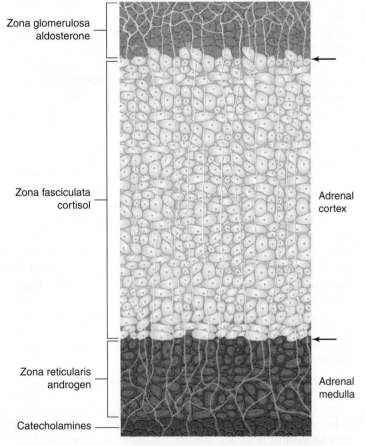

■ FIGURE 15-4   The anatomical layers of the adrenal gland indicating the sources of selected hormones from the cortex and medulla.

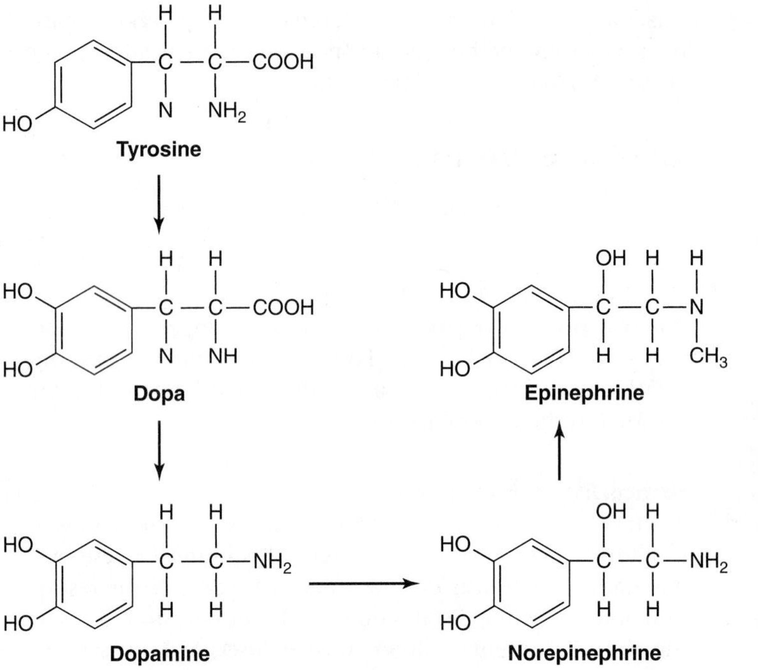

Zona glomerulosa aldosterone

Zona fasciculata cortisol

Adrenal cortex

Zona reticularis androgen

Adrenal medulla

Catecholamines

■ FIGURE 15-5   Biosynthetic pathway for major catecholamines. They are derived from the amino acid tyrosine.

The catecholamines play important roles as neurotransmitters both in the CNS and in the peripheral sympathoadrenal medullary system. Norepinephrine is primarily a neurotransmitter. Both epinephrine and norepinephrine influence the vascular system, whereas epinephrine affects metabolic process such as carbohydrate metabolism. Catecholamines initiate their respective biological action via their interaction with two different types of specific cell-membrane receptors: the α-adrenergic and β-adrenergic receptors. The affinities of epinephrine and norepinephrine for cell-membrane receptors are different, which results in opposing physiological effects. Norepinephrine primarily interacts with α-adrenergic receptors, whereas epinephrine interacts with both α-adrenergic and β-adrenergic receptors.[4]

Stimulation of either α- or β-adrenergic receptors initiate a myriad of body responses. For example, a signal to the β-adrenergic receptors may lead to:

• vasodilation;
• stimulation of insulin release;
• increased cardiac contraction rate;
• relaxation of smooth muscle in the intestinal tract stimulation of renin releases, which enhance sodium resorption from the kidney; and
• enhanced lipolysis and bronchodilation by relaxation of smooth muscles in bronchi.

Initiation of α-adrenergic receptor actions results in vasoconstriction, decreased insulin secretion, sweating, and stimulation of glycogenolysis in the liver and skeletal muscle, leading to an increase in blood-glucose concentration.

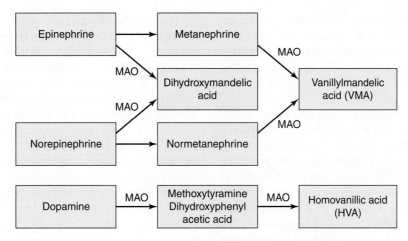

■ FIGURE 15-6   Metabolism of adrenal medullary hormones.

The catecholamines are degraded rapidly once released into circulation. The metabolism of catecholamines is shown in Figure 15-6 ■. Epinephrine and norepinephrine are metabolized by the enzyme catechol-O-methyltransferase to metanephrine and normetanephrine, respectively. They are then converted to vanillylmandelic acid (VMA) by the enzyme monoamine oxidase (MAO). The catecholamines may also be converted directly by MAO to dihydroxymandelic acid and then to VMA. Only a small fraction of catecholamines (2%) remained unmetabolized and excreted into the urine as free catecholamines.

 ### Checkpoint! 15–5

Identify three hormones produced by the adrenal medulla.

CLINICAL APPLICATION

Measurement of catecholamines and their metabolites is useful for assessment of pheochromocytoma and neuroblastomas. These are examples of catecholamine-producing tumors of the neurochromaffin cells. Other examples of catecholamine-secreting neurochromaffin tumors include paragangliomas and other autonomic and genetic disorders that will not be covered in this chapter. Measurements of urinary catecholamines and metabolites are useful in the diagnosis of adrenal medullary diseases; the metanephrines are most specific.

CLINICOPATHOLOGIC CORRELATIONS

Pheochromocytomas

A pheochromocytoma affects the neurochromaffin cells in the adrenal medulla are rare (<0.1%), and afflicted patients are labored with hypertension. It is curable if correctly diagnosed and treated, and fatal if not. The classical picture of catecholamine excess consists of substained or paroxysmal hypertension, weight loss, sweating, headache, palpitations, and anxiety.

Catecholamine elevation may be induced by stress, upright posture, exercise, hypoglycemia, cold, and mental states of anxiety and anger. Because the release of catecholamines may be sporadic, the measurement of intermediate metabolites tends to give results that are more definitive.

Biochemical tests for reliable detection of this rare, but serious disorder should include a combination of measurements of catecholamines and catecholamine metabolites. Quantitative measurement of VMA, a metabolite of both epinephrine and norepinephrine, in urine is useful but lacks the diagnostic sensitivity to make it an initial test for pheochromocytoma. Current recommendations are to measure urine or plasma catecholamines and urinary metanephrines (see Figure 15-6). False positive tests and confounders are issues that need to be addressed when interpreting urinary metanephrine results. These may occur because of inappropriate sampling conditions such as failure to fast and failure of the patient to rest for 20 minutes in a supine position. Certain medications may also cause false results—for example, tricyclic antidepressants and phenoxybenzamine.

NEUROBLASTOMA

Neuroblastoma is a malignant neoplasm of neural crest origin. It is characterized by the overproduction of catecholamines and their metabolites. Close to 50% of neuroblastoma cases occur in children before age 3, accounting for 6–10% of childhood cancer. Hypertension is uncommon in neuroblastoma, and symptoms relate primarily to tumor mass. Laboratory results showing a functional catecholamine-producing tumor is necessary in the diagnosis. Most patients have elevated urinary VMA and homovanillic acid (HVA) at time of diagnosis.

SPECIMEN COLLECTION AND STORAGE

Collection and storage of specimens, either plasma or urine, is critical to the reliability and interpretation of test results. A 24h urine collection is commonly requested by the clinician instead of plasma samples because of the more stringent patient-preparation requirements recommended for plasma samples. Timed urine collections do present potential problems. For example, there may be issues regarding specimen reliability, the difficulty in collecting specimens from pediatric patients, and the effects of a patient's diet, physical activity, and changes in posture.

The stability of catecholamines in urine collections has historically been a problem but due to extensive assay revisions, these issues have been resolved. The general recommendation is that catecholamines in urine samples should be preserved with hydrochloric acid to maintain an acid pH. Aliquots of specimens can be stored frozen over extensive periods at –80 °C to reduce autooxidation and deconjugation. Similarly, plasma samples can be collected in tubes containing heparin or EDTA as an anticoagulant and stored on ice before centrifugation at 4 °C, and prolonged storage of plasma fraction at –80 °C.[5]

ADRENAL CORTICAL HORMONES

The adrenal cortex secretes **adrenocorticosteroid** hormones and adrenal androgens. All steroids have the chemical structure of three six-sided rings forming the phenanthrene nucleus, to which is attached the cyclopentane ring (see Figure 15-7 ■). The steroids have a remarkable diversity of biological effects that depend on the nature of the chemical modifications of the basic steroid nucleus. The modifications include unsaturation of the C—C bonds in the ring and the attachment of hydroxyl, ketone, or other groups to specific carbon atoms. Steroids isolated from the adrenal gland include the corticosteroids and the adrenal androgens. The corticosteroids, consisting of **mineralocorticoids** and **glucocorticoids**, are the most important groups of adrenal steroids both physiologically and quantitatively.

Glucocorticoids

Cortisol is the major glucocorticoid in humans; more than 80% is naturally bound to cortisol-binding globulin (CBG). It is synthesized in the zona fasciculata and reticularis of the human adrenal cortex. Cortisol strongly influences carbohydrate, lipid, and protein metabolism. In its role as a glucocorticoid, cortisol promotes gluconeogenesis, increases the deposition of liver glycogen, and reduces glucose use and uptake by peripheral tissues such as muscle and fat.

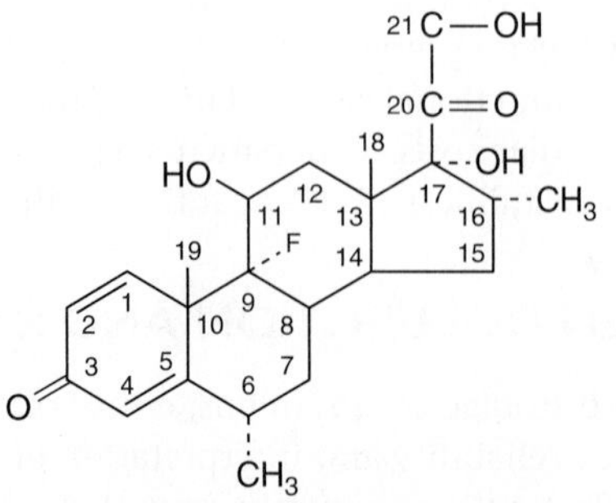

■ FIGURE 15-7   Structure of a C_{21} adrenocorticosteroid molecule.

The increase in gluconeogenesis is the result of several factors: (1) activation of enzymes in this pathway, including phosphoenolpyruvate carboxykinase and glucose-6-phosphatase; (2) an increase in muscle protein breakdown; and (3) an inhibition of amino acid uptake and protein synthesis in muscle, skin, and bone. Glucocorticoids also activate lipolysis and the release of free fatty acids into the circulation. They can stimulate adipocyte differentiation, promote lipogenesis through the activation of enzymes such as lipoprotein lipase, and increase expression of leptin. Other clinically important effects of glucocorticoids include their ability to suppress the immune system, their capacity to act as antiinflammatory agents, and their possession of antiallergic properties.

Mineralocorticoids

Aldosterone secretion is regulated predominantly by the renin-angiotensin system and enhanced by potassium and modestly by ACTH. In hypovolemic states, rennin, a proteolytic enzyme secreted by renal cells, acts on its substrate to produce angiotensin I, which is rapidly converted to angiotensin II in the lung by angiotensin-converting enzyme. Angiotensin II stimulates aldosterone production in the zona glomerulosa of the adrenal cortex.

Aldosterone acts on the distal tubule to increase sodium retention and volume repletion, thus shutting off the stimulus for renin release. Unlike cortisol, aldosterone is only weakly bound to CBG. Aldosterone concentrations are affected by dietary sodium, potassium, and posture.

Adrenal Androgens

The adrenal glands produce and secrete androgens, progesterone, and estrogen, all of which are produced in the gonads as well. Adrenal androgens are produced in the zona fasciculata or reticularis from a precursor compound, 17α-hydroxypregnenolone. Three examples of specific adrenal androgens are dehydroepiandrosterone (DHEA), testosterone, and androstenedione.

Metabolism of Cortisol and Adrenal Androgens

Steroids are extensively transformed and conjugated in the liver via the cytochrome P450 metabolizing enzyme system. The kidneys also serve an important role in steroid metabolism. The kidney excretes approximately 90% of conjugated steroids released by the liver, and nearly 50% of cortisol found in the urine is in the form of tetrahydrocortisol and tetrahydrocortisone.

Cortisol is tightly bound to CBG, and only approximately 2% of cortisol is excreted unchanged in the urine, which results in a slow rate of metabolism (half-life ~100 minutes). In the liver, cortisol is metabolized via several enzymatic reactions, yielding catabolites that are conjugated in the liver with glucuronic acid and sulfur. An inactive product formed by the catabolism of cortisol is 17-hydroxycorticosteroid (17-OHCS), which is measurable in urine samples.[6]

Adrenal androgens are also metabolized via a series of several complex reactions that produce metabolites. A principle metabolite dehydroepiandrosterone–sulfate (DHEA-S) is formed in the adrenal cortex or by sulfokinases in the liver and kidney from DHEA and excreted by the kidney. Testosterone, androstenedione and DHEA are metabolized to a group of steroids known as 17-ketosteroids (17-KS). These metabolites are excreted primarily in the urine.

The following factors influence the metabolism of steroids.[6]

- **Liver disease.** Patients with chronic liver disease will develop reduced cortisol clearance and produce urinary metabolites. Acute alcohol ingestion will elevate cortisol levels in blood.

- **Kidney disease.** Patients with chronic renal failure may have normal plasma levels of total and free cortisol and ACTH. The plasma concentrations of conjugated metabolites or cortisol are elevated, and renal excretion is reduced.

- **Thyroid disease.** Patients with hyperthyroidism will have elevated secretion and metabolism of cortisol without significant changes in blood concentrations. Conversely, in hypothyroidism, both secretion and metabolism of cortisol is reduced without significant changes in blood cortisol levels.

- **Stress.** Patients with acute stress from trauma events or illness have elevated concentrations of adrenal glucocorticoids, cortisol and corticosterone, and the adrenal androgens DHEA and DHEA-S (DHEA with a sulfate ester). Chronic stress-related illness results in elevated plasma cortisol levels but suppressed DHEA, DHEA-S, and androstenedione levels.

- **Age.** In disease-free individuals, the production of glucocorticoids changes with age. In newborns, the production of cortisone is higher than cortisol. A few days after birth, cortisol production increases over that of cortisone and production patterns become similar to those of adults. The secretion of adrenal androgen begins to rise about age 9, peaks at about 25, and declines after 30.[7]

- **Estrogen therapy.** Estrogen-containing medications tend to raise plasma concentrations of CBG and thus influence total cortisol blood concentrations. Free cortisol blood levels are not affected, and the clearance of cortisol and its production rate and concentrations of urinary metabolites are below normal in patients on estrogen therapy.

- **Nutrition.** Obese individuals tend to have elevated levels of conjugated metabolites of cortisol in their urine, which may suggest the presence of Cushing's syndrome. Clinicians may request the measurement of free cortisol measured in urine, which is not influenced by obesity to screen for Cushing's syndrome. Plasma concentrations of total and free cortisol and urinary free cortisol excretion rates in obese individuals are normal. Obesity can also present with elevated DHEA, DHEA-S, and androstenedione production and clearance rates.

- **Drugs.** Drugs that can induce the hepatic hydroxylase enzymes systems may produce reduced urinary excretion rates of cortisol metabolites, whereas plasma cortisol concentrations and urinary free excretion rates of cortisol are not affected. Examples of drugs that produce these changes are phenytoin, phenobarbital, aminoglutethimide, and rifampin.

Hypothalamus–Pituitary–Adrenal Axis

The relationship of the hypothalamus to the pituitary gland and its target organ, the adrenal gland, is based on the principle of feedback control. The hypothalamus secretes CRH, and the posterior pituitary secretes ACTH, a polypeptide containing 39 amino acids. Elevated blood levels of cortisol, an adrenal cortex hormone, reduces the synthesis and release of CRH from the hypothalamus, resulting in decreased secretion of pituitary ACTH. This mechanism is an example of negative feedback control (see Figure 15-8 ■). This ultimately causes reduced stimulus of the adrenal gland, resulting in decreased secretion of cortisol. Conversely, with low levels of blood cortisol, both the pituitary and the hypothalamus are stimulated to secrete greater amounts of ACTH and CRH. This results in an increased stimulus of the adrenal gland to secrete cortisol. Emotional stress may also act as a stimulus to enhance cortisol secretion.

Patterns of ACTH and Cortisol Secretion

ACTH is secreted in brief episodic bursts that cause sharp rises in plasma cortisol concentration as outlined in Box 15-2 ■. Cortisol and ACTH levels are lowest at 4 a.m. and rise early in the morning, peak at about 8 a.m., and gradually fall over the

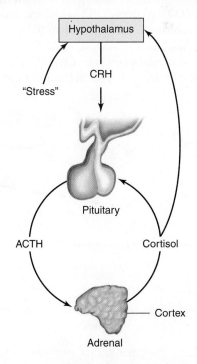

■ **FIGURE 15-8** The hypothalamic–pituitary–adrenal axis and the negative-feedback mechanism associated with cortisol and ACTH.

Time (hours)	0800	1200	1600	2000	2400	0400
cortisol (μg/dL)	15	8	14	5	10	4
ACTH (pg/mL)	120	60	80	50	70	50

course of the day, with a nadir around midnight. Diurnal variation of cortisol is lost in adrenal disorders such as adrenal adenoma and Cushing's syndrome (see Figure 15-9 ■).

Clinicopathologic Correlations

Adrenal Hypofunction Adrenal insufficiency can be classified as primary, secondary, or tertiary. Primary adrenal insufficiency, also known as **Addison's disease**, results from progressive destruction or dysfunction of the adrenal glands by a local disease process or systemic disorder. Because the entire gland is affected, all classes of adrenal steroids are deficient. Complete glucocorticoid deficiency can manifest in a variety of ways, including fatigue, weakness, weight loss, hypoglycemia, hyperpigmentation, hypotension, and hyperkalemia.

In secondary and tertiary adrenal insufficiency, diminished cortisol production can result from destructive events in the hypothalamic–pituitary that result in a decreased ability to secret ACTH (secondary) or CRH (tertiary). Suppression of hypothalamic–pituitary–adrenal function by chronic administration of pharmacological dosages of glucocorticoids is the most common cause of tertiary adrenal insufficiency. It decreases CRH synthesis and secretion from the hypothalamus and blocks its tropic and secretory actions on the pituitary corticotropes. Synthesis of ACTH by the anterior pituitary decreases. In the absence of ACTH stimulation, the adrenal zona fasciculata and reticularis atrophy and can no longer secrete cortisol. The clinical features of secondary and tertiary adrenal insufficiency are similar to those of primary insufficiency, except that hyperpigmentation is not evident and the magnitude of hypertension is less severe. Plasma levels of ACTH and cortisol drawn at 8 a.m. may be normal to decreased in secondary and tertiary adrenal insufficiency.

For the diagnosis of adrenal insufficiency, inadequate cortisol production must first be demonstrated. A morning sample is the best to use because cortisol values are normally higher from diurnal variation. The patient must not be under stress, which would cause cortisol elevation.

An ACTH stimulation test, which measures cortisol response to a dose of a commercially available ACTH analog (cosyntropoin or Synacthen), should be performed for virtually all patients suspected of adrenal insufficiency. A normal response is >20 μg/dL of cortisol. An impaired response confirms adrenal insufficiency.

✓ **Checkpoint! 15–6**

Explain the cause of Addison's disease.

Metyrapone Stimulation

The metyrapone testing that assesses the pituitary reserve is useful for differential diagnosis of adrenal insufficiency. Metyrapone inhibits 11β-hydroxylase (shown in Figure 15-10 ■), causing cortisol deficiency, which stimulates ACTH secretion. Testing protocol requires that a specified

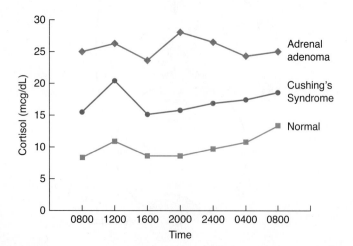

■ **FIGURE 15-9** Changes in cortisol levels over time in a patient without disease (normal), with adrenal adenoma and Cushing's disease.

■ **FIGURE 15-10** Metyrapone stimulation test. Metyrapone inhibits 11-hydroxylase, which catalyzes the step immediately preceding cortisol syntesis.

amount of metyrapone be administered to a patient at midnight, followed by blood samples drawn at 8 a.m. Plasma is tested for cortisol, 11-deoxycortisol, and ACTH. Other protocols may require testing of urine samples at 8 a.m. for 17-OHCS and 17-KS. Increased 11-deoxycortisol (>7 μg/dL) and ACTH (>150 pg/mL) with a decreased cortisol level (<3 μg/dL) implies a normal pituitary–adrenal axis. Subnormal results or a lack of response may be seen in patients with secondary adrenal insufficiency or pituitary or hypothalamic diseases combined with inadequate enzyme blockage or with Cushing's syndrome caused by adrenal tumors or nonendocrine ACTH-secreting tumors. This test is not useful if the clinician suspects primary adrenal insufficiency. An ACTH stimulation test should be performed first in this condition.[8]

ADRENAL HYPERFUNCTION

Adrenal hyperfunction is associated with either Cushing's disease, which is ACTH dependent, or Cushing's syndrome, which is ACTH independent. An elevated serum cortisol is the starting point in establishing a diagnosis of Cushing's. It may be achieved by the following.

• **Single blood sampling for cortisol.** The morning sample may be normal, but the loss of diurnal variation is a clue to the diagnosis, and the evening sample should be increased.

• **A 17-OHCS test.** A urinary 17-hydroxy corticosteroid test measures the cortisol metabolites in urine. It is not as specific and sensitive as either serum or urine cortisol measurements.

• **Urinary free cortisol.** Although free cortisol excretion represents only 1% of the cortisol secreted each day, it provides a valid index of glucocorticoid secretion. Urinary free cortisol excretion results from glomerular filtration of plasma free cortisol and is therefore an index of integrated 24h plasma-free cortisol. Urinary cortisol is a more sensitive indicator of increased cortisol because the increase is rapid as soon as plasma cortisol exceeds the binding capacity of CBG.

Dexamethasone Suppression Test

The overnight dexamethasone suppression test is used to document hypersecretion of the adrenocortical hormones. The integrity of the cortisol feedback mechanism can be tested by administering dexamethasone, a potent glucocorticoid, and estimating the suppression of ACTH secretion by measuring serum or urine cortisol level. Urinary free cortisol measurements are requested to provide the clinician with information for determining the appropriate diagnosis.

A low-dose (1.0 mg) dexamethasone test is used initially to document true hypersecretion of cortisol (because of Cushings, for example). Larger doses (4 mg) of dexamethasone are then used to establish the differential diagnosis of an ACTH-secreting pituitary adenoma. In the low-dose dexamethasone test, plasma levels of cortisol are suppressed to about 2 μg/dL or lower in normal subjects but not in patients with Cushing's (8 a.m. plasma sample >10 μg/dL). Plasma cortisol levels of >2 μg/dL may be found also in patients with stress, obesity, infection, or acute illness. Once glucocorticoid excess is confirmed, initiate a differential study to identify the cause.[8]

✓ **Checkpoint! 15–7**

Explain the purpose of the metyrapone and dexamethasone suppression tests.

CONGENITAL ADRENAL HYPERPLASIA

Congenital adrenal hyperplasia (CAH) refers to syndromes caused by inherited enzymatic defects in one of the four enzymes required for cortisol biosynthesis. A common feature is decreased negative feedback inhibition of cortisol on pituitary ACTH, resulting in adrenal hyperplasia.

The clinical manifestations of CAH are consequences of the deficient synthesis of cortisol and, in some cases, aldosterone. The most common cause of CAH is 21-hydroxylase deficiency followed by 11-hydroxylase deficiency. Figure 15-11 illustrates the biochemical pathway for the formation of steroids in a patient with CAH caused by a deficiency of 21-hydroxylase.[9] Mineralocorticoid deficiency leads to salt and water wasting and dehydration with hypotension, hyponatremia, and hyperkalemia. Excess pituitary synthesis of ACTH and related precursor peptides, unchecked by negative feedback, may cause hyperpigmentation of the skin and mucous membranes through the action of MSH on melanocytes.

Blockage of the synthesis of aldosterone and cortisol would result in elevated levels of urinary 17-KS and urinary 17-hydroxyprogesterone. Those would be desirable laboratory tests to perform for diagnosis.

PRIMARY ALDOSTERONISM

There are two main causes of autonomous hypersecretion of aldosterone: (1) an adrenal aldosterone-producing adenoma

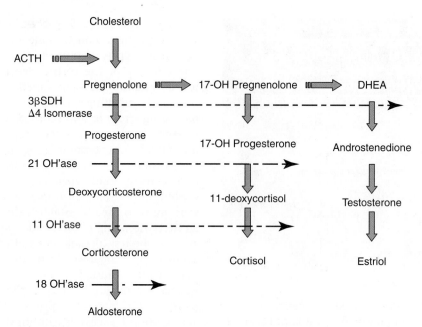

■ **FIGURE 15-11** Pathway of steroid formation in patients with congenital adrenal hyperplasia as a result of 21-hydroxylase deficiency (21 OH'ase). 3βSDHΔ4 isomerase, 3β-hydroxysteroid dehydrogenase-isomerase.

(APA) or **Conn's Syndrome** (about 70% of cases) and (2) an idiopathic hyperaldosteronism (IHA) caused by bilateral adrenocortical hyperplasia (about 30% of cases). The occurrence of adrenal carcinoma is rare. Primary hyperaldosteronism is an uncommon cause of hypertension that accounts for 0.1–0.5% of all cases of hypertension. Hypokalemia is the most consistent laboratory finding along with elevated urinary aldosterone and suppressed plasma renin activity. Urinary potassium measurement is a cost-effective test to establish hyperaldosteronism, which should be elevated in a hypokalemic patient.

The differential diagnosis of APA from IHA is important because APA is responsive to surgery, whereas drug therapy is usually the treatment for IHA. Determination of 18-hydroxy-corticosterone may help because values are higher with APA and lower with IHA. Posture stimulation (the effects of standing versus laying down) is also helpful because most patients with APA show declining values on standing and IHA responds to normal postural rise. The most reliable means, nonetheless, is adrenal bilateral vein catheterization and measuring serum aldosterone.

SECONDARY ALDOSTERONISM

Aldosterone secretion from the adrenal gland is not autonomous, but adrenal secretion of aldosterone is responding to secondary causes, often to the production and release of renin from the kidney. This may be caused by sodium loss, decreased renal perfusion, renal artery stenosis, or vascular volume depletion. The hallmark of secondary aldosteronism is elevated urinary aldosterone in the presence of elevated plasma renin activity.

▶ REPRODUCTIVE SYSTEM

PHYSIOLOGY

The reproductive function and pregnancy are regulated by a variety of hormones synthesized and secreted by the testes, ovaries, adrenals, pituitary, hypothalamus, and placenta. Several of the significant hormones released from these tissue sources are described in the following sections.

HORMONES IN THE OVARY AND TESTIS

Sex steroids are synthesized from cholesterol in the ovaries and testes and adrenal glands. Once in the blood, the sex hormones become reversibly protein bound to **sex-hormone binding globulin (SHBG)**, CBG, and albumin. Only 1–2% of the sex steroids that circulate in blood are free or unbound to protein.

Estrogens
Estrogens are a group of C_{18} sterol compounds, with estradiol identified as the principle hormone produced by ovaries. Estrone and estriol are mostly metabolites of ovarian and glandular conversion. Estrogens promote the secondary sexual characteristic changes in the female and are responsible for follicular phase changes in the uterus. Estrogen deficiency results in irregular and incomplete development of the endometrium.

Progesterone
Progesterone is a steroid hormone with 21 carbons that induces secretory activities of the endometrium for implantation. It is the dominant hormone responsible for the luteal

phase of the menstrual cycle. Progesterone deficiency results in the failure of an embryo to implant.

Androgens

Adrenal androgens include any of the C_{19} steroids synthesized from cholesterol by the .Androgens stimulate or control the development and maintenance of masculine characteristics in by binding to . The principal sex hormone is testosterone. Testosterone is primarily secreted by the male and female ovaries, although small amounts are also secreted by the . In women, most testosterone is produced from the metabolism of androstenedione, which is also a member of the androgens. Another important androgen, DHEA, is the primary precursor of natural estogens.

Testosterone

Androgen synthesis continues into advanced age. Excess production in women leads to hirsutism, which is abnormal abundant androgen-sensitive terminal hair growth; more severe cases would lead to defeminization and virilization as shown in Figures 15-14 and Figure 15-15.

REGULATION OF REPRODUCTIVE SYSTEMS AND PREGNANCY

Follicle-Stimulating Hormone and Luteinizing Hormone

The hypothalamus secretes GnRH that is responsible for the secretion of both gonadotropins, LH and FSH, from the anterior pituitary as shown in Figures 15-12 ■ and Figure 15-13 ■. LH binds to the Leydig cell receptors to enhance conversion of cholesterol to testosterone, and FSH activates the seminiferous

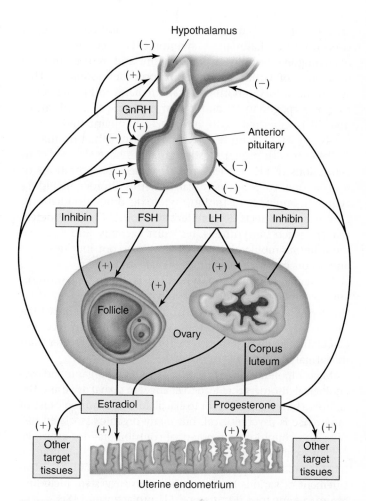

■ FIGURE 15-13   Hormonal regulation of reproduction in females.

tubules for sperm production. LH and FSH each consists of two subunits that share the same subunit but have different subunits that confer their functional specificities. In the male, LH induces Leydig cells to synthesize testosterone. In the female, the regulatory process is cyclic. In the follicular phase, FSH stimulates the series of pituitary, ovarian, and uterine changes that occur. Estradiol is secreted to restore the endometrium. A surge of GnRH stimulates LH and FSH secretion at the end of the follicular phase. LH in turn stimulates the secretion of progesterone and estradiol in the luteal phase to prepare the endometrium for implantation should fertilization occur. A more detailed explanation with appropriate illustrations can be found in the reference cited.[10]

Male Reproductive Disorders

There are many disorders associated with the male reproductive system. Several abnormalities arise from the hypothalamus and pituitary, and others are located in the testes. A few disorders result from defects in androgen activity, impotence, and gynecomastia. Clinical laboratory tests are available to assist the clinician in assessing many of these disorders.

Defects in the hypothalamus or pituitary prevent normal gonadal stimulation. Examples of causes for these conditions include panhypopituitarism, hypothalamic syndromes, hyperprolactinemia, malnutrition, and GnRH deficiency; all

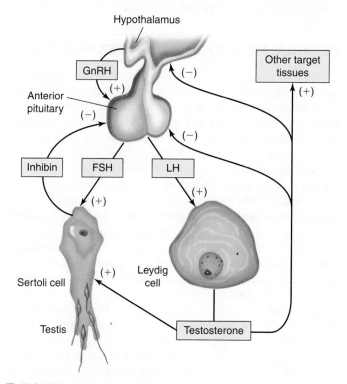

■ FIGURE 15-12   Hormonal regulation of reproduction in males.

are associated with decreased testosterone and gonadotropin concentration. Kallmann's syndrome is a common form of hypogonadotropic hypogonadism and results from a deficiency of GnRH during embryonic development. This syndrome occurs in both males and females and is characterized by hypogonadism and loss of the sense of smell (anosmia). It is inherited as an autosomal dominant trait. Blood levels of FSH, LH, and testosterone are lower than normal.

Patients with primary testicular failure have elevated concentrations of LH and FSH and decreased concentrations of testosterone. There are many causes of this disorder, including acquired (mumps, castration, cytotoxic drugs, irradiation), chromosome defects (Klinefelter's syndrome, 47, XXY), defective androgen biosynthesis, testicular agenesis, and selective seminiferous tubular disease. Aging is also a contributing factor and occurs in a high percentage of males older than age 60.

Testicular feminization syndrome is the most common and severe defect in androgen activity. It is characterized by the presence of female habitus and development of breast tissue. The defect is thought to be with the androgen receptor. Circulating levels of testosterone is the same as or greater than in normal men.

Male impotence can be caused by a variety of organic and psychological disorders that lead to changes in sexual drive and the ability to have an erection or to ejaculate. The common cause of impotence is psychological, but many organic associated diseases and conditions are potential contributors—for example, cardiovascular diseases, diabetes mellitus, hypertension, uremia, hypogonadism, hyperthyroidism, and hypothyroidism. Measurement of various hormones has been suggested, including serum free and total testosterone, LH, prolactin, and TSH.

Gynecomastia is a benign growth of glandular breast tissue in men and is found in males of varied ages. Gynecomastia is associated with an increase in the estrogen:androgen ratio. It may develop from iatrogenic causes, hyperthyroidism, germ cell tumors that produce human chorionic gonadotropin, and estrogen-producing tumors of the adrenal glands, the testes, or the liver. An iatrogenic cause reflects a disease or illness that results from physician care—for example, drug-induced gynecomastia can result from such patient medications as calcium channel blockers, cancer chemotherapeutic agents, and histamine-2 receptor blockers.[11] Patients with suspected gynecomastia where history and physical examination point to no specific disorder—testicular failure, for example—measurement of human chorionic gonadotropin (hCG), plasma estradiol, testosterone, and LH may prove useful. Measurement of prolactin is only useful as an indirect association with gynecomastia.

NORMAL MENSTRUAL CYCLE

The menstrual cycle consists of a series orchestrated events involving the hypothalamus, pituitary, and ovaries. Each cycle consists of two phases—*follicular* and *luteal*—during which several hormones interact through feedback mechanisms to initiate functional and structural changes in the ovaries, uterus, cervix, and vagina. This entire cycle results in the creation of an ovum for possible implantation and fertilization.

The follicular phase begins just after the previous luteal phase ends and terminates at ovulation. Follicle-stimulating hormone levels are elevated early in the menstrual cycles and decline until ovulation. Secretion of LH increases around the middle of the follicular phase. Estrogen secretion commences just before ovulation, initiates a feedback signal to the hypothalamus, and causes the LH surge.

The luteal phase begins about day 14 and is characterized by an increase in production of progesterone and estrogen from the corpus luteum and a reduction of circulating FSH and LH. If ovulation does not occur, then the corpus luteum does not develop and progesterone levels decrease. If pregnancy occurs, hCG is available to maintain the corpus luteum and progesterone levels rise.

Several hormones play a key role in menstrual cycle.

- GnRH triggers the surge of LH that precedes ovulation.
- FSH initiates the growth and maturation of a group of ovarian follicles.
- LH seeks out ovarian follicle receptors and initiates a signal to enhance differentiation of the theca cells (form the membrane sheath around the follicle) and the production of progesterone by the developing corpus luteum.
- Estradiol enhances the effects of FSH on a maturing follicle through changes in FSH receptors of the follicular cells, but it suppresses pituitary FSH and LH release during the follicular phase through negative feedback.
- Progesterone levels increase significantly during the midcycle LH surge and ovulation. Progesterone is believed to stimulate the ovulation peak of FSH and promote the growth of secretory endometrium, which is required for the implantation of a fertilized ovum.[12]

FEMALE REPRODUCTIVE DISORDERS

Primary amenorrhea is the absence of menstruation by age 16. A large number of women with this disorder fail to develop secondary sex characteristics. Primary amenorrhea has many causes, including a group of genetically related conditions. Turner's syndrome (55 X karyotype) and pure gonadal dysgenesis (either 46 XX or XY karyotype) are two examples of specific ovarian disorders. Disorders that are not genetically related involve the pituitary, thyroid, adrenals, and uterus.

Secondary amenorrhea occurs when there was previous menstruation, but then menstruation ceased. Oligomenorrhea is infrequent menstruation that occurs less than nine times per year. There are many causes of secondary amenorrhea and pregnancy is considered a common cause, therefore pregnancy should be ruled out first before further assessment of the patient. Secondary amenorrhea due to iatrogenic causes involve ingestion of a myriad of pharmacological drugs such as antipsychotic, antidepressant, antihypertensive, drugs with estrogenic activity and drugs with ovarian toxicity. The causes of primary and secondary amenorrhea overlap significantly thus the causes listed above for primary amenorrhea may apply to secondary amenorrhea.

Characteristic features of a patient with an excess of androgens include excess hair on the face, chest, abdomen, and thighs as well as acne and obesity. Amenorrhea may result from an excess of androgens in patients with adult-onset CAH, corticotropin-dependent Cushing's syndrome, or polycystic ovary syndrome (PCOS).

Polycystic ovary syndrome occurs in a small percentage of premenopausal women and is believed to be caused by a disorder of the hypothalamus. PCOS is clinically described as hyperandrogenism with chronic anovulation in women without underlying diseases of the adrenal or pituitary glands.[13] This syndrome is characterized by infertility, hirsutism, obesity, and irregular vaginal bleeding. PCOS is associated with polycystic ovaries, but this is not essential for the diagnosis. Women with PCOS have a greater frequency of both hyperinsulinism and insulin resistance. Women with PCOS may develop acanthosis nigricans (dark pigmentation of the spiny layer of the skin). Blood levels of FSH are low, and LH levels are high in PCOS. Serum androstenedione and testosterone concentration are also elevated.

Hirsutism is defined as excessive growth of body hair in women and children. The causes of hirsutism include disorders of the ovaries, adrenal gland, iatrogenic factors, various endocrine disorders, and familial occurrences. Laboratory tests that may be useful for the clinician's assessment of a patient with possible hirsutism include measurements of serum total or free testosterone and DHEA-S. Elevations in testosterone levels indicate either an adrenal or an ovarian source, whereas elevations of DHEA-S levels suggest an adrenal origin of androgens.

LABORATORY TESTS FOR ASSESSMENT OF MALE AND FEMALE REPRODUCTIVE DISORDERS

The majority of hormone precursors, active hormones, and metabolites of hormones can be measured by immunoassay techniques, chromatography (liquid or gas), and mass spectrometry. Specimen types appropriate for testing include plasma, serum, amniotic fluid, and saliva. Quantitative measurement of progesterone, FSH, LH, prolactin, hCG, estradiol, estriol, and testosterone are routinely available in many clinical chemistry laboratories because the methods do not require extensive preanalytical sample preparation and can be automated. Many health-care facilities have fertility clinics, high-risk pregnancy clinics, and obstetrics–gynecology services that incorporate assays to measure these hormones. A brief description of a select group of hormones is discussed here.

Evaluation of Semen

Semen analysis consists of determining ejaculate volume, pH, sperm count, motility, and forward progression. Analysis should be performed within an hour after collection. Semen analysis is not a test for infertility but is considered an important laboratory test in the evaluation of male fertility.

Testosterone

Testosterone circulates in three different forms: nonprotein bound or free form, loosely bound, and tightly bound. Albumin is the principle binding protein for the loosely bound form, and SHBG is for the tightly bound form. A *total testosterone* measurement includes (1) free testosterone, (2) albumin-bound testosterone, and (3) SHBG-bound testosterone. *Bioavailable testosterone* includes circulating free testosterone and albumin-bound testosterone. The SHBG bound testosterone is not biologically active, whereas the free testosterone is available for binding to target cell receptors. Albumin-bound testosterone is available for binding target cell receptors because testosterone can dissociate from albumin and rapidly diffuse into target cells.

Laboratory methods are currently available for estimating levels of total testosterone, SHBG, free testosterone, loosely bound testosterone, bioavailable testosterone, testosterone precursors, and metabolites in blood (e.g., dihydrotestosterone, DHT, and DHEA) or urine (e.g., 17-KS). The majority of assays are immunoassays that use a variety of detector molecules and preanalytical sample-preparation techniques. For a comprehensive review of these assays and their respective clinicopathological correlation, refer to the reference cited.[14]

Mathematical models exist to estimate SHBG-bound testosterone. Studies have shown that SHBG correlates with the percentage of free testosterone. Based on the findings, both percent free testosterone[15] and the concentration of free testosterone[16] in mol/L can be estimated using the equations cited.

Estrogens

Several specific estrogen compounds are measurable in blood and urine samples. They include estrone (E_1), estradiol (E_2), total and free estriol (E_3), and unconjugated estetrol (E_4). Estradiol is the most active of endogenous estrogens. Measuring estradiol is useful in conjunction with the measurement of additional gonadotropins for evaluating menstrual and fertility problems in adult females. Measurement of estradiol is also useful in the assessment of gynecomastia, menstrual cycle irregularities, and sexual maturity in females. The predominant urinary estrogen during pregnancy is estriol. Free estriol measurements are more specific than total estriol for monitoring the output of the fetoplacental unit at the time a sample is secured. Multiple serial sampling over time is recommended for accurate assessment of this purpose. Circulating levels are decreased if the fetus is compromised. Measurement of unconjugated estriol (uE_3) is routinely used instead of total estriol for identifying a fetus with Down syndrome. Estrone is a more potent estrogen than estriol but is less potent than estradiol. It is the major circulating estrogen after menopause. The measurement of estrone is limited to diagnosis of postmenopausal bleeding, the menstrual dysfunction caused by extraglandular estrone production. Estetrol is produced exclusively during pregnancy by the fetal liver. During human pregnancy, estretrol is detectable from week nine to delivery. Estetrol reaches the maternal blood circulation via the placenta. The role of estetrol in humans if not known, and its clinical usefulness is undetermined.

✓ Checkpoint! 15-8

Identify the fundamental approach to the assessment of disorders of the reproductive systems.

▶ THE THYROID GLAND

BIOLOGICAL FUNCTION

The thyroid hormones have a far-reaching effect on the body habitus. They control the basal metabolic rate and calorigensis through increased oxygen consumption in tissue. Thyroid hormones contribute to neural development, normal growth, and sexual maturation in mammals. Other actions include stimulation of protein synthesis and carbohydrate metabolism, stimulation of adrenergic activity with increased heart rate and myocardial contractility, increased synthesis and degradation of cholesterol and triglycerides, increased requirements for vitamins, increased calcium and phosphorus metabolism, and enhanced sensitivity of adrenergic receptors to catecholamines.[17]

PHYSIOLOGY

The normal thyroid gland weighs 20–25 grams and is histologically composed of closely packed follicles. The follicles contain colloid as its major constituent, with thyroglobulin (Tg) as the main storage site of the thyroid hormones. In response to the stimulation of thyroid-stimulating hormone from the pituitary, uptake of iodide by the gland becomes active. The source of the iodine is primarily diet, and it is ingested in the form of iodide. The follicular cells of the thyroid concentrate iodide several folds greater than normal plasma levels by means of an energy-dependent pump mechanism. Iodide is oxidized at or near the thyroid gland membrane to a more reactive iodine and within seconds is bound to the tyrosine residues in Tg, resulting in the formation of monoiodotyrosine (MIT) and diiodothyronine (DIT). Two MIT residues couple in the synthesis of thyroxine (T_4), while one MIT residue couples with one DIT residue to form triiodothyronine (T_3). The products of this process—T_4, T_3, and reverse T_3 (rT_3)—are shown with their respective iodination substitutions in Figure 15-14 ■.

The **hypothalamic–pituitary–thyroid axis** is shown in Figure 15-15 ■. TRH stimulates the pituitary for TSH secretion; as a tropic hormone, TSH stimulates the thyroid gland to produce T_3 and T_4. When enough of the glandular hormones are synthesized, a feedback message goes to the pituitary to inhibit the TSH secretion; more specifically, feedback goes to the hypothalamus to stop the stimulation of TSH. In the thyroid gland, T_4 is the major hormone secreted completely from the gland. Only 20% of T_3 is glandular; the rest is derived outside the thyroid gland through deiodination of T_4, which takes place in the liver, kidney, and muscle. This has implications on the clinical utility of T_3, as will be delineated

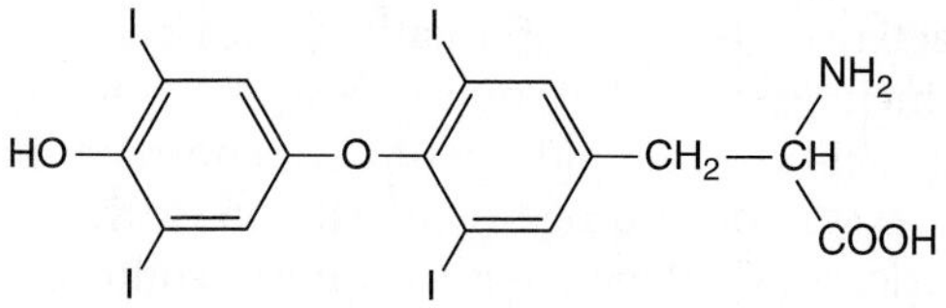

Thyroxine (T4)

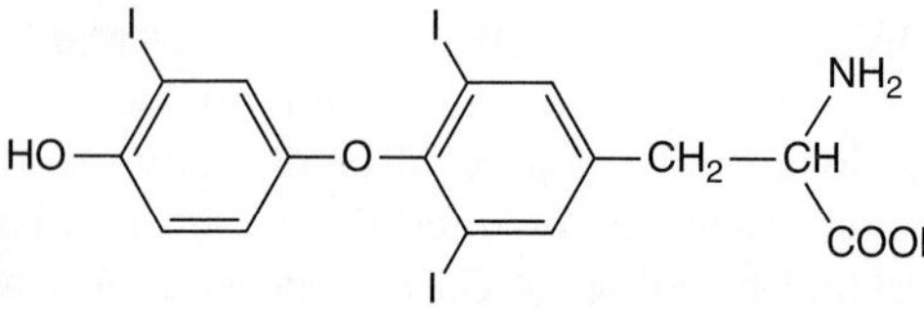

Triiodothyronine (T3)
(active)

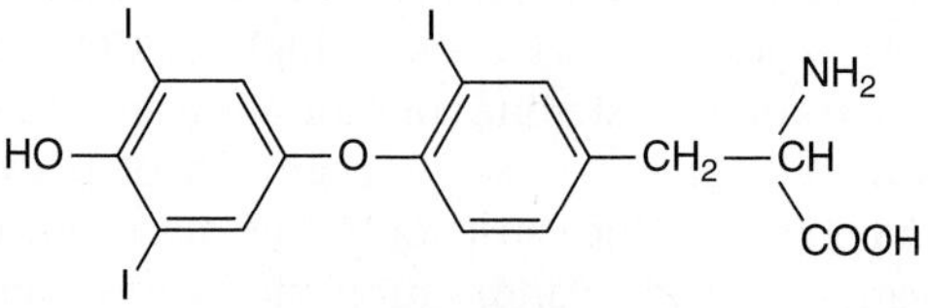

Reverse triiodothyronine (rT3)
(inactive)

■ **FIGURE 15-14**   Molecular structures of three thyroid hormones: T_4, T_3, and rT_3, the inactive metabolite of T_4.

later. Peripheral deiodination may also produce rT_3, which is biologically inactive.

Thyroxine and T_3 circulate in blood primarily complex to proteins such as thyroid binding globulin, thyroid-binding pre-albumin (TBPA), transthyretin (TTR), and serum albumin. The reaction is reversible, with the unbound fraction possessing biological activity. The interaction between free thyroid hormones (FT_4) and their binding proteins [in this example, it is TBG (T_4—TBG)], conforms to a reversible binding equilibrium (K') typified by the mass action relationship:[18]

$$FT_4 + TBG \xleftrightarrow{K} T_4 \cdot TBG$$

(0.03% biologically active)   (99.97% biologically inactive)

$$K = \frac{[T_4 \cdot TBG]}{[FT_4][TBG]}$$

$$[FT_4] = \frac{[T_4 \cdot TBG]}{K[TBG]}$$

where

K = association constant for either T_4(2×10^{10} M^{-1}) or T_3(5×10^8 M^{-1})

[TBG] = concentration of unoccupied binding sites on TBG

[$T_4 \cdot$ TBG] = concentration of T_4-occupied binding sites on TBG

[FT_4] = concentrations of free or unbound T_4

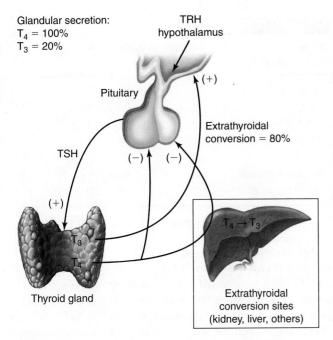

Glandular secretion:
T_4 = 100%
T_3 = 20%

TRH
hypothalamus

Pituitary

(+)

Extrathyroidal
conversion = 80%

TSH

(−)   (−)

(+)

T_3

T_4

Thyroid gland

$T_4 \rightarrow T_3$

Extrathyroidal
conversion sites
(kidney, liver, others)

■ FIGURE 15-15 Schematic of the hypothalamic-pituitary thyroid axis.

The relationship of these factors has significant diagnostic implications. A significant increase in the concentration of either FT_4 or any of the binding proteins (in this example, TBG) drives this reaction to the right, increasing serum $[T_4 \cdot TBG]$. Because 99.96% of the TT_4 is bound to TBG, the value of $[T_4 \cdot TBG]$ is considered equivalent to that of TT_4. Hyperthyroid diseases—for example, Graves' disease—produce an increase in FT_4, whereas pregnancy, estrogen therapy, hepatitis, chronic heroin abuse, and acute intermittent porphyria (AIP) may produce a primary increase in TBG. In both examples, $[T_4 \cdot TBG]$ is increased; in Graves' disease, however, the patient is ill and requires treatment; in the other examples, the patient is euthyroid. Therefore, it is important to differentiate between changes in $[T_4 \cdot TBG]$ that result from primary changes in $[FT_4]$ (e.g., hyperthyroidism or hypothyroidism) and those that are caused by primary changes in $[TBG]$.[19]

As much as 99.96% of T_4 and 99.6% of T_3 are protein bound. Because only the free moiety is metabolically active, changes in TBG concentration interfere with total T_4 and T_3 concentrations but not with the free fraction. Factors resulting in increased TBG include pregnancy, estrogen therapy, and AIP. Conditions resulting in decreased TBG include androgens, glucocorticoids, nephrotic syndrome, active acromegaly, major illness, cirrhosis, surgical stress, and drugs such as salicylates, phenytoin, phenylbutazone, and diazepam.

 Checkpoint! 15–9

What form of the thyroid hormones is metabolically active?

THYROID-FUNCTION TESTING

Thyroid-Stimulating Hormone (TSH)

Currently, TSH is the primary and most important test in the assessment of thyroid dysfunction in both hypothyroid and hyperthyroid patients. This has been made possible with greatly improved sensitivity and specificity in TSH assays. Quantitative immunoassays for TSH have undergone modifications that have incrementally decreased analytical sensitivity since their inception in the 1960s.[20] Currently, high-sensitivity TSH (hsTSH) assays are capable of detecting the hormone at <0.001 mU/L and are able to assist clinicians in differentiating **euthyroid** from hyperthyroid and hypothyroid states (see Box 15-3 ■). Table 15-2 ✪ illustrates the incremental improvements in diagnostic utility of several generations of TSH assays produced since the 1960s. These hsTSH assays have largely replaced the TRH stimulation test, which is now used only when the diagnosis of hyperthyroidism is equivocal.

A primary abnormality of thyroid function may be excluded in patients with TSH values within the reference interval of 0.5–4.8 mU/L. TSH values of <0.1 mU/L and >4.8 mU/L are indicative of hyperthyroidism and hypothyroidism, respectively (see Table 15-2). Both diagnoses should be confirmed with assays of circulating thyroid hormone—free T_4(FT_4) or free T_3(FT_3)—with elevated results for hyperthyroidism and subnormal results for hypothyroidism. TSH values between 0.1 mU/L and 0.49 mU/L are considered borderline. The diagnosis should be confirmed with FT_4 or FT_3. If still equivocal, TRH stimulation testing should be performed. See "Thyrotropin-Releasing Hormone Stimulation" later in the chapter for more discussion.

✪ TABLE 15-2			
Sensitive TSH Assays Used for Evaluating Patients Who May Have Hyper- or Hypothyroidism			
<0.1 mU/L	0.1–0.49 mU/L	0.5–4.8 mU/L	>4.8 mU/L
hsTSH Assay*			
Hyperthyroid suspect	Borderline thyroid status	No thyroid dysfunction	Hypothyroid suspect
Confirm with FT_4 or FT_3	Do FT_4, FT_3, or TRH simulation	No further testing	Confirm with FT_4

*Reference interval = 0.5–4.8 mU/L.

TOTAL THYROXINE

In circulation, T_4 is extensively (99.97%) bound to the plasma proteins (TBG ~ 60–75%, TTR/TBPA ~ 15–30%, albumin ~10%). The earliest method measured the amount of protein-bound iodine in T_4. Serum total thyroxine (TT$_4$) may also be measured by competitive-protein-binding assay using TBG as the binder. Serum TT$_4$ is now routinely determined using chemiluminescent immunoassay technique. The bound T_4 is released from its carrier proteins with a blocking agent, 8-anilino-1-naphthalene sulfonic acid. The method is sensitive and amenable to automation. The reference interval for adult males is 4.6–10.5 μg/dL; for females, it is 5.5–11.0 μg/dL.

Since TT$_4$ concentration is dependent on changes in thyroid binding protein concentrations, which frequently occur unrelated to thyroid diseases, the measurement of TT$_4$ as a test of thyroid function has been replaced in several medical protocols by FT$_4$ measurement.

Free Thyroxine

Because FT$_4$ is independent of TBG concentration, it is more closely correlated with the thyroid status of the patient. The reference method for measuring FT$_4$ begins with equilibrium dialysis of a patient's specimen to which a known tracer amount of iodinated T_4 has been added.[21] The sample is assayed for TT$_4$ by immunoassay, and percentage FT$_4$ is calculated using equation 15-1. The procedure is accurate but technically demanding.[22]

$$\%FT_4 = TT_4 \times \% \text{ tracer } T_{4(\text{dialyzed})} \quad (15.1)$$

When T_4 is synthesized from the thyroid gland, it binds with the plasma protein just discussed to form T_4–protein complex. This reaction is reversible, but its equilibrium goes to the right. As a result, 99.97% of T_4 is protein bound, which is biologically inactive; only the remaining 0.03% is FT$_4$, the biologically active moiety. To accurately assay this minute amount of FT$_4$ [relative index (RI) = 0.9–2.3 ng/dL] was a challenge. Testing is now routinely performed by immunometric assay techniques using primarily chemiluminescent assays. As shown in the following equations, the antiserum (Ab) used in the assay to generate T_4—Ab and T_4*—Ab must interact with the unbound T_4 without disturbing the endogenously established bound–free equilibrium. FT$_4$ is increased in hyperthyroidism and decreased in hypothyroidism.

$$T_4* + Ab$$
$$+$$
$$\text{Protein} + T_4 \leftrightarrow \text{Protein} - T_4$$
$$\updownarrow$$
$$T_4 - Ab + T_4* - Ab$$

where

Ab = antibody

T_4* = labeled T_4

T_4 = serum or blood levels of T_4

☼ TABLE 15-3

Thyroid-Function Test Results in Various Clinical Conditions*

Clinical conditions	T_4	T_3	rT_3	FT_4	FT_4I	TSH
Graves' disease	I	I	NA	I	I	D, U
Toxic multinodular goiter	I	I	NA	I	I	D, U
Toxic adenoma	I	I	NA	I	I	D, U
T_3 toxicosis	N	I	NA	N	N	D, U
Nonthyroid disease	N, D	N, D	N, I	N, I	N, D	N
Primary hypothyroidism	D	D, N	NA	D	D	I

*I, increase; D, decrease; U, undetectable; NA, not applicable; N, normal

Free Thyroxine Index

The free thyroxine index (FT$_4$I) can be estimated using TT$_4$ and T_3 resin uptake (RT$_3$U). RT$_3$U is *not* synonymous with serum T_3. It measures the degree of unsaturation of TBG. To perform the test in the laboratory, radioactive T_3 binds to free TBG in serum, and the resin takes up the excess radioactive T_3.

FT$_4$I is calculated using equation 15-2 and is an indirect assessment of active thyroxine status. In hyperthyroidism, FT$_4$I is elevated (both TT$_4$ and RT$_3$U are increased); in hypothyroidism, FT$_4$I is decreased (both TT$_4$ and RT$_3$U are decreased). A normal FT$_4$I is expected in euthyroid patients with either a decreased TBG (low TT$_4$ and high RT$_3$U) or an increased TBG (high TT$_4$ and low RT$_3$U). Table 15-3 ☼ summarizes these results.

$$FT_4I = TT_4 \times RT_3U \quad (15.2)$$

SERUM TRIIODOTHYRONINE

Serum triiodothyronine (T_3) is determined by immunoassay using a specific antiserum that does not cross-react with T_4 and a blocking agent to release bound T_3 from carrier proteins. The reference interval for total T_3 varies with age; for adults (20–50 years old), it is 70–204 ng/dL. T_3 is a potent thyroid hormone. It is much less tightly bound to serum proteins than T_4 (0.3% of T_3 exists in the free state). T_3 measurement is helpful in confirming the diagnosis of hyperthyroidism. However, in patients with hyperthyroidism coexisting with nonthyroidal illness (NTI), serum T_3 levels may be within RI or low. T_3 levels are not useful in the diagnosis of hypothyroidism because they do not reflect glandular secretion and often are normal in hypothyroidism. T_3 levels are affected by age, starvation, illness, stress, and drugs.

✓ Checkpoint! 15–10

Briefly describe what is being measured or calculated for the following:

Total serum thyroxine

Total serum triiodothyronine

Free thyroxine

Free thyroxine index

THYROTROPIN-RELEASING HORMONE STIMULATION

In the thyrotropin-releasing hormone stimulation test, the patient is infused with TRH to produce a rise in TSH response that is at a maximum between 20 and 40 minutes. TRH levels revert toward baseline at 60 minutes. In Figure 15-16 ■, an exaggerated response is shown in patients with primary hypothyroidism with elevated baseline. For patients with hypothalamic hypothyroidism, basal TSH is inappropriately low for the degree of thyroid-hormone insufficiency, and the TSH response is small and delayed. Finally, in patients with pituitary hypothyroidism, the TSH response is flat, as in hyperthyroidism.[23]

The TRH stimulation test is useful in patients who show equivocal signs and symptoms of thyroid dysfunction and thyroid-function test results are not clearly diagnostic. Since the advent of the sensitive TSH assay, this procedure is less frequently used.

REVERSE T_3

Reverse T_3 is a metabolite with minimal metabolic activity produced by 5-deiodination of T_4. The RI for rT_3 in adults as measured by radioimmunoassay is 10–38 ng/dL. In certain clinical situations, rT_3 levels are found to vary reciprocal to T_3 (i.e., when T_3 is elevated, rT_3 is decreased; conversely, when T_3 is decreased, rT_3 is elevated). A case in point is in patients with NTI in whom T_3 levels are decreased in conjunc-

tion with elevated values of rT_3. Serum T_3 levels are decreased because of a block in the 5-deiodinase enzyme that converts T_4 to T_3 in peripheral tissues. Patients with hypothyroidism without other complications will have a lower than normal T_3 and rT_3. This test strategy is used clinically to differentiate NTI (T_3 and rT_3) from hypothyroidism (T_3 and rT_3).

THYROGLOBULIN ANTIBODIES

Thyroglobulin antibodies (TgAbs) are directed against the thyroglobulin protein located within the thyroid colloid. Several assays are available to estimate the amount of TgAb, including passive hemagglutination, immunoassays, and radioassays. Agglutination assays are still used but are quickly being replaced with immunoassays. The agglutination assays provide a Tg antibody titer; in normal patients, the titers are less than 1:100. Higher titers are seen in patients with Hashimoto's thyroiditis and Graves' disease; titers >1:1600 are common.

MICROSOMAL ANTIBODIES

Antimicrosomal antibodies are directed against a protein component of thyroid cell microsomes. Assays for the estimation of microsomal antibodies are similar to those for TgAb. The passive hemagglutination assays provides a titer for antibody activity, and for nondiseases individuals the titer is <1:100. In patients with Hashimoto's thyroiditis and Graves' disease, the titer is much higher.

The principal autoantigenic component of thyroid microsomes is the enzyme thyroid peroxidase. This enzyme is currently the preferred biomarker for clinical use in the management of patients with suspected autoimmune disease of the thyroid. Several immunoassays are available, and the reference cutoff point for currently used sensitive chemiluminescent immunoassays is <9 U/mL; a titer assays <1:100.

RADIOACTIVE IODINE UPTAKE

Because of the availability of sensitive and specific assays for thyroid hormones and TSH, radioactive iodine uptake (RAIU) is no longer recommended as a screening test for thyroid function. Although increased RAIU is indicative of hyperthyroidism, decreased RAIU is useful for establishing diagnosis of subacute thyroiditis and thyrotoxicosis factitia. It is also helpful in assessing the treatment dose of radioactive iodine-131 (^{131}I) needed for specific patients.

CLINICOPATHOLOGIC CORRELATIONS

Evaluation of the patient should include a thorough history and physical examination.

The most frequently encountered thyroid diseases relate to either the excessive or inadequate production of thyroid hormones, namely, hyperthyroidism and hypothyroidism. All types of thyroid pathology may be subtle in their initial

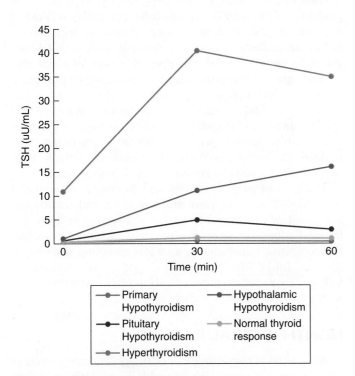

■ FIGURE 15-16 Typical responses for a TRH stimulation test where a quantity of TRH is given to a patient intravenously and serum TSH is measured at 0, 30 and 60 minutes.

clinical presentation, thus a high index of suspicion is required to make an early diagnosis. Thyroid-function tests are essential for ruling out or confirming the presumptive diagnoses of these conditions. When a sensitive serum TSH assay is used together with a valid serum FT_4 assay, a sensitive and specific assessment of thyroid status can usually be established from the general relation between the two hormones.

HYPOTHYROIDISM

Hypothyroidism is defined as a deficiency of thyroid activity. It is a common disorder that occurs in mild or severe forms in 2–15% of the population. Women are afflicted more often than men, and both sexes are affected more frequently with increasing age. Signs and symptoms include bradycardia, hoarseness, cold sensitivity, dry skin, and muscle weakness. Primary hypothyroidism is due to destruction or ablation of the thyroid gland, which fails to secrete an adequate amount of thyroid hormone. Secondary hypothyroidism results from pituitary or hypothalamic dysfunction when the thyroid gland is not stimulated to produce thyroid hormones. The most common cause of hypothyroidism is Hashimoto thyroiditis. This is an autoimmune disease associated with goiters, or enlargements of the thyroid gland. Several conditions can lead to the formation of a goiter, including biosynthetic defects, iodine deficiency, autoimmune disease, and nodular disease. In Hashimoto thyroiditis, a defect in hormone synthesis results in an increase in TSH production, which stimulates the growth of the thyroid gland.

Measurements of T_4 and TSH with the use of hsTSH assays are usually sufficient to arrive at the correct diagnosis. Serum TSH levels should be elevated in patients with primary hypothyroidism. In the absence of hypothalamic pituitary disease, illness, or drugs, TSH is an accurate indicator of thyroid-hormone status and the adequacy of thyroid-hormone replacement.

THYROID DISORDERS IN PREGNANCY

The most common thyroid problem, hypothyroidism, or a deficiency of T_4, affects 2–5% of pregnant women. It has been associated with diminished intellectual capacity and developmental delay in children and miscarriage, preterm delivery, preeclampsia, and placental abruption in mothers. There is also an ongoing debate among laboratorians, obstetricians, and endocrinologists over whether all pregnant women should be screened for TSH, which affects T_4 levels, and thyroid peroxidase antibody (TPOAb) or only those with existing disease or who are at high risk of disease.

Guidelines for managing thyroid disease in pregnancy published by the Endocrine Society do not recommend universal thyroid-disease screening for pregnant women.[24] The main reason is a lack of recommendation on what to do with patients having positive TPOAb and normal TSH. However, it is suggested that a new upper limit for the TSH reference range be established when running TSH tests for pregnant women because an existing new study indicates that a small TSH increase is a risk factor. Thus, a lowered TSH upper limit should help identify more hypothyroid women.

HYPERTHYROIDISM

Primary hyperthyroidism is far more frequent than secondary hyperthyroidism. Classical signs and symptoms include heat intolerance, tachycardia, weight loss, weakness, emotional liability, and tremor. The most common single cause is Graves' disease, an autoimmune disease caused by circulating antibodies to the TSH receptor. Other causes of hyperthyroidism include toxic adenoma and multinodular goiter. These are mostly benign tumors formed as a result of thyroid tissue secreting excessive amount of thyroid hormones and being unresponsive to normal feedback mechanism. Subacute thyroiditis is another cause of hyperthyroidism because of the inflammation of the thyroid gland.

TSH values from sensitive TSH assays are expected to be subnormal in conjunction with elevated T_4, FT_4, or FT_3. In a patient suspected of hyperthyroidism with normal or borderline high T_4 values, if an FT_4 test is unavailable, then T_3 measurement should be made to look for increased $T_3:T_4$ ratio. If still inconclusive, look for TSH unresponsiveness to TRH stimulation.

NEONATAL HYPOTHYROIDISM

Neonatal hypothyroidism is decreased thyroid-hormone production in a newborn. In rare cases, no thyroid hormone is produced. If the baby is born with the condition, it is called *congenital* hypothyroidism. Hypothyroidism in the newborn may be caused by a missing or abnormally developed thyroid gland, failure to stimulate the thyroid because of a defective pituitary gland, or defective or abnormal formation of thyroid hormones. Incomplete development of the thyroid is the most common defect and occurs in about 1 out of every 3,000 births. Females are affected twice as often as males. Early diagnosis is advantageous because most of the effects of hypothyroidism are easily reversible. Replacement therapy with thyroxine is the standard treatment of hypothyroidism.

The goal of early detection and treatment of neonatal hypothyroidism is to eliminate severe mental retardation associated with thyroid-hormone deficiency in early infancy. Nationwide screening programs are now available. Protocol usually includes T_4 and TSH assays on filter-paper disk. An initial filter-paper blood spot T_4 measurement is followed by a measurement of TSH in filter-paper specimens with low T_4 values.[25]

NONTHYROIDAL ILLNESS

Abnormalities of circulating TSH or thyroid-hormone levels in the absence of underlying thyroid disease may occur in patients with severe acute illnesses such as infections as well as with myocardial infarction, trauma, surgery, and malignancy.

Patients with NTI typically present with low serum T_3 and high levels of rT_3. Total T_4 and FT_4 will progressively diminish, with total T_4 showing a drastic decrease during the severe stages of the illness. The magnitude of the fall in T_3 usually correlates with the severity of the illness. Conversion of T_4 to T_3 via peripheral deiodination becomes impaired, leading to increased rT_3. These findings mimic those of hypothyroidism to a varying degree and are often called *euthyroid sick syndrome*. This low T_3 state is adaptive: It can be induced in normal individuals by fasting. To rule out hypothyroidism and to confirm NTI, rT_3, or TSH should be performed.[26]

The diagnosis of NTI depends on previous history of thyroid disease and the severity of thyroid-function test results, the time course of the patient's acute illness medications that may affect thyroid function or thyroid-hormone levels, and measurements of rT_3, FT_4, and TSH.

Treatment of T_4 replacement does not appear to help, but most authorities recommend monitoring the patient's thyroid-function tests during recovery, without administering thyroid hormone, unless there is historic or clinical evidence suggestive of hypothyroidism. Because clinical presentations and outcomes in NTI are highly variable, more clinical trials are needed to ascertain if intervention is beneficial.

SUBCLINICAL THYROID DISEASES

Subclinical hypothyroidism is defined as a serum TSH concentration above the statistically defined upper limit of the reference range when serum FT_4 concentration is within its reference range. Subclinical hyperthyroidism is defined as a serum TSH concentration below the statistically defined lower limit of the reference range when serum FT_4 and T_3 concentrations are within their reference ranges. Hence, subclinical hyperthyroid and hypothyroid diseases are both laboratory diagnoses. The prevalence of subclinical hypothyroidism is about 4–8%, and may be as high as 20% in women older than 60 years, whereas subclinical hyperthyroidism is approximately 2% of the population.

The enhanced sensitivity of TSH assays has allowed screening and detection of mild subclinical thyroid diseases. In subclinical hyperthyroidism (Figure 15-17 ■), suppressed TSH values are accompanied by T_3 and FT_4 values within RI, while in subclinical hypothyroidism, minimally increased TSH values are accompanied by T_3 and FT_4 values within RI.

Most national organizations recommend against routine screening of asymptomatic patients. Although subclinical hypothyroidism is associated with progression to overt disease, there is insufficient evidence to suggest that treatment is beneficial. Similarly, treatment is not recommended with subclinical hyperthyroidism, although subnormal levels of TSH (<0.1 μU/mL) are associated with progression to overt hyperthyroidism, atrial fibrillation, reduced bone mineral density, and cardiac dysfunction.[27,28,29]

SUMMARY

The human body's endocrinology systems are extremely complex and highly interactive. There are literally hundreds and diseases, conditions, and syndromes associated with their dysfunction. Parallel to this, the number of clinical laboratory and nonlaboratory diagnostic tests available to assess endocrine dysfunctions is just as numerous. Therefore, because one textbook cannot cover such a voluminous amount of material, this chapter has provided an overview of topics relevant to students of clinical laboratory science. Student are encouraged to use reference clinical chemistry books to augment and increase their knowledge base of the topics presented in this chapter.

The chapter specifically examined the endocrine glands and appropriate tests commonly encountered in most hospital-based laboratories, physician office laboratories, reference clinical laboratories, and similar facilities. Several major endocrine organs and glands have been discussed, including the pituitary, the thyroid, adrenals, and reproductive. Other endocrine organs and glands not include here may be discussed in other chapters of this textbook (e.g., parathyroid, gastric).

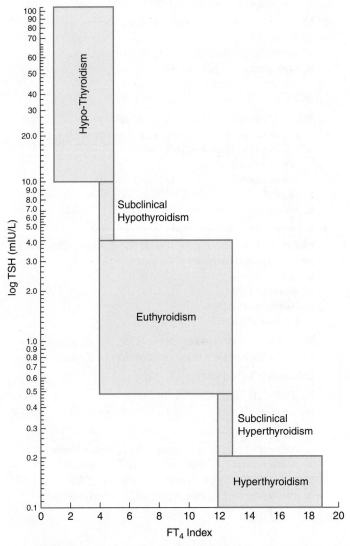

■ FIGURE 15-17 Sub-clinical assessment of thyroid conditions based upon serum TSH concentrations and calculated FT_4 index.

REVIEW QUESTIONS

LEVEL I

1. Acromegaly is associated with: (Objective 11)
   a. growth hormone deficiency.
   b. growth hormone excess.
   c. hypercorticolism.
   d. hypeprolactinemia.

2. The adrenal cortex releases: (Objective 8)
   a. corticosteroids.
   b. mineralosteroids.
   c. androgens.
   d. All of the above.

3. Which of the following class of compounds are derived from tyrosine and includes epinephrine, norepinephrine, and dopamine? (Objective 2)
   a. steroids
   b. androgens
   c. catecholamines
   d. estrogens

4. Thyroid hormones are derived from which of the following amino acids? (Objective 2)
   a. phenylalanine
   b. methionine
   c. tyrosine
   d. histidine

5. Serum thyroid-stimulating hormone (TSH) levels are decreased in: (Objective 11)
   a. primary hyperthyroidism.
   b. primary hypothyroidism.
   c. primary parathyroidism.
   d. hyperpituitarism.

6. TSH is secreted by the: (Objective 4)
   a. anterior pituitary.
   b. posterior pituitary.
   c. hypothalamus.
   d. thyroid gland.

7. Which of the following pituitary hormones regulates renal free water excretion to maintain homeostasis? (Objective 12)
   a. oxytocin
   b. antidiuretic hormone (ADH)
   c. glucagon
   d. cortisol

8. Where in the human body are the adrenal glands located? (Objective 6)
   a. on the anterior section of the esophagus
   b. at the base of the brain
   c. on top of the kidneys
   d. near the thymus gland

LEVEL II

1. A patient with suspected nonthyroidal illness (NTI) would present with which of the following laboratory results? (Objective 8)
   a. increased rT_3, decreased serum T_3, and increased TT_4 as the disease progresses
   b. decreased rT_3, decreased serum T_3, and increased TT_4 as the disease progresses
   c. increased rT_3, increased serum T_3, and increased TT_4 as the disease progresses
   d. increased rT_3, decreased serum T_3, and decreased TT_4 as the disease progresses

2. Free thyroxine assays measures: (Objective 11)
   a. total thyroxine levels.
   b. only the bound thyroxine level.
   c. only the unbound thyroxine level.
   d. the sum of the bound and unbound thyroxine.

3. Metabolites of the adrenal medulla hormones include: (Objective 6)
   a. methoxytryramine dihydroxyphenyl acetic acid.
   b. vanillymandelic acid (VMA).
   c. homovanillic acid (HVA).
   d. All of the above.

4. What would the expected serum cortisol level be after a dexamethasone suppression test was performed on a patient with Cushing's disease? (Objective 4)
   a. above normal
   b. below normal
   c. normal
   d. no change

5. A patient with Graves' disease would present with which of the following laboratory results for a serum TSH value derived by an immunoassay? (Objective 8)
   a. normal TSH
   b. elevated TSH
   c. below normal TSH
   d. None of the above.

6. What would be the predicted results for the following laboratory tests for a patient with hypothyroidism? (Objective 11)

   Laboratory Tests (serum)

   (1) Total T_4 (TT_4)

   (2) FT_4

   (3) TSH
   a. below normal TT_4, below normal FT_4, below normal TSH
   b. below normal TT_4, elevated normal FT_4, elevated TSH
   c. elevated TT_4, below normal FT_4, elevated TSH
   d. below normal TT_4, below normal FT_4, elevated TSH

REVIEW QUESTIONS (continued)

LEVEL I

9. Tetraiodothyronine describes which thyroid hormone? (Objective 7)
   a. T_4 (thyroxine)
   b. T_3
   c. MIT
   d. reverse T_3

10. The principle method used to measure hormone levels in blood in most clinical laboratories is: (Objective 13)
    a. gas chromatography
    b. thin layer chromatography
    c. electrophoresis
    d. immunoassays

11. All of the following proteins transports thyroid hormones *except:* (Objective 5)
    a. thyroxin-binding globulin
    b. gamma globulin
    c. albumin
    d. thyroxine-binding pre-albumin

12. What would the expected serum cortisol level be after an ACTH stimulation test is performed on a patient with Addison's disease? (Objective 11)
    a. above normal
    b. below normal
    c. normal
    d. no change

LEVEL II

7. Cushing's syndrome is characterized by: (Objective 3)
   a. excess secretion of pituitary ACTH.
   b. adrenal insufficiency.
   c. corticosteroid excess (in blood).
   d. low plasma levels of cortisol.

8. Pheochromocytoma is a benign or malignant tumor arising from: (Objective 7)
   a. bile caniculi.
   b. trophoblastic cells.
   c. neurochromaffin cells in the adrenal medulla.
   d. follicular cells of the thyroid.

9. Which of the following laboratory tests would be appropriate for a patient with a neuroblastoma? (Objective 7)
   a. TT_4, FT_4, and TSH
   b. cortisol and ACTH
   c. urinary VMA and HVA
   d. total serum testosterone

PEARSON myhealthprofessionskit™

Use this address to access the interactive Companion Website created for this textbook. Simply select "Clinical Laboratory Science" from the choice of disciplines. Find this book and log in using your user name and password.

REFERENCES

1. Hershman JM. Principles of clinical endocrinology. In Hershman JM (Ed.), *Endocrine pathophysiology: A patient-oriented approach* (Philadelphia: Lea & Febiger, 1977): 1–2.

2. Demers LM, Vance ML. Pituitary function. In Burtis CA, Ashwood ER, Bruns DE (Eds.), *Tietz textbook of clinical chemistry and molecular diagnostics*, 4th ed. (St Louis: Saunders, 2006): 1992–1993.

3. Ibid., 1994–1996.

4. Pudek MR. Adrenal hormones and hypertension. In Kaplan LA, Pesce AJ, Kazmierczak SC (Eds.), *Clinical chemistry: Theory, analysis, correlation*, 4th ed. (St. Louis: Mosby, 2003): 881–882.

5. Rosano TG, Eisenhofer G, Whitley RJ. Catecholamines and serotonin. In Burtis CA, Ashwood ER, Bruns DE (Eds.), *Tietz textbook of clinical chemistry and molecular diagnostics*, 4th ed. (St Louis: Saunders, 2006): 1042–1055.

6. Demers LM. The adrenal cortex. In Burtis CA, Ashwood ER, Bruns DE (Eds.), *Tietz textbook of clinical chemistry and molecular diagnostics*, 4th ed. (St Louis: Saunders, 2006): 2012–2014.

7. Sizonenko PC, Paunier L. Hormonal changes in puberty: III. Correlation of plasma dehydroepiandrosterone, testosterone, FSH and LH with stages of puberty and bone age in normal boys and girls and in patients with Addison's disease or hypogonadism or with premature or late adrenarche. *J Clin Endocrinol Metab* (1975) 41: 894–904.

8. Demers LM. The adrenal cortex. In Burtis CA, Ashwood ER, Bruns DE (Eds.), *Tietz textbook of clinical chemistry and molecular diagnostics*, 4th ed. (St Louis: Saunders, 2006): 2018–2019.

9. Ibid., 2028–2029.

10. Webster RA. Reproductive function and pregnancy. In McPherson RA, Pincus MR (Eds.), *Henry's clinical diagnosis and management by laboratory methods*, 21st ed. (Philadelphia: Saunders Elsevier, 2007): 366–367.

11. Thompson DF, Carter JR. Drug-induced gynecomastia. *Pharmacotherapy* (1993) 13, 1: 37–45.

12. Carr BR. Disorders of the ovary and female reproductive tract. In Williams RH, Foster DW, Kronenberg HM, et al. (Eds.), *William*

textbook of endocrinology, 9th ed. (Philadelphia: WB Saunders, 1998): 751–817.

13. Lewis V. Polycystic ovary syndrome: A diagnostic challenge. *Obstet Gynecol Clin North Am* (2001) 28: 1–20.

14. Haymond S, Gronowski AM. Reproductive related disorders. In Burtis CA, Ashwood ER, Bruns DE (Eds.), *Tietz textbook of clinical chemistry and molecular diagnostics*, 4th ed. (St. Louis: Saunders, 2006): 2097–2138.

15. Nanjee MN, Wheeler MJ. Plasma free testosterone: Is an index sufficient? *Ann Clin Biochem* (1985) 22 (pt. 4): 387–390.

16. Vermeulen A, Stoica T, Verdonck L. The apparent free testosterone concentrations: An index of adrogenicity. *J Clin Endocrinol Metab* (1971) 33: 759–767.

17. Demers LM, Spencer C. The thyroid: Pathophysiology and thyroid function testing. In Burtis CA, Ashwood ER, Bruns DE (Eds.), *Tietz textbook of clinical chemistry and molecular diagnostics*, 4th ed. (St Louis: Saunders, 2006): 2053–2054.

18. Barrett EJ. The thyroid gland. In Boron WF, Boulpaep EL (Eds.), *Medical physiology: A cellular and molecular approach*, 2nd ed. (Philadelphia: Saunders, 2009): 1046–1047.

19. Demers L, Spencer C. The thyroid: Pathophysiology and thyroid function testing. In Burtis CA, Ashwood ER, Bruns DE (Eds.), *Tietz textbook of clinical chemistry and molecular diagnostics*, 4th ed. (St Louis: Saunders, 2006): 2073–2075.

20. Spencer CA. Clinical utility and cost-effectiveness of sensitive thyrotropin assays in ambulatory and hospitalized patients. *Mayo Clin Proc* (1988) 63: 1214–1218.

21. Demers L, Spencer C. The thyroid: pathophysiology and thyroid function testing. In Burtis CA, Ashwood ER, Bruns DE (Eds.), *Tietz textbook of clinical chemistry and molecular diagnostics*, 4th ed. (St Louis: Saunders, 2006): 2075–2076.

22. Nelson J, Tomei R. Direct determination of free thyroxine in undiluted serum by equilibrium dialysis/radioimmunoassay. *Clin Chem* (1988) 34: 1737–1744.

23. Marshall WJ, Bangert SK. *Clinical chemistry*, 6th ed. (New York: Mosby–Elsevier, 2008): 168–169.

24. Abalovich M, Amino N, Barbour LA, et al. Management of thyroid dysfunction during pregnancy and postpartum: An Endocrine Society Clinical Practice Guideline. *J Clin Endo & Metab* (2007) 92 (8 suppl): s1–s47.

25. U.S. Preventive Services Task Force. Screening for thyroid disease: Recommendation statement. *Ann Intern Med* (2004) 140, 2: 125–127.

26. Brent GA, Hershman JM. Effects of nonthyroidal illness on thyroid function tests. In Van Middleworth L (Ed.), *The thyroid gland: A practical clinical treatise* (Chicago: Year Book Medical, 1986): 86–87.

27. Surks MI, Ortiz E, Daniels GH, et al. Subclinical thyroid disease: Scientific review and guidelines for diagnosis and management. *JAMA* (2004) 291, 2: 228–238.

28. Surks MI. Subclinical thyroid dysfunction: A joint statement on management from the American Association of Clinical Endocrinologists, the American Thyroid Association, and the Endocrine Society. *J Clin Endocrinol Metabolism* (2005) 90: 586–587.

29. Demers LM, Spencer CA. Laboratory medicine practice guidelines: Laboratory support for the diagnosis and monitoring of thyroid disease. *Clin Endocrin* (2003) 58: 138–140.

16

Gastrointestinal Function

■ OBJECTIVES—LEVEL I

Following successful completion of this chapter, the learner will be able to:

1. Review the gross anatomy of the gastrointestinial (GI) tract from the mouth to the anus.

2. Outline the functions of each significant component of the GI tract.

3. Identify three examples of GI regulatory peptides.

4. Define the following terms: *peptic ulcer*, *gastrinoma*, and *protein-losing enteropathy*.

5. Explain the principal pathological condition associated with each of the following GI tract disorders: Zollinger–Ellison's syndrome, peptic ulcer, celiac disease, protein-losing enteropathy, lactase deficiency, and carcinoid tumors.

6. Identify five nonclinical laboratory diagnostic tests or procedures used to assess patients with disorders of the GI tract.

■ OBJECTIVES—LEVEL II

Following successful completion of this chapter, the learner will be able to:

1. Contrast the chemistry of gastrin, secretin, cholecystokinin, and vasoactive intestinal polypeptide.

2. Explain the function of the following GI regulatory peptides: cholecystokinin, gastrin, secretin, vasoactive intestinal polypeptide, and glucose-dependent insulinotropic peptide.

3. Predict whether the fluid concentrations of the following analytes will be decreased, normal, or increased for each of the associated disorders:

Analyte	Disorder
Gastrin	Zollinger–Ellison's syndrome
Basal acid output	Peptic ulcer disease
Serum albumin	Protein-losing enteropathy
Glucose	Lactase deficiency

4. Explain the principal laboratory tests available to assess diseases of the GI tract.

5. Discuss the pathology of GI tract disorders.

KEY TERMS

Bile acids

Carcinoid tumors

Celiac disease

Cholecystokinin

Crohn's disease

Enterochromaffin cells

Gastrin

Gastrinoma

G cells

Helicobacter pylori

Lactose tolerance test

Neuroendocrine tumor (NET)

Peptic ulcer

Secretin

Vasoactive intestinal
   polypeptide

Zollinger–Ellison's syndrome

 A CASE IN POINT

A 37-year-old women developed renal colic at the age of 23 years. One year later she was diagnosed with hyperparathyroidism, and a three-gland parathyroidectomy was performed when she was 25 years old. In her 25th year she was diagnosed with multiple endocrine neoplasia type 1 (MEN1). It was later discovered that she had a family history of MEN1. One year later she began to develop symptoms that included abdominal pain, diarrhea, nausea, vomiting, and weight loss.

Clinical Laboratory Results

Analytes	Results	Reference interval/ cutoff value
Basal acid output (BAO)	56 mEq/hr	<10 mEq/hr
Gastrin (fasting serum)	555 pg/mL	<100 pg/mL

Nonclinical diagnostic tests:

Upper GI endoscopy procedure, biopsy of a duodenal nodule, adnominal imaging studies, and somatostatin receptor scintigraphy.

Results of these procedures:

All of these procedures revealed duodenal gastrinoma and pancreatic lesions.

Issues and Questions to Consider

1. Based on the history provided, identify clinical laboratory tests that may provide valuable information and help clinicians determine a correct diagnosis.

2. Indicate other nonclinical laboratory tests or procedures that, in combination with the clinical laboratory tests, will provide clinicians with enough information to make the correct diagnosis.

3. What is the most probable diagnosis?

WHAT'S AHEAD

▶ Review of the anatomy and function of the gastrointestinal tract

▶ Description of gastrointestinal regulatory peptides

▶ Discussion of gastrointestinal disorders

▶ Overview of clinical laboratory assessment of gastrointestinal disorders

---

▶ ANATOMY

The major organs of the gastrointestinal tract (GI) include the stomach, the small and large intestines, the pancreas, and the gall bladder. Chapter 17 of this textbook is devoted to the pancreas; therefore, in this chapter only succinct references will be made to this organ where appropriate.

The GI tract is approximately 10 meters in length and extends from the mouth to the anus. Separating the organs mentioned above are specialized, independently controlled sphincters that serve to compartmentalize the gut. The gut wall is arranged in layers that contribute to the functional activities in each area. Layers of mucosal cells serve as a barrier to luminal contents or as a site of transfer of fluids and nutrients. Gut smooth muscle facilitates propulsion from one area to the next.

The stomach is composed of three major regions: the cardiac, the body, and the pyloric regions, as shown in Figure 16-1 ■. In the cardiac region, which includes the fundus, specialized epithelial cells secrete mucus and group II pepsinogens. Also within the fundus are several types of endocrine-secreting cells.

The region identified as the body of the stomach contains a diverse population of cells, including (1) surface epithelial cells, which secrete mucus; (2) parietal cells, which secrete hydrochloric acid and intrinsic factor (IF); (3) the chief cells (also zymogen or peptic cells), which secrete groups I and II pepsinogen; (4) enterochromaffin cells, which secrete serotonin; and (5) other types of endocrine-secreting cells. The pyloric region consists of the antrum, the pyloric canal, and the sphincter. Cells within the pyloric region secrete mucus, group II pepsinogen, serotonin, gastrin, and other hormones but no hydrochloric acid.

The requirements of the gut and body are facilitated through interactions with other organ systems. The panceaticobiliary duct delivers bile and enzymes into the duodenum. An abundant vascular supply is affected by GI tract activity. Nerves located in the gut wall assisting in controlling propulsion and fluid regulation. Lymphatic conduits facilitate gut immune activities. Involuntary control activities for each region are modulated via extrinsic neural inputs.

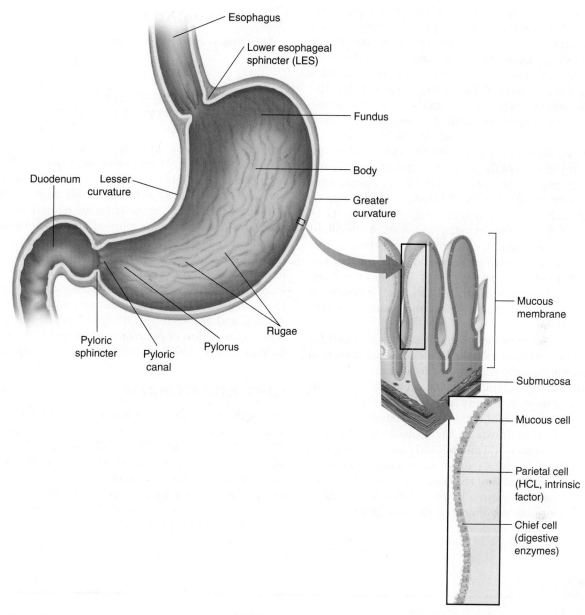

Esophagus

Lower esophageal
sphincter (LES)

Fundus

Body

Greater
curvature

Duodenum   Lesser
curvature

Mucous
membrane

Submucosa

Mucous cell

Parietal cell
(HCL, intrinsic
factor)

Chief cell
(digestive
enzymes)

Pyloric
sphincter

Pyloric
canal

Pylorus

Rugae

■ **FIGURE 16-1**   Anatomy of the stomach and intestines.

▶ FUNCTIONS OF THE GASTROINTESTINAL TRACT

The GI tract is designed to serve two primary functions: assimilation of nutrients and elimination of waste. The process begins in the mouth, where food is mixed with salivary amylase and delivered to the GI tract. Processed foods (bolus) are propelled by the esophagus into the stomach, and the lower esophageal sphincter prevents oral reflux of gastric contents. The esophageal mucosa has a protective squamous tissue that does not permit a large amount of diffusion or absorption. Esophageal propulsions are directed away from the mouth

and are in concert with relaxation of the upper and lower esophageal sphincter upon swallowing.

In the stomach, continuous processing of the food, including mixing with pepsin and acid, helps reduce the solid components into a homogenous mixture. Gastric acid also sterilizes the upper gut. The proximal stomach serves a storage function by relaxing to accommodate the meal. In the distal portion of the stomach, phasic contractions propel solid food residue against the pylorus, where it is repeatedly propelled proximally for further mixing before it is emptied into the duodenum. In the final phase of stomach digestion, intrinsic factor is secreted for vitamin B_{12} absorption.

Food entering the intestines passes through the intestinal mucosa, which is made of villi that provide a large surface area for absorption of nutrients. Absorption of nutrients is facilitated by specialized enzymes and transporter substances. Pancreatic juices and bile enter the proximal duodenum to facilitate digestion. Pancreatic juices contain the enzyme for carbohydrate, protein, and fat digestion along with bicarbonate to optimize the pH and promote activation of these enzymes. Digestion of lipids in the intestine is accomplished by secretion of bile from the gall bladder. The proximal intestine rapidly absorbs nutrient breakdown products and most minerals, whereas the ileum absorbs vitamin B_{12} and bile acids. The small intestine plays a role in waste elimination by eliminating bile, (which contains byproducts of erythrocyte degradation, toxins, metabolized and unaltered medications, and cholesterol). Indigestible food substances and sloughed enterocytes are moved into the colon by intestinal motor functions. The ileocecal junction represents the terminal end of the small intestine. A sphincter prevents the coloileal reflux and maintains small-intestinal sterility.

Waste material enters the colon, where it is prepared for controlled elimination. Stool is dehydrated by colonic mucosa cells, which significantly decrease the fecal volume that will exit the body via the rectum. Fermentation of undigested carbohydrates and short-chain fatty acids occurs in the lumen of the colon by a large mass of bacterial colonization. This stage of digestion of food stuffs requires days, whereas time through the esophagus is seconds and minutes and passage through the stomach and small intestine requires a few hours. Motor control in the colon facilities slow fecal desiccation, where mixing absorption of fluids progresses. In the distal colon, peristaltic contractions cause the movement of digested material to be expelled in the stool.

✓ Checkpoint 16–1

Identify two primary functions of the GI tract.

▶ GASTROINTESTINAL REGULATORY PEPTIDES

The gut is the largest endocrine organ in the body and is a major target for many hormones that are released locally and from other sites. Table 16-1 ✪ lists examples of hormones that are either released by the gut or act on it. GI regulatory hormones such as cholecystokinin are released from the gut mucosa whereas others are released from other tissue sources (e,g,m pancreatic islets). Several of the peptides (e.g., vasoactive intestinal polypeptide [VIP] and somatostatin) are found in enteric nerves and the central nervous systme (CNS) and serve as neuroedocrine control substances of the gut. Overall the GI regulatory peptides influence motility, secretion, digestion, and absorption in the gut. They also regulate bile flow and secretion of pancreatic hormones and affect vascular wall tonicity, blood pressure, and cardiac output.

CHOLECYSTOKININ

Cholecystokinin (CCK) is a linear polypeptide with multiple molecular forms. Examples of these forms include CCK-33, CCK-8, CCK-39, and CCK-58. All molecular forms of CCK consist of five C-terminal amino acid that are identical to those of gastrin and are necessary, together with a sulfate tyrosyl residue, for physiological activity. Preprocholecystokinin, the precursor compound for all forms of CCK, is a 115-amino-acid-long polypeptide.[1]

✪ TABLE 16-1

Characteristics of GI regulatory peptides

GI Peptide	Molecular weight/number of amino acids	Principal location	Principal actions
Cholecystokinin	3918/33	Duodenum and jejunum Enteric nerves	Stimulates gallbladder contraction, intestinal motility, secretion of pancreatic enzymes, insulin glucagon
Secretin	3056/27	Duodenum and jejunum	Stimulates pancreatic secretion of bicarbonate, enzymes and insulin; reduces gastric and duodenal motility; inhibits gastrin release and gastric acid secretion
Gastrin	Depends on which form	Gastric antrum and duodenum	Stimulate the secretion of gastric acid, pepsinogen, intrinsic factor, and secretin; stimulate intestinal mucosal growth and increase gastric and intestinal motility
Vasoactive intestinal polypeptide	3326/28	Enteric nerves	Relaxes smooth muscles of gut, circulatory system, and genitourinary system; increases water and electrolyte secretion; stimulates the release of hormones from pancreas, gut, and hypothalamus
Glucose-dependent insulinotropic peptide	4976/42	Duodenum and jejunum	Stimulates insulin release; inhibits gastric acid, pepsin, and gastrin secretion; reduces gastric and intestinal motility; increases fluid and electrolyte secretion form small intestine

Cholecystokinin resides in the I cells of the upper small intestinal mucosa. Release of CCK from cells is caused by several factors, including the presence of mixtures of polypeptides, amino acids (e.g., tryptophan and phenylalanine), fatty acids with chains of nine or more carbons, and gastric acid entering the duodenum. Thus CCK concentrations increase following ingestion of a mixed meal containing proteins, carbohydrates, and lipids.

The physiological functions of CCK are numerous. It regulates gallbladder contractions and increases small intestinal motility. Because the structure of CCK is similar to gastrin, it has a mild stimulatory effect on gastric HCL and pepsinogen secretion, motility, and pancreatic bicarbonate secretion. Other functions include stimulation of pancreatic growth, relaxation of the sphincter of Oddi, and stimulation of secretion from the duodenal glands. Also, release of CCK from the GI tract aids in regulating an individual's appetite.

GASTRIN

Several molecular forms of gastrin exist in blood and tissues. They include big gastrin (G34), little gastrin (G17), mini gastrin (G-14), and smaller amounts of G-71, G-52, and G-6. Gastrin is formed after cleavage of a single precursor, preprogastrin, which contains 101 amino acids.[1]

The endocrine cells (**G cells**) of the antral mucosa are the primary producers and storage site of gastrin. Secondary sites include G cells of the proximal duodenum and delta cells of the pancreatic islets. Gastrin that is released by G cells enters the circulatory system and is transported through the liver to the parietal cells of the fundus of the stomach, where it stimulates the secretion of gastric acid. In the gastric mucosa, gastrin serves to promote the secretion of gastric pepsionogens and IF. It also stimulates release of secretin by the small intestinal mucosa, as well as secretion of pancreatic bicarbonate, enzymes, and hepatic bile.

Gastrin release is stimulated in response to antral distention and by amino acids, peptides, and polypeptides from partially digested proteins in the stomach. Other stimuli of gastrin release are alcohol, caffeine, insulin-induced hypoglycemia, infusion of calcium, and vagal stimulation caused by smelling, tasting, chewing, and swallowing food.

SECRETIN

Secretin is a linear polypeptide containing 27 amino acids. Its structure is similar to glucagon, VIP, glucose-dependent insulinotropic peptide (GIP), peptide histidine-isoleucine (PHI), and growth hormone-releasing hormone (GHRH). Only the intact polypeptide form of secretin is biologically active.[1]

The mucosal granular S cells located primarily in the duodenum but also throughout the small intestine contain the majority of secretin that is released to aid in digestion. Release of secretin occurs when gastric HCL comes into contact with S cells. This interaction is modulated based on the acidity of the gastric fluids that come into contact with S cells.

A principal physiological role of secretin is stimulating the pancreas to secrete an increased volume of juice with high bicarbonate contents. Other roles include stimulating bicarbonate and water secretion from the liver and other glands; changing gallbladder concentration and increasing hepatic bile flow; reducing stimulation of insulin secretion; promoting active stimulation of parathyroid hormone (PTH); releasing pancreatic enzymes and pepsinogen by the chief cells of the stomach; reducing gastric and duodenal motility; and inhibiting normal gastrin secretion and thus gastric acid secretion.

VASOACTIVE INTESTINAL POLYPEPTIDE

Vasoactive intestinal polypeptide is a linear polypeptide consisting of 28 amino acids and is structurally similar to secretin, GIP, and glucagon. The principal source of VIP is the nervous system and gut, but it is also present throughout the body. However, VIP is not found in the mucosal endocrine cells of the GI tract, as are secretin and other GI hormones. It is believed that VIP is a neurotransmitter located in peripheral and central nervous tissue and that VIP-containing nerve fibers are found throughout the GI tract.[1]

The mechanism that results in the release of VIP into circulation is not well understood. It is not known whether VIP is released during digestion, but its secretion is increased by vagal stimulation. The half-life of VIP is about one minute, and most of the hormone is inactivated by a single passage through the liver.

Physiological actions of VIP are shared with other similar polypeptide hormones. As a neurotransmitter it causes vasodilatation and relaxation of the smooth muscles of the circulatory and genitourinary systems and the gut. Other actions of VIP include increasing water and electrolyte secretion from the pancreas and gut; releasing hormone from the pancreas, gut, and hypothalamus; inhibiting gastrin and gastric acid secretion; and stimulating lipolysis, glycolysis, and bile flow. The duration of action for VIP is short because of its rapid degradation.

GLUCOSE-DEPENDENT INSULINOTROPIC PEPTIDE (GASTRIC INHIBITORY POLYPEPTIDE)

Glucose-dependent insulinotropic peptide (GIP) is a linear peptide consisting of 42 amino acids. The N terminus closely resembles glucagon and secretin, but the C terminus (17 amino acids) is not common to any other known intestinal hormone.

In the body GIP is produced and released by K cells located in the duodenal and jejuna mucosa. Plasma levels of GIP are increased by oral administration of glucose and triacylglycerols and by intraduodenal infusion of solutions containing a mixture of amino acids. Food substances will only increase stimulation of GIP if they are absorbed by the intestinal mucosa.

The physiological action of GIP includes the following:

- Stimulating insulin secretion in the presence of hyperglycemia

- Increasing intestinal fluid and electrolyte secretion

- In supraphysiological concentrations, inhibiting gastric acid, pepsin, and gastrin secretions.

The insulinotropic effect of GIP is presumed to be the most significant function, and thus the hormone is referred to as "glucose-dependent insulinotropic peptide" to denote the more accurate description of its physiological action.[1]

 ## Checkpoint 16–2

Identify five examples of how GI regulatory peptides influence GI tract functions.

► GASTROINTESTINAL TRACT DISORDERS

Gastrointestinal tract diseases or disorders occur due to abnormalities within or outside the gut and range in severity from conditions that produce mild symptoms and no long-term morbidity to those with serious symptoms and adverse outcomes. GI tract diseases can be isolated to a single organ or manifest themselves at several body sites. These diseases are characterized by alterations in nutrient assimilation, waste evacuation, or in the activity supporting these main functions.

ZOLLINGER–ELLISON'S SYNDROME

Impaired digestion and absorption include several conditions associated with the stomach, intestine, biliary tree, and pancreas. Gastric hypersecretory conditions such as **Zollinger–Ellison's syndrome (ZES)** fall within this grouping.

Zollinger–Ellison described a syndrome in 1955 that was characterized by fulminate peptic ulcers, large amounts of gastric hypersecretion, and non–β-islet-cell tumors of the pancreas. Other endocrinopathies may be present, including hypergastrinemia, diarrhea, and steatorrrhea.

Gastrinomas are a causative agent of ZES. A **gastrinoma** is a neuroendocrine tumor (NET) that secretes gastrin, and the hypergastrinemia that develops causes gastric acid hypersecretion. The chronic gastric acid hypersecretion leads to growth of the gastric mucosa with a concomitant increase in the number of parietal cells and proliferation of gastric hydrochloric acid cells. The excess gastric acid usually results in peptic ulcer diseases (PUDs) and diarrhea.[2]

About 25% of gastrinomas are part of the multiple endocrine neoplasia type 1 syndrome (MEN-1). MEN-1 is an autosomal dominant predisposition to tumors of the parathyroid glands, anterior pituitary, and pancreatic islet cells. This syndrome is characterized by tumors or hyperplasia in pancreatic islets, parathyroids, and pituitary glands.[3]

PEPTIC ULCER

Peptic ulcer is a disorder in which there is a defect in the gastrointestinal mucosa that extends through the musculais mucosa. Peptic ulcers persist as a function of the acid or peptic activity in gastric juice. PUD is an important cause of morbidity in humans.

Peptic ulcer disease is associated with two factors: *Helicobacter pylori* infection and the consumption of nonsteroidal anti-inflammatory drugs (NSAIDs) such as aspirin. A significant number of patients with PUD are infected with *H. pylori*. *H. pylori* infection predominantly affects the gastric mucosa, with the antrum usually being the most densely colonized areas. The pathogenesis of ulcer formation involves the following:[3]

- Increased gastric acid secretion

- Gastric metaplasia

- Immune response

- Mucosal defense mechanisms

An acute *H. pylori* infection induces a short period of hypochlorhydria, whereas in chronic infections the body tends to increase basal acid output (BAO). *H. pylori* eradication reduces basal and stimulated acid output significantly.

Gastric metaplasia refers to the presence of gastric epithelium in the first portion of the duodenum. This atypical condition may be a response of the mucosa to excessive acid exposure. In addition to acid hypersecretion, impaired duodenal bicarbonate secretion induced by *H. pylori* may also contribute to the low duodenal luminal pH. *H. pylori* infection in areas of gastric metaplasia may weaken the mucosa, making it more susceptible to acid injury.

The immune response in PUD is brought about by the presence *H. pylori* and may play a role in peptic ulcer formation. This response includes increased production of inflammatory cytokines (e.g., interleukins-1, -6, and -8 (IL-1, IL-6, IL-8), and tissue necrosis factor-alpha. *H. pylori* also stimulate a β-cell response consisting of immunoglobulins G and A (IgG and IgA) antibodies, which are produced locally in the gastroduodenal mucosa and systemically.

H. pylori may downregulate several principal mucosal defense factors, examples of which are as follows:[3]

- Decreased amounts of epidermal growth factor (EGF) and transforming growth factor-alpha (TGF-α) inhibit gastric acid production.

- *H. pylori* may decrease the amount of mucosal bicarbonate production.

- Increased release of proteases from *H. pylori* tend to degrade normally protective mucosal glycoproteins.

Patients ingesting large amounts of NSAIDs can harm the gastric mucosa. The mechanism for many of the NSAID compounds occurs because they are carboxylic acid derivatives. As a result, they are not ionized at the acidic pH found in the gastric lumen and therefore can be absorbed across the gastric mucosa. When the drug moves from the acidic environment of the gastric lumen into the pH-neutral mucosa, it is ionized and trapped momentarily in epithelial cells, where it can damage these cells.

 Checkpoint 16–3

What two factors are associated with peptic ulcers?

LACTASE DEFICIENCY

A common intestinal maldigestion syndrome is lactase deficiency. In this condition there is an increased production of gas and diarrhea after consumption of dairy product but there is no adverse effect on survival. Congenital lactase deficiency is a rare disorder in which lactase activity in the mucosa is low or undetectable at birth. Symptoms develop as soon as milk is taken; stool (feces) samples will have a low pH and contain glucose produced by bacterial action on undigested lactose. A definitive diagnosis is usually deferred until after maturation of the lactase synthesis system has occurred. Differential diagnosis requires performance of an oral glucose tolerance test in conjunction with the lactose tolerance test.

Acquired lactase deficiency occurs later in life and is characterized by a diminished expression of enzyme activity. Symptoms include abdominal discomfort, bloating, and/or diarrhea after consumption of one or two glasses of milk or a large portion of ice cream or yogurt.

 Checkpoint 16–4

Which sugar is found in the feces of a patient who has lactase deficiency?

BILE SALT MALABSORPTION

Bile salt malabsorption is a disorder that leads to chronic diarrhea when there is ileal disease (e.g., Crohn's disease) or after resection of the terminal ileum. It may also occur following cholecystectomy or irritable bowel syndrome.

Bile acids (e.g., cholic acid and tauracholic acid) are synthesized in the liver and flow into the lumen of the small bowel via the gallbladder. Bile acids exist as taurine or glycine conjugates (salts). Their major functions are to act as surface-active agents (forming micelles) and to facilitate digestion of triglycerides and absorption of cholesterol and fat-soluble vitamins. Most bile salts are reabsorbed in the terminal ileum, with little absorption occurring in the proximal small bowel. Bile salts return to the liver in the portal circulation (i.e., enterohepatic circulation) and are resecreted into the bile. Only about 10% of bile acid is lost in the feces.

PROTEIN-LOSING ENTEROPATHY

Protein-losing enteropathy (PLE) is a condition in which large amounts of serum proteins pass into the bowel lumen and ultimately into the feces. This condition is the result of a wide range of GI tract disorders. Protein-losing enteropathy may be associated with the following:[4]

- Inflammation or ulceration of a segment of the small or large bowel (e.g., Crohn's disease and ulcerative colitis) or stomach
- Diseases in which the intestinal lymphatic system is obstructed or where there is elevated lymphatic pressure (e.g., lymphoma and Whipple's disease)
- Disorders associated with altered immune status (e.g., systemic lupus erythematosus) and some food allergies

Several conditions, such as celiac disease, bacterial overgrowth, and Crohn's disease (discussed below), affect digestion and absorption more diffusely and may produce anemia, dehydration, electrolyte disorders, or malnutrition. Biliary obstruction from stricture or neoplasm may impair fat digestion. Chronic pancreatitis or pancreatic cancer can result in impaired release of pancreatic enzymes and may lead to profound malnutrition.

CROHN'S DISEASE

Crohn's disease is described as a condition of unknown etiology that is characterized by transmural inflammation of the GI tract. It may involve the entire GI tract from mouth to anus. Examples of diverse involvement include the following: (1) small bowel involvement, usually in the distal ileum, which may lead to ileitis; (2) development of ileocolitis that involves both the ileum and colon; (3) disease that is limited to the colon; (4) predominant involvement of the mouth or gastroduodenal area; and (5) perianal disease.[5]

The clinical manifestations of Crohn's disease are variable because of the transmural involvement and variability of the extent of disease. Hallmarks of Crohn's disease include fatigue, prolonged diarrhea with abdominal pain, weight loss, fever, and, in some patients, gross bleeding. In children there may be growth retardation.

BACTERIAL OVERGROWTH

Bacterial overgrowth refers to bacterial colonization of the upper small bowel and usually occurs as a consequence of other abnormalities, such as motility or structural conditions of the small intestine. Examples of types of bacteria that colonize in this location are *Escherichia coli* and *Bacteroides* species. These bacteria deconjugate and dehydroxylate bile salts, leading to conjugated bile salt deficiency, which causes fat malabsorption. The bacteria may also metabolize vitamin B_{12}, leading to vitamin B_{12} deficiency. The usual symptoms of bacterial overgrowth are abdominal pain, diarrhea, and steatorrhea.[6]

Checkpoint 16–5

What is the most notable characteristic of Crohn's disease?

OTHER GUT-RELATED CONDITIONS

Dysregulation of gut secretion can lead to GI diseases. Examples include ZES, discussed previously; G-cell hyperplasia; retained antrum syndrome; and duodenal ulcers. Some individuals may develop conditions where there is little or no gastric acid (e.g., atrophic gastritis or pernicious anemia). Certain acute bacterial or viral infectious agents, such as *Giardia* or *Cryptosporidi*a, cause hypersecretory conditions that result in diarrhea.

Another group of GI diseases involve alteration in gut transit. This condition can be due to mechanical obstruction. Esophageal occlusions results from acid-induced stricture or a neoplasm. PUD or gastric cancer can result in gastric outlet obstruction. Obstruction of the small intestine commonly results from adhesions but may also occur with Crohn's disease, radiation, or drug-induced strictures. The most common cause of colon obstruction is colon cancer, but inflammatory strictures can develop in patients with inflammatory bowel disease, after certain infections, or with some drugs.

The movement or propulsion of digested food substances may be hindered or retarded along the GI tract. The cause is usually a disorder of gut motor function. A condition called *achalasia* is characterized by impaired esophageal body peristalsis and incomplete lower esophageal sphincter relaxation. A symptomatic delay in gastric emptying of solids or liquid meals secondary to impaired gastric motility is termed *gastroparesis*. Constipation is produced by diffusely impaired colonic propulsion or by outlet abnormalities, such as rectal prolapse, intussusceptions, or failure of anal relaxation upon attempted defecation. Rapid propulsion is less common than delayed or slowed transit.[7]

CELIAC DISEASE

Several inflammatory GI tract diseases are a result of modified gut immune function. One example is **celiac disease** (also called nontropical sprue or celiac sprue). Celiac disease occurs in genetically predisposed individuals as a result of an inappropriate T-cell-mediated immune response to the dietary ingestion of gluten-containing grains such as wheat. Gluten is a complex group of proteins present in wheat that forms a sticky mass when dough is washed with water and the starch is removed. Most of the proteins that are toxic to the small bowel mucosa of subjects with celiac disease contain large amounts of glutamine. The major toxic protein of wheat is the gliadins, and similar types of proteins (the hordeins and secalins) are found in barley and rye, respectively. The gliadins are a large family of proteins accounting for about half of the wheat protein.[8]

The presence of these toxic proteins in cereal leads to intestinal epithelial damage and release of tissue transgluaminase. The enzyme will cross link and produce gliadin-gliadin or gliadin-enzyme complexes, which expose new antigenic epitopes that bind to HLA DQ2 molecules on the antigen-presenting cells. This results in an immune response by gut-derived T cells.[9]

The clinical presentation of an individual with celiac disease is varied. Most of the diagnosis is made in adulthood. In the classical presentation of celiac disease, presenting in infancy up to the age of about two years, one finds individuals with failure to thrive, abdominal distention, and diarrhea. Most adults present with nonspecific symptoms such as mild iron deficiency and anemia, diabetes, thyroid dysfunction, and other presenting symptoms.

Ischemic damage from impaired blood flow may affect different regions of the GI tract. The most common ischemic conditions occur in the intestines and colon. These usually result from arterial embolus, arterial thrombosis, venous thrombosis, hypoperfusion from dehydration, sepsis, hemorrhage, or reduced cardiac output. The net effect can be mucosal injury, hemorrhage, or possible perforation.

Checkpoint 16–6

Identify the immunologic condition that results in celiac disease.

GI MALIGNANCIES

Malignant degeneration of GI tract tissue is possible. In the United States, colorectal cancer ranks third highest for new cases of cancer, just behind breast and prostate cancer. Colorectal cancer usually presents itself after the age of 50 years. Worldwide, gastric cancer is prevalent, especially in certain Asian regions. Patients who consume large amounts of alcohol or use tobacco develop may esophageal cancer along with chronic acid reflux. Other cancers (e.g., small intestine, anal, pancreatic, and biliary) are less frequent but can have dire consequences.

CARCINOID TUMORS

Carcinoid tumors are rare but are the most common GI NET. The age distribution ranges from the twenties to the nineties, with peak occurrence between the ages of 50 and 70 years. Clinical features vary, from asymptomatic to episodes of flushing and diarrhea. Carcinoid tumors can appear in several sites throughout the body, including the lungs, stomach, small intestine, appendix, colon, and rectum.

The term *carcinoid tumor* is applied to this group of tumors because they appear morphologically different and clinically less aggressive than the more common intestinal adenocarcinoma. Carcinoids arise from **enterochromaffin cells** of the GI tract. *Enterochromaffin* is the term given to cells that can be stained with potassium chromate (chromaffin) due to the presence of serotonin in the cell. Carcinoid tumors appear as well-circumscribed, round, submucosal lesions, and the cut surface appears yellow due to lipid content.[10]

Classifications of carcinoid tumors are based on their origin from the embryonic division of the alimentary tract. The classifications include tumors of the foregut (including the lungs, bronchi, and stomach), midgut (including the small intestine, appendix, and proximal colon), and hindgut (including the distal colon, rectum, and genitourinary tract). The tumors are further elucidated based on their morphologic pattern, silver staining affinity, and clinical behavior.[10]

The clinical features of carcinoid tumors depend on their origin. Most patients with carcinoid tumors are asymptomatic, and the tumor is usually detected and diagnosed during endoscopy, surgery, or autopsy.

Symptoms of foregut tumors vary with the site. Gastric carcinoids may be present with PUD, abdominal pain, or bleeding. Midgut tumors usually produce abdominal pain that may be vague and nonspecific. Some obstruction is seen in a small population of patients. Finally, hindgut carcinoid tumors are usually nonsecretory. When symptoms occur they include changes in bowel habit, obstruction, or bleeding.[10]

✓ Checkpoint 16–7

Where do carcinoid tumors originate?

▶ LABORATORY ASSESSMENT OF GI TRACT DISORDERS

Diagnostic tests for GI tract disorders are provided by the clinical laboratory and other diagnostic testing departments. In many cases both are utilized. Several clinical laboratory tests have become obsolete (e.g., tests for fecal fats) and have been replaced by newer technologies, such as mass spectroscopy and nonclinical laboratory technologies and procedures. Examples of nonclinical laboratory tests and procedures are listed in Table 16-2 ✪.

The selection of clinical laboratory tests appropriate for any specific GI tract disorder is somewhat complicated and difficult to present in a textbook format. Therefore, this discussion focuses on several of the disorders presented earlier in the chapter. A table listing GI tract disorders and laboratory tests useful for patient assessment is presented in Box 16-1 ■. A detailed discussion of this material is presented in *Harrison's Principles of Internal Medicine* (17th edition) and a convenient clinical guide to laboratory tests edited by Alan Wu.[10,11]

ZOLLINGER–ELLISON'S SYNDROME

Assessment of patients suspected of having ZES may include fasting serum gastrin concentration, secretin stimulation test, and gastric acid secretion test. Both the gastric acid secretion test and the secretin stimulation test are not widely used

---

 TABLE 16-2

Laboratory tests used to assess patients with GI disorders

Gastrointestinal Disorders	Laboratory Tests
Lactase deficiency	Oral lactose tolerance test
Gastrinoma	Gastrin
Mucosal blood loss	Serum iron, TIBC
Small intestinal diseases • Gastric • Pancreatic	Vitamin B_{12}
Inflammatory conditions • Leukopenia • Viremic illness	Leukocyte count, sedimentation rate
Severe vomiting/diarrhea	Electrolytes, acid–base chemistries, urea nitrogen
Pancreaticobiliary/liver	Pancrolauryl test, fecal elastase, liver function tests
Endocrine-caused GI symptoms	Thyroid tests, cortisol, calcium
Intraabdominal malignancies	CA 19-9, CEA, alpha-fetoprotein
Peptic ulcer disease (PUD)	*Helicobacter pylori*, serum IgA, IgG, CagA
Celiac disease	Gliadin antibodies (IgA & IgG), IgA-antitissue transglutaminase antibody, IgA-endomysial antibody
Carcinoid tumors (gut)	5-hydroxyindolacetic acid
Zollinger–Ellison's syndrome	Basal acid output, gastrin
Protein-losing enteropathy	alpha-1-antitrypsin

---

in medicine today and have been replaced by improved computed tomography (CT) scanning techniques or somatostatin-receptor imaging using somatostatin analog octreotide.[12]

Serum gastrin tests require a fasting sample. In blood the principal forms of gastrin are G-34 (big gastrin), G-17 (little gastrin), and G-14 (mini gastrin). Each of these polypeptides circulates in nonsulfated (I) or sulfated (II) forms. The immunoassays should be designed to detect all forms of gastrin to reduce the chance of false-negative results.

BOX 16-1 Nonclinical laboratory tests and procedures used to assess patients with GI disorders

▶ Endoscopy
▶ Colonoscopy
▶ Sigmoidscopy
▶ Endoscopic retrograde cholangio pancreaticography (ERCP)
▶ Endoscopic ultrasound
▶ Contrast radiography
▶ Computed tomography
▶ Magnetic resonance imaging
▶ Scintigraphy
▶ Histology (biopsy)
▶ Manometric techniques
▶ Radiography
▶ Breath test using radioisotopes

Gastrin values follow a circadian rhythm (lowest 3:00 to 7:00 a.m. highest during the day) or fluctuate in relation to meals. Most patients with ZES have fasting gastrin levels >500 pg/mL (reference interval [RI] <100 pg/mL) and high basal gastric acid output (BAO). Determination of BAO requires titration of a gastric juice sample with sodium hydroxide to determine the amount of free HCL produce during a given amount of time, usually an hour. Patients with ZES will produce more than 25 mmol/hr and have an elevated serum fasting gastrin level.[11]

✓ Checkpoint 16–8

In the following sentence, choose the correct adjective to describe a patient with ZES: "Serum gastrin levels will be (increased or decreased) and BAO will be (increased or decreased)."

PEPTIC ULCERS

A large percentage of patients with duodenal ulcers are infected with *Helicobacter pylori* **(H. pylori)**; thus eradication of the organism leads to healing of the ulcer and reduction in relapse rates. Infection with the organism leads to increases in both basal and meal-stimulated serum gastrin concentrations (mainly G-17). BAO is also increased.

H. pylori produces urease, and hydrolysis of endogenous urea to bicarbonate and ammonia creates a better environment for the survival of the organism in the stomach. Mammalian cells do not hydrolyze urea, and in 1984 it was shown that "gastric urease" was associated with the presence of *H. pylori*.[13] The urea breath test and direct urease test on gastric biopsy samples are based on the ability of *H. pylori* to rapidly hydrolyze urea.

There are several diagnostic tests for *H. pylori*, including invasive and noninvasive tests. The invasive tests are usually performed on gastric mucosal biopsy samples whereas the noninvasive tests use breath, blood, saliva, or feces for specimens. The breath test uses radio-labeled carbon or carbon dioxide. Whole-blood tests are available using a variety of point-of-care analyzers to measure specific IgG antibodies.

Serum measurements of IgA, IgG, and cytotoxin-associated gene A (CagA) proteins are available to determine whether *H. pylori* has colonized or not. Also, *H. pylori* antigen detected in stool by enzyme-linked immunosorbent assay (ELISA) is a highly sensitive, specific test to identify active infections in symptomatic adult patients, not just exposure as determined by serological antibody tests. The antigen test can be used to monitor response during and after therapy.

CELIAC DISEASE

Assessment of patients for celiac disease is supported by tests performed on serum samples. The most sensitive and specific tests are IgA-antitissue transglutaminase antibody (TGA) and IgA-endomysial (EMA) antibody. Testing for IgA-antigliadin antibody is not used routinely due to lack of standardization

of most immunoassays and unacceptable analytical sensitivity and specificity. The use of older laboratory tests (e.g., D-xylose absorption and fat malabsorption) is no longer warranted, such tests are not discussed in this chapter.

✓ Checkpoint 16–9

Identify the two principal immunoglobulins associated with tests for PUD.

LACTASE DEFICIENCY

Lactase deficiency can be assessed in patients using the oral **lactose tolerance test**. In this procedure the patient is given a solution of lactose, usually 50 g in 200 mL water. Multiple blood samples are drawn over a period of time (usually 2 hours) and the amount of glucose or galactose is measured. To exclude lactase deficiency, the increase above baseline for venous plasma glucose must be >30 mg/dL.

The lactose breath hydrogen test is another approach to assessing lactase deficiency. It is simple to perform, noninvasive, and has a sensitivity and specificity reputed to be better than that of absorption tests.[14] This procedure is based on the fact that hydrogen is not an end product of mammalian metabolism, and breath hydrogen is derived from bacterial metabolism in the intestinal tract. Normally when lactose is ingested, the disaccharide will split into its constituent monosaccharides (galactose and glucose) and be absorbed. Patients with lactase deficiency will not absorb disaccharides, which will pass into the large bowel. In additioni, bacterial metabolism will produce hydrogen that is absorbed into the systemic circulation and exhaled in the breath.

A breath hydrogen analysis can be performed using an electrochemical hydrogen monitor. A patient is given a dose of lactose (50 g). Breath hydrogen samples are acquired at baseline and at 30-minute intervals after the ingestion of lactose for up to three hours. The postlactose and baseline values are compared. Cutoff values may vary depending on laboratories, but a breath hydrogen value of 10 ppm (10 µL/L) is normal and values of >20 ppm (20 µL/L) are considered diagnostic of lactose malabsorption.[14]

✓ Checkpoint 16–10

Answer true or false: A patient whose glucose value during a lactose tolerance test is 20 mg/dL higher than the fasting baseline result is suspected of having lactase deficiency.

BACTERIAL OVERGROWTH

Assessment of patients with bacterial overgrowth includes (1) intubation with aspiration of jejuna contents and demonstration of a bacterial count of >10^7 organism/mL and >10^4 anaerobes/mL[15]; (2) hydrogen breath tests using either lactulose or glucose as substrates; and (3) therapeutic trials of antibiotics. Another procedure used less frequently uses ^{14}C-xylose or ^{14}C-glycocholic acid. In patients with bacterial overgrowth, the bacteria deconjugate the glycocholic acid to produce ^{14}C-glycine that is absorbed and metabolized with a concomitant increase in breath $^{14}CO_2$. An early increase in the marker indicates either bacterial overgrowth or rapid transit to the large bowel, where the normal colonic flora releases ^{14}C-glycine.[16]

BILE SALT MALABSORPTION

There are three different procedures used to diagnosed bile salt malabsorption. The first involves a whole-body scanning technique after a patient has been given an oral dose of synthetic radioactive bile acid selenohomocholyltaurine (SeHCAT). The test is interpreted as positive for bile salt malabsorption if the retention of the administered dose is less than 10%.[17]

A simpler alternative to the SeHCAT is measurement of 7α-hydroxy-4-cholesten-3-one. The principle is based on findings that the fasting serum levels of this intermediary between cholesterol and taurocholic acid reflect the activity of hepatic cholesterol 7α-hydroxylase and thus the rate of bile acid synthesis. Bile acid malabsorption is associated with increased serum concentration of 7α-hydoxylase as hepatic synthesis increases to maintain the pool of circulating bile acids.[18] This sterol intermediary compound can be measured by tandem mass spectrometry.

PROTEIN-LOSING ENTEROPATHY

The diagnostic workup for PLE usually begins with discovery of hypoalbuminemia in a patient. Clinicians must begin to rule out causes (e.g., renal disease or possible malabsorption). Once these conditions are ruled out with the assistance of a battery of clinical laboratory tests, the physician can focus on PLE as the reason for the below-normal serum albumin level. One approach to determine a diagnosis of PLE is to measure fecal ^{51}Cr-albumin. Albumin labeled with the radioisotope

^{51}Cr is administered by intravenous injection. This technique is not used frequently for several reasons, including unavailability of testing equipment for 51Cr and the fact that it is an invasive technique. Other imaging techniques using ^{99m}Tc-human serum albumin or ^{99m}Tc-dextran that are used to determine the specific site of the protein loss are available.

A laboratory test, such as fecal clearance of α-1-antitrypsin (AT), can be used as a marker for PLE. AT is a relatively small (~51 kDa) glycoprotein that is synthesized in the liver and is normally present in the serum at a concentration of about 2 g/L. It is a protease inhibitor and thus is resistant to degradation by proteolytic enzymes in the GI tract. The fecal clearance of AT correlates with protein loss measured by ^{51}Cr techniques.[19] AT is normally present in the excreted stool, primarily complex to pancreatic trypsin and elastase. In patients with PLE, fecal clearance of AT increases and the amount of protein loss also increases.

SUMMARY

Disorders associated with GI function occur at varying frequencies ranging from rare or infrequent cases of Zollinger–Ellison's syndrome to common ailments such as peptic ulcer disease. Many disorders are genetically related, whereas some occur as a result of modifications in diets, stress, and fecal-oral or oral-oral transmission (*Helicobacter pylori*).

Several GI tract disorders were discussed to a level commensurate with readers'needs. These disorders can be complex and incorporate multiple body organs. It can be difficult to describe something as simple as a gastric ulcer because of the many external influences that participate in this disorder. Similarly, it may be difficult to describe the assessment of these GI tract disorders. Readers should understand that the clinical laboratory role in patient assessment of GI tract disorders is much less than it was two decades ago. However, for many GI disorders, clinical laboratory data in combination with other nonclinical diagnostic testing provide clinicians with all of the information they require to arrive at a diagnosis.

REVIEW QUESTIONS

LEVEL I

1. Which of the following are two primary functions of the GI tract? (Objective 2)
   a. anaerobic respiration and bicarbonate excretion
   b. assimilation of nutrients and elimination of waste
   c. release of hypothalamic factors and maintenance of blood pressure
   d. detoxification of drugs and hydroxylation of drug metabolites

2. Cholecystokinin (CCK) is located in which of the following cells? (Objective 1)
   a. I cells from the upper small intestinal mucosa
   b. B cells from the pancreas
   c. G cells from the gut
   d. S cells from the salivary glands

LEVEL II

1. A gastrinoma is classified as which of the following? (Objective 5)
   a. neuroendocrine tumor (NET)
   b. pituitary adenoma
   c. glioma
   d. hepatoma

2. Which of the following laboratory analytes should be tested for a patient with suspected lactose intolerance? (Objective 4)
   a. mannose
   b. sodium
   c. glucose
   d. lactate

REVIEW QUESTIONS (continued)

LEVEL I

3. *Big*, *little*, and *mini* are all descriptors of which of the following peptides? (Objective 3)
   a. parathyroid hormone
   b. secretin
   c. cholecystokinin
   d. gastrin

4. Which of the following is the principal physiological role of secretin? (Objective 2)
   a. stimulate the gut to secrete all of its contents
   b. stimulate the intestine to move its contents to the large bowel
   c. stimulate the pancreas to secrete an increased volume of juice with high bicarbonate contents
   d. stimulate the release of gastrin from the antral mucosa

5. Which of the following factors are associated with peptic ulcer disease (PUD)? (Objective 5)
   a. *Staphylococcus aureas* and *Bacteriodies*
   b. *Helicobacter pylori* and consumption of nonsteroidal anti-inflammatory drugs (NSAIDs)
   c. *Helicobacter pylori* and consumption of cocaine
   d. *Escherichia coli* and consumption of the steroid cortisol

6. Zollinger–Ellison's syndrome (ZES) is associated with which of the following conditions? (Objective 5)
   a. hyposecretion of IgA and IgM
   b. overactive large bowel
   c. gastric hypersecretion
   d. excessive loss of protein

7. Which of the following GI disorders is characterized by a modified gut immune function? (Objective 5)
   a. Zollinger–Ellison's syndrome
   b. carcinoid tumor
   c. achalasia
   d. celiac disease

LEVEL II

3. A patient with suspected protein-losing enteropathy (PLE) would have which of the following laboratory results? (Objective 3)
   a. elevated serum protein and albumin
   b. decreased serum protein and albumin
   c. decreased serum protein and increased albumin
   d. no change in serum protein or albumin

4. Which of the following GI disorders is associated with transmural inflammation? (Objective 5)
   a. celiac disease
   b. bile salt malabsorption
   c. Crohn's disease
   d. Zollinger–Ellison's syndrome

5. Which of the following clinical laboratory tests is diagnostically useful for patients with peptic ulcer disease (PUD)? (Objective 4)
   a. serum measurement of IgA, IgG, and cytotoxin-associated gene A (CagA)
   b. serum measurement of IgM and IgE
   c. basal acid output and IgM
   d. alpha-1-antitrypsin

6. Which of the following is a potentially useful clinical laboratory tests for bile salt malabsorption? (Objective 4)
   a. measurement of 1,25-dihydroxycholecalciferol
   b. measurement of cytotoxin-associated gene A (CagA)
   c. measurement of cholesterol
   d. measurement of 7α-hydroxy-4-cholesten-3-one

7. Carcinoid arises from which of the following cells? (Objective 5)
   a. G cells of the gut
   b. enterochromaffin cells of the GI tract
   c. parafollicular cells the thyroid gland
   d. islet cells of the pancreas

PEARSON myhealthprofessionskit™

Use this address to access the interactive Companion Website created for this textbook. Simply select "Clinical Laboratory Science" from the choice of disciplines. Find this book and log in using your user name and password.

REFERENCES

1. Miller LJ. Gastrointestinal hormones and receptors. *In*: Yamada T (ed.). *Textbook of gastroenterology*, 4th ed. Philadelphia: Lippincott Williams and Wilkins, 2003: 48–77.

2. Barakat MT, Meeran K, Bloom SR. Neuroendocrine tumors. *Endocrine-Related Cancer* (2004) 11: 1–18.

3. Jensen RT. Endocrine tumors of the gastrointestinal tract and pancreas. *In*: Fauci AS, Kasper DL, Braunwald E et al. (eds.). *Harrison's principles of internal medicine*, 17th ed. New York: McGraw-Hill, 2008: 2348–50.

4. Kim KE. Protein losing enteropathy. *In*: Feldman M, Friedman LS, Sleisenger MH (eds.). *Sleisenger and Fortran's gastrointestinal and liver diseases*, 7th ed. Philadelphia: W. B. Saunders, 2002: 446–52.

5. Peppercorn MA. Clinical manifestations, diagnosis and natural history of Crohn's disease in adults. *In*: Rose BD (ed.). *UpTo Date*. Waltham, MA, 2008. http://www.utdol.com (accessed June 1, 2008).

6. Kumar PH, Clark ML. Malabsorption and weight loss. *In*: Bloom S (ed.). *Practical gastroenterology*. London: Marin Dunitz, 2002: 371082.

7. Hasler WL, Owyang C. Approach to the patient with gastrointestinal disease. *In*: Fauci AS, Kasper DL, Braunwald E et al. (eds.). *Harrison's principles of internal medicine*, 17th ed. New York: McGraw-Hill, 2008: 1832–35.

8. Shan L, Molberg O, Parrot I et al. Structural basis for gluten intolerance in celiac sprue. *Science* (2002) 297: 2275–79.

9. Mowat AM. Celiac disease—a meeting point for genetics, immunology, and protein chemistry. *Lancet* (2003) 361: 1290–92.

10. Jensen RT. Endocrine tumors of the gastrointestinal tract and pancreas. *In*: Fauci AS, Kasper DL, Braunwald E et al. (eds.). *Harrison's principles of internal medicine*, 17th ed. New York: McGraw-Hill, 2008: 2352–57.

11. Wu Alan HB. *Tietz clinical guide to laboratory tests*, 4th ed. St. Louis, MO: W. B. Saunders, 2006.

12. Lamberts SWJ, Hofland LJ, Nobels FRE. Neuroendocrine tumors markers. *Frontiers in Neuroendocrinology* (2001) 22: 309–39.

13. Langenberg ML, Tytgat GNJ, Schipper MEI et al. Campylobacter-like organism in the stomachs of patients and healthy individuals. *Lancet* (1984) i: 1348–50.

14. Karcher RE, Truding RM, Stawick LE. Using a cutoff of <10 ppm for breath hydrogen testing: A review of 5 years' experience. *AnnClinLabSci* (1999) 29: 1–8.

15. Flourie B, Turk J, Lemann M et al. Breath hydrogen in bacterial overgrowth. *Gastroenterology* (1989) 96: 1225–26.

16. Stotzer PO, Kilander AF. Comparison of the 1-gram (14) C-D-xylose breath test and the 50-gram hydrogen glucose breath test for diagnosis of small intestinal bacterial overgrowth. *Digestion* (2000) 61(3): 165–67.

17. Hardt PD, Hauenschild A, Nalop J, Marzeion AM, Porsch–Ozcurumez M, Luley C, Sziegoleit A, Kloer HU. The commercially available ELISA for pancreatic elastase 1 based on polyclonal antibodies does measure an as yet unknown antigen different from purified elastase 1. Binding studies and clinical use in patients with exocrine pancreatic insufficiency. *Zeitschrift fur Gastroenterologie* (2003) 41(9): 903–906.

18. Eusufzai S, Axelson M, Angelin B, Einarsson K. Serum 7 alpha-hydroxy-4-cholesten-3-one concentrations in the evaluation of bile acid malabsorption in patients with diarrhea: correlation to SeHCAT test. *Gut* (1993),34(5): 698–701.

19. Florent C, L'Hirondel C, Desmazures C, Aymes C, Bernier JJ. Intestinal clearance of alpha 1-antitrypsin. A sensitive method for the detection of protein-losing enteropathy. *Gastroenterology* (1981) 81(4): 777–80.

17

Pancreas

CHAPTER OUTLINE

■ OBJECTIVES—LEVEL I

Following successful completion of this chapter, the learner will be able to:

1. Review the location and anatomy of the pancreas.

2. Identify the islets of Langerhans and the major cells found in the islets of Langerhans.

3. Summarize the endocrine and exocrine functions of the pancreas.

4. Explain the major invasive test for assessing exocrine pancreatic function secretin–cholecystokinin (CCK).

5. Summarize the most common noninvasive tests for assessing pancreatic exocrine insufficiency: pancreatic elastase-1, pancreatic chymotrypsin, pancreatic serum enzymes, breath test (C-mixed triglyceride test), urinary amylase, fecal fat, phospholipase A_2, NBT-PABA, and fecal elastase.

6. Review the two major tests for monitoring the endocrine function of the pancreas: insulin and C-peptide.

7. Summarize briefly diabetes mellitus, the major endocrine pancreatic disease.

8. List the two primary causes of acute pancreatitis.

9. Outline Ranson's indicators of severity in acute pancreatitis.

10. Briefly review the etiology and prognosis of chronic pancreatitis.

11. Summarize the etiology of cystic fibrosis.

■ OBJECTIVES—LEVEL II

Following successful completion of this chapter, the learner will be able to:

1. Briefly explain CA 19-9, the major tumor marker for colorectal and pancreatic cancer.

2. Compare and contrast acute and chronic pancreatitis.

3. Summarize the major complications of acute and chronic pancreatitis.

4. Compare and contrast pancreatic neoplasms, including ductal adenocarcinoma, insulinomas, glucagonomas, and somatostatinomas.

KEY TERMS

Acini

Acute pancreatitis

Adult respiratory distress syndrome (ARDS)

Ampulla of Vater (hepatopancreatic ampulla)

Amylase

Breath tests

CA 19-9

C-mixed-chain triglyceride test

C-peptide

Chronic pancreatitis

Enterokinase

Fecal elastase

Fecal fat

Glucagon

Glucagonoma

Insulin

Insulinoma

Islets of Langerhans

Lipase

NBT-PABA test

Pancreatic chymotrypsin

Pancreatic elastase-1

Pancreatic neoplasm

Pancreatitis

Phospholipase A_2 (PLA$_2$)

Pseudocyst

Ranson's indicators of severity in acute pancreatitis

Secretin–cholecystokinin (CCK) test

Somatostatinoma

Trypsin

Zymogen

Zymogen granules

ⓔ A CASE IN POINT

John L., a 50-year-old man, was seen in the emergency room with severe epigastric pain. He also complained of nausea and vomiting and had a history of alcoholism. The following chemistry tests were ordered

Chemistry Tests	Results	Reference Range
Sodium	143	135–145 mEq/L
Potassium	3.9	3.6–5.0 mEq/L
Chloride	106	98–107 mEq/L
CO_2	31.0	24.0–34.0 mEq/L
Glucose	95	70–99 mg/dL
Bilirubin, total	1.4	0.2–1.0 mg/dL
AST	90	5–40 IU/L
ALP	200	30–157 IU/L
Protein	6.8	6.0–8.4 g/dL
BUN	40	6–20 mg/dL
The following day		
Calcium	8.1	8.5–10.5 mg/dL
Albumin	3.9	3.5–5.0 g/dL
ALT	100	5–40 IU/L
Amylase	838	10–110 IU/L
Lipase	2500*	31–186 IU/L
GGT	335	13–86 IU/L

*Lipemic sample; results verified

CBC: Normal, hematocrit stable

Issues and Questions to Consider

1. Circle or highlight the abnormal laboratory values.
2. What two organ systems appear to be involved? Support your answer.
3. (A) John's profile is indicative of what condition?
   (B) Briefly describe the pathogenesis that is involved (ie, the origin and development of disease).
4. (A) What are the two most common causes or etiologies of this condition?
   (B) List three other less common causes.
5. List two factors that increase the risk of this condition.
6. Which two enzymes are critical in evaluating the status of the exocrine function of the pancreas?
7. Is hypocalcemia associated with this condition? Explain why or why not.
8. Discuss two reasons for the elevated liver enzymes (AST, ALP, and GGT).
9. How many of Ranson's indicators does John have, if any?
10. What is his prognosis?

WHAT'S AHEAD

▶ An overview of pancreatic anatomy and physiology

▶ A review of the methodologies most commonly used in the assessment of pancreatic function.

▶ A summary of pancreatic diseases, including acute and chronic pancreatitis, diabetes mellitus, and cystic fibrosis.

▶ A discussion of pancreatic neoplasms, including ductal adenocarcinoma, insulinomas, glucagonomas, and somatostatinomas.

▶ INTRODUCTION

The pancreas is a vital organ responsible for the synthesis of many enzymes and hormones that serve critical functions in the human body. Pancreatic disease, including acute and chronic pancreatitis, is a common cause of hospitalization. Other disorders that involve the pancreas are diabetes mellitus and cystic fibrosis. The clinical laboratory is responsible for performing valuable screening tests for pancreatitis, including amylase and lipase.

► PANREATIC ANATOMY AND PHYSIOLOGY

The pancreas is a pyramid-shaped organ located behind the peritoneum between the stomach and the duodenum. It is an elongated structure divided into three sections with a *head* located within the curvature of the duodenum, a *body*, and a *tail*. It is approximately 15 cm long (~5 inches) and weighs between 85 and 100 grams Figure 17-1 ■ illustrates its location and approximate size.

The pancreas functions as both an exocrine and endocrine gland. As an exocrine gland, it contains **acini** (from the Latin *acinus*, which means grape or berry) These small clusters of glandular epithelial cells make up 98% of the pancreatic mass. Acini produce pancreatic juice, which is carried to the duodenum through a duct system. They form lobules that are separated by thin septa, and clusters of acini are connected by small intercalated ducts leading to intralobular ducts that leave the lobules to join interlobular ducts between the lobules. The main pancreatic duct joins the common bile duct and drains into the gall bladder.

Enzymes are stored in **zymogen granules** in an inactive form called **zymogens**. The number of zymogen granules increases in the fasting state and decreases following a meal when they are released in the digestive process. The pancreatic juice contains pancreatic enzymes, including amylase and lipase, which are discussed in Chapter 9, Enzymes. Approximately 1500 mL of pancreatic juice is secreted each day in response to stimulation by gastric acid and products of digestion. The pancreatic duct joins the common bile duct from the liver and gallbladder and enters the duodenum as a single duct called the **ampulla of Vater** (or **hepatopancreatic ampulla**). On entering the duodenum, **enterokinase**, a duodenal peptidase, cleaves a small amino acid chain from the proenzymes—for example, converting protrypsin to the active form **trypsin**. The active trypsin begins a cascade or chain reaction that activates the remaining proenzymes.

The pancreas synthesizes 22 digestive enzymes, including 15 proteases that act on proteins Trypsin, chymotrypsin, and elastase are the three major proteases discussed later in this chapter. Other enzymes catalyze lipids and carbohydrates. Lipase, phospholipase A_2, and cholesterol esterase digest various lipids. Amylase catalyzes the catabolism of complex carbohydrates.

The endocrine portion of the pancreas consists of the **islets of Langerhans**, which make up approximately 1% of the pancreas and produce hormones that are secreted into the circulatory system. The pancreas contains approximately 1 million islets of Langerhans distributed among the ducts and acini. Each islet comprises alpha, beta delta, and F cells *Alpha cells* constitute about 20% of the islet cells and synthesize **glucagon**, a hormone that increases glucose concentration. *Beta cells*, which constitute 75% of the islet cells, produce proinsulin, which is converted to the active form insulin before it is secreted. *Delta* or *D cells* make up approximately 5% of the islet cells and secrete somatostatin, which is identical to the growth-hormone–inhibiting factor secreted by the pituitary *F cells* constitute the remainder of the pancreatic islet cells and secrete pancreatic polypeptide. Pancreatic polypeptide inhibits somatostatin secretion, contraction of the gallbladder, and secretion of amylase and lipase by the pancreas. Insulin and glucagon are discussed in Chapter 6, Carbohydrates.

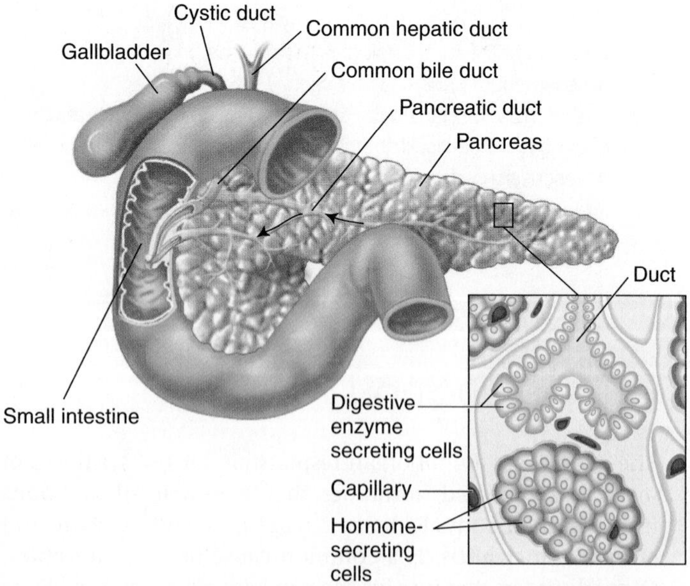

✓ Checkpoint! 17–1

1. *Inactive forms of enzymes which are stored in the pancreas are called _____.*
2. *What hormone is secreted by the following cells in the islets of Langerhans?*
   a. Alpha cells: _____
   b. Beta cells: _____
   c. Delta cells: _____

► ASSESSMENT OF EXOCRINE PANCREATIC FUNCTION

The exocrine function of the pancreas can be evaluated by several laboratory tests. Pancreatic insufficiency is most often caused by cystic fibrosis in children and chronic pancreatitis in adults. Two major categories of pancreatic function tests are invasive and noninvasive. The most common invasive test is the **secretin–cholecystokinin (CCK) test**, which requires gastroduodenal intubation. Invasive tests are uncomfortable and time consuming. Noninvasive tests do not require intubation, are simpler, and less expensive but do not have the sensitivity and specificity of secretin–CCK, especially for mild

Cystic duct
Gallbladder
Common hepatic duct
Common bile duct
Pancreatic duct
Pancreas
Duct
Small intestine
Digestive enzyme secreting cells
Capillary
Hormone-secreting cells

■ **FIGURE 17-1**   Anatomy of the pancreas
From Colbert, Ankey, and Lee, *Anatomy and Physiology for Health Professions* (Upper Saddle River, NJ: Prentice Hall, 2007), p. 389.

pancreatic insufficiency. Examples of noninvasive tests include pancreatic elastase-1, pancreatic chymotrypsin, and pancreatic serum enzymes Box 17-1 ■ has a more extensive list of invasive and noninvasive (direct and indirect) tests. This chapter will concentrate on the most routine and clinically significant tests.

SECRETIN–CHOLECYSTOKININ

The secretin-CCK test is based on the principle that the secretion of pancreatic juice and the output of bicarbonate are directly related to the functional mass of pancreatic tissue Secretin and CKK are normally released by the enteroendocrine cells of the duodenum during a meal. Secretin stimulates the secretion of pancreatic juice and bicarbonate, but stimulation of the secretion of pancreatic enzymes (eg, amylase and lipase) is not consistent CCK or ceruletide stimulates pancreatic enzyme secretion and provides a more complete assessment than using secretin alone.

Following an overnight fast, the patient is intubated with a gastroduodenal tube into the duodenum using X-ray to guide the process. A baseline sample of the stomach and duodenal contents is collected One unit (U) of secretin/kg body weight/hour is given intravenously, and 30 mg ceruletide/kg body weight/hour is given simultaneously during the second hour.

Pancreatic juice is collected in 15-minute intervals, and volume, pH, bicarbonate, and enzymes are measured Procedures are not standardized among laboratories, but it is recommended that doses of secretin and CCK and ceruletide are adequate to provide maximal pancreatic stimulation, and these are given for at least 60 to 90 minutes to increase diagnostic accuracy. Decreased volume of pancreatic juice and increased enzymes provide evidence of pancreatic obstruction.

Other pancreatic disorders, including cystic fibrosis and chronic pancreatitis, are characterized by low bicarbonate and enzyme levels.[1]

PANCREATIC ELASTASE-1

Pancreatic elastase-1 is a pancreatic-specific protease synthesized by the pancreatic acinar cells with other digestive enzymes and found in pancreatic juice. It is not catabolized in the intestine, and its concentration in feces is five to six times greater than in pancreatic juice. Fecal concentration can be quantified through an enzyme-linked immunosorbent assay (ELISA) assay, which uses two monoclonal antibodies specific for the human enzyme.[2]

The use of pancreatic enzymes supplements does not interfere with this test; therefore, patients are not required to stop substitution therapy before providing a stool sample for analysis. The major advantage of pancreatic elastase-1 is that it is noninvasive, unlike the gold standard secretin-CCK test described in the previous section. Pancreatic elastase-1 is highly sensitive in evaluating moderate to severe pancreatic damage, but its sensitivity is lessened in mild disease. Decreased levels of pancreatic elastase-1 are associated with chronic pancreatitis, alcoholism, and cystic fibrosis.

PANCREATIC CHYMOTRYPSIN

Pancreatic chymotrypsin is almost completely digested during its passage through the intestine in adults. Residual activity of chymotrypsin in feces is stable for several days at room temperature. Correlation between chymotrypsin levels in duodenal contents and stool are poor when duodenal chymotrypsin is measured after stimulation with secretin-CCK.

In patients with normal pancreatic function, falsely low results in approximately 12% of patients may be explained by a large stool, resulting in less enzyme per gram of feces; inadequate food intake; obstruction of the bile duct; or partial gastrectomy or mucosal disease. Fecal chymotrypsin is an extremely sensitive test for steatorrhea; like duodenal chymotrypsin, its sensitivity decreases with increased damage to the pancreas. It is used to monitor patients with pancreatic insufficiency.

PANCREATIC SERUM ENZYMES

Amylase and lipase are the pancreatic enzymes routinely measured to assess pancreatic function. Amylase is synthesized by the salivary glands and the pancreas and digests starch into smaller carbohydrate groups (dextrins and maltose). Amylase is added to a chemistry profile to rule out acute or chronic pancreatitis. The primary laboratory finding in acute pancreatitis is an amylase level elevated 10 to 20 times the upper limit of normal; a level three times the upper level is diagnostic.

Lipase hydrolyzes triglycerides into monoglycerides, alcohol, and fatty acids. Amylase and lipase provide valuable information useful in differentiating acute pancreatitis from other intraabdominal conditions, acute from chronic pancreatitis, and the stage of pancreatitis. The reader is referred to "Digestive and Pancreatic Enzymes" in Chapter 9, Enzymes, for a detailed discussion of these two enzymes.

BREATH TESTS

Breath tests have been developed to evaluate fat absorption. However, most of them are unable to differentiate between pancreatic and nonpancreatic causes of malabsorption. For example, one study of cystic fibrosis (CF) patients suggested that continuing fat malabsorption in CF patients receiving enzyme-replacement therapy does not result from insufficient lipolytic enzyme activity but from incomplete solubilization of long-chain fatty acids in the intestine, reduced mucosal uptake of long-chain fatty acids, or both.[3]

The **C-mixed-chain triglyceride test** evaluates intraluminal pancreatic lipase activity. The substrate is 1,3-distearyl, 2(carboxyl-C13) octanoyl glycerol, which contains long-chain fatty acids in positions 1 and 3 and the ^{13}C-labeled octanoic acid in position 2. It is given orally to fasting patients with a "standard meal" of toast and butter.

Breath samples are collected for 5 hours, and the exhaled $^{13}CO_2$ is expressed as a percentage of the administered dose. The test is based on the rationale that before ^{13}C-octanate or ^{13}C-octanoyl monoglyceride can be absorbed or metabolized, the stearyl groups must be hydrolyzed by the activity of pancreatic lipase. Decreased secretion of pancreatic lipase will result in the reduction in the amount of ^{13}C-label absorbed and subsequently metabolized to CO_2.[3]

URINARY AMYLASE EXCRETION

Urinary amylase remains elevated longer than serum amylase or serum lipase. The amylase creatinine clearance ratio compares the renal clearance of amylase to the clearance of creatinine on the same urine and serum. See Chapter 9, Enzymes, for more information.

FECAL FAT

Fecal fat is performed to determine if the steatorrhea results from a pancreatic or intestinal dysfunction. Steatorrhea, an excess of fat in the stool, is characteristic of some malabsorption syndromes. It is often obvious on macroscopic examination with distinctive pale, frothy, and foul-smelling feces. In qualitative fecal fat procedures, the fat is examined microscopically by staining with an oil-soluble dye such as Sudan III, Sudan IV, or Oil Red O.

Four types of lipids are found in feces: neutral fats (triglycerides), fatty acid salts (soaps), fatty acids, and cholesterol. In qualitative fecal fat procedures, two slides are examined. The first slide, in which only neutral fats are stained, is a suspension of the stool examined microscopically for the presence of fat droplets stained orange or red, depending on the dye. More than 60 droplets/high-power field (hpf) is indicative of steatorrhea.

The specimen on the second slide is mixed with acetic acid and heated to release the fatty acids by hydrolysis of the soaps and neutral fats. The number and size of droplets must be considered. A normal stool can contain as many as 100 droplets ranging in size from 6 to 75μ. The number of fat droplets can be used to approximate the percent fat content of the stool. Normally, fat excretion should result in fewer than 10 droplets/hpf.

Quantitative fecal fat measures fecal fat excretion in g per day; a normal fecal fat excretion is less than 6 g per day. The patient ingests a diet of 100 g of fat per day for 6 days followed by the collection of stool specimens over a 3-day period. A number of procedures are used to determine the fat content of an aliquot of the 72h stool specimen.

Decreased fat absorption or increased fecal fat may result from acute or chronic pancreatitis, pancreatic cancer, cystic fibrosis, cholelithiasis, celiac disease, biliary cancer, Crohn's disease, and Whipple's disease. Decreased fecal fat does not determine the cause of the steatorrhea. Fecal fat is normal until 85% to 90% of the pancreatic acini are destroyed; therefore, pancreatic exocrine function must be almost lost before fecal fat will identify pancreatic insufficiency.[1]

PHOSPHOLIPASE A₂

Phospholipase A₂ (PLA₂) has been associated with the pathogenesis of acute pancreatitis. Although the exact mechanism is not known, phospholipase A_2 is released into the serum during an acute attack. The activity of PLA_2 is greater in severe, necrotizing, acute pancreatitis than the milder edematous form of acute pancreatitis. Acute pancreatitis can be complicated by an infection with necrotic tissue, endotoxemia, or sepsis, which will be discussed later. Increased PLA_2 is also associated with pancreatic cancer and various inflammatory conditions, including sepsis, infections, and postoperative states.

PLA_2 has been implicated in the pulmonary and renal damage associated with acute pancreatitis. It is useful in the diagnosis of acute pancreatitis and can be used to monitor patients following an attack.[4]

NBT-PABA TEST (BENTIROMIDE)

The **NBT-PABA** test of pancreatic function is based on the hydrolysis of a synthetic tripeptide, N-benzoyl-1-tyrosyl-ρ-aminobenzoic acid (NBT-PABA) or bentiromide by chymotrypsin. It is administered orally with a test meal to stimulate pancreatic secretion. The NBT-PABA is hydrolyzed by chymotrypsin to release PABA, which is absorbed by the intestine and metabolized in the liver to PABA glucoronide and PABA acetylate. These two substances are excreted by the kidney. In pancreatic insufficiency, the decreased chymotrypsin levels result in less peptide being hydrolyzed as well as less chromagen in the urine and serum. The PABA levels are directly proportional to the chymotrypsin levels or activity.[5]

The specificity of the test was significantly improved by adopting high-performance liquid chromatography instead of the previous colorimetric procedures. However, NBT-PABA is not readily available, and the test has a lower diagnostic sensitivity and specificity for pancreatic insufficiency than does pancreatic elastase-1 discussed earlier.

FECAL ELASTASE

Fecal elastase is an important noninvasive test to assess pancreatic insufficiency. In children with cystic fibrosis, it can differentiate those with from those without pancreatic insufficiency. Low fecal elastase ($<200 \mu g/g$ of stool) in infants older than 4 weeks of age is indicative of cystic fibrosis. Fecal elastase has been found to be superior to fecal chymotrypsin to evaluate pancreatic exocrine function in CF patients. A disadvantage of chymotrypsin is the interference of the enzyme-substitution therapy with the colorimetric reaction. Enzyme supplements should be discontinued for 3 days before the test. The fecal elastase procedure is not influenced by the enzyme-substitution therapy.[6]

Fecal elastase is measured using a monoclonal antibody and is the test of choice for both diagnostic sensitivity and specificity and patient comfort. It has a high sensitivity for detecting moderate to severe chronic pancreatitis. Although it is more sensitive than other tests for detecting mild pancreatitis, fecal elastase may be negative in mild disease. It does, however, have a high sensitivity and high predictive value for discriminating between pancreatic and nonpancreatic diarrhea.[7] Fecal elastase may give false positives in nonpancreatic diseases such as malnutrition, inflammatory bowel disease, and chronic diarrhea. Liquid stools may also decrease its accuracy.

> ✓ **Checkpoint! 17–2**
>
> 1. *Which test is the gold standard for evaluating pancreatic exocrine function?*
> 2. *What is measured in the pancreatic juice collected in the secretin–CCK test?*
> 3. *What is the major advantage of using pancreatic elastase-1 over secretin–CCK to assess pancreatic function?*
> 4. *The substrate 1,3-distearyl, 2(carbonzyl-C13) octanyl glycerol is used in which pancreatic function test?*

▶ ASSESSMENT OF ENDOCRINE PANCREATIC FUNCTION

INSULIN

Insulin is a small peptide hormone secreted by the beta cells of the islets of Langerhans. It is the only hormone that decreases blood glucose by decreasing glycogenolysis and increasing glycolysis, glycogenesis, and lipogenesis. Insulin increases the cell membrane's permeability to glucose by binding with the insulin receptors on the cell membrane.

Insulin is stored in the inactive form, proinsulin, which is converted to insulin and C-peptide. Insulin and C-peptide levels are important in investigating the presence of insulinomas and in ruling out factitious hypoglycemia. See Chapter 8, Amino Acids and Proteins, for further information.

C-PEPTIDE

Proinsulin, the inactive form of insulin, is cleaved on activation into a 31-amino acid connecting (C)-peptide and insulin. C-peptide serves no known biological function but appears to protect the correct protein structure of insulin. Therefore, on activation proinsulin is cleaved into equimolar amounts of insulin and C-peptide. C-peptide is removed from circulation by the kidneys and excreted in the urine.

Measurement of C-peptide is primarily used to evaluate fasting hypoglycemia as discussed in Chapter 6, Carbohydrates. In patients with suspected insulinomas or factitious hypoglycemia, the measurement of both insulin and C-peptide is important in the diagnosis. Other clinical uses of C-peptide measurement are classification of diabetes mellitus as well as to obtain insurance coverage for insulin pumps, regulate insulin levels in diabetics, and monitor therapy following pancreatectomies and pancreas transplants.

SWEAT CHLORIDE

The analysis of electrolytes in sweat (eg, sweat chloride) is performed to confirm the diagnosis of CF. The sweat test can be divided into three parts: (1) sweat stimulation by pilocarpine iontophoresis, (2) collection of sweat onto appropriate medium (eg, gauze, patch, capillary tube), and (3) the qualitative or quantitative measure of sodium, chloride, or osmolality. The most common methods are chloride measurement and chloride conductivity. Sweat chloride is described in more detail in Chapter 12, Body Water and Electrolyte. Homeostasis. CF is discussed later in this chapter in "Pancreatic Diseases".

▶ TUMOR MARKERS

CA 19-9

CA 19-9 is a marker for colorectal and pancreatic carcinoma, and it has been approved by the FDA to monitor patients with pancreatic cancer. CA 19-9 is a glycolipid synthesized by pancreatic and biliary ductal cells and also gastric, colon, endometrial and salivary epithelia.[9] Patients who are Lewis null phenotype (Le a– b–), about 5% of the population, do not express CA 19-9 because the antigen requires the Lewis gene product 1,4-fucosyl transferase.

In addition to pancreatic and colorectal cancer, elevated CA 19-9 is also associated with hepatobiliary, gastric, hepatocellular, and breast cancers with prevalence from 80% in pancreatic cancer to 15% in breast cancer. Increased levels have also been found in pancreatitis and other benign gastrointestinal diseases but at far lower levels than with neoplasms.

▶ PANCREATIC DISEASES

In adults, the three major exocrine pancreatic disorders are acute pancreatitis, chronic pancreatitis, and pancreatic carcinoma. Other conditions associated with pancreatic disease are diabetes mellitus and cystic fibrosis.

ACUTE PANCREATITIS

Pancreatitis is an inflammation of the pancreas that occurs as a result of autodigestion. Certain proteases—including trypsinogen, proelastase, phospholipase A_2, and chymotrypsin—are normally stored in zymogen granules and secreted by the pancreas as inactive zymogens. They are released from the acinar cells into the pancreatic duct and secreted into the duodenum, where trypsinogen is activated to trypsin and the other zymogens are activated to the active forms of the other enzymes. The conversion of trypsinogen to trypsin appears to be the critical early step because trypsin can activate most of the other pancreatic enzymes, including phospholipase A_2 and proelastase. If activated in the pancreas, phospholipase A_2 attacks phospholipids in the walls of membranes; once activated, elastase digests the walls of blood vessels, which results in hemorrhaging. The activation of chymotrypsin also results in edema and vascular damage.

In **acute pancreatitis**, the enzymes are prematurely activated in the pancreas and begin digesting the pancreatic cells. Acute pancreatitis occurs suddenly, lasts for a short period of time, and usually resolves. The pancreas often becomes edematous and enlarged, and steatorrhea and malabsorption are symptoms of acute pancreatitis. Acute pancreatitis can progress to necrosis and hemorrhaging and can become life threatening, although most cases recover uneventfully. Mortality varies with the etiology, development of possible complications, and the number and severity of other medical conditions that can aggravate the situation.[8,10]

The two major etiologies or triggering mechanisms for acute pancreatitis, accounting for 75% of cases, are alcohol and biliary tract disease or obstructive liver disease. Alcohol has a direct toxic effect on the pancreas as well as the liver. Prolonged alcohol intake of more than 100 g/day for 3 to 5 years may cause the precipitation of pancreatic enzymes within the small pancreatic ductules. Other factors that increase the risk of acute pancreatitis are smoking and diets high in fat and protein.[8,10]

Ductal obstruction may result in premature activation of pancreatic enzymes, and an alcoholic binge may trigger an attack of acute pancreatitis in these patients. The biliary obstruction is often a gallstone in gallbladder disease. In cases of acute pancreatitis, 25% result from other causes, including surgery near the pancreas, drugs, and unknown

✪ TABLE 17-1

Causes of Acute Pancreatitis

Cause	Example
Drugs	• Angiotensin-converting enzyme inhibitors • Asparaginase furosemide • Sulfa drugs • Valproic acid • Aminosalicylates
Infectious	• Coxsackie B virus • Rubella cytomegalovirus • Mumps
Inherited	• Multiple known gene mutations, including a small percentage of cystic fibrosis patients
Mechanical or structural	• Gallstones • Trauma • Pancreatic or periampullary cancer • Sphincter of Oddi stenosis
Metabolic	• Hypertriglyceridemia • Hypercalcemia (including hyperparathyroidism)
Toxins	• Alcohol • Methanol
Other	• Pregnancy • Postrenal transplant

Source: Merck Manual: Professional (www.merck.com/mmpe/print/sec02/ch015/ch015b.html), accessed May 21, 2007.

etiology (idiopathic) Causes of acute pancreatitis are summarized in Table 17-1 ✪.

Symptoms of acute pancreatitis are upper-right abdominal pain, often after a heavy meal or alcoholic binge. It is a severe knife-like pain associated with nausea and vomiting. Vomiting with a decrease in pain is a hallmark symptom of acute pancreatitis. Fever and rapid pulse may also be present in acute pancreatitis, and severe cases may result in dehydration and hypotension. The mortality rate of an uncomplicated attack is less than 5%; however, a hemorrhagic attack raises the mortality rate to 50% to 90%. A wide continuum occurs from a mild, self-limiting edematous form to a full-blown necrotizing, hemorrhagic pancreatitis.[8]

Ranson's indicators of severity in acute pancreatitis are helpful in determining the severity of an acute pancreatitis attack (see Box 17-2 ■). Calcium levels are decreased because of the fat necrosis and release of free fatty acids that occurs when pancreatic lipase is released by the inflamed acinar cells. Calcium binds with the free fatty acids to form soaps, and the parathyroid cannot respond quickly enough to compensate for the decreased calcium. The decreased albumin levels also contributes to the low calcium because approximately 50% of calcium is protein bound, predominantly to albumin. Mortality prediction for fewer than three signs of Ranson's indicators is

1%, three to four signs equals 15%, and five to six signs is 40%. With more than six signs, the mortality rate approaches 100%.[8,10]

Treatment to support vital body functions and prevent complications depends on the severity of the attack. Four major goals are to (1) provide supportive care, (2) decrease and prevent further local pancreatic necrosis and the inflammatory process, (3) recognize and treat complications, and (4) prevent future attacks. The patient is hospitalized so fluids can be replaced intravenously. A critical step is to make the patient *nil per os* (NPO)—allow nothing by mouth—to effect pancreatic rest. Pain and nausea can be controlled by using moderate doses of intravenous analgesics and antiemetics. On release from the hospital, the patient is advised not to drink alcohol or eat large meals.[8]

CHRONIC PANCREATITIS

Chronic pancreatitis is defined as irreversible damage to the pancreatic tissue with evidence of inflammation and fibrosis. It usually occurs after repeated bouts of acute pancreatitis and results in the pancreatic cells being replaced with scar tissue. Pancreatic ducts become calcified, resulting in malabsorption. In chronic pancreatitis, both the endocrine and exocrine functions of the pancreas are destroyed. In acute pancreatitis, the pancreas is normal before the attack, and the changes are reversible after the attack. In chronic pancreatitis, the pancreas is abnormal before the attack, and the changes are permanent.[1]

In population studies, males are affected more commonly than females (6.7 vs 3.2 per 100,000).[1] Symptoms of chronic pancreatitis include intermittent or constant pain. However, 10% to 20% of patients experience no pain and present with diabetes, jaundice, anorexia, malabsorption, and weight loss.

The most common cause of chronic pancreatitis is years of alcohol abuse, but it may be triggered by one acute attack if the pancreatic ducts are damaged. Only 5% to 15% of heavy drinkers experience an attack of acute pancreatitis; therefore, other predisposing factors must be involved. Additional causes of chronic pancreatitis are cystic fibrosis, surgery, hyperlipidemia, hypercalcemia, and glucagonomas, but 25% of cases are idiopathic[1] Box 17-1 lists conditions associated with acute pancreatitis that can progress to chronic pancreatitis.

A new mnemonic system called TIGAR-O has been developed to classify etiologic factors associated with chronic pancreatitis. In this scheme,

T = toxic metabolic,

I = idiopathic;

G = genetic,

A = autoimmune,

R = recurrent and severe acute pancreatitis, and

O = obstructive

See the references and suggested readings for more detailed discussions.[14]

Chronic pancreatitis patients often exhibit abdominal pain, although some do not experience any pain. The pain may get worse when eating or drinking, and it may become constant and disabling. Other symptoms are nausea, vomiting, weight loss, and fatty stools. Weight loss occurs because even though a patient's appetite and diet are normal, the pancreas does not secrete the pancreatic enzymes to digest the food, so the nutrients are not absorbed.

Complications of acute pancreatitis include **adult respiratory distress syndrome (ARDS)**, cardiac complications, metabolic complications, gastrointestinal bleeding, and pancreatic infection and abscess. Systemic complications, especially shock, ARDS, and multiorgan failure are the most common causes of death from acute pancreatitis during the first week of illness. The most common complications of chronic pancreatitis are pseudocyst formation and obstruction of the duodenum and common bile duct.

ARDS is a serious complication of acute pancreatitis. It may be caused by the autodigestion of pulmonary capillaries by activated pancreatic enzymes, including phospholipase A_2. Enzymes in the systemic circulation may overwhelm antiprotease defenses such as α_1-antitrypsin in the host. Other organs, including the kidneys and parathyroids, can also be attacked.

A variety of cardiac complications including congestive heart failure (CHF), myocardial infarction, and cardiac arrhythmias may occur in severe acute pancreatitis. Metabolic complications associated with pancreatitis are hypocalcemia, hyperglycemia, and hyperlipidemia. Calcium replacement is usually not required unless the patient's ionized calcium is decreased or the patient exhibits signs of tetany (neuromuscular instability). Calcium levels should be monitored because hypocalcemia can be a marker of severe pancreatitis as discussed in Ranson's indicators earlier in the previous section.

Hyperglycemia, another of Ranson's indicators, can suggest a poor prognosis. It usually does not require treatment unless the glucose levels increase above 200 mg/dL Hyperlipidemia is associated with acute pancreatitis as an etiological factor and as a consequence. Patients with acute pancreatitis may develop a moderate elevation of serum triglycerides (300–400 mg/dL) as a consequence of their condition. Hypertriglyceridemia can also be a cause of acute pancreatitis when the serum triglycerides are more than 1000 mg/dL. Triglyceride levels drop rapidly when the patient is NPO.

Infected pancreatic necrosis, the most serious form of pancreatic infection, occurs in 1% to 4% of patients with pancreatic infection and in 15% to 30% of those with pancreatic necrosis. The mortality rate in necrotizing pancreatitis is approximately 10%, but this rises to 30% if an infection occurs. This usually develops in the second or third week of an acute attack, and the infecting organisms are from the intestinal area—for example *Klebsiella*, *Escherichia coli*, and *Enterococcus*.[1]

Gastrointestinal bleeding from peptic ulcers, pseudoaneurysms, or varices from splenic vein thrombosis may be a complication of acute pancreatitis. Splenic varices may bleed and can be managed by a splenectomy if required. Monitoring of the patient's hematocrit to detect hemorrhage is also one of Ranson's indicators.

A **pseudocyst** is a collection of pancreatic juice that is enclosed by a wall of fibrous or granulation tissue that usually occurs 2 weeks after the initial symptoms. Pseudocysts are a complication in 1% to 8% of acute pancreatitis cases. Peripancreatic fluid collections are usually amorphous and not encapsulated, and they typically resolve spontaneously. Pseudocysts develop in approximately 10% of patients with chronic pancreatitis as a result of ductal disruptions in which the cyst is enclosed by a wall of fibrous or granulation tissue. Pseudocysts can be single or multiple, large or small, and located inside or outside of the pancreas. Asymptomatic pseudocysts less than 6 cm are usually safely monitored.[1] They can become infected, leading to abscess formation as discussed earlier. Pseudocysts in chronic pancreatitis are generally mature at the time of diagnosis, and therapy can be initiated if indicated.

Bile and duodenal obstruction develops in 5% to 10% of patients with chronic pancreatitis. Elevated liver function tests (LFTs), including hyperbilirubinemia, are suggestive of bile duct obstruction. This is most commonly seen in patients with dilated bile ducts from inflammation and fibrosis in the head of the pancreas or a pseudocyst. Surgical or endoscopic drainage of the pseudocyst may help to relieve the obstruction.[1]

A pseudoaneurysm affecting blood vessels near the pancreas such as the splenic, hepatic, and gastroduodenal arteries is another complication. Surgery for bleeding pseudoaneurysms is challenging and has a high rate of morbidity and mortality.[3]

DIABETES MELLITUS

Diabetes mellitus (DM) is the major endocrine pancreatic disorder. Diabetes mellitus is typically the result of islet cell destruction, which is categorized as type 1 (insulin dependent) or insulin resistant type 2 (noninsulin dependent) In type 1 diabetes mellitus, the beta cells of the islets of Langerhans are destroyed by a cellular mediated autoimmune response, which results in an absolute lack of insulin Type 2 diabetes mellitus is associated with insulin resistance and dyslipidemia; it is a relative rather than an absolute insulin deficiency. A detailed discussion of diabetes, including laboratory findings, is found in Chapter 6, Carbohydrates.

In rare cases, chronic pancreatitis or drugs toxic to the pancreas can lead to diabetes. An increase in serum amylase is common in uncontrolled diabetes and diabetic ketoacidosis; it is normally the result of an increase in salivary amylase.

CYSTIC FIBROSIS

Cystic fibrosis (CF) is one of the most common autosomal recessive diseases in people of northern European descent. In the United States, the incidence is approximately 1 in 3200, and the frequency of carriers of the gene is 1 in 29. CF involves many systems, including pulmonary, gastrointestinal, and reproductive organs.

Morbidity and mortality in CF is related to mucus accumulation, recurrent infection with unusual pathogens (eg, *Pseudomonas aeruginosa*), and increased inflammation in the lungs. In the past, CF resulted in death in early childhood, however, the Cystic Fibrosis Foundation now approximates the median survival age at 32 years and ranging from 7 months to 74 years. The increased survival age can be attributed to organ transplantation, improved nutrition, and new drug therapies.

The United States Cystic Fibrosis Foundation has three criteria for the diagnosis of CF: (1) the presence of one of the characteristic phenotypic pictures, (2) a history of CF in a sibling, and (3) laboratory evidence of a cystic fibrosis transmembrane conductance regulator protein abnormality as documented by an abnormal sweat chloride or identification of two CF mutations.

Pancreatic insufficiency is present at birth in approximately 65% of patients with CF, and an additional 15% develop it during infancy and early childhood. The remaining 20% who never develop pancreatic insufficiency have a better prognosis with fewer complications.

Fecal pancreatic elastase-1 is a reliable test for pancreatic insufficiency in infants 2 weeks and older as well as in older children at the time of diagnosis. It can also be used to detect pancreatic insufficiency in CF patients who were pancreatic sufficient. Sweat chloride testing is discussed in Chapter 12, Body Water and Electrolyte Homeostasis.

▶ NEOPLASMS

Pancreatic cancer is the fourth leading cause of death from cancer in both males and females in the United States. Approximately 95% of **pancreatic neoplasms** are exocrine pancreatic cancers: Two-thirds are found in the pancreatic head and one-third in the pancreatic body and tail. The remaining 5% of malignant tumors are islet cell tumors, including insulinomas and somatostatinomas described later in this chapter.[9]

Risk factors for pancreatic cancer are tobacco smoking, age, and predisposing medical conditions. Tobacco smoking has been implicated as a cause in approximately 30% of pancreatic cancers, and it is associated with a dose response: It is higher in two-pack-a-day smokers than in light smokers. Smoking is the strongest environmental factor correlated to pancreatic cancer. Risk of pancreatic cancer increases exponentially with age. The median age at diagnosis is 69 years in whites and 65 years in blacks, and it is rare in persons younger than 45 years. Predisposing medical conditions include long-standing chronic pancreatitis and diabetes mellitus. Abrupt onset of diabetes in a previously healthy, nonobese middle-aged or older patient with no family history of diabetes can be the first sign of pancreatic cancer. Significant weight loss is also a characteristic feature of pancreatic cancer.[9,11]

Ductal adenocarcinoma is the most common pancreatic cancer comprising 80% of pancreatic neoplasms. Only 2% of tumors of the exocrine pancreas are benign. Cystic neoplasm of the pancreas accounts for less than 5% of pancreatic tumors.[9]

▶ INSULINOMAS

Insulinomas, tumors of the pancreatic islets, are the most common type of tumor associated with hypoglycemia but one of the rarest causes of hypoglycemia. The incidence of insulinomas in the United States is 1 to 4 per million, with women slightly more prone than men with a ratio of 3:2. Insulinomas are neuroendocrine tumors of the pancreatic islet cells that produce excessive amounts of insulin. These small, benign tumors are difficult to locate but can be confirmed by transabdominal ultrasound and triple-phase spiral computed tomography. The most efficient method is intraoperative ultrasonography with careful mobilization of the pancreas by a surgeon experienced with insulinoma surgery. Insulinomas are benign in 90% of cases; however, they can be malignant in patients with multiple endocrine neoplasms (MENs).[11,12]

Insulinomas are diagnosed by the following laboratory findings: low plasma glucose, inappropriately elevated levels of insulin and C-peptide, and, to a lesser degree, high levels of proinsulin. Insulin is reported as the ratio of insulin to glucose with glucose in mg/dL and insulin in U/L. A ratio above 0.3 indicates inappropriate insulin production, too much insulin for the glucose level. Another ratio of insulin to C-peptide is used to detect the presence of exogenous insulin. The normal ratio of insulin to C-peptide is 1.0 A ratio less than 1.0 indicates the presence of exogenous (injected) insulin. See Chapter 6, Carbohydrates, for further information on hypoglycemia and insulinomas.[11,12]

▶ GLUCAGONOMAS

Glucagonomas, tumors that produce glucagon, are usually found in the pancreas and are characterized by hyperglycemia, weight loss, and a peculiar skin rash, necrolytic migratory erythema (NME). NMEs are described as painful, scaly, and erythematous patches. They may also be present in pseudoglucagonomas associated with inflammatory bowel disease, pancreatitis, malabsorption syndromes, and other malignancies in the absence of a glucagonoma. The pathogenesis of an NME has not been determined, but four major proposed etiologies are glucagon excess, nutritional deficiencies, inflammatory mediators, and liver disease.[15] Glucagonomas are usually found in patients 40 and older.

Diagnosis is determined by measuring plasma glucagon. The upper limit of normal is 200 ρg/mL, but patients with glucagonomas have levels above 1000 ρg/mL. Although pancreatitis, starvation, renal failure, and other conditions can result in elevated glucagon, the levels in these patients are rarely above 500 ρg/mL. Most glucagonomas are large at the time of diagnosis and are extremely aggressive.

▶ SOMATOSTATINOMAS

Somatostatinomas are tumors of the pancreas or intestine that secrete excess somatostatin. Somatostatin regulates and inhibits the release of hormones by many neuroendocrine cells in the brain, pancreas, and gastrointestinal tract. It inhibits gastric motility and secretion of gastric acid, and it blocks the endocrine and exocrine functions of the pancreas. Somatostatin has a marked effect on gastrointestinal transit time, intestinal motility, and absorption of nutrients from the duodenum. Somatostatin has been associated with inhibition of pancreatic proteolytic-enzyme secretion and gallbladder motility. Somatostatinomas are usually found in adults 50 and older and are more common in women than men.[13]

Clinical features are glucose intolerance (diabetes mellitus), gallbladder disease, and diarrhea often associated with steatorrhea. Diarrhea, described as 3 to 10 foul-smelling stools per day and steatorrhea of 20 to 76 g of fat per 24h, is common in pancreatic somatostatinomas. In one study, 75% of patients with pancreatic somatostatinomas had diabetes mellitus. Symptoms of pancreatic somatostatinomas are different than intestinal stomatostatinomas. Pancreatic somatostatinomas are usually large and malignant and found in the head of the pancreas, where they are responsible for bile duct obstruction. Eighty percent of pancreatic somatostatinomas are metastatic at diagnosis, which may be because of the late diagnosis. Symptoms are less pronounced than with insulinomas and glucagonomas, so the tumor often is not detected until the patient has extremely high levels of the hormone.[13]

SUMMARY

The pancreas, a pyramid-shaped organ located between the stomach and duodenum, has both exocrine and endocrine functions. The exocrine function of the pancreas consists of enzymes produced in the inactive form called *zymogens* that are stored in zymogen granules in the acinar cells of the pancreas. The pancreas synthesizes 22 digestive enzymes, including 15 proteases, enzymes (lipase, phospholipase A_2, cholesterol esterase) that catalyze lipids, and amylase, which catalyzes the digestion of complex carbohydrates. The endocrine function is located in the islets of Langerhans in which the alpha cells produce glucagon and the beta cells produce insulin.

The exocrine function is evaluated by several tests. The most common invasive test is secretin–cholecystokinin (CCK), which requires a gastroduodenal tube. Following stimulation by secretin and CKK, pancreatic juice is collected and the volume, pH, bicarbonate, and enzymes are measured Pancreatic elastase-1, pancreatic chymotrypsin, pancreatic serum enzymes, breath tests (C-mixed-chain triglyceride), urinary amylase excretion, fecal fat, phospholipase A_2, NBT-PABA, and fecal lipase are other tests of the exocrine function of the pancreas. Assessment of endocrine function is measured mainly through insulin and C-peptide measurements.

Pancreatic diseases include acute and chronic pancreatitis, diabetes mellitus, and cystic fibrosis. Pancreatitis is an inflammation of the pancreas that occurs as a result of autodigestion Zymogens, including trypsinogen and chymotrypsinogen, are prematurely activated in the pancreas and attack the walls of membranes and blood vessels. The two main etiologies of acute pancreatitis are alcohol and biliary tract disease or obstructive liver disease. Other causes are drugs, infections, inherited (genetic), mechanical or structural, metabolic, toxins, and idiopathic. Mortality varies with the etiology, but most cases of acute pancreatitis recover uneventfully.

Chronic pancreatitis is the irreversible damage to the pancreatic tissue with evidence of inflammation and fibrosis. In chronic pancreatitis, the endocrine and exocrine functions of the pancreas are destroyed. The most common cause is alcohol abuse. Other conditions associated with chronic pancreatitis are surgery, hyperlipidemia (hypertriglyceridemia), cystic fibrosis, and glucagonomas.

Complications of acute pancreatitis include adult respiratory distress syndrome (ARDS), cardiac complications, metabolic complications, gastrointestinal bleeding, and pancreatic infection and abscess. The most common complications of chronic pancreatitis are pseudocyst formation and mechanical obstruction of the duodenum and common bile duct. Adult respiratory distress syndrome (ARDS), a serious complication of acute pancreatitis, is caused by autodigestion of pulmonary capillaries by activated pancreatic enzymes, including phospholipase A_2. A variety of cardiac complications such as congestive heart failure (CHF), myocardial infarction, and cardiac arrhythmias may occur in severe acute pancreatitis. Metabolic complications include hypocalcemia, hyperglycemia, and hyperlipidemia. A pseudocyst is a collection of pancreatic juice that is enclosed by a wall of fibrous or granulation tissue—which can be single or multiple, large or small—located inside or outside of the pancreas.

Pancreatic cancer is the fourth leading cause of death from cancer in both males and females in the United States. Ductal adenocarcinoma is the most common pancreatic neoplasm, comprising 80% of pancreatic neoplasms. Only 2% of tumors of the exocrine pancreas are benign. Cystic neoplasm of the pancreas accounts for less than 5% of pancreatic tumors. Risk factors are tobacco smoking, age, and predisposing medical conditions (eg, chronic pancreatitis and diabetes mellitus).

Insulinomas, tumors of the pancreatic islets, are the most common type of tumor associated with hypoglycemia but one of the rarest causes of hypoglycemia. Insulinomas are benign in 90% of cases; they can, however, be malignant in patients with multiple endocrine neoplasms. Insulinomas are diagnosed by the following laboratory findings: low plasma glucose, inappropriately elevated levels of insulin and C-peptide, and, to a lesser degree, high levels of proinsulin.

Glucagonomas, tumors that produce glucagon, are usually found in the pancreas and are characterized by hyperglycemia, weight loss, and a peculiar skin rash, necrolytic migratory erythema (NME). Somatostatinomas are tumors of the pancreas or intestine that secrete excess somatostatin. Somatostatin regulates and inhibits the release of hormones by many neuroendocrine cells in the brain, pancreas, and gastrointestinal tract. Clinical features are glucose intolerance (diabetes mellitus), gallbladder disease, and diarrhea often associated with steatorrhea.

REVIEW QUESTIONS

LEVEL I

1. The main pancreatic duct joins the common bile duct from the liver and drains into the: (Objective 1)
   a. gallbladder.
   b. colon.
   c. duodenum.
   d. stomach.

LEVEL II

1. Which of the following is *not* a possible complication of pancreatitis? (Objective 3)
   a. hyperlipidemia
   b. hypoglycemia
   c. hypocalcemia
   d. congestive heart failure

REVIEW QUESTIONS (continued)

LEVEL I

2. The cells responsible for the production of pancreatic juice are: (Objective 3)
   a. beta cells.
   b. alpha cells.
   c. islet cells.
   d. acinar cells.

3. Zymogen granules contain: (Objective 2)
   a. gastrin.
   b. digestive enzymes.
   c. amylase.
   d. secretin.

4. Insulin is synthesized by which of following: (Objective 2)
   a. acinar cells.
   b. alpha cells of the islets of Langerhans.
   c. beta cells of the islets of Langerhans.
   d. delta cells of the islets of Langerhans.

5. A physician suspects a patient has pancreatitis. Which test would be most indicative of this disease? (Objective 5)
   a. creatinine
   b. LD
   c. amylase
   d. CK

6. Which of the following tests is used to evaluate the exocrine secretory function of the pancreas? (Objective 4)
   a. quantitative fecal fat
   b. amylase
   c. lipase
   d. secretin–CCK test

7. Which of the following is the test of choice for a noninvasive test to assess pancreatic insufficiency? (Objective 5)
   a. fecal elastase
   b. amylase
   c. secretin–CCK test
   d. phospholipase A_2

8. In the C-mixed-chain triglyceride breath test, which of the following is measured? (Objective 5)
   a. ^{13}C-octonate
   b. $^{13}CO_2$
   c. CO_2
   d. All of the above

9. The NBT-PABA test measures the activity of which of the following pancreatic enzymes? (Objective 5)
   a. trypsin
   b. chymotrypsin
   c. amylase
   d. lipase
   e. elastase

LEVEL II

2. Pseudocysts are a result of: (Objective 3)
   a. high levels of bicarbonate.
   b. activated enzymes.
   c. pancreatic juice and debris.
   d. digestive products.

3. Which of the following conditions is *not* associated with chronic pancreatitis? (Objective 2)
   a. alcoholism
   b. hyperlipidemia
   c. surgery
   d. hypocalcemia

4. Necrolytic migratory erythema (NME) is associated primarily with chronic pancreatitis and is more commonly found in: (Objective 3)
   a. acute pancreatitis.
   b. glucagonomas.
   c. somatostatinomas.
   d. insulinomas.
   e. chronic pancreatitis.

LEVEL I

10. The product measured in the NBT-PABA test is: (Objective 5)
    a. *N*-benzoyl-1-tyrosyl
    b. ρ-aminobenzoic acid
    c. phospholipase A_2
    d. elastase

11. Fecal fat is increased in all of the following *except:* (Objective 5)
    a. cystic fibrosis
    b. acute pancreatitis
    c. hepatitis
    d. Crohn's disease

12. Which of the following are the most common cause(s) of acute pancreatitis? (Objective 8)
    a. gallstones
    b. abdominal trauma
    c. acute alcohol ingestion or drug use
    d. chronic alcohol use and biliary tract disease
    e. hypertriglyceridemia

13. The critical step precipitating the pathogenesis of acute pancreatitis is the premature release of: (Objective 8)
    a. amylase
    b. lipase
    c. elastase
    d. trypsin

14. In acute pancreatitis, serum enzymes that increase significantly include: (Objective 5)
    1. creatine phosphate
    2. amylase
    3. aldolase
    4. lipase
    a. Only 1, 2, and 3 are correct.
    b. Only 1 and 3 are correct.
    c. Only 2 and 4 are correct.
    d. All are correct.

15. Which of the following does not meet Ranson's indicators of severity in acute pancreatitis? (Objective 9)
    a. BUN: 55 mg/dL
    b. glucose: 250 mg/dL
    c. calcium: 84 mg/dL
    d. AST: 270 mg/dL

16. Chronic pancreatitis is more commonly found in: (Objective 10)
    a. men
    b. women
    c. men and women age 60 and older.
    d. individuals with a normal pancreas before an acute attack.

LEVEL II

REFERENCES

1. Obideen K. Chronic pancreatitis. Emedicine (www.emedicine.com/MED/topic1721.htm), accessed May 30, 2007.

2. DiMagno MJ, DiMagio EP. Chronic pancreatitis. *Curr Opin Gsatroenterol* (2006) 22, 5: 487–497.

3. Kalivianakis M, Minich DM, Bifleveld C, et al. Fat malabsorption in cystic fibrosis patients receiving enzyme replacement therapy is due to impaired intestinal uptake of long-chain fatty acids. *Am J Clin Nutr* (1999) 69: 127–340.

4. Nevalainen TJ. Serum phospholipases A_2 in inflammatory diseases. *Clin Chem* (1993) 39, 12: 2453–2458.

5. Scharpe S, Illano L. Two indirect tests of exocrine pancreatic function evaluated. *Clin Chem* (1987) 33, 1: 5–12.

6. Walkowiak J, Herzig K-H, Stuzykala K, et al. Fecal elastase-1 is superior to fecal chymotrypsin in the assessment of pancreatic involvement in cystic fibrosis. *Pediatrics* (2002) 110, 1: e-7.

7. Stein J, Jung M, Sziegoleit A, et al. Immunoreactive elastase 1: Clinical evaluation of a new noninvasive test of pancreatic function. *Clin Chem* (1996) 43, 2: 222–226.

8. Merck Manual Professional. Acute pancreatitis (www.merck.com/mmpe/print/sec02/ch015/ch015b.htmL), accessed May 21, 2007.

9. Sun W, Haller D. Pancreatic, gastric and other gastrointestinal cancers: Pancreatic cancer. ACP Online @002 (www.medscape.com/voewartoc;e/534510print), accessed May 31, 2007.

10. National Digestive Diseases Information Clearinghouse. Pancreatitis (www.digestive.niddk.nih.gov/ddiseases/pubs/pancreatitis/index.htm), accessed May 21, 2007.

11. Erickson RA. Pancreatic cancer. Emedicine (www.emedicne.com/MED/topic1712.htm), accessed June 28, 2007.

12. John Hopkins Pancreas Web Site. Islet cell/endocrine/tumors of the pancreas (http://pathology2.jhu.edu/pancrease/endocrineislet.cfm), accessed June 5, 2007.

13. Vinik A. Chapter 8: Somatostatinomas (August 2, 2004) (www.endotext.org/guthormones/guthomone8/guthormone8.htm), accessed June 5, 2007.

14. DiMagno MJ, DiMagno EP. Chronic pancreatitis. *Curr Opin Gastroenterol* (2005) 21, 5: 544–554.

15. Tierney EP, Badger J. Etiology and pathogenesis of necrolytic migratory erythema. *Medscape Gen Med* (2004) 6, 3 (accessed June 5, 2007).

Suggested Readings

1. Draganov P, Toskes PP. Chronic pancreatitis. *Curr Opin Gastroenterol* (2002) 18, 5: 558–562.

2. Dragnov P, Forsmark C. Diseases of the pancreas. *ACP Med* (2006) (www.medscscape.com/viewarticle/532779), accessed May 30, 2007.

3. Duffy JP, Reber HA. Pancreatic neoplasms. *Curr Opin Gastroenterol* (2003) 19, 5: 458–466.

4. Meng Q, Luxton G. Recurrent confusion and seizures in an adult male. *Lab Med* (2004) 35, 8: 478–483.

5. Service FJ. Hypoglycemia: Conditions that cause hypoglycemia. ACP Medicine Online (2002), accessed June 4, 2007.

6. Xydakis AM, Gagel RF. A man with low blood glucose and coarse facial features. *Medscape Diab & Endocrin* (2003) 5, 1 (accessed June 5, 2007).

18

Cardiac Function

■ OBJECTIVES—LEVEL I

Following successful completion of this chapter, the learner will be able to:

1. Explain the inflammatory response associated with atherosclerosis.

2. Define *acute coronary syndrome* (*ACS*).

3. List five factors that define an ideal cardiac biomarker.

4. Identify two biomarkers used to evaluate each of the following events associated with vascular inflammation:
   a. Proinflammatory cytokine release
   b. Plaque destabilization
   c. Plaque rupture
   d. Acute-phase reactant response
   e. Ischemia
   f. Necrosis

5. Define *hs-CRP* relative to cardiac usefulness.

6. Identify the clinical usefulness of the following cardiac biomarkers:
   a. Lipoprotein (a)
   b. Lipoprotein-associated phospholipase A2
   c. Glycogen phosphorylase isoenzyme BB
   d. Omega-3 fatty acids
   e. Matrix metalloproteinases
   f. Placental growth factor
   g. Oxidized low-density lipoprotein (LDL)
   h. Myeloperoxidase
   i. Cardiac troponin I and T
   j. Brain-type natriuretic peptide and NT-proBNP
   k. Ischemia-modified albumin (IMA)

7. Discuss the advantages of point-of-care testing (POCT) for cardiac biomarkers.

8. Discuss the temporal relationship and concentration of each the following relevant to acute myocardial infarction (AMI):
   a. Myoglobin
   b. CK-MB
   c. Cardiac troponin I

■ OBJECTIVES—LEVEL II

Following successful completion of this chapter, the learner will be able to:

1. Distinguish between biomarkers used for each of the stages resulting in an acute coronary syndrome (ACS), including
   a. Inflammation
   b. Ischemia
   c. Necrosis

2. Explain the differences between cTnT and cTnI and include the following:
   a. Function
   b. Specificity

3. Explain the difference between BNP and NT-ProBNP and include the following:
   a. Function
   b. Specificity

4. Explain the principle of the cobalt binding assay (CBA) used to measure IMA.

5. Identify potential sources of interferences both analytically and physiologically for the following biomarkers:
   a. Troponins
   b. BNP and NT-proBNP

6. Identify the limitations of the following biomarkers relative to various disease states:
   a. CK-MB
   b. Myoglobin
   c. Cardiac troponin I and T

KEY TERMS

Acute coronary syndrome (ACS)
Angiogenesis
Arteriosclerosis
Atherosclerosis
Atheroma

Biomarker
Coronary artery disease
 (CAD)
Cytokines
Endothelium

Ischemia
Necrosis
Risk factor
Temporal
Troponin

 A CASE IN POINT

A 35-year-old male presented to the emergency department complaining of chest pain. The chest pain started at about 11:30 p.m. The pain was "kind of sharp" and pressure was continuous. The patient said that he had no shortness of breath, sweating, or nausea. The physician ordered an electrocardiogram, blood chemistries, and a complete blood count (CBC). The laboratory results are shown below.

Tests	Patient Results	Reference Values
Blood Chemistries:		
Troponin I	16.2 ng/mL	<1.5 ng/mL
Total creatine kinase	1,581 U/L	20–200
CK-MB	9.3 µg/L	<5
CK-index	0.6%	<4
Sodium	138 mmol/L	136–145
Potassium	3.6 mmol/L	3.5–5.1
Chloride	106 mmol/L	98–107

Bicarbonate	23 mmol/L	23–29
Creatinine	1.1 mg/dL	0.62–1.10
BUN	22 mg/dL	6–20
Glucose	132 mg/dL	74–100
Magnesium	1.6 mg/dL	1.6–2.6
Phosphorus	2.8 mg/dL	2.5–4.5
Calcium	8.8 mg/dL	8.6–10.2
CBC		
WBC	7.0×10^9/L	4–10
RBC	5.0×10^{12}/L	4.6–6.1
Hemoglobin	13.7 g/dL	13.5–18
Hematocrit	38.7%	41–53
Platelet count	234×10^9/L	150–400
Neutrophils	60%	45–66
Lymphocytes	20%	10–23
EKG results		

ST elevation over the precordial leads, which indicates that the patient had an acute myocardial infarction.

Issues and Questions to Consider

1. Identify specific diseases or conditions that would result in an elevated serum troponin I.

2. What is the current acceptable emergency department (ED) medical protocol for diagnosing an acute myocardial infarction based on (World Health Organization [WHO], American College of Cardiology [ACC], European Society of Cardiology [ESC])?

3. Identify additional diagnostic tests that might be useful for making a diagnosis and establishing a treatment plan.

The physician ordered serial measurements of troponin I and an echocardiogram. The echocardiogram is a procedure that uses sound waves to show motions of the heart (e.g., chamber wall motions).

Laboratory results of serial troponin I (ng/mL) measurements are as follows:

Day	Time (hour)	cTnI Concentrations
Day 1	1337	16.2
Day 1	1948	11.9
Day 2	0339	11.0
Day 2	1220	10.9
Day 2	2020	7.6

Results of echocardiogram: negative for wall motion abnormality.

4. Do the results for serum troponin I follow a reasonable pattern of concentration versus time?

5. Is it reasonable to assume that this patient had an acute myocardial infarction?

WHAT'S AHEAD

▶ Discussion of the anatomy and function of the heart
▶ Description of the events leading to acute coronary syndrome and cardiac dysfunction
▶ Discussion of cardiac diseases
▶ Description of cardiac biomarkers used to evaluate patients with heart-related problems
▶ Identification of assays used to measure cardiac biomarkers

The focus of this chapter is on the cardiac blood biomarkers used to evaluate the processes associated with cardiovascular dysfunction. A **biomarker** is a biochemical feature or facet that can be used to measure the progress of disease or the effects of treatment. Even though the total number of deaths from cardiac disease has steadily declined for several decades, there remains a significant number of Americans who suffer from cardiac diseases. Knowledge of the physiochemical attributes of each biomarker presented along with methodologies and clinical usefulness will allow the laboratorian to properly address issues relating to each biomarker.

The word **atherosclerosis** originates from the Greek word for gruel, which is *Athero*, and refers to the massive accumulation of lipids in vascular lesions. Thus for decades atherosclerotic vascular disease (ASVD) was considered to be a bland lipid storage diseases. In 1976 a paper published in the new *England Journal of Medicine* by R. Ross proposed the idea that ASVD was in fact the result of an inflammatory process.[1] He proposed the idea that ASVD begins with an initial insult to the endothelial cells and subsequent dysfunction via the deleterious effects of known cardiac **risk factors**, such as oxidized LDL, infection, hyperglycemia, hypertension, hyperhomocysteinemia, and smoking.

The proponents of the significance of vascular inflammation did not discount the importance of the contribution of lipids to the development of ASVD. They believed that ASVC was the result of both processes. **Acute coronary syndrome (ACS)** is described as a continuum of clinical signs and symptoms ranging from unstable angina (chest pain) to non-Q-wave (refers to a unique EKG pattern) acute myocardial infarction (AMI) and Q-wave AMI.[2] The focus of medicine in the past has been to monitor a patient's lipids; recommend treatment modalities for hyperlipidemia; and, when a patient has a heart attack (AMI), diagnose, treat, and provide prophylaxis for the remainder of the patient's life. The laboratory has always played a key role by providing laboratory tests that help clinicians care for their patients.

The laboratory will continue to play an important role in patient care by providing laboratory tests that will supply information relevant to vascular inflammation and allow clinicians to take a more proactive role in their patients' well-being. For example, cytokines such as interleukin-6 and tissue necrosis factor-α (TNF-α) can provide information about the proinflammatory stage. **Cytokines** are locally acting autocoid polypeptides that mediate vasoconstriction by interacting with phospholipaseC–linked receptors. These cytokines are released early in the inflammatory response preceding an ACS. Measurement of high-sensitivity C-reactive protein (hs-CRP), an acute-phase reactant, can provide information regarding risk stratification for a patient. Monitoring the blood levels of myeloperoxidase (MPO) may provide the clinician with insight into the extent of progress of plaque formation. Finally, the ischemic state of ACS can be evaluated by measuring ischemia-modified albumin (IMA).

In this chapter several diseases of the heart relevant to clinical laboratory testing will be presented. A discussion of the laboratory tests utilized by the clinician for diagnosing and monitoring these diseases will follow. Specific test parameters (e.g., expected values, diagnostic sensitivity, and specificity) will be included where appropriate.

▶ THE HEART

An adult heart weighs about 2 kg and is about the size of a fist. It lies diagonally in the mediastinum, which is an area above the diaphragm and between the lungs. The wall of the heart has three layers: a pericardium, myocardium, and endocardium. The pericardium is a double-walled sac that encloses the heart. This pericardial sac prevents displacement of the heart during acceleration or deceleration and serves as a physical barrier between the heart and lungs. Pain receptors and mechanoreceptors are located in the pericardium and elicit reflex changes in blood pressure and heart rate.

The thickest layer of the heart wall, the myocardium, is made up of cardiac muscle and is anchored to the heart's fibrous skeleton. Myocardium thickness varies among the chambers of the heart and is dependent on the amount of resistance the muscle must overcome to pump blood from that chamber. Connective tissue and squamous cells make up the internal lining of the myocardium and are called the endocardium. This lining is continuous with the **endothelium**, which has a single cell thickness and lines all the arteries, veins, and capillaries of the body. This creates a continuous, closed circulatory system.

The heart has four chambers: the left and right atria, and the left and right ventricles. The atria are smaller than the ventricles and have thinner walls. The ventricles have a thicker myocardial layer and contribute most of the weight of the heart. An image of the heart is shown in Figure 18-1 ■.[3]

Arteriosclerosis, or hardening of the arteries (especially the coronary arteries), results from the deposition of tough, rigid collagen inside the vessel wall and around the **atheroma**, which is described as fatty degeneration or thickening of the walls of the larger arteries. The large arteries of the heart include the left and right coronary arteries, the left anterior descending artery, and the circumflex artery.

Atherosclerosis refers to the most common form of arteriosclerosis affecting the arterial blood vessels and is caused by the formation of multiple plaques within the coronary arteries. Coronary arteries supply the cardiac tissue with blood containing oxygen and other nutrients. If these coronary vessels are occluded and the blood supply is reduced, then myocardial tissue supplied by the vessels begins to lose function. This process is called cardiac **ischemia**. If cardiac ischemia is allowed to continue, cells will begin to die; this cell death is called **necrosis**.

The cardiac cycle characterized by a continuous rhythmic repetition depends on the transmission of electrical impulses throughout the myocardium. A cardiac cycle results in the pumping action of the heart and consists of contraction and relaxation of the myocardial layer of the heart wall. During relaxation or diastole, the ventricles fill with blood. This process is followed by systole or contraction, which propels the blood out of the ventricles and into the systemic circulation.

The electrical impulses or signals travel through the myocardium via a conduction system consisting of specialized cells that enable the heart to generate and transmit electrical potentials without stimulation from the nervous system. The electrical conductivity through the atria and ventricles creates a characteristic voltage pattern referred to as an electrocardiogram (ECG, EKG), as shown in Figure 18-2 ■.

The electrical events that allow the atria and ventricle muscle cells to contract and relax follow an ordered pattern that begins with the electrical discharge of the atria at the sinoatrial node (SA). The cells are depolarized, leaving the inside of the cell positive or less negative in respect to the exterior surface of the cell. The electrical charge continues to the atrioventricular node (AV), which causes the ventricular muscle cells to depolarize. This is represented in the ECG as the QRS wave. Finally, the left ventricle repolarizes (T wave).

Abnormalities in electrical conductivity through the heart create abnormal wave patterns, which are interpreted by the clinician. The most significant parts of the waveform relevant to AMI are the Q wave, the ST segment, and the T wave. ECG changes will be discussed later in this chapter.

The cells of cardiac muscle (myocardium) and of muscle that make voluntary movement possible (skeletal muscle) are similar in structure, function, and microscopic appearance. Differences between cardiac and skeletal muscles reflect heart function. For example, cardiac myocytes are arranged in branching networks throughout the myocardium that make up the myofibril, whereas skeletal muscle cells tend to be arranged in parallel units throughout the length of the muscle. The attachment of myocytes creates the striated appearance of cardiac muscle.

Cardiac muscle contraction is the result of a series of electrical, chemical, and mechanical processes. Integral to this process are molecules of **troponin** and tropomyosin. These

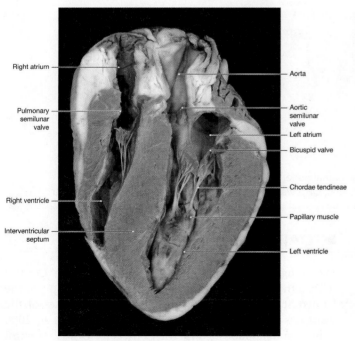

Right atrium — Aorta
Pulmonary semilunar valve — Aortic semilunar valve
— Left atrium
— Bicuspid valve
— Chordae tendineae
Right ventricle — Papillary muscle
Interventricular septum — Left ventricle

■ **FIGURE 18-1** The anterior view of a human heart.

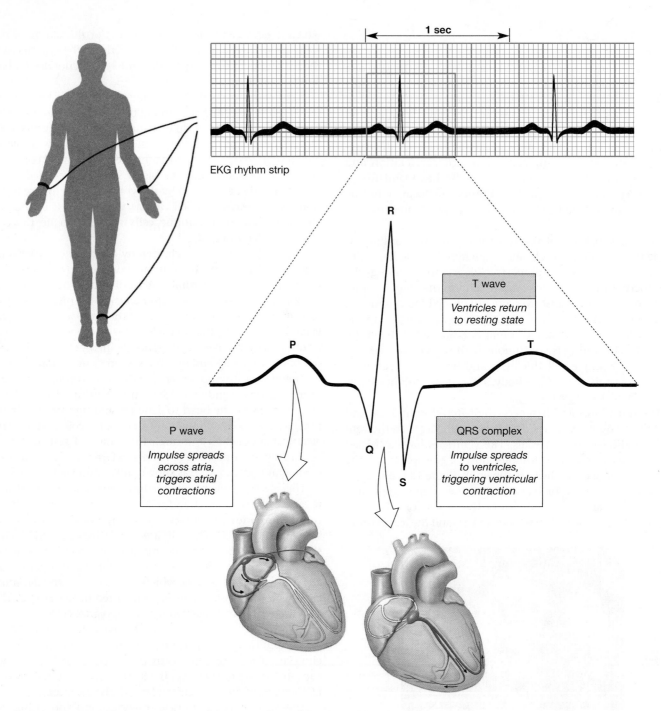

1 sec

EKG rhythm strip

R

T wave

Ventricles return to resting state

P

T

P wave

Impulse spreads across atria, triggers atrial contractions

Q

S

QRS complex

Impulse spreads to ventricles, triggering ventricular contraction

■ **FIGURE 18-2** Electrocardiogram of the human heart. SA—sinoatrial node, AV—atrioventricular node, LV—left ventricle. P wave is caused by the depolarization of the atrial muscle tissues, QRS wave corresponds to depolarization of ventricular muscle, and T wave corresponds to ventricular repolarization.

two compounds associate to form the troponin–tropomyosin complex. The troponin complex consist of three components: (1) troponin T, which aids in the binding of the troponin complex to actin and tropomyosin; (2) troponin I, which inhibits the ATPase of actomyosin; and (3) troponin C, which has binding sites for calcium ions.

Cardiac troponin will be discussed in more detail in relationship to cardiac diseases later in this chapter.

► CARDIAC DISEASES

According to data from the Centers for Disease Control (CDC), the age-adjusted death rate per 100,000 people in the United States (both males and females) for diseases of the heart has declined from about 560 in 1960 to 211 in 2006 (based on data published by the Centers for Disease Control in the National Vital Statistics Report [NVSR] in) 2009.[4] The

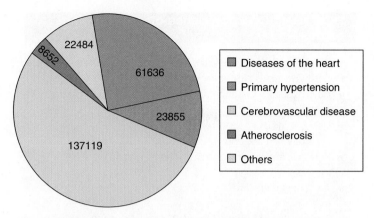

■ FIGURE 18-3   Number of deaths associated with major cardiovascular diseases for males and females according to CDC data for 2005.

actual number of individuals (both males and females) who died of major cardiovascular diseases (CVDs) was 823,746. A distribution of the number of deaths within this category is shown Figure 18-3 ■ Interestingly, ischemic heart disease, which includes AMIs, accounts for nearly 68% of the deaths in this category. Thus, although the total number of individuals with heart-related diseases is declining, there is still a large number of Americans suffering from heart disease. American Heart Association (AHA) data based on a survey from 2005 revealed that the prevalence of CVD in Americans (both male and female) is approximately 80.7 million, and another 8.1 million have had an AMI.[5]

A major emphasis in heart disease research is the discovery and evaluation of biomarkers that may be released into the systematic circulation in the early stages of vascular dysfunction. For example, studies are currently in progress to evaluate the significance of proinflammatory cytokines such as interleukin-6 (IL-6), TNF-α, and others released during the inflammatory phase of atherosclerosis. Other research is ongoing to identify biomarkers of ischemia, such as IMA, which could give clinical insight into a patient's condition before it worsens and ultimately results in an ACS. Biomarkers of cardiac cell necrosis have been available for decades. Examples of cardiac cell necrosis include creatine kinase-MB (CK-MB) isoenzyme and, more recently, cardiac troponin I (cTnI) and cardiac troponin T (cTnT). Theses biomarkers have made a significant impact on reducing the overall death rate as a result of cardiac disease.

Clinical laboratory tests are also available to assist the clinician in evaluating patients' suffering from other types of cardiac diseases (e.g., congestive heart failure [CHF]). Traditionally the laboratory was not a part of the diagnostic protocols for CHF until studies showed that brain type natriuretic peptide (BNP) and N-terminal (NT-ProBNP) correlated to the severity of CHF. Currently tests for BNP or NT-ProBNP are requested routinely on patients being "worked up" for shortness of breath often associated with CHF.

Coronary artery disease (CAD), also called coronary heart disease (CHD) and atherosclerotic heart disease (AHD),

results from the accumulation of atheromatous plaques within the walls of the arteries that supply the myocardium. CAD can be classified as chronic CAD, ACS. and sudden death, from lowest to highest risk, respectively. CAD may present clinically from asymptomatic conditions to unexpected cardiac collapse. Chronic CAD is second to coronary atherosclerosis leading to diminished oxygen supply and a stable pattern of coronary ischemia.[6]

▶ CARDIAC BIOMARKERS

Cardiac biomarkers (e.g., cTnI, cTnT, and CK-MB) can be detected and measured because they are released into the systemic circulation via disruption in the sarcolemma. The release of cardiac biomarkers is influenced by several factors:

- Cytosolic enzymes
- Subcellular location
- Molecular mass
- Plasma clearance
- Concentration gradients

The key to better understanding the clinical utility of cardiac biomarkers is knowledge of the process of plaque development, which may have deleterious effects on the myocardium. The current hypothesis—which can now be challenged and studied because current methods and instrumentation can measure analytes, especially protein substances—is that ACS begins with disruption in the one-cell thick-endothelium. The inflammatory process begins when endothelial cells undergo inflammatory activation and there is an increased expression of selectin, vascular cell adhesion molecule-1 (VCAM-1), and intercellular adhesion molecule-1 (ICAM-1), the latter of which promotes adhesion of monocytes. The following are additional examples of proteins that induce adhesion of monoctyes:

- Proinflammatory cytokines (e.g., IL-6, IL-1, TNF-α)
- Acute-phase protein (e.g., CRP)
- Protease-activated receptor signaling (PARS)
- Oxidized low-density lipoprotein (ox-LDL) uptake
- Soluble CD 40 ligand (sCD40L) interactions with CD40 receptor molecules

Once monocytes aggregate, they migrate inward to the intima layer of the vessel. The monocytes begin to modify lipoproteins (e.g., LDL), which gives rise to lipid-laden macrophages or foam cells that are characteristic of early atherosclerotic plaque development.

Within the atheroma, the foam cells begin to secrete proinflmamatory cytokines that serve to promote and maintain leukocyte adhesion. Mast cells, T cells, and dendrite cells also present themselves in the area of the athermanous plaque and may be incorporated into it. If the risk factors inducing endothelial dysfunction and inflammation remain, the atheroma will progress from a fatty streak to a larger and more complex plaque. Fatty streaks are a description of what is seen in tissue when lipids are present. These streaks look similar to the 'marbling effect" in beef steaks.

The remaining process that eventually leads to plaque destabilization and rupture involves smooth cell migration toward the intima and continued release of cytokines. Enlargement of the plaque within an artery can result in lumen obstruction, which causes a reduction of blood flow and results in a clinical symptom such as angina or chest pain.[7] Biochemical markers associated with the events leading to ACS are shown in Figure 18-4 ■.[8] In summary, the development of

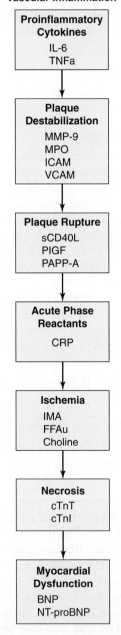

Vascular Inflammation

■ FIGURE 18-4   Progression of CVD to cell death or cardiac dysfunction and associated biomarkers.

BOX 18-1   Characteristics of ideal cardiac biomarkers.

► Smaller markers are released faster from injured tissues.
► Soluble cytoplasmic marker is preferred to structural markers.
► Have absolute cardiac tissue specificity and should not exist in other tissues.
► Useful to differential between reversible (ischemia) and irreversible (necrosis) damage.
► Releases from the myocardium should be complete following injury.
► Amount of marker released should be in direct proportion to the size of injury (infarct sizing).
► Remain elevated long enough (12 to 24 hr) to be detected in the serum of the "late presenter." This individual is described as a person who experiences a cardiac-related event such as angina or heart attack but delays going to the hospital for evaluation.
► For risk stratification, there should be a correlation between outcome and the presence or absence of a marker in serum or the degree of elevation of the marker above "normal."
► Should be cleared rapidly to allow diagnosis of recurrent injury.
► Should be useful for monitoring of reperfusion and reocclusion.
► Assays should be relatively easy and quick to perform.

many cardiac diseases begins with vascular inflammation and progresses to plaque development, plaque destabilization, plaque rupture, and ischemia and terminates with cell death or necrosis.

A biomarker for cardiac disease must possess specific characteristics to make it valuable to clinicians. Every biomarker has at least one salient feature that makes it potentially useful for patient care. Some biomarkers (e.g., cTnI) have several characteristics that make them valuable to clinicians in providing optimum care for the patient. A list of the characteristics of an ideal cardiac marker is provided in Box 18-1 ■.[9] The greater the number of these characteristics a biomarker has, the more useful it is to the clinician.

Each cardiac biomarker provides the clinician with a unique purpose that spans the breadth of clinical usefulness. For example, a biomarker may be used to assist a physician in diagnosing a cardiac-related problem. Many biomarkers are used for evaluating risk stratification. **Risk stratification** can be described as the layering or gradation of **risk factors**, which are described as elements or constituents that may put someone's health in peril (e.g., smoking, high blood pressure, and high blood cholesterol).

The following is an example of an application of these two terms. If a person smokes, overeats, and has diabetes (three risk factors) he has a much higher risk of developing CHD and therefore would be placed in a higher risk strata (risk stratification) than a person who just smokes. Measuring certain biomarkers may be useful for monitoring the progress of a patient or evaluating the prognosis relevant to a specific disease.

Laboratory assays for cardiac biomarkers are primarily immunoassays using a variety of labels, including chemiluminescence compounds, flurophores, enzymes, and other techniques (e.g., chromatography, electrophoresis, and enzyme-linked immunosorbent assay [ELISA] have also been used). Methods for measuring cardiac biomarkers have been incorporated into POCT devices, automated analyzers, and nonautomated analyzers. In the future, more sophisticated techniques, including molecular diagnostics and mass spectroscopy, will replace many of the methods currently used.

Parameters that require attention before implementing laboratory assays for cardiac biomarkers include the following:

- Analytical imprecision
- Detection limits
- Calibration characteristics
- Assay specificity
- Assay standardization
- Preanalytical issues
- Appropriate reference intervals

For each of the following cardiac biomarkers, a discussion will be provided that includes a description of the biomarker, identifications of assays to measure the biomarker, and the clinical usefulness of such assays.

TROPONINS

Myofibrils are cylindrical elements of skeletal muscle containing smaller filaments called myofilaments. Each of these myofibrils is made up of repeating units or sarcomeres. The repeating arrangement of sarcomeres imparts a striped or striated appearance to the muscle tissue. Both skeletal and cardiac muscles are referred to as striated muscle. Troponin is a regulatory protein of the myofibril.

Troponin is a heterotrimer consisting of troponin T (which binds tropomyosin), troponin I (which binds to actin and inhibits contraction), and troponin C (which contains four calcium binding sites and regulates contraction). The subunits exist as a number of isoforms. All of the isoforms are distributed between cardiac muscle and slow- and fast-twitch skeletal muscle. There are two major isoforms of troponin C, and they are found in human heart and skeletal muscle. Isoforms of cardiac-specific troponin T (cTnT) and cardiac-specific troponin I (cTnI) have been identified and are the products of specific genes. cTnT is a 37-kDa protein encoded by a different gene than encodes skeletal muscle isoform. An eleven-amino-acid-terminal residue gives this marker unique cardiac specificity. cTnI is a 22-kDa protein that regulates striated muscle contraction and is not expressed in skeletal muscle, whereas cTnT is. Three isoforms of cTnI exist, with only one located within cardiac myocytes. The cTnI isoform has a post-translation 31-amino-acid residue on the amino terminal end compared with skeletal muscle TnI, which confers unique cardiac

specificity. In humans, cTnT isoform expression has been demonstrated in patients with muscular dystrophy, polymyositis, and end-stage renal disease (ESRD).

The location of troponin is along the myofibril. Each troponin heterotrimer interacts with a single tropomyosin molecule, which in turn interacts with seven actin monomers. The troponin complex also interacts directly with actin filaments. Thus, the coordinated interaction among troponin, tropomyosin, and actin allows actin–myosin interactions to be regulated by changes in calcium ion concentration.[10]

✓ Checkpoint! 18–1

Differentiate cTnI from cTnT.

Cardiac Troponin Assays

Cardiac troponin I assays are available in many immunoassay formats. Until recently there was not a material that could be used as a primary reference to standardize all cTnI immunoassays. Therefore, correlation among the different assays was difficult and the correlation statistics were marginal. The NIST has developed SRM number 2921—Human Cardiac Troponin Complex, which is available for use by laboratories interested in assessing their cTnI assays.

Another confounding issue with cTnI assays is that analyte concentrations between reagent systems tend not to agree due to the different epitopes recognized by the different antibodies used in individual assay platforms. Therefore, any comparisons between different cTn assays should be avoided because troponin circulates in various forms (see below).[11]

Circulating forms of cTnI include the following:[12]

- free form
- bound as a two-unit binary complex (cTnI–cTnC)
- bound as three-unit tertiary complexes (cTnT–cTnI–cTnC)

Currently assays using monoclonal antibodies are being developed to be more specific for the circulating forms of cTnI.

There is only one Food and Drug Administration (FDA)–approved electrochemiluminescence immunoassay (ECLIA) for cTnT. The manufacturer of this assay has licensed its technology to several other manufacturers to be incorporated into their assays platforms and to develop POCT devices. The ECLIA is in its third generation, and there is no interference with skeletal muscle TnT as was found in the first generation of ELISA. Thus, the assay used to measure cTnT levels has cardiac specificity equivalent to that of assays for cTnI.[13]

Troponin is the biomarker of choice for assessing cardiac injury in patients with renal failure, including those with ESRD receiving long-term dialysis. It has been shown that asymptomatic ESRD patients are more likely to have elevated levels of cTnT than cTnI even with the newer third-generation assays, which are not supposed to detect cross-reacting isoforms from skeletal muscle.[13]

cTnI and cTnT are very good biomarkers for cardiac dysfunction because they possess many of the qualities associated with an ideal biomarker. These qualities include the following:

- They are specific for cardiac tissue.
- They have a high level of diagnostic specificity and sensitivity.
- They possess early release kinetics after an AMI.
- They remain elevated for a long interval of time.
- They have very low to undetectable concentrations in serum from patients without cardiovascular disease.
- There are relatively few interfering substances.

Unfortunately, during the past few years, several substances have been identified that negatively affects select assays for cTnI. Several examples are listed below along with their impact on cTnI assays.[14]

- False positive
  - Heterophile antibodies
  - Rheumatoid factor

- False negative
  - Bilirubin
  - Hemoglobin
  - Circulating cTnI autoantibodies
  - Interfering factor (IF)

Laboratorians should be aware that cardiac troponin might be elevated in diseases and conditions other than ACS. Several of these are listed in Box 18-2 ■.[15,16]

 Checkpoint! 18–2

List five conditions other than ACS in which cTnI or cTnT may be elevated in blood.

The rationale for requesting troponin measurements by clinicians has evolved over a 40-year period. The World Health Organization (WHO) initially proposed the diagnostic criteria for AMI, which includes the following:[17]

- History of characteristic chest pain
- Diagnostic changes in the EKG
- Changes in serum enzyme levels

The WHO criteria were later modified by the European Cardiology Society in collaboration with the American College of Cardiology, which recommended the following:[18]

- Promote troponin to a pivotal role.
- Relegate CK-MB to a secondary role.
- Eliminate the need for total CK measurement.

These guidelines are routinely used for differentiating AMI from unstable angina. Thus the routine use and monitoring of the concentrations of troponin in patients was encouraged to assist clinicians in the following:

- Diagnosing AMI.
- Establishing a high-risk profile based on elevated blood levels in the appropriate clinical stetting. The guidelines

BOX 18-2   Elevated troponin levels in conditions other than ACS.

- Trauma:
  - Contusion
  - Ablation
  - Pacing
  - Cardioversion
  - Endomyocardial biopsy
  - Cardiac surgery
- Nonatherosclerotic ischemia (e.g., due to cocaine or other sympathomimetic agent or coronary vasospasm)
- Hypertension
- Hypotension, often with arrhythmias
- Cerebrovascular accident (stroke)
- Rhabdomyolysis (cardiac injury)
- Post-op noncardiac surgery
- Renal failure
- Diabetes mellitus
- Hypothyroidism
- Myocarditis
- Pulmonary embolism
- Sepsis
- Inflammatory disease (e.g., myocarditis, parvovirus B19)
- Percutaneous coronary intervention without complications
- Burns (especially if total body surface area affected is >30%)
- Vital exhaustion
- Transplant-related vasculopathy
- Severe asthma
- Drug toxicity (e.g., adriamycin, 5-fluorouracil, herceptin, and snake venoms)
- Amyloidosis
- Heart failure (e.g., congestive heart failure or CHF)
- Aortic valve disease
- Cocaine-induced rhabdomyolysis
- Chest injury from motor vehicle accident
- Pulmonary conditions, diseases, and syndromes
- Hypereosinophilic syndrome—myeloproliferative type (mIHES)

recognize that neither the clinical presentation nor the EKG provide adequate clinical sensitivity and specificity.

The result of the recommendations of these organizations was the establishment of an evaluation protocol for patients presenting to the emergency departments with chest pain.[19] If a patient presents to the emergency department (ED) with chest pain, an EKG is performed. If the patient has an abnormal electrical cardiac rhythm (e.g., elevation of ST segment, called ST segment elevation myocardial infarction [STEMI]; ST-segment depression [NSTEMI]; or inverted T wave), the patient will be managed accordingly. For example, a patient with STEMI will be given aspirin, and blood will be drawn for cardiac testing, coagulation testing, and a complete blood count (CBC). The important point for laboratory staff to realize is

TABLE 18-1

Rise and fall (temporal) patterns of cardiac markers.

Enzyme	Starts to Rise (hours)	Peaks (hours)	Returns to Normal (days)
Total CK	4–6	24	3–4
CK-MB	4	18	2
Myoglobin	1–3	12	1
Troponin T	4–6	10–24	10
Troponin I	4–6	10–24	4

that almost every patient entering an ED complaining of chest pain will have cardiac biomarkers requested by the clinician; therefore, knowledge of these biomarkers is important.

In addition to the aforementioned protocols, the clinician continues to monitor blood levels of cardiac-related biomarkers over time to determine their **temporal** patterns (which relates to time) versus concentration or activity. These temporal patterns are characteristic of the markers and are shown in Table 18-1 ✪. Other biomarkers shown in Table 18-1 will be discussed later in this chapter.

> ✓ **Checkpoint! 18–3**
>
> *Identify five characteristics of cTnT and cTnI that make them useful biomarkers for acute myocardial infarction.*

CREATINE KINASE MB ISOENZYME

There are three cytosolic and one mitochondrial isoenzyme of creatine kinase (CK). The cytosolic isoenzymes are dimers composed of two monomers subunits designated as M (muscle) and B (brain). Isoenzyme CK-BB is found in high activity in the brain. The cytosolic isoenzymes are designated as CK-MM (CK-1), CK-MB (CK-2), and CK-BB (CK-3). CK-MM isoenzyme is present in both heart and skeletal muscle. CK-MB isoenzyme is more prevalent in myocardium, which contains about 10–20% of the total creatine kinase (TCK) activity compared to 2–5% in skeletal muscle.

Isoforms (subtypes) of CK isoenzymes exist and can be detected in serum. At least three CK-MM isoforms and at least four CK-MB isoforms have been identified. The isoforms of CK-MB have been studied, and attempts have been made to use these isoforms for clinical use. When CK-MB tissue isoform is released into circulation, carboxypeptidase cleavage of the CK-B carboxy-terminal lysine residue gives rise to a B-chain negative product and a product devoid of lysines on both chains. Only the CK-MB2 and CK-MB1 have been used for diagnostic purposes.

Laboratory assays for CK-MB have evolved during the past several decades from electrophoresis procedures, to immunoinhibition and immunoprecipitation assays, to current methodologies using immunoassays that incorporate monoclonal anti-CK-MB antibodies. These new generation assays have greatly improved on analytical and clinical sensitivity and specificity.

Quantitation of CK-MB activity can be determined by measuring isoenzyme activity. The units for isoenzyme activity are units per liter or international units per liter (U/L or IU/L). Isoenzyme measurement based on activity may be influenced by inhibitors, assay temperature, interference from other enzymes and drugs, prolonged storage or inadequate preservatives, pH, and ionic concentrations used in their analysis.[20] Another means of quantitating CK-MB is to use mass assays. The units for mass assays are micrograms/L (μg/L), and the mass assays are shown to provide more reliable results.

The mass concentration of CK-MB can be determined by sandwich-type immunoassays using antibody labels such as acridinium esters, fluorescein, and enzymes. A standard curve prepared from known mass units of CK-MB is used for quantitation of unknown samples.

The isoenzyme of CK that is of primary clinical interest for cardiac disease is CK-MB. This isoenzyme has been used for the evaluation of cardiac-related conditions such as AMI since the early 1980s. There are limitations to the use of CK-MB for evaluating patients with AMI. For example, CK-MB can be elevated in several diseases not related to the heart. Also, CK-MB is found in other tissues sources (e.g., skeletal muscle, brain, and lung). Therefore, if these tissues sources are diseased and the cells destroyed, the CK-MB present will leave the cell and enter the blood. Box 18-3 ■ provides a partial listing of non-cardiac-related conditions where CKMB may be elevated.

CK-MB concentrations in blood following an AMI rise and fall in a fairly predictable manner. CK-MB increases above reference interval in about 4–6 hours following the onset of chest pain due to cardiac dysfunction. Peak levels occur at approximately 24 hours and return to normal within 48–72 hours. Many laboratories offer a %CK-MB relative index or %CK-MB that may aid clinicians in evaluating their patients for possible AMI. The % relative index is determined by dividing the CK-MB *mass* by the total CK (TCK) *activity* and multiplying by 100. An increased relative index or %CK-MB usually means that the heart is the source of the CK-MB in serum. If the TCK is normal, then the neither of the two calculated parameters should be used because they may yield falsely elevated values. CK-MB determinations are being replaced by

---

BOX 18-3 Non-cardiac-related causes of elevated CK-MB isoenzyme.

► Severe skeletal muscle injury following trauma or surgery
► Chronic muscle disease in patient with muscular dystrophy, end-stage renal disease, or polymyositis
► Healthy people who exercise and participate in physical activity such as long-distance running
► Extensive rhabdomyolysis
► Early dermatomyositis

cTnI and cTnT for assessment of AMI. This process is currently in transition; thus many laboratories still perform both measurements.

 Checkpoint! 18–4

Why is CK-MB less useful than cTnI or cTnT as a biomarker for AMI?

MYOGLOBIN

Myoglobin is a heme protein approximately 17.8 kDa that is located in the cytoplasm of both cardiac and skeletal muscle cells. Myoglobin is the first cardiac biomarker to appear in circulation that may be used for assessment of ACS. Diagnostic sensitivity of myoglobin is greater than 90%, with serial sampling beginning at presentation and then 2 to 6 hours later.[9] However, the diagnostic specificity of myoglobin suffers because its amino acid sequence is the same in both cardiac and skeletal muscle. Thus a patient with significant skeletal muscle injury will present with elevated blood levels of myoglobin. Studies have shown that diagnostic specificity of myoglobin for cardiac-related diseases (especially AMI) ranges from 60 to 95%. Patients with renal insufficiency will have elevated myoglobin levels due to decreased clearance of myoglobin.

An appropriate use of myoglobin as a biomarker for cardiac injury is to make the test available within 30 minutes of the request and to use the result along with other biomarkers, such as CK-MB and cTn.[21]

 Checkpoint! 18–5

Identify a significant advantage and disadvantage of measuring serum myoglobin in patients complaining of chest pain with a suspected AMI.

BNP AND NT-ProBNP

Natriuretic peptides (NPs) are secreted to regulate fluid volume, blood pressure, and electrolyte balance. They are active in both the central and peripheral nervous systems. There are several NPs, including the following:

• Atria natriuretic peptide (ANP) is a 28-amino-acid peptide found in the atria of the heart.

• C-type natriuretic peptide (CNP) is a 22-amino-acid peptide found in the endothelium.

• Brain natriuretic peptide (BNP) is a 32-amino-acid peptide found in the ventricles of the heart.

• Urodilatin is a 32-amino-acid peptide found in the renal distal convoluted tubule cells.

The structures of each NP are each characterized by a 17-amino-acid ring and a closed ring with a disulfide bond between two cysteine residues.[22]

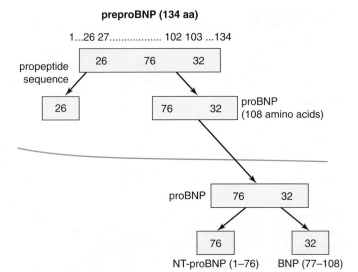

■ **FIGURE 18-5**   Overview of the biochemical formation of BNP and NT-proBNP.

ANP and BNP are released in response to atrial and/or ventricular stretch from volume overload. Thus blood NPs will be increased in diseases with volume overload, including renal and liver diseases and some endocrine diseases (e.g., Cushing's disease). BNP is a hormone originally isolated from porcine brain tissue. In humans BNP is produced in the brain, but the main source of circulating BNP is the heart ventricles.

The major circulating forms of BNP are NT-proBNP, whose function is unknown, and C-terminal BNP (a physiologically active form). BNP is cleared by neutral endopeptidase degradation; by receptor-mediated clearance; and to a small degree via the kidneys, which also secrete BNP. NT-proBNP is cleared predominantly by the kidneys. Thus NP-proBNP will be more sensitive to changes in renal function. The NPs currently measured to evaluate heart-related conditions (especially CHF) are BNP and NT-proBNP. The biochemical sequence of events leading to the formation of BNP and NT-proBNP is shown in Figure 18-5 ■. Initially, BNP is synthesized as a 134-amino-acid precursor *preproBNP*. PreproBNP is cleaved to a 26-amino-acid propeptide sequence and proBNP with 108 amino acids in the cytoplasm. Pro BNP is cleaved into the 76-amino-acid NT-proBNP and the biologically active 32-amino-acid BNP (also referred to as BNP-32) either within the cell or after secretion by the enzyme corin.[22]

BNP and NT-proBNP Assays

The first assay made available for clinical use was marketed along with its own nonautomated bench-top analyzer. The technology was made available to other instrument manufacturers, who developed larger instrument platforms in an effort to automate the analysis for faster turnaround times.

The decision whether to measure BNP or NT-proBNP requires knowledge of their respective biological and physiological attributes. NT-proBNP has a longer half-life; $t_{1/2}$ is 60 to 120 minutes versus ~20 minutes for BNP. The molar concentration of NT-proBNP in circulation is several folds higher than BNP. NT-proBNP has a longer in-vitro stability (~48 hours) than BNP (~4 hours) at room temperature. BNP is not dependent on renal or hepatic clearance in the absence of heart failure or cardiac-associated left ventricle hypertrophy (LVH). NT-proBNP as mentioned above is predominantly cleared by the kidneys. The specimen requirement for the Biosite BNP assay is whole blood or ethylenediaminetetraacetic acid (EDTA) plasma versus serum or EDTA plasma for NT-proBNP. Finally, the compound nesiritide used in the preparation of Natrecor (Scios Inc., Fremont CA), which is a recombinant human BNP used for treatment of CHF, may interfere with BNP assays, whereas there is no interference with NT-proBNP assays.[22]

BNP and NT-proBNP and Congestive Heart Failure

Before physicians accepted BNP as a biomarker for CHF, the diagnosis or medical workup for patients with suspected CHF included a patient history, physical findings, and diagnostic tests such as X-rays, magnetic resonance imaging (MRI), and echocardiograms. Thus the clinical laboratory was not involved in the diagnostic protocol until it was shown that BNP could provide useful information for clinical evaluation of patients with possible CHF.

Heart failure is a complex clinical syndrome derived from ventricular dysfunction (acute or chronic), where the venous return to the heart is normal but the heart is unable to pump sufficient blood to meet the body's metabolic needs and normal filling pressure.[23]

Congestive heart failure is characterized by a build-up of blood in the veins and other body fluids in tissue. It is a clinical syndrome with complex and variable signs and symptoms that include dyspnea or shortness of breath (SOB), increased fatigue, and edema. Also, jugular venous distention and ascites are usually present.

There are several potential blood biomarkers of heart failure, but the two that provide the most promise for the clinicians are BNP and NT-proBNP. Examples of other potential blood biomarkers of heart failure are listed in Table 18-2 ⊗.[24]

The focus of clinical laboratory testing for CHF using BNP or NT-proBNP testing is as follows:

• Determine the cause of clinical symptoms.

• Estimate the degree of severity of CHF.

• Estimate the risk of disease progression and rise.

• Screen for a less symptomatic disease.

A study published in 2002 provided correlation data of BNP values with New York Hospital Association (NYHA) classification of heart failure.[25] BNP was measured in selected subjects representing the four classes of heart failure, with class I being least severe and class IV the most severe form of heart failure. The data revealed an almost linear increase of blood BNP levels with increasing severity of heart failure. The results of this study provided clinicians with significant

⊗ TABLE 18-2

Potential biomarkers of heart failure.

Natriuretic Peptides (amino acid)	Molecular Type	Size (number of amino acids)	Tissue Source
N-terminal proANP	Peptide	98	Heart atria
Pro-ANP (99-126)	Peptide	28	Heart atria
Pro-ANP (80-96)	Peptide	17	Heart atria
Pro-ANP (26-55)	Peptide	30	Heart atria
BNP	Peptide	32	Heart ventricles
Pro-BNP	Peptide	108	Heart ventricles
NT-ProBNP	Peptide	76	Heart ventricles
Cardiotonic Steroids			
DLIF	Steroid-like	NA	Adrenal cortex
OLF	Steroid-like	NA	Adrenal cortex/hypothalamus
Cytokines			
TNF-α	Protein	157	Endothelial cells
IL-6	Protein	184	Endothelial cells

DLIF: digoxin-like immunoreactive factor; OLF: Ouabain-like factor.

information to justify the use of BNP as a blood biomarker for CHF patients.

Blood BNP and NT-proBNP have provided clinicians, especially emergency department physicians, with a useful tool for evaluating patients with signs and symptoms of heart failure. A general consensus is that BNP or NT-proBNP should be performed according to the following guidelines:

1. To confirm the diagnosis of CHF in patients with a suspected diagnosis of CHF but presenting with ambiguous clinical features or confounding pathology etiologies such as chronic obstructive pulmonary disease (COPD)

2. To improve the diagnostic accuracy for detecting heart failure.

3. To aid in ruling out CHF when the concentration of BNP or NT-proBNP is within normal limits[26]

BNP measurements are currently being evaluated for prognosis and risk stratification of CHF and ACS. Because BNP is released by intact cells, including those that are not ischemic, measuring BNP may in the future be useful in detecting pathophysiological consequences of ischemia and cell necrosis.[27]

Several studies have focused on the use of BNP for screening the general population, especially those patients considered to be at risk for CHF. The results of the studies are encouraging, but because more variables need to be evaluated, the use of BNP or NT-proBNP for population-based screening is not recommended at this time.[28]

The circulating concentrations of cardiac NPs are regulated by several physiologic factors, such as circadian variations, age, gender, exercise, body posture, eating habits (e.g., sodium intake). Also, cardiac NPs may be elevated or decreased in conditions not involving the heart. Several examples of these are listed in Box 18-4 ■.[29]

 Checkpoint! 18–6

Identify three differences between NT-proBNP and BNP that should be considered before deciding which assay to select.

LACTATE DEHYDROGENASE ISOENZYMES

Isoenzymes of lactate dehydrogenase (LD), a 180-kDa protein, have been used in the past to evaluate patients with cardiac diseases, especially AMI. Specifically, LD-1 and LD-2 were used as cardiac biomarkers. These isoenzymes are no longer used to evaluate patients with suspected AMI; therefore, only a brief review will be presented here.

The highest activities of LD are found in liver, skeletal muscle, heart, kidney, and erythrocytes. Electrophoresis techniques have revealed the existence of five isoenzymes composed of two distinct types. These types are designated *M* for muscle and *H* for heart. The five isoenzymes are composed of four subunit peptides (tetramers) with the following designations: LD-1 (H4), LD-2 (H3M1), LD-3 (H2M2), LD4 (H1M3), and LD-5 (M4). LD-1 is found in high concentrations in heart, kidney (cortex), and erythrocytes. LD-5 is found in highest concentration in liver and kidney. LD-2, LD-3, and LD-4 are distributed in these and other tissues throughout the body. LD is not a tissue-specific enzyme; thus it is elevated in a variety of diseases.

C-REACTIVE PROTEIN (ALSO hs-CRP)

Measurement of CRP, a nonspecific, acute-phase reactant, has been useful as a diagnostic marker of inflammation or infection for decades. CRP is a pentamer or pentraxin that undergoes structural modification, forming monomeric subunits. These monomeric forms are considered to be proinflammatory and proatherogenic.[30]

High-sensitivity C-reactive protein (hs-CRP) is a term applied to the analytical test for CRP that allows for detection of smaller concentrations of serum CRP and is used to assess the relative to risk for cardiac disease. Analytical sensitivity of hs-CRP differs significantly from the conventional quantitative assays for CRP used to indicate inflammation in the body. For example, CRP in acute and chronic infections can be as high as 500 mg/L, whereas CRP in cardiac-related conditions (especially early onset) may only be in the range of 2–3 mg/L. Therefore, the analytical sensitivity for hs-CRP must be much lower (e.g., 1.0 mg/L) and the measuring range should be significantly lower (e.g., 1–90 mg/L) than conventional CRP assays.

Recently the FDA's OIVD (Office of In Vitro Diagnostics) published an announcement that defines the different CRP assays. Three assays, designated as conventional CRP, hs-CRP, and cardiac CRP (cCRP), have been defined based on their intended use. *cCRP* is used to designate tests intended to identify individuals at risk for future coronary vascular disease (CVD) and *hs-CRP* for those tests in which there is no evidence of efficacy in CVD risk stratification. No hs-CRP assay has been allowed to have the *cCRP* designation without supporting clinical studies. The designation *cCRP* is intended solely to identify in vitro devices (IVDs) that meet the performance specifications.[31]

There are numerous automated assays for measuring hs-CRP. These include both nephlometric assays and immunoassays. At first reagent manufacturers developed unique and specific assays for hs-CRP along with a separate assay for conventional CRP. Recently assays have been developed that allow users to measure CRP for both analytical ranges using one single-reagent system. These combination or wide-range assays make more efficient and economical use of reagents kits for the laboratory. An example of a wide-range assay is Equal Diagnostics' (Exon, PA) CRP ultra WR immunoturbidimetric assay. Assay sensitivity is down to 0.05 mg/L and the linear range is up to 320 mg/L. For POCT application, Cholestech Corporation (Hayward, CA) has developed an hs-CRP assay for its LDX analyzer. The measurement requires a fingerstick specimen, and results are available in 6 minutes.

hs-CRP is an independent marker of risk and may be used as part of a global coronary risk assessment in adults without known CVD. Also, patients with persistently unexplained marked elevation hs-CRP (>10 mg/L) after repeated testing should be evaluated for noncardiac vascular etiologies.[32]

Atheromatous plaques in diseased arteries typically contain inflammatory cells. Multiple prospective studies have demonstrated that baseline CRP is a good marker for future cardiovascular events. The CRP level is a stronger predictor of cardiovascular events than low-density lipoprotein (LDL). Because of the individual variability in hs-CRP, two separate measurements are recommended to reliably assess an individual's risk level. In patients with stable coronary disease or ACS, hs-CRP measurement may be useful as an independent marker for assessing likelihood of recurrent events, including death, AMI, or restenosis after percutaneous coronary intervention (PCI).

Measuring blood levels of hs-CRP provides better risk assessment than measuring any other biomarker. Measuring both hs-CRP and the total cholesterol:HDL-C ratio slightly improves the risk assessment for AMI.[33]

 Checkpoint! 18–7

What is the clinical usefulness of hs-CRP?

ISCHEMIA-MODIFIED ALBUMIN

Ischemia-modified albumin (IMA) is produced when circulating albumin comes in contact with ischemic tissue in the heart or other organs. During ischemia, the N-terminus of albumin is altered, probably through a series of chemical reactions involving free radical damage by compounds such as nitric oxide. This altered albumin, termed *ischemia-modified albumin*, is produced continually during ischemia, which means its blood concentrations rise quickly and remain elevated during an ischemic event.

A laboratory test that has been cleared by the FDA is available to measure IMA. The test is referred to as the albumin cobalt binding (ACB) test (ACB, Ischemia Technologies Inc., Denver CO) and is available on several automated chemistry analyzers currently in use. The principle of the ACB assay is based on the concept that when cobalt is added to serum it does not bind to the NH2 terminus of IMA, thus leaving more free cobalt to react with the reagent dithiothreitol, which forms a darker color in samples from patients with ischemia. Increased serum IMA levels indicate a reduced metal binding capacity of albumin associated with cardiac ischemia and the production of oxygen free radical species.

The cutoff for a positive ACB is above 75 U/mL and should be viewed with caution if the serum albumin is <2.0 g/dL or >5.5 g/dL or the blood lactate or ammonia concentrations are elevated.[36] Increased serum IMA values are also found in patients with cancer, infections, end-stage renal disease, liver disease, and brain ischemia; this may limit its clinical usefulness.

 Checkpoint! 18–8

Explain the principle of the albumin cobalt binding test (ACB) used to measure IMA.

MYELOPEROXIDASE

Myeloperoxidase (MPO) is a hemoprotein of approximately 140 kDa. It is stored in azurophilic granules of polymorphonuclear neutrophils and macrophages. MPO catalyzes the conversion of chloride anion and hydrogen peroxide to hypochlorous acid during the release of reactive oxygen species in neutrophils and macrophages. During inflammatory conditions, MPO is released into the extracellular fluid and general circulation. Macrophages secrete matrix metalloproteinases (MMPs) and metal independent MPO, which degrade the collagen layer that protects atheromas from erosion or abrupt rupture. As a result, plaques that have been highly infiltrated with macrophages have a thin fibrous cap and are vulnerable to erosion or rupture, leading to ACS. Thus, MPO is a better biomarker of plaque instability.

The clinical usefulness of MPO is summarized below. Elevated blood levels of MPO are associated with the following processes:

- Oxidation of LDL cholesterol followed by the uptake of Ox-LDL by macrophages producing foam cells. The accumulation of foam cells leads to unstable plaque formation.
- Rendering of HDL "dysfunctional" in human atheroma.
- Destabilization and rupture of the atherosclerotic plaque surface.
- Consumption of nitric oxide, which leads to vasoconstriction and endothelial dysfunction.

MPO and Risk of CAD

Significantly higher levels of MPO were found in patients with angiographically demonstrated CAD.[35]

MPO and Risk of ACS

MPO may give prognostic information independent of myocardial necrosis in patients with known ACS.[36]

MPO and Chest Pain

MPO appears useful in both short- and long-term risk stratification of patients who present with chest pain, and it may help identify cTn-negative patients at risk for major adverse cardiac events.[37]

MPO Assays

Several assays are available to measure MPO relative to atherosclerotic heart disease, including the following:

- Calbiochem, a brand of EMD Biosciences, Inc. InnoZyme MPO activity assay kit (ELISA) (EMD, a subsidiary of Merck Germany)
- Assay Designs Inc. (Ann Arbor, MI) Human Myeloperoxidase TiterZyme EIA Kit
- Zen MPO ELISA serum assay, Invitrogen (Carlsbad, CA)

Recently the FDA has approved the CardioMPO, (PrognostiX, Inc., Cleveland, OH), which is an enzyme immunoassay intended for quantitative determination of MPO in human plasma and is to be used in conjunction with clinical history, ECG, and cardiac biomarkers to evaluate patients presenting with chest pain that are at risk for major adverse cardiac events, including AMI.

MPO immunoassays are currently being developed for automated analyzers. Two examples are as follows:

- Abbott Diagnostics (Abbott Park, IL) ARCHITECT MPO assay is a chemiluminescent immunoassay using the CHEMIFLEX technology that incorporates acridinium as the label.

- Dade Behring, Inc. (Deerfield, IL) Dimension RxL is a heterogeneous immunoassay using a chromogenic substrate with an enzyme label.

POCT devices are currently being developed to measure MPO. Biosite Inc. (San Diego, CA) has developed a device and method to measure MPO called the Triage MPO Test that is used along with other cardiac biomarkers to assess patients who present to the ED with chest pain.

 Checkpoint! 18-9

What is the clinical usefulness of MPO in patients who present to the ED with chest pain?

CYTOKINES

Several cytokines, especially interleukin-6 (IL-6), are known to contribute to the pathogenesis of the inflammatory response and atherosclerosis. IL-6 is produced by monocytes, fibroblasts, and lymphocytes and has numerous pathophysiological functions. It is one of the major cytokines that is produced during sepsis and other catabolic processes. IL-6 also serves a prominent role in the genesis of the acute-phase response, including synthesis of CRP by the liver. Finally, IL-6 has been shown to be an independent risk factor for future AMIs and is elevated in patients with ACS.

Flow cytommetry and ELISA methods are available for determining cytokine levels in blood. A novel approach to measuring several cytokines has been developed that provides a profile format. Luminex (Austin, TX) Multi Analyte Profiling system is a flow cytometer–based instrument that allows multiple cytokines to be assayed simultaneously in a single sample by multiplex fluoroimmunoassay. Cytokines IL-2, IL-4, IL-6, IL-8, IL-10, IL-12, TNF-α, and interferon-γ are included in this profile. The consensus is that cytokine profiles might have a role in differentiating patients with CAD and AMI from those with chest pain due to other disorders.[38]

PREGNANCY-ASSOCIATED PLASMA PROTEIN A

Pregnancy-associated plasma protein A (PAPP-A) is a glycoprotein (~200 kDa) synthesized by the syncytiotrophoblast and is typically measured during pregnancy to detect Down syndrome. PAPP-A is a zinc binding metalloproteinase and is an insulin-like growth factor (IGF)–dependent protein that is potentially a proatherosclerotic molecule through its role in disrupting the integrity of the atheroma's protective cap.[39] PAPP-A in pregnancy differs from PAPP-A associated with atherosclerotic plaques. In pregnancy, PAPP-A circulates as an heterotetrameric complex consisting of two PAPP-A subunits covalently bound to two subunits of the proform of eosinophil major basic protein (proMBP). PAPP-A present in human fibroblasts and released during atherosclerotic plaque disruption is a homodimeric active form, uncomplexed with proMBP.

PAPP-A has been found in unstable plaques from a patient who died suddenly of cardiac causes and has been found in patients with unstable angina and AMIs. Thus PAPP-A measurement may be valuable for detecting unstable ACS in patients without increased concentration of biomarkers of necrosis, such as cTn, thereby potentially identifying high-risk patients whose unstable clinical situation might otherwise remain undiagnosed.[40]

Immunoassays for determining PAPP-A cannot differentiate the two forms, thus making it difficult to measure PAPP-A as a cardiac marker. There are a few immunoassays currently on the market that measure total PAPP-A (which includes both free and complex forms):

- Diagnostic System Laboratory (Beckman Coulter)

- Kryptor (Brahms AG)

- Aio! (Innotrac Diagnostics Oy Turku, Finland), which measures mostly PAPP-A/proMPB complexes

OXIDIZED LDL

Oxidized LDL (Ox-LDL) is the oxidized form of LDL and is found primarily in the fatty plaques that may line the walls of coronary arteries in patients with atherosclerosis. Cardiovascular specialists believe that Ox-LDL is contributory to the inflammation of the fatty plaques and elicits an immune response.[41]

Studies have shown that circulating levels of Ox-LDL are strongly associated with angiographically documented CAD in patients 60 years of age or younger. Data show that the atherogenicity of lipoprotein (a) (Lp a) may be mediated in part by association with proinflammatory Ox-LDL (as phospholipids).[41]

Ox-LDL is not one homogeneous entity but consists of multiple chemical and immunogenic modifications of the lipid and apoB-100 on LDL. The designation *Ox-LDL* actually represents many different types of Ox-LDL. Most researchers now designate different types of Ox-LDL using antibodies specific to epitopes on apoprotein, such as apoB. A specific example of a type of Ox-LDL is Ox-LDL-E06, which is used to measure Ox-LDL epitopes on apoB.

There are currently specific ELISA assays developed based on murine monoclonal antibodies that recognize various epitopes of Ox-LDL. Both the Mercodia Inc. (Winston Salem, NC) Ox-LDL ELISA solid-phase two-site enzyme immunoassay and the new Ox-LDL competitive ELISA using one antibody incorporate the specific antibody 4E6. The 4E6 antibody is directed against a conformational epitope in the apoB-100 moiety of LDL that is generated as a consequence of aldehyde substitution of the lysine residues of apoB-100. These two assays are FDA approved.

PLACENTAL GROWTH FACTOR

Placental growth factor (PlGF) is a 50- kDa heterodimer consisting of 149 amino acids and has high homology with vascular endothelial growth factor (VEGF). There are two forms of PlGF: PlGF-1 and PlGF-2. The principal difference is that PlGF-2 has within its structure a highly basic 21-amino-acid moiety inserted at the carboxyl end. PGlF is one of a family of platelet-derived proteins that functions as a potent chemo attractant for monocytes and is involved in the regulation of VEGF.[42]

PlGF appears to function in the early stages of the inflammatory process, which includes the recruitment of circulating macrophages into atherosclerotic lesions, stimulation of smooth muscle cell growth, and up-regulation of both TNF-α and monocyte chemotactic protein (MCP-1) by macrophages.[43] PlGF may serve as a strong candidate biomarker for plaque instability, myocardial ischemia, and patient prognosis in the continuum of ACS.

A serum/plasma ELISA for PlFG marketed as Quantikine Human PlGF immunoassay is available from R & D systems, Inc. (Minneapolis, MN). This assay is a quantitative sandwich immunoassay whose concentration is proportional to the intensity of the colored chromogen, tetramethylbenzidine.

MATRIX METALLOPROTEINASE-9

Matrix metalloproteinases (MMPs) are a class of 24 endopeptidases that are physiologic regulators of the extracellular matrix. They are found in most human tissues. In the heart, matrix metalloproteinase participates in vascular remodeling, plaque instability, and ventricular remodeling after cardiac injury. A zinc-dependent form, MMP-9, also known as gelatinase, B is able to accomplish the following:

• Generate angiostatin via cleavage of plasminogen
• Increase its affinity for collagen
• Interact with intercellular adhesion molecule-1
• Function as an anti-inflammatory protein by processing IL-1β from its precursor and by reducing IL-2 response.[44]

In vascular tissue, several MMPs, including MMP-9, are localized at the shoulder of a plaque, which is described as the area that is thinner and prone to rupture. It is thought that some degree of MMP production is necessary as part of the reparative process after acute cardiac injury.[45]

Much interest has been generated in developing an assay for measurement of MMP-9 in patients with ACS. Several studies have revealed a potential use for MMP-9 measurements to evaluate patients with CAD and after acute coronary events. In tissue, Pro-MMP-9 (92 kDa) is cleaved to an active free form (82 kDa) that is elevated in cardiovascular disease.

Immunoassays have been developed to measure MMP-9s but are not designed to exclusively measure the active form associated with CVD. The assays do measure various amounts of MMP-9, Pro-MMP-9 (inactive form), and the pro MMP-9/TIMP-1 complex (tissue inhibitor of MMP-1).[46]

Two examples of assays currently available and their specificities are the Quantikine Human MMP-9 (total) sandwich enzyme immunoassay technique using a monoclonal antibody specific for MMP-9 (R & D Systems Inc., Minneapolis, MN) and the Biotrak immunofluorescent ELISA (GE Healthcare, Life Science, Biotrak Amersham, Sweden). The Quantikine assay measures human active and Pro-MMP-9, which equates to a total MMP-9 measurement. The Biotrak Immunostat measures endogenous active MMP-9, Pro-MMP-9 (inactive form), and the ProMMP-9/TIMP-1 complex.

HEART-TYPE FATTY ACID BINDING PROTEIN

Fatty acid binding protein (FABP) binds long-chain fatty acids reversibly and noncovalently. FABPs are small (15 kDa) intracellular proteins that are abundantly produced in tissues with active fatty acid metabolism, including the heart, liver, and intestine. Human H-FABP is an acidic protein consisting of 132 amino acid residues and is located in the myocardium. Human heart FABP is released into the circulation when the myocardium is injured. The isoform of H-FABP is produced in cardiomyocytes, skeletal muscle, kidney, brain, placenta, and lactating mammary glands.[47]

Several immunoassays are available to measure H-FABP, and some have been adapted to automated immunoassay analyzers (e.g., the Roche COBAS MIRA Plus analyzer). There are also several immunosensors in development that readily detect H-FABP in serum or plasma. These immunosensors may be incorporated into automated and nonautomated instrument designs.

The clinical usefulness of plasma measurement of H-FABP includes the following:

• Early marker of AMI
• Urinary marker of AMI
• Marker of myocardial injury after cardiac surgery
• Detector of coronary reperfusion
• Tool for infarct sizing

The release of H-FABP from myocytes is rapid and closely resembles the release kinetics of myoglobin. Application of H-FABP as a sensitive early marker for myocardial injury has been investigated. Diagnostic utility of serum measurements of H-FABP for AMI was as good as or better than measurement of myoglobin. H-FABP performed better than or similar to myoglobin. Several studies have been conducted that support the other clinical uses of H-FABP, but the data do not suggest routine measurement of H-FABP at this time.[47]

UNBOUND FREE FATTY ACID

Free fatty acids (FFAs) are bound mostly to albumin, and a small amount of the total FFA is unbound (FFA$_u$). Studies have shown that FFA$_u$ and not FFA may provide a sensitive guide to the underlying pathophysiology of CHD. The mechanisms that initiate and maintain increased blood levels of FFA$_u$ after ischemia have not been fully elucidated. Data suggest that in patients with ischemic symptoms, monitoring blood levels of FFA$_u$ may provide an early indication of cardiac ischemia.[48]

A fluorescent assay has been developed by FFA Sciences (San Diego, CA) to measure FFA$_u$. A probe using a recombinant fatty-acid-binding protein labeled with a fluorescent tag *acrylodan labeled intestinal fatty acid binding protein* (ADIFAB) is used to bind any FFA in the sample. If FFA$_u$ is present, the complex fluoresces green, and if FFA$_u$ is not present in the sample, a blue fluorescent color develops. This assay has also been modified using ADIFAB2 and adapted to a hand-held device, the FFA Picofluor (Turner BioSystems Inc., Sunnyvale, CA).

OMEGA-3 FATTY ACIDS

Omega-3 fatty acids are obtained from several dietary sources, including seafood, nuts, and plant oils. These food sources contain various amounts of fatty acids (e.g., eicosapentaenoic acid [EPA], docosahexaenoic acid [DHA], and α-linolenic acid [ALA]) that have been shown to reduce the number of deaths from heart disease. Research suggests that omega-3 fatty acids can reduce CVD risk in the following manner:[49]

- Decrease risk for arrhythmias
- Decrease risk for thrombosis
- Decrease triglycerides and remnant lipoprotein levels
- Decrease rate of growth of the atherosclerotic plaque
- Improve endothelial function
- Slightly lower blood pressure

Continued research into the benefits of omega-3 fatty acids in reducing heart disease has produced additional findings. For example, in a community-based sampling, omega-3 fatty acids were independently associated with lower levels of proinflammatory markers (e.g., IL-6, IL-1ra, TNF-α, CRP) and higher levels of anti-inflammatory markers, including soluble IL-6r, IL-10, and TGF-β, independent of confounders (e.g., age and gender). The findings support the concept that omega-3 fatty acids may be beneficial in patients affected by diseases characterized by active inflammation.[50] Another study group concluded from a multicenter case-control pilot study that omega-3 fatty acids are a negative risk factor for myocardial infarction.[51]

Although routine measurement of omega-3 fatty acids is not conducted in most clinical laboratories, assays are available in most large reference laboratories. The most widely used methods require a gas chromatograph and are somewhat labor intensive.

LIPOPROTEIN-ASSOCIATED PHOSPHOLIPASE A2 (Lp-PLA2)

Lp-PLA2 is a calcium-independent serine lipase that is associated with LDL in human plasma and is distinct from other phospholipases such as cPLA2 and sPLA2.[52] Lp-PLA2 is produced by macrophages and is expressed in higher concentrations in atherosclerotic lesions than in normal vascular tissues. It is widely believed that oxidation of LDL plays a role in the development and progression of atherosclerosis. Lp-PLA2 participates in the oxidative modification of LDL by hydrolyzing oxidized phosphatidylcholines and generating lysophosphatidylcholine and oxidized free fatty acids, both of which are potential proinflammatory products that contribute to the formation of atherosclerotic plaques.[53]

A retrospective case study of hypercholesterolemia in men in the West of Scotland Coronary Prevention Study (WOSCOPS) revealed a two-fold greater risk of CHD in subjects in the highest quintile level of Lp-PLA2 concentrations, compared to the lowest quintile. Also, the risk of CHD associated with Lp-PLA2 was shown to be independent of LDL and other markers of inflammation, such as CRP, white cell counts, and fibrinogen. The author's conclusion was that elevated levels of Lp-PLA2 appear to be a strong risk factor for CHD.[54]

Recent studies have advanced the work of the WOSCPS group. For example, one study concluded that both CRP and Lp-PLA2 may be complementary in identifying middle-aged individuals with high CHD risk and low LDL-C.[55] Also, this group concluded that Lp-PLA2 activity is a new and independent predictor of ischemic stroke in the general population. Another study provided further evidence of the independent role of Lp-PLA2 in the prediction of CHD and indicated that the effect of Lp-PLA2 on CHD is independent of a subject's total cholesterol level and markers of inflammation.[56]

A commercial test is available to measure Lp-PLA2 in human plasma. The diaDexus PLAC Test (San Francisco, CA) is a sandwich enzyme immunoassay that uses two highly specific monoclonal antibodies for the direct measurement of Lp-PLA2 concentrations in plasma.

LIPOPROTEIN (a)

Lipoprotein (a) has many properties in common with LDL. It consists of a unique plasminogen-like glycoprotein, apo (a), which is linked by a disulphide linkage to apolipoprotein B-100, which is anchored in a lipid-rich LDL-like core.

Apo (a) is a highly glycosylated, hydrophilic apoprotein that somewhat resembles plasminogen in that it contains an extended kringle domain and a carboxyl terminal serine protease domain. Kringles are autonomous structural domains found throughout the blood-clotting and fibrinolyic proteins. Kringle domains are believed to play a role in binding mediators (e.g., membranes, other proteins or phospholipids) and in the regulation of proteolytic activity. Kringle domains are characterized by a triple loop, 3-disulfide bridge structure whose conformation is defined by a number of hydrogen bonds and small pieces of antiparallel beta-sheets.

Apolipoprotein (a) gene is located on chromosome *6q279* and is responsible for Lp (a) synthesis. Lp (a) has homology with LDL and plasminogen and is under genetic control.[57] It migrates in the pre-beta region in electrophoresis separation.

Lp (a) particle diameter is 25–30 nm with a density in the range of 1.040–1.130.

Lipoprotein (a) is a lipoprotein whose physiological functions are not completely known. The roles of Lp (a) are thought to be as follows:

- Respond to tissue injury and vascular lesions
- Prevent infectious pathogens from invading cells
- Promote wound healing

Lp (a) assays have been available for many years. The assays can be categorized as follows:

1. Methods based on Lp (a) associated lipids
   - Lipoprotein electrophoresis
   - Ultracentrifugation: beta-quantification (co isolated with LDL)
   - Electrophoresis with enzymatic cholesterol detection: Lp (a)-C
   - Lysine affinity chromatography: Lp (a)-C
2. Methods based only on apo (a) detection or apo (a) and apo B) detection
   - Radioimmunoassay (RIA)
   - Radial immunodiffusion (RID)
   - Electroimmunoassay
   - ELISA
   - Immunoturbidimetry
   - Immunonephelometry

Since its discovery by Kare Berg over 40 years ago, Lp (a) has been subject to controversy and debate regarding its role in humans and specifically in CHD. Novel discoveries about the structure and function of this molecule, such as its effects on plaque stability and the tissue factor pathway, suggest new potential targets for drug design and therapy.[58]

Lp (a) is believed to promote atherosclerosis by a number of separate but related mechanisms, which are as follows:[53]

- Adhesion molecules are increased in the presence of Lp (a).
- Leukocyte adhesion and migration is increased.
- Lp (a) induces vascular endothelial cells to produce monocyte chemotactic protein (MCP), which has been shown to be a potent chemoattractant for monocytes.
- Oxidized Lp (a) induces reduction of nitric oxide (NO).
- Lp (a) mediates extracellular matrix degradation and plaque rupture.
- Lp (a) binds to endothelial cells, macrophages, fibroblasts, platelets, and the subendothelial matrix.
- Lp (a) inhibits clot formation at the site of injury.
- Lp (a) delivers cholesterol to the site (Lp (a) is made up of approximately 40% cholesterol).
- Lp (a) competes with plasminogen for binding sites and thus interferes with clot lysis and increases the risk of AMI.
- Lp (a) promotes smooth cell proliferation.

Medical protocols for screening patients for Lp (a) suggest that patients be evaluated for the following before being tested for Lp (a):[57]

- Family history of premature CVD
- Family history of hyperlipidemia
- Established CVD with normal routine lipid profile
- History of recurrent arterial stenosis

ADIPONECTIN

Adiponectin (also apM1, Arcp-30, AdiopoQ, and GBP-28) belongs to the family of adipokines. Specifically, it is a protein hormone with a molecular mass of 28 kDa and contains 244 amino acids. It is produced exclusively by adipocytes and is a modulator of lipid and glucose metabolism. Adiponectin influences the body's response to insulin and also has anti-inflammatory effects on the cells lining the walls of blood vessels.[59]

High blood levels of adiponectin are associated with reduced risk of heart attacks. Low blood concentrations of adiponectin are found in individuals who are obese and who are at increased risk of heart attack. Recent studies suggest that adiponectin may have antiatherogenic and anti-inflammatory properties. It is suggested that "plasma adiponectin levels may predict cardiovascular events years in advance in a population without diagnosed cardiovascular disease."[60]

The association of adiponectin with atherogenic dyslipidemia has been studied. Adiponectin levels in patients with left ventricular dysfunction or association with markers of systemic inflammation have recently been investigated. Researchers have found that serum adiponectin is associated with NT-proBNP concentrations but not with markers of systemic inflammation in patients with manifest CHD.[61]

There are several ELISA assays available to measure adiponectin, including the following:

- Panomics Affymetrix (Freemont, CA) human adiponectin in kit form
- Linco Research Inc. (St. Charles, MO) human adiponectin assay
- B-Bridge International, Inc. (Mountainview, CA) human ELISA kit
- BioVendor Laboratory Medicine, Inc. (Candler, NC) competitive ELISA and sandwich immunoassays
- Mercodia, Inc. (Winston-Salem, NC) human serum or plasma ELISA
- R & D Systems Quantikine (Minneapolis, MN) human adiponectin/Acrp30 immunoassay (currently for research only)

LEPTIN

Leptin is a protein hormone with important effects in regulating body weight, metabolism, and reproductive function. Leptin is ~16 kDa in mass, contains 167 amino acids, and is encoded by the obese (ob) gene. It is expressed predominantly by adipocytes and is related to amount of body fat. Leptin is also associated with increased heart rate, blood pressure, and sympathetic neural activity. Leptin is also thought to be an independent risk factor for CVD.[59]

Leptin has been linked to CRP, a robust predictor of risk of CHD. Studies have shown that elevated levels of leptin may signal an increased risk of vascular events. Leptin was found to have an independent protective association in patients with unstable CAD. The protective mechanisms include (1) pro **angiogenesis** of leptin, (2) coronary artery vasodilation, and (3) activation of endothelial nitric oxide production.[62] In patients with established CAD, lower serum concentrations of leptin are associated more with unstable CAD than stable CAD.[63]

Plasma concentrations of leptin can be determined by radioimmunoassay and ELISA assays (Linco Research Inc., St Charles, MO). Leptin analysis is part of a Human Adiptocyte Panel provided by Linco. LINCO*plex* kits are designed for use with xMAP Technology and Luminex Instrumentation. This quantitative technique is based on a laser-activated fluorescent-labeled micro-bead platform.

ADHESION MOLECULES

The adherence of leukocytes to the vascular endothelium is one of the earliest events in atherosclerosis. The process of leukocyte adhesion is mediated by cellular adhesion molecules (CAMs), which are expressed on the endothelial surface in response to atherogenic stimuli. There are several CAMs that are involved in this process, including the following:

- P-selectin
- E-selectin
- ICAM-1 (also CD54)
- VCAM-1 (also CD106)
- Platelet endothelial adhesion molecule 1 (PECAM-1)

Both ICAM-1 and VCAM-1 are expressed within atherosclerotic lesions, and elevated plasma levels of soluble forms of these molecules suggest a role in plaque disruption.[64]

R & D Systems, Inc. (Minneapolis, MN) manufactures an ELISA procedure to measure both ICAM and VCAM. The assay

✪ TABLE 18-3

POCT devices currently on the market along with manufacturers' performance claims.

Device	Cardiac Marker	Manufacturer's Claim	Time (min)
[1]Roche Cardiac T Rapid Assay	cTnT	Myocardial damage detected	12
Roche Cardiac M Rapid Assay	Myoglobin	Above normal range	8
[2]Cardiac Stratus	CK-MB Myoglobin cTnI	Aid in diagnosis of cardiac ischemia	15
[3]Alpha Dx POINT-OF-NEED system	CK mass CKMB CK-MB index Myoglobin cTnI	Aid in diagnosis of AMI	18
[4]Stratus CS STAT	CK-MB Myoglobin cTnI	Aid in diagnosis of AMI and cTnI for risk stratification	14 min to first result
[5]Triage Cardiac Panel	CK-MB Myoglobin cTnI	Aid in diagnosis of AMI	15
TriageCardioProfilER	CK-MB cTnI Myoglobin BNP	Aid in diagnosis of AMI and CHF	15
TriageMPO	MPO	Rapid assessment of patients with chest pain	15
[6]RAMP (Response Biomedical)	cTnI myoglobin CK-MB	Aid in diagnosis of AMI	15

[1]Roche Diagnostics; [2]Spectral Diagnostics, Toronto, Canada; [3]Mountainview, CA; [4]Flurometric analyzer, Dade Behring, Inc., Deerfield IL; [5]Biosite Diagnostics; [6]Burnaby, British Columbia, Canada

uses a monoclonal antihuman ICAM-1 or monoclonal antihuman VCAM-1 reagent in the procedure. These assays are for research only and are not available for clinical laboratory use.

MULTIPLE BIOMARKERS

A popular approach to evaluating patients with cardiac dysfunction is to measure several biomarkers as a group or panel. The earliest use of multiple biomarkers for cardiac disease occurred in the 1960s. Clinicians used a combination of the enzymes AST, CK, and LD. They relied on their unique temporal relationship to total enzyme concentration. The enzymes in this group have characteristic rise and fall patterns after a cardiac event (e.g., AMI). This grouping of enzymes changed during the next decades, but the principle remains the same. Currently the proteins used in such a manner include CK-MB, Troponin I or T, and myoglobin.

▶ POCT FOR CARDIAC BIOMARKERS

The advantages and disadvantages of POCT are well documented. POCT for cardiac biomarkers in emergency departments do provide some unique advantages to central laboratory testing. These advantages include the following:

- Rapid identification of AMI
- Rapid exclusion of ACS
- Rapid stratification of cardiac patients
- Rapid identification and exclusion of CHF

For POCT to work effectively, the laboratory staff and analyzer must provide the following to clinicians in the emergency room:

- POCT device should be fail-safe and turnkey so that the ED staff has only to place the sample tube in the analyzer and push a button.
- A POCT device performs the tests that are actually needed to produce a decreased length of stay.
- Laboratory maintains control of the POCT device and all quality assurance, compliance, proficiencies, competencies, and charges.

- Maintenance of the POCT device is easy so that minimally trained ED staff can perform simple daily routine duties.
- Compliance and proper control is assured through automatic errors and lockouts.

There are several POCT devices on the market today with various tests menus available that provide measurement of cardiac biomarkers. Examples of these devices along with some salient features of each are presented in Table 18-3 ✪.

SUMMARY

There are many reasons why the death rate for cardiac disease has been declining since the 1960s. Improvements in laboratory testing are certainly close to the top of the list. A tremendous amount of research and development has led to the discovery of cardiac biomarkers that are available to clinicians to support their care and treatment of patients. Specific biomarkers for cardiac necrosis have been developed that are both specific and sensitive and have been used as a first line of diagnostic tests for patients suspected of having a myocardial infarction. Cardiac biomarkers are also available that are widely used to evaluate cardiac-related diseases that traditionally relied on nonclinical laboratory tests (e.g., the use of brain natriuretic peptide [BNP] for patients with congestive heart failure and ischemia-modified albumin for cardiac ischemia).

A shift in paradigm has occurred in relation to the process of atherosclerosis and has led to the development of unique markers that are able to evaluate the deterioration of cardiac vessels long before an acute event occurs. Significant discoveries led the medical community to view atherosclerotic heart disease as not just a lipid deposition process but also an inflammatory process. The current approach to evaluating patients with atherosclerosis is to investigate the process "downstream" or toward the beginning and not at the very end, where a major cardiac event occurs.

This chapter presented the current cardiac biomarkers, including potentially useful biomarkers that are currently under investigation. In addition, detailed discussion was presented on clinical utility, assays parameters, and preanalytical considerations. The inflammatory process relevant to the cardiovascular system was reviewed and biomarkers that test for each stage of the process were identified and discussed.

REVIEW QUESTIONS

LEVEL I

1. Which of the following factors describes an ideal cardiac biomarker? (Objective 3)
   a. Smaller markers are released faster from injured tissues.
   b. It has absolute cardiac tissue specificity and should not exist in other tissues.
   c. Release from the myocardium should be complete following injury.
   d. All of the above

LEVEL II

1. Natrecor (Nesiritide) is a drug that will interfere with the measurement of which of the following? (Objective 6)
   a. BNP
   b. cTnI
   c. CK-MB
   d. myoglobin

REVIEW QUESTIONS *(continued)*

LEVEL I

2. Cardiac troponin I and T are useful biomarkers for patients presenting to the emergency department with chest pain because: (Objective 8)
   a. they remain elevated in blood on average longer than 7 days.
   b. they remain elevated in blood longer than 1 day.
   c. they remain elevated in blood longer than 1 hour.
   d. they remain elevated in blood longer than 30 seconds.

3. How many hours post–myocardial infarction does myoglobin first appear in the blood? (Objective 8)
   a. 1–3
   b. 4–8
   c. 12–24
   d. 24–36

4. BNP is elevated in the blood of patients with which of the following? (Objective 6)
   a. brain tumors
   b. prostate cancer
   c. left-sided heart failure
   d. liver disease

5. The acute-phase reactant C-reactive protein is useful as a cardiac marker because of which of the following? (Objective 5)
   a. It is considered to be anti-inflammatory and antiatherogenic.
   b. It is considered to be proinflammatory and proatherogenic.
   c. It is found in high concentrations in the left ventricle.
   d. It is useful for diagnosing aortic aneurysm.

6. *Ischemia-modified albumin* (*IMA*) is a term that reflects which of the following? (Objective 6)
   a. the total albumin concentration in blood
   b. the amount of albumin released after an AMI
   c. the altered C-terminus of albumin that occurs during cardiac ischemia
   d. the altered N-terminus of albumin that occurs during cardiac ischemi.

7. Myeloperoxidase (MPO) is a marker for which of the following? (Objective 6)
   a. cardiac necrosis
   b. left ventricular heart failure
   c. plaque instability
   d. obesity

8. Which of the following biomarkers is thought to contribute to the pathogenesis of the inflammatory response and atherosclerosis? (Objective 1)
   a. BNP
   b. cytokines, especially IL-6 and IL-8
   c. PAPP-A
   d. cTnT

LEVEL II

2. Which of the following represents the "active" form of BNP? (Objective 3)
   a. PreproBNP
   b. ProBNP
   c. NT-proBNP
   d. BNP

3. Which of the following is a limitation to the interpretation of CK-MB relative to AMI? (Objective 7)
   a. It is released into circulation too long after an AMI.
   b. Its mass is too large to be measured by current laboratory technology.
   c. It is structurally similar to cTnI.
   d. CK-MB is not cardiac specific.

4. Which of the following is a limitation of cTnT in evaluating patients with suspected AMI? (Objective 7)
   a. cTnT is also found in high concentration in the prostate.
   b. Patients with renal disease (e.g., end-stage renal disease) may have high cTnT.
   c. Its molecular structure is similar to myoglobin.
   d. Plasma levels of cTnT do not begin to rise until 24 hours post-AMI.

5. Which of the following accurately reflects structural and functional differences between TnI and TnT? (Objective 2)
   a. cTnI has a higher mass.
   b. cTnI isoforms have been expressed in patients with muscular dystrophy and end-stage renal disease.
   c. cTnI binds to actin and inhibits muscle contraction and cTnT does not.
   d. cTnI is expressed in skeletal muscle and cTnT is not.

6. Which of the following characteristics of NT-proBNP makes it more advantageous to measure than BNP? (Objective 3)
   a. higher endogenous levels
   b. shorter half-life
   c. shorter in vitro stability
   d. its molecular weight

REVIEW QUESTIONS *(continued)*

LEVEL I

9. Which of the following is a unique characteristic of H-FABP? (Objective 6)
   a. It is produced by adipocytes and is a modulator of lipid and glucose metabolism.
   b. It is linked to C-reactive protein.
   c. The release of H-FABP into the blood closely resembles that of CK-MB.
   d. The release of H-FABP into the blood closely resemble that of myoglobin.

LEVEL II

PEARSON
myhealthprofessionskit™

Use this address to access the interactive Companion Website created for this textbook. Simply select "Clinical Laboratory Science" from the choice of disciplines. Find this book and log in using your user name and password.

REFERENCES

1. Ross R, Glossmet JA. The pathogenesis of atherosclerosis. *NEnglJMed* (1976) 295:365–77.

2. Theroux P, Fuster V. Acute coronary syndromes: Unstable angina and non-Q-wave myocardial infarction. *Circulation* (1998) 97(12): 1195–1206.

3. McCance KL. Structure and function of the cardiovascular and lymphatic systems. *In*: Huether SE, McCance KL (eds.). *Understanding pathophysiology*, 3rd ed. St Louis, MO: Mosby, 2004:597–609.

4. Heron M, Hoyert DL, Murphy SL et al. Deaths: Final data for 2006. *CDC National Vital Statistics Reports* (2009) 57(14):32–34.

5. American Heart Association. *Heart disease and stroke statistics—2008 update*. Dallas, TX: American Heart Association, 2008.

6. Kim MC, Kini AS, Fuster V. Definitions of acute coronary syndromes. *In*: Fuster V, Alexander RW, O'Rourke RA (eds.). *Hurst's the heart*, 11th ed. New York: McGraw-Hill, 2004:1215–20.

7. Szmitko PE, Wang CH, Weisel RD et al. New markers of inflammation and endothelial cell activation. Part 1. *Circulation* (2003) 108:1917–23.

8. Apple FS, Wu A, Mair J et al. Future biomarkers for detection of ischemia and risk stratification in acute coronary syndrome. *ClinChem* (2005) 51(5):810–24.

9. Azzazy H, Christenson RH. Cardiac markers of acute coronary syndromes: Is there a case for point of care testing? *ClinBiochem* (2002) 35:13–37.

10. Kamisago M, Sharma SD, DePalma SR et al. Mutations in sarcomere protein genes as a cause of dilated cardiomyopathy. *NEnglJMed* (2000) 343:1688–96.

11. Apple FS. Analytical issues for cardiac troponin. *Progress in Cardiovas Diseases* (2004) 47(3):189–95.

12. Katrukha AG, Bereznikova AV, Filatov VL et al. Troponin I is releasesed in bloodstream of patients with acute myocardial infarction not in free form but as complex. *ClinChem* (1997) 43:1379–85.

13. Babuin L, Jaffe AS. Troponin: The biomarker of choice for the detection of cardiac injury. *CMAJ* (2005) 173(10):1191–99.

14. Eriksson S, Halenius H, Pulkki K et al. Negative interference in cardiac troponin I immunoassays by circulating troponin autoantibody. *ClinChem* (2005) 5:803–4.

15. Jaffe AS, Katus HA. Acute coronary syndrome biomarkers: The need for more adequate reporting. *Circulation* (2004) 110:104–6.

16. Pearlman ES. An unusual cause of an elevated cardiac troponin-I concentration. *ClinChem* (2006) 52(6): suppl: A19–A20.

17. Hypertension and coronary heart disease: Classification and criteria for epidemiological studies. *WHO Tech Support Ser* (1959) 168:3–28.

18. The Joint European Society of Cardiology/American College of Cardiology Committee. Myocardial infarction redefined—a consensus document of the Joint European Society of Cardiology/American College of Cardiology. *EurHeartJ* (2000) 21:1502–13.

19. Kamineni R, Alpert J. Acute coronary syndromes: initial evaluation and risks stratification. *Progress in Cardiovas Diseases* (2004) 46(5):379–92.

20. Eisenberg PR, Shaw D, Schaab C et al. Concordance of creatine kinase MB activity and mass. *ClinChem* (1989) 35:440–43.

21. Newby LK, Storrow AB, Gibler WB et al. Bedside multimarker testing for risk stratification in chest pain units: The chest pain evaluation by creatine kinase-MB, myoglobin and troponin I (CHECKMATE) study. *Circulation* (2001) 103:1832–44.

22. Winter WE, Elin RJ. The role and assessment of ventricular peptides in heart failure. *Clin in Lab Med* (2004) 24:235–74.

23. Francis GS, Tang HW, Sonnenblick EH. Pathophysiology of heart failure. *In*: Fuster V, Alexander RW, O'Rourke RA (eds.). *The heart*. 11th ed. New York: McGraw-Hill, 2004:697–99.

24. Jortani S, Prabhu, SD, Valdes R. Strategies for developing biomarkers of heart failure. *ClinChem* (2004) 50(2):265–78.

25. Maisel AS, Krishnaswamy P, Nowak RM et al. Rapid measurement of B-type natriuretic peptide in the emergency diagnosis of heart failure. *NEnglJMed* (2002) 347:1671–72.

26. Cowie MR, Jourdain P, Maisel A et al. Clinical application B-type natriuretic peptide (BNP) testing. *EurHeartJ* (2003) 24:1710–18.

27. Jernberg T, Lindahl B, Siegbahn A et al. N-terminal pro-brain natriuretic peptide in relation to inflammation, myocardial necrosis, and the effect of an invasive strategy in unstable coronary artery diseases. *JAmCollegeCardiology* (2003) 42:1909–16.

28. Vasan RS, Benjamin EJ, Larson MG et al. Plasma natriuretic peptides for community screening for left ventricular hypertrophy and systolic dysfunction: The Framingham Heart Study. *JAMA* (2002) 288:1252–59.

29. Clerico A, Emdin M. Diagnostic accuracy and prognostic relevance of the measurement of cardiac natriuretic peptides: A review. *ClinChem* (2004) 50: 33–50.

30. Verma S, Szmitko PE, Yeh E. C-reactive protein—structure affects function. *Circulation* (2004) 109:1914–17.

31. US Food and Drug Administration. Review criteria for assessment of C-reactive protein (CRP), high sensitivity C-reactive protein, and cardiac C-reactive protein (cCRP) assays. 2005. http://www.fda.gov/cdrh/ovid/guidance/1246.pdf (accessed November 2005).

32. Person TA, Nensah GA, Alexander RW et al. Markers of inflammation and cardiovascular disease: Application to clinical and public health practice: A statement for healthcare professional from the Centers for Disease Control and Prevention and the American Heart Association. *Circulation* (2003) 107:499–511.

33. Ridker PM. Evaluating novel cardiovascular risk factors: Can we better predict heart attacks? *AnnInterMed* (1999) 130:933–37.

34. Sinha MK, Gaze DC, Tippins JR et al. Ischemia modified albumin is a sensitive marker of myocardial ischemia after percutaneous coronary intervention. *Circulation* (2003) 107:2403–05.

35. Zhang R, Brennan ML, Fu X et al. Association between myeloperoxidase levels and risk of coronary artery disease. *JAMA* (2001) 286:2136–42.

36. Baldus S, Heeschen C, Meinetz T et al. Myeloperoxidase serum levels predict risk in patients with acute coronary syndromes. *Circulation* (2003); 108:1440–45.

37. Brennan ML, Penn MS, Van Lente F et al. Prognostic value of myeloperoxidase in patients with chest pain. *NEnglJMed* (2003); 349:1595–604.

38. Martin TB, Anderson JL, Muhlestein JB et al. Risk factor analysis of plasma cytokines in patients with coronary artery diseases by a multiplexed fluorescent immunoassay. *AmJClinPath* (2006) 125: 906–13.

39. Lawrence JB, Oxvig C, Overgaard MT et al. The insulin-like growth factor (IGF)-dependent IGF binding protein-4 protease secreted by human fibroblasts is pregnancy-associated plasma protein-A. *ProcNatlAcadSciUSA* (1999) 96:3149–53.

40. Bayes-Genis A, Conover CA, Overgaard MT et al. Pregnancy-associated plasma protein A as a marker of acute coronary syndromes. *NEnglJMed* (2001) 345:1022–29.

41. Tsimikas S, Brilakis ES, Miller ER et al. Oxidized phospholipids, Lp (a) lipoprotein, and coronary artery diseases. *NEnglJMed* (2005) 353(1):46–57.

42. Maglione D, Guerriero V, Viglietto G et al. Isolation of a human placenta cDNA coding for a protein related to the vascular permeability factor. *ProcNatlAcad SciUSA* (1991) 88:9267–71.

43. Tjwa M, Luttun A, Autiero M et al. VEGF and PlGF: Two pleiotropic growth factors with distinct roles in development and homeostasis. *CellTissueRes* (2003) 314:5–14.

44. Visse R, Nagase H. Matrix metalloproteinases and tissue inhibitors of metalloproteinases; Structure, function and biochemistry. *CirRes* (2003) 92:827–39.

45. Thompson MM, Squire IB. Matrix metalloproteinase-9 expression after myocardial infarction: Physiological or pathological? *CardiovascRes* (2002) 54:495–98.

46. Gavin P, Nilsson L, Carstensen J et al. Circulating matrix metalloproteinase-9 is associated with cardiovascular risk factors in a middle-aged normal population. Public Library of Science (Plos ONE) 2008;3(3): e1774. Doi:10.1371/journal.pone.0001774. http://www.ncbi.nlm.nih.gov/pmc/issues/161333/ (accessed 1/26/2010)

47. Azzazy H, Pelsers M, Christenson RH. Unbound free fatty acids and heart-type fatty acid-binding protein: Diagnostic assays and clinical applications. *ClinChem* (2006) 52(1):19–29.

48. Richieri GV, Ogata RT, Kleinfeld AM. The measurement of free fatty acid concentration with the fluorescent probe ADIFAB: A practical guide for the use of the ADIFAB probe. *MolCellBiochem* (1999) 192:87–94.

49. Kris-Etherton PM, Harris WS, Apple LJ. Omega-3 fatty acids and cardiovascular disease. New recommendations from the AHA. *ArteriosclerThrombVasBiol* (2003) 23:151–52.

50. Ferrucci L, Cherubini A, Bandinelli S et al. Relationship of plasma polyunsaturated fatty acids to circulating inflammatory markers. *JClinEndo Metab* 2006; 91(2): 439–46.

51. Oda Eiji, Hatada K, Katoh K et al. A case-control pilot study on n-3 polyunsaturated fatty acid as a negative risk factor for myocardial infarction. *International Heart Journal* (2005) 46(4):583–91.

52. Caslake MJ, Packard CJ, Suckling KE et al. Lipoprotein-associated phospholipase A2, platelet-activating factor acetylhydrolase: A potential new risk factor for coronary artery disease. *Atherosclerosis* (2000) 150(2): 413–19.

53. Macphee CH. Lipoprotein-associated phospholipase A2: A potential new risk factor for coronary artery disease and a therapeutic target. *Review Current Opinion in Pharmacology* (2001) 1(2):121–25.

54. Packard CJ, O'Reilly DS, Caslake MJ et al. Lipoprotein-associated phospholipase A2 as an independent predictor of coronary heart disease. West of Scotland Coronary Prevention Study. *NEnglJMed* (2000) 343(16):1148–55.

55. Ballantyne CM, Hoogeveen RC, Bang H et al. Lipoprotein-associated phospholipase A2, high-sensitivity C-reactive protein, and risk for incident coronary heart disease in middle-aged men and women in the Atherosclerosis Risk in Communities (ARIC) study. *Circulation* (2004) 109(7):837–42.

56. Oei HH, van der Meer IM, Hofman A et al. Lipoprotein-associated phospholipase A2 activity is associated with risk of coronary heart disease and ischemic stroke: The Rotterdam Study. *Circulation* (2005) 111(5):570–75.

57. Hackam DG, Anand SS. Emerging risk factors for atherosclerotic vascular diseases: A critical review of the evidence. *JAMA* (2003) 290(7):932–40.

58. Berg K. A new serum type system in man—the Lp system. *ActaPathol MicrobiolScand* 1963; 59:369–82.

59. Haque WA, Garg A. Adipocyte biology and adipocytokines. *ClinLabMed* (2004) 24:217–34.

60. Pischon T, Girman CJ, Hotamisligil GS et al. Plasma adiponectin levels and risk of myocardial infarction in men. *JAMA* (2004) 291(14):1730–37.

61. von Eynatten M, Hamann A, Twardella D et al. Relationship of adiponectin with markers of systemic inflammation, atherogenic dyslipidemia, and heart failure in patients with coronary heart disease. *ClinChem* (2006) 52:853–59.

62. Shamsuzzaman A, Winnicki M, Wolk R et al. Independent association between plasma leptin and C-reactive protein in healthy humans. *Circulation* (2004) 109: 2181–85.

63. Wolk R, Berger P, Lennon RJ et al. Body mass index: A risk factor for unstable angina and myocardial infarction in patients with angiographically confirmed coronary artery disease. *Circulation* (2003) 108(18):2206–11.

64. Pradhan A, Rifai N, Ridker P. Soluble intercellular adhesion molecule-1, soluble vascular adhesion molecule-1, and the development of symptomatic peripheral arterial diseases in men. *Circulation* (2002) 106(7): 820–25.

19

Liver Function

■ OBJECTIVES—LEVEL I

Following successful completion of this chapter, the learner will be able to:

1. Diagram a hepatic lobule and identify the major vessels and cell types.

2. Review major liver functions and list examples of each category.

3. Summarize the main steps of bilirubin metabolism from the breakdown of hemoglobin to excretion as urobilin.

4. Differentiate conjugated and unconjugated bilirubin, including composition and solubility in water and alcohol.

5. Review the clinical significance of bilirubin, including levels of total, direct, and indirect bilirubin.

6. Define jaundice, and identify and list examples of the three major categories of jaundice.

7. Explain the enzyme deficiency or metabolic defect involved in Crigler–Najjar, Gilbert, Dubin–Johnson, and Rotor syndromes.

8. Identify type of virus, route of transmission, at risk populations, incubation period, and recovery rate for the following types of viral hepatitis: A, B, C, and D.

9. Briefly examine the progression in alcoholics from alcoholic fatty liver to alcoholic hepatitis to alcoholic cirrhosis.

10. Review the Jendrassik–Grof methodology, including reagents and the direct bilirubin procedure.

11. Summarize the clinical significance of increased ammonia.

12. Briefly outline other liver function tests: enzymes, albumin, urinary and fecal urobilinogen, and prothrombin time.

13. Review common ammonia methodologies.

■ OBJECTIVES—LEVEL II

Following successful completion of this chapter, the learner will be able to:

1. Briefly summarize cirrhosis and the Child–Pugh and MELD systems for predicting prognosis in cirrhosis patients.

2. Describe three inherited metabolic liver diseases: hereditary hemochromatosis, Wilson disease, and α_1-antitrypsin deficiency.

3. Compare and contrast the changes that occur in nonalcoholic fatty liver diseases from steatosis to nonalcoholic steatohepatitis.

4. Examine the three most common cholestatic liver diseases: primary biliary cirrhosis, primary sclerosing cholangitis, and mechanical obstruction of the bile ducts.

KEY TERMS

α_1-antitrypsin (AAT)
Alcoholic cirrhosis
Alcoholic fatty liver
Alcoholic hepatitis (AH)
Alcoholic liver disease (ALD)
Ammonia
Bile canaliculi
Child–Pugh system
Cholestasis
Chronic hepatitis
Cirrhosis
Crigler–Najjar syndrome (CNS)
Delta bilirubin
Direct bilirubin
Dubin–Johnson (DJ) syndrome
Gilbert syndrome
Glutamate dehydrogenase

Hepatic jaundice
Hepatocytes
Hereditary hemochromatosis
   (HH)
Indirect or unconjugated bilirubin
Jaundice
Jendrassik–Grof
Kupffer cells
Liver lobule
Mechanical obstruction of the
   bile ducts
Model for end-stage liver
   disease (MELD)
Nonalcoholic fatty liver
   disease (NAFLD)
Nonalcoholic steatohepatitis
   (NASH)

Portal hypertension
Portal vein
Posthepatic jaundice
Prehepatic jaundice
Primary biliary cirrhosis (PBC)
Primary sclerosing cholangitis
   (PSC)
Reye's syndrome
Rotor syndrome
Space of Disse
Uridyl diphosphate (UDP)–glucoronyl
   transferase
Urobilin
Urobilinogen
Viral hepatitis
Wilson disease (WD)

℮ A CASE IN POINT

Tom N., a 60-year-old man, came into the emergency room complaining of fatigue and overall flulike symptoms. He was mildly jaundiced with icteric sclera.

Liver Profile

	2/15	2/22	Reference Range
Bilirubin, total	2.2	2.4	0.2–1.2 mg/dL
Bilirubin, direct	1.19		0.00–0.40 mg/dL
AST	564	735	5–40 IU/L
ALP	250		30–157 IU/L
ALT	373		5–40 IU/L
Albumin	2.1	1.9	3.5–5.0 g/dL
Total protein	6.5	6.4	6.0–8.4 g/dL
Ammonia	32	48	20–80 µg/dL
Folic acid	17.8		2.9–15.6 ng/dL
Vitamin B$_{12}$	377		180–710 pg/mL

Hepatitis Serum Panel

IgM anti-HAV:	Nonreactive
HbsAg:	Nonreactive
IgM anti-HBc:	Nonreactive
Anti-HCV:	Reactive

Issues and Questions to Consider

1. Circle or highlight the abnormal results.
2. What organ system is involved?
3. List six probable explanations for this chemistry profile.
4. What is the most likely explanation for these results?
5. What is the most likely diagnosis on this patient?

6. List five main etiologic factors (causes) for this infection.
7. List six groups that have higher than normal risk of infection (high-risk groups).
8. Describe the pathogenesis (i.e., what happens?).
9. How can the physician differentiate between active and chronic forms of this infection?

WHAT'S AHEAD

▶ An overview of liver anatomy and function
▶ A discussion of bilirubin metabolism
▶ A summary of pathological hepatic conditions
▶ A review of liver function tests

▶ INTRODUCTION

Liver function is critical for survival, playing an extremely vital role in the metabolism, digestion, detoxification, and elimination of substances from the body. Most plasma proteins except for immunoglobulins and hemoglobin are synthesized in the liver. Portal circulation from the intestine ensures that all nutrients from the digestion of food with the exception of fats pass through the liver before entering the general circulation to the rest of the body. As discussed in Chapter 6, Carbohydrates, the liver is responsible for the storage of carbohydrates in the form of glycogen to provide energy between meals. The liver is unique in that it is the only organ in mammals that has the ability to regenerate. Following damage from liver disease, as little as 25% of the remaining liver can regenerate into a whole organ.

▶ LIVER ANATOMY

The liver is the largest organ in the upper right quadrant of the body beneath the diaphragm and weighs 1400 to 1600 g, making up approximately 2.5% of total body weight. It is divided into two main lobes, with the right lobe six times larger than the left, as well as two smaller lobes on the posterior and inferior surfaces of the right lobe. The liver has a rich blood supply, accounting for 30% of the blood pumped through the heart per minute.

LIVER LOBULE

The **liver lobule** is the basic microscopic unit, 1–2 μ in size and containing four to six portal triads. The liver is also unique in that it has a dual blood supply. The **portal vein**, which originates in the gastrointestinal tract provides 70% of the total blood volume; the *hepatic artery* from the aorta contributes the remaining 30%. The *hepatic vein* drains blood from the liver to the inferior vena cava. Each portal triad contains a branch of the (1) portal vein, (2) hepatic artery, and (3) bile duct, hence the name *triad*. Figure 19-1 ■ illustrates a liver lobule.

Blood flows from the portal vein and hepatic artery through the *sinusoids* toward the *central vein*, which merges with the hepatic vein and is responsible for draining blood from the liver. The sinusoids are capillary channels between the rows of hepatic parenchymal cells called **hepatocytes** that make up 70% of the liver's mass. The sinusoids are also lined with two types of cells: endothelial cells and macrophages called **Kupffer cells**. Kupffer cells are active phagocytes that engulf bacteria, old red blood cells (RBCs), toxins, and cellular debris from the blood flowing through the sinusoids. They contain enzymes stored in the lysosomes that break down foreign particles such as bacteria. A good analogy for Kupffer cells is a vacuum cleaner that cleans the debris from the blood as it flows through the liver. The two primary cells are hepatocytes and Kupffer cells, with the endothelial cells not as important to liver function. The lobular structure is illustrated in more detail in Figure 19-2 ■.

Each hepatocyte has a large surface area that is in contact with both nutrient uptake from the sinusoids and an outlet system, the **bile canaliculi**, that carries secretions and excretions away from the hepatocytes. The nutrient uptake occurs in the **space of Disse** between the endothelial cells and hepatocytes. The bile in the bile canaliculi and blood in the sinusoids flow in opposite directions.

✓ Checkpoint! 19–1

1. *Blood leaves the liver through the* _____ .
2. *The portal triad of the liver consists of what three structures?*
3. _____ _____ *remove toxins from blood flowing through the sinusoids.*

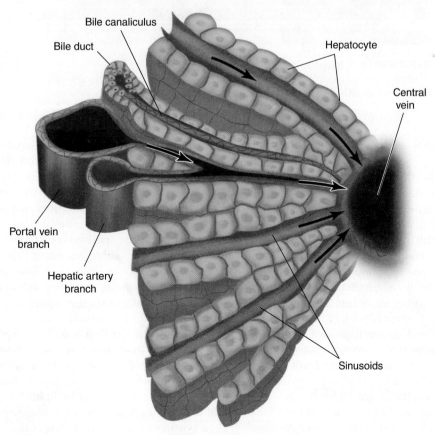

Bile canaliculus
Bile duct
Hepatocyte
Central vein
Portal vein branch
Hepatic artery branch
Sinusoids

■ **FIGURE 19-1**   The flow of blood and bile in a liver lobule.

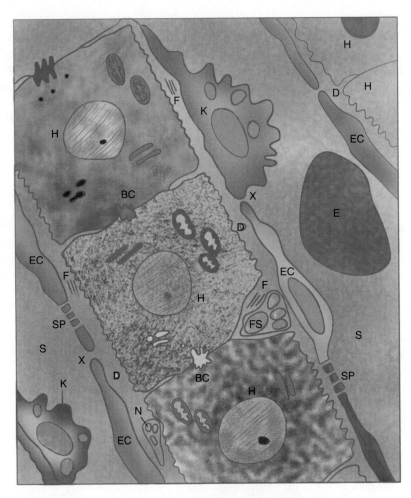

■ **FIGURE 19-2**   Hepatic lobule.
H, hepatocyte; EC, endothelial cell; S, sinusoid; D, space of Disse; K, Kupffer cell;
BC, bile canaliculi

► LIVER FUNCTION

Liver function is divided into five major categories: metabolism, excretion, detoxification, storage, and immunologic. The liver is responsible for more than 500 separate and critical activities; in fact, a patient with a completely nonfunctional liver would die within 10 hours. The liver is also unique in that is has full regenerative powers within normal limits until 75% of the hepatocytes have been destroyed.

The liver has a major role in carbohydrate, protein, lipid and fat, bilirubin, and hormone metabolism and synthesis. As described in Chapter 6, Carbohydrates, the liver is the major processor of dietary and endogenous carbohydrates. Review Chapter 6 for a discussion of glycolysis, gluconeogenesis, and glycogenolysis. The liver plays an important role in maintaining stable plasma glucose by its ability to store glucose as glycogen and to release it as required for energy.

The liver is also the site of synthesis of most plasma proteins except gamma globulins and hemoglobin. Proteins are synthesized in the rough endoplasmic reticulum of the hepatocyte and released into the hepatic sinusoids. Amino acids are delivered to the liver from the portal vein and used to synthesize proteins to meet the body's needs. The liver takes free fatty acids from the diet and breaks them down to acetyl-CoA and produces *endogenous cholesterol* for cell membranes. Another role in lipid and fat metabolism and synthesis is the synthesis of lipoproteins to make insoluble lipids more soluble, allowing them to be transported to other parts of the body. The liver synthesizes endogenous lipids in response to excess carbohydrates as well as to the normal intake of dietary lipids. See Chapter 7 (Lipids, Lipoproteins, and Cardiovascular Disease) for a detailed discussion of the liver's role in lipid metabolism.

One of the main *excretion* and *secretory* functions of the liver is the excretion of bile acids, cholesterol, and bilirubin through the bile ducts to the hepatic duct for storage in the gallbladder. Bile acids solubilize lipids for intestinal absorption and play an important role in their digestion and absorption. Urea synthesis from ammonia is critical in the excretion of nitrogen-containing metabolic products of protein catabolism.

The liver is critical in the *detoxification* of many products, including bilirubin, alcohol, and many drugs. Bilirubin is conjugated to direct bilirubin, which is water soluble and excretable through bile. Alcohol is converted to acetalaldehyde by alcohol dehydrogenase followed by conversion to acetate and finally CO_2 and H_2O. Many drugs are detoxified in the liver—phenobarbitol, for example. Drugs that are particularly hepatotoxic are monitored by the physician by periodic liver function tests (LFTs). The liver also plays an important role in the regulation of plasma hormone concentrations by catalyzing thyroid, steroid, and other hormones.

The liver is the primary site for glycogen storage. It also stores iron for the production of hemoglobin, copper, and fat-soluble vitamins A, D, E, and K. The storage of these various substances is crucial to maintain their presence in normal levels in the body. Immunologic roles include phagocytosis of bacteria and other substances by Kupffer cells and the secretion of IgA, among other roles in humoral defenses.

Checkpoint! 19–2

List the five major categories of liver function and an example of each.

▶ BILIRUBIN METABOLISM

The liver is the only organ that has the capacity to eliminate waste products—for example, heme from the breakdown of hemoglobin (old RBCs) from the body. Approximately 80% of the 200 to 400 mg of bilirubin formed daily is produced from hemoglobin released from old red blood cells. The remaining 20% is formed from enzymes and other products that contain heme, including cytochromes, myoglobin, and catalase.

Degradation of RBCs occurs in the reticuloendothelial system (spleen, liver, and bone marrow). The globin portion of hemoglobin is returned to the amino acid pool for reuse. Iron is attached to transferrin for transport to iron stores and recycling into hemoglobin. The porphyrin ring is converted to *biliverdin* (green bilirubin) by the action of heme oxygenase. A second enzyme, biliverdin reductase, reduces biliverdin into **indirect** or **unconjugated bilirubin**:

$$\text{Heme} \xrightarrow{\text{Heme Oxygenase}} \text{Biliverdin} \xrightarrow{\text{Biliverdin reductase}}$$
$$\text{Indirect and unconjugated bilirubin} \quad (19.1)$$

Ninety percent of the indirect or unconjugated bilirubin circulates in the bloodstream bound to albumin. Indirect bilirubin is insoluble in water, so it travels bound to a carrier protein, mainly albumin. The remaining 10% does not bind to albumin and can cross cell membranes. This unbound or free indirect bilirubin has a special affinity for brain and nervous tissue and in large amounts can cause brain damage called *kernicterus*, especially to developing or immature brain tissue (e.g., infants).

Two nonalbumin proteins isolated from the liver cell cytoplasm, called Y and Z proteins, are responsible for the intracellular binding and transport of bilirubin. When bilirubin reaches the liver and flows through the sinusoidal spaces in the liver lobule, albumin is removed and the bilirubin is bound to ligandin, a carrier protein. The ligandin–bilirubin complex is transported to the hepatocytes, where bilirubin is conjugated in the microsomes, which are particles within the endoplasmic reticulum. The enzyme **uridyl diphosphate (UDP)–glucoronyl transferase** binds two glucoronic acid molecules to the bilirubin molecule. This bilirubin, which is called *conjugated* or *direct bilirubin* (bilirubin diglucoronide), is now water soluble and can cross lipid cell membranes.

Three types of conjugated or direct bilirubin are found in the plasma: bilirubin monoglucoronide (one glucoronic acid), bilirubin diglucoronide (two glucoronic acids), and delta bilirubin. **Delta bilirubin** is conjugated bilirubin bound through a covalent bond with albumin; it is seen only in cases of significant hepatic obstruction. Four types of bilirubin are found in serum: one unconjugated or indirect bilirubin and three conjugated or direct bilirubins.

The bilirubin can now be excreted into the bile canaliculi and eliminated from the body. It proceeds from the bile canaliculi to larger bile ducts through the hepatic ducts and into the gallbladder for storage between periods of food digestion. When signaled by the digestive tract, the gallbladder constricts and expels the bile through the common bile duct into the duodenum.

Bacterial action in the duodenum reduces bilirubin to **urobilinogen**, which comprises urobilinogen, stercobilinogen, and mesobilinogen, all of them colorless. They are then oxidized to red-brown **urobilin** and excreted in the stool. Most urobilinogen and urobilin is excreted in the feces, but 20% is reabsorbed through the portal circulation and recycled through the liver for a return trip to the gallbladder and intestines. Some urobilinogen (2% to 5%) is excreted in the urine and is a normal constituent of urine. Figure 19-3 ■ illustrates the major steps in bilirubin formation and excretion.

Checkpoint! 19–3

1. *What is the difference in chemical structure between unconjugated or indirect bilirubin and conjugated or direct bilirubin?*
2. *Which bilirubin is water soluble and can be excreted in the urine?*

▶ PATHOLOGICAL HEPATIC CONDITIONS

Jaundice is a condition characterized by yellow discoloration of the skin, sclera, and mucous membranes. It is most commonly caused by elevated bilirubin levels and may be the first and only symptom of liver disease. Jaundice is not clinically apparent until bilirubin is increased to 2–3 mg/dL, and often yellow sclera are the only sign of liver problems. Conjugated bilirubin is accountable for more symptoms of

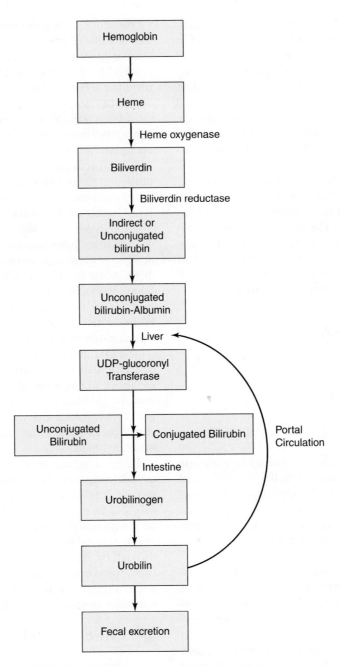

■ FIGURE 19-3   Bilirubin Metabolism.

jaundice than unconjugated because it is more easily absorbed into the tissues.

Jaundice can be divided into three categories: prehepatic, hepatic, and posthepatic jaundice. **Prehepatic jaundice** is the result of increased production of bilirubin by the body. Any process or disease resulting in increased hemolysis and excessive destruction of RBCs leads to increased bilirubin levels and, in some cases, prehepatic jaundice. Prehepatic jaundice results in increased total serum bilirubin and unconjugated bilirubin because the excess bilirubin reaching the liver cannot be excreted. Unconjugated bilirubinemia occurs when the

liver is unable to process the bilirubin load. Examples of prehepatic jaundice include hemolytic disease of the newborn, congestive heart failure, hemolytic anemias, transfusion reactions, and exposure to some chemicals. Hemolytic disease of the newborn occurs when the fetus's RBCs are destroyed by maternal antibodies when the mother has been previously sensitized by a transfusion or an earlier pregnancy. Congestive heart failure causes impaired circulation so that bilirubin does not reach the liver to be conjugated and eliminated.

Hepatic jaundice is the result of a problem occurring in the liver. Gilbert, Crigler–Najjar, Dubin–Johnson, and Rotor

BOX 19-1 Prehepatic, Hepatic and Posthepatic Jaundice

Prehepatic Jaundice

► Hemolytic anemias
► Exposure to chemicals
► Disease state (e.g., some cancers)
► Drugs coating red cells (autoimmune hemolytic anemia)
► Transfusion reaction
► Hemolytic disease of the newborn
► Congestive heart failure

Hepatic Jaundice

► Gilbert syndrome
► Crigler–Najjar syndrome
► Dubin–Johnson syndrome
► Rotor syndrome
► Cirrhosis
► Viral hepatitis
► Alcoholic liver disease
► Drug-induced liver disease
► Hepatocellular carcinoma
► Toxic liver injury
► Neonatal physiologic jaundice

Posthepatic Jaundice

► Common bile duct stones, gallbladder stones (most common)
► Cancer of the bile ducts, pancreas, or ampulla of Vater
► Bile duct stricture or stenosis

syndromes are examples of genetic defects in some part of bilirubin metabolism. Examples of other causes of hepatic jaundice are cirrhosis, viral hepatitis, and alcoholic liver disease. Neonatal physiologic jaundice is the most common cause of unconjugated hyperbilirubinemia. It is usually seen as a temporary elevation of bilirubin in newborns mainly because of an immature liver that cannot metabolize and excrete bilirubin completely. UDP-glucoronyl transferase is not fully developed at birth, especially in premature infants, and the liver usually does not become functional until around the third day. Neonatal physiologic jaundice is often treated with ultraviolet light by a bilirubin lamp.

Posthepatic jaundice or obstructive jaundice is triggered by blockage of the flow of bile from the liver. Common bile duct stones or gallbladder stones are the most common causes of posthepatic jaundice. Other conditions include cancer of the bile ducts or pancreas as well as bile duct stricture or

stenosis. Box 19-1 ■ offers a more complete list of conditions associated with prehepatic, hepatic, and posthepatic jaundice.

DISORDERS OF BILIRUBIN METABOLISM

Gilbert syndrome is the least serious of inherited unconjugated hyperbilirubinemias and the most common. It is found in 2–13% of the population, and males are affected more often than females. Gilbert has to do with a problem involving active transport of bilirubin by ligandin through the hepatocyte cell membrane to the microsome. Activity of UDP-glucoronyl transferase is reduced by 20–50%, and levels of unconjugated bilirubin are only slightly increased. Symptoms include jaundice and vague complaints of fatigue. It is a mild, benign condition, and no treatment is necessary.

Crigler–Najjar syndrome (CNS) is the rarest of unconjugated hyperbilirubinemias. It is an inherited condition, an autosomal recessive trait, whereby children are born with a decreased or absolute lack of UDP-glucoronyl transferase. Children with type 1 CNS usually die within the first year of life. Most have irreversible brain damage, kernicterus, because of the toxicity of unconjugated bilirubin in the central nervous system. Type 1 is caused by a mutation of one of the five exons (a coding region in a gene) that code for UDP-glucoronyl transferase. This leads to severe jaundice because unconjugated bilirubin is not water soluble and cannot be excreted in the urine. Treatment options include liver transplant or gene replacement (gene-repair therapy). In type 2, the milder form, patients have no symptoms other than jaundice.

Dubin–Johnson (DJ) syndrome is a chronic and benign condition that produces an obstructive liver disease resulting in impaired biliary excretion of conjugated bilirubin. The liver's uptake, processing, and storage of bilirubin are normal; only the action of removal of bilirubin from the hepatocyte and excretion in bile is defective. The bilirubin cannot be transported from the hepatic microsome to the bile canaliculi to be excreted. Symptoms include mild jaundice, hepatomegaly, and dark urine in some individuals. It is a mild disorder with an excellent prognosis and a normal life expectancy, and it requires no treatment.

Rotor syndrome, closely similar to DJ, is also a benign conjugated hyperbilirubinemia. A reduction in the concentration or activity of intracellular binding protein (e.g., ligandin) may be the specific defect. It is not progressive, and the only abnormality is the elevated conjugated bilirubin and total bilirubin. It is inherited and less common than DJ, but it requires accurate diagnosis to distinguish it from more serious liver diseases. These conditions are illustrated in Figure 19-4 ■.

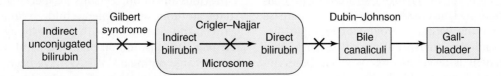

■ FIGURE 19-4 Inherited Hyperbilirubinemias.

TABLE 19-1

Bilirubin and Urobilinogen in Prehepatic, Hepatic, and Posthepatic Jaundice

	Prehepatic	Hepatic	Posthepatic
Total bilirubin	N to ↑	↑	↑
Conjugated bilirubin	N to ↑	↑ (Except CNS)	↑
Unconjugated bilirubin	↑	N or ↑	↑
Urine urobilinogen	↑	N or ↑	↓
Fecal urobilinogen	N or ↑	N or ↑	↓↓
Urine bilirubin	N	↑ (Pos)	↑ (Pos)

N, normal; ↑, increased; ↓, decreased

Table 19-1 ✪ summarizes the bilirubin concentrations, including conjugated and unconjugated, in serum, urine, and stool in prehepatic, hepatic, and posthepatic jaundice.

✓ Checkpoint! 19-4

Identify the following as prehepatic, hepatic or posthepatic jaundice.

1. *Viral hepatitis:* _____
2. *Congestive heart failure:* _____
3. *Bile duct obstruction:* _____
4. *Cirrhosis:* _____
5. *Hemolytic disease of the newborn:* _____

REYE'S SYNDROME

Reye's syndrome is an acute and often fatal childhood syndrome that is presumably caused by a virus (e.g., chicken pox, influenza, varicella). The child starts with an upper respiratory tract infection and begins to improve after 2 to 4 days. This is followed by an abrupt onset of vomiting and diarrhea that leads to delirium and progresses in some patients to coma and terminal respiratory arrest. Aspirin has been found to precipitate Reye's syndrome in more than 90% of cases, hence the warning on children's aspirin.[1]

The age distribution is 5 to 15 years, although younger and much older patients have been diagnosed with Reye's syndrome. Encephalopathy and fatty degeneration of the liver are hallmarks of this condition. Hepatomegaly is often associated with disturbed consciousness and seizures.[1]

Treatment includes reduction of protein intake to decrease the production of ammonia (discussed later in this chapter). Suppression of colonic bacteria with antibiotics also decreases the production of ammonia. Exchange transfusion with dialysis is another option. Diuretics may be used to increase the production of urine. Drugs—steroids, for example—are administered to minimize brain swelling and seizures. As a last resort, the patient is taken to surgery, where holes are drilled in the skull to allow room for the brain swelling to prevent further brain injury.

TOXIC HEPATITIS

As discussed earlier, the liver is the metabolic center of the body where a multitude of organic compounds are catabolized or altered for use by the body. The portal circulation brings ingested material, including all oral drugs, first to the liver. Some drugs and their metabolites, however, are toxic to the liver (e.g., acetaminophen), so physicians order routine LFTs to monitor patients who are on such hepatotoxic medications.

VIRAL HEPATITIS

Viral hepatitis is a worldwide disease of serious proportion that is accompanied by hepatocellular inflammation, injury, and necrosis of the hepatocyte. It can be caused by viruses, bacteria, parasites, drugs, chemicals, and toxins. Viral infections are the most common cause of acute hepatitis.[2] This section will provide a brief overview of viral hepatitis, including hepatitis A, B, C, D, and E. For further information about hepatitis and hepatitis testing, consult the excellent immunology and serology textbooks.

HEPATITIS A

Hepatitis A (HAV), formerly called *infectious hepatitis*, is transmitted through contamination of water and food supplies with feces infected with HAV virus. It comprises about 20% of viral hepatitis infections. HAV is rarely if at all transmitted through blood or blood products. It is found in areas with poor personal hygiene, contaminated water supplies, or poor public sanitation. Approximately one-third of the population in the United States shows evidence of past infection with HAV.

HAV is a picovirus, a single-stranded RNA virus, that replicates in the hepatocytes and is excreted through the bile into the gastrointestinal tract. This is how it contaminates water and food supplies. HAV has a short incubation period— 28 days (ranging from 15 to 50 days)—and a short course of acute symptoms lasting 2 to 3 weeks. Symptoms are flulike and fairly mild and nonspecific (e.g., anorexia, fever, nausea, and vomiting). HAV is often a subclinical disease with patients anicteric, showing no signs of jaundice. Practically all patients recover within 3 to 4 months, and there is no carrier state or chronic form of this infection.[3]

HAV is identified by finding IgM anti-HAV antibody in the serum during the acute stage. IgM antibodies are detectable 5 to 10 days before the onset of symptoms, peak within 1 week, and disappear at 8 weeks. IgG anti-HAV antibodies peak at 1 to 2 months and persist for years, indicating a previous infection. Persons at special risk in the United States are homosexual men, children and staff in day care centers, and any facility with people living in crowded quarters (e.g., prisons, military facilities, and psychiatric institutions). Large outbreaks of HAV have been associated with a single source such as an infected food handler or a contaminated water supply.[3]

HEPATITIS B

Hepatitis B (HBV), formerly called *serum hepatitis*, comprises about 60% of viral hepatitis. HBV is transmitted by blood or blood products through three major routes: parenteral, sexual, and perinatal. The parenteral route is by means of contaminated blood or blood products, including transfusion, accidental puncture with contaminated needles, hemodialysis, coagulation factor concentrates (hemophiliacs), and IV drug users. The sexual route involves transmission through sex by homosexuals and people with multiple sex partners. The perinatal route is transmission by an infected mother to the fetus.[4]

HBV belongs to a new class of viruses called Hepadna viruses. It is a double-shelled particle called the *Dane particle*, which consists of a core with a number of proteins and envelope and surface proteins. The outer surface structure contains hepatitis B surface antigen (HBsAg), the inner core contains hepatitis B core antigen (HBcAg) and has a single molecule of partially double-stranded DNA. Hepadna viruses are unusual among DNA viruses in that they replicate from an RNA template using reverse transcriptase.

HBV has a long incubation of 1 to 6 months, with a range of 6 to 26 weeks. Liver disease associated with hepatitis B infection can result in a wide spectrum of conditions ranging from subclinical (30% of persons have no obvious symptoms) to acute but self-limited to fulminant hepatitis. About 10% of HBV patients do not recover completely and develop chronic hepatitis; the remaining 90% recover completely. Some recover but remain carriers of the live virus.[5]

Testing for HBV consists of measuring various antigens and antibodies depending, on the stage of infection. HBsAg is the first to be detected early in the process and the most widely used marker in the acute or active phase of the HBV infection. Other antigens that can be identified later in the infection are HBcAg and HBeAg, which is closely associated with the core antigens. Anti-HBs follows the disappearance of HBsAg. Other antibodies, including anti-HBc and anti-HBe, can also be measured. Anti-HBc is the most commonly detected antibody.[6]

The hepatitis B vaccine is highly effective and recommended for all people exposed to blood or blood products (e.g., medical personnel). Routine vaccination of infants to 18-year-olds is recommended by the Centers for Disease Control and Prevention.[5]

HEPATITIS C

Hepatitis C was formerly identified by ruling out HAV and HBV and thus was called *non-A, non-B hepatitis* before the actual virus was initially identified in 1989 and fully recognized two years later. It is spread primarily through percutaneous contact with infected blood or blood products. Injectable drug abuse is the most common risk factor for hepatitis C virus (HCV) infection.

HCV is a small, enveloped, single-stranded RNA virus of the *Flaviviridae* family with an incubation period of 6 to 15 weeks.[7]

Some 80% of persons with HCV have no signs or symptoms.[8] HCV is the most common cause of posttransfusion hepatitis (90%). In 85% of individuals, the infection can progress to chronic hepatitis, cirrhosis, or carcinoma. HCV is the major cause of chronic hepatitis and associated with evidence of chronic liver injury in most cases. In 15–20% of chronic HCV patients, cirrhosis develops within 20 to 30 years. This may progress to hepatocellular cancer (HCC), with HCV being the most common risk factor for developing HCC.[7]

HCV testing is available through enzyme immunoassays, Western blot, or polymerase chain reaction (PCR) amplification. Anti-HCV antibody is the principal screening test for HCV.[7]

HEPATITIS D

Hepatitis D virus (delta hepatitis or HDV) is caused by an RNA virus that lacks a protein coat. The virus cannot reproduce unless HBV is present. HDV is spread primarily by direct contact of HBsAg carriers with HDV- or HBV-infected individuals. Apparently, HDV uses excess HBsAg as a protein coat and becomes infective. HDV consists of a single-stranded circular RNA coated in HBsAg. Co-infection with HBV and HDV results in a more serious prognosis, including death, so HBV-positive patients should be screened for HDV. Hepatitis D is a severe and rapidly progressive liver disease. Some who recover progress to chronic hepatitis, leading to cirrhosis or hepatic cancer. Testing involves antibodies to HDV and IgM-anti-HDV.

HEPATITIS E

Hepatitis E is a 34-nm, single-stranded unenveloped RNA virus. It is considered to be rare in the United States, although a few cases have been identified. Like HAV, it is transmitted through contaminated food and water.[9] Immunological tests have been developed to detect the antibody to the *open reading frame 2* (ORF2) antigen. The clinical progression of the disease is similar to HAV. It infects younger people, has a self-limited course, and there is no chronic infection.

CHRONIC HEPATITIS

Chronic hepatitis is defined as a chronic inflammation of the liver that persists for at least 6 months. It is characterized by continuing inflammation of hepatocytes, which is confirmed by elevated liver enzymes that can be accompanied by hepatocyte regeneration and scarring. The most common causes of chronic hepatitis are hepatitis B, hepatitis C, autoimmune disease, Wilson disease, α_1-antitrypsin deficiency, and idiopathic. Chronic hepatitis B and hepatitis C will be covered in this section; the others will be reviewed later in this chapter.

Chronic hepatitis B is the most important cause of chronic hepatitis. Hepatitis B usually does not cause pathological changes to hepatocellular cells, and the cellular damage is from an immune-mediated attack against the hepatocyte. In chronic hepatitis, the immune response is incomplete and the virus is not totally eliminated from the infected cells. It continues to replicate and damage newly regenerated cells. Chronic HBV is diagnosed with the patient's history, including HBsAg, anti-HBs, anti-HBc, and HBV DNA levels.[10]

Chronic hepatitis B can be classified into two categories: replicative and nonreplicative. In chronic replicative HBV, viral DNA is found in the hepatocytes and viral particles are released into the plasma. Approximately 3% to 5% of chronic replicative hepatitis B patients convert to nonreplicative every year. In the nonreplicative form, the viral DNA is integrated into the host cell DNA and the viral load or circulating viral particles is low and often undetectable. This condition is often referred to as the HBV *carrier state*.

Chronic hepatitis C develops in about half of acute hepatitis C infections. If HCV is detected 6 months following the initial exposure, it will not resolve spontaneously. Approximately 20% of hepatitis C patients will progress to cirrhosis within 20 years. The factors that tend to be associated with progression are age 40 or over at diagnosis, male gender, alcohol abuse, and immunosuppression. Chronic HCV is diagnosed with the following tests: anti-HCV and HCV RNA by PCR.

Chronic hepatitis C is characterized by fluctuating alanine aminotransferase (ALT) levels in comparison to HBV infection where ALT increases at the time of seroconversion and values do not fluctuate. It is common for patients with chronic HCV to have several normal ALTs between two elevated ALTs.

ALCOHOLIC LIVER DISEASE

Hepatic injury resulting from long-term consumption of alcohol is a common problem. Alcohol is catabolized to acetalaldehyde by alcohol dehydrogenase, and acetalaldehyde becomes a hepatocyte toxin when concentrations become too high. A woman may be at higher risk because her liver has reduced levels of alcohol dehydrogenase. Injury progresses from alcoholic fatty liver to alcoholic hepatitis to alcoholic cirrhosis.[11]

Risk factors for **alcoholic liver disease (ALD)** include the following.

- **History and magnitude of alcohol consumption.** Alcoholic liver disease does not occur in all individuals with a history of chronic daily alcohol consumption. However, there seems to be a threshold below which it does not occur: 40 g/day in men (~14–27 drinks per week) and 10 g/day in women (~7–13 drinks/week). Daily ethanol consumption exceeding 40–80 g/day for males and 20–40 g/day for females for 10–12 years will almost certainly lead to ALD. Daily drinking appears to be riskier than intermittent drinking.
- **Hepatitis B or C infection.** Patients with hepatitis are more likely to have increased severity of symptoms.
- **Gender.** Women have lower levels of alcohol dehydrogenase in their gastric mucosa, resulting in higher levels of blood alcohol. Family, twin, and adoption studies have indicated heritability ranging from 50% to 60%.[12]
- **Genetic factors.** A genetic predisposition to alcoholism has been identified.

- **Nutritional status.** Poor nutritional status is common in alcoholics. Although poor nutrition may contribute to the progression of alcoholic liver disease, adequate nutrition does not protect an individual from liver disease.

Alcoholic fatty liver follows 6 months to a year of moderate consumption and results in few lab abnormalities. Hepatomegaly may be noted on physical examination. Fatty infiltration of the liver with fat collected in vacuoles would be found on biopsy. It is a benign, reversible disease; abstinence from alcohol leads to a complete recovery.[12]

Alcoholic hepatitis (AH) occurs with alcohol consumption over a longer period of time. AH has a wide variety of symptoms, including hepatomegaly, vomiting, jaundice, ascites, fever, abdominal pain, and peripheral neuritis. Signs and symptoms include malnutrition, abnormal liver function tests, leukocytosis, and increased levels of acute-phase reactant proteins. Prognosis depends on the severity and type of damage, and mortality ranges from 2% to 27%. Aspartate aminotransferase (AST) levels are often more than twice ALT, and the AST is usually less than eight times the upper limit of normal. Alcohol depletes vitamin B_6-dependent pyridoxyl-5-phosphate, an essential precursor in aminotransferase synthesis. For further information, see "Enzymes in Liver Disease" in Chapter 9, Enzymes.

Alcoholic cirrhosis is the result of the most severe damage and leads to an irreversible scarring process. Patients are usually more than 40 years old because it takes years for alcohol to produce that amount of damage to liver. Only 10% to 15% of heavy drinkers develop cirrhosis, which further suggests a genetic link. A liver biopsy is needed for confirmation. The mortality rate 2 years after diagnosis is 25%. Cirrhosis in general, not only alcoholic, is one of the 10 leading causes of death in the United States.

The prognosis of alcoholic liver disease is much better than that for other forms of liver disease. The primary treatment is abstinence from alcohol. New approaches for drug treatment have shown great promise. Liver transplantation is the best option for those who progress to end-stage liver disease.[12]

✓ **Checkpoint! 19–5**

1. *Briefly describe the symptoms and pathology of Reye's syndrome.*
2. *Categorize each of the following as hepatitis A, B, C, or D.*
   a. *Requires co-infection with hepatitis B: _____*
   b. *Transmitted through contaminated food and water supplies:*
   _____
   c. *Formerly called non-A, non-B hepatitis: _____*
3. *Do all alcoholics develop cirrhosis? Explain why or why not.*

CIRRHOSIS

Cirrhosis is defined as a chronic liver disease characterized pathologically by fibrosis and the conversion of normal liver architecture into structurally abnormal lobules. The progression of liver damage from minimal change to cirrhosis can occur over weeks to years. Hepatitis C may take as long as 40 years to progress to cirrhosis. Clinical symptoms of cirrhosis are caused by the loss of functioning hepatocytes and increased resistance to hepatic blood flow or portal hypertension.[13]

The cause of cirrhosis cannot be determined based on the histology. Viral hepatitis B, HCV, alcohol, hemochromatosis, Wilson disease, and other causes result in similar histological changes. Virtually all chronic liver diseases can progress to cirrhosis. Hepatitis C has emerged as the nation's leading cause of both chronic hepatitis and cirrhosis.[13]

Complications associated with cirrhosis include portal hypertension, varices, edema, ascites, bruising, itching, and hepatic encephalopathy. **Portal hypertension** is caused by blockage of blood flow in the portal vein by scarring, resulting in increased pressure within the vein. It is a result of increased portal venous inflow and increased resistance to portal blood flow. Varices are formed in the stomach and esophagus as a result of portal hypertension. Varices are enlarged blood vessels with thin walls that are more likely to burst or rupture with increased pressure (portal hypertension), leading to bleeding problems in the stomach or esophagus. Bleeding varices are a life-threatening problem that require immediate medical attention.[13]

Edema and ascites are caused by fluid retention in the legs and abdomen because of portal hypertension, changes in hormone levels, and decreased albumin. Ascites is the accumulation of fluid in the peritoneal cavity. Bruising occurs as the result of decreased levels of coagulation factor proteins synthesized in the liver. Increased levels of bilirubin may result in intense itching. Hepatic encephalopathy is caused by increased toxins, including ammonia in the bloodstream because of decreased hepatic function.[13]

Cirrhosis is classified into three categories: (1) micronodular, (2) macronodular, and (3) mixed, based on the histology and gross appearance of the liver. The earliest laboratory abnormalities are (1) increased prothrombin time, (2) decreased platelet count, (3) a decrease in the albumin to globulin ratio to < 1, and (4) a DeRitis ratio (AST/ALT) > 1.

A variety of staging systems have been developed to predict prognosis in cirrhosis. The most common of these was the **Child–Pugh system** summarized in Table 19-2 .

⊗ TABLE 19-2

Child–Pugh System for Classifying Severity of Cirrhosis

Feature	One Point	Two Points	Three Points
Encephalopathy	None	Grade 1	Grade 2–3
Ascites	None	Slight	Moderate–severe
Albumin (g/dL)	>3.5	2.5–3.5	<2.5
Prothrombin time (> control)	0–2	2–4	≥6
Bilirubin (mg/dL)	<2	2–3	>3

Source: Haranath SP. Perioperative management of the patient with liver disease. eMedicine (www.emedicine.com/med/topic3167.htm), accessed July 2, 2007.

A score <7 points is class A, 7–9 points class B, and >9 points class C.[14] A new system called **model for end-stage liver disease (MELD)** appears to be superior for predicting short term survival. Three blood tests: bilirubin, prothrombin time as international normalized ratio (INR) and creatinine, are used to determine the MELD score. The MELD equation is

$$3.8 \times \log_e (\text{bilirubin mg/dL}) + 11.2 \times \log_e (\text{INR}) + 9.6 \log_e (\text{creatinine mg/dL})$$

Several variations of this equation have been developed,[15] and several Internet sites have automatic calculators to assist in the calculation—all you have to do is plug in the numbers.[1] It is used to identify seriously ill patients who are candidates for liver transplantation to ensure that available organs are directed to transplant candidates based on the severity of liver disease or medical urgency rather than length of time on the waiting list. Scores range from 6 to 40, with 6 being the least sick patient and 40 the sickest. Categories predict risk of death in 3 months: <10 = low risk, 10–20 = moderate risk, and >20 = high risk.

> **✓ Checkpoint! 19–6**
>
> *John has a grade-1 encephalopathy with slight ascites; albumin is 3.6, prothrombin is prolonged 3 seconds, and his bilirubin is 5.0 mg/dL. What is his Child–Pugh score? What class of cirrhosis is he—A, B, or C?*

METABOLIC LIVER DISEASE

Three inherited liver diseases that present as chronic hepatitis are hemochromatosis, Wilson disease, and α_1-antitrypsin deficiency. These genetic disorders should be ruled out along with the viral causes of chronic hepatitis.

Hereditary hemochromatosis (HH) is an autosomal recessive disorder of iron metabolism that is characterized by excessive iron absorption and accumulation in tissue. The HH gene on chromosome 6 is called the *HFE gene*, and it codes for a transmembrane protein. It binds the transferrin–transferrin receptor complexes, which stops iron absorption.[16] Several mutations are associated with HH, but in North America more than 90% are homozygous for a single-point mutation that inserts a tyrosine instead of cysteine at residue 282.

HH is diagnosed by testing for excess iron. Transferrin saturation (plasma iron divided by the total iron binding capacity, or TIBC) greater than 45% or unsaturated iron binding capacity less than 155 μg/dL has a 100% positive predictive value (PV+) for HH homozygous C282Y mutation.[17] HH is characterized by a class triad of symptoms: cirrhosis, diabetes mellitus, and bronzing of the skin. The skin pigment is not iron, but melanin. Other symptoms have also been associated with HH, including abdominal pain, arthritis, and hypogonadism.[17]

Treatment is most commonly phlebotomy to remove excess iron and stimulate erythropoiesis to deplete iron stores. Phlebotomy is performed weekly until iron deficiency develops.

The patient is then monitored with regular phlebotomy to maintain normal plasma iron indices.[17]

Wilson disease (WD) is an autosomal recessive disorder of copper metabolism. It involves a mutation in the copper-transporting ATPase (ATP7B), which moves copper into bile for excretion.[18] A deficiency of this enzyme results in accumulation of copper in the tissues, including the liver. WD is also associated with low levels of ceruloplasmin, a copper-containing enzyme. The mutated ATPase results in less than a full complement of copper in apoceruloplasmin and a reduced half-life. The plasma concentrations of ceruloplasmin are also decreased. For further information see ceruloplasmin in the α_2-globulin section in Chapter 8, Amino Acids and Proteins.

The most important serine protease inhibitor is α_1-**antitrypsin (AAT)**. A deficiency of AAT is one of the most common genetically lethal diseases in caucasians and associated with lung and liver disease. AAT deficiency is also associated with hepatitis in newborns. In the 20% of AAT-deficient newborns who develop hepatitis, there is a 35% mortality rate in the first year. Those who survive the first year show evidence of diminished liver disease, which usually resolves by age 12. In adults, the correlation between AAT deficiency and liver disease is not as strong. In fact, contradictory studies have found an increased frequency of liver disease in patients with homozygous versus heterozygous PiZ variant gene responsible for AAT. No clear-cut determinations can be made at this point. However, a few studies suggest that AAT deficiency may increase the risk of liver damage from other factors such as HCV. In adults, AAT deficiency is associated with early-onset emphysema. AAT is also discussed in more detail in Chapter 8, Amino Acids and Protein.

CHOLESTATIC LIVER DISEASE

Cholestasis is defined as the stoppage or obstruction of the flow of bile from intrahepatic causes, obstruction of the bile duct by gallstones, or any process that blocks the bile duct such as cancer. The most common cholestatic diseases are (1) primary biliary cirrhosis, (2) primary sclerosing cholangitis, and (3) mechanical obstruction of the bile ducts.

Primary biliary cirrhosis (PBC) is a rare autoimmune disorder that targets intrahepatic bile ducts.[19] Its prevalence is 2 to 8 per 100,000, with a median age of onset at 50 years and a female-to-male ratio of 6:1. It is often associated with other autoimmune processes (e.g., Sjogren syndrome).

PBC patients usually present with an asymptomatic elevation of ALP but may also have features of cholestasis or fatigue. Aminotransferases are elevated in 50% of cases but are usually less than two times the upper limit of normal. The hallmark of PBC is the presence of antimitochondrial antibodies in the serum of 80–95% of patients. These antibodies target different components of the mitochondria, especially enzymes.[19] PBC patients typically show a slow progression to portal hypertension and often do not progress to cirrhosis.

Primary sclerosing cholangitis (PSC) is a chronic inflammatory disease of unknown origin marked by inflammation and obliteration of the intrahepatic and extrahepatic bile ducts. PSC differs from PBC in that PSC is found predominantly in males and has a younger median age of onset of 30 years. It is associated with ulcerative colitis with irritable bowel disease, which precedes PSC in 70–80% of patients.[20]

The patient's presentation is similar to PBC, with elevated ALP and other biliary tract enzymes found in asymptomatic patients. Serum aminotransferase levels may be elevated to several times normal. Transplantation is the main treatment for end-stage PSC, resulting in a high rate of long-term survival.[20] It does, however, tend to increase the severity of ulcerative colitis when present.

Mechanical obstruction of the bile ducts is the most common cause of cholestatic liver disease. Gallstones in the common bile duct or tumors in the head of the pancreas or duodenum are the predominant causes of mechanical obstruction of the bile ducts. Other causes include bile duct strictures, PSC, and extrinsic compression of the bile ducts by enlarged lymph nodes.

Extrahepatic obstruction often results in jaundice, and elevation of biliary tract enzymes is common but not found in all cases. Transient increases in AST and ALT are more common with choledocholithiasis (gallstones in the common bile duct) than with other causes of extrahepatic obstruction.

Mechanical obstruction of the bile ducts can also occur in the liver; this is known as intrahepatic cholestasis. This is rarely associated with jaundice, although direct bilirubin may be elevated. Common conditions related to intrahepatic obstruction are (1) granulomatous disease (e.g., sarcoidosis), (2) tumors, and (3) infiltrative processes (e.g., lymphomas and leukemia).

NONALCOHOLIC FATTY LIVER DISEASE AND NONALCOHOLIC STEATOHEPATITIS

Nonalcoholic fatty liver disease (NAFLD) is used to define a wide spectrum of disorders from fatty liver alone (steatosis) to nonalcoholic steatohepatitis, an inflammatory or fibrosing disease that can lead to progressive fibrosis and cirrhosis in 10% to 25% of individuals. Steatosis is the accumulation of fat in the liver with no inflammation or scarring. It is a benign condition that by itself does not cause any significant liver damage.[21] NAFLD is the hepatic component of the metabolic syndrome (obesity, type 2 diabetes mellitus, insulin resistance, dysplipidemia, and hypertension) described in Chapter 7 (Lipids, Lipoproteins, and Cardiovascular Disease).

Steatosis can progress to **nonalcoholic steatohepatitis (NASH)**. NASH is a necroinflammatory liver disease associated with fat accumulation in the liver without inflammation or scarring. Patients with nonalcoholic steatohepatitis have no history of alcohol abuse and do not have AST levels greater than ALT [DeRitis ratio (AST/ALT) <1]. It is more common in women than men and is often associated with diabetes or obesity. The threshold of less than 20 g of ethanol daily has been set as the maximum alcohol intake for the diagnosis of NAFLD, and individuals whose alcohol intake is greater than 20 g of ethanol daily are classified under alcoholic liver disease (e.g., alcoholic hepatitis).

NAFLD is defined as fat accumulation greater than 5–10% of the liver's mass. Fat accumulation has been linked with insulin resistance. The frequency of NAFLD in the United States is 20% and NASH 2–3%, which makes these conditions as common a cause of chronic hepatitis as chronic HCV. Obesity and insulin resistance are key components of hepatic inflammation and fibrosis. Insulin resistance is an antecedent in the accumulation of hepatocellular fat, and is a strong predictor of NAFLD.[21]

NASH leads to fibrosis and the irreversible scarring of the liver and is not a benign condition. It can lead to the most severe stage, which is cirrhosis. The symptoms of NAFLD and NASH are identical and highly nonspecific, and most patients have no symptoms. Others experience vague, right upper-quadrant abdominal pain, possibly because of stretching of the capsule surrounding the liver in hepatomegaly. The cirrhosis stage, which occurs later in patients 50 to 60 years of age, can have the typical symptoms of compensated or decompensated cirrhosis. In compensated cirrhosis, the liver is still able to function or cope with the damage; LFTs, including albumin, bilirubin, and sometimes even AST and ALT are normal. In decompensated cirrhosis, the liver damage cannot be overcome, and there is a significant risk of life-threatening complications, including the formation of abnormal veins called varices and hepatic encephalopathy.[21]

Laboratory diagnosis of NASH and NAFLD is not currently available. Increased levels of AST and ALT can differentiate NASH from other causes of NAFLD, but it does not correlate with the degree of necroinflammatory damage. The major treatment is lowering body weight and fat content, which has been shown to lower ALT values. In one study, a 1% decrease in weight was associated with an 8% decrease in ALT activity.[21] Monitoring and regulating diabetes mellitus, use of lipid-lowering medications, and pharmacologic treatment of insulin resistance in symptomatic patients are also suggested treatments for NAFLD. Studies are being conducted to determine the processes responsible for progression from steatosis to NASH to cirrhosis.

LIVER FUNCTION TESTS

Bilirubin

Bilirubin measurement is the cornerstone of a liver function profile. Total bilirubin and direct bilirubin concentrations are critical in the diagnosis of liver-related disorders. Classical bilirubin methodologies began with Ehrlich's reaction in 1883. He found that reacting urine with diazotized sulfanilic acid (sulfanilic acid and sodium nitrite) yielded a red or blue pigment with Ehrlich-reactive substances, including bilirubin. Van der Bergh applied Ehrlich's reaction to bilirubin using alcohol as an accelerator to speed up the reaction. Evelyn–Malloy modified it to a quantitative technique using 50% methanol.

Jendrassik–Grof is the method of choice in today's instruments. **Direct bilirubin** reacts with a diazo reagent—diazotized sulfanilic acid—composed of sulfanilic acid in hydrochloric acid and sodium nitrite. Following an incubation period that varies with the instrument, a solution of ascorbic acid is added that stops the reaction by destroying the excess diazotized sulfanilic acid. Alkaline tartrate and caffeine are added to the tube, alkalinizing the solution and resulting in the formation of blue-green azobilirubin. The reaction is read at 600 nm.

A second aliquot of serum, used to measure total bilirubin, is pretreated with caffeine–sodium benzoate. The caffeine, an accelerator and dissociating reagent, promotes the displacement of bilirubin from albumin and increases the reaction rate of indirect bilirubin. Sodium benzoate buffers the solution and may also play a role in the dissociation of indirect bilirubin from albumin. Hydrochloric acid and diazotized sulfanilic acid are added next. The diazo reaction is stopped by ascorbic acid, and alkaline tartrate is added to form the blue-green azobilirubin. Note that the same reagents are used in both total and direct bilirubin but in a different order.

Unconjugated bilirubin is not measured directly but can be calculated using the total bilirubin and direct bilirubin:

$$\text{Unconjugated (indirect) bilirubin} = \text{Total bilirubin} - \text{Direct (conjugated) bilirubin} \qquad (19.2)$$

Jendrassik–Groff is less sensitive than Evelyn–Malloy and other bilirubin procedures to variations in pH, protein, and hemoglobin concentration, which may result in a falsely decreased bilirubin in those methodologies. Bilirubin is light and temperature sensitive. It is photoxidized, so specimens should be shielded from light, which can decrease bilirubin by 10% within 30 minutes.

Reference Ranges

Total bilirubin	0.2–1.0 mg/dL
Conjugated (direct) bilirubin	<0.8 mg/dL
Unconjugated (indirect)	<0.2 mg/dL
Conjugated hyperbilirubinemia	>50% of the total bilirubin is conjugated
Unconjugated hyperbilirubinemia:	>80% of the total bilirubin is unconjugated

SERUM ENZYMES

Serum enzymes most often included in a panel of liver function tests are the aminotransferases (AST and ALT), alkaline phosphatase (ALP), lactate dehydrogenase (LD), and gamma glutamyl transferase (GGT). Using a combination of the enzymes as well as other LFTs, physicians can categorize or narrow down which liver disease is the most likely cause of the problem.

AST, ALT, and ALP are the most useful tests in that they allow differentiation of hepatocellular disease from obstructive liver disease or cholestatic disease. See Chapter 9,

Enzymes, specifically the liver enzyme section, for further information.

SERUM ALBUMIN

Serum albumin levels are important in determining the chronicity and severity of the liver disease. In acute liver disease, the albumin levels are often normal or within the reference range. A decreased albumin is indicative of chronic liver disease, although it can be present in severe, acute liver disease. Serial measurements of albumin levels can provide evidence of increasing liver failure if the levels continue to drop over weeks and months.

URINARY AND FECAL UROBILINOGEN

Urine urobilinogen levels increase when there is an increase in secretion of bilirubin from the gallbladder to the intestinal tract. This occurs when a patient is recovering from hepatitis or cirrhosis or has a decreased liver clearance (e.g., portal hypertension).

In cholestasis, the biliary excretion of bilirubin is impaired and the urinary urobilinogen is decreased. The decrease in urinary urobilinogen is not measurable or detectable by current reagent strip methodologies. A decrease in fecal urobilinogen, characterized by clay- or chalk-colored feces, is caused by the decrease in bilirubin reaching the intestinal tract. The decrease in bilirubin results in a decrease in fecal urobilinogen and urobilin that normally gives feces its characteristic brown color. Urinary and fecal urobilinogen provide little additional information required in the diagnosis of liver disease.

PROTHROMBIN TIME

Prothrombin time (PT) is often elevated in liver disease. Serial measurements of PT can be used to differentiate between cholestasis and severe hepatocellular disease. To determine the cause of the elevated PT, a repeat PT 4 hours following injection with vitamin K will provide diagnostic evidence. If the PT is corrected, then the prolonged PT was the result of malabsorption of vitamin K in cholestasis. If the PT is not corrected, then hepatocellular disease is the diagnosis.

AMMONIA

Ammonia is the product of amino acid and protein catabolism. Ammonia and amino acids are urea precursors. The rate of hepatic urea synthesis is dependent on exogenous intake of nitrogen as well as endogenous protein catabolism. The major source of ammonia is the gastrointestinal tract, where it is derived from the action of bacteria on contents of the colon as well as hydrolysis of glutamine in the large and small intestines. The primary source of ammonia production is the small intestine.

Normally, most ammonia in the portal vein is metabolized to urea in the Krebs–Henselheit urea cycle by hepatocytes during the first pass through the liver. In liver dysfunction, the liver is unable to convert excess ammonia into urea because of the low levels of enzymes of the urea cycle.

Ammonia is neurotoxic and can cause encephalopathy as discussed in the section on Reye's syndrome. Ammonia is present in serum in minute amounts, 10–70 μg/dL.

AMMONIA CLINICAL SIGNIFICANCE

Ammonia is elevated in advanced liver disease and renal failure. Hepatic conditions resulting in increased ammonia are impending hepatic coma, advanced liver disease, acute viral hepatitis, Reye's syndrome, and chronic hepatitis. Inherited deficiencies of enzymes in the Krebs–Henselheit urea cycle are the major cause of hyperammonemia in infants.

AMMONIA METHODOLOGIES

Glutamate dehydrogenase, an enzymatic procedure, is the method of choice.

$$\text{2-Oxoglutarate} + NH_4^+ + NADPH \xrightarrow{\text{Glutamate dehydrogenase}}$$
$$\text{Glutamate} + NADP^+ + H_2O \qquad (19.3)$$

A decrease in absorbance as NADPH is oxidized to NADP$^+$ is measured at 340 nm. Enzymatic reactions offer increased specificity and speed of reaction.

A second methodology, the Berthelot reaction, uses a cation-exchange resin, and the ammonia is eluted with NaCl. Sodium hypochlorite (NAOCl) and phenol (Berthelot's reaction) are then added to the eluate. Berthelot's reaction cannot be automated, a major disadvantage.

Ammonia is particularly susceptible to contamination, resulting in increased values. For this reason, several precautions should be observed when performing this procedure. No smoking by phlebotomist or patient. Poor venipunctures (e.g., short draws), probing for veins, and drawing from a heparin lock will also result in erroneous results. Metabolism of nitrogenous constituents in the blood will result in increased ammonia; therefore, specimens should be placed on ice immediately and centrifuged as soon as possible.

The reference range is 19–60 μg/dL with the enzymatic method.

LIVER TRANSPLANTS

When the liver becomes so damaged from scarring that it stops functioning and complications cannot be controlled, a liver transplant is the only option. In the liver-transplant procedure, the diseased organ is removed and replaced by a healthy liver. About 80% to 90% of patients survive liver transplantation. The health of the patient before the procedure, the disease that necessitated the transplant, and the presence of complicating factors play a role in the prognosis.

Several advances in liver transplantation have resulted in more patients on the transplant list receiving livers. A reduced-size liver transplant allows an adult size liver to be transplanted into a pediatric patient. In split-liver transplants, the larger right lobe is transplanted into an adult and the smaller left lobe in a pediatric patient, resulting in two patients receiving livers from one donor. Live liver transplants are a relatively new procedure when the part of the donor's liver is transplanted and the donor's liver regenerates back to its normal size within 6 to 8 weeks of surgery.[15]

Organ rejection is always a concern, and most transplant recipients have at least one incident of rejection. Survival rates have increased in the last few years because of more effective immunosuppressive drugs such as cyclosporine, tracrolimus, and prednisone. The status of the recipient's liver is monitored through liver function tests, including AST, ALT, GGT, ALP, and prothrombin time.

SUMMARY

The liver plays a critical role in the metabolism, digestion, detoxification, and elimination of substances from the body. It is unique in that it is the only organ in mammals that has the ability to regenerate.

The liver is also the only organ that eliminates heme from hemoglobin released from old red blood cells. Following degradation of RBCs in the reticuloendothelial system, the porphyrin ring is converted to biliverdin by heme oxygenase and reduced by biliverdin reductase to indirect or unconjugated bilirubin. In the liver, the indirect bilirubin is conjugated with two glucoronic acid molecules by UDP–glucoronyl transferase to conjugated or direct bilirubin, which is water soluble and can cross cell membranes.

Bilirubin is stored in the gallbladder until needed. During a meal, it is signaled by the digestive tract, constricts, and releases bile through the common bile duct into the duodenum. The bilirubin is reduced to urobilinogen by bacterial action in the duodenum and then oxidized to urobilin, which is responsible for the red-brown color of feces.

Jaundice is characterized by discoloration of the skin, sclera, and mucous membranes because of elevated bilirubin. It is categorized into prehepatic, hepatic, and posthepatic jaundice. Prehepatic jaundice is the result of increased bilirubin production from increased hemolysis or excess destruction of red cells. Examples of prehepatic jaundice include hemolytic anemias, congestive heart failure, transfusion reaction, and hemolytic disease of the newborn. Hepatic jaundice is a problem involving the liver. Cirrhosis, viral hepatitis, alcoholic liver disease, and genetic conditions including Crigler–Najjar, Dubin–Johnson, and Gilbert syndromes are examples of hepatic jaundice. Posthepatic jaundice is obstructive liver disease caused by blockage of the flow of bile from the liver—for example, common bile duct stones, gallstones, and cancer.

Viral hepatitis is a worldwide disease of serious proportion that is characterized by hepatocellular injury, necrosis, and inflammation of the hepatocytes. Hepatitis A, B, C, D, and E viruses are the most common in the United States. They can be differentiated through antigen and antibody testing, characteristics of the disease (e.g., incubation period), and medical history. Hepatitis A is transmitted through the oral–fecal route and contaminated food and water supplies. It has a short incubation period and is identified by finding IgM–anti-HAV during the acute period and IgG–anti-HAV antibodies that persist for years. Hepatitis B is transmitted through blood or blood products through three major routes: perenteral, sexual, and perinatal. Antigens and antibodies are used to detect the presence and stage of the infection. Hepatitis C is the most prevalent cause of posttransfusion hepatitis and the most common cause of chronic hepatitis. Hepatitis D is caused by an RNA virus that requires the presence of HBV to become infective. Co-infection with HBV and HDV has a profoundly serious prognosis. Hepatitis D is a severe and rapidly progressive liver disease.

Chronic hepatitis in defined as a chronic inflammation of the liver that persists for at least 6 months. The most common causes of chronic hepatitis are hepatitis B, hepatitis C, autoimmune, Wilson disease, α_1-antitrypsin deficiency, and idiopathic. Chronic hepatitis C is the most common cause.

Alcoholic liver disease progresses from alcoholic fatty liver to alcoholic hepatitis to alcoholic cirrhosis, depending on risk factors. Examples of risk factors include history and magnitude of alcohol consumption, hepatitis B or C infection, gender, genetic factors, and nutritional status. Alcoholic fatty liver, described as fatty infiltration of the liver with fat collected in vacuoles, is a benign, reversible condition with a prognosis of complete recovery on abstinence from alcohol. Alcoholic hepatitis results in more serious liver damage, and prognosis depends on the severity and type of damage, with a mortality ranging from 2% to 27%. Alcoholic cirrhosis is the most severe, with irreversible scarring of the liver. The mortality rate 2 years after diagnosis is 25%.

Cirrhosis is defined as a chronic liver disease characterized pathologically by liver scarring with loss of normal hepatic architecture and areas of ineffective regeneration. Viral hepatitis B, HCV, hemochromatosis, Wilson's disease, and other conditions must be ruled out to determine the exact cause. The earliest laboratory abnormalities are (1) increased prothrombin time, (2) decreased platelet count, (3) decrease in albumin to globulin ratio to <1, and (4) DeRitis ratio (AST/ALT) >1. A variety of staging systems have been developed to predict prognosis in cirrhosis. The most common of these are the Child–Pugh system and the model for end-stage liver disease (MELD).

Nonalcoholic fatty liver disease (NAFLD) involves a wide spectrum of disorders from only fatty liver (steatosis) to nonalcoholic steatohepatitis to cirrhosis. Steatosis is a benign condition that does not cause any significant liver damage, but it can progress to nonalcoholic steatohepatitis (NASH). NASH can lead to fibrosis and irreversible scarring of the liver, resulting in cirrhosis.

Bilirubin measurement is the cornerstone of the liver function profile. Total and direct bilirubin are critical tests as well as serum enzymes (AST, ALT, ALP, and GGT). Serum albumin, urinary and fecal urobilinogen, prothrombin time, and ammonia are often included in a liver function panel.

REVIEW QUESTIONS

LEVEL I

1. Bilirubin is a metabolic product of: (Objective 3)
   a. hemoglobin catabolism.
   b. protein catabolism.
   c. nucleoprotein catabolism.
   d. carbohydrate catabolism.

2. The phagocytic cells that line the sinusoidal space in the liver are called: (Objective 1)
   a. Kupffer cells.
   b. hepatocytes.
   c. renal epithelial.
   d. endothelial cells.

3. The liver's blood supply is provided by the: (Objective 1)
   a. hepatic artery.
   b. hepatic vein.
   c. portal artery.
   d. portal vein.
   e. Both a and d.

4. In the liver bilirubin is converted to: (Objective 3)
   a. urobilinogen.
   b. biliverdin.
   c. bilirubin–albumin complex.
   d. bilirubin diglucoronide.

5. Which of the following may be classified as being a function of the liver? (Objective 2)
   a. detoxification of drugs
   b. excretion of bile acids
   c. metabolism of glucose
   d. synthesis of proteins
   e. All of the above.

6. As a result of excessive accumulation of bilirubin in the blood, certain body areas such as the skin and sclera take on a yellow-pigmented appearance. The condition that is characterized by this yellow pigmentation is known as: (Objective 6)
   a. jaundice.
   b. hemolysis.
   c. kernicterus.
   d. lipemia.
   e. cholestasis.

7. Which of the following groups would be at greater than normal risk of getting hepatitis A? (Objective 8)
   a. drug addicts
   b. dialysis patients
   c. medical personnel
   d. day care workers

8. The type of hepatitis spread by the fecal–oral route is: (Objective 8)
   a. type A.
   b. type B.
   c. delta.
   d. type C.

LEVEL II

1. The inability of bilirubin to be transported from the microsomal region to the bile canaliculi in the liver is characteristic of: (Objective 7)
   a. Gilbert syndrome.
   b. Crigler–Najjar syndrome.
   c. prematurity.
   d. Dubin–Johnson syndrome.
   e. Rotor syndrome.

2. The biochemical defect involved in Gilbert disease is: (Objective 7)
   a. inhibition of glucoronyl transferase activity by a compound in the mother's serum.
   b. defective transport of direct bilirubin from the hepatocyte into the bile canaliculi.
   c. defective transport of bilirubin from the plasma (sinusoidal space) to the hepatocyte.
   d. inability to conjugate bilirubin.

3. In which of the following conditions does decreased activity of glucoronyl transferase cause an increase in the unconjugated bilirubin concentration? (Objective 7)
   a. viral hepatitis
   b. Rotor's syndrome
   c. Dubin–Johnson syndrome
   d. Crigler–Najjar syndrome

4. The antibody initially increased in response to hepatitis A infection is: (Objective 8)
   a. HAV antibody, IgM.
   b. HAV antibody, IgG.
   c. HAV antibody, IgA.
   d. *Escherichia coli*, protein C.

5. A clay-colored stool is frequently associated with an obstructive type of jaundice. The dark brown color of a normal stool results from the presence of intestinal: (Objective 6)
   a. bilirubin.
   b. porphobilinogen.
   c. urobilin.
   d. delta-aminolevulenic acid.

REVIEW QUESTIONS (continued)

LEVEL I

9. The hepatitis B virus: (Objective 8)
   a. may be transmitted by transfusion.
   b. cannot be transmitted by transfusion.
   c. cannot be transmitted by IV drugs.
   d. contains no Dane particles in the plasma.

10. Which of the following causes 90% of posttransfusion hepatitis? (Objective 8)
    a. HAV
    b. HBV
    c. HCV
    d. HIBG

11. Hepatitis C differs from hepatitis A and hepatitis B because it: (Objective 8)
    a. has a highly stable incubation period.
    b. is associated with a high incidence of icteric hepatitis.
    c. is associated with a high incidence of chronic carrier state.
    d. is seldom implicated in cases of posttransfusion hepatitis.

12. If a total bilirubin is 3.1 mg/dL and the conjugated bilirubin is 2.0 mg/dL, then the unconjugated bilirubin is: (Objective 10)
    a. 1.1.
    b. 2.2.
    c. 4.2.
    d. 5.1.

13. Which of the following statements regarding bilirubin metabolism is true? (Objective 10)
    a. Bilirubin is inhibited by salicylates.
    b. Bilirubin is excreted only as conjugated bilirubin.
    c. Bilirubin undergoes photooxidation when exposed to daylight.
    d. Bilirubin excretion is decreased by barbiturates.

14. Direct bilirubin reacts with diazo reagent in aqueous media because of: (Objective 10)
    a. the addition of OH groups.
    b. fewer NH_2 groups.
    c. glucoronide conjugation.
    d. additional COOH groups.
    e. protein conjugation.

15. Direct bilirubin: (Objective 10)
    a. reacts with Ehrlich's diazo reagent only in the presence of methanol.
    b. reacts with Ehrlich's diazo reagent in aqueous solution.
    c. is calculated by subtracting the direct from the indirect bilirubin.
    d. is the same as total bilirubin.
    e. must be hydrolyzed by acetic acid before reacting with Ehrlich's diazo reagent.

LEVEL II

REVIEW QUESTIONS *(continued)*

LEVEL I

16. When measuring bilirubin, the purpose of adding caffeine–sodium benzoate or methanol to the reaction mixture is to: (Objective 10)
    a. accelerate the reaction with conjugated bilirubin.
    b. accelerate the reaction with unconjugated bilirubin.
    c. destroy excess diazo reagent.
    d. shift the wavelength absorbed by azobilirubin.

17. In the Jendrassik–Grof bilirubin methodology, what converts purple azobilirubin to blue-green azobilirubin measured at 600 nm? (Objective 10)
    a. hydrochloric acid
    b. caffeine
    c. methanol
    d. alkaline tartrate

18. Specimens for ammonia should be: (Objective 13)
    a. kept on ice as soon as drawn.
    b. kept at room temperature.
    c. kept at 37°C.
    d. kept at 25°C.

19. Ammonia measurement is helpful in the diagnosis of all of the following *except*: (Objective 13)
    a. Reye's syndrome.
    b. impending hepatic coma.
    c. end-stage liver disease.
    d. kidney failure.

20. An 8-year-old boy was recovering from chicken pox when he suddenly suffered a relapse. He eventually went into a coma. An ammonia level was performed that was extremely elevated. What is a possible diagnosis? (Objective 6)
    a. hepatitis
    b. cirrhosis
    c. Reye's syndrome
    d. renal failure

LEVEL II

PEARSON myhealthprofessionskit™

Use this address to access the interactive Companion Website created for this textbook. Simply select "Clinical Laboratory Science" from the choice of disciplines. Find this book and log in using your user name and password.

REFERENCES

1. National Reye's Syndrome Foundation (www.reyessyndrome.org), accessed June 30, 2008.
2. Liang S-L. An overview of current practice in hepatitis C testing. *MLO* (2008) 40, 6: 14–19.
3. Fiore AE. Hepatitis A transmitted by food. *Clin Infect Dis* (2004) 38: 705–715.
4. Hepatitis B Foundation. Transmission (www.hepb.org/hepb/transmission.htm), accessed June 1, 2008.
5. Centers for Disease Control and Prevention. Hepatitis B fact sheet (www.CDC.gov/hepatitis), accessed June 1, 2008.
6. Centers for Disease Control and Prevention. Hepatitis B serology (www.cdc.gov/ncidod/diseases/hepatitis/b/Bserology.htm), accessed July 1, 2008.
7. Worman HJ. *The hepatitis B sourcebook* (New York: McGraw-Hill, 2002).
8. Centers for Disease Control and Prevention. Viral hepatitis C fact sheet (www.cdc.gov/ncidod/diseases/hepatitis/c/fact.htm), accessed July 1, 2008.
9. Centers for Disease Control and Prevention. Viral hepatitis C fact sheet (www.cdc.gov/ncidod/diseases/hepatitis/e/fact.htm), accessed July 1, 2008.

10. Malet PF. Chronic hepatitis. In Dale DC, Federman DD (Eds.), *ACP Medicine* (New York: WebMD Publishing, 2006).

11. Gramenzi A, Caputo F, Biselli M, Kuria F, et al. Alcoholic liver disease-pathophysiological aspects and risk factors. *Aliment Pharmacol Ther* (2006) 24, 8: 1151–1161.

12. Tome S, Lucey MR. Current management of alcoholic liver disease. *Aliment Pharmacol Ther* (2004) 19, 7: 707–714.

13. Wolf DC. Cirrhosis. eMedicine (www.emedicne.com/med/TOPIC3183.HTM), accessed June 10, 2008.

14. Haranath SP. Perioperative management of the patient with liver disease. eMedicine (www.emedicine.com/med/TOPIC3167.HTM), accessed June 9, 2008.

15. Durand F, Valla D. Assessment of prognosis of chronic hepatitis (www.medscape.com/viewarticle/572659), accessed July 10, 2008.

16. National Digestive Diseases Information Clearinghouse. Hemochromatosis. (www.digestive.niddk.nih.gov/diseases/pubs/hemochromatosis/ index.htm), accessed July 10, 2008.

17. Iron Disorder Institute. Hemochromatosis (www.irondisorders.org/Disorders/Hemochromatosis.asp), accessed July 10, 2008.

18. Das S, Das K, Ray K. Wilson's disease: An update. *Nat Clin Pract Neurol* (2006) 2, 9: 482–493.

19. Pyrsopoulos N. Primary biliary cirrhosis. eMedicine (www.emedicine.com/med/TOPIC223.HTM), accessed July 10, 2008.

20. Gillis L. Primary sclerosing cholangitis. eMedicine (www.emedicine.com/ped/TOPIC1895.HTM), accessed July 10, 2008.

21. Mendler M. Fatty liver: Nonalcoholic fatty liver disease (NAFLD) and nonalcoholic steatohepatitis (NASH) (www.medicinenet.com/script/maine.asp?articlekey=1909&pf=3&page=1), accessed November 26, 2006.

Suggested Readings

1. Anaizi M. The drug monitor, MELD (www.thedrugmonitor.com/meld.htmL), accessed June 9, 2008.

2. Balistreri WF. Fatty liver disease: A major cause of obesity-related morbidity and mortality (www.medscape.com/viewprogram/4224_pnt), accessed October 26, 2006.

3. Blaistreri WF. Nonalcoholic fatty liver disease: New insight into a major cause of obesity-related morbidity and mortality (December 15, 2005) (www.medscape.com/viewprogram/4837_pnt), accessed November 26, 2006.

4. Dufour R. Laboratory approach to acute and chronic hepatitis. *MLO* (September 2003) 35, 9: 10–16.

5. Falck-Ytter Y, McCullough AJ, et al. Clinical features and natural history of nonalcoholic steatosis syndromes. *Sem Liver Dis* (2001) 21, 1: 17–26.

6. Liver Disease. Model for end-stage liver disease: The MELD score (www.liverdisease.com/meld._hepatitis.htmL), accessed October 25, 2006.

7. Mayo Clinic. Cirrhosis (www.mayoclinic.com/health/cirrhosis/DS00373), accessed July 5, 2007.

8. MDCalc. MELD score (model for end-stage liver disease). *Clin Calc* (www.mdcalc.com/meld), accessed October 25, 2006.

9. National Digestive Diseases Information Clearinghouse. Cirrhosis of the liver (http://digestive.niddk.nih.giv/ddiseases/pubs/cirrhosis/), accessed July 6, 2007.

10. Tran TT. Issues in cirrhosis and liver transplantation (www.medscape.com/viewprogram/4224_pnt), accessed November 26, 2006.

20

Iron, Porphyrins, and Hemoglobin

■ OBJECTIVES—LEVEL I

Following successful completion of this chapter, the learner will be able to:

1. Explain the biochemistry of iron in humans.
2. Explain how iron is transported in the human body.
3. Outline the metabolism of iron and iron-containing compounds.
4. Cite examples of specific diseases associated with iron deficiency and iron overload.
5. Identify methods used to measure iron in serum or plasma.
6. Identify types of instrumentation used to measure iron, porphyrins, and porphobilinogen.
7. Identify examples of types of specimens used for laboratory assessment of iron, porphyrins, and hemoglobin.
8. List the two classes of porphyrias and outline specific porphyrias within each class.
9. Diagram the metabolic pathway of heme.
10. Draw the basis structure of a porphyrin.
11. Define the following terms: *porphyrins*, *porphyrias*, *ferritin*, *transferrin*, *heme*, *hemin*, *hematin*, and *hemoglobin*.

■ OBJECTIVES—LEVEL II

Following successful completion of this chapter, the learner will be able to:

1. Correlate laboratory results with selected diseases associated with iron disorders.
2. Correlate laboratory results with selected porphyrias.
3. Explain the clinical significance of ferritin and transferrin.
4. Summarize the chemistry of porphyrins.
5. Outline the eight enzymatic reactions involved in biosynthesis of heme.
6. Distinguish the porphyrias discussed in this chapter.
7. Predict the outcome of a deficiency of any enzyme used in the biosynthesis of heme.
8. Define the negative feedback loop associated with heme.
9. Define the primary, secondary, tertiary, and quaternary structures of hemoglobin.
10. Identify clinical features and symptoms for selected porphyrias and iron-related diseases.
11. Explain the principle of selected methods used to measure iron and porphyrin-related compounds.

KEY TERMS

Anemia
Apotransferrin
Ferritin
Hematin
Heme

Hemin
Hemochromatosis
Hemoglobin
Hemosiderin
Photosensitivity

Porphyrins
Porphyrias
Porphyrinogens
Transferrin

 A CASE IN POINT

A 40-year-old male Caucasian presented with the following medical scenario:

The patient was not feeling well for some time. He noted a gradual onset of fatigue, decreased libido, and erectile dysfunction (ED). He also complained of dry mouth and polyuria and noted some loss of muscle mass. He did not have shortness of breath, a cough, a fever, night sweats, or visual changes. The patient had developed arthralgias in his ankles. The patient did not show signs of hepatomegaly. His skin was tanned (hyperpigmented).

The patient had a history of pulmonary sarcoidosis that was treated successfully. The patient drank 5 to 12 alcoholic beverage per week and never smoked. His father died of heart disease at the age of 46, and his brother is healthy at this time. The patient's vital signs and physical evaluation were all relatively normal.

The results of clinical laboratory tests are shown below: (*Note*: Hematology test results were all within normal limits.)

Serum Chemistries	Results (conventional units)	Reference Interval (conventional units)
Sodium	144	136–145 mEq/L
Potassium	4.4	3.5–5.1 mEq/L
Chloride	101	98–107 mEq/L
Bicarbonate	7	23–29 mEq/L
Anion gap	36	6–10 mEq/L
Glucose	300	74–100 mg/dL
Creatinine	1.0	0.9–1.3 mg/dL
Glycated hemoglobin	7.1	<6.0 %
Urea	11.2	6–20 mg/dL
Albumin	4.1	3.3–4.1 g/dL
Akaline phosphatase	125	40–115 U/L
Alanine aminotranferase	125	10–55 U/L
Aspartate aminotransferase	97	10–40 U/L
Bilirubin		
Total	0.8	0.0–1.0 mg/dL
Direct	0.3	0.0–0.4 mg/dL
Iron	197	30–160 μg/dL
Iron binding capacity	202	228–428 μg/dL
Transferrin saturation	97	<45%
Ferritin	4890	30–300 ng/mL
Testosterone	146	270–1070 ng/dL
Luteinizing hormone	1.2	2.1–12.0 U/L (males)
Follicle-stimulating hormone	0.5	1.0–12.0 U/L (males)
Estradiol	6	10–50 pg/mL

Issues and Questions to Consider

1. What laboratory tests, signs, symptoms, and physical findings are most striking with regard to this patient?
2. What disease, condition, or syndrome may be present based on the history, physical findings, and laboratory data?
3. What are the key clinical laboratory tests that provide the clinician with important diagnostic information?
4. What is the course of action for this patient?

WHAT'S AHEAD

▶ Discussion of the biochemistry and physiology of iron, hemoglobin, and porphyrins
▶ Discussion of the clinical relevance of iron-related disorders
▶ Explanation of the abnormalities of porphyrin metabolism
▶ Description of the analytical methods for measuring iron and porphyrins
▶ Explanation of physical and structural configuration of the heme and globin components of hemoglobin

▶ IRON

Iron is an essential element for the function of all cells and is normally present in small quantities in most cells of the body, in plasma, and in other extracelluar fluids. The body conserves its iron supply and protects itself from free iron, which is highly toxic in that it participates in chemical reactions that generate free radicals such as singlet oxygen or hydroxyls. A free radical is devoid of one electron and is therefore unstable and will react quickly with other molecules that are trying to capture the needed electron to gain stability. When it succeeds in capturing this electron, it leaves the other molecules in an unstable state. This process continues in a cascade fashion and tends to disrupt the function of the cells. Thus, sophisticated physiological mechanisms are available within body tissues to maintain an appropriate concentration of iron for physiological functions while at the same time conserving this element and handling it in a way that toxicity is avoided.

BIOCHEMISTRY

Iron in the body is distributed into several different compounds (e.g., hemoglobin, myoglobin, enzymes) and body sites that are used for transfer and storage. The approximate

Body distribution of iron in both males and females.		
	Iron	Content (mg)
	Males, ~80 kg	Females, ~60 kg
Hemoglobin	2500	1700
Myoglobin/enzymes	500	300
Transferrin iron	3	3
Iron stores	600–1000	0–300

amounts of iron distributed for males and females are summarized in Table 20-1 ❂.

The storage forms of iron are **ferritin** and **hemosiderin**. Ferritin consists of a protein shell surrounding an iron core, and hemosiderin is formed when ferritin is degraded in secondary lysosomes. The outer shell of ferritin is composed of apoferritin with an interior ferric oxyhydroxide (FeCOOH)x crystalline core. There are approximately 24 subunits in apoferritin that are comprised of either L (light) or H (heavy) ferritin chains. The amount of light and heavy chains varies from tissue to tissue. For example, in serum, ferritin consists primarily of light chains and is designated H_0L_{24} whereas ferritin in tissues has variable amounts of primarily heavy chains and fewer light chains. Ferritin incorporates iron in the (3^+) or ferrous state and, through enzymatic reactions involving flavin mononucleotide (FMN), converts the iron to the ferric or (2^+) states. The ferric iron leaves the ferritin particles and is distributed to nearly all cells in the body.

Ferritin in cells (e.g., liver and macrophages of the marrow) provides a reserve of iron readily available for formation of hemoglobin and other heme proteins to (discussed later in this chapter). Liver injury and other pathologies result in the release of relatively large amounts of ferritin into iron-deficient plasma.

Hemosiderin is partially deproteinized ferritin and is insoluble in aqueous solutions. Iron is only slowly released from hemosiderin due in part to the large aggregates that it forms. Similar to ferritin, hemosiderin is found predominantly in cells of the liver, spleen, and bone marrow.

Iron is required for several cellular enzymes and coenzymes as either an integral part of the molecule or as a cofactor. Examples include peroxidase, cytochromes, aconitase, ferrodoxin, and many enzymes found in the Krebs cycle.

A small quantity of iron, ~80 mg, is located in the labile pool. The labile pool is a compartment that has no definitive anatomical location but has been described based on kinetic measurement with radiolabeled iron.[1]

✓ Checkpoint! 20–1

Describe the structural composition of ferritin.

TRANSPORT

Apotransferrin is a protein in plasma that transports iron from one organ to another. It is a β_1-globulin with a molecular mass of 75 kDa and consists of two binding sites, one for Fe^{3+} and the other for bicarbonate (HCO_3^-). The apotransferrin–Fe^{3+} complex is termed **transferrin**. The mechanism believed to result in the movement of iron from outside the cell to inside is called the transferrin cycle and is diagrammed in Figure 20-1 ■.[2] Transferrin binds to the transferrin captor cells and becomes a part of the interior of the cell. The iron is released from transferrin and is incorporated into heme or stored in the form of ferritin. The transferrin is then transported back to the cell surface, is released from the cytosol, and is available to bind to another iron atom.

METABOLISM

The metabolic cycle for iron in humans is outlined in Figure 20-2 ■. Iron absorbed from diet or released from stores circulates in the plasma bound to transferrin. Most of the iron is transported to the erythroid marrow and interacts with the transferrin receptors for uptake in the cells, as discussed previously. The iron incorporated into the hemoglobin structure subsequently enters the circulation as new red cells are released from the bone marrow. The iron is then part of the red cell mass and will not become available for reutilization until the red cell dies.

The average life span of erythrocytes is 120 days. Thus about 1% of red cells turn over each day under normal conditions. At the end of its life span, the red cell is recognized as senescent by the cells of the reticuloendothelia (RE) system, and the cell undergoes phagocytosis. Once inside the RE cell, the hemoglobin from the ingested red cell is broken down, the globins and other proteins are retuned to the amino acid pool, and the iron is sent back to the surface of the cell, where it is available to circulating transferrin.

Checkpoint! 20–2

Explain the mechanism for transporting iron into the cell using apotransferrin–Fe^{3+}.

CLINICAL SIGNIFICANCE

Disorders of iron metabolism reflect either a deficiency or excess of body sources of iron. Several other diseases exist in which abnormal distribution of iron contributes to either a primary or secondary cause. Examples include hyperferritinemia with cataracts, aceruloplasminemia, neuroferritinopathy, and atransferrinemia. A partial listing of disorders and laboratory measurements used to evaluate these diseases is shown in Table 20-2 ❂.

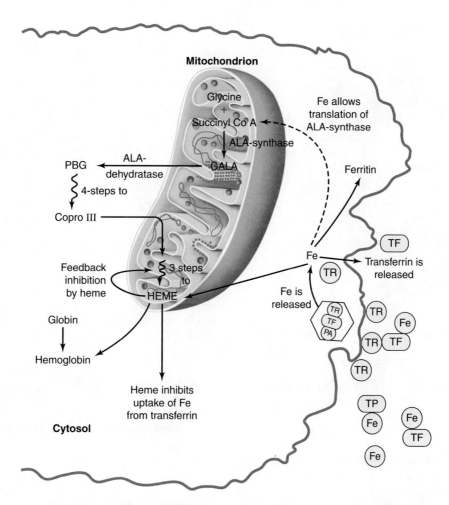

FIGURE 20-1   Transport of iron from outside the cell and into the cytosol for heme synthesis and storage. The reactions utilized to synthesize heme are presented in detail in Figure 20-4. (Key: TF, transferrin, TR, transferrin receptors, Fe, iron)

IRON DEFICIENCY

Iron deficiency states are more prevalent in humans than are iron excess or overload. Iron deficiency occurs primarily in children, young women, and older people but is present in people of all ages and social strata. The primary cause of iron deficiency in children is an iron-deficient diet; and in adults iron deficiency is usually the result of chronic blood loss or childbearing.

The pathological condition that develops when iron intake falls below the amount required for red blood cell production and iron reserves become depleted is termed **anemia**. There are several types of anemias, and laboratory tests of iron status can distinguish the various types. For a more detailed discussion of anemia, the reader is directed to the references listed at the end of the chapter.

IRON OVERLOAD

Hemosiderosis

Hemosiderosis is a condition characterized by an iron overload without associated tissue injury. It occurs locally at sites of bleeding or inflammation and can be present in individuals who have been given large amounts of iron, either as iron medication or in blood transfusions. Evaluation of patients presenting with possible hemosiderosis requires examination of tissue for stainable iron (hemosiderin), measurement of serum iron, measurement of total iron binding capacity (TIBC), and calculation of transferrin saturation (see section below).

Hemochromatosis

Hemochromatosis (primary) is a genetically related disease in which the body accumulates excess amounts of iron; it is one of the most common genetic diseases in humans. The symptoms of hemochromatosis include a triad of bronzing of the skin, cirrhosis, and diabetes. Other conditions include cardiomyopathies, arrhythmias, and endocrine deficiencies.

Secondary hemochromatosis is usually the result of problems with administration or absorption of iron. Patients whose treatment modality for anemia includes iron supplementation via intravenous (IV) infusion may accumulate an excess amount of iron. The most common causes of secondary hemochromatosis are thalassemia major and acquired myelodysplastic states.

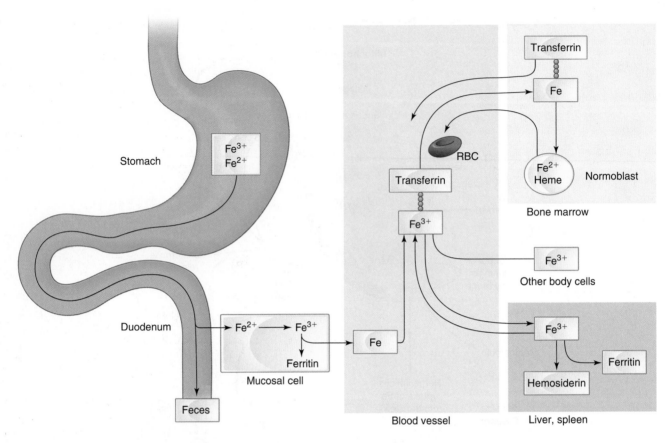

■ **FIGURE 20-2**   Internal iron exchanges.

Several forms of hemochromatosis exist and include the more common *hereditary form, juvenile form, African iron overload*, and *ferroportin deficiency*. Hereditary hemochromatosis results from hereditary abnormalities of proteins that regulate iron hemostasis. It has been linked to the human leukocyte antigen (HLA) loci on chromosome 6. Juvenile hemochromatosis is a rare disorder that resembles hereditary hemochromatosis clinically, but occurs at a much earlier average. These young individuals usually develop endocrine and cardiac dysfunction.

Assessment of patients with signs and symptoms of iron overload includes clinical laboratory tests for serum iron, TIBC, and transferrin saturation, as well as histological techniques to look for iron in tissues. Other clinical laboratory tests that may provide clinicians with useful information include a complete blood count (CBC), blood glucose, hemoglobin A_{1c}, and liver function tests.

Management of patients with hemochromatosis includes iron depletion monitoring and dietary changes. Iron depletion diminishes the symptoms and signs of most all the

⊗ **TABLE 20-2**

Relative concentrations of laboratory analytes in disease.

	Serum Iron (μg/dL)	TIBC (μg/dL)	Transferrin Saturation (%)	Serum Ferritin (μg/L)
Iron deficiency anemia	Dec	Inc	Dec	dec
Normochromic, of chronic diseases	Dec	Dec or Norm	Dec or Norm	Norm or slightly Inc
Thalassemia major	Inc	Dec	Inc	N/A
Sideroblastic anemia	Inc	N	Inc	Inc
Hemochromatosis	Inc	Dec	Inc	Inc
Hemosiderosis	Inc	Dec	Inc	Inc

Note: Dec, decreased; Inc, increased; Norm, normal

complications due to iron overload. Most therapeutic regimens involve weekly therapeutic phlebotomy, which removes iron. Patients must have their hemoglobin and serum ferritin measured throughout therapy. Dietary recommendations usually involve the reduction of red meats or organ meat and avoidance of iron supplements and alcohol consumption. The patient should avoid vitamin C supplements also because it tends to increase intestinal iron absorption.

 Checkpoint! 20–3

Define hemochromatosis *and include symptoms, secondary causes, and appropriate assessment of patients.*

FERRITIN

Ferritin is present in blood in very low concentrations, usually <250 ng/mL. It is considered an acute-phase protein and roughly reflects the body iron content. Circulating ferritin is iron poor and is largely apoferritin. The plasma ferritin concentration declines early in the progression of iron deficiency and usually before any noticeable changes occur in blood hemoglobin concentration, erythrocyte size, or serum iron concentration. Therefore, measurement of serum ferritin concentration serves as a good indication of iron deficiency irrespective of any concurrent diseases. Plasma ferritin is elevated in numerous chronic diseases and conditions, including rheumatoid arthritis, renal disease, heart disease, lymphomas, and leukemia. Plasma ferritin is also increased in patients with iron storage disease and is used to evaluate the effectiveness of phlebotomy therapy.

TRANSFERRIN

Measurement of plasma transferrin is useful in the differential diagnosis of hypochromic microcytic anemia and for monitoring treatment. Patients with iron deficiency will have an increased level of transferrin, but the protein is less saturated with iron. If the patient has anemia due to a failure to incorporate iron into erythrocytes (as found in patients with chronic inflammation), the transferrin level may be normal or low, but the protein is normally saturated with iron. In iron overload conditions the transferrin is normal, but saturation is above normal. High levels of transferrin are also seen in pregnancy, estrogen therapy, and hyperestrogenism.

Transferrin is described as a negative acute-phase reactant protein; therefore, concentrations will be decreased in inflammation or malignancy. Patients with liver disease will present with low levels due to decreased synthesis. Protein-losing enteropathies and nephritic syndrome also result in low blood levels of transferrin.

ANALYTICAL METHODS

Serum Iron

Measuring serum iron using spectrophotometry is routinely performed in clinical chemistry laboratories. The assays are easily adaptable to most automated analyzers. The serum iron concentration reflects the Fe^{3+} bound to serum transferrin and does not include the iron contained in serum as free hemoglobin. Methods for quantitative analysis of serum iron involve three fundamental steps: (1) releasing iron from transferrin by decreasing the pH of the serum using an acid, (2) reducing Fe^{3+} to Fe^{2+} using an acid such as ascorbic acid, and (3) complexing Fe^{2+} with a chromogen. Examples of chromogens widely used in serum iron assays include bathophenanthroline, tripyridyl triazine, and ferrozine.

Unsaturated Iron Binding Capacity and Total Iron Binding Capacity

Normally only about one-third of the iron binding sites of transferrin are occupied by Fe^{3+}. The additional amount of iron that can be bound is termed *unsaturated* or *latent* iron binding capacity (UIBC). The sum of the serum iron and UIBC represents the TIBC. Thus TIBC is a measure of the maximum iron concentration that transferrin can bind.

Serum UIBC and TIBC are measured by the addition of sufficient Fe^{3+} to saturate iron binding sites on transferrin. The excess Fe^{3+} is removed by adsorption with a compound (e.g., magnesium carbonate or sodium hydrogen carbonate) and the assay for iron content is then repeated. The TIBC can be determined from this second measurement.

Transferrin and Transferrin Saturation

Serum transferrin can be determined using immunoassays. Several Food and Drug Administration (FDA)–approved applications are available for use on automated immunochemistry analyzers. Also, serum transferrin concentration may be estimated from the TIBC using the following relationship:

$$\text{Serum transferrin (g/L)} = 0.007 \times \text{TIBC } (\mu g/dL) \quad (20.1)$$

Serum **transferrin saturation** is calculated using the following equation:

$$\text{Serum transferrin saturation (\%)} = 100 \times \text{serum iron/TIBC}$$

$$(20.2)$$

Ferritin

Serum concentrations of ferritin can be measured by a variety of immunoassay techniques, including enzyme-linked immunosorbent assay (ELISA), chemiluminescent immunoassay (CLIA), immunoradiometric assay (IRMA), and fluorrescent immunoassay (FIA). The reagents are available in kit form from several manufactures and can be applied to most automated immunoassay analyzers.

Reference Interval

The RIs for iron-related analytes are highly variable and dependent on many factors. Examples of conditions known to affect serum iron concentration, TIBC, and transferrin saturation include the following:

- Diurnal variation
- Menstruation
- Pregnancy
- Ingestion of iron
- Progesterone-like oral contraceptives
- Hepatitis
- Acute and chronic inflammation

A difference of 20 to 35% may exist between commercially available methods. Therefore it is recommended that each laboratory establish a RI for its method.

 Checkpoint! 20–4

Identify the three fundamental steps in procedures for measuring iron in serum.

▶ PORPHYRINS

The **porphyrins** are a group of compounds that contain four monopyrrole rings connected by methene bridges to form a tetrapyrrole ring, as shown in Figure 20-3 ■. Porphyrins were named from the Greek root for "purple" (porphyra) and owe this color to the conjugated double-bond structure of the tetrapyrrole ring. Note the alternate double and single bonds within the pyrroles. There are many porphyrin compounds, but only a few are of clinical interest, as shown in Table 20-3 ✪ along with their substituent groups. Variation in the arrangement of the same substituents around the peripheral position of the tetrapyrrole ring gives rise to porphyrin isomers, which are identified using Roman numerals (e.g., protoporphyrin IX).

A **porphyrinogen** is the reduced form of a porphyrin and differs by the absence of six hydrogens (Figure 20-3). Porphyrinogens are unstable in vitro and quickly oxidized to the corresponding porphyrins. In the cell, porphyrinogens are stable due to the lower oxygen content and tend to form intermediates in the heme biosynthesis pathway.

Porphyrins are able to chelate metals because there are four nitrogen atoms in the center of the tetrapyrrole ring. Protoporphyrins that chelate iron are termed **heme**. *Ferroheme* refers to Fe^{2+} complex, and *ferricheme* is Fe^{3+}. Ferricheme associated with hydroxide is known as **hematin**, and if chloride is the counter ion, ferricheme is referred to as **hemin**.

Porphyrins are purple, as stated earlier, whereas porphyrinogens are colorless due to a lack of conjugated double bonds. Porphyrins possess a unique spectral property in that they absorb electromagnetic radiation (EMR) near 400 nm (referred to as the Soret band). This band represents a strong absorption in the blue region of the EMR and is unique to heme protein. When porphyrins are illuminated with 400 nm

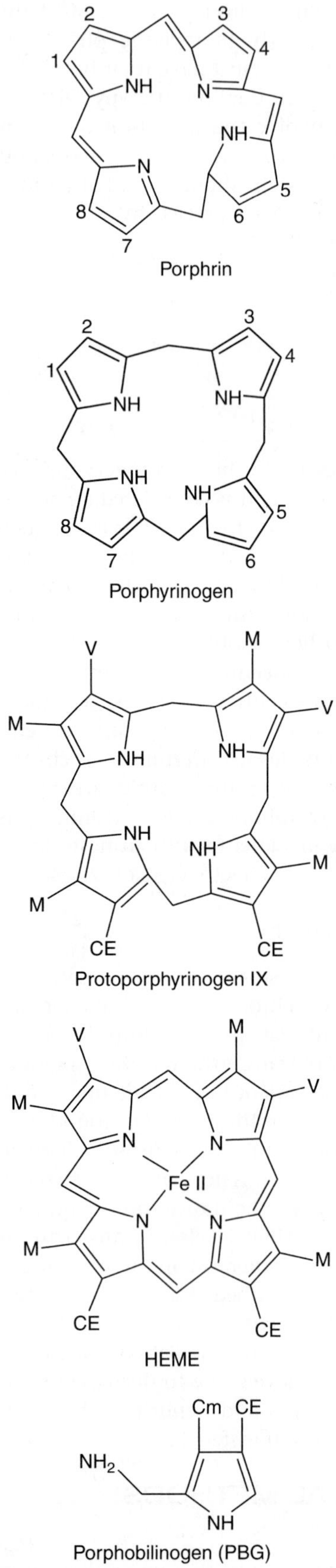

Porphrin

Porphyrinogen

Protoporphyrinogen IX

HEME

Porphobilinogen (PBG)

■ **FIGURE 20-3** Structures of selected porphyrins.

✪ TABLE 20-3

Porphyrins of clinical interest and their substituent groups that occupy the peripheral positions 1–8.

Position	1	2	3	4	5	6	7	8
Uroporphyrin-I	CM	CE	CM	CE	CM	CE	CM	CE
Uroporhyrin-III	CM	CE	CM	CE	CM	CE	CE	CM
Coproporphyrin-III	M	CE	M	CE	M	CE	CE	M
Coproporphyin-I	M	CE	M	CE	M	CE	M	CE
Protoporphyrin	M	V	M	V	M	CE	CE	M

Key: CM, carbonmethyl ($-CH_3COOH$); CE, carbonethyl ($-CH_2CH_2COOH$);
M, methyl ($-CH_3$); E, ethyl ($-CH_2CH_3$); V, vinyl ($-CH_2=CH_2$)

light, they produce an orange-red fluorescence near the 500–650 nm region of the electromagnetic spectrum (EMS). Both the absorbance and fluorescent spectral qualities vary depending on the substituent groups around the tetrapyrrole and the metal chelated to the nitrogen atoms.

The solubility properties of porphyrins and porphyrinogens are important both in vivo and in vitro. The ability of porphyrins to solubilize in either an aqueous or lipid medium dictates the route of excretion from the human body. Water-soluble porphyrins are excreted via the kidney into urine, and non-water-soluble porphyrins are excreted in the feces via the biliary tract. Analytical methods, especially those requiring extraction techniques for measuring porphyrins, rely on the differences in solubility of the various porphyrins. The porphyrins are only slightly soluble in water, as are the porphyrinogens. Porphyrins can be made to be more water soluble as the number of carboxyl groups increases. The analytical extraction techniques used to assay porphyrins and porphyrinogens require procedural steps that include extraction into acidified organic solvent first followed by a second or back extraction into an aqueous acid.

✓ Checkpoint! 20–5

Briefly describe the structural configuration of a basic porphyrin without any specific substituent groups.

HEME BIOSYNTHESIS

Heme biosynthesis involves eight enzymatic reactions that take place in cell mitochondria and cytoplasm. These eight reactions and associated **porphyrias** are shown in Figure 20-4 in a flow diagram format. Porphyrias (see below) are diseases caused by inherited defects in the genes required to create the enzymes found in the pathway for heme biosynthesis, or by conditions such as lead toxicity (see Chapter 22,

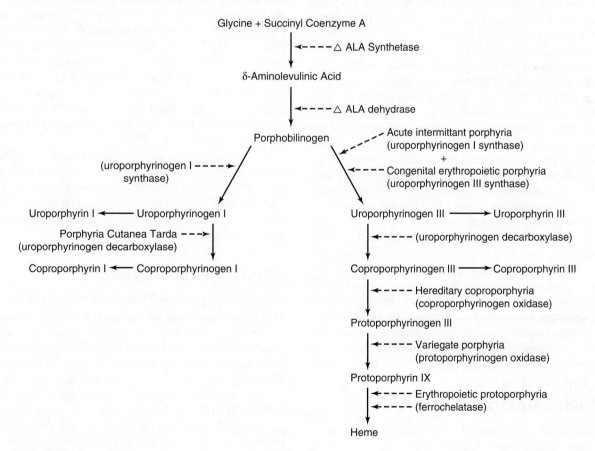

■ FIGURE 20-4   Porphyrias and eight steps to heme synthesis.

"Toxicology") that affect the enzymatic activity in individuals with normal heme synthesis. For example, if a gene defect involves the enzyme hydroxymethylbilane synthase, the individual will develop acute intermittent porphyria (AIP) and present with signs and symptoms associated with that disease. Also, the enzyme deficiency will result in an accumulation of porphobilinogen (PBG) in biological fluids, including blood, urine, and feces, that can be detected and measured by laboratory tests and provide useful diagnostic information to the clinician.

Heme biosynthesis begins in the mitochondria with chemical bonding of glycine and succinyl Co A in the presence of 5-aminolevulinate synthase (ALAS). The next four reactions occur in the cytoplasm and begin with the second enzyme, 5-aminolevulinic dehydratase (ALAD), catalyzing the condensation of two molecules of aminolevulinic acid (ALA) to form PBG. Hydroxymethylbilane synthase (HMB-synthase; also known as PBG-deaminase) catalyzes the head-to-tail condensation of four PBG molecules by a series of deaminations to form the linear tetrapyrrole, hydroxymethylbilane (HMB). Uroporphyrinogen III synthase (URO-synthase) catalyzes the rearrangement and rapid cyclization of HMB to form the asymmetric, physiologic, octacarboxylate porphyrinogen, uroporphyrinogen III (URO III).

The fifth enzyme in the pathway, uroporphyrinogen decarboxylase (URO-decarboxylase), catalyzes the sequential removal of the four carboxyl groups from the acetic acid side chains of URO III to form coproporphyrinogen III (COPRO III), a tetracarboxylate porphyrinogen. This compound then enters the mitochondrion, where COPRO-oxidase, the sixth enzyme, catalyzes the decarboxylation of two of the four propionic acid groups to form the two vinyl groups of protoporphyrinogen IX (PROTO IX), a decarboxylate porphyrinogen. Next, PROTO-oxidase oxidizes PROTO IX to protoporphyrin IX by the removal of six hydrogen atoms. The product of the reaction is a porphyrin (oxidized form), in contrast to the preceding tetrapyrrole intermediates, which are porphyrinogens (reduced forms). Finally, ferrous iron is inserted into PROTO IX to form heme, a reaction catalyzed by the eighth enzyme in the pathway, ferrochelatase (also known as heme synthetase or protoheme ferrolyase).[3]

Heme is necessary for a variety of hemoproteins, such as **hemoglobin**, myoglobin, respiratory cytochromes, and the cytochrome P450 enzymes. Hemoglobin synthesis in erythroid precursor cells accounts for approximately 85% of daily heme synthesis in humans. Hepatocytes account for most of the rest, primarily for synthesis of cytochromes, which are abundant in the liver endoplasmic reticulum and turn over more rapidly than many other hemoproteins, such as the mitochondrial respiratory cytochromes.

✓ **Checkpoint! 20–6**

Distinguish between the mitochondrial and cytoplasmic reactions used in the biosynthetic pathway of porphyrins and heme.

REGULATION OF HEME BIOSYNTHESIS

Regulation of heme synthesis differs in the two major heme-forming tissues, the liver and erythrocyte. In the liver, "free" heme is involved in a negative feedback loop in which heme inhibits the activity of ferrochelatase and acquisition of iron from transferrin. The decrease in iron acquisition leads to a decrease in iron uptake into the cell, with subsequent decrease in δ-ALA and heme production, as shown in Figure 20-1.

In the erythrocyte, unique regulatory mechanisms allow for the production of the very large amounts of heme, which is necessary for hemoglobin synthesis. The response to stimuli for hemoglobin synthesis occurs during cell differentiation, leading to an increase in cell number. The erythroid-specific ALA-synthase is expressed at higher levels, and erythroid-specific control mechanisms regulate other pathway enzymes as well as iron transport into erythroid cells.

CLASSIFICATION OF THE PORPHYRIAS

The porphyrias can be classified as either *hepatic* or *erythropoietic*, depending on whether the heme biosynthetic intermediates that accumulate arise initially from the liver or developing erythrocytes, or as *acute* or *cutaneous*, based on their clinical manifestations. Several clinically significant porphyrias are listed in Table 20-4 ⊙ with their biochemical and clinical features. Four of the five hepatic porphyrias—acute intermittent porphyria (AIP), hereditary coproporphyria (HCP), variegate porphyria (VP), and ALA-dehydratase porphyria (ADP)—manifest themselves during adult life and present with acute attacks of a neurologic nature and elevated levels of one or both of the porphyrin precursors, ALA and PBG. These are classified as *acute porphyrias*. The fifth hepatic disorder, porphyria cutanea tarda (PCT), presents with blistering skin lesions. HCP and VP also may have cutaneous manifestations similar to PCT.

The erythropoietic porphyrias—congenital erythropoietic porphyria (CEP) and erythropoietic protoporphyria (EPP)—are characterized by elevations of porphyrinogens in bone marrow and erythrocytes and present with cutaneous **photosensitivity**. The skin lesions vary in severity, physical appearance, and time of onset.

HEPATIC PORPHYRIAS

ALA-Dehydratase-Deficient Porphyria

ALA-dehydratase-deficient porphyria (ADP) is a rare autosomal recessive acute hepatic porphyria caused by a severe deficiency of ALA-dehydratase activity. To date, there are only a few documented cases, some in children or young adults, in which specific gene mutations have been identified. These affected individuals had <10% of normal ALA-dehydratase activity in erythrocytes, but their clinically asymptomatic parents and heterozygous relatives had about half-normal levels of activity and did not excrete increased levels of ALA. The frequency of ADP is unknown, but the frequency of

⭐ TABLE 20-4

Biochemical and clinical features of human porphyrias.

[2]Porphyria	Enzyme Defect	Inheritance	Abdominal Pain, Neurological Dysfunction (NV)	Photosensitivity, Cutaneous Lesions (CP)	Tissue Expression
Hepatic porphyrias					
ADP	ALAD	AR	YES	NO	Liver
AIP	HMBS	AD	YES	NO	Liver
PCT	UROD	AD	NO	YES	Liver
HCP	CPO	AD	YES	YES	Liver
VP	PPOX	AD	YES	YES	Liver
Erythropoietic porphyrias					
CEP	UROS	AR	NO	YES	Erythroid cells
EPP	FECH	AD[1]	NO	YES	Erythroid cells

[1]Polymorphism in intron 3 of wild-type allele affects levels of enzyme activity and clinical expression.

[2]See text for acronyms relevant to porphyrias.

Abbreviations: AD, autosomal dominant; ALAD, 5′-aminolevulinic acid; AR, autosomal recessive; CP, cutaneous photosensitivity; CPO, coproporphyrinogen oxidase; FECH, ferrochelatase; HMB, hydroxymethylbilane; NV, neurovisceral; PPOX, protoporphyrinogen oxides; UROD, uroporphyrinogen decarboxylase; UROS, uroporphyrinogen III synthase.

heterozygous individuals with <50% normal ALA-dehydratase activity was ~2% in a study in Sweden.[4] Because there are multiple causes for deficient ALA-dehydratase activity, it is important to confirm the diagnosis of ADP by mutation analysis.

Acute Intermittent Porphyria

Acute intermittent porphyria is an autosomal dominant disease resulting from the half-normal levels of HMB-synthase activity. The disease is widespread but is especially common in Scandinavia and Great Britain. Clinical expression is highly variable, and activation of the disease is often related to environmental or hormonal factors, such as drugs, diet, and steroid hormones. Attacks can be prevented by avoiding known precipitating factors. Rare homozygous dominant AIP also has been described in children (see below).

Because the neurovisceral symptoms rarely occur before puberty and are often nonspecific, a high index of suspicion is required to make the diagnosis. The disease can be disabling but is rarely fatal. Abdominal pain, the most common symptom, is usually steady, poorly localized, and may result in cramps. Ileus, abdominal distention, and decreased bowel sounds are common. However, increased bowel sounds and diarrhea may occur. Abdominal tenderness, fever, and leukocytosis are usually absent or mild because the symptoms are neurologic rather than inflammatory.

Peripheral neuropathy may develop due to axonal degeneration (rather than demyelinization) and primarily affects motor neurons. Significant neuropathy does not occur with all acute attacks; abdominal symptoms are usually more prominent. Motor neuropathy affects the proximal muscles initially, more often in the shoulders and arms. The course and degree of involvement are variable and sometimes may be focal and involve cranial nerves.

Porphyria Cutanea Tarda

Porphyria cutanea tarda (PCT) is a skin disorder that does not usually appear until adulthood. It is the most common type of porphyria and is caused by a partial deficiency of uroporphyrinogen decarboxylase. Some cases of the disease are familial and inherited as an autosomal dominant trait, but most cases are sporadic and most likely represent an acquired deficiency of the hepatic enzyme. Symptoms include fragile skin, blister formation, and hyperpigmentation. The disease is usually dominant until some form of liver dysfunction develops, such as iron overload or alcoholic liver disease. Patients on estrogen therapy may activate the skin lesion.

Hereditary Coproporphyria

Hereditary coproporphyria (HCP) is characterized by a partial deficiency of the enzyme CPO, which causes CPO III to accumulate in the cytosol. Patients may experience acute attacks, photosensitivity, and skin lesions.

Variegate Porphyria

Patients with variegate porphyria (VP) experience acute neurological attacks and/or sensitivity of the skin to sunlight and mechanical trauma. The enzymatic defect is a partial deficiency of PPOX. Both protoporphyrinogen IX and coproporphyrinogen III accumulate in the body, which results in photosensitivity and cutaneous lesions. The disease is most common among South African whites.

▶ THE ERYTHROPOIETIC PORPHYRIAS

CONGENITAL ERYTHROPOIETIC PORPHYRIA

Congenital erythropoietic porphyria (EPP), also known as Günter's disease, is an autosomal recessive disease. It is due to the reduced activity of URO-synthase and the resultant accumulation of uroporphyrin I and coproporphyrin I isomers. CEP is associated with hemolytic anemia and cutaneous lesions.

This disease begins in early infancy and is characterized by severe cutaneous photosensitivity. The skin may become brittle. Some individuals develop bullae, and vesicles are prone to rupture and infection. Skin sections may become hypo- and hyperpigmented, and excess hair on the face and extremities is characteristic. Secondary infection of the cutaneous lesions can lead to disfigurement of the face and hands. Porphyrin compounds are deposited in teeth and in bones, with the teeth appearing reddish-brown and fluorescing on exposure to long-wave ultraviolet light. Hemolysis is probably due to the marked increase in erythrocyte porphyrins and leads to splenomegaly.

ERYTHROPOIETIC PROTOPORPHYRIA

Erythropoietic protoporphyria (EPP) is an inherited disorder resulting from the partial deficiency of ferrochelatase activity, the last enzyme in the heme biosynthetic pathway (Figure 20-4). Children develop EPP more frequently than adults. A ferrochelatase activity in a patient with EPP is as low as 15–25% in lymphocytes and cultured fibroblasts. Protoporphyrin accumulates in bone marrow reticulocytes and then appears in plasma, is taken up in the liver, and is excreted in bile and feces. Protoporphyrins transported to the skin cause nonblistering photosensitivity.

This form of skin photosensitivity, which differs from that of other porphyrias, usually begins in childhood. It usually consists of pain, redness, and itching that occurs within minutes of sunlight exposure. The patient develops substantial elevations in EPP, which only occurs in patients with genotypes that result in ferrochelatase activities below ~35% of normal. Redness, swelling, burning, and itching can develop shortly after sun exposure. The primary source of excess protoporphyrin is the bone marrow reticulocytes. Erythrocyte protoporphyrin is free (not complexed with zinc) and is mostly bound to hemoglobin.

LABORATORY ASSESSMENT OF PORPHYRIAS

Clinical laboratory tests are important for proper assessment of patients with suspected porphyrias. In many cases the clinical features presented to the clinician are nonspecific, thus making diagnosis more difficult. Diagnosis of porphyrias depends on demonstration of specific patterns of overproduction of heme precursors, as shown in Table 20-5 ✪. It is incumbent on the clinician to request the correct intermediate compounds to be measured and to submit the appropriate specimen for laboratory analysis.

The diagnostic strategy for assessment of porphyrias will vary and depends on whether the patient exhibits an acute condition or is in remission. Another significant factor is

✪ TABLE 20-5

Clinical laboratory tests results for hepatic and erythropoietic porphyrias.

Porphyria	Urine PBG/ALA	Urine Porphyrins	Fecal Porphyrins	Erythrocyte Porphyrins
Hepatic porphyrias				
ADP	ALA	Copro III	Not increased	Zn-proto
AIP	PBG > ALA	Increased uro	Normal	Normal
PCT	Not increased	Uro, Hepta	Isocopro, Hepta	Normal
HCP	PBG > ALA	Copro III; Uro	Copro III	Normal
VP	PBG > ALA	Copro III, Uro	Proto IX > copro III, X-porphyrin	Normal
Erythropoietic porphyrias				
CEP	Not increased	Uro I, copro I	Copro I	Zn-proto, proto, copro I, Uro I
EPP	Not increased	Not increased	Proto	Proto

Abbreviations: Uro, uroporphyrin; Zn-proto, zinc-protoporphyrin; Hepta, heptacarboxylate; Copro III, Coproporphyrin III.

whether the symptoms present neurovisceral or cutaneous photosensitivity. A brief overview of diagnostic approaches to acute and cutaneous porphyrias is presented here, and readers are encouraged to broaden their knowledge by reading the relevant sections in the references listed at the end of this chapter.

Three strategies are used when testing patients with acute neurovisceral symptoms: (1) investigating the acute attack, (2) diagnosing the cause, and (3) investigating possible acute porphyrias when patients are in remission.

Patients who present with acute attacks require measurement of urinary PBG. The PBG level should be elevated. If a screening method is used, it should be confirmed by a specific quantitative assay (see the discussion of analytical methods below). Urinary ALA may also be measured, and the relative concentrations of each are compared, as shown in Table 20-5.

 Checkpoint! 20–7

State which porphyrins are elevated in urine for the following hepatic porphyrias: ADP, AIP, and VP.

The cause of porphyrias in a patient usually involves a distinction between the acute porphyrias, and it is appropriate to measure porphyrins in feces. If the total fecal porphyrin is normal, then VP and HCP are excluded and the patient must have AIP. If the total fecal porphyrin is increased, porphyrins should be fractionated by high-performance liquid chromatography (HPLC). Patients with HCP will have elevated urine coproporphyrin III and a normal to slight elevation of protoporphyrin IX. In VP, protoporphyrin IX is elevated, with smaller increases in coproporphyrin.

 Checkpoint! 20–8

State which porphyrins are elevated in urine for the following diseases: CEP and EPP.

For patients in remission, a retrospective study after a patient has fully recovered may provide useful information to the clinician. Usually the first step is to have the physician request the laboratory to quantitate urinary PBG. Also, a fecal porphyrin measurement will serve to exclude HCP. A fluorescent assay will determine if the patient has VP. If all of the laboratory tests are negative, then the symptoms presented by the patient are most likely *not* due to a porphyrias.

A patient who presents with cutaneous symptoms requires a different strategy from that of acute porphyrias. For suspected EPP, the test request should be the measurement of whole blood (erythrocytes) porphyrin using a sensitive fluorometric method. If the test is normal, then EPP can be excluded as the cause of the symptoms. If the result is increased, then the clinician needs to determine if the increase is caused by free protoporphyrin, as in EPP, or ZPP, as in iron deficiency and lead toxicity. If the patient has skin lesions, then there are four possibilities, as shown in Table 20-5:

HCP, VP, PCT, and EPP. A urine and fecal specimen is required for measurement of porphyrins. If tests are normal, then the clinician may rule out a porphyria.

 Checkpoint! 20–9

List the three strategies for assessment of porphyrias.

ANALYTICAL METHODS

Laboratory testing for porphyrias include measuring PBG, ALA, and urinary, fecal, plasma, and blood porphyrins. All specimens must be protected from exposure to light (due to the photosensivity of porphyrins). Specimens can be collected in brown containers and low actinic glassware where appropriate. Urine samples for testing must be fresh and can be random or 24-hour collections. No preservatives are necessary for 24-hour collections. Dilute urine specimens are not usually acceptable. Fecal specimens should be from 5 to 10 grams in weight. Blood specimens can be whole blood or plasma collected in heparin, ethylenediaminetetracetic acid (EDTA), or potassium EDTA blood tubes, depending on the specific analyte to be measured.

Porphobilinogen Assays

Many clinical laboratories have discontinued the use of qualitative PBG screening tests such as the Watson–Schwartz test and have elected to send specimens to a larger reference laboratory for a quantitative PBG assay. Several spectrophotometric assays are available that provide reliable results. Most colorimetric methods for PBG are based on the reaction of Ehrlich's reagent (dimethylaminobenzaldehyde in acidic solution) with the α-methene carbon of the pyrrole ring to produce a magenta or purple product. Porphyrins do not contain any α-methene bridges and thus will not react. Potential interfering substances (e.g., urobilinogen) can be removed before analysis using ion exchange chromatography.[5]

Laboratory techniques, including HPLC[6] and liquid chromatography and tandem mass spectrometry (LC-MSMS),[7] are available that can improve the sensitivity of the measurement. The LC-MSMS technique allows for detection of mild elevations of PBG that can occur between acute events in AIP. Mild PBG elevation of ~5 mg in 24 hours can be observed in affected yet asymptomatic postpubertal family members of patients with diagnosed AIP. Measuring PBG has become useful in family studies to identify asymptomatic individuals who can be counseled to avoid substances and situations known to result in acute episodes.

Aminolevulinic Acid

The specimen of choice for laboratory measurement of ALA is urine. Specimens can be random or 24-hour collections. A 24-hour collection should be stabilized by addition of an acid (e.g., hydrochloric acid). Analytical techniques include GC and spectrophotometry. A separation step is usually recommended to separate PBG from ALA using ion exchange chromatography.

The ALA is then converted into an Ehrlich's reaction pyrrole by condensation with a reagent such as acetylacetone.[8] Chemical interferences can occur with compounds such as ammonia, glucosamine, and penicillin.

Porphyrins

Analysis of porphyrins is appropriate in several specimen types, including urine, feces, whole blood, and plasma. Clinical laboratories equipped with spectrophotometers or fluorometers can provide the clinician with qualitative and possibly quantitative assays that are more sensitive and specific than screening assays. The spectrophotometric methods involve a wavelength spectrum analysis of acidified sample. This technique involves monitoring the absorbance of a sample while varying the wavelength of the spectrophotometer. A positive result is interpreted as the presence of a Soret band (peak) in the 400-nm region of the EMS.[9]

Another approach to porphyrin analysis is fractionation of the porphyrin compounds based on their different solubilities. The variations of solubility are often the result of the different order of β-substituent groups on the tetrapyrrole rings. A widely used technique for analyzing the fractions produced after chemical manipulation of the samples is HPLC. Fluorescent detection produces superior sensitivity and specificity over absorbance spectroscopy.

 Checkpoint! 20–10

Identify four types of specimens that can be used for qualitative and quantitative analysis of porphyrins.

► HEMOGLOBIN

The free heme formed in the mitochondria (Figure 20-1) relocates into the cytoplasm, where it can function in the following manner: (1) inhibit the uptake of iron from transferrin (see above), (2) inhibit ferrochelatase (see above), or (3) chemically attach itself to globin to form hemoglobin.

The globin proteins are derived from gene clusters consisting of (but not inclusive), α-, β-, δ-, ζ-globin genes. The end product of each unique genetic process is various globin "chains" subunits. These subunits are available to be incorporated into protein structures and to confer unique biochemical and physiological properties to the protein.

Hemoglobin is globular protein with a diameter of 6.4 nm and a molecular weight of ~64.5 kDa. It consists of four subunits: two alpha and two beta chains. Each globin chain is looped about itself to form a pocket or cleft in which the heme group resides, as depicted in Figure 20-5 ■. The heme moiety is located within this pocket by an attachment of its iron atom to the imidazole group of a nearby histidine residue. A distal histidine that is in close proximity to the iron appears to swing in and out of this position to permit the passage of oxygen into and out of the hemoglobin molecules.[10]

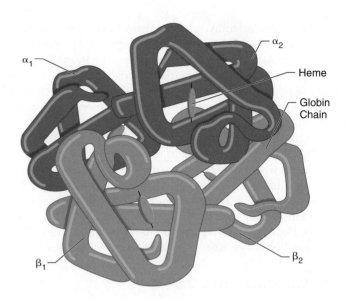

■ FIGURE 20-5   Model of hemoglobin.

 Checkpoint! 20–11

Identify the three pathways that are available to "free" heme that relocates to the cytoplasm.

GLOBIN STRUCTURE OF NORMAL HEMOGLOBIN

In normal human adults, hemoglobin is composed of two normal alpha and two normal beta polypeptide chains and is represented as $\alpha_2\beta_2$. Most of the adult hemoglobin in whole blood is hemoglobin A, with a smaller fraction being hemoglobin A_2, and the rest is comprised of other hemoglobin variants.

The primary structure of globin consists of amino acid chains with varying lengths. The amino acid terminal of the β-chain is the site of attachment of glucose ($HgbA_{1c}$), urea, and salicylate. The carboxy terminal amino acid of the β-chain is tyrosine and can function as a salt bridge. There are no disulphide bonds. A majority of the secondary structure of hemoglobin is α and non-α polypeptide chains arranged in helices, some of which from nonhelical turns. The β-chains are arranged into eight helices. Consistent with many other globular proteins, the tertiary structure of hemoglobin refers to the arrangement of the helices in a three-dimensional, twisted structure. The heme group, located in the cavity between helices, is attached to histidine residue in each β-chain. This attachment is essential to maintain the secondary and tertiary structures of the globin chains. The quaternary structure of hemoglobin results from the attachment for the four globin chains to each other. The molecule is held together via strong $\alpha_1\beta_1$ and $\alpha_2\beta_2$ dimeric bonds. Stability of the structure is maintained through the tetrameric $\alpha_1\beta_1$ and $\alpha_2\beta_2$ bonds. Several physiological effects, including the oxygen dissociation curve and the Bohr effect (see Chapter 13, "Blood Gases, pH, and Acid–Base Balance")are a result of shifting, rotation, and sliding in the quaternary structure.

Checkpoint! 20–12

Explain the structural configuration of the hemoglobin molecule and include the primary, secondary, tertiary, and quaternary components.

MODIFIED HEMOGLOBINS

Carboxyhemoglobin

Carboxyhemoglobin (also called carbon monoxide hemoglobin) is formed when carbon monoxide attaches to the molecule instead of oxygen (see Chapter 22, "Toxicology"). The chemical bond is extremely strong, and the half-life is 4 to 5 hours at sea level.

Methemoglobin

Normally the iron in heme is in the reduced ferrous state (Fe^{2+}). In an alkaline medium, the iron is oxidized to the ferric state (Fe^{3+}) by toxic agents. Several examples of toxic agents include nitrates, aniline dyes, chlorates, quinones, phenacetin, procaine, benzocaine, and lidocaine. The oxidation reaction converts the heme to hematin and the hemoglobin to methemoglobin. Patients with elevated methemoglobin show signs of cyanosis due to the inability of methemoglobin to reversibly bind oxygen. In normally functioning cells, methemoglobin is reduced to hemoglobin by the reduced form of the NADH (nicotinamide adenine dinucleotide plus hydrogen)–cytochrome reductase system.

Sulfhemoglobin

Sulfhemoglobin is produced by the reaction of sulfur-containing compound with heme to form an irreversible chemical modification of hemoglobin by the introduction of sulfur in one or more of the porphyrin rings. A common cause of sulfhemogboin production is exposure to drugs such as phenacetin and sulfonamide. Cyanosis develops due to the inability of sulfhemogboin to transport oxygen.

A discussion of hemoglobinopathies and analytical methodologies will not be included in this chapter. There are several references listed in the suggested readings at the end of this chapter that will provide the reader with in-depth material on these topics.

SUMMARY

This chapter focused on iron and the compounds that incorporate it within their structures. These compounds include porphyrins and hemoglobin. Several clinical diseases are associated with both increased and decreased levels of total body iron. Laboratory tests play a significant role in diagnosis and management of patients with iron-related diseases. The laboratory technologist should be familiar with the clinical correlation between laboratory tests and diseases.

Porphyrias are rare in the United States. Most laboratories are not equipped to perform tests that will provide the clinician with information about diagnosis and management of a patient. A working knowledge of porphyrins and porphyrias is worthwhile so that if a patient does present to a health-care facility, the laboratory staff will know how to process the specimens for physician-requested tests.

A brief discussion of hemoglobin was presented with an emphasis on its relationship with iron. There are many reference texts, including those in clinical chemistry and hematology, that provide voluminous amounts of information on hemoglobinopathies and methodologies if the reader wants to explore the subject in greater depth than presented in this chapter.

REVIEW QUESTIONS

LEVEL I

1. Which of the following is the storage form of iron? (Objective 2)
   a. transferrin
   b. ferritin
   c. albumin
   d. hemin

2. The following statement best describes which of the terms listed below? (Objective 4) *The pathological condition that develops when iron intake falls below the amount required for red blood cell production and iron reserves become depleted.*
   a. anemia
   b. hemochromatosis
   c. porphyria
   d. acute intermittent porphyria

LEVEL II

1. Patients with hemochromatosis usually present to their clinician with which of the following triad of symptoms? (Objective 10)
   a. yellowing of the sclera of the eyes, shortness of breath, and chest pain
   b. bronzing of the skin, cirrhosis, and diabetes
   c. polyuria, polydypsia, and polyphagia
   d. abdominal pain, chest pain, and grayish-blue lips

2. Which of the following laboratory test groups best reflects a patient with primary hemochromatosis? (Objective 1)
   a. serum iron, increased; serum TIBC, increased; transferrin saturation, decreased
   b. serum iron, increased; serum TIBC, increased; transferrin saturation, increased
   c. serum iron, decreased; serum TIBC, decreased; transferrin saturation, decreased
   d. serum iron, increased; serum TIBC, decreased; transferrin saturation, increased

REVIEW QUESTIONS (continued)

LEVEL I

3. Hemochromatosis is a disease that is best described as which of the following? (Objective 4)
   a. a disease characterized by an iron overload without associated tissue injury
   b. a disease characterized by an accumulation of porphyrins in the blood
   c. a genetically related disease in which the body accumulates excess amounts of iron
   d. a genetically related disease in which the body develops an iron deficiency

4. Which of the following is an example of a chromogen used to measure iron in serum? (Objective 5)
   a. bathophenanthroline
   b. picric acid
   c. Erhlich's reagent
   d. diphenylcarbazone

5. A porphyrin is a compound that contains how many pyrrole rings? (Objective 10)
   a. one
   b. two
   c. four
   d. five

6. Protoporphyrins that chelate iron are referred to as which of the following? (Objective 11)
   a. hemoglobin
   b. hemin
   c. hemosiderin
   d. heme

7. The biosynthesis of heme begins with which two compounds? (Objective 9)
   a. alanine and glycine
   b. glycine and succinyl CoA
   c. glycerin and acetyl-CoA
   d. lysine and globulin

8. Which enzyme is deficient in a patient with acute intermittent porphyria? (Objective 8)
   a. hydroymethylbilane synthase
   b. aminolevulinic acid synthase
   c. ferrochelatase
   d. coproporphyrinogen oxidase

9. The globin portion of hemoglobin in normal adult humans is composed of which of the following? (Objective 9)
   a. two deltas, two betas ($\delta_2\beta_2$)
   b. three betas, three deltas (β_3, $\delta3$)
   c. two alphas, two betas ($\alpha_2\beta_2$)
   d. two sigmas, two gammas ($\sigma_2\gamma_2$)

10. Which of the following terms is synonymous with *transferrin*? (Objective 11)
    a. Lipotransferrin-Fe^{2+}
    b. Prototransferrin-Fe^{3+}
    c. Apotransferrin-Fe^{2+}
    d. Apotransferrin-Fe^{3+}

LEVEL II

3. A patient with ALA-dehydratase-deficient porphyria (ADP) may present with which of the following clinical features? (Objective 10)
   a. severe chest pain, shortness of breath, and rapid heart rate (tachycardia)
   b. photosensitivity and cutaneous lesion
   c. abdominal pain and neurological dysfunction
   d. bulging eyes, goiter in the neck, and heat intolerance

4. The addition of sufficient Fe^{3+} to saturate the iron bind site of transferrin is referred to as which of the following? (Objective 3)
   a. total iron binding saturation
   b. unsaturated iron binding capacity
   c. serum transferrin saturation
   d. total iron index

5. What is the required oxidation state of iron that enables it to complex with a chromogen in most spectrophotometric assays? (Objective 11)
   a. +1
   b. +2
   c. +3
   d. −3

6. What is the unique quality that many porphyrin compounds possess and that allows for their detection within body tissues and in a test tube? (Objective 11)
   a. They emit gamma radiation and can be detected using a radiation detector.
   b. They absorb heat and can be detected using a thermal sensor.
   c. They fluoresce and can be detected by directing EMR at a specific wavelength on tissue or in test tubes.
   d. They emit an odor that is characteristic of the particular type of porphyrin.

7. Which of the following porphyrias best reflects the following description of a patient's symptoms and laboratory results? (Objective 2)

   A 35-year-old male presented to his clinician with alcoholic liver disease, hyperpigmentation, blister formation, and fragile skin. The results of the liver function tests were abnormal. The patient's urine PBG/ALA was not increased. Urine porphyrins analysis revealed an elevation of uroporphyrinogen and heptacarboxylate.
   a. ALA-dehydratase-deficient porphyria
   b. acute intermittent porphyria
   c. porphyria cutanea tarda
   d. erythropoietic protoporphyria

REFERENCES

1. Kakhlon O, Cabantchik ZI. The labile iron pool: Characterization, measurement, and participation in cellular processes (1). *Free Radical Biology & Medicine* (2002) 33(8):1037–46.

2. Lebron JA, Bennett MJ, Vaughn DE et al. Crystal structure of the hemochromatosis protein HFE and characterization of its interaction with transferrin receptor. *Cell* (1998) 93(1):111–23.

3. Desnick RJ, Astrin KH. The porphyrias. *In*: Fauci AS, Kasper DL, Braunwald et al (eds.). *Harrison's principles of internal medicine*, 17th ed, New York: McGraw-Hill, 2008.

4. Maruno M, Furuyama K, Akagi R et al. Highly heterogeneous nature of delta-aminolevulinate dehydratase (ALAD) deficiency in ALAD porphyria. *Blood* (2001) 97:2972–78.

5. Buttery JE, Stuart S. Measurement of porphobilinogen in urine by a simple resin method with use of a surrogate standard. *Clinical Chemistry* (1991) 37(12):2133–36

6. Jamani A, Pudek M, Schreiber WE. Liquid-chromatographic assay of urinary porphobilinogen. *Clinical Chemistry* (1989) 35(3):471–75.

7. Ford RE, Magera MJ, Kloke KM et al. Quantitative measurement of porphobilinogen in urine by stable-isotope dilution liquid chromatography-tandem mass spectrometry. *Clinical Chemistry* (2001) 47(9):1627–32.

8. Buttery JE, Stuart S, Pannall PR. An improved direct method for the measurement of urinary delta-aminolevulinic acid. *Clinical Biochemistry* (1995) 28(4):477–80.

9. Deacon AC, Elder GH. ACP Best Practice No 165: Front line tests for the investigation of suspected porphyria. *Journal of Clinical Pathology* (2001) 54(7):500–507.

10. Winter WP, Yodh J. Interaction of human hemoglobin and its variants with agar. *Science* (1983) 221(4606):175–78.

Suggested Readings

1. Higgins T, Beutler, E Doumas BT. Hemoglobin, iron and bilirubin. *In*: Burtis CA, Ashwood ER, Bruns DE (eds.). *Teitz textbook of clinical chemistry and molecular diagnostics*, 4th ed. Philadelphia: W. B. Saunders, 2006:1165–1208.

2. Vajpayee N, Graham SS, Bem S. Basic examination of blood and bone marrow. *In*: McPherson RA, Pincus MR (eds.). *Henry's clinical diagnosis and management by laboratory methods*, 21st ed. Philadelphia: Saunders Elsevier, 2007:457–65.

3. McKenzie SB. *Clinical laboratory hematology*. Upper Saddle River, NJ: Prentice Hall, 2004.

21

Therapeutic Drug Monitoring (TDM)

■ OBJECTIVES—LEVEL I

Following successful completion of this chapter, the learner will be able to:

1. Identify the four principle biological events associated with pharmacokinetics.
2. Identify the factors that influence drug absorption.
3. List examples of conjugation compounds.
4. Discuss two mechanisms associated with drug excretion.
5. List four parameters that affect changes in dosage of drugs.
6. List an example of a specific drug from a given therapeutic category.
7. Identify examples of additional laboratory tests that may be requested to evaluate organ function(s) in patients taking prescribed medications.
8. Contrast chemical, generic, and trade name nomenclature for drugs.
9. Identify appropriate specimens for selected therapeutic drugs.
10. Identify two factors that significantly affect steady state.
11. Name the sources of selected drugs.

■ OBJECTIVES—LEVEL II

Following successful completion of this chapter, the learner will be able to:

1. Describe the first-pass effect that occurs in the liver and indicate how it influences bioavailability.
2. Compare and contrast enteral and parenteral routes for drug administration.
3. Distinguish free versus bound drug.
4. Correlate high drug concentration with the amount of bound versus free drug.
5. Compare phase I– and phase II–type reactions.
6. Compare and contrast first-order and zero-order enzyme kinetics as they relate to the metabolism of drugs.
7. Describe the cytochrome P450 system relative to drug metabolism.
8. Discuss the significance of sampling time in monitoring therapeutic drug levels.
9. Describe how a given drug initiates its in vivo effect.
10. Compare an agonist-type drug interaction with an antagonist-type drug interaction.
11. Describe how compounds can interfere or cross-react with selected therapeutic drugs.
12. Identify health conditions associated with toxic levels of selected therapeutic drugs.
13. Discuss the metabolites of selected therapeutic drugs.
14. Explain treatment modalities for selected therapeutic drugs.

KEY TERMS

Agonist	First-pass effect	Phase II reactions
Antagonist	Free drug (unbound drug)	Receptor
Bioavailability	Half-life (t½)	Steady state
Biotransformation	Maintenance dose	Subtherapeutic
Bound drug	Peak drug concentration	Therapeutic range
Cytochrome P450	Pharmacodynamics	Trough drug level
Efficacy	Pharmacokinetics	
Elimination half-life	Phase I reactions	

 A CASE IN POINT

A 39-year-old male was seen in the emergency department following a possible grand mal seizure. The patient informed the physician that he had been taking therapeutic doses of phenobarbital and phenytoin for several months. The doctor ordered drug levels for each compound. A phlebotomist drew a blood sample in a red-top tube that contained serum separator gel. Results of the laboratory tests are shown below:

Test	Result	Therapeutic range
Phenobarbital	4.0 μg/mL	15–30 μg/mL
Phenytoin	3.5 μg/mL	10–20 μg/mL

The physician questioned the laboratories results and asked for a repeat analysis.

Issues and Questions to Consider

1. Are these results consistent with the patient's history?
2. Are the laboratory results incorrect? If so, what is the source of the error?
3. Does the laboratory need to repeat the analysis?
4. What course of action should the laboratory follow to prevent future questioning of these assays?

WHAT'S AHEAD

▶ A discussion of pharmacokinetics and pharmacodynamics.
▶ Preanalytical considerations for therapeutic drug monitoring.
▶ A detailed discussion of several therapeutic drugs routinely measured in clinical laboratories.
▶ Mechanism of action.
▶ Absorption, metabolism, and excretion.
▶ Therapeutic and toxicity issues for each drug.
▶ Discussion of the therapeutic and toxic levels of the drugs presented.

▶ INTRODUCTION

Therapeutic drug monitoring (TDM) comprises a significant number of tests performed in the clinical chemistry laboratory. The introduction of nonradioisotopic homogeneous immunoassays such as enzyme multiplied immunoassay technique (EMIT) and fluorescent polarization immunoassay allowed the laboratorian to measure therapeutic drugs in a timely manner without sacrificing sensitivity. Increased requests by clinicians led to laboratories measuring more than a dozen therapeutic drugs on a routine and stat basis. Laboratories had significant influence on the implementation of TDM programs, with laboratory staff working closely with pharmacies, nursing staffs, and clinicians to address preanalytical and postanalytical issues relating to sample procurement, pharmacology, and interpretation of test results. The menu of drugs continues to increase and now includes immunosuppressant drugs such cyclosporine A and tacrolimus (FK506).

The goal of therapeutic drug monitoring is to provide information to the clinician that will establish optimal therapeutic drug regimens for individual patients. The factors influencing dosage, blood concentrations, and toxicity for many of the therapeutic drugs commonly measured in the clinical laboratory will be discussed.

▶ PHARMACOKINETICS

Pharmacokinetics involves the dynamics associated with the movement of drugs across cell membranes. Four principle biological events govern the pharmacokinetics of drugs: absorption, distribution, metabolism, and excretion. Figure 21-1 ■ shows the interrelationship of these events.

ABSORPTION

Drug absorption involves the rate at which a drug leaves its site of administration and the extent to which this occurs. Factors that affect rate of absorption are the area of the absorbing capillary membranes and the solubility of the substance in the interstitial fluid.

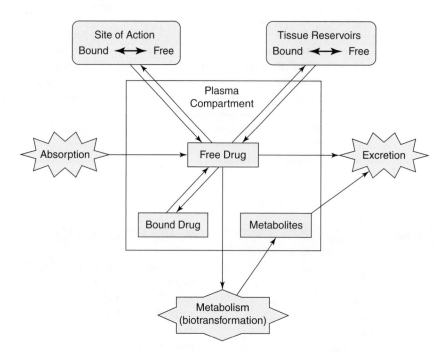

■ **FIGURE 21-1**   Interrelationship of absorption, distribution, metabolism (biotransformation), and excretion of a drug.

Drugs are absorbed from the gastrointestinal (GI) tract by either passive diffusion or active transport. The major force for passive absorption of a drug is the concentration gradient across a membrane separating two body compartments. Thus, a drug moves from a region of high concentration to one of low concentration. Most drugs move between compartments by passive diffusion. Lipid-soluble drugs readily move across most biological membranes, and water-soluble drugs gain access to cells through aqueous channels. Active transport requires specific carrier proteins that span the membrane. Active transport is energy dependent and driven by the hydrolysis of adenosine triphosphate (ATP). The active-transport process is capable of moving drugs against a concentration gradient. Therefore, drugs are moved from an area of low concentration to one of high concentration via active transport.

The pH of a drug affects its absorption. A drug that is uncharged tends to pass through membranes more readily than a charged drug. Therefore, the effective concentration of the permeable form of each drug at its absorption site is determined by the relative concentrations of the charge and uncharged forms. The ratio of the two forms is determined by the pH at the site of absorption and by the strength of the weak acid or base, defined as its pKa.

Several physical factors influence the absorption of drugs. Blood flow is an important physical factor at the absorption site. The faster the blood flows, the more quickly the drug arrives at its site for possible absorption. Any condition such as shock that tends to slow blood flow will increase the time it takes a drug to reach its site. The total surface area available for absorption is another factor and affects the amount of drug

absorbed. The intestine has a surface rich in microvilli and a total surface area a thousand times greater than the stomach; thus, absorption of the drug across the intestine is more efficient. Finally, the amount of contact time at the absorptive surface is a factor that serves a significant role in the overall absorption of a drug. If a drug moves too quickly through the GI tract, as is the case in a patient with severe diarrhea, the drug is not absorbed well. Conversely, any function that slows the transports of a drug from the stomach to the intestine delays the rate of the drug's absorption. Food taken with a drug both dilutes the drug and slows gastric emptying.

Bioavailability indicates the fractional extent to which a dose of drug reaches its site of action. Bioavailability of a drug is more useful to the clinician than absorption because it takes into account anatomical, physiological, and pathological factors. For example, if a drug is given orally, it must be absorbed first from the stomach and intestine. This process may be limited by the characteristics of the dosage form, the drug's physiochemical properties, or both. Next, the drug passes through the liver, where metabolism or biliary excretion may occur before it reaches the systemic circulation. This is referred to as the **first-pass effect** through the liver. A fraction of the administered and absorbed dose of drug will be inactivated or diverted before it can reach the systemic circulation and be distributed to its site of action. Bioavailability will be markedly reduced if the metabolic or excretory capacity of the liver for the drug is compromised.

The choice of the route of drug administration must be made with a good understanding of these conditions. There are two major routes of drug administration: enteral and parenteral. Several specific methods of drug administration are listed in Box 21-1 ■. Oral (enteral) ingestion is the most common means of drug administration. It is safe, convenient, and economical. Some drugs are absorbed from the stomach, although the duodenum is often the major site of entry to the systemic circulation because of its larger absorptive surface. Most drugs absorbed from the GI tract enter the portal circulation and encounter the liver before they are distributed through general circulation (see Figure 21-2 ■). First-pass metabolism by the intestine or liver limits the efficacy of many drugs when taken orally.

Some drugs may be destroyed in the stomach by its acid. This usually occurs if there is food in the stomach. As noted, food in the stomach delays gastric emptying time so that the drug is in the stomach for a longer period of time. To avoid this problem, certain drug preparations have an enteric coating that protects the drug from the acid environment and may prevent gastric irritation.

There are disadvantages to oral administration of a drug, including limited absorption of some drugs because of their physical characteristic such as solubility, emesis or vomiting, destruction by digestive enzymes, and low gastric pH that may inactivate the drug.

Controlled-release preparations of orally administered drugs are designed to produce slow, uniform absorption of the drug for 8 hours or longer. The advantages of this type of preparation are a reduction in the frequency of the drug's administration, maintenance of therapeutic effect overnight, and fewer undesired effects by eliminating the peaks in drug concentration that usually occur after administration of immediate-release dosage forms.

Sublingual (under the tongue) administration of a drug—nitroglycerin, for example—may be advantageous. Nitroglycerin is nonionic and has high lipid solubility. The drug is absorbed rapidly because of the close proximity of blood vessels to membrane surfaces. Even though the surface area under the tongue is much smaller, only a few molecules of nitroglycerin are needed to elicit the desired effects.

Rectal administration of a drug provides an alternate route if the individual is unconscious or vomiting. Approximately 50% of the drug that is absorbed from the rectum will bypass the liver, thereby reducing the hepatic first-pass metabolism. Some disadvantages of rectal administration of a drug include irritation of the rectal mucosa and irregular absorption.

The major routes of parenteral administration are intravenous (IV), subcutaneous and intramuscular (IM). Drugs may be injected directly into arteries (intra-arterial) or spinal subarachnoid spaces (intrathecal). Parenteral administration is used for drugs that are poorly absorbed from the GI tract and for compounds such as insulin that are unstable in the GI tract. Patients who are unconscious or require treatment with a rapid onset of action may need an injectable form of the drug.

Absorption from subcutaneous and IM sites into the plasma occurs by simple diffusion along the gradient or barrier from where the drug was injected. The rates of absorption are limited by the area of the absorbing capillary membranes and by the solubility of the substance in the interstitial fluid.

Drugs administered into the systemic circulation may be subject to first-pass elimination in the liver before distribution to the rest of the body. First-pass elimination may take a drug out of circulation by one of the following mechanisms:

1. partitioning into the lipids located in liver tissue,

2. filtration, and

3. elimination as a volatile substance.

IV-administered drugs avoid the GI tract and thus the first-pass metabolism by the liver. This route allows for a

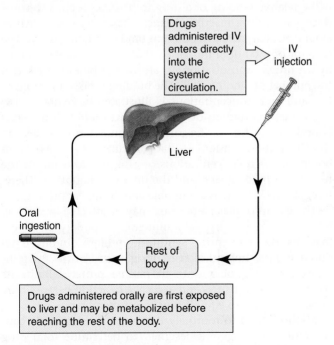

Drugs administered IV enters directly into the systemic circulation.

IV injection

Liver

Oral ingestion

Rest of body

Drugs administered orally are first exposed to liver and may be metabolized before reaching the rest of the body.

■ FIGURE 21-2 First-pass metabolism of orally administered drugs versus intravenous injectable drugs.

rapid effect and a maximal degree of control over the circulating levels of the drug.

IM drug preparations can be aqueous solutions or specialized depot preparations in ethylene glycol or peanut oil. An example of a depot preparation is an oil-based ester of estradiol and estrone suitable for IM injection. Absorption of drugs in aqueous solution is fast, whereas depot preparations are absorbed more slowly.

Subcutaneous administration of a drug is similar to that of IM injections except that subcutaneous injections are usually placed into the fatty tissue and not necessarily into the muscle. The absorption of drugs subcutaneously is slightly slower than the IV route. Examples of drugs administered subcutaneously include solids such as silastic capsules containing the contraceptive levonorgestrel, programmable mechanical pumps that can be implanted to deliver insulin in some diabetic patients, heparin, vaccines, and epinephrine.

Some drugs are administered by inhalation—for example, treatments for asthma and anesthetics gases. Inhalation provides the rapid delivery of a drug across the large surface area of the mucous membranes of the respiratory tract and pulmonary epithelium. The effect is as rapid as IV injection.

There are also topical applications of compounds to mucous membranes on various parts of the body; these are used primarily for their local effects. Topical applications can be applied to the skin for absorption through the dermis. Topical drugs include tetracain and lidocaine, which provide anesthesia to the localized area. There are several examples of controlled-release topical patches that have become popular in an effort to assist patients to stop smoking. Ophthalmic drugs (atropine) are topically applied to the eye to treat a variety of eye-related problems.

 Checkpoint! 21–1

What factors affect the rate of absorption of a drug?

DISTRIBUTION

The distribution of an administered drug into interstitial and intracellular fluids depends on several physiological factors and the specific physiochemical properties of the individual drug. The rate of delivery and the potential amount of drug distributed into tissues depends on blood flow, capillary permeability, binding of drugs to proteins, and tissue volume.

Initially, most of the drug administered goes to the liver, kidneys, brain, and other well-perfused organs, followed by delivery to muscles, skin, fat, and viscera. Diffusion of a drug into the interstitial fluid occurs rapidly because of the highly permeable construction of the capillary endothelial membrane.

The tissue distribution of a drug depends on the following:

- partitioning of drug between blood and the particular tissue,
- lipid solubility, and
- pH gradients between intracellular and extracellular fluid.

A Dose is less than available binding sites

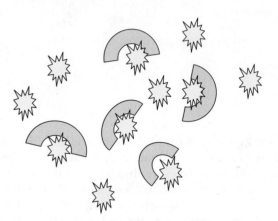

B Dose is greater than available binding sites

■ FIGURE 21-3 The binding of drugs to proteins is saturable and a nonlinear process. A. The dose of drug is less than the number of binding sites; B. the dose of drug is greater than the number of binding sites.

An important determinant of blood and tissue partitioning is the relative binding of a drug to plasma proteins (**bound drug**) and tissue macromolecules. Plasma proteins such as albumin and α_1-acid glycoprotein bind circulating drugs. The amount of total drug in plasma that is bound is determined by the drug's concentration, its affinity for binding sites, and the number of binding sites. The binding process for drugs of low and high concentrations is illustrated in Figure 21-3 ■. For low concentrations of drug, the fraction of a drug that is bound is a function of the concentration of binding sites and the dissociation constant of that particular drug. The fraction bound for high concentrations of drug is a function of the number of binding sites and the drug concentration. Therefore, plasma binding is a saturable and nonlinear process.

The extent of plasma binding may be affected by disease-related factors. In hypoalbuminemia secondary to severe liver disease or nephritic syndrome, protein binding is reduced and the amount of **free drug** or **unbound drug** increases. Binding of a drug to plasma proteins limits its concentration in tissues and at its site of action because only unbound drug is in equilibrium across membranes.

Methods used to routinely measure the therapeutic concentration of drugs are designed to determine total drug concentration. Total drug concentration represents the amount of free drug and bound drug, although the form

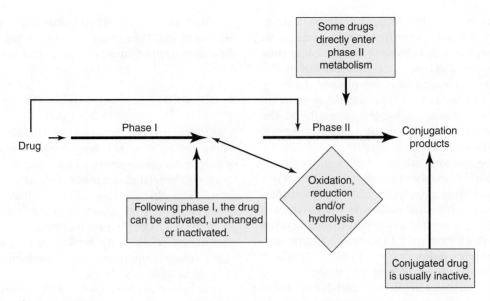

■ FIGURE 21-4   Biotransformation of drugs illustrating possible outcomes during phase I and phase II reactions.

that elicits the desired pharmacological effect is the free form not the bound fraction.

 Checkpoint! 21–2

What is an important determinant of blood and tissue partitioning of a drug?

METABOLISM (BIOTRANSFORMATION)

The metabolism of a drug into more hydrophilic metabolites is essential for the elimination of these compounds from the body and the cessation of their biological activity. **Biotransformation** reactions generate more polar, inactive metabolites that are readily excreted from the body. On occasion, metabolites with potent biological activity or toxic properties are created. For example, procainamide, a drug with antiarrhythmic effects, is metabolized to an active metabolite, *N*-acetylprocainamide (NAPA) with similar potency to procainamide

Drug biotransformation reactions are classified as either phase I (functionalization) or phase II (conjugation)–type reactions. **Phase I reactions** introduce or expose a functional group to the parent compound. Examples of functional groups include hydroxyls, methyl, and amines. The products of phase I reactions may be rapidly excreted into the urine, or they can react with endogenous compounds to form highly water-soluble conjugates. A **phase II reaction** leads to the formation of a covalent linkage between a functional group on the parent compound with an endogenously derived glucuronic acid, sulfate, glutathione, amino acid, or acetate group. The polar conjugates formed are generally inactive and are excreted rapidly into the urine and feces.

The metabolism of a drug usually involves reactions using enzymes. All reactions catalyzed by enzymes follow

Michaelis–Menten kinetics (see Chapter 9, Enzymes). Reactions are either first-order or zero-order kinetics. The interactive pathways available for drug molecules are shown in Figure 21-4 ■. In most clinical applications, the concentration of the drug is much less than the Michaelis–Menten constant, and the rate of drug metabolism is directly proportion to the concentration of free drug. This is an example of a first-order kinetic-type enzyme reaction. With some drugs, the doses may be very large (e.g., aspirin and ethanol) so that their concentration is larger than the Michaelis–Menten constant. The enzyme becomes saturated by a high free drug concentration, and the rate of metabolism remains constant over time. This represents zero-order kinetics. The Michaelis–Menten plot shown in Figure 21-5 ■ illustrates the area of the graph at low and high drug concentrations.

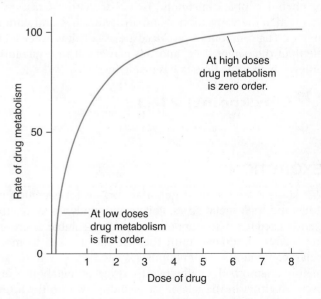

■ FIGURE 21-5   The effects of drug dosages on the rate of metabolism follow first- or zero-order kinetics.

The enzymes involved in the biotransformation of drugs are found mainly in the liver. Most tissues have some metabolic activity, and organs such as those of the gastrointestinal tract, kidney, and lungs also have the capacity to metabolize drugs. Most drug metabolic activities takes place in the endoplasmic reticulum and the cytosol. The enzyme systems involved in phase I reactions are located primarily in the endoplasmic reticulum. Phase II conjugation enzymes are mainly located in the cytosol.

A major enzyme system that metabolizes drugs is the **cytochrome P450** (CYP) monooxygenase system. This heme-thiolate protein functions as a terminal oxidase in a multicomponent electron-transfer chain that introduces a single atom of molecular oxygen into the substrate, with the other atom being incorporated into water. Cytochrome P450 catalyzes many reactions in the human body—deamination, N-hydroxylation, and aromatic and side-chain hydroxylation, among others.

Several hydrolytic enzymes are used to metabolize drugs. The alcohol and amine groups exposed following hydrolysis of esters and amides are suitable substrates for conjugation reaction. Epoxide hydrolase, a detoxification enzyme, hydrolyzes highly reactive arene oxides generated from CYP oxidation reaction to inactive, water-soluble metabolites.

During the most predominant conjugation reaction, glucuronidation, uridine diphosphate glucuronosyltransferase catalyzes the transfer of glucuronic acid to aromatic and aliphatic alcohol, carboxylic acids, amines, and free sulphydryl groups. These reactions occur for both exogenous and endogenous compounds to form O-, N- and S-glucuronides. The increased water solubility of a glucuronide conjugate promotes its elimination into the urine or bile.

Sulfation and acetylation reactions are also important conjugation reactions. Cytosolic sulfation involves the catalytic transfer by sulfotransferase or inorganic sulfur from activated 3'-phophoadenosoine-5'-phosphsulfate to the hydroxyl group of phenol aliphatic alcohols. Two N-acetyltransferases are involved in the acetylation of amines, hydrazines, and sulfonamides. The compounds produced are usually less water soluble than the parent drug, and this may result in crystalluria unless a high urine flow rate is maintained.

> **Checkpoint! 21–3**
>
> *Compare phase I– and phase II–type reactions.*

EXCRETION

The kidneys are the most important organs for excretion of drugs and their metabolites. Polar compounds are more efficiently excreted than nonpolar ones. Lipid-soluble drugs are not readily eliminated until they are metabolized to more polar compounds. Compounds excreted in the feces are mainly unabsorbed. Orally ingested drugs or metabolites are excreted either in the bile or secreted directly into the intestinal tract and, subsequently, not reabsorbed. Drugs excreted in breast milk are significant because the excreted drugs are potential sources of unwanted pharmacological effects in the nursing infant. Pulmonary excretion is important mainly for the elimination of anesthetic gases and vapors.

▶ CLINICAL PHARMACOKINETICS

A relationship exists between the pharmacological effects of a drug and an attainable concentration of the drug in blood or plasma, so the concentration of drug in the systemic circulation will be related to the concentration of drug at its site of action. The pharmacological effects that result may be the desired effect, a toxic effect, or possibly an effect that is not predicted. Clinical pharmacokinetics attempts to establish a quantitative relationship between dose and effect that provide a means to interpret the measurement of concentrations of drugs in biological fluids.

Four important parameters affect changes in the dosage of drugs for patients: *clearance, volume of distribution, elimination half-life,* and extent and rate of *bioavailability.* These factors must be taken into consideration by the clinician so that an appropriate drug level can be maintained in the patient.[1]

CLEARANCE

Clearance (CL) is a measure of the body's efficiency in eliminating drugs. Plasma clearance is defined as the volume of plasma from which all drug appears to be removed in a given time (e.g., mL/min). Clearance equals the amount of renal plasma flow multiplied by the extraction ratio (see below.) Because these factors normally do not vary over time, CL is constant.

Two quantifiable parameters are associated with renal clearance of drugs: extraction ratio and extraction rate. Extraction ratio, shown in equation 21.1, represents the decline of drug concentration in the plasma from the arterial to the venous side of the kidneys. Drugs enter the kidneys at a concentration (C_1) and exit the kidneys at concentration (C_2):

$$\text{Extraction ratio} = C_2/C_1 \qquad (21.1)$$

The excretion rate (mg/min) is a function of clearance (mL/min) times plasma drug concentration (mg/mL). The elimination of a drug follows first-order kinetics, and the concentration of drug in plasma drops exponentially with time.

Total body clearance of a drug is the sum of the clearances from the various body organs that metabolize and eliminate drugs. Total body clearance (CL_{total}) can be calculated by using equation 21.2:

$$CL_{total} = CL_{hepatic} + CL_{rena}1 + CL_{pulmonary} + CL_{other} \qquad (21.2)$$

It is not practical to measure and sum these individual clearances. Total clearance of a drug can be derived from studies of the steady-state characteristic of a drug.

The underlying principles associated with the clearance of a drug are similar to those of renal clearance of solutes. For example, creatinine clearance (see Chapter 11, Nonprotein

Nitrogen and Renal Function) is defined as the rate of elimination of creatinine in urine relative to its concentration in plasma. CL of a drug is its rate of elimination by all routes normalized to the concentration (C) of a drug in some biological fluid, as shown in equation 21.3:

$$CL = \text{rate of elimination}/C_1 \quad (21.3)$$

Therefore, when clearance is constant, the rate of drug elimination is directly proportional to drug concentration. Clearance of a drug does not indicate how much drug is being removed but rather the volume of biological fluid such as blood or plasma from which a drug would have to be completely removed to account for elimination.

VOLUME OF DISTRIBUTION

Volume of distribution (V_d) is a measure of the apparent space the body has available to contain the drug. It relates to the amount of drug in the body to the concentration of drug in the blood or plasma. This volume does not essentially refer to a physiological volume but to the fluid volume that would be required to contain the entire drug in the body at the same concentration as in the blood or plasma.

A drug's volume of distribution reflects the extent to which it is present in extravascular tissues. Thus, volume of distribution is equal to the amount of drug in the body divided by concentration of the drug in blood or plasma. For example, the plasma volume of a typical 70-kg man is 3 liters (L), blood volume is about 5.5 L, extracellular fluid volume outside the plasma is about 12 L, and the volume of total body water is approximately 42 L. The relationship of V_d to total amount of drug in the body and plasma drug concentration is shown in equation 21.4.

$$V_d = D/C \quad (21.4)$$

where

V_d = volume of distribution

D = total amount of drug in the body

C = plasma concentration

Example

What is the volume of distribution for a 70-kg male who receives a dose of a drug that results in a total amount of drug in the body of 20 μg and the plasma concentration is 0.75 ng/mL?

Answer:

$$V_d = D/C$$
$$V_d = 20\ \mu g/0.75\ ng/mL$$
$$V_d = 26.7\ L$$

ELIMINATION HALF-LIFE

Half-life ($t_{1/2}$) is the time required for an amount of drug in blood to decline to one-half its measured value and determines (1) the time taken to reach steady state on multiple

dosing, (2) the time taken to reach a new steady state when dosage is altered, and (3) the time taken to eliminate most the drug from the body when dosing is terminated. In general, it takes five to seven half-lives to achieve steady state. If dosing is terminated, most of the drug will have been eliminated from the body with the exception of possible metabolites with longer half-lives.

Specifically, **elimination half-life** is the time it takes for the plasma concentration or the amount of drug in the body to be reduced by 50%. It is now believed that half-life is a derived parameter that changes as a function of both clearance and volume of distribution (see equation 21.5).

$$\text{Half-life} = 0.693 \times V_d/CL \quad (21.5)$$

When a patient has a condition that alters the half-life of a drug, an adjustment in dosage is required. It is important to be able to predict in which patients a drug is likely to have a longer half-life. The following conditions may increase the half-life of a drug:

1. decreased renal blood flow resulting from heart disease, shock, or hemorrhaging;

2. the addition of a second drug that may displace the first drug from its carrier protein such as albumin and thereby increase the volume of distribution of the drug; and

3. a decrease in metabolism because of hepatic insufficiency (cirrhosis) or because another drug inhibits its biotransformation.

 Checkpoint! 21–4

A patient is diagnosed with alcoholic liver disease with an accompanying arrhythmia. The patient requires a daily dose of an antiarrhythmic drug. Will the elimination half-life of the arrhythmic drug be shorter or longer than that in a patient without alcoholic liver disease?

STEADY STATE

Steady state of a drug is defined as the time required for an amount of drug in blood to decline to one-half its measured value and is dependent on the dosing rate and clearance. In equation 21.6, the steady-state concentration will eventually be achieved when a drug is administered at a constant rate.

$$\text{Dosing rate (mg/min)} = CL(mL/min) \times C_{ss}(mg/mL) \quad (21.6)$$

where

CL = clearance

C_{ss} = the concentration at steady state

In addition, from equation 21.6, if C_{ss} is known, then the rate of clearance of a drug by the patient will determine the rate at which the drug should be administered.

At this point, drug elimination (equation 21.3) will equal the rate of drug availability. This concept also extends to drugs given by repeated dosing (e.g., 100 mg of drug every 4 hours).

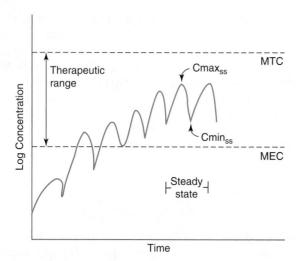

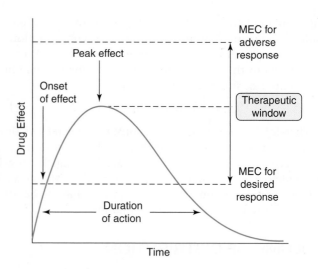

■ FIGURE 21-6   Steady state after seven half-lives. Multiple identical doses of a drug are administered once each half-life. MEC, minimum effective concentration; MTC, minimum toxic concentration; Cmax$_{ss}$, minimum steady-state concentrations.

■ FIGURE 21-7   Sequential characteristics of drug effect and relationship to the therapeutic window.

Between each dosage, the concentration of drug rises and falls. At each steady state, the entire cycle is repeated identically; thus, equation 21.6 stills applies for repeated doses. The major difference is that C_{ss} is now described as the average drug concentration during an interdose interval. In Figure 21-6 ■, a plot of repeated dosing is derived by measuring plasma drug concentrations over an interval of time. Depending on the drug given, a steady-state concentration represented by an average C_{ss} will be attained. It may take four to seven half-lives to achieve a steady-state concentration.

EXTENT AND RATE OF BIOAVAILABILITY

Bioavailability is defined as the fraction of drug dose reaching the systemic circulation following administration by any route. Bioavailability depends on (1) administered dose, (2) the fraction of the dose that is absorbed, and (3) the fraction of dose that escaped any first-pass elimination.

A clinician that administers a drug by a route that is subject to first-pass loss (enteral) must take into consideration bioavailability (F) when making decision regarding dosing or dosing rates. Therefore, equation 21.6 would need to include F, which results in equation 21.7:

$$F \times dosing\ rate = CL \times C_{ss} \qquad (21.7)$$

Rate of absorption does not usually influence steady-state concentration but still may affect the decision by the clinician regarding drug therapy. For example, if a drug is absorbed rapidly (e.g., given IV bolus*) and has a small volume of distribution, the initial drug concentration will be high. The concentration will become lower as the drug is distributed to its final or larger volume. If the same drug is given slowly (e.g., by slow IV infusion) and absorbed more slowly, then it will be

*IV bolus: a concentrated mass of an intravenous drug given rapidly.

distributed while it is being given and peak concentration will be lower and will occur later. Many drugs are designed and prepared to be released and absorbed in a slow and sustained rate. This type of preparation produces a less-fluctuating plasma *concentration × time* profile during the dosage interval compared to more immediate-release formulation.

DOSAGE REGIMENS

When a drug is administered, its effects usually show a characteristic sequence of events as presented in Figure 21-7 ■. A plot of time versus drug effect from a single dose yields a curve with several significant points. A lag period exists before the drug concentration exceeds the minimum effective concentration (MEC) for the desired effect.

Following the onset of response, the intensity of the effect increases as the drug continues to be absorbed and distributed. Eventually, a **peak drug concentration** is reached. Following the peak level of a drug, elimination of the drug results in a decline in the effects of the drug until the drug concentration falls below MEC. MEC for adverse response represents a toxic level of the drug. The goal of the clinician is to determine a drug concentration within the therapeutic window. Drug response below MEC for desired effect will be **subtherapeutic.**[2]

MAINTENANCE DOSE

For most clinical applications, drugs are administered as a series of repetitive doses or as a continuous infusion. This serves to maintain steady-state concentration of the drug associated with the therapeutic window. The **maintenance dose** required is calculated based on the clinician's desired target plasma concentration, patient clearance, and bioavailability of the drug.

LOADING DOSE

A loading dose of a drug usually results in the attainment of target drug concentration in a shorter period of time. An example of when a clinician may want to achieve a target dose quickly would be in the treatment of an arrhythmia using the drug lidocaine. The half-life of lidocaine is about 1–2 hours. Arrhythmia represents a cardiac dysrhythmia that can be life threatening. The clinician and patient thus cannot wait 4–8 hours to achieve a steady-state level, so administering a loading dose of lidocaine to a patient is part of the protocols of most coronary care units.

THERAPEUTIC DRUG MONITORING

The goal of TDM is to provide the patient with the optimum dosage of drug to effect the best outcome. This can be achieved by monitoring or measuring the concentration of drug in a plasma sample on a periodic basis. The process of TDM involves several medical professionals, including the ordering clinician, clinical laboratorian, clinical pharmacologist, and the nurse involved in medication delivery.

The clinical utility of measuring concentrations of drug at steady state is to reevaluate and adjust CL/F for the patient being treated. Rearranging equation 21.7 by substituting CL/F, we have equation 21.8:

$$CL/F \text{ (patient)} = \text{dosing rate}/C_{ss} \text{ (measured)} \quad (21.8)$$

The new value for CL/F can be used to modify the maintenance dose to achieve the desired target concentration.

Several factors need to be addressed when implementing therapeutic drug monitoring. An important issue that requires attention is the time of sampling for measurement of drug concentrations. If the sample is taken any time during dosing (also known as *random sampling*), this may help in assessing drug toxicity. A problem with this sampling protocol is that there could be a significant amount of interindividual variability in sensitivity to the drug. This will make the interpretation of drug levels more difficult for the clinician.

The sample could be drawn shortly after administration of the drug. A problem could arise in evaluating the drug level because of variable rates of distribution and altered pharmocdynamics. Therefore, acquiring a sample from a patient shortly after a drug is given may produce results that are not useful or are even misleading.

If sampling is for adjusting dosage, then it is recommended that the patient have blood drawn *just before the next dose*. This places the sampling well after the previous dose and should provide the clinician with a reliable result.

Another significant issue regarding the timing of sampling is when to begin the maintenance dosage regimen. Remember that steady state is usually achieved after five to seven half-lives have elapsed. If sampling is done too soon after dosage begins, it may not accurately reflect steady state and the clearance of the drug. Conversely, for toxic drugs, if sampling is delayed until steady state is confirmed, the damage may have already occurred. A current practice is to sample the patient after about two to three half-lives, assuming no loading dose has been given. Based on the results of laboratory testing, the dosage regimen can then be adjusted.

COMPLIANCE

The issue of compliance is a problematic for the clinician. Compliance addresses the issue of whether the patient is actually taking the drug or is taking it as prescribed. The success of the therapeutic drug regimen established for the patient depends on total compliance by the patient. Noncompliance of a dosing schedule is a major reason for therapeutic failure. This occurs more frequently in the long-term treatment of disease using antihypertensive, antiretroviral, and anticonvulsant drugs. A prominent reason for therapeutic failure is missed doses by the patient. Successful therapeutic programs are ones in which communication and understanding between patient and physician is at a maximum.

 Checkpoint! 21–5

In general, when is a sample to be drawn for the purpose of adjusting dosage?

▶ PHARMACODYNAMICS

Analysis of drug action is important so that mechanisms of the chemical or physical interactions between drug and target cell can be determined. Also, characterization of sequence and scope of action of each drug can be derived from analysis. This total analysis of the **pharmacodynamics** of drugs may provide the basis for both the rational therapeutic use of a drug and the design of new and better therapeutic agents.

MECHANISM OF ACTION

The effect of a drug results from its interaction with macromolecular components of the organism. The interactions alter the function of the component and initiate the biochemical and physiological changes that are characteristic of the response to the drug. **Receptors** are the components of the organism that the drug is thought to interact with. The drug binds to the receptor to initiate its effects. In equation 21.9, the binding of drug to an appropriate receptor forms a drug–receptor complex. The formation of a drug–receptor complex leads to a biologic response or effect. Proteins form the majority of receptors. Examples of protein receptors are those for hormones, growth factors, neurotransmitters, enzymes, and proteins involved in transport processes and structural proteins. An example of a nonprotein receptor is a nucleic acid that binds cancer chemotherapeutic agents.

$$\text{Drug} + \text{Receptor} \leftrightarrow \text{Drug} \times \text{Receptor complex} \rightarrow \text{Effect} \quad (21.9)$$

Drugs may bind to receptors by forming chemical bonds: that is, ionic, hydrogen, hydrophobic, van der Waals, and covalent. In most interactions of drugs and receptors, multiple types of bonding are used. The type of bonding may result in a drug's prolonged effect, or it may render the bond irreversible.[3]

Not all drug interactions are mediated by receptors of macromolecular proportions. Some drugs bind small molecules or ions—for example, the neutralization of gastric acid by a base compound found in antacids. Another example is mesna (sodium 2-mercapto ethane sulfanate), which is a free radical scavenger that is rapidly eliminated by the kidneys. It binds reactive metabolites associated with some cancer chemotherapeutic agents and thus minimizes their effects on the urinary tract.

Drugs that bind to physiological receptors and mimic the regulatory effects of the endogenous signaling compounds are termed **agonists**. Drugs are **antagonists** if they bind to receptors without regulatory effect. The binding of an antagonist compound blocks the binding of the endogenous agonist. Antagonist drug interactions can produce useful effects even though they inhibit the action of an agonist. Thus, an agonist compound may compete for binding to receptors with antagonist compounds. An example of this type of competition is shown in Figure 21-8 . Acetylcholine (an agonist) competes with atropine and scopolamine (both are antagonist) for binding sites on the muscarinic receptor. The binding of either antagonist to muscarinic receptors will inhibit all muscarinic-related functions.

The overall effectiveness and efficiency of a drug and receptor interaction can be defined and measured. The probability that a drug molecule will interact with its receptor to form the drug–receptor complex is measurable. This determination represents a drug's *affinity* for a receptor. The measurement of biological effectiveness of a drug–receptor complex is called *intrinsic activity*.

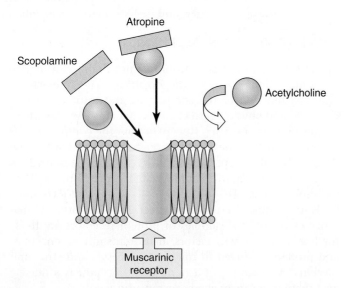

■ **FIGURE 21-8** The competitive nature of acetylcholine, an agonist drug, and antagonist drugs scopolamine and atropine for binding sites to muscarinic receptors.

Potency of a drug refers to the least amount of drug required to produce the maximum effect. For example, if two analgesic (pain relief) drugs, morphine and dihydrocodeine, were compared, it would be shown that morphine is more potent than dihydrocodeine. Thus, it would take less morphine to obtain maximal effect.

Finally, it is important to know whether a drug has the ability to produce the desired therapeutic effect. A measure of this ability of a drug is called **efficacy**. Two drugs may have the same efficacy but differ in potency as can be seen in the preceding example of morphine and dihydrocodeine.

The conclusions drawn from each measured parameter cited earlier come from experimental data that are used to derive dose–response or dose–effect relationships.

> ✓ **Checkpoint! 21–6**
>
> *A drug's effect on a host results from the formation of what type of complex?*

FACTORS THAT INFLUENCE DRUG EFFECTS AND DRUG DOSAGE

There are many factors that modify the effects of drugs. Some of these factors result in qualitative differences in a drug's effects. For example, the development of a drug allergy may result in a discontinuation of the drug's use. Other factors may produce a quantitative change in the usual effects of the drug and can be compensated for by appropriate adjustment of dosage. Several of these factors are shown in Box 21-2 ■.

Drug Toxicity

A majority of drugs have a toxic effect that may be minor in nature. However, some drugs do produce a serious and at times fatal effect. Some of these effects appear quickly, whereas others develop after a prolonged medication interval. Some toxic effects occur only in certain patients or only when multiple drugs are taken in combination.

Drugs that are given for therapeutic use may also result in drug-induced complications. Examples of these are drug

BOX 21-2 Examples of Factors that Determine the Relationship Between Prescribed Drug Dosage and Drug Effect

► Medication errors and patient compliance
► Body weight
► Age
► Routes of administration
► Time of administration
► Rate of elimination
► Tolerance
► Physiological variables
► Pathological factors
► Genetic factors

allergies, blood dyscrasias, heptotoxicity and nephrotoxicity, teratogenic effects, behavioral toxicity, drug dependence, and drug addictions. These complications or adverse effects can be avoided if drugs are used more carefully and more wisely. Physicians can avoid using a toxic drug if a less toxic one will suffice. Minimizing the use of concurrent or multiple medications also may reduce toxicity.

▶ PREANALYTICAL CONSIDERATIONS

SPECIMENS AND COLLECTION TUBES

Serum is the specimen of choice for the majority of therapeutic drugs measured in the clinical laboratory today. Plasma samples can be used for the measurement of some drugs, so consult the appropriate references to confirm that plasma is an acceptable substitute. Whole blood may also be used for selected therapeutic drug measurements—cyclosporine A, for example. Analysis of drugs in whole blood is often complicated by the need for preanalytical treatment of the specimen, which direct affects the turnaround time and may result in greater imprecision.

PEAK AND TROUGH SAMPLE

Many drugs are sampled at their peak time or level. Knowledge of peak drug concentrations is important not only to the clinician but also to the laboratorian. A **trough drug level** is often requested for antibiotic medications such as the aminoglycosides. A trough levels represents the minimum drug concentrations just before the next dose. The person giving the medication and the personnel drawing the specimens should carefully document peak and trough levels. Failure to document this information may lead to inappropriate therapeutic intervention such as overdosing or underdosing the patient.

▶ SPECIFIC DRUG GROUPS (THERAPEUTIC CATEGORIES)

OVERVIEW

In the following sections, several categories of therapeutic drugs will be identified and discussed. A brief discussion of the naming of drugs is presented in the next section. Specific drugs in each category that are often measured in the clinical chemistry laboratory are also discussed. Information about therapeutic use and unique chemical features are presented for each drug listed. In addition, the mechanism of action, absorption, metabolism, and excretion of each drug are covered.

NAMING OF DRUGS

There is a system of nomenclature for drugs. Laboratory personnel should be familiar with the names for each drug

measured in their respective laboratories. Drugs may be referred to by their chemical name. This form of nomenclature specifies the chemical structure of the drug and uses standard chemical terminology. For example, N-acetyl-para-aminophenol is the chemical name for acetaminophen or Tylenol. Most health-care professionals use the generic or nonproprietary name of a drug. An example of a generic name is acetaminophen. Finally, the proprietary or trade name for a drug is a registered trademark belonging to a particular drug manufacturer and used to designate a drug product marketed by that manufacturer. An example of a trade name for acetaminophen is Tylenol (McNeil Pharmaceuticals).

Several of the drugs discussed in this chapter will have additional names listed that are commonly used by nurses, laboratory staff, and clinicians. For a complete list of other generic or proprietary names, consult the annual edition of a *Physicians' Desk Reference* (PDR).[4]

▶ ANTIBIOTICS

If taken literally, antibiotics are substances produced by various species of microorganism (e.g., bacteria, fungi, and actinomycetes) that suppress the growth of other microorganism. The definition has been extended to include synthetic antimicrobial agents (e.g., quinolones and sulfonamides). There are hundreds of antibiotic compounds that differ markedly in their physical, chemical, and pharmacological properties.

AMINOGLYCOSIDES (AMIKACIN, GENTAMICIN, AND TOBRAMYCIN)

The aminoglycosides consist of two or more amino sugars joined by a glycoside linkage to a hexose nucleus. This hexose or amiocyclitol is either streptidine (found in streptomycin) or 2-deoxystreptamine (all other aminoglycosides). These compounds are therefore aminoglycosidic aminocyclitols, but they are usually referred to as aminoglycosides. The aminoglycoside family of drugs is distinguished by the amino sugars attached to the aminocyclitol.

Mechanism of Action
The aminoglyoside antibiotics are rapidly bactericidal, and this function is dependent on concentration. This killing effect begins with the aminoglycoside diffusing through aqueous channels formed by porous porin in the outer membrane of gram-negative bacteria to enter the periplasmic space. Transport of aminoglycosides across the cytoplasmic membrane depends on electron transport. Once inside the cell, an aminoglycoside binds to polysomes and interferes with protein synthesis by causing premature termination of translation of messenger RNA. The aberrant proteins produced may be inserted into the cell membrane, leading to altered permeability and further stimulation of aminoglycoside transport.[5] This progressive disruption of the cell envelope, as well as other vital cell processes, explains the lethal action of aminoglycosides.

Absorption, Metabolism, and Excretion

The aminoglycosides are highly polar cations and are poorly absorbed from the gastrointestinal tract. Less than 1% of a dose is absorbed after either oral or rectal administration. All of the aminoglycosides are absorbed rapidly from intramuscular sites of injection. Peak concentrations in plasma occur after 30 to 90 minutes. These same levels are seen 30 minutes after completion of an intravenous infusion of an equal dose over a 30-minute period.

The polar nature of aminoglycosides markedly affects their distribution in tissues. They are largely excluded from most cells, the central nervous system, and the eye. There is negligible binding of aminoglycosides to plasma albumin with the one exception: streptomycin.

The aminoglycosides are excreted almost entirely by glomerular filtration, and concentration in the urine may approach 50 to 200 µg/mL. A large percentage of parentally administered doses are excreted unchanged during the first 24 hours, with most of this appearing in the first 12 hours. The *half-life (t½)* of the aminoglycosides in plasma is between 2 and 3 hours in patients with normal renal function.

Toxicity

Tissue concentrations of aminoglycosides are low, although high concentrations are found in the renal cortex and in the endolymph and perilymph of the inner ear. This appears to cause the nephrotoxicity and ototoxicity caused by these drugs. The level of aminoglycosides in bile is about 30% that of plasma as a result of active hepatic secretion.

All aminoglycosides have the potential to produce reversible and irreversible vestibular, cochlear, and renal toxicity. Appropriate use of these drugs is complicated by these often serious adverse effects.

Ototoxicity occurs when plasma levels of aminoglycosides are too high. The drugs accumulate in the perilymph and endolymph of the inner ear.[6] Ototoxicity is largely irreversible and results from progressive destruction of vestibular or cochlear sensory cells. These sensory cells are highly sensitive to damage by aminoglycosides.

Mild renal impairment may occur in many patients receiving aminoglycosides for more than several days. This impairment is usually reversible. The toxicity occurs when the aminoglycoside accumulates and is retained in the proximal tubular cells. After several days, there is a defect in renal concentration ability, mild proteinuria, and the appearance of hyaline and granular casts.[7] The glomerular filtration rate is reduced. The most common finding is a slight rise in plasma creatinine. The impairment of renal function is almost always reversible because the proximal tubular cells have the capacity to regenerate.

 Checkpoint! 21–7

Identify two renal function tests that should be monitored in patients treated with an aminoglycoside.

AMIKACIN [ALSO AMIKIN AND CEFTRIAXONE (CHILDREN)]

Amikacin is the preferred agent for initial treatment of serious nosocomial gram-negative bacillary infections in hospitals, where resistance to gentamicin and tobramycin have become a significant problem. It is the resistance to aminoglycoside-inactivating enzymes that gives Amikacin a broad spectrum of activity against gram-negative bacilli. This includes most strains of *Serratia*, *Proteus*, and *Pseudomonas aeruginosa*. It is active against nearly all strains of *Klebsiella*, *Enterobacter*, and *Escherichia coli* that are usually resistant to gentamicin and tobramycin. In addition, amikacin is also effective against most gram-positive anaerobic bacteria.

GENTAMICIN (ALSO GARAMYCIN)

Gentamicin is isolated from *Micromonospora purpurea*, a gram-positive bacterium. Gentamicin is a compound used for the treatment of many serious gram-negative bacillary infections. In most instances, it is the aminoglycoside of first choice because of its reliable activity against most gram-negative aerobes. Emergence of resistant microorganisms in some individuals has become a serious issue and may limit the future use of this compound. Gentamicin is often administered in combination with penicillin to treat serious gram-negative microbial infections, especially those from *Pseudomonas aeruginosa*, *Enterobacter*, *Klebsiella*, and *Serratia* species.

TOBRAMYCIN (ALSO NEBCIN)

Tobramycin may be given either intramuscularly or intravenously. The antimicrobial activity and pharmacokinetic properties of tobramycin are similar to gentamicin. Indications for use of tobramycin are also similar for those for gentamicin. Tobramycin is superior for treatment of *P. aeruginosa*. and it should also be used concurrently with an antipseudomonal β-lactam antibiotic such as one of the cephalosporins.

Tobramycin, unlike gentamicin, shows poor activity in combination with penicillin against many strains of *Enterococci*. Most strains of *Enterococcus facecium* are highly resistant to tobramycin. Further, tobramycin is ineffective against mycobacteria.

VANCOMYCIN

Vancomycin is a tricyclic glycopeptide antibiotic produced by the microorganism *Amycolatopsis orientalis* (formerly *Nocardia orientalis*) and is unrelated to other commercially available antibiotics. Vancomycin is used principally for the treatment of severe infections caused by gram-positive bacteria in patients who cannot receive or who have failed to respond to penicillin and cephalosporins or for the treatment of gram-positive bacterial infections that are resistant to β-lactams and other anti-infective agents. Examples of such bacteria include *Streptococci*, *Corynebacteria*, *Clostridia*, *Listeria*, and *Bacillus* species.

Mechanism of Action

Vancomycin is bactericidal and appears to bind to the bacterial cell wall, causing blockage of glycopeptide polymerization. This results in immediate inhibition of cell-wall synthesis and secondary damage to the cytoplasmic membrane. Magnesium, manganese, calcium, and ferrous ions reduce the degree of adsorption of vancomycin to the cell wall, but the in vivo importance of this interaction is unknown.

Absorption, Metabolism, and Excretion

Vancomycin hydrochloride is usually not appreciably absorbed from the GI tract. Limited data suggest, however, that clinically important serum concentrations of the drug may follow enteral or oral administration of vancomycin in some patients with colitis, particularly those who also have renal impairment. Following IV administration of 1 g of the drug to adults, plasma concentrations have been reported to average 25 ug/mL at 2 hours. Serum vancomycin concentrations are higher in patients with renal dysfunction than in those with normal renal function, and toxic serum concentrations may result.

Parentally administered vancomycin is excreted primarily by glomerular filtration. More than 80% of a single IV dose is excreted within 24 hours. Specific urinary concentrations of the drug have not been determined. Only small amounts of vancomycin are excreted in the bile following IV administration. Oral doses are excreted primarily in the feces. Vancomycin is only minimally removed by hemodialysis or peritoneal dialysis, including continuous ambulatory peritoneal dialysis. The drug is substantially removed by hemofiltration.

Toxicity

Ototoxicity and nephrotoxicity are the most serious adverse effects of parenteral vancomycin therapy. To date, these adverse effects have not been reported in patients receiving vancomycin orally. The incidences of ototoxicity and nephrotoxicity have not been well established, but clinical experience to date suggests that these adverse effects occur relatively infrequently. Ototoxicity and nephrotoxicity are most likely to occur in patients who (1) have renal impairment, (2) are receiving high-dose IV vancomycin for prolonged periods, or (3) are receiving other ototoxic or nephrotoxic drugs. Ototoxicity and nephrotoxicity have been associated with serum or blood vancomycin concentrations of >80 μg/mL. Vancomycin may cause damage to the auditory branch of the eighth cranial nerve. Permanent deafness has occurred. Rarely have vertigo, dizziness, and tinnitus been reported. Tinnitus may precede the onset of deafness and necessitates discontinuance of the drug. Deafness may progress despite cessation of vancomycin therapy.

Vancomycin-induced nephrotoxicity may result in transient elevations of serum urea nitrogen or creatinine concentrations, urine albumin levels, and the appearance of hyaline or granular casts in urine. Fatal uremia has occurred. Rarely has the drug been associated with acute interstitial nephritis.

Checkpoint! 21–8

Why should the clinician monitor both peak and trough levels of aminoglycosides and vancomycin antibiotic compounds?

▶ ANTIEPILEPTIC DRUGS

Antiepileptic drugs—also known as antiseizure or anticonvulsant drugs—are formulated to best control seizures in an individual patient while minimizing unwanted side effects. Side effects include impairment of the central nervous system (CNS), aplastic anemia, liver disease, and death. These drugs are often prescribed in combination (e.g., phenobarbital and phenytoin).

PHENOBARBITAL (LUMINAL)

Phenobarbital and phenobarbital sodium are used principally for routine sedation or to relieve anxiety and provide sedation preoperatively. Because several hours are required to achieve maximal effects, phenobarbital is not generally used orally as a hypnotic. Phenobarbital may also be used to assist patients who are in the process of drug withdrawal from persistent use of barbiturate or nonbarbiturate hypnotics.

The drug is also used in the prophylactic management of epilepsy. The usefulness of parenteral phenobarbital sodium in terminating acute seizure episodes is limited by its relatively slow onset of action.

Mechanism of Action

Anticonvulsant effects of barbiturate derivatives are multiple and rather nonselective. The principal mechanism of action is believed to be reduction of monosynaptic and polysynaptic transmission, which results in decreased excitability of the entire nerve cell. Barbiturates also increase the threshold for electrical stimulation of the motor cortex.

Absorption, Metabolism, and Excretion

About 70% to 90% of an oral dose of phenobarbital is absorbed slowly from the GI tract. Following oral administration of phenobarbital, peak blood concentrations are reached in 8–12 hours and peak brain concentrations in 10–15 hours. Because phenobarbital has a long plasma half-life (2–6 days), 3 to 4 weeks of therapy may be required to achieve steady-state plasma concentrations unless a loading dose is given. Subcutaneous or IM subcutaneous administration of phenobarbital sodium results in a slightly slower onset of action. The duration of action of parenterally administered phenobarbital sodium is usually 4–6 hours but may be 6–10 hours following hypnotic doses.

In vitro, 20–45% of phenobarbital in the blood is bound to plasma proteins. The liver hydroxylates phenobarbital to form an inactive metabolite, *p*-hydroxyphenobarbital. Phenobarbital is a potent inducer of the enzymes involved in the metabolism of other drugs, but there is no conclusive evidence that phenobarbital accelerates its own metabolism.

Approximately 25% of a dose is excreted unchanged in urine, and about 75% of the drug is excreted in urine as the *p*-hydroxy metabolite and its glucuronide and sulfate conjugates. Alkalinization of the urine or increases in urinary flow substantially elevate the rate of excretion of unchanged phenobarbital. Unmetabolized drug may accumulate in patients with oliguria or uremia.

PHENYTOIN (ALSO DIPHENYLHYDANTOIN, DILANTIN, OR DIPHENYLAN)

Phenytoin is used mainly in the prophylactic management of tonic–clonic (grand mal) seizures and partial seizures with complex symptomatology (psychomotor seizures). The drug is also effective in controlling autonomic seizures. The drug is often administered concomitantly with phenobarbital or other anticonvulsants. Phenytoin is not recommended for the treatment of pure absence (petit mal) seizures because it may increase the frequency of these seizures; however, phenytoin may be useful in conjunction with succinimide or oxazolidinedione anticonvulsants in the management of combined absence and tonic–clonic seizures.

Phenytoin is a diphenyl-substituted hydantoin. It has much lower sedative properties than compounds with alkyl substituents at the 5-carbon position. A 5-phenyl or other aromatic substituent appears essential for activity against generalized tonic–clonic seizures.

Mechanism of Action

Most authorities agree that the principal mechanism of action of the hydantoin anticonvulsants, particularly phenytoin, is limitation of seizure propagation by reduction of post-tetanic potentiation (PTP). PTP is described as the excitatory postsynaptic potential created after a brief period of high-frequency (tetanic) activation that leads to an increased release of transmitters by the presynaptic terminals. Phenytoin acts by reducing the passive influx of sodium ions or by increasing the efficiency of the sodium pump. These mechanisms may reduce excess accumulation of intracellular sodium during tetanic stimulation.

Absorption, Metabolism, and Excretion

Phenytoin and its sodium salt are usually completely absorbed from the GI tract. Bioavailability may vary enough among oral phenytoin sodium preparations of different manufacturers to result in toxic serum concentrations or a loss of seizure control (subtherapeutic serum concentrations.) *Prompt* phenytoin capsules are rapidly absorbed and generally produce peak serum concentrations in 1.5 to 3 hours, whereas *extended* phenytoin sodium capsules are more slowly absorbed and generally produce peak serum concentrations in 4 to 12 hours. When phenytoin sodium is administered IM, absorption may be erratic; this may result from

crystallization of the drug at the injection site because of the change in pH.

Following oral administration, the plasma half-life of phenytoin averages about 22 hours and ranges from 7 to 42 hours in individual patients. The plasma half-life of phenytoin in humans following IV administration ranges from 10 to 15 hours.

The major route of metabolism of phenytoin is oxidation by the liver to the inactive metabolite 5-(*p*-hydroxyphenyl)-5-phenylhydantoin (HPPH). HPPH goes into the enterohepatic circulation and is excreted in urine via glomerular filtration and tubular secretion, mainly as the glucuronide. Approximately 60–75% of the daily dose of the drug is excreted in this form. Minor metabolites also appear in urine. In therapeutic doses, approximately 1% is excreted unchanged in urine; in toxic doses, as much as 10% of the ingested drug may be excreted unchanged by the kidneys.

VALPROIC ACID (ALSO DEPAKENE)

Valproic acid is a carboxylic acid derivative (*n*-dipropylacetic acid). There are many formulations of this compound, but only two will be presented here. Valproic acid (Depakene), valporate sodium (Depacon) and divalproex sodium (Depakote) is used alone or with other anticonvulsants (e.g., ethosuximide) in the prophylactic management of simple and complex absence (petit mal) seizures. The drugs also may be used in conjunction with other anticonvulsants in the management of multiple seizure types that include absence seizures. Valproic acid has been used as and is considered by some clinicians to be a drug of choice for management of other generalized seizures, including primary generalized tonic–clonic seizures, atypical absence, myoclonic, and atonic seizures.

Divalproex sodium is used in the treatment of manic episodes associated with bipolar disorder; valproic acid and valproate sodium have also been used. The term *valproic acid* will be used in the following discussion because there are only minor differences in the pharmacokinetics of the formulations, and all forms of the drug circulate in plasma as valproic acid. Divalproex sodium is used in the prophylaxis of migraine headaches, with or without associated aura; valproic acid and sodium valproate have also been used. Valproic acid has been used as an adjunct to antipsychotic drugs in the symptomatic management of schizophrenia in patients who fail to respond sufficiently to an adequate trial of the antipsychotic agent alone.

Mechanism of Action

The mechanism of action of valproic acid is not known. Effects of the drug may be related, at least in part, to increased brain concentrations of the inhibitory neurotransmitter gamma-aminobutyric acid (GABA). Animal studies have shown that valproic acid inhibits GABA transferase and succinic aldehyde dehydrogenase, enzymes that are important for GABA catabolism. Results of one study indicate the drug inhibits neuronal activity by increasing potassium conductance. In animals, valproic acid protects against seizures induced by electrical stimulation as well as those induced by pentylenetetrazol.[8]

Absorption, Metabolism, and Excretion

Following oral administration, valproate sodium is rapidly converted to valproic acid in the stomach. Valproic acid is rapidly and almost completely absorbed from the GI tract. Absorption of the drug is delayed but not decreased by administration with meals. Following oral administration of divalproex sodium extended-release tablets, divalproex sodium dissociates into valproic acid in the GI tract.

Peak plasma concentrations of valproic acid are usually attained 1–4 hours following a single oral dose of the acid or the sodium salt, 3–5 hours following a single oral dose of divalproex sodium, and 7–14 hours following oral administration of multiple doses of divalproex sodium extended-release tablets.

Valproic acid has an elimination half-life of 5–20 hours (average 10.6 hours). Elimination half-lives in the lower portion of the range are usually observed in patients receiving other anticonvulsants concomitantly. Plasma half-lives of as long as 30 hours may be seen in cases of valproate sodium overdose.

Valproic acid is metabolized principally in the liver. Valproic acid metabolites are excreted in urine; 30–50% of an administered dose is excreted as glucuronide conjugates. Less than 3% of an administered dose is excreted in urine unchanged. The major metabolite in urine is 2-propyl-3-ketopentanoic acid. Several minor metabolites also appear in the urine. Liver disease impairs the ability to eliminate valproic acid and plasma half-life may increase from 12 to 18 hours.

CARBAMAZEPINE (ALSO TEGRETOL AND CARBATROL)

Carbamazepine is an iminostilbene that is characterized as a tricyclic tertiary amine with a carbamyl group at carbon number five position. It is structurally related to the family of tricyclic antidepressants such as amitriptyline and imipramine. Carbamazepine is used as both an anticonvulsant and for the relief of pain associated with trigeminal neuralgia (tic douloureux). This disorder affects the fifth cranial nerve and results in facial neuralgia. It also may be used for various psychiatric disorders. Carbamazepine is used in adults and children in the prophylactic management of partial seizures with complex symptomatology (psychomotor or temporal lobe seizures), generalized tonic–clonic seizures, and mixed seizure patterns. Carbamazepine may be administered concomitantly with other anticonvulsants—for example, phenytoin, phenobarbital, or primidone. Carbamazepine has been used in the symptomatic management of the acute phase of schizophrenia as an adjunct to therapy with an antipsychotic agent in patients who fail to respond to an adequate trial of the antipsychotic agent alone.

Carbamazepine has been used for the management of aggression or uncontrolled rage outbursts or loss of control (dyscontrol) in patients with or without an underlying seizure disorder. Alcohol withdrawal syndrome has also been treated with carbamazepine. It has also been used to relieve neurogenic pain and to control seizures in a variety of conditions, including "lightning" pains of tabes dorsalis, a condition that results from the destruction of the dorsal columns in the spinal cord, which is normally responsible for postural stability. The drug has also been used for its antidiuretic effects in the management of neurohypophyseal diabetes insipidus.

Mechanism of Action

The pharmacological actions of carbamazepine are similar to those of the hydantoin-derivative anticonvulsants. The anticonvulsant activity of carbamazepine, like phenytoin, principally involves limitation of seizure propagation by reduction of PTP of synaptic transmission.

Absorption, Metabolism, and Excretion

Carbamazepine is slowly absorbed from the GI tract. Following chronic oral administration of carbamazepine tablets, peak plasma concentration is reached in 4.5 hours. Carbamazepine is widely distributed in the body and is about 77–90% bound to plasma protein.

The metabolic fate of carbamazepine has not been completely elucidated. A major metabolic pathway is oxidation by microsomal enzymes in the liver (principally cytochrome P-450 isoform 3A4) to form carbamazepine-10,11-epoxide (CBZ-E). CBZ-E is almost completely metabolized to trans-10,11-dihydroxy-10,11-dihydrocarbamazepine (trans-CBZ-diol) and excreted in urine mainly in the unconjugated form. Carbamazepine and its metabolites are excreted in urine. Only about 1–3% of the drug is excreted in urine unchanged.

ETHOSUXIMIDE (ALSO ZARONTIN)

Ethosuximide is a succinimide-derivative anticonvulsant. Succinimide derivatives also have a five-member ring structure similar to the hydantoins. The difference is that the imino nitrogen at position 3 in the hydantoin structure is replaced with a methylene group, which is mono- or di-substituted with a methyl, ethyl, or phenyl group.

Ethosuximide is generally considered the drug of choice in the management of absence seizures. The drug is usually ineffective in the management of partial seizures with complex symptomatology (psychomotor seizures) or tonic–clonic seizures. When used alone in mixed seizures, ethosuximide may increase the frequency of tonic–clonic seizures. When patients with absence seizures also have tonic–clonic seizures, other anticonvulsants such as phenytoin or phenobarbital must be used in combination with ethosuximide.

Mechanism of Action

Oxazolidinedione and succinimide derivatives elevate the seizure threshold in the cortex and basal ganglia and reduce synaptic response to low-frequency repetitive stimulation. They have no appreciable effect on PTP.

Absorption, Metabolism, and Excretion

Ethosuximide is absorbed from the GI tract. Following oral administration of a single dose, peak blood concentrations are reached within 4 hours. However, about 4 to 7 days of therapy at usual dosage are required to achieve steady-state plasma concentrations.

The plasma half-life of ethosuximide is about 60 hours in adults and about 30 hours in children. Ethosuximide is excreted slowly in urine. Approximately 20% of a dose is excreted unchanged, and as much as 50% may be excreted in urine as the hydroxylated metabolite or conjugated with glucuronic acid. Small amounts of unchanged drug are also excreted in bile and feces.

PRIMIDONE (ALSO MYSOLINE)

Primidone is a structural analog of phenobarbital in which a methylene group has replaced the carbonyl group at position 2. Primidone is used mainly in the prophylactic management of partial seizures with complex symptomatology (psychomotor seizures), and some clinicians consider it the drug of choice. Primidone is also useful in the prophylactic management of other partial seizures such as those with autonomic symptoms, akinetic seizures, and tonic–clonic seizures, particularly those refractory to other anticonvulsant therapy. Primidone is often used concomitantly with other anticonvulsants, especially phenytoin or phenobarbital. Some clinicians, however, do not recommend the concurrent administration of primidone and phenobarbital because of possible increased sedation.

Mechanism of Action

Primidone resembles phenobarbital in many antiseizure effects, although primidone is much less potent than phenobarbital in antagonizing seizures induced by pentylenetetrazol. Most anticonvulsive effects of primidone are attributed to both the drug and its active metabolites, primarily phenobarbital.

Absorption, Metabolism, and Excretion

Approximately 60–80% of an oral dose of primidone is absorbed from the GI tract. Following oral administration, peak serum concentrations are reached in about 4 hours. Serum half-life of primidone is about 21 hours.[2]

Primidone is slowly metabolized by the liver and slowly excreted in urine as phenylethylmalonamide (PEMA), phenobarbital, and *p*-hydroxyphenobarbital. Both PEMA and phenobarbital possess anticonvulsant activity; however, PEMA has only weak anticonvulsant properties and is more toxic than primidone. High serum phenobarbital concentrations during chronic administration of primidone may be the result of enzyme induction, differences in the serum half-life of the two drugs, or both. During chronic therapy, approximately 15–25% of an oral dose of primidone is excreted in urine unchanged. Approximately 15–25% of an oral dose of the drug is metabolized to phenobarbital, and 50–70% is excreted in urine as PEMA.

 Checkpoint! 21–9

Describe how a false positive phenobarbital result measured by immunoassay could occur if a patient was also taking primidone.

TOPIRAMATE (ALSO TOPAMAX)

Topiramate is a sulfamate-substitute monosaccharide with a structure distinctly different from other anti-epileptic drugs. It is a derivative of the naturally occurring monosaccharide D-fructose. It is used to treat partial seizure in adults when used in addition to other drugs.

Mechanism of Action

Topiramate reduces voltage-gated sodium currents in cerebellar granule cells and may act on the inactivated state of the channel in a manner similar to that of phenytoin. Also, topiramate enhances postsynaptic gamma-aminobutyric acid (GABA-receptor) currents and also limits activation of the α-amino-3-hydroxy-5-methyl-4-isoxazol propionic acid (AMPA)-kainate-subtypes(s) of glutamate receptor. It also is a weak carbonic anhydrase inhibitor.

Absorption, Metabolism, and Excretion

Topiramate is rapidly absorbed after oral administration with a bioavailability of 80%. It is less than 20% bound and is mostly excreted unchanged in the urine. The rest is metabolized by hydroxylation, hydrolysis, and glucuronidation. The half-life of topiramate is about 20 to 30 hours.

 Checkpoint! 21–10

How does the structure of topiramate, which has anti-epileptic effects, differ from other anti-epileptic drugs?

► ANTIHYPERTENSIVE DRUGS

Antihypertensive agents are prescribed to patients in an effort to reduce their blood pressure.

SODIUM NITROPRUSSIDE (ALSO NITROPRESS)

Nitroprusside is a nitrovasodialator used for short-term control of severe hypertension. It is also useful in improving cardiac function in patients with left ventricular failure. It reduces both ventricular filling pressure and systemic and arterial resistance. It has a rapid onset of action, within 2–5 minutes. It is commonly used in intensive-care settings for rapid control of hypertension and for the management of acutely decompensated heart failure. Knowledge of the drugs benefits have been known since 1850s but it was not until the 1950s that physicians began using it on a regular basis.

Mechanism of Action

Nitroprusside is metabolized by blood vessels to its active metabolite, nitric oxide. Nitric oxide activates guanylyl cyclase, which leads to the formation of cyclic guanosine monophosphate (GMP) and vasodilatation.[9] Nitroprusside dilates both arterioles and venules. Venous pooling and reduced arterial impedance contribute to the hemodynamic effects of administration of nitroprusside.

Absorption, Metabolism, and Excretion

Sodium nitroprusside is unstable and decomposes under strong alkaline conditions. It will also decompose when exposed to light. To be effective, the drug is given by continuous intravenous infusion. Onset of action is within 30 seconds. Peak hypotensive effect occurs within 2 minutes.

Metabolism of nitroprusside occurs in smooth muscle. The compound is reduced to cyanide and then nitric oxide.[10] Cyanide is further metabolized by liver rhodanase to thiocyanate (SCN⁻) and is eliminated mostly in the urine. The average elimination half-life for thiocyanate is 3 days in patients with normal renal function. Patients with renal insufficiency may take much longer to completely eliminate the compound.

Checkpoint! 21–11

Why should thiocyanate be measured in blood on a routine basis if a patient is taking sodium nitroprusside to treat hypertension?

▶ ANTINEOPLASTIC DRUGS

Neoplasms are described as any new growth of tissue in which the growth is uncontrolled and progressive. Malignant neoplasms show a greater degree of anaplasia or loss of differentiation of cells and have the properties of invasion and metastasis. Antineoplasic drugs are a groups of compounds used for cytotoxic antitumor therapy. Their mechanisms and sites of action vary, but one primary concern with all of these drugs is the potential for harmful side effects. These compounds also have narrow **therapeutic range**, so it is important to measure these drugs in an effort to minimize their unpleasant effects.

METHOTREXATE (ALSO AMETHOPTERIN, RHEUMATREX, TREXALL)

Methotrexate is an antifolate or folic acid analog that inhibits dihydrofolate reductase. It also directly inhibits the folate-dependent enzymes of de novo purine and thymidylate synthesis. This makes methotrexate an effective antineoplastic agent for management of certain types of cancers.

Methotrexate is often used in the management of lymphoblastic leukemia in children. It can induce remission and consolidation if given in high doses. During methotrexate infusion, high steady-state levels are associated with a lower leukemia relapse rate.[11] In adults, methotrexate alone rarely is effective in the treatment of acute myeloblastic leukemia; remissions are short, with relapses common and resistance develops rapidly. It is useful for treatment and prevention of leukemic meningitis in children and adults. The intrathecal administration of methotrexate has been used for treatment or prophylaxis of meningeal leukemia or lymphoma and for treatment of meningeal carcinomatosis.

Women with choriocarcinoma and related trophoblastic tumors may be treated with methotexate.[12] It is often given in conjunction with dactinomycin (Cosmegen, Merk & Co., West Point, PA), which is an actinomycin antibiotic compound. In contrast to uterine choriocarcinoma, testicular choriocarcinomas are usually resistant to methotrexate alone. Methotrexate has been used alone or, more commonly, in combination chemotherapy for the treatment of breast cancer.

Mechanism of Action

Methotrexate has a high affinity for dihydrofolate reductase and prevents the formation of tetrahydrofolate. This produces an intracellular deficiency of certain folate coenzymes and a larger accumulation of the toxic inhibitory substrate, dihydrofolate polyglutamate. The one-carbon transfer reaction for the de novo synthesis of purine nucleotides and thymidlylate cease, with the subsequent interruption of the synthesis of DNA and RNA.

Absorption, Metabolism, and Excretion

Methotrexate is readily absorbed from the gastrointestinal tract at low doses. Higher doses are absorbed incompletely and are administered intravenously. After intravenous administration, the drug disappears from plasma in a triphasic fashion. The first phase is a rapid distribution of the drug, followed by a second phase that reflects renal clearance. The third phase has a half-life of approximately 8–10 hours. If a patient is in renal failure, then toxic levels may occur and adversely affect the marrow and gastrointestinal tract.

Approximately 50% of methotrexate is bound to plasma proteins. Of the drug absorbed, about 90% is excreted unchanged in the urine within 50 hours. Methotrexate is not metabolized to a large extent in humans. In high-dose therapy, though, metabolites may accumulate and are nephrotoxic.[13] Methotrexate is retained in the form of polyglutamates for long periods, especially in the kidneys and liver.

Therapeutic Levels

The serum concentration of methotrexate is usually monitored at 24, 48, and 72 hours if necessary. The times at which the drug is monitored and the toxic concentrations may vary with different dosing regimens. Therapeutic levels vary, depending on whether the patient is on low-dose or high-dose therapy. Therapeutic monitoring of high-dose methotrexate is essential to guide the timing and amount of Leucovorin (Xanodyne Pharmacal, Florence, KY), a reduced derivative of folic acid, that a patient may receive to be "rescued" from the toxic effects of the drug.[14]

Toxicity

The primary toxic effects of methotrexate for a patient are to the bone marrow and intestinal epithelium. These patients may be at risk for spontaneous hemorrhage or life-threatening infections. Additional toxic effects include alopecia, dermatitis, intestinal pneumonitis, nephrotoxicity, defective oogenesis or spematogenesis, and teratogenesis.

The toxic concentrations of methotrexate at various sampling intervals are as follow:

24 hours	≥5.0 μmol/L
21 hours	≥0.5 μmol/L
72 hours	≥0.05 μmol/L

▶ ANTIPSYCHOTIC (PSYCHOTHERAPEUTIC) DRUGS

Lithium carbonate or citrate has been used for decades for the treatment of mania and recurrences of mania and depression in bipolar disorders. Close control of serum levels of lithium is important because of the narrow therapeutic range. Other psychotropic drugs used to treat mania include benzodiazepines, divalproex sodium and carbamazepine.

LITHIUM (ALSO ESKALITH AND LITHOBID)

The use of lithium carbonate in psychiatry for the treatment of mania in Australia dates back to 1949.[15] It was not until 1970 that physicians in the United States began using lithium for this purpose. The reason for the delayed use in the United States was the concern about the use of lithium chloride to treat various cardiac-related conditions that led to several deaths. The lithium-containing compounds used as mood-stabilizing agents in the United States include lithium carbonate and lithium citrate.[16] Alternatives to lithium salts for treatment of manic states are the anticonvulsants carbamazepine and valproic acid.[17]

Within the past two decades, lithium compounds have been widely used to treat manic-depressive states. A narrow therapeutic window made it necessary to monitor this drug in patients on lithium therapy. Lithium is the lightest of the alkali metals. It is a group IA metal with an atomic weight of 6.94 g. The salts of this monovalent cation share some characteristics with those of sodium and potassium. It can be assayed by atomic absorption and ion-selective electrodes.

Mechanism of Action

In normal individuals, therapeutic concentrations of lithium ion have practically no discernible psychotropic effects. Lithium is not a depressant, sedative, or euphoriant. The exact mechanism of action as a mood-stabilizing agent remains unknown.

Absorption, Metabolism, and Excretion

Lithium is absorbed almost completely from the gastrointestinal tract. Complete absorption occurs within 8 hours, with peak concentration in plasma achieved within 2-4 hours after an oral dose. Elimination occurs via the kidneys. About 95% of a single dose of lithium will be eliminated. The elimination half-life averages 20–24 hours.

Therapeutic Levels

Therapeutic ranges for lithium in blood are 0.5–1.5 mmol/L. If plasma samples are submitted to the laboratory, verify that the anticoagulant used does not contain lithium salts. Becton Dickinson (Franklin Lakes, NJ) makes several types of Vacutainer tubes that contain mixtures of inorganic salts and heparin. The lithium-containing Vacutainer tubes includes BD Vacutainer PST and BD Vacutainer Heparin tubes, which are spray coated with lithium heparin.

Toxicity

Toxicity is related to the serum concentration of lithium and its rate of rise following administration. Symptoms of acute toxicity include vomiting, profuse diarrhea, coarse tremor, ataxia, and convulsions. Blood levels greater than 1.5 mmol/L are considered toxic in patients.

▶ BRONCHODILATORS (ANTI-ASTHMATIC COMPOUNDS)

Asthma is described as a disease associated with airway inflammation, airway hyperactivity, and acute bronchoconstriction. There are several treatment modalities for asthma, including inhaled glucocorticoids, β-adrenergic receptor agonists, and leukotriene-modifying drugs. These compounds tend to relax the airways smooth muscles and promote bronchodilation, or they can inhibit airway inflammation.

THEOPHYLLINE (ALSO AMINOPHYLLINE AND UNIPHYL)

Theophylline is used for the treatment of asthma. In the United States, theophylline has been relegated to a third-line treatment. The reasons for this decreased use include the drug's narrow therapeutic window, the need to monitor its levels, and its modest benefit to patients. Theophylline is most often used in patients whose asthma may be difficult to control. Widely used theophylline preparations include Aminophylline (Watson Laboratories, Corona CA) and anhydrous theophylline, Uniphyl (Purdue Pharma, Stamford, CT.)

Theophylline is classified as a methylxanthine and is structurally similar to two other methylated xanthines, caffeine and theobromine. Xanthine itself is a dioxypurine and is structurally related to uric acid. The methylated xanthines are alkaloids found in plants widely distributed geographically. Structurally, caffeine is 1,3,7-trimethylxanthine; theophylline is 1,3-dimethylxanthine, and theobromine is 3,7-dimethylxanthine.

Mechanism of Action
Theophylline inhibits cyclic nucleotide phosphodiesterase enzymes. Phosphodiesterase enzymes catalyze the breakdown of cyclic AMP and cyclic GMP to 5'-AMP and 5'-GMP, respectively. Inhibition of phosphodiesterase enzymes will lead to an accumulation of cyclic AMP and cyclic GMP, thereby increasing the signal transduction through these pathways. Theophylline is also a competitive antagonist at adenosine receptors. Adenosine can cause bronchoconstriction in asthmatics.

In the lungs, theophylline effectively relaxes airway smooth muscle and is classified as a bronchodilator. The bronchodilating effects result from a combination of both adenosine receptor antagonism and phosphodiesterase enzymes inhibition.

Absorption, Metabolism, and Excretion
Preparations of theophylline administered in liquid or uncoated tablets are rapidly and completely absorbed. Theophylline is distributed into all body compartments and is bound to plasma proteins and at therapeutic concentrations; the protein binding is about 60%.[18]

Theophylline is eliminated primarily by metabolism in the liver. In premature infants, the rate of elimination is quite slow. Plasma half-lives may be as long as 36 hours, and extensive conversion of theophylline to caffeine occurs. This conversion is an important metabolic pathway in pre-term infants. In children, the half-life averages about 3.5 hours, and the adult half-life may be 8–9 hours.[19]

Toxicity
Toxic doses of theophylline may occur from rapid intravenous infusions of 500-mg doses of theophylline preparations such as Aminophylline or as a result of repeated administration by either oral or parenteral routes. Symptoms of toxicity include headache, palpitation, dizziness, nausea, hypotension, and precordial pain. In addition, restlessness, agitation, and emesis may occur at plasma concentrations of more than 20 μg/mL. Plasma concentrations greater than 20 μg/mL are considered toxic and should be reported to the clinician as soon as possible.

▶ CARDIOACTIVE DRUGS

Cardioactive drugs include compounds that have an effect on the heart. These drugs affect a wide range of cardiac functions, including electrical conduction, muscle contractility, cardiac pressures, and the flow of blood through the heart.

DIGOXIN (ALSO LANOXIN)
Digoxin is a cardiac glycoside containing a steroid nucleus with an unsaturated lactone at the C17 (carbon) position. Glycosidic residues are located at C3. The cardiac glycosides are used to treat heart failure. They have a positive inotropic effect on failing myocardium and efficacy in controlling the ventricular rate response to atrial fibrillation. They also modulate sympathetic nervous system activity.

Mechanism of Action
The influence of digoxin or digitalis glycoside on the myocardium involves both a direct action on cardiac muscle and the specialized conduction system and indirect actions on the cardiovascular system through the influence on the autonomic nervous system. The indirect action mediated by the autonomic nervous system involves a vagus nerve–like action that is responsible for the effects of digitalis on the sinoatrial and atrioventricular (AV) nodes. Another indirect action is sensitization of the baroreceptors, which results in increased carotid sinus nerve activity and increased sympathetic response to increases in mean arterial pressure. The pharmacological consequences of these direct and indirect effects are:

- an increase in the force and velocity of myocardial systolic contraction,
- a slowing of heart rate, and
- decreased conduction velocity through the AV node.

Absorption, Metabolism, and Excretion
Oral bioavailability for most digoxin tablets averages 70% to 80%. Liquid-filled capsules have a higher bioavailability than do tablets, so dosage must be adjusted if a patient is switched from one form to another. Parenteral digoxin is available for intravenous administration. Digoxin can be administered IM, but this route of administration is rarely justified because it frequently causes severe local irritation, and IV administration produces more rapid, predictable effects.

In patients with normal renal function, the elimination half-life of digoxin is 36–48 hours. Digoxin is excreted almost entirely unchanged, with a clearance rate that is proportional to the glomerular filtration rate. The principal tissue reservoir is skeletal muscle, not adipose tissue, so digoxin dosage should be based on lean or ideal body weight.

Spironolactone (Aldactone, Pfizer, New York) is a synthetic steroid that acts as a diuretic, and it can cause false-positive digoxin results with some immunoassays.[20] These misleading results may cause the clinician to administer insufficient therapy, or they may result in unnecessary alarms about toxicity. Negative interference has also been reported with immunoassays used to measure digoxin.[21] In this report, the drug digitoxin was administered to patients and the laboratory assayed blood samples for digoxin. The effects of a false-negative or falsely low result can be more serious than a false positive. Two things may happen as a result of negative interference: (1) Toxic concentration may remain unidentified, and (2) intoxication might occur if therapy is based on these misleading values.

 Checkpoint! 21–12

Identify several serious health conditions that may arise if a patient's digoxin level is 3.0 ng/mL (therapeutic range is 0.5–1.5 ng/mL).

LIDOCAINE (ALSO XYLOCAINE)

Lidocaine hydrochloride is an amide-type local anesthetic that is also used as a class IB anti-arrhythmic agent. Lidocaine hydrochloride is used parenterally for the acute treatment of ventricular arrhythmias that occur following myocardial infarction or during cardiac manipulative procedures such as cardiac surgery or cardiac catheterization. It was previously considered the drug of choice by the American Heart Association and other experts for advanced cardiovascular life support (ACLS) in the treatment of ventricular ectopy such as premature ventricular contractions and ventricular tachycardia associated with acute myocardial ischemia or infarction.

Lidocaine continues to be used as a first-line agent in the treatment of ventricular tachyarrhythmias and remains an acceptable choice in current ACLS guidelines for defibrillation–refractory ventricular fibrillation or pulseless ventricular tachycardia.

Mechanism of Action
The cardiac actions of lidocaine appear to be similar to those of phenytoin. Lidocaine is considered a class I (membrane-stabilizing) anti-arrhythmic agent. Lidocaine controls ventricular arrhythmias by suppressing automaticity in the His–Purkinje system and by suppressing spontaneous depolarization of the ventricles during diastole.

Absorption, Metabolism, and Excretion
Although lidocaine hydrochloride is absorbed from the GI tract, it passes into the hepatic portal circulation, and only about 35% of an oral dose reaches systemic circulation unchanged. Plasma lidocaine concentrations of approximately 1–5 μg/mL are required to suppress ventricular arrhythmias. Toxicity has been associated with plasma lidocaine concentrations greater than 5 μg/mL. Following IV administration of a bolus dose of 50–100 mg of lidocaine hydrochloride, the drug has an onset of action within 45 to 90 seconds and a duration of action of 10 to 20 minutes. Therapeutic plasma concentrations are achieved in 30 to 60 minutes after the start of a continuous infusion of 60–70 μg/kg per minute when no loading dose is given.

Binding of lidocaine to plasma proteins is variable and concentration dependent. The drug is approximately 60–80% bound to plasma proteins at concentrations of 1–4 μg/mL. Lidocaine is partially bound to alpha-1-acid glycoprotein (α-1-AGP),

and the extent of binding to α-1-AGP depends on the plasma concentration of the protein.

Approximately 90% of a parenteral dose of lidocaine is rapidly metabolized in the liver by de-ethylation to form monoethylglycinexylidide (MEGX) and glycinexylidide (GX) followed by cleavage of the amide bond to form xylidine and 4-hydroxyxylidine, which are excreted in urine. Less than 10% of a dose is excreted unchanged in urine. MEGX and GX are pharmacologically active and may also cause CNS toxicity in some patients.

PROCAINAMIDE (ALSO PROCAN AND PROCANBID)

Procainamide is an analog of procaine, a local anesthetic. Loading and maintenance intravenous infusions are used in the acute therapy of many supraventricular and ventricular arrhythmias. It may be given as a long-term oral treatment.

Mechanism of Action
Procainamide blocks open sodium channels, with an intermediate time constant of recovery from the blocking action. It prolongs cardiac action potentials by blocking outward potassium currents. Procainamide effects include decreased automaticity, increased refractory periods, and slow conduction.

A major metabolite of procainamide is N-acetyl procainamide. This metabolite lacks the sodium channel–blocking activity of the parent drugs but is equipotent in prolonging action potential.[22]

Absorption, Metabolism, and Excretion
Procainamide is rapidly eliminated by both renal excretion of unchanged drug as well as by hepatic metabolism. The half-life of procainamide is 3–4 hours. The major pathway for hepatic metabolism is conjugation by N-acetyl transferase to form N-acetyl procainamide. The elimination half-life of NAPA is 6–10 hours via renal excretion. The metabolite is not significantly converted back to procainamide.

Procainamide and NAPA are usually administered in slow-release formulation because of the rapid elimination half-lives. In patients with renal failure, both compounds can accumulate to potentially toxic plasma concentrations.[23] Therefore, in patients with compromised renal function, the dosage and frequency of drug administration should be reduced. It may also be necessary to monitor the plasma concentrations of both compounds in these patients. Because the parent drug and metabolite exert different pharmacological effects, the practice of using the sum of their concentrations to guide therapy may be inappropriate.

QUINIDINE (ALSO QUINAGLUTE AND QUINIDEX)

Quinine, an alkaloid obtained from the bark of the cinchona tree and the levorotatory isomer of quinidine, is an antimalarial agent. Quinidine is a class IA antiarrhythmic agent that exhibits antimalarial activity. It is used primarily as prophylactic therapy to maintain normal sinus rhythm after conversion of atrial fibrillation or flutter by other methods. The drug also is used to prevent the recurrence of paroxysmal atrial fibrillation, paroxysmal atrial tachycardia, paroxysmal AV junctional rhythm, paroxysmal ventricular tachycardia, and atrial or ventricular premature contractions.[24]

Mechanism of Action

In brief, quinidine blocks sodium ion (Na$^+$) channels and multiple cardiac potassium ion (K$^+$) currents. The blocking action on Na$^+$ and K$^+$ channels results in an increased threshold for excitability and decreased automaticity. Therefore, electrical conduction is slowed throughout the heart. The results of the blocking of K$^+$ and Na$^+$ channels are increased durations of the QRS and QT intervals on EKG strips.

By slowing electrical conduction in the heart muscle and prolonging the effective refractory period, quinidine can prevent arrhythmias, atrial flutter, atrial fibrillation, and paroxysmal supraventricular tachycardia.

Absorption, Metabolism, and Excretion

Quinidine salts are almost completely absorbed from the GI tract. The amount of drug that reaches the circulation after oral administration of quinidine depends on the amount of drug metabolized on the first pass through the liver. The polygalacturonate derivative is slowly dissociated in the GI tract and absorbed as free quinidine. In patients with congestive heart failure, the rate and extent of quinidine absorption is reduced, but these patients have higher plasma concentrations of quinidine because of a decreased volume of distribution.

Quinidine is rapidly distributed into all body tissues except the brain. Quinidine is concentrated in the heart, the liver, the kidneys, and skeletal muscle. Quinidine also is distributed into erythrocytes, where it is bound to hemoglobin.

Approximately 80% of quinidine is bound to plasma albumin at therapeutic plasma concentrations; this binding is decreased in the presence of hypoalbuminemia from various causes, including cirrhosis. Quinidine generally has a plasma half-life of 6 to 8 hours in healthy individuals, but the half-life may range from 3 to 16 hours or longer.

Quinidine is metabolized in the liver, principally via hydroxylation to 3-hydroxyquinidine and 2-quinidinone. Some metabolites have antiarrhythmic activity. Approximately 10–50% of a dose is excreted in urine by glomerular filtration as unchanged drug within 24 hours.

Toxicity

An overdose of quinidine has produced ataxia, respiratory depression or distress, apnea, vomiting, diarrhea, severe hypotension, syncope, and anuria. Atypical electrocardiogram changes may also occur. Other cardiac-related problems might occur such as ventricular arrhythmias, extrasystoles, heart block, and heart failure. Ultimately. coma and death may result.

▶ IMMUNOSUPPRESSIVE DRUGS

Immunosuppressive drugs are used to reduce the immune response in organ transplantation and autoimmune disease. The major classes of drugs used in transplantations are (1) calcineurin inhibitors, (2) glucocorticoids, and (3) antiproliferative and antimetabolic compounds. These drugs have been used successfully in treating conditions such as acute immune rejection, organ transplants, and severe autoimmune diseases.

CYCLOSPORINE A (ALSO NEORAL, SANDIMMUNE, AND SangCya)

Cyclosporine A (CSA) is an 11-amino acid cyclic peptide of fungal origin (*Beauveria nivea*) with strong immunosuppressive actions.[25] All of the amide nitrogens are either hydrogen bonded or methylated. A single D-amino acid is found at position 8 (D-alanine). Because cyclosporine is lipophilic and strongly hydrophobic, it must be made soluble for clinical administration.

Mechanism of Action

Cyclosporine inhibits the production of interleukin-2 (IL-2) by helper T cells thereby blocking T-cell activation and proliferation (amplification of immune response). It is effective in both prevention and treatment of ongoing acute rejection.[26] The current model for the mechanism of action of CSA suggests that, in the T-cell cytoplasm, CSA binds to a specific binding protein called *immunophilin* that is actually a cis-trans isomerase. The CSA-immunophilin complexes (cyclophilin) in turn bind to and block a phosphatase called *calcineurin*. The latter is required for the translocation of the nuclear factor of activated T cells (NFAT) from the cytosol to the nucleus, where it would normally bind to and activate enhancers or promoters of certain genes. In the presence of CSA, the cytosolic activation factor is unable to reach the nucleus, and the transcription of IL-2 (and other early activation factors) is strongly inhibited. As a result, T cells do not proliferate, secretion of gamma interferon is inhibited, and no major histocompatibility complex (MHC) also known as human leukocyte antigen class II antigens are induced, and no further activation of the macrophages occurs.[27]

Absorption, Metabolism, and Excretion

Absorption of oral CSA is incomplete, erratic, and dependent on food and bile, but less so with the microemulsion (Neoral), which has an absorption that may be 20–50% higher than the regular emulsion preparation (Sandimmune). CSA is 90% bound to proteins and is distributed extensively outside the vascular compartment. CSA is metabolized extensively by liver cytochrome P450–IIIA (CYP3A) system, and at least 25 metabolites have been identified in human bile, feces, blood, and urine.[28] CSA and metabolites are eliminated mainly through the bile into the feces, with only about 5–6% being excreted in the urine.[29] The half-life for CSA is about 12 hours (range 6–24h).

Therapeutic Levels

Several factors contribute to the determination of optimal whole-blood (specimen of choice) levels of cyclosporine. These include the complexity of the clinical state, individual differences in sensitivity, the immunosuppressive and nephrotoxic effects of cyclosporine, the coadministration of other immunosuppressants, the type of transplant, and the time post-transplant. Individual whole-blood cyclosporine values should not be used as the only indicators for making changes in the treatment regimen. Patients should be thoroughly evaluated clinically before treatment adjustments are made.

Most clinicians currently base monitoring on trough concentrations of the drug. It is important that the sampling time for a given patient is standardized and that consideration be given to the effect of once versus twice daily dosing of cyclosporine. Laboratory assays may measure unchanged cyclosporine or total drug (both cyclosporine and metabolites). Clinicians should be familiar with the forms of the drug that their laboratories are measuring. Because there are so many factors that affect blood levels of cyclosporine, most laboratories do not publish a therapeutic range. They will leave the interpretation of the result to the ordering physician.

Toxicity

The toxic effects of cyclosporine are varied. They include nephrotoxicity, hypertension, hyperkalemia, hypomagnesemia, and hyperuricemia. Symptoms such as headache, abdominal pain, tremors, leg cramps, and convulsions may arise.

 Checkpoint! 21–13

Identify the specimen of choice for measuring cyclosporine A.

TACROLIMUS (ALSO FK 506 AND PROGRAF)

Tacrolimus was discovered in Japan in 1984. It is a macrolide of fungal origin (produced by *Streptomyces tsukubaensis*).[30] It is a potent, selective anti–T cell immunosuppressive drug. It inhibits the production of IL-2 by helper T cells, thereby blocking activation and proliferation.[31] Tacrolimus is used to prevent the rejection of liver and renal allografts. The structure of tacrolimus is described as a hemiketal diketoamide incorporated into a 23-member ring. It is shown to be 10 to 100 times more potent than cyclosporine A.

Mechanism of Action

The current model for the mechanism of action of tacrolimus (or cyclosporine) suggests that, in the T-cell cytoplasm, tacrolimus binds to a specific binding protein called *immunophilin*, which is actually a cis-trans isomerase. The tacro–immunophilin complex in turn binds to and blocks a phosphatase called *calcineurin*. The latter is required for the NFAT from the cytosol to the nucleus, where it would normally bind to and activate enhancers or promoters of certain genes. In the presence of tacrolimus, the cytosolic activation factor is unable to reach the nucleus, and the transcription of IL-2 (and other early activation factors) is strongly inhibited. As a result, T cells do not proliferate, secretion of gamma interferon is inhibited, no MHC class II antigens are induced, and no further activation of the macrophages occurs.[27]

Absorption, Metabolism, and Excretion

Oral bioavailability is variable at about 30%. Absorption is reduced by food. The time to peak concentration is from 1 to 4 hours. Tacrolimus is poorly and incompletely absorbed in the GI tract because of its low water solubility. Protein binding is about 77%, mostly with α-1-glycoprotein. It is found primarily in the erythrocyte. Tacrolimus is metabolized by liver via the CYP3A system into at least eight metabolites.[27] Excretion is through the biliary circulation. The half-life is 4–40 hours, with an average being 12 hours.

Therapeutic Levels

The complexity of the clinical state and individual differences in sensitivity to the immunosuppressive effects of tacrolimus may cause different requirements for optimal whole-blood level. Patients should be thoroughly evaluated clinically before treatment adjustments are made. Because there are so many factors that affect blood levels of tacrolimus, most laboratories do not publish a reference range. They will leave the interpretation of the result to the ordering physician.

Toxicity

Side effects and signs of toxicity are similar to that of cyclosporine and include nephrotoxicity, hyperkalemia (or hypokalemia), and hypomagnesemia.[31] Symptoms such as headache, insomnia, abdominal pain, and weakness may occur.

▶ ANALYTIC TECHNIQUES FOR MEASURING THERAPEUTIC DRUGS

One of the earliest nonradioisotopic immunoassays used for therapeutic drug monitoring was EMIT (Syva Corp., Palo Alto, CA). Introduced in the early 1970s, this assay is a homogeneous immunoassay requiring no physical separation of bound antigen complexes from free or unbound antigens. The initial reagents were adaptable to a variety of photometric platforms and were subsequently adapted to highly automated instrumentation. This assay gained popularity in the laboratory because it provided shorter turnaround times without loss of sensitivity or specificity. Many more methodologies have since been developed to measure therapeutic drugs. A detailed description of many of these assays and accompanying instruments are described in other chapters in this text—for example, Chapter 2 (Instrumentation) and Chapter 5 (Immunoassays).

SUMMARY

The drugs discussed in this chapter represent the commonly measured compounds in clinical chemistry laboratories. The principle pharmacological effect or purpose of each drug is identified. Preanalytical points of interest are presented when appropriate (for example, whether peak and trough levels are necessary). Knowledge of drug toxicity is important and gives the analyst insight into the effects of toxic concentrations of drugs on the patient. Many times additional laboratory tests (e.g., renal function tests) need to be requested by the clinician to further evaluate the effects of a drug on a specific organ. This information has been presented in the text when appropriate.

Familiarization with the structure of each drug may help in questions related to the specificity of the assay. The addition of drug synonyms and trade names also helps the laboratorian communicate with nurses and physicians. Many times nurses or clinicians use names of drugs familiar to them when they question laboratory staff. An example is the drug valproic acid. The nursing staff may call the laboratory and ask whether it measures Depakene.

Therapeutic drug monitoring, as we know it today, has been practiced in nearly every clinical laboratory since the early 1970s. Several new drugs will be added to the test menu in these laboratories in the future. Knowledge about and understanding of the pharmacokinetics and pharmacodynamics of each drug allows the laboratorian and the clinician an opportunity to provide the patient with the best possible care.

REVIEW QUESTIONS

LEVEL I

1. Which of the following represents the four principal biological events governing drug disposition? (Objective 1)
   a. absorption, distribution, mechanism of action, and excretion
   b. absorption, distribution, biotransformation, and excretion
   c. absorption, distribution, body fat content, and biotransformation
   d. distribution, biotransformation, excretion, and pharmacodynamics

2. Which of the following does not influence the absorption of a drug? (Objective 2)
   a. cardiac output.
   b. tissue volume.
   c. availability of cytochrome P450 (CYP).
   d. pH.

3. Tylenol (McNeil Pharmaceuticals) is an example of which type of naming for drugs? (Objective 8)
   a. chemical
   b. generic
   c. trade
   d. nonproprietary

4. Which of the following drugs have antiepileptic effects and is structurally different from other commonly used antiepileptic drugs? (Objective 6)
   a. phenobarbital
   b. phenytoin
   c. ethosuximide
   d. topiramate

5. Most drugs are excreted in which form? (Objective 4)
   a. nonhydrated
   b. nonpolar
   c. polar
   d. as the parent compound

LEVEL II

1. First-pass effect occurs in which organ of the human body? (Objective 1)
   a. liver
   b. kidney
   c. lungs
   d. gallbladder

2. Phase I reactions that occur during biotransformation of a drug: (Objective 5)
   a. introduce or expose a functional group (e.g., hydroxyl) into the parent compound.
   b. lead to the formation of a covalent linkage between a functional group and the parent compound with glucuronic acid.
   c. involve an enzyme and substrate.
   d. result in the creation of an ionized form of the drug.

3. The major enzyme system that metabolizes a drug in phase I–type reactions is: (Objective 7)
   a. cytochrome Y450.
   b. cytochrome P650.
   c. apolipoprotein P450.
   d. cytochrome P450.

4. Which of the following statements best describes first-order drug metabolism? (Objective 6)
   a. The enzyme becomes saturated by a high free drug concentration and the rate of metabolism remains constant over time.
   b. The rate of drug metabolism is directly proportion to the concentration of free drug.
   c. Polar compounds are more efficiently excreted than nonpolar ones.
   d. The binding of a drug to plasma proteins limits its concentration in tissues and at its site of action.

REVIEW QUESTIONS (continued)

LEVEL I

6. Which of the following analytes would be useful to measure in a patient taking a nephrotoxic drug (e.g., aminoglycosides)? (Objective 7)
   a. creatinine
   b. cardiac troponin I
   c. aspartate aminotransferase
   d. amylase

7. Which of the following are examples of drugs derived from fungi? (Objective 11)
   a. digoxin and lidocaine
   b. phenobarbital and dilantin
   c. cyclosporine A
   d. sodium nitroprusside and lithium

LEVEL II

5. A drug agonist is described as: (Objective 10)
   a. a drug that binds to receptors without regulatory effect.
   b. a drug that binds to mitochondria without regulatory effect.
   c. a drug that binds to receptors and mimics the regulatory effects of the endogenous signaling compounds.
   d. a drug that binds to mitochondria and mimics the regulatory effects of the endogenous signaling compounds.

6. When should a trough level for drug measurement be drawn? (Objective 8)
   a. after the next dose
   b. just before the next dose
   c. when the drug has reached its optimal level
   d. 60 minutes after the IV infusion

7. Which of the following is considered to be the most predominant conjugation reaction during phase II–type metabolism? (Objective 5)
   a. hydroxylation
   b. sulfoxidation
   c. oxidation
   d. glucuronidation

8. When the concentration of a drug (digoxin, for example) is measured by the most routinely used immunoassay, which of the following is actually measured? (Objective 3)
   a. total drug concentration (i.e., free drug plus bound drug)
   b. only the free drug concentration.
   c. only the bound drug concentration.
   d. total drug concentration minus free drug concentration.

9. Evaluating renal function tests (creatinine and urea nitrogen, for example) should be done if a patient is on long-term aminoglycoside therapy because aminoglycosides are: (Objective 12)
   a. ototoxic.
   b. nephrotoxic.
   c. cardiotoxic.
   d. hepatotoxic.

10. Which of the following compounds, if taken along with digoxin, may interfere with certain immunoassay methodologies? (Objective 11)
    a. spironolactone
    b. procainamide
    c. phenobarbital
    d. acetaminophen

11. Toxic levels of methotrexate may result in: (Objective 12)
    a. hypotension.
    b. ototoxicity.
    c. spontaneous hemorrhage or life-threatening infection.
    d. bone disease marrow abnormalities.

REVIEW QUESTIONS *(continued)*

LEVEL I

LEVEL II

12. Sodium nitroprusside, a nitrovasodilator, metabolizes to: (Objective 13)
    a. nitrosoamine.
    b. ferricnitride.
    c. thiocyanate.
    d. sodium thiohydroxide.

13. Which of the following drugs is commonly used to treat lymphoblastic leukemia in children? (Objective 14)
    a. digoxin
    b. phenobarbital
    c. tacrolimus
    d. methotrexate

PEARSON myhealthprofessionskit™

Use this address to access the interactive Companion Website created for this textbook. Simply select "Clinical Laboratory Science" from the choice of disciplines. Find this book and log in using your user name and password.

REFERENCES

1. Brody TM. Introduction and definitions. In Brody TM, Larner J, Minnoman KP (Eds.), *Human pharmacology: Molecular to clinical*, 3rd ed. (New York: Mosby, 1998): 3–8.

2. Winter JC. Dose-effect relationships, interactions and therapeutic index. In Smith CM, Reynard AM (Eds.), *Textbook of pharmacology* (Philadelphia: W.B. Saunders, 2000): 9–15.

3. Buxton ILO. Pharmacokinetics and pharmacodynamics: The dynamics of drug absorption, distribution, action and elimination. In Brunton LL, Lazo JS, Parker KL (Eds.), *Goodman and Gilman's: The pharmacological basis of therapeutics*, 11th ed. (New York: McGraw Hill, 2006): 18–20.

4. *Physician's desk reference*, 61st ed. (Montvale, NY: Thomson PDR, 2008) (published annually).

5. Busse HJ, Wostmann C, Bakker EP. The bactericidal action of streptomycin: Membrane permeabilization caused by the insertion of mistranslated proteins into the cytoplasmic membrane of *Escherichia coli* and subsequent caging of the antibiotic inside the cells due to degrading of these proteins. *J Gen Microbiol* (1992) 138: 551–561.

6. Tran Ba Huy P, Meulemans A, Wassef M, et al. Gentamicin persistence in rat endolymph and perilymph after a two-day constant infusion. *Antimicrob Agent's Chemother* (1983) 23: 344–346.

7. Schentag JJ, Jusko WJ. Renal clearance and tissue accumulation of gentamicin. *Clin Pharmacol Ther* (1977) 22: 364–370.

8. Phillips NI, Fowler LJ. The effects of sodium valproate on gamma-aminobutyrate metabolism and behavior in naïve and ethanolamine-O-sulphate pretreated rats and mice. *Biochem Pharmacol* (1982) 31: 2257–2261.

9. Murad F. Cyclic guanosine monophosphate as a mediator of vasodilatation. *J Clin Invest* (1986) 78: 1–5.

10. Bates JN, Baker MT, Guerra R Jr, et al. Nitric oxide generation from nitroprusside by vascular tissue. Evidence that reduction of the nitroprusside anion and cyanide losses are required. *Biochem* (1991) 42: S157–S165.

11. Borsi JD, Moe PJ. Systemic clearance of methotrexate in the prognosis of acute lymphoblastic leukemia in children. *Cancer* (1987) 60: 3020–3024.

12. Hertz R. Folic acid antagonist: Effects on the cell and the patient. Clin staff conference at NIH. *Ann Intern Med* (1963) 59: 931–956.

13. Messmann R, Allegra CJ. Antifolates. In Chabner BA, Longo DL (Eds.), *Cancer chemotherapy and biotherapy: Principles and practice*, 3rd ed. (Philadelphia: Lippincott Williams, and Wilkins, 2001).

14. Ackland SP, Schilsky RL. High-dose methotrexate: A critical reappraisal. *J Clin Oncol* (1987) 5: 2017–2031.

15. Cade JF. Lithium salts in the treatment of psychotic excitement. *Med J Australia* (1949) 2: 349–352.

16. Davis JM, Janicak PG, Hogan DM. Mood stabilizers in the prevention of recurrent affective disorders: a meta analysis. *Acta Psychiatric Scand* (1999) 100: 406–417.

17. Post RM. Psychopharmacology of mood-stabilizers. In Buckley PF, Waddington JL (Eds.), *Schizophrenia and mood disorders: The new drug therapies in clinical practice* (Boston: Butterworth-Heinemann, 2000): 127–154.

18. Hendeles L, Weinberger M. Improved efficacy and safety of theophylline in the control of airway hyperreactivity. *Pharmacol Ther* (1982) 18: 91–105.

19. Roberts RJ. *Drug therapy in infants: Pharmacological principles and clinical experience* (Philadelphia: W.B. Saunders, 1984).

20. Pleasants RA, Williams DM, Porter RS, et al. Reassessment of cross-reactivity of spironolactone metabolites with four digoxin immunoassays. *Ther Drug Monit* (1989) 11: 200–204.

21. Datta P, Dasgupta A. Bi-directional (positive/negative) interference in a digoxin immunoassay: Importance of antibody specificity. *Ther Drug Monit* (1998) 20: 352–357.

22. Dangman KH, Hoffman BF. In vivo and in vitro antiarrhythmic and arrhythmogenic effects of N-acetyl procainamide. *J Pharmacol Exp Ther* (1981) 217: 851–862.

23. Drayer DE, Lowenthal DT, Woosley RL, et al. Accumulation of N-acetylprocainamide, an active metabolite of procainamide, in patients with impaired renal function. *Clin Pharmacol Ther* (1977) 22: 63–69.

24. Grace AA, Camm J. Quinidine. *N Engl J Med* (1998) 338: 35–45.

25. Borel JF, Feurer C, Gubler HU, et al. Biological effects of cyclosporine A: A new antilymphocytic agent. *Agns Action* (1976) 6: 468–475.

26. Kahan BD. Cyclosporine. *N Engl J Med* (1989) 32: 1725–1738.

27. Schreiber SL, Crabtree GR. The mechanism of actions of cyclosporine A and FK506. *Immunol Today* (1992) 13: 136–142.

28. Fahr A. Cyclosporine clinical pharmacokinetics. *Clin Pharmacokin* (1993) 24: 472–495.

29. Christians U, Sewing KF. Cyclosporine metabolism in transplant patients. *Pharmacol Ther* (1993) 57: 291–345.

30. Goto T, Kino T, Hatanaka H, et al. Discovery of FK-506, a novel immunosuppressant isolated from *Streptomyces tsukubaenisis*. *Transplant Proc* (1987) 19: 4–8.

31. Plosker GL, Foster RH. Tacrolimus: A further update of its pharmacology and therapeutic use in the management of organ transplantations. *Drugs* (2000) 59: 323–389.

22

Toxicology

■ OBJECTIVES—LEVEL I

Following successful completion of this chapter, the learner will be able to:

1. List several examples of toxic substances measured in clinical laboratories.
2. Identify methods used to measure selected toxic substances.
3. Identify types of instrumentation used to measure toxic substances in clinical laboratories.
4. List six substances that are frequently used as adulterants in urine specimens for drug abuse testing.
5. List several examples of classes of drugs that are included in urine-drugs-of-abuse screening procedures.
6. Identify several sources of lead that may result in high blood levels of lead.
7. Identify the acidic or ketone metabolites of the following compounds:
   a. Ethanol
   b. Methanol
   c. Ethylene glycol
   d. Isopropyl alcohol
   e. Salicylate

■ OBJECTIVES—LEVEL II

Following successful completion of this chapter, the learner will be able to:

1. Distinguish between drug screening and drug confirmatory methods.
2. Explain the toxic effects of selected compounds.
3. Explain the mechanism of action of selected toxic substances.
4. Describe the chemistry of selected toxic substances.
5. Explain the physiochemical impact on toxic levels of selected drugs.
6. Discuss the physiochemical relationship of lead and free erythrocyte protoporphyrin.
7. State the toxic levels of selected drugs and toxic substances in biological fluids.
8. Explain the clinical significance of measuring selected drugs of abuse and toxic substances.

KEY TERMS

Agonist
Analgesic
Antagonist
Antidepressant
Anxiolytic
Carboxyhemoglobin
Chain of custody

Drug abuse screen
Drug confirmation
Hepatotoxicity
Hypnotics
Opiates
Opioid
Poisoning

Salicylism
Sedative
Tolerance
Toxicant
Toxicology

A CASE IN POINT

A 54-year-old male was admitted to the emergency department (ED). The patient was conscious but his level of consciousness (LOC) was diminished. He was neither very alert nor coherent in responding to verbal questioning. He was responsive to pain stimulus. He did not have "alcohol breath" or a "fruity odor" upon exhalation. The patient admitted that his vision was slightly blurred and that he was seeing double (diplopia). Slight nystagmus was evident. Patient experienced multiple episodes of emesis. Additional symptoms included cephalalgia, slurred speed, and unsteady gait. Lips and fingernails were bluish in color.

Vital signs were as follows:

Pulse = 110 beats per minute	normal: ~80
Blood pressure = 100/74 mmHg	normal: 120/80
Respirations = 28 per minute	normal: 12–16

Results of the initial laboratory tests were as follows:

Serum Chemistries	Results (conventional units)	Reference Interval (conventional units)
Sodium	135 mEq/L	136–145
Potassium	4.5 mEq/L	3.5–5.1
Chloride	108 mEq/L	98–107
Carbon dioxide	7 mEq/L	23–29
Anion gap	20 mEq/L	6–10
Glucose	162 mg/dL	74–100
Creatinine	1.5 mg/dL	0.9–1.3
Urea nitrogen	17 mg/dL	6–20
Calculated osmolal	275 mOsm/kg	282–300
Ethanol	<10 mg/dL	N/A
Acetaminophen	<2.5 μg/mL	N/A
Salicylate	<2.8 μg/mL	N/A

Hematology Results

WBC	$11 \times 10^3/\mu m^3$	$4.0–11.0 \times 10^3$
RBC	$4.3 \times 10^6/\mu m^3$	$4.70–6.10 \times 10^6/\mu m^3$
Hemoglobin	11 g/dL	13.0–18.0
Hematocrit	34.9%	39–50
Platelets	$110 \times 10^3/\mu m^3$	140–440

Urinalysis

Color: amber		
Appearance: hazy		clear
pH = 6		5–6
Specific gravity = 1.028		1.002–1.030

All other dipstick results are negative

Microscopic analysis:

Birefringent octahedral, envelope-shape calcium oxalate crystals

Note: N/A = "not applicable."

Issues and Questions to Consider

1. What course of action should the clinician pursue?
2. What is the possible origin of the calcium oxalate crystals in the urine?
3. What additional laboratory tests should be considered?

The ED physician exposed a urine specimen provided by the patient to a Wood's lamp and the specimen emitted a yellow-green color (i.e., it glowed). The physician suspected the presence of a chemical substance in the urine that may be ethylene glycol. Additional laboratory tests were requested and included the following:

Serum Chemistries	Results (conventional units)	RI or Cutoff Values (conventional units)
Serum osmolality (Using freezing-point depression osmometry)	372 mOsm/kg	275–295 mOsm/kg
Osmol gap	97 mOsm/kg	5–10 mOsm/kg
Troponin I	<0.05 μg/L	<0.05 μg/L
Volatiles:		
Ethylene glycol	190 mg/dL	neg
Methanol	<1.5 mg/L	<1.5 mg/L
Isopropanol	none detected	none detected

Drug Abuse Urine (DAU) screen: negative for seven classes of abused drugs

Issues and Questions to Consider

4. Explain the cause of the increased osmolal gap.
5. Why did the urine emit a yellow-green color when irradiated with ultraviolet light using a Wood's lamp?
6. Is the ethylene glycol concentration representative of a toxic dose?
7. Why is ethylene glycol toxic to the human body?
8. What is the treatment for ethylene glycol ingestion?

WHAT'S AHEAD

This chapter will focus on analytes measured on a routine basis in many clinical laboratories that are permitted to perform toxicology testing procedures. These procedures are designed to screen for toxicological substances and require confirmation of all positive and questionable results. Laboratories that perform toxicology testing may also provide quantitative test procedures (e.g., ethanol, salicylate, and acetaminophen). Following this discussion, the chapter discusses the analytes that are measured in larger laboratories (e.g., reference labs that have methods and analyzers to measure more esoteric analytes, including ethylene glycol, methanol, and toxic metals). Testing modalities include confirmatory procedures that require more sophisticated instrumentation than commonly found in clinical laboratories, including gas chromatography—mass spectrometry (GCMS), liquid chromatrography—mass spectrometry (LCMS), and liquid chromatography—mass spectrometry—mass spectrometry (tandem) (LC-MSMS). The pharmacology, toxic effects, and analytical considerations of the drugs presented will be discussed.

Toxicology and all the terms associated with it have been defined and described in various ways as far back as the 16th century. Paracelsus (1453–1541), who is known as the father of modern toxicology because he was the first to explain the dose–response relationship of toxic substances, stated that "all things are poison and not without poison, only the dose makes a thing not a poison."

▶ INTRODUCTION

For the purpose of this textbook, the Greek derivation of **toxicology** (*toxicos logos*) is defined as "the science of poisons, including their source, chemical composition, action, tests, and antidotes." This definition embodies the scope of medical and clinical toxicology. A **poison** is defined by *Webster's Dictionary* as "any agent which, when introduced into the animal organism, is capable of producing morbid, noxious, or deadly effects upon it." A **toxicant** "is an agent or a substance that acts as a poison." Finally, **poisoning** is defined as the "damaging physiological effects of ingestion, inhalation, or other exposure to a range of pharmaceuticals, illicit drugs, and chemicals, including pesticides, heavy metals, gases/vapors and common household substances, such as bleach and ammonia."[1]

Medical toxicology is a term used to describe a broad, diverse discipline in medicine that encompasses clinical, regulatory, and research components. The clinical component focuses on the medical aspects of direct patient care resulting from admission to hospital-based facilities, usually an emergency department. This is when the clinical laboratory has an opportunity to offer its laboratory services either internally or via external reference laboratories. Here is where technologists, under the appropriate laboratory toxicology license or permit, are able to use their knowledge and experience to perform, interpret, and release laboratory results to the patient's caregivers.

✪ TABLE 22-1

Partial listing of the total number of human exposures reported by the AAPCC

Year	No. of Participating Centers	Human Exposures Reported	Exposures per Thousand Population
1983	16	251,012	5.8
1984	47	730,224	7.3
⋮	⋮	⋮	⋮
1990	72	1,713,462	8.9
⋮	⋮	⋮	⋮
2000	63	2,168,248	8.0
⋮	⋮	⋮	⋮
2004	62	2,438,644	8.3
2005	61	2,424,180	8.3
2006	61	2,403,539	8.0
2007	61	2,482,041	8.1

The need for toxic substance testing continues to grow, according to data published by the *American Association of Poison/Control Centers* (AAPCC) with data compiled using the *Toxic Exposure Surveillance System* (TESS).[2] The number of human exposures to substances reported to TESS by over 60 poison control centers in the United States has increased steadily up to 2004 and showed a slight decrease for 2005 and 2006. There was a significant increase in the number of human exposures reported for the year 2007. These data are shown in Table 22-1 ✪.[3] The number of fatalities reported for the over two million exposures was 1597. Examples of substances most frequently involved in human exposures are listed in Table 22-2 ✪.[4] Exposure to drugs classified as **analgesics** (drugs that relieve pain), such as aspirin and acetaminophen, and also sedative/hypnotics/antipsychotic comprise nearly 21% of the total number of exposures. Examples of specific substances that may result in clinical toxicological testing are given in Table 22-3 ✪.[5] Compounds, including

✪ TABLE 22-2

Partial listing of substances most frequently involved in human exposures based on data from AAPCC

Substance	Number	Percent per Total Number of Exposures
Analgesic	309,431	12.5
Sedatives/hypnotics/antipsychotics	154,602	6.2
Cold and cough preparations	111,222	4.5
Antidepressants	98,898	4.0
Alcohols	82,432	3.3
Stimulants and street drugs	46,143	1.9
Anticonvulsants	43,080	1.7

TABLE 22-3

Examples of toxic substances and their sources that may result in laboratory test requests from hospital emergency departments

Toxins	Source(s) (includes substance by itself or included in mixtures)	No. of Reportable Exposures
Alcohols and glycols:		
Isopropanol	Rubbing alcohol, wall/floor tiles, cleaners, glass cleaners	~10,500
Ethylene glycol	Antifreeze, automotive, boat, aircraft	~600
Ethanol	Ethanol, rubbing alcohol, wall/floor/tiles, cleaners mouthwash	~63,500
Methanol	Methanol, automotive products	~2100
Fumes/gases/vapors		
Carbon monoxide	Car exhaust, fires, faulty furnaces, hot water heaters	~15,500
Heavy metals		
Lead	Batteries, paints	~3000
Insecticides/ pesticides/herbicides		
Organophosphates/ carbamates	Insect control, fungus, algaecide products	~9000

Information provided by AAPCC.

ethanol, carbon monoxide, and hemoglobin, are measured routinely in clinical laboratories whereas ethylene glycol and methanol test requests are often sent to reference laboratories.

Clinical laboratories offering toxicology testing at any level have experienced an increased volume of testing. Data such as those shown in the aforementioned tables support this trend. The demands and pressures brought on the clinical laboratory by hospital emergency room departments and by hospital mental health facilities have caused clinical laboratories to provide increased toxicology test menus and reduced turnaround times. Many clinical laboratories have included confirmation procedures for their screening tests and include the use of point-of-care testing (POCT) (e.g., biosensors) drug testing devices.

▶ REQUIREMENTS OF TOXICOLOGY TESTING

Many states require an additional permit to perform toxicology testing. These permits allow the laboratory to perform toxicology testing for specific medical purposes. For example, a forensic permit requires the laboratory to provide **chain of custody** from the time a specimen is drawn to the time the results are released to a clinician. The results of the laboratory test(s) are often used in medico-legal cases. An additional requirement of a laboratory that tests for toxic substances with

a forensic permit is to provide **drug confirmation** of all positive screening tests. The confirmation method must have the attributes of a definitive or reference method such as GC-MS or LC-MS.

The toxicology permits most often used in hospital-based laboratories include only qualitative testing on emergency department patients and do not include confirmation procedures for positive screening results. Larger hospital laboratories in urban areas may obtain a permit that allows them to perform both qualitative **abused drug** screening and provide confirmatory methods for all positive results. **Drug abuse screens** are tests that qualitatively identify the presence of one or more drugs or classes of drugs.

▶ GROUP I ANALYTES

The following compounds are commonly measured in clinical laboratories and will be presented first. This group of analytes will be followed by a discussion of toxic substances that are not routinely offered in most clinical laboratories.

ACETAMINOPHEN

Acetaminophen (also paracetamol) is a nonsteroidal anti-inflammatory drug (NSAID) that has three major effects:

- Anti-inflammatory effects (i.e., modification of the inflammatory reaction)
- Analgesic effects (i.e., reduction of certain types of pain)
- Antipyretic effects (i.e., lowering of raised body temperature)

The mechanisms of action for NSAIDs can be summarized as follows:[6]

- Anti-inflammatory effects are a result of the decrease in vasodilator prostaglandins (e.g., prostaglandin E2 [PGE2] and prostacyclin), which leads to less vasodilatation and, indirectly, less edema. However, the response by inflammatory cells is not reduced.
- Pain reduction (i.e., analgesic) effects result from a decrease in prostaglandin production and thus less sensitization of nociceptive nerve endings (which are nonmyelinated C fibers with low conduction velocities) to inflammatory mediators such as bradykinin and 5-hydroxytryptamine. Relief of headaches is most likely a result of decreased prostaglandin-mediated vasodilatation.
- Antipyretic effects result in part from a decrease in mediator prostaglandins (which are generated in response to the inflammatory pyrogen interleukin-1), which are responsible for elevating the hypothalamic set point for temperature control, thus causing fever.

Acetaminophen is given orally and is nearly completely absorbed. Peak plasma concentrations are attained in 30 to 60 minutes, and the drug is inactivated in the liver by conjugation to glucuronic acid or sulphate. The plasma half-life of acetaminophen with therapeutic doses is 2 to 4 hours, but with toxic doses it may be increased to between 4 and 8 hours.

Side effects with therapeutic doses are few, though allergic skin reactions may occur. Regular usage of acetaminophen in large doses over a long period of time may increase the risk of kidney damage.

Toxic doses, described as two to three times the maximum therapeutic dose, cause a serious and potentially fatal **hepatotoxicity**. Hepatotoxicity results from the saturation of enzymes catalyzing the normal conjugation reaction and results in the drug being metabolized by mixed function oxidases. This reaction produces a toxic metabolite, N-acetyl-p-benzoquinoneimine, which is normally inactivated by conjugation with glutathione. If glutathione is depleted, the toxic intermediate accumulates and reacts with the nucleophilic constituents in the cell. This causes necrosis in the liver and also in the kidney tubules.

The initial symptoms of acetaminophen toxicity are nausea and vomiting. Hepatotoxicity is a delayed manifestation that occurs 24–48 hours later. Treatment begins with gastric lavage followed by oral administration of activated charcoal. Liver damage can be reduced by giving an agent (e.g., acetylcysteine) intravenously (IV) or administering methionine orally, both of which will increase glutathione formation in the liver.

Measurement of serum acetaminophen concentration is important in assessing the severity of intoxication. A nomogram, shown in Figure 22-1 ■, that relates serum acetaminophen concentration and hours after ingestion is often used to estimate the probability of hepatic toxicity.[7] Interpretation using this nomogram is incumbent on the following qualifications:

- Blood samples should not be obtained earlier than 4 hours after ingestion to ensure that absorption is complete.

- The nomogram applies only to acute and not chronic ingestion.

- The nomogram is not useful if the time of ingestion is unknown or is considered unreliable.

- Serial determination of serum levels is warranted if extended-release medication has been ingested.

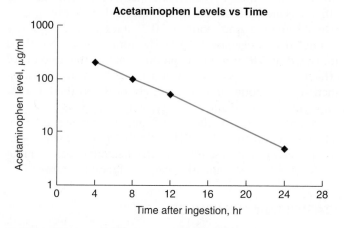

■ **FIGURE 22-1** Rumack and Matthew plot to determine clinical significance of blood levels of acetaminophen.

Several photometric assays are available to measure acetaminophen. One significant factor is whether the photometric assay measures only parent compound or parent compound plus metabolite(s). Select the assay that detects only parent compound for best results.

There is an enzymatic assay that uses arylacylamide amidohydrolase and that catalyzes hydrolysis of acetaminophen (but not conjugate) to p-aminophenol and acetate. This method has been adopted by several instrument manufacturers and provides the laboratory with a fast, reliable, and specific assay to quantitate acetaminophen in serum or plasma.

Chromatographic assays, including gas and liquid chromatography, coupled with a mass spectrometer are very accurate and are considered to be reference methods. These assays are not widely used by clinical laboratories due to the cost, labor, and skill level required.

Several different immunoassays are available, including enzyme multiplied immunoassay technique EMIT, fluorescent polarization immunsoassay (FPIA), and particle enhanced turbidimetric inhibition immunoassay (PETINIA), for measuring acetaminophen in blood. These assays have the advantage of providing reliable, specific, and sensitive measurements on automated analyzers.

✓ **Checkpoint! 22–1**

What is the mechanism for acetaminophen-induced hepatotoxicity?

SALICYLATE

Salicylate (also called aspirin, and acetylsalicylic acid) is a NSAID and is the most commonly consumed drug in the world. It is reputed to be the consummate anti-inflammatory drug, but within the past several years acetylsalicylic acid has been recommended for use in an increasing number of other conditions, including the following:

- Cardiovascular disorders
- Colonic and rectal cancer
- Alzheimer's disease
- Radiation-induced diarrhea

Acetylsalicylic acid is a weak acid chemically derived by the reaction of salicylate with a strong acid such as sulfuric acid followed by the addition acetic anhydride. Acetylsalicylic acid is largely un-ionized in the acid environment of the stomach, and thus its absorption is increased. Most absorption occurs in the ileum because of the extensive surface area of the microvillus. Acetylsalicylic acid is hydrolyzed by esterases in the plasma and the tissues—particularly in the liver—yielding salicylate.

Nearly 25% of the salicylate is oxidized; some is conjugated to form glucuronide sulphate before excretion and about 25% is excreted unchanged. The plasma half-life of aspirin will depend on the dose, but the duration of action in not directly related to the plasma half-life because of the irreversible nature of the action of the drug.

Unwanted effects of salicylate are associated with several disorders. **Salicylism** can occur with repeated ingestion of fairly large doses of salicylate. It is a syndrome associated with tinnitus (a high-pitched buzzing noise in the ears), vertigo, decreased hearing, and occasionally nausea and vomiting.

There is an association between aspirin intake and *Reye's syndrome*. This syndrome is a rare disorder in children. Reye's syndrome usually follows an acute viral illness and involves both the liver and central nervous system (CNS). An encephalopathy may develop, and the syndrome has a 20–40% mortality outcome. It is not entirely clear to what extent aspirin is implicated in its causation, but it is suggested that the drug not be given to children.[6]

Salicylate poisoning, as a result of accidental ingestion or knowingly ingesting large amounts of the drug to commit suicide, causes several metabolic changes in the body. Most notably are changes in the body's acid–base balance and electrolyte status. The sequence of events resulting from high-dose salicylate ingestion begins with salicylate uncoupling oxidative phosphorylation mainly in skeletal muscle, leading to increased oxygen consumption and thus increased production of carbon dioxide. Respiration is stimulated by the change in carbon dioxide levels and the direct action on the respiratory center. Hyperventilation results, which leads to respiratory alkalosis; this alkalosis is normally compensated by renal mechanisms involving increased bicarbonate excretion. Larger doses (>200 μg/mL) of salicylate can cause depression of the respirator center, which leads to retention of carbon dioxide and thereby increases plasma carbon dioxide. Because this is concurrent with a reduction in plasma bicarbonate, an uncompensated respiratory acidosis will occur. A metabolic acidosis can result from the accumulation of metabolites, including pyruvic, lactic, and acetoacetic acid as well as the acid load associated with the salicylate itself.

Hyperpyrexia (excessive elevation of body temperature >107.6°C, also called high fever) is often present due to the increased metabolic rate, and dehydration may follow from repeated vomiting.

Toxic doses of salicylate result in disturbance of hemostasis due to the action of salicylate on platelet aggregation. The effect of toxic doses on the CNS is, initially, stimulation with excitement, but eventually coma and respiratory depression develop. A potentially hazardous situation may arise in a patient on warfarin therapy who takes aspirin. Aspirin causes an increase in the effect of warfarin, partly by displacing it from plasma proteins and partly because of its effect on platelets, which results in an interference with hemostatic mechanisms.

Several methods based on the work of Trinder are available for measuring blood salicylate levels.[8] In these methods, salicylate reacts with Fe^{3+} to form a colored complex that is measured at 540 nm. The assay is affected by interference due to endogenous background, salicylate metabolites; endogenous compounds; structurally related drugs such as diflunisal (difluorophenyl salicylate); and azide present in commercially prepared control material. A comparative study involving 115 patients' specimens was conducted using photometric assays, including the Trinder reaction, and produced results that agree very closely with those of a reference high-performance liquid chromatography (HPLC) procedure.[9]

There are other methods used for salicylate quantitation, including fluorescent polarization immunoassay and an enzymatic assay using salicylate hydroxylase-mediated photometric procedures. This enzyme method has been adapted to automated chemistry analyzers currently in use.[10]

✓ Checkpoint! 22–2

Explain the mechanism associated with the metabolic acidosis that develops and identify three acids that accumulate in the blood of a patient who has overdosed on salicylate.

ETHANOL

The main effects of ethanol are on the CNS, where its depressant action resembles those of the volatile anesthetic such as ether. Ethanol is purely a depressant at the cellular level, though it increases impulse activity in some parts of the CNS. The main theories of ethanol action are as follows:[11]

- Enhancement of gamma-aminobutyric acid (GABA)–mediated inhibition, similar to the action of benzodiazepines (see below)
- Inhibition of Ca^{2+} energy through voltage-gated calcium channels
- Inhibition of *N*-methyl-D-aspartic acid (NMDA) receptor function

The effects of acute ethanol intoxication in humans are well known and include slurred speech, uncoordinated body movements, increased self-confidence, and euphoria. Intellectual performance and sensory discrimination show increased impairment by ethanol. Ethanol has deleterious effects on other body tissues and systems, as summarized in Box 22-1 ■.[12]

In many clinical laboratories, blood ethanol concentrations are measured by enzymatic analysis. In this method, ethanol is measured by oxidation to acetaldehyde with nicotinamide adenine dinucleotide (NAD^+), a reaction catalyzed by alcohol dehydrogenase (ADH). The formation of NADH is measured at 340 nm and is proportional to the amount of ethanol in the specimen. Interference from isopropanol, methanol, acetone, and ethylene glycol is less than 1% for most currently used automated systems using ADH.

Serum or plasma is the most common specimen used for ethanol determination by ADH methods. Urine and saliva samples produce reliable results using ADH methods but are not routinely submitted as a sample type to the clinical laboratory.

CARBON MONOXIDE

Carbon monoxide (CO) is a colorless, odorless, and tasteless gas produced as a byproduct of incomplete combustion of

BOX 22-1 The far-reaching effects of ethanol on the human body

▶ The cardiovascular effects of ethanol produce cutaneous vasodilatation, central in origin, which causes a warm feeling but actually increases heat loss.

▶ Salivary and gastric secretion is increased by ethanol consumption. Heavy consumption causes damage directly to the gastric mucosa, resulting in chronic gastritis, which may lead to GI bleeding.

▶ Ethanol increases the output of adrenal steroid hormones by stimulating the anterior pituitary gland to secrete adrenocorticotrophic hormone (ACTH).

▶ Antidiuretic hormone (ADH) secretion is inhibited by ethanol, and patients develop diuresis.

▶ Oxytocin secretion is also inhibited, resulting in delayed parturition at term.

▶ Chronic male alcoholics are often impotent and show signs of feminization, which is associated with impaired testicular steroid synthesis.

▶ Liver damage is one of the most serious long-term consequences of excessive ethanol consumption.

▶ Moderate drinking reduces mortality associated with coronary heart disease; this may be due to its effect on lipoproteins (e.g., raises blood levels of HDL-C).

▶ Ethanol may also protect against ischemic heart disease by inhibiting platelet aggregation.

▶ Excessive alcohol intake during pregnancy is associated with fetal alcohol syndrome (FAS), which results in a variety of abnormal features exhibited by neonates.

▶ Ethanol is rapidly absorbed, principally from the stomach, and most of it is cleared by first-pass hepatic metabolism. Ethanol is quickly distributed throughout the body water, where the rate of its redistribution depends mainly on the blood flow to individual tissues. About 90% of ethanol is metabolized and 5–10% is excreted unchanged in expired air and in urine. Metabolism of ethanol in the liver involves successive oxidations, first to acetaldehyde by the enzyme alcohol dehydrogenase and then to acetic acid using aldehyde dehydrogenase. Refer to Figure 22-2. The intermediate metabolite acetaldehyde is a reactive and toxic compound, and this may contribute to the hepatotoxicity.

carbon-containing substances. Exogenous sources and approximate numbers of exposures reported by TESS are shown in Table 22-3.[5]

The lungs rapidly absorb CO, which combines with hemoglobin at 200 to 240 times greater affinity than does oxygen. Small amounts of CO are metabolized by oxidation to carbon dioxide. Approximately 85% of absorbed CO combines with hemoglobin, and the remainder is attached to myoglobin and blood proteins. Carbon monoxide is eliminated primarily through the lungs. In acute CO poisoning, the half-life of **carboxyhemoglobin (COHb)** is from about 26 to 146 minutes.[13] The term *carboxyhemoglobin* is synonymous with *carbon monoxide hemoglobin*.

Carbon monoxide toxicity arises from impaired oxygen delivery and use and leads to cellular hypoxia, dysfunction, and death. The brain and heart are particularly susceptible to CO toxicity because they contain vessels with high metabolic activity. CO replaces oxygen on the hemoglobin molecule, leading to relatively functional anemia. The human body requires a consistent amount of oxygen per unit of blood, and impairment of oxyhemoglobin formation by CO can result in cellular hypoxia.

When CO enters tissues such as blood, it replaces oxygen on the hemoglobin molecule. Thus the presence of CO impairs and decreases oxyhemaglobin formation and results in cellular hypoxia. An elevated carboxyhemoglobin level impairs the release of oxygen from hemoglobin by increasing the affinity of hemoglobin from oxygen binding sites. The results of this CO substitution are an allosteric modification of the hemoglobin molecule and impaired oxygen release. The effect on the oxygen dissociation curve is a shift to the left, which decreases the release of oxygen in the tissue.

Other effects of CO on body tissues include the following:

• CO binds to cytochrome oxidase, and if there is reduced oxygen delivery to the brain, then the effects of CO may be enhanced.

• Myocardial myoglobin can be saturated with CO three times higher than skeletal muscle, thereby diminishing the normal functions of myoglobin.

Overall effects on the cardiovascular system are myocardial depression and hypotension, both of which may cause ischemia and enhanced tissue hypoxia induced by impaired oxygen delivery.

Carbon monoxide can cause oxidative stress, including brain lipid peroxidation via activation and release of oxygen radicals by neutrophils. Also, peroxynitrate is formed via interaction of CO with nitric oxide (NO) released by platelets; the release of peroxynitrate can facilitate endovascular oxidative stress following CO exposure.

Typically, emergency room departments are presented with patients suffering from carbon monoxide exposure due to the following:

• Accidental poisoning via faulty furnaces or gas stoves

• Remaining in an automobile that is running for a prolonged period of time with all of the windows and doors closed

• Smoke inhalation due to a fire

• Attempted suicide via gas stoves or automobiles

In normal, healthy individuals in an environment relatively free of carbon monoxide, their percent carboxyhemoglobin levels range from nearly zero to three. Individuals who smoke may have levels between 5 and 10%. These individuals are often short of breath with moderate exertion. Critical levels of carboxyhemoglobin occur at greater than 60%.[14]

Treatment for exposure to CO usually begins with first aid care by emergency medical personnel and includes airway management, ventilation, and administration of 100% oxygen. Once the patient arrives at a hospital, hyperbaric oxygen therapy may be administered if necessary. Hyperbaric oxygen increases the amount of dissolved oxygen in plasma and enhances the elimination of carboxyhemoglobin. This treatment modality is usually reserved for acute CO poisoning, in which carboxyhemoglobin levels exceed 20%.

Diagnostic tests are useful for confirming CO intoxication and assessing the degree of end organ injury. Carboxyhemoglobin levels are usually determined by cooximetry using a stand-alone cooximeter or an oximeter incorporated into a blood-gas analyzer. A cooximeter uses the principles of absorption spectroscopy to measure the various hemoglobin compounds found in blood (e.g., oxyhemoglobin, carboxyhemoglobin, reduced hemoglobin, methemoglobin, and sulfhemoglobin). The specimen, usually whole blood, is hemolyzed in the oximeter and all of the aforementioned hemoglobin compounds are released if present. The absorbance of each hemoglobin compound is measured, and the microprocessor calculates the percent of each hemoglobin compound. The laboratory usually reports percent carboxyhemoglobin.

Several additional laboratory tests may be requested by the physician, including arterial blood gases, electrolytes, lactic acid (in severe CO exposure), total creatine kinase (TCK), lactate dehydrogenase (LD), and myoglobin (for muscle necrosis). An electrocardiogram (EKG) may also be requested to evaluate cardiac function.

 Checkpoint! 22–3

Identify three sources of exposure to carbon monoxide that may produce toxic levels in blood.

DRUGS OF ABUSE URINE (DAU) SCREEN

Screening urine for drugs of abuse by immunoassay has been available to the clinical laboratory since the early 1970s, when Syva Corp. (Palo Alto, CA) introduced the EMIT DAU assay. Clinical laboratories no longer had to rely on chromatographic methods that were labor intensive and time consuming. Since that time, thousands of laboratories have been permitted to screen for DAUs with or without confirmatory techniques available onsite.

The drugs or classes of drugs routinely tested in clinical laboratories include the following:

• Amphetamines
• Barbiturates
• Benzodiazepines
• Canabinoids
• Cocaine
• Methadone
• Opiates
• Phencyclidine
• Tricyclic antidepressants

There are several significant factors with which technologists should be familiar before screening for DAUs:

1. Most drug screening methods are immunoassays and are calibrated at established cutoff concentrations. Specimens yielding instrument responses greater than the cutoff value are considered positive. A value lower than the cutoff is considered negative.

2. Cutoff values are not the same as assay detection limits. The recommended cutoff value is higher than the detection limit but low enough to detect drug use within a reasonable time frame.

3. Immunoassays are not always specific for the drug tested. Many immunoassays are developed to allow detection of drugs within a particular class of compounds. For example, the antibody used in the barbiturate screening assay is produced against secobarbital but will cause a response with many different barbiturates at certain concentrations. Therefore, all positive drugs responses must be confirmed by a second method using a completely different measurement principle.

 Checkpoint! 22–4

List four significant factors with which technologists should be familiar in reference to drug abuse screening methods.

AMPHETAMINES

Amphetamine and amphetamine-like compounds (listed below) are described as psychomotor stimulants that act by releasing monoamines from nerve terminals in the brain. Noradrenaline and dopamine are the most important mediators in this connection, but 5-hydroxytrypamine release also occurs.

Examples of amphetamine-like compounds include the following:

• Dextroamphetamine
• Methamphetamine
• Methylphenidate
• Methylenedioxymethamphetamine (MDMA) or ecstasy
• Fenfluramine

β-phenylethylamine is considered the parent compound of this class of drugs and consists of a benzene ring and an ethylamine side chain. The structure allows substitutions to be made on the aromatic ring, the α- and β-carbon atoms, and the terminal amino group to yield a variety of compounds with similar pharmacological activity.[15]

The pharmacological effects of these drugs include the following:

- Locomotor stimulation
- Euphoria and excitement
- Anorexia

Amphetamine-type compounds also have peripheral sympathomimetic actions. Sympathomimetics are a class of drugs whose effects mimic those of stimulated sympathetic nervous systems. They tend to increase cardiac output, dilate bronchioles, inhibit gastrointestinal (GI) motility, and constrict blood vessels, thereby raising blood pressure.

Amphetamines are readily absorbed from the GI tract and freely penetrate the blood–brain barrier. They are also readily absorbed from the nasal mucosa and are often taken by "snorting." Amphetamine is primarily excreted unchanged in the urine, and the rate of excretion is increased when the urine is made more acidic. Plasma half-life of amphetamine varies from about 5 hours to between 20 and 30 hours, depending on urine flow and urinary pH.

A primary use for amphetamines is in the treatment of attention deficit/hyperactivity disorder (ADHD), principally in children. Methylphenidate (Ritalin and others) is a piperdine derivate that is structurally related to amphetamine and is widely prescribed to patients with ADHD.

Amphetamines can be used to treat narcolepsy, a condition in which the patient suddenly and unpredictably falls asleep. They have also been used as appetite suppressants in the treatment of obesity.

Unwanted side effects of amphetamine use include hypertension, insomnia, tremors, risk of exacerbating schizophrenia, and risk of dependence. Sudden deaths have occurred in "ecstasy" users. The drug can induce a condition resembling "heat stroke" and associated with muscle damage and renal failure. It can also cause inappropriate secretions of antidiuretic hormone, leading to thirst, overhydration, and hyponatremia (low plasma levels of sodium).

BARBITURATES

Barbituric acid is 2,4,6-trioxohexahydropyrimidine. It lacks central depressant activity, but the presence of alkyl or aryl groups at position 5 confers sedative-hypnotic and other activities. Other substitutions at carbons 2, 3 and nitrogen number 3 produce the myriad of barbiturate-like compounds, many of which are shown in Table 22-4 ✪ along with their half-lives.[16]

Barbiturates are a class of drugs that form the largest group of **hypnotics** and **sedatives**. A hypnotic drug produces drowsiness and facilitates the onset and maintenance of a state of sleep. Sedatives decrease activity, moderate excitement, and calm the recipient. The barbiturates reversibly depress the activity of all excitable tissues. Their pharmacologic effects on the CNS are profound. They can produce all degrees of depression of the CNS, ranging from mild sedation

✪ TABLE 22-4

Partial listing of frequently prescribed barbiturates, their respective half-lives, and therapeutic use(s)

Nonproprietary Name (trade name)	Half-Life (h)	Therapeutic Uses
Amobarbital (AMYTAL)	10–40	Insomnia, preoperative sedation
Butabarbital (BUTISOL)	35–50	Insomnia, preoperative sedation
Mephobarbital (MEBARAL)	10–70	Seizure disorders, daytime sedation
Methohexital (BREVITAL)	3–5	Induction and maintenance of anesthesia
Pentobarbital (NEMBUTAL)	15–50	Insomnia, preoperative sedation
Phenobarbital (LUMINAL)	80–120	Seizure disorders, daytime sedation
Secobarbital (SECONAL)	15–40	Insomnia, preoperative sedation
Thiopental (PENTOTHAL)	8–10	Induction and/or maintenance of anesthesia

to general anesthesia. Certain barbiturates (e.g., phenobarbital and mephobarbital) have selective anticonvulsant activity. The antianxiety properties of barbiturates are inferior to those exerted by the benzodiazepines.

Hypnotic doses of barbiturates increase the total sleep time and alter the stages of sleep in a dose-dependent fashion. Barbiturates decrease sleep latency, the number of awakenings, and the duration of rapid eye movement (REM) and slow-wave sleep.

Barbiturates also affect the following:

- Central nervous system
- Respiration
- Cardiovascular system
- GI tract
- Liver
- Kidney

Barbiturates are usually administered orally for sedative-hypnotic use. They are absorbed rapidly, and the onset of action varies from 10 to 60 minutes depending on the formulation. Some preparations containing specific barbiturates are given intramuscularly (IM) or rectally.

Barbiturates are distributed widely, and they readily cross the placenta. Highly lipid-soluble barbiturates used to induce anesthesia undergo redistribution after IV injection. With few exceptions, the barbiturates are nearly completely metabolized/or conjugated in the liver and proceed to excretion via the kidneys. The most important biotransformation of barbiturates occurs with the oxidation of radicals at C5. Oxidation results in the formation of alcohols, ketones, phenols, or carboxylic acids, all of which may appear in the urine.

The elimination of barbiturates is more rapid in young people than in the elderly and infants, and half-lives are increased during pregnancy. Patients with chronic liver diseases often increase the half-life of the biotransformation of barbiturates.

Overdosing with barbiturates is of great concern along with the fact that they can induce a high degree of tolerance and dependence. Barbiturates also induce the synthesis of hepatic cytochrome P450 and conjugating enzymes, thus increasing the rate of metabolic degradation of many other drugs and potentially increasing the number of problematic drug interactions. They also precipitate attacks of acute porphyria in susceptible individuals.

Large doses of barbiturates cause death from respiratory and cardiovascular depression, which is why they are not used on a large scale now as hypnotic agents or **anxiolytic** medication (i.e., medication whose purpose is to reduce anxiety). Nevertheless, poisoning with barbiturates is a significant clinical problem, and death occurs in a few percent of cases. Many of the cases are a result of suicide, but some are from accidental poisoning of children. Lethal doses of barbiturate vary, but severe poisoning is likely to occur when more than 10 times the full hypnotic dose has been ingested at once. If alcohol or other depressant drugs are also ingested, the concentration of barbiturates that can cause death is lower.

BENZODIAZEPINES

In 1961, chlordiazepoxide, one of the first benzodiazepines compounds, was synthesized by accident in the laboratories of Hoffman La Roche. Since then, drugs classified as benzodiazepines represent the most widely prescribed drugs due to their principal pharmacological effects. Examples of drugs included in this classification and their characteristics are shown in Table 22-5 ✪.[17]

The basic structure of benzodiazepines consists of a seven-membered ring fused to an aromatic ring, with four main substituent groups that can be modified without loss of activity. The many compounds included in this group of drugs have similar pharmacological actions, though some degree of selectivity has been shown. For example, clonazepam shows anticonvulsant activity with less marked sedative effects. Benzodiazepines act selectively on GABA receptors, which mediate fast inhibitory synaptic transmission throughout the CNS. Benzodiazepines enhance the response to GABA. The ultimate pharmacologic effects include the following:

- Reduction of anxiety and aggression
- Sedation and induction of sleep (hypnotic)
- Reduction of muscle tone and coordination
- Anticonvulsant effect
- Anterograde amnesia (new events are not transferred to long-term memory)

Benzodiazepines are almost completely absorbed when taken orally and usually exhibit peak plasma concentrations within an hour. They bind strongly to plasma proteins, and their high lipid solubility causes them to accumulate gradually in body fat.

Benzodiazepines are completely metabolized and are excreted as glucuronide conjugates in the urine. Several benzodiazepines are converted to active metabolites (Table 22-6 ✪) with long half-lives, which accounts for the tendency of many benzodiazepines to produce cumulative effects and a long hangover when given at regular intervals.

Several unwanted effects of benzodiazepines are as follows:[18]

- Benzodiazepines in acute overdose are considerably less dangerous than other anxiolytic/hypnotic drugs. Individuals who overdose with benzodiazepines experience prolonged sleep without serious depression of respiration or cardiovascular function. However, in the presence of other CNS depressants, particularly alcohol, benzodiazepine can cause severe, even life-threatening, respiratory depression.

- Side effects during therapeutic use include drowsiness, confusion, amnesia, and impaired coordination. These effects will impact a person's manual skills, such as driving performance.

- **Tolerance**, or the gradual escalation of the dose needed to produce the required effect, occurs with all benzodiazepines, as does *dependence*, which is their primary downside.

✪ TABLE 22-5

Drugs classified as benzodiazepines

Nonproprietary Name (trade name)	Active Metabolite	Overall Duration of Action	Therapeutic Uses
Triazolam (HALCION), Midazolam (VERSED)	Hydroxylated derivative	Ultra-short (<6 h)	Insomnia Preanesthetic, intraoperative medication
Lorazepam (ATIVAN), Oxazepam (SERAX), Temazepam (RESTORIL), Lormetazepam (LORETAM)	No	Ultra-short (~4 h)	Hypnotic, anxiolytic
Alprazolam (XANAX)	Hydroxylated derivative	Medium (24 hr)	Anxiolytic, antidepressant
Nitrazepam (MOGADON)	No	Medium	Hypnotic, anxiolytic
Diazepam (VALIUM), Chlordiazepoxide (LIBRIUM)	Nordazepam	Long (24–48 h)	Anxiolytic, muscle relaxant Diazepam used IV as anticonvulsant
Flurazepam (DALMANE)	Desmethylflurazepam	Long	Anxiolytic
Clonazepam (KLONOPIN)	No	Long	Anticonvulsant, anxiolytic (especially mania)

❂ TABLE 22-6

A partial listing of opiates and opiate agonist compounds that are often abused		
Nonproprietary Name (trade name)	Half-Life (h) and/or Comments	Unwanted Effects
Morphine sulphate (AVINZA)	3–4	Sedation, respiratory depression, constipation, nausea, vomiting, tolerance and dependence
Diamorphine (Heroine)	Rapidly metabolized to 6-monoacetylmorphine and then to morphine. Acts more rapidly than morphine, metabolized to morphine	Same as morphine
Hydromorphone (DILAUDID)	2–4	Same as morphine
Methadone (DOLOPHINE)	>24	Same as morphine
Pentazocine lactate (TALWIN)	2–4	Dysphoria
Fentanyl (ACTIQ)	1–2	Same as morphine
Codeine phosphate	Acts as a precusor drug, metabolized to morphine and other active opioids	Mainly constipation
Hydrocodone (VICODIN, LORCET)	Pain relief and cough suppressant	Same as morphine
Oxycodone HCL (OXYCONTIN)	~3.5	Same as morphine
Naloxone (NARCAN)	Opioid antagonist	
Meperidine, also pethidine (DEMEROL)	~3.5	Nausea, vomiting, tremor, seizures
Propoxyphene HCL (DARVON)	1–2	Similar to other opioids and also the following: EKG changes, ototoxicity, delusions, hallucinations, seizure, and confusion
Tramadol (ULTRAM)	4–6	Dizziness and convulsions

CANNABINOIDS

The hemp plant, *Cannabis sativa,* contains the active substance Δ^9-tetrahydrocannabinol (THC). Marijuana is the name given to the dried leaves and flower heads, prepared as a smoking mixture; hashish is the extracted resin. These substances have been used for centuries as intoxicant preparations and medicines to treat various diseases.

Cannabis extracts contain several related compounds, called cannabinoids, that are insoluble in water. The most abundant cannabinoids are THC, its precursor cannabidiol, and cannabinol. THC is the most pharmacologically active and abundant cannabinoid. The metabolite, 11-hydroxy-THC, is more active than THC and contributes to the pharmacological effect.

THC acts mainly on the CNS, producing a mixture of depressant and psychotomimetic effects. Subjective effects in humans include the following:

- A feeling of relaxation and well-being, similar to the effect of ethanol, but without the accompanying aggression
- A feeling of heightened sensory awareness, with sounds and sights seeming more awesome and intense

Several effects that can be measured in humans are as follows:

- Impairment of short-term memory
- Analgesia
- Increased appetite
- Impairment of motor coordination
- Catalepsy (retention of fixed unnatural postures)

Several peripheral effects of cannabis are as follows:

- Tachycardia (increased heart rate)
- Vasodilatation
- Bronchodilatation
- Reduction of intraocular pressure

The pharmacokinetic aspects of cannabis are slower to develop than with other drugs taken for similar effects, such as lysergic acid diethylamide (LSD). Cannabis, taken by smoking or by IV, takes about 1 hour to fully develop these effects and last for 2 to 3 hours. A small amount is converted to 11-hydroxy-THC. It is partly conjugated and undergoes enterohepatic recirculation. THC and its metabolites are highly lipohilic and therefore are sequestered in body fat, and excretion continues for several days after a single dose.

Immunoassays urine screening tests are developed to detect metabolites of THC. The principale urinary metabolite of THC is 11-Nor-Δ^9-tetrahydrocannabinol-9-carboxylic acid (THC-COOH) and its glucuronide conjugate. THC is slowly released from tissue storage sites; thus urine may test positive for THC metabolite (>20 ng/mL cutoff THC-COOH) for 2 to 5 days after last marijuana use. A positive screening result for metabolites of THC by immunoassay is confirmed by GCMS analysis in urine specimens.

COCAINE

Cocaine is found in the leaves of a South American shrub, coca. The leaves have been chewed by natives of South America for their psychotropic effects for thousands of years. The natives knew about the numbing effect they produce on the mouth and tongue. Cocaine has been isolated and used as a local anesthetic for surgical procedures. Unfortunately, cocaine has become one of the most abused drugs in Western countries in the past several decades.

Cocaine is an ester of benzoic acid, the complex alcohol 2-carbomethoxy, 3-hydroxy-tropane. Because of cocaine's toxicity and addictive properties, several synthetic substitutes have been produced (e.g., procaine, lidocaine, bupivacaine, and tetracaine) as alternatives for local anesthesia.

The pharmacological effects of cocaine lie in its ability to inhibit catecholamine uptake by the noradrenalin and dopamine transporters, thus enhancing the peripheral effects of sympathetic nerve activity and producing marked psychomotor stimulant effects. The psychomotor stimulant effect produces euphoria, garrulousness, increased motor activity, and a magnification of pleasure, similar to the effects of amphetamine. With excessive dosage, tremors and convulsion, followed by respiratory and vasomotor depression, may occur. Peripheral sympathomimetic actions lead to tachycardia, vasoconstriction, and an increase in blood pressure. Also, body temperature may increase due to the increased motor activity coupled with reduced heat loss.

Cocaine is readily absorbed by several routes. Illegal preparations using hydrochloride salt could be given by nasal inhalation or by IV. The latter route produces an intense and immediate euphoria, whereas nasal inhalation produces a less dramatic sensation. Cocaine in the free base form, referred to as "crack" is a popular street form of the drug. Crack cocaine can be smoked, thus giving an effect similar to IV methods.

The toxic effects of cocaine are usually exhibited in abusers of the drug. Primary adverse effects include cardiac dysrhythmias and coronary or cerebral thrombosis. Also, slowly developing damage to the myocardium can occur, leading to heart failure.

Cocaine can impair brain development in utero. Brain size is reduced in babies exposed to cocaine in pregnancy, with an increased incidence of neurological and limb malformation.

The qualitative screening test for cocaine in urine is designed to detect the metabolite benzoylecgonine. Immunoassays are widely used for the initial screening tests with cutoff concentrations for benzoylecgonine of approximately 300 ng/mL. Benzoylecgonine excretion can be detected for 1 to 3 days following cocaine use. A positive screening result for benzoylecgonine by immunoassay is confirmed by GCMS analysis in urine specimen.

METHADONE

Methadone (trade name DOLOPHINE) is a long-acting μ-receptor agonist (i.e., it activates the μ-receptors) with pharmacological properties qualitatively similar to those of morphine. Thus it is classified as a narcotic analgesic drug or opiate (see below). Other significant attributes of methadone include its efficacy when administered orally, its extended duration

of action in suppressing withdrawal symptoms in physically dependent individuals, and its tendency to show persistent effects with repeated administration. The effects of methadone on the respiratory system can be detected for more than 24 hours after a single dose, and, on repeated administration, marked sedation is seen in some patients. Methadone causes mitosis or constriction of the pupils in most patients. This effect may persist for many hours. The effects on bowel motility, cough, changes in biliary tone, and the secretion of pituitary hormones are similar to those of morphine.[19]

Methadone is absorbed from the GI tract and can be detected in plasma within 30 minutes of oral ingestion. Peak plasma concentrations are reached in about 4 hours. Following a therapeutic dose, approximately 90% of methadone is bound to plasma protein. Following a subcutaneous or intramuscular injection, methadone reaches peak concentration in the brain within 1 to 2 hours, and this correlates with the intensity and duration of analgesia.

Biotransformation of methadone occurs in the liver. The major metabolites, a result of N-demethylation and cyclization to form pyrrolidines and pyrroline, are excreted in the urine and bile along with small amounts of unchanged drug. The half-life of methadone is approximately 15–40 hours.

Methadone is widely used as a drug to treat morphine and diamorphine addicts. In the presence of methadone, given at regular oral doses, an injection of morphine does not cause the normal euphoria, and the lack of an abstinence syndrome or withdrawal makes it possible to reduce an addict's need or desire for morphine or diamorphine.

OPIATES

Opioid is a term that applies to any substance, whether endogenous or synthetic, that produces morphine-like effects that are blocked by antagonists (compounds that combine at the same receptor site without causing activation) such as naloxone hydrochloride (NARCAN). The term opiate refers to synthetic, morphine-like drugs with nonpeptidic structures. Opium is an extract of the poppy Papaver somniferum that has been used for centuries as an agent to produce euphoria, analgesia, and sleep and tp prevent diarrhea. In the 17th century, residents of Great Britain took the drug orally as "tincture of laudanum," and doing so conveyed a certain social status at the time. In the mid-1800s, the hypodermic needle was invented and was the chief conveyance of the drug for humans and may have led to opiates' status as a drug of dependence.

There are many alkaloids related to morphine contained within opium. The basic structure of morphine is that of a benzylisoquinoline alkaloid with two additional ring closures. Chemical variants of the morphine molecule have been derived by substitution at one or both of the hydroxyl groups or a nitrogen atom.

Morphine analogues are compounds that structurally resemble morphine and are often synthesized from it. Examples of morphine analogues include diamorphine (heroin), codeine, nalophine, and naloxone. Another group of morphine-like drugs is the synthetic derivatives. This group of drugs includes pethidine (meperidine in the United States), fentanyl, sufentanil, methadone, dextropropoxyphen, and pentazocine. Examples of these compounds and other morphine-like drugs and their half-lives are listed in Table 22-6.[20,21]

The mechanism of action for this group of compounds involves opioid receptors that belong to the family of G-protein-coupled receptors and inhibit adenylate cyclase, thus reducing intracellular cyclic adenosine monophosphate (cAMP) content. Pharmacological terms associated with these mechanisms include (1) **agonist**, defined (pharmacologically) as a compound such as morphine that leads to activation of the receptor and usually elicits a response by cell or host; and (2) **antagonist**, which is a substance (e.g., naloxone) that binds to a receptor and thus does not allow an agonist such as morphine to bind to the receptor and result in a lack of response by cells or hosts to morphine.[17]

Morphine-like compounds have significant impact on the CNS and the GI tract, with many effects of lesser importance on other human systems. Examples of morphine's effects on the CNS include the following:

- Analgesia
- Euphoria
- Respiratory depression
- Depression of cough reflex
- Nausea and vomiting
- Pupillary constriction

Morphine affects many parts of the GI tract by increasing tone and reducing motility, which results in constipation and may cause a significant problem in certain individuals. The resulting delay in gastric emptying can retard the absorption of other drugs. Also, the gall bladder may contract and the biliary sphincter may constrict, which will result in increased pressure in the biliary tract. A rise in intrabiliary pressure can cause a transient increase in the concentration of amylase and lipase in the plasma.

Examples of unwanted effects of morphine-like compounds are shown in Table 22-6. Acute overdose results in coma and respiratory depression; constricted pupils are a sign of overdose. Treatment for overdose may include administration of naloxone by IV. Administration of naloxone also serves as a diagnostic test, because failure to respond to naloxone indicates a cause, other than opioid poisoning, of the patient's comatose state.

Immunoassays for opiates are designed primarily to detect morphine and codeine at a cutoff concentration equivalent to approximately 300 ng/mL of morphine. Cross reactivity with oxycodone and oxymorphone is very low. False-positive results have been reported in patients ingesting compounds such as dextromethorphan, diphenhydroamine, and ephedrine/pseudoephedrine. In general, urine specimens test positive for 1 to 3 days following morphine or codeine use

at a cutoff of 300 ng/mL. A positive screening test for opiates is confirmed by GCMS analysis of urine specimens.

PHENCYCLIDINE

Phencyclidine (PCP) (brand name SERNYL, street name "angel dust"), is a piperidine and is grouped with drugs referred to as psychotomimetic (also called psychedelic or hallucinogenic drugs) that affect thought, perception, and mood without causing marked psychomotor stimulation or depression. Notable characteristic features of this group of drugs are that thoughts and perceptions tend to become distorted and dreamlike, rather than being just sharpened or dulled, and the change in mood is more complex than a simple shift in the direction of euphoria or depression. Psychotomimetic drugs are further categorized into two broad groups:

- Drugs with a chemical resemblance to known neurotransmitters (e.g.,LSD)
- Drugs unrelated to monoamine neurotransmitters (e.g., PCP)

Initially PCP was intended as an IV anesthetic drug but was found to produce a period of disorientation and hallucinations in many patients following recovery of consciousness. The primary concern now with PCP is as a drug of abuse.

The pharmacological effects of PCP are similar to other psychotomimetic drugs but also include analgesia. There have been reported incidences of "bad trips" similar to those of LSD users. PCP's mode of action at the cellular level is not well understood. Significant binding sites occur on neuronal membranes—in particular, the frontal cortex and hippocampus, which are located on the underside of the brain. There are two binding sites for PCP: (1) the σ-receptor recognized by various opioids, and (2) glutamate-operated ion channels, which are blocked by PCP and ketamine (see below). The σ-receptor is thought to mediate the effects of dysphoria and hallucinations produced by certain opiates, and this may account for the psychotomimetic effects of PCP.[22]

TRICYCLIC ANTIDEPRESSANTS

Tricyclic antidepressants (TCAs) are closely related in structure to the phenothiazines (see below) and were initially produced as possible antipsychotic drugs. **Antidepressant** drugs serve to alleviate depression. Imipramine, a TCA, was not helpful in treating schizophrenia but was effective in treating depression. This led to formulation of other TCAs, such as clomipramine. TCAs differ in structure from phenothiazines principally in the incorporation of an extra atom (sulphur) into the central ring, which twists the structure so that the molecule is no longer planar, like phenothiazine. A partial listing of several TCAs and their mechanism of action is shown in Table 22-7 ✪.[23]

The primary mechanism of action of TCAs is to block the uptake of amines by nerve terminals for binding sites on the transport protein. Most TCAs inhibit noradrenalin and 5-hydroxytryptamine uptake by brain synaptosomes to a

⊙ TABLE 22-7

A select list of tricyclic, tetracyclic antidepressants and their mechanisms of action

Nonproprietary Name (trade name)	Mechanism of Action
Amitriptyline (ELAVIL)	Norepinephrine reuptake inhibitor
Chomipramine (ANAFRANIL)	Norepinephrine reuptake inhibitor
Doxepin (ADAPIN, SINEQUANT)	Norepinephrine reuptake inhibitor
Imipramine (TOFRANIL)	Norepinephrine reuptake inhibitor
Trimipramine(SURMONTIL)	Norepinephrine reuptake inhibitor
Amoxapine (ASENDIN)	Norepinephrine reuptake inhibitor
Desipramine (NORPRAMIN)	Norepinephrine reuptake inhibitor
Maprotiline (LUDINOMIL)	Norepinephrine reuptake inhibitor
Nortriptyline (PAMELOR)	Norepinephrine reuptake inhibitor
Citalopram (CELEXA)	Seratonin reuptake inhibitor
Escitalopram (LEXAPRO)	Seratonin reuptake inhibitor
Fluoxetine (PROZAC)	Seratonin reuptake inhibitor
Paroxetine (PAXIL)	Seratonin reuptake inhibitor
Sertraline (ZOLOFT)	Seratonin reuptake inhibitor

similar degree but have much less effect on dopamine uptake. Also, TCA affects one or more types of neurotransmitter receptors, including muscarinic acetylcholine receptors, histamine receptors, and 5-hydroxytryptamine receptors.[24]

Tricyclic antidepressants are quickly absorbed when administered orally and bind strongly to plasma albumin. They bind to extravascular tissues, which accounts for their generally large distribution volumes and low rates of elimination.

There are two main routes of metabolism of TCAs in the liver:

• N-demethylation, where, for example, imipramine is converted to desmethylimipramine and amitriptyline to nortriptyline

• Ring hydroxylation

Methods for measuring TCAs are available on most automated immunochemistry analyzer platforms. Immunoassays for quantitative TCAs in serum and urinary TCA qualitative screening are often provided by clinical chemistry laboratories. The quantitative measurement is useful to clinicians whose patients are taking the drug for therapeutic purposes. Confirmatory assays are available using thin layer chromatography (TLC), gas chromatography (GC), HPLC, and GCMS.

 Checkpoint! 22–5

Identify six classes of abused drugs that a clinical laboratory can screen for and indicate at least one pharmacological effect associated with each group.

LEAD

Lead is a metal found in many industrial processes and products. Several sources of lead are shown in Table 22-8 ⊙.[25] A common use of lead is was as an additive to paints and coating products. The use of lead in paints was banned in the United States around 1977. Unfortunately, lead-based paints still adorn houses and other buildings constructed before 1977 and provide a source of lead exposure, especially for children.

The daily intake of lead is 5–15 μg/day across all age groups. Blood lead levels (BLLs) may be higher in children due to ingestion of and contact with soil. Children can also ingest lead at home by coming into contact with paints (chips), soil, dust, food, cosmetics, art materials, and toys. The major exposure of children to lead is via paint chips and dust.

Studies carried out by the Centers for Disease Control and Prevention (CDC) show that the levels of lead in the blood of U.S. children have decreased steadily over the past several years. The steady decline in BBLs is due in large part to the efforts of the government to ban lead from gasoline, residential paint, and solder used for food cans and water pipes. However, about 310,000 U.S. children between the ages of 1 and 5 years are believed to have BLLs equal to or greater than 10 μg/dL. This BLL is considered to unacceptable, and efforts are under way to find ways to reduce BLLs by 2010.[26]

The data provided by the 2007 Annual Report of the AAPCC show that the number of exposures reported to the agency was approximately 1900 for children under the age of 6 years and nearly 750 for adults over the age of 19.[5]

Most human exposure to lead occurs through ingestion or inhalation. Lead exposure in the general population (including children) occurs mainly through ingestion, although inhalation also contributes to lead accumulation and may be the primary contributor for workers in lead-related occupations. Most of the lead inhaled is absorbed into the body, whereas 20–70% of ingested lead is absorbed (except in children, who absorb higher amounts than adults). Once lead is absorbed into the body, it may be stored for long periods of time in mineralizing tissues such as teeth and bones and then released into the blood stream. For example, in times of calcium stress (e.g., pregnancy, lactation, osteoporosis, or calcium deficiency) lead is readily released from storage tissue sites. Most of the lead that is absorbed into the body is excreted either by the kidneys (in urine) or through biliary clearance and ultimately in the feces. In adults lead is excreted by the kidneys at a rate of approximately 30 μg/day, which represents 50–60% of an absorbed fraction of lead. Adults retain only about 1% of absorbed lead, but children tend to retain more than adults. In infants from birth to 2 years, approximately one-third of the total amount of lead to which the infant is exposed is absorbed.

Absorbed lead that is not excreted is exchanged primarily among three compartments: (1) blood; (2) soft tissue, which includes liver, kidneys, lungs, brain, spleen, muscle, and heart; and (3) mineralizing tissues (bones and tissues).

TABLE 22-8

Sources of lead exposure

Occupational Lead Exposures

Lead mining, roofing, smelting and manufacturing industry employees	Printers	Gas station attendants	Bridge reconstruction workers
Plumber, pipe fitters	Police officers	Plastic manufacturers	Glass manufacturers
Battery manufacturers	Auto repairs	Steel welders or cutters	Shipbuilders

Environmental Lead Exposure

Lead-containing paint	Soil/dust near lead industries, roadways	Plumbing leachate (from pipes and solder)	Lead gasoline
Homes painted with lead based paints	Ceramic ware		

Hobbies and Related Activities

Glazed pottery making	Painting	Car or boat repairs	Target shooting at firing ranges
Stained glass making	Preparing lead shot or fishing sinkers	Home remodeling	Lead soldering

Other Potential Sources

Folk remedies	Moonshine whiskey	Tobacco smoking	Gasoline "huffing"(purposeful inhalation of vapors)
Cosmetics			

Blood transports only a small fraction of the total lead body burden. Blood serves as the initial compartment of absorbed lead and distributes it through the body, making it available to other tissues or for excretion. The half-life of lead in adult human blood is about 28 days. Approximately 99% of lead in blood is bound to erythrocytes; the rest resides in blood plasma. Knowledge of lead distribution in the blood and cells is important because the BLL is frequently used to measure lead exposure. The less sensitive free erythrocyte protoporphyrin (FEP) assay is the also used to evaluate the effects of lead on the metabolism of porphyrins. Both BLL and FEP do not measure total body burden, but rather they are more reflective of recent or ongoing exposures.

Clinically relevant BLLs and recommended courses of action for caregivers are shown in Table 22-9 ☆.[27, 28] The BLLs have undergone modifications since the 1980s in an effort to provide patients (especially children) with timely treatment modalities.[29]

Hematological effects of lead provide a mechanism for the initial evaluation and monitoring of lead intoxication. Lead inhibits the body's ability to make hemoglobin by interfering with several enzymatic steps in the heme pathway. Specifically, lead decreases heme biosynthesis by inhibiting aminolevulinic acid dehydrase and ferrochelatase activity. Ferrochelatase, which catalyzes the insertion of iron into protoporphyrin IX, is sensitive to the presence of lead. A decrease in the activity of this enzyme may result in an increase of the substrate, erythrocyte protoporphyrin (EP), in the red blood cells. A decrease in iron results in zinc replacing the iron and forming erythrocyte zinc protoporphyrin (ZPP), which remains elevated for the life of the red cell.

Lead can induce two types of anemia, often accompanied by basophilic stippling of the erythrocytes. Acute, high-level lead exposure has been associated with *hemolytic anemia*. In chronic lead exposure, lead induces *hypochromic microcytic anemia* by both interfering with heme biosynthesis and diminishing red blood cell survival.

Treatment modalities will be determined by the BLL and organs or tissues affected. For example, if a patient experiences seizures, she will need antiepileptic drug therapy. If the BLL is in the range of 11–44 μg/dL, the clinician may not initiate chelation therapy, but if the BLL is greater than 45 μg/dL, then chelation therapy should be started. For symptomatic patients without encephalopathy and for asymptomatic patients with elevated BLLs requiring chelation, the use of British anti-Lewis (BAL) or succimer (2,3-dimercaptosuccinic acid, DMSA) with or without calcium disodium edetate (CaNa2-EDTA) is recommended and should be administered in the emergency department.[30]

The current biomarker for assessment of lead exposure is venous blood lead with or without measured FEP (also referred to as ZPP). Blood lead measurements are commonly measured by anodic stripping voltammetry (ASV) or graphite furnace atomic absorption spectroscopy (see Chapter 2, "Instrumentation"). These techniques have been available in laboratories for many years, with ASV used more often due to ease of operation compared to atomic absorption spectroscopy.

ESA Inc., a subsidiary of Magellan Biosciences Inc. (Chelmsford, MA), has developed a hand-held, portable POCT lead analyzer, the LeadCare Childhood Blood Lead Testing device, which is categorized by the Clinical Laboratory Improvement Act (CLIA '88) as a moderately complex test. The measuring principle is ASV and is similar to ESA Inc.'s benchtop model 3010B Lead Analyzer.

A second lead analyzer designed by ESA Inc. for POCT purposes has been approved by the Food and Drug Administration (FDA) and categorized as a waived test by CLIA '88. The

☼ TABLE 22-9

Blood lead concentration and suggested management. Updates include laboratory tests recommendations (see references 9 and 10)

Blood Lead Level (BLL, μg/dL)	Recommendations
<10	Children with this degree of exposure are not considered to have lead poisoning

10–14	Lead education
	    Dietary
	    Environmental
	Follow-up blood lead monitoring
	Proceed according to actions for 20–44 μg/dL if
	• a follow-up blood lead concentration in this range at least 3 months after initial venous test; or
	• blood lead concentration increases

20–44	Lead education
	    Dietary
	    Environmental
	Follow-up blood lead monitoring
	Complete history and physical examination
	Lab work
	    Hemoglobin or hematocrit
	    Iron status
	Environmental investigation
	Lead hazard reduction
	Neurodevelopmental monitoring
	Abdominal radiography (if particulate lead ingestion is suspected) with bowel decontamination if indicated

45–69	Lead education
	    Dietary
	    Environmental
	Follow-up blood lead monitoring
	Complete history and physical examination
	Lab work
	    Hemoglobin or hematocrit
	    Iron status
	    FEP or ZPP
	Environmental investigation
	Lead hazard reduction
	Neurodevelopmental monitoring
	Abdominal radiography (if particulate lead ingestion is suspected) with bowel decontamination if indicated
	Chelation therapy

≥70	Hospitalize and commence chelation therapy Proceed according to actions for 45–69 μg/dL

LeadCare II POCT device requires no manual calibration or refrigeration. It is designed for physician offices, health clinics, and outreach screening for whole blood levels.

Free erythrocyte protoporphyrin is measured by front-surface fluorometry (see Chapter 2, "Instrumentation") using a hematofluorometer. Two examples of hematofluorometers currently used in clinical laboratories are as follows:

• AVIV (Lakewood, NJ)

• ProtoFluor Z (Helena Laboratories, Beaumont, TX)

The compound actually measured in whole blood is zinc protoporphyrin and not FEP. To measure FEP the analyst must perform an extraction to isolate the "free" form. *Free* means that the chelated porphyrins (e.g., heme) are not measured. Doing so would require additional time and expense. Therefore, clinical laboratories routinely choose to use an extraction-free method that is simple to perform and requires only a drop of whole blood.[31]

 Checkpoint! 22–6

Explain the mechanism for the adverse effects of lead intoxication.

▶ GROUP II ANALYTES

This group of analytes represents those measured in laboratories equipped to provide qualitative and/or quantitative analysis of toxic substances less commonly requested by clinicians.

METHANOL

Methanol (also called methyl alcohol, wood alcohol) finds wide application for use in antifreeze solutions (gas line and windshields), fuel, photocopy fluids, and other products, as listed in Table 22-3.[4] It is also used in "home-brews," as a substitute for ethanol; for attempted suicide; and as an ethanol denaturant (to render ethanol solutions unsafe to drink).

Methanol is metabolized using the same enzymes as used in ethanol metabolization. Competition between methanol and ethanol for ADH forms the basis of the use of ethanol as an antidote in methanol poisoning. The first oxidation step in the metabolism of methanol produces formaldehyde, instead of acetaldehyde as with ethanol. Formaldehyde is more reactive than acetaldehyde and reacts rapidly with proteins, causing the inactivation of enzymes involved in the tricarboxylic acid cycle. Formaldehyde, which disappears from the circulation with a half-life of less than one minute, is then converted to formic acid (Figure 22-2 ■). Formic acid, unlike acetic acid from ethanol, cannot be utilized in the TCA cycle and is likely to cause tissue damage if produced in high concentrations. Formic acid production, and derangement of the TCA cycle, also produces severe acidosis. Complete elimination of formic acid from the body requires formic acid to combine with tetrahydrofolate with subsequent oxidization to carbon dioxide.[32]

Conversion of methanol to aldehyde occurs not only in the liver but also in the retina and is catalyzed by the dehydogenase enzymes responsible for retinol–retinal conversion. Formation of formaldehyde in the retina accounts for one of the main toxic effects of methanol, namely blindness, which can occur after ingestion of as little as 10 g of methanol.

Clinical presentation of patients with suspected methanol ingestion varies, especially if co-ingestion of ethanol occurred. Methanol itself is a CNS depressant, and like other alcohols,

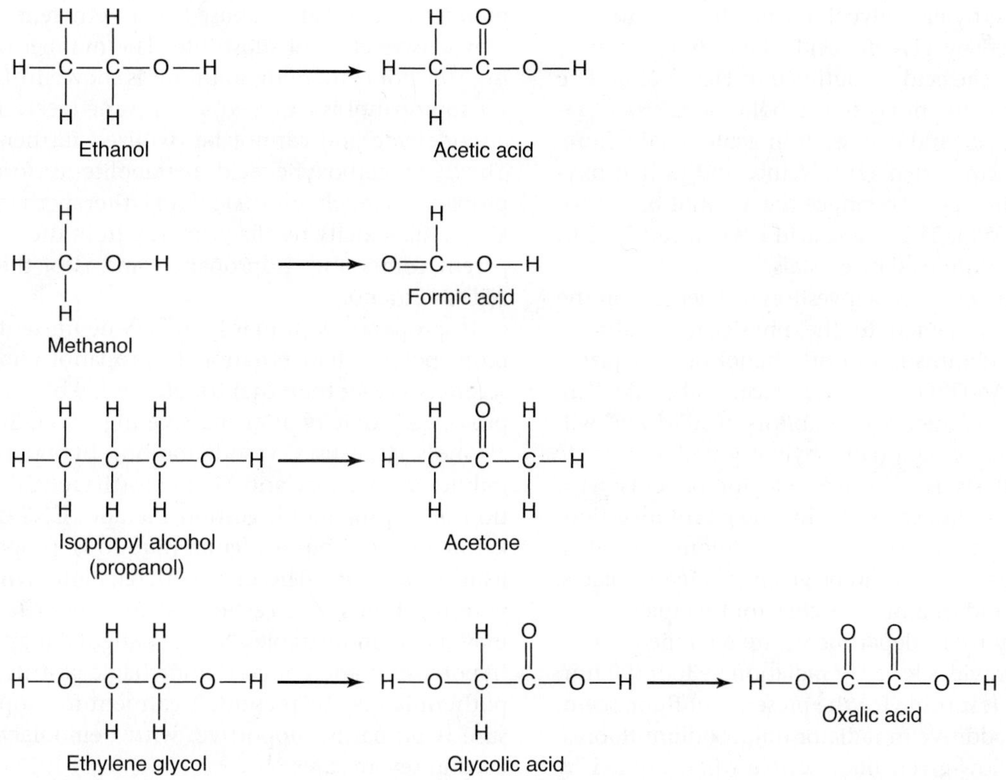

FIGURE 22-2 Structure of the *ol*'s (alcohols) and their respective metabolic products.

produces vasodilatation, hypotension, and reduced cardiac output. Approximately 0.5–40.0 hours postingestion of methanol only, the patient may exhibit mild disinhibition, sedation, or ataxia. After several more hours, patients may complain of headache, vomiting, vertigo, or abdominal pain. Patients usually complain of central scotomata (an island-like blind spot in the visual field), blurred vision, tunnel vision, or diplopia. More severely intoxicated patients who present late to the ED may have coma, seizure, blindness, GI hemorrhage, and pancreatitis.

Methanol poisoning is common, and it is treated by administration of large doses of ethanol, which acts to retard methanol metabolism by competition for alcohol dehydogenase. This is often done in conjunction with hemodialysis to remove unchanged methanol, which has a small volume of distribution.

Diagnostic test protocols should include a quantitative blood methanol measurement using GC analysis. Because most hospital-based clinical laboratories do not offer methanol determinations, the clinician can use other clinical laboratory tests, including pH, blood gasses, anion gap, electrolytes, and osmolal gap, to assess whether the patient has ingested methanol.

✓ Checkpoint! 22–7

What are two treatment modalities for methanol intoxication?

ETHYLENE GLYCOL

Ethylene glycol (1,2-ethandiol) is a colorless, odorless, sweet-tasting, viscous fluid. It is used in de-icing solutions, brake fluid, and other fluids used in boats and aircrafts. The number of exposures reported to TESS is shown in Table 22-3.[5] Ingestion of ethylene glycol itself is nontoxic, but the metabolites produced can cause tissue destruction and metabolic toxicity. Ethylene glycol presents itself in three clinical stages, as follows:[33]

Stage 1 (30 minutes–12 hours): Inebriation, nystagmus, paralysis, and seizures, also nausea and vomiting. Calcium oxalate crystals may form and appear in the urine.

Stage 2 (12–24 hours): Cardiopulmonary symptoms with tachycardia and hypertension, anion gap metabolic acidosis with hyperventilation, hypoxia, congestive heart failure (CHF).

Stage 3 (>24 hours): Renal phase characterized by acute tubular necrosis and renal failure. Oliguria, anuria, hematuria, and proteinuria may occur.

Death can be the ultimate outcome following ethylene glycol ingestion. The delayed onset of toxicity after ethylene glycol ingestion, attributable to the biotransformation and accumulation of the toxic metabolites, resembles that of methanol ingestion, although the latency period is generally shorter.

Once ingested, ethylene glycol is metabolized to glycoaldehyde, followed by glycolic acid, then glyoxylic acid, and ultimately to oxalic acid, as outlined in Figure 22-2. The glycolic acid (glycolate) induces metabolic acidosis, negatively impacts the CNS, and can result in acute renal failure. Glyoxylic acid is converted to glycine and α-hydroxy-β-ketoadipic acid and can also reduce the amount of adenosine triphosphate (ATP). The oxalic acid created can lead to the formation of calcium oxalate crystals.

Treatment for ethylene glycol ingestion will depend on the clinical conditions presented to the physician. Treatment modalities include administration of ethanol or fomepizole (4-methylpyrazole, ANTIZOL, Jazz Pharmaceuticals, Palo Alto, CA), which serve as competitive inhibitors of ADH and will block further biotransformation of ethylene glycol to its toxic products. Hemodialysis is also available for patients with more severe (>100 mg/dL) cases of ethylene glycol ingestion. Patients who have confounded chronic conditions (e.g., alcoholism or vitamin deficiency) can be given tetrahydrofolates, such as pyridoxine and thiamine, as cofactor therapies.

Many emergency room departments are equipped with a Wood's lamp that provides long UV radiation (~360–390 nm) that can be used to test urine for the presence of fluorescein, which serves as an additive in radiator fluid. Sodium fluorescein produces a yellow-green fluorescence when excited by the radiation emitted from a Wood's lamp. If the patient tests positive using this technique, the physician will begin a treatment regime for the patient. Samples of blood and urine can be sent to a laboratory for GC analysis of ethylene glycol and its major metabolite, glycolic acid.

The clinical laboratory can serve as a valuable resource to the patient and clinician who is assessing the patient by providing routine laboratory tests, including but not limited to CBC, electrolytes, glucose, urea nitrogen, creatinine, urine and serum osmolality using the freezing-point depression method, calculated serum osmolality, osmolal gap, anion gap, serum calcium, quantitative ethanol, acetaminophen, salicylate, and blood gases. A urinalysis and microscopic examination can be performed to detect the presence of protein, blood, and calcium oxalate crystals. Also, a urine pH will be useful.

The presence of an increased osmolal gap suggests the possibility of toxic alcohol (including ethylene glycol) ingestion. A low blood pH may reflect an increase in glycolate. Also, the anion gap may be increased, indicating a metabolic acidosis with an elevated anion gap. Renal function should be monitored by measuring urea nitrogen and creatinine. If a patient presents with cyanosis or symptoms of methemoglobinemia, then a test for blood methemoglobin level should be performed. Some patients may develop hypocalcemia; therefore, it is useful to monitor total serum calcium and ionized calcium.

ISOPROPANOL

Isopropanol (2-isopropanol) is the principal compound in rubbing alcohol (70% v/v) and is also found in several commercial products. It is also abused like ethylene glycol as an inexpensive ethanol substitute. The number of exposures to isopropanol-containing solutions is shown in Table 22-3.[5]

Isopropanol is oxidized to a ketone (acetone) rather than an aldehyde and cannot be oxidized further (Figure 22-2). Thus, no carboxylic acid metabolite is formed and isopropanol is much less toxic than either methanol or ethylene glycol. Its toxicity results primarily from the properties of the parent compound, isopropanol, and is of the same magnitude as ethanol.

Isopropanol is primarily a CNS depressant. It is slightly more potent than ethanol. Isopropanol elimination from serum is slower than that for ethanol, which accounts for its prolonged toxicity after massive ingestion. Similar to other alcohols, it causes vasodilatation, hypotension, gastritis, pulmonary edema, and GI hemorrhage. Clinical presentation of isopropanol ingestion includes CNS depression and ketonemia without acidemia (due to isopropanol's metabolism to acetone). The onset of signs and symptoms occurs within 2 hours of ingestion. A fruity odor (acetone breath) may be demonstrable. Some patients may present with hypotension and coma. Hemorrhagic gastritis and mild hypothermia may be present. Treatment for isopropanol exposure is primarily supportive, with hemodialysis procedures used in severe cases.[34]

Diagnostic tests for patients who present to the emergency department with possible isopropanol ingestion include quantitative analysis of isopropanol by GC or GCMS, osmolal gap, serum and urine ketones, glucose to exclude hypoglycemia, ethanol, and renal function tests. Serum creatinine may be falsely elevated due to the acetone present.

PHENOTHIAZINES

Phenothiazines have a tricyclic structure, with two benzene rings linked by sulfur and a nitrogen atom. The chemically related thioxanthenes have a carbon in place of the nitrogen at position 10, with the R1 moiety linked through a double bond. There are also several other heterocyclic compounds that exhibit similar effects to that of phenothiazine. Several examples of phenothiazine and chemically related drugs are shown in Table 22-10 ☺.[35]

Phenothiazines and chemically related drugs are used to treat psychotic disorders, including schizophrenia, the manic phase (manic-depressive) of bipolar illness, acute idiopathic psychotic illnesses, and other conditions marked by severe agitation. These disorders have in common major disturbances in reasoning, often with delusions and hallucinations. Antipsychotic drugs are also useful alternatives to electroconvulsive therapy (ECT) in severe depression with psychotic features and may be used on occasion to assist in managing patients with psychotic disorders associated with delirium or dementia.

Overdosing with phenothiazines is uncommon, and death rarely results. If death does occur, usually it is a result of concomitant use with other drugs and alcohol. Adverse

TABLE 22-10

Tricyclic antipsychotic agents, including phenothiazine and chemically related agents

Phenothiazines Nonproprietary Name (trade name)	Thioxanthenes Nonproprietary Name (trade name)	Other Heterocyclic Nonproprietary Name (trade name)
Chloropromazine HCL (THORAZINE)	Chlorprothixene (TARACTAN)	Aripiprazole (ABILIFY)
Mesoridazine besylate (SERENTIL)	Thiothixene HCL (NAVANE)	Clozapine (CLOZARIL)
Thioridazine HCL (MELLARIL)		Haloperidol; haloperidol decanoate (HALDOL)
Fluphenazine HCL (PERMITIL)		Olanzapine (ZYPREXA)
Perphenzine (TRILAFON)		Quetiapine fumarate (SEROQUEL)
Trifluorperazine HCL (STELAZINE)		Risperidone (RISPERDAL)

effects of phenothiazines may involve cardiovascular, central and autonomic nervous, and endocrine systems. Other significant effects include seizure, agranulocytosis, cardiac toxicity, and pigmentary degeneration of the retina.

Phenothiazines are almost completely metabolized (>99%) by the liver to metabolites, some of which are pharmacologically active. The correlation of dose, serum concentration, and pharmacological effect is not precise; thus quantitative measurement in overdose conditions is usually not necessary. A qualitative assessment of phenothiazines in urine is sufficient to verify ingestion for patients presenting with appropriate signs and symptoms. Qualitative testes using Toxi Lab (Varian, Inc., Palo Alto, CA) or Forrest reagent[14] may be sufficient.

ORGANOPHOSPHATE AND CARBAMATE COMPOUNDS

Organophosphates and carbamates belong to a class of compounds referred to as anticholinesterase agents. The function of cholinergic agents such as acetylcholinesterase (AChE) is to terminate the action of acetylcholine (ACh) at the junctions of the various cholinergic nerve endings with their effector organs or postsynaptic sites. Compounds that inhibit AChE are anticholinesterase (anti-ChE) agents. They cause ACh to accumulate in the vicinity of cholinergic nerve terminals and thus are capable of producing effects equivalent to excessive stimulation of cholinergic receptors throughout the central and peripheral nervous systems. These anti-ChE agents are used extensively in agricultural insecticides, in pesticides, and as potential chemical warfare "nerve gas." Coincidentally, several compounds in this class of agents are used therapeutically and others that cross the blood–brain barrier have been approved or are in clinical trials for treatment of Alzheimer's disease.[36]

The basic structure of organophosphates consists of a phosphate radical with various functional groups substituted for any or all of the oxygen atoms. Several examples of specific organophosphate compounds are shown in Table 22-11.[36]

Carbamate-containing insecticides include carbaryl ethienocarb (Sevin) and carbofuran (Furadan). The basic structural component of carbamates is an esterified form of carbamic acid with substitutions occurring on the nitrogen atom.

Occupational exposure occurs commonly on skin surfaces and via inhalation into the lungs, whereas oral ingestion is a common route in cases of nonoccupational poisoning. Toxic exposure is a primary concern for children because the developing nervous system may be particularly susceptible to specific agents. The effects of acute exposure to anti-ChE compounds are characterized by muscarinic and nicotinic signs and symptoms. Muscarinic symptoms include bronchospasms or laryngeal spasms that may lead to a compromised airway. Nicotinic actions at the neuromuscular junction of skeletal muscle consist of fatigue and generalized weakness, involuntary twitching, and eventually severe weakness and paralysis.

Ocular manifestations include marked miosis, ocular pain, conjuctival congestion, diminished vision, and ciliary spasm. Changes in respiratory function are characterized by rhinorrhea (thin watery discharge from the nose) and hyperemia (presence of an unusual amount of blood) of the upper respiratory tract, tightness in the chest, wheezing respiration, and increased bronchial secretion. GI symptoms occur after ingestion and include anorexia, nausea, vomiting, abdominal cramps, and diarrhea. Measurement of cholinesterases is appropriate for individuals who have been exposed to insecticides listed in Table 22-11. Organophosphate-containing

TABLE 22-11

Partial listing of chemicals classified as organophosphates, including their proprietary names and use

Common Name (trade name)	Comments
DFP; isoflurophate	Potent, irreversible inactivator
Tabun; Ethyl N-dimethylphosphoramidocyanidate	Extremely toxic "nerve gas"
Sarin (GB) isopropyl methylphosphonofluoridate	Extremely toxic "nerve gas"
Soman (GD)Pinacolyl methylphosphonofluoridate	Extremely toxic "nerve gas"
Paraoxon (MINTACOL) O,O-diethyl O-(4-nitrophenyl)-phosphate	Active metabolite of parathion
Malaoxon, O,O-dimethyl S-(1,2-dicarboxyethyl)-phosphorothioate	Active metabolite of malathion
Parathion, O,O-diethyl O-(4-nitrophenyl)-phosphorothioate	Agricultural insecticide
Diazinon, O,O-diethyl O-(2-isopropyl-6-methyl-4-pyrimidinyl) phosphorothioate	Insecticide used for gardening and agriculture
Malathion, O,O-dimethyl S-(1,2-dicarbethoxyethyl) phosphorodithioate	Used as an insecticide

compounds inhibit ChE activity. If enough of the toxic substance is absorbed to inactivate the entire AChE of nervous tissue, death will result. A 40% drop in ChE activity occurs before the first symptoms develop, and a drop of 80% is required before neuromuscular effects become apparent.

Laboratory assays are available to measure serum butyrylcholinesterase (pseudocholinesterase) activity and are useful for diagnosing acute ingestion or for monitoring chronic exposure. The best definitive approach for determining organophosphate or carbamate insecticide exposure is to measure their urinary metabolites using GCMS.

GAMMA-HYDROXYBUTYRATE (GHB)

Gamma-hydroxybutyrate (GHB) and its metabolic precursors, gamma-butyrolactone (GBL) and 1, 4-butanediol (BDL), have become popular drugs of abuse in adolescents and adults. These substances have been implicated in several celebrity overdoses, and GHB has been popularized as a "date-rape drug" by the media. GHB is a CNS depressant that is structurally similar to the inhibitory neurotransmitter γ-aminobutyric acid (GABA). These compounds cause deep but short-lived coma; the coma often lasts only 1 to 4 hours. GHB is claimed to produce a variety of desirable effects, including improved athletic performance, sleep, sexual prowess, mood elevation, and euphoria.

Once ingested, GBL and BDL are rapidly metabolized to GHB. GBL in the presence of lactonase is converted to GHB. BDL via alcohol dehydrogenase and aldehyde dehydrogenase is also converted to GHB. GHB is rapidly absorbed, with peak plasma concentrations occurring 20–60 minutes after oral administration. The onset of action is fairly rapid, often within 15 minutes. GHB has a half-life of 0.3–1.0 hour and a volume of distribution of 0.4 L/kg. Less than 5% of the dose is eliminated unchanged in the urine. Approximate detection times are 6–8 hours in blood and 12 hours in urine.[37]

GHB is a relatively fast-acting depressant. Within approximately 15 minutes of administration, an individual may begin to experience drowsiness, dizziness, euphoria, nauseas, and visual disturbances. The person may also become unconscious. The duration of action is about 3 hours. Some users present with combative or aggressive behavior. Emergence delirium, characterized by myoclonic jerking, confusion, and agitation, is usually transient and lasts about 30 minutes.

Management of acute GHB intoxication focuses on alleviating symptoms and providing support airway management, mechanical ventilation, and prevention of aspiration, and measures to counter bradycardia are commonly required. Extubation following 2–6 hours of mechanical ventilation is common. Spontaneous recovery from GHB overdose is common usually within a few hours. Patients often have no recollection of having taken GHB due to the anterograde amnesia associated with the drug.

Analysis of specimens in suspected cases of GHB intoxication involve measurement of GHB due to the efficient transformation of its precursors BDL and GBL in vivo. In antemortem samples, metabolic precursors like BDL and GBL are usually undetectable. GBL is not detected in blood or urine following administration of GHB, indicating that lactonization of GHB does not occur in vivo.

In vitro conversion of GHB to GBL has been widely used for toxicological analyses. GBL is easier to analyze using conventional methods of extraction and GC compared with GHB, which is considerably more polar and less volatile. Despite its low molecular weight (86 Da), GBL has been analyzed directly using GCMS. The mass-to-charge ratios are low following mass analysis; therefore, an additional extract that does not undergo acidification should also be analyzed to determine whether intact GBL is present in the sample.

KETAMINE (KETALAR AND KETAJECT)

Ketamine is a nonbarbiturate anesthetic compound structurally similar to PCP and was developed in the 1960s to replace PCP as a surgical anesthetic drug. Effects of the drug include sedation, immobilization, amnesia, analgesia, and a trancelike cataleptic feature, wherein the patient seems to be "dissociated" from his environment but not actually asleep. Ketamine is used as a short-term anesthetic for painful diagnostic or surgical procedures in children. Ketamine is an abused drug that can result in death. An overdose of ketamine can produce sedation and respiratory depression. It may also result in tachycardia, muscle rigidity, hypertension, and delirium.

Therapeutic dosing of ketamine for induction of anesthesia is ~1–4.5 mg/kg IV administration over 1 minute, or it may be given IM to adults. The dose for abuse is ~1–2 mg/kg IM or IV. The usual adult abuse dose is ~100–200 mg IM or subcutaneously. Fatal doses of the drug have occurred but are not common.[38]

Oral administration of ketamine (tablet form) is not the most effective route of administration. Abusers usually snort the drug, smoke it, inject it, or on occasion (as with date rape incidents) mix it in drinks. Ketamine undergoes first-pass metabolism, in which it undergoes hepatic N-demethylation by microsomal cytochrome P450 biotransformation to the active metabolite norketamine. Ketamine is also hydroxylated and conjugated with glucuronic acid followed by excretion of metabolites in the urine. Ketamine is a noncompetitive antagonist of N-methyl-D-aspartate receptors. The cerebral cortex and limbic system are the primary sites of action. Ketamine causes electrophysiological dissociation between the limbic and cortical system. It also inhibits the reuptake of catecholamine, which causes an increase in sympathetic activity, hypertension, and tachycardia.

Clinical presentation and treatment vary depending on the dose of ketamine and whether the patient has taken other drugs that are either administered therapeutically or used illegally. Patients can present with amnesia and cardiovascular and CNS effects. Blood concentrations of ketamine are not usually helpful in managing patients who have overdosed. Ketamine can result in a false-positive PCP urine drug screen. Definitive methods for measuring ketamine are GC or GCMS, and therefore ketamine is not usually measured in a hospital-based clinical chemistry laboratory because they are not equipped with this type of instrumentation. Larger reference laboratories have assays to detect and quantitate blood and urine ketamine levels.

Checkpoint! 22–8

Gamma hydroxybutyrate and ketamine are commonly referred to as date rape drugs, club drugs, and rave drugs. Identify several adverse effects of these drugs on humans.

LYSERGIC ACID DIETHYLAMIDE (LSD)

Lysergic acid diethylamide (LSD) is classified as a hallucinogenic drug due to its ability to produce distortions of reality. Chemically, LSD is classified as an indolamine and is a potent drug with a high affinity for serotonin 5HT-2 receptors (5-hydroxytryptamine). LSD affects primarily the cerebral cortex and locus ceruleus. The locus ceruleus is a nucleus in the brain stem responsible for physiological responses to stress and panic. The nucleus contains norepinephrine and is found in the upper pons, below the cerebellum in the caudal midbrain.

The route of administration is usually oral, but the drug can be smoked or injected. Onset of symptoms is ~30–90 minutes, and symptoms peak ~2–4 hours. Duration of activity is ~6–8 hours. Metabolism of LSD is by hepatic hydroxylation, and serum half-life is 2.5 hours.

Individuals using LSD experience a variety of clinical effects that vary among users. A common effect is altered visual perception, and many users may be hypersensitive to sound. Rapid mood swings may occur, ranging from euphoria to fear. Serious effects of LSD use include anxiety, despair, suicidal thoughts, and panic (these serious effects are often referred to as a "bad trip" and may also include feelings of insanity, loss of control, or impending death).

Signs and symptoms reflect sympathomimetic effects and include the following:

- Mydriasis
- Flushing
- Tremor
- Hyperthermia
- Tachycardia
- Increased blood pressure
- Diaphoresis
- Ataxia

Severe toxicity can result in coma, seizures, and respiratory arrest. Another important concern regarding the use of LSD is drug interaction. Patients who are taking certain medications and use LSD may experience serious side effects. For example, patients taking the antidepressant drug fluoxetine (Prozac) and lithium (for manic depression) may develop seizures after taking LSD. Another example involves patients who currently use LSD and begin taking serotonin reuptake inhibitor drugs such as sertraline (Zoloft) or paroxetine (Paxil). These drug combinations may result in the exacerbation of LSD-like panic and visual symptoms.

Treatment for LSD exposure begins with the caregiver providing a nonthreatening environment and possibly benzodiazepine for anxiety or agitation. Thus a sample for drug abuse testing may be positive for benzodiazepine in these patients.

Urine drug abuse screening for LSD is not available in most clinical chemistry laboratories. However, several nonradioisotopic immunoassays are available, including the following: (1) Microgenics (Freemont, CA) CEDIA (cloned enzyme donor immunoassay), (2) Dade Behring (Deerfield, IL) EMIT II LSD assay, and (3) Roche (Nutley, NJ) Online KIMS (kinetic interaction of microparticle in solution).

Historically, blood radioimmunoassay (RIA) techniques have been used to provide a sensitive and specific assay for quantitative measurement of LSD. These assays have been replaced by nonradioisotopic techniques (e.g., GCMS and LCMS), which are used for confirmation of urine screening procedures.

URINE SPECIMEN VALIDITY

Specimens submitted for drug screening should be tested for possible adulteration, dilution, or substitution. An adulterated urine specimen can be described as a specimen containing a substance that is not a normal constituent or containing an endogenous substance at a concentration that is not a normal physiological concentration. A *dilute* urine specimen is a specimen with creatinine and specific gravity values that are lower than expected for human urine. A *substituted* urine specimen is a specimen with creatinine and specific gravity values that are so diminished or so divergent that they are not consistent with normal human urine.

The entire list of criteria for establishing whether a specimen has been adulterated, diluted, or substituted is presented in the references cited.[39,40,41] A brief description of these criteria is presented below.

A specimen is considered to have been adulterated if it meets specific criteria established by Substance Abuse and Mental Health Administration (SAMHSA), which includes in part the following:

1. pH less than 3 or greater than or equal to 11.
2. Nitrite concentration greater than or equal to 500 μg/mL.
3. The presence of chromium (VI).
4. The presence of halogens (e.g., bleach, iodine, fluoride).
5. The presence of glutaraldehyde.
6. The presence of pyridine (pyridinium chlorochromate).
7. The presence of a surfactant.
8. The presence of any other adulterant not specified in 3 through 7 is verified using an initial test on the first aliquot and a different confirmatory test on the second aliquot.

Urine specimens can be adultered by simply adding common household chemicals, such as vinegar or ammonia solutions, that may interfere with the drug testing methods. There are also several commercially available drug testing

BOX 22-2 Examples of adulterants available to individuals to use to modify urine specimens submitted for drug testing. These substances are designed to remove unwanted substances or to interfere with chemical reactions

- Mary Jane's SuperClean 13
- Klear
- Whizzies
- UrinAid
- Urine Luck
- Stealth
- Instant Clean ADD-IT-IVE

adulterant products that can be used to remove unwanted substances from the urine or interfere with the chemical reactions of the urine adulteration test strips (discussed below). Examples of adulterants available on the market for sale to individuals are shown in Box 22-2 ■.

A urine specimen is reported as substituted when the creatinine concentration is less than 2 mg/dL and the specific gravity is ≤1.0010 or ≥1.0200 on both the initial and confirmatory creatinine tests (i.e., the same colorimetric test may be used to test both aliquots) and on both the initial and confirmatory specific gravity tests (i.e., a refractometer is used to test both aliquots) on two separate aliquots.

A urine specimen is reported dilute when the creatinine concentration is ≥2 mg/dL but ≤20 mg/dL and the specific gravity is ≥1.0010 but ≤1.0030 on a single aliquot. Laboratories can test for urine specimen integrity using urine adulteration test strips. These strips provide a quick and easy way to screen urine specimens before analysis. Examples of two commercially available test strips are as follows:

1. Intect 7 (Bioscan Screening Systems, Inc., Smyrna, TN), which tests for the following:

   Creatinine
   Nitrite
   Glutaraladehyde
   pH
   Specific gravity

Bleach
Pyridinium chlorochromate

2. AdultaCheck 6 (Diagnostics Inc., Asheville, NC), which tests for the following:

   Oxidants
   Creatinine
   Nitrite
   Glutaraldehyde
   pH
   Chromate

The seven tested substances in the Intect 7 and the six in the AdultaCheck 6 test strip can reveal individuals' attempts to dilute or adulterate their urine specimen. A positive pyridinium chlorochromate may indicate the presence of an oxidant added to the specimen by the individual. A positive nitrite, glutaraldehyde, or bleach indicates an individual's attempt to foil drug tests by interfering with the reagents used, confounding result interpretation or influencing components of the drug testing device.

SUMMARY

Toxicology testing varies significantly in laboratories throughout the United States. Many laboratories perform no toxicology testing at all, whereas many provide a limited number of tests for the emergency department, including tests for alcohol, acetaminophen, and salicylate. Some laboratories provide a large menu of toxicology tests and services. Toxicology testing is an example of a controlled service provided to caregivers. Each state provides and regulates the types of testing and analytes to be tested using state-issued permits.

Several examples of analytes detected by laboratories not involved in forensic toxicology were discussed in this chapter. The primary emphasis was on drug testing in laboratories providing patient care for emergency departments. Reference was made to drug testing in reference laboratories and /or larger point-of-care laboratories or group practices.

The focus of discussion for most analytes or drug groups included a brief description of pharmacology, methods, instrumentation, and limitations, if any.

REVIEW QUESTIONS

LEVEL I

1. Salicylate intoxication can result in a metabolic acidosis due to the accumulation of which of the following acids? (Objective 7)
   a. salicylic acid
   b. lactic acid
   c. pyruvic acid
   d. all of the above

LEVEL II

1. Toxic doses of acetaminophen can cause serious damage to which of the following organs? (Objective 2)
   a. lungs
   b. heart
   c. liver
   d. brain

REVIEW QUESTIONS *(continued)*

LEVEL I

2. Most photometric methods used to measure salicylate concentration in serum are based on which of the following methods? (Objective 2)
   a. Rosalki
   b. Trinder
   c. Henry
   d. Karmen

3. Which of the following enzymes is used to provide a specific measurement of ethanol in serum? (Objective 2)
   a. alcohol dehydrogenase
   b. pseudocholinesterase
   c. alcohol dehydratase
   d. alcohol synthetase

4. Carboxyhemoglobin can be measured by which of the following techniques? (Objective 2)
   a. Betherlot reaction
   b. anodic stripping voltammetry
   c. a PO_2 electrode
   d. cooximetry

5. Ethylene glycol metabolizes to which of the following acids? (Objective 7)
   a. acetic acid
   b. oxalic acid
   c. gamma-aminobutyric acid
   d. Formic acid

6. Which of the following instruments can be used to measure toxic substances in blood or urine? (Objective 3)
   a. osmometer
   b. refractometer
   c. GCMS
   d. electrophoresis system

7. Lead inhibits which of the following enzymes? (Objective 3)
   a. pseudocholinesterase and gamma-glutamylhydroxylase
   b. 5'-nucleotidase and ferricochelatase
   c. aminolevulinic acid dehydrase and ferrochelatase
   d. delta aminolevulinic acid dehyroxylase and ferrochelatase

8. Which of the following classes of compounds may be included in drugs-of-abuse screening (DAU)? (Objective 5)
   a. barbiturates
   b. alcohols
   c. sterols
   d. prostaglandins

9. Which of the following is an example of a definitive or reference method that is suitable for confirming positive drugs-of-abuse testing (DAU)? (Objective 2)
   a. immunoassays
   b. HPLC
   c. GCMS
   d. electrophoresis

LEVEL II

2. Measurement of serum acetaminophen concentrations is important for which of the following? (Objective 8)
   a. assessment of the severity of methanol intoxication
   b. prognosis for patient with lung cancer
   c. assessment of liver damage
   d. presumption diagnosis of acute pancreatitis

3. Initially, salicylate poisoning results in which of the following? (Objective 2)
   a. hyperventilation
   b. cardiac arrest
   c. liver disease
   d. metabolic alkalosis

4. Toxic doses of salicylate can adversely affect which of the following? (Objective 5)
   a. white blood cell count
   b. red blood cell count
   c. platelet aggregation
   d. lymphocyte count

5. The pharmacological effects of ethanol at the cellular level are accurately described as which of the following actions? (Objective 2)
   a. antidepressant
   b. depressant
   c. vasodilator
   d. antiasthmatic

6. Carbon monoxide affinity for hemoglobin is: (Objective 4)
   a. lower than oxygen's affinity for hemoglobin.
   b. 200–250 times greater than oxygen's affinity for hemoglobin.
   c. about the same as oxygen;s affinity for hemoglobin.
   d. negliable.

7. Which of the following statements provides an accurate description of drug screening methods? (Objective 1)
   a. They provide accurate quantitative measurements of drug concentrations in biological specimens.
   b. They are highly specific for one particular drug within a class of drugs.
   c. They provide qualitative information on a particular class of drugs.
   d. Only a definitive or reference technique such as GCMS can be used to screen for drugs in biological specimens.

8. Reye's syndrome is associated with which of the following substances? (Objective 8)
   a. ethanol
   b. cocaine
   c. salicylate (aspirin)
   d. phenothiazines

REVIEW QUESTIONS *(continued)*

LEVEL I

LEVEL II

9. The mechanism of action of organophosphates, such as Malathion, with acetylcholinesterase is best described by which of the following statements? (Objective 4)
   a. Organophosphates enhance the production of acetyl-cholinesterase.
   b. Organophosphates inhibit acetylcholinesterase.
   c. Organophosphates facilitate the production of acetylcholine via interaction with acetylcholinesterase.
   d. Organophosphates cause a depletion of acetylcholine in the vicinity of cholinergic nerve terminals.

10. Gamma hydroxybutyrate (GHB) ingestion by humans results in which of the following physical/mental effects? (Objective 5)
    a. drunken stupor
    b. a deep but short-lived coma
    c. hallucinations
    d. ocular manifestations, including marked miosis, ocular pain, and conjuctival congestion

PEARSON myhealthprofessionskit™

Use this address to access the interactive Companion Website created for this textbook. Simply select "Clinical Laboratory Science" from the choice of disciplines. Find this book and log in using your user name and password.

REFERENCES

1. Osterhoudt KC. The lexiconography of toxicology. *JMedToxic* (2006) 2(1):1–3.

2. Bronstein AC, Spyker DA, Cantilena LR et al. 2007 Annual Report of the American Association of Poison Control Centers' National Poisoning and Exposure Database. Clinical Toxicology: The Official Journal of the American Academy of Clinical Toxicology & European Association of Poisons Centres & Clinical Toxicologists. *Clinical Toxicology* (2008) 46:930–31.

3. Bronstein AC, Spyker DA, Cantilena LR et al. 2007 Annual Report of the American Association of Poison Control Centers' National Poisoning and Exposure Database. Clinical Toxicology: The Official Journal of the American Academy of Clinical Toxicology & European Association of Poisons Centres & Clinical Toxicologists. *Clinical Toxicology* (2008) 46:931–32.

4. Bronstein AC, Spyker DA, Cantilena LR et al. 2007 Annual Report of the American Association of Poison Control Centers' National Poisoning and Exposure Database. Clinical Toxicology: The Official Journal of the American Academy of Clinical Toxicology & European Association of Poisons Centres & Clinical Toxicologists. *Clinical Toxicology* (2008) 46:944–945.

5. Bronstein AC, Spyker DA, Cantilena LR,et al. 2007 Annual Report of the American Association of Poison Control Centers' National Poisoning and Exposure Database. Clinical Toxicology: The Official Journal of the American Academy of Clinical Toxicology & European Association of Poisons Centres & Clinical Toxicologists. *Clinical Toxicology* (2008) 46:1009–1015.

6. Rang HP, Dale MM, Ritter PK et.al. *Pharmacology*, 5th ed. New York: Churchill Livingstone, 2003:251–52.

7. Rumack BH, Matthew H. Acetaminophen poisoning and toxicity. *Pediatrics* (1975) 55:871–76.

8. Trinder P. Rapid determination of salicylate in biological fluids. *BiochemJ* (1954) 57:301–303.

9. Jarvie DR, Heyworth R, Simpson D. Plasma salicylate analysis: A comparison of colorimetric, HPLC and enzymatic techniques. *AnnClinBiochem* (1987) 4:364–73.

10. Morris HC, Overton PD, Ramsay JR et.al. Development and validation of an automated, enzyme-mediated colorimetric assay of salicylate in serum. *ClinChem* (1990) 36:131–35.

11. Rang HP, Dale MM, Ritter PK et.al. *Pharmacology*, 5th ed. New York: Churchill Livingstone, 2003:603–605.

12. Rang HP, Dale MM, Ritter PK et.al. *Pharmacology*. 5th ed. New York: Churchill Livingstone, 2003:605–606.

13. Weaver LK. Carbon monoxide. *In*: Dart RC (ed.). *Medical toxicology*, 3rd ed. Philadelphia: Lippincott Williams & Wilkins, 2004:1146–152.

14. Porter WH. Clinical toxicology. *In*: Burtis CA, Ashwood ER, Bruns DE (eds.). *Tietz textbook of clinical chemistry and molecular diagnostics*, 4th ed. Philadelphia: W. B. Saunders, 2006:1296–98.

15. Rang HP, Dale MM, Ritter PK et al. *Pharmacology*, 5th ed. New York: Churchill Livingstone, 2003:585–90.

16. Charney DS, Mihic SJ, Harris RA. Hypnotics and sedative. *In*: Brunton LL, Lazo JS, Parker KL (eds.). *Goodman and Gilman's the pharmacological basis of therapeutics*, 11th ed. New York: McGraw-Hill, 2006:415–16.

17. Charney DS, Mihic SJ, Harris RA. Hypnotics and sedative. *In*: Brunton LL, Lazo JS, Parker KL (eds.) *Goodman and Gilman's the pharmacological basis of therapeutics*, 11th ed. New York: McGraw-Hill, 2006:410–11.

18. Rang HP, Dale MM, Ritter PK et al. *Pharmacology*, 5th ed. New York: Churchill Livingstone, 2003:522–24.

19. Gutstein HB, Huda Akil. Opioid analgesics. *In*: Brunton LL, Lazo JS, Parker KL (eds.). *Goodman and Gilman's the pharmacological basis of therapeutics*, 11th ed. New York: McGraw-Hill, 2006:572–74.

20. Gutsein HB, Akil H. Opioid analgesics. *In*: Brunton LL, Lazo JS, Parker KL (eds.). *Goodman and Gilman's the pharmacological basis of therapeutics*, 11th ed. New York: McGraw-Hill, 2006:579–82.

21. Rang HP, Dale MM, Ritter PK et al. *Pharmacology*, 5th ed. New York: Churchill Livingstone, 2003:580–81.

22. Rang HP, Dale MM, Ritter PK et al. *Pharmacology*, 5th ed. New York: Churchill Livingstone, 2003:592–93.

23. Baldessarini RJ. Drug therapy of depression and anxiety disorders. *In*: Brunton LL, Lazo JS, Parker KL (eds.). *Goodman and Gilman's the pharmacological basis of therapeutics*, 11th ed. New York: McGraw-Hill, 2006:432–35.

24. Rang HP, Dale MM, Ritter PK et al. *Pharmacology*, 5th ed. New York: Churchill Livingstone, 2003:542–44.

25. Agency for Toxic Substances and Disease Registry. 2005. Toxicological profile for lead. Atlanta, GA: U.S. Department of Health and Human Services, Public Health Service, Agency for Toxic Substances and Disease Registry. http://www.atsdr.cdc.gov (accesssed January 15, 2010).

26. Silbergeld EK. Preventing lead poisoning in children. *AnnReview PublicHealth* (1997) 18:187–210.

27. Centers for Disease Control. Preventing Lead Poisoning in Young Children: A statement by the centers for Disease Control. 1985. Atlanta: US Department of Health and Human Services, Public Health Service, 1985; DHHS 99-2230.

28. Committee on Environmental Health. Policy statement. *Pediatrics* (2005) 116: 1036–1046.

29. Centers for Disease Control and Prevention. *Managing elevated blood lead levels among young children: recommendations from the advisory committee on childhood lead poisoning prevention.* Atlanta, GA: Centers for Disease Control and Prevention; 2002. Available at www.cdc.gov/nceh/lead/CaseManagement/caseManagement_main.htm (accessed January 15, 2010).

30. Klaassen CD. Heavy metals and heavy-metal antagonists. *In*: Brunton LL, Lazo JS, Parker KL (eds.). *Goodman and Gilman's the pharmacological basis of therapeutics*, 11th ed. New York: McGraw-Hill, 2006:1758–59.

31. Blumberg WE, Eisinger J, Lamola AA et al. Zinc protoporphyrin level in blood determined by a portable hematofluorometer: A screening device for lead poisoning. *JClinLabMed* (1977) 89(4):712–23.

32. Silvilotti MLA. Toxic alcohols and their derivatives. *In*: Dart RC (ed.). *Medical toxicology*, 3rd ed. Philadelphia: Lippincott Williams & Wilkins, 2004:1217–20.

33. Jolliff HA, Sivilotti MLA. Ethylene glycol. *In*: Dart RC (ed.). *Medical toxicology*, 3rd ed. Philadelphia: Lippincott Williams & Wilkins, 2004:1223–30.

34. Sivilotti ML. Toxic alcohols and their derivatives. *In*: Dart RC (ed.). *Medical toxicology*, 3rd ed. Philadelphia: Lippincott Williams & Wilkins, 2004:1215–16.

35. Baldessarini RJ. Drug therapy of depression and anxiety disorders. *In*: Brunton LL, Lazo JS, Parker KL (eds.). *Goodman and Gilman's the pharmacological basis of therapeutics*, 11th ed. New York: McGraw-Hill, 2006:463–66.

36. Taylor P. Anticholinesterase Agents. *In*: Brunton LL, Lazo JS, Parker KL (eds.). *Goodman and Gilman's the pharmacological basis of therapeutics*, 11th ed. New York: McGraw-Hill, 2006:206–07.

37. Thai D, Dyer JE, Jacob P et al. Clinical pharmacology of 1, 4-butanediol and gamma-hydroxybutyrate after oral 1, 4-butanediol administration to healthy volunteers. *ClinPharm&Therap* (2007) 81(2):178–84.

38. Evers AS, Crowder CM, Balser JR. Genera anesthetics. *In*: Brunton LL, Lazo JS, Parker KL (eds.). *Goodman and Gilman's the pharmacological basis of therapeutics*, 11th ed. New York: McGraw-Hill, 2006:351–52.

39. From the Mandatory Guidelines for Federal Workplace Drug Testing Programs published in the *Federal Register* on April 13, 2004 (59 FR 29908), effective June 9, 1994.

40. From the Mandatory Guidelines for Federal Workplace Drug Testing Programs published in the *Federal Register* on April 13, 2004 (69 FR 19644), effective November 1, 2004.

41. United States Department of Transportation, *Federal Register* 49 CFR Part 40.

23

Trace Elements

■ OBJECTIVES—LEVEL I

Following successful completion of this chapter, the learner will be able to:

1. List several examples of trace elements.

2. Identify methods used to measure selected trace elements.

3. Identify types of instrumentation used to measure trace elements.

4. Cite biological uses of selected trace elements.

5. State food and other sources of selected trace elements.

6. Identify examples of types of specimens used for laboratory assessment of trace elements.

7. Name 10 essential trace elements.

8. Define the following terms: trace elements, ultra trace elements, chelating agents, and metalloproteins.

9. Name specific organs affected by the presence of trace elements.

10. Identify routes of expose for selected trace elements.

■ OBJECTIVES—LEVEL II

Following successful completion of this chapter, the learner will be able to:

1. Correlate laboratory results with diseases associated with trace elements.

2. Explain the mechanism of toxicity of selected trace elements.

3. Summarize the chemistry of selected trace metals.

4. Identify treatment modalities used after exposure to a toxic amount of trace elements.

5. Select the proper diagnostic tests to use for patients exposed to selected trace elements.

6. Describe symptoms associated with exposure to selected trace elements.

KEY TERMS

Alopecia	Essential trace elements	Trace elements
Argyria	Metalloenzyme	Succimer
British anti-Lewisite (BAL)	Metalloprotein	Ultratrace elements
Chelating agent	Prion protein	Wilson disease

 A CASE IN POINT

A 25-year-old male presented to the hospital emergency department (ED) complaining of headaches, nausea, dizziness, and episodes of repetitive vomiting. The patient was conscious and denied having chest pain, shortness of breath, abdominal pain, diarrhea, and hematemesis. He also stated that he did not smoke, drink, or use illicit drugs.

While being interviewed further, the patient did reveal to the clinician that he had attempted suicide in the past and had a history of depression. He was currently taking olanzapine (Zyprexa, Eli Lilly Co.), which is an antipsychotic and benzodiazepine derivative, and venlafaxine (Effexor, Wyeth Laboratory), an antidepressant medication, on a daily basis.

The patient also disclosed to the clinician that he had purchased more than 1 gram of the organic form of arsenic trioxide (Trisenox) in liquid form from an Internet auction site for less than $100. Trisenox is a chemotherapeutic agent used to treat various cancers such as leukemia. The drug was shipped to him in an unmarked clear plastic Ziploc bag. He dissolved the drug in water and consumed it about 12 hours before his ED admission.

A physical exam including head, eye, ear, nose, and throat (HEENT) revealed nothing abnormal, and his vital signs were normal except for an increased heart rate (tachycardia) of 112 beats/minute. An electrocardiogram revealed a sinus tachycardia, and a chest X-ray was unremarkable. His kidney–ureter–bladder X-ray, which also included the abdomen, showed a high-density material within the distal area of the stomach.

Results of the initial laboratory tests are as follow.

Serum Chemistries	Results (Conventional Units)	Reference Interval (Conventional Units)
Sodium	135	136–145 mEq/L
Potassium	4.5	3.5–5.1 mEq/L
Chloride	108	98–107 mEq/L
Carbon dioxide	7	23–29 mEq/L
Anion gap	20	6–10 mEq/L
Glucose	162	74–100 mg/dL
Creatinine	1.5	0.9–1.3 mg/dL
Urea nitrogen	17	6–20 mg/dL
Calculated osmolality	275	282–300 mOsm/Kg
Ethanol	<10 mg/dL	N/A
Acetaminophen	<2.5 μg/mL	N/A
Salicylate	<2.8 μg/mL	N/A

Urine Drug Abuse Screen

Opiates	Negative
Methadone	Negative
Amphetamines	Negative
Barbiturates	Negative
Benzodiazepines	Positive
Cocaine	Negative
Phencyclidine	Negative

Hematology Results

WBC	11×10^9/L	$4.0–11.0 \times 10^9$/L
RBC	4.3×10^{12}/L	$4.70–6.10 \times 10^{12}$/L
Hemoglobin	11.0	13.0–18.0 g/dL
Hematocrit	34.9	39–50%
Platelets	110×10^9/L	$140–440 \times 10^9$/L

Urinalysis

Color amber

Appearance hazy

pH = 6

Specific gravity = 1.028

All other dipstick results are negative

Microscopic analysis: Nothing abnormal

Issues and Questions to Consider

1. What additional clinical laboratory tests should the physician consider?

2. Are the patient's signs and symptoms consistent with any trace-element poisoning?

3. What laboratory instrumentation is used to test for trace-element content in biological samples?

The results of additional laboratory tests follow.

	Patient's Results	RI/Cutoff Values
Serum arsenic concentration	<2.0	2–23 μg/L
24-hr urine arsenic concentration	9950	<100 μg/g creatinine

Other Trace-Element Whole-Blood Panel

Mercury	<1	1.0–59 μg/L
Thallium	<1	<5 μg/L
Lead	<1	<25 μg/dL
Hair arsenic concentration	6.5	<1 μg/g

Issues and Questions to Consider

4. Why is the serum arsenic concentration so low?

5. Why is the urine arsenic concentration so high?

6. What are some pharmacokinetic issues that may be relevant to this patient?

7. Identify possible treatment modalities associated with arsenic poisoning.

WHAT'S AHEAD

▶ Biochemistry of trace elements.
▶ Physiology of trace elements.
▶ Dietary features of trace elements.
▶ Biological functions of trace elements.
▶ Clinical implications of deficiencies and toxic amounts of trace elements in body tissues.
▶ Laboratories roles and instrumentation used for trace-element analysis.

▶ INTRODUCTION

Trace elements are in reference to a group of elements that include mostly metals (for example, zinc) that serve a biological role in humans as well as in animals. The word *trace* was originally used because quantitation with the analytical methods available at the time was not possible. Trace elements can also be described as elements that are present in μg/dL in body fluids and mg/kg in tissues. Another designation commonly used for elements found at ng/dL or μg/kg is **ultratrace elements**.

Trace elements may be classified as **essential trace elements**, possibly essential, and not essential to humans. A listing of elements within each class is shown in Table 23-1 ✪. An element is considered essential when the signs and symptoms that appear when a diet is deficient in the element are uniquely reversed when an adequate supply of the particular trace element is present.[1] Ten elements are recognized as essential with copper, iodine, iron, selenium, and zinc being associated with well-characterized deficiency states. Several elements (e.g., chromium, fluorine, and manganese) do not have clearly definitive biochemical roles, yet signs of deficiency have been documented.[2] Other elements that are included in the possibly essential category have been shown to have essential biochemical roles in some animal species and are likely to be in humans as well.[3]

Conclusive characterization of trace or ultratrace elements is often difficult, especially when attempting to evaluate whether or not they are essential. This is because of (1) the widespread distribution of these elements in the environment and food supply, which often causes contamination of the testing systems, and (2) only minute amounts are needed to support physiologic processes. A summary of the roles and conditions associated with deficiencies and toxicities are shown in Table 23-2 ✪.

✪ TABLE 23-1

Essential, Possibly Essential, and Nonessential Classification of Trace Elements[4]

Essential	Possibly Essential	Not Essential
Chromium	Arsenic	Aluminum
Cobalt	Boron	Antimony
Copper	Lithium	Bismuth
Fluorine	Nickel	Germanium
Iodine	Silicon	Mercury
Iron	Vanadium	Silver
Manganese		Thallium
Molybdenum		Titanium
Selenium		
Zinc		

✪ TABLE 23-2

The Role of Essential Trace Elements and Associated Clinical Disorders

Element	Role (s)	Deficiency	Toxicity
Aluminum	Not an essential trace element Not clear	Not significant*	May affect bone formation and remodeling May be associated with Alzheimer's disease Can lead to dialysis encephalopathy or dialysis dementia
Antimony	Not an essential trace element Not clear	Not significant*	Pneumoconiosis Affects heart Dermatitis Spontaneous abortions and premature births Lymphocytosis and reduction in leukocytes and platelets
Arsenic	Possibly an essential trace element Not clear	Not significant*	Known carcinogen GI tract involvement Cardiac effects Encephalopathy Lung injury Renal failure

✪ TABLE 23-2 (continued)

Element	Role (s)	Deficiency	Toxicity
Beryllium	Not necessary for human health	Not significant*	Chronic Be disease Significant lung involvement
Cadmium	No known biological role	Not significant*	Nephrotoxicity Chronic emphysema Cancer Osteomalacia
Chromium	Metabolism of glucose	Impaired glucose tolerance	Occupational: renal failure, dermatitis, pulmonary cancer
Cobalt	Component of vitamin B_{12}	Vitamin B_{12} deficiency anemia	Developmental defects Male sterility Testicular atrophy
Copper	Cofactor for oxidase enzymes	Anemia Growth retardation Defective keratinization and pigmentation of hair Hypothermia Degenerative changes in aortic elastin Osteopenia Mental deterioration	Nausea and vomiting Diarrhea Hepatic failure Tremor Mental deterioration Hemolytic anemia Renal dysfunction
Fluorine	Prevents tooth decay	Increased dental caries	Fluorosis and mottled enamel
Manganese	Required for glycoprotein and proteoglycan synthesis	Impaired growth and skeletal development, reproduction, lipid and carbohydrate metabolism; upper body rash	General: neurotoxicity, Parkinson-like symptoms Occupational: encephalitis-like syndrome, Parkinson-like syndrome, psychosis, pneumoconiosis
Mercury	None found	Not significant	Brain involvement Gastroenteritis Interstitial pneumonitis Cardiac problems Renal tubular dysfunction
Molybdenum	Component of sulfite and xanthenes oxidases	Severe neurologic abnormalities	Reproductive and fetal abnormalities
Nickel	May be essential for life but unclear	Not significant	Dermatitis Pulmonary, hepatic, and neurologic dysfunction
Selenium	Component of glutathione peroxidase and iodinothyroneine-5′-deiodinase	Cardiomyopathy Heart failure Striated muscle degeneration	General: Alopecia, nausea, vomiting, abnormal nails, emotional liability, peripheral neuropathy, lassitude, garlic odor to breath, dermatitis Occupational: Lung and nasal carcinomas, liver necrosis, pulmonary inflammation
Silver	Not clear	Not significant	Argyria Hemopoiesis Cardiac enlargement Degeneration of the liver and destruction of renal tubules Growth retardation
Thallium	Not clear	Not significant	Alopecia (hair loss) Peripheral neuropathy Renal failure Seizures
Zinc	Cofactor of more than 300 metalloenzymes	Growth retardation Reduced taste and smell Alopecia Dermatitis Diarrhea Immune dysfunction Failure to thrive Gonadal atrophy Congenital malformations	General: Reduced copper absorption, gastritis, sweating, fever, nausea, vomiting. Occupational: Respiratory distress, pulmonary fibrosis

*Not significant: No known clinical symptoms associated with low levels of trace elements in blood or tissues.[5]

A useful website that summarizes several parameters associated with many trace elements discussed in this chapter is provided by the Agency for Toxic Substances and Disease Registry (ATSDR; www.atsdr.cdc.gov).

► TRACE ELEMENTS

ALUMINUM

Aluminum (Al, atomic number 13, atomic weight 26.98) is a metallic element present in bauxite or clay. It is incorporated into antacids and astringents. Antacids—aluminum hydroxide [Al $(OH)_3$] and aluminum carbonate [Al_2 $(CO_3)_3$], for example—are used to counteract stomach acidity. Astringents such as alum (potassium aluminum sulfate [KAl $(SO_4)_2 \cdot 12$ H_2O] is used to shrink or constrict body tissues.

Normally, the daily dietary intake of 5–10 mg of aluminum is completely excreted. Aluminum is activity filtered from the blood by the glomerulus of the kidney. A patient with compromised renal function may lose the ability to filter aluminum from the blood and develop aluminum toxicity.

Unfiltered aluminum accumulates in blood and becomes bound to proteins such as transferrin and is distributed throughout the body. Two principal tissue sites tend to accumulate aluminum: bone and brain. Aluminum replaces calcium in bone near the mineralization site. The result of this replacement is an interruption of normal osteoid formation. This interruption leads to an altered feedback mechanism involving blood calcium levels with parathyroid hormone (PTH) and 1,25-dihydroxyvitamin D. The aluminum binds to calcium binding sites in the parathyroid gland, which causes an abnormal physiological response by the gland. The resulting biochemical profile, which can be diagnostic for aluminum toxicity in patients with compromised renal function, is a decreased serum PTH level and increased serum aluminum level.

Normal serum concentrations of aluminum are <5.41 μg/L (<0.2 μmol/L). Serum levels >100 μg/L (3.7 μmol/L) represent possible toxicity, and serum levels >200 μg/L (7.4 μmol/L) usually show clinical symptoms of toxicity.[6] The symptoms of toxicity reflect which organs or tissues are most affected. For example, aluminum usually accumulates in patients with renal failure and massive accumulation from long-term intermittent dialysis against high-aluminum dialysate leads to dialysis encephalopathy and vitamin D–resistant osteomalacia.

Aluminum toxicity has also been linked to the following: (1) oral exposure from aluminum-containing pharmaceutical products such antacids, (2) Alzheimer's disease resulting from the accumulation of aluminum in the neurofibrillary tangles of patients with the disease, (3) dialysis dementia, (4) Hodgkin's disease, (5) pregnancy, and (6) cystic fibrosis.[7]

Laboratory assessment of aluminum-related disorders is accomplished by measuring the aluminum concentration in serum or plasma. Special attention must be made to the type of container or tube used for collecting the blood. The tubes must be free of aluminum, and care should be taken to avoid the use of aluminum-containing stoppers. Most rubber stoppers contain a significant amount of aluminum silicate, which can produce falsely elevated aluminum concentrations.

Analysis of aluminum in biological samples can be accomplished using several different instrumentation techniques; the choice depends on the type of specimen. For example, serum, plasma, and urine specimens can be measured using atomic absorption spectroscopy (AAS) or inductively coupled plasma mass spectrometry (ICP-MS). Hair specimens are measured by AAS or neutron activation analysis (NAA). Tissue samples can be measured using mass spectroscopy (MS) techniques.

 Checkpoint! 23–1

Identify three examples of medicinal preparations that are commonly used by the public that contain aluminum-based compounds.

ANTIMONY

Antimony, (Sb, atomic number 51, atomic weight 121.58) is in the same group on the periodic table as arsenic and therefore shares many similar chemical, physical, and toxicological properties. It is classified as a metalloid because it reacts with both metals and nonmetals. In nature, antimony is found in the mineral stibnite (Sb_2S_3). Antimony is used for industrial manufacturing and is added to metals such as tin, lead, and copper to increase their hardness. Another use of antimony, in the form of antinomy oxychloride ($Sb_6O_6C_{14}$), is in the preparation of flame-retardant materials. The current medical use of antimony is in the treatment of Leishmaniasis (kala-azar), schistosomiasis, and bilharziasis, which are all examples of parasitic diseases.[8] Antimony has been used on rare occasions as an aversive therapy for substance abuse.[9] The compound would cause the substance abuser to avoid using the illicit drug.

Antimony is not an essential element. Exposure usually occurs with inhalation of antimony dust over a long period of time. It may be ingested or introduced transcutaneously. Absorption from the gastrointestinal (GI) tract begins immediately on ingestion, and the oral bioavailability of antimony ions ranges from 15% to 50%. Ingestion of antimony is usually followed by an episode of vomiting, which facilitates the removal of the metal from the body.

Distribution and excretion of antimony depends on the oxidation state of antimony. Antimony is incorporated into erythrocytes, liver, kidneys, thyroid, and adrenals. Excretion of antimony is via bile and urine. It is eliminated slowly in most forms and more rapidly in the trivalent oxidation state.

Workplace exposure to antimony dust over a period of years leads to pneumoconiosis or pneumonitis (inflammation of the lungs). Symptoms include metallic taste, cough, difficulty breathing, headache, and dizziness in concert with vomiting, diarrhea, and intestinal spasms.

Acute and chronic exposure to antimony may also result in cardiovascular, renal, hematologic, dermatologic, and neurologic effects in individuals. An important point to remember is that exposure to metallic antimony may also mean exposure to other toxic metals such as lead and arsenic.

Treatment of antimony exposure includes decontamination, supportive care, and possibly chelation therapy. Because of the limited number of cases of antimony intoxication and its concomitant exposure to other metals, chelation therapy and other treatment modalities have not been studied in detail.

Clinical assessment of a patient with a history, signs, and symptoms of antimony intoxication may require laboratory measurement of biological specimens. The specimen of choice for acute exposure is a 24-hour urine collection in acid-washed, metal-free containers. A 24-hr urine-collection result greater than 1.0 mg/L is considered toxic using AAS. Plasma samples collected in heparinized container are also acceptable. The reference interval (RI) for plasma antimony by NAA is 0.033–0.070 μg/dL (2.71–5.75 nmol/L).[4] Blood and urine antimony levels can also be determined by AAS and ICP-MS.

ARSENIC

Arsenic (As, atomic weight 33, atomic weight 74.92) exists in nontoxic and toxic forms. Two examples of nontoxic organic forms of arsenic are arsenobetaine and arsenocholine. The nontoxic forms of arsenic are found in many foods. Food sources of arsenic include shellfish, cod, and haddock. Arsenic is also found elsewhere, and some examples of these sources are shown in Box 23-1 ■. Following ingestion of arsenobetaine and arsenocholine, both metabolites undergo rapid renal clearance to become concentrated in the urine. The organic forms of arsenic are completely excreted within 1 to 2 days of ingestion.

Arsenic is most noted for its toxic effects on humans. Toxic forms of arsenic are the inorganic species As^{3+}, As^{5+}, and metabolites monomethylarsine (MMA) and dimethylarsine (DMA). Detoxification of arsenic occurs in the liver, where As^{5+} is reduced to As^{3+} and then methylated to MMA and DMA.[10] The results of the detoxification process are the presence of As^{3+} and As^{5+} in urine shortly after ingestion. These metabolites are present in significant amounts 24 hours after ingestion. Peak concentrations of As^{3+} and As^{5+} in urine occur after approximately 10 hours and return to normal approximately 20 to 30 hours after ingestion. After ingestion, the urine metabolite concentrations of arsenic normally peak at about 40–60 hours and return to baseline 6–20 days.[11] The half-life of organic arsenic in blood is 4–6 hours, whereas the metabolites are 20–30 hours. Arsenic appears in serum for only a short amount of time after ingestion and rapidly disappears into the body's phosphate pool. Elevated levels of

arsenic in serum can be detected for only a few hours (<4) after ingestion.

There are three mechanisms associated with arsenic toxicity in humans[9]:

1. Arsenic binds to dihydrolipoic acid, which is a cofactor for pyruvate dehydrogenase. Without this cofactor, pyruvate cannot proceed to acetyl coenzyme A, which is the first step in gluconeogenesis.

2. Arsenic competes with phosphate for reaction with ADP and results in the formation of the lower-energy ADP rather than ATP.

3. Arsenic also binds with any hydrated sulfhydryl group on proteins and distorts the three dimensional configuration of the protein, which results in a loss of protein activity.

BOX 23-1 Sources of Arsenic

Inorganic

Manufacturing

- ▶ Brass and bronze
- ▶ Ceramics and glass
- ▶ Computer chips
- ▶ Dyes and paints
- ▶ Electron microscopy
- ▶ Fireworks (Chinese)
- ▶ Coal
- ▶ Herbicides
- ▶ Metallurgy
- ▶ Mining
- ▶ Rodenticides
- ▶ Semiconductors (gallium arsenide)
- ▶ Smelting
- ▶ Soldering
- ▶ Wood preservatives

Medicines and Drugs

- ▶ Chemotherapy
- ▶ Depilatories
- ▶ Herbals
- ▶ Homeopathy remedies
- ▶ Kelp
- ▶ Moonshine ethanol
- ▶ Opium

Miscellaneous

- ▶ Well water
- ▶ Contaminated foods and candies (e.g., licorice)

Organic

- ▶ Melarsoprol
- ▶ Parasitic therapy
- ▶ Seafood

Arsenic is also a known carcinogen and has been associated with increased risk of lung, skin, and bladder cancer from water contaminated with arsenic. The Environmental Protection Agency and World Health Organization recommend a level of less 10 parts per billion (0.001 mg/dL) for drinking water.[12] A causal association also has been established between exposure to environmental tobacco smoke and lung cancer with an RI in the order of 1.2.[13] Arsenic also interferes with the activity of several enzymes of the heme biosynthetic pathway.[14]

The antidotes available in the United States for treating arsenic toxicity are British anti-Lewisite [BAL (2,3-dimercaptopropanol)] and succimer (2,3-dimercaptosuccinic acid). British anti-Lewisite is the initial drug used to treat severe arsenic toxicity and is a dimercaprol that functions as a chelating agent to reduce sulfhydryl groups. Succimer is an oral hydrophilic analog of BAL and the chelating agent of choice for subacute and chronic arsenic toxicity.[15]

Arsenic can be measured in several different samples types, including, blood, hair, tissues, and other biological fluids. Serum is not a suitable specimen because arsenic is elevated for only a short time after administration or ingestion. In serum, arsenic binds to proteins and rapidly disappears into the phosphate pool. Adult RI for arsenic in whole-blood specimens is 2–23 μg/L (0.03–0.31 μmol/L). Adults excrete approximately 0–120 μg/day in their urine.

Laboratory determination of arsenic levels in biological fluids and tissues is accomplished using high-performance liquid chromatography (HPLC) followed by ICP-MS, HPLC followed by hydride-generation atomic-fluorescence spectrometry or AAS. For MS applications, urine is the specimen of choice because arsenic is excreted predominantly by the kidney, where it becomes concentrated.

BERYLLIUM

Beryllium (Be, atomic number 4, atomic weight 9.0) is an alkaline earth metal found in Earth's crust. It is not essential for human health and is poisonous. Beryllium alloy and ceramics are used in dental appliances, golf clubs, wheelchairs, circuit-board production, satellites, spacecraft manufacturing, and weapons as a neutron modulator.

Humans can be exposed to beryllium via food, drinking water, and industrial processes, so the major routes of exposure by which beryllium enters the body are inhalation and ingestion. Beryllium compounds that are inhaled are cleared by the lungs. Soluble beryllium compounds are absorbed more quickly than less-soluble compounds such as beryllium oxide. A large percentage of absorbed beryllium accumulates in the skeleton. Clearance of beryllium by the kidneys is slow. A significant negative effect of beryllium in humans is its ability to inhibit several enzyme systems, including alkaline phosphatase, acid phosphatase, hexokinase, and lactate dehydrogenase.

Most cases of beryllium exposure are the chronic type and occur in industry. An acute exposure may result from an industrial explosion or accident and usually results in chemical pneumonitis. Chronic exposure to beryllium may result in the development of a condition known as *chronic beryllium disease* or *berylliosis*. This disease can lead to a progressive and potentially fatal respiratory illness. Chronic beryllium disease is characterized by the formation of noncaseating granulomas in the lung.[16]

An assay has been developed that provides useful diagnostic information to the clinician for patients suspected of beryllium exposure. The *beryllium lymphocyte proliferation test* (BeLPT) uses a sample of blood that measures beryllium sensitization, which represents an "allergic" reaction to beryllium. The test is highly specific, and a positive result rules out the presence of any other metal in the sample.[17] This test is also recommended by OSHA for exmployees who are suspected of exposure to a significant amount of beryllium (see www.osha.gov/SLTC/beryllium/be_proliferation_test.html).

Clinical laboratory assays are available to measure beryllium concentrations in biological fluids and tissues. Techniques available include fluorescent spectroscopy, AAS, and ICP-MS. A 24-hr urine collection is the specimen of choice for quantitative measure of beryllium. The cutoff value for toxicity is >20 μg/L (>2.22 μmol/L). Lung specimens are also acceptable, and a patient with beryllium disease can have beryllium value in the range of 4 to 45,700 μg/kg dry weight of tissue (0.33–5073 μmol/kg).

 Checkpoint! 23–2

Name the disease associated with toxic levels of beryllium.

CADMIUM

Cadmium (Cd, atomic number 48, atomic weight 112.4) is a transition metal in group IIB of the periodic table. It readily oxidizes to a divalent ion, Cd^{2+}. Most naturally occurring cadmium exist as cadmium sulfide (CdS). Cadmium sulfide, cadmium oxide (CdO), and other cadmium-containing compounds are refined to produce elemental cadmium, which is used for industrial purposes. Cadmium alloys (with zinc and lead) are used extensively in solder and brazing rods. Industrial applications include electroplating and the production of nickel-based rechargeable batteries. It is also used as a common pigment in organic--based paints.

Exposure to cadmium can result from one of the following settings: environmental, occupational, or hobby work. Environmental exposure occurs through the consumption of foods grown in areas of cadmium contamination. For example, soil and water near facilities involved in mining or refining of ores can become contaminated. Environmental exposure also occurs in smokers because of soil contamination where the tobacco was grown.

There are many examples of occupational and hobby exposures to cadmium: welders, solderers, jewelry workers, and auto-repair mechanics who spray paint cars and car parts. The source of cadmium is cadmium oxide fumes and dust. Workers and hobbyists who use cadmium-containing alloys must use protective breathing apparatus or incorporate the use of appropriate ventilation systems to remove the cadmium-laden dust and fumes.

The biological role of cadmium in humans is unknown. The bioavailability of elemental cadmium has not been determined. Cadmium salts that are orally ingested have minimal bioavailability. Inhaled cadmium fumes from cadmium oxide are readily bioavailable, so most cases of cadmium intoxication and subsequent injury to body organs originate from inhalation.

Once cadmium enters the body, it is taken into the bloodstream and bound to proteins (e.g., α_1-macroglobulin and albumin). It next goes to the liver and kidney, where it is complexed with metallothionein (MT). Metallothionein binds to cadmium and functions to hold the complex for slow release from the liver to the circulatory system. Circulating Cd–MT is filtered by the glomerulus. A significant amount is reabsorbed and concentrated in proximal tubular cells. The slow release of cadmium from MT-complex hepatic cells accounts for its long biological half-life of nearly 10 years.

The toxicological impact of cadmium in humans manifests itself in four principle tissue sites: (1) kidney, (2) bone, (3) lung, and (4) GI tract. In the kidney, renal damage develops over years. Proteinuria is a common finding in these patients. Cadmium also produces hypercalcuria via damage to the proximal tubules. Bone diseases, including osteomalacia, result from an abnormality in calcium–phosphorus homeostasis because of renal proximal tubular dysfunction. Cadmium toxicity often results in acute cadmium pneumonitis. This disease is characterized by the presence of an infiltrate that is visible on chest radiography. Finally, ingested cadmium salts can be caustic and result in GI tract inflammation. This may lead to episodes of vomiting and abdominal pain that may lead to GI hemorrhage, neurosis, and perforation.

Management of patients exposed to cadmium begins with treatment of presenting complication (e.g., acid–base imbalance, renal dysfunction, abnormalities found on the chest X-ray) followed by GI tract decontamination procedures. Administration of activated charcoal or chelating agents may provide comfort for the patient.

Diagnostic testing of cadmium concentrations in blood usually is of limited value in cases of acute exposure. Blood and urine specimens are acceptable and can be measured by AAS and ICP-MS. The RI for whole-blood cadmium in nonsmokers is 0.6–3.9 μg/L (5.3–34.7 nmol/L). Toxic levels in whole blood are 100–3000 μg/L (0.9–26.7 μmol/L) . RIs for urine levels are 0.5–4.7 μg/L (4.4–41.8.0 nmol/ L) or 0.0–3.0 μg/g creatinine (0.0–26.69 nmol/g).

✓ Checkpoint! 23–3

State the cause of the very long biological half-life (~10 years) of cadmium.

⭐ TABLE 23-3

Industrial Uses of Chromium-Containing Compounds

Barium chromate	$BaCrO_4$	Safety matches Paint pigment
Calcium chromate	$CaCrO_4$	Batteries
Chromic acid	H_2CrO_4	Electroplating
Chromic fluoride	CrF_3	Mordant in dye industry
Chromic oxide	Cr_2O_3	Metal plating Wood treatment
Chromium picolinate	$C_{18}H_{12}CrN_3O_6$	Nutritional supplements
Lead chromate	$PbCrO_4$	Yellow pigment for paints and dyes
Potassium dichromate	$K_2Cr_2O_7$	Oxidizer Leather tanning Porcelain painting

CHROMIUM

Chromium (Cr, atomic number 24, atomic weight 51.99) is a transition metal found naturally in various crystal materials and used in many industrial products. Examples of the uses of chromium for industrial applications are shown in Table 23-3 ⭐. It exists with valances of Cr^{3+} or Cr^{6+} in biological substances. Chromium is an essential trace element that augments the action of insulin. Tissue damage can occur with hexavalent chromium (Cr^{6+}) because of its strong oxidant properties. However, the toxic Cr^{6+} ion is usually reduced to Cr^{3+} during contact with various foods and gastric contents. The structure of the bioactive ion Cr^{3+} associated with insulin is thought to be an octahedral chromium complex. This complex contains two molecule of nicotinic acid with four coordination sites linked to glutamic acid, glycine, and cysteine.[18]

Dietary sources of chromium include processed meats, whole-grain products, green beans, broccoli, and some spices. The estimated daily intake in the United States is 20–30 μg/day.

Adequate intake (AI) is 35 μg Cr/day for males and 25 μg Cr/day for females (dietary or supplement).[19] Patients who are on short-term total parenteral alimentation (TPN) are supplemented with 10–20 μg Cr/day. Those on long-term TPN will usually receive a sufficient amount of chromium via contamination of this supplement.[20]

Intestinal absorption of chromium is low, so fecal output is primarily unabsorbed dietary chromium. Once chromium is absorbed, it is bound to plasma transferrin with the same affinity as iron. It usually settles in the liver, spleen, heart, bone, and other soft tissues. The urinary excretion rate is ~0.3 μg/day, depending on intake. Resistive exercise and running tend to increase urine chromium excretion.

Chromium has been shown to enhance the action of insulin receptors on cell membranes by binding with chromodulin. Chromodulin is a low-molecular-weight intracellular octapeptide that binds four Cr^{3+} ions and then moves to sites on the cell membrane near insulin receptors.[21] The biological mechanisms that allows chromodulin to affect insulin is as follows: (1) Inactive insulin receptors on cell membranes are converted to an active form by binding circulating insulin; (2) binding of insulin serves to stimulate the movement into cells of chromium bound to plasma transferrin; (3) intercellular chromium will bind to apoLMWCr (low-molecular-weight chromium), converting it to an active form that functions to bind to the insulin receptors and results in kinase activity; and (4) as plasma glucose and insulin fall to normal levels, the LMWCr factor is released from the cell to terminate its effect.[22]

Chromium deficiency is characterized by insulin resistance, glucose intolerance, weight loss, and possible neurological deficits. Relatively few cases have been reported of patients exhibiting chromium deficiency, but the United States Food and Nutrition Board has designated chromium an essential trace element.[19]

The clinical importance of chromium resides in its influence on insulin. Studies have shown that >15% of adults ages 40 to 74 may have impaired glucose tolerance as a result of decreased dietary or supplemental intake of chromium.[19] Poor nutrition resulting in diminished body chromium levels is thought to be a cardiovascular risk factor.[23] Studies have shown that proper body levels of chromium raised high-density lipoprotein cholesterol and decreased insulin levels.[24]

Chromium toxicity principally comes from the Cr^{6+} ion. It is a carcinogen, and industrial exposure to fumes and dust containing this metal ion is associated with lung cancer, skin ulcers, and dermatitis. Contaminated soil near leather tanning facilities and dye products manufacturers pose an environmental risk.[25] A widely used dietary supplement containing chromic picolinate has been associated with renal and hepatic damage when used in high doses.[26] Treatment of acute chromium poisoning include GI decontamination if indicated and supportive care. Use of N-acetylcysteine (NAC) and chelating agents has not proven to be efficacious for humans.

The role of laboratories in assessing chromium status in a patient is limited. Assays are prone to problems with contamination from numerous sources—for example, stainless steel needles used for blood drawing. The best techniques require the use of an MS or AAS, which are not available in most clinical laboratories. For this reason, the samples need to be sent to a reference laboratory.

 Checkpoint! 23–4

Identify a specific function of chromium relative to insulin receptors on human cell membranes.

COBALT

Cobalt (Co, atomic number 27, atomic weight 58.93) is an essential element in vitamin B_{12} and is discussed in Chapter 24, Vitamins. It does not have any other function in humans.

Intestinal microflora cannot use cobalt to synthesis physiologically active cobalamin. Free (nonvitamin B_{12}) cobalt does not interact with the body vitamin B_{12} pool. Quantitative analysis of vitamin B_{12} is the best means to assess the nutritional status of a patient, and measurement of blood or urine cobalt is not usually pursued.

Cobalt is rare although widely distributed in the environment. It is found in metal alloys that have high melting points and are resistant to oxidation. Exposure to cobalt usually occurs during the production and machining of metal alloys and can lead to interstitial lung disease. Severe complications of acute cobalt exposure are cardiomyopathy and renal failure. Chronic exposure results in pulmonary edema, allergy, nausea, vomiting, hemorrhage, and thyroid abnormalities.[27]

Patients with acute cobalt poisoning require aggressive therapy. Clinicians must critically evaluate a patient's signs, symptoms, and history because cobalt toxicity usually presents multiorgan and multisystem involvement. Target tissues and systems include cardiac, endocrine, hematopoietic, GI, and neurological systems. Treatment modalities may include GI decontamination and chelation therapy using $CaNa_2$–EDTA, NAC, or succimer.

Measuring urinary cobalt is the preferred method to assess patients who have been exposed to cobalt. Cobalt is quantified in biological samples by AAS or ICP-MS. The RI for serum cobalt is 0.11–0.45 μg/L (1.9–7.6 nmol/L), and urine is 1–2 μg/L (17–34 nmol/L).

 Checkpoint! 23–5

Which vitamin requires the presence of cobalt within its structure to function properly?

COPPER

Copper (Cu, atomic number 29, atomic weight 63.24) exist as Cu^{1+} and Cu^{2+} in human tissues. It is associated with several **metalloproteins** and is essential for selective reduction–oxidation reactions. In biological material, copper is complexed with proteins, peptides, and other organic moieties. Copper exists in the free form but at minute concentrations. It is an integral element in several industrial preparations used as fungicides and algicides as shown in Table 23-4 ⊙.

Dietary sources of copper include meats (liver and kidney), shellfish, nuts, whole-grain cereals, and cocoa products. Lower amounts of copper are found in milk products and white meats. Dietary intake in U.S. adults is 1–1.6 mg/day.[17]

Most dietary copper is absorbed in the small intestines and a small fraction in the stomach. Copper absorption is influenced by the presence of zinc, molybdate, sodium, and amino acids. A large fraction of the absorbed copper is transported to the liver via portal circulation, where it is bound to albumin and incorporated into hepatocytes. In the hepatocytes, copper is bound to enzymes—for example, ceruloplasmin-producing

⭐ **TABLE 23-4**

Commercial Uses of Copper-Containing Compounds

Copper octanoate	$Cu[CH_3(CH_2)_6COO]_2$	Fungicide
Copper triethanolamine	$Cu[(HOCH_2CH_2)_3N]_2$	Algicide
Cupric arsenite	$CuHAsO_3$	Insecticide Wood preservative
Cupric hydroxide	$Cu(OH)_2$	Fungicide
Cupric oxide	CuO	Flux Polishing agents
Cupric sulfate	$CuSO_4$	Fungicide Plant-growth regulator

cuproenzyme and other proteins. These copper-containing complexes enter the peripheral circulation and locate in various tissues and organs. More than 60% of total body copper content is located in skeleton and muscle, but the liver serves as the principle organ for copper homeostasis.

The majority of copper released from the liver and into the circulation is in the form of ceruloplasmin. This copper-containing enzyme is an acute-phase reactant and increases during infection and after tissue injury. Ceruloplasmin is elevated in pregnancy and during use of oral contraceptives. This results in an increase in blood copper levels.

A small fraction of absorbed copper is excreted via bile into feces. A patient with cholestatic liver disease may present with an increase in copper accumulation because of reduced excretion. Small amounts of copper are also lost in urine and sweat. Normal urinary output of copper is about 60 μg/day.

Copper is an important essential trace element because of its association with several enzymes and inclusion in other proteins. Several specific functions of copper are discussed below.

Copper and iron are incorporated into cytochrome c oxidase which is located on the external surface of mitochondrial membranes. Cytochrome c catalyzes a four-electron reduction of molecular oxygen, which creates a high-energy proton gradient across the internal mitochondrial membrane. This mechanism results in the production of intracellular energy.

The copper-containing enzyme protein lysine 6-oxidase is essential for stabilization of extracellular matrixes—for example, cross-linking collagen and elastin in bone formation. This enzyme is associated with tissues found in the aorta and the dermis, as well as in fibroblast and the cytoskeletons of many other cells.

Copper-containing enzymes are important in iron metabolism. The enzymes ferroxidase I, ferroxidase II, and hephaestin (in erythrocytes) serve to oxidize ferrous iron to ferric iron. This reaction facilitates the binding of Fe^{3+} into transferrin and eventually into hemoglobin.

Copper is also used as a cofactor for certain enzyme systems. An example of a specific enzyme is dopamine monoxygenase (DMO). In this reaction, copper serves as a cofactor; ascorbate is the electron donor. DMO catalyzes the conversion of dopamine to norepinephrine, a central nervous system (CNS) neurotransmitter. Another enzyme, monoamine oxidase, is a copper-containing enzyme that catalyzes the catabolism of serotonin in the brain and is involved in the metabolism of the catecholamines.

Prion protein (PrP) binds Cu^{2+} and may be involved in copper regulation within the brain.[28] PrP has been identified as a metalloprotein capable of binding multiple copper ions. An increase in Cu^{2+} results in an increase in the level of PrP–PrP interactions. The binding of Cu^{2+} to PrP serves to act as a switch that induces PrP–PrP interaction in a reversible manner.[29]

Creutzfeldt–Jakob diseases, kuru, and mad cow disease are all examples of transmissible spongiform encephalopathies (TSEs). TSEs arise from the conversion of PrP (C) to PrP (Sc). PrP (C) binds to Cu^{2+} in the section of the protein referred to as the *octarepeat region*.[30] The complete mechanism and impact of Cu^{2+} in this biochemical process is currently under investigation.

Melanocytes contain the enzyme tyrosinase, a copper-containing enzyme that catalyzes the synthesis of melanin. Melanin is an organic polymer and the primary determinate of skin color. Tyrosinase converts L-dopa to melanin biopigments pheomelanin and eumelanin.

Deficiencies of copper can arise in several conditions associated with malnourished infants, premature infants, inappropriate nutritional support of adults or children, and malabsorption syndrome. Examples of symptoms that may develop include anemia, growth retardation, defective keratinization and pigmentation of hair, hypothermia, osteopenia, changes in aortic elastin, and mental deterioration.

Menkes syndrome is an inborn error of metabolism caused by a defective gene that regulates the metabolism of copper in the body. It is a rare condition (1/100,000 live births) in which the mutation is X linked and usually occurs in infants at 2 to 3 months. This syndrome is characterized by loss of previously normal development, hypotonia, seizures, and failure to thrive. Physical changes are also apparent and involve hair and facial features along with neurological abnormalities. Plasma copper and ceruloplasmin levels will be lower than normal.

Wilson disease is an inherited autosomal recessive trait characterized by a defect in the metabolism of copper, with copper depositing in the liver, brain, kidney, cornea, and other tissues. The incidence of Wilson disease is 1/30,000 live births. Both plasma copper and ceruloplasmin will be decreased. The nonceruloplasmin bound fraction of copper is increased, allowing copper to be deposited in the skin, eyes, brain, and kidneys.

Aceruoplasminemia is a result of a gene defect that is characterized by an absence of the protein ceruloplasmin and a progressive neurodegenerative condition that results in tremors. The liver loses its ability to synthesize ceruloplasmin. Additional symptoms that develop in the fourth or fifth decade of life include retinal damage, secondary iron overload that results in an excess amount of iron deposition in tissue, and insulin-dependent diabetes.[31]

Increases in plasma copper have been associated with cardiovascular disease. Studies have shown that increased plasma copper levels are a positive risk factor for developing cardiovascular disease. The mechanism involves both copper and ceruloplasmin and their response to the inflammation of arteries found in atherosclerosis.

Copper toxicity is a serious condition usually resulting from human exposure to copper-containing substances—for example, pesticides, marine antifouling paints, and wood preservatives. Severe GI discomfort results from ingestion of any of these sources of copper and results in significant irritation of the epithelial layer of the GI tract, hemolytic anemia, centrilobular hepatitis with jaundice, and renal damage. Toxic levels of copper also interfere with absorption of zinc, resulting in zinc deficiency (see the section on zinc toward the end of this chapter).[32] Patients with copper toxicosis usually present with symptoms similar to Wilson disease and which include hepatocellular damage and changes in mood behavior in some patients because of copper deposition in central neurons.

Treatment for acute copper poisoning requires aggressive supportive care. Antiemetic therapy, fluid and electrolyte correction, and normalization of vital signs are important first steps followed by chelation therapy, if necessary. Patients may be given IM injections of BAL, or D-penicillamine (Cuprimine), or both. D-penicillamine is a structurally distinct metabolite of penicillin and is an orally bioavailable monothiol chelator. The D-penicillamine–copper complex undergoes rapid renal clearance in patients with normal functioning kidneys.

Laboratory assessment of both copper and ceruloplasmin and acute-phase reactants should be included by the clinician in cases in which there is a suspicion of copper excess or depletion, and it should be interpreted in concert with clinical and drug information.

The reference interval for plasma copper in adults is 70–140 μg/dL (11.0–22.0 μmol/L), and values may vary between laboratories. Women of childbearing age or who are pregnant will usually have higher plasma levels. A plasma copper level below 50 μg/dL (7.85 μmol/L) in adults and 30 μg/dL (4.7 μmol/L) in infants usually indicates copper deficiency. Urine copper output is normally less than 60 μg/day (0.942–μg/day) with values >200 μg/day (3.14 μg/day) found in patients with Wilson disease.

✓ Checkpoint! 23–6

What are the expected blood levels (increase, decrease, or normal) of plasma copper and ceruloplasmin in a patient with Wilson disease?

MANGANESE

Manganese (Mn, atomic number 25, atomic weight 54.94) is usually found bound to proteins in the Mn^{2+} or Mn^{3+} oxidation state. In humans, manganese is required for formation of connective and bony tissue, in growth and reproductive functions, and in lipid and carbohydrate metabolism. The Mn^{2+} form is paramagnetic and can be detected in tissue by magnetic resonance imaging (MRI).

Nonfood sources of manganese include (1) anticorrosive in steel alloys, (2) pigment in paints and glazes, (3) binding agents in red brick, and (4) cleaning agents for glassware commonly used in laboratories. Individuals working in an environment in which these materials and processes are present are potential candidates for exposure.

Food sources of manganese include whole-grain foods, nuts, leafy vegetables, soy products, and teas. The average intake for U.S. adults is about 2 mg/day. The Food and Nutrition Board has set an AI level for adults at 2.3 mg/day for males and 1.8 mg/day for females. A tolerable upper intake limit of 11 mg/day was set for adults.[17]

The mechanism of absorption of manganese from the intestines is similar to that of iron. Absorption increases at low dietary intakes and decreases at higher intakes. A diet high in iron, calcium, magnesium, phosphates, fiber, phytic acid, and tannins from tea can reduce the absorption of manganese.

Most of the absorbed manganese is transported in portal blood to the liver bound to albumin and then exported to other tissue bound to transferrin and α_2-macroglobulin. Manganese is excreted mostly via bile into feces, with urine output being extremely low.

Manganese serves as an important constituent in several **metalloenzymes** that function to facilitate specific enzymatic reactions (e.g., a nonspecific enzyme activator). Manganese ions can be replaced by Mg^{2+}, Co^{2+}, and other cations during the activations of some enzymes. The following are several specific enzymes that require manganese:[33]

- superoxide dismutase,
- pyruvate carboxlyase,
- arginase, and
- glycosyl transferases.

A person eating a natural diet should not have an issue with manganese deficiency, although an individual who is deprived of manganese under special conditions may develop problems with bone demineralization and impaired growth. Humans exposed to large quantities of dust containing the metal via manufacturing and mining processes may develop symptoms because of manganese deposition in the substantia nigra of the brain. Symptoms include Parkinson-like neurodegenerative disorder.

Patients with severe liver disease may develop signs of neurotoxicity because of a failure to excrete manganese in bile. Manganese deposits in a subcortical region of the brain and results in decreased neurotransmitter production and symptoms of postsystemic hepatic encephalopathy.[34]

Patients who exhibit signs and symptoms of manganese toxicity are evaluated for blood and urine manganese levels. The RI for serum or plasma in adults is 0.4–1.1 μg/L (7.28–20.0 nmol/L). Urine concentrations measured in acid-washed, metal-free containers is 0.5–9.8 μg/L (9.1–178 nmol/L). Whole blood and serum manganese in combination with brain MRI scans and neurological assessment are used to detect excessive exposure to manganese. Methods for measurement of manganese include AAS and ICP-MS.

✓ **Checkpoint! 23–7**

State a specific example of the importance of manganese in human cells.

MERCURY

Mercury (Hg, atomic number 80, atomic weight 200.59) is found naturally in small amounts as elemental silver-colored liquid (quicksilver), inorganic salts such as mercury sulfide (cinnabar) and mercurous chloride (calomel), and organic compounds (methylmercury and dimethylmercury). Several examples of nonoccupational and occupation exposure to mercury are shown in Box 23-2 . Over the centuries, mercury has been used for medicinal purposes—for example, treating syphilis.

Mercury compounds are organized into three groups (elemental, inorganic, and organic) that differ with respect to toxicodynamics and toxicokinetics. For example, mercuric oxide (HgO) represents the elemental form, mercuric chloride ($HgCl_2$) the inorganic mercury salts form, and short-chain alkyl-mercury compounds the organic forms. Each group presents unique clinical features of mercury exposure.

Elemental mercury is absorbed primarily via inhalation of vapors. Other routes of exposure include slow absorption following aspiration, subcutaneous deposition, and direct intravenous embolization. Volatilization of elemental mercury increases as the temperature increases. The creation of mercury vapor increases the risk for inhalation by individuals using mercury. Elemental mercury is negligibly absorbed from a normally functioning gut and is considered nontoxic when ingested.

After any form of mercury is absorbed, it distributes widely to all tissues, predominantly the kidneys, liver, spleen, and CNS. The initial distributive pattern into the CNS of elemental and organic mercury differs from that of the inorganic salts because of their greater lipid solubility.

The peak levels of elemental mercury are delayed in the CNS as compared to the organs, and significant accumulation in the CNS usually occurs following an acute, intense exposure to elemental mercury vapor. Elemental mercury is converted to the charged mercuric cation within the CNS, thereby facilitating local accumulation of the metal. Mercury toxicity depends on its oxidation initially to the mercurous ion (Hg^+) and then to the mercuric ion (Hg^{2+}) by the enzyme catalase.

The principle route of absorption for inorganic mercury salts is the GI tract. Some of the ingested inorganic mercury salts are absorbed after dissociation of ingested soluble divalent mercuric salts such as mercuric chloride. These

BOX 23-2 Nonoccupational and Occupation Sources of Mercury Compounds

Nonoccupational

- ▶ Antiseptics
- ▶ Calomel teething powders
- ▶ Dental amalgams
- ▶ Diuretic laxatives
- ▶ Thermometers
- ▶ Button batteries
- ▶ Chemistry sets
- ▶ Lightbulbs (fluorescent)
- ▶ Self-injections
- ▶ Preservatives

Occupational

Elemental

- ▶ Bronzers
- ▶ Ceramic workers
- ▶ Chlorine workers
- ▶ Dentists
- ▶ Electroplaters
- ▶ Jewelers
- ▶ Mercury refiners
- ▶ Paint makers
- ▶ Photographers

Salts

- ▶ Dye makers
- ▶ Explosives workers
- ▶ Fireworks makers
- ▶ Fur processors
- ▶ Laboratory workers
- ▶ Taxidermists

Organic

- ▶ Drug makers
- ▶ Embalmers
- ▶ Histology technicians
- ▶ Seed handlers

mercuric salts are also absorbed across skin and mucous membranes from dermal applications of mercurial ointments and powders containing mercuric chloride.

A large concentration of mercuric ions is found in the kidney, particularly within the renal tubules. Almost no mercury is found as free mercuric ions. In blood, mercuric ions are found within the erythrocytes and bound to plasma proteins in approximately equal proportions. Plasma concentrations are greatest immediately following inorganic mercury exposure, with a rapid decrease as distribution to other tissues occurs. Penetration of the blood–brain barrier is poor because of the low lipid solubility. Slow elimination and prolonged exposure contribute to an accumulation within the CNS of mercuric ions.

Organic mercury compounds are primarily absorbed from the GI tract. Short-chain alkyl compounds are approximately 90% absorbed from the gut. Aryl and long-chain alkyl compounds have greater than 50% GI absorption.[35] Once absorbed, the aryl and long-chain alkyl mercury compounds that possess a labile carbon–mercury bond is cleaved, releasing the inorganic mercuric ion. The short-chain organic mercury compounds possess relatively stable carbon–mercury bonds that persist through the absorptive phases. Because it is a lipophilic bond, they readily distribute across all tissues, including the blood–brain barrier and placenta. This leads to neurologic degeneration that develops in prenatally exposed infant with mercury poisoning.

Mercuric ions are excreted through the kidney by both glomerular filtration and tubular secretion. In the GI tract, they transfer across gut mesenteric vessels into feces. A small amount is reduced to elemental mercury vapor and volatilized from skin and lungs.

The elimination of organic mercury compounds is predominantly fecal. Enterohepatic recirculation contributes to its half-life of about 70 days versus 30–60 days for elemental and inorganic mercury salts.

Clinical symptoms of acute elemental mercury inhalation occur within hours of exposure and consist of chills, fever, cough, and shortness of breath. GI symptoms include nausea, vomiting, and diarrhea. Other symptoms include a metallic taste, dysphasia, salivation, weakness, headaches, and visual disturbance. Patients may have interstitial pneumonitis and both patchy atelectasis and emphysema. Some symptoms may progress to acute lung injury, respiratory failure, and death.

Acute ingestion of mercuric salts produces a characteristic spectrum from severe irritation to caustic gastroenteritis. Patients develop a grayish discoloration of mucous membranes and a metallic taste. Abdominal pain, nausea, and vomiting also can occur. Severe acute mercuric-salt ingestion results in hemorrhagic gastroenteritis and massive fluid loss, resulting in shock and acute tubular necrosis.

Treatment modalities include chelation therapy with BAL or succimer, hemodialysis, and combinations of the two therapies. Other chelating agents used include 2,3-dimercapto-1-propanesulphonate, N-acetyl-d, l-penicillamine, and D-penicillamine.

Measurement of mercury in blood, urine, and hair is used to determine exposure. Hair analysis is done in forensic and toxicology laboratories associated with law-enforcement agencies. Clinical toxicology laboratories equipped with ICP-MS have the capability to measure mercury levels in blood and urine specimens.

✓ Checkpoint! 23–8

Identify at least two examples of specific treatment modalities available for patients who have been poisoned with mercury.

MOLYBDENUM

Molybdenum (Mo, atomic number 42, atomic weight 95.94) is a component of a few metalloenzymes that are important in many biochemical reactions. It can have several oxidation states, but the most stable in biological systems is Mo^{6+}, as in molybdate (MoO_4^{2-}).

Dietary sources of molybdenum include peas, lentil, beans, grains, and nuts. The average dietary intake is 76–109 μg/day for U.S. adults. Molybdenum is absorbed mainly as molybdate. In whole blood, molybdenum is bound to red cell proteins; in plasma, it is transported complexed with α_2-macroglubilin. Urinary output reflects the dietary intake of molybdenum.

Specific enzymes that require molybdenum as metallocofactors include sulfite oxidase, xanthine dehydrogenase, and aldehyde oxidase. Sulfite oxidase is the most significant of this group in relation to human health. This enzyme catalyzes the last step in the degradation of sulfur amino acids, oxidizing sulfite to sulfate and transferring electrons to cytochrome c.[36]

A deficiency of molybdenum is extremely rare in individuals who are healthy and consume a normal diet. Few cases have been reported that describe molybdenum deficiency. There is a rare recessive inherited disease that results from a defect in the biosynthesis of molybdenum cofactor that can result in early childhood death.

Excess intake of molybdenum creates a copper deficiency by inhibiting copper absorption through formation of an insoluble thiomolybdate–copper complex. It has been proposed that ammonium tetrathiomolybdate be given to patients with Wilson disease.[37]

Laboratory assessment of molybdenum deficiency is limited because of the minimal levels found in serum or plasma. Instead, urine is a practical specimen to evaluate because urinary output is responsive to increased or decreased input. The laboratory may measure urate or sulfite in urine, which reflects changes in the metabolism of sulfur and purines. Instrumentation available to provide the sensitivity required include NAA and ICP-MS.

The RI of serum molybdenum is 0.1–3.0 μgL (1.0–31.3 nmol/L); for a 24-hr urine collection, the interval is 40–60 μg/day using ICP-MS analysis. The urine RI may vary because of variations in recent dietary intake.

NICKEL

Nickel (Ni, atomic number 28, atomic weight 58.7) is found naturally in soil, volcanic ash, freshwater, and saltwater. Nickel is used in the production of metal alloys and nickel-based rechargeable batteries and as a catalyst in the hydrogenation of oils. Specific nickel-containing compounds of interest include (1) nickel carbonyl [Ni (CO)$_4$], which is used in nickel refining and petroleum processing and is an extremely deadly liquid form of nickel; (2) nickel oxides and sulfates and metallic nickel, which are associated with combustion incinerator and smelting processes; and (3) metallic nickel, which is found in drinking water, usually at a level of below 20 μg/L. Elevated levels may occur as a result of corrosion and leaching of nickel alloys from plumbing fixtures such as valves and faucets.

Dietary intake of nickel comes from foods such as nuts, legumes, cereal, and chocolate. Nickel is not considered an essential element (but possibly essential), so there are no dietary requirements established by any agency.

Nickel can enter the body via the skin, lungs, and GI tract. The degree of absorption depends on the solubility of the nickel compounds in water. Once in the body, nickel exists primarily as the divalent cation. Inhalation of nickel-containing compounds results in an accumulation of nickel in the lungs, where about 25% is absorbed into the bloodstream. The remainder is swallowed, expectorated, or deposited in the upper respiratory tract.

Following oral ingestion of nickel, a portion is absorbed. Most remains in the gut, and the rest is excreted in the feces. Serum peak levels of nickel occur approximately 1.5–3 hours after ingestion. There is evidence that food reduces the absorption of nickel.

Nickel compounds applied to the skin are capable of quickly penetrating inward with subsequent absorption into the circulation. No evidence exists to show that the absorbed nickel reaches the deep layers or into the bloodstream. Once nickel is absorbed, it is bound to proteins (e.g., nickeloplasmin, albumin, α_2-macroglobulin) and distributed throughout the body. Nickel can cross the placenta and may accumulate in breast milk, thus increasing exposure to a nursing child. It is also found in the lungs, the thyroid, the adrenals, the kidneys, the liver, and other tissues.

Nickel is excreted in the feces and urine. Individuals may normally excrete as much as 10 μg/day, whereas workers in refineries may excrete as much as 800 μg/L. Patients with nickel carbonyl poisoning can have urine levels between 100 and 2500 μg/L. Normal fecal levels can average 260 μg/day.

The severity of acute and chronic exposure to nickel depends on the specific compound and route of exposure. The most common disorders associated with acute exposure to nickel are an allergic dermatitis. A common mode for local nickel dermatitis is nickel transfer from jewelry, garments, wristwatches, body piercings, and other items that come into contact with the skin.

The most prevalent cause of acute nickel toxicity is exposure to nickel carbonyl compounds. Symptoms may develop quickly or be delayed. Examples of symptoms include acute lung injury, intestinal pneumonitis, myocarditis, neurological symptoms, possible leukocytosis, and cerebral edema. Patients who undergo dialysis may develop nickel toxicity because of contaminated leachate from the inside of nickel-plated tanks.

Treatment of nickel toxicity should consist of the immediate removal of the source material followed by decontamination. The clinician can then begin to treat the symptoms presented in an attempt to remove any nickel-containing

compounds from the patient. This can be accomplished by hemodialysis or chelation therapy.

Laboratory assessment may include measuring nickel levels in serum, plasma, and feces. Instrumentation techniques include AAS and ICP-MS. Reference intervals for serum and plasma nickel are 0.14–1.00 μg/L (2.4–17.0 nmol/L). The RI for workers in refineries is 3–11 μg/L (50–187 nmol/L).

SELENIUM

Selenium (Se, atomic number 34, atomic weight 78.96) is a nonmetal with various chemical forms and valences. It is an essential element for humans and a component of the enzyme glutathione peroxidase. Selenium is thought to be closely associated with vitamin E in its functions.

In the periodic table of elements, selenium is grouped with sulfur (group VI), thus giving it a bioinorganic chemistry. Selenocysteine is a biologically active compound in which selenium is substituted for sulfur in cysteine. This active compound is considered to be the twenty-first amino acid and is eventually incorporated into proteins such as glutathione peroxidase-1 by a specific three-nucleotide codon, uracil–guanine–adenine (UGA).[38]

Selenomethionine is synthesized by plants but not humans. It is biologically identical to methionine, thus sharing the same metabolic pathways. Selenomethionine and selenium provide about half the total dietary intake. Both compounds are made available for selenocysteine synthesis when the methionine pathways catabolize selenomethionine. Dietary sources of selenium compounds selenate, selenite, selenocysteine, and selenomethionine are metabolized largely via selenide.

Human dietary sources of selenium are from the selenomethionine synthesized from plants that take up the element from the soil but do not use it. The soil content of selenium is highly variable, so the amount humans receive will vary from region to region and soil type to soil type. Wheat and other cereal products are a good source of selenium. Avererage dietary intake is about 80 to 220 μg/day.

The various forms of dietary selenium are absorbed in the intestines. Selenium is distributed to the liver, lungs, kidney, and muscles. The concentration of selenium in whole blood, serum, or plasma is related to dietary intake. Nearly half of the total plasma selenium is present as the protein selenoprotein P, with about 10 atoms of selenium per molecules. About 30% of selenium is present as glutathione peroxidase (GSHPx-3). Urinary excretion varies widely, depending on the geographical origins of the food.

Selenium specific selenoprotein abound in humans. There may be more than 30 biologically active selenoproteins in humans that play a significant role in many biochemical reactions. Examples include:[39]

- glutathione peroxidases (classical GPx1, gastrointestinal GPx2, plasma GPx3, and phospholipid hydroperoxide GPx4);
- thioredoxin reductase;

- selenoprotein P, which makes up about 60% of selenium in plasma and contains 10 Se atoms per molecules as selenocysteine;
- iodothyronine deiodinase, which catalyzes the removal of one iodine from thyroxine (T_4) to form the active thyroid hormone triiodothyronine (T_3);
- seleno-phosphate synthetase;
- sperm capsule selenoprotein GPx4; and
- selenoprotein W.

Selenium is an important trace element for the proper maintenance of health in humans. It is, for example, important for healthy immune responses. Research has shown that selenium has a protective effect against some forms of cancer—for example, prostate, colon, and lung cancers. Low blood selenium concentrations have been associated with increased cardiovascular disease mortality, and selenium has been shown to regulate the inflammatory mediators in asthma.[37]

Deficiencies of selenium are associated with several disorders, including the following:

- Keshan disease,
- Kashin–Beck disease,
- nutritional depletion in hospital patients,
- thyroid function,
- immune function,
- reproductive disorders,
- mood disorders,
- inflammatory conditions,
- cardiovascular disease,
- cancer chemoprevention, and
- viral virulence.

Readers may broaden their knowledge and understanding of selenium-related deficiencies by reading the reference cited.[40]

Symptoms of selenium toxicity vary, depending on method and site of exposure. General symptomatology includes **alopecia**, emotional liability, peripheral neuropathy, lassitude, dermatitis, and garlic breath. Dermatitis and tissue sensitivity results from acute exposure to selenium dioxide, which is converted to selenious acid or selenium oxychloride and is, in turn, converted to hydrochloric acid.

Occupational exposure can result in lung and nasal carcinomas, liver necrosis, and pulmonary inflammation. Severely poisoned patients develop weakness, increased creatine kinase, and renal insufficiency. In addition, caustic esophageal and gastric burns, acute myocardial infarction, metabolic acidosis, and multiorgan failures may occur in patients exposed to high amounts of selenium compounds.

Treatment of selenium toxicity is usually focused on the symptoms presented by the patient to the clinician. For skin exposure, an application of a topical ointment may be effective. Patients who have inhaled fumes contaminated with selenium compounds may find relief by inhaling fumes from sponges soaked with ammonium hydroxide. Decontamination procedures should be implemented as soon as possible to remove the sources of contamination. To date, there are no proven antidotes for selenium toxicity.

The laboratory assessment for selenium exposure includes measurement of plasma selenium or GSHPx-1 and GSHPx-3. Blood levels for selenium can be determined using carbon furnace AAS. Enzymatic assays are available for measuring GSHPx-1 and GSHPx-3. Urinary selenium output reflects recent exposure on intake, so measuring selenium levels in urine is not usually performed because of the preanalytic effects of recent food intake and hydration status of the patients.

The adult RI for serum selenium is 60–160 μg/L (0.76–2.03 μmol/L); in children, the levels are lower. A 24-hr urine concentration for adults is 7–160 μg/L (0.09–2.03 μmol/L), and a result >400 μg/L (5.08 μmol/L) represents a serious toxicity.

✓ Checkpoint! 23–9

List three examples of the importance of selenium in the proper maintenance of health in humans.

SILVER

Silver (Ag, atomic number 47, atomic weight 107.86) in the ionized form is able to bond with other elements and compounds to form complex ions and salts. Because of its microcidal effects, silver has been used in medicine as a bactericidal agent. The antibacterial activity of silver is related to its direct binding to biotic molecules and its disruption of H^+ activity and thus pH balance. Silver ions bind to electron donor groups of proteins (amines, phosphates, carbonyls) to inhibit enzyme activity, which results in protein denaturation and precipitation.[41] Silver also inserts itself into DNA structures and does not destroy the double helix, thus inhibiting fungal DNAse.[40] Interactions of silver ions within bacterial cell membranes that cause protein leakage from the membrane result in loss of energy and cell death.[42]

Humans are exposed to minute amounts of silver on a daily basis. Silver is ubiquitous to the environment and is deposited through manufacturing processes that include photographic, plating, inks and dyes, porcelain, preparation of germicides, antiseptics, and analytical reagents.

Exposure is usually occupational in nature via skin, mucous membranes, and inhalation. Another common form of exposure is by use of medicinal silver-containing products as shown in Table 23-5 .

Silver that is absorbed into the body is transported by globulins in blood and stored primarily in skin and liver. It is excreted in the bile and eliminated in the feces and urine.

A deficiency of silver is not an issue, but overexposure is. The most significant effect of silver overexposure or ingestion in humans is **argyria**. Patients with argyria (derived from the Greek word for silver *argyros*) exhibit a permanent

⊗ TABLE 23-5		
Silver-Containing Products Used for Medicinal Purposes		
Product Name	**Route of Administration**	**Medicinal Use**
Silver nitrate (1%)	Ophthalmic	Prevention of gonorrheal ophthalmia neonatorum
Silver nitrate (10%)	Cutaneous	In podiatry for corns, calluses, impetigo vulgaris, plantar warts and papillomatous growths
Silver sulfadiazine	Cutaneous	Antimicrobial treatment for wound infections for patients with second- and third-degree burns

bluish-gray discoloration of skin, mucous membranes, and nails because of silver deposits in those tissues. Argyria is associated with growth retardation, hemopoisesis, cardiac enlargement, liver degeneration, and renal tubule destruction. Treatment for argyria includes the use of (1) topical hydroquinone and (2) supplementation with selenium, vitamin E, or sulfur compounds. Accidental ingestion of silver nitrate causes corrosive damage to the GI tract, abdominal pain, diarrhea, vomiting, shock, convulsions, and death. The amount of silver necessary to cause argyria is about >1 g of soluble silver salts.

Laboratory assessment of patients exposed to silver are central to two situations: (1) monitoring burn patients treated with silver sulfadiazine and (2) monitoring patients treated with silver-containing nasal decongestants. The normal concentration for serum silver by NAA is <2 μg/L (<18.5 nmol/L). Silver is primarily excreted in the feces; a small amount is found in urine. A normal 24-hr urine collection will produce <1 μg/day (<9.3 nmol/day) measured by AAS.

 Checkpoint! 23–10

Identify a significant finding involving the skin that is associated with patients having argyria.

THALLIUM

Thallium (Tl, atomic number 81, atomic weight 204.37) is commonly found in granite, shale, volcanic rock, and pyrites. It is a by-product of lead smelting and is used to make sulfuric acid. Thallium has been used in alloys as an anticorrosive and in optical lenses, artist paints, tungsten-filament lamps, and fireworks. It is also a metal commonly found in rodenticides. Environmentally, thallium poses a problem because it is a waste product of coal combustion and in the manufacturing of cement.

Exposure can occur via three routes: inhalation of dust, ingestion (accidental swallowing of rodenticides), and absorption through intact skin. Thallium is rapidly absorbed

following all routes of exposure. Distribution of thallium is uneven throughout the body, with highest concentrations found in the large and small intestines, liver, kidney, heart, muscle, and brain.

The primary mechanism of thallium elimination is secretion into the intestine. It is excreted primarily via the feces and urine. Thallium is filtered by the glomerulus and partially reabsorbed in the tubules.

The mechanism of thallium toxicity includes competition with potassium because the two elements have similar ionic radii. Thus, thallium ions can migrate where potassium tends to migrate (i.e., central and peripheral nervous systems and hepatic and muscle tissues). Thallium can also inhibit DNA synthesis, binds to sulfhydryl groups on proteins in neural axons, and concentrates in renal tubules cells to cause necrosis.[43]

Individuals exposed to sufficient quantities of thallium may develop conditions associated with GI tract, cardiovasculature, respiratory apparatus, kidneys, central and peripheral nervous systems, and skin. Patients may also develop alopecia, which may be permanent if the exposure is great enough.

Treatment of thallium exposure includes the following options: (1) decontamination, (2) potassium supplementation, (3) chelation therapy, (4) administration of Prussian blue antidote (a Food and Drug Administration approved treatment modality),[44] (5) extracorporeal drug removal (hemodialysis), (6) gastric lavage, (7) whole bowel irrigation, and (8) administration of activated charcoal therapy.

Laboratory assessment of thallium toxicity includes measurement of thallium in biological fluids. Normal whole-blood levels in specimens collected in sodium–heparin tubes and measured by AAS is <0.5 μg/dL (<24.5 nmol/L). Toxic concentrations are >1.0 μg/dL (48.9 nmol/L). Normal 24-hr urine-collection concentrations are <2.0 μg/L (9.8 nmol/L); toxic is >1.0 mg/L (4.9 μmol/L).

ZINC

Zinc (Zn, atomic number 30, atomic weight 65.39) is second to iron as the most abundant trace element in the body. Zinc has been associated with several clinical disorders and thus is important in human nutrition.

Zinc is found in many food sources and is bound to proteins. In humans, the bioavailability of dietary zinc depends on the digestion of these proteins to release zinc and bind with amino acids, peptides, phosphates, and other ligands within the intestinal tract. High amounts of zinc are found in red meat and fish, with lesser amounts in white meat and flesh from young animals. Wheat germ and whole bran are good sources of zinc. The median intake in the United States is about 14 mg/day for men and 9 mg/day for women.[17]

Dietary sources of zinc are absorbed by the intestines. Several substances tend to lower the intestinal absorption of zinc, including phytate, calcium, iron, dietary fiber, and a constituent in beans. The absorbed zinc is transported to the liver by portal circulation, where active incorporation into metalloenzymes and plasma proteins such as albumin and α_2-macroglobulin occurs. Less than 1% of the total body content of zinc in found in plasma, and thus plasma levels are only 80–120 μg/dL. A larger portion of zinc, ~80%, is bound to albumin; the rest is tightly bound to α_2-macroglobulin. The zinc on albumin is in equilibrium with amino acids such as histidine and cysteine and may be important in cellular uptake.

The amount of zinc in an adult body is approximately 2–2.5 g and is present in the cells of all metabolically active tissue and organs. Nearly 55% is found in muscle and about 30% in bone. Zinc is also present in the prostate, semen, and the retina. Zinc found in erythrocytes is in the form of carbonic anhydrase, so it is about 10 times higher than in plasma.[45]

Zinc is also released from the body in feces and includes unabsorbed dietary zinc and zinc resecreted into the gut. Urine output of zinc is approximately 0.5 mg/day. Higher concentrations of zinc in urine occur during certain chronic illnesses and conditions, resulting in muscle breakdown.

Approximately 300 zinc metalloenzymes span all of the enzyme categories. Examples of significant metalloenzymes include carbonic anhydrase, alkaline phosphatase, RNA and DNA polymerases, alcohol dehydrogenase, and thymidine kinase. Zinc also plays a vital role in protein and nucleic acid synthesis. The structural stability of proteins is often attributed to the presence of zinc. The mechanism of zinc-related reactions is discussed in detail in various texts.[43]

Zinc deficiency is associated with several conditions—most notably, a reduction in growth. This reduction in overall growth is usually attributed to a diminished intake of dietary zinc. It often occurs in countries where a cereal-based diet high in phytate and fiber but low in animal protein is common.[46] Fortunately, the stunting in growth is easily reversed with a diet supplemented with zinc.

A condition known as acrodermatitis enteropathica (AE) is characterized by periorificial and acral dermatitis, alopecia, and diarrhea. It is an autosomal recessive inborn error that affects zinc absorption from the intestinal mucosa. Patients with this disorder have lower zinc levels than normal (<30 μg/dL). Symptoms are reversed with a diet supplemented with zinc. This often serves as the diagnostic criteria for AE.[47]

Patients who are on parenteral nutrition may develop zinc deficiency if their solutions are not supplemented. A patient who has a surgical procedure done may often develop a zinc deficiency because of increased zinc loss from the intestinal tract via diarrhea and in urine from catabolism of muscles during periods of negative nitrogen balance. Other symptoms associated with zinc deficiency in these patients are mental depression, dermatitis, delayed wound healing, and alopecia that develop during the anabolic period of weight regain when there is insufficient zinc in the nutritional regimen to support tissue repair.[48]

Studies have shown that zinc depletion can impair human immunity.[49] It also has a direct effect on the GI tract, thereby increasing the severity of enteric infections.[50] Signs and symptoms include increased incidence of infection, diarrhea, and defects in carbohydrate use and skin lesions. In patients with zinc deficiency, there is a decrease in the activity of serum thymulin, a thymus-specific hormone that is involved in T-cell function, and the ratio of Th1 and Th2 helper cells becomes significantly different. The lytic activity of natural killer cells also decreases. Zinc is also necessary for intracellular binding of tyrosine kinase to the T-cell receptors CD-4 and CD-8. These receptors are required for T-lymphocyte activation. These changes result in an impairment of cell-mediated immunity and may be the basis for increased infection rates seen in zinc-deficient states.[47]

Zinc may have a role in the synthesis and actions of several hormones. The reaction mechanism may involve zinc-transcription factors. Patients with low circulating levels of testosterone, free thyroxin (T_4), insulin-like growth factor (IGF-1), and thymulin may have decreased blood levels of zinc.[47]

Toxicity associated with ingestion of zinc-contaminated diet is abdominal pain, diarrhea, nausea, and vomiting. Larges doses for examples 225 mg of zinc can induce vomiting and other GI disorders. Increased levels can also result in copper depletion by causing intestinal blockage of intestinal absorption. The U.S. Nutrition Board recommends the tolerable upper level of intake for adults at 40 mg/day.[17]

Treatment for acute oral zinc toxicity is primarily supportive. The patient should be well hydrated and given antiemetic therapy and either H_2 receptor antagonist or proton pump inhibitors to alleviate abdominal discomfort. Chelating agents such as BAL in combination with $CaNa_2$–EDTA may also be effective in certain clinical conditions.[51]

Laboratory assessment of zinc deficiency includes measurement of plasma zinc. Plasma zinc determinations are minimally affected by dietary zinc intake and are subject to several influences as discussed below, but remain the most widely used laboratory test to confirm severe deficiency and to monitor adequacy of zinc supplementation. Although there are no definitive laboratory procedures to clearly identify patients with borderline zinc depletion, the clinical and biochemical responses to zinc supplementation are used to determine a presumption diagnosis and borderline zinc-depleted states.

The specimen of choice for zinc determinations is plasma. Plasma specimens are preferred to serum because of the possibility of zinc contamination from erythrocytes, platelets, and leukocytes during clotting and centrifugation. Plasma zinc concentrations can be measured by AAS.

Blood levels of zinc will vary, depending on the type of specimen. The zinc concentration in serum is usually 5–15% higher than plasma because of osmotic fluid shifts from the blood cells when various anticoagulants are used. Plasma zinc levels are affected by circadian and postprandial fluctuations. For example, zinc concentration decreases after food intake and is higher in the morning that in the evening. A reference interval for clinical guidance is 80–120 μg/dL. A fasting morning value of plasma zinc below 70 μg/dL (<10.7 μmol/L) on more than one occasion requires further testing. A result below 30 μg/dL (4.59 μmol/L) is suggestive of zinc deficiency. Urine zinc excretion is in the range of 150–1200 μg/day (2.3–18.4 μmol/day).[52]

 Checkpoint! 23–11

Indicate the specimen of choice for measuring zinc in patients suspected of being deficient in zinc. Also state the reason (s) why this is the specimen of choice.

▶ MINIMAL RISK TRACE ELEMENTS

The following trace elements represent minimal risks to individuals who may be deficient or toxic; exposures usually occur at an extremely low frequency, so only a brief dissuasion of their clinical significant will be presented in this chapter. More detailed information on their toxicity can be found in *Goldfrank's Toxicological Emergencies* as cited in the reference section at the end of this chapter.

BORON

The exact role of boron in humans has not been clearly elucidated. It is present in fruits, leafy vegetables, nuts, and legumes. Most of the dietary boron is absorbed as boric acid, B (OH)$_3$, and transported in blood and readily excreted by the kidney.

Boron in the form of boric acid is used for medicinal purposes to treat arthritis. It is used as an antiseptic and preservative. Boron compounds are also found in eye-drop solutions. Adverse effects of boron compounds include dermatitis, GI upset, renal and hepatic toxicity, seizure, coma, and death.

FLUORINE

Fluorine is used in dentistry for prophylactic therapy to help prevent the formation of dental caries. The fluoride ion can exchange for hydroxyls in the crystal structure of apatite, a major component of skeletal bone and teeth. This process tends to stabilize the regenerating tooth surfaces.

A conditions known as *fluorosis* can develop, usually in children, when an excess amount of fluoride is deposited on the teeth. Fluorosis can also occur on bone tissue. Patients with fluorosis experience a slower rate of tooth development, yellowing of teeth, white spots, and pitting or mottling of enamel.

PLATINUM

Several chemotherapeutic agents contain platinum. For example, cis-platin (*cis*-dichlorodiammineplatinum dihydrate) and its analogs. These compounds may be nephrotoxic, depending on their relative concentrations in blood. Measuring blood levels of platinum is not commonly requested on patients receiving cis-platin therapy. If a patient has compromised renal function, however, measuring blood levels of platinum may help identify whether an underlying condition is the cause of the renal dysfunction.

SILICON

Silicon is one of the most abundant elements in the environment. Two forms of silicon are of medical interest: (1) amorphous oxides found in asbestos and (2) methylated polymers of silicon.

Silica combined with other minerals are referred to as *silicates*, the form found in asbestos. Inhaling asbestos containing dust leads to deposition of asbestos fibers in the pulmonary alveoli. This condition is called *asbestosis*. Another condition, *silicosis*, includes a variety of pulmonary diseases associated with the inhalation of crystalline silica oxide (SiO$_2$), or quartz. This occurs in workers who remove and pulverize rock or granite in quarries.

Individuals who undergo breast augmentation are often filled with silicone breast implants. The pathological concern is the development of a disease or condition associated with connective tissue that may possibly be triggered by the presence of silicone in the tissue. Silicone may induce a response from polymorphonuclear cells and macrophages that bind small particles of silicon and transport them to lymph nodes, where they can accumulate.[53]

Detailed discussion of other trace elements such as lead, iron, and iodine appear in Chapters 22 (Toxicology), 20 (Iron, Porphyrins, and Hemoglobin), and 15 (The Endocrine System). Additional information is available in *Goldfrank's Toxicological Emergencies*.

SUMMARY

Although testing for trace and ultratrace elements is not a part of most hospital-based facilities, it is important to possess a working knowledge of a select group of these elements. The occurrence of diseases and conditions associated with either a deficiency or toxic amounts is relatively rare in the United States. Outbreaks of diseases are sporadic and are usually localized if they do occur. Mass screenings of employees working in facilities where these trace elements are present is rarely done, so there are an insufficient number of test requests for most laboratories to justify purchasing the sophisticated and expensive instrumentation required to measure these trace elements in biological fluids.

For those laboratories that require knowledge of these trace elements, the focus should be on specimen requirements, preanalytical factors, collection containers, and methodologies. Because these aspects of the analytes are so variable, this chapter emphasized a review of the chemistry, sources, metabolism, function, and pathology states.

REVIEW QUESTIONS

LEVEL I

1. Which of the following trace elements is not classified as essential? (Objective 7)
   a. silver
   b. chromium
   c. fluorine
   d. cobalt

2. Which of the following instrumentation techniques is used to quantitate trace elements in biological fluids? (Objective 3)
   a. atomic absorption spectrometry
   b. osmometry
   c. reflectometry
   d. enzyme immunoassay

3. Which of the following substances if ingested can result in an excessive amount of aluminum in blood and tissue? (Objective 5)
   a. wine
   b. antacids and astringents
   c. aspirin
   d. citrus fruit

4. Which of the following is an example of a trace element? (Objective 1)
   a. calcium
   b. potassium
   c. cobalt
   d. magnesium

5. Which of the following techniques is used to measure trace elements? (Objective 2)
   a. refractometry
   b. reflectometry
   c. osmometry
   d. atomic absorption spectroscopy

6. Which of the following organs is most affected in patients with berylliosis? (Objective 9)
   a. urinary bladder
   b. heart
   c. lungs
   d. brain

7. Aluminum is commonly used in: (Objective 4)
   a. baking soda.
   b. antacids.
   c. saline.
   d. radiator fluid.

8. A metalloenzyme is described as: (Objective 8)
   a. a trace element that is associated with an enzyme to serves as an essential component or cofactor.
   b. a trace element that is associated with an enzyme to serves as a nonessential component or cofactor.
   c. a trace element that is associated with an enzyme to serves as a catalyst for the enzyme reaction.
   d. a trace element that is associated with an enzyme to serves as an inhibitor of an enzyme reaction.

LEVEL II

1. Which of the following is the correct test to use for patients exposed to beryllium? (Objective 5)
   a. beryllium lymphocyte propagation test (BeLPT)
   b. beryllium lymphocyte proliferation test (BeLPT)
   c. beryllium lipoprotein test (BeLPT)
   d. belgium leukocyte proliferation test (BeLPT)

2. Chromium deficiency is most associated with which of the following conditions? (Objective 1)
   a. insulin resistance and glucose intolerance
   b. heart diseases
   c. chronic obstructive pulmonary disease
   d. acute renal failure

3. A patient with Wilson disease will typically have: (Objective 1)
   a. decreased plasma copper and increased plasma ceruloplasmin.
   b. increased plasma copper and increased plasma ceruloplasmin.
   c. increased plasma copper and decreased plasma ceruloplasmin.
   d. decreased plasma copper and plasma ceruloplasmin.

4. A patient with Menkes syndrome will have which of the following laboratory results? (Objective 1)
   a. decreased blood arsenic
   b. increased blood zinc
   c. increased blood copper
   d. decreased blood copper

5. Which of the following sets of laboratory results characterizes a patient with aceruoplasminemia? (Objective 1)
   a. increased blood ceruloplasmin, increased copper deposition in the liver, and increased iron deposition in tissues
   b. decreased blood ceruloplasmin, decreased copper deposition in the liver, and increased iron deposition in tissues
   c. decreased blood ceruloplasmin, increased copper deposition in the liver, and increased iron deposition in tissues
   d. decreased blood ceruloplasmin, increased copper deposition in the liver, and increased zinc deposition in tissues

6. Excess intake of molybdenum creates a deficiency of: (Objective 2)
   a. lead
   b. zinc
   c. copper
   d. boron

7. Which of the following statement about selenium is false? (Objective 3)
   a. It is present in several selenoproteins in humans.
   b. It is not important for proper maintenance of health in humans.
   c. It is a nonmetal.
   d. Selenocysteine is a biologically active compound in humans.

REVIEW QUESTIONS (continued)

LEVEL I

9. Which of the following represents the primary route of exposure to mercury? (Objective 10)
   a. inhalation of vapors
   b. ingestion of mercury-containing medications
   c. contact with the skin of elemental mercury (e.g., mercury-containing thermometers)
   d. direct injection of mercury from contaminated needles

LEVEL II

8. Which of the following terms describes a patient with argyria? (Objective 6)
   a. Patients with argyria exhibit a copper or copper tone discoloration of the skin.
   b. Patients with argyria exhibit mottling or blotching of the skin.
   c. Patients with argyria exhibit a permanent bluish-gray discoloration of the skin.
   d. Patients with argyria exhibit a yellow appearance in the skin, especially in the sclera of the eyes.

9. Antimony is used to treat patients with: (Objective 4)
   a. lupus erythematosus.
   b. pneumonia.
   c. schistosomiasis.
   d. diabetes.

10. Which of the following compounds functions as a chelating agent used as a treatment modality for trace-element exposures? (Objective 4)
   a. sodium heparin
   b. sodium EDTA
   c. copper sulfate
   d. British anti-Lewisite (BAL)

11. Chromium in the blood binds: (Objective 3)
   a. chromoglobulin.
   b. chromodulin.
   c. glucuronide.
   d. microalbumin.

12. Which of the following proteins incorporates copper into its structure? (Objective 3)
   a. albumin
   b. creatine kinase
   c. haptoglobin
   d. ceruloplasmin

PEARSON myhealthprofessionskit

Use this address to access the interactive Companion Website created for this textbook. Simply select "Clinical Laboratory Science" from the choice of disciplines. Find this book and log in using your user name and password.

REFERENCES

1. Mertz W. *Trace elements in human and animal nutrition*, 5th ed. (New York: Academic Press, 1987).
2. Mertz W. The scientific and practical importance of trace elements. *Philos Trans R Soc Lond Biol Sci* (1981b) 294: 9–18.
3. Milne DP. Laboratory assessment of trace elements and mineral status. In Bogden JD, Keevay LM (Eds.), *Laboratory assessment of trace elements and mineral* (Totowa, NJ: Humana Press, 2000): 69–90.
4. O'Dell BL, Sunde RA. *Handbook of nutritionally essential mineral elements* (New York: Marcel Dekker, 1997).
5. Russell RM. Vitamins and trace mineral deficiency and excess: Introduction. In Kasper DL, Braunwald E, Fauci AS, et al. (Eds.), *Harrison's principles of internal medicine*, 16th ed. (Philadelphia: McGraw-Hill, 2005): 411.
6. Wu AHB. *Tietz clinical guide to laboratory tests*, 4th ed. (St. Louis: Saunders Elsevier, 2006): 92–95.
7. Moyer TP, Burritt MF, Butz J. Toxic metals. In Burtis CA, Ashwood ER, Bruns DE (Eds.), *Tietz textbook of clinical chemistry and molecular diagnostics*, 4th ed. (Philadelphia: WB Saunders, 2006): 1374–1376.

8. Carrio J, de Colmenares M, Riera C, et al. Leishmania infantum: Stage specific activity of pentavalent antimony related with the assay conditions. *Exp Parasitol* (2000) 95: 209–214.

9. Tarabar AF, Khan Y, Nelson LS. Antimony toxicity from the use of tartar emetic for the treatment of alcohol abuse. *Veterin & Hum Toxicol* (2004) 46, 6: 331–333.

10. Zakharyan RA, Sampayo-Reyes A, Healy SM, et al. Human monomethylarsonic acid reductase is a member of the glutathione-S-transferase superfamily. *Chem Res Toxicol* (2001) 14: 1051–1057.

11. Moyer TP. Testing for arsenic. *Mayo Clin Proc* (1993) 68: 1210–1211.

12. Environmental Protection Agency. National primary drinking water regulation. Arsenic and clarifications to compliance and new source contaminants monitoring. Proposed rules. 40 CFR parts 141 and 142. *Fed Reg* (2000) 65: 63027–63035.

13. Boffetta P, Nyberg F. Contribution of environmental factors to cancer risk. *Br Med Bull* (2003) 68: 71–94.

14. Gardner DE, Crapo JD, Massaro EJ (Eds.). *Toxicology of the lung* (New York: Raven Press, 1988).

15. Muckter H, Liebl B, Reichl FX, et al. Are we ready to replace dimercaprol (BAL) as an arsenic antidote? *Hum & Exper Toxicol* (1997) 16, 8: 460–465.

16. Deubner CD, Goodman MG, Iannuzzi J. Variability, predictive value, and uses of the beryllium blood lymphocyte proliferation test (BLPT): Preliminary analysis of the ongoing workforce survey. *Appl Occup Environ Hyg* (2001) 16: 521–526.

17. Lebeau J. Use of the beryllium lymphocyte proliferation test (BeLPT) for screening. *J Occ and Envir Med* (2007) 49, 3: 357–358.

18. Offenbacker GE, Pi-Sunyer XF, Stoecker BJ. Chromium. In O'Dell BL, Sunder RA (Eds.), *Handbook of nutritionally essential mineral elements* (New York: Marcel Dekker, 1997): 389–412.

19. Food and Nutrition Board IOM. *Dietary reference intakes for vitamin A, vitamin K, arsenic, boron, chromium, copper, iodine, iron, manganese, molybdenum, nickel, silicon, vanadium and zinc* (Washington, DC: National Academy Press, 2002).

20. Anderson RA. Chromium and parenteral nutrition. *Nutr* (1995) 11: 83–86.

21. Vincent JB. The biochemistry of chromium. *J Nutr* (2000) 130: 715–718.

22. Clodfelder BJ, Emmanuelle J, Hepburn DD, et al. The trail of chromium (III) in vivo from the blood to the urine: The roles of transferrin and chromodulin. *J Biol Inorg Chem* (2001) 6: 608–617.

23. Mertz W. Trace minerals and atherosclerosis. *Fed Proc* (1982) 41: 2807–2812.

24. Riales R, Albrink MJ. Effect of chromium chloride supplementation on glucose tolerance on serum lipids including high-density lipoprotein of adult men. *Am J Clin Nutr* (1981) 34: 2670–2678.

25. U.S. Department of Health and Human Services. Draft toxicological profile for chromium. Agency for Toxic Substances and Disease Registry. ATSDR 2008; 2–5.

26. Cerulli J, Grabe DW, Gauthier I, et al. Chromium picolinate toxicity. *Ann Pharmacother* (1998) 32: 428–431.

27. Nemery B, Lewis CP, Demedts M. Cobalt and possible oxidant-mediated toxicity. *Sci Total Environ* (1994) 150: 57–64.

28. Burns CS, Aronoff-Spencer E, Legname G, et al. Copper coordination in the full-length, recombinant prion protein. *Biochem* (2003) 42: 6794–6803.

29. Kenward AG, Bartolotti LJ, Burns CS. Copper and zinc promote interactions between membrane-anchored peptides of the metal binding domain of the prion protein. *Biochem* (2007) 46, 14: 4261–4271.

30. Millhauser GL. Copper and the prion protein: Methods, structures, function, and disease. *Ann Rev Phys Chem* (2007) 58: 299–320.

31. Harris ZL. Aceruloplasminemia. *J Neurol Sci* (2003), 207: 108–109.

32. Sandstead HH. Requirements and toxicity of essential trace elements, illustrated by zinc and copper. *Am J Clin Nutr* (1995) 61: 621S–624S.

33. Liang WJ, Johnson D, Jarvis SM. Vitamin C transport systems of mammalian cells. *Mol Membr Biol* (2001) 18: 87–95.

34. Hazell AS. Butterworth RF. Hepatic encephalopathy: An update of pathophysiologic mechanism. *Proc Soc Exp Biol Med* (1999) 222: 99–112.

35. Sue Young-Jin. Mercury. In Flomenbaum NE, Goldfrank LR, Hoffman RS, et al. (Eds.), *Goldfrank's toxicological emergencies*, 8th ed. (New York: McGraw-Hill, 2006): 1337.

36. Johnson JL. Molybdenum. In O'Dell BL, Sunde RA (Eds.), *Handbook of nutritionally essential mineral elements* (New York: Marcel Dekker, 1997): 413–438.

37. Brewer GJ, Hedera P, Kluin KJ, et al. Treatment of Wilson disease with ammonium tetrathiomolybdate III: Initial therapy in a total of 55 neurologically affected patients and follow-up with zinc therapy. *Arch Neurol* (2003) 60: 379–385.

38. Hatfield DL, Gladyshev VN. How selenium has altered our understanding of the genetic code. *Mol Cell Biol* (2002) 22: 3565–3576.

39. Brown KM, Arthur JR. Selenium, selenoproteins and human health: A review. *Pub Health Nutr* (2001) 4 (2B): 593–599.

40. Shenkin A, Baines M, Path FRC, et al. Vitamins and trace elements. In Burtis CA, Ashwood ER, Bruns DE (Eds.), *Tietz textbook of clinical chemistry and molecular diagnostics*, 4th ed. (Philadelphia: WB Saunders, 2006): 1134–1136.

41. Fung MC, Bowen DL. Silver products for medical indications: Risk-benefit assessment. *J Toxicol–Clin Toxicol* (1996) 34, 1: 119–126.

42. Dibrov P, Dzioba J, Gosink KK, et al. Chemiosmotic mechanism of antimicrobial activity of Ag^+ in Vibrio cholerae. *Antimicrob Agents & Chemotherapy* (2002) 46, 8: 2668–2670.

43. Gardner DE, Crapo JD, Massaro EJ (Eds.), *Toxicology of the lung* (New York: Raven Pres, 1988).

44. Thompson DF, Callen ED. Soluble or insoluble Prussian blue for radiocesium and thallium poisoning? *Ann Pharmacother* (2004) 38, 9: 1509–1514.

45. Chesters JK. Zinc. In O'Dell BL, Sunde RA (Eds.), *Handbook of nutritionally essential mineral elements* (New York: Marcel Dekker, 1997): 185–230.

46. Prasad AS. Zinc deficiency. *BMJ* (2003) 326: 409–410.

47. Barnes PM, Moynahan EJ. Zinc deficiency in acrodermatitis enteropathica: Multiple dietary intolerance treated with synthetic diets. *Proc R Soc Med* (1973) 66: 327–329.

48. Kay RG, Tasman-Jones C, Pybus J, et al. A syndrome of acute zinc deficiency during total parenteral alimentation in man. *Ann Surg* (1976) 183: 331–340.

49. Prasad AS. Effects of zinc efficiency on immune functions. *J Trace Elem in Exper Med* (2000) 13: 1–20.

50. Sturniolo GC, Mestriner C, D'Inca RD. Trace elements and mineral nutrition in gastrointestinal disease. In Bogen JD, Klevay LM (Eds.), *Clinical nutrition of the essential trace elements and minerals* (Totowa, NJ: Humana Press, 2000): 289–307.

51. Barceloux DG. Zinc. *J Toxicol Clin Toxicol* (1999) 37: 279–292.

52. Shenkin A, Baines M, Path FRC, et al. Vitamins and trace elements. In Burtis CA, Ashwood ER, Bruns DE (Eds.), *Tietz textbook of clinical chemistry and molecular diagnostics*, 4th ed. (Philadelphia: WB Saunders, 2006): 1141.

53. Shons AR, Schubert W. Silicone breast implants and immune diseases: An overview. *Ann Plast Surg* (1992) 28: 491–501.

24

Vitamins

■ OBJECTIVES—LEVEL I

Following successful completion of this chapter, the learner will be able to:

1. List the fat-soluble vitamins.
2. Identify methods used to measure selected vitamins.
3. Identify types of instrumentation used to measure vitamins.
4. Cite biological uses of selected vitamins.
5. State food sources of selected vitamins.
6. Identify examples of types of specimens used for laboratory assessment of vitamins.
7. Identify selected vitamins by both their common and trivial chemical names.
8. Define the following terms: *functional assay*, *direct assay*, *hypervitaminosis*, and *hypovitaminosis*.
9. Associate selected vitamins with disease.

■ OBJECTIVES—LEVEL II

Following successful completion of this chapter, the learner will be able to:

1. Correlate laboratory results with disease(s) associated with selected vitamins.
2. Explain the function(s) of selected vitamins.
3. Summarize the chemistry of selected vitamins.
4. Outline the mechanisms of vitamin absorption, metabolism, and excretion.
5. Correlate symptoms of hypervitaminosis and/or hypovitaminosis in humans.

KEY TERMS

Antioxidant

Carotenoids

Direct assay

Dry beriberi

Flavin

Functional assay

Hypervitaminosis

Hypovitaminosis

Nyctalopia

Pellagra

Pernicious anemia

Rickets

Scurvy

Tocopherol

Wernicke–Korsakoff syndrome

Wet beriberi

 A CASE IN POINT

The patient described in this case has multiple-organ involvement; therefore, the clinician requested a significant number of diagnostic tests, including clinical laboratory tests, to rule in and rule out various diseases and conditions.

A 65-year-old woman who is a chronic alcoholic suddenly collapsed at home and was transported to the emergency department (ED). Her medical history includes hypothyroidism, chronic renal impairment, and osteopenia. The patient complained of mild epigastric pain, with occasional radiation to the right costal margin. She denied chest pain, headache, or vertigo. She had shortness of breath (dyspnea) on examination. Her medications consisted of furosemide (a diuretic), potassium, captopril (ACE inhibitor), and thyroxine. She has smoked for about 30 years.

Physical examination revealed a tired and lethargic-looking woman. Her weight was 130 pounds and her height was 61 inches. Her blood pressure was 110/70 lying down and 90/60 sitting down, with a pulse rate of 102 beats per minute. Other significant findings were crepitus (crackling or popping sound) in her chest, marked cardiomegaly, and pulmonary edema.

Issues and Questions to Consider

1. Identify several clinical observations relevant to the patient's history and physical examination.

2. List examples of diagnostic laboratory and nonlaboratory (e.g., X-ray) tests that may provide the clinician with useful data to support clinical findings.

The results of clinical laboratory tests and additional diagnostic tests are as follows:

Serum Chemistries	Results (Conventional units)	Reference Interval (RI) (Conventional units)
Sodium	144	136–145 mEq/L
Potassium	4.4	3.5–5.1 mEq/L
Chloride	101	98–107 mEq/L
Bicarbonate	7	23–29 mEq/L
Anion gap	36	6–10 mEq/L
Glucose	100	74–100 mg/dL
Creatinine	2.5	0.9–1.3 mg/dL
Urea	11.2	6–20 mg/dL
Lactic acid	88.3	1.8–18 mg/dL

Calcium	8.9	8.7–10.0 mg/dL
Phosphate	3.1	0.26–4.2 mEq/L
Magnesium	1.2	1.5–2.3 mg/dL
Albumin	3.1	3.3–4.1 g/dL
Gamma glutamytransferase	389	1–25 U/L
Ethanol	Not detected	Undetectable
Alkanine phosphase	250	40–115 U/L
Alanine amino transferase	1880	0–55 U/L
Asparate amino transferase	7542	5–34 U/L
Lipase	125	<120 U/L
Troponin I	0.48	<0.01 ng/mL
B-type natriuretic peptide	458	<20 pg/mL
pH	7.10	7.35–7.45
Serum thiamine	0.18	0.21–0.43 μg/dL
Erythrocyte transketolase	0.60	0.75–1.30 U/g Hb

Other nonclinical laboratory tests:

Left ventricular ejection Fraction	32%	>50%

Issues and Questions to Consider

3. Based on all of the information provided, including laboratory tests, identify possible medical conditions affecting this patient.

4. What is the relationship between the increased serum BNP and the decreased thiamine levels?

5. What type of assay is erythrocyte transketolase?

6. Explain why erythrocyte transketolase is elevated.

WHAT'S AHEAD

► Biochemistry of vitamins

► Physiology of vitamins

► Dietary sources of vitamins

► Biological functions of vitamins

► Clinical implications of deficiencies and toxic amounts of vitamins in body tissues

► Laboratories' role in assessing vitamin status

► Direct and functional assays for ascertaining vitamin status

► ASSESSMENT OF VITAMIN STATUS

Clinical evaluations of patients with suspected vitamin deficiency (or excess) often includes the performance of **functional assays** (indirect) or **direct assays**. A summary of functional and direct assays is presented in Table 24-1 ✪.

A functional assay is an indirect measure of nonvitamin substances using one of several techniques: (1) increased or decreased activity of reactions, (2) cell response to inhibitors, or (3) activators that will reflect alteration in vitamin concentrations.

Direct assays are methods that can detect and quantitate specific vitamins, and possibly their respective metabolites, using any of several methodologies and instruments. Examples of such methods include immunoassays and colorimetric, fluorescent, and separation techniques. Types of instrumentation used for vitamin analysis include spectrophotomers, fluorometers, gas and liquid chromatographs, and mass spectrometers. Most clinical laboratories do not offer assays to measure vitamins because of the low frequency of **hypervitaminosis** or **hypovitaminosis**, expensive instrumentation (e.g., gas chromatography [GC], liquid chromatography [LC], mass spectrometry [MS]), high level of technical skills required, and long assay times. The two exceptions are vitamin B_{12} and folic acid, which are measured primarily by immunoassays and the tests for which are easily adaptable to instruments found in most clinical chemistry laboratories.

Several analytical factors must be considered when assessing the vitamin status of an individual: reference interval, sample type, and apparent concentration of the vitamin in body fluids. There are published reference intervals for each vitamin, and these can be found in many sources. Appropriate interpretation of a patient's vitamin concentration and nutritional status in health and disease must be based on knowledge of factors that may influence the vitamin levels (e.g., ethnicity, geographic location, and seasonal variations).

Most direct assays are performed on blood samples, including plasma, serum, or whole blood. For some vitamin determinations, serum or plasma provides a reliable assessment of patient's status (e.g., vitamin B_{12}) whereas for other vitamins whole blood may be the specimen of choice (e.g., folate).

Factors that confound the interpretation of vitamin levels in plasma include the following: (1) There may be a lack of correlation between the amount of nutrient in the plasma compartment with the amount in the intracellular compartment of most body tissues; (2) there may be changes in the binding proteins in plasma as a result of disease; and (3) there may be variation in acute-phase reactant (APR) concentration in a patient with infection or inflammatory disease.[1]

✪ TABLE 24-1		
Summary of both functional (indirect) and direct assays for the evaluation of vitamin status		
Vitamin	**Functional Assays**	**Direct**
B_1 (thiamine)	Erythrocyte trans-ketolase activation	Thiamine concentration in blood as TPP or TDP
B_2 (riboflavin)	Measure FAD-dependent glutathione reductase activity in freshly lysed erythrocytes	Measure riboflavin in erythrocytes or urinary riboflavin
B_6 (pyridoxine)	Measure activity of erythrocytes	Measure PLP
B_{12}	(1) Urine and serum concentration of methylmalonic acid (2) Plasma homocysteine (3) Deoxyuridine suppression test (4) Vitamin B_{12} absorption test	Serum concentration of vitamin B_{12}
B_3 (niacin)	None available	(1) Measure two urine metabolites, N'methyl nicotinamide and N'-methyl-2-pyridone-5'carboxyamide (2) Determine ratio of NAD/NADP in erythrocyte and plasma tryptophan
C	No useful test	Plasma, urine, serum, tissue, or cell levels of vitamin
Biotin	Measure biotin in samples by microbiological assays	Urine concentration of biotin
Folate	None available	Serum and whole-blood folate
Pantothenic acid	None available	Whole blood and urine vitamin levels
A	Relative dose response	Plasma vitamin level, retinol binding protein using radial immunodiffusion or nephelometry
E	(1) Protection of erythrocyte hemolysis on addition of peroxides (2) Inhibition of lipid peroxidation products	(1) Serum α-tocopherol (2) In tissue (lymphocytes, platelets, or red blood cells [RBCs])
K	(1) Prothrombin clotting time (PT) (2) Immunoassay of γ-carboxy prothrombin or undercarboxylated prothrombin (2) PIVKA-II (3) Plasma undercarboxylated osteocalcin	Plasma phylloquinone

TPP, thiamine pyrophosphate; TDP, thiamine dipyrophosphate; FAD, flavin adenine dinucleotide; PLP, pyridoxal-5'-phosphate; RBP, retinol binding protein; PIVKA, protein induced by vitamin K absence or antagonism

☼ TABLE 24-2

Function and clinical findings of vitamins

Vitamin	Primary Function	Primary Clinical Condition and Symptoms
B₁ (thiamine)	Coenzyme for cleaving carbon–carbon bonds	Beriberi
B₂ (riboflavin)	Cofactor for oxidation, reduction reactions	Seborrhea, magenta tongue
B₃ (niacin)	Coenzyme for oxidation and reduction reactions	Pellagra: pigmented rash of sun-exposed areas, reddish tongue
B₆ (pyridoxine)	Cofactor for enzymes of amino acid metabolism	Glossitis, seborrhea, neuropathy, microcytic anemia
B₁₂	Coenzyme for methionine synthase and L-methylmalonyl CoA mutase	Megaloblastic anemia, dementia, increased homocysteine and methylmalonic acid
Folate	Coenzyme for one carbon transfer in nucleic acid and amino acid metabolism	Megaloblastic anemia, depression, increased homocysteine
C	Participation as a redox ion in several biological oxidation and hydrogen transfer reactions	Scurvy: petechiae, ecchymosis, inflamed and bleeding gums
A	Formation of rhodopsin and glycoproteins	Nightblindness, follicular hyperkeratosis, xerophthalmia
E	Antioxidants	Peripheral neuropathy, skeletal muscle atrophy, retinopathy
K	Cofactor for posttranslational carboxylation of many proteins and clotting factors	Elevated prothrombin time, bleeding

Tissue samples are rarely used for vitamin assessment. An exception is the need for a liver biopsy, which can prove helpful in Wilson's disease. More frequently, selected types of cells from blood samples rather than tissue samples may provide the clinician with information that is more useful. An example may be the assessment of thiamine status in a patient using erythrocyte transketolase activation analysis (see below).

Urine samples are not useful in some conditions because most vitamins are not under homeostatic control, and their excretion may be a direct reflection of consumption rather than active retention relative to whole-body deficiency. For example, high levels of excretion of a water-soluble vitamin may indicate ingestion of large amounts of vitamin supplements.

The assessment of vitamin status in an individual usually begins with the patient presenting signs and symptoms of deficiency or, less frequently, excess of a vitamin. Patients may develop characteristics clinical findings relevant to the vitamin status at the time of presentation. A summary of clinical findings and diseases is presented in Table 24-2 ☼. The reader is advised to refer to this table throughout the chapter for additional information relevant to the vitamins discussed.

 Checkpoint! 24–1

Distinguish between a functional assay and direct assay used to assess vitamin status in humans.

► FAT-SOLUBLE VITAMINS

VITAMIN A (RETINOL)

Vitamin A (retinol) is a fat-soluble vitamin that is found in plants and animals. The actions or effects of vitamin A are primarily through hormone-like receptors. These physiological effects are diverse within the human body and are not exclusive to the eye. Analogs of vitamin A possess a unique effect on epithelial differentiation and thus have been used as therapeutic agents in the treatment of a variety of dermatological conditions and are being evaluated for use in cancer chemotherapy.

Retinoid is the term used for the chemical compound retinol and other closely related, naturally occurring derivatives; several examples are shown in Box 24-1 ■. Retinoid functions by binding to specific nuclear receptors and modulating gene expression. Retinoids also include structurally related synthetic analogs that may not have retinol-like activity.[2]

Carotene (provitamin A) is a purified plant pigment and is an important source of vitamin A. β-carotene is the most active form of **carotenoid** found in plants. Retinal is a primary alcohol and is present in esterified form in the tissues of animal and saltwater fish (primarily livers). Several *cis-trans* isomers exist because of the unsaturated carbons in the retinal side chain.

Preformed vitamin A (e.g., retinyl ester) is found in animal-derived foods such as fish oils, liver, and other organ

BOX 24-1 Examples of retinoid/vitamin A family of compounds

► ALL-*trans*-retinol
► ALL-*trans*-14-hydroxyretrorentinol
► ALL-*trans*-retinal
► ALL-*trans*-retinoic acid
► 9-*cis*-retinoic acid
► 11-*cis*-retinal
► 13-*cis*-retinal
► 13-*cis*-retinoic acid

meats. Nonanimal sources include full cream milk, butter, and fortified margarines. Good sources of provitamin A carotenoids are yellow to orange pigmented fruits and vegetables and green leafy vegetables. Other sources include carrots, pumpkins, tomatoes, lettuces, grapefruit, and apricots.

Once preformed vitamin A is ingested, it is emulsified and formed into micelles by the action of bile salts (e.g., cholic acid, taura cholic acid) before being transported into the intestinal cell. The retinyl esters are moved across the mucosal membrane and hydrolyzed to retinol within the cell for re-esterification by cellular retinol-binding II and packaged in chylomicra. The chylomicrons then enter the mesenteric lymphatic system and pass into the systematic circulation. Some ingested retinoid is converted into retinoic acid in the intestinal cells.[3]

Carotenoids, in micellular form, are absorbed into the duodenal mucosal cells by passive diffusion. Once inside the cell, most of the β-carotene is converted to retinol, which is oxidized to retinal and further oxidized to retinoic acid. The newly created retinyl esters then pass with chylomicrons via the lymphatic system to the liver and are taken up by the parenchymal cells. In the liver, retinol is conjugated with retinol-binding protein and transthyretin and returned to the circulation, or stored as esters in the stellate cells (Kupffer cells). Movement of retinol to the tissue is controlled by the availability of the vitamin A protein complex in circulation.

Retinoic acid formed in the intestinal mucosal cells is bound to albumin and transported via the portal vein. Retinoic acid is quickly metabolized in tissue, such as liver, to produce polar catabolites, which are conjugated and then excreted in the urine. A small portion of retinoic acid is excreted in the bile as a glucuronic acid conjugate.

Retinoic acid forms when the alcohol moiety has been oxidized. It shares some but not all the actions of retinol. In certain species, retinoic acid is ineffective in restoring visual or reproductive function whereas retinol is effective. Retinoic acid has been shown to promote growth and control differentiation and maintenance of epithelial tissue in vitamin A–deficient animals.[3]

Vitamin A is essential for several reasons, including the following: (1) the proper functioning of the retina, (2) growth and differentiation or epithelial tissues, (3) growth of bone, and (4) reproduction and embryonic development. Vitamin A, together with other carotenoids, facilitates immune functions, diminishes the consequences of some infectious diseases, and may protect against the development of certain malignances. Thus retinoids are currently being evaluated for cancer prophylaxis and for treating various premalignant conditions. Vitamin A and its analogs are used to treat several skin diseases, consequences associated with aging, and problems associated with prolonged exposure to the sun.

Vitamin A compounds are necessary for proper vision. Photoreception is accomplished by two types of specialized retinal cells: rods and cones. Rods are sensitive to light of low intensity, and cones act as receptors of high-intensity light and are responsible for color vision.

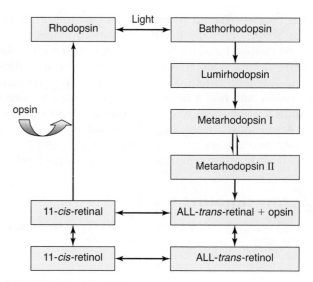

■ FIGURE 24-1   Vitamin A and vision.

The visual cycle illustrated in Figure 24-1 ■ is initiated by the absorption of a photon of light. These photons cause photodecomposition of rhodopsin through a series of unstable conformational states, leading to the isomethylation of 11-cis-retinal to the ALL-trans form and subsequent dissociation of the opsin moiety.

The chromophore of both rods and cones is 11-cis retinal. The holoreceptor in rods is rhodopsin—a combination of the protein opsin and 11-cis retinal attached as a prosthetic group. The three different types of cone cells (red, green, and blue) contain individual, related photoreceptor proteins that respond to light of different wavelengths.

Vitamin A is important in the functional and structural integrity of epithelial cells throughout the body. It serves to induce and control epithelial differentiation in mucus secretion or keratinizing tissues. Basal epithelial cells are stimulated to produce mucus in the presence of retinal or retinoic acid. A large concentration of retinoid leads to the production of a thick layer of mucin, the inhibition of keratination, and the display of goblet cells.

Individuals who lack sufficient amounts of vitamin A have a reduced number of goblet mucous cells, which are replaced by basal cells that tend to proliferate. This leads to an increase in stratification and keratination of the epithelium. Normal secretions are suppressed, resulting in increased irritation and infection. When this process happens in the cornea, degenerative changes occur that lead to poor dark adaptation or night blindness (**nyctalopia**). The severe form of this condition is called hyperkeratination (xerophthalmia) and may lead to the formation of small gray plaques with foamy surfaces (Bitot's spots). These lesions are reversible with vitamin A administration. More serious effects of deficiencies are termed *keraomalacia*, which may lead to permanent blindness.

Direct assays such as high-performance liquid chromatography (HPLC) are useful but not an ideal indicator vitamin A status because vitamin A levels do not decline until liver stores become critically depleted. However, indirect assays may be useful. The relative dose–response test used to assess vitamin A total-body stores is an example of an indirect assay.[4] This assay requires two blood samples to be collected, one before and one 5 hours after a physiological dose of vitamin A is administered. Patients who are deficient in vitamin A will show a rapid, large, and sustained rise in serum retinal concentration; in contrast, in vitamin A–sufficient patients there is a slower, more shallow rise in retinal levels.

 Checkpoint! 24–2

Briefly describe the role of vitamin A deficiency in the development of nyctalopia.

VITAMIN D

A brief discussion of vitamin D is presented here. The physiological actions, regulatory functions, and assessment of vitamin D are discussed in Chapter 14, "Mineral and Bone Metabolism." For a more extensive discussion of vitamin D, the reader is directed to the references listed at the end of this chapter.[5]

Vitamin D deficiency is associated with **rickets**. Rickets is described as a disorder of primarily vitamin D and also of calcium and phosphorous that results in soft bone-tissue formation. Rickets, although a rare condition, often manifests itself in children. Principae causes of rickets include (1) a deficiency of vitamin D, calcium, or phosphorus that may result from malabsorption of vitamin D; (2) diminished dietary intake of vitamin D, calcium, or phosphorus; (3) lack of exposure to sunlight; and (4) genetic factors transferred to offspring.

Dietary sources of vitamin D include irradiated foods and commercially prepared milk. Some vitamin D is found in butter, egg yolk, liver, salmon, sardines, and tuna. Thus rickets in children or osteoporosis in adults can be minimized though dietary supplementation with vitamin D, calcium, and phosphorus.

VITAMIN E

Vitamin E belongs to a family of compounds that are two radical (2R) stereoisomers of α-**tocopherol**. Natural vitamin E exists in eight different forms that include four tocopherol and four tocotrienols. The α-tocopherol form is presumed to be the most common, biologically active form and has the highest bioavailability. Tocopherol and tocotrienol are viscous oils at room temperature, soluble in fat solvents, and insoluble in aqueous solutions.

Food sources with significant amounts of vitamin E include vegetable oil, nuts, sunflower seeds, and wheat germ. In addition, sunflower oil, safflower oil, corn oil, and soybeans contain vitamin E. Meats, vegetables, and fruits contain only small amounts of vitamin E. Alpha-tocopherol is the form of vitamin E widely used to supplement food products.

Vitamin E is absorbed from the small intestines and mixed with bile. Absorbed vitamin E is secreted in chylomicrons along with triglyercides and cholesterol. A portion of chylomicron-bound vitamin E is transported and delivered to the peripheral tissue using lipoprotein lipase as a catalyst. Chylomicron remnants travel to the liver, where α-tocopherol is incorporated into very low density lipoprotein (VLDL), which enables continued distribution of α-tocopherol throughout the body aided (by low density lipoprotein cholesterol [LDL-C] and high density lipoprotein cholesterol [HDL-C]). Vitamin E is excreted via the bile, made water soluble (via tocopheroinic acid and its β-glucuronide conjugate), and excreted in the urine, as carboxyethyl hydroxychromans.[6]

Vitamin E functions is a chain-breaking **antioxidant** as well as a pyroxyl radical scavenger that protects LDL and polyunsaturated fats in membranes from oxidation. Through this action vitamin E helps protect the body against the damaging effects of free radicals. Other functions of vitamin E, though controversial and not established in clinical studies, include (1) enhancing immune function, (2) blocking formation of nitrosamine (a known carcinogen), and (3) delaying development of coronary artery disease (as a result of limited oxidation of LDL-C).

Deficiency of vitamin E in humans is rare. Three conditions that may result in vitamin E–deficient states in humans are as follows:

- Premature birth
- Malabsorption of dietary fat
- Rare disorders of fat metabolism

In premature and low-birth-weight infants, placental transfer of biological substances may be reduced, which could lead to a deficiency of vitamin E. Also, these infants have insufficient amounts of adipose tissue, which serves as a storage site of vitamin E.[7] These infants often develop edema and anemia due to the shortened life span of erythrocytes and the fragile membranes of erythrocytes.

Another condition that is not categorized as a true deficiency or toxicity yet is associated with vitamin E and may afflict premature infants (and has been linked to high oxygen therapy) is retinopathy of prematurity (formerly retrolental fibroplasias). This condition is characterized by abnormal development of blood vessels in premature infants, especially the blood vessels in the retina of the eye (which may lead to retinal detachment and eventual blindness). *High oxygen therapy* refers to a treatment modality that consists of supplying the patient with higher than normal amounts of oxygen. Vitamin E supplementation for premature infants has been shown to reduce the number cases of retinopathy of prematurity.

Vitamin E deficiency can be caused by severe or prolonged malabsorptive diseases, such as celiac disease, or it may follow resection of the small intestine. Children with cystic fibrosis or prolonged cholestasis may develop vitamin E deficiency, which could lead to areflexia and hemolytic anemia. Abetalipoproteinemia, a rare genetic disorder in children, may lead to vitamin E deficiency due to an inappropriate amount of vitamin E absorption or transport. Signs and symptoms of vitamin E deficiency include peripheral neuropathy, skeletal myopathy, opthalmoplegia, and pigmented retinopathy.

Toxicity of vitamin E is usually a result of excess dietary supplementation. High doses of vitamin E may reduce platelet aggregation and thus interfere with vitamin K metabolism. Therefore, patients taking warfarin compounds should not oversupplement their diets with vitamin E.

Evaluation of vitamin E status may include both functional and direct assays. Functional methods include protection of erythrocytes hemolysis on addition of peroxide,[8] or inhibition of lipid peroxidation products.[9] Vitamin E concentrations in serum, erythrocytes, lymphocytes, or platelets may be determined by direct assays such as HPLC.

 Checkpoint! 24–3

Describe the role of vitamin E as an antioxidant.

VITAMIN K

There are two naturally occurring substances associated with vitamin K activity: vitamin K_1 and vitamin K_2. Vitamin K_1 (phylloquinone) is 2-methyl-3-phytyl-1, 4-naphthoquinone and can be found in plants. Vitamin K_2 is actually a series of compounds (the menaquinones). Synthesis of menaquinone occurs in gram-positive bacteria (within the human intestine).

Dietary sources of vitamin K (phylloquinone) are green vegetables, plant oils, and margarines. The menaquinones are found in cheese, eggs, and other milk products.

Vitamin K absorption into the intestines depends on the compound's solubility. Bile salts facilitate the intestinal absorption of phylloquinone and the menaquinones primarily via the lymph. Following absorption, phylloquinone is incorporated into chylomicrons and correlates with triglyceride and lipoproteins levels. Both phylloquinone and menaquinone are localized in the liver.

Phylloquinone is metabolized to polar metabolites, which are excreted in the bile and urine. Urinary metabolites are produced by removing carbon atoms on the side chain at carbon 3, yielding carboxylic acids. Carboxylic acid is conjugated with glucuronic acid before excretion.

A physiological function of vitamin K is to promote clotting of blood. Vitamin K serves as a cofactor to vitamin K–dependent carboxylase. Carboxylase is an enzyme necessary for the posttranslational conversion of specific glutamyl residues in target proteins to γ-carboxyglutamyl (Gla) residues. Thus vitamin K facilitates formation of the Gla proteins prothrombin, proconvertin plasma thromboplastin, and Stuart factor. The enzyme γ-glutamylcarboxylase couples the oxidation of the reduced hydroquinone form of vitamin K_1 or K_2 to γ-carboxylation of glucose residue on vitamin-dependent proteins such as prothrombin. The product of this reaction is an epoxide of vitamin K and γ-carboxyglutamate (Gla) residues in vitamin K–dependent precursor proteins found in the endoplasmic reticulum. The reaction sequence is shown in Figure 24-2 ■.

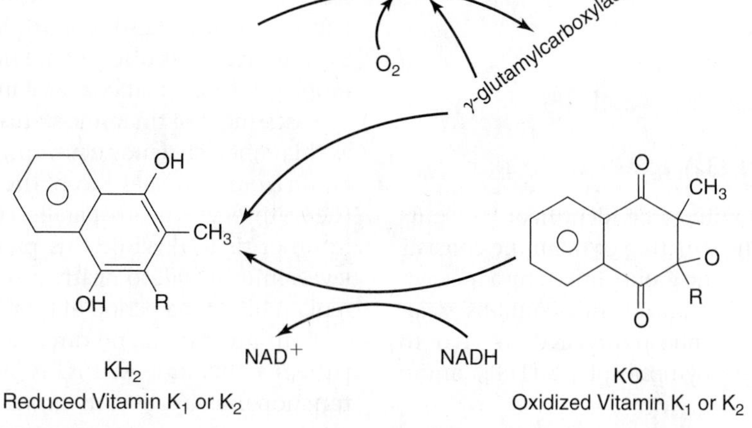

■ FIGURE 24-2 Reaction of vitamin K with prothrombin. The enzyme γ-glutamyl carboxylase couples the oxidation of the reduced form (KH_2) of vitamin K_1 or K_2 to γ-carboxylation of decarboxyprothrombin. An epoxide of vitamin K (KO) is formed in concert with prothrombin.

Vitamin K stimulates biosynthesis of the bone Gla protein osteocalcin and matrix Gla protein. Thus evidence exists that vitamin K plays a role in maintenance of the adult skeleton and prevention of osteoporosis. Low concentration of vitamin K is associated with deficits in bone mineral density and fractures.

Vitamin K deficiency is uncommon in adults. There are risks for depletion deficiencies associated with malabsorption states, including bile duct obstruction, chronic pancreatitis, and liver diseases. Patients taking coumarin anticoagulants (e.g., warfarin) and antibiotics (e.g., cephalosporin) may be at risks for developing vitamin K deficiency. Other at-risk patients include those hospitalized and receiving poor nutrient intakes or receiving total parenteral nutrition (TPN) with insufficient amounts of vitamin K supplementation.

A disease of the newborn known as hemorrhagic disease can develop due to any of the following conditions: (1) poor placental transfer of vitamin K, (2) hepatic immaturity leading to inadequate synthesis of coagulation proteins, and (3) low vitamin K content of breast milk.

Vitamin K concentration in normal subjects is very low, and thus functional assays such as prothrombin time (PT) have been used to assess vitamin K status. Another approach for the assessment of vitamin K status with respect to PT is to measure des-γ-carboxyprothrombin or undercarboxylated prothrombin and PIVKA-II (protein induced by vitamin K absence) by immunoassay.[10] A high concentration of PIVKA-II indicates vitamin K deficiency.

Direct assays for measurement of plasma phylloquinone concentrations are a good approach to assessing vitamin K status. Methods using HPLC with electrochemical or fluorometric detection provide acceptable precision and detection limits.

 Check point! 24-4

Identify the enzyme that catalyzes the following two reactions nearly simultaneously: (1) decarboxylates decarboxy prothrombin to prothrombin, and (2) oxidizes vitamin K_1 or K_2 (KH_2) to oxidized vitamin K_1 or K_2 (KO).

▶ WATER-SOLUBLE VITAMINS

THIAMINE (VITAMIN B₁)

Thiamine was the first B vitamin to be identified; hence its designation as vitamin B_1. The structure of thiamine consists of a pyrimidine ring with an amino group, linked by a methylene bridge to a thiazole ring. The thiazole ring contains an alcohol side chain that can be phosphorylated in vivo to produce esters (e.g., thiamine pyrophosphate [TPP], mono and triphosphates).

Food sources of thiamine include yeast, pork, legumes, beef, whole grains, and nuts. Rice that has been milled and polished contain very little thiamine. Therefore, cultures whose diets rely heavily on milled rice develop more cases of thiamine deficiency than in societies whose diets are not rice dependent. Thiaminases are compounds that tend to destroy or degrade the vitamin and are found in tea, coffee (caffeinated and decaffeinated), raw fish, and certain shellfish.

The majority of thiamine absorption occurs in the proximal small intestine by a saturable process and passive diffusion. Once absorbed, thiamine undergoes intracellular phosphorylation to pyrophosphate, whereas a portion remains in the free form. Thiamine is carried by the portal blood to the liver. Free thiamine resides in the plasma, whereas TPP is the primary cellular component. Nearly half the body stores of thiamine are located in the skeletal muscles, and the rest in the heart, kidneys, and nervous tissue.

Thiamine functions as the coenzyme TPP, which is necessary for the essential decarboxylation reactions catalyzed by the enzymes pyruvate and α-ketoglutarate dehydrogenase. TPP also serves as a coenzyme for a transketolase reaction that mediates the conversion of hexose and pentos phosphates.

Thiamine deficiency usually is a result of poor dietary intake. In Western countries, alcoholism and chronic illness, such as cancer, are the principae causes of thiamine deficiency. Alcohol interferes directly with the absorption of thiamine and with the synthesis of TPP.

Most individuals in early stages of thiamine deficiency develop anorexia and nonspecific symptoms (e.g., irritability). Chronic thiamine deficiency results in beriberi, which is classified as *wet* or *dry*. In either case, patients often complain of pain and paresthesias. Patients with **wet beriberi** exhibit cardiovascular symptoms resulting from impaired myocardial energy metabolism. Patients may present with tachycardia, enlarged heart, congestive heart failure (CHF), peripheral edema, and peripheral neuritis. Patients with **dry beriberi** often present with peripheral neuropathy of the motor and sensory systems. The legs are affected the most, and patients have difficulty rising from a squatting position.

Alcoholics with thiamine deficiency present with a unique set of signs and symptoms. They tend to develop central nervous system complications known as Wernicke's encephalopathy (also **Wernicke–Korsakoff syndrome**). Symptoms of this disorder are horizontal nystagmus, opthalmoplegia, cerebral ataxia, and mental impairment.

Assessment of thiamine status is predicated on the fact that as thiamine deficiency develops, there is a rapid loss of the vitamin from all tissues except the brain. The decrease of TPP in the erythrocytes nearly parallels the decrease of this coenzyme in other tissues. While this process is continuing, thiamine levels in urine fall to nearly zero and the urinary metabolites remain high for a period of time before decreasing.

Both functional and direct assays are available to assess a patient's thiamine status. A widely used enzyme for the functional assay is transketolase. In red cells, transketolase in the presence of magnesium (metallocofactor) and TPP (coenzyme) catalyzes two unique substrates to products, α, β-Dihydroxyethyl-TPP and thiamine pyrophosphate (TPP).

As thiamine decreases, the transketolase activity diminishes. The TPP depletion or effect measures the extent of depletion of the transketolase enzyme for coenzyme by assaying enzyme activity before and after TPP supplementation.

Direct assays involve measuring plasma, erythrocytes, or whole-blood concentrations of thiamine. Separation techniques such as HPLC provide reliable results that reflect body stores of thiamine.

 Checkpoint! 24–5

Identify the two types of beriberi and explain the principal differences between the two types.

RIBOFLAVIN (VITAMIN B₂)

Riboflavin serves an important role in the metabolism of carbohydrates, fats, and proteins by participating as a respiratory coenzyme and an electron donor. Enzymes that incorporate **flavin** adenine dinucleotide (FAD) or flavin mononucleotide (FMN) as prosthetic groups are referred to as flavoenzymes (e.g., monoamine oxidase [MAO], glutathione reductase, and succinic acid dehydrogenase).

Dietary sources of riboflavin are milk, other dairy products, enriched bread and cereal, legumes, lean meat, fish, eggs, and broccoli. Riboflavin deficiency is almost always due to reduced dietary intake. Riboflavin is photosensitive, and foods containing riboflavin must be protected from light.

In food riboflavin exists as a complex of food protein with coenzymes FMN and FAD. In the gut the coenzymes are released from the protein. Riboflavin is primarily absorbed in the proximal small intestine by a saturable transport mechanism. Flavins are transported in blood loosely attached to albumin and tightly attached to several immunoglobulins. Once inside the cell, flavin enters into various redox reactions manifested within each cell.

Riboflavin and its coenzyme derivatives contribute to several biochemical reactions that ultimately result in energy production. For example, the coenzyme forms are involved in the electron transfer reaction necessary for cellular respiration. Flavin proteins have a role in drug metabolism via the cytochrome P450 system and lipid metabolism.

Deficiencies associated with riboflavin are manifested primarily by lesions of the mucocutaneous surface of the mouth and skin. Other conditions associated with riboflavin deficiency include cornmeal vascularization and anemia.

Assessment of riboflavin status can be accomplished via a functional assay incorporating FAD-dependent glutathione reductase or direct assays measuring urine riboflavin excretion and measuring riboflavin in plasma.

Direct assays of riboflavin in blood include HPLC using fluorescence detection after protein precipitation and capillary zone electrophoresis with laser-induced fluorescence detection.

NIACIN (VITAMIN B₃)

Niacin is the term for nicotinic acid, and its amide is nicotinamide. Both niacin and nicotinamide are converted to the redox coenzymes nicotinamide-adenine dinucleotide (NAD)$^+$ and nicotinamide-adenine dinucleotide phosphate (NADP)$^+$, respectively. Structurally, niacin or nicotinic acid is pyridine-3-carboxylic acid, and this basic compound produces derivatives with significant functions.

Niacin in the form of NAD and NADP is found in high amounts in beans, milk, lean meats, cereal and whole grains, yeast, liver, poultry, and eggs. A smaller amount of niacin is found in canned salmon and some leafy vegetables. Flours are enriched with the "free" or non-coenzyme form of niacin and therefore are a good dietary source of niacin.

Nicotinic acid and nicotinamide are absorbed from the stomach and small intestine. The amino acid tryptophan can be converted to niacin, thus providing an endogenous form of the vitamin. The urinary excretion products of niacin include 2-pyridone and 2-methylnicotinmide.

Cofactors of niacin are important in several oxidation and reduction reactions in the body. These coenzymes function along with dehydrogenases and catalyze reactions that convert alcohols to aldehydes, or ketones; hemiacetals to lactones; aldehydes to acid; and certain amino acids to keto acids. In addition, NAD and NADP are active in adenine diphosphate-ribose transfer reactions required for DNA repair and calcium mobilization. Nicotinic acid used as a pharmaceutical agent has an antiatherogenic phenotype. Nicotinic acid can lower triglycerides, raise HDL-C, and shift LDL-C particles to a less atherogenic phenotype.[11]

Individuals who become deficient of niacin develop **pellagra**. This syndrome occurs in people whose diet consists primarily of corn-based foods in locales such as China, Africa, and India. In North America, pellagra frequently occurs in alcoholics, in patients with carcinoid tumors (e.g., tumores of the adrenal glands), and in patients with Hartnup's disease (a congenital defect of intestinal and kidney absorption of tryptophan). Patients who develop pellagra present with a variety of signs and symptoms, including the following:

- Loss of appetite
- Generalized weakness and irritability
- Abdominal pain and vomiting
- Bright red tongue (glossitis)
- Skin rash ("Casal's necklace")
- Diarrhea
- Vaginitis
- Esophagitis
- Depression
- Dementia

There are no blood biochemical markers currently available for assessing niacin deficiency. Urinary measurement of two metabolites, N'-methylnicotinamide and N'-methyl-2-pyridone-5-carboxamide, is available. In normal adults, approximately 20–30% of niacin is excreted as methylnicotinamide and 40–60% as pyridone. An excretion ratio of pyridone to methylnicotinamide less than 1.0 is indicative of niacin deficiency.[12] Capillary electrophoresis and HPLC techniques are available for analyzing these metabolites.

 Checkpoint! 24–6

What condition is characterized by Casal's necklace and a bright red tongue?

PYRIDOXINE (VITAMIN B₆)

Pyridoxine (pyridoxol) (PN), pyridoxamine (PM), and pyridoxal (PL) are the three natural forms of vitamin B₆ that are converted to 5′-phosphates derivatives. The derivatives are required for the synthesis, catabolism, and interconversion of amino acids. Further metabolism of the derivatives results in the formation of pyridoxamine-5′-phosphate (PMP) and pyridoxal-5-phosphate (PLP, P-5′-P). A large number of vitamin B₆–dependent reactions require the coenzyme PLP.

Vitamin B₆ is found in plants as pyridoxine, whereas animal tissues contain 5′-pyridoxal phosphate and pyridoxamine phosphate. The animal form is more bioavailable than the plant form. Food sources rich in vitamin B₆ are legumes, nuts, wheat bran, and meats. Vitamin B₆ is also found in lesser amounts in all other food groups.

Intestinal absorption of PLP and PMP differs from absorption of pyridoxine-5′-glycoside. PLP and PMP are first hydrolyzed by intestinal alkaline phosphatase, whereas pyridoxine-5-glucose is less effectively hydrolyzed by glycosidase within the cells. The unphosphorylated forms are transported to the liver via the portal vein. Once the vitamin enters the cell, it is phosphorylated and subsequently participates in over 100 enzyme reactions related to amino acid metabolism. Vitamin B₆ is an important contributor to the synthesis of heme and several neurotransmitters. It is also required for the metabolism of glycogen, lipids, steroids, and sphingoid bases and for the conversion of tryptophan to niacin.

Individuals deficient in vitamin B₆ can develop a wide range of symptoms depending on age, current medications, concurrent disease, nutritional status, and alcohol intake. Vitamin B₆ deficiency usually results in skin conditions such as seborrhea. Peripheral neuropathy, confusion, and depression may also occur. Patients may develop anemia because low levels of vitamin B₆ tend to diminish hemoglobin synthesis.

Biochemical assessment of vitamin B₆ can be made by both functional and direct assays of either the vitamin or metabolites. Two functional assays used are as follows: (1) measuring the activity of red cell aspartate or alanine amininotransferase and its activation coefficient on incubation with PLP, and (2) measuring urinary tryptophan metabolites (e.g., xanthurenic acid) following an oral load of L-tryptophan.

Direct assay by HPLC of plasma PLP, plasma 4-pyridoxic acid (4-PA), or urine 4-PA using fluorescence detection is available. Other approaches to testing vitamin B₆ include using a homogenous nonradioactive recombinant enzymatic method for PLP.

 Checkpoint! 24–7

Identify two specific liver function enzymes that use pyridoxal-5′-phosphate as a coenzyme.

VITAMIN B₁₂ (CYANOCOBALAMIN)

Vitamin B₁₂ is an essential dietary vitamin. A deficiency of vitamin B₁₂ results in impaired DNA synthesis in any cell in which chromosomal replication and division are taking place. Tissues with the greatest rate of cell turnover show the most dramatic changes. This is especially true with the hematopoietic system. An early sign of vitamin B₁₂ deficiency is megaloblastic anemia. This type of anemia is characterized by an increase in the number of abnormal macrocytic erythrocytes, and the patient becomes anemic. This disease is called **pernicious anemia**.

The structural configuration of vitamin B₁₂ includes three major portions: (1) a planar group or corrin nucleus that is a porphyrin-like ring structure with four reduced pyrrole rings linked to a central cobalt atom with a substantial number of substituent groups; (2) a 5, 6-dimethylbenzimidazolyl nucleotide, which is bound to the cobalt atom by one of its imidazol nitrogens and whose 2′-ribose carbon is linked with a ester of aminoisopropanol and proprionic acid to the corrin right; and (3) a variable R group attached to the cobalt atom. The R groups may be, for example, –CN (thus cyanocobalamin), –OH (thus hydroxocobalamin), –CH₃ (thus methylcobalamin) and –5′-deoxyadenosyl (thus 5′-deoxyadenosylcobalamin).[13] The terms *vitamin B₁₂* and *cyanocobalamin* are used interchangeably to represent the group of cobamides active in humans.

The active coenzymes methylcobalamin and 5-deoxyadenosylcobalamin are necessary for cell growth and replication. Methylcobalamin is required for the converse of homocysteine to methionine and its derivative *S*-adenosylmethionine, as shown in reaction one Figure 24-3 ■. A second reaction sequence involves folate as 5-methyltetrahydrofolate (discussed later in the chapter). The methyl groups contributed by methyltetrahydrofolate are used to form methylcobalamin, which then acts as a methyl group donor for the conversion of homocysteine to methionine. Reaction three in Figure 24-3 illustrates the reaction involving the second coenzyme, 5-deoxyadenosylcobalamin, that is required for the isomerization of L-methylmalonyl CoA to succinyl CoA.[13]

Primary exogenous sources of vitamin B₁₂ for humans are certain organisms that grow in soil, sewage, water, or the intestinal lumen of animals that produce the vitamin.

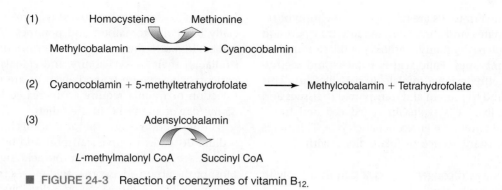

(1) Homocysteine → Methionine

Methylcobalamin → Cyanocobalmin

(2) Cyanocoblamin + 5-methyltetrahydrofolate → Methylcobalamin + Tetrahydrofolate

(3) Adensylcobalamin

L-methylmalonyl CoA → Succinyl CoA

 FIGURE 24-3 Reaction of coenzymes of vitamin B_{12}.

Most vegetables do not contain vitamin B_{12} unless they are contaminated with such microorganisms. Thus animals must synthesize the vitamin in their own alimentary tract or ingest animal products containing vitamin B_{12}.

Dietary vitamin B_{12} is released from food and salivary binding proteins in the presence of gastric acid and pancreatic proteases and is subsequently bound to gastric intrinsic factor (IF). In the ileum, the vitamin B_{12}–IF complex interacts with a receptor on the mucosal cell surface and is actively transported into circulation. Optimum transport of vitamin B_{12} also requires the presence of sodium bicarbonate (for appropriate pH) and bile.

Absorbed vitamin B_{12} is bound to transcobalamin II, a plasma beta-globulin, for transport to tissues. This bond complex is quickly cleared from the plasma and is primarily distributed to the hepatic parenchyma cells. In normal adults, approximately 90% of the body's stores of vitamin B_{12} are in the liver. Vitamin B_{12} is stored as the active coenzyme to be used in metabolic reactions previously discussed.

The clinical impact of vitamin B_{12} deficiency is on the hematopoietic and nervous systems. Cells associated with hematopoiesis have high turnover rates, as do mucosal and cervical epithelium. All of these cells have high requirements for vitamin B_{12}. Vitamin B_{12} deficiency results in abnormal DNA replication. The process begins once a hematopoietic stem cell begins to enter a prescribed series of cell divisions. A defect in chromosome replication results in an inability of maturing cells to complete nuclear division, whereas cytoplasmic maturation continues at fairly normal rate. The results of this defect are an increased production of morphologically abnormal cells and cell death during maturation. A microscopic examination of samples from marrow and peripheral blood samples will reveal the outcome of this process. The microscopist will observe a highly abnormal maturation of erythrocytes (megaloblastic erythropoiesis), many cell fragments, poikilocytes, macrocytes, and abnormal-looking cells in the peripheral blood cells. The mean red cell volume increases, and pancytopenia may be present.[14]

Vitamin B_{12} deficiency can irreversibly damage the nervous system. A progressive swelling of myelinated neurons, demyelization, and neuronal cell death may be present in the spinal column and cerebral cortex. The results of this type of damage to the nervous system may include paresthesias of the hands and feet, unsteadiness, decreased deep tendon reflexes, and several other neurologic symptoms.

Indirect assays assess the functional adequacy of vitamin B_{12}. Several assays are available and include urinary and serum concentrations of methylmalonic acid, plasma homocysteine, the deoxyuridine suppression test, and the vitamin B_{12} absorption test. Other tests include cytochemical staining of red blood cell precursors and a test for IF blocking antibodies.

Several assays are in use for measuring vitamin B_{12} directly in blood. Examples of test methods include the following:

- Competitive protein binding (CPB) assay using ^{57}Co-labeled cobalamin
- Immunometric methods using solid-phase separation by immobilizing the IF binder on beads or magnetic particles
- Chemiluminescent immunoassays using acridium ester for a label to produce the light signal

✓ Checkpoint! 24–8

Identify the origin of the chemical term cyanocobalamin.

FOLIC ACID

Folic acid and *folate* are generic terms for a group of compounds that function as coenzymes that process one-carbon units. Folic acid is derived from pteroic acid, which has attached to it one or more glutamic acid moieties. Pteroic acid includes a pteridine ring joined to a *p*-aminobenzoic acid residue (PteGlu). Pteroylglutamic acid is formed when one molecule of L-glutamic acid is attached to the *p*-amino benzoic acid structure. Pteroylglutamic acid can be reduced to dihydrofolic acid, which is biologically active. Several forms of folic acid are normally present in serum and other body fluids, with the principal form being 5-methyltetrahydrofolate.

Food sources of folate are liver, spinach, other dark leafy vegetables, legumes, and orange juice. Some cereals are fortified with 25–100% of the recommended daily allowance of folic acid.

Folates contained in foods are primarily in the form of reduced polyglutamates, and absorption requires transport and the action of a pteroylglutamyl carboxypeptidase found in mucosal cell membranes. Folic acid is reduced and methylated to form methyltetrahydrofolate ($CH_3H_4PteGlu$). This compound is bound to protein and transported to tissues and the liver. In the liver, $CH_3H_4PteGlu$ is reduced and transported to bile and enters the enterohepatic cycle. This helps to provide a consistent source of folate along with dietary intake.

In the cells, folic acid congeners play a significant role in intracellular metabolism, as follows:[15]

- Conversion of homocysteine to methionine
- Conversion of serine to glycine
- Synthesis of thymidylate
- Metabolism of histidine
- Synthesis of purines
- Utilization or generation of formate

A deficiency of folate may result from (1) an insufficient number of intestinal microorganism (stomach sterilization), (2) inadequate intestinal absorption (e.g., following surgical resection or in celiac disease), (3) decreased dietary intake (as may occur in alcoholics), (4) increased demand (e.g., during pregnancy), and (5) administration of anti–folic acid drugs (e.g., methotrexate) or anticonvulsant drugs (e.g., phenobarbital), which tend to increase demand for folate.[16]

The major clinical manifestation of folate deficiency is megaloblastic anemia. This disease is characterized by the presence of large, abnormally nucleated erythrocytes in the bone marrow. The patient's serum folate will be decreased, and plasma homocysteine will be increased.

Assessment of folate status in patients may be reliably determined using direct assays for folic acid in whole blood, serum, and erythrocytes. Competitive binding protein assays using radioactive nuclides for labels are available, but most clinical laboratories use immunoassays that incorporate nonradioactive labels. For example, chemiluminescent or enzyme immunoassays and HPLC techniques have been developed.

 Checkpoint! 24–9

List three conditions that may result in a deficiency of folic acid.

VITAMIN C

The term *vitamin C* refers to a group of molecules that exhibit antiscorbutic properties in humans and includes both ascorbic acid and its oxidized form, dehydroascorbic acid (DHA). Ascorbic acid serves as a reducing agent in many important hydroxylation reactions in the body.

Food sources of vitamin C have been well publicized since the research of Linus Pauling was popularized in the 1970s.[17]

Vitamin C is found in citrus fruits, green vegetables (especially broccoli), tomatoes, and potatoes. Vitamin C is consumed in both the natural and synthetic form. Both are similar in their bioavailability. Individuals who smoke and patients who require hemodialysis or are under stress from infection or trauma require an increased amount of nearly 35 mg/day of vitamin C in their diet.

The absorption of ascorbic acid occurs by a combination of sodium-dependent active transport and by simple diffusion. On average, nearly 90% of ascorbic acid ingested is absorbed. Most of the absorbed ascorbic acid moves from the intestinal cells into blood by diffusion and is distributed throughout the body, where it is found in most tissues. Ascorbic acid can be converted to DHA, which is reduced to ascorbate, and in plasma vitamin C exists predominantly as the ascorbic ion. Excretion of unchanged ascorbate occurs within approximately 24 hours. The oxidized form of ascorbic acid, DHA, is further degraded to oxalic acid for excretion in urine.

Ascorbic acid and its oxidized product DHA are both biologically active. Vitamin C has several functions, including antioxidant activity, promotion of nonheme iron absorption, biosynthesis of carnitine, and conversion of dopamine to norepinephrine. Other functions include involvement in connective tissue metabolism and cross-linking, and acting as a component of drug-metabolizing enzyme systems.

Scurvy is a disease characterized by a deficiency of vitamin C. In the United States, this disease is most prevalent in the poor and elderly and in alcoholics, who ingest lower amounts (<10 mg/day) of vitamin C per day. Symptoms of scurvy include bleeding into the skin (petechiae, ecchymosis), bleeding and irritated gums, and bleeding elsewhere (e.g., joints, the peritoneal cavity, the pericardium, and the adrenal glands).

Vitamin C is one of the few vitamins in which consumption of large doses may cause serious problems. The symptoms of hypervitaminosis include abdominal pain, diarrhea, and nausea. Blood levels of ALT, lactate dehydrogenase (LD), and uric acid may be elevated. Chronic, high-dose supplementation of vitamin C can lead to kidney stones.

Assessment of vitamin C status is primarily accomplished by direct methods. Quantitative analysis of plasma, urine, or tissue samples for ascorbic acid or total vitamin C is available to clinicians. Spectrophotometric assays, HPLC, and gas chromatography-mass spectrometry (GC-MS) techniques are used for these analyses.

 Checkpoint! 24–10

List three symptoms of scurvy.

PANTOTHENIC ACID

Pantothenic acid is a component of coenzyme A (CoA) and in nature it is synthesized by most microorganisms and plants from pantoic acid. Pantoic acid is modified by addition of cysteamine at the C-terminal end and phosphorylation at carbon 4 to form 4'-phosphopantetheine, which serves as a covalently attached prosthetic group of acyl carrier proteins. This structure attaches to ribose-3'-phosphate and adenine to form CoA.

Food sources of pantothenic acid include legumes, broccoli, sweet potatoes, molasses, whole-grain cereals, and animal sources (e.g., liver, beef, and poultry). Other good food sources include egg yolk, skimmed milk, kidney, liver, and yeast.

Pantothenic acid within dietary CoA compounds and 4'-phosphopeantetheine are hydrolyzed by enzymes in the intestinal lumen to dephospho-CoA, phosphopantetheine, and pantethine. These forms are further hydrolyzed to pantethenic acid. Most is absorbed as pantethenic acid, enters circulation, and is carried by proteins and taken up by cells.

Deficiencies associated with pantothenic acid are relatively rare in the United States. Pantothenic acid is widely available in foods, and thus deficiency does not pose a problem for humans. Direct assays are available to measure pantothenic acid in whole blood or urine. The urine assays best reflect dietary intake. Analytical techniques include immunoassays, GC, and GC-MS.

SUMMARY

The material presented in this chapter focused on the vitamins of clinical interest. They are categorized as fat- and water soluble based on their chemical properties in solution. Although most clinical laboratories are not equipped with the instrumentation required for vitamin analysis, it is important for clinicians to understand the vitamins discussed in this chapter.

A review of the chemistry and functions of each vitamin was included in this chapter. Dietary sources and distribution of vitamins and excretion of metabolites were included. Reference was made to both functional (indirect methods) and direct methods of assessing vitamin status in humans.

REVIEW QUESTIONS

LEVEL I

1. Beriberi is associated with a deficiency of which of the following vitamins? (Objective 9)
   a. niacin
   b. thiamine
   c. riboflavin
   d. folate

2. Night blindness is associated with which of the following vitamins? (Objective 9)
   a. vitamin C
   b. vitamin A
   c. vitamin B_1
   d. vitamin E

3. Which of the following vitamins is a fat-soluble vitamin? (Objective 1)
   a. vitamin D
   b. vitamin C
   c. vitamin B_{12}
   d. vitamin B_2

4. Which groups of individuals are most susceptible to vitamin E–deficient states? (Objective 9)
   a. premature infants
   b. young adults
   c. middle-aged adults
   d. senior citizens

5. Which of the following instruments can be used to identify and quantitate vitamins in blood? (Objective 3)
   a. refractometer
   b. reflectometer
   c. atomic absorption spectrometer
   d. mass spectroscopy

LEVEL II

1. Which of the following drugs may interfere with vitamin K metabolism? (Objective 4)
   a. digoxin
   b. warfarin
   c. barbiturates
   d. theophylline

2. Which of the following statements best describes the difference between wet beriberi and dry beriberi? (Objective 5)
   a. A patient with wet beriberi tends to sweat profusely whereas a patient with dry beriberi does not.
   b. A patient with wet beriberi will have an increased pro-thrombin clotting time whereas a patient with dry beriberi will have a normal pro-time.
   c. A patient with wet beriberi will show signs of peripheral neuritis whereas a patient with dry beriberi will show signs of cardiac failure.
   d. A patient with wet beriberi will show signs of cardiac failure whereas a patient with dry beriberi will show signs of peripheral neuritis.

3. Vitamin B_{12} participates in which of the following biochemical conversions? (Objective 3)
   a. L-aspartate + 2-oxoglutarate $\longleftrightarrow$ Glutaric acid + oxaloacetate
   b. L-methylmalonyl CoA $\longleftrightarrow$ succinyl-CoA Homocysteine $\longrightarrow$ methionine
   c. L-lactate + NAD^+ $\longleftrightarrow$ Pyruvate + NADH + H^+ CK
   d. Creatine phosphate + ADP $\longrightarrow$ creatine ATP

REVIEW QUESTIONS (continued)

LEVEL I

6. Which of following functions is associated with vitamin C? (Objective 4)
   a. participates as a cofactor for enzymes of amino acid metabolism
   b. cleaves carbon–carbon bonds.
   c. scavenges physiologically important reactive oxygen species and nitrogen species
   d. participates in controlling bleeding

7. Which of the following statements describes the function of vitamin E in humans? (Objective 4)
   a. Vitamin E functions as a chain-breaking antioxidant.
   b. Vitamin E functions to maintain the integrity of the eyes.
   c. Vitamin E functions to promote bone growth.
   d. Vitamin E functions to promote clotting of blood.

8. Measuring the concentration of vitamin C in plasma by HPLC is an example of which type of assessment? (Objective 8)
   a. functional
   b. direct
   c. indirect
   d. overt

LEVEL II

4. Folate deficiency can result from which of the following? (Objective 4)
   a. the absence of intestinal microorganisms
   b. poor intestinal absorption
   c. administration of drugs (e.g., methotrexate, anticonvulsants)
   d. all of the above

5. Vitamin A is a fat-soluble vitamin for which of the following reasons? (Objective 3)
   a. Vitamin A goes into aqueous-based solutions.
   b. Vitamin A is a polar compound.
   c. Vitamin A goes into organic solvent–based solutions.
   d. Vitamin A can be extracted into aqueous reagents.

6. A patient with megaloblastic anemia would have which of the following test results? (Objective 1)
   a. elevated blood levels of vitamin B_{12} and folic acid
   b. elevated blood levels of vitamin B_{12} and decreased folic acid
   c. decreased blood levels of vitamin B_{12} and decreased folic acid
   d. normal blood levels of vitamin B_{12} and folic acid

7. NAD and NADP belong to which family of vitamins? (Objective 3)
   a. niacin
   b. vitamin A
   c. vitamin C
   d. vitamin D

8. A deficiency of vitamin K can result in which of the following? (Objective 1)
   a. an increase in prothrombin time
   b. a decrease in prothrombin time
   c. a normal prothrombin time
   d. decreased clotting time

PEARSON
myhealthprofessionskit™

Use this address to access the interactive Companion Website created for this textbook. Simply select "Clinical Laboratory Science" from the choice of disciplines. Find this book and log in using your user name and password.

REFERENCES

1. Shenkin A. Trace elements and inflammatory response: Implications for nutritional support. *Nutrition* (1995) 11(1 Suppl):100–105.

2. Evans TR, Kaye SB. Retinoids: Present role and future potential. *BrJCancer* (1999) 80:1–8.

3. Napoli JL. A gene knockout corroborates the integral function of cellular retinol-binding protein in retinoid metabolism. *Nutrition Reviews* (2000) 58(8):230–36.

4. Loerch JD, Underwood BA, Lewis KC. Response of plasma levels of vitamin A to a dose of vitamin A as an indicator of hepatic vitamin A reserves in rats. *J of Nutrition* (1979) 109(5):778–86.

5. Endres DB, Rude RK. Mineral and bone metabolism. *In*: Burtis CA, Ashwood ER, Bruns DE (eds.). *Tietz textbook of clinical chemistry and molecular diagnostics*, 4th ed. Philadelphia: W. B. Saunders, 2006:1920–26.

6. Schultz M. Leist M. Petrzika M. et al. Novel urinary metabolite of alpha-tocopherol, 2, 5, 7, 8-tetramethyl-2(2'-carboxyethyl)-6-hydroxychroman, as an indicator of an adequate vitamin E supply? *American J of Clin Nutrition* (1995) 62(6 Suppl):1527S–1534S.

7. Bieri JG. Evarts RP. Effect of plasma lipid levels and obesity on tissue stores of alpha-tocopherol. *Proceedings of the Society for Experimental Biol Medicine* (1975) 149(2):500–502.

8. Farrell PM, Bieri JG, Fratantoni JF et al. The occurrence and effects of human vitamin E deficiency. A study in patients with cystic fibrosis. *J of Clin Investigation* (1977) 60(1):233–41.

9. Van Gossum A. Shariff R. Lemoyne M. Kurian R. Jeejeebhoy K. Increased lipid peroxidation after lipid infusion as measured by breath pentane output. *American J of Clin Nutrition* (1988) 48(6):1394–99.

10. Blanchard RA, Furie BC, Kruger SF et al. Immunoassays of human prothrombin species which correlate with functional coagulant activities. *J of Lab & Clini Med* (1983) 101(2):242–55.

11. Ito MK. Niacin-based therapy for dyslipidemia: Past evidence and future advances. *American J of Managed Care* (2002) 8(12 Suppl): S315–22.

12. Sauberlich HE. *Laboratory tests for the assessment of nutritional status*, 2nd ed. Boca Raton, FL: CRC Press, 1999.

13. Kaushansky K, Kipps TJ. Hematopoietic agents. *In*: Brunton LL, Lazo JS, Parker KL (eds.). *Goodman & Gilman's the pharmacological basis of therapeutics*, 11th ed. New York: McGraw-Hill, 2006:1454–56.

14. Kaushansky K, Kipps TJ. Hematopoietic agents. *In*: Brunton LL, Lazo JS, Parker KL (eds.). *Goodman & Gilman's the pharmacological basis of therapeutics*, 11th ed. New York: McGraw-Hill, 2006:1456–1457.

15. Shankin A, Baines M, Fell GS et al. Vitamins and trace elements. *In*: Burtis CA, Ashwood ER, Bruns DE (eds.). *Tietz textbook of clinical chemistry and molecular diagnostics*, 4th ed. Philadelphia: W. B. Saunders, 2006:1110–11.

16. Lindenbaum J, Allen RH. Clinical spectrum and diagnosis of folate deficiency. *In*: Bailey LB (ed.). *Folate in health and diseases*. New York: Marcel Dekker, 1996:43–74.

17. Pauling L. Evolution and need for ascorbic acid. *Proceedings of the National Acad of Sciences* (1970) 67(4):1643–48.

25

Molecular Diagnostics

■ OBJECTIVES—LEVEL I

Following successful completion of this chapter, the learner will be able to:

1. List the three components of a nucleotide.
2. Describe the basic steps of transcription and translation.
3. Using absorbance values, calculate the DNA purity and yield of a sample.
4. Discuss the basic principles of gel electrophoresis of DNA.
5. Define the basic principles of restriction endonucleases.
6. Determine the melting temperature of a short DNA sequence.
7. Explain the basic principles of the following techniques: liquid-phase hybridization, in situ hybridization, fluorescent in situ hybridization (FISH), Southern blot, Western blot, traditional polymerase chain reaction (PCR), real-time PCR, and the Sanger method of DNA sequencing.
8. List clinical applications of the methods mentioned in objective 7.

■ OBJECTIVES—LEVEL II

Following successful completion of this chapter, the learner will be able to:

1. Identify the structures of DNA, RNA, adenine, guanine, cytosine, thymine, and uracil.
2. Compare and contrast the chemical structures of DNA and RNA.
3. Given a sequence of nucleotide bases in a DNA strand, deduce the sequence of nucleotide bases in the complementary RNA strand and protein product.
4. Discuss factors that may lead to reduced purity and yield following a DNA isolation procedure.
5. Compare and contrast agarose and polyacrylamide as matrices for gel electrophoresis.
6. Given a specific restriction endonuclease and DNA sequence, determine the number and size of the resulting fragments.
7. Discuss factors that affect the ability of hybridization to occur.
8. Explain the basic principles of the following techniques: liquid-phase hybridization, in situ hybridization, FISH, Southern blot, Western blot, DNA microarrays, traditional PCR, real-time PCR, the Sanger method of DNA sequencing, and pyrosequencing.
9. Select an appropriate assay from those listed in objective 8 for a particular clinical application.

KEY TERMS

Amplicon	Gene	Primers
Anticodon	Hybridization	Probe
Chromosome	In situ hybridization	Promoter
Codon	Intron	Restriction endonucleases
DNA (deoxyribonucleic acid)	Melting temperature (T_m)	RNA (ribonucleic acid)
DNA polymerase	Microarray	Stringency
DNA sequencing	Molecular diagnostics	Thermocycler
Exons	Mutation	Transcription
Fluorescent in situ	Nucleotides	Translation
hybridization (FISH)	Polymerase chain reaction (PCR)	Western blot

A CASE IN POINT

A 30-year-old male presented to his physician with fatigue, fever and chills, night sweats, diarrhea, and a weight loss of 10 lbs. His white blood count was elevated, at 11,000/μL, with 85% granulocytes, 11% lymphocytes, and 4% monocytes. Lymphocyte subset analysis revealed a decreased percentage and number of CD4-positive cells (15%, absolute number of 181/μL). Tests for HIV antibody were positive. The patient was started on antiretroviral therapy.

Issues and Questions To Consider

After reading the patient's case history, consider:

1. What is the patient's diagnosis based on his clinical findings and laboratory results?

2. What are the benefits of molecular testing in the care of this patient?

3. What specific molecular methods could be used to monitor this patient, and what are the principles of each?

WHAT'S AHEAD

▶ Discussion of basic concepts necessary for the understanding of molecular testing.
▶ Overview of the principles and clinical applications of major molecular methods used in the clinical laboratory.

▶ INTRODUCTION

NUCLEIC ACID STRUCTURE AND FUNCTION

It is remarkable to think that every physical characteristic of our bodies—hair color, eye color, and height, to name a few—as well as the basic functions needed for our survival, such as respiration and digestion, are all made possible by genetic codes contained in molecules called *nucleic acids*. Nucleic acids are complex, acidic macromolecules that were first identified in the nuclei of cells by chemist Frederick Miescher in the 1860s. There are two main types of nucleic acids: DNA and RNA.

DNA, or **deoxyribonucleic acid**, contains the genetic information needed by cells to synthesize all the proteins necessary to maintain their structure, carry out their essential functions, and replicate to form new cells. DNA is found in all living organisms and in most viruses. In cells from higher-level organisms, or eukaryotes, DNA is found in distinct threadlike structures called **chromosomes**, which are located mainly in the nucleus and are separated from the rest of the cell by the nuclear membrane. A small amount of DNA is also found in the mitochondria, which contain a single circular chromosome. In contrast, simpler organisms such as bacteria (i.e., the prokaryotes) do not have a nucleus, and their DNA is found on a single "chromosome" within the cell proper. Some bacteria also contain a small circular double-stranded DNA called a *plasmid*. Plasmids are located outside of the main chromosome and often contain DNA that confers special properties to the cell such as resistance to antibiotics.

The structure of DNA was determined in the 1950s by James Watson and Francis Crick, who analyzed physical and chemical data generated by other scientists to construct a model.[1,2] The significance of this discovery and its potential applications were recognized by the scientific community, and Watson and Crick received the Nobel Prize for Medicine and Physiology in 1962 for their work, along with Maurice Wilkins, who provided them with X-ray crystallography results of the DNA molecule.

The major features of DNA structure, as explained in the Watson–Crick model, are illustrated in Figures 25-1 through 25-4 and are summarized below:[1–4]

• **Nucleotides** are the building blocks of DNA. A nucleotide has three components: (1) a phosphate group, (2) a five-carbon sugar called *deoxyribose*, and (3) a nitrogen-containing, or nitrogenous, base shown in Figure 25-1 ■. Note that the base is attached to the 1 prime (1′) carbon of the sugar, whereas the phosphate is attached to the 5′ carbon by covalent bonds. The phosphate group gives DNA its acidic properties and an overall negative charge.

• Each nucleotide contains one of four possible nitrogenous bases: adenine (A), guanine (G), cytosine (C), or thymine (T).

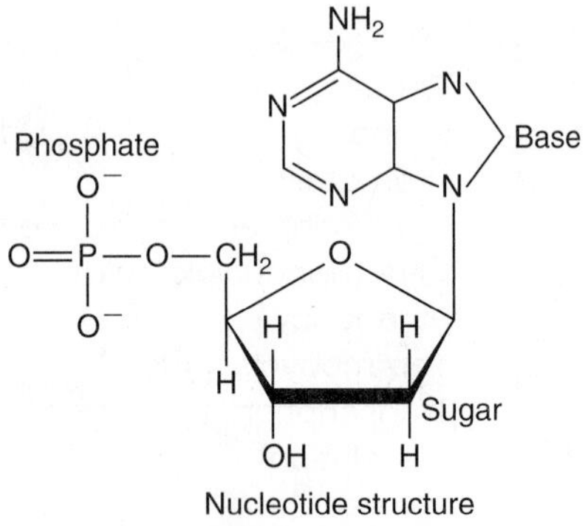

FIGURE 25-1 Nucleotide structure.

Adenine and guanine are classified as *purines* because they have a double-ring structure, whereas cytosine and thymine are classified as *pyrimidines* because of their single-ring structure as illustrated in Figure 25-2 ■.

- The nucleotides are attached to each other by covalent bonds between the phosphate group of one nucleotide and the sugar molecule of the next nucleotide in the sequence to form a polynucleotide chain. By convention, the sequence of nucleotides is written in the 5′ to 3′ direction—for example, 5′ATCGAACAGTAC3′.

- DNA is composed of two chains of nucleotides that run in opposite directions to each other, with one chain running in the 5′ to 3′ direction, and the other in the 3′ to 5′ direction. These two chains are said to be *antiparallel* as seen in Figure 25-3 ■.

- In the three-dimensional structure of DNA, its two nucleotide chains are coiled around each other to form a double helix. Alternating sugar and phosphate groups are on the exterior of the helix, forming the backbone of each chain. The nucleotide bases are in the interior of the helix and are paired to bases of the opposite chain by hydrogen bonds.

- To maintain the regularity of the helical structure, it is necessary for the larger purines on one strand to pair with the smaller pyrimidines on the opposite strand, and for base pairing to be highly specific. Adenine on one chain will only pair with thymine on the other chain. A and T are held together by two hydrogen bonds. Guanine will only pair with cytosine, and these two bases are held together by three hydrogen bonds as shown in Figure 25-4 ■. Each A–T and G-C duo is referred to as a *base pair*.

- Because of these specific pairing rules, the sequence of bases in the second chain of a DNA molecule is determined by the sequence of bases in the first chain. For example, if the base sequence in strand 1 is 5′ATTCGA3′, then the sequence in strand 2 must be 3′TAAGCT5′. The two strands are said to be complementary to each other, and strand 2 is referred to as *complementary DNA* (cDNA).

When DNA replicates itself during the formation of new cells, it is necessary for this base complementarity to be precisely maintained. Otherwise, even a single base change can result in a mutation that can lead to disease or even death. DNA synthesis occurs by a semiconservative mechanism in which the double-stranded molecule of DNA separates into

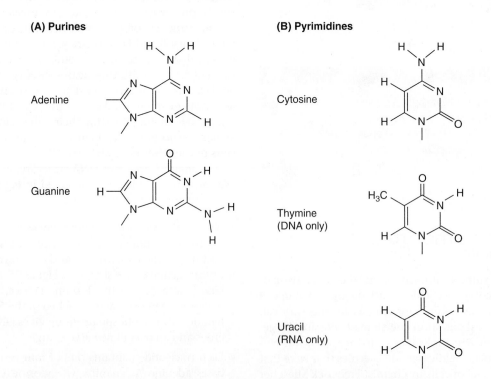

FIGURE 25-2   Chemical structure of nucleotide bases: A. purines and B. pyrimidines.

Deoxyribonucleic Acid (DNA)

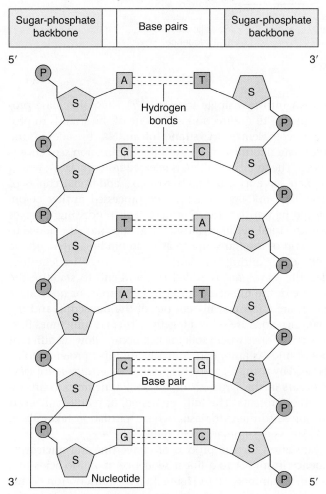

■ FIGURE 25-3   Antiparallel DNA strands.

■ FIGURE 25-4   A–T pairing and G–C pairing.

two single strands; each single strand serves as a template for the synthesis of a complementary DNA strand. In addition to the template strands to be copied, DNA synthesis also requires the presence of the four deoxynucleotide triphosphates (dNTPs) [i.e., the building blocks of DNA: deoxyadenosine triphosphate (dATP), deoxyguanosine triphosphate (dGTP), deoxythymidine triphosphate (dTTP), and deoxycytidine triphosphate (dCTP)], the enzyme **DNA polymerase** (which forms the phosphodiester bonds between adjacent nucleotides in the growing chain), and **primers** (short DNA or RNA sequences complementary to the ends of each template strand that are necessary for binding of the polymerase).

The second type of nucleic acid, **RNA** or **ribonucleic acid**, is present in both the nucleus and the cytoplasm. RNA plays a major role in protein synthesis and serves as the source of genetic information in RNA viruses. There are four types of RNA: messenger RNA (mRNA), transfer RNA (tRNA), ribosomal RNA (rRNA), and small RNAs. Messenger RNA plays a key role in the process of **transcription**, where the genetic information contained in DNA is transferred to a complementary nucleotide sequence in mRNA. After further structural modifications, the mRNA is transported from the nucleus to the cytoplasm, where its genetic code is translated into the specific amino acid sequence of a protein. The second type of RNA, tRNA, plays an important role in the process of **translation** by carrying specific amino acids to the appropriate position on the protein strand being synthesized. The third type of RNA, rRNA, is a major component of the ribosomes, the cellular organelles that serve as the site of protein synthesis. Several classes of small, noncoding RNA molecules have been discovered more recently and are believed to play a role in regulating transcription and translation.[5]

The structure of RNA is chemically similar to that of DNA, with three major differences: (1) RNA usually exists as a single-stranded molecule, whereas DNA is typically double stranded; (2) RNA contains the sugar ribose instead of the deoxyribose found in DNA; and (3) RNA contains the nitrogenous base uracil instead of thymine (see Figure 25-2). Like thymine, uracil binds to adenine during nucleic acid synthesis. Thus, for the DNA sequence, 5' ATTCGA 3', the complementary strand of mRNA would be 3' UAAGCU 5'.

✓ Checkpoint! 25–1

Identify the four types of RNA.

▶ GENE STRUCTURE AND EXPRESSION

Although DNA is the primary carrier of genetic information, all genetically determined features of an organism, or traits, are determined directly through the actions of proteins. Proteins have many essential functions in the body, serving as structural components of cells, enzymes that carry out the

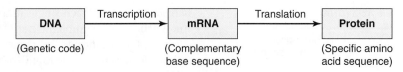

FIGURE 25-5 Flow of genetic information.

chemical reactions in cells, messengers and receptors that transmit signals between cells, antibodies that play a role in the immune system, and transporters of atoms and small molecules throughout the body. Proteins are large, complex molecules that are made up of smaller units called *amino acids*. There are 20 different types of amino acids, and hundreds to thousands of these combine together in a particular sequence that determines the three-dimensional structure and function of the resulting protein.

The amino acid sequence of a protein is coded for by a specific nucleotide sequence in DNA. However, DNA, which resides in the nucleus, is not directly translated into proteins, which are synthesized in the cytoplasm. The intermediate molecule mRNA, which travels from the nucleus to the cytoplasm, is required for this process. The central dogma of molecular biology states that genetic information flows from DNA to RNA to protein as illustrated in Figure 25-5 ■. The first step in this process, in which DNA is converted into a complementary sequence of nucleotide bases in mRNA, is called *transcription*. The second step in the process, translation, involves conversion of the genetic information contained in mRNA into a linear sequence of amino acids in a protein.

The process of protein synthesis begins in the DNA with a fundamental unit of heredity called a **gene**. A gene is a specific sequence of nucleotides located in a particular position on a chromosome that provides cells with the instructions needed to synthesize a specific protein or a particular type of RNA (e.g., rRNA). The typical gene is between 1,000 and 4,000 nucleotides long, although many smaller and larger genes exist. The basic structure of a eukaryotic gene is illustrated in Figure 25-6 ■. The gene begins at the 5' end with a **promoter** region, which is necessary for the initiation of transcription, and ends at the 3' end with a terminator sequence that specifies the end of transcription.[1,3,4] In between the promoter and terminator regions are the sequences called **exons** that ultimately code for the amino acid sequence of the protein. In most genes, the exons are not continuous but are interrupted by one or more noncoding regions called **introns**.

To begin the process of transcription, the enzyme RNA polymerase binds to the promoter region of the gene, separating the DNA into two single strands.[1,3,4,6] The polymerase proceeds in a 3' to 5' direction along one of the strands to produce a complementary strand of mRNA by adding the appropriate ribonucleotides until the termination sequence is reached. The resulting mRNA is released from the DNA, which can then serve as a template to make additional copies of mRNA. The mRNAs must then be processed further before they can be translated into protein. This processing involves three structural modifications: (1) a guanosine cap is added to the 5' end of the transcript to assist in binding the molecule to ribosomes during translation, (2) a string of adenosines called the *poly-A tail* is added to the 3' end to stabilize the structure of the molecule, and (3) the introns, or noncoding regions, are enzymatically cut out of the transcript, and the remaining exons are spliced together. There is sometimes flexibility in the sites where splicing can occur, allowing different types of mRNAs (and, consequently, multiple proteins) to be produced by a single gene. This process, called *alternative splicing*, occurs in the synthesis of antibodies, neurotransmitters, and other proteins. The fully processed, or mature mRNA, is transported to the cytoplasm, where translation takes place. These steps are summarized in Figure 25-7 ■.

Translation, or the process of converting the nucleotide sequence in mRNA to a linear sequence of amino acids in a protein, is summarized in Figure 25-8 ■. Transcription occurs on organelles called *ribosomes* that are present freely in the cytoplasm of cells or are attached to a system of membrane channels in the cytoplasm known as the *endoplasmic reticulum*. Ribosomes are composed of a small subunit that contains binding sites for mRNA and tRNA and a large subunit on which amino acids are joined together to form the newly synthesized protein chains. Both subunits are composed of rRNA and proteins.

To begin translation, the mRNA strand associates with ribosomes in the groove between the small and large subunits.[1,3,4,6] The ribosome moves along the mRNA, reading the genetic code within the molecule, which is made up of units called **codons**. Each codon, which consists of three nucleotides, codes for one amino acid. There are 64 different codons (four nucleotide bases occurring as triplet combinations, or 4_3) but only 20 amino acids, so more than one codon can specify a particular amino acid. This genetic code and its

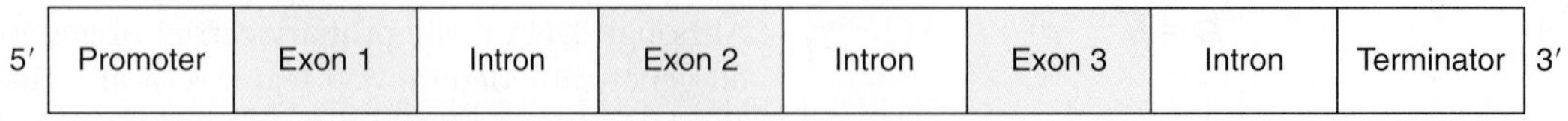

5'	Promoter	Exon 1	Intron	Exon 2	Intron	Exon 3	Intron	Terminator	3'

FIGURE 25-6 Basic structure of the eukaryotic gene.

Sample DNA Sequence:

3′ T A C T C G C T T A A C C C G G A T T A G A C G A T G . . . 5′

↓ Transcription

Initial mRNA:

5′ A U G A G C G A A U U G G G C C U A A U C U G C U A C . . .

(contains exons and introns)

RNA Transcription and Processing
DNA:

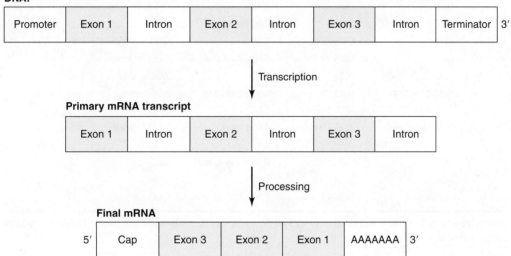

| Promoter | Exon 1 | Intron | Exon 2 | Intron | Exon 3 | Intron | Terminator | 3′ |

↓ Transcription

Primary mRNA transcript

| Exon 1 | Intron | Exon 2 | Intron | Exon 3 | Intron |

↓ Processing

Final mRNA

| 5′ | Cap | Exon 3 | Exon 2 | Exon 1 | AAAAAAA | 3′ |

■ FIGURE 25-7   Steps of transcription.

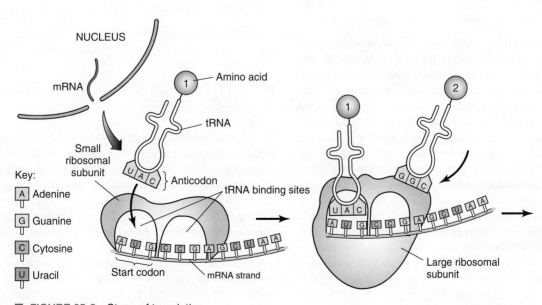

■ FIGURE 25-8   Steps of translation.

⊗ TABLE 25-1

Codons and Their Corresponding Amino Acids

First Position		Second Position of Codon				Third Position
		U	C	A	G	
U		UUU	UCU	UAU	UGU	U
		UUC Phenylalanine	UCC	UAC Tyrosine	UGC Cysteine	C
		UUA	UCA	UAA Ter (end)	UGA Ter (end)	A
		UUG Leucine	UCG Serine	UAG Ter (end)	UGG Tryptophan	G
C		CUU	CCU	CAU	CGU	U
		CUC	CCC	CAC Histidine	CGC	C
		CUA	CCA	CAA	CGA	A
		CUG Leucine	CCG Proline	CAG Glutamine	CGG Arginine	G
A		AUU	ACU	AAU	AGU	U
		AUC Isoleucine	ACC	AAC Asparagine	AGC Serine	C
		AUA	ACA	AAA	AGA	A
		AUG Methionine	ACG Threonine	AAG Lysine	AGG Arginine	G
G		GUU	GCU	GAU	GGU	U
		GUC	GCC	GAC Aspartic acid	GGC	C
		GUA	GCA	GAA	GGA	A
		GUG Valine	GCG Alanine	GAG Glutamic acid	GGG Glycine	G

Interpretation: Codons are read as the nucleotide in the left column, then the row at the top, and then the right column. Note the three termination (ter) codons: UAA, UAG, and UGA.

corresponding amino acids are summarized in Table 25-1 ⊗. It is universal; in other words, the same code is used by all organisms, from bacteria to humans.

Example 25-1

An example of a genetic code contained in mRNA and its corresponding amino acid sequence would be:

mRNA: A U G   G C C   U C C   C G C   U A A
Protein: methionine alanine serine arginine stop

Translation always begins with the codon AUG, which specifies the amino acid methionine (Met). Met is carried to the mRNA on a ribosome by a specific tRNA, which is a cloverleaf-shaped molecule that has a specific amino acid (e.g., Met) attached to one end and a three-nucleotide sequence called an **anticodon** attached to another site. The anticodon nucleotide sequence is complementary to a specific codon on mRNA. For example, the anticodon UAC on tRNA corresponds to the codon AUG on mRNA. When the anticodon on a tRNA binds to its complementary codon on mRNA, it brings the appropriate amino acid, which attaches to a specific site on the ribosome. As the ribosome moves along the mRNA, the second tRNA travels to the site, bringing its amino acid, which binds to an adjacent site on the ribosome. A peptide bond forms between adjacent amino acids, the first amino acid is released from the ribosome, and the second amino acid shifts its position to the site that was occupied by the first. A third tRNA brings its amino acid, which attaches to the ribosome and forms a covalent bond with the second amino acid. This process continues and the protein chain grows until the ribosome reaches a stop or termination codon. The three stop codons—UGA, UAA, and UAG—are not associated with any amino acid, causing protein synthesis to terminate. The peptide chain formed is then released from the ribosome and can be recycled to read another mRNA strand.

✔ **Checkpoint! 25–2**

Describe the process of transcription and translation.

▶ INTRODUCTION TO MOLECULAR DIAGNOSTICS

As you can see from our discussion so far, it is extremely important that accuracy of the genetic code be maintained throughout DNA replication, transcription, and translation in order to produce the correct sequence of amino acids needed for a functional protein. Even a single base change resulting in the substitution of one incorrect amino acid can cause a nonfunctional protein to be produced or can prevent synthesis of the protein altogether, with severe clinical consequences. For example, sickle-cell anemia is caused by a single nucleotide substitution in DNA that results in a change in codon 6 of the mRNA for beta globin protein: the normal GAG, which codes for the amino acid glutamic acid, is converted to GUG, which specifies valine, resulting in production of the abnormal hemoglobin S, which has a reduced ability to carry oxygen and causes the red blood cells to sickle.[1]

This substitution is an example of a **mutation**, a permanent change in the nucleotide sequence or arrangement of DNA that can be passed on from parents to their offspring for many generations. Mutations can involve small changes, such as the one in our example, or large changes such as duplication or deletion of a chromosome segment. The average mutation rate for humans has been estimated to be 1×10^{-5} to 1×10^{-6} per gene, or one mutation per 100,000 to 1,000,000 copies of a gene.[1] This suggests that in a room full of 100 people, 10 are likely to be carrying a mutated gene. However, not all of these mutations are harmful.

Detection of mutations associated with human disease is the heart of **molecular diagnostics**. Knowledge of these mutations has expanded enormously in recent years as a result of the Human Genome Project, a 13-year, $3-billion project in which the nucleotide sequence of the entire set of human genes, or human genome, was determined.[7,8] By analyzing the sequence information, scientists have learned that humans possess about 25,000 genes and that 99.9% of the genome is identical in all people.[9] Research is now focusing on the differences in the 0.1% of nucleotide sequences among humans and how some of these may be associated with disease. Current research is also focused on the protein products produced by human genes. As these findings grow and newer technologies become available, the field of molecular diagnostics continues to expand.

Molecular diagnostics involves the use of nucleic acids as analytes in clinical assays to determine the presence of specific diseases or the likelihood that individuals could develop these diseases. Molecular tests have many applications in clinical medicine.[1,10] First, they can be used as diagnostic tests to confirm or rule out the presence of a specific disease in individuals who may be symptomatic or asymptomatic. They can also be used to detect specific mutations that can predict future development of certain genetic diseases in asymptomatic individuals. Molecular tests can be used to identify asymptomatic carriers of a mutation and to calculate the likelihood of these carriers passing on the mutation to their offspring. They can be performed on fetal samples obtained by amniocentesis or other procedures to determine whether a fetus carries an undesired gene, or on early embryos obtained during in vitro fertilization. Finally, molecular tests are performed routinely on newborns to identify specific genetic disorders early in life.

Molecular tests have been applied to a variety of medical fields, including genetics, microbiology, hematology, oncology, transplantation, and forensics. We will discuss some of the important clinical applications of molecular tests in this chapter. As we do this, we will look at the basic principles of many of the commonly used tests, beginning with the isolation of nucleic acids from clinical samples and the detection of nucleic acids by electrophoresis. We will then proceed to discuss more specialized tests such as nucleic acid amplification assays, blotting assays, hybridization assays, and microarrays.

 Checkpoint! 25-3

Nucleic acids can be used as analytes in clinical assays to determine the presence of specific diseases or the likelihood that individuals could develop these diseases. This describes which type of diagnostic assay?

► NUCLEIC ACID ISOLATION

The first step involved in molecular tests is to extract DNA or RNA from the sample and purify it by separating it from other components of the cell. A variety of samples have been used for this purpose, including whole blood, mononuclear cells, bone marrow aspirates, tissue culture cells, bacterial cells, and solid tissues (the latter must first be homogenized to form individual cells).[11] The most common sample used in the clinical laboratory for nucleic acid isolation is whole peripheral blood, so we will focus on the processing of this sample in our discussion.

Many methods have been developed to isolate nucleic acid from cells, including salt extraction, phenol and chloroform extraction, column chromatography, resins, and magnetic bead separation.[12] A variety of commercial kits have been developed for this purpose and have several advantages over standard research procedures: They can be done more rapidly, process multiple samples simultaneously, require less handling of potentially hazardous materials, and use minimal amounts of toxic organic solvents. Many of the commercial kits are based on salt extraction or column chromatography.

Methods for isolation of DNA vary in specific details but have four basic steps in common: (1) collection of the sample; (2) lysis of the cell and nuclear membranes to release DNA; (3) removal of membrane fragments, protein, and RNA; and (4) concentration of the purified DNA.[12] For collection of whole-blood samples, the anticoagulant ethylenediaminetetraacetic acid (EDTA) is routinely used, although acid citrate dextrose or sodium citrate are also acceptable. Heparin is not recommended because it can interfere with the results of some of the molecular methods that follow the DNA isolation such as **polymerase chain reaction (PCR)**. Whole blood can be stored at 2–25°C for as long as 24 hours before use. After collection of the blood, it should be centrifuged to collect the buffy coat, a rich source of the nucleated white blood cells. Buffy coat or mononuclear cells can be used immediately or stored at –70°C for as long as 1 year before testing. The membranes of these cells are then lysed with a mild detergent such as sodium dodecyl sulphate (SDS). For salt-extraction methods, the nuclei are pelleted and treated with a proteolytic enzyme such as protein K to digest proteins and an RNase enzyme to destroy RNA. In column purification methods, the sample is centrifuged through spin columns to separate the nucleic acid from other cell components. Separation can occur on the basis of size of the molecules or binding of the negatively charged DNA to an anionic resin attached to the beads of the column; unwanted components can then be washed from the column and the DNA eluted in a small volume of water or buffer. Following either method, the DNA can be concentrated by precipitation in a solution of sodium chloride and ethanol and then resuspended in water or ethanol.

After the ethanol is added, the sample should be handled gently to avoid sheering the DNA—that is, breaking it into smaller fragments. RNA isolation can be accomplished by similar methods, but it requires more careful laboratory technique to avoid contamination with RNase enzymes, which are ubiquitous in the environment and could destroy RNA if present. These techniques include cleaning laboratory gloves, the laboratory bench, pipettors and other equipment with an RNase decontamination solution; using nuclease-free water; and using RNase-free filtered pipette tips, microcentrifuge tubes, and disposable plasticware.[13,14]

Purity of the isolated nucleic acid can be estimated by spectrophotometry readings at 260 nm and 280 nm, which can be used to determine the absorbance ratio as follows:[15–17]

$$\text{DNA or RNA Purity} = \frac{\text{Absorbance @ 260 nm}}{\text{Absorbance @ 280 nm}} \quad (25.1)$$

A ratio of exactly 2.0 is characteristic of highly pure nucleic acid, but ratios of 1.6 to 2.0 are considered acceptable for many applications. A ratio of <1.6 indicates contamination of the sample with protein and often a need to reprocess the sample before use.

The 260 nm reading can also be used to estimate the DNA or RNA concentration in reasonably pure samples by using the following standard concentration formulas:[13,15,18]

$$1\ A_{260} \text{ unit of double-stranded DNA} = 50\ \mu g/mL$$
$$1\ A_{260} \text{ unit of RNA} = 40\ \mu g/mL \quad (25.2)$$

To calculate the yield of double-stranded DNA in $\mu g/mL$, the absorbance value at 260 would be multiplied by 50. To determine the yield of RNA, the absorbance value at 260 would be multiplied by 40. Factors that influence DNA yield from a sample include the patient's white blood-cell count, the age and storage conditions of the sample, inadequate harvesting of the buffy coat, and technologist technique.

✓ **Checkpoint! 25–4**

Identify the four basic steps used to isolate DNA.

▶ **GEL ELECTROPHORESIS**

Gel electrophoresis is a method that has been traditionally used to detect and analyze nucleic acids after they have been isolated. The method involves separating large molecules such as proteins and nucleic acids by how much they migrate in an electrical field. The rate of migration depends on physical characteristics of the molecules—namely, their mass and charge. Because all nucleic acids contain a negatively charged

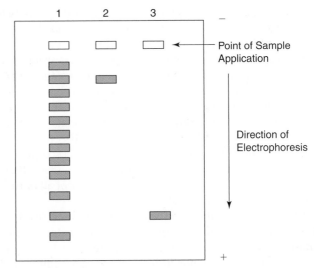

- Lane 1 contains a DNA ladder, a sample containing a mixture of DNA fragments of known lengths.
- Lane 2 contains a sample of DNA with a high molecular weight.
- Lane 3 contains a sample of DNA with a low molecular weight.

■ **FIGURE 25-9**   Gel electrophoresis.

phosphate backbone, they will migrate toward the positive pole, or anode. Separation of nucleic acids therefore depends only on their mass.[12] Larger DNA molecules take longer to travel through the sieve created by the particles within the gel matrix; smaller DNA molecules travel faster through the sieve. Thus, the mobility of DNA molecules through the sieve is inversely proportional to their molecular weight, with larger molecules remaining close to the point of sample application, and smaller molecules traveling farther from their point of application. Figure 25-9 ■ diagrams this technique.

Two types of matrices are used for electrophoresis of nucleic acids: agarose and polyacrylamide. Agarose, a polysaccharide extracted from seaweed, is prepared by dissolving powdered agarose in boiling water; on cooling, it solidifies to form a gel. Agarose is typically used in concentrations of 0.5% to 2.0% and can separate DNA fragments ranging from about 200 to 50,000 base pairs (bps) in size. The main advantage of agarose is its ability to separate a large range of DNA molecules.[19]

Polyacrylamide is a cross-linked polymer prepared from the liquid or powder form of the chemical, acrylamide. Acrylamide is a potent neurotoxin and must be handled with gloves and, if being weighed out as a powder, a mask as well. Polyacrylamide is typically prepared in concentrations of 3.5% to 20% and can separate DNA fragments ranging from about 10 to 1000 bp in size. The major advantage of polyacrylamide gel electrophoresis (PAGE) is that it has a high degree of resolving power and is capable of distinguishing nucleic acid fragments that differ by only a single base pair.[19]

The amount of voltage and the type of electrophoresis buffer also influence the mobility of DNA fragments in the gel. Increasing the voltage will increase the migration of fragments in the gel (particularly the larger fragments) but decrease the resolution. It is generally recommended that the voltage not exceed 5 volts/cm, where the centimeter value represents the distance between the two electrodes. The two types of electrophoresis buffers used most commonly are tris-acetate-EDTA (TAE) and tris-borate-EDTA (TBE). These buffers provide ions to support conductivity and establish a basic pH for the system, which allows the DNA to remain in a deprotonated state. For electrophoresis to be conducted properly, the correct concentration of buffer must be used.

Horizontal electrophoresis is typically used for agarose gels, whereas a vertical assembly is routinely used for PAGE. To run a standard agarose gel, melted agarose is poured into a casting tray containing a sample comb that outlines the wells into which the samples will be placed. The gel is allowed to solidify at room temperature and then the comb is carefully removed. Before loading DNA samples into their appropriate wells, the samples are mixed with a loading buffer that contains (1) glycerol or sucrose to add density to the sample and facilitate its placement into the wells and (2) a tracking dye that helps monitor the progress of the electrophoresis. The most commonly used tracking dyes are bromphenol blue, which migrates at the same rate as a DNA fragment of 300 bp, and xylene cyanol, which migrates at the same rate as a DNA fragment of 4000 bp.[19]

When the electrophoresis has been completed, the DNA fragments will appear as bands in which position corresponds to size. Their length in base pairs can be determined by comparison to a commercially available DNA ladder, a sample containing a mixture of DNA fragments of known lengths. The bands are visualized by staining the gel with a dye such as ethidium bromide, silver, or Coomassie Blue. Ethidium bromide is most commonly used because of its high sensitivity. It is a fluorescent dye that intercalates between the bases of nucleic acids; staining can be visualized by exposing the gel to ultraviolet light. It is a known mutagen, however, so it must be handled with gloved hands and disposed of with special precautions.

Variations in electrophoresis can be made to accommodate special situations. For example, capillary electrophoresis, performed at an ultahigh voltage in thin capillary tubes, is used routinely in DNA sequencing. Pulsed field gel electrophoresis, which involves alternating the direction of the current over a gradient of time intervals, is used to separate the largest DNA fragments from bacterial pathogens during the epidemiological analysis of infectious disease outbreaks.[20] Denaturing gel electrophoresis, which uses chemicals such as urea or formamide to denature secondary structures that can interfere with migration patterns, is used most often in the electrophoresis of RNA.[21]

 Checkpoint! 25–5

What is the major advantage of PAGE?

▶ RESTRICTION ENDONUCLEASES

Often the total DNA composition of a cell is too large to study in its entirety, and DNA must be cut into fragments for easier analysis. **Restriction endonucleases**, also referred to as *restriction enzymes*, have become essential tools in molecular biology for this reason. These enzymes, which occur naturally in bacteria as a defense against foreign DNA, act as molecular scissors by recognizing specific nucleotide sequences and cutting the DNA at those sites.[22]

Example 25-2

As an example, the enzyme EcoRI, which is derived from *Escherichia coli* bacteria, recognizes the sequence GAATTC and cuts both strands of a double-stranded piece of DNA between the G and the A every time this sequence occurs as follows:

5′ . . . G^AATTC . . . Fragment 1 . . . G^AATTC . . . Fragment 2 . . . G^AATTC . . . 3′

3′ . . . CTTAA^G . . . Fragment 1 . . . CTTAA^G . . . Fragment 2 . . . CTTAA^G . . . 5′

Thus, if the GAATTC sequence occurred three times in a strand of DNA, the EcoRI enzyme would cut the strand three times, generating two fragments of a predictable size.

The mixture of fragments generated is called a *restriction digest*. If a mutation occurred in this DNA that altered a GAATTC sequence, a different number of fragments would result. For example, if the mutation destroyed one of the GAATTC sequences, then the enzyme could only cut the DNA two times, and one fragment would result. On the other hand, if the mutation added an additional base to an AATTC sequence to create a new GAATTC site, then the enzyme would cut the DNA four times and produce three fragments. Some of the fragments created will be of a different length than those from the original DNA that lacks the mutation.

These variations in fragment length and number can be detected by their corresponding band patterns in gel electrophoresis and are known as *restriction fragment length polymorphisms* (RFLPs); these are illustrated in Figure 25-10 ■. RFLP bands can be detected by their reaction with complementary probes in Southern blotting (see the discussion under "Blotting Techniques" later in the chapter) or by amplifying the sample DNA by the PCR to detectable levels before gel electrophoresis (also under "Blotting Techniques"). Applications of RFLP analysis include the detection of genetic diseases, paternity testing, and forensic analyses (DNA fingerprinting). RFLP patterns are inherited, and a particular pattern may be linked to the presence of a disease-related gene in a given family. Thus, RFLP patterns can serve as markers to identify the presence of disease-related genes in family members in cases where the gene structure has not been identified. This type of testing is known as *linkage analysis*.[15]

Hundreds of different restriction endonucleases have been discovered, and multiple enzymes are used in the analyses just

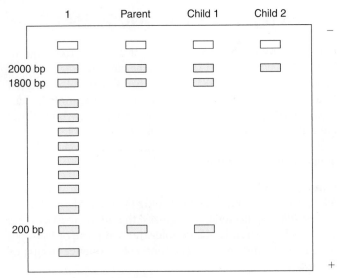

- Lane 1 contains the DNA ladder.
- Lane 2 contains a DNA sample from a parent. Three bands are produced when the parent's DNA is treated with a restriction enzyme. The 2000-bp band represents DNA from one of the parent's chromosomes that does not contain a restriction site for the enzyme and is not cut. The 1800- and 200-bp bands represent DNA from the other chromosome of the parent that contains the restriction site, so the 2000-bp piece of DNA is cut to produce the 1800- and 200-bp fragments.
- Lane 3 contains a DNA sample from child 1, who is heterozygous for the restriction site, like the parent.
- Lane 4 contains a DNA sample from child 2, who lacks the restriction site.

■ **FIGURE 25-10**   Restriction fragment length polymorphism (RFLP).

mentioned.[23] Examples of these enzymes and the sequences they recognize are listed in Table 25-2 ✪. By convention, these enzymes are named by using the first letter of the genus of the organism from which they are derived, followed by the first two letters of the species name and then a Roman numeral if more than one restriction enzyme has been isolated from that organism.[12] For example, the enzyme, EcoRI is the first restriction enzyme that was isolated from the bacterium *E. coli.*

Restriction enzymes typically recognize sequences that are 4–6 bps—and sometimes 8 bps—in length; these are referred to as *4-cutters*, *6-cutters*, and *8-cutters*, respectively. Most restriction enzymes recognize *palindromes*, or sequences that read the same in the 5' to 3' direction on one strand of the DNA as in the 3' to 5' direction on the opposite strand (for example, the GAATTC sequence recognized by EcoRI—see above). Some enzymes, such as EcoRI, create unpaired bases at the 5' or 3' end of each strand, referred to as "sticky ends." The enzyme EcoRI, for example, generates the sticky ends 5'-G, and –AATTC-3' on one strand and 3'-CTTAA and -G-5' on the other strand). These unpaired sequences protrude as an overhang and can be joined to a piece of DNA from another source that has been cut with the same enzyme and has the same sticky ends. This technique is used in cloning genes,

✪ TABLE 25-2

Examples of Restriction Endonucleases

Enzyme	Source	Recognition Site
Alu I	*Arthrobacter luteus*	AG^CT
BamHI	*Bacillus amyloliquefaciens H*	G^GATCC
EcoRI	*Escherichia coli*	G^AATTC
HindIII	*Haemophilus influenzae*	A^AGCTT
PstI	*Providencia stuartii*	CTGCA^G
SmaI	*Serratia marcescens*	CCC^GGG
XmaI	*Xanthomonas malvacearum*	C^CCGGG

allowing human genes to be inserted into bacterial plasmids, for example. Other restriction enzymes cut a recognized sequence symmetrically to create blunt ends—for example, the enzyme SmaI, which cuts in the center of the sequence CCC^GGG.

▶ HYBRIDIZATION, MELTING TEMPERATURES, AND STRINGENCY

As we discussed previously, DNA normally occurs as a double-stranded molecule that has an interior consisting of nucleotide bases bound together by hydrogen bonds. Treating DNA with heat or an alkaline solution breaks these hydrogen bonds, separating the double helix into two single strands. This process is called *denaturation*.

The temperature required to denature 50% of a double-stranded nucleic acid is referred to as its **melting temperature**, or T_m. The T_m of a nucleic acid is influenced by the base composition of the molecule because guanine–cytosine pairs are joined by three hydrogen bonds and are more difficult to break apart than adenine–thymidine pairs, which are bound by only two hydrogen bonds. The T_m of DNA strands 30 or fewer base pairs in length can be calculated by the following formula:

$$T_m = (2°C \times \text{number of A–T pairs})$$
$$+ (4°C \times \text{number of G–C pairs}) \quad (25.3)$$

Example 25-3
It follows then, for example, that the T_m of a DNA strand consisting of three A–T pairs and 6 G–C pairs would equal $(2 \times 3) + (4 \times 6) = 6 + 24 = 30°C$.

Strands of DNA longer than 30 bp require a more complex formula to calculate the T_m:[24]

$$T_m = 81.5 + 16.6\,[\log10(Na^+)] + 0.41(\%\,G–C)$$
$$- (500 \times \text{number of base pairs}) - 1.0\,(\%\,\text{mismatch}) \quad (25.4)$$

In the laboratory, we can identify a target sequence within a single strand of DNA by its ability to bind to a **probe** that contains a complementary nucleic acid sequence. This process of joining two complementary strands of nucleic acids together to form a double-stranded, hybrid molecule is

A T A C G C A A A T T C C T G G G A T C A T G G A C C G C A A A Sample DNA
| | | | | | | | | template
T A T G C G T T T Complementary
                   Probe

■ **FIGURE 25-11**   Hybridization.

called **hybridization** and is shown in Figure 25-11 ■. The probes consist of single stranded DNA or RNA sequences labeled with a fluorescent or chemiluminescent dye, an enzyme, or less frequently, a radioactive isotope.

Hybridization is an integral step in hybridization assays, hybrid capture assays, blotting techniques, and microarrays. Each technique will be discussed in the following section.

First, however, we examine the concept of **stringency**, or the factors that affect the ability of hybridization to occur. Two major factors that affect stringency are the temperature and salt concentration of the buffer used in the hybridization reaction. Increasing the temperature and lowering the salt concentration results in high stringency, where specific binding of the probe to its homologous sequence is enhanced by encouraging dissociation of mismatched sequences. In contrast, decreasing the temperature and increasing the salt concentration results in low stringency by enabling the probe to bind to sequences with lower homology. These factors are typically applied during the washing steps in the procedure.

▶ HYBRIDIZATION ASSAYS

Hybridization assays have been developed in three major formats; (1) liquid-phase hybridization, (2) solid-phase hybridization, and (3) in situ hybridization.

In liquid-phase hybridization, the target nucleic acid and probe are free to interact in an aqueous solution. This format allows for rapid duplex formation with shortened assay times. Liquid-phase hybridization assays that detect rRNA from a variety of bacterial pathogens in clinical specimens or primary cultures have been popular in clinical microbiology laboratories. Detection of rRNA from bacterial or fungal cultures is commercially available as a liquid-phase hybridization format known as the *hybridization protection assay*.[25–27] In solid-phase hybridization, the target DNA is isolated, denatured into single strands, and pipetted onto a solid surface such as nitrocellulose or a nylon membrane to produce a *dot blot* or a *slot blot*.[25,27,28] Multiple samples can be pipetted onto different areas of the membrane and analyzed simultaneously. This method can be used to distinguish between alleles that differ by a single nucleotide by using allele-specific oligonucleotide (ASO) probes, which are complementary to the normal and mutant alleles and incubated with the target DNA under conditions of high stringency. The technique is illustrated in Figure 25-12 ■. This method can be used to probe for mutations characteristic of genetic diseases such as sickle-cell anemia.

For **in situ hybridization**, labeled probes are hybridized with tissues, cells, or chromosomes that have been fixed onto a glass slide.[29,30] Use of tissue has the advantage of being able to detect the presence or absence of DNA or mRNA in its native location in the context of the tissue architecture. Detection of mRNA also allows information to be obtained on the expression of genes by the cells within the tissue.

In situ hybridization that uses fluorescent-labeled probes is called **fluorescent in situ hybridization**, or **FISH**. FISH is most commonly applied to the analysis of DNA in chromosomes. Although the traditional method to identify chromosomal abnormalities has been to construct a karyotype, or photograph, of stained chromosomes from a single cell arranged by size, banding pattern, and centromere position,

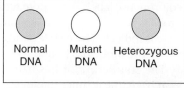

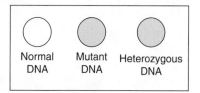

| Normal DNA | Mutant DNA | Heterozygous DNA | | Normal DNA | Mutant DNA | Heterozygous DNA |

Incubated with probe for Normal DNA   Incubated with probe for Mutant DNA

■ **FIGURE 25-12**   ASO dot plot.
The membrane on the left has been incubated with a probe that detects a normal DNA sequence. Note that there is reactivity with the sample DNA that is homozygous for the normal sequence (circle on left) and the DNA sample that is heterozygous normal or mutant (right circle), but not with the sample that is homozygous for the mutant sequence (middle circle). The membrane on the right has been incubated with a probe that detects a mutant DNA sequence. Note that there is reactivity with the sample DNA that is homozygous for the mutant sequence (middle circle) and the DNA sample that is heterozygous normal or mutant (right circle), but not with the sample that is homozygous for the normal sequence (left circle).

the resolution of this method is not sufficient to detect small molecular changes. FISH attains the resolution needed through the use of specific probes that bind to complementary sequences of DNA on the chromosomes of interest. Suitable specimens for FISH analysis include amniotic fluid, peripheral blood cells, bone marrow cells, paraffin-embedded tissue sections, or other preparations from solid tumors.[11,31] Three major classes of probes have been used in FISH, depending on the specific application: (1) locus-specific or unique-sequence probes that bind to part of a specific gene on the chromosome of interest, (2) satellite probes that bind to repetitive sequences in the centromeres or telomeres of the chromosome, and (3) whole-chromosome probes, which are collections of smaller probes that can be used to "paint" an entire chromosome by binding to different sequences along the length of the chromosome.[11,32] Following incubation with the probe, unbound probes are removed by washing. The slides are viewed under a fluorescent microscope, and fluorescence is analyzed with the help of computer-aided image-processing systems.[11,33] A FISH technique is illustrated in Figure 25-13 ■.

The clinical applications of FISH have expanded rapidly because of advances in sensitivity, the types of probes available, improved speed of analysis with interphase chromosomes, simultaneous multitarget visualization, and the capability of automated data collection and analysis.[11,34] These applications include the following:[11,31,33]

- detection of microdeletions,
- detection of abnormalities in chromosome numbers,
- analysis of tumor cells,
- monitoring of bone marrow transplant patients,
- screening for chromosomal mosaicism, and
- genetic analysis of embryos.

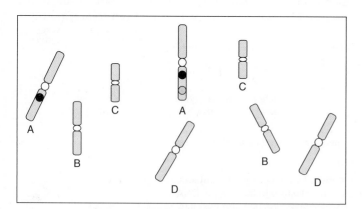

■ FIGURE 25-13 Fluorescent in situ hybridization (FISH). Hypothetical metaphase spread of chromosomes incubates with two probes, a control probe for chromosome A (black circle), and a probe for a specific normal sequence within chromosome A (gray circle). Note that one of the chromosomes in the pair does not hybridize with the second probe, indicating that a deletion has occurred.

✓ **Checkpoint! 25–6**

Identify four clinical applications of FISH assays.

▶ **BLOTTING TECHNIQUES**

Blotting techniques involve transfer of nucleic acids or proteins that have been separated by gel electrophoresis onto a solid support membrane for analysis.[15,35] Blotting techniques to analyze DNA are called *Southern blots*. Those for analysis of RNA are called *Northern blots*, and those for analysis of proteins, **Western blots**.

The Southern blot was named after British scientist Edwin Southern, who developed the method in the 1970s.[36,37] The Southern blot is used to identify a specific DNA sequence within a complex mixture of genes such as the entire genome of an organism.[35] To achieve this, the sample DNA is first cut with restriction endonucleases to produce smaller fragments, which are electrophoresed on a gel as illustrated in Figure 25-14 ■. Next, the double-stranded DNA fragments are denatured into single strands by treatment with a strong base. The single-stranded DNA fragments are transferred onto a solid support membrane, which is usually made from nitrocellulose or nylon, to facilitate the hybridization step that follows. Transfer to the membrane has been traditionally accomplished by capillary action through a stack of dry papers placed in between the gel and the membrane in the presence of a high salt buffer, but newer methods involving a vacuum apparatus or electrophoresis can speed up this process.[11] The membrane is then treated with ultraviolet light or heat in order to immobilize the DNA and prehybridized by treatment with a buffer containing Denhardt solution and salmon sperm DNA, which block nonspecific binding of the probe to the membrane surface. The membrane is then hybridized with a labeled single-stranded DNA probe complementary to the desired sequence. Whereas the original method used ^{32}P-labeled nucleotides, nonisotopic probes with chemiluminescent or colorimetric labels have more recently been developed. Excess probe is removed by washing, and the presence of the desired gene is indicated by visible bands representing the fragment(s) where the probe has bound.

The Southern blot has had numerous applications in research, clinical microbiology, genetics, hematology, and oncology.[38] Although many of these have been replaced by PCR-based methods in recent years because of the ease of performance and more rapid turnaround times for results, Southern blot hybridization is still used in situations involving large DNA fragments that may not amplify well or fragment sizes that are unique to individual patients.

In 1979, Alwine and Stark developed a method to transfer RNA from a gel to a solid support, and they jokingly named it "Northern blot."[37] The Northern blot involves separation of RNA fragments by gel electrophoresis, with subsequent transfer to a solid support membrane and hybridization with a

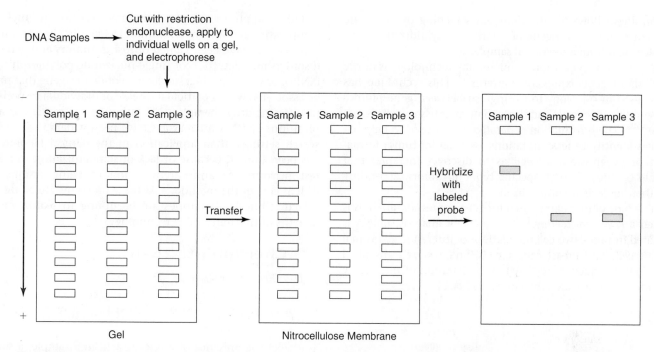

Cut with restriction endonuclease, apply to individual wells on a gel, and electrophorese

DNA Samples →

Sample 1   Sample 2   Sample 3

Sample 1   Sample 2   Sample 3

Sample 1   Sample 2   Sample 3

Transfer →

Hybridize with labeled probe →

Gel

Nitrocellulose Membrane

■ **FIGURE 25-14**   Southern blot. In sample 1, the probe has hybridized to a piece of DNA that has a higher molecular weight than that of samples 2 and 3, indicating that there is a genetic difference in the region of DNA recognized by the probe in sample 1 as compared to the other samples.

specific probe.[12] This method has been used primarily as a research tool to detect gene expression in cells through the presence of specific mRNA.

When a blotting method for detection of proteins was developed, it was named *Western blot* in keeping with the "geographic" terminology.[39] In the Western blot method, the proteins of interest are denatured and separated on the basis of their molecular weight by sodium dodecyl sulfate polyacrylamide gel electrophoresis (SDS-PAGE).[39] Larger proteins will travel slower through the gel and remain closer to the point of application, while proteins having a lower molecular weight will travel faster and migrate farther away from the point of application. The proteins are then blotted onto a membrane (preferably polyvinylidene difluoride, or PVDF, which has a high protein-binding capacity) using simple diffusion by capillary action, vacuum-assisted flow, or electrophoretic elution. Instead of using a molecular probe as do Southern and Northern blots, Western blots use antibodies to identify the protein targets. A primary antibody that has specificity for the protein of interest is incubated with the membrane. Following incubation with the primary antibody, the membrane is washed and incubated with a secondary antibody that has specificity for the primary antibody. The secondary antibody contains a label, most commonly an enzyme such as horseradish peroxidase or alkaline phosphatase. On addition of a colorimetric substrate, insoluble colored products precipitate at the locations on the membrane where the secondary antibody has bound, producing visible bands that represent proteins of a specific molecular weight. Alternatively, chemiluminescent substrates can be

used that emit light at the sites of immune complex formation, detectable by exposure of the blot to an X-ray film.

The Western blot has had numerous research and clinical applications. For many of its clinical applications, patient serum is used as the primary antibody and labeled antihuman immunoglobulin as the secondary antibody. In this way, patient antibodies to specific proteins present in various pathogens can be detected. Because of its high level of specificity, the Western blot has been used as a confirmatory test for antibodies to HIV, hepatitis C, HTLV-I, and *Borrelia burgdorferi*.[40]

✓ **Checkpoint! 25-7**

In the Western blot technique what type of molecule is used to identify the protein targets?

▶ **DNA MICROARRAYS**

Another molecular method that relies on hybridization is **microarray** technology. A DNA microarray consists of an orderly arrangement of specific gene sequences that have been immobilized onto precise locations of a small solid support, such as a glass microscope slide, nylon membrane, or silicon chip.[41] These arrays are produced by robotic printing of known DNA sequences or the actual synthesis of short DNA sequences called *oligonucleotides* onto thousands of miniaturized spots on the solid support.[42,43] This revolutionary technology allows for simultaneous screening of the entire human genome for a gene of interest by hybridization with nucleic acid from a biological sample.

A common application of microarray technology is in the analysis of gene-expression patterns.[44] This technique has been used in oncology to distinguish between gene profiles present in cancer cells and those in normal tissue of the same type.[42,45] The information generated can be used to help scientists identify genetic mutations essential for tumor formation or progression, as well as to discover therapies that would be effective with specific types of tumors, depending on their molecular composition.

The type of microarray used for gene expression analysis is called a *cDNA microarray*.[41,43] In a cDNA analysis, mRNA is isolated from the two cell populations of interest (e.g., normal breast cells and breast cancer cells) and reverse-transcribed into cDNA sequences (cDNAs). An illustration of microarray technology is provided in Figure 25-15 ■.

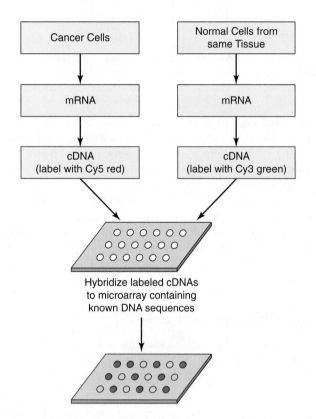

Scan for fluorescence:

- Spots containing more red fluorescence indicate higher gene expression in cancer cells.
- Spots containing more green fluorescence indicate higher gene expression in normal cells.
- Yellow spots indicate that gene expression is about equal in both populations.

■ **FIGURE 25-15** Microarray technology in the detection of gene expression.

Miller, Linda E Tumor Marker. Inservice Reviews in Clinical Laboratory Science Vol. 18 (No. 12) October 2005

Other applications of microarray technology include comparative genomic hybridization to identify chromosomal regions that have been gained or lost in tumors and genetic disorders and detection of single-nucleotide polymorphisms (SNPs), or variations in a single nucleotide in a gene that may be associated with a particular disease or individual responses to different drug therapies.[43,46] Although the numerous applications of DNA microarrays have been widely used in research settings, their application in the clinical laboratory has been limited because of lack of standardization in both test performance and interpretation of being results, as well as cost of the method.[47] As these problems are being resolved, clinical laboratories are beginning to realize more benefits of this remarkable technology.

✓ **Checkpoint! 25–8**

Describe a DNA microarray.

▶ **AMPLIFICATION TECHNIQUES**

To detect the presence of a nucleic acid in a sample, a sufficient number of DNA or RNA molecules must be present. This is not always the case with clinical samples. For this reason, nucleic acid amplification techniques have been developed to increase the sensitivity of molecular tests despite the presence of a small number of nucleic acid molecules in the original sample. Amplification can be accomplished by one of three ways: (1) target amplification, in which millions of copies of a specific DNA sequence are produced in vitro; (2) probe amplification, which involves amplification of synthetic probes that bind to the target sequence; and (3) signal amplification in which a signal generated from a probe bound to the target sequence is increased.[15,27] We will discuss examples of methods that fall under each category below.

Target amplification methods include polymerase chain reaction, nucleic acid sequence–based amplification (NASBA), transcription-mediated amplification (TMA), and strand-displacement amplification (SDA). The first amplification method to be developed was the PCR, which has become the workhorse of the molecular biology laboratory. PCR was developed in 1983 by chemist Kary Mullis, who envisioned the method while driving through the moonlit mountains of California.[48] This method, which has the power to generate millions of copies of a specific DNA sequence from a single copy, has revolutionized biomedical research and clinical laboratory testing and earned Mullis the Nobel Prize in 1993. Clinical applications of PCR are too numerous to be discussed at length in this chapter but include detection of genetic mutations in inherited disorders (e.g., cystic fibrosis, Duchenne muscular dystrophy, Huntington's disease, hemochromatosis) and malignancies (e.g., leukemias, lymphomas, breast cancer, colon cancer), detection of infectious diseases for which the organism cannot be cultured or for which culture results are not rapidly available (e.g., *Mycobacterium tuberculosis*, HIV, hepatitis C), HLA typing, paternity testing, and forensic analyses.[49,50]

The basic steps of the PCR method mimic the steps of DNA replication in vivo and may be found in the cited reference cited.[51]

The traditional PCR method involves three basic steps: (1) denaturation, (2) annealing, and (3) extension as shown in Figure 25-16 ■. In the denaturation step, double-stranded DNA in the sample is heated to 94–96°C to separate the molecule into two single strands, each of which will serve as a template for the synthesis of new DNA. This step typically requires 20–60 seconds. The reaction is then cooled to 50–70°C,

and the annealing step is conducted, whereby the primers will bind to their complementary regions on each template strand (20–90 seconds). Next, the reaction is heated to 68–72°C for 10–60 seconds, allowing each of the two DNA strands to be extended as DNA polymerase adds deoxynucleotide triphosphates to the 3' ends of the primers attached to each strand. The exact incubation times and temperatures for each step will vary, depending on the nature of the template and type of PCR. This sequence of three steps is referred to as a *cycle*. Multiple cycles are conducted by an automated

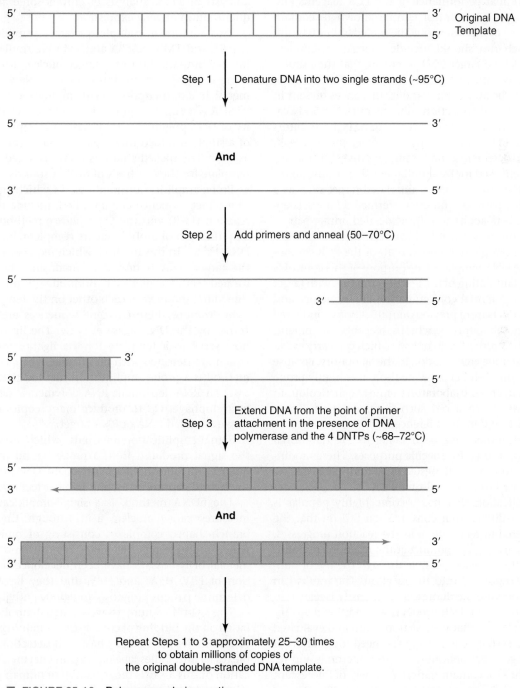

■ FIGURE 25-16   Polymerase chain reaction.

thermocycler that can be programmed to carry the reaction through the series of required incubation steps. With each cycle, the number of DNA copies is doubled so that, after n cycles, approximately $2n$ copies are produced. The PCR product, called an **amplicon**, typically consists of more than 1 million copies of the original DNA generated after 25–30 cycles. The amplicon can be detected by gel electrophoresis in which a band of a specific size is visualized after ethidium bromide staining, hybridization assays using specific probes, enzyme-linked immunosorbent assay (ELISA), or DNA sequencing.

To ensure optimal performance of the PCR reagents and thermocycler, a positive control sample known to contain the gene sequence of interest should be included in each run. In addition, each run should include a negative template control that lacks the target DNA to ensure that the primers are not binding to unintended DNA sequences. The laboratorian also needs to be aware that various substances present in the test sample, blood-collection tube, or nucleic acid isolation procedure can inhibit the PCR reaction. PCR inhibitors include substances in urine, heparin, heme, proteinase K, SDS and other detergents, and some fixatives. These inhibitors can be checked for by simultaneously running an internal control that consists of an unrelated target such as a gene essential for basic cell function (termed a "housekeeping gene"—e.g., beta-actin) and the associated primer pair.

Another important control to include with each run is a negative control sample that contains all of the PCR components except the DNA template. The purpose of this control is to check for contamination of the reaction mixture with target DNA. Sources of potential contamination are numerous and include PCR products from previous amplifications conducted in the laboratory, laboratory benchtops, reagents, equipment, and the even laboratorian's skin or hair, which can carry extraneous DNA to an undesired location in the laboratory. Because the PCR is an exquisitely sensitive method, it is highly prone to contamination, and the laboratorian must be meticulous in performing measures to avoid such contamination. These measures can be found in the references.[15,52]

Many modifications have been made to the traditional PCR that have been used for specific purposes. These modifications include hot-start PCR, nested PCR, multiplex PCR, reverse transcriptase PCR, and real-time PCR.[51,53–57]

A PCR modification that has become highly popular is real-time PCR. It differs from conventional PCR in that the product is detected in real time or as the reaction progresses, in contrast to analyzing the product at the reaction endpoint. Real-time detection is made possible through the use of fluorescent dyes incorporated into the reaction. The results can be quantitative because the fluorescence intensity is related to the concentration of the PCR product. An additional advantage of real-time PCR is that the reaction is run and evaluated in a closed-tube system, eliminating the need for analysis of the product by gel electrophoresis or other methods and reducing the chance for contamination. A variety of fluorescent chemistries have been employed in real-time PCR tests, depending on the commercial manufacturer of the method.[58]

Some of the more common chemistries include DNA binding dyes, hybridization probes, hydrolysis probes, and hairpin-shaped probes.[59,60]

With any of these fluorescent chemistries, fluorescence will increase as the PCR product accumulates. The PCR cycle number at which the fluorescent signal becomes detectable is called the *threshold cycle* and is inversely related to the amount of template present at the start of the reaction.[59] A real-time PCR amplification plot serves as the basis for quantitative real-time PCR. Real-time PCR has been widely applied to a variety of analyses, including detection of pathogens, analysis of SNPs, analysis of chromosome aberrations, and quantitation of gene expression.[60]

Two additional examples of target amplification tests are NASBA and TMA. NASBA and TMA are similar methods that involve amplification of a target nucleic acid, usually RNA, using enzymes, reverse transcriptase, RNase, and RNA polymerase in a transcription-based amplification system.[15,27,61,62]

SDA is a target amplification test that is based on the ability of DNA polymerase to initiate DNA replication at the site of a single-stranded nick created by a restriction endonuclease.[61,62] The nicked strand is then displaced and serves as a template for the synthesis of further strands.

Probe-amplification methods, in which probes that bind to a target sequence are amplified, include the ligase chain reaction (LCR) and the Qβ replicase method. LCR involves amplification of probes that are complementary to the target DNA.[15,61,62] In this method, which requires a thermal cycler, the target DNA is heat denatured, and the single strands formed are subsequently hybridized with probes that are designed to bind next to each other on the template. Following hybridization, these adjacent sequences are connected together by the DNA ligase enzyme. The ligated product can now serve as a template for the ligation of more probes, which are detected with an enzyme label. The Qβ replicase method is a probe-amplification method that uses Qβ replicase, an RNA-dependent RNA polymerase derived from the bacteriophage, Qβ to produce many copies of RNA from a single stranded DNA or RNA target.[15,62]

Signal-amplification methods, which involve increasing the signal produced from a probe bound to the target sequence, include the branched chain DNA (bDNA) method, the hybrid-capture assay, and Invader technology.

The bDNA method, is a signal-amplification assay that measures target nucleic acid through the formation of branch-shaped complexes consisting of multiple probes to which multiple reporter molecules can bind.[27,57,61,63] Commercial bDNA assays have been developed for the quantitation of HIV, HCV, and HBV, and they have been used to determine patient prognosis and guide antiviral therapy.[57,61]

The hybrid capture assay is a signal-amplification system based on the binding of complexes containing the target DNA to antibodies on a solid phase and detection of these complexes by antibodies labeled with an enzyme. A popular application of this assay is the detection of human papilloma virus (HPV), which has been associated with the development of cervical cancer, in a hybrid capture assay.[27,64,65]

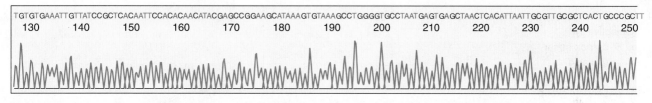

■ FIGURE 25-17   Dideoxynucleotide structure.

■ FIGURE 25-18   Sample electropherogram of DNA sequencing.

The Invader Assay, developed by Third Wave Technologies, Inc., is an isothermal signal-amplification method that uses a primary probe, an "invader" probe, a reporter probe, and the enzyme cleavase to generate a large number of reporter probes from a single probe bound to a DNA target.[27,62,66]

Products using this technology have been developed to detect genetic disorders such as cystic fibrosis, coagulation disorders such as Factor V Leiden deficiency, and infectious diseases such as HPV.[66]

✓ Checkpoint! 25–9

What is the purpose of nucleic acid amplification?

▶ DNA SEQUENCING

DNA sequencing provides the most definitive knowledge about the composition of a gene by determining the precise order of the nucleotides in the DNA molecule. Sequencing methods have found a variety of applications in the research and clinical laboratories, including the detection of genetic mutations, genotyping of microorganisms, identification of mutations associated with antimicrobial resistance, identification of human haplotypes and polymorphisms, and determination of the entire genomic sequences of humans and other organisms studied in the Human Genome Project and beyond.[62,67,68] Two methods of DNA sequencing that have been widely used are the Maxim–Gilbert method and the Sanger method, while pyrosequencing is a more recent method that is gaining popularity. The basic principles of each method are discussed in this last section of the chapter.

The Maxim–Gilbert method is a chemical sequencing method that was developed by Allan Maxim and Walter Gilbert in the late 1970s.[69,70] Although this method has the advantages of being able to analyze DNA of a completely unknown sequence and of analyzing the original DNA strand rather than a copy of the strand, it is not practical for the analysis of long sequences and employs several hazardous chemicals.[69] For these reasons, it has been largely replaced by the Sanger method.

The Sanger, or dideoxy chain-termination method, is the most widely used method for sequencing DNA because of its ease of performance and automated format.[69,71] In this method, a copy of the DNA sequence to be analyzed is synthesized in a PCR-like process using the original DNA template, specific primers, DNA polymerase, the four dNTPs, and 2'3'dideoxynucleotide triphosphates (ddNTPs) of each base.[69,70,72] The ddNTPs are very similar in structure to their corresponding dNTPs, except that they lack an OH group at the 3' carbon of the deoxyribose sugar, as shown in Figure 25-17 ■. Therefore, they are incorporated into the growing strand of DNA along with the dNTPs, but DNA chain synthesis is terminated wherever a ddNTP is located because a phosphodiester bond cannot be formed when the next nucleotide is added. This results in the synthesis of DNA fragments of varying length that can be resolved by size using capillary gel electrophoresis. Before the reaction is performed, each ddNTP is labeled with a fluorescent dye of a different color. The dyes are excited by a laser in the automated sequencer, generating an electropherogram that displays a series of fluorescent peaks from which the nucleotide sequence can be determined. A sample electropherogram of a DNA sequence is shown in Figure 25-18 ■.

Pyrosequencing is a method that uses light-generating chemical reactions to analyze short DNA sequences. Pyrosequencing has been applied to the detection of mutations and SNPs, bacterial genotyping through analysis of rRNA, genotyping of viruses, and identification of expressed genes.[67]

SUMMARY

In the field of molecular diagnostics, laboratory tests are performed to detect changes in the normal DNA sequence called *mutations*, which are associated with specific diseases. These tests have been applied to a variety of medical fields, including genetics, microbiology, hematology, oncology, transplantation, and forensics.

This chapter has provided the reader with an introduction to molecular diagnostics. A discussion of basic concepts necessary for the understanding of molecular testing was presented first and then followed by a review of the principles and clinical applications of major molecular methods used in the clinical laboratory. Several specific techniques are discussed, including gel electrophoresis, hybridization, blotting, amplification, and DNA sequencing.

A review of this chapter should provide the reader with fundamental information to be able to move to the next phase of education in molecular diagnostic testing. Mastery of this material will allow the reader to investigate methods and techniques at a more advanced level.

REVIEW QUESTIONS

LEVEL I

1. Which of the following is not a component of a nucleotide? (Objective 1)
   a. 5′ carbon sugar
   b. amino acid
   c. nitrogenous base
   d. phosphate group

2. Which of the following is a step in the process of translation? (Objective 2)
   a. RNA polymerase binds to the promoter sequence.
   b. Introns are removed from the transcript.
   c. A string of As are added to the 3′ end of the mRNA.
   d. Anticodons on the tRNAs bind to their complementary codons on mRNA.

3. How much DNA would be contained in 200 μl of a sample that has an A_{260} of 0.40 and an A_{280} of 0.20? (Objective 3)
   a. 2 μg
   b. 4 μg
   c. 10 μg
   d. 20 μg

4. Which of the following is true about gel electrophoresis of DNA? (Objective 4)
   a. DNA of smaller molecular weight will travel far from the point of application.
   b. DNA migrates toward the cathode.
   c. Increasing the voltage increases the resolution of DNA fragments separated in the gel.
   d. DNA electrophoresis requires polyacrylamide as a matrix.

5. Restriction endonucleases function by: (Objective 5)
   a. linking two pieces of DNA from different sources.
   b. adding a phosphate group onto one end of the DNA.
   c. cutting the DNA at specific nucleotide sequences.
   d. breaking the DNA randomly to create fragments of manageable size.

6. What is the melting temperature (T_m) of the following DNA sequence? (Objective 6)

   5′ TAACCTATGCGA 3′
   3′ ATTGGATACGCT 5′
   a. 12°C
   b. 24°C
   c. 34°C
   d. 38°C

LEVEL II

1. The following figure represents the structure of: (Objective 1)

   a. adenine.
   b. guanine.
   c. cytosine.
   d. thymine.

2. RNA differs from DNA in that RNA: (Objective 2)
   a. contains thymine instead of uracil.
   b. contains an H group instead of an OH group at its 2′ carbon.
   c. is normally single-stranded.
   d. is used to carry the genetic information of most viruses.

3. The protein product derived from the DNA sequence below contains which of the following amino acid sequences? (Objective 3)

   DNA: 3′ TACTTTCGCGGAACT 5′
   a. tyrosine–phenylalanine–arginine–glycine–threonine
   b. tyrosine–phenylalanine–arginine–glycine
   c. methionine–lysine–alanine–proline
   d. methionine–lysine–alanine–proline–tryptophan

4. Suppose a laboratory got 2 μg of DNA from a clinical sample, but the DNA isolation procedure stated that 10 μg was optimal. The laboratory's result can be explained by all of the factors below except: (Objective 4)
   a. a low patient white blood cell count.
   b. inadequate harvesting of the buffy coat from the sample.
   c. storage of the whole-blood sample at 22°C for 12 hr before use.
   d. contamination of the sample with protein.

5. Suppose a laboratory wanted to separate small DNA fragments that were 10 bps apart in size. The optimal technique to use for the separation is: (Objective 5)
   a. agarose gel electrophoresis.
   b. polyacrylamide gel electrophoresis.
   c. pulse-field gel electrophoresis.
   d. denaturing gel electrophoresis.

REVIEW QUESTIONS (continued)

LEVEL I

7. Real-time PCR differs from traditional PCR in that: (Objective 7)
   a. primers are not necessary.
   b. the amplicon is detected as the reaction progresses.
   c. the amplicon is detected on a special fluorescent gel.
   d. the reaction requires hybridization probes.

8. All of the following are clinical applications of FISH except for: (Objective 8)
   a. detection of chromosome microdeletions.
   b. detection of oncogenes in tumor cells.
   c. monitoring patients with sex-mismatched bone marrow transplants for engraftment.
   d. detecting DNA from tumor-causing viruses in cytologic specimens.

LEVEL II

6. Which of the following fragments would be generated when the following sequence is cut by SmaI? (Objective 6)

5′ TACCCCGGGGGGCAATTCCCGGGAGATTCCCGGGAACTC 3′

   a. one 3-bp fragment, two 11-bp fragments, and one 13-bp fragment
   b. one 4-bp fragment, one 10-bp fragment, one 11-bp fragment, and one 13-bp fragment
   c. two 19-bp fragments
   d. one 6-bp fragment, one 8-bp fragment, one 11-bp fragment, one 13-bp fragment

7. More nonspecific binding of a probe to DNA sequences in a sample is encouraged when the: (Objective 7)
   a. size of the probe is increased.
   b. salt concentration is decreased.
   c. temperature is decreased.
   d. temperature is increased.

8. Labeled antibodies are used as probes to detect proteins in the: (Objective 8)
   a. hybrid capture assay.
   b. Northern blot.
   c. Southern blot.
   d. Western blot.

9. A suitable method to use for quantitation of HIV in patient blood is: (Objective 9)
   a. bDNA.
   b. DNA sequencing.
   c. SDA.
   d. Southern blot.

PEARSON
myhealthprofessionskit

Use this address to access the interactive Companion Website created for this textbook. Simply select "Clinical Laboratory Science" from the choice of disciplines. Find this book and log in using your user name and password.

REFERENCES

1. Cummings MR. *Human heredity: Principles and issues*, 7th ed. (Belmont, CA: Thomson Brooks/Cole, 2006): 178–193, 198–213, 324–329.

2. Watson JD, Crick FHC. Molecular structure of nucleic acids: A structure for deoxyribose nucleic acid. *Nature* (1953), 171: 737–738.

3. Barnum SR. *Biotechnology: An introduction* (Belmont, CA: Thomson Brooks/Cole, 2005): 25–55.

4. Thieman WJ, Palladino MA. *Introduction to biotechnology* (San Francisco: Pearson Benjamin Cummings, 2004): 23–49.

5. Finnegan EJ, Matzke MA. The small RNA world. *J Cell Science* (2003) 116: 4689–4693.

6. Nussbaum RL, McInnes RR, Willard HF. *Thompson & Thompson genetics in medicine*, 6th ed. (Philadelphia: WB Saunders, 2001): 17–32.

7. International Human Genome Sequencing Consortium. Initial sequencing and analysis of the human genome. *Nature* (2001) 409: 860–921.

8. Ventor JC, et al. The sequence of the human genome. *Science* (2001) 291, 5507: 1304–1351.

9. U.S. Department of Energy. Early insights from the human DNA sequence (www.ornl.gov/sci/techresources/Human_Genome/publicat/primer2001/4.shtml), accessed June 4, 2007.

10. Csako G. Present and future of rapid and/or high-throughput methods for nucleic acid testing. *Clin Chim Acta* (2006) 363: 6–31.

11. Mahon CR, Tice D. *Clinical laboratory immunology* (Upper Saddle River, NJ: Prentice Hall, 2006): 217–231, 233–251, 253–276.

12. Dale JW, von Schantz M. *From genes to genomes* (New York: John Wiley & Sons, 2002): 5–20, 31–39, 41–47, 228–229.

13. Connolly MA, Clausen PA, Lazar JG. RNA purification. Ch. 10 in Dieffenbach CW, Dveksler GS (Eds.), *PCR primer: A laboratory manual*, 2nd ed. (Cold Spring Harbor, NY: Cold Spring Harbor Laboratory Press, 2003): 117–133.

14. Sambrook J, Russell DW. *Molecular cloning: A laboratory manual*, 3rd ed. (Cold Spring Harbor, NY: Cold Spring Harbor Laboratory Press, 2001): 7.82–7.83.

15. Buckingham L, Flaws ML. *Molecular diagnostics: Fundamentals, methods, & clinical applications* (Philadelphia: FA Davis, 2007): 65–93, 94–120, 227–229.

16. Held PG. Nucleic acid purity assessment using A260/A280 ratios. *Biotek Application Notes* (2001) (www.biotek.com/products/tech_res_detail.php?id=43), accessed June 5, 2007.

17. Sambrook J, Fritsch EF, Maniatis T. *Molecular cloning: A laboratory manual*, 2nd ed. (Cold Spring Harbor, NY: Cold Spring Harbor Laboratory Press): E5.

18. Protocol Online. RNA quantitation and purity determination. Adapted from Qiagen RN, Easy kit instruction (2005) (www.protocol-online.org/cgi-bin/prot/page.cgi?g=print_page/3538.html), accessed June 28, 2006.

19. Bowen R. Gel electrophoresis of DNA and RNA (www.vivo.colostate.edu/hbooks/genetics/biotech/gels/index.html), accessed December 14, 2006.

20. Goering RV. Pulsed-field gel electrophoresis. In Persing DH (Ed.), *Molecular microbiology: Diagnostic principles and practice* (Washington, DC: ASM Press, 2004): 185–196.

21. National Diagnostics. DNA and RNA electrophoresis (http://nationaldiagnostics.com/articles.php/tPath/1_3?osCsid=21a15109140ec22197537a44afa52f17), accessed December 14, 2006.

22. Simmer M, Secko D. Restriction endonucleases: Molecular scissors for specifically cutting DNA. *Sci Creat Q* (2006): 1–6.

23. New England Biolabs. Restriction endonucleases (www.neb.com/nebecomm/products/category1.asp?2), accessed June 7, 2007.

24. Wetmur JG. DNA probes: Applications of the principles of nucleic acids hybridization. *Crit Rev Biochem & Mol Biol* (1991) 26: 227–259.

25. Li J, Hanna BA. DNA probes for culture confirmation and direct detection of bacterial infections: A review of technology. Ch. 2 in Persing DH, et al. (Eds.), *Molecular microbiology: Diagnostic principles and practice* (Washington, DC: ASM Press, 2004): 19–36.

26. GenProbe. Core technologies overview (www.gen-probe.com/sci_tech/core.htm), accessed May 8, 2007.

27. Podzorski RP. Introduction to molecular methods. Ch. 5 in Detrick B, Hamilton RG, Folds JD (Eds.), *Manual of molecular and clinical laboratory immunology*, 7th ed. (Washington, DC: ASM Press, 2006): 26–51.

28. Strachan T, Read AP. *Human molecular genetics*, 2nd ed. (New York: John Wiley & Sons, 2000): 95–118.

29. GeneDetect.com. In situ hybridization (www.genedetect.com/insitu.htm), accessed May 10, 2007.

30. Wilkinson DG. *In situ hybridization: A practical approach*, 2nd ed. (Oxford, UK: Oxford University Press, 1998): 1–21.

31. Yale Department of Genetics. Cytogenetics test information (http://info.med.yale.edu/genetics/cytogenetics/cytogenetics_testInfo.php), accessed July 18, 2007.

32. National Human Genome Research Institute. Fluorescence in situ hybridization (FISH) (www.genome.gov/10000206), accessed May 11, 2007.

33. Waters JJ, Barlow AL, Gould CP. Demystified . . . FISH. *Br Med J* (1998) 51, 2: 62–70.

34. Levsky JM, Singer RH. Fluorescence in situ hybridization: Past, present, and future. *J Cell Sci* (2003) 116: 2833–2838.

35. Mama Ji's Molecular Kitchen (http://lifesciences.asu.edu/resources/mamajis/), accessed May 15, 2007.

36. Southern EM. Detection of specific sequences among DNA fragments separated by gel electrophoresis. *J Mol Biol* (1975) 98: 503–517.

37. Southern EM. Blotting at 25. *Trends Biochem Sci* (2000) 25: 585–588.

38. Tenner KS, O'Kane DJ. Clinical application of Southern blot hybridization with chemiluminescence detection. *Methods in Enzymology* (2000) 305: 450–466.

39. Kurien BT, Scofield RH. Protein blotting: A review. *J Immunol Methods* (2003) 274: 1–15.

40. Stevens CD. *Clinical immunology and serology*, 2nd ed. (Philadelphia: FA Davis, 2003): 305–306, 336–337, 357–358.

41. National Center for Biotechnology Information. Microarrays: Chipping away at the mysteries of science and medicine (http://ncbi.nlm.nih.gov/About/primer/microarrays.html), accessed May 18, 2007.

42. Lobenhofer EK, Bushel PR, Afshari CA, Hamadeh HK. Progress in the application of DNA microarrays. *Environmental Health Perspectives* 2001; 109 (9): 881–891.

43. Geschwind DH. DNA microarrays: translation of the genome from laboratory to clinic. Lancet Neurology 2003; 2: 275–282.

44. Stoughton RB. Applications of DNA microarrays in biology. *Ann Rev Biochem* (2005) 74: 53–82.

45. Pusztai L, Ayers M, Stec J, Hortobagyi GN. Clinical applications of cDNA microarrays in oncology. *Oncologist* (2003) 8: 252–258.

46. Oostlander AE, Meijer GA, Ylstra B. Microarray-based comparative genomic hybridization and its applications in human genetics. *Clin Genet* (2004) 66: 488–495.

47. King HC, Sinha AA. Gene expression profile analysis by DNA microarrays: Promise and pitfall. *JAMA* (2001) 286, 18: 2280–2288.

48. Mullis K. The unusual origin of the polymerase chain reaction. *Scient Am* (1990) 262: 56–65.

49. Kim Y, Flynn TR, Donoff RB, et al. The gene: The polymerase chain reaction and its clinical application. *J Oral Maxillofac Surg* (2002) 60: 808–815.

50. Ma TS. Applications and limitations of polymerase chain reaction amplification. Chest 1995; 108(5): 1393–1404.

51. Baumforth KRN, Nelson PN, Digby JE, et al. Demystified . . . The polymerase chain reaction. *Molec Pathol* (1999) 52, 1: 1–10.

52. Mifflin TE. Setting up a PCR laboratory. Ch. 1 in Dieffenbach CW, Dveksler GS (Eds.), *PCR primer: A laboratory manual*, 2nd ed. (Cold Spring Harbor, NY: Cold Spring Harbor Laboratory Press, 2003): 5–14.

53. Tzanakaki G, Tsopanomichalou M, Kesanopoulos K, et al. Simultaneous single-tube PCR assay for the detection of *Neisseria meningitides*, *Haemophilus influenzae* type b and *Streptococcus pneumoniae*. *Clin Microbiol & Infect* (2005) 11, 5: 386–390.

54. Mylonakis E, Paliou M, and Rich JD. Plasma viral load testing in the management of HIV infection. *Am Fam Physician* (2001) 63, 3: 483–490.

55. Centers for Disease Control. Guidelines for laboratory test result reporting of human immunodeficiency virus type 1 ribonucleic acid determination. *MMWR* (2001) 50 (RR-20): 1–12.

56. Scott JD, Gretch DR. Molecular diagnostics of hepatitis C virus infection. *JAMA* (2007) 297, 7: 724–732.

57. Weikersheimer PB. Viral load testing for HIV. *Lab Med* (1999) 30, 2: 102–108.

58. Wong ML, Medrano JF. Real-time PCR for mRNA quantitation. *Biotechniq* (2005) 39: 75–85.

59. BioRad. Real-time PCR applications guide, *Bio-Rad Labs* (2005): 1–18.

60. Kubista M., et al. The real-time polymerase chain reaction. *Molec Aspects Med* (2006) 27: 95–125.

61. Hayden RT. In vitro nucleic acid amplification techniques. In Persing DH, et al. (Ed.), *Molecular microbiology: Diagnostic principles and practice* (Washington, DC: ASM Press, 2004): 43–69.

62. Wolk DL, Mitchell S, Patel R. Principles of molecular microbiology testing methods. *Infect Dis Clinics North Am* (2001) 15, 4: 1157–1204.

63. Collins ML, et al. A branched DNA signal amplification assay for quantification of nucleic acid targets below 100 molecules/mL. *Nucleic Acids Res* (1997) 25, 15: 2979–2984.

64. Digene Corporation. Hc2 HPV DNA test (www.digene.com/pdf/L2290P.I,%20hc2%20HPV%20DNA%20Test%20US.pdf), accessed May 30, 2007.

65. Vernick JP, Steigman CK. The HPV DNA virus hybrid capture assay. *MLO* (2003): 8–15.

66. Third Wave Technologies. About the invader chemistry (www.twt.com/invader_chemistry/invader.com), accessed on May 24, 2007.

67. Ahmadian A, Ehn M, Hober S. Pyrosequencing: History, biochemistry, and future. *Clin Chim Acta* (2006): 83–94.

68. Amos J, Grody W. Development and integration of molecular genetic tests into clinical practice: The U.S. experience. *Expert Rev Molecular Diagnost* (2004) 4, 4: 465–477.

69. Alphey L. *DNA sequencing: From experimental methods to bioinformatics* (New York: Springer-Verlag, 1997): 1–25.

70. Maxam A, Gilbert W. A new method for sequencing DNA. *Proc Natl Acad Sci USA* (1977) 74, 2: 560–564.

71. Sterky F, Lundeberg J. Sequence analysis of genes and genomes. *J Biotech* (2000) 76, 1: 1–31.

72. Sanger F, Nicklen S, Coulson AR. DNA sequencing with chain-terminating inhibitors. *Proc Natl Acad Sci USA* (1977) 74, 12: 5463–5467.

LIST OF APPENDIXES

A. U.S. CUSTOMARY SYSTEM OF MEASUREMENT OF LENGTH, AREA, LIQUID, AND MASS

B. CONVERSIONS FROM U.S. CUSTOMARY TO METRIC UNITS

C. METRIC PREFIXES FOR THE BASIC UNITS OF MEASUREMENT

D. CONVERSION AMONG DIFFERENT MEASUREMENTS WITHIN THE METRIC SYSTEM

E. BASE UNITS OF LE SYSTÈME INTERNATIONAL D' UNITÉS

F. DERIVED UNITS OF LE SYSTÈME INTERNATIONAL D' UNITÉS

G. COMMON ACIDS AND BASES USED IN LABORATORIES

H. pH VALUES AND THEIR CORRESPONDING HYDROGEN AND HYDROXYL MOLAR CONCENTRATIONS

I. NOMOGRAM FOR THE DETERMINATION OF BODY SURFACE AREA OF CHILDREN AND ADULTS

J. BODY MASS INDEX (BMI) TABLE

K. REFERENCE INTERVALS FOR SELECTED ANALYTES

L. PERIODIC TABLE OF ELEMENTS

M. ANSWERS TO CASE STUDY QUESTIONS

N. ANSWERS TO CHECKPOINTS

O. ANSWERS TO REVIEW QUESTIONS

U.S. Customary System of Measurement of Length, Area, Liquid, and Mass

Length and Area

12 inches	1 foot
3 feet	1 yard
1760 yards	1 mile
5280 feet	1 mile
1 square foot	144 square inches
1 square yard	9 square feet
43,560 square feet	1 acre
1 square mile	640 acres

Liquid

2 tablespoons	1 fluid ounce
8 fluid ounces	1 cup
2 cups	1 pint
2 liquid pints	1 liquid quart
4 liquid quarts	1 gallon

Mass

Troy	
12 ounces	1 pound
5760 grains	1 pound

Apothecaries

8 drams	1 ounce
12 ounces	1 pound

Avoirdupois

16 drams	1 ounce
16 ounces	1 pound
2000 pounds	1 short ton
2240 pounds	1 long ton

APPENDIX B

Conversions from U.S. Customary to Metric Units

Convert from U.S. to Metric	Conversion Factor (Multiply U.S. Unit by)	Metric Unit
Inch	25.4	Millimeter
Inch	2.54	Centimeter
Yard	0.91	Meter
Mile	1.61	Kilometer
Fluid ounce	29.6	Milliliter
Pound	0.453	Kilogram
Pint	0.474	Liter
Gallon	3.79	Liter

APPENDIX C

Metric Prefixes for the Basic Units of Measurement

Prefix	Abbreviation
Femto	f
Pico	p
Nano	n
Micro	μ
Centi	c
Deci	d
Deca	da
Hector	H
Kilo	K
Mega	M

APPENDIX D
Conversion Among Different Measurements Within the Metric System

Unit	Abbreviation	Comparable Unit
1 kilometer	km	1000 meters
1 deciliter	dL	0.1 or 10^{-1} liters
10 deciliter	dL	1 liter
10 centimeters	cm	1 decimeter
1 millimeter	mm	0.001 or 10^{-3} liters
10 millimeters	mm	1 centimeter
1 microgram	μg	0.000001 or 10^{-6} grams
1000 microgram	μg	1 milligram
1 nanometers	nm	0.000000001 or 10^{-9} meters
1 picograms	pg	0.00000000001 or 10^{-12} grams
1 femtoliter	fL	0.00000000000001 or 10^{-15} liters
1 cm^2		100 mm^2
1 m^2		1,000,000 mm^2

APPENDIX E

Base Units of Le Système International d' Unités

Quantity	Name	Symbol
Length	Meter	M
Mass	Kilogram	Kg
Time	Second	S
Electric current	Ampere	A
Thermodynamic temperature	Kelvin	K
Amount of substance	Mole	Mol
Luminous intensity	Candela	Cd

APPENDIX F

Derived Units of Le Système International d' Unités

Quantity	Name	Symbol
Area	Square meter	m^2
Volume	Cubic meter	m^3
Wave number	1 per meter	m^{-1}
Concentration (of amount of substance)	Mole per cubic meter	mol/m^3
Pressure	Pascal	Pa
Radionuclide activity	Becquerel	Bq
Celsius temperature	Degree Celsius	°C
Luminance	Candela per square meter	cd/m^2
Specific volume	Cubic meter per kilogram	m^3/kg
Electrical potential	Volt	V
Electrical charge	Coulomb	C
Force	Newton	N

Appendix G

Common Acids and
Bases Used in Laboratories

Acid or Base	Specific Gravity	% by Weight (w/w)	MW	Approximate Molarity	Approximate Normality	Approximate mL required to make a 1 L of 1 N solution
Acetic Acid	1.06	99.5	60.05	17.6	17.6	57
Ammonium hydroxide	0.880	29	17.03 (NH_3)	14.78	14.8	67
Hydrochloric Acid	1.19	37	36.46	12.1	12.1	83
Nitric acid	1.42	70	63.02	15.7	15.7	64
Phosphoric acid	1.7	85	98.00	14.7	44.1	23
Potassium hydroxide, saturated	1.55	~50	56.11	14	14	71
Sodium Hydroxide, saturated	1.50	~50	40.00	19	19	53
Sulfuric acid	1.84	96	98.08	18	36	28

pH Values and Their Corresponding Hydrogen and Hydroxyl Molar Concentrations

pH	Hydrogen Ion Concentration (Molar)	Hydroxyl Ion Concentration (Molar)
0	1.0×10^0	1.0×10^{-14}
1	1.0×10^{-1}	1.0×10^{-13}
2	1.0×10^{-2}	1.0×10^{-12}
3	1.0×10^{-3}	1.0×10^{-11}
4	1.0×10^{-4}	1.0×10^{-10}
5	1.0×10^{-5}	1.0×10^{-9}
6	1.0×10^{-6}	1.0×10^{-8}
7	1.0×10^{-7}	1.0×10^{-7}
8	1.0×10^{-8}	1.0×10^{-6}
9	1.0×10^{-9}	1.0×10^{-5}
10	1.0×10^{-10}	1.0×10^{-4}
11	1.0×10^{-11}	1.0×10^{-3}
12	1.0×10^{-12}	1.0×10^{-2}
13	1.0×10^{-13}	1.0×10^{-1}
14	1.0×10^{-14}	1.0×10^0

Nomogram for the Determination of Body Surface Area of Children and Adults

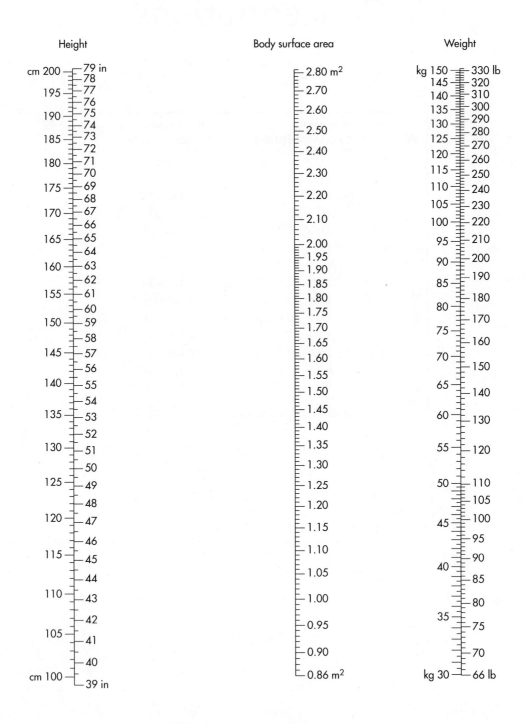

Body Mass Index (BMI) Table

The number at the top of the column is the BMI at that height and weight. Pounds have been rounded off.

BMI	19	20	21	22	23	24	25	26	27	28	29	30	31	32	33	34	35
							Body Weight (lb)										
Height (in.) 58	91	96	100	105	110	115	119	124	129	134	138	143	148	153	158	162	167
59	94	99	104	109	114	119	124	128	133	138	143	148	153	158	163	168	173
60	97	102	107	112	118	123	128	133	138	143	148	153	158	163	168	174	179
61	100	106	111	116	122	127	132	137	143	148	153	158	164	169	174	180	185
62	104	109	115	120	126	131	136	142	147	153	158	164	169	175	180	186	191
63	107	113	118	124	130	135	141	146	152	158	163	169	175	180	186	191	197
64	110	116	122	128	134	140	145	151	157	163	169	174	180	186	192	197	204
65	114	120	126	132	138	144	150	156	162	168	174	180	186	192	198	204	210
66	118	124	130	136	142	148	155	161	167	173	179	186	192	198	204	210	216
67	121	127	134	140	146	153	159	166	172	178	185	191	198	204	211	217	223
68	125	131	138	144	151	158	164	171	177	184	190	197	203	210	216	223	230
69	128	135	142	149	155	162	169	176	182	189	196	203	209	216	223	230	236
70	132	139	146	153	160	167	174	181	188	195	202	209	216	222	229	236	243
71	136	143	150	157	165	172	179	186	193	200	208	215	222	229	236	243	250
72	140	147	154	162	169	177	184	191	199	206	213	221	228	235	242	250	258
73	144	151	159	166	174	182	189	197	204	212	219	227	235	242	250	257	265
74	148	155	163	171	179	186	194	202	210	218	225	233	241	249	256	264	272
75	152	160	168	176	184	192	200	208	216	224	232	240	248	256	264	272	279
76	156	164	172	180	189	197	205	213	221	230	238	246	254	263	271	279	287

(Continued)

BMI	36	37	38	39	40	41	42	43	44	45	46	47	48	49	50	51	52
								Body Weight (lb)									
Height (in.) 58	172	177	181	186	191	196	201	205	210	215	220	224	229	234	239	244	248
59	178	183	188	193	198	203	208	212	217	222	227	232	237	242	247	252	257
60	184	189	194	199	204	209	215	220	225	230	235	240	245	250	255	261	266
61	190	195	201	206	211	217	222	227	232	238	243	248	254	259	264	269	275
62	196	202	207	213	218	224	229	235	240	246	251	256	262	267	273	278	284
63	203	208	214	220	225	231	237	242	248	254	259	265	270	278	282	287	293
64	209	215	221	227	232	238	244	250	256	262	267	273	279	285	291	296	302
65	216	222	228	234	240	246	252	258	264	270	276	282	288	294	300	306	312
66	223	229	235	241	247	253	260	266	272	278	284	291	297	303	309	315	322
67	230	236	242	249	255	261	268	274	280	287	293	299	306	312	319	325	331
68	236	243	249	256	262	269	276	282	289	295	302	308	315	322	328	335	341
69	243	250	257	263	270	277	284	291	297	304	311	318	324	331	338	345	351
70	250	257	264	271	278	285	292	299	306	313	320	327	334	341	348	355	362
71	257	265	272	279	286	293	301	308	315	322	329	338	343	351	358	365	372
72	265	272	279	287	294	302	309	316	324	331	338	346	353	361	368	375	383
73	272	280	288	295	302	310	318	325	333	340	348	355	363	371	378	386	393
74	280	287	295	303	311	319	326	334	342	350	358	365	373	381	389	396	404
75	287	295	303	311	319	327	335	343	351	359	367	375	383	391	399	407	415
76	295	304	312	320	328	336	344	353	361	369	377	385	394	402	410	418	426

Source: www.nhlbi/nih.gov/guidelines/obesity/bmi_tbl.htm (accessed June 2009)

Reference Intervals for Selected Analytes

Reference intervals (RIs; also known as a *reference range* or a *normal range*) represent data derived from apparently normal healthy individuals that are used as comparators by clinicians to assess the health status of their patients. Several factors can influence RIs as outlined in Chapter 4 (Laboratory Operations). Four factors that have been extensively evaluated and are often cited in RI tables are age, gender, specimen type, and methods that provide the user with specific information appropriate to their patients and laboratory testing facility. Therefore, it is imperative that an RI obtained from any source should not be used as an absolute indicator of health and disease.

The selected reference intervals presented here are used to provide general information about ranges commonly used; they were obtained from highly reputable sources. All of the RIs included are for adults (>18 years); where significant differences exist, they are separated by gender. Each analyte is expressed in both conventional and recommended SI units when applicable. The conversion factor used is also indicated. Some of the RIs were taken from chapters in this textbook; others were selected from Wu AHB, *Tietz Clinical Guide to Laboratory Tests*, 4th ed. (Philadelphia: W.B. Saunders, 2006) and McPherson PM, *Henry's Clinical Diagnosis by Laboratory Methods*, 21st ed. (Philadelphia: Saunders Elsevier, 2007)

Analyte	Specimen	Conventional Units	Conversion Factor (×)	Recommended SI Units
Adrenocorticotropic hormone	P	p/mL Adult (7–10 a.m.) 9–52	0.22	pmol/L 2–11
Alanine aminotransferase (ALT)	S	U/L 6–37	0.017	μKat/L 0.102–0.630
Albumin	S	g/dL 3.5–5.0	10	g/L 35–50
Aldostersone	P	ng/dL Adults: Supine: 3–16 Upright: 7–30	0.0277	mmol/L 0.08–0.61 0.19–0.83
Alkaline phosphatase (ALP)	S	U/L 30–90	0.017	mKat/L 0.51–1.53
Ammonia	P	μg/dL 19–60	0.587	mmol/L 11–35
Amylase, serum	S	U/L 30–100	0.017	μKat/L 0.51–1.7
Amylase, urine	U	U/L 170–2000 U/hr 1–17	0.017	μKat/L 2.89–34.0
Anion gap	S, P	mEq/L $Na^+ - (Cl^- + HCO_3)$: 7–16 $Na^+ + K^+) - (Cl^- + HCO_3)$: 10–20	1.0	mmol/L 7–16 10–20
Aspartate aminotransferase (AST)	S	U/L 5–30	0.017	μKat/L 0.085–0.510
Base excess	WB	mEq/L −1 to +3	1.0	mmol/L −1 to +3

(Continued)

Analyte	Specimen	Conventional Units	Conversion Factor (×)	Recommended SI Units
Bicarbonate	WB	mEq/L 22–26	1.0	mmol/L 22–26
Bilirubin, total	S	mg/dL 0.2–1.0	17.1	μmol/L 3.4–17.1
Bilirubin, conjugated (direct)	S	mg/dL <0.8	17.1	μmol/L <13.6
Bilirubin, unconjugated (indirect)	S	mg/dL <0.2	17.1	μmol/L <3.4
Calcium, ionized	WB	mg/dL 4.60–5.08	0.25	mmol/L 1.15–1.27
Calcium, total	S	mg/dL 8.60–10.0	0.25	mmol/L 2.15–2.50
Carbon dioxide, total	WB	mEq/L 23–29	1.0	mmol/L 23–29
Carbon dioxide, partial pressure	WB	mmHg Adult: Male: 35–48 Female: 32–45	0.133	kPa 4.66–6.38 4.26–5.99
Carboxyhemoglobin	WB	% Nonsmokers: 0.5–1.5 Toxic: >20		
Chloride	S, P	mEq/L 98–107	1	mmol/L 98–107
Cholesterol, total	S	mg/dL <200	0.0259	mmol/L <5.18
Cholinesterase	S	U/mL Male: 4.0–7.8 Female: 3.3–7.6	1	kU/L 4.0–7.8 3.3–7.6
Chorionic gonadotropin	S	mU/mL Males and nonpregnant females: <5.0	1	U/L <5.0
Cortisol, total	S	μg/dL Adults: 8 a.m.: 5–23 8 p.m.: <50% of 8 a.m.	27.6	Mmol/L 138–635 <50%
Creatine kinase (CK)	S	U/L Male: 46–180 U/L Female: 15–171 U/L		
Creatine kinase relative index	S	% <3		
Creatinine	S	mg/dL Male: 0.9–1.2 Female: 0.6–1.1	88.4	μmol/L 80–106 53–97
Creatinine clearance	S	ml/min/1.73 m^2 Males: 97–137 Females: 88–128	0.00963	mL/sec/m^2 0.93–1.32 0.85–1.23
Ferritin	S	ng/mL Adult: Male: 20–250 Female: 10–120	1.0	μg/mL 20–250 10–120
Free thyroxine index $FTI = TT_4 \times THBR$	S	Adult: 4.2–13.0		

(Continued)

Analyte	Specimen	Conventional Units	Conversion Factor (×)	Recommended SI Units
Glucose	S	mg/dL 74–106	0.0555	mmol/L 4.1–5.9
γ-Glutamyltransferase (GGT)	S	U/L Males: 10–34 Females: 9–22	0.017	μKat/L 0.17–0.59 0.15–0.37
High-density lipoprotein cholesterol	S	mg/dL Low: <40 High: ≥60	0.0259	mmol/L <1.04 ≥1.55
Iron	S	μg/dL Adult: Male: 65–175 Female: 50–170	0.179	μmol/L 11.6–31.3 9.0–30.4
Iron saturation (% transferrin saturation)	S	% Adult: Male: 20–50 Female: 15–50		
Iron binding capacity (TIBC)	S, P	mg/dL 250–425	0.179	mmol/L 44.8–76.1
Lactate dehydrogenase	S	U/L 140–280		
L-lactate (lactic acid)	P	mg/dL Venous: 4.5–19.8 Arterial: 4.5–14.4	0.111	mmol/L 0.5–2.2 0.5–1.6
Lead	WB	μg/dL Child: <10 Adult: <25	0.0483	μmol/L <0.48 <1.21
Lipase	S	U/L 13–60	0.017	μKat/L 0.22–1.00
Low-density lipoprotein cholesterol (LDL-C)	S	mg/dL Optimal: <100	0.0259	mmol/L <2.85
Magnesium	S	mg/dL 1.6–2.6	0.4113	mmol/L 0.66–1.07
Osmolality	S, P, U	mOsm/kg Serum: 275–295 Urine: 50–1200		
Oxygen, partial pressure (PO_2)	WB	mmHg Arterial: 95–100	0.133	kPa 12.64–13.3
Oxygen saturation	WB	% 94–98		
pH	WB	Arterial: 7.35–7.45 Venous: 7.32–7.43		
Phosphorus, Inorganic	S, P (heparin)	mg/dL 2.5–4.5	0.323	mmol/L 0.81–1.45
Potassium	S, P	mEq/L 3.5–5.1	1.0	mmol/L 3.5–5.1
Protein, total	S	g/dL 6.4–8.3	10	g/L 64–83
Sodium	S, P	mEq/L 136–145	1.0	mmol/L 136–145
Thyroid-stimulating hormone (TSH)	S	μU/mL 0.4–4.2	1.0	mU/L 0.4–4.2
Thyroxine, free (FT_4)	S	ng/dL 0.8–2.7	12.9	pmol/L 10–35

(Continued)

Analyte	Specimen	Conventional Units	Conversion Factor (×)	Recommended SI Units
Thyroxine (T$_4$)	S	μg/dL 4.6–12.0	12.9	nmol/L 59–155
Transferrin	S	mg/dL 200–360	0.01	g/L 2.0–3.6
Triglycerides	S	mg/dL Recommended cut point: <150	0.0113	mmol/L 1.70
Triiodothyronine (T$_3$), total	S	ng/dL 70–204	0.0154	nmol/L 1.08–3.14
Urea nitrogen	S, P	Urea nitrogen, mg/dL 6–20	0.357	Carbamide (urea), mmol/L 2.1–7.1
Uric acid	S, P	mg/dL Adult: Male: 3.5–7.2 Female: 2.6–6.0	59.48	μmol/L 208–428 155–357

Abbreviations: S, serum; P, plasma; U, urine; WB, whole blood

APPENDIX L

Periodic Table of Elements

Periodic Table of the Elements

Representative [main group] elements												Representative [main group] elements						

Transition metals

1	IA																	VIIIA
	1 H 1.0079	IIA											IIIA	IVA	VA	VIA	VIIA	2 He 4.003
2	3 Li 6.941	4 Be 9.012											5 B 10.813	6 C 12.011	7 N 14.007	8 O 15.999	9 F 18.998	10 Ne 20.180
3	11 Na 22.990	12 Mg 24.305	IIB	IVB	VB	VIB	VIIB		VIIIB		IB	IIB	13 Al 26.982	14 Si 28.086	15 P 30.974	16 S 32.066	17 Cl 35.453	18 Ar 39.948
4	19 K 39.098	20 Ca 40.078	21 Sc 44.956	22 Ti 47.88	23 V 50.942	24 Cr 51.996	25 Mn 54.938	26 Fe 55.845	27 Co 58.933	28 Ni 58.933	29 Cu 63.566	30 Zn 65.39	31 Ga 69.723	32 Ge 72.61	33 As 74.922	34 Se 78.96	35 Br 79.904	36 Kr 83.8
5	37 Rb 85.468	38 Sr 87.62	39 Y 88.906	40 Zr 91.224	41 Nb 92.906	42 Ma 95.94	43 Tc 98	44 Ru 101.07	45 Rh 102.906	46 Pd 106.42	47 Ag 107.868	48 Cd 112.411	49 In 114.82	50 Sn 118.71	51 Sb 121.76	52 Te 127.60	53 I 126.905	54 Xe 131.29
6	55 Cs 132.905	56 Ba 137.327	57 La 138.906	72 Hf 178.49	73 Ta 180.948	74 W 183.84	75 Re 178.49	76 Os 190.23	77 Ir 192.22	78 Pt 195.08	79 Au 196.907	80 Ag 200.59	81 Tl 204.383	82 Pb 207.2	83 Bi 208.980	84 Po 209	85 At 210	86 Rn 222
7	87 Fr 223	88 Ra 226.025	89 Ac 227.028	104 Rf 261	105 Db 262	106 Sg 263	107 Bh 264	108 Hs 265	109 Mt 266	110 Uun 268	111 Uuu 272	112 Uub 227		114		116		118

Rare earth elements

Lanthanides	58 Ce 140.115	59 Pr 140.908	60 Nd 144.24	61 Ph 143	62 Sm 150.36	63 Eu 151.964	64 Gd 157.25	65 Tb 158.925	66 Dy 162.5	67 Ho 164.93	68 Er 167.26	69 Tm 168.934	70 Yb 173.04	71 Lu 174.967
Actinides	90 Th 232.038	91 Pa 231.038	92 U 230.029	93 Np 237.048	94 Pu 244	95 Am 243	96 Cm 247	97 Bk 247	98 Cf 251	99 Es 252	100 Fm 257	101 Md 258	102 Na 259	103 Lr 262

APPENDIX M

Answers to Case Study Questions

CHAPTER 1

Case Summary: The laboratory supervisor discovered the presence of a shift in quality-control values for serum creatinine in both levels of serum-based quality-control (QC) material. The quality-control data was consistently lower than previous results. The quality-control material is lyophilized (freeze dried) and requires reconstitution with 5.0 mL of the diluent provided by the manufacturer. An investigation into the cause of this shift in quality-control results was initiated.

Question

1. Identify several causes of a shift in quality-control data.

Explanation

Quality-control shifts may be caused by any of the following:

a. instrument malfunctions (e.g., light source, pipettors)
b. changes in environmental conditions (e.g., room temperature and humidity)
c. improperly prepared quality-control material
d. problems with instrument or method calibration

Question

2. Indicate the type of pipette that is suitable for reconstituting quality-control material.

Explanation

Pipettes suitable for preparing quality-control material include class A volumetric glass pipettes, and calibrated volumetric pipettes that are not designated class A.

Question

3. Is the procedure and techniques required to use pipettes the same for every pipette?

Explanation

No, the procedures for using pipettes are not the same. The technologist should read the accompanying direction for properly operating any pipette device. In this case, the quality-control shift resulted from the improper use of the 5.0-mL glass class A volumetric pipette. The technologist thought that all glass pipettes are blow-out type pipettes and proceeded to blow out the entire contents of the pipette. The pipette that was used did not have a single- or double-etched marking at the aspiration end of the pipette, so more diluent was delivered than required and the results were lower because of a slight dilution effect.

CHAPTER 2

Case Summary: A staff technologist discovered that the quality-control results for photometric assays—including tests for glucose, urea nitrogen, and creatinine—were out of range. Further investigation revealed the presence of a shift in quality-control results that resulted in a QC failure according to Westgard rules. The shift in QC was below the mean for all photometric assays.

The technologist initiated the laboratory QC protocol for out-of-control situations. None of the procedures (including reassaying the quality-control serum and recalibrating the assay) resolved the problem.

Question

1. Identify components within the spectrometer that may result in a shift in QC results to below the mean.

Explanation

Polychromatic light source, monochromator, cuvettes, or detector

Question

2. Identify specific photometric function checks that can be completed and which may provide evidence that a component is not functioning properly.

Explanation

Wavelength calibration and absorbance check

Question

3. Which function check has the highest probability of being the one that failed and is responsible for the downward shift of QC results?

Explanation

Absorbance check

Question

4. Provide a solution for the problem that would result in an upward movement of the QC data toward the mean.

Explanation

A failing light source is a common problem with spectrophotometers, so change the light source first.

CHAPTER 3

Case Summary: The supervisor of a laboratory has been given the task of updating the core laboratory, including processing, instrumentation in chemistry, hematology, and urinalysis.

Question
1. What are several options available to accomplish this reorganization of the laboratory?

Explanation
Options include the following:

a. purchase separate analyzers for chemistry, hematology, and urinalysis and continue with manual processing of specimens;
b. purchase a stand-alone automated processing unit and separate analyzers for chemistry, hematology, and urinalysis;
c. purchase a stand-alone automated processing unit and integrated modular system to automated chemistry (including immunoassays) as well as a separate hematology analyzer and urinalysis analyzer; and
d. purchase a fully integrated modular system to incorporate general chemistries, immunochemistries, hematology, and specimen processing, and also purchase a fully automated urinalysis system to include the ability to perform a urine microscopic analysis.

Question
2. What factors should be considered before making a final decision?

Explanation
Factors to consider before making final decision include the following:

a. finances;
b. space;
c. utilities;
d. number of specimens to test;
e. analytes that need to be measured;
f. throughput rates;
g. staff experience, expertise, training concerns, and number of staff required to operate and maintain the equipment;
h. service contracts; and
i. turnaround-time demands.

 CHAPTER 4

Case Summary: A member of the laboratory staff was asked to clean out a closet storage area in which chemicals had been stored and discovered a bottle of picric acid.

Question
1. What is the NFPA 704-M placard designation for health hazard, fire and flammability, reactivity, and special hazard?

Explanation
For picric acid:

Fire and flammability	Red diamond: 0
Health hazard	Blue diamond: 2
Reactivity	Yellow diamond: 2
Special or specific hazard	White diamond: None

Question
2. What is the physical appearance of the chemical?

Explanation
The physical appearance of the chemical is described as dried and yellow in color, and the label looks old, indicating that the chemical was acquired many years earlier.

Question
3. Where can we find information regarding the safety issues for this chemical?

Explanation
The laboratory staff should find the material safety data sheet for the chemical either in the laboratory or from an online source.

Question
4. What information about the chemical is important for laboratory workers to focus on?

Explanation
The following information is significant.
An emergency overview shows the following:

- The greatest hazard associated with these solutions occurs when picric acid is allowed to dry completely: It may detonate. It also may be absorbed through the skin, so any areas of contact should be washed with plenty of water. If it is ingested, dilution with water is essential.
- There is an aspiration hazard, so do not induce vomiting. Call a physician if necessary.
- Target organs are the eyes, skin, liver, and blood.
- Eye contact may result in irritation, redness, pain, and tearing.
- Inhalation may cause pulmonary irritation.
- May be absorbed through the skin and cause irritation. All allergic reactions and symptoms are similar to inhalation. Will stain areas of contact. Stains may be removed using reagent alcohol or dilute ammonium hydroxide solutions.
- Ingestion may cause symptoms similar to inhalation.
- Chronic effects and carcinogenicity: None.

When handled properly by qualified personnel, the product does not present a significant healthy or safety hazard. Alterations of its characteristic by concentration, evaporation, addition of other substances, or other means may present hazards not specifically addressed here; these must be evaluated carefully by the user.

Question

5. What specific steps should be taken to properly handle, use, and dispose of this chemical?

Explanation

Specific steps to handle, use, and dispose of picric acid include the following.

- Wear gloves, a laboratory coat, and proper eye protection.
- Transfer picric acid from its original container using appropriate engineering controls.
- Keep away from heat and do not shock container.
- Work in an adequately ventilated room.
- Add acid to water.
- Do not allow the picric acid to dry out completely through prolonged exposure to the air.
- Dispose of any source of picric acid within a reasonable amount of time, especially if the reagent will not be reused.
- Contact any state or federal agency for proper instructions on disposing of the chemical.

CHAPTER 5

Case Summary: A 21-year-old female (gravida 0, para 0) presents to the emergency department complaining of nausea and abdominal pain over the past couple of weeks. The patient states that she shows vaginal spotting, is sexually active, admits to drug and alcohol use, and has an intrauterine device. An attending physician conducts a physical examination that reveals lower leg edema and a palpable lower abdominal mass. The physician orders an ultrasound and a urine pregnancy test followed by a serum quantitative β-hCG measurement. (*Gravida* is the total number of times a women has been pregnant and *para* indicates the number of viable births.)

Diagnostic Test Results

Results of the ultrasonography	The ultrasound revealed a "snowstorm" appearance that may be caused by the presence of a molar pregnancy (for example, a hydaditiform mole).
Urine qualitative pregnancy test	Negative
Serum quantitative β-hCG	700 IU/L (RI = <0.5–2.9 IU/L)

Question

1. What is the clinical significance of a serum quantitative β-hCG value of 700 IU/L?

Explanation

The concentration of serum β-hCG for a normally progressing pregnancy at 4 weeks (since last missed period) is in the range of 5–100 U/L; at 5 weeks (last missed period), it is in the range of 200–3000 U/L. In normal intrauterine pregnancies, serum hCG rapidly increases, with a doubling time of 1.5 days during weeks 2–5. In abnormally progressing pregnancies, serum hCG concentrations increase more slowly or may

decrease. Therefore, the concentration of hCG may be reasonable for either a normally progressing pregnancy or an abnormally progressing pregnancy. In this case, the clinician thought that the result of the serum β-hCG was too low in light of the fact that the snowstorm appearance from the ultrasound procedure usually indicates the possibility of an abnormally progressing pregnancy.

Question

2. Why is the urine qualitative pregnancy test negative, whereas the serum quantitative β-hCG is 700 IU/L?

Explanation

The discrepant results between the laboratory's urine qualitative and serum quantitative pregnancy tests are disturbing. Several reasons could account for such a discrepancy and include human error, assay limitations, analytical error, and methodological differences.

Question

3. What methodologies were used for both the qualitative and quantitative tests?

Explanation

Immunoassays may be affected by a phenomenon referred to as the *hook effect* or *high-dose hook effect*. The hook effect arises when the amount of patient antigen exceeds the amount of antibody present in the reagent system. The result of the measurement is a falsely low value of measured analyte. In this case, both the qualitative and quantitative assays were affected by the hook effect, thereby yielding a false-negative urine qualitative assay and a spuriously low serum quantitative assay.

Question

4. What could account for the questionable results of the urine qualitative pregnancy test?

CHAPTER 6

Case Summary: Heidi, an 18-year-old white woman, was brought to the hospital emergency room in a comatose state. Her roommates stated that she had been nauseated earlier in the day. On physical examination, it was noted that she was breathing deeply and rapidly, her breath had a fruity odor, and her skin and mucous membranes were dry.

Heidi's physician received the following laboratory data.

Laboratory Results

		Reference Range
Na+	128	135–145 mmol/L
K+	5.7	3.4–5.0 mmol/L
pH	7.12	7.35–7.45
Serum glucose	750	70–99 mg/dL
Serum acetone	3+	Neg

(Continued)

Urinalysis		
Color/appearance	Pale yellow/clear	
Glucose	4+	Neg
Ketones	2+	Neg

Question

1. Identify all abnormal laboratory values.

Explanation

Heidi's abnormal results are decreased sodium and pH (acidosis). Chemistry results above the reference range are an extremely high serum glucose, elevated potassium, and a positive serum acetone. Abnormal urinalysis results are glucose (4+) and ketones (2+).

Question

2. On the basis of Heidi's history, clinical findings, and laboratory data, Heidi would be classified as having what type of hyperglycemia? Why?

Explanation

Heidi would be diagnosed as a type 1 diabetic because of the high glucose and the presence of ketones, and her condition would be classified as diabetic ketoacidosis (DKA). Type 1 diabetics are often, but not always, juvenile onset, prone to ketosis, and insulin dependent.

Question

3. Which laboratory findings are *most* valuable in establishing the diagnosis?

Explanation

The very high glucose and positive ketones are the most significant findings.

Question

4. Explain the presence of glucose in the urine.

Explanation

Heidi's serum glucose (750 mg/dL) exceeds the renal threshold of 160–180 mg/dL, or the amount of glucose that can be reabsorbed by the proximal convoluted tubules in the kidney. Therefore, her kidneys are unable to reabsorb the excess glucose and it spills into the urine.

Question

5. What type of antibodies (autoimmune) are often found in type 1 diabetics?

Explanation

Autoantibodies in type 1 diabetes mellitus are islet cell cytoplasmic antibodies, insulin autoantibodies, glutamic acid decarboxylase antibodies, and autoantibodies to tyrosine phosphatases IA-2 and IA-2β.

Question

6. What is causing the fruity odor on her breath?

Explanation

Ketones (acetone) being released in the expired air are responsible for the fruity odor on her breath. The excess production of ketone bodies is associated with diabetic ketoacidosis, which is probably the cause of her coma.

Question

7. The nitroprusside chemical reaction with glycine, used to indicate a positive acetone reaction, reacts with which ketone bodies?

Explanation

Sodium nitroprusside and glycine react with acetoacetic acid and acetone but not with β-hydroxybutyric acid.

Question

8. Why are the ketone bodies increased in diabetes mellitus?

Explanation

The body is metabolizing fat for energy but does not have enough acetyl CoA to metabolize it all the way to CO_2 and H_2O, so ketones accumulate. In other words, ketones are formed as a result of incomplete fat metabolism.

Question

9. Heidi is prescribed a daily regimen of insulin. Which of the laboratory procedures discussed in this chapter would be of most value in determining the degree of glucose control over a 2-month period? What would be Heidi's target value?

Explanation

Glycosylated hemoglobin would provide an index of her control over the previous 2–3 months. The American Diabetes Association recommends an HbA_{1c} target value of less than 7%.

CHAPTER 7

Case Summary: Robert R., a 47-year-old male, went to his physician for an "annual" physical that had been postponed for more than 3 years. He had started his own business 4 years earlier and had been extremely busy getting it established. Robert's medical history indicated he was a nonsmoker, and his father and grandfather had histories of myocardial infarctions (MIs) before age 55. Because he had 'not seen this patient in a few years, the physician decided to order routine screening tests.

Chemistry Results

		Reference Range
Sodium	143	136–145 mEq/L
Potassium	4.6	3.6–5.0 mEq/L
Chloride	109	101–111 mEq/L
CO_2	29.0	24.0–34.0 mEq/L

(Continued)

		Reference Range
Glucose	95.0	70–99 mg/dL
BUN	16	7–24 mg/dL
Creatinine	1.0	0.5–1.2 mg/dL
Bilirubin total	0.7	0.2–1.2 mg/dL
AST	32	5–40 IU/L
ALP	80	30–157 IU/L
Protein	7.5	6.0–8.4 g/dL
Albumin	4.6	3.5–5.0 g/dL
Calcium	8.5	8.5–10.5 mg/dL

Miscellaneous Chemistry

		Reference Range
Cholesterol	355	Recommended (Desirable): <200 mg/dL

Urinalysis

Macroscopic		Reference Range
Color	Yellow	Colorless to amber
Appearance	Clear	Clear
Specific Gravity	1.014	1.001–1.035
pH	6.0	5–7
Protein	Neg	Neg
Glucose	Neg	Neg
Ketones	Neg	Neg
Bilirubin	Neg	Neg
Blood	Neg	Neg
Urobilinogen	Normal	Normal
Nitrite	Neg	Neg
Leukocyte esterase	Neg	Neg

Microscopic: Not indicated

Question

1. Circle or highlight the abnormal result(s).

Explanation

The only abnormal result is the elevated cholesterol.

Question

2. List eight secondary conditions or disorders associated with the abnormal result(s) in question 1.

Explanation

Some of the more common conditions associated with hypercholesterolemia are the following:

a. obesity and diet;
b. alcohol;
c. drugs such as steroids, thiazides, anticonvulsants, beta blockers, and certain oral contraceptives;
d. diabetes mellitus;
e. hypothyroidism;

f. renal disease (e.g., chronic renal failure, nephrotic syndrome);
g. acute or transient conditions such as burns, hepatitis, acute trauma (e.g., surgery), myocardial infarction, and bacterial and viral infections.

Question

3. Which conditions can the physician rule out with Robert's medical history, physical exam, and current laboratory results?

Explanation

The following can be ruled out with Robert's medical history, physical exam, and current laboratory results:

a. diabetes mellitus (glucose is within reference range),
b. renal disease such as chronic renal failure and nephrotic syndrome (BUN and creatinine are within reference range and urinalysis is "normal"),
c. acute or transient conditions,
    i. burns (medical history is negative),
    ii. hepatitis (AST, bilirubin, and ALP are within reference range),
    iii. acute trauma such as surgery (medical history is negative),
    iv. myocardial infarction (asymptomatic: no symptoms),
    v. bacterial and viral infections (asymptomatic), and
d. hypothyroidism (asymptomatic but may require further questioning or testing).

Question

4. What is the most probable cause of the abnormal result in this patient?

Explanation

The most probable cause is diet or obesity.

The following week, a lipid profile was performed on a 12-hour fasting specimen.

Lipid Profile

Cholesterol	350	Recommended (desirable): <200 mg/dL
High-density lipoprotein cholesterol	40	Recommended (desirable): >60 mg/dL
Triglycerides	340	Recommended (desirable): <150 mg/dL

Question

5. Given the above information, what is Robert's low-density lipoprotein cholesterol (LDL-C)? Based on his LDL-C, is he at high or moderate risk of coronary heart disease (CHD) or is he within the recommended (desirable) range?

Explanation

Mike's LDL-C can be calculated using the Friedewald formula:

$$\text{LDL-cholesterol} = \text{Total cholesterol} - (\text{HDL} + \text{triglycerides}/5)$$
$$= 350 - (40 + 340/5)$$
$$= 242 \text{ mg/dL}$$

His LDL-cholesterol would put him in the high risk category.

Question

6. (A) If Robert's triglycerides were 450 mg/dL, could the LDL–C be calculated? Why or why not? (B) What would the next step be?

Explanation

(A) The LDL–C cannot be calculated if the triglycerides are more than 400 mg/dL; the Friedewald formula is not valid. Triglycerides/5 is *not* a valid indicator of very-low-density lipoprotein (VLDL) cholesterol when the triglycerides are extremely elevated because these samples may also contain chylomicrons, chylomicron remnants, or abnormal VLDLs, which have higher triglyceride:cholesterol ratios than normal VLDL.
(B) The LDL–C would have to be determined directly after separation from the HDL–C through immunochemical separation, chemical precipitation of HDL, or ultracentrifugation.

Question

7. List eight risk factors associated with CHD as determined by the National Cholesterol Education Program (NCEP) Adult Treatment Panel.

Explanation

Eight risk factors associated with coronary heart disease as determined by the NCEP Adult Treatment Panel are:

1. age (>45 for men; ≥ 55 or premature menopause for women),
2. family history of CHD,
3. current cigarette smoking,
4. hypertension (BP ≥ 140/90 mm Hg or taking antihypertensive medications),
5. LDL–C concentration ≥ 160 mg/dL with two or fewer risk factors,
6. LDL–C concentration 130–159 mg/dL with two or more risk factors,
7. HDL concentration < 35 mg/dL, and
8. diabetes mellitus.

Question

8. How many other risk factors in question 7 does Robert currently have given the information provided? Is he at high risk for CHD?

Explanation

Robert has three risk factors: age, family history, and high LDL–C.

Question

9. What is the follow–up testing and treatment decision recommended by the NCEP Adult Treatment Panel?

Explanation

The NCEP Adult Treatment Panel recommends clinical evaluation, including family history of CHD, and dietary and drug therapy. The goal is to bring the LDL–C to <130 mg/dL.

CHAPTER 8

Case Summary: Judy, a 40-year-old woman with a past history of kidney infections, was seen by her physician because she had been feeling lethargic for a few weeks. She also complained of decreased frequency of urination and a bloated feeling. The physician noted periorbital swelling and general edema, including a swollen abdomen.

Urinalysis

Macroscopic

Color:	Yellow
Appearance:	Cloudy or frothy
Specific gravity:	1.022
pH:	7.0
Protein:	3 + (500 mg/dL) (SSA: 4+)
Glucose:	Neg
Ketones:	Neg
Bilirubin:	Neg
Blood:	Neg
Urobilinogen:	Normal
Nitrite:	Neg
Leukocyte Esterase:	Neg

Microscopic

WBCs:	0–3/HPF
RBCs:	0–1/HPF
Epithelial cells:	Rare squamous/HPF
	Rare renal tubular epithelial/HPF
Casts:	0–3 Hyaline/LPF
	0–1 Renal tubular epithelial/LPF
	0–1 Granular/LPF
	0–1 Waxy/LPF
	0–1 Fatty/LPF
Other:	Occasional oval fat bodies

Chemistry

		Reference Range
Protein:	5.0 g/dL	6.0–8.4 g/dL
Albumin:	2.4 g/dL	3.5–5.0 g/dL
Cholesterol:	370 mg/dL	<200 mg/dL
BUN:	33 mg/dL	7–24 mg/dL
Creatinine:	2.1 mg/dL	0.5–1.2 mg/dL

Question

1. Circle or highlight the abnormal value(s) or discrepant result(s) in the urinalysis.

Explanation

The abnormal macroscopic urinalysis results are as follow.

Appearance: Cloudy/Hazy
Protein: 3+

The abnormal microscopic results are as follow.

Casts: 0–1 Renal tubular epithelial/LPF
       0–1 Granular/LPF
       0–1 Waxy/LPF
       0–1 Fatty/LPF
       Occasional oval fat bodies

Discrepant results: None.

The abnormal chemistry results are decreased total protein and albumin, and increased cholesterol, BUN, and creatinine.

Question

2. What type of disease or condition would be characterized by the urinalysis and chemistry results?

Explanation

The urinalysis is consistent with nephrotic syndrome.

Question

3. What urinalysis result(s) led to your probable diagnosis?

Explanation

Marked proteinuria and lipiduria (oval fat bodies and fatty casts) are hallmarks of nephrotic syndrome.

Question

4. Are the abnormal chemistry tests consistent with the probable diagnosis? Why or why not?

Explanation

Nephrotic syndrome is characterized by hypoalbuminemia and hyperlipidemia (elevated cholesterol). Hyperlipidemia is inversely proportional to the albumin concentration: The lower the albumin levels fall, the higher the lipid values. Liver synthesis of all proteins increases to compensate for the increased loss of protein in the urine, and large proteins (α_2-macroglobulin) that are retained are increased in the blood. Hypoalbuminemia stimulates the liver to synthesize LDLs and VLDLs. The minimally increased BUN and creatinine are indicative of slightly decreased renal function.

Question

5. Discuss the physiological cause of the edema.

Explanation

The edema is caused by the hypoalbuminemia, which leads to reduced plasma osmotic pressure. This reduced osmotic pressure allows fluids to leak from blood vessels to the interstitial spaces, resulting in edema.

Question

6. What is Judy's albumin:globulin (A:G) ratio? Is it within the reference range? Is it consistent with the probable diagnosis?

Explanation

Judy's A:G ratio is 0.92, which is below the reference range. Normal albumin levels are 60% of the total protein. A low A:G ratio is consistent with a diagnosis of nephrotic syndrome.

Question

7. Describe what you would expect to see in this patient's protein electrophoresis.

Explanation

The patient's protein electrophoresis would be characterized by a marked increase in α_2- and β-globulins and a marked decrease in albumin and γ-globulins, especially IgG.

Question

8. Which specific proteins would be decreased? Which specific proteins would be increased?

Explanation

In nephritic syndrome, the smaller molecular weight proteins such as albumin and γ-globulins will be decreased in the serum but increased in the urine. Larger proteins, including α_2-macroglobulin, would be retained.

 CHAPTER 9

Case Summary: Doreen T., a 60-year-old woman, was seen in the emergency room complaining of moderate to severe chest pain. She had experienced substernal pain for the previous 6 to 7 weeks with dyspnea on exertion. The pain, however, had become more frequent and severe with constant pain and pressure for the last 2 to 3 days. Doreen appeared anxious and complained of weakness, sweating, and nausea. Her blood pressure was 110/66.

Doreen was given Nitrostat 1/150 grain PRN for pain, and Inderal was administered. She was also taking digoxin. The ECG was performed and revealed an atrial flutter and the possibility of a true posterior infarct and lateral ischemia

Question

1. (A) After reading Doreen's initial patient history, what chemistry profile would the ER physician order on this patient? (B) What tests are included in this profile in your laboratory and what are the collection times?

Explanation

(A) The initial chemistry order would include a cardiac profile that includes troponin I or troponin T, CK-MB, total CK, and possibly myoglobin. (B) The current recommendation to rule out acute myocardial infarction is the cardiac profile (CK-MB and troponin) at: (1) admission, 6–9 hours; and (2) 12–24 hours (Burtis & Ashwood, 2006).

Question

2. What laboratory results in Tables 1 and 2 are abnormal?

⊙ TABLE 1

Chemistry Results

	Day 1	Day 4	Day 6	Day 9	Day 11	Reference Range or Desired Range
Sodium	136	130	139	143	153	135–145 mEq/L
Potassium	3.7	3.0	4.3	3.9	3.8	3.6–5.0 mEq/L
Chloride	94	103	107	113	114	98–107 mEq/L
CO_2		25.0	24.0	23	30.0	24.0–34.0 mEq/L
Anion Gap		2.0	8.0	7.0	9.0	10–20 mmol/L
Glucose	319	519	379	310	234	70–99 mg/dL
BUN	23	53	81	99	79	7–24 mg/dL
Creatinine	0.9	1.4	2.3	2	1.9	0.5–1.2 mg/dL
Calcium	9.7					8.5–10.5 mg/dL
Magnesium		1.1	1.7	1.9	2.0	1.3–2.5 mEq/L
Digoxin			2.60	1.12		0.80–2.00 ng/ml
Cholesterol	350					<200 mg/dL
Triglyceride	275					<150 mg/dL
Bilirubin	0.2					0.2–1.2 mg/dL
AST	76					5–40 IU/L
ALP	84					30–157 IU/L
Protein	7.3					6.0–8.4 g/dL
Albumin	4.1					3.5–5.0 g/dL
TSH				0.72		0.3– 3.0 ulU/ml

⊙ TABLE 2

Cardiac Profile

	Day 1 20:30	Day 2 5:06	Day 2 12:25	Day 2 20:35	Day 3 12:15	Day 4 7:15	Reference Range
CK	668	1383	3461	3743	2117	973	24–170 IU/L
CK-MB	47.1		146.6	93.0	24.8	12.1	0.0–3.8 ng/ml
Troponin I	36.6	184.0	5745.0	926.1			0.0–0.4 ng/ml

CK-MB Reference Range		Troponin I Reference Range	
0–3.8:	Normal	0–0.4 ng/ml:	No evidence of myocardial injury
3.9–10.4:	Borderline	0.5–2.0 ng/ml:	Mild elevation, possible myocardial injury
>10.4:	Significantly elevated	>2.0 ng/ml:	Significantly elevated, consistent with myocardial injury

Explanation

In Table I, the following laboratory values are abnormal: increased glucose, BUN, creatinine, AST and slightly decreased chloride on admission, and a decreased anion gap. In Table 3 CK, CK-MB, and troponin I are elevated.

Question

3. The laboratory results in Tables 1 and 2 indicate what condition?

Explanation

Laboratory results are indicative of an acute myocardial infarction (AMI).

Question

4. Heart muscle contains which CK isoenzyme(s)?

Explanation

Heart muscle contains both CK-MB and CK-MM isoenzymes.

Question

5. How many hours after a myocardial infarction would you find an elevated CK and CK-MB? How long would they remain elevated?

Explanation

CK is elevated 4 to 6 hours after an acute myocardial infarction, peaks 10 to 24 hours following infarction, and remains elevated 2 to 4 days. CK-MB rises 4–8 hours after infarction, peaks 12–24 hours, and returns to normal within 48 to 72 hours. CK-MB levels do not remain persistently elevated and may return to normal levels 24 hours after a minor infarct.

Question

6. Define CK RI. Calculate the CK-MB RI for all specimens with total CK and CK-MB.

Explanation

CK-MB RI compares the CK-MB mass concentration to the total CK activity as a percentage.

$$CK\text{-}MB\ RI = \frac{CK\text{-}MB\ (mass)}{Total\ CK\ (activity)} \times 100\%$$

For mass assays values exceeding 6% (method dependent) are indicative of a myocardial infarction (a myocardial source of CK-MB). The relative index represents a disproportionately high concentration of CK-MB consistent with cardiac necrosis.

$$Day\ 1\ (20{:}30) = \frac{47.1}{668} = 7.1\%$$

$$Day\ 2\ (12{:}25) = \frac{146.6}{3461} = 4.2\%$$

$$Day\ 2\ (20{:}35) = \frac{93.0}{3743} = 2.5\%$$

$$Day\ 3\ (12{:}15) = \frac{24.8}{2117} = 1.2\%$$

$$Day\ 4\ (07{:}15) = \frac{12.1}{973} = 1.2\%$$

Question

7. What is the cause of the elevated AST?

Explanation

AST is elevated following a myocardial infarction, but it is not specific for damage to the myocardium. Liver disease and skeletal muscle disease also cause an increase in AST.

Question

8. What other medical problem or condition does this patient have? What laboratory values support your decision?

Explanation

Kathy also has diabetes mellitus as indicated by the increased serum glucose and positive glucose and acetone in the urinalysis report.

Question

9. Do patients with the condition you described in the previous questions have a higher risk of myocardial infarction or stroke than the general population? Why or why not?

Explanation

Diabetics are at higher risk for MIs or strokes than the general population. Elevated triglycerides and decreased high-density lipoproteins (HDLs) contribute to the risk of AMI.

 CHAPTER 10

Case Summary: A 38-year-old woman presented for biopsy and excision of a recently discovered lump in her right breast. Along with her routine preoperative blood work, her physician requested a blood test for CA27.29. Subsequent requests for CA27.29 levels were made for her 2-week postop follow-up visit and at 1 year and 2 years following her surgery. Results are shown in the following table. All testing was performed by the same laboratory using the same immunoassay instrumentation and reagent systems. Reference range for CA27.29 using this test system is 0–40 U/mL. Between-run precision of the assay is 10% at a level of 30 U/mL and 5% at a level of 90 U/mL.

Sample	Result (U/mL)
Preop	602
2 weeks postop	38
1 year postop	41
2 years postop	60

Question

1. Why did the values fall after surgery?

Explanation

The postsurgery results for CA 27.29 were much lower than the preop results because the source of CA 27.29 was removed via surgery.

Question

2. Which value(s) indicate significant changes from the previous value?

Explanation

The result that reflects a significant change from the previous value is the 2-year postop result of 60 U/mL. An increase in CA27.29 postop may indicate the recurrence of the malignant neoplasia.

Question

3. How would interpretation of these results be affected if the laboratory had changed instrument manufacturers during the time period represented?

Explanation

If the instrument and method changed during the testing of this patient over the time period indicated, then the interpretation may change because of a change in cutoff value or a change in antibody specificity that may result in lower or higher values than previous results.

CHAPTER 11

Case Summary: Dave C., an 80-year-old male, was admitted through the emergency room with difficulty breathing, coughing, and chest pain.

Chemistry Results

	2/18	2/19	Reference Range
Sodium	141	140	136–145 mEq/L
Potassium	4.4	4.7	3.6–5.0 mEq/L
Chloride	107	106	101–111 mEq/L
CO_2	27.0	30.0	24.0–34.0 mEq/L
Anion gap	5.0	5.0	10–20 mmol/L
Glucose	95	112	70–99 mg/dL
Bilirubin total	0.3		0.2–1.2 mg/dL
Troponin I	<0.3	<0.3	0.0–0.4 ng/ml
AST	10		5–40 IU/L
ALP	42		30–157 IU/L
Protein	6.5		6.0–8.4 g/dL
BUN	48	49	7–24 mg/dL
Creatinine	1.7	1.8	0.5–1.2 mg/dL
Calcium	8.8		8.5–10.5 mg/dL
Albumin	3.6		3.5–5.0 g/dL
Digoxin	1.06	0.82	0.00–2.00 ng/mL
CK-MB	3.1	3.2	*

*0–3.8 ng/mL = normal; 3.9–10.4 = borderline;
>10.4 = significantly elevated

Question

1. What are Dave's abnormal test results?

Explanation

Dave's abnormal chemistry results are elevated BUN and creatinine.

Question

2. Do the laboratory results rule out a myocardial infarction?

Explanation

Lab results including normal CK-MB and troponin I provide no evidence of a myocardial infarction.

Question

3. What is Dave's BUN:creatinine ratio? What is a normal BUN: creatinine ratio?

Explanation

The BUN:creatinine ratio is BUN ÷ creatinine = 48 ÷ 1.7=28.2 on 2/18 and 49 ÷ 1.8 = 27.2 on 2/19. A normal BUN:creatinine ratio is between 12:1 and 20:1.

Question

4. What type of azotemia does this indicate?

Explanation

A significantly elevated BUN with a normal or slightly elevated creatinine is characteristic of prerenal azotemia.

Question

5. (A) Where does this type of azotemia originate? (B) List five conditions associated with this type of azotemia.

Explanation

(A) Prerenal azotemia results from a decreased renal blood flow. Decreased blood flow to the kidneys leads to less BUN filtered and excreted in the urine and a higher serum BUN.
(B) Conditions associated with prerenal azotemia include: congestive heart failure (CHF), shock, hemorrhage, dehydration, a high protein diet, increased protein catabolism in fever, major illness, and stress. Heart failure is fairly common (10:1000 people), and the incidence increases with age.

Question

6. What is the most likely diagnosis in this case?

Explanation

The most probable diagnosis in this case based on presenting symptoms and laboratory results is CHF. He is also on digoxin, which provides further evidence.

CHAPTER 12

Case Summary: A 42-year-old Caucasian male presented to the emergency department complaining of malaise, tiredness ("just no energy") and mild pain described as coming from "my bones." The physician requested an initial set of laboratory tests based on a complete history and physical examination. Results of the initial laboratory tests are shown in the following table.

Serum Chemistries

	Results (Conventional Units)	Reference Interval (Conventional Units)
Sodium*	122 mmol/L	136–145
Potassium*	3.5 mmol/L	3.5–5.1
Chloride	101 mmol/L	98–107
Total carbon dioxide	20 mmol/L	23–29
Anion gap	1 mmol/L	7–16
Glucose	125 mg/dL	74–106
Creatinine	0.8 mg/dL	0.9–1.3
Urea nitrogen	8 mg/dL	6–20
Calculated osmolality	254 mOsm/Kg	N/A
Calcium	6.0 mg/dL	8.8–10.2

(Continued)

	Results (Conventional Units)	Reference Interval (Conventional Units)
Phosphate	1.3 mg/dL	2.4–4.4
Magnesium	1.8 mg/dL	1.6–2.6
Triglyceride	300 mg/dL	<150
HDL-C	15 mg/dL	27–67
LDL-C	50 mg/dL	87–186
Total protein	12.5 g/dL	6.4–8.4
Albumin	1.0 g/dL	3.5–5.2
A:G	0.09 mg/dL	1.0–2.2

*measured using an indirect measuring ISE

Hematology Results

WBC count	$12.5 \times 10^3/\mu L$	$4.0–11.0 \times 10^3/\mu L$
RBC count	$3.0 \times 10^6/\mu L$	$4.70–6.10 \times 10^6/\mu L$
Hemoglobin	11.0 g/dL	13.0–18.0
MCV	93.0 fL	82.0–99.0
MCH	33.0 pg	27.0–34.0
MCHC	36.0%	32.0–36.0
Platelet count	200,000 μL	140,000–440,000 μL
Neutrophils	68%	39–78
Bands	2%	0–12
Lymphocytes	20%	15–56
Monocytes	4%	2–14
Plasma cells	6%	0–1
Anisocytosis	Mildly elevated	
Rouleaux	Markedly elevated	

Question
1. What is (are) this patient's most striking laboratory result(s)?

Explanation
This patient has several abnormal test results for a variety of analytes. The most significant abnormal tests are:

- a peripheral blood smear with 6% plasma cells and marked Rouleaux;
- markedly decreased serum sodium and albumin concentrations;
- markedly decreased calculated anion gap, osmolal, and A:G ratios; and
- markedly increased total protein.

Question
2. What disease(s) or condition(s) could account for these abnormal results?

Explanation
An elevated total protein with a large number of plasma cells suggests a hyperproteinemia-related disease such as multiple myeloma. Also,

the technologist should be aware of the extremely low serum sodium and that high protein concentrations may result in spuriously low sodium values.

Question
3. Why is the patient's anion gap decreased?

Explanation
The anion gap is low because of the low serum sodium value.

Question
4. What additional test(s) should be ordered to support presumptions made in question 2?

Explanation
The clinician should request a serum protein electrophoresis, quantitative immunoglobulins to support presumptive diagnosis of multiple myeloma, and serum osmolality because of the hyponatremia. Also, the laboratory technologist should repeat the electrolytes, especially the sodium, and use a direct-measurement procedure.

Case Summary: Based on the results of the initial laboratory tests, history, and physical, the clinician requested additional laboratory tests to confirm the diagnosis of multiple myeloma. The laboratory staff also completed confirmation tests of the patient's specimen to rule out the possibility of pseudohyponatremia. The results of the additional laboratory tests are shown in the following table.

Serum Chemistries

	Results (Conventional Units)	RI (Conventional Units)
Measured serum osmolality	310 mOsm/kg	275–295
Osmolal gap	56 mOsm/kg	
SPE	IgG kappa monoclonal gammopathy	
IgG	5000 mg/dL	700–1,600
IgA	56 mg/dL	70–400
IgM	39 mg/dL	40–230

Laboratory Confirmation Tests*

Sodium	141 mmol/L
Potassium	3.5 mmol/L
Chloride	100 mmol/L
Total CO_2	21 mmol/L
Anion gap	20 mmol/L
Recalculated osmolal	291 mOsm/kg
Recalculated osmolal gap	18 mOsm/kg

*Sodium and potassium were assayed using a direct ISE.

Question

5. Was the laboratory's course of action appropriate? Explain?

Explanation

The serum protein electrophoresis (SPE) technique produced an abnormal amount of paraprotein of the IgG kappa type. The quantitative IgG also was markedly elevated. The measured serum osmolality by freezing-point depression osmometry is elevated. The sodium concentration was measured by a direct ISE, and the result was within the reference range. A recalculated osmolality was within reference interval, and the osmolal gap was significantly reduced.

Question

6. What is the explanation for the falsely low sodium value?

Explanation

A falsely low serum sodium value can occur because of an abnormally high concentration of proteins or lipids in the sample. This is the result of using an indirect method to measure sodium in serum or plasma and is referred to as the *electrolyte exclusion effect*.

CHAPTER 13

Case Summary: A 42-year-old male Caucasian enters the emergency department of a local hospital. He is confused, disoriented, sleepy, and combative, and his breath has a fruity odor and smells like acetone. The emergency room staff establishes intravenous and intraarterial lines. An emergency room physician requests blood samples to be drawn for glucose, electrolytes, serum acetone, and arterial blood gas. Results of the laboratory tests are shown in the following table.

Routine Chemistries		
Tests	Results	Reference Intervals
Sodium	125 mmol/L	135–145
Potassium	3.4 mmol/L	3.5–5.1
Chloride	90 mmol/L	98–107
Total CO_2	10 mmol/L	22–28
Anion gap	25 mmol/L	7–16
Glucose	600 mg/dL	74–106
Urea nitrogen (BUN)	40 mg/dL	6–20
Creatinine	3.5 mg/dL	0.7–1.3
Acetone	positive	Negative
Arterial Blood Gases (ABGs)		
pH	7.30	7.35–7.45
PCO_2	25 mmHg	35–45
Bicarbonate (HCO_3^-)	16 mmol/L	18–23
Base excess (deficit)	−2	(−2)–(2)

Question

1. What are the most striking abnormal test results?

Explanation

Sodium, potassium, chloride, and bicarbonate are below normal. Glucose is markedly elevated, urea nitrogen and creatinine are elevated, pH is acidic, anion gap is elevated, acetone is positive, and pCO_2 is below normal.

Question

2. Is this patient in acidosis or alkalosis?

Explanation

The blood pH value is below normal, therefore the patient is in acidosis.

Question

3. Is the acid–base imbalance metabolic or respiratory?

Explanation

The acidosis is metabolic in nature because the bicarbonate is below normal, hence primary bicarbonate deficit, and the PCO_2 is below normal.

Question

4. What is the possible cause of the acid–base imbalance based on the history provided and laboratory findings?

Explanation

The cause of the metabolic acidosis is a result of diabetes mellitus (markedly elevated glucose and positive acetone due to the presence of ketones. This is a good example of a patient with diabetic ketoacidosis (DKA) and kidney dysfunction as shown by the elevation in both urea nitrogen and creatinine. The blood sodium and potassium are below normal because of polyuria and co-excretion of these cations with acetoacetate and β-hydroxybutyric acid (ketones).

Case Summary: A second set of ABGs were drawn and the results are shown in the following table.

pH	7.34
PCO_2	34 mmHg
HCO_3^-	18 mmol/L
Base excess (deficit)	−1 mmol/L

Question

5. Why have these results begun to change?

Explanation

The body is beginning to compensate for the low blood pH.

Question

6. Specifically what is the body doing to cause these results to shift?

Explanation

The lungs attempt to compensate for the reduced pH by stimulating the respiratory mechanism and causing hyperventilation, which results in the elimination of H_2CO_3 as CO_2, a decrease in PCO_2, and consequently a decrease in $cdCO_2$. The kidney increases the excretion of acid and conservation of base (increases the rate of Na^+-H^+ exchange), increases ammonia formation, increases reclamation of HCO_3^-, and secretes H^+ and Cl^-.

Question

7. What modified condition does this patient have?

Explanation

This patient has partially compensated metabolic acidosis.

Case Summary: A final set of ABGs were drawn from the patient, and the results are shown in the following table.

pH	7.45
PCO_2	40 mmHg
HCO_3^-	22 mmol/L
Base excess	0 mmol/L

Question

8 What has happened to this patient by this point in time?

Explanation

The patient's ABGs have returned to normal.

Question

9 What is the patient's condition now?

Explanation

The patient is considered fully compensated at this time.

CHAPTER 14

Case Summary: A 55-year-old Caucasian female underwent a routine physical that was unremarkable in all aspects. Her clinician requested several laboratory tests, and the results of key tests are shown in the following table.

Question

1. Which test(s) are outside the RI?

Explanation

The serum total carbon dioxide level is slightly decreased, creatinine is slightly increased, total serum calcium is increased, and serum intact parathyroid hormone (PTH) is significantly increased.

Question

2. What are some plausible explanations for these abnormal results?

Explanation

This patient is hypercalcemic, and a myriad of causes for this clinical condition are highlighted in the text. The slight increase in creatinine could be a random occurrence for the patient, but the clinician may want to investigate this further to exclude a secondary renal condition. A low total carbon dioxide level is consistent with a mild form of metabolic acidosis. An increase in PTH can lead to decreased urinary hydrogen-ion excretion and an increase in bicarbonate excretion at the level of the renal tubules, thereby resulting in a mild acidosis.

A significant elevation of PTH represents a serious clinical condition involving the parathyroid gland. The challenge for the clinician is to determine the cause of the increased PTH. The usual approach is to distinguish between hypercalcemia caused by malignancy versus hyperparathyroidism. In most cases of malignancy-associated hypercalcemia, the production of PTH by the parathyroid gland is suppressed, so serum PTH levels are low. Patients with hypercalcemia resulting from primary hyperparathyroidism (PHPT) will present with elevated total serum calcium and serum PTH.

Question

3. What is the probable diagnosis for this patient?

Explanation

The most likely diagnosis is PHPT because of the presence of an adenoma on one or more parathyroid glands.

Question

4. How can this laboratory facilitate the management of this patient?

Tests	Patient's Results	RI (Conventional Units)	RI (SI Units)
Sodium	142 mEq/L	136–145 mEq/L	136–145 mmol/L
Potassium	4.0 mEq/L	3.8–5.0 mEq/L	3.8–5.0 mmol/L
Chloride	108 mEq/L	98–110 mEq/L	98–110 mmol/L
TCO_2	23 mEq/L	24–30 mEq/L	24–30 mmol/L
Glucose	96 mg/dL	70–110 mg/dL	3.9–6.1 mmol/L
Urea nitrogen	22 mg/dL	8–23 mg/dL	2.9–8.2 mmol/L
Creatinine	1.3 mg/dL	0.6–1.2 mg/dL	53–106 μMol/L
Calcium, total	11.5 mg/dL	9.2–11.0 mg/dL	2.3–2.7 mmol/L
PTH, intact	110 ng/L	10–65 ng/L	10–65 pg/mL

Explanation

The clinician will have a Tc-99m sestamibi scintigraphy procedure performed and a surgical neck exploration completed. This surgical procedure will serve to localize and remove any abnormal parathyroid gland(s). The laboratory can provide intraoperative PTH testing, which facilitates the surgical management of this patient. A partial protocol for testing during neck exploration surgery is shown in Table 14-12.

Question

5 What is the outcome of this patient?

Explanation

The patient had postsurgical calcium testing done to ensure that the serum total calcium levels were within the reference interval and that all of the tumors were removed during the surgical procedure. The patient's total calcium returned to normal, and she was discharged.

CHAPTER 15

Case Summary: A 21-year-old white female involved in a motor vehicle accident is admitted to the emergency department. The patient had multiple injuries to soft tissue, fracture of ankle, pneumothorax, subarachnoid hemorrhage, and contusion of brain. CT scan of the abdomen revealed a mass near the kidney. Fourteen days after the accident, the patient developed severe abdominal pain during a physical therapy session. Her family revealed that the patient had a previous history of hypertension. When questioned, the patient admitted to having attacks of intermittent headache, palpitation, and anxiety some months earlier, each lasting for a few minutes. Another CT scan showed the same mass in the same area. Laboratory results are shown in the following table.

Tests	Laboratory Results	Reference Interval
Plasma epinephrine	300 pg/mL	20–97 pg/mL
Plasma morepinephrine	10,285 pg/mL	125–310 pg/mL
Metanephrine	3850 μg/day	0.3–0.9 μg/day
VMA	15.5 mg/day	1.5–7.5 mg/day
Sodium	141 mEq/L	136–142 mEq/L
Potassium	4.3 mEq/L	3.5–5.0 mEq/L
Cortisol (a.m.)	16 μg/dL	6–21 μg/dL

Question

1. Based on the patient's symptoms and available history, what is your presumptive diagnosis? What tests would you order to confirm your diagnosis?

Explanation

One should suspect pheochromocytoma. Order catecholamines and metabolites: epinephrine, norepinephrine, metanephrine, and vanillylmandelic acid (VMA) in urine. Other less likely disorders are primary aldosteronism and Cushing's disease. Order sodium and potassium to rule out primary aldosteronism and order cortisol to rule out Cushing's.

Question

2. Comment on the relative usefulness of catecholamines and metabolites in the diagnosis of this patient's condition.

Explanation

Sample collection is extremely important. For example, 24h urine may be more superior to blood sampling because single collection may miss the time of paroxysmal attack. Assays of plasma epinephrine and norepinephrine are specific and reliable in the diagnosis of pheochromocytoma and also allow assessment of tumor location, although they are somewhat difficult to perform. Urine metanephrine is simpler to perform and usually gives valid results. Urinary VMA is nonspecific and is not a definitive diagnostic test.

Question

3. What would you recommend as the best management for this entity?

Explanation

Surgical removal of the tumor is the method of choice.

Question

4. What is the prevalence of this condition as a cause of hypertension? How often is the tumor malignant? What are the associated familiar disorders?

Explanation

About 0.1%. Between 70% and 90% of the tumors are located within the adrenals, more often on the right than the left. In about 10–20% of cases, tumors are bilateral, and about 10% are extra-adrenal neoplasm. About 5–10% of the tumors are malignant.

Associated familial disorders include the following.

1. Multiple endocrine neoplasia (MEN) type IIa, also known as Sipple's syndrome: There is the concurrence of pheochromocytoma, medullary carcinoma of the thyroid, and parathyroid hyperplasia or adenoma.
2. MEN IIb: There is the concurrence of pheochromocytoma, medullary carcinoma of the thyroid, and mucosal neuromas.

CHAPTER 16

Case Summary: A 37-year-old women developed renal colic at age 23. One year later, she was diagnosis with hyperparathyroidism, and a three-gland parathyroidectomy was performed when she was 25 years old. In her 25th year, she was diagnosed with MEN type 1 (MEN1). It was later discovered that she had a family history of MEN1. One year later, she began to develop symptoms, including abdominal pain, diarrhea, nausea, vomiting, and weight loss.

Clinical Laboratory Results

Analytes	Results	Reference Interval/Cutoff Value
Basal acid output (BAO)	56 mEq/Hr	<10 mEq/Hr
Gastrin (fasting serum)	555 pg/mL	<100 pg/mL
Secretin test	*971 pg/mL	*<200 pg/mL

*Increases with secretin

Nonclinical Diagnostic Tests

The following tests were performed: upper GI endoscopy procedure, biopsy of a duodenal nodule, adnominal imaging studies, and somatostatin receptor scintigraphy.

Results of These Procedures

All of these procedures revealed duodenal gastrinoma and pancreatic lesions.

Question

1. Based on the history provided, identify clinical laboratory tests that may provide valuable information to the clinician in determining a correct diagnosis.

Explanation

Fasting serum gastrin, secretin test, and gastric fluid for a basal acid output (BAO) tests may be revealing.

Question

2. Indicate other nonclinical laboratory tests or procedures that, in combination with the clinical laboratory tests, will provide the clinical with enough information to make the diagnosis.

Explanation

Also perform endoscopy, imaging studies, and scintigraphy.

Question

3. What is the most probable diagnosis?

Explanation

MEN-1 Zollinger–Ellison syndrome is the most probable condition.

ⓔ CHAPTER 17

Case Summary: John L., a 50-year-old man, was seen in the emergency room with severe epigastric pain. He also complained of nausea and vomiting and had a history of alcoholism. The following chemistry tests were ordered.

Chemistry Tests

		Reference Range
Sodium	143	135–145 mEq/L
Potassium	3.9	3.6–5.0 mEq/L
Chloride	106	98–107 mEq/L
CO_2	31.0	24.0–34.0 mEq/L
Glucose	95	70–99 mg/dL
Bilirubin, total	1.4	0.2–1.0 mg/dL
AST	90	5–40 IU/L
ALP	200	30–157 IU/L

Protein	6.8	6.0–8.4 g/dL
BUN	40	6–20 mg/dL
The Following Day		
Calcium	8.1	8.5–10.5 mg/dL
Albumin	3.9	3.5–5.0 g/dL
ALT	100	5–40 IU/L
Amylase	838	10–110 IU/L
Lipase	2500*	31–186 IU/L
GGT	335	13–86 IU/L

Lipemic sample; results verified
CBC: normal, hematocrit stable

Question

1. Circle or highlight the abnormal laboratory values.

Explanation

The following chemistry results are elevated: amylase, lipase, AST, ALP, and γGT. Calcium level is slightly decreased, and bilirubin in slightly elevated.

Question

2. What two organ systems appear to be involved? Support your answer.

Explanation

The liver and pancreas are the two organ systems that appear to be involved. The pancreas is evidenced by the increased amylase and lipase. Liver involvement is established by the increased liver enzymes: AST, ALT, ALP, and GGT.

Question

3. (A) John's profile is indicative of what condition? (B) Briefly describe the pathogenesis involved (origin and development of disease).

Explanation

(A) The markedly elevated amylase (AMY) and lipase (LPS) are indicative of acute pancreatitis. Acute pancreatitis is estimated to be the cause of 1 out of 500 acute admissions to the hospital and is most prevalent in the 50 and over age group.

(B) Acute pancreatitis is an acute response to tissue necrosis caused by digestive enzymes (proteolytic and lipolytic) released from exocrine pancreatic cells. In the pancreas digestive enzymes are stored in the inactive form as zymogen granules. Free enzymes are not normally present in the pancreas; they are only activated on entering the duodenum. In acute pancreatitis the enzymes are activated prematurely in the pancreas leading to autodigestion of pancreatic tissue. The activation of protrypsin to trypsin leads to a proteolytic attack on the gland's structure and vessels.

Question

4. (A) What are the two most common causes or etiologies of this condition? (B) List three other less common causes.

Explanation

(A) The two main etiologies of acute pancreatitis are (1) biliary tract disease or obstructive liver disease (e.g., cholelithiasis—gallstones leading to cholecystitis, which spreads to the pancreas) and (2) alcoholism. They account for 75% of all cases in the United States.

(B) Other less common causes of acute pancreatitis include the following:

- drugs (e.g., ACE inhibitors, asparaginase, furosemide, sulfa drugs, valproate),
- infections (e.g., Coxsackie B virus, cytomegalovirus, mumps),
- inherited genes (e.g., multiple known gene mutations, including a small percentage of cystic fibrosis patients),
- mechanical or structural problems (e.g., trauma, pancreatic or peri-ampullary cancer, sphincter of Oddi stenosis),
- metabolism (e.g., hypertriglyceridemia and hypercalcemia, including hyperparathyroidism),
- toxins (e.g., alcohol, methanol), and
- other (e.g., pregnancy, postrenal transplant).

Question

5. List two factors that increase the risk of this condition.

Explanation

Two factors that increase the risk of this condition are smoking and diets high in fat and protein.

Question

6. Which two enzymes are critical to evaluate the status of the exocrine function of the pancreas?

Explanation

AMY and LPS are the two most common enzymes used to evaluate the exocrine function of the pancreas.

Question

7. Is hypocalcemia associated with this condition? Why or why not?

Explanation

Hypocalcemia is associated with acute pancreatitis. The elevated lipase increases fat digestion and necrosis, resulting in increased free fatty acids in abdominal adipose tissue. The fatty acids bind calcium as they form free fatty acid salts, decreasing the levels of serum calcium.

Question

8. Discuss two reasons for the elevated liver enzymes (AST, ALP, and GGT).

Explanation

Liver enzymes are elevated in acute pancreatitis and liver disease. GGT is elevated in all types of liver disease but highest in intrahepatic or posthepatic biliary obstruction, including gallstone obstruction. In acute or chronic pancreatitis, enzyme activity may be 5 to 15 times the upper limit of normal if it is associated with hepatobiliary obstruction. In acute pancreatitis, the AST and ALT levels are slightly to moderately elevated.

Question

9. How many of Ranson's indicators does John have, if any?

Explanation

John does not have any of Ranson's indicator of severity of acute pancreatitis.

Question

10. What is his prognosis?

Explanation

John's prognosis is excellent. With fewer than three risk factors, the mortality rate is less than 1%.

CHAPTER 18

Case Summary: A 35-year-old male presented to the emergency department complaining of chest pain. The chest pain started at about 11:30 p.m. The pain was "kind of sharp," and pressure was continuous. The patient said that he had no shortness of breath, sweating, or nausea. The physician ordered an electrocardiogram, blood chemistries, and a complete blood count (CBC). The laboratory results are shown in the following table.

Blood Chemistries

Tests	Patient Results	Reference Values
Troponin I	16.2 ng/mL	<1.5 ng/mL
Total creatine kinase	1581 U/L	20–200
CK-MB	9.3 μg/L	<5
CK-index	0.6%	<4
Sodium	138 mmol/L	136–145
Potassium	3.6 mmol/L	3.5–5.1
Chloride	106 mmol/L	98–107
Bicarbonate	23 mmol/L	23–29
Creatinine	1.1 mg/dL	0.62–1.10
BUN	22 mg/dL	6–20
Glucose	132 mg/dL	74–100
Magnesium	1.6 mg/dL	1.6–2.6
Phosphorus	2.8 mg/dL	2.5–4.5
Calcium	8.8 mg/dL	8.6–10.2

(Continued)

CBC		
WBC	7.0×10^9/L	4–10
RBC	5.0×10^{12}/L	4.6–6.1
Hemoglobin	13.7 g/dL	13.5–18
Hematocrit	38.7%	41–53
Platelet count	234×10^9/L	150–400
Neutrophils	60%	45–66
Lymphocytes	20%	10–23

EKG Results

ST elevation over the precordial leads which indicates that the patient had an acute myocardial infarction.

Question

1. Identify specific diseases or conditions that would result in an elevated serum troponin I?

Explanation

The following conditions may elevate serum troponin I:

- trauma
- contusion
- ablation
- cardiac pacemaker response
- cardioversion
- endomyocardial biopsy
- cardiac surgery
- hypertension
- hypotension (often with arrhythmias)
- cerebrovascular accident
- rhabdomyolysis (cardiac injury)
- postop noncardiac surgery
- renal failure
- diabetes mellitus
- myocarditis
- pulmonary embolism
- sepsis
- inflammatory disease (e.g., myocarditis, parvovirus B19)
- percutaneous coronary intervention without complications
- drug toxicity (e.g., adriamycin, 5-fluorouracil, Herceptin, and snake venoms)
- heart failure (e.g., congestive heart failure)
- aortic valve disease
- cocaine-induced rhabdomyolysis
- chest injury from motor vehicles accident
- pulmonary conditions, diseases, and syndromes

Question

2. What is the current acceptable emergency department medical protocol [World Health Organization (WHO), American College of Cardiology (ACC), European Cardiology Society (ECS)] for diagnosing and treating an acute myocardial infarction?

Explanation

For WHO:

- history of characteristic chest pain
- diagnostic changes in the EKG
- changes in serum enzyme levels

For ACC and ECS:

- promote troponin to a pivotal role
- relegate CK-MB to a secondary role
- eliminate the need for total CK measurement

Question

3. Identify additional diagnostic tests that might be useful for making a diagnosis and establishing a treatment plan.

Explanation

Serial CK-MB determinations may be helpful as well as an echocardiogram.

The physician ordered serial measurements of troponin I and an echocardiogram. The echocardiogram is a procedure that uses sound waves to show motions of the heart (for example, chamber wall motions).

Laboratory results of serial troponin I (ng/mL) measurements are shown in the following table.

Day	Time (hour)	cTnI Concentrations
Day 1	1337	16.2
Day 1	1948	11.9
Day 2	0339	11.0
Day 2	1220	10.9
Day 2	2020	7.6
Results of echocardiogram: negative for wall motion abnormality.		

Question

4. Do the results for serum troponin I follow a reasonable pattern of concentration versus time?

Explanation

Yes, serum troponin I levels begin to rise 4–6 hours after chest pain and return to normal within 4–10 days.

Question

1. Is it reasonable to assume that this patient had an acute myocardial infarction?

Explanation

Yes, it is reasonable to assume the patient had an AMI because the EKG and serum levels of troponin were indicative of an AMI.

CHAPTER 19

Case Summary: Tom N., a 60-year-old man, came into the emergency room complaining of fatigue and overall flulike symptoms. He was mildly jaundiced with icteric sclera.

Liver Profile

	2/15	2/22	Reference Range
Bilirubin, total	2.2	2.4	0.2–1.2 mg/dL
Bilirubin, direct	1.19		0.00–0.40 mg/dL
AST	564	735	5–40 IU/L
ALP	250		30–157 IU/L
ALT	373		5–40 IU/L
Albumin	2.1	1.9	3.5–5.0 g/dL
Total Protein	6.5	6.4	6.0–8.4 g/dL
Ammonia	32	48	20–80 μg/dL
Folic Acid	17.8		2.9–15.6 ng/dL
Vitamin B$_{12}$	377		180–710 pg/ml

Hepatitis Serum Panel

IgM anti-HAV		Nonreactive
HbsAg		Nonreactive
IgM anti-HBc		Nonreactive
Anti-HCV		Reactive

Question

1. Circle or highlight the abnormal results.

Explanation

Tom's abnormal chemistry results are increased total bilirubin, direct bilirubin, AST, ALT, and ALP, and decreased albumin. Abnormal CBC results include decreased WBC, RBC, Hgb, Hct, and platelet count and increased MCH.

Question

2. What organ system is involved?

Explanation

Liver disease is the most likely explanation for the abnormal results.

Question

3. List six probable explanations for this chemistry profile.

Explanation

The following categories of liver disease are probable causes:

a. acute viral hepatitis (HAV , HBV, HBV, HDV, HEV)
b. chronic hepatitis
c. cirrhosis
d. alcoholic liver disease
e. primary biliary cirrhosis
f. hepatic tumors or cancer

Question

4. What is the most likely explanation for these results?

Explanation

The most likely explanation is hepatitis.

Question

5. What is the most likely diagnosis on this patient?

Explanation

The serology results indicate hepatitis C. The fatigue, flulike symptoms and jaundice are consistent with viral hepatitis.

Question

6. List five main etiologic factors (causes) for this infection.

Explanation

Five main etiologic factors for HCV infection are:

1. blood products or transfusion (32.3%),
2. intravenous drug abuse (16.0%),
3. sexual partner with HCV infection (4.2%),
4. body piercing or tattooing (3.6%), and
5. professional contact with HCV-contaminated person or material (medical personnel) (3.3%).

Unknown factors constitute 40.6%.

Question

7. List six groups that have higher than normal risk of infection (high-risk groups).

Explanation

Among groups at increased risk of HCV are:

1. drug users who share needles and syringes;
2. people who undergo skin-penetrating procedures such as tattooing, body piercing, and nail manicuring;
3. people who received blood or blood products before 1992;
4. health-care workers (exposure);
5. kidney dialysis patients;
6. hemophiliacs;
7. military personnel; and
8. prisoners.

Question

8. Describe the pathogenesis (i.e., what happens?).

Explanation

Acute hepatitis results in acute injury to the hepatocytes, which leads to marked elevations of AST and ALT, slight elevations in ALP, and normal or slightly decreased albumin. An increased prothrombin time (not performed on this patient) and a decrease in albumin indicates a more serious disease and prognosis.

Chronic infection occurs in 50–80% of all cases and is often asymptomatic. Chronic hepatitis C is responsible for 40% of all chronic liver disease among nearly 40 million people in the United States. It is also accountable for 20–30% of all liver transplantations and more than 8,000 deaths annually.

Question

9. How can the physician differentiate between active and chronic forms of this infection?

Explanation

The physician can distinguish based on knowledge of clinical history (time from the onset of symptoms). A previously positive anti-HCV test with more than 6 months between tests is considered a change from acute to chronic infection.

CHAPTER 20

Case Summary: A 40-year-old male Caucasian presented with the following medical scenario. The patient was not feeling well for some time. He noted a gradual onset of fatigue, decreased libido, and erectile dysfunction (ED). He also complained of dry mouth and polyuria and noted some loss of muscle mass. He did not have shortness of breath, a cough, a fever, night sweats, or visual changes. The patient had developed arthralgias in his ankles. No hepatomegaly. His skin was tanned (hyperpigmented).

He had a previous history of pulmonary sarcoidosis that was treated successfully. The patient drank 5–12 alcoholic beverages per week and never smoked. His father died of heart disease at the age of 46, and his brother is healthy at this time. The patient's vital signs and physical evaluation were all relatively normal.

The results of clinical laboratory tests follow. (*Note:* Hematology test results were all within normal limits.)

Serum Chemistries	Results (Conventional Units)	RI (Conventional Units)
Sodium	144.0	136–145 mEq/L
Potassium	4.4	3.5–5.1 mEq/L
Chloride	101	98–107 mEq/L
Bicarbonate	7	23–29 mEq/L
Anion gap	36	6–10 mEq/L
Glucose	300	74–100 mg/dL
Creatinine	1.0	0.9–1.3 mg/dL
Glycated hemoglobin	7.1	<6.0 %
Urea	11.2	6–20 mg/dL
Albumin	4.1	3.3–4.1 g/dL
ALP	125	40–115 U/L
ALT	125	10–55 U/L
AST	97	10–40 U/L
Bilirubin		
Total	0.8	0.0–1.0 mg/dL
Direct	0.3	0.0–0.4 mg/dL

Iron	197	30–160 µg/dL
Iron-binding capacity	202	228–428 µg/dL
Transferrin saturation	97	<45%
Ferritin	4890.0	30–300 ng/ML
Testosterone	146	270–1070 ng/dL
Luteinizing hormone	1.2	2.1– 12.0 U/L (males)
Follicle-stimulating hormone	0.5	1.0–12.0 U/L (males)
Estradiol	6	10–50 pg/mL

Question

1. What laboratory tests, signs, symptoms, and physical findings are most striking with regard to this patient?

Explanation

The patient admitted to feeling fatigue and having decreased libido, ED, and polyuria. His skin was tanned (hyperpigmented). He also had symptoms associated with arthralgias. Significant laboratory tests included elevated glucose, glycosylated hemoglobin, iron, ferritin, transferrin saturation, iron-binding capacity, liver function, and sex hormone (which had lower than normal values).

Question

2. What disease, condition, or syndrome may be present based on the history and physical and laboratory data?

Explanation

This patient has hereditary (primary) hemochromatosis. The most common form of hereditary hemochromatosis is an autosomal recessive disorder associated with a mutation of the HFE gene on chromosome 6. Clinical symptoms include fatigue, malaise, weakness, arthralgias, loss of libido (impotence), and diabetes (polyuria). Signs of the disease are hepatomegaly, hyperpigmentation, testicular atrophy, hypogonadism, and arthritis. The patient's liver function tests were abnormal and without hepatomegaly.

Question

3. What key clinical laboratory tests provide the clinician with important diagnostic information?

Explanation

Clinical laboratory tests that are important for the diagnosis of hereditary hemochromatosis include total serum iron, iron-binding capacity, ferritin, and transferrin saturation.

Question

4. What is the course of action for this patient?

Explanation

Therapeutic management for patients with hereditary hemochromatosis includes a reduction in body iron and diet modifications. Regular therapeutic phlebotomy can remove approximately 200 mg of iron per unit of blood. Dietary restrictions of red meat, iron supplements, and alcohol consumption are also appropriate.

 CHAPTER 21

Case Summary: A 39-year-old male was seen in the emergency department following a possible grand mal seizure. The patient informed the physician that he had been taking therapeutic doses of phenobarbital and phenytoin for several months. The doctor ordered drug levels for each compound. A phlebotomist drew a blood sample in a red-top tube that contained serum separator gel. Results of the laboratory tests are shown in the following table.

Test	Result	Therapeutic Range
Phenobarbital	4.0 μg/ml	15–30 μg/ml
Phenytoin	3.5 μg/ml	10–20 μg/ml

The physician questioned the laboratory's results and asked for a repeat analysis.

Question

1. Are these results consistent with the patient's history?

Explanation

There are several questions regarding this patient's condition and the drug levels obtained by the laboratory that need to be answered in an effort to determine whether the results are consistent with the patient's history. Has the patient been taking his medication as prescribed by the physician? Is the dosage correct? Does the patient have any other underlying condition that may affect absorption and distribution of the drug?

Question

2. Are the laboratory results incorrect? If so, what is the source of the error?

Explanation

If the drug levels are an accurate reflection of the patient at that time, then laboratory error should be considered. What are possible sources of errors with these particular tests? Is there a problem with the analyzer—for example, improper calibration, operation, or quality control? There may be a problem with the particular methodology? Is there a problem in the preanalytical phase the analysis?

Question

3. Does the laboratory need to repeat the analysis?

Explanation

The case stated that the phlebotomist drew the sample into a red-top tube that contained serum separator gels. It has been shown that the concentration of phenobarbital and phenytoin in samples drawn in this type of tube may be affected by the presence of the gel material.

Question

4. What course of action should the laboratory follow to prevent future questioning of these assays?

Explanation

The laboratory should change its blood drawing requirements to exclude the use of serum separator tubes for the measurement of phenobarbital and phenytoin. The assays should not be repeated on the original tubes but a new sample should instead be obtained in a plain red-top tube.

 CHAPTER 22

Case Summary: A 54-year-old male was admitted to the emergency department. The patient was conscious, but his level of consciousness was diminished. He was neither alert nor coherent to oral questioning. He was responsive to pain stimulus. He did not have alcohol breath or a fruity odor on exhalation. The patient admitted that his vision was slightly blurred and that he was seeing double (diplopia). Slight nystagmus was evident. Patient experienced multiple episodes of emesis. Additional symptoms included cephalalgia, slurred speed, and unsteady gait. Lips and fingernails were bluish in color.

Vital signs are shown in the following table.

Pulse = 110 beats per minute	normal: ~80
Blood pressure = 100/74 mmHg	normal: 120/80
Respirations = 28 per minute	normal: 12–16

Results of the initial laboratory tests are shown in the following table:

Serum Chemistries	Results (Conventional Units)	RI (Conventional Units)
Sodium	135 mEq/L	136–145
Potassium	4.5 mEq/L	3.5–5.1
Chloride	108 mEq/L	98–107
Carbon dioxide	7 mEq/L	23–29
Anion gap	20 mEq/L	6–10
Glucose	162 mg/dL	74–100
Creatinine	1.5 mg/dL	0.9–1.3
Urea nitrogen	17 mg/dL	6–20
Calculated Osmolal	275 mOsm/Kg	282–300
Ethanol	<10 mg/dL	N/A
Acetaminophen	<2.5 μg/mL	N/A
Salicylate	<2.8 μg/mL	N/A
Hematology Results		
WBC	$11 \times 10^3/\mu m^3$	$4.0–11.0 \times 10^3$
R BC	$4.3 \times 10^6/\mu m^3$	$4.70–6.10\ 10^6/\mu m^3$

(Continued)

Hemoglobin	11 g/dL	13.0–18.0
Hematocrit	34.9%	39–50
Platelets	$110 \times 10^3/\mu m^3$	140–440

Urinalysis

Color	Amber	
Appearance hazy	clear	
pH = 6	5–6	
Specific gravity = 1.028	1.002–1.030	
All other dipstick results are negative		

Microscopic analysis:

Birefringent octahedral, enveloped-shaped calcium oxalate crystals

Serum Chemistries	Results (Conventional Units)	RI or Cutoff Values (Conventional Units)
Serum osmolality	372 mOsm/Kg	275–295 mOsm/Kg
(Using freezing-point depression osmometry)		
Osmol gap	97 mOsm/Kg	5–10 mOsm/Kg
cTnl	<0.05 μg/L	<0.05 μg/L
Volatiles		
Ethylene glycol	190 mg/dL	Neg
Methanol	<1.5 mg/L	<1.5 mg/L
Isopropanol	None detected	None detected
DAU screen	Negative for seven classes of abused drugs	

Question

1. What course of action should the clinician pursue?

Explanation

The clinician suspects the patient may have ingested an abused drug or volatiles other than ethanol based on the initial group of laboratory tests. The clinician tested the patient's urine for the presence of ethylene glycol by subjecting the specimen to ultraviolet light using a Wood's lamp.

Question

2. What is the possible origin of the calcium oxalate crystals in the urine?

Explanation

The urine oxalate crystals originate from the metabolic product oxalic acid. The identification of these crystals supports the diagnosis of ethylene glycol poisoning but does not confirm ethylene glycol exposure because there are several other reasons for finding oxalate crystals in the urine.

Question

3. What additional laboratory tests should be considered?

Explanation

The clinical laboratory does not offer additional volatile substance testing and must send out specimens to a reference laboratory. Therefore, the clinician orders volatile testing (reference laboratory); the following tests will be done in the hospital laboratory:

- Measured serum osmolality by freezing-point depression osmometry
- Drug of abuse urine [DAU] screen

The ED physician exposed a urine specimen provided by the patient to a Wood's lamp, and the specimen emitted a yellow-green color (i.e., it glowed). The physician suspected the presence of a chemical substance in the urine that might be ethylene glycol. Additional laboratory tests were requested and included the following:

Question

4. Explain the cause of the increased osmolal gap.

Explanation

The measured serum osmolal will be increased in the presence of volatiles (e.g., methanol, ethylene glycol, ethanol, and isopropanol) using freezing-point depression osmometer, not a vapor pressure osmometer.

Question

5. Why did the urine emit a yellow-green color when irradiated with ultraviolet light using a Wood's lamp?

Explanation

Many manufacturers of radiator fluid add fluorescein, a fluorescent dye, that will produce yellow-green color when subjected to ultraviolet light (e.g., from a Wood's lamp.)

Question

6. Is the ethylene glycol concentration representative of a toxic dose?

Explanation

Yes. Toxic concentrations of ethylene glycol are defined as >20.0 mg/dL, and "potentially fatal" concentrations are defined as >50.0 mg/dL.

Question

7. Why is ethylene glycol toxic to the human body?

Explanation

Ethylene glycol itself is relatively nontoxic but is metabolized into toxic metabolites (e.g., oxalic acid.)

Question

8. What is the treatment for ethylene glycol ingestion?

Explanation

Depending on the patient's condition, treatments may include but are not limited to hemodialysis or the administration of ethanol or fomepizol.

CHAPTER 23

Case Summary: A 25-year-old male presented to the hospital emergency department (ED) complaining of headaches, nausea, dizziness, and episodes of repetitive vomiting. The patient was conscious and denied having chest pain, shortness of breath, abdominal pain, diarrhea, and hematemesis. He also stated that he did not smoke, drink, or use illicit drugs.

While being interviewed further, the patient did reveal to the clinician that he had attempted suicide in the past and had a history of depression. On a daily basis, he was currently taking olazpine (Zyprexa, Eli Lilly Co.), which is an antipsychotic and benzodiazepine derivative drug; and venlafaxine (Effexor, Wyeth Laboratory), an antidepressant medication.

The patient also disclosed to the clinician that he had purchased more than 1 gram of the organic form of arsenic trioxide (Trisenox) in liquid form from an Internet auction site for less than $100. Trisenox is a chemotherapeutic agent used to treat various cancers such as leukemia. The drug was shipped to him in an unmarked clear plastic Ziploc bag. He dissolved the drug in water and consumed it about 12 hours before his ED admission.

A physical exam including head, eye, ear, nose, and throat revealed nothing abnormal, and his vital signs were normal except for an increased heart rate (tachycardia) of 112 beats/minute. Electrocardiogram revealed a sinus tachycardia; chest X-ray was unremarkable. His kidney–ureter–bladder X-ray, which included the abdomen, showed a high-density material within the distal area of the stomach.

Results of the initial laboratory tests are shown in the following table.

Serum Chemistries	Results (Conventional Units)	RI (Conventional Units)
Sodium	135	136–145 mEq/L
Potassium	4.5	3.5–5.1 mEq/L
Chloride	108	98–107 mEq/L
Carbon dioxide	7	23–29 mEq/L
Anion gap	20	6–10 mEq/L
Glucose	162	74–100 mg/dL
Creatinine	1.5	0.9–1.3 mg/dL
Urea nitrogen	17	6–20 mg/dL
Calculated osmolal	275	282–300 mOsm/Kg
Ethanol	<10 mg/dL	N/A
Acetaminophen	<2.5 μg/mL	N/A
Salicylate	<2.8 μg/mL	N/A
Urine drug abuse screen		
Opiates	Negative	
Methadone	Negative	
Amphetamines	Negative	
Barbiturates	negative	
Benzodiazepines	Positive	
Cocaine	Negative	
Phencyclidine	Negative	

Hematology Results		
WBC	11×10^9/L	$4.0–11.0 \times 10^9$/L
RBC	4.3×10^{12}/L	$4.70–6.10 \times 10^{12}$/L
Hemoglobin	11.0	13.0–18.0 g/dL
Hematocrit	34.9	39–50%
Platelets	110×10^9/L	$140–440 \times 10^9$/L
Urinalysis		

Color amber
Appearance hazy
pH = 6
Specific gravity = 1.028
All other dipstick results are negative
Microscopic analysis: Nothing abnormal

Question

1. What additional clinical laboratory tests should be considered by the physician?

Explanation

Trace-element testing should be done, including arsenic and a heavy metal panel (lead, thallium, mercury).

Question

2. Are the patient's signs and symptoms consistent with any trace-element poisoning?

Explanation

A significant amount of vomiting is consistent with certain trace-element poisoning.

Question

3. What laboratory instrumentation is used to test for trace-element content in biological samples?

Explanation

High-performance liquid chromatography should be followed by inductively coupled plasma–mass spectrometry or hydride-generation atomic–fluorescence spectrometry or atomic absorption spectroscopy.

The results of additional laboratory tests are shown in the following table.

	Patient's Results	RI/Cutoff Values
Serum arsenic concentration	<2.0	2–23 μg/L
24-hr urine arsenic concentration	9,950	<100 μg/g creatinine
Other trace elements whole-blood panel:		
Mercury	<1	1.0–59 μg/L
Thallium	<1	<5 μg/L
Lead	<1	<25 μg/dL
Hair arsenic concentration	6.5	<1 μg/g

Question

4. Why is the serum arsenic concentration so low?

Explanation

Serum arsenic can be low because of a delay in intestinal absorption or the anticholinergic effects of olanzapine. Arsenic trioxide also has a low water solubility, which makes GI absorption less efficient than dissolved arsenic.

Question

5. Why is the urine arsenic concentration so high?

Explanation

Urine levels of arsenic were markedly elevated because of the delayed absorption that resulted in large amounts of arsenic being eliminated via the kidneys. In addition, urinary excretion of arsenic is intermittent, so a 24h urine is the urine specimen of choice.

Question

6. What pharmacokinetic issues may be relevant to this patient?

Explanation

Normally, arsenic is readily absorbed by the GI tract and is dependent on which form or compound has been ingested. For example, arsenic trioxide (trivalent form) is less absorbed than more soluble trivalent and pentavalent compounds.

Question

7. Identify possible treatment modalities associated with arsenic poisoning.

Explanation

The treatment options available to the patient include (1) whole bowel irrigation with polyethylene glycol electrolyte solution, (2) colonic irrigation, (3) and administration of chelating agents such as BAL and succimer.

CHAPTER 24

Case Summary: This patient has multiple organ involvement, so a significant number of diagnostic tests, including clinical laboratory tests, were requested by the clinician to rule in and rule out various diseases and conditions.

A 65-year-old woman, a chronic alcoholic, suddenly collapsed at home and was transported to the emergency department. Her medical history included hypothyroidism, chronic renal impairment, and osteopenia. The patient complained of mild epigastric pain, with occasional radiation to the right costal margin. She denied chest pain, headache, or vertigo. She had shortness of breath (dyspnea) on examination. Her medications consisted of furosemide (a diuretic), potassium, captopril (an ACE inhibitor), and thyroxine. She had smoked for about 30 years.

Physical examination revealed a tired and lethargic-looking woman. Her weight was 130 pounds and her height 61 inches. Her blood pressure was 110/70 lying down and 90/60 sitting down with a pulse rate of

102 beats per minute. Other significant findings were crepitus (crackling or popping sound) in her chest and marked cardiomegaly with pulmonary edema.

Question

1. Identify several clinical observations relevant to the patient's history and physical symptoms and status.

Explanation

Among the observed symptoms and conditions were:

- dyspnea
- chronic alcoholic cirrhosis
- hypothyroidism
- osteopenia
- renal impairment
- tired and lethargic elderly woman
- hypotensive pulmonary edema

Question

2. List examples of diagnostic laboratory and nonlaboratory (e.g., X-ray) tests that may provide the clinician with useful data to support his or her clinical findings.

Explanation

The following laboratory tests may be of value to the clinician:

- ethanol
- BNP
- cardiac troponin
- serum thiamin
- serum red blood cell transketolase activity
- electrolytes

The following nonlaboratory tests also may be helpful:

- EKG
- cardiac
- chest X-ray
- transthoracic echocardiogram
- blood pressure

The results of clinical laboratory tests and additional diagnostic tests are shown in the following table.

Serum Chemistries	Results (Conventional Units)	RI (Conventional Units)
Sodium	144	136–145 mEq/L
Potassium	4.4	3.5–5.1 mEq/L
Chloride	101	98–107 mEq/L
Bicarbonate	7	23–29 mEq/L
Anion gap	36	6–10 mEq/L
Glucose	100	74–100 mg/dL
Creatinine	2.5	0.9–1.3 mg/dL

(Continued)

Urea	11.2	6–20 mg/dL
Lactic acid	88.3	1.8–18 mg/dL
Calcium	8.9	8.7–10.0 mg/dL
Phosphate	3.1	0.26–4.2 mEq/L
Magnesium	1.2	1.5–2.3 mg/dL
Albumin	3.1	3.3–4.1 g/dL
GGT	389	1–25 U/L
Ethanol	ND	Undetectable
ALP	250	40–115 U/L
ALT	1880	0–55 U/L
AST	7542	5–34 U/L
Lipase	125	<120 U/L
Troponin I	0.48	<0.01 ng/mL
BNP	458	<20 pg/mL
pH	7.10	7.35 –7.45
Serum thiamine	2.7	<25.3 mg/dL
Serum red blood cell transketolase activity	34	<5%

Other nonclinical laboratory tests:

Left ventricular ejection fraction	32%	>50%

Question

3. Based on all of the information provided, including laboratory tests, identify possible medical conditions affecting this patient.

Explanation

Possible conditions affecting the patient include the following:

- congestive heart failure
- thiamine deficiency
- cardiac "wet" beriberi
- advanced age
- malnourishment
- alcoholism

Question

4. What is the relationship between the increased serum BNP and the decreased thiamine levels?

Explanation

High BNP and lower left ventricular ejection fraction is consistent with CHF. Thiamin deficiency remains an under-recognized cause of congestive cardiac failure

Question

5. What type of assay is serum red cell transketolase?

Explanation

This is a functional assay.

Question

6. Explain why it is elevated.

Explanation

As thiamine concentrations decreases, the transketolase activity diminishes.

CHAPTER 25

Case Summary: A 30-year-old male presented to his physician with fatigue, fever and chills, night sweats, diarrhea, and a weight loss of 10 lbs. His white blood count was elevated at 11,000/μL, with 85% granulocytes, 11% lymphocytes, and 4% monocytes. Lymphocyte subset analysis revealed a decreased percentage and number of CD4-positive cells (15%, absolute number of 181/μL). Tests for HIV antibody were positive. The patient was started on antiretroviral therapy.

Question

1. What is the patient's diagnosis based on his clinical findings and laboratory results?

Explanation

Based on the patient's clinical symptoms, positive HIV antibody test result, and decreased number of CD4-positive cells below 200/μL, the patient can be classified as having AIDS.

Question

2. What are the benefits of molecular testing to the care of this patient?

Explanation

Molecular testing has had an important role in monitoring AIDS patients and guiding their physicians in prescribing the most effective therapies.

Question

3. What specific molecular methods could be used to monitor this patient? What are the principles of each?

Explanation

Molecular tests called *viral load tests* measure the levels of HIV RNA in the patient's blood; these are performed periodically to determine whether the patient's therapy is effective in decreasing the amount of virus harbored by the patient, as reflected by a decrease in viral load, or whether a change in therapy is indicated.

Molecular methods that have been developed to measure HIV viral loads.

APPENDIX N
Answers to Checkpoints

Chapter 1

Checkpoint! 1-1
Identify which type of glass pipette would be the best to use to reconstitute lyophilized, serum–based, quality-control material.

Answer
Class A volumetric TD pipette.

Checkpoint! 1-2
What is the value in SI units for a blood glucose of 100 mg/dL?

Answer
Glucose value (SI units, mmol/L) =

$$100 \text{ mg/dL} \times 10 \text{ dL/1 L} \times 1 \text{ g/1000 mg}$$
$$\times 1 \text{ mol glucose/180 g glucose} \times 1000 \text{ mmol/L mol}$$

Glucose value (SI units) = 5.5 mmol/L

Or

Glucose value (SI units, mmol/L) =

$$100 \text{ mg/dL} \times 0.055 \text{ (conversion factor)} = 5.5 \text{ mmol/L}$$

Checkpoint! 1-3
Convert the following:

1. *98°F to °C*
2. *10°C to °F*
3. *1°C to Kelvin*

Answer
1. $T_c = (T_F - 32)/1.8$
   $T_c = (98 - 32)/1.8$
   $T_c = 36.7°C$
2. $T_F = 1.8 T_c + 32$
   $T_F = 1.8(10) + 32$
   $T_F = 50°F$
3. $T_k = T_c + 273$
   $T_k = 1 + 273$
   $T_k = 274K$

Checkpoint! 1-4
The partial pressure of oxygen (PO_2) reported in a European journal of medicine article is 13 kPa. What is the PO_2 in mmHg?

Answer

$$13 \text{ kPa} \times \frac{760 \text{ mmHg}}{101.33 \text{ kPa}} = 97.5 \text{ mmHg}$$

Or use the conversion factor of 0.133:

$$\frac{13 \text{ kPa}}{0.133} = 97.7 \text{ mmHg}$$

Checkpoint! 1-5
1. *Convert 100 milligrams/deciliter to milligrams/liter.*
2. *Convert 2.5 quarts to liters.*
3. *Convert 140 milliequivalents of sodium/liter to milligrams of sodium/ deciliter.*

Answer
1. $(100 \text{ mg/dL})(10 \text{ dL/L}) = (100 \text{ mg})(10 \text{ dL})/(1 \text{ dL})(1 \text{ L}) = 1000 \text{ mg/L}$
   Or

$$100 \text{ mg/dL} = 100 \text{ mg/100 mL}$$
$$(100 \text{ mg/100 mL})(1000 \text{ mL/L}) = 1000 \text{ mg/L}$$

2. 1 quart = 0.95 L

$$(2.5 \text{ quarts})(0.95 \text{ L/quart}) = 2.4 \text{ L}$$

3. GMW sodium = 22.9 = GEW

$$(140 \text{ mEq Na}^+/\text{L})(1 \text{ L/10 dL})(1 \text{ Eq/1000 mEq}) = 0.014 \text{ Eq/dL}$$
$$(0.014 \text{ Eq/dL})(22.9 \text{ g Na}^+/1 \text{ GEW Na}^+)(1000 \text{ mg/1 g})$$
$$= 320.6 \text{ mgNa}^+/\text{dL}$$

Checkpoint! 1-6
Describe the proper preparation of a 1 ml to 10 ml dilution of a serum sample with saline.

Answer
Pipette 1.0 ml of serum and add 9.0 ml of saline for a total volume of 10 ml.

Checkpoint! 1-7
A patient's glucose result is "flagged" with the comment "Results exceed the upper limit of linearity." The method's upper limit of linearity is 700 mg/dL. The technologist prepares a 1 to 2 dilution of the serum specimen with saline and reanalyzes the diluted sample. A result, 400 mg/dL, is displayed. What is the concentration of the glucose in this patient's serum sample?

Answer

$$400 \text{ mg/dL} \times 2(\text{dilution factor}) = 800 \text{ mg/dL}$$

Checkpoint! 1-8
The molecular mass of albumin, a protein found in significant amounts in the human body, is 66.438 kDa. How many grams does this represent?

Answer

$$(66,438 \text{ Daltons})(1.66024 \times 10^{-24}) = 1.10 \times 10^{-19} \text{ g}$$

Checkpoint! 1-9

Determine 1 mole of sucrose ($C_{12}H_{22}O_{11}$) in daltons and grams.

Answer

Calculate molecular weight:

Step 1: Determine the atomic weight of each element.
   Atomic weight of Carbon = 12 Da
   Atomic weight of Hydrogen = 1 Da
   Atomic weight of oxygen = 16 Da

Step 2: Calculate the total atomic weight of the compound:
   12 carbon atoms $\times$ 12 Da = 144 Da
   22 hydrogen atoms $\times$ 1 Da = 22 Da
   16 oxygen atoms $\times$ 16 Da = 176 Da

Therefore, the total atomic weight of the compound in Daltons is 342 Da, so 1 mole of sucrose is equivalent to 342 Da or 342 g.

Checkpoint! 1-10

How many grams of NaCl are required to prepare 1 liter of a 0.5-M solution? The GMW of NaCl is 58.5.

Answer

$$\text{GMW} \times \text{M} = \text{g/L}$$
$$58.5 \times 0.5 = 29.25 \text{ g/L}$$

Weigh out 29.25 grams of NaCl and transfer it to a 1-L volumetric flask. Add water to the 1-L mark on the flask.

Checkpoint! 1-11

1. *What is the gram equivalent weight of $Ca(OH)_2$(GMW = 74)?*
2. *What is the GEW of H_2SO_4(GMW = 98)?*
3. *How many milliliters of concentration H_2SO_4 (specific gravity 1.84, percent purity 96.2%) are required to prepare 1 liter of a one normal solution?*

Answer

1. Gram equivalent weight = 74/2 = 37
   (1 mole of Ca $(OH)_2$ contains 2 equivalents of hydroxyls)
2. Gram equivalent weight. = 98/2 = 49
   (1 mole H_2SO_4 contains 2 equivalents of hydrogen)
3. Step 1: compute GEW H_2SO_4 = 98/2 = 49
   Step 2: Compute the number of grams of H_2SO_4 in one liter = 98 g/L
   Step 3: Compute number of grams of H_2SO_4 per milliliter of solution = SG $\times$ % assay = 1.84 $\times$ 96.2 = 1.77
   Step 4: Compute numbers of milliliters of concentrated H_2SO_4 required preparing 1 liter of a one normal solution.
$$\frac{1.77 \text{ g}}{1 \text{ ml}} = \frac{49 \text{ g}}{\text{ml}}$$
$$X = 27.6 \text{ ml}$$

Checkpoint! 1-12

How many grams of NaOH are required to prepare a 2.00 molal solution?

Answer

Step 1: Determine the GMW of NaOH = 40 g

Step 2: Compute the number of grams of NaOH required preparing the solution by using the following formula:

$$\text{Molality} = \frac{\text{Gram of solute/GMW of solute}}{1.0 \text{ kg of solvent}}$$

$$2 \text{ Molal} = \frac{X \text{ g of NaOH}/40.00 \text{ g}}{\text{kg solvent}}$$

$$X = 80 \text{ g of NaOH}$$

Checkpoint! 1-13

How many grams of solid NaOH are required to prepare 100 ml of a 10% solution in water?

Answer

$$100 \text{ ml} \times 10 \text{ g NaOH}/100 \text{ mL} = 10 \text{ g of NaOH}$$

Checkpoint! 1-14

A buffer solution requires that 3.00 grams of anhydrous Na_2HPO_4 be dissolved into 100 mL of reagent-grade water. The laboratory has $Na_2HPO_4 \cdot 7 H_2O$. How much of the $Na_2HPO_4 \cdot 7 H_2O$ should be used to prepare the solution?

Answer

$$141.96 \text{ GMW } (Na_2HPO_4) = 3.00 \text{ g } (Na_2HPO_4)$$
$$268.10 \text{ GMW } (Na_2HPO_4 \cdot 7 H_2O) \, X \text{ g } (Na_2HPO_4 \cdot 7 H_2O)$$
$$(141.96)(X) = (3.00)(268.10)$$
$$X = 5.66 \text{ g } Na_2HPO_4 \cdot 7 H_2O$$

Checkpoint! 1-15

Determine the enzyme activity (concentration) of LDH in a patient specimen using a spectrophotometric assay. The method uses NADH as the reaction indicator. The reaction volume is 300 μL, and the sample volume is 50 μL. Absorbance readings are taken at 30-second intervals for 2.5 minutes. The light path for the spectrophotometer is 1.0 cm, and the delta absorbance for the sample is 0.020.

Answer

Two approaches can be used. One is to determine an enzyme factor and multiply it by the delta absorbance per minute (method 1). The second approach is to use modification of the formula shown above (method 2).

Method 1

$$[1/(\varepsilon) \, (d)] \, [10^{-6}] \, [1/T] \, [V_t/V_s] = \text{enzyme factor}$$
$$[1/(6.22 \times 10^3 \text{ mol}^{-1} \text{ cm}^{-1})(1 \text{ cm})](10^6)(1/1 \text{ min})(300 \, \mu\text{L}/50 \, \mu\text{L})$$
$$= \text{enzyme factor}$$

Combining the equation:

$$\frac{300 \times 10^3}{311} = 965$$

Therefore:

$$\text{Enzyme factor} = 965$$
$$\text{U/L LD} = \Delta \text{ A/min} \times 965$$
$$\text{U/L LD} = 0.020 \times 965$$
$$19.3 \text{ U/L LD activity}$$

Method 2

Using a modification of the enzyme factor equation –

IU/L LD activity = $[\Delta A/(6.22 \times 10^3 \text{ mol}^{-1} \text{ cm}^{-1})(1 \text{ cm})]$
$$[10^{-6}] [1/T] [V_t/V_s]$$

IU/L LD activity = $0.020/(6.22 \times 10^3 \text{ mol}^{-1} \text{ cm}^{-1})(1 \text{ cm})]$
$$[10^{-6}][1/1 \text{ min }][0.3 \text{ mL}/0.05 \text{ mL}]$$

IU/L LD activity = $(3.2154 \times 10^{-6})(10^6)(1/1 \text{ min })(6)$
$$19.3 \text{ U/L LD activity}$$

Chapter 2

Checkpoint! 2-1

What is the wavelength in nanometers for EMR having a frequency of 1.58×10^{15} Hz?

Answer

$$c = \nu\lambda$$
$$\lambda = c/\nu$$
$$\lambda(\text{nm}) = \frac{3.00 \times 10^8 \text{ m/s}}{1.58 \times 10^{15} \text{ Hz}}$$
$$\lambda \text{ (nm)} = 190$$

Checkpoint! 2-2

What is the photon energy (E) in eV of EMR of (a) 190 nm? (b) 520 nm?

Answer
a. $E = hc/\nu$

$$E = \frac{(6.626 \times 10^{-34} \text{ J} \cdot \text{s})(3.0 \times 10^8 \text{ m/s})}{190 \text{ nm } (10^{-9} \text{ m/nm})}$$

$$E = 1.046 \times 10^{-18} \text{ J}$$

To convert to eV, use the conversion of 1 J = 6.24×10^{18} eV. Therefore, the number of eV_s is 6.53.

b. $E = hc/\nu$

$$E = \frac{(6.626 \times 10^{-34} \text{ J} \cdot \text{s}) (3.0 \times 10^8 \text{ m/s})}{520 \text{ nm } (10^{-9} \text{ m/nm})}$$

$$E = 3.82 \times 10^{-19} \text{ J or 2.38 eV}$$

Checkpoint! 2-3

What wavelength should be selected for a filter that will be used to measure a solution that appears purple in color?

Answer

A solution that appears purple absorbs photons in the green region of the EMS; therefore, the wavelength of the filter should be between 500 and 560 nm.

Checkpoint! 2-4

What is the absorbance of a solution with a transmittance of 10%?

Answer

$$A = 2 - \log \%T = 2 - \log 10\% = 2 - 1 = 1$$

Checkpoint! 2-5

What is the molar absorptivity of a 9.8×10^{-6} M solution of bilirubin dissolved in methanol having an absorbance of 0.600 when measured in a 1.0-cm cuvet at 435 nm?

Answer

$$A = abc$$
$$A = A/bc = 0.600/1 \text{ cm} \times 9.8 \times 10^{-6} \text{ molar}$$
$$A = 60,700 \text{ M}^{-1} \text{cm}^{-1}$$

Checkpoint! 2-6

What is the concentration in g/L of a solution of uric acid with an absorbance of 0.250, an absorptivity of 0.0625 Lg^{-1} cm^{-1} at 570 nm, and cuvette path length of 1 cm?

Answer

$$A = abc$$
$$c = A/cb = 0.250/(0.0625 \text{ L g}^{-1}\text{cm}^{-1}) (1 \text{ cm})$$
$$c = 4.0 \text{ g/L}$$

Checkpoint! 2-7

What is the concentration of creatinine in a serum sample if given the following?

$A_t = 0.140$
$A_s = 0.125$
$C_s = 0.90$ mg/dL

Answer

$$CT = A_T/A_S \times C_S$$
$$C_T = 0.140/0.125 \times 0.9 \text{ mg/dL}$$
$$C_T = 1.0 \text{ mg/dL}$$

Checkpoint! 2-8

What is the concentration of glucose in an unknown serum sample derived from the standard curve produced from the data shown in the following table?

Standard (mg/dL)	Absorbance
50	0.100
100	0.200
250	0.500
500	1.000
Sample:	
Patient	0.250

Answer
135 mg/dL

Checkpoint! 2-9

What would a laboratory's course of action be if its spectrophotometer failed a stray light check?

Answer
To reduce the effects of stray light, every effort must be made to reduce the room light or any light that is directed to the spectrometer.

Examples of action include pulling down shades or blinds on windows, dimming any room lights, and moving sources of light from lamps or other lighting fixtures.

Checkpoint! 2-10
How many detectors does a double-beam-in-space spectrophotometer have?

Answer
Two

Checkpoint! 2-11
What is the source of EMR required for a metal to be measured in a sample by atomic absorption spectroscopy?

Answer
The element to be measured is coated on the cathode of hollow cathode lamp. When the lamp is charged with an appropriate voltage, the cathode heats up and the metal emits characteristic EMR.

Checkpoint! 2-12
True or false? The excitation wavelength of a fluorophore has a higher energy value and a shorter wavelength than the emission wavelength.

Answer
True

Checkpoint! 2-13
Explain the mechanism of light production by chemiluminescence.

Answer
Chemiluminescence is the production of light through a chemical reaction.

Checkpoint! 2-14
What does the line of demarcation seen within the refractometer after a sample is applied to the glass surface represent?

Answer
It represents the critical angle created by light refracted as it goes through the sample.

Checkpoint! 2-15
If the number of solute particles in an aqueous solution is increased, would the change in freezing-point temperature increase or decrease?

Answer
It would decrease.

Checkpoint! 2-16
Which electrode would albumin migrate to if the buffer pH is 8.6 and its pI is 4–5.8?

Answer
Albumin will migrate to the positive electrode because albumin's overall net electronic charge is negative.

Checkpoint! 2-17
What is the difference between a split and a splitless injector?

Answer
The split injector is used to introduce only a small amount of the sample vapor onto the analytical column. The splitless injector is designed to transfer the entire sample to the column.

Checkpoint! 2-18
Compare these two solvent delivery methods for liquid chromatography: isocratic and gradient elution.

Answer
Isocratic LC uses one mobile phase. Gradient LC involves the use of two or more mobile phases that are automatically programmed to pump for a specific interval of time.

Checkpoint! 2-19
Identify the two measured parameters of a compound using a mass spectrometer.

Answer
The masses of the ionized molecules (or atoms) and the relative abundances of each.

Checkpoint! 2-20
Identify the term used to describe the injection of a sample cell suspension into the center of a rapidly flowing stream or sheath.

Answer
Hydrodynamic focusing

Checkpoint! 2-21
Identify the major components of a biosensor.

Answer
Biocatalyst, transducer, and electronics

Checkpoint! 2-22
List several advantages for patients and caregivers when using POCT.

Answer
Reduced turnaround time
Improvements to patient management
Small sample volumes
Laboratory tests are performed at bed side

Chapter 3

Checkpoint! 3-1
Identify several areas of concern in the laboratory that were improved using automated chemistry analyzers.

Answer
- turnaround times
- cost
- staff safety
- laboratory errors
- increase in the number of specimens

Checkpoint! 3-2
The preanalytical stage of laboratory testing remains the greatest source of laboratory error. Identify several examples of preanalytical tasks that may lead to laboratory error.

Answer
- transcription mistakes
- mislabel specimens
- incorrect blood drawing tube

- pouring one specimen into another specimen
- incorrect storage of specimen aliquots

Checkpoint! 3-3

Identify two techniques that can be used in automated analyzers to reduce carryover of samples.

Answer

Carryover can be reduced by:

- aspirating a wash solution in between each pipetting,
- back flushing the probe using a wash solution, or
- using disposable plastic pipette tips to transfer samples.

Checkpoint! 3-4

Indicate whether the following situation represents a closed- or an open-reagent analyzer:

A staff technologist is interested in adding a new assay to the laboratory's chemistry analyzer. The technologist contacts the instrument manufacturer and is told that the assay is not currently available and that no competitor's reagents can be used on this chemistry analyzer.

Answer

It is a closed reagent analyzer.

Checkpoint! 3-5

Give three examples of how liquid solutions are warmed in automated analyzers.

Answer

Liquid solutions can be warmed by:

- using an elongated cuvette path length and a fluorocarbon oil incubation bath,
- using a Peltier thermal electric module, or
- using water.

Checkpoint! 3-6

Identify two advantages of integrated modular systems over single-batch, discrete analyzers.

Answer

Integrated modular systems allow for multiple platform analysis, incorporation of multiple numbers of similar analyzers, and inclusion of preanalytical modules.

Checkpoint! 3-7

List five examples of detectors that can be found in automated clinical chemistry analyzers.

Answer

Five types of detectors are:

1. photometers,
2. ion-selective electrodes,
3. nephelometers,
4. fluorometers, and
5. luminometers (for chemiluminescence).

Chapter 4

Checkpoint! 4-1

A laboratory technologist was asked to determine the descriptive statistics for the data presented in the following table. that follows. The data represent replicate measurements of sodium in serum samples. Twenty measurements were completed on one sample. Determine the statistics listed for the data shown in the following table.

- *arithmetic mean*
- *standard deviation*
- *variance*
- *percent coefficient of variation*
- *median*
- *mode*
- *range*

Data Number	Result	Data Number	Result
1	140	11	139
2	141	12	138
3	139	13	140
4	140	14	140
5	140	15	141
6	138	16	139
7	142	17	139
8	141	18	141
9	141	19	140
10	140	20	140

Answer

Arithmetic mean = 139.95
Standard deviation = 1.05
Variance = 1.10
Coefficient of variation = 0.75%
Median = 140
Mode = 140
Range = 4.0 (138 – 142)

Checkpoint! 4-2

A normal population study for uric acid in serum samples reveals that the mean of the group of samples is 6.0 mg/dL and the standard deviation is 0.5 mg/dL. What is the probability of a serum uric acid value of 6.9 or greater occurring in this series?

Answer

$$Z = x - \mu/\sigma$$
$$Z = 6.9 - 6.0/0.5 = 1.8$$

From a table of the area under the tails of a normal curve, the Z value of (1.8) is 0.36. Therefore, the probability is 0.036 or 3.6%.

Checkpoint! 4-3

A clinical laboratory is required to correlate its new biosensor method for whole-blood potassium concentrations with results obtained from its existing method, which uses an ion-selective electrode analyzer. The following data were obtained on paired samples. (Note: The number of sample pairs has been reduced below the recommended number for comparison of methods to simplify the problem.)

1. *Determine the following statistics from the following data.*

   - *bias*
   - *standard deviation of the difference*

- standard error of the mean of differences
- p-value

2. Assuming a $t_{\alpha = 0.05}$ two-tailed distribution, should the null hypothesis, H0: $\mu1 = \mu2$, be accepted or rejected?

Sample Number	K$^+$ mEq/L Biosensor Technique	K$^+$ mEq/L ISE Technique
1	4.5	4.6
2	4.0	4.0
3	3.9	4.0
4	3.8	3.9
5	5.0	5.1
6	5.5	5.7
7	3.0	3.2
8	3.4	3.6
9	3.1	3.3
10	4.2	4.2
11	4.4	4.7
12	5.2	5.3
13	5.1	5.0
14	3.7	3.9
15	4.4	4.5
16	3.2	3.3
17	2.8	3.0
18	6.0	6.3
19	5.8	6.2
20	4.6	4.6

Answer

Bias = −0.14
Standard deviation of the differences = 0.310
Standard error of the mean of differences = 0.069
p value = 0.000

Therefore, the null hypothesis cannot be accepted!

Checkpoint! 4-4
Calculate the following from the data presented in Checkpoint! 4.3:

- linear regression by least squares, including slope and intercept;
- correlation coefficient; and
- coefficient of determination.

Answer
Linear regression:

Slope = −0.011
Intercept = 0.97
Correlation coefficient = 0.992
Coefficient of determination = 0.986

Checkpoint! 4-5
The mean and standard deviation for creatinine levels I and II quality-control pool are shown in the following table below.

Creatinine	Level I	Level II
Mean	0.8	6.0
Standard deviation	0.1	0.3

The technologists in the chemistry laboratory recorded the following quality-controls values for a 10-day period and are shown in the following table.

	Level I	Level II
Day 1	0.9	6.1
Day 2	1.15	6.2
Day 3	0.8	5.9
Day 4	0.7	6.7
Day 5	0.7	6.8
Day 6	0.8	5.5
Day 7	0.9	6.0
Day 8	1.3	6.5
Day 9	0.8	6.2
Day 10	0.9	6.1

Using the data presented, determine the following:

1. all Westgard rule violations,
2. the type of error associated with each rule violation, and
3. a course of action for each rule violation.

Answer
Day 2 level I shows a 1_{2s} rule violation resulting from random error. This is a warning rule, and the patient results can be reported.

Day 4 level II shows a 1_{2s} rule violation caused by random error. This is a warning rule, and the patient results can be reported.

Day 5 level II shows a 2_{2s} rule violation because of systematic error. The patient results cannot be reported, and the technologist must rerun the same vial of control material.

Day 8 level I shows a 1_{3s} rule violation from random error. The patient results cannot be reported, and the technologist must rerun the same vial of control material.

Chapter 5

Checkpoint! 5-1
Distinguish immunogen from antigen.

Answer
An immunogen is any chemical substance capable of inducing an immune response. The term immunogen is used when referring to material capable of eliciting antibody formation when injected into a host. The term antigen is used for any material capable of reacting with an antibody without necessarily being capable of inducing antibody formation.

Checkpoint! 5-2
Which of the following types of immunoassays—homogeneous or heterogeneous immunoassays—requires a physical separation of bound from free fractions?

Answer

Homogeneous immunoassays do not require a physical separation of bound fractions from free fractions.

Checkpoint! 5-3

How is fluorescent electromagnetic radiation produced in the microparticle enzyme immunoassay method?

Answer

The compound that creates fluorescent light in the MEIA is methylumbelliferone (MU), which is created by the cleavage of the substrate 4-methylumbelliferylphosphate (MUP) by alkaline phosphatase.

Checkpoint! 5-4

In the EMIT assay, which form—bound or free—is left to react with a substrate?

Answer

Free or unbound fraction

Checkpoint! 5-5

What type of technology is used to engineer the creation of large polypeptides of β-galactosidase in the CEDIA?

Answer

Recombinant DNA technology

Checkpoint! 5-6

Explain why a positive result in a fluorescent polarization immunoassay produces a low polarized signal.

Answer

If the sample is positive for a compound (e.g. a drug), then more patient drug will bind to the antibody and result in a greater amount of free unbound reagent drug labeled with fluorescein. Free unbound reagent drug labeled with fluorescein results in a lower net polarization.

Checkpoint! 5-7

What chemicals are used to facilitate the production of EMR in a chemiluminescent immunoassay that incorporates the label acridinium ester?

Answer

Hydrogen peroxide, acid, and strong alkali are used to initiate the light-emitting reaction from acridinium.

Checkpoint! 5-8

Identify the element and its electronic transition state that is responsible for triggering chemiluminescence in the LOCI.

Answer

Single oxygen ($^1\Delta_g O_2$)

Checkpoint! 5-9

Identify the test protocol used to determine the presence of hook effect in a patient sample measured by an immunoassay

Answer

Dilution protocol to test the linearity

Chapter 6

Checkpoint! 6-1

1. *Explain the difference in the linkage of two monosaccharides to form a disaccharide that results in a reducing carbohydrate and the linkage that*

forms a nonreducing carbohydrate. Name one reducing disaccharide and one disaccharide that is a nonreducing carbohydrate.
2. *A five-carbon monosaccharide is called a _____.*
3. *A glucose molecule with the hydroxyl group on the left on the next-to-last carbon atom is a(n) _____-glucose.*

Answer

1. In the formation of disaccharides, if the linkage between the two monosaccharides is between the aldehyde and ketone groups of both sugars, the disaccharide has no free aldehyde or ketone group and is a nonreducing sugar (sucrose, for example). If the linkage is between the aldehyde or ketone group of one sugar and the hydroxyl group of the second monosaccharide, there is a potentially free aldehyde or ketone group and the disaccharide is a reducing sugar (for example, glucose).
2. pentose
3. L

Checkpoint! 6-2

1. *Two hormones regulating glucose metabolism are secreted by the pancreas. The hypoglycemic hormone is _____, and the hyperglycemic hormone is _____.*
2. *List two glands, other than the pancreas, that regulate glucose metabolism and the hormones they secrete.*
3. *A patient's insulin level is twice her C-peptide level (in moles). Provide a brief explanation for this scenario.*

Answer

1. Insulin is the hypoglycemic hormone, and glucagon is the hyperglycemic hormone.
2. Gland possibilities and hormones include the following:

Gland	Hormone
Adrenal medulla	Epinephrine
Adrenal cortex	Glucocorticoids (cortisol)
Anterior pituitary	ACTH
Anterior pituitary	Growth hormone
Thyroid	Thyroxine and triiodothyronine
Pancreas	Somatostatin

3. Insulin and C-peptide are produced in equimolar amounts from endogenous insulin. A patient whose insulin level is twice her C-peptide level indicates an exogenous source: injected insulin. Injected or exogenous insulin will result in high insulin without an increase in C-peptide.

Checkpoint! 6-3

1. *Identify the following conditions or symptoms as type 1 or type 2 diabetes mellitus.*
   a. *ketosis prone _____*
   b. *non–insulin-dependent _____*
   c. *obesity _____*
   d. *autoantibodies _____*
2. *List three risk factors for diabetes mellitus.*

Answer

1. a. type 1
   b. type 2
   c. type 2
   d. type 1
2. Family history, obesity, age, racial or ethnic group, and abnormal glucose tolerance are risk factors for diabetes mellitus.

Checkpoint! 6-4

1. *List the three components of Whipple's triad.*
2. *Linda has a glucose level of 50 mg/dL but no symptoms (no weakness, shaking, or dizziness). Does she suffer from hypoglycemia?*

Answer

1. Whipple's triad is defined as having signs and symptoms of hypoglycemia, documentation of low plasma glucose at the time the patient is experiencing symptoms, and alleviation of symptoms with the ingestion of glucose and an increase in plasma glucose.
2. Linda is not hypoglycemic. The low blood glucose must be associated with symptoms.

Checkpoint! 6-5

1. *John, a diabetic, has a glycosylated hemoglobin of 8%. According to ADA, is this acceptable? Approximately what would be his average glucose level?*
2. *Why is microalbuminuria usually present at the time of diagnosis in type 2 diabetics, whereas it takes several years to develop in type 1?*

Answer

1. John's glycosylated hemoglobin is abnormal. ADA recommends a maximum of 7% HbA_{1c}. An 8% Hba_{1c} is approximately an average glucose of 180 mg/dL.
2. In type 2 diabetics, it is usually present at the time of diagnosis because the kidneys have already been damaged because the diabetes has been present for a few years before diagnosis. In 80% of type 1 diabetics, urinary albumin excretion increases at a rate of 10–20% per year, with development of clinical proteinuria in 10 to 15 years.

Chapter 7

Checkpoint! 7-1

1. *Differentiate between exogenous cholesterol and endogenous cholesterol.*
2. *Why is it difficult to lower cholesterol by only limiting dietary cholesterol?*
3. *Which lipid class is responsible for 95% of fat stored in tissue?*

Answer

1. Exogenous cholesterol is cholesterol originating from outside the body or dietary cholesterol. It is absorbed in the diet, bile, intestinal secretions, and cells. Endogenous cholesterol is produced by the liver and is regulated by feedback from the levels of exogenous cholesterol.
2. Regulating diet only affects the exogenous cholesterol. The body responds to lower exogenous cholesterol by increasing the production of endogenous cholesterol. Therefore, lowering exogenous cholesterol can only lower total cholesterol to a certain point before the synthesis of endogenous cholesterol is increased.
3. Triglycerides are responsible for 95% of fat stored in tissue.

Checkpoint! 7-2

1. *Which of the following apolipoproteins is described by the following?*
   a. *Major protein found in HDL:_____*
   b. *Associated with a high risk of cardiovascular disease:_____*
   c. *Associated with chylomicron remnants and renal failure:_____*
   d. *Activates lipoprotein lipase:_____*

Answer

1. a. Apo A-1
   b. Apo B-100
   c. Apo B-48
   d. Apo C-II

Checkpoint! 7-3

1. *Which lipoprotein has the lowest ratio of lipid to protein (the highest percentage of protein)?*
2. *The exogenous pathway involves primarily which lipoprotein?*
3. *Chylomicron remnants are catabolized and channeled in what three pathways?*
4. *Which lipoprotein class is described as*
   a. *the "good" lipoprotein: _____*
   b. *containing almost exclusively Apo B-100: _____*
   c. *the transporter of endogenous triglycerides from the liver to muscle and adipose cells: _____*

Answer

1. HDL has the lowest ratio of lipid to protein.
2. The exogenous pathway is primarily associated with chylomicrons.
3. Chylomicron remnants are used to synthesize very low density lipoproteins, are released to form bile acids, or are stored as cholesteryl esters.
4. a. HDL
   b. LDL
   c. VLDL

Checkpoint! 7-4

1. *Mary has diabetes mellitus and is hypothyroid. Would she be at higher or lower risk of cardiovascular disease? Explain.*
2. *An opaque or lipemic plasma would indicate elevated levels of which lipid? What would be the approximate level?*
3. *Lp(a) contains one molecule of which two proteins?*

Answer

1. Mary is definitely at higher risk of cardiovascular disease with both diabetes and hypothyroidism, which are significant risk factors for CVD.
2. An opaque or lipemic plasma would indicate grossly elevated triglycerides of more than 600 mg/dL.
3. Lp(a) molecules contain one molecule of Apo B-100 and one molecule of apo (a).

Checkpoint! 7-5

1. *Bill's lipid profile is*

   Total cholesterol: 290 mg/dL
   TG: 140 mg/dL
   HDL-C: 40 mg/dL
   What is his calculated LDL-C?

2. *List the five criteria for the clinical diagnosis of metabolic syndrome.*
3. *Describe dyslipidemia as defined for the metabolic syndrome.*

Answer

1. LDL-C = Total cholesterol – (HDL-C + Tg/5)
   290 mg/dL – (40 mg/dL + 140 mg/dL/5) = 222 mg/dL
2. Five criteria for the clinical diagnosis of metabolic syndrome are elevated waist circumference, elevated triglycerides, reduced HDL-C, elevated blood pressure, and elevated fasting glucose.

3. Dyslipidemia in metabolic syndrome is defined as low HDL, high LDL-C, increased sdLDL, and high triglycerides.

Chapter 8

Checkpoint! 8-1
1. Define amphoteric.
2. At a pH above its isoelectric point a protein carries a _____ (positive or negative) charge.
3. A protein with three polypeptide chains are arranged to forms its _____ structure.
4. Conjugated proteins that contain cholesterol and triglycerides are called _____.

Answer
1. Amphoteric describes a molecule that contains two ionizable sites: a proton-accepting group (NH_2), and a proton-donating group (COOH).
2. negative
3. quaternary
4. lipoproteins

Checkpoint! 8-2
1. Most plasma proteins are synthesized in the _____
2. List three major protein functions.
3. Identify the aminoacidopathy associated with the following:
   a. ochronosis _____
   b. "mousy" urine odor _____
   c. formation of kidney stones _____
   d. deficiency of homogentisic acid oxidase _____

Answer
1. liver.
2. Major protein functions include the following:

Maintenance of colloidal osmotic pressure and water distribution	Enzymes
	Peptide hormones
	Coagulation
Structural (e.g., hair nails)	Hemoglobin
Transport molecule (e.g., transferrin, bilirubin)	Antibodies

3. a. alkaptonuria
   b. phenylketonuria
   c. cystinuria
   d. alkaptonuria

Checkpoint! 8-3
1. Identify the protein described by the following:
   a. early-onset emphysema _____
   b. spina bifida _____
   c. sensitive indicator of intravascular hemolysis _____
   d. Wilson disease _____
2. The most common cause of hypoalbuminemia is _____.
3. Hyperalbuminemia is most commonly found in _____.

Answer
1. a. α_1-antitrypsin
   b. alpha-fetoprotein
   c. haptoglobin
   d. ceruloplasmin
2. increased catabolism; tissue damage and inflammation
3. dehydration

Chapter 9

Checkpoint! 9-1
1. The apoenzyme and the cofactor or coenzyme that form the catalytically active unit is called the_____.
2. The actual place in the enzyme where the substrate is converted to product is the _____ _____ .
3. Nicotinamide adenine dinucleotide (NAD^+) is an example of a(n) _____.
4. How do enzymes affect the energy of activation of a reaction?
5. Define zero-order kinetics.
6. List four factors that influence enzyme reactions.

Answer
1. holoenzyme
2. active site
3. coenzyme
4. Enzymes lower the energy of activation (EA) of the reaction. Less energy is required to energize one mole of the substrate to form the activated complex (enzyme–substrate complex).
5. Zero-order kinetics is a plateau that is reached where the reaction rate is independent of the substrate concentration. All of the enzyme is bound to substrate, therefore the enzyme cannot work any faster. Addition of more substrate will not result in a faster reaction.
6. Factors that influence enzyme reactions are:
   a. substrate concentration,
   b. enzyme concentration,
   c. temperature,
   d. pH,
   e. cofactors, activators or coenzymes, and
   f. inhibitors.

Checkpoint! 9-2
1. CK-3 (MM) is found primarily in what tissue?
2. List the CK isoenzyme associated with the following:
   a. tumor-associated marker _____
   b. muscular dystrophy _____
   c. cerebrovascular accident _____
   d. rhabdomyolysis _____
   e. myocardial infarction _____
3. John's CK-2 is 30 $\mu g/L$, and total CK is 400 U/L. What is his relative index, and what does it indicate?
4. Describe the "flipped" pattern in LD electrophoresis and what condition(s) are associated with it.
5. Bill has a negative or normal CK-MB at 8 hours. Can his physician confirm or rule out a myocardial infarction?

Answer
1. muscle
2. a. CK-1 (BB)
   b. CK-3 (MM)
   c. CK-1 (BB)
   d. CK-3 (MM)
   e. CK-2 (MB)
3. $\frac{30\ \mu g/L}{400\ U/L} = 7.5\%$

   John's relative index indicates a myocardial infarction. An RI more than 6% is indicative of an MI.

4. In the flipped pattern in LD electrophoresis, LD1 is greater than LD2, whereas in normal individuals LD2 is greater than LD1.
5. Yes, the physician could rule out an MI. A negative CK-MB at 6 to 8 hours has a 95% predictive value ruling out an MI.

Checkpoint! 9-3
1. *Which enzyme is most specific for hepatic disease: AST or ALT?*
2. *Linda's AST is 190 U/L, and her ALT is 90 U/L. What is the De Ritis ratio, and what does it indicate?*
3. *Tom is a 15-year-old boy with an ALP of 160 U/L.*
   a. *Should his physician be concerned?*
   b. *What is the most likely explanation?*
4. *The primary sources or the tissues richest in the following enzymes are:*
   a. *AST* _____ _____ _____
   b. *ALT* _____
   c. *ALP* _____ _____
5. *Bill's AST is 50 U/L, and his ALP is 250 U/L. Is his problem hepatic or obstructive liver disease?*

Answer
1. ALT is most specific for hepatic disease. AST can also be elevated in heart and muscle disease.
2. Linda's De Ritis ratio is $\frac{190\ U/L}{90\ U/L} = 2.1$

   A De Ritis ratio greater than 2 points toward alcoholism or alcoholic hepatitis.
3. Tom's physician should not be unduly concerned with the elevated ALP. A physiologic increase in ALP is normal in growing children especially before or during a growth spurt.
4. AST: liver, cardiac, and muscle
   ALT: liver
   ALP: liver (hepatocytes lining the biliary tract) and osteoblasts (bone).
5. Bill slightly elevated AST and moderately elevated ALP indicates a biliary tract problem or obstructive liver disease.

Checkpoint! 9-4
1. *Robert has an elevated ALP and an elevated GGT. What is the most likely source?*

Answer
Robert's elevated ALP and GGT point to the liver as the source.

Checkpoint! 9-5
1. *Mary has an elevated amylase and a normal lipase. What is the most likely source of the amylase? What are other possible explanations?*
2. *Amy has an amylase creatinine clearance ratio of 2%. What is the most likely condition associated with the low ACCR?*
3. *Which of the following pancreatic enzymes is elevated first in acute pancreatitis: serum amylase, serum lipase, or urine amylase? Which would remain elevated the longest?*

Answer
1. Mary's elevated amylase and normal lipase rule out the pancreas as the source; the most likely explanation would be salivary amylase (e.g., parotitis, mumps). Other possible explanations would be intraabdominal conditions such as perforated peptic ulcer, appendicitis, or ectopic pregnancy.
2. An ACCR of 2% would be associated with macroamylasemia.

3. Serum lipase would be elevated first, and urine amylase would remain elevated the longest.

Chapter 10

Checkpoint! 10-1
Define the following terms: diagnostic sensitivity, diagnostic specificity, predictive value of a positive test, and predictive value of a negative test.

Answer
Diagnostic sensitivity, as related to tumor markers, is a measure of how often the assay system detects the biomarker when the disease is present (i.e., positivity in disease).

Diagnostic specificity is used to describe the probability that a laboratory test will be negative in the absence of disease.

The predictive value of a positive test relates the number of true positives for a test to the total number of positive tests.

The predictive value for a negative test relates the true negative to the total number of negative tests.

Checkpoint! 10-2
Why is knowledge of analytical specificity important for quantitative β-hCG assays when used as a tumor markers assay?

Answer
In many cases, the malignant neoplasia of the patient produces varying amounts of different molecular forms of hCG, so one should be familiar with which assays detect which molecular forms.

Checkpoint! 10-3
Briefly describe the rise and fall pattern of serum alpha fetal protein in a normal progressing pregnancy.

Answer
During pregnancy, maternal AFP levels increase from 12 weeks of gestation to a peak of about 500 µg/L during the third trimester. The fetal AFP reaches a peak of 2 g/dL at 14 weeks and then declines to about 70 mg/dL at term.

Checkpoint! 10-4
Explain the differences among tPSA, fPSA, and cPSA.

Answer
Total PSA represents the sum of all the molecular forms present in a sample. Free PSA is the uncomplexed form, and the complexed PSA is the molecular form that is complexed to several different proteins.

Checkpoint! 10-5
Explain why PCA3 may be a better biomarker for prostate cancer than tPSA.

Answer
PCA3 can add specificity to a diagnostic algorithm for prostate cancer in men with a negative biopsy but elevated serum tPSA.

Checkpoint! 10-6
What is the relationship of serum HER-2/neu levels to treatment with Herceptin (trastuzmab)?

Answer
If the serum HER-2/neu level is high while the patient is on Herceptin therapy, then the outcome usually results in a shorter treatment response.

Checkpoint! 10-7

What is the principal clinical usefulness of CA 125 measurement in patients with ovarian cancer?

Answer

CA 125 measurement is most often used to monitor recurrence of disease and for differential diagnosess of pelvic masses in post-menopausal women.

Checkpoint! 10-8

What is the clinical usefulness of serum carcinoembryonic antigen measurements?

Answer

The clinical usefulness of CEA is primarily in monitoring therapy.

Checkpoint! 10-9

On a molecular level, CA 19-9 is associated with which blood group antigen?

Answer

Lewis a (Le^a)

Checkpoint! 10-10

Identify the factor detected by the BTA test.

Answer

Human complement-related H factor

Checkpoint! 10-11

Which biomarker is reported to be the most sensitive for non–small cell lung carcinoma?

Answer

Neuron-specific enolase

Chapter 11

Checkpoint! 11-1

1. *A journal article reports a urea nitrogen of 10 mg/dL. What would be the equivalent urea concentration?*
2. *Classify the following as prerenal, renal, or postrenal azotemia.*
   a. *Dehydration _____*
   b. *Glomerulonephritis _____*
   c. *Congestive heart failure _____*
   d. *Nephrolithiasis _____*
   e. *Shock _____*
3. *A BUN:CR ratio of 15 with a moderately elevated BUN and creatinine can be classified as prerenal, renal, or postrenal azotemia?*

Answer

1. 10 mg/dL urea nitrogen $\times$ 2.14 (conversion factor) = 21.4 mg/dL urea
2. a. prerenal
   b. renal
   c. prerenal
   d. postrenal
   e. prerenal
3. A "normal" BUN:creatinine ratio with elevated BUN and creatinine would be classified as renal azotemia.

Checkpoint! 11-2

1. *Why is creatinine clearance the most widely used test for estimating the glomerular filtration rate?*
2. *List three reasons why creatinine is described as a good indicator of the glomerular filtration rate.*
3. *What is the most common source of error in calculating the creatinine clearance?*
4. *A creatinine clearance was ordered on an obese patient with kidney disease.*

   Weight: 350 lb
   Height: 5'40"
   24-hour urine volume: 1850 mL
   Plasma creatinine: 6.5 mg/dL
   Urine creatinine: 120 mg/dL
   BSA: 2.30m^2
   What is this patient's creatinine clearance?

Answer

1. Creatinine clearance is the most widely used test because it is easily measured and extensive data are available.
2. Creatinine is a good indicator of glomerular filtration because it is freely filtered by the glomeruli, is not reabsorbed by the tubules, and is released in the plasma at a constant rate.
3. The most common source of error in creatinine clearance calculations is the completeness of the 24 h collection.
4.

$$\frac{UV}{P} \times \frac{1.73 \text{ m}^2}{BSA} = \frac{120 \text{ mg/dL} \times \dfrac{1850 \text{ ml/24 h}}{1440 \text{ ml/24 h}}}{6.5 \text{ mg/dl}} \times \frac{1.73 \text{ m}^2}{2.30 \text{ m}^2}$$

$$= 17.8 \text{ ml/min}$$

Checkpoint! 11-3

1. *Which renal disease is associated with each of the following?*
   a. *Bacterial infection of the renal tubules: _____*
   b. *Circulating antigen–antibody complexes: _____*
   c. *Bladder infection: _____*
2. *What chronic disease is the leading cause of renal failure?*

Answer

1. a. acute pyelonephritis
   b. acute glomerulonephritis
   c. cystitis
2. diabetes mellitus

Chapter 12

Checkpoint! 12-1

Compare and contrast serum versus plasma.

Answer

Electrolyte composition of serum versus plasma is not significant except for potassium. Potassium concentrations are about 8% lower in plasma than in serum. Serum does not have fibrinogen; serum is what remains after blood clots, and plasma is what remains after cells and other particulates are separated from the aqueous phase of a whole-blood sample.

Checkpoint! 12-2

Identify examples of unmeasured anions that could cause an increase in the anion gap.

Answer

Examples of unmeasured anions are proteins, organic acids, sulfates, and phosphates.

Checkpoint! 12-3

Given the following data, calculate the serum osmolality using equation 12.3 and osmolal gap equation 12.7. Also list two examples of substances that can result in an increased osmolal gap.

Sodium	135 mEq/L
Potassium	4.5 mEq/L
Chloride	108 mEq/L
Carbon dioxide	7 mEq/L
Anion gap	20 mEq/L
Glucose	162 mg/dL
Creatinine	1.5 mg/dL
Urea nitrogen	17 mg/dL
Serum osmolality*	320 mOsm/kg

Measured by freezing-point depression osmometry.

Answer

Calculated serum osmolality is 285 mOsm/kg, therefore the osmolal gap is determined by subtracting the measured serum osmolality from the calculated osmolality. Therefore, measured osmol – calculated osmol = 320 – 285 = 35 mOsm/kg. Volatiles such as ethanol, methanol, isopropanol, and ethylene glycol can cause elevations of measured serum osmolality.

Checkpoint! 12-4

Explain the body's response via the renin–angiotensin–aldosterone system to a situation of water deprivation.

Answer

In times of water deprivation (also dehydration), plasma sodium concentration increases (hypernatremia) and both total and intracellular water decrease. A low effective circulating blood volume can be sensed by baroreceptors located in the carotid sinus and aortic arch; this signals medullary control centers in the brain to increase sympathetic outflow to the juxtaglomerular cells, thereby increasing release of renin into the circulatory system. Renin converts angiotensinogen to angiotensin I. Angiotensin I is converted to angiotensin II in the lung and kidneys. Angiotensin II is a potent vasoconstrictor. Also angiotensin II stimulates aldosterone secretion by the adrenal cortex, thirsting behavior, and ADH secretion. Aldosterone stimulates sodium reabsorption in the distal nephron; as a result of this sodium reabsorption, the body retains water. The thirsting mechanism also increases water intake.

Checkpoint! 12-5

Describe the response of ANP in a patient with CHF and severe hypertension.

Answer

ANP relaxes venous capacitance of blood vessels by suppressing sympathetic nervous system activity. The increase in venous pressure is reduced. ANP also increases vascular permeability and promotes natriuresis and diuresis. Glomerular filtration rate is increased, the renin–angiotensin system is suppressed, and tubular Na^+ reabsorption is inhibited. In the brain, ANP inhibits salt appetite, water intake, and secretion of ADH and corticotrophin.

Checkpoint! 12-6

Indicate several conditions that would result in depletional-type hyponatremia.

Answer

Depletional-type hyponatremia may be caused by several factors:

- Na^+ loss along with water,
- Na^+ loss via the kidney because of osmotic diuresis or adrenal insufficiency,
- excess edema, and
- SIADH.

Checkpoint! 12-7

Identify at least three techniques used to measure Na^+ in body fluids.

Answer

Methods for measuring Na^+ in body fluids include:

- both direct and indirect ISEs,
- flame-emission spectroscopy,
- atomic-absorption spectroscopy, and
- spectrophotometry.

Checkpoint! 12-8

Plasma potassium concentration may be elevated or decreased in acid–base disturbances, depending on the conditions associated with the patient. Identify the plasma K^+ levels as either hyperkalemia or hypokalemia in metabolic acidosis and metabolic alkalosis. Also, briefly discuss the mechanism for these acid–base conditions.

Answer

Metabolic acidosis is often associated with hyperkalemia because potassium moves out of the cell and into the extracellular water space. Metabolic alkalosis is a condition characterized by hypokalemia. Potassium ions move from the extracellular water space and into the cell.

Chapter 13

Checkpoint! 13-1

What is the calculated pH for a patient sample if the measured bicarbonate concentration is 22 mmol/L and pCO_2 is 57 mmHg?

Answer

Substituting into the following equation used to calculate hydrogen-ion concentration:

cH^+ nmol/L = 24.1 × 57 mmHg/22 mmol/L cH^+ = 62.4 nmol/L

To convert 62.4 nmol/L to pH, calculate the following:

$$pH = -\log cH^+$$
$$pH = -\log (62.4 \text{ nMol/L})$$

Convert 62.4 nmol/L to mol/L = 6.24 × 10^{-8} M/L

$$pH = -\log 6.24 \times 10^{-8}$$
$$pH = 8 - \log 6.24$$
$$pH = 8 - 2.39$$
$$pH = 7.2$$

Checkpoint! 13-2

Which buffer is the major buffer of blood and why?

Answer

Hemoglobin

Checkpoint! 13-3

Compare and contrast external and internal convection systems and include anatomical, physiological, and mechanical features

Answer

The external convection system consists of the lungs, airway, and respiration muscles. It serves to maximize gas exchange by continuously supplying bulk-phase water or oxygen to the external surface of the gas-exchange barriers. The internal convection system is the circulatory system. It maximizes the flow of oxygen and carbon dioxide across the gas-exchange barriers by delivering blood that has low PO_2 and high PCO_2 to the inner surface of the barriers.

Checkpoint! 13-4

Atmospheric oxygen needs to enter the body and be distributed to cells. Briefly outline the processes involving the respiratory apparatus that serve to accomplish this end.

Answer

1. Air needs to be moved from the outside to the inside of the body.
2. The body needs to carry O_2 and CO_2 in the blood. Erythrocytes serve this purpose.
3. The body needs a surface for gas exchange. The alveoli serve this purpose.
4. The circulatory system (internal convection system) moves gases throughout the body via systemic and pulmonary vessels.
5. The body needs a mechanism to locally regulate ventilation and perfusion.
6. The body needs a mechanism to centrally regulate ventilation.

Checkpoint! 13-5

What is the approximate percent hemoglobin saturation if a sample of whole blood has a PO_2 of 30 mmHg?

Answer

The approximate percentage is 50 (use hemoglobin–oxygen dissociation curve in Figure 13-4.

Checkpoint! 13-6

Predict whether the hemoglobin–oxygen dissociation curve will shift to the right or left for each of the following:
1. *Increased blood pH.*
2. *Increased 2,3-DPG.*

Answer

1. An increase in blood pH results in a shift to the left (Figure 13-6).
2. An increase in 2,3-DPG results in a shift to the right (Figure 13-4).

Checkpoint! 13-7

Carbon dioxide needs to be removed from cells, carried through the blood, and removed from the body. Identify the five different forms in which CO_2 is carried in the blood.

Answer

CO_2 is carried as:

1. bicarbonate (HCO_3^-),
2. carbonate (CO_3^{2-}),
3. carbonic acid (H_2CO_3),
4. dissolved CO_2, and
5. carbamino compounds.

Checkpoint! 13-8

Explain how peripheral and central chemoreceptors respond to changes in blood PCO_2.

Answer

The peripheral chemoreceptors are mainly sensitive to PO_2 but will respond to high PCO_2. The central chemoreceptors are sensitive only to increases in arterial PCO_2. The signals produce an increase in alveolar ventilation that tends to return the PO_2, PCO_2, and pH to normal.

Checkpoint! 13-9

Name the three most significant measured arterial blood-gas parameters that are required to evaluate acid–base status of a patient.

Answer

The three most significant measures are pH, PCO_2, and HCO_3^-.

Checkpoint! 13-10

Identify the specimen of choice for blood-gas determinations.

Answer

The specimen of choice for arterial blood gases is heparinized whole blood drawn from an artery.

Checkpoint! 13-11

Explain why calibration of a blood-gas analyzer is so important.

Answer

Calibration of a blood-gas analyzer is important because the calibration curve or data is used to generate all other results for quality-control and patient samples. If the calibration is not correct, then the results of all other measurements will be unacceptable.

Checkpoint! 13-12

Identify several sources of preanalytical errors and indicate briefly an example of a negative effect on patient care.

Answer

Sources of errors include:
1. introducing air into the syringe, which will tend to increase the PO_2 result;
2. delays in transport, which will allow the gases to continue to move in and out of the erythrocytes and other phases within the sample; and
3. not keeping specimens cold during transport, which will allow the continued movement of gases and other constituents between the phases of the sample.

Chapter 14

Checkpoint! 14-1

Identify the three forms of calcium that exist in plasma. Which form is the biologically active form?

Answer

Calcium in plasma exists in three distinct forms: free or ionized calcium, complexes with a variety of anions, and bound to plasma protein. The biologically active form is ionized calcium.

Checkpoint! 14-2

Identify the three major organs involved in both calcium and phosphorus homeostasis.

Answer
Kidney, bone, and small intestine.

Checkpoint! 14-3
Provide examples for each of the following: (1) two metallochromatic indicators used for measuring total serum calcium, (2) the ammonium complex widely used to measure serum inorganic phosphorus, and (3) two compounds used to measure serum total magnesium.

Answer
1. orthocresolphthalein complexone (CPC or OCPC) and arsenazo III;
2. molybdate;
3. any two of the following: calmagite, methylthymol blue, formazan dye, mango or xylidyl blue, chlorophosphan aso II, and arsenazo.

Checkpoint! 14-4
Indicate the principal effect of PTH on the kidneys, intestine, and bone.

Answer
Kidneys: Simultaneous reduction in reabsorption of sodium, phosphorus, calcium and bicarbonate ions in the proximal tubule and the enhanced reabsorption of calcium at the distal tubule.

Intestine: Absorption of dietary calcium is indirect; PTH stimulates the renal synthesis of the active vitamin D metabolite, $1,25(OH)_2D_3$, which in turn acts as a regulator of intestinal calcium absorption.

Bone: Bone resorption to restore calcium concentration in ECF. The end result of PTH action on bone is true bone resorption and not simply demineralization. PTH bone resorption is mediated by increased activity of osteoclasts. Increased conversion of osteoprogenitor cells to osteoclasts occurs as a consequence of more prolonged PTH stimulation.

Checkpoint! 14-5
Predict the serum concentration of total calcium and PTH in a patient with PHPT.

Answer
Total serum calcium is elevated and serum PTH is elevated.

Checkpoint! 14-6
Describe the difference between bone modeling and remodeling.

Answer
During bone modeling (endochrondral ossification), the calcified cartilage matrix serves as a site for deposition of mineral in the form of primary spongiosa. Mineralized crystals develop into woven, compact bone. The calcified cartilage and woven bone is removed and replaced by woven bone alone. This modeling is accomplished by the resorption of bone on the endosteal surface and the formation of periosteal bone.

Bone remodeling is a process that involves formation of bone on surfaces previously containing bone. There are two phases to remodeling: (1) Bone breakdown is started by the osteoclasts that adhere to the bone surface and create a secondary lysosome with an acid pH to dissolve the hydroxyapatite mineral. (2) The rebuilding phase depends on the recruitment of osteoblasts that lay down an osteoid and become mineralized in a tightly regulated fashion. *Bone turnover* refers to the amount of bone renewed during the bone remodeling process.

Checkpoint! 14-7
Identify two reasons why measurement of pyridinoline and deoxypyridinoline provides an advantage over measurement of urinary hydroxyproline for assessment of bone disorders.

Answer
Measurement of pyridinoline and deoxypyridinoline provides an advantage over urinary hydroxyproline because they are not influenced by dietary intake and are unaffected by the degradation of newly synthesized collagen.

Chapter 15

Checkpoint! 15-1
Identify in which class each of the following hormones belongs: epinephrine, insulin, and testosterone.

Answer
Epinephrine is derived from amino acids. Insulin is a polypeptide. Testosterone is a steroid.

Checkpoint! 15-2
The circulating levels of thyroxin in a 42-year-old female patient are below the normal range. Explain how the body attempts to increase the blood levels of T_4 in this patient using negative feedback.

Answer
When circulating levels of thyroxine (T_4) are low, the hypothalamus rapidly senses the decline in hormone output and increases production of hypothalamus-based thyroid-releasing hormone (TRH), which enters the portal circulation in the brain to stimulate pituitary hormone synthesis and the secretion of thyroid-stimulating hormone (TSH) to reestablish normal hormone output. This is the *positive-feedback loop.*

Checkpoint! 15-3
Distinguish between a stimulation test and a suppression test.

Answer
A stimulation test evaluates the secretory reserve of the gland when testing for hypofunction. A suppression test evaluates a hyperfunctioning gland by demonstrating its inability to suppress excessive hormone production.

Checkpoint! 15-4
Why is prolactin a sensitive indicator of pituitary dysfunction?

Answer
Prolactin is the hormone most frequently produced in excess by pituitary tumors; it is also the first hormone to become deficient from infiltrative disease or tumor compression of the pituitary.

Checkpoint! 15-5
Identify three hormones produced by the adrenal medulla.

Answer
Epinephrine, norepinephrine, and dopamine

Checkpoint! 15-6
Explain the cause of Addison's disease.

Answer
Addison's disease results from progressive destruction or dysfunction of the adrenal glands by a local disease process or systemic disorder.

Checkpoint! 15-7
Explain the purpose of the metyrapone and dexamethasone suppression tests.

Answer

The dexamethasone suppression test (DST) is used to document hypersecretion of the adrenocortical hormones. The metyrapone testing that assesses the pituitary reserve is useful for differential diagnosis of adrenal insufficiency.

Checkpoint! 15-8

Identify the fundamental approach to the assessment of disorders of the reproductive systems.

Answer

Disorders of reproductive systems may generally be viewed from the standpoints of hormone deficiency or excess as well as of primary (gonad) or secondary (pituitary) origin.

Checkpoint! 15-9

What form of the thyroid hormones is metabolically active?

Answer

The free forms of the thyroid hormones are metabolically active.

Checkpoint! 15-10

Briefly describe what is being measured or calculated for the following:

> *Total serum thyroxine*
> *Total serum triiodothyronine*
> *Free thyroxine*
> *Free thyroxine index*

Answer

Total serum thyroxine: Current assays measure the sum of both bound and free forms of thyroxine.

Total serum triiodothyronine: Current assays measure the sum of both bound and free forms of triiodothyronine.

Free thyroxine: Current assays measure the free (unbound forms) of thyroxine.

Free thyroxine index: The index is a calculated parameter based on the measurements of total thyroxine and T_3 resin uptake ($FT_4I = TT4 \times RT_3U$).

Chapter 16

Checkpoint! 16-1

Identify two primary functions of the GI tract.

Answer

The GI tract assimilates nutrients and eliminates waste.

Checkpoint! 16-2

Identify five examples of how GI regulatory peptides influence GI tract functions.

Answer

Overall, GI regulatory peptides influence motility, secretion, digestion, and absorption in the gut. They also regulate bile flow and secretion of pancreatic hormones and affect vascular wall tonicity, blood pressure, and cardiac output.

Checkpoint! 16-3

What two factors are associated with peptic ulcers?

Answer

The two main factors associated with peptic ulcers are *Helicobacter pylori* infection and the consumption of nonsteroidal antiinflammatory drugs (NSAIDs).

Checkpoint! 16-4

Which sugar is found in the feces of a patient who has lactase deficiency?

Answer

Glucose

Checkpoint! 16-5

What is the most notable characteristic of Crohn's disease?

Answer

Transmural inflammation of the GI tract

Checkpoint! 16-6

Identify the immunologic condition that results in celiac disease.

Answer

Celiac disease is caused by the inappropriate T-cell–mediated immune response to the dietary ingestion of gluten-containing grains such as wheat.

Checkpoint! 16-7

Where do carcinoid tumors originate?

Answer

Carcinoid tumors originate in the enterochromaffin cells of the GI tract.

Checkpoint! 16-8

A patient with ZES will have an increased *or* decreased *serum gastrin level and an* increased *or* decreased *BAO?*

Answer

Increased, increased

Checkpoint! 16-9

Identify the two principal immunoglobulins associated with tests for PUD.

Answer

IgA and IgG

Checkpoint! 16-10

Answer true or false: A patient whose glucose value during a lactose tolerance test is 20 mg/dL higher than the fasting baseline result is suspected of having lactase deficiency.

Answer

False

Chapter 17

Checkpoint! 17-1

1. *Inactive forms of enzymes that are stored in the pancreas are called _____.*

2. *What hormone is secreted by the following cells in the islets of Langerhans?*
   a. *Alpha cells:* _____
   b. *Beta cells:* _____
   c. *Delta cells:* _____

Answer

1. zymogens
2. a. glucagon
   b. insulin
   c. somatostatin

Checkpoint! 17-2

1. *Which test is the gold standard for evaluating pancreatic exocrine function?*
2. *What is measured in the pancreatic juice collected in the secretin–CCK test?*
3. *What is the major advantage of using pancreatic elastase-1 over secretin–CCK to assess pancreatic function?*
4. *The substrate 1,3-distearyl, 2(carbonzyl-C13) octanyl glycerol is used in which pancreatic function test?*

Answer

1. Secretin–cholecystokinin (secretin–CCK)
2. Volume, pH, bicarbonate, and enzymes are measured in the pancreatic juice collected in the secretin–CCK test.
3. Pancreatic elastase-1 is noninvasive and does not require intubation.
4. C-mixed-chain triglyceride test

Checkpoint! 17-3

1. *Which enzyme activation is the critical first step in acute pancreatitis?*
2. *What is the most common cause of chronic pancreatitis?*
3. *Using Ranson's indicators of severity in acute pancreatitis indicate whether the following would or would not meet the criteria. (Answer Yes or No.)*
   a. *BUN: 40 mg/dL* _____
   b. *Glucose: 210 mg/dL* _____
   c. *Calcium: 8.4 mg/dL (24h)* _____
   d. *AST: 310 U/L* _____

Answer

1. The activation of tyrpsinogen to trypsin is the critical early step because trypsin can activate most of the other pancreatic enzymes.
2. Alcoholism is the most common cause of chronic pancreatitis.
3. a. No
   b. Yes
   c. Yes
   d. Yes

Checkpoint! 17-4

1. *List and briefly describe four complications of pancreatitis.*
2. *Why are calcium levels decreased in acute pancreatitis?*
3. *Discuss two explanations for hyperlipidemia in acute pancreatitis.*

Answer

1. Complications of acute pancreatitis include adult respiratory distress syndrome (ARDS), cardiac complications, metabolic complications, gastrointestinal bleeding, and pancreatic infection and abscess.
   a. ARDS is a serious complication of acute pancreatitis. It may be caused by autodigestion of pulmonary capillaries by activated pancreatic enzymes, including phospholipase A_2.
   b. A variety of cardiac complications, including congestive heart failure (CHF), myocardial infarction, and cardiac arrhythmias may occur in severe acute pancreatitis.
   c. Metabolic complications include hypocalcemia, hyperglycemia, and hyperlipidemia.
   d. Gastrointestinal bleeding from peptic ulcers, pseudoaneurysms, or varices from splenic vein thrombosis may be a complication of acute pancreatitis.
   e. Hyperglycemia is another of Ranson's indicators, which can suggest a poor prognosis.
   f. Hyperlipidemia is associated with acute pancreatitis as an etiologic factor and as a consequence.
2. Calcium levels are decreased as a consequence of the fat necrosis that occurs in acute pancreatitis. This leads to the release of free fatty acids that occurs when pancreatic lipase is released by inflamed acinar cells and calcium binds with the free fatty acids to form soaps.
3. Hypertriglyceridemia can be an etiologic factor or occur as a consequence of acute pancreatitis. It can precipitate acute pancreatitis in patients with triglycerides of more than 1000 mg/dL or it can occur as a consequence of acute hepatitis associated with triglyceride levels of less than <500 mg/dL.

Chapter 18

Checkpoint! 18-1

Differentiate cTnI from cTnT.

Answer

cTnI:

- binds to actin and inhibits contraction,
- weighs 22 kDa, and
- is not expressed in skeletal muscle

cTnT:

- binds tropomyosin,
- weights 37 kDa, and
- is expressed in skeletal muscle.

Checkpoint! 18-2

List five conditions other than ACS where cTnI or cTnT may be elevated in blood.

Answer

1. trauma
2. nonatherosclerotic ischemia
3. hypertension
4. cerebrovascular accident
5. renal failure

Checkpoint! 18-3

Identify five characteristics of cTnT and cTnI that make them useful biomarkers for acute myocardial infarction.

Answer

1. They are specific for cardiac tissue.
2. They have high diagnostic specificity and sensitivity.
3. They possess early release kinetics after an AMI.
4. They remain elevated for a long interval of time.
5. They have extremely low to undetectable values in serum from patients without cardiovascular disease.

Checkpoint! 18-4

Why is CK-MB less useful than cTnI or cTnT as a biomarker for AMI?

Answer

CK-MB is less useful because it is not as cardiac specific as cardiac troponin I. CK-MB is found in numerous tissues throughout the body other than the heart.

Checkpoint! 18-5

Identify a significant advantage and disadvantage of measuring serum myoglobin in patients complaining of chest pain with a suspected AMI.

Answer

Advantage: Myoglobin is released early in the blood following an AMI and rises quickly above normal concentrations in such events.

Disadvantage: Myoglobin is found in many other tissues throughout the body, including skeletal muscle, so it is not highly specific test for a possible AMI.

Checkpoint! 18-6

Identify three differences between NT-proBNP and BNP that should be considered before deciding which assay to select.

Answer

Three major differences are:
1. half-lives,
2. molar concentrations, and
3. in vitro stability.

Checkpoint! 18-7

What is the clinical usefulness of hs-CRP?

Answer

Because hs-CRP is an independent marker of risk, it can be used in risk assessment of adults without known cardiovascular disease.

Checkpoint! 18-8

Explain the principle of the albumin cobalt binding test (ACB) used to measure ischemia modified albumin.

Answer

Cobalt is added to serum and does not bind to the NH_2 terminus of ischemia-modified albumin, thus leaving more free cobalt to react with the reagent dithiothreitol and form a darker color in samples from patients with ischemia.

Checkpoint! 18-9

What is the clinical usefulness of MPO in patients who present to the ED with chest pain?

Answer

MPO is useful for both short- and long-term risk stratification and may help identify troponin-negative patients at risk for major adverse cardiac events.

Chapter 19

Checkpoint! 19-1

1. *Blood leaves the liver through the* _____.
2. *The portal triad of the liver consists of what three structures?*
3. _____ *remove toxins from blood flowing through the sinusoids.*

Answer

1. hepatic vein
2. portal vein, hepatic artery, and bile duct make up the portal triad.
3. Kupffer cells

Checkpoint! 19-2

List five major categories of liver function and an example of each.

Answer

Among the five categories possible are the following:
1. Metabolism or synthesis: carbohydrates, amino acids and proteins, lipids and fats, bilirubin, and hormones
2. Excretion and secretory: bile acids, cholesterol, and bilirubin
3. Hematologic
4. Detoxification: bilirubin conjugation, alcohol, and drugs
5. Storage: glycogen, copper, and vitamins A, D, E, and K
6. Immunologic: phagocytosis, secretion of IgA

Checkpoint! 19-3

1. *What is the difference in chemical structure between unconjugated or indirect bilirubin and conjugated or direct bilirubin?*
2. *Which bilirubin is water soluble and can be excreted in the urine?*

Answer

1. Conjugated bilirubin has two glucoronic acid molecules added to the unconjugated molecule.
2. Conjugated bilirubin is water soluble, and unconjugated bilirubin is not.

Checkpoint! 19-4

Identify the following as prehepatic, hepatic, or posthepatic jaundice.
1. *Viral hepatitis:* _____
2. *Congestive heart failure:* _____
3. *Bile duct obstruction:* _____
4. *Cirrhosis:* _____
5. *Hemolytic disease of the newborn:* _____

Answer

1. hepatic
2. prehepatic
3. posthepatic
4. hepatic
5. prehepatic

Checkpoint! 19-5

1. *Briefly describe the symptoms and pathology of Reye's syndrome.*
2. *Categorize each of the following as hepatitis A, B, C, or D.*
   a. *Requires co-infection with hepatitis B:* _____
   b. *Transmitted through contaminated food and water supplies:* _____
   c. *Formerly called non-A, non-B hepatitis:* _____
3. *Do all alcoholics develop cirrhosis? Explain why or why not.*

Answer

1. Reye's syndrome is an acute and often fatal childhood syndrome that follows a viral infection such as mumps or measles. The child appears to recover, but then there is an abrupt onset of vomiting, diarrhea leading to delirium, and possibly coma and terminal respiratory arrest. Encephalopathy and fatty degeneration of the liver are hallmarks of Reye's syndrome.

2. *a.* HDV
   b. HAV
   c. HCV
3. No. Only 10% to 15% of alcoholics develop cirrhosis, which suggest a genetic link or multiple factors.

Checkpoint! 19-6

John has a grade-1 encephalopathy with slight ascites; albumin is 3.6, prothrombin is prolonged 3 seconds, and his bilirubin is 5.0 mg/dL. What is his Child–Pugh score? What class of cirrhosis is he—A, B, or C?

Answer

John's Child–Pugh score is 9, which is category B.

Chapter 20

Checkpoint! 20-1

Describe the structural composition of ferritin.

Answer

Ferritin consists of a protein shell surrounding an iron core; hemosiderin is formed when ferritin is degraded in secondary lysosomes. The outer shell of ferritin is composed of apoferritin with an interior ferric oxyhydroxide (FeCOOH) $\times$ crystalline core. There are approximately 24 subunits in apoferritin in either L (light) or H (heavy) ferritin chains.

Checkpoint! 20-2

Explain the mechanism for transporting iron into the cell using apotransferrin–Fe^{3+}.

Answer

Transferrin binds to the transferrin captor cells and becomes a part of the interior of the cell. The iron is released from transferrin, and apotransferrin is then transported back to the cell surface, ready to transport another transferrin molecule to the cell's interior.

Checkpoint! 20-3

Define hemochromatosis and include symptoms, secondary causes, and appropriate assessment of patients.

Answer

Hemochromatosis (primary) is a genetically related disease in which the body accumulates excess amounts of iron; it is one of the most common genetic diseases in humans. It has a distinctive triad of symptoms: bronzing of the skin, cirrhosis, and diabetes. Other conditions include cardiomyopathies, arrhythmias, and endocrine deficiencies. Secondary hemochromatosis is usually the result of problems with the administration and absorption of iron.

Checkpoint! 20-4

Identify the three fundamental steps for measuring iron in serum.

Answer

The three fundamental steps: (1) Iron is released from transferrin when the pH of the serum is decreased by using an acid; (2) Fe^{3+} is reduced to Fe^{2+} using, for example, ascorbic acid; and (3) Fe^{2+} is complex with a chromogen.

Checkpoint! 20-5

Briefly describe the structural configuration of a basic porphyrin without any specific substituent groups.

Answer

The porphyrins are a group of compounds that contain four monopyrrole rings connected by methene bridges to form a tetrapyrrole ring.

Checkpoint! 20-6

Distinguish between the mitochondrial and cytoplasmic reactions used in the biosynthetic pathway of porphyrins and heme.

Answer

Mitochondrion reactions
  ALA synthase

$$\text{Succinyl CoA} + \text{Glycine} \rightarrow \text{Delta-aminolevulinic acid} \quad (1)$$

Cytoplasm reactions
  PBG synthase

$$2 \text{ ALA} \rightarrow \text{Porphobilinogen (PBG)} \quad (2)$$

Hydroymethylbilane synthase

$$\text{PBG} \rightarrow \text{hydroxymethylbilane} \quad (3)$$

Uroporphyrinogen III synthase

$$\text{Hydroxymethylbilane} \rightarrow \text{Uroporphyrinogen III} \quad (4)$$

Uroporphyrinogen decarboxylase

$$\text{Uroporphyrinogen III} \rightarrow \text{coproporphyrinogen III} \quad (5)$$

Mitochondrion
  Coproporphyrinogen oxidase

$$\text{Coproporphyrinogen III} \rightarrow \text{Protoporphyrinogen IX} \quad (6)$$

Protoporphyrinogen oxidase

$$\text{Protoporphyrinogen IX} \rightarrow \text{Protoporphyrin IX} \quad (7)$$

Ferrochelatase and Iron (III)

$$\text{Protoporphyrin IX} \rightarrow \text{Heme} \quad (8)$$

Checkpoint! 20-7

State which porphyrins are elevated in urine for the following hepatic porphyrias: ADP, AIP and VP.

Answer

  ADP: coproporphyrin III
  AIP: uroporphyrin
  VP: coproporphyrin III and uroporphyrin

Checkpoint! 20-8

State which porphyrins are elevated in urine for the following diseases: CEP and EPP.

Answer

  CEP: uroporphyrin I and coproporphyrin I
  EPP: none

Checkpoint! 20-9

List the three strategies for assessment of porphyries.

Answer

The three strategies are (1) investigating the acute attack, (2) diagnosing the cause, and (3) investigating possible acute porphyrias when the patients are in remission.

Checkpoint! 20-10

Identify four examples of types of specimens that can be used for qualitative and quantitative analysis of porphyrins.

Answer

Four types of specimens are (1) urine, (2) serum, (3) feces, and (4) whole blood.

Checkpoint! 20-11

Identify the three pathways that are available to "free" heme that relocates to the cytoplasm.

Answer

The three pathways for heme include (1) inhibiting the uptake of iron from transferrin (see above), (2) inhibiting ferrochelatase (see above), and (3) chemically attaching itself to globin to form hemoglobin.

Checkpoint! 20-12

Explain the structural configuration of the hemoglobin molecule and include the primary, secondary, tertiary and quaternary components.

Answer

- The primary structure of globin consists of amino acid chains with varying lengths.
- A majority of the secondary structure of hemoglobin is α and non-α polypeptide chains arranged in helices, and some form nonhelical turns.
- Tertiary structure of hemoglobin refers to the arrangement of the helices into a three-dimensional, twisted structure.
- The quaternary structure of hemoglobin results from the attachment of the four globin chains to each other.

Chapter 21

Checkpoint! 21-1

What factors affect the rate of absorption of a drug?

Answer

The factors that affect a drug's absorption rate include the area of the absorbing capillary membrane and the solubility of the substance in interstitial fluid.

Checkpoint! 21-2

What is an important determinant of blood and tissue partitioning of a drug?

Answer

Relative binding of drugs to plasma proteins

Checkpoint! 21-3

Compare phase I- and phase II-type reactions.

Answer

Phase I reactions introduce functional groups on the parent compound, and phase II reactions develop covalent bonds with functional groups on the parent compound and endogenously derived compounds such as glucuronic acid.

Checkpoint! 21-4

A patient is diagnosed with alcoholic liver disease with an accompanying arrhythmia. The patient requires a daily dose of an antiarrhythmic drug. Will the elimination half-life of the arrhythmic drug be shorter or longer than that in a patient without alcoholic liver disease?

Answer

The elimination half-life will be longer because if the liver is not functioning properly—that is, metabolizing drugs—then the compound will stay in the body for a longer period of time.

Checkpoint! 21-5

In general, when is a sample to be drawn for the purpose of adjusting dosage?

Answer

Just before the next dose

Checkpoint! 21-6

A drug's effect on a host results from the formation of what type of complex?

Answer

Drug receptors

Checkpoint! 21-7

Identify two renal function tests that should be monitored in patients treated with an aminoglycoside.

Answer

Urea nitrogen and creatinine to monitor renal function

Checkpoint! 21-8

Why should the clinician monitor both peak and trough levels of aminoglycosides and vancomycin antibiotic compounds?

Answer

Patients may develop ototoxicity, nephrotoxicity, or other significant diseases if the concentrations of these drugs are allowed to reach toxic levels.

Checkpoint! 21-9

Describe how a false positive phenobarbital result measured by immunoassay could occur if a patient was also taking primidone.

Answer

Primidone metabolizes to phenobarbital. If a patient is on both of these medications, then the assay for phenobarbital will measure the entire phenobarbital present.

Checkpoint! 21-10

How does the structure of topiramate, which has anti-epileptic effects, differ from other anti-epileptic drugs?

Answer

Topiramate contains a sugar or monosaccharide moiety within its structure. This is not true of any other anti-epileptic drug.

Checkpoint! 21-11

Why should thiocyanate be measured in blood on a routine basis if a patient is taking sodium nitroprusside to treat hypertension?

Answer

Patients receiving sodium nitroprusside as treatment for hypertension will metabolize the compound to thiocyanate (SCN^-), which is toxic to the body.

Checkpoint! 21-12

Identify several serious health conditions that may arise if a patient's digoxin level is 3.0 ng/mL (therapeutic range is 0.5–1.5 ng/mL).

Answer

Toxic levels of digoxin can produce cardiac arrhythmias (e.g., ventricular arrhythmia or progressive bradydysrhythmia) that may be life threatening. Also, hyper- or hypokalemia (high and low blood levels of potassium) may result.

Checkpoint! 21-13
Identify the specimen of choice for measuring cyclosporine A.

Answer
Cyclosporine should be measured using whole blood. Heparin or EDTA samples may be used for most whole-blood assays.

Chapter 22

Checkpoint! 22-1
What is the mechanism for acetaminophen-induced hepatotoxicity?

Answer
Hepatotoxicity results from the saturation of enzymes catalyzing the normal conjugation reaction and results in acetaminophen being metabolized by mixed-function oxidases. This reaction produces a toxic metabolite that causes necrosis in the liver.

Checkpoint! 22-2
Explain the mechanism associated with the metabolic acidosis that develops and identify three acids that accumulate in the blood of a patient who has overdosed with salicylate.

Answer
Carbon dioxide accumulates in the blood because of increased production and retention. A reduction in plasma bicarbonate also develops along with the accumulation of metabolites of pyruvic, lactic, and acetoacetic acid, which leads to a metabolic acidosis.

Checkpoint! 22-3
Identify three sources of exposure to carbon monoxide that may produce toxic levels in blood.

Answer
Three sources of exposure to carbon monoxide are:

• accidental poisoning through faulty furnaces or by gas stoves,
• remaining in an automobile with the engine running for a prolonged period of time with all of the windows and doors closed,
• smoking inhalation because of a fire, and
• attempted suicide using gas stoves or automobiles.

Checkpoint! 22-4
List four significant factors that the technologist should be familiar with in reference to drug abuse screening methods.

Answer
1. detection limit
2. specificity
3. cutoff values
4. interpreting positive versus negative results

Checkpoint! 22-5
Identify six classes of abused drugs that a clinical laboratory can screen for and indicate at least one pharmacological effect associated with each group.

Answer
Classes of abused drugs include the following:

1. amphetamines: psychomotor stimulants
2. barbiturates: sedatives or hypnotics
3. benzodiazepines: antianxiety medications
4. cocaine: psychomotor stimulant
5. opiates: analgesics
6. phencyclidine: psychotomimetic
7. tricyclics: antidepressants and antipsychotics

Checkpoint! 22-6
Explain the mechanism for the adverse effects of lead intoxication.

Answer
Lead inhibits the body's ability to make hemoglobin by interfering with several enzymatic steps in the heme pathway. Specifically, lead decreases heme biosynthesis by inhibiting aminolevulinic acid dehydratase and ferrochelastase activity. Ferrochelatase, which catalyzes the insertion of iron into protoporphyrin IX, is quite sensitive to the presence of lead. A decrease in the activity of this enzyme results in an increase of the substrate, erythrocyte protoporphyrin (EP), in the red blood cells.

Checkpoint! 22-7
What are two treatment modalities for methanol intoxication?

Answer
1. administration of larges doses of ethanol
2. hemodialysis

Checkpoint! 22-8
Gamma hydroxybutyrate and ketamine are commonly referred to as date-rape drugs, club drugs, and rave drugs. Identify several adverse effects of these drugs on humans.

Answer
These compounds can cause deep but short-lasting comas, often only 1 to 4 hours. Gamma hydroxybutyrate (GHB) is claimed to produce a variety of desirable effects, including improved athletic performance, sleep, sexual prowess, mood elevation, and euphoria. It is a relatively fast-acting depressant: Within about 15 minutes of administration, drowsiness, dizziness, euphoria, nauseas, visual disturbance, and unconsciousness may ensue. The duration of action is about 3 hours. Some users present with combative or aggressive behavior.

The effects of ketamine are, sedation, immobilization, amnesia, analgesia, and a trance-like cataleptic feature in which the patient seems to be "dissociated" from his or her environment but not actually asleep. Abusers usually snort, smoke, or inject the drug or on occasion, as with date-rape incidents, mix it in drinks.

Chapter 23

Checkpoint! 23-1
Identify three examples of medicinal preparations that are commonly used by the public that contain aluminum-based compounds.

Answer
Two examples of antacids are aluminum hydroxide and aluminum carbonate. An example of an astringent is potassium aluminum sulfate.

Checkpoint! 23-2
Name the disease associated with toxic levels of beryllium.

Answer
Berylliosis or chronic beryllium disease

Checkpoint! 23-3
State the cause of the very long biological half-life (~10 years) of cadmium.

Answer
The slow release of cadmium from metallothioenin complex

Checkpoint! 23-4
Identify a specific function of chromium relative to insulin receptors on human cell membranes.

Answer
Chromium enhances the action of insulin receptors on human cell membranes.

Checkpoint! 23-5
Which vitamin requires the presence of cobalt within its structure to function properly?

Answer
Vitamin B_{12}

Checkpoint! 23-6
What are the expected blood levels (increase, decrease, or normal) of plasma copper and ceruloplasmin in a patient with Wilson disease?

Answer
Plasma copper and ceruloplasmin is decreased.

Checkpoint! 23-7
State a specific example of the importance of manganese in human cells.

Answer
Manganese is a constituent of several important metalloenzymes that facilitate specific enzyme reactions.

Checkpoint! 23-8
Identify at least two examples of specific treatment modalities available for patients who have been poisoned with mercury.

Answer
Among possible examples are:
1. chelating agents such as BAL, succimer, DMPS, or NAP;
2. hemodialysis; and
3. a combination of both chelating agents and hemodialysis.

Checkpoint! 23-9
List three examples of the importance of selenium in the proper maintenance of health in humans.

Answer
Among possible examples are the following:
1. Selenium is important for healthy immune responses.
2. Selenium has a protective effect against some forms of cancer (e.g., prostate, colon, and lung).
3. Selenium is associated with increased cardiovascular disease mortality.
4. Selenium has been shown to regulate the inflammatory mediators in asthma.

Checkpoint! 23-10
Identify a significant finding involving the skin that is associated with patients having argyria.

Answer
Permanent bluish-gray discolorization of skin caused by silver deposits in tissue

Checkpoint! 23-11
Indicate the specimen of choice for measuring zinc in patients suspected of being deficient in zinc. Also state the reason(s) why this is the specimen of choice.

Answer
Plasma specimens are preferred to serum because of the possibility of zinc contamination from erythrocytes, platelets, and leukocytes during clotting and centrifugation.

Chapter 24

Checkpoint! 24-1
Distinguish between a functional assay and a direct assay used to assess vitamin status in humans.

Answer
Functional assays are indirect measures of nonvitamin substance, increased or decreased activity of reactions, cells, responses to inhibitors or activators that reflect alterations in vitamin concentrations.

Direct assays are described as methods that can detect and quantitate specific vitamins and possibly their respective metabolites using any of several instrumentation principles.

Checkpoint! 24-2
Briefly describe the role vitamin A deficiency in the development of nyctalopia.

Answer
Individuals who lack enough vitamin A have a reduced number of goblet mucous cells, which are replaced by basal cells that tend to proliferate. This leads to an increase in stratification and keratinization of the epithelium. Normal secretions are suppressed, resulting in increased irritation and infection. When this process happens in the cornea, degenerative changes occur that lead to poor dark adaptation or night blindness (nyctalopia).

Checkpoint! 24-3
Describe the role of vitamin E as an antioxidant.

Answer
Vitamin E compound act as a chain-breaking antioxidant and is a pyroxyl radical scavenger that protects LDL and polyunsaturated fats in membranes from oxidation. Through this action, vitamin E helps protect the body against the damaging effects of free radicals.

Checkpoint! 24-4
Identify the enzyme that catalyzes the following two reactions nearly simultaneously: (1) decarboxylates decarboxy prothrombin to prothrombin, and (2) oxidizes vitamin K_1 or K_2 (KH_2) to oxidized vitamin K_1 or K_2 (KO).

Answer
Gamma-gluamlcarboxylase

Checkpoint! 24-5
Identify the two types of beriberi and explain the principal differences between the two types.

Answer
The two types of beriberi are wet and dry. The principal difference between them relates to the symptoms a patient experiences. Patients with wet beriberi exhibit cardiovascular symptoms resulting from impaired myocardial energy metabolism. Patients may present with tachycardia, enlarged heart, CHF, peripheral edema, and peripheral

neuritis. Patients with dry beriberi often present with peripheral neuropathy of the motor and sensory systems.

Checkpoint! 24-6

What condition is characterized by presence of Casal's necklace and a bright red tongue?

Answer

Pellagra

Checkpoint! 24-7

Identify two specific liver function enzymes that use pyridoxal-5′-phosphate as a coenzyme.

Answer

Alanine aminotransferase (ALT) and aspartate aminotransferase (AST)

Checkpoint! 24-8

Identify the origin of the chemical term cyanocobalamin.

Answer

A variable R group attached to the cobalt atom is able to bind a –CN functional group, thus cyanocobalamin.

Checkpoint! 24-9

List three conditions that may result in a deficiency of folic acid.

Answer

Possible conditions that may lead to folic acid deficiency include (1) an insufficient number of intestinal microorganism (stomach sterilization), (2) inadequate intestinal absorption (e.g., surgical resection or celiac disease), (3) decreased dietary intake (alcoholics), (4) increased demand (e.g., during pregnancy), and (5) administration of antifolic acid drugs (e.g., methotrexate) or anticonvulsant drugs (e.g., phenobarbital), which tend to increase demand for folate.

Checkpoint! 24-10

List three symptoms of scurvy.

Answer

Symptoms of scurvy include bleeding into the skin (petechiae, ecchymosis), bleeding and irritated gums, and bleeding elsewhere such as joints, the peritoneal cavity pericardium, and the adrenal glands.

Chapter 25

Checkpoint! 25-1

Identify the four types of RNA.

Answer

The four types of RNA are messenger RNA, transfer RNA, ribosomal RNA, and small RNAs.

Checkpoint! 25-2

Describe the process of transcription and translation.

Answer

In the transcription process, the enzyme RNA polymerase binds to the promoter region of the gene, separating the DNA into two single strands. The polymerase proceeds in a 3′ to 5′ direction along one of the strands to produce a complementary strand of mRNA by adding the appropriate ribonuclotides until the termination sequence is reached. The resulting mRNA is released from the DNA, which can then serve as a template to make additional copies of mRNA.

In the translation process, the nucleotide sequence in mRNA is converted to a linear sequence of amino acids to create a protein.

Checkpoint! 25-3

Nucleic acids can be used as analytes in clinical assays to determine the presence of specific diseases or the likelihood that individuals could develop these diseases. This describes which type of diagnostic assay?

Answer

Molecular diagnostic assays

Checkpoint! 25-4

Identify the four basic steps used to isolate DNA.

Answer

Step 1 Collect sample
Step 2 Lyse cells and nuclear membranes to release DNA
Step 3 Remove membrane fragments, proteins, and RNA
Step 4 Concentrate the purified DNA

Checkpoint! 25-5

What is the major advantage of PAGE?

Answer

The major advantage of PAGE is its high degree of resolving power and its ability to distinguish nucleic acid fragments that differ by only a single base pair.

Checkpoint! 25-6

Identify four clinical applications of FISH assays.

Answer

The following possibilities can be chosen:

detection of microdeletions
detection of abnormalities in chromosome number
analysis of tumor cells
monitoring patients with sex-mismatched bone marrow transplants
screening for chromosomal mosaicism
genetic analyses of embryos

Checkpoint! 25-7

In the Western blot technique, what type of molecule is used to identify the protein targets?

Answer

Antibodies

Checkpoint! 25-8

Describe a DNA microarray.

Answer

A DNA microarray consist of an orderly arrangement of specific gene sequences that have been immobilized onto precise locations of a small solid support such as silicon chips, nylon membrane, or glass microscope slides.

Checkpoint! 25-9

What is the purpose of amplification?

Answer

Amplification is necessary to detect the presence of a nuclei acid in a sample, especially in cases where insufficient numbers of DNA or RNA molecules are present.

APPENDIX O
Answers to Review Questions

CHAPTER 1
Level I
1. B
2. B
3. D
4. B
5. C
6. D
7. D
8. B

Level II
1. C
2. B
3. D
4. C

CHAPTER 2
Level I
1. B
2. D
3. B
4. A
5. D
6. C
7. D
8. C
9. A
10. D
11. C
12. A

Level II
1. B
2. A
3. C
4. D
5. A
6. C

CHAPTER 3
Level I
1. B
2. A
3. D
4. B
5. B
6. D
7. A
8. C

Level II
1. A
2. A
3. D
4. C

CHAPTER 4
Level I
1. B
2. D
3. C
4. B
5. A
6. C
7. C
8. A
9. C

Level II
1. C
2. A
3. A
4. C
5. B
6. C
7. A
8. D
9. C
10. B
11. C
12. C
13. B
14. A
15. B
16. A

CHAPTER 5
Level I
1. C
2. A
3. D
4. C
5. A

Level II
1. B
2. D
3. C
4. B
5. A

CHAPTER 6
Level I
1. C
2. C
3. C
4. D
5. D
6. B
7. C
8. B
9. B
10. D
11. D
12. C
13. B
14. D
15. A
16. C
17. C
18. D
19. B
20. A
21. D
22. C
23. B

Level II
1. A
2. D
3. D
4. C
5. C

CHAPTER 7
Level I
1. A
2. A
3. D
4. C
5. C
6. B
7. B
8. B
9. A
10. A
11. E
12. A
13. A
14. C
15. E
16. A

17. A
18. B
19. B
20. A
21. A
22. B
23. D

Level II
1. C
2. B

CHAPTER 8
Level I
1. C
2. E
3. B
4. B
5. C
6. B
7. B
8. C
9. B
10. C
11. B
12. D
13. C
14. A
15. D
16. A
17. C
18. A
19. C
20. A

Level II
1. B
2. D
3. C
4. C
5. A

CHAPTER 9
Level I
1. E
2. B
3. C
4. A
5. C
6. B
7. B
8. A
9. A
10. A
11. D
12. D
13. B
14. B
15. B
16. C

17. B
18. C
19. B
20. B
21. D
22. D

Level II
1. A
2. D
3. C

CHAPTER 10
Level I
1. A
2. D
3. C
4. A
5. C
6. D
7. A
8. B
9. B

Level II
1. C
2. D
3. A
4. C
5. A
6. A
7. C
8. A
9. B

CHAPTER 11
Level I
1. B
2. D
3. C
4. E
5. E
6. A
7. D
8. C
9. A
10. B
11. A
12. B
13. D
14. D
15. A
16. D
17. A
18. A
19. B
20. D
21. D

Level II
1. A
2. A
3. B
4. D

CHAPTER 12
Level I
1. C
2. D
3. B
4. C
5. A
6. B
7. C
8. A
9. C
10. B
11. B
12. C
13. A

Level II
1. D
2. B
3. A
4. C
5. C
6. A
7. D
8. C
9. D
10. D
11. B
12. C
13. C

CHAPTER 13
Level I
1. C
2. D
3. D
4. C
5. B
6. C
7. B
8. C

Level II
1. B
2. A
3. B
4. C
5. A
6. B
7. C
8. D
9. A
10. C
11. B
12. B

CHAPTER 14
Level I
1. C
2. A
3. C
4. A
5. B
6. D
7. C
8. B
9. A

Level II
1. A
2. C
3. B
4. A
5. D
6. C
7. B
8. A
9. A
10. B
11. D

CHAPTER 15
Level I
1. B
2. D
3. C
4. C
5. A
6. A
7. B
8. C
9. A
10. D
11. B
12. B

Level II
1. A
2. C
3. D
4. A
5. C
6. D
7. A
8. C
9. C

CHAPTER 16
Level I
1. B
2. A
3. D
4. C
5. B
6. C
7. D

Level II
1. A
2. C
3. B
4. C
5. A
6. D
7. B

CHAPTER 17
Level I
1. C
2. D
3. B
4. C
5. C
6. D
7. A
8. B
9. B
10. B
11. C
12. D
13. D
14. C
15. C
16. A

Level II
1. B
2. C
3. D
4. B

CHAPTER 18
Level I
1. D
2. A
3. A
4. C
5. B
6. D
7. C
8. B
9. D

Level II
1. A
2. D
3. D
4. B
5. C
6. A

CHAPTER 19
Level I
1. A
2. A
3. E
4. D

5. E
6. A
7. D
8. A
9. A
10. C
11. C
12. A
13. C
14. C
15. B
16. B
17. D
18. A
19. D
20. C

Level II
1. D
2. C
3. D
4. A
5. C

CHAPTER 20
Level I
1. B
2. A
3. C
4. A
5. C
6. D
7. B
8. A
9. C
10. D

Level II
1. B
2. D
3. C
4. B
5. B
6. C
7. C

CHAPTER 21
Level I
1. B
2. C
3. C
4. D
5. C
6. A
7. C

Level II
1. A
2. A
3. C
4. B

5. C
6. B
7. D
8. A
9. B
10. A
11. D
12. C
13. D

CHAPTER 22
Level I
1. D
2. B
3. A
4. D
5. B
6. C
7. C
8. A
9. C

Level II
1. C
2. C
3. A
4. C
5. B
6. B
7. C
8. C
9. B
10. B

CHAPTER 23
Level I
1. A
2. A
3. B
4. C
5. D
6. C
7. B
8. A
9. A

Level II
1. B
2. A
3. D
4. D
5. C
6. C
7. B
8. C
9. C
10. D
11. B
12. D

CHAPTER 24
Level I
1. B
2. B
3. A
4. A
5. D
6. C

7. A
8. B

Level II
1. B
2. D
3. B
4. D
5. C
6. C
7. A
8. B

CHAPTER 25
Level I
1. B
2. D
3. B
4. A
5. C
6. C
7. B
8. D

Level II
1. B
2. C
3. C
4. C
5. B
6. D
7. C
8. D
9. A

Glossary

▶ A

Absolute specificity Enzyme that catalyzes only one specific substrate and one reaction.

Absorption, drug Transfer of drug from its site of administration to bloodstream.

Accuracy Closeness of the agreement between the measured value of an analyte to its "true" value; accuracy is also defined as the estimate of nonrandom, systematic bias between samples of data or between a sample of data and the true population value.

Acidemia Condition of decreased pH of blood.

Acidosis Pathological condition characterized by increased acidity of blood because of accumulation of acids or excessive loss of bicarbonate; fluid hydrogen in concentration increases, lowering pH.

Acid phosphatase Group of hydrolases with maximal pH between 4.0 and 5.0; frees inorganic phosphate from an organic phosphate monoester. The highest concentration is found in the prostate gland; for this reason, it was historically a screening test for prostate cancer.

Acini Small clusters of glandular epithelial cells that make up 98% of pancreatic mass; produce pancreatic juice, which is carried to duodenum through duct system.

Acridinium ester Acridinium is a compound used as a label for chemiluminescence immunoassays; during the immunoassay, acridinium is converted to a sulfonyl acridinium ester, which is oxidized with alkaline hydrogen peroxide in the presence of a detergent and produces a rapid flash of light at 429 nm; this label possesses high specific activity and can be used to label both antibodies and haptens.

Activators Include metallic ions—for example, Fe^{+2}, Ca^{+2}, Mg^{+2}, Zn^{+2}, Cu^{+2}, and Mn^{+2}; also called *inorganic cofactors,*

Active site The portion of the enzyme where the substrate molecule fits and forms temporary bonds.

Acute coronary syndrome (ACS) Continuum of clinical signs and symptoms ranging from unstable angina (chest pain) to non–Q wave (represents EKG changes) AMI and Q-wave AMI.

Acute glomerulonephritis (AGN) Renal disease characterized by rapid onset of symptoms that indicate damage to glomeruli; most often associated with children and young adults following group A streptococcal infection but also associated with systemic lupus erythematosus, subacute bacterial endocarditis, various forms of vasculitis, and other conditions. Symptoms include rapid onset, fever, malaise, nausea, oliguria, hematuria, and proteinuria.

Acute pancreatitis Condition when pancreatic enzymes trypsinogen and chymotrypsin are prematurely activated in pancreas and begin digesting pancreatic cells; symptoms include epigastric pain, nausea, vomiting, and elevated levels of amylase and lipase; alcohol abuse and obstruction of pancreatic duct by gallstones are most common causes of acute pancreatitis.

Acute-phase reactions When both major α_1 proteins (α_1-antitrypsin and α_1-glycoprotein), α_2-band proteins (ceruloplasmin and haptoglobin), β-globulins, C3, C4, and C-reactive protein are increased and albumin is decreased.

Acute pyelonephritis Condition most often seen in women as result of untreated cases of cystitis or lower urinary tract infections; an ascending infection; usually does not cause permanent damage to renal tubules.

Acyl-cholesterol acyltransferase (ACAT) Enzyme that converts free cholesterol to esterified cholesterol for storage.

Addison's disease Primary adrenal insufficiency caused by progressive destruction of adrenal gland because of infectious diseases such tuberculosis or infiltration by neoplastic tissue.

Adenohypophysis Anterior glandular lobe of pituitary gland.

Adenosine diphosphate (ADP) Coenzyme composed of adenosine and two molecules of phosphoric acid that is important in intermediate cellular metabolism.

Adenosine triphosphate (ATP) Adenosine-derived nucleotide containing high-energy phosphate bonds used to transport energy to cells for biochemical processes, including muscle contraction and enzymatic metabolism, through its hydrolysis to ADP; hydrolyzed to AMP when incorporated into DNA or RNA.

Adrenal cortex Outer portion of adrenal gland; produces several steroid hormones.

Adrenal medulla Inner portion of both adrenal glands that produces catecholamines such epinephrine and norepinephrine.

Adrenergic symptoms Symptoms associated with reactive hypoglycemia, including sweating, nervousness, faintness, palpitations, and hunger.

Adrenocorticosteroids Group of steroids secreted by the adrenal cortex, including cortisol and aldosterone.

Adult respiratory distress syndrome (ARDS) Serious complication of acute pancreatitis; may be caused by autodigestion of pulmonary capillaries by activated pancreatic enzymes, including phospholipase A_2; inflammation of lung parenchyma leads to impaired gas exchange.

Affinity Thermodynamic quantity defining the energy of interaction of a single antibody binding site and its corresponding epitope on the antigen.

A:G ratio. *See* Albumin:globulin ratio.

Agonist Drug that has affinity for receptor and intrinsic activity at receptor.

Alanine aminotransferase (ALT) An intracellular enzyme involved in amino acid metabolism; catalyzes the transfer of an amino group from alanine to α-oxoglutarate; present in high concentrations in the liver and one of the analytes measured in a liver function profile.

Albumin Synthesized in liver; makes up approximately 60% of total serum protein; chief biological function is to maintain plasma colloidal osmotic pressure.

Albumin:globulin ratio (A:G ratio) Ratio of albumin to globulin proteins in serum; reference range for ratio is approximately 1.0–1.8, with albumin levels normally higher than globulins.

Albumin methodologies Laboratory assay methods based on binding of albumin with anionic dyes such as bromcresol green and bromcresol purple.

Alcoholic cirrhosis Most severe of alcoholic liver diseases, causing damage that leads to irreversible scarring; occurs in approximately 10–15% of individuals who drink heavily for more than 10 years; characterized pathologically by fibrosis and conversion of normal liver architecture into structurally abnormal lobules; associated complications include portal hypertension, varices, edema, ascites, bruising, itching, and hepatic encephalopathy.

Alcoholic fatty liver Fatty infiltration of the liver with fat collected in vacuoles that would be found on biopsy; benign, reversible disease with abstinence from alcohol leading to complete recovery.

Alcoholic hepatitis (AH) Liver disease that occurs with excess alcohol consumption over a year; has wide variety of symptoms, including hepatomegaly, vomiting, jaundice, ascites, fever, abdominal pain, and peripheral neuritis; prognosis depends on severity and type of damage, and mortality ranges from 2% to 27%.

Alcoholic liver disease (ALD) Hepatic injury caused by long-term consumption of alcohol; alcohol is catabolized to acetalaldehyde by alcohol dehydrogenase, and acetalaldehyde becomes hepatocyte toxin when concentrations become too high.

Aldehyde Molecule containing a terminal carbonyl group.

Aldolase (ALD) Enzyme present in the liver, skeletal, and heart muscle; important in the metabolic pathway converting glycogen to lactic acid; elevated in certain muscle diseases and hepatitis.

Aldose Monosaccharide produced when an aldehyde is at the end position of a carbon chain (first or last).

Aliquot A known amount of a sample such as blood or urine; a process to divide a solution into aliquots.

Alkalemia Condition of increased blood pH.

Alkaline phosphatase (ALP) Enzyme with an optimal pH of 9.0–10.0; frees inorganic phosphate from an organic phosphate monoester; present in highest concentration in the cells of the biliary tract and osteoblasts, so elevated levels are present in obstructive liver disease and various bone disorders.

Alkalosis Condition resulting in increased blood alkalinity because of accumulated of alkalies or reduction of acids.

Alkaptonuria A rare inherited disease that results from the deficiency of the enzyme homogentisic acid oxidase in the catabolic pathway of tyrosine; deficiency leads to a buildup of homogentisic acid in the tissues of the body.

Allosteric sites Sites on an enzyme other than the active site.

Alopecia Loss of hair especially on head from any of several causes, including exposure to toxic substances such as selenium, endocrine disorders, tumors, and treatment modalities such as radiation.

α_1-acid glycoprotein (AAG) Major glycoprotein that increases during inflammation; elevated levels are found in rheumatoid arthritis, cancer, pneumonia, and other conditions.

α_1-antichymotrypsin (ACT) Serine proteinase that catalyzes chymotrypsin and mast cell chymase.

α_1-antitrypsin (AAT) Major α_1-globulin making up approximately 90% of α_1-proteins; a glycoprotein, serum trypsin inhibitor, and acute-phase reactant with antiprotease activity; deficiency of AAT results in emphysema or chronic obstructive pulmonary disease in adults and cirrhosis in adults or children.

α_1-antitrypsin deficiency Deficiency in most important serine protease inhibitor; one of most common genetically lethal diseases in Caucasians and associated with lung and liver disease; deficiency is associated with hepatitis in newborns and early-onset emphysema in adults.

α_2-macroglobulin (AMG) One of the largest plasma proteins; involved in primary or secondary inhibition of enzymes in the complement, coagulation, and fibrinolytic pathways.

Alpha-fetoprotein Principal fetal protein (fetal albumin-like protein) in maternal serum used to screen for the antenatal diagnosis of neural tube defects, including spina bifida and anencephaly.

Alveoli Air sacs of lungs

Amino acids Molecule containing amino group (NH_2), carboxyl group (COOH), hydrogen, and R group (radical or side chain) with formula $RCH(NH_2)COOH$.

Ammonia Product of amino acid and protein catabolism that is converted to urea in liver by Krebs–Henselheit urea cycle; markedly elevated in cases of impending hepatic coma and Reye's syndrome.

Amperometry Measurement of current flow produced by oxidation–reduction reaction.

Amphoteric Containing two ionizable sites: proton-accepting group (NH_2) and proton-donating group (COOH).

Ampholyte When both COOH and NH_2 groups on amino acid become ionized NH_3^+.

Amplicon Product of PCR.

Ampulla of Vater Single duct formed where pancreatic duct joins common bile duct from liver and gallbladder and enters duodenum; also called *hepatopancreatic ampulla*.

Amylase (AMY) Class of enzymes that catalyze the hydrolysis of starch; two main sources of amylase are the pancreas and salivary glands, so serum levels are elevated in pancreatic and salivary gland disorders.

Amylase creatinine clearance ratio (ACCR) Ratio that compares the clearance of amylase to creatinine on the same urine and serum; elevated in acute pancreatitis (>8%) because the renal clearance of AMY is greater than that of CR.

Analbuminemia Absence of albumin, a genetic autosomal recessive trait.

Analgesic Class of drugs that relieve pain without causing loss of consciousness.

Anderson's disease Type IV glycogen storage disease; associated with glycogen brancher enzyme deficiency in liver, brain, heart, skeletal muscles, and skin fibroblasts.

Androgen Class of sex hormones that includes testosterone, which is responsible for developing masculine traits.

Anemia Pathological condition that develops when iron intake falls below amount required for red blood cell production and iron reserves become depleted.

Anencephaly Congenital absence of the brain and cranial vault; condition is incompatible with life.

Angiogenesis Development of blood vessels.

Anion gap Difference between sum of measured cations and sum of measured anions.

Antagonist Drug that has affinity for receptor but is devoid of intrinsic activity.

Anterior pituitary Lobe of pituitary gland that lies on anterior side; source of growth hormone, ACTH, thyroid-stimulating hormone, follicle-stimulating hormone, and luteinizing hormones.

Antibody Immunoglobulin that binds specifically to an antigen or hapten.

Anticodon Unit of three nucleotides in tRNA that pairs with specific codon on mRNA during process of translation.

Antidepressant Class of drugs that prevent, cure, or alleviate mental depression; tricyclic and tetracyclic antidepressants are commonly used by patients for this purpose.

Antigen Substance capable of reacting with antibody without necessarily being capable of inducing antibody formation.

Antioxidant Agent that prevents or inhibits oxidation; naturally occurring or synthetic substance that helps protect cells from damaging effects of oxygen free radicals.

Anxiolytic Refers to drug that is prescribed to treat symptoms of anxiety (antianxiety medication).

Apnea Temporary cessation of breathing

Apoenzyme The heat labile protein portion of an enzyme.

Apo(a) Unique plasminogen-like glycoprotein linked by disulphide linkage to apolipoprotein B-100, which is anchored in lipid-rich LDL-like core.

Apolipoproteins Proteins found as integral part of outer shell of lipoprotein molecules; main function is metabolizing and transporting lipoproteins; designated as Apo AI, Apo AII, Apo B48, Apo B 100, Apo CI, ApoC II, Apo C III, and Apo E.

Apoprotein Conjugated proteins without their nonprotein groups or ligands.

Apotransferrin Protein in plasma that transports iron from one organ to another; β_1-globulin with molecular mass of 75kDa, consisting of two binding sites, one for Fe^{3+} and one for bicarbonate (HCO_3^-).

Argyria Symptom occurring in humans exposed to silver; characterized by permanent bluish-gray discoloration of skin, mucous membranes, and nails from silver deposits in those tissues; associated with growth retardation, hemopoiesis, cardiac enlargement, liver degeneration, and renal tubule destruction.

Arithmetic average Similar to mean, or average number produced when all values are totaled and then divided by total number of items.

Arteriosclerosis Hardening of arteries, especially coronary arteries; results from deposition of tough, rigid collagen inside vessel wall and around atheroma.

Aspartate aminotransferase (AST) One of a group of enzymes that catalyze interconversion of amino acids and α-oxoacids by transfer of amino groups; catalyzes deamination of aspartate to oxalacetate; highest levels found in liver, muscle, and brain; its measurement is included in liver function profile.

Atherogenic Causing formation of fatty degeneration or thickening of walls of large arteries that occurs in atherosclerosis.

Atheroma Fatty degeneration or thickening of walls of larger arteries.

Atherosclerosis Most common form of arteriosclerosis; affects arterial blood vessels and is caused by formation of multiple plaques within coronary arteries.

Atomic mass unit (amu) Unit of mass used by chemists and physicists for measuring masses of atoms and molecules; equivalent to 1/12th the mass of most common atoms of carbon: carbon-12 atoms (6 protons and 6 neutrons); 1 mole of substance contains just as many elementary entities (atoms, molecules, ions, or other kinds of particles) as atoms in 12 grams of ^{12}C.

Avidity Overall strength of binding of antibody to antigen; includes sum of binding affinities of all individual binding sites on the antibody.

Avogadro's number Actual number of "elementary entities" in 1 mole; careful measurement has determined it to be approximately 6.02×10^{23}; named for Italian chemist and physicist Amedeo Avogadro (1776–1856).

Azotemia Increased level of blood urea and other NPN compounds in blood.

 B

Bar code Series of parallel lines or squares of varying thickness, printed in a fashion to represent numbers or numbers and letters and which can be read by automated analyzers.

Barometric pressure Barometric pressure of all gases on face of Earth; at sea level [1 atmosphere (atm)], approximately 760 mmHg (torr).

Basal levels Standard low levels of activity of an organism, organ, or function, as during total rest.

Base excess Difference between titratable acids and bases of blood sample and normal blood sample at pH of 7.4, pCO$_2$ of 40 mmHg, and temperature of 37°C.

Batch analysis When group of samples is prepared for analysis and single test is performed on each sample in group.

Beer's law Concentration of substance is directly proportional to amount of light energy absorbed.

Benign Disease or condition that is not malignant, recurrent, or progressive.

β_2-microglobulin Low-molecular-weight protein on cell membrane of most nucleated cells; found in especially high levels in lymphocytes; makes up common light chain of class I major histocompatability complex antigens found in all nucleated cells.

Bias Difference between two means or the mean difference; also presence of nonrandom events (for example, estimating serum cholesterol levels in apparently normal healthy individuals without knowing subjects were on cholesterol-lowering medications); also described as lack of accuracy.

Bidirectional interface Computer interface that allows laboratory information system to simultaneously transmit or download information and receive uploaded information such as specimen accession numbers, tests to be performed, and test results from instrument.

Bile acids Include cholic acid and taurocholic acid, which are synthesized in liver and flow into lumen of small bowel via gallbladder; exist as salts; major functions are to act as surface-active agents, form micelles, and facilitate digestion of triglycerides, absorption of cholesterol, and fast-soluble vitamins.

Bile canaliculi   Small canal that carries secretions and excretions away from hepatocytes to larger ducts, which convey bile to gallbladder.

Bioavailability   Indicates fractional extent of drug dose that is absorbed, escapes any first-pass elimination, and ultimately reaches site of action.

Biomarker   Biochemical feature or facet that can be used to measure progress of disease or effects of treatment.

Biotransformation   Series of chemical modification of a compound—for example, a drug—that occur within body (as by enzymatic activity).

Biuret reaction   Commonly used protein methodology that occurs when solution of protein is treated with cupric divalent ions in moderately alkaline medium and forms violet-colored chelate that absorbs light at 540 nm; peptide bond forms between cupric ion and carbonyl oxygen and amide nitrogen atoms of peptide bond.

Blood-borne pathogen   Pathogenic microorganism present in human blood that can cause disease in humans.; include, but are not limited to, hepatitis B virus and human immunodeficiency virus.

Blood urea nitrogen (BUN)   Nitrogen found in urea as distinguished from nitrogen in other blood proteins; serum BUN levels are measured to screen for renal function.

Bohr effect   Effects of pH on hemoglobin–oxygen affinity; as H^+ concentration in tissues increases, affinity of hemoglobin for oxygen decreases, permitting unloading of oxygen.

Bond specificity   Enzyme that catalyzes reaction with certain type of bond.

Bone alkaline phosphatase   Isoenzyme of alkaline phosphatase that serves as marker for bone formation and is found in osteoblasts.

Bone modeling   Process of forming bone; calcified cartilage matrix serves as site for deposition of mineral in form of primary spongiosa; mineralized crystals develop into woven and compact bone; calcified cartilage and woven bone is removed and replaced by woven bone alone; modeling accomplished by resorption of bone on endosteal surface and formation of periosteal bone; also called *endochrondral ossification*.

Bone remodeling   Process that involves formation of bone on surfaces previously containing bone. Two phases: (1) Bone breakdown is started by osteoclasts that adhere to bone surface and create secondary lysosome with an acid pH to dissolve hydroxyapatite mineral; and (2) rebuilding phase depends on recruitment of osteoblasts that lay down an osteoid and become mineralized in tightly regulated fashion.

Bone resorption   Process occurring in bone where osteoclasts are broken down and release minerals such as calcium and phosphorus.

Bone turnover   Amount of bone renewed during bone remodeling process.

Bound drug   Pharmacological compound that exists in blood attached to protein or lipid.

Branched DNA (bDNA)   Signal amplification method that measures target nucleic acid through formation of branched-shaped complexes consisting of multiple probes bound to multiple reporter molecules.

Breath tests   Tests that have been developed to evaluate fat absorption; detect specific substance in exhalate that helps explain metabolic changes—for example, malabsorption in intestine.

British anti-Lewisite (BAL)   Antidote (2,3-dimercaptopropanol) available in United States for treating arsenic toxicity; initial drug used to treat severe arsenic toxicity; dimercaprol that functions as chelating agent to reduce sulfhydryl groups.

Bromcresol green   Anionic dye of triphenylmethane family with four bromine atoms that binds to albumin and is common albumin methodology.

Bromcresol purple   Anionic dye with two bromine atoms that binds to albumin and is common albumin methodology.

Buffer   Compound that tends to preserve solution's original hydrogen-ion concentration (pH).

Bulk reagents   Large quantity of reagents or solutions used for most analyses or for flushing and priming; ready for use with little or no preparation.

BUN:creatinine (BUN:CR) ratio   Ratio that can be used to distinguish among three major types of azotemia—prerenal, renal, and postrenal.

 C

CA 19-9   Marker for colorectal and pancreatic carcinoma; glycolipid synthesized by pancreatic and biliary ductal cells and also gastric, colon, endometrial, and salivary epithelia.

Calibration   Procedure used to ensure accuracy of test system.

C-mixed-chain triglyceride test   Test that evaluates intraluminal pancreatic lipase activity; substrate is 1,3-distearyl, 2(carboxyl-C13) octanoyl glycerol, which contains long-chain fatty acids in positions 1 and 3, ^{13}C-labeled octanoic acid in positions 1 and 3, and ^{13}C-labeled octanoic acid in position 2; labeled ^{13}C can be detected in exhaled CO_2, with amount recovered an indirect measure of lipolysis within small intestine.

Carbamino compound   Compound with structure R–NH–COO$^-$; also form in which CO_2 is carried in blood.

Carbohydrate antigen (CA)   Group of tumor marker antigens that contain a major carbohydrate component found on the surface of cells or secreted by cells (e.g., CA 15-3, CA 19-9, and CA-125).

Carbohydrates   Group of organic compounds (including sugars, glycogen, and starches) that contain only carbon, oxygen, and hydrogen; ratio of hydrogen to oxygen is usually 2 to 1, and general formula is $C_x(H_2O)_y$.

Carbon dioxide, partial pressure (PCO_2)   Pressure exerted by carbon dioxide in total pressure of gases in atmosphere; mole fraction of carbon dioxide gas times total pressure; for example, partial pressure of oxygen, PCO_2, is fraction of carbon dioxide gas times barometric pressure.

Carbonic anhydrase   Enzyme that catalyzes reaction between CO_2 and water to form carbonic acid.

Carboxyhemoglobin   Hemoglobin combined with carbon monoxide; also termed *carbon monoxide hemoglobin* (COHb).

Carcinogen   Agent that causes cancer.

Carcinoid tumors   Rare tumors that can appear in several sites throughout body, including lungs, stomach, small intestine, appendix, colon, and rectum; clinical features vary fro asymptomatic to episodes of flushing and diarrhea.

Carotenoids   Group of compounds structurally similar to β-carotene (provitamin A); occur naturally in vegetables and fruits.

Carryover Transport of quantity of analyte or reagent from one specimen to another and contaminating subsequent specimen.

Catecholamines Class of biogenic amines having sympathomimetic action; aromatic portion of molecule is catechol, and aliphatic portion is an amine; examples include dopamine, norepinephrine, and epinephrine.

Celiac disease Disease occurring in genetically predisposed individuals as result of inappropriate T-cell–mediated immune response to dietary ingestion of gluten-containing grains such as wheat; also called *nontropical sprue* or *celiac sprue*.

Ceruloplasmin (Cp) Principal copper-containing protein in plasma, comprising 95% of the total serum copper; primary role seems to be in plasma redox reactions, where it can be an oxidant or antioxidant, depending on various factors such as presence of ferric ions and ferritin binding sites; serum levels increase in Wilson disease.

Chain of custody Chronological documentation or paper trail that shows the seizure, custody, control, transfer, analysis, and disposition of physical or electronic evidence.

Chelating agent In toxicology, compound that can enclose or grasp toxic substance and make it non-native and therefore non-toxic.

Chemical hygiene program An OSHA-mandated program designed to provide laboratory staff with information necessary to handle chemicals.

Chemiluminescence Chemical reaction, usually including oxidation, in which one product is light energy.

Child–Pugh system Staging system to predict prognosis in cirrhosis; factors used are presence and severity of encephalopathy and ascites, albumin levels, prothrombin time, and bilirubin; each factor graded 1 to 3; score <7 points is class A, 7–9 points class B, and >9 points class C.

Chloride shift Exchange of Cl^- in serum for HCO_3^- in erythrocytes in peripheral tissues as response to PCO_2 of blood.

Cholecystokinin Linear polypeptide, 33 amino acids, with multiple molecular forms that consist of five C-terminal amino acids identical to those of gastrin; secreted by upper intestinal mucosa and found in central nervous systems; causes gallbladder contraction and release of pancreatic exocrine enzymes and affects other gastrointestinal functions.

Cholestasis Stoppage or obstruction of bile flow because of intrahepatic causes, obstruction of bile duct by gallstones, or any process that blocks bile duct; most common cholestasis diseases are primary biliary cirrhosis, primary sclerosing cholangitis, and mechanical obstruction of bile ducts.

Cholesterol Sterol that contains 27 carbon atoms and four fused rings (A, B, C, and D) called a *perhydrocyclopentanophenathrene nucleus*; found in animal tissues as well as in egg yolks, various oils, and fats in brain, spinal cord, liver, and kidney; endogenous cholesterol synthesized in liver, and exogenous cholesterol found in diet; major part of gallstones.

Cholesterol oxidase Enzyme that catalyzes hydrolysis of cholesteryl esters.

Cholinesterase (CHE) Enzyme that catalyzes hydrolysis of choline esters such as acetylcholinesterase.

Chromosome Threadlike structure in nucleus that contains sequence of genes and associated proteins.

Chronic glomerulonephritis End stage of persistent glomerular damage with irreversible loss of renal tissue and chronic renal failure; 80% of patients have had some other form of glomerulonephritis, and 20% are unrecognized or have subclinical symptoms.

Chronic hepatitis Chronic inflammation of liver that persists for at least 6 months and can be accompanied by hepatocyte regeneration and scarring; characterized by continuing inflammation of hepatocytes and confirmed by elevated liver enzymes; most common causes are hepatitis B and C, autoimmune disorders, Wilson disease, α_1-antitrypsin deficiency, and idiopathic.

Chronic pyelonephritis Causes permanent scarring of renal tubules and can lead to renal failure; most common cause is vesicoureteral reflux nephropathy; from 10% to 15% of chronic pyelonephritis patients progress to chronic renal failure and end-stage renal disease.

Chronic pancreatitis Irreversible damage to pancreatic tissue with evidence of inflammation and fibrosis; usually occurs after repeated bouts of acute pancreatitis and results in pancreatic cells being replaced with scar tissue.

Chylomicrons Largest and least dense (<0.94) of lipoprotein classes; formed from lipids absorbed in intestines and responsible for transporting dietary or exogenous fat, mostly triglycerides, from intestines to liver and peripheral cells.

Chymotrypsin (CHY) Serine proteinase that hydrolyzes peptide bonds connecting hydroxyl groups of tryptophan, leucine, tyrosine, or phenylalanine.

Circadian rhythm Events that occur at approximately 24h intervals such as rise and fall of hormone (cortisol) concentrations.

Cirrhosis Chronic liver disease characterized pathologically by fibrosis and conversion of normal liver architecture into structurally abnormal lobules; clinical symptoms caused by loss of functioning hepatocytes and increased resistance to hepatic blood flow or portal hypertension; although associated with alcohol abuse, many causes are not related to alcohol—for example, hepatitis B and C, hemochromatosis, and Wilson disease.

Clearance tests Rate at which kidneys remove substance from plasma or blood or quantitative expression of rate at which substance is excreted by kidneys in relation to concentration of same substance in plasma (usually expressed as ml cleared per minute).

Closed-reagent analyzer System in which operator can only use instrument manufacturer's reagents.

Clot detector Device capable of detecting specimens that have clotted through use of pressure transducer; as pressure transducer comes into contact with sample, analyzer measures difference in pressure between air and surface of sample; if pressure created is greater than specified cutoff value, then sample is not aspirated.

Codes of Federal Regulations (CFR) Codification of general and permanent rules published in *Federal Register* by executive department and agencies of federal government; rules are divided into 50 titles that represent broad areas subject to federal regulations.

Codon Unit of three nucleotides in mRNA that codes for an amino acid.

Coefficient of variation (CV) Way of expressing standard deviation; defined as 100 times standard deviation divided by mean; expressed as percentage; also referred to as *relative standard deviation*.

Coenzymes Organic cofactors that commonly have structure related to vitamins (usually vitamin B); include nicotinamide adenine dinucleotide (NAD^+) and pyridoxal-5-phosphate (P-5'-P).

Cofactors   Organic or inorganic compounds required for enzyme function.

Collagens   Proteins that provide structural framework of bones and cartilages and provide shape most of their biomechanical properties (e.g., resistance to pressure, torsion, and tension).

Colligative properties   Properties of solutions that depend on number of solute particles present but not on chemical properties of solute; four colligative properties are osmosis, freezing point, boiling point, and vapor pressure.

Compensation   In blood-gas physiology, refers to body's actions to correct or adjust blood gases and pH in an attempt to bring values back to normal.

Competitive immunoassay   Immunoassay in which all reactants are mixed together either simultaneously or sequentially; when reactants reach equilibrium, bound and free fractions are physically separated; amount of bound fraction is estimated and used to calculate unlabeled antigen concentration.

Competitive inhibition   In enzyme reactions, when inhibitor is similar to normal substrate and competes with it for binding or active site of enzyme.

Concentration gradient   Difference in concentrations between two solutions separated by barrier.

Conjugated protein   Proteins that have nonprotein groups attached to them that provide certain characteristics to protein (e.g., lipoproteins).

Connecting peptide (C-peptide)   Peptide in proinsulin that connects α and β chains of insulin; activated insulin forms equimolar amounts of C-peptide and insulin.

Conn's syndrome   Rare cause of hypertension; also known as *primary hyperaldosteronism*; symptoms include muscle weakness, polyuria, hypertension, hypokalemia, and alkalosis; associated with abnormally high rate of aldosterone secretion by adrenal cortex.

Cori's disease   Type III glycogen storage disease; caused by glycogen debrancher enzyme deficiency in liver, muscles, and some blood cells.

Convection   Loss or gain of gas by transferring it from one environment to another.

Coronary artery disease (CAD)   Condition in which coronary arteries narrow because of atherosclerosis; most common cause of death in industrialized countries.

Coronary heart disease (CHD)   Pathologic condition in which fatty substance and plaque build up and narrow walls of coronary arteries, leading to atherosclerosis; leading cause of death in United States for men and women.

Cortisol   Adrenocorticosteroid secreted by cortex in response to ACTH from anterior pituitary.

Counterregulatory hormones   Group of hormones that regulate insulin to ensure that glucose levels do not become too low; examples include glucagon, epinephrine, and cortisol.

C-reactive protein   Acute-phase reactant and nonspecific indicator of bacterial or viral infection, inflammation, and tissue injury or necrosis.

Creatine kinase (CK)   Cytoplasmic and mitochondrial enzyme that catalyzes reversible phosphorylation of creatine by ATP; present in highest concentrations in heart, muscle, and brain.

Creatininase (creatinine amidohydrolase)   Enzyme that breaks down creatinine to creatine.

Creatinine (CK)   Product formed by breakdown of creatine phosphate; provides energy for muscle contraction; elevated serum levels are found in renal disease.

Creatinine clearance (CrCl)   Laboratory test for measuring kidney's glomerular filtration rate; compares creatinine levels in blood and urine, taking into consideration patient's body surface area and urine volume; decreased CrCl indicates renal insufficiency.

Crigler–Najjar syndrome (CNS)   Inherited condition caused by autosomal recessive trait; children are born with decreased or absolute lack of UDP-glucoronyl transferase, resulting in unconjugated hyperbilirubinemia with brain damage because of kernicterus (deposits of bilirubin in brain); in most severe form, death may occur within 15 months.

Crohn's disease   Condition of unknown etiology characterized by transmural inflammation of GI tract.

Cumulative trauma disorder (CTD)   Disorders associated with overloading of particular muscle groups from repeated use or constrained posture; disorders typically develop over periods of weeks, months, or years and may involve tendons, nerve entrapment, muscles, or blood vessels.

Cushing's syndrome   Condition characterized by increased concentration of adrenal glucocorticoids hormone in bloodstream.

Cystatin C   Single-chain, nonglycosylated, low-molecular-weight protein synthesized by all nucleated cells; acts as cysteine protein inhibitor; superior to creatinine for detection of renal disease and especially useful in detecting mild to moderate impairment of renal function.

Cystinuria   Hereditary disease characterized by excretion of large amounts of cystine, arginine, ornithine, and lysine in urine that leads to development of cystine calculi.

Cytochrome P450   Terminal electron acceptor and binding site of drugs; generic term that denotes family of cytochromes that are immunologically and biochemical distinct.

Cytokines   Locally acting autocoid polypeptide mediators that interact with receptors that are phospholipase C–linked and mediate vasoconstriction.

▶ D

Dalton   Used as unit of molar mass, where 1 Da equals 1g/mol, thus 1 Da is equivalent to $1.660538782 \times 10^{-27}$ kg.

Degree(s) of freedom (df, DF)   Number of data points affecting statistical analysis.

Dehydration   Process of dehydrating, especially abnormal depletion of body fluids.

Delta bilirubin   Conjugated bilirubin bound through covalent bond with albumin; present only in cases of significant hepatic obstruction.

Denaturation   Separation of double-stranded DNA molecule into two single strands by treatment that disrupts hydrogen bonds between strands; disruption of bonds holding secondary, tertiary, or quaternary structures of protein, resulting in loss of activity and also functional and structural characteristics of protein molecule.

Density   Amount of matter (weight) per unit volume of substance.

Deoxyribonucleic acid (DNA)   Nucleic acid that serves as primary carrier of genetic information in cells and most viruses.

Depot preparation   Formulation of drug that is suitable for introduction into body; usually represents appropriate maintenance dose.

De Ritis ratio   Ratio of AST to ALT; helpful in determining cause of liver disease; in alcoholic liver disease and cirrhosis, AST > ALT, so ratio of AST:ALT is >1.0; viral hepatitis, acute inflammatory disease, and obstructive liver disease are associated with ALT > AST, so ratio of AST:ALT is < 1.0.

Descriptive statistics   Numerical or graphical summary of data, including mean, range, variability, and distribution; most commonly used statistical computations in clinical laboratory.

Dextro (D)   Isomer in which hydroxyl group is on right part of molecule; most sugars are D isomers.

Diabetes mellitus   Group of chronic metabolic diseases characterized by hyperglycemia because of defects in insulin production (type 1), insulin action (type 2), or both; complications include retinopathy, nephropathy, neuropathy, and hyperlipidemia leading to increased risk of myocardial infarctions and strokes.

Diabetic ketoacidosis (DKA)   Condition affecting type 1 diabetics because of tendency to produce high levels of ketones when glucose levels are out of control; may lead to severe hyperglycemia.

Diabetic nephropathy   Diabetic complication that accounts for 12,000 to 24,000 cases of blindness each year; some 45% of type 1 diabetes mellitus patients develop condition within 15 to 20 years of diagnosis.

Diagnosis   Act of identifying disease from its signs and symptoms.

Diagnostic sensitivity   Measure of how often assay system detects biomarker when disease is present.

Diagnostic specificity   Describes probability that laboratory test will be negative in absence of disease.

Dialysis   Process in which larger macromolecules are separated from low-molecular-weight compounds by rates of diffusion through semipermeable membrane; may be only treatment option available for patients with acute renal failure or end-stage renal disease when kidneys can no longer excrete body's waste products.

Diffusion   Tendency of molecules of substance (gas, liquid, or solid) to move from region of high concentration to region of lower concentration.

Direct assay   Term that applies to test that requires mixing of sample and reagents followed by measurement of product that is directly proportional to analyte concentrations.

Direct or **conjugated bilirubin**   Bilirubin conjugated by UDP-glucoronyl transferase that adds two glucuronic acid molecules; water soluble and excreted in urine.

Disaccharide   Carbohydrate formed by interaction of two monosaccharides with loss of water molecule; three of most common disaccharides are maltose (two glucose molecules), lactose (glucose and galactose), and sucrose (glucose and fructose).

Discordant results   Term commonly used to describe laboratory results that do not agree.

Discrete testing   Term applied to instruments that compartmentalize each sample reaction.

Distillation (water)   Process by which liquid is vaporized and condensed to purify or concentrate substance or separate volatile substances from less volatile substances.

Distribution   Division of drug into various components of body that include interstitial and intracellular fluids.

Diuresis   Secretion and passage of large quantities of urine; often occurs in diabetes mellitus.

Diurnal rhythm   Pertains to daily rhythms; hormone concentrations, for example, change during a 24h interval (i.e., plasma concentration of cortisol at 8 p.m. is approximately 50% of plasma concentration at 8 a.m.); other hormones that show diurnal rhythms are GH, ACTH, TSH, PTH, and catecholamines.

DNA polymerase   Enzyme that catalyzes synthesis of complementary DNA strand using DNA template, primers, and nucleotides.

DNA sequencing   Method used to determine exact order of nucleotides in DNA molecule.

Drug abuse screen   Test procedure designed to screen for groups of abused drugs, including opiates, cocaine, barbiturates, amphetamines, cannabinoids, and benzodiazepines.

Drug confirmation   Protocol used to verify results of drug abuse screen test procedure; most laboratories reassay any screen result that is positive or questionable; confirmation assay must be different method than screening method.

Drug-induced hypoglycemia   Most common form of hypoglycemia; accounts for more than 50% of patients who are hospitalized for hypoglycemia.

Dry beriberi   Chronic condition associated with deficiency of vitamin B_1 (thiamine); patients often present with peripheral neuropathy of motor and sensory systems; legs are most affected, and patients have difficulty rising from squatting position.

Dubin–Johnson (DJ) syndrome   Chronic and benign condition that produces obstructive liver disease resulting in impaired biliary excretion of conjugated bilirubin and conjugated bilirubin retained in hepatocytes.

 E

Efficacy   Maximal effect or response of drug.

Electrolytes   Major ions of body fluids; constitute majority of osmotically active particles; cation electrolytes include sodium and potassium; anion electrolytes include chloride and bicarbonate.

Electromagnetic radiation (EMR)   Energy emanating from source that consists of both Maxwell's waves and streams of particles called *photons*; photons are exchanged whenever electrically charged subatomic particles interact.

Elimination   Excretion or removal of drug from body.

Elimination half-life   Time required for plasma concentration to decline by one-half when elimination is first order.

Embden–Meyerhoff pathway   Glycolysis of glucose into pyruvate or lactate with or without oxygen present; principal means of energy production in humans; glucose is broken down into two 3-carbon molecules, lactate or pyruvate, with net gain of two ATP molecules (four ATP molecules are produced, but two are needed to catalyze reaction).

Endocrinopathies Conditions such as Cushing's syndrome, hyperthyroidism, glucagonoma, and acromegaly associated with secondary diabetes.

Endogenous cholesterol Cholesterol produced by liver and made from simpler molecules, particularly acetate; 500 to 1000 mg daily is produced daily by body.

Endothelium Form of squamous epithelium consisting of flat cells that line blood and lymphatic vessels, heart, and various other body cavities.

End-stage renal disease Last stage of chronic kidney disease.

Energy of activation (EA) Amount of energy required to energize 1 mole of substrate to form activated complex.

Engineered control Safety equipment that isolates or removes bloodborne pathogen hazard from workplace; preferred method for controlling hazards.

Enteral Refers to drugs taken orally or by mouth; if swallowed, drug enters stomach and may pass into intestine (enteric).

Enterochromaffin cells Cells that can be stained with potassium chromate (chromaffin) because serotonin is present in cells; these tumors appear as well-circumscribed, round submucosal lesions, and cut surface appears yellow because of lipid content of tumors.

Enterokinase Duodenal peptidase in duodenum that cleaves small amino acid chain from proenzymes—for example, converting protrypsin to active form of trypsin.

Enzyme–substrate (ES) complex When enzyme binds with substrate; provides free energy required for reaction (no additional external energy is required).

Epinephrine Catecholamine produced by adrenal glands; stimulated in "fight or flight" syndrome and secreted during periods of physical or emotional stress.

Epitope Molecular region on surface of antigen capable of eliciting immune response and of combining with specific antibody produced by such a response.

Ergonomics Study of problems of people in adjusting to their environment.

Essential amino acid Amino acid not synthesized by body that must be ingested in diet; in humans, 8 to 10 essential amino acids (valine, leucine, isoleucine, phenylalanine, tryptophan, methionine, threonine, and lysine) required by adults; infants require two additional amino acids (arginine and histidine).

Essential fatty acids Fatty acids body cannot synthesize; must be present in diet (e.g., linolenic and linoleic acids).

Essential trace elements Element considered essential when signs and symptoms that appear from dietary deficiency of element are uniquely reversed when adequate supply of particular trace element is present.

Euthyroid Having normally functioning thyroid gland.

Exhalation process of breathing out.

Exogenous cholesterol Cholesterol absorbed from diet, bile, intestinal secretions, and cells.

Exon One of several regions in gene that codes for parts of protein; separated by introns.

Exposure control plan Plan mandated by OSHA to help prevent accidental exposure of laboratory personnel to bloodborne pathogens.

External convection Mechanism in humans that maximizes gas exchange by continuously supplying atmospheric air to external surface of gas-exchange barrier (e.g., alveoli), thereby maintaining high external PO_2 and low external PCO_2; average human inspires about 4.0 liters of fresh air per minute with alveolar PO_2 of about 100 mmHg and alveolar PCO_2 of about 40 mmHg.

Extracellular water (ECW) Water external to cell membranes.

▶ F

Fasting hypoglycemia Occurs after overnight fast or fast lasting longer than 8 hours; in infancy or childhood, can be caused by various liver-enzyme deficiencies.

Fatty acids Simplest form of lipids; has chemical formula RCOOH and general formula C_nH_{2n+1}.

Fecal elastase Noninvasive test used to assess pancreatic insufficiency; fecal elastase cleaves amino acids from proteins in presence of trypsin.

Fecal fat Test to determine quantity of lipids in timed stool specimen; used to evaluate chronic diarrhea and confirm diagnosis of malabsorption or pancreatic dysfunction.

Ferritin Storage form of iron; consists of protein shell surrounding iron core; hemosiderin is formed when ferritin is degraded in secondary lysosomes; outer shell composed of apoferritin with interior ferric oxyhydroxide (FeCOOH) × crystalline core.

First-order kinetics Reaction involving enzyme and drug where rate of drug elimination is proportionate to its concentration (i.e., the higher the concentration, the greater the amount of drug eliminated per unit of time); area of enzyme-reaction curve with substrate concentration versus velocity, where velocity is directly proportional to substrate concentration.

First-pass effect When drug is metabolized in gut wall and liver before it reaches systemic circulation.

5'-nucleotidase (5'-NT, NTP) Hydrolyzes phosphate group from nucleoside-5'-phosphates—for example, 5'-adenosine monophosphate (AMP); microsomal and membrane-associated enzyme found in variety of tissues, most specifically in liver tissue.

Flavin Group of yellow water-soluble pigments that include riboflavin, flavin adenine dinucleotide, and flavin mononucleotide.

Fluorescent in situ hybridization (FISH) In situ hybridization method that uses fluorescent-labeled probes to identify specific DNA sequences in chromosomes.

Fluorescein Fluorescent compound used as label for immunoassays; absorbs EMR at 492 nm and reemits EMR at 520 nm; one widely used application that incorporates fluorescein-labeled antigen is fluorescent polarization immunoassay (FPIA), which measures therapeutic and toxic drugs.

Fluorescence Light (EMR) emitted by atom or molecule after absorption of photon; energy is at longer wavelength (less energy) than absorbed light and usually emitted in less than 10^{-8} sec.

Fraction of inspired air (FIO₂) Percentage of oxygen in ambient air, which is equal to 0.2093 (20.93%).

Free drug Represents fraction of drug that is not bound to proteins or other nonprotein compounds; also referred to as *unbound* drug.

Friedewald formula Used to calculate LDL from total cholesterol, HDL, and triglycerides; LDL cholesterol (LDL) = (Total cholesterol)

(HDL-C + TG/5); not recommended for patient triglycerides of more than 400 mg/dL.

Functional assay   A functional assay is an indirect measure of non-vitamin substances using one of several techniques: (1) increased or decreased activity of reactions, (2) cell response to inhibitors, or (3) activators that will reflect alteration in vitamin concentrations.

Fusion of the β–γ bands   Also called *bridging;* common abnormal protein electrophoresis pattern that results from fast moving γ-globulins that prevent resolution of β- and γ-globulins; cirrhosis is most common cause of β–γ bridging.

▶ **G**

Galactosemia   Autosomal recessive trait characterized by lack of an enzyme, galactose-1-phosphate uridyl transferase, which prevents breakdown of galactose and results in increased levels of galactose in serum; screened for by testing urine for nonreducing substances (galactose) followed by confirmation by testing for actual enzyme.

γ-globulin (IgG)   Immunoglobulin or humoral antibody; IgG antibodies are produced in response to antigens of most bacteria and viruses.

Gamma glutamyl transferase (GGT)   Membrane-associated enzyme that transfers γ-glutamyl group from glutathione and other γ-glutamyl peptides to amino acids or small peptides to form γ-glutamyl amino acids and cysteinyl-glycine; elevated in various conditions involving hepatic damage, especially hepatobiliary, and may also indicate alcohol or drug toxicity.

Gastrin   Group of peptide hormones secreted by gastrointestinal mucosa cells; stimulates stomach parietal cells to produce hydrochloric acid.

Gastrinoma   Neuroendocrine tumor that secretes gastrin; resulting hypergastrinemia causes gastric-acid hypersecretion.

G cells   Endocrine cells of antral mucosa that produce and store gastrin.

Gene   Unit of DNA that codes for a protein.

Gestational diabetes   Abnormal glucose concentrations discovered for first time during pregnancy; occurs in 1% to 5% of pregnancies because of changes in glucose metabolism and hormonal changes (insulin resistance).

Gilbert syndrome   Least serious and most common of inherited unconjugated hyperbilirubinemias; problem involves defect in active transport of bilirubin by ligand through hepatocyte cell membrane to microsome; mild, benign condition that does not require treatment.

Globular protein   Compact, folded, and coiled protein chains that are relatively soluble.

Globulin   Group of proteins that make up approximately 38% of total protein in serum; help to control plasma osmotic pressure in capillaries, have an important role in body's immune response, and serve as carrier proteins for various substances in plasma.

Glomerular filtration   Filtration occurring in glomerulus, which is first part of nephron.

Glomerular filtration rate (GFR)   Rate of urine formation as plasma passes through glomeruli of kidney.

Glomerulus   Coil of approximately 40 capillary loops referred to as *capillary tuft* located within Bowman's capsule (initial section of nephron).

Glucagon   Hormone produced by α-cells of islets of Langerhans of pancreas; increases blood glucose (hyperglycemic agent) and stimulates liver to release stored glycogen as glucose.

Glucagonomas   Tumors that produce glucagon; usually found in pancreas and characterized by hyperglycemia, weight loss, and peculiar skin rash (necrolytic migratory erythema).

Glucocorticoids   Group of steroid hormones secreted by adrenal cortex that have multiple physiological effects, including regulation of carbohydrate metabolism; corticol is major hormone in group.

Gluconeogenesis   Formation of glucose from noncarbohydrate sources such as amino acids, glycerol, or lactate; occurs during long-term fasting.

Glucose-6-phosphatase deficiency   Type I and most common form of glycogen storage disease; also known as *von Gierke disease.*

Glucose oxidase   Enzyme that catalyzes oxidation of glucose to gluconic acid and hydrogen peroxide (H_2O_2).

Glutamate dehydrogenase   Preferred enzymatic procedure for measuring ammonia. Glutamate dehydrogenase catalyzes the conversion of 2-oxoglutarate in the presence of NADPH and ammonium to glutamate and $NADP^+$.

Glycerol esters   Common alcohols found in human metabolism; three-carbon molecule containing three hydroxyl groups.

Glycerol kinase   Enzyme that catalyzes transfer of phosphate group from ATP to glycerol to form glycerol phosphate.

Glycogen storage diseases (GSD)   Several inherited diseases characterized by abnormal storage and accumulation of glycogen in tissues, especially liver; classified according to enzyme deficiency that is responsible.

Glycogen   Polysaccharide that is major storage form of glucose in liver and muscles.

Glycogenesis   Conversion of glucose to glycogen for storage; often occurs after a heavy meal.

Glycogenolysis   Breakdown of glycogen to form glucose and other intermediate products; regulates glucose levels between meals.

Glycolysis   Conversion of glucose or other hexoses into three-carbon molecules (lactate or pyruvate).

Glycoprotein   Compounds consisting of simple protein and carbohydrates, which make up less than 4% of total weight.

Glycosylated hemoglobin   Formed by glycation of glucose to N-terminal valine of one or both β chains of hemoglobin A, eventually forming HbA_{1c}; used as index of patient's average blood glucose over 2- to 3-month period.

Goiter   Enlargement of thyroid gland that may be caused by deficiency of iodine in diet, thyroiditis, tumors, or hyperfunction or hypofunction.

Gonadotropins   Gonad-stimulating hormones—for example, follicle-stimulating hormone and luteinizing hormone.

Gout   Inborn error of metabolism found predominantly in men 30 to 50 years old and associated with overproduction of uric acid and essential hyperuricemia; symptoms include arthritis (pain, inflammation of joints), nephropathy, and nephrolithiasis.

Gram molecular weight (GMW)   Alternate designation to *mole* because 1 mole of chemical compound is same number of grams as molecular weight of molecule of that compound measured in atomic mass units.

Group specificity   Enzyme that catalyzes substrates with similar structural groups—for example, phosphatases, which catalyze reactions containing organic phosphate esters.

Growth hormones   Hormones that stimulate growth in humans and other animals and are antagonistic to insulin.

► **H**

Haldane effect   Inverse relationship between CO_2 content and PO_2.

Half-life ($t_{1/2}$)   Time required for amount of drug in blood to decline to one-half its measured value.

Hapten   Chemical determinant that has ability to stimulate synthesis of antibody specific for hapten when conjugated to immunogenic carrier.

Haptoglobin   An α_2-globulin, mucoprotein, and acute-phase reactant that binds free hemoglobin in plasma; increases in inflammatory conditions and decreases in hemolytic disorders.

Haworth projection   Formed when aldehyde or ketone reacts with alcohol group on carbon 5, resulting in symmetrical ring structure.

Hazard identification system (704-M)   Identification system developed by National Fire Protections Agency to properly label containers storing hazardous material; covers nine classes of hazardous materials—for example, explosives, flammables, and compressed gases.

Heat of fusion   Energy required to melt solid to liquid.

Helicobacter pylori   Bacterium found in mucous layer of stomach; various strains all secrete (1) proteins that cause inflammation of mucosa and (2) enzyme urease, which produces ammonia from urea; some strains also produce toxins that injure gastric cells.

Hematin   Nonprotein portion of hemoglobin molecule referred to as *ferricheme* (Fe^{3+}) and associated with hydroxide.

Heme   Iron-containing nonprotein portion of hemoglobin molecules in which iron is in ferrous (2+) state.

Hemin   Brownish-red crystalline salt of heme formed when hemoglobin is heated with glacial acetic acid and sodium chloride; iron is present in ferric (Fe^{3+}) state.

Hemochromatosis   Genetically related disease in which body accumulates excess amounts of iron; symptoms of hemochromatosis include the triad of skin bronzing, cirrhosis, and diabetes; other conditions include cardiomyopathies, arrhythmias, and endocrine deficiencies.

Hemodialysis (HD)   Traditional and most common method of dialysis in which synthetic membrane in machine outside body filters toxic products from blood before it returns to circulation.

Hemoglobin   Oxygen-carrying, heme-containing protein contained in red blood cells and formed by developing erythrocyte in bone marrow; conjugated protein containing four heme groups and globin; has property of reversible oxygenation.

Hemopexin   Protein that removes heme from circulation; when red blood cells are destroyed, hemopexin transports heme to liver, where it is catabolized by reticuloendothelial system.

Hemosiderin   Intracellular storage form of iron in which iron particles are contained in granules consisting of ferric hydroxides, polysaccharides, and proteins.

Henderson–Hasselbalch equation   Equation that defines relationships among pH, bicarbonate, and partial pressure of dissolved carbon dioxide gas.

Hepatic jaundice   Result of problem occurring in liver; causes include cirrhosis, viral hepatitis, and alcoholic liver disease as well as genetic defects such as Gilbert, Crigler–Najjar, Dubin–Johnson, and Rotor syndromes.

Hepatocytes   Parenchymal liver cells that make up 70% of the liver's mass.

Hepatopancreatic ampulla.   *See* Ampulla of Vater.

Hepatotoxicity   Any compound (for example, a drug) that injures or destroys hepatocytes; such compounds tend to saturate enzymes catalyzing normal conjugation reaction and result in drug being metabolized by mixed-function oxidases.

Hereditary hemochromatosis (HH)   Autosomal recessive disorder of iron metabolism characterized by excessive iron absorption and accumulation in tissue; treatment involves therapeutic phlebotomies until patient's serum iron levels are below normal.

Heterogeneous immunoassay   Immunoassay that requires physical separation of free-labeled antigen from antibody-bound labeled antigen before signal generated by label is measured.

Hexokinase   Enzyme that catalyzes phosphorylation of glucose by ATP to form glucose-6-phosphate and ADP.

Hexose monophosphate shunt   Oxidizes glucose to ribose and CO_2.

High-density lipoprotein (HDL)   Cholesterol scavenger that removes cholesterol from tissues, esterifies it, and carries it to liver for disposal; has low lipid-to-fat ratio, more protein than lipids, and more protein than VLDLs and LDLs; called "good" lipoprotein because it provides protection against coronary heart disease.

Holoenzyme   Apoenzyme and cofactor or coenzyme that forms catalytically active unit.

Homogeneous immunoassay   Type of immunoassay that does not require physical separation of bound and free-labeled antibody or antigen; activity of label attached to antigen is modified directly by antibody binding, with amount of modification proportional to concentration of antigen or antibody being measured.

Homogentisic acid   Product of autosomal recessive condition and alkaptonuria.

Hook effect   Effect created in an antigen–antibody reaction when concentration of analyte begins to exceed amount of antibody present; dose–response curve will flatten (plateau) and may become negatively sloped with further increase, exhibiting hook-like feature of reaction curve.

Hormone   Chemical substance that has specific regulatory effect on activities of certain organ, organs, or cell types.

Hybrid capture   Signal-amplification method based on detection of complexes containing target DNA by enzyme-labeled antibodies on solid phase.

Hybridization   Process of joining two complementary single strands of nucleic acids to form double-stranded hybrid molecule.

Hydrolases   Group of enzymes that catalyze hydrolytic (addition of water) cleavage of compounds.

Hydroxyapatite Crystal lattice of bone composed of calcium, phosphorus, and hydroxide.

Hyperalbuminemia Condition associated with increase in albumin, which is usually artifactual; relative hypoalbuminemia due to decrease in plasma volume.

Hypercalcimia Abnormally high plasma concentration.

Hypercapnia Increased amount of carbon dioxide in blood.

Hypercholesterolemia Excessive amount of cholesterol in blood; possible risk factor for cardiovascular disease.

Hyperkalemia Abnormally high plasma potassium concentration.

Hyperlipidemia Increased level of lipids in blood.

Hypernatremia Abnormally high plasma sodium concentration.

Hyperparathyroidism Condition caused by excessive amounts of parathyroid hormone in body.

Hyperproteinemia Condition associated with positive nitrogen balance; dietary nitrogen intake is greater than excretion or loss of nitrogen, which occurs mainly in urine; usually associated with hemoconcentration or dehydration, resulting in decreased plasma volume.

Hypertriglyceridemia Increased level of triglycerides in blood; possible risk factor for cardiovascular disease.

Hyperuricemia Serum or plasma uric acid concentration >7.0 mg/dL in men and > 6.0 mg/dL in women; causes divided into four categories (increased dietary intake, overproduction of uric acid, underexcretion of uric acid, and specific enzyme defects).

Hypervitaminosis Abnormally high plasma level of vitamin.

Hypnotics Agents that cause insensitivity to pain or inhibit reception of sensory impressions in cortical centers of brain, thus causing partial or complete unconsciousness; pertains to sleep or hypnosis.

Hypoalbuminemia Decreased albumin levels with most common cause being increased catabolism because of tissue damage and inflammation; found in people who are malnourished and individuals with chronic liver, kidney, and pancreatic diseases.

Hypocapnia Decreased amount of carbon dioxide in blood.

Hypocholesterolemia Abnormally low amount of cholesterol in blood.

Hypoglycemia Abnormally low plasma glucose level, usually defined as below 50 mg/dL in men and below 45 mg/dL in women, accompanied by symptoms such as nervousness, sweating, intense hunger, trembling, weakness, palpitations, and trouble speaking.

Hypoproteinemia Decreased levels of protein in blood; condition related to negative nitrogen balance or when excretion of nitrogen exceeds intake or synthesis of protein; most common cause is increase in plasma water volume or hemodilution, which results in decrease in concentration of all proteins.

Hypothalamic–pituitary–thyroid axis Three-gland complex that makes up feedback mechanisms for release or suppression of hormones located within glands.

Hypothalamus Region of brain that controls immense number of bodily functions; located in middle of base of brain and encapsulating ventral portion of third ventricle.

Hypotriglyceridemia Decreased level of triglycerides in blood.

Hypovitaminosis Abnormally low plasma vitamin concentration.

Hypovolemia Diminished blood volume.

Hypoxemia Decreases oxygen tension (concentration) in blood.

Hypoxia Decreased oxygen concentration in inspired air.

► **I**

Immunogen Chemical substance capable of inducing an immune response; term used when referring to material capable of eliciting antibody formation when injected into host.

Immunoglobulins (Igs) Plasma polypeptides that bind with antigenic proteins and serve as one of body's defenses against disease; comprise two heavy chains and two light chains; five types:

(1) **immunoglobulin A (IgA)** Molecular weight 160,000; 10–15% of Igs
(2) **immunoglobulin D (IgD)** Molecular weight 184,000; <1% of serum Igs
(3) **immunoglobulin E (IgE)** Molecular weight 180,000; extremely low concentrations ($0.3 \mu g/mg$)
(4) **immunoglobulin G (IgG)** Molecular weight 150,000; most abundant Ig in serum (70–75% of Igs)
(5) **Immunoglobulin M (IgM)** Largest Ig (molecular weight 900,000); 5–10% of total Igs

Impaired fasting glucose (IFG) Fasting plasma glucose $\geq$ 100 mg/dL but < 126 mg/dL, which is in gray area between normal glucose levels and diabetes.

Impaired glucose tolerance (IGT) Fasting plasma glucose $\geq$ 100 mg/dL but < 126 mg/dL; 2-h values are > 140 mg/dL but <200 mg/dL; glucose levels are in gray area between normal and diabetes.

Indirect or unconjugated bilirubin Bilirubin in serum that is fat soluble but not water soluble and is bound to albumin as carrier protein.

Inferential statistics Process used to attempt to draw conclusions about something based on partial information.

Inhalation Drawing air into lungs; also called *inspiration*.

Inhibitors Various substances that can decrease enzymatic reaction rates.

In situ hybridization Method in which labeled probes are hybridized to tissues, cells, or chromosomes to detect presence or absence of specific nucleotide sequence in its native location.

Insulin Pancreatic enzyme secreted by beta cells of the islets of Langerhans; controls metabolism of sugars and is principal hormone responsible for decreasing blood glucose.

Insulinoma Tumor of pancreatic islets; most common type of tumor associated with hypoglycemia but one of rarest causes of hypoglycemia.

Insulinopenia Decrease in insulin production.

Internal convection Specialized mechanism in circulatory system that maximizes flow of O_2 and CO_2 across gas-exchange barriers by delivering blood that has low PO_2 and high PCO_2 to barrier's inner surface; *perfusion* is process of delivering blood to lungs.

International unit (IU) Measure of enzyme activity; defined as amount of enzyme that will catalyze reaction of 1 μmol of substrate per minute (μm/min) under defined conditions such as temperature, pH, activators, and substrate; $1U = 1.67 \times 10^{-8}$ katal.

Interstitial fluid   Fluid that lies between and bathes tissue cells and constitutes portion of physiological extracellular water; includes extravascular and extracellular water into which ions and small molecules diffuse freely from plasma.

Intraoperative PTH (iPTH)   In surgical management of primary hyperthyroidism, iPTH assays have shown to improve success of parathyroid gland surgery; have been used by many surgeons to detect decreases in plasma PTH levels after all hypersecreting tissue has been excised.

Intrinsic activity   Measure of biological effectiveness of drug–receptor complex.

Introns   Noncoding region in DNA that is removed from mRNA transcript by splicing.

Invader assay   Signal amplification method that uses primary probe, invader probe, reporter probe, and cleavase to generate large number of reporter probes from single probe bound to DNA target.

Ionizing radiation   Sources that emit particles (e.g., alpha, beta, neutrons) and electromagnetic (X-rays, gamma rays) radiation.

Ion-selective electrode   Membrane-based electrochemical transducer capable of responding to specific ion; potential difference or electron flow created by selectively transferring ion to be measured from sample solution to membrane phase.

Ischemia   Local and temporary deficiency of blood supply because of obstruction of circulation; if process occurs in heart vessel, may lead to angina pectoris and, if untreated, to myocardial infarction.

Islets of Langerhans   Endocrine portion that makes up approximately 1% of pancreas and produces hormones that are secreted into circulatory system; approximately 1 million islets in pancreas distributed among ducts and acini. Each islet contains three types of cells: alpha, beta, and delta.

Isoelectric point (pI)   Hydrogen-ion concentration (pH) at which amino acid or protein has no net charge and positive charges equal negative charges.

Isoenzymes   Various forms of enzyme found in particular tissue; catalyze same reaction but forms from different tissue are specific and may be separated from each other.

Isohydric shift   Series of reactions in red blood cells in which CO_2 is taken up and oxygen is released without production of excess hydrogen ions.

Isomerases   Group of enzymes that catalyze interconversion of isomers—for example, phosphohexose isomerase converts α-glucose to β-glucose.

▶ J

Jaffe reaction   Reaction between creatinine and picric acid in alkaline medium, yielding red-orange compound that is creatinine and picrate ion.

Jaundice   Condition characterized by yellow discoloration of skin, sclera, and mucous membranes; most commonly caused by elevated bilirubin; symptom of many liver diseases—for example, obstructive liver disease (gallstones), hepatitis, and cirrhosis.

Jendrassik–Grof   Method of choice in measuring bilirubin.

▶ K

Kalium   Latin for *potassium*.

Ketone   Substance with a carbonyl group (C = O) attached to two carbon atoms formed by incomplete catabolism of fatty acids from lipid stores; acetone, acetoacetic acid, and β-hydroxybutyric acid are three major ketones.

Ketonemia   Presence of excess ketones in blood; causes characteristic fruity breath odor of diabetics in ketoacidosis.

Ketonouria   Presence of excess ketones in urine.

Ketose   Monosaccharide form of ketone.

Kupffer cells   Active phagocytes in liver that engulf bacteria, old red blood cells, toxins, and cellular debris from blood flowing through sinusoids.

▶ L

Lab-on-a-chip   Microchip design to represent total microanalysis system (μTAS) that incorporates sample preparation, separation, detection, and quantification on microchip surface.

Laboratory information system (LIS)   Group of microprocessors and computers connected together to provide management and processing of information throughout laboratory.

Lactate dehydrogenase (LD)   Oxidoreductase and hydrogen-transfer enzyme that catalyzes oxidation of L-lactate to pyruvate with NAD^+ as hydrogen acceptor; present in many tissues, including heart, muscle, liver, and red cells.

Lactose tolerance test   Test designed to assess patients for possible lactase deficiency; patients are given loading dose of lactose, and serum glucose is measure over interval of time; to exclude lactase deficiency, increases above baseline for venous plasma glucose must be >30 mg/dL.

Lecithin-cholesterol acytransferase (LCAT)   Enzyme that converts free cholesterol into cholesteryl esters that are incorporated into core of lipoprotein—for example, HDL.

Lesch–Nyhan syndrome   X-linked genetic disorder caused by deficiency of hypoxanthine-guanine phosphoribosyl transferase, which produces elevated uric acid; characterized by mental retardation, abnormal muscle movements, and behavioral problems such as pathological aggressiveness and self-mutilation.

Levo (L)   Isomer in which hydroxyl group is on left part of molecule.

Ligase chain reaction (LCR)   Probe amplification technique based on binding of two adjacent probes to single-stranded target DNA sequence.

Ligases   Group of enzymes that catalyze joining of two molecules coupled with hydrolysis of pyrophosphate bond in ATP or similar compound.

Light amplification by stimulated emission of radiation (LASER)   Monochromatic, highly intense, coherent light in ultraviolet, infrared, and visible regions of electromagnetic spectrum.

Limit of detection   Represents lowest concentration or quantity of an analyte that significantly exceeds measurement of blank sample.

Linearity   Quality or state of being linear.

Lipase (LPS)   Pancreatic enzyme that catalyzes breakdown of triglycerides into glycerol and fatty acids.

Lipids   Class of organic compounds that are actually or potentially esters of fatty acids; soluble in organic solvents and nearly insoluble in water; include true fats such as triglycerides, phospholipids, cerebrosides, and sterols (cholesterol).

Lipoprotein (a) [Lp (a)]   Lipoprotein particle that resembles LDL with addition of carbohydrate-rich protein apo (a) that is bound to apo B-100, with each molecule containing one molecule of apo (a) and one molecule of apo B-100; contributes to obstruction of blood vessels in atherosclerosis.

Lipoprotein lipase (LPL)   Enzyme found on surface of endothelial cells and responsible for hydrolyzing triglycerides to glycerol and free fatty acids located on surface of chylomicrons and VLDLs.

Lipoproteins   Micelles composed of cholesterol, phospholipids, and triglycerides bound to simple proteins; categories are chylomicrons, very-low-density lipoproteins (VLDLs), intermediate-density lipoproteins (IDLs or LDL1), low-density lipoproteins (LDLs), and high-density lipoproteins (HDLs).

Liquid-level sensor   Electrical transducer designed to respond to presence of liquid in container.

Liver lobule   Basic microscopic unit of hexagonal structure consisting of hepatic cells arranged in spoke-like plates around central vein; sinusoids are between cellular plates; portal triad at periphery consists of hepatic artery, portal vein, and bile duct.

Loading dose   One or series of doses that may be given at onset of therapy in effort to achieve target concentration.

Local area network (LAN)   Digital computer network that serves localized area; usually limited to less than three miles.

Low-density lipoprotein (LDL)   Class of lipoproteins that are principal carriers of cholesterol, accounting for approximately 70% of total cholesterol in plasma; principal cause of atherosclerosis.

Luminescence   Emission of light or radiant energy when electron returns from excited or higher energy level to lower energy level.

Lyases   Group of enzymes that catalyze removal of groups from substrates without hydrolysis (addition of water), leaving double bonds in end product.

▶ **M**

Macroamylasemia   Presence of macroamylase in blood; formed when amylase binds to IgA or IgG, thereby limiting its excretion in urine; benign condition (not correlated with any disease).

Maintenance dose   Amount of drug needed to maintain steady-state concentration.

Malignant   Disease or condition that is described as becoming worse or tending toward or threatening to produce death; term often associated with cancers.

Maple syrup urine disease (MSUD)   Genetic disorder named for characteristic maple syrup or burnt sugar odor of affected individual's urine; caused by absence of or extremely low levels of α-ketoacid decarboxylase enzyme, which results in abnormal metabolism of three essential amino acids: leucine, isoleucine, and valine.

Mass-to-charge ratio (m/z)   Ratio of mass (m) of ion in atomic mass units and charge (z); ratio is obtained by dividing atomic or molecular mass by number of charges that ion bears.

Material Safety Data Sheet (MSDS)   Document that consists of 16 sections and includes identification, composition, hazards, first aid, firefighting concerns, and proper safe handling that are relevant to chemical being used.

Mean   Value determined by summing all data and dividing sum by number of data represented by n or N.

Mechanical obstruction of the bile ducts   Most common cause of cholestatic liver disease; gallstones in common bile duct or tumors in head of pancreas or duodenum are predominant causes of mechanical obstruction; other causes include bile duct strictures, PSC, and extrinsic compression of bile ducts by enlarged lymph nodes.

Median   Middle value or 50th percentile value in sample set when data are rank ordered by magnitude.

Melting temperature (T_m)   Temperature required to denature double-stranded DNA or RNA molecule.

Messenger RNA (mRNA)   Type of RNA that carries genetic code for particular protein from DNA in cell's nucleus to ribosome in cytoplasm during protein synthesis.

Metabolic acidosis   Pathological condition that leads to accumulation of acid, which lowers bicarbonate concentration and decreases pH.

Metabolic alkalosis   Pathological condition that leads to accumulation of base, which raises bicarbonate concentration and increases pH.

Metabolic syndrome   Group of interrelated metabolic risk factors that appear to promote development of atherosclerotic cardiovascular disease; most commonly recognized metabolic risk factors are atherogenic dyslipidemia (low HDL, high LDL, and elevated triglycerides), elevated blood pressure, and elevated blood glucose.

Metabolism   Biotransformation of drug for eventual elimination from body.

Metalloenzymes   Enzymes that contain metal ions in their structures.

Metalloproteins   Proteins that have metal ion attached to amino acid residue—for example, ceruloplasmin, which contains copper.

Metastases   Change in location of disease or transfer from one organ or part to one or more others (from singular *metastasis*); term used often in reference to malignant tumors (e.g., tumor that originates in prostate but moves to upper GI tract).

Michaelis–Menten constant (Km)   Substrate concentration in moles per liter when initial velocity is ½ V max.

Microalbumin   Excretion of minute amounts of albumin in urine, too small to be detected by reagent dipsticks; screening test used to detect early diabetic nephropathy.

Microalbuminuria   Excretion of minute amounts of albumin in urine; undetectable by urinalysis reagent strips; using urine albumin, defined as 30–300 mg albumin on two 24h urine samples or albumin-to-creatinine ratio of 30:300.

Microarray   Orderly arrangement of DNA molecules immobilized onto precise locations on solid support; used to identify specific gene sequences through their ability to bind to complementary labeled probes.

Middleware   Software that allows laboratory to connect existing LIS and instrumentation to facilitate automated information and performs tasks not currently done with laboratory's existing hardware and software.

Mineralocorticoids Steroid hormones secreted by adrenal cortex that stimulate resorption of sodium and excretion of potassium in kidneys; aldosterone is major mineralocorticoid in humans.

Mixed acid-base disorder Occurrence of more than one acid–base disorder simultaneously, in which blood pH may be low, high, or within reference interval.

Mode Data point that occurs most frequently in data array.

Model for end-stage liver disease (MELD) System for predicting short-term survival for patients with end-stage liver disease; three blood tests (bilirubin, prothrombin time as international normalized ratio, and creatinine) used to determine MELD score.

Modeling Formation of bone on sites where it has not been before.

Molarity (M) Unit equal to number of moles of solute per liter of solution (solvent).

Molecular diagnostics Detection of mutations associated with human disease.

Monochromator Spectroscope modified for selective transmission of narrow band of spectrum or device used to isolate certain wavelength or range of wavelengths.

Monosaccharide Simple sugar—for example, glucose, fructose, and galactose.

Monounsaturated fatty acid Fatty acid that contains only one double bond between carbon atoms.

Mucoprotein Proteins linked with large, complex carbohydrates (>4% of total weight).

Multiple myeloma Cancer of plasma cells in bone marrow.

Mutation Permanent change in nucleotide sequence or arrangement of DNA that is transmitted from parents to offspring.

► **N**

Natriuresis Excretion of large quantities of sodium in urine.

Natrium Latin for *sodium.*

Natriuretic peptides Includes family of proteins that function to regulate water and sodium metabolism; examples include atrial natriuretic peptide and brain-type natriuretic peptide.

NBT-PABA test Test of pancreatic function based on hydrolysis of synthetic tripeptide N-benzoyl-1-tyrosyl-ρ-aminobenzoic acid (NBT-PABA) or bentiromide by chymotrypsin; NBT-PABA is hydrolyzed by chymotrypsin to release PABA, which is absorbed by intestine and metabolized in liver to PABA glucoronide and PABA acetylate; PABA levels are directly proportional to chymotrypsin levels or activity.

Necrosis Death of cells, tissues, or organs.

Negative feedback Control mechanism used by body to regulate hormone levels; specifically, product of second gland causes decrease in hormone released from first gland, resulting in diminished stimulation of second gland; for example, if concentration of thyroxine (second gland) in blood drops to abnormally low level, then pituitary (first gland) responds by increasing secretion of thyroid-stimulating hormone, which stimulated thyroid gland to increase releases of thyroxine.

Neoplasia Term used to indicate new growth; unrestricted growth of cells resulting in cancer.

Nephrons Structural and functional units of kidney; approximately 1 million to 1.5 million nephrons in each kidney; each one consists of glomerulus within Bowman's capsule, proximal convoluted tubule, loop of Henle, and distal convoluted tubule.

Nephrotic syndrome Condition characterized by increased glomerular permeability to protein, resulting in massive loss of protein in urine; urinalysis characterized by marked proteinuria, hematuria, and sediment containing urinary fat droplets, oval fat bodies, renal tubular cells, and renal epithelial, waxy, and fatty casts; acute symptoms are shock and decreased renal blood flow.

Nernst equation Equation that correlates chemical energy and electric potential of galvanic cell or battery; derived by Nobel Prize recipient Walther H. Nernst.

Nervous system damage Condition present in 60% to 70% of diabetics; includes impaired sensation in feet or hands, carpal tunnel syndrome, and other nerve problems.

Neuroglycopenic Symptoms of fasting hypoglycemia, including confusion, inappropriate behavior, visual disturbance, stupor, seizures, and coma.

Nonnormal Asymmetrical or skewed distribution of data.

Nonparametric Data that do not follow normal Gaussian distribution.

Neuroendocrine tumor (NET) Abnormal mass affecting nervous and endocrine systems as integrated functioning mechanism; specifically, gastroenteropancreatic or gastroenteropancreatic neuroendocrine tumors (GEP-NETs; examples include carcinoid tumors and gastrinomas.

Neurohypophysis Posterior part of pituitary gland; extension of central nervous system that produces hormones oxytocin and antidiuretic hormone.

Nominal wavelength Wavelength in nanometers at peak transmittance.

Nonionizing radiation Sources that emit electromagnetic radiation ranging from extremely low frequency to ultraviolet; examples include incandescent lamps and microwaves.

Nonalcoholic fatty liver disease (NAFLD) Wide spectrum of disorders from fatty liver alone (steatosis) to nonalcoholic steatohepatitis; hepatic component of metabolic syndrome (obesity, type 2 diabetes mellitus, insulin resistance, dysplipidemia, and hypertension).

Nonalcoholic steatohepatitis (NASH) Necroinflammatory liver disease, associated with fat accumulation in liver without inflammation or scarring; patients have no history of alcohol abuse and do not have AST levels greater than ALT [De Ritis ratio (AST/ALT) <1].

Noncompetitive inhibition Enzyme inhibition when inhibitor is structurally different from substrate and binds to allosteric site on enzyme molecule.

Nonionizing radiation Electromagnetic radiation ranging from extremely low frequency to ultraviolet.

Nonlinear process Results of all interactions do not produce straight lines (linear).

Nonprotein nitrogen (NPN) Products of catabolism of proteins and nucleic acids that contain nitrogen but are not part of protein molecule; urea (BUN) is largest component of NPN.

Normality (N) Number of gram equivalents of solute per liter of solution; dependent on type of reaction involved (e.g., acid–base, oxidation).

Northern blot   Technique used to identify mRNA that has been separated by electrophoresis and blotted onto membrane by hybridization with specific probe.

Nucleic acid sequence based amplification (NASBA)   Target-amplification technique that uses reverse transcriptase, RNase, and RNA polymerase to make multiple copies of RNA sequence.

Nucleoprotein   Combination of simple protein and nucleic acids (DNA, RNA).

Nucleotides   Structural units of DNA or RNA, consisting of five-carbon sugars (either deoxyribose or ribose), nitrogenous bases, and phosphate groups.

Nyctalopia   Inability to see in faint light or at night; synonym for night blindness; may be result of vitamin A deficiency.

 O

Occupational and Safety Health Administration (OSHA)   Agency of U.S. Department of Labor; mission is prevention of work-related injuries, illnesses, and occupational fatalities by issuing and enforcing standards for workplace safety and health.

Ochronosis   Darkening of skin, urine, and connective tissues of body because of excess homogentisic acid in patients with alkaptonuria.

Oncofetal antigen (protein)   Glycoprotein (70-kDa) normally produced by fetal yolk sac and fetal hepatocytes; decreases to low or undetectable levels after birth; can appear in some forms of cancer.

Oncogene   Gene having potential to cause normal cell to become cancerous.

Oncoprotein   Protein that is coded for by viral oncogene that has been integrated into genome of eukaryotic cell and involved in regulation or synthesis of proteins linked to tumorigenic cell growth.

Online data entry   Automatic data transmission from laboratory analyzer to laboratory information system.

Open-reagent analyzer   System in which reagents other than those from instrument manufacturer can be used.

Opiate   Synthetic morphine-like drug with nonpeptidic structure.

Opioid   Term applied to any substance, whether endogenous or synthetic, that produces morphine-like effects that are blocked by antagonists such as naloxone hydrochloride (Narcan).

Oral glucose tolerance test (OGTT)   Screening test for diabetes mellitus when glucose levels are measured after patient is given glucose drink containing 75 g glucose and serial blood samples are drawn to measure rise and fall of glucose levels; no longer recommended by American Diabetes Association.

Osmolal gap   Calculated parameter derived by subtracting calculated osmolality from measured osmolality.

Osmolality   Measurement of number of moles of particles per kilogram of water.

Osmolarity   Osmotic concentration expressed as osmoles or milliosmoles of solute per liter of solvent.

Osmometry   Measurement of osmolality of an aqueous solution such as serum, plasma, or urine.

Ossification   Process of bone formation.

Osteoblasts   Cells responsible for formation of bone; arranged as layer of contiguous cells that are cuboidal in their active state.

Osteocalcin   Propeptide (75 amino acids in length) synthesized by mature osteoblasts, odontoblasts, and hypertrophic chondrocytes; most protein found in circulation reflects osteoblast activity because number of osteoblasts exceeds number of odontoblasts.

Osteoclasts   Promote bone resorption; found on growth surfaces of bone.

Osteocytes   Cells found within bony matrix; derived from osteoblasts that have encased themselves within bone.

Osteoporosis   Condition characterized by reduction in bone mass, leading to fractures with minimal trauma.

Osteoprotegerin   Cytokine that protects bone from osteoclastic activity by binding RANKL.

Oxidoreductases   Group of enzymes that catalyze oxidation--reduction reactions.

Oxygen, partial pressure (PO_2)   Pressure exerted by oxygen as fraction of total pressure of gases in atmosphere; mole fraction of oxygen gas times total pressure; for example, the partial pressure of oxygen, PO_2, is fraction of oxygen gas times barometric pressure.

Oxygen saturation (O_2 sat or sO_2)   Fraction of total hemoglobin in form of HbO_2 at defined PO_2.

 P

P50 PO_2 for given blood sample at which hemoglobin of blood is half-saturated with O_2.

Paget's disease   Localized, nonmetabolic bone diseases characterized by osteoclastic bone resorption followed by replacement of bone in random fashion; also called *osteitis deformans*.

Pancreatic amylase   Pancreatic enzyme that completes digestion of starch and glycogen to limit dextrins and maltose.

Pancreatic chymotrypsin   Pancreatic enzyme secreted into duodenum that hydrolyses proteins to peptones and amino acids; fecal chymotrypsin is extremely sensitive test for steatorrhea, but like duodenal chymotrypsin, sensitivity decreases with increased damage to pancreas and thus a test for pancreatic insufficiency.

Pancreatic elastase-1   Pancreatic-specific protease synthesized by pancreatic acinar cells with other digestive enzymes and found in pancreatic juice; cleaves amino acids from proteins when trypsin is present.

Pancreatic neoplasm   Tumor or growth in pancreas.

Pancreatitis   Inflammation of pancreas that occurs as result of autodigestion.

Paramagnetic particle   Particulate material coated with iron oxide to allow its attraction to magnetic force; such particles are used to facilitate separation of antibody bound from free fractions in immunoassay.

Parametric   Data that follow normal Gaussian distribution.

Parathyroid hormone (PTH)   Polypeptide hormone secreted by parathyroid glands in response to hypocalcemia that increases calcium in blood by increasing bone resorption, increasing renal reabsorption of calcium, and increasing the synthesis of 1,25-dihydroxyvitamin D.

Parent drug   Drug originally taken into body and eventually catabolized into several metabolites in liver.

Parenteral Administration of drug by injection—for example, intravenous (IV), intramuscular (IM), and subcutaneous.

Partial pressure (*P* or *p*) Pressure exerted by gas, whether alone or mixed with other gases.

Pascal MKS unit (SI) of pressure and equal to 1 newton meter squared ($1 \, Nm^2$), which is equal to $1 \, kg \, m^{-1}s^{-2}$.

Peak drug concentration Highest concentration of drug encountered in given dose interval; usually achieved soon after next dose is given.

Pellagra Systemic nutritional wasting disease caused by a deficiency of vitamin B_3 (niacin); condition occurs in people eating corn-based foods in such regions and countries as Africa, China, and India; in North America, frequently occurs in alcoholics, patients with carcinoid tumors (adrenal glands), and patients with Hartnup's disease (congenital defect of intestinal and kidney absorption of tryptophan).

Pentose phosphate pathway Oxidizes glucose to ribose and CO_2.

Peptic ulcer Disorder in which defect in gastrointestinal mucosa extends through muscularis mucosa; may persist as function of acid or peptic activity in gastric juice.

Peptide Molecule containing two or more amino acids.

Peptide bond Bond that links amino acids in proteins; a molecule of water is split between the carboxyl group of one amino acid and the amino group of another, and a covalent bond is formed.

Peptide Compound containing two or more amino acids.

Perfusion Passing of fluid through spaces.

Peritoneal dialysis (PD) Dialysis in which peritoneal wall acts as dialysis membrane and dialysate is introduced and removed through gravity; two types are available (continuous ambulatory peritoneal dialysis and continuous cycling peritoneal); not as effective as hemodialysis.

Pernicious anemia Megaloblastic anemia caused by failure to absorb vitamin B_{12}.

Personal protective equipment (PPE) Specialized clothing or equipment worn by laboratory worker for protection against hazard.

pH Negative logarithm of hydrogen-ion concentration (H^+); degree of acidity or alkalinity of a substance expressed in pH values.

Pharmacodynamics Fundamental action of drug on biochemical and physiological level.

Pharmacokinetics Dynamic factors that affect fate of drug during its passage through body; factors include rate of absorption, distribution, metabolism (biotransformation), and excretion.

Phase I reaction One of two types of reactions that occur during drug's metabolism or biotransformation; functional group (for example, a methyl group) is attached to the drug being metabolized; also known as *functionalization* reaction.

Phase II reaction One of two types of reactions that occur during drug's metabolism or biotransformation; leads to formation of covalent linkage between functional group on parent compound with endogenously derived glucuronic acid, sulfate, glutathione, amino acid, or acetate; also known as *conjugation* reactions.

Phenylketonuria (PKU) Inborn error of metabolism, an autosomal recessive trait, that results in inability to metabolize essential amino acid phenylalanine to tyrosine; phenylalanine accumulates in body

and results in severe neurological defects (mental retardation) in infant if not recognized and treated.

Phospholipase A_2 (PLA$_2$) Enzyme associated with pathogenesis of acute pancreatitis; although exact mechanism is unknown. PA$_2$ is released into serum during acute attack; especially useful in diagnosis of acute pancreatitis and used to monitor patients following an attack.

Phospholipids Diglycerides containing phosphorous (e.g., lecithin, cephalin); constitute most of lipid portion of cell membranes.

Phosphotungstic acid (PTA) Uric acid methodology; blue color (tungsten blue) develops when PTA is reduced by uric acid in an alkaline medium; also called *Carraway*.

Photosensitivity Abnormal reactivity of skin to sunlight.

Pituitary gland Elliptically shaped gland located at base of brain in sella turcica and attached by stalk to hypothalamus; divided into anterior (adenohypophysis), intermediate, and posterior (neurohypophysis) sections; anterior and posterior lobes produce variety of hormones.

***p*K$_a$** Negative logarithm of dissociation constant and measure of strength of interaction of compound with proton.

Plasma Liquid component of blood in which cells are suspended; differs from serum in containing fibrinogen and related compounds that are removed from serum when blood clots.

Poison Any agent capable of producing morbid, noxious, or deadly effects when introduced into animal organisms.

Poisoning Damaging physiological effects of ingestion, inhalation, or other exposure to range of pharmaceuticals, illicit drugs, and chemicals, including pesticides, heavy metals, gases and vapors, and common household substances such as bleach and ammonia.

Polydipsia Excessive thirst; symptom that may indicate dehydration or hyperglycemia (diabetes mellitus).

Polymerase chain reaction (PCR) Target-amplification technique that employs repeated cycles of denaturation, annealing of primers, and extension to generate multiple copies of specific DNA fragment.

Polyphagia Increased appetite; eating large amounts of food.

Polysaccharides Group of complex carbohydrates composed of more than 20 monosaccharides that are usually insoluble in water; most common polysaccharides are starch, glycogen, and cellulose.

Polyunsaturated fatty acids Fatty acids that have more than one double bond between carbon atoms.

Polyuria Increased secretion and discharge of urine; 24h urine volume of more than 2 liters.

Pompe's disease Type II glycogen storage disease; also known as *acid maltase deficiency*; divided into two forms based on age of onset—infantile and juvenile and adult forms.

Population Universe of values or attributes, such as all of fasting serum cholesterol levels of all apparently healthy males in United States.

Porphyrias Group of primarily inherited metabolic disorders that result from partial deficiencies of enzymes of heme biosynthesis that cause increased formation and excretion of porphyrins.

Porphyrinogen Reduced form of a porphyrin; differs by absence of six hydrogens; unstable in vitro and quickly oxidized to corresponding porphyrin; stable in cell because of lower oxygen content; tends to form intermediates in heme biosynthesis pathway.

Porphyrins Group of compounds that contain four monopyrrole rings connected by methene bridges to form tetrapyrrole ring; named from Greek root for *purple* (porphyra); owe their color to conjugated double-bond structure of tetrapyrrole ring.

Portal hypertension Increased pressure in portal vein because of obstructive liver disease; seen in cirrhosis; results in splenomegaly and ascites.

Portal vein Vein that originates in gastrointestinal tract and provides 70% of liver's total blood volume.

Posterior pituitary Lobe of pituitary that lies in posterior section of gland; source of oxytocin and antidiuretic hormone.

Postrenal azotemia Azotemia occurring after urine has left kidney; results from obstruction of urine flow through kidneys, bladder, or urethra.

Potency Overall effectiveness of drug to produce its maximal effect.

Precision Ability to produce same value for replicate measurements of same sample; also described as *random variation* in population of data.

Prehepatic jaundice Result of increased production of bilirubin; results in increased total serum bilirubin and unconjugated bilirubin because excess bilirubin reaching liver cannot be excreted.

Prerenal azotemia Azotemia that occurs before kidney, usually because of decreased renal blood flow.

Primary biliary cirrhosis (PBC) Rare autoimmune disorder that targets intrahepatic bile ducts and is often associated with other autoimmune processes; hallmark is presence of antimitochondrial antibodies in serum of 80–95% of patients; symptoms include jaundice, pruritus, and fatigue.

Primary gout Condition associated with overproduction and essential hyperuricemia; inborn error of metabolism found predominantly in men 30 to 50 years of age; 7 times more common in men than women.

Primary hyperparathyroidism (PHPT) Caused by disorder of calcium metabolism in relationship to excessive PTH release.

Primary sclerosing cholangitis (PSC) Chronic inflammatory disease of unknown origin marked by inflammation and obliteration of intrahepatic and extrahepatic bile ducts; found predominantly in males; has median age of 30 years at onset.

Primary structure Protein structure determined by sequence and specific order of amino acids in polypeptide chain.

Primer Short strand of DNA or RNA that is complementary to region of DNA template that will serve as starting point for DNA replication.

Prion protein (PrP) Infectious disease protein that causes degeneration of central nervous system; prion disease is disorder of protein conformation; most common in humans is Creutzfeldt–Jakob disease.

Probe Single-stranded nucleic acid that is used to identify complementary nucleotide sequence in target DNA or RNA molecule.

Prognosis Prediction of course and end of disease as well as chance for recovery.

Promoter Region at 5′ end of DNA, where enzyme RNA polymerase binds to begin transcription.

Prostaglandins Derivatives of fatty acids comprising 20 C atoms, including 5-carbon cyclopentane ring.

Prosthetic groups The non–amino acid components of conjugated protein.

Protein electrophoresis Movement of charged particles through medium in which they dispersed because of changes in their charges of electrical potential; proteins are negatively charged because of buffer pH (8.6) and thus migrate to anode.

Protein-energy malnutrition (PEM) Malnutrition from inadequate protein intake; people at risk for PEM are elderly, hospitalized, or nursing home patients and those with chronic illnesses (diabetes, arthritis), increased nutritional losses, open wounds, burns, and malabsorption (gastrointestinal protein-losing diseases).

Proteinuria Increase in urinary protein resulting from renal or systemic disease.

Pseudocyst Collection of pancreatic juice that is enclosed by wall of fibrous or granulation tissue that usually occurs 2 weeks after initial symptoms of acute pancreatitis; pseudocysts can be single or multiple, large or small, and located inside or outside of pancreas.

Pyelonephritis Inflammation of kidney and renal pelvis, usually caused by bacterial infection that has ascended from bladder (cystitis); urinalysis is characterized by pyuria, significant bacteria, low specific gravity, and alkaline pH.

Pyridinium Cross-linking in bone tissue involves hydroxylation of telopeptide lysine residues that give rise to primarily keto-imine forms of cross-links; reaction results in formation of pyridinium and pyrrolic cross-links on maturation; two specific pyridinium cross-links are pyridinoline and an analog named *deoxypyridinoline*.

Pyrosequencing DNA sequencing method based on sequential addition of phosphorylated dNTPs to sequencing reaction and emission of light on binding of each nucleotide to DNA template.

▶ **Q**

Quality control Procedures for monitoring and evaluating quality of analytical testing process of each method to ensure accuracy and reliability of patient test results and reports.

Quaternary structure Protein structure that describes arrangement of two or more polypeptide chains to form protein; only proteins with two or more polypeptide chains have quaternary structure.

Qβ Replicase Probe-amplification technique that uses RNA-dependent RNA polymerase to produce multiple copies of RNA from single-stranded DNA or RNA target.

▶ **R**

Random-access testing Analyzing specimens without regard to their initial order.

Range Measure of spread or variation in set of data.

Ranson's indicators of severity in acute pancreatitis List of laboratory tests (e.g., glucose, lactate dehydrogenase, aspartate aminotransferase, BUN, and calcium) and values that are helpful in determining severity of acute pancreatitis attack; mortality prediction increases with number of positive indicators.

Reactive hypoglycemia Form of hypoglycemia that occurs within 2 to 4 hours after eating because of delayed and exaggerated increase in plasma insulin; can be predictor of early-onset type 2 diabetes mellitus.

Receptor   Represents drug's site of action; specific macromolecule where initial molecular event occurs with drug.

Receptor activator of nuclear factor (NF-kB) ligand (RANKL)   Specific protein substance that is major stimulator of both differentiation of preosteoclasts to osteoclasts and activities of mature osteoclasts; expressed by osteoblasts, bone marrow stromal cells, and activated T cells.

Reference interval   Interval between and including two reference limits; consists of upper and lower reference limits, which are determined by several different techniques, including ±2 standard deviations about mean, percentiles (e.g., 2.5 and 97.5), and confidence limits.

Reflectometer   Filter photometer that measures quantity of light reflected by liquid sample that has been dispensed onto nonpolished surface.

Relative centrifugal force (RCF)   Weight of particle in centrifuge relative to its normal weight expressed as some number times gravity (g); term describes force required to separate two phases being separated by centrifugation.

Relative index (RI)   Calculated by dividing the CK-MB (CK-2) mass ($\mu g/L$) by total CK activity (U/L); ratios greater than 5 indicate cardiac source, and ratios between 3 and 5 are a gray zone that requires serial determinations to diagnose or rule out myocardial infarction.

Relative hyperproteinemia   Hyperproteinemia in which protein concentration is usually normal but dissolved in less plasma.

Relative hypoproteinemia   Form of hypoproteinemia that occurs in water intoxication, salt-retention syndromes, and as result of massive IV infusions and administration of volume expanders (e.g., dextran).

Releasing factor   Peptide synthesized by hypothalamus and released into portal circulation to affect pituitary hormone synthesis and secretion.

Remodeling   Formation of bone on surfaces previously containing bone.

Renal azotemia   Azotemia associated with kidney disease—for example, glomerulonephritis, nephrotic syndrome, and acute renal failure.

Renal clearance   Rate at which kidneys remove substance from plasma or blood or quantitative expression of rate at which substance is excreted by kidneys in relation to concentration of same substance in plasma; usually expressed as mL cleared/min.

Resistivity   Electrical resistance in ohms measured between opposite faces of 1.00-cm cube of aqueous solution at specified temperature.

Resolution   In chromatography, ability of chromatographic system to separate multicomponent mixtures of chemical compounds.

Respiratory acidosis   Pathological condition that leads to accumulation of carbon dioxide, which raises PCO_2 and decreases pH.

Respiratory alkalosis   Pathological condition that leads to excessive elimination of carbon dioxide, which lowers PCO_2 and increases pH.

Respiratory apparatus   Systems used to transport oxygen and carbon dioxide throughout body.

Restriction endonuclease   Enzyme that recognizes specific nucleotide sequence and cuts DNA molecule wherever that sequence occurs.

Restriction fragment length polymorphism (RFLP)   Genetic variations in fragment length and number that result from treating DNA from different individuals with restriction enzymes.

Retention time   Time elapsed from injection of sample into chromatograph to recording of peak maximum of component in chromatogram.

Reverse osmosis   Process in which water is forced through semipermeable membrane that acts as molecular filter.

Revolutions per minute   Unit for expressing number of complete rotations of rotor occurring per minute.

Reye's syndrome   Acute, often fatal, childhood syndrome presumably caused by virus; hallmarks are encephalopathy and fatty infiltration of liver; correlation was discovered between Reye's syndrome and aspirin that prompted pharmaceutical companies to place warnings label on aspirin bottles.

Rf   Ratio used in thin-layer chromatography that represents distance compound migrates from origin to center of "spot" or separated zone divided by distance traveled by solvent from origin, represented as solvent front.

Ribonucleic acid (RNA)   Nucleic acid located in nucleus and cytoplasm of cells that serves important role in protein synthesis; also serves as primary carrier of genetic information in some viruses.

Ribosomal RNA (rRNA)   Type of RNA that is structural component of ribosome.

Rickets   Condition associated with vitamin D deficiency, as well as calcium and phosphorous, that results in soft bone tissue formation; although rare, often manifests itself in children.

Risk factors   Chemical, psychological, physiological, or genetic elements thought to predispose individual to development of disease.

Risk stratification   Statistical process by which quality of care can be assessed independently of patient-specific presentation.

Rotor syndrome   Benign autosomal recessive trait resulting in conjugated hyperbilirubinemia; caused by reduction in concentration or activity of intracellular binding protein such as ligandin; not progressive; only abnormality is elevated conjugated bilirubin and total bilirubin.

▶ S

Salicylism   Syndrome associated with repeated ingestion of fairly large doses of salicylate; symptoms include tinnitus (high-pitched buzzing noise in ears), vertigo, decreased hearing, and occasionally nausea and vomiting.

Salivary amylase   Enzyme produced by salivary glands; catabolizes polysaccharides to intermediate-sized glucosans (limit dextrins) and maltose; inactivated by acid pH when it arrives in stomach.

Sandwich immunoassay   Noncompetitive technique (also referred to as *immunometric assay*) that uses label attached to second antibody; antigen must have two binding sites to complex both antibodies.

Saturable   Capable of being saturated.

Saturated fatty acid   Fatty acid that contains alkyl chain without double bond between carbon atoms.

Screening test   Test that is quick, easy to perform, and frequently used to separate samples into those considered presumptively positive for specified analyte (e.g., drugs and hormones) and those considered negative.

Scurvy Disease characterized by severe ascorbic acid deficiency; symptoms include spongy gums with loosening teeth, weakened capillary beds, and defective cartilage synthesis.

Secondary diabetes mellitus Diabetes form associated with secondary conditions or *endocrinopathies* such as Cushing's syndrome, hyperthyroidism, glucagonoma, and acromegaly, as well as certain drugs or chemicals.

Secondary structure Protein structure determined by interaction of adjacent amino acids; affected by winding of polypeptide chain, formation of hydrogen bonds between NH and CO groups of peptide bonds, and occasional disulfide bonds.

Secretin Basic peptide of 27 amino acid residues with structure similar to glucagon; located in S cells of mucosa of duodenum and jejunum; functions to inhibit smooth muscle contraction and decrease gastric acid secretion.

Secretin–cholecystokinin (CCK) test Direct test of pancreatic function that assesses both endocrine and exocrine functions of pancreas; bicarbonate and trypsin in duodenal juices are measured to determine pancreatic function.

Secretory IgA Second form of IgA found in secretions, including tears, sweat, saliva, and milk as well as gastrointestinal and bronchial secretions.

Sedatives Agents that decrease activity, moderate excitement, and calm their recipients; hypnotic drug produces drowsiness and facilitates onset and maintenance of state of sleep; barbiturates reversibly depress activities of all excitable tissues.

Sella turcica Depression in upper surface of sphenoid bone in which pituitary gland rests.

Serum Clear liquid that separates from blood on clotting.

Sex-hormone binding globulin (SHBG) Carrier protein for steroid hormones produced in liver; testosterone and dihydrotestosterone circulate primarily bound to SHBG in men.

Simple protein Protein composed only of amino acids (albumin, for example).

Skewness Data that are not symmetrical (asymmetrical).

Solid phase Material commonly used to facilitate separation of bound from free fractions in immunoassay; in this technique, antibody immobilized onto solid-phase material via covalent bonding or physical adsorption through noncovalent interactions; examples of solid-phase material are gel particles made of agaros or polyacrylamide, plastic beads, and particles coated with iron oxide.

Somatostatins Hormones secreted by delta cells of islets of Langerhans in pancreas; do not appear to directly affect carbohydrate metabolism but do indirectly inhibit secretion of insulin, growth hormone, glucagon, and other counterregulatory hormones.

Somatostatinoma Tumors of pancreas or intestine that secrete excess somatostatin.

Southern blot Technique used to identify DNA that is separated by electrophoresis and blotted onto membrane by hybridization with specific probe.

Space of Disse Space between endothelial cells and hepatocytes where nutrient uptake occurs.

Spectral bandwidth Range of wavelengths at point halfway between baseline and peak; term often referred to as *half-power point* or *full width at half maximum* (FWHM).

Spectrometer Instrument that provides information about intensity of radiation as function of wavelength or frequency; most use grating monochromator to disperse light into spectrum.

Spina bifida Congenital defect in walls of spine that allows protrusion of spinal cord or meninges.

Standard deviation Measure of dispersion of group of values around mean.

Staging Process of diagnosis in which patholgist determines position of cancer in progressive cycle of phases.

Steady state Represents point when total amount of drug in body does not change over multiple loading-dose intervals; rate of drug input equals rate of elimination; time to reach steady state is about four to five half-lives.

Stereoisomers Molecules that have same empirical or chemical formula but have mirror-image structural formulas.

Stereospecificity Enzyme that catalyzes reactions with only certain optical isomers.

Steroids Organic compound containing in its chemical nucleus the perhydrocyclopentanophenanthrene ring structure include groups of compounds related to sterol, including vitamin D, bile acids, certain hormones, and glucosides of digitalis.

Stimulation test Pertains to challenging procedures used to document hyposecretion of hormones—for example, adrenocortical hormones.

Strand displacement amplification (SDA) Target-amplification technique based on ability of DNA polymerase to initiate DNA replication at site of single-stranded nick created by restriction endonuclease.

Stringency Factors affecting ability of hybridization to occur.

Subtherapeutic When plasma drug concentration is below therapeutic level.

Succimer Antidote (2,3-dimercaptosuccinic acid) used to treat arsenic exposure; oral hydrophilic analog of BAL and chelating agent of choice for subacute and chronic arsenic toxicity.

Suppression test Procedure that tests organ's ability to decrease release of analyte after patient is given specific medication; for example, dexamethasone suppression test is used to assess patients for Cushing's syndrome.

Susceptibility to infection Condition seen in diabetics, who are more likely to require amputation of lower limbs because of resulting gangrene, hyperglycemia, and poor circulation.

Syndrome Group of symptoms, signs, laboratory results, and physiological disorders that are linked by common anatomical, biochemical, or pathological history.

Syndrome of inappropriate ADH secretion (SIADH) Condition in which inappropriate antidiuretic hormone secretion produces hyponatremia, hyopvolemia, and elevated urine osmolality.

Système Internationale d'Unités Internationally accepted system of measurement.

▶ T

Telopeptide Portion of amino acid sequence of protein that is removed in maturation of protein; examples are N- and C-terminal telopeptides of procollagen, which contain cross-linking sites involved in development of quaternary structure and are then proteolytically removed by procollagen peptidases.

Temporal Pertaining to time.

Teratogen Substance such as thalidomide and diethylstilbestrol that acts preferentially on embryo at precise stages of development, thereby leading to possible anomalies and malformations.

Tertiary structure Protein structure described by the way in which protein chain folds back on itself to form three-dimensional structure.

Test menu Listing of all tests or analytes that an instrument is capable of measuring; can include as many as 150 analytes.

Therapeutic range Relationship between desired clinical effect of drug and its concentration in plasma.

Therapeutic window Range between minimum effective therapeutic concentration of dose and minimum toxic concentration of dose; reflects concentration range that provides efficacy without unacceptable toxicity.

Thermistor Transducer that converts changes in temperature (heat) to resistance.

Thermocouple Sensor that consists of two dissimilar metals joined at one end; when junction of two metals is heated or cooled, voltage is produced that is calibrated to temperature.

Thermocycler Instrument that automates timed, temperature-controlled cycles of denaturation, primer annealing, and extension in PCR.

Throughput Numbers of tests performed per hour.

Thyroxine Thyroid hormone stimulated by thyroid-stimulating hormone (TSH) from anterior pituitary.

To contain (TC) pipette Referred to as *rinse-out pipettes* because they must be refilled or rinsed out with appropriate solvent after initial liquid is drained from pipette; contain exact amount of liquid that must be completely transferred for accurate measurement.

To deliver (TD) pipette Pipettes designed to drain by gravity; must be held vertically with tip placed against side of container and without touching liquid; stated volume is obtained when draining stops; this type of pipette should not be blown out.

Tocopherol Class of chemical compounds, many of which have vitamin E activities; fat soluble and function as antioxidants; common form of tocopherol, α-tocopherol, is commonly added to food products.

Tolerance Gradual escalation of needed dose of drug to produce required effect.

Tonic drive Refers to the stimulus that results in muscle contraction or tension.

Torr Pressure exerted by 1 mm of mercury at certain specified conditions, notably 0°C.

Total body water All water within body, both inside and outside cells, including that contained in gastrointestinal and genitourinary systems.

Total laboratory automation (TLA) Use of automated devices and robots to perform all phases of clinical laboratory testing; comprehensive TLA system can perform such functions as specimen labeling, identification, transportation, specimen introduction to and removal from analyzers, and pipetting.

Toxic Means poisonous to body; occurs when drug's plasma concentration is higher than body can tolerate; may lead to deleterious effects.

Toxicant Agent or substance that acts like poison.

Toxicology Study of poisons, including their sources, chemical compositions, actions, tests, and antidotes.

Trace elements Group of elements that includes mostly metals—for example, zinc—that serve biological role in humans as well as animals; word *trace* was originally used because quantitation with analytical methods then available was not possible; can also be described as elements that are present in $\mu g/dL$ in body fluids and mg/kg in tissues.

Transcription Process during which DNA is converted into complementary sequence of nucleotide bases in mRNA.

Transcription-mediated amplification (TMA) Target-amplification technique that uses reverse transcriptase with RNase activity and RNA polymerase to make multiple copies of RNA sequence.

Transducer Device that converts one form of energy to another; photodetectors, for example, are transducers that convert EMR into electron or photocurrent that is passed to readout circuit.

Transfer RNA (tRNA) Type of RNA that carries specific amino acid to site on ribosome during protein synthesis.

Transferases Group of enzymes that catalyze transfer of group other than hydrogen between two substrates.

Transferrin (TRF) Major component of β-globulins and principal plasma protein for transport of ferric iron (Fe^{3+}) from intestine, where it is absorbed by apotransferrin to red cell precursors in bone marrow or to liver, bone marrow, or spleen for storage.

Translation Process of converting nucleotide sequence in mRNA to linear sequence of amino acids in proteins.

Transthyretin (TTR) Prealbumin protein that binds with thyroxine and triiodothyronine (thyroid hormones) and retinol (vitamin A) and serves as transport protein.

Triglyceride Most common glycerol ester in plasma, comprising glycerol and three fatty acids; neutral fat that combines with proteins to form lipoproteins; stored in adipose tissue for energy.

Trinder reaction Reaction in which glucose oxidase catalyzes oxidation of glucose to gluconic acid and hydrogen peroxide (H_2O_2).

Trioses Smallest carbohydrates, having only three carbon atoms.

Troponin Polypeptide involved in muscle function; three distinct polypeptides—troponin T (responsible for tropomyosin-binding activity), troponin I (binds to actin and inhibits activity of actomyosin ATPase), and troponin C (facilitates calcium-binding activities of muscle contraction); with tropomyosin, three polypeptide forms create complex that regulates actin and myosin interaction and muscle contraction.

Trough drug level Represents minimum drug concentration in plasma or blood measured before next dose is given.

Trypsin (TRY) Proteolytic enzyme in intestine that catalyzes peptide bonds in partly digested proteins.

Tubular reabsorption Process by which renal tubule substances filtered by glomerulus are reabsorbed as filtrate passes through tubules.

Tubular secretion Passage of substances from peritubular capillaries into tubular filtrate; serves two major functions of (1) eliminating waste products not filtered by glomerulus and (2) regulating acid–base balance in body through secretion of 90% of hydrogen ions excreted by kidneys.

Tumor Swelling or enlargement of tissue; abnormal mass.

Turbidimetry Measurement of reduction in light transmission caused by particle formation.

Type I collagen Term *collagen* usually refers to type I collagen, most common collagen in vertebrates, making up as much as 90% of skeleton and found throughout body (e.g., skin, tendons, ligaments, intervertebral disks, arteries, and granulation tissues); type I collagen molecules form D-periodic (D = 67 nm, the characteristic axial periodicity of collagen) cross-striated fibrils in extracellular space, giving tissues their mechanical strength and providing major biomechanical lattice for cell attachment and anchorage of macromolecules.

Type 1 diabetes mellitus Accounts for about 5% to 10% of diabetes; previously called *juvenile-onset diabetes* and *insulin-dependent diabetes*; most commonly diagnosed in childhood and adolescence but has been found in adults.

Type 2 diabetes mellitus Most common and milder form of diabetes; accounts for 90% to 95% of diabetics; previously called *adult-onset* and *non–insulin-dependent* diabetes; affected individuals usually do not require insulin to maintain adequate glucose control.

▶ **U**

Ultratrace elements Trace elements present in elements at ng/dL or μg/kg level of concentration.

Unbound drug Drug not attached to proteins; also termed *free drug*.

Uncompetitive inhibition In enzyme reaction, when inhibitor binds to ES complex to form enzyme–substrate inhibiting complex that does not yield product.

Unit test reagents Amount of reagent present on chemistry analyzer for performance of single test.

Universal precautions Practice described by National Institute for Occupational Safety and Health; every clinical laboratory should treat all human blood and other potentially infectious material as if they were known to contain infectious agents such as hepatitis B virus, HIV, and other bloodborne pathogens.

Urea Major nitrogen-containing metabolic product of protein catabolism in humans; formed from exogenous protein (protein in diet) or endogenous protein from breakdown of cells in body; synthesis of urea carried out exclusively by hepatic enzymes of Krebs and Henselheit urea cycle.

Urease Enzyme that catalyzes hydrolysis of urea into carbon dioxide and ammonia; initial reaction in clinical laboratories for the measurement of BUN levels.

Urea reduction ratio (URR) Simplest calculation for determining adequacy of dialysis; determines percentage fall in urea during dialysis session.

Uremia Increased level of urea or BUN in blood.

Uricase Uric acid methodology that is more specific than PTA, does not require protein-free filtrate, and is used by 99% of labs; uricase catalyzes oxidation of uric acid to allantoin.

Uridyl diphosphate (UDP)–glucuronyl transferase Enzyme synthesized by hepatocytes that binds two glucuronic acid molecules to indirect bilirubin molecule to form direct bilirubin.

Urobilin Brown pigment formed by oxidation of urobilinogen that gives characteristic brown color to feces.

Urobilinogen Colorless product of bilirubin caused by action of intestinal bacteria.

▶ **V**

Vasoactive intestinal polypeptide (VIP) Linear polypeptide consisting of 28 amino acids that is structurally similar to secretin, gastric inhibitory polypeptide (GIP), and glucagon; principal source is nervous system and gut; not found in mucosal endocrine cells of GI tract; believed to be neurotransmitter located in peripheral and central nervous tissue with VIP-containing nerve fibers through GI tract.

Very-low-density lipoprotein (VLDLs) Class of lipoproteins that are primarily transporters of endogenous triglycerides from liver to muscles and adipose cells; contain greater lipid-to-protein ratio and are least dense of lipoproteins.

Viral hepatitis Worldwide disease of serious proportion accompanied by hepatocellular inflammation, injury, and necrosis of hepatocyte; most common viruses associated with viral hepatitis are hepatitis A, B, C, and D.

Vitamin D metabolites Vitamin D is metabolized to its main circulating form (25-hydroxyvitamine D) and then to its biologically active form (1, 25-dihydroxyviatmin D), which regulates calcium and phosphate metabolism; 25-hydroxyviatmin D reflects vitamin D nutritional status.

V max Occurs when substrate concentration is high enough that all enzyme molecules are bound to substrate and all active sites are engaged.

Von Gierke disease Type I and most common form of glycogen storage disease; also known as *glucose-6-phosphatase deficiency*.

▶ **W**

Wavelength (λ) Linear distance between any two equivalent points on successive wave; nanometer (nm) (10^{-9} m) is widely used unit for wavelength in visible spectrum.

Wet beriberi Condition resulting from chronic vitamin B_1 (thiamine) deficiency; patients will exhibit cardiovascular symptoms resulting from impaired myocardial energy metabolism and may present with tachycardia, enlarged heart, CHF, peripheral edema, and peripheral neuritis.

Wernicke–Korsakoff syndrome Condition associated with chronic alcoholics who present with thiamine deficiency and exhibit unique set of signs and symptoms; tend to develop central nervous system complications known as Wernicke's encephalopathy; symptoms of disorder are horizontal nystagmus, opthalmoplegia, cerebral ataxia, and mental impairment.

Western blot   Technique used to identify proteins that have been separated by electrophoresis and blotted onto membrane by their reactions with specific antibodies.

Wet beriberi   Chronic thiamine deficiency; patients exhibit cardiovascular symptoms resulting from impaired myocardial energy metabolism and may present with tachycardia, enlarged heart, congestive heart failure, peripheral edema, and peripheral neuritis.

Whipple's triad   Three factors used to diagnose hypoglycemia: symptoms known or likely to be caused by hypoglycemia, low plasma glucose measured at time of symptoms, and relief of symptoms when glucose is raised to normal.

Wilson disease (WD)   Rare autosomal recessive trait in which ceruloplasmin levels are reduced and dialyzable copper concentration is increased; involves mutation in copper-transporting ATPase (ATP7B) that moves copper into bile for excretion; copper accumulates in various organs (e.g., brain, liver, kidney, and cornea); associated with degenerative changes in the brain, cirrhosis, splenomegaly, involuntary movements, psychic disturbances, and progressive weakness and emaciation.

Work cell   Combination of specimen manager with instruments or consolidated instruments of chemistry and immunoassay reagents that provide broad spectrum of analytical tests.

Workstation   Designated area in which limited number of specific tasks are completed; for example, processing workstation serves as area where specimens requiring laboratory testing are received.

▶ **X**

Xenobiotic   Substance that is foreign to living organism and usually harmful.

▶ **Z**

Zero-order kinetics   Reaction involving enzyme and drug where rate of change of plasma concentration of drug is independent of plasma drug concentration; reaction rate is based only on enzyme concentration; enzyme measurements in laboratory are measured when reaction is in zero-order kinetics.

Zollinger–Ellison syndrome   Condition caused by tumor (gastrinoma) of pancreatic islet cells that results in overproduction of gastric acid; excess gastric acid may cause ulceration of esophagus, stomach, duodenum, and jejunum, resulting in hypergastinemia, diarrhea, and steatorrhea.

Zymogen   Protein that becomes enzyme; exists in inactive form antecedent to active enzyme and is stored in zymogen granules of pancreas.

Zymogen granule   Granule in pancreas that contains zymogens (precursor enzymes) before they are secreted.

Sodium homeostasis, 274–275
Sodium homeostasis disorders, 275
Sodium measurement, 275–276
Sodium nitroprusside, 464–465
Sodium vapor lamps, 28
Solid-phase hybridization, 547
Solid-phase material, 127
Solution concentrations, 14
Solutions, percent, 16
Somatostatinomas, pancreatic, 385–386
Somatotropin, 341
Southern, Edwin, 548
Southern blots, 545, 548, *549*
Space of Disse, 415
SPE. *See* Serum protein electrophoresis (SPE)
Spearman's correlation coefficient, 88
Special reagent water (SRW), 3
Specific gravity, 15
  in urinalysis, 256
Specificity
  diagnostic, 226
  of test, 226
Specimen manager, 71
Specimens
  archiving, 66
  automated processing of, 66
  blood-gas
    preanalytical handling, 301
    types of, 300–301
  collection and storage of, 345–346
  labeling, 65
  new assays and type of, 89
  retrieving, 66
  sorting, 65
  therapeutic drug measurement, 459
  transport of, 64–66
  urine, 495–496, 524
Spectral bandwidth, 28
Spectrograde, 3
Spectrometer, 33
Spectrophotometer, 27, *27*
Spectroscope, 33
Spectroscopy, quality assurance in, 32–34
Specular reflectance, 34
Sphingosine derivatives, 163
Spina bifida, 186
Spironolactone, 467
Split/splitless injector, 47
Splitting the samples, 65
Sprinkler systems, 112
Squamous cell carcinoma antigen (SCCA), 239
SRMs. *See* Standard reference materials (SRMs)
SRW. *See* Special reagent water (SRW)
SSA. *See* Sulfosalicylic acid (SSA)

Stability, 93
Staging of germ cell tumors, alpha fetal protein and, 232
Stand-along processing system, 66
Standard deviation of differences, 84
Standard deviation of the duplicates, 84
Standard deviation *(s)*, 81
Standard error of the estimate $(S_{y \cdot x})$, 87–88
Standard error of the mean, 84
Standard error of the mean of differences, 84–85
Standard hydrogen electron (SHE), 41
Standard reference materials (SRMs), 3
Statistical applications, of reference intervals, 98–99
Statistics. *See also* Laboratory statistics
  defined, 79
  descriptive, 79
  inferential, 79, 82–84
  symbols, 80
Stat samples, 9
Steady state, 455–456, *456*
Steatorrhea, 380
Stereoisomer specificity, 203
Stereospecificity, 203
Steroids, 337, *337*, 338, 346
Sterol derivatives, 163
Stimulation tests, 341
Stoke's shift, 54
Stomach, 364, *365*
Storage, of specimens, 345–346
Strand-displacement amplification (SDA), 550, 552
Stray light, 33
StreamLAB, *71*
Streptavidin-coated microparticles, 135, 137
Streptococci, 460
Streptococcus pneumoniae infections, 188
Streptomyces tsukubaensis, 470
Stress, steroid metabolism and, 347
Stringency, 547
Stroke, CK-BB and, 206
Strong acids, 17
Strong bases, 17
Student's *t*-test, 83
Subclinical thyroid diseases, 359
Subcutaneous drug administration, 452
Sublingual drug administration, 451
Substance Abuse and Mental Health Administration (SAMHSA), 495
Substituted urine specimen, 495
Substrate concentration, 204
Substrate-labeled fluorescent immunoassay (SLFIA), 123, 131, *132*
Subtherapeutic effect, 456

Succimer, 506
Succinyldicholine (Suxemethonium), 218
Sufentanil, 487
Sulfation, 454
Sulfhemoglobin, 445
Sulfosalicylic acid (SSA), 193
Sulfur chemiluminescence detector (SCD), 48
Suppression test, 341
Surface-enhanced laser desorption ionization (SELDI), 51, *51*
Surface-enhanced laser desorption ionization-time of flight (SELDI-TOF), 227
Surveys, 105
Sweat chloride, 279, 381
Swinging-bucket centrifuge, 8–9, *9*
Symbols, population and samples statistic, 80
SYNCHRON LX 1725 Clinical System, 68, 69
Syndrome of inappropriate antidiuretic hormone secretion (SIADH), 250, 343
Synthesis
  cholesterol, 163
  protein, 182–183
Syringe devices, 68
Systematic errors, 91
  sources of, 101, 103
Systemic lupus erythematosus (SLE), 188
SYVA Company, 123
Syva Corp., 482

▶ T

Tacrolimus, 449, 470
TAG-22. *See* Tumor-associated glycoprotein (TAG-22)
Tandem mass spectrometer (MSMS), 52
Tanzer-Gilvarg reaction and, 207
Targeted Diagnostics Technologies and Therapeutics Inc., 237
Target values, 100
Taring, 7
Tartrate-resistant acid phosphatase, 327
TBW. *See* Total body water (TBW)
TCA. *See* Trichloroacetic acid (TCA)
TCC. *See* Transitional cell carcinoma (TCC)
TCD. *See* Thermo conductivity detector (TCD)
TDM. *See* Therapeutic drug monitoring (TDM)
Tecan FE500, 66